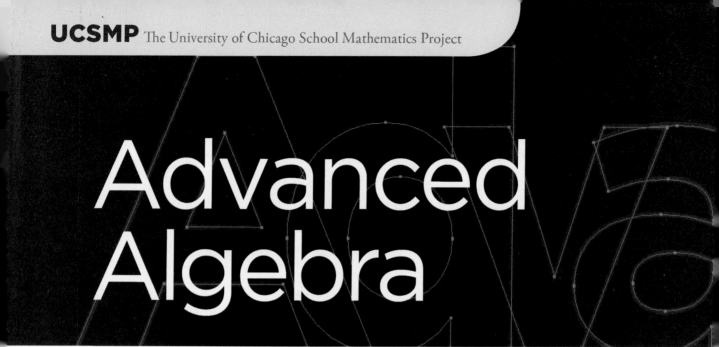

UCSMP The University of Chicago School Mathematics Project

Advanced Algebra

Authors

James Flanders

Marshall Lassak

Jean Sech

Michelle Eggerding

Paul J. Karafiol

Lin McMullin

Neal Weisman

Zalman Usiskin

Director of Evaluation

Denisse R. Thompson

UChicagoSolutions

Authors

3rd EDITION AUTHORS

James Flanders
Researcher and Technology Integration
Center for Elementary Mathematics and Science
Education, The University of Chicago

Marshall Lassak
*Associate Professor of Mathematics and
Computer Science*
Eastern Illinois University, Charleston, IL

Jean Sech *Mathematics Teacher*
Mattawan High School, Mattawan, MI

Michelle Eggerding *Mathematics Teacher*
Schaumburg High School, Schaumburg, IL

Paul J. Karafiol *Mathematics Teacher*
Walter Payton College Prep High School, Chicago, IL

Lin McMullin *Mathematics Teacher (retired)*
Burnt Hills-Ballston Lake High School, Burnt Hills, NY

Neal Weisman *Mathematics Teacher*
Oak Park/River Forest High School, Oak Park, IL

Zalman Usiskin *Professor of Education*
The University of Chicago

AUTHORS OF EARLIER EDITIONS

Sharon L. Senk
Associate Professor of Mathematics
Michigan State University, East Lansing, MI

Natalie Jakucyn *Mathematics Teacher*
Glenbrook South High School, Glenview, IL

Denisse R. Thompson
Assistant Professor of Mathematics Education
University of South Florida, Tampa, FL

Steven S. Viktora
Chairman, Mathematics Department
New Trier High School, Winnetka, IL

Nils P. Ahbel *Mathematics Teacher*
Kent School, Kent, CT

Suzanne Levin
UCSMP

Marcia L. Weinhold *Mathematics Teacher*
Kalamazoo Area Mathematics and Science Center,
Kalamazoo, MI

Rheta N. Rubenstein *Mathematics Department Head*
Renaissance High School, Detroit, MI

Judith Halvorson Jaskowiak *Mathematics Teacher*
John F. Kennedy High School, Bloomington, MN

Gerald Pillsbury
UCSMP

http://ucsmp.uchicago.edu/secondary/overview

UChicagoSolutions

Copyright © 2016 by The University of Chicago

Printed in the United States of America.

Send all inquiries to:
UChicagoSolutions
1427 E. 60th Street
Chicago, IL 60637
uchicagosolutions.com

ISBN 978-1-943237-16-6
ISBN 1-943237-16-6

1 2 3 4 5 6 7 8 9 RRDW 21 20 19 18 17 16

UCSMP EVALUATION, EDITORIAL, AND PRODUCTION

Director of Evaluation
Denisse R. Thompson
Professor of Mathematics Education
University of South Florida, Tampa, FL

Coordinator of School Relations
Carol Siegel

Executive Managing Editor
Clare Froemel

Production Coordinator
Benjamin R. Balskus

Editorial Staff
Kathryn Rich, Gary Spencer, Carlos Encalada,
Evan Jenkins, Currence Monson, Yan Yan Wang,
Nathaniel Loman

Evaluation Consultant
Sharon L. Senk, *Professor of Mathematics*
Michigan State University, East Lansing, MI

Evaluation Assistants
Allison Burlock, Julian Owens, Sophia Zhang,
Zhuo Zheng, Gladys Mitchell, Shravani Pasupneti,
Alex Yablon

Production Assistants
Paul Campbell, Gretchen Neidhardt, Sara Mahoney,
Nurit Kirschenbaum, Don Reneau, Elizabeth Olin,
Loren Santow

Technology Assistant
Luke I. Sandberg

Since the first two editions of *Advanced Algebra* were published, millions of students and thousands of teachers have used the materials. Prior to the publication of this third edition, the following teachers and schools participated in evaluations of the trial version during 2006–2007.

Clay Brown
Greenwood High School
Greenwood, Arkansas

Kathy Coskey
Glenbrook South High School
Glenview, Illinois

Audra Spicer
New Trier High School
Winnetka, Illinois

Craig Flietstra
Unity Christian High School
Hudsonville, Michigan

Catherine Feuerstein
Westfield High School
Westfield, New Jersey

Michael Buescher
Hathaway Brown School
Shaker Heights, Ohio

Cathy Buckingham
Nolan Catholic High School
Ft. Worth, Texas

Tami Wittkopf
Kewaskum High School
Kewaskum, Wisconsin

The following schools participated in field studies in 1993–1994, 1987–1988, 1986–1987, or 1985–1986 as part of the first edition or the second edition research.

Brentwood School
Los Angeles, California

Boulder High School
Boulder, Colorado

Hernando High School
Brooksville, Florida

Lassiter High School
Marietta, Georgia

Taft High School
Chicago, Illinois

Whitney Young High School
Chicago, Illinois

Kenwood Academy
Chicago, Illinois

Steinmetz Academic Center
Chicago, Illinois

Glenbrook South High School
Glenview, Illinois

Thornton Fractional South High School
Lansing, Illinois

Mt. Zion High School
Mt. Zion, Illinois

Rich South High School
Richton Park, Illinois

Lake Park West High School
Roselle, Illinois

Argo Community High School
Summit, Illinois

Shawnee Mission NW High School
Shawnee, Kansas

Framingham High School
Framingham, Massachusetts

Renaissance High School
Detroit, Michigan

Pontotoc High School
Pontotoc, Mississippi

Sentinel High School
Missoula, Montana

West Genessee High School
Camillus, New York

Lake Oswego High School
Lake Oswego, Oregon

Springfield High School
Springfield, Pennsylvania

Hanks High School
El Paso, Texas

We wish to acknowledge the generous support of the **Amoco (now BP) Foundation** and the **Carnegie Corporation of New York** in helping to make it possible for the first edition of these materials to be developed, tested, and distributed, and the additional support of the **Amoco (now BP) Foundation** for the second edition.

We wish also to acknowledge the contribution of the text *Advanced Algebra with Transformations and Applications*, by Zalman Usiskin (Laidlaw, 1975), to some of the conceptualizations and problems used in this book.

UCSMP The University of Chicago School Mathematics Project

The University of Chicago School Mathematics Project (UCSMP) is a long-term project designed to improve school mathematics in Grades K–12. UCSMP began in 1983 with a 6-year grant from the Amoco Foundation. Additional funding has come from the National Science Foundation, the Ford Motor Company, the Carnegie Corporation of New York, the Stuart Foundation, the General Electric Foundation, GTE, Citicorp/Citibank, the Exxon Educational Foundation, the Illinois Board of Higher Education, the Chicago Public Schools, from royalties, and from publishers of UCSMP materials.

From 1983 to 1987, the director of UCSMP was Paul Sally, Professor of Mathematics. Since 1987, the director has been Zalman Usiskin, Professor of Education.

UCSMP *Advanced Algebra*

The text *Advanced Algebra* has been developed by the Secondary Component of the project, and constitutes the core of the fifth year in a seven-year middle and high school mathematics curriculum. The names of the seven texts around which these years are built are:

- *Pre-Transition Mathematics*
- *Transition Mathematics*
- *Algebra*
- *Geometry*
- *Advanced Algebra*
- *Functions, Statistics, and Trigonometry*
- *Precalculus and Discrete Mathematics*

Why a Third Edition?

Since the second edition, there has been a general increase in the performance of students coming into middle and high schools due, we believe, to a combination of increased expectations and the availability of improved curricular materials for the elementary and middle school grades. On the other hand, there has also been a substantial increase in the percent of students expected to be successful in *Advanced Algebra*. In many places, all students are expected to learn the material of this course, more than double the percent of students taking this course a generation ago. As a consequence, these materials are designed to take into account the wide range of backgrounds and knowledge students bring to the classroom.

These increased expectations for the performance of all students in high schools and the increased levels of testing that have gone along with those expectations require a broad-based, reality-oriented, easy-to-comprehend approach to mathematics. UCSMP third-edition materials were written to accommodate these factors.

The writing of the third edition of UCSMP is also motivated by the recent advances in technology both inside and outside the classroom, coupled with the widespread availability of computers with Internet connections at schools and at homes.

A factor supporting the continued importance of the entire UCSMP curriculum is the increase in the number of students taking a full course in algebra before ninth grade. With the UCSMP curriculum, these students will have four years of mathematics beyond algebra before calculus and other college-level mathematics. UCSMP is the only curriculum so designed. Thousands of schools have used the first and second editions and have noted success in student achievement and in teaching practices.

Research from these schools shows that the UCSMP materials really work. Many of these schools have made suggestions for additional improvements in future editions of the UCSMP materials. We have attempted to utilize all of these ideas in the development of the third edition.

UCSMP *Advanced Algebra*–Third Edition

All lessons are fresh for this edition but the overall structure and mathematical prerequisites are the same as in previous editions. Many features are retained or enhanced in this third edition. There is **wider scope**, including significant amounts of geometry employed to motivate, justify, extend, and otherwise enhance the algebra. The use of transformations is particularly important because transformations are functions and they allow all graphs to be considered as geometric sets of points. This feature enables this text to be particularly beneficial in any further study of algebra and functions. A **real-world orientation** has guided both the selection of content and the approaches allowed the student in working out problems, because being able to do mathematics is of little use to an individual unless he or she can apply that content. We ask students to **read mathematics**, because students must read to understand mathematics in later courses and must learn to read technical matter in the world at large. The use of **new and powerful technology** is integrated throughout, with *computer algebra systems* and *dynamic geometry systems* assumed throughout the materials.

Four dimensions of understanding are emphasized: Skill in drawing, visualizing, and following algorithms; understanding of Properties, mathematical relationships, and proofs; Using algebraic ideas in realistic situations; and Representing algebraic concepts with graphs or other diagrams. We call this the SPUR approach: **S**kills, **P**roperties, **U**ses, **R**epresentations.

The **lessons** show why the content is important and explain how ideas are related to each other, and include fully-developed examples that often show multiple worked-out solutions and checks. Each question set begins with the **Covering the Ideas** questions that demonstrate the student's knowledge of the overall concepts of the lesson. **Applying the Mathematics** questions go beyond lesson examples. The **Review** questions relate either to previous lessons in the course or to content from earlier courses. The **Exploration** questions ask students to explore ideas related to the lesson.

The **book organization** is designed to maximize the acquisition of both skills and concepts. At the end of each chapter, a carefully focused Progress Self-Test and a Chapter Review, each keyed to SPUR objectives, are used to solidify performance of skills and concepts from the chapter so that they may be applied later with confidence. Finally, to increase retention, important ideas are reviewed in later chapters.

New instructional features for this edition include: a **Big Idea** highlighting the key concept of each lesson; **Mental Math** questions at the beginning of lessons to sharpen "in your head" skills; **Activities** in virtually every lesson to develop concepts and skills; **Guided Examples** that provide partially completed solutions to encourage active learning; and **Quiz Yourself** **(QY)** stopping points to periodically check understanding as you read.

Comments about these materials are welcomed. Please address them to:

The University of Chicago School Mathematics Project
http://ucsmp.uchicago.edu
ucsmp@uchicago.edu
773-702-1130

▷ Contents

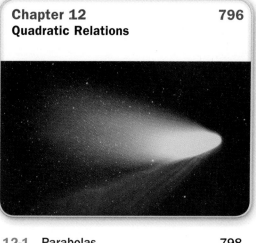

To the Student: Getting Started

Welcome to *Advanced Algebra*! We hope you enjoy this book and find it useful; it was written for you.

Advanced Algebra contains much of the mathematics that educated people around the world use in conversation, and that most colleges want or expect you to have studied. Familiar ideas, such as properties of numbers, graphs, expressions, equations, and inequalities, appear throughout the book. In addition, you will study many topics that may be new to you, including matrices, logarithms, trigonometry, and conic sections. Throughout the course you will use the broad ideas of function and mathematical modeling to help organize ideas.

The name *Advanced Algebra* indicates that the content of this book is related to what you learned in first-year algebra. However, this book is not necessarily more difficult. Some questions may be harder, but you know a lot more than you did then. In particular, you know much more geometry, and you have had a year's more practice with algebra.

Studying Mathematics

A goal of this book is to help you learn mathematics on your own, so that you will be able to deal with the mathematics you see in newspapers and magazines, on television, on any job, and in school. Everything in the book is written with the expectation that you will read it, learn from it, and enjoy that experience.

Not all students learn the same way. You need to find a way that works best for you. Here are a few suggestions.

1. **Mathematics is not a spectator sport.** You can watch an athlete, actor, juggler, or dancer perform, but you cannot replicate their performance without trying, failing, and improving through repeated attempts.

2. **Reading mathematics takes practice.**

▶ *Read slowly and thoughtfully*, paying attention to each word, graph, table, and symbol. Look up the meaning of any word you do not understand. Use the **QY** (Quiz Yourself) questions to help you check your understanding as you read. The answers to QY questions are found at the end of each lesson.

▶ *Work examples* yourself as you follow the steps in the text. **Guided Examples** are in most lessons. You should write these examples in a notebook so that you can refer to them throughout the course. Parts of the solutions to examples are written in this font to suggest what to write on your homework paper as a solution to a problem. Many lessons also have **Activities**. You should do these activities to help you understand the mathematics of the lesson.

▶ *Draw graphs, diagrams, and tables* by hand or with technology when following a complicated example.

3. **Writing mathematics is an important skill.** Writing is a tool for communicating your solutions and

thoughts to others, and can help you understand mathematics, too. Writing good explanations takes practice. Use solutions to the examples in each lesson to guide your writing.

4. **Be persistent.** Not all questions can be answered immediately. If you are struggling with a problem, don't give up! Read the lesson again. Read the question again. Look for examples. If you can, go away from the problem and come back to it a little later.

5. **Develop good study habits.** Many students have found study groups to be helpful. Get the phone numbers or email addresses of a few students in your class, so you can share thoughts about what you are learning. And perhaps most important, ask your teacher for help with important ideas that are still not clear to you after you have worked on them.

Tools Needed for This Book

To be a successful student in this course, you will need lined paper, sharp pencils, erasers, a ruler, and graph paper.

You need technology for use while reading the lessons, doing homework, participating in class, and taking tests. The technology needs to be able to:

1. deal with arithmetic operations, the numbers π and e, square roots, reciprocals (x^{-1} or $\frac{1}{x}$), powers (x^y or $x^{\wedge}y$), logarithms (log and ln), and trigonometric functions (cos, sin, and tan);

2. graph functions;

3. make lists and tables of values for equations, and calculate statistics on the data in these tables (table generator);

4. manipulate algebraic expressions and solve symbolic equations (a computer algebra system (CAS));

5. construct and manipulate geometric objects (a dynamic geometry system (DGS));

6. operate on matrices; and

7. generate sequences recursively (spreadsheet).

A few calculators have all seven of the above features. Often they are called CAS calculators, though they do much more. Such a calculator will be particularly useful in future mathematics courses and in many courses in college.

Altogether this technology allows you to solve realistic problems without having to do tedious computations, and it enables you to understand better many mathematics ideas. You will begin using technology in Lesson 1-3.

Using Technology

In many places in this book, you have a choice as to whether

> paper-and-pencil (✐) or
> calculator or computer (🖳)

is the most appropriate technology to solve a problem. Here is some advice.

▶ Whether you use ✐ or 🖳, take care to copy the problem correctly.

▶ Do not be content with blindly following a paper-and-pencil procedure or mindlessly pushing buttons on a calculator to solve a problem. Always learn additional ways to do calculations so that you can use them to check your first way. *It is not a good idea to check a problem by merely repeating the same steps you used to solve the problem in the first place.* If you made an error the first time, you are likely to make the same error when you check your work.

- Use the appropriate technology. Avoid using ✐ or ▭ on a problem you should be able to do in your head. If a problem can be solved quickly with a ✐ procedure, try not to use ▭. Do not necessarily use ✐ on a problem that is more appropriate solved with the aid of ▭.

- When using ▭, record what you did in some way so that you can go back and repeat the process if it was a good one, or change the process if it did not work out.

Getting Off to a Good Start

The questions that follow are designed to help you become familiar with *Advanced Algebra*.

We hope you join the hundreds of thousands of students who have enjoyed earlier editions of this book. We wish you much success.

Questions

COVERING THE IDEAS

1. Name four mathematical topics that *Advanced Algebra* includes. Indicate which of these topics are new to you.

2. What tools other than paper and pencil are needed for this course?

3. What is a CAS?

4. Identify two strategies that you might use to improve your reading comprehension.

5. What should you do if you cannot answer an assigned question?

6. Why is it not a good idea to check an answer to a problem by going through the same steps you used to solve the problem?

KNOWING YOUR TEXTBOOK

In 7–10, answer the questions by looking at the Table of Contents, the lessons and chapters of the textbook, or material at the end of the book.

7. Refer to the Table of Contents. What lesson would you read to learn about matrices for reflections?

8. Suppose you are working in Lesson 1-8.
   a. Where can you find the answers to the *Guided Example* in the lesson?
   b. On what page can you find answers to check your work on the Questions?
   c. Answers to which Questions are given?

9. In the *Vocabulary* sections at the end of each chapter, why are some terms marked with an asterisk(*)?

10. Look at a *Self-Test* at the end of a chapter. What should you do after taking the Self-Test?

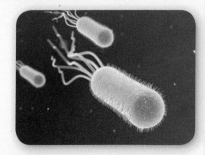

Chapter

1 Functions

Contents

Gold holds a special significance in almost all cultures around the world. It has long symbolized wealth and power, and it has always held great economic value. Unlike most other metals, gold is often found alone in the form of nuggets in a wide range of geographic areas, so it was likely the first metal used by early human beings. Since its discovery, gold has seen a wide range of uses, from ornamentation, to currency (the first gold coins were created in Turkey around 700 BCE), to modern scientific, technological, and medical applications.

Gold was once used to define the value of currency. In 1879, the United States dollar was set so that $20.67 was equal to one troy ounce of gold. It remained at this exact value until 1933, when President Roosevelt devalued the dollar as part of an attempt to

combat the Great Depression. Since then, the price of gold has varied.

In 1976, one troy ounce of gold was worth $145.10. By 1980, the price had more than quadrupled, to $641.20. By 1985, it had dropped to about half of its 1980 value. The price of gold at the end of each year from 1960 to 2005 is displayed in the graph below.

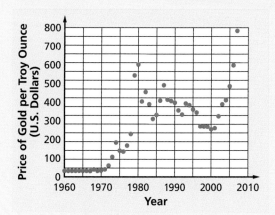

Each point on the graph represents a pair of numbers: the year and the closing price of gold in that year. From this graph it is difficult to predict the future value of gold. However, when a graph has a pattern, you can often describe its values with a formula, and you may be able to estimate future values with some confidence.

This graph represents a *function*. Functions exist whenever the value of one or more variables determines the value of another variable. The study of functions is basic to the study of all mathematics. In this chapter, you will see how to describe functions with words, tables, graphs, and symbols in order to solve a wide variety of problems.

Lesson

1-1

The Language of Algebra

Vocabulary

variable

algebraic expression

expression

algebraic sentence

evaluating an expression

order of operations

vinculum

equation

formula

▶ **BIG IDEA** Algebra is a language with expressions and sentences. There are precise rules for evaluating algebraic expressions so the meaning and values of expressions are unambiguous.

The language of algebra uses numbers and variables. It lets you describe patterns and relationships between quantities. A **variable** is a symbol that can be replaced by any one of a set of numbers or other objects. When variables stand for numbers and when numbers and variables are combined using the operations of arithmetic, the result is called an **algebraic expression**, or simply an **expression**. For instance, the expression $\frac{s^2 \sqrt{3}}{4}$ for the area of an equilateral triangle with side length s uses the variable s and the number $\frac{\sqrt{3}}{4}$. The expression for the volume of a cone with radius r and height h is $\frac{1}{3}\pi r^2 h$. That expression involves the variables r and h and the numbers π and $\frac{1}{3}$. Both expressions involve the operations of multiplication and powering.

Mental Math

a. What is the area of a square with perimeter 28 inches?

b. What is the area of a rectangle with length 6 cm and perimeter 20 cm?

c. What is the width of a rectangle with length 12 ft and area 108 ft²?

d. What is the length of a rectangle with width 25 mm and perimeter 130 mm?

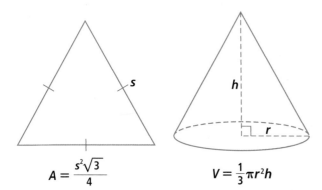

$$A = \frac{s^2\sqrt{3}}{4} \qquad\qquad V = \frac{1}{3}\pi r^2 h$$

An **algebraic sentence** consists of expressions related with a verb in symbolic form.
Common verbs are shown in the table at the right.

Examples of algebraic sentences are $A = \pi r^2$, $A \approx 3.14 \cdot r^2$, $a + b = b + a$, and $3x + 9 < 22$.

Symbol	Meaning
$=$	is equal to
$<$	is less than
$>$	is greater than
$\leq$	is less than or equal to
$\geq$	is greater than or equal to
$\neq$	is not equal to
$\approx$	is approximately equal to

Writing Expressions and Sentences

From your earlier study of algebra, you have gained experience writing expressions and sentences, modeling real situations, and evaluating expressions or sentences. In Example 1 below, part of the solution is written **using this typestyle**. This style is used to indicate what you might write on your homework paper as the solution to the problem.

Example 1

Joseph has a collection of 1,380 comic books and buys 17 new titles every month. If Joseph continues collecting comic books in this way, how many comic books will he have after *m* months?

Solution Make a table. Beginning with 1,380 comic books, in each month there will be an increase to Joseph's collection of 17 issues.

The first American comic book was *The Adventures of Mr. Obadiah Oldbuck* by Rudolphe Töpffer, published in 1842.

Months from Now	Number of Comics
1	$1380 + 1 \cdot 17$
2	$1380 + 2 \cdot 17$
3	$1380 + 3 \cdot 17$
4	$1380 + 4 \cdot 17$

Notice in this table that the arithmetic in the right column is not carried out. This makes the pattern easier to see. The number in the left column, which gives the number of months, is always in a particular place in the expression in the right column. You should see the following pattern.

$$m \qquad\qquad 1380 + m \cdot 17$$

Because of the Commutative Property of Multiplication, $m \cdot 17 = 17m$. So, after *m* months, Joseph will have $1380 + 17m$ comic books.

Check Pick a value for *m* not in the table and substitute it in $1380 + 17m$. We pick $m = 5$, indicating 5 months from now, and get $1380 + 17m = 1380 + 17 \cdot 5 = 1,465$ comic books. Then calculate the number of comics 5 months from now using the table. The table shows that in 4 months, Joseph would have $1380 + 4 \cdot 17 = 1,448$ comics. Add 17 for the fifth month to get $1448 + 17 = 1,465$. It checks.

You could also describe the situation in Example 1 with the sentence $C = 1380 + 17m$, where *C* is the number of comic books after *m* months.

See Quiz Yourself 1 at the right.

Quiz Yourself (QY) questions are designed to help you follow the reading. You should try to answer each Quiz Yourself question before reading on. The answer to the Quiz Yourself is found at the end of the lesson.

▶ **QUIZ YOURSELF 1**

In the situation of Example 1, how many comic books would Joseph have after 6 years?

Evaluating Expressions and Formulas

Substituting numbers for the variables in an expression and calculating a result is called **evaluating the expression**. In the expression $1380 + 17m$ in Example 1, we multiplied the value of m by 17 and then added 1380. We were using the standard rules for **order of operations** to evaluate the expression.

Rules for Order of Operations

1. Perform operations within parentheses or other grouping symbols from the innermost group out.

2. Within grouping symbols, or if there are no grouping symbols:
   a. Take powers from left to right.
   b. Multiply and divide in order from left to right.
   c. Add and subtract in order from left to right.

GUIDED

Example 2

A Guided Example is an example in which some, but not all of the work is shown. You should try to complete the example before reading on. Answers to Guided Examples are in the Selected Answers section at the back of this book.

Find the value of $\dfrac{-b - \sqrt{b^2 - 4ac}}{2a}$ **when** $a = 3$, $b = -1$, **and** $c = -4$.

Solution

Step 1 Substitute: $\dfrac{-b - \sqrt{b^2 - 4ac}}{2a} = \dfrac{? - \sqrt{?^2 - 4 \cdot ? \cdot ?}}{2 \cdot ?}$

Step 2 In both the fraction and the radical symbol ($\sqrt{\ }$), the bar (—) is a grouping symbol. (The bar is called a **vinculum**.) The square root vinculum is inside the fraction vinculum, so work inside the square root first.

Compute the power and then do the multiplications followed by the subtraction. (Watch the sign!)

$= \dfrac{? - \sqrt{?}}{2 \cdot ?}$

Step 3 Now compute the square root and subtract in the fraction's numerator. Then multiply in the fraction's denominator.

$$= \frac{?}{?}$$

Step 4 You may wish to rewrite the fraction in lowest terms.

$$= \underline{\quad?\quad}$$

 See Quiz Yourself 2 at the right.

▶ QUIZ YOURSELF 2

Evaluate
$88 - 16 \div 2 \cdot 3^{6-4}$.

An **equation** is a sentence stating that two expressions are equal. A **formula** is an equation stating that a single variable is equal to an expression with one or more different variables on the other side. The single variable on one side of a formula is said to be written *in terms of* the other variables. Below are some examples.

$$x = \frac{b \pm \sqrt{b^2 - 4ac}}{2a} \qquad \text{both an equation and the Quadratic Formula}$$

$$A = \pi r^2 \qquad \text{both an equation and a formula}$$

$$y = 3x + 4 \qquad \text{both an equation and a formula}$$

$$a + b = b + a \qquad \text{an equation that is not a formula}$$

$$x = 15 \qquad \text{an equation that is not a formula}$$

Formulas are useful because they express important ideas with very few symbols and can be easily applied to many situations.

Example 3 shows how to evaluate an expression with more than one variable taken from a real situation. It also illustrates how to work with units in evaluating expressions.

GUIDED

Example 3

A kilowatt-hour is one kilowatt of power used for one hour.

a. What does it cost to have a 60-watt bulb turned on for 33 hours at a cost of 9.53¢ per kilowatt-hour?

b. Give a formula for the cost in terms of the three numbers in the problem.

Solution

a. $\text{Cost} = 60 \text{ watts} \cdot \underline{\ ?\ } \text{ hours} \cdot \underline{\ ?\ } \dfrac{\text{cents}}{\text{kilowatt-hour}}$

You need to change 60 watts to kilowatts. Since 1 kilowatt = 1000 watts, $1 = \dfrac{1 \text{ kilowatt}}{1000 \text{ watts}}$. This fraction is a *conversion factor*.

Thus, $60 \text{ watts} = 60 \text{ watts} \cdot \dfrac{1 \text{ kilowatt}}{1000 \text{ watts}} = \underline{\ ?\ } \text{ kilowatts}$.

$\text{Cost} = \underline{\ ?\ } \text{ kilowatts} \cdot \underline{\ ?\ } \text{ hours} \cdot \underline{\ ?\ } \dfrac{\text{cents}}{\text{kilowatt-hour}}$

(continued on next page)

The Language of Algebra **9**

Notice that the units are multiplied and divided as if they were numbers.

$$= \underline{\quad ? \quad} \; \frac{\text{kilowatts} \cdot \text{hours} \cdot \text{cents}}{\text{kilowatt} - \text{hour}} = \underline{\quad ? \quad} \text{ cents}$$

b. Let p = the number of watts in the bulb. Let t = the number of hours. Let u = the (unit) cost in cents per kilowatt-hour. Then Cost = __?__.

Questions

COVERING THE IDEAS

These questions cover the content of the lesson. If you cannot answer a Covering the Ideas question, you should go back to the reading for help in obtaining an answer.

1. Refer to page 5. The price of shares of stock in gold-mining companies tends to go up as the price of gold goes up. In what year between 1975 and 2000 would it have been best to put your money in gold stocks for five years?

2. In your own words, describe the difference between an equation and a formula.

3. In your own words, describe the difference between an expression and an equation.

4. a. Name all the variables in $\pi\left(\frac{d}{2}\right)^2$.
   b. Classify $\pi\left(\frac{d}{2}\right)^2$ as an equation, formula, expression, or sentence. Explain your answer.

5. Give an example of an algebraic expression not found in the reading.

6. Give an example of an algebraic sentence not found in the reading.

7. Consider the sentence $c^2 = a^2 + b^2$.
   a. Is this sentence an equation? Why or why not?
   b. Is this sentence a formula? Why or why not?

8. Evaluate $\frac{-b + \sqrt{b^2 - 4ac}}{2a}$ when $a = -5$, $b = 30$, and $c = -25$.

9. Paula has collected 6 years of back issues of the magazine *Nature Today and Tomorrow*. *Nature Today and Tomorrow* prints 51 issues per year. If Paula reads two issues per day, how many issues will Paula have left to read after m months? Consider that the average month has 30 days.

10. **a.** Evaluate $12 \div 3 \cdot 5^{(4-2)} + 76$.

   b. Indicate the order in which you applied the five operations in Part a.

11. Refer to Example 3. Find the cost of operation of a central air conditioner that uses 3.5 kilowatts and runs for 12 hours a day for one week at a cost of 9.53¢ per kilowatt-hour.

APPLYING THE MATHEMATICS

These questions extend the content of the lesson. You should study the examples and explanations if you cannot answer the question. For some questions, you can check your answers with the ones in the Selected Answers section at the back of this book.

12. **a.** The expression $\frac{6^2 - 7}{5^2}$ is not equivalent to the expression $6 \cdot 6 - 7 \div 5 \cdot 5$. Why not? Use the order of operations to justify your answer.

   b. Insert parentheses into the second expression to make it equivalent to the first expression.

13. **a.** If a person owns C comic books and buys 31 new comics every month, how many comic books will this person have after t years?

   b. If a person owns C comic books and buys b new comics every month, how many comic books will this person have after t years?

14. Yuma used 1,040 ft of fence to enclose the rectangular pasture shown below. One side borders a river where there is already a thick hedge. That side needed no fencing.

   a. Let x be the width of the pasture as labeled. Write an expression for L, the length of the pasture, in terms of x.

   b. Write an expression for the area of the pasture in terms of L and x.

   c. Write an expression for the area of the pasture in terms of x only.

   d. Suppose Yuma wants the pasture to enclose at least 60,000 square feet. Write a sentence relating your answer in Part c to the area the fence must enclose.

In 15–17, evaluate each expression to the nearest tenth when $x = 7.2$, $y = \sqrt{3}$, and $z = -2$.

15. $\dfrac{10x}{y^4 - z^3}$

16. $\dfrac{x}{y^2} + z$

17. $y - x^2 - z$

In May 2008, NASA sent Phoenix to Mars, looking for evidence of life.

18. The formula $d = \frac{1}{2}gt^2$ tells how to find d, the distance an object has fallen during time t, when it is dropped in free fall from near the Earth's surface. The variable g represents the acceleration due to gravity. Near the Earth's surface, $g = 9.8 \frac{m}{sec^2}$.

   a. About how far will a rock fall in 5 seconds if it is dropped close to the Earth's surface?

   b. About how far will a rock fall in 5 seconds if it is dropped near the surface of the moon, where $g = 1.6 \frac{m}{sec^2}$?

   c. Looking at the results of Parts a and b, notice that the smaller g-value on the moon resulted in the rock falling a shorter distance. What conjecture might you make about the g-value on Mars in relation to Earth, if a rock dropped close to the surface of Mars fell 46.25 m in 5 seconds?

REVIEW

Every lesson contains review questions to practice ideas you have studied earlier.

In 19–23, tell which expression, (a) $x + y$, (b) $x - y$, (c) $y - x$, (d) xy, (e) $\frac{x}{y}$, or (f) $\frac{y}{x}$, correctly answers the given question. **(Previous Course)**

19. You download x files in y minutes. What is your rate of download in files per minute?

20. Toy cars are made x times the size of the actual car they represent. If the original car is y feet long, what is the length of the related toy car?

21. You had x dollars, but after paying for lunch you have y dollars left. How much did lunch cost?

22. Mindy gave Jian x marbles. Then Jian lost some of his marbles. If he has y marbles left, how many of his marbles did Jian lose?

23. Destinee walked x mph for y hours. How many miles did she walk?

24. Write examples of situations different from those in this lesson that lead to each expression you did not use as an answer in Questions 19–23. **(Previous Course)**

25. **Multiple Choice** Which sentence correctly relates the angle measures in the figure below? (**Previous Course**)

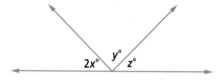

A $x + y + z = 180$

B $x = 90 - y - z$

C $2x + y + z = 180$

D none of these

26. What name is given to the polygon that is the shape of a stop sign? (**Previous Course**)

27. **a.** Solve $2x = 4x + 18$ for x.

    b. Check your work. (**Previous Course**)

EXPLORATION

These questions ask you to explore topics related to the lesson. Sometimes you will need to use references found in a library or on the Internet.

28. The graph of gold prices on page 5 looks distinctly different before and after 1971. Describe this difference in your own words. Do research on the Internet or at a library to find out what event happened to affect gold prices, and explain why this event caused the pattern of the graph to change.

29. Silver is another economically important metal. Using the information from the Internet or a library, make a graph of silver prices since 1960 similar to the graph of gold prices on page 5. Does your graph of silver prices follow the same patterns as the graph of gold prices?

QUIZ YOURSELF ANSWERS

1. 2604

2. 16

Lesson

1-2

Relations and Functions

Vocabulary

relation

function

dependent variable

independent variable

input, output

mathematical model

▶ **BIG IDEA**   Functions are the mathematical models of relationships between two variables.

You saw a graph of the yearly closing prices of gold at the beginning of this chapter. The closing prices for 1966 to 1975 are given in the following table.

Year y	Closing Price p (in dollars)
1966	35.4
1967	35.5
1968	43.5
1969	35.4
1970	37.6
1971	43.8
1972	65.2
1973	114.5
1974	195.2
1975	150.8

Mental Math

Rewrite the expression as the power of a single variable, if possible.

a. $x^7 \cdot x^2$

b. $\dfrac{a^7}{a^2}$

c. $d^8 + d^4$

d. $n^{16} \cdot \dfrac{n^3}{n^5}$

Two graphs of these data are shown below.

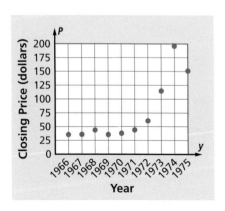

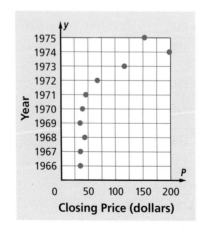

The graph on the left on the previous page shows the ordered pairs (y, p) in the set {(1966, 35.4), (1967, 35.5), (1968, 43.5), (1969, 35.4), (1970, 37.6), (1971, 43.8), (1972, 65.2), (1973, 114.5), (1974, 195.2), (1975, 150.8)}. The graph on the right shows the ordered pairs (p, y) in the set {(35.4, 1966), (35.5, 1967), (43.5, 1968), (35.4, 1969), (37.6, 1970), (43.8, 1971), (65.2, 1972), (114.5, 1973), (195.2, 1974), (150.8, 1975)}.

Each of these sets of ordered pairs and their graphs shows the relationship between the variables y and p. In general, a **relation** is any set of ordered pairs. Any correspondence or pairing between two variables can be written as a set of ordered pairs. In the relation (y, p) graphed on the left, no two ordered pairs have the same first coordinate. Each value of y relates to *one and only one* value of p. A relation having this property describes a *function*.

Definition of Function

A **function** is a set of ordered pairs (x, y) in which each first component x of the pair is paired with exactly one second component y.

In the relation (y, p), the variable p is the **dependent variable** because finding its value depends on knowing the value of y. The variable y is the **independent variable**. Sometimes the independent variable values of a function are called **inputs** and the dependent variable values are called **outputs**.

When the relationship between two quantities is a function, we say that the dependent variable (the second component) **is a function of** the independent variable (the first component). In our gold example, the price of gold at the end of the year is a function of the year.

We also say that the function **maps** each value of the independent variable onto the corresponding value of the dependent variable. Because of this property, every function is a relation, and we can reword the definition of function given earlier.

Definition of Function (reworded)

A **function** is a relation in which no two ordered pairs have the same first component x.

Not every relation is a function. Consider the relation with the ordered pairs (p, y). Both (35.4, 1966) and (35.4, 1969) are in this relation. That is, if the closing price of gold is 35.4, the year could be either 1966 or 1969. Because of this ambiguity, y is not a function of p.

The table and list of ordered pairs in the situation about gold prices are examples of *mathematical models* of the yearly gold prices. A **mathematical model** for a real situation is a description of that situation using the language and concepts of mathematics. Often the situation has to be simplified for the mathematical model. In the case of gold, since the price of gold is always changing, we used the closing price of the end of the year. Sometimes models are created to describe properties of mathematical objects.

Example 1

In a right triangle, the sum of the measures of the two acute angles is 90 degrees. Suppose x is the degree measure of one acute angle, and y is the degree measure of the other, as shown in the drawing at the right.

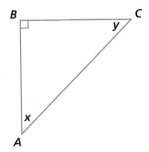

You can write an algebraic sentence to describe this situation.

$$x + y = 90$$

Now you can subtract x from both sides of this equation to get y alone on one side.

$$y = 90 - x$$

In this question, which of x or y is the dependent variable and which is the independent variable? Explain your answer.

Solution In this formula, y is the dependent variable because finding its value depends on knowing the value of x. This means that the independent variable is x.

A function can be described in words, by a formula, or in a graph or table.

Example 2

The percentage P of the adult U.S. population aged 25 and over who had earned a 4-year college degree by the year Y is given in this table.

Year (Y)	1986	1988	1990	1992	1994	1996	1998	2000	2002	2004	2006
Percent (P)	19	20	21	21	22	24	24	26	27	28	28

a. Is Y a function of P? Why or why not?

b. Is P a function of Y? Why or why not?

Solution

a. Think: If Y is a function of P, then for each value of P there should be only one value of Y. So Y is not a function of P because $P = 21$ is paired with two different years, $Y = 1990$ and $Y = 1992$.

b. P is a function of Y because each year is paired with only one value of P. That is, there was a unique percentage of graduates each of these years.

🛑 **See Quiz Yourself at the right.**

In Example 2, *P* is a function of *Y*, so *Y* is the independent variable and *P* is the dependent variable.

▶ **QUIZ YOURSELF**

Are there other ordered pairs that show *Y* is not a function of *P*? If so, what are they?

Questions

COVERING THE IDEAS

1. In your own words, what is a function?

2. a. What is the difference between a function and a relation?
   b. Give an example of a relation that is not a function.

3. Give the definition and an example of a mathematical model.

4. Consider the table below.

s	−10	−5	0	5	10
t	100	25	0	25	100

   a. Is *t* a function of *s*? Explain your answer.
   b. Is *s* a function of *t*? Explain your answer.

5. Recall the formula for the circumference of a circle, $C = 2\pi r$. Is *C* a function of *r*? Justify your answer.

In 6–8, a relation is graphed. Is the relation a function? How can you tell?

6.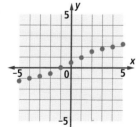

7.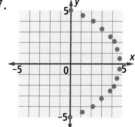

8.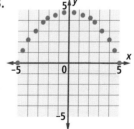

APPLYING THE MATHEMATICS

9. Let D be the degree measure of each interior vertex angle in a regular polygon with n sides. The formula $D = \frac{180(n-2)}{n}$ gives D as a function of n.

   a. Identify the dependent variable. Explain your answer.

   b. Find the number of degrees in one of the interior angles of a regular hexagon.

   c. Draw a regular hexagon. Check that your answer to Part b is correct either by using a protractor, or by giving a logical argument based on the Triangle-Sum Theorem, which states that the sum of the measures of the angles of a triangle is $180°$.

10. The table below shows data about some of the costs of owning a vehicle. $Y =$ year, $T =$ the average cost (cents) per mile of owning and operating the vehicle, and $G =$ the average cost (cents) per mile of gasoline for the vehicle. (The data assume an average of 15,000 vehicle-miles per year.)

   a. Is T a function of Y? Explain your answer.

   b. Is Y a function of G? Again, explain your answer.

$Y =$ Year	$T =$ Average Cost per Mile to Own and Operate (in cents)	$G =$ Average Cost per Mile for Gasoline (in cents)
1985	23.2	5.6
1990	33.0	5.4
1991	37.3	6.6
1992	38.8	5.9
1993	38.7	5.9
1994	39.4	5.6
1995	41.2	5.8
1996	42.6	5.6
1997	44.8	6.6
1998	46.1	6.2
1999	47.0	5.6
2000	49.1	6.9
2001	51.0	7.9
2002	50.2	5.9
2003	51.7	7.2
2004	56.2	6.5
2005	52.2	9.5
2006	52.2	8.9

Source: American Automobile Association

In **11** and **12**, find the value of the dependent variable for the given value of the independent variable.

11. $p = 3x^2, x = -5$

12. $d = -6b^2 + 2b - 8, b = 5.6$

13. Recall from geometry that the *unit circle* is the circle with center $(0, 0)$ and radius 1. A graph of the unit circle is given at the right. Is the unit circle a graph of a function? Explain your answer.

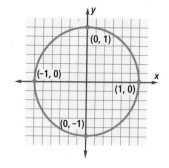

REVIEW

A lesson reference following a review question indicates a place where the idea of the question is discussed.

14. Multiple Choice Which of the following equations are formulas? There may be more than one correct answer. **(Lesson 1-1)**

A $A = \pi r^2$

B $Ax + By = C$

C $E = mc^2$

D $(x - a)(x + a) = x^2 - a^2$

In **15–20**, state the answer in terms of the variables given. **(Previous Course)**

15. Farid bought m blank CDs for a total of b dollars. What was his cost per CD?

16. Butter normally costs n dollars per pound. If butter is on sale for s dollars per pound, how much will you save?

17. How many different outfits can be made from K skirts and L shirts?

18. A bank pays you I interest on P dollars of savings. How much do you have altogether?

19. If a recipe for m people requires g eggs, how many eggs will a recipe for n people require?

20. If a website has h hits now and is getting m hits an hour, how many hits will it have in all, t hours from now?

EXPLORATION

21. The word *function* has several meanings beside the definition given in this lesson. Discuss the similarities and differences between the mathematical definition of function and its nonmathematical definitions.

In 2007, the average per capita egg consumption in the United States was 253.8 eggs.

Lesson
1-3
Function Notations

Vocabulary

$f(x)$ notation

argument of a function

value of a function

mapping notation

▶ **BIG IDEA**   A function f may be written as a set of ordered pairs $(x, f(x))$, described in $f(x)$ notation by an equation $y = f(x)$, or defined by the mapping $f: x \rightarrow y$.

Consider these three equations:

$$y = 90 - x,$$
$$y = 1.8x + 32,$$
$$y = \sqrt{25 - x^2}.$$

The first equation relates the degree measures x and y of the two acute angles in a right triangle. The second equation converts a temperature x in degrees Celsius into a temperature y in degrees Fahrenheit. The third equation gives the length y of one leg of a right triangle with another leg of length x and hypotenuse of length 5. Each formula is easy to use, but when they are all together on the page, all the x's and y's can be confusing. To avoid this confusion, you can use **$f(x)$ notation.** If A names the angle function, T names the temperature conversion function, and L names the formula for leg length, you can write:

$$A(x) = 90 - x \qquad A(x) \text{ is read "}A \text{ of } x.\text{"}$$
$$T(x) = 1.8x + 32 \qquad T(x) \text{ is read "}T \text{ of } x.\text{"}$$
$$L(x) = \sqrt{25 - x^2} \qquad L(x) \text{ is read "}L \text{ of } x.\text{"}$$

Now each function can be simply referred to by its name: A, T, or L. Any letter or string of letters (such as ABS or $SQRT$) or letters and numbers (such as $T1$) can name a function.

Mental Math

a. How much would 15 cans of tuna cost if each can costs 59 cents?

b. How much would 15 frozen dinners cost if each dinner costs $5.90?

c. How much would 1.5 pounds of mixed nuts cost at a price of $5.90 per pound?

d. How much would 4.5 pounds of mixed nuts cost at a price of $5.90 per pound?

$f(x)$ notation

The symbol $f(x)$ is read "f of x". It does *not* mean f times x. This notation is attributed to Leonhard Euler (pronounced "oiler"), an extraordinary Swiss mathematician. In 1770, Euler wrote what is considered the most influential algebra book of all time, *Vollständige Anleitung zur Algebra (Complete Introduction to Algebra).* In honor of him, $f(x)$ notation is also called "Euler's notation."

Leonhard Euler (1707–1783)

You can combine $f(x)$ notation with a descriptive name for the independent variable. If you replace x with C for degrees Celsius, function T above is described by

$$T(C) = 1.8C + 32.$$

The letter C is the independent variable, or input, and $T(C)$ names the dependent variable, or output. T is the name of the function which multiplies the input C by 1.8 and adds 32.

You may also think of the function as

$$T(\ ) = 1.8(\ ) + 32.$$

The same variable, number, or expression is placed in the ( ) wherever the ( ) appears. What is written in the parentheses is called the **argument of the function**; the result is called the **value of the function**.

Example 1

If $T(C) = 1.8C + 32$, evaluate $T(20)$.

Solution Substitute 20 for C in the equation $T(C) = 1.8C + 32$.

$T(20) = 1.8(20) + 32$
$T(20) = 36 + 32$
$T(20) = 68$

Check Look at a thermometer with both Celsius and Fahrenheit scales to see that $20°C = 68°F$.

 See Quiz Yourself 1 at the right.

> **QUIZ YOURSELF 1**
>
> In the function A on page 20, calculate $A(32)$.

Computer Algebra Systems (CAS)

In this course, you will often use a *computer algebra system* (CAS). Computer algebra systems have many uses. They allow you to define and find values of functions and expressions, simplify and solve algebraic equations and inequalities, and perform operations with algebraic expressions. Computer algebra systems use $f(x)$ notation to define functions and $f1, f2, f3$, and so on to name the functions to be graphed. Once defined, you can use these names on the home screen to find values of the function such as $f1(17)$ or $f3(\pi)$. The following Activity shows how to define and evaluate functions on a CAS.

Activity

MATERIALS CAS

Let f be the function that gives the length of one leg of a right triangle with a hypotenuse of 5 in terms of x, the length of the other leg.

Step 1 Find an equation for y in terms of x.

Step 2 Put your calculator in **REAL** (not complex) mode. Define the function f on your CAS using the equation you found in Step 1. The entry is made on the command line and the CAS may write the expression in a different form for easier reading. Here is what one calculator shows.

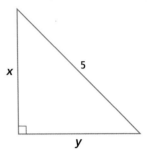

Define $f(x) = \sqrt{5^2 - x^2}$ Done

Step 3 Use your CAS to evaluate $f(4)$, $f(2.95)$, $f(\pi)$, and $f(3 \cdot c)$. One CAS shows the following results.

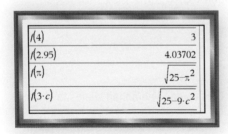

$f(4)$	3
$f(2.95)$	4.03702
$f(\pi)$	$\sqrt{25 - \pi^2}$
$f(3 \cdot c)$	$\sqrt{25 - 9 \cdot c^2}$

Step 4 Use your CAS to evaluate $f(6)$. Write a brief explanation of the result.

 See Quiz Yourself 2 at the right.

▶ QUIZ YOURSELF 2

Use your CAS to find the values of $f(2)$ and $f(-2)$.

Mapping Notation for Functions

In the Activity, the function f is described by $f(x) = \sqrt{5^2 - x^2}$. Another notation, called **mapping notation**, is sometimes used for functions. Mapping notation uses a colon (:) and an arrow ($\rightarrow$). The function from the Activity would be written in mapping notation as follows:

$$f{:}x \rightarrow \sqrt{5^2 - x^2}\ .$$

This is read, "the function f maps x onto $\sqrt{5^2 - x^2}$." For $f(4) = 3$, in mapping notation we write

$$f{:}4 \rightarrow 3.$$

Example 2

Let $f: x \rightarrow \sqrt{5^2 - x^2}$. Evaluate $f(3)$ and $f(r) - f(3)$.

Solution 1 Substitute 3 for x.

$$f(3) = \sqrt{5^2 - 3^2} = 4$$

Substitute r for x. $\sqrt{5^2 - x^2} = \sqrt{5^2 - r^2} = \sqrt{25 - r^2}$

$$f(r) = \sqrt{5^2 - r^2} = \sqrt{25 - r^2}$$

So $f(3) = 4$ and $f(r) - f(3) = \sqrt{25 - r^2} - 4$.

Solution 2 Use a CAS. Define $f(x)$. Then evaluate $f(3)$ and $f(r) - f(3)$, as shown at the right.

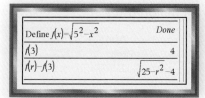

Questions

COVERING THE IDEAS

In 1–3, how is each read?

1. $f(x)$
2. $T: a \rightarrow a^2 + 3a + 4$
3. $A(x) = \frac{1}{2}x(5 - x)$

In 4–8, suppose $f(x) = \sqrt{25 - x^2}$. Evaluate the expression. If necessary, round to the nearest hundredth.

4. $f(0)$
5. $f(\sqrt{2})$
6. $f(-5)$
7. $6 \cdot f\left(-\frac{1}{2}\right)$
8. $f(a) - f(b)$

In 9–11, use the function with equation $T(C) = 1.8C + 32$.

9. Evaluate $T(86)$.
10. Write this function in mapping notation.
11. What would you enter into your CAS command line to define the function T?
12. Is the value of a function the dependent or the independent variable?

In 13–14, let $g: x \rightarrow 12 - 2x$.

13. **Fill in the Blank**

    a. $g: 12 \rightarrow$ __?__
    b. $g: -3 \rightarrow$ __?__

14. Rewrite the function g using $f(x)$ notation.

APPLYING THE MATHEMATICS

In 15 and 16, let $V(r) = \frac{4}{3}\pi r^3$.

15. **a.** What is the argument of the function V?
    b. What is the value of the function V when $r = 6$?
    c. What is the value of the function V when the argument is $3R$?

16. What argument of the function V produces a value of $\frac{4}{3}\pi$?

17. Define $g(x) = 2^x$ and $h(x) = x^2$ on your CAS.

    a. Complete the following table. (When necessary, round to the nearest hundredth.)

X	h(x)	g(x)	h(x) − g(x)
−4	16	0.06	15.94
−2	?	?	?
−0.5	?	?	?
0.5	?	?	?
2.5	?	?	?
3	?	?	?
3.5	?	?	?
4.5	?	?	?
10	100	?	?

    b. Is $h(x) > g(x)$ for all values of x? Explain based on the results of your table.

    c. The g and h functions in your CAS remain stored in memory until you replace them with new definitions or delete them. Find out how to check the variable and function contents of your CAS memory.

    d. Are g and h stored as functions in your CAS memory? How can you tell?

    e. Delete the h function from your CAS memory. What happens now when you use your CAS to evaluate $h(8)$?

In 18–20, use the table below, in which x is the year and $p(x)$ is the average price (in cents) of one gallon of unleaded gasoline during the year x.

x	1997	1998	1999	2000	2001	2002	2003	2004	2005	2006
$p(x)$	123.4	105.9	116.5	151.0	146.1	135.8	159.1	188.0	229.5	258.9

Source: http://www.economagic.com

18. **a.** Evaluate $p(2006) - p(2005)$.
    b. Explain in words what you have just calculated.
    c. What does the answer to Part a mean about the price of gasoline?

19. For what values of x is $p(x) > 150$?

20. Solve $p(x) = 116.5$ for x.

21. Given that $g: x \rightarrow |-x| + (x - 2)^3$, evaluate.
    a. $g(4)$ **b.** $g(-4)$
    c. $g(2.376)$ **d.** $g(4b)$

REVIEW

22. Is y a function of x? Why or why not? (**Lesson 1-2**)

x	0	1	−1	2	−2
y	0	1	1	8	8

23. Amy has three sisters and two brothers. Suppose she writes out the relation of all pairs (x, y) where x and y are siblings of hers and x is y's brother. Is this relation a function? (**Lesson 1-2**)

24. Berto is saving money for a car. He has $657 in his account. He plans on saving $50 each week. How much will he have (**Lesson 1-1**)
    a. after 11 weeks?   **b.** after n weeks?

25. **a.** Give an equation for the horizontal line through $(-4, 6)$.
    b. Give an equation for the vertical line through $(-4, 6)$.
    c. Write equations for the horizontal line and the vertical line that intersect at (h, k). (**Previous Course**)

EXPLORATION

26. Leonhard Euler contributed to all branches of mathematics. Various theorems, formulas, and functions are named after him. Research Euler's *totient function* and write a paragraph about it.

QUIZ YOURSELF ANSWERS

1. $A(32) = 58$

2. $f(2) \approx 4.58$; $f(-2) \approx 4.58$

Lesson

1-4

Graphs of Functions

Vocabulary

real function

domain of a function

set-builder notation

range of a function

trace

standard window

natural numbers

counting numbers

whole numbers

integers

real numbers

rational numbers

irrational numbers

▶ **BIG IDEA**   Because a *real function* is a set of ordered pairs of real numbers, it can be graphed on the coordinate plane.

A **real function** is a function whose independent and dependent variables stand only for real numbers. You can graph a real function on a rectangular coordinate graph. If $y = f(x)$, then the coordinates of points on the graph have the form (x, y) or $(x, f(x))$.

An Example from Driver's Education

When a driver attempts to stop a car, the distance d the vehicle travels before stopping is a function of the car's speed x and can be modeled by the equation $d = x + \frac{x^2}{20}$. If we give the name C to the function giving the stopping distance, then $C(x) = x + \frac{x^2}{20}$. Several ordered pairs of this function C are shown in the table below.

Speed (mph) = x	0	10	20	30	40	50	60	70
Car's Stopping Distance (ft) = C(x)	0	15	40	75	120	175	240	315

On the following page are two graphs relating the data above. The graph at the left is a graph of the function C for values of x from 0 to 70. All the points on it are of the form $(x, C(x))$. To find the value of $C(60)$ from the graph, start at 60 on the x-axis. Read up to the curve and then across to find the value on the y-axis. So, $C(60) = 240$. Notice that this agrees with the value in the table above.

The graph at the right includes the graph of a second function S. S maps the speed x of an SUV (sport utility vehicle) onto its stopping distance $S(x)$ at that speed. Using function notation helps distinguish the y-coordinates of the graphs when more than one function is displayed.

Mental Math

Refer to △*ABC* below. Estimate to the nearest integer.

a. *BC*

b. the area of △*ABC*

c. the perimeter of △*ABC*

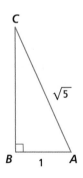

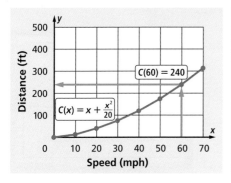

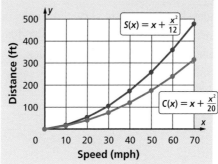

Skid marks on this road were caused by heavy braking.

Domain and Range of a Function

For the functions C and S on the previous page, the situation and table determine what values of the independent and dependent variables to include in the graph. The set of allowable values for the independent variable is called the **domain of the function**. Allowable speeds for both C and S are from 0 to 70 mph, so the domain for each is $\{x \mid 0 \leq x \leq 70\}$. This notation, called **set-builder notation**, is read "the set of all x such that 0 is less than or equal to x and x is less than or equal to 70." Some people write $\{x : 0 \leq x \leq 70\}$ for set-builder notation.

The **range of a function** is the set of values for the dependent variable that can result from all possible substitutions for the independent variable. According to the table, when x has values from 0 to 70, the values of $C(x)$ range from 0 to 315. So the range of C is $\{y \mid 0 \leq y \leq 315\}$. For these values of x, the graph shows that values of $S(x)$ range from 0 to about 480. So the range of S is $\{y \mid 0 \leq y \leq 480\}$.

When graphing a function on a graphing utility, you need to consider its domain and range in order to get an appropriate picture. The *window* of the graph must be large enough for you to see what you want to see, but not so large that the graph is too tiny. In this Activity you will learn how to set your grapher's window and use its TRACE feature to further examine the functions C and S.

Activity

MATERIALS CAS or graphing calculator

Step 1 Locate your grapher's place for entering equations to be graphed.

Step 2 Enter $x + \frac{x^2}{20}$ for the first function and $x + \frac{x^2}{12}$ for the second function. A possible display is shown at the right.

(continued on next page)

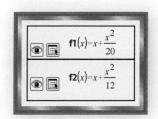

Step 3 Set the viewing window large enough to show the graphs of both C and S. Because the range of S is larger, use its domain and range for the window. Your graphs should now look similar to the ones shown at the right.

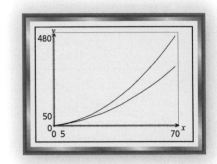

Step 4 The **TRACE** feature on your grapher allows you to estimate input and output values, much as you would on a graph drawn by hand. Trace along the graphs within the window, as shown. The numbers on the screen are the approximate x- and y-coordinates of the point on the graph where the **TRACE** marker is currently placed. Which ordered pair on the graph of function S appears at the right edge of the window?

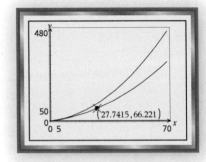

Step 5 The upper limit of allowable speeds in the table for C is 70 because that is the maximum speed limit in many parts of the U.S. However, some states have limits up to 75 mph. Continue tracing the C and S functions off the right side of the viewing window to estimate $C(75)$ and $S(75)$. Then define a new window which allows you to see allowable values of $C(x)$ and $S(x)$ over the domain $\{x \mid 0 \le x \le 75\}$.

Step 6 Your grapher has a default window, often called the **STANDARD WINDOW**, that shows all four quadrants at a reasonably close scale. On many calculators, the default window is where $\{x \mid -10 \le x \le 10\}$, $\{y \mid -10 \le y \le 10\}$; and the x-scale and the y-scale are both 1. This means that both the horizontal (x) axis and the vertical (y) axis are viewable from -10 to 10, with 1 unit between tick marks. Find and use the standard window to graph C and S, as shown at the right.

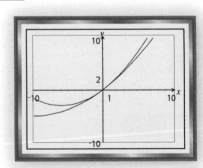

Step 7 Do all the ordered pairs displayed in this window of the graphs make sense in the situations of car and SUV speeds? Explain your answer.

Step 8 Trace along the graphs of the *C* and *S* functions beyond the left edge of the standard window. You will see that the graphs of these functions are in Quadrants I, II, and III. Set the viewing window as specified below.

$$-40 \leq x \leq 40 \ \ x \text{ scale} = 5$$
$$-40 \leq y \leq 40 \ \ y \text{ scale} = 5$$

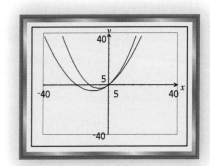

Step 9 Trace along the graphs of *C* and *S* again to estimate the minimum value of each function.

Negative speeds and negative distances are not reasonable measures. However, when you examine the graphs of the *C* and *S* functions without regard to stopping distances, you see that the functions are defined for negative values of *x* and *y*. When examining functions outside of real situations, it is common practice to identify the largest possible domain and its associated range.

Since there exist minimum values for $C(x)$ and $S(x)$, the biggest range for *C* is about $\{y|\ y \geq -5\}$ and for *S* it is about $\{y|\ y \geq -3\}$. But even if you imagine the graphs extending forever to the right and left, there do not seem to be minimum or maximum values for *x*. So *x* can be any real number, and we say that the domain of both *C* and *S* is the set of all *real numbers*, or the set of all reals.

Some sets of numbers that you are probably familiar with are frequently used as domains and often appear as ranges.

The set of **natural numbers** or **counting numbers** {1, 2, 3, 4, 5, ...}
The set of **whole numbers** {0, 1, 2, 3, 4, 5, ...}
The set of **integers** {..., –3, –2, –1, 0, 1, 2, 3, ...}
The set of **real numbers** (those numbers that can be represented by decimals)

> Samples: 0, 1, –7, 35 million, 2.34, π, $\sqrt{5}$

The set of **rational numbers** (those numbers that can be represented as ratios of the form $\frac{a}{b}$, where *a* and *b* are integers and $b \neq 0$)

> Samples: 0, 1, –7, $\frac{2}{3}$, $1\frac{9}{11}$, $-\frac{34}{10}$, 0.0004, $-9.6\overline{18}$, $\sqrt{16}$

The set of **irrational numbers** (real numbers that are not rational)

> Samples: π, $\sqrt{5}$, $\sqrt{10}$, e

The diagram at the right shows how each set of numbers relates to the others. A number in any set is also in any set in the path above it. For example, an integer is also a rational number and a real number, but not all integers are whole numbers or natural numbers.

🛑 **See Quiz Yourself at the right.**

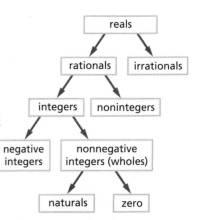

> ▶ **QUIZ YOURSELF**
>
> Give an example of a negative rational number that is not an integer.

Questions

COVERING THE IDEAS

In 1–3, refer to the graphs of functions C and S at the beginning of this lesson.

1. Give their common domain.

2. Give the range of S.

3. Explain how to estimate C(45) from the graph.

In 4–6, refer to the Activity.

4. Use your grapher's TRACE feature to estimate C(55).

5. Assume you are in a state with a maximum speed limit of 55 mph. Define a window which shows all allowable values of C(x) and S(x) over the domain $\{x \mid 0 \leq x \leq 55\}$.

6. What is the standard window on your grapher?

In 7 and 8, the graph of a function is given. From the graph, determine the function's domain and range. Use set-builder notation.

7.
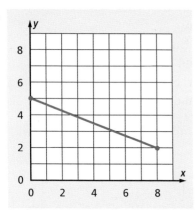

8.
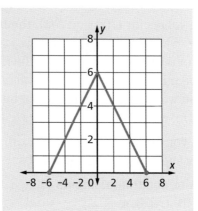

In 9–11, identify each number as an integer, a rational number, an irrational number, or a real number. A number may belong to more than one set.

9. -97

10. $\frac{23}{47}$

11. $-\sqrt{18}$

12. Name a rational number that is not an integer.

13. Name a real number that is an integer and not positive.

14. Name a real number that is an irrational number between 0 and 1.

APPLYING THE MATHEMATICS

15. The graph at the right shows the relationship between the age of a grapefruit tree and the diameter of its trunk. Let $A(d)$ be the age of a grapefruit tree whose diameter is d inches.

    a. Estimate the range of A.

    b. Estimate $A(10)$ from the graph.

    c. What diameters correspond to a tree aged between 10 and 20 years?

    d. Sketch a graph of this relation with age on the x-axis and diameter on the y-axis. Is this the graph of a function?

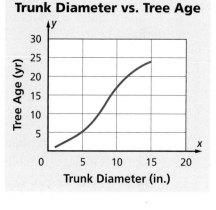

Trunk Diameter vs. Tree Age

16. The volume V of a sphere with radius of length r is given by the formula $V = \frac{4}{3}\pi r^3$.

    a. State the domain for r. (*Hint*: Are there any numbers that do not make sense for the value of the radius?)

    b. Find the volume of a beach ball with a 15 cm radius. Round your answer to the nearest tenth.

In 17 and 18, graph the function in a standard window on your grapher, then answer the questions.

17. $f(x) = 5$

    a. What is $f(15)$?

    b. State the domain and range of f.

18. $g(x) = -2x^2 - 7$

    a. What is the maximum value of g?

    b. State the domain and range of g.

19. On your grapher or CAS, graph the function $f(x) = \frac{4}{1+x}$.

    a. Use your grapher to estimate $f(0)$. Check your answer using algebra.

    b. For what value of x is $f(x) = \frac{1}{2}$? Use your grapher and check your answer using algebra.

20. Refer to Question 17 in Lesson 1-3, where you were asked to compare functions with equations $g(x) = 2^x$ and $h(x) = x^2$.

a. Graph g and h together on your grapher in the window defined below.

xmin	xmax	xscale	ymin	ymax	yscale
−5	5	1	−2	15	1

b. In this window, sometimes $h(x) > g(x)$ and sometimes $g(x) > h(x)$. Your grapher has a ZOOM feature that allows you to get a magnified view of certain regions of your graph. Use ZOOM and TRACE to approximate values of x for which $h(x) > g(x)$.

21. A formula for the height h of an object in free fall t seconds after its release from a height of 25 feet is $h = -16t^2 + 25$.

a. Graph this equation on your grapher. Sketch the graph. (Most graphers use x for the independent variable, and $y1$, $y2$, etc. for the dependent variables, so you must substitute x for t and y for h when entering this equation.)

b. Use the TRACE feature on your grapher to estimate the height of the object after 1.1 seconds.

In an egg drop, contestants build contraptions to protect an egg to be dropped from a certain height. The winning contraption will protect the egg from breaking.

REVIEW

22. Suppose that for all x, $d(x) = 2x^3 - x + 3$. Evaluate. **(Lesson 1-3)**

a. $d(2)$ b. $d(-2)$ c. $d(\pi r)$

23. Use a CAS to find $S(100)$ when $S(n) = \dfrac{n(n + 1)}{2}$. **(Lesson 1-3)**

24. Let n = the number of sides of a polygon and $f(n)$ = the number of its diagonals. The figures below show that $f(4) = 2$ and $f(5) = 5$. Find $f(3)$ and $f(6)$. **(Lesson 1-3)**

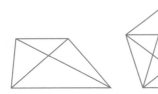

In 25 and 26, solve and check. (Previous Course)

25. $\frac{1}{3}r - 27 = 60$ 26. $2x = \frac{3}{2}x + 4$

EXPLORATION

27. Find a coordinate graph in a newspaper or magazine. Does it represent a function? Why or why not? If it is a function, identify the domain and range of the function.

QUIZ YOURSELF ANSWER

Answers vary. Sample: $-\frac{4}{5}$

Lesson 1-5

Using Graphs and Tables of Functions

▶ **BIG IDEA** The graph of a function can tell a story and from it you can learn great deal about the function.

When a graph is plotted in the (x, y) coordinate plane, x is the independent variable, and y is the dependent variable. Time is often an independent variable, as in Example 1.

GUIDED

Example 1

Frank Furter rode his bicycle to Chuck Roast's house to plan a back-to-school picnic. The boys then rode a short distance to the park and then back to Frank's house. The graph below models Frank's trip where t is time in minutes and d is Frank's distance from home in miles.

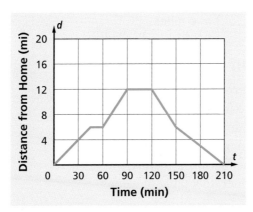

Use the graph of Frank's bicycle trip to answer the following questions.

a. Which variable is a function of the other variable? Express this relationship in $f(x)$ notation. (Give the function the name f.)

b. State the domain and range of the function in this situation.

c. Estimate Frank's distance from home after 60 minutes, and write your answer using $f(x)$ notation.

d. Estimate how much time it took Frank to first reach 12 miles from home. Write your answer using $f(x)$ notation.

e. What is the significance of the left horizontal line segment? Estimate its endpoints, and describe it using $f(x)$ notation.

(continued on next page)

(continued on next page)

Mental Math

Suppose Berta drove 280 miles in four hours.

a. What was her average speed?

b. Berta drove for another 30 minutes at the same speed. How far has she traveled in all?

c. Berta stopped for a half-hour for lunch, then drove another 210 miles at the same speed. How long did she drive after lunch?

d. If Berta started her trip at 9:30 A.M., at what time did she arrive at her destination?

Solution

a. The time t is on the horizontal axis, and the distance d is on the vertical axis. So, the distance from home d is a function of the length of time t since Frank left home.

In $f(x)$ notation, __?__ $= f($ __?__ $)$.

b. The domain is the set of possible values of the independent variable.

The domain is $\{t|$ __?__ $\leq t \leq$ __?__ $\}$ minutes.

The range is the set of values attained by the dependent variable.

The range is $\{d|$ __?__ $\leq d \leq$ __?__ $\}$ miles.

c. When $t = 60, d \approx$ __?__. The distance is about __?__ miles.

In $f(x)$ notation, $f($ __?__ $) \approx$ __?__ miles.

d. The leftmost point on the graph at $d = 12$ is $(90, 12)$. It took about __?__ minutes for Frank to ride 12 miles away from home.

In $f(x)$ notation, $f($ __?__ $) = 12$ miles.

e. From about $t =$ __?__ to about $t =$ __?__ minutes, Frank's distance remained constant at about 6 miles. This shows that Frank stayed at Chuck's house for about 15 minutes. One explanation is that it might take time for Chuck to get ready to leave for the park.

In $f(x)$ notation, for __?__ $\leq t \leq$ __?__, $f($ __?__ $) = 6$ miles.

Creating and Reading Tables on a Grapher

Sometimes it is helpful to represent the ordered pairs of a function in a table. In previous courses, you may have found values for tables by substituting values into a formula. For example, if $y = 15 - 4x$ and $x = -2$, then $y = 15 - 4(-2) = 23$, and the ordered pair is $(-2, 23)$. In the previous lesson, you used your grapher's TRACE feature to find ordered pairs. Your grapher can also generate a table of values automatically.

Activity

MATERIALS CAS or graphing calculator

Suppose a puddle of water is evaporating and its depth $D(t)$ in inches after t days is given by the formula $D(t) = 6 - \frac{1}{16}t^2$.

Step 1 Enter the equation into your grapher.

Step 2 Set up a table to start at $x = -5$ and to generate values in increments of 1. If your grapher needs an end value, use $x = 5$.

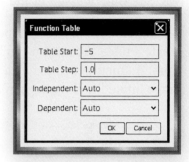

Step 3 Your table should be similar to the one to the right. If necessary, you can see more values in the table by scrolling with the up and down arrows on your grapher.

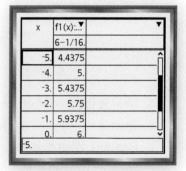

Step 4 The values in the x column represent which variable in this problem?

The values in the $f(x)$ column represent which variable in this problem?

Do all of the ordered pairs in the table realistically model the puddle of water situation? Why or why not?

Step 5 Determine the water depth after 6 days. Write your answer in $D(t)$ notation.

Step 6 When is the water depth 5.75 inches? Write your answer in $D(t)$ notation.

Recall that a function *can* have two or more ordered pairs with the same second coordinate. For instance, in the table from the Activity, when $f(x) = 5$, then $x = -4$ or 4. However, only $x = 4$ models a real number of days.

Unlike graphs in which smooth or *continuous* curves show the ordered pairs at *all* values of the domain and range, tables give a *discrete* view of a function that shows only some of the ordered pairs. For example, your table in the Activity only shows the ordered pairs for integer values of *x*. To find more pairs, you can change the table increment in the table setup.

 See Quiz Yourself at the right.

A function does not need to be defined by a formula in order to create a table. A function can be defined simply as a set of ordered pairs. A function defined by a set of ordered pairs has a domain and range limited to the values in the ordered pairs.

▶ **QUIZ YOURSELF**

Find two other ordered pairs for the puddle function *D* by changing the table increment to 0.5.

Example 2

In 2000, the Centers for Disease Control and Prevention (CDC) found the mean height *h* in centimeters for girls in the U.S. at various ages *a* where $h = f(a)$. Some of these data are given in the table below.

a	2	4	6	8	10
h	85	101	115	127	138

a. What is the domain of the function defined by this table?

b. What is the range of this function?

c. Find $f(6)$. What does this represent?

d. For what value of *a* does $f(a) = 138$?

Solution

a. The domain is the set of values the independent variable *a* can have. The domain is {2, 4, 6, 8, 10}.

b. The range is the set of possible values of the dependent variable *h*. The range is {85, 101, 115, 127, 138}.

c. $f(6) = 115$. At age 6, girls in the U.S. have an average height of 115 cm.

d. $f(a) = 138$ when $a = 10$.

Questions

COVERING THE IDEAS

In 1–3, refer to the graph in Example 1.

1. How far from home was Frank when he ended his trip? How does the graph show this?

2. Estimate $f(75)$ and explain what it means.

3. What is the maximum value of $f(t)$?

In 4–6, refer to the following table of values for the function with equation $s = f(t)$.

s	−3	−2	0	4	9
t	9	4	0	16	81

4. What is the domain of this function? What is the range?
5. Find each of the following:
    a. $f(16)$ b. $f(4)$
6. For what value of t does $f(t) = 9$?

In 7 and 8, refer to the Activity.

7. Change the table increment to 0.2.
    a. What is the depth of the water after 3.6 days?
    b. When does the depth reach 4.04 inches?
8. Can you find an increment to show all the ordered pairs of D? Why or why not?
9. Use your grapher to make a table of values for $f(x) = 9 - x^2$ that begins at $x = 1$ and has increments of 0.5. List the first five ordered pairs in the table.

APPLYING THE MATHEMATICS

10. The graph below shows the distances $M(t)$ and $P(t)$ that Maria and Pia traveled during a 40-km bike race.

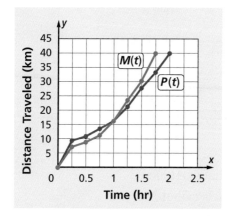

    a. Estimate $P(0.5) - M(0.5)$. What does this quantity represent?
    b. Find a value of t such that $M(t) = P(t)$. What is happening at this moment?
    c. Do M and P have the same domain? Why or why not?

In 11–13, a set of ordered pairs is given in a table. Is the set a function? If it is, give its domain and range. If it is not, explain why not.

11.

x	y
-3	-0.1
-2	-0.06
-1	-0.03
0	0
1	0.03
2	0.06
3	0.1

12.

x	y
0	-6
1	-4
2	-2
3	0
2	2
1	4
0	6

13.

x	y
-6	0
-4	1
-2	2
0	3
2	2
4	1
6	0

In 14–16, refer to the table and graph below. They give information about the number of farms and the average size (in acres) of these farms across the United States from 1995 to 2005.

Year	1995	1996	1997	1998	1999	2000	2001	2002	2003	2004	2005
Number of Farms (thousands)	2,200	2,190	2,190	2,190	2,190	2,170	2,160	2,160	2,130	2,110	2,100

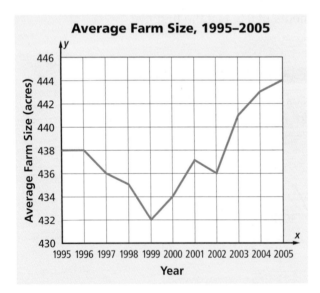

Average Farm Size, 1995–2005

14. If the average size function is called A, what is $A(2000)$?

15. If the function giving the number of farms is called N, find $N(2002)$.

16. a. Calculate $A(2004) \cdot N(2004) - A(1996) \cdot N(1996)$.

    b. Explain in words what the computation in Part a represents.

17. For the equation $s = 4 - r^2$:

a. Find the values of s which correspond to the given values of r.

r	–3	–2	–1	0	1	2	3
s	?	?	?	?	?	?	?

b. Plot (r, s) on a coordinate graph for each of the values in the table. Is s a function of r? Explain.

c. Plot (s, r) on a coordinate graph for each of the values in the table. Is r a function of s? Explain.

REVIEW

In 18 and 19, use this information. In the United States many teachers grade tests on a scale of 1–100. In France it is common to grade tests on a scale of 1–20 (with 20 being the highest). Suppose G is a function that converts a grade x in the American system to a grade $G(x)$ in the French system. (Lesson 1-4)

In 2007, the high school graduation rate in France was 85%.

18. Write an expression for $G(x)$.

19. Give a reasonable domain and range for the function G.

20. Write out the elements of the set $\{S | S$ is a state in the United States, and S does not share a border with any other state$\}$. (Lesson 1-4)

21. Consider all ordered pairs of the form $(x, 2x - \pi x^2)$. Do these ordered pairs describe a function? If they do, describe this function using mapping notation. If they do not, explain why not. (Lessons 1-3, 1-2)

22. The formula for the surface area of a sphere with radius R is $A = 4\pi R^2$.

a. Write the function mapping R onto A in $f(x)$ notation.

b. Which is the dependent variable and which is the independent variable? (Lessons 1-3, 1-2)

23. The radius of the planet Neptune is approximately 24,750 kilometers. Estimate its surface area to the nearest million square kilometers. (Previous Course)

EXPLORATION

24. Think about your trips to and from school yesterday. Draw a graph similar to the graph in Example 1 to model your trip. What is your independent variable? What is your dependent variable? Are there any horizontal lines on your graph? If so, what do they represent? If not, why not?

QUIZ YOURSELF ANSWER

Answers vary. Samples: (1.5, 5.8594); (2.5, 5.6094).

Lesson

1-6

Solving Equations

▶ **BIG IDEA** By solving $y = f(x)$ for x you can find the value of x that leads to a particular value of y.

Suppose you are studying a function f. Two key questions often arise.

1. Given a value of x, what is the value of $f(x)$?
2. Given a value of $f(x)$, what is the value of x?

In the previous lesson, you learned how to answer these questions using graphs and tables. If the function is described with an equation, then answering the first question usually involves evaluating an expression. Answering the second question usually involves solving an equation. Example 1 illustrates these two situations. The expressions and equations are easy enough that you should be able to work with them by hand, but we also show how to use a CAS on the same questions since a CAS is useful for more complicated problems.

Mental Math

a. How many outfits can you make with 4 t-shirts and 5 pairs of pants, if one of the pairs of pants cannot be worn with two of the shirts?

b. How many outfits can you make with 4 blouses, 5 skirts, and 6 pairs of shoes?

c. How many outfits can you make with 4 shirts, x pairs of pants, y pairs of shoes, and 7 ties?

Example 1

In the board game Monopoly®, Boardwalk is the most expensive property. The one-time cost of the property plus a hotel is $1,400, while the rent earned from another player landing on it is $2,000. If $n =$ the number of players landing on Boardwalk, then $f(n) = 2000n - 1400$ is the owner's total profit after n players have landed on the property.

a. What will be the owner's profit after 4 people land on Boardwalk?

b. Is it possible for the owner to make exactly $10,000 in profit?

Solution 1

a. The profit is given by $f(4)$.
   $f(n) = 2000n - 1400$
   So, $f(4) = 2000 \cdot 4 - 1400 = 6600$.
   The owner's profit will be $6,600.

b. The question asks you to find n when $f(n) = 10,000$. Substitute for $f(n)$ and solve the equation.

$$2000n - 1400 = 10,000$$
$$2000n - 1400 + 1400 = 10,000 + 1400 \quad \text{Add 1400 to both sides.}$$
$$2000n = 11,400$$
$$n = 5.7 \qquad \text{Divide both sides by 2000.}$$

The property owner needs 5.7 people to land on Boardwalk. So it is not possible to earn exactly $10,000 in profit.

Solution 2 Use a CAS.

a.

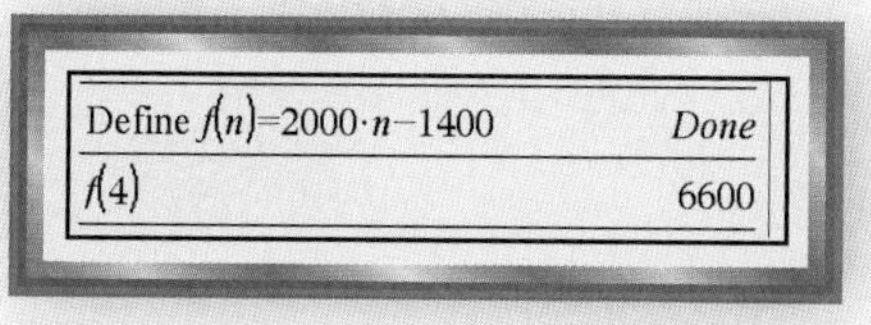

Define $f(n) = 2000 \cdot n - 1400$	*Done*
$f(4)$	6600

b.

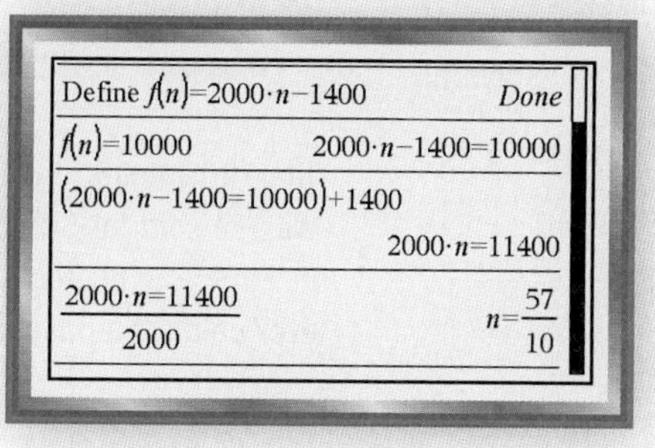

Define $f(n) = 2000 \cdot n - 1400$	*Done*
$f(n) = 10000$	$2000 \cdot n - 1400 = 10000$
$(2000 \cdot n - 1400 = 10000) + 1400$	
	$2000 \cdot n = 11400$
$\dfrac{2000 \cdot n = 11400}{2000}$	$n = \dfrac{57}{10}$

$\frac{57}{10}$ is not an integer, so it is not possible for an owner to earn exactly $10,000 in profit.

STOP **See Quiz Yourself 1 at the right.**

Recall that when you subtract (or add the opposite of) a number from each side of an equation, you are applying the Addition Property of Equality to find an *equivalent,* but simpler, equation. There are many other properties that are frequently applied when solving equations. The properties of real numbers used in algebra are listed on page S1 at the back of the book. Skim this page now, and refer to it whenever you need to check the meaning or name of a particular property. The next three examples explore how one of these properties, the Distributive Property, can be used to solve equations.

▶ **QUIZ YOURSELF 1**

Can the owner make exactly $12,000 in profit? Explain.

Distributive Property

For all real numbers a, b, and c,
$$c(a + b) = ca + cb.$$

Using the Distributive Property

In Example 2, the Distributive Property is used to solve an equation containing parentheses.

Example 2

Darius had $500 to spend at the baseball card convention. He decided to spend all his money buying the rookie cards of his two favorite players. At the end of the season, Darius found that the first card increased in value by 21%, while the second card decreased in value by 7%. He spent d dollars on the first card, so he spent $(500 - d)$ dollars on the second card.

So,

$$T(d) = 1.21d + 0.93(500 - d)$$

gives the total value of the two cards. If Darius's cards were worth $550 at the end of the season, how much did he spend on each card?

Solution 1 You are given that $T(d) = 550$, and asked to find d. Solve an equation.

$$1.21d + 0.93(500 - d) = 550$$

$1.21d + \underline{\ ?\ } - \underline{\ ?\ } = 550$	Distribute the 0.93.
$\underline{\ ?\ } + \underline{\ ?\ } = 550$	Add like terms.
$\underline{\ ?\ } = \underline{\ ?\ }$	Subtract 465 from each side.
$d \approx \underline{\ ?\ }$	Divide each side by 0.28.
$500 - \underline{\ ?\ } = \underline{\ ?\ }$	Find the cost of the other card.

For the cards to be worth __?__ at the end of the season, Darius needed to spend about __?__ on the first card and about __?__ on the second card.

Solution 2 Enter the equation in a CAS. The CAS automatically simplifies the equation to get one of the steps in Solution 1. You can then solve this equation.

The 1909 T206 Honus Wagner sold for $2.8 million in 2007, the highest price ever paid for a baseball card.

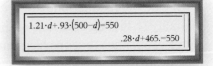

$1.21 \cdot d + .93 \cdot (500 - d) = 550$
$.28 \cdot d + 465. = 550$

Clearing Fractions in Equations

Often, an equation modeling a real situation will involve fractions. If you want to clear an equation of fractions for simplicity, multiply each side of the equation by a common multiple of the denominators. If there is more than one term on either side of the equation, you will then need to apply the Distributive Property. Example 3 illustrates this procedure.

Example 3

In game 5 of the 2006 NBA Finals between Miami and Dallas, Dirk Nowitzki scored $\frac{1}{5}$ of Dallas's points and Josh Howard scored $\frac{1}{4}$ of Dallas's points. The rest of the Dallas Mavericks scored a total of 55 points. How many points did Dallas score?

Solution Write an equation to model the situation, then solve it.

If P = the total number of points scored by Dallas in the game, then Dirk scored $\frac{1}{5}$ of P, Josh scored $\frac{1}{4}$ of P, and the rest of the team scored 55 points. So, $P = \frac{1}{5}P + \frac{1}{4}P + 55$.

To clear the fractions in this equation, multiply both sides by a common multiple of the denominators 4 and 5. One common multiple is 20.

$20P = 20\left(\frac{1}{5}P + \frac{1}{4}P + 55\right)$	
$20P = 4P + 5P + 1100$	Distribute the 20.
$20P = 9P + 1100$	Combine like terms.
$11P = 1100$	Subtract $9P$ from each side.
$P = 100$	Divide each side by 11.

So, the Dallas Mavericks scored 100 points, with Dirk scoring $\frac{1}{5}(100) = 20$ points and Josh scoring $\frac{1}{4}(100) = 25$ points.

Dirk Nowitzki of the Dallas Mavericks

Check Use the SUCH THAT command on your CAS to verify that the computed solution is true. On the CAS pictured here this function is represented by a vertical bar |.

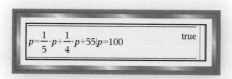

$$p = \frac{1}{5} \cdot p + \frac{1}{4} \cdot p + 55 \,|\, p = 100 \qquad \text{true}$$

In words, this display reads "p equals one-fifth p plus one-fourth p plus 55 such that p equals 100 is true", so it checks. Sometimes "such that" is read "with."

Opposite of a Sum Theorem

From the Distributive Property and the fact that $-1 \cdot x = -x$ for all x, you can deduce that

$$-(a + b) = -1 \cdot (a + b) = -1 \cdot a + -1 \cdot b = -a + -b = -a - b.$$

This result is the Opposite of a Sum Theorem.

Opposite of a Sum Theorem

For all real numbers a and b,
$$-(a + b) = -a + -b = -a - b.$$

The next example applies this theorem in solving an equation.

GUIDED

Example 4

Suppose $g(a) = 3a - (7 - 5a)$. For what value of a is $g(a) = 12$?

Solution

$12 = 3a - (7 + \underline{\ ?\ })$	Write the subtraction in parentheses as a sum.
$12 = 3a - \underline{\ ?\ } + \underline{\ ?\ }$	Opposite of a Sum Theorem
$12 = \underline{\ ?\ } - 7$	Add like terms.
$\underline{\ ?\ } = \underline{\ ?\ }$	Add 7 to each side.
$\underline{\ ?\ } = a$	Divide each side by the coefficient of a.

Check To check your answer, calculate the value $g(a)$ using the value you calculated for a. The result should be 12.

 STOP See Quiz Yourself 2 at the right.

> ▶ **QUIZ YOURSELF 2**
>
> True or False?
> $-(-(-2 - 3)) = -2 + 3$

Questions

COVERING THE IDEAS

In 1 and 2, suppose a bathtub contains 60,000 cubic inches of water. If water can be drained from the tub at a rate of 800 cubic inches per second, then $w(t) = 60,000 - 800t$ represents the volume of water left in the tub after draining for t seconds.

1. What volume of water will be in the tub after 18 seconds?

2. After how many seconds will the tub be empty?

In 3 and 4, assume Jamila has a collection of 200 nickels and dimes. If n is the number of nickels in her collection, then $c(n) = 0.05n + 0.10(200 - n)$ represents the face value amount of money her collection is worth.

3. Evaluate $c(122)$. What does the answer mean in the context of this problem?

4. Solve $c(n) = 19.15$. What does the answer mean in the context of this problem?

In 5–7, an equation is given. Solve the equation and check your solution using the SUCH THAT command on a CAS.

5. $4y + 50 = 5y + 42$

6. $2z + 2 = 2 - 6z$

7. $4 = \frac{6}{x}$

In 8 and 9, identify a common multiple of the denominators and solve the equation.

8. $\frac{h}{6} + \frac{h}{10} = 1$

9. $\frac{1}{8}x + \frac{2}{3}x = 5$

10. **Fill in the Blank** According to the Opposite of a Sum Theorem, $-(3 - 9y) = $ ___?___.

In 11 and 12, an equation is given.

a. Solve each equation.

b. Check your work.

11. $3x - (x + 1) = 7$

12. $5n - (9 - 5n) = 9$

APPLYING THE MATHEMATICS

In 13–15, use this information. A bowler's *handicap* is a bonus given to some bowlers in a league. The handicap h is a function of A, the bowler's average score, and is sometimes determined by the formula $h(A) = 0.8(200 - A)$, when $0 < A < 200$.

13. What is the handicap for a bowler whose average is 135?

14. If a bowler has a handicap of 28, what is the bowler's average?

15. What is the domain of the function h?

In 16–19, a. solve the equation by hand, and
b. enter the equation into a CAS and solve the resulting equation.

16. $0.05x + 0.12(50000 - x) = 5{,}995.8$

17. $\frac{1}{3}t + \frac{1}{4}t + 6 = t$

18. $\frac{2}{3}p + \frac{1}{2} = p - 3$

19. $0.05z + 0.1(2z) + 0.25(100 - 3z) = 20$

20. Gloria owns a video rental store. Two-fifths of the store items are comedies, one-eighth are horror films, one-fourth are dramas, and the remaining 270 are video games.

    a. How many items are there in total?

    b. How many comedies are there?

    c. How many horror films are there?

21. Suppose $f(n) = \frac{n - 1}{n + 2}$.

    a. What is $f(18)$?

    b. If $f(n) = \frac{46}{49}$, find n.

In September 1895, the first meeting of the American Bowling Congress met in New York City. The ABC standardized dimensions for bowling balls, pins, and lanes.

Source: www.bowlingmuseum.com

22. **Multiple Choice** Which of the following sentences are equivalent to $3(x - 7) = \frac{5}{2}$? There may be more than one correct answer.

    A $3x - 21 = \frac{5}{2}$ B $3x - 21 = \frac{5}{21}$ C $x - 7 = \frac{5}{6}$

REVIEW

In 23–26, refer to the graph below. In the graph, $x =$ the year, $i(x) =$ the value in millions of dollars of imports into the United States, and $E(x) =$ the value of exports from the United States, also in millions of dollars. **(Lessons 1-3 and 1-4)**

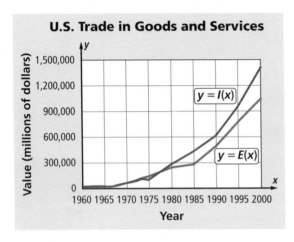

23. Estimate $I(1995)$.

24. In what unit is the dependent variable of function E measured?

25. In what year(s) was $E(x) > \$500,000$ million?

26. A negative *balance of trade* exists when a country's imports are greater than its exports. Write a few sentences comparing the balance of trade for the period 1960–1980 to the balance of trade for the period 1980–2000.

27. A cylindrical soft drink can has a lateral area of about 241 cm². If the radius of the can is 3.2 cm, approximate the can's height to the nearest tenth of a cm. **(Previous Course)**

EXPLORATION

28. Veronica was asked to find the product of two given numbers. By mistake, she added instead of multiplying. Yet she got the right answer! What two different numbers might have been given?

29. The graph for Questions 23–26 only displays the values of U.S. imports and exports through the year 2000. Research on the Internet to find more recent import and export data. What can you say about trends in imports, exports, or balance of trade by looking at the more recent data?

Lesson

1-7

Rewriting Formulas

Vocabulary

equivalent formulas

▶ **BIG IDEA** Every formula defines one variable in terms of other variables; by using equation-solving properties you can manipulate a formula so that any one of the variables is defined in terms of the rest of the variables.

The volume V of a cylindrical barrel is given by the formula $V = \pi r^2 h$, where r is the barrel's radius and h is the barrel's height. This formula gives the volume V in terms of r and h. If you have values for r and h, you can use this formula to calculate V. For instance, if a barrel has radius 30 cm and height 100 cm, then

$$V = \pi (30 \text{ cm})^2 (100 \text{ cm})$$

$$= 90{,}000\pi \text{ cm}^3$$

$$\approx 283{,}000 \text{ cm}^3.$$

In some situations it is useful to solve a formula for one of the other variables. Example 1 illustrates how to convert a formula to a more useful form.

Mental Math

The 180-member marching band is going on a field trip. Suppose one bus can transport 50 students.

a. How many buses are needed to transport the band members?

b. One chaperone for every 10 students is going on the trip. Now how many buses are needed?

c. The instruments take up 15 seats on one of the buses. Now how many buses are needed?

Example 1

Justin Case is an engineer for a barrel manufacturing company. He wants to design a cylindrical barrel to contain 500,000 cm^3 of liquid and is considering several different radii and heights. To do this, it is useful to solve the volume formula for h in terms of r.

a. Solve the formula $V = \pi r^2 h$ for h in terms of V and r.

b. Find the height of a barrel with radius 30 cm that will contain 500,000 cm^3 of liquid.

(continued on next page)

Solution 1

a. Solve for h in terms if V and r.

$$V = \pi r^2 h \quad \text{Original formula}$$

$$\frac{V}{\pi r^2} = \frac{\pi r^2 h}{\pi r^2} \quad \text{Divide both sides by } \pi r^2.$$
$$\text{(You can do this because } r \neq 0.)$$

$$\frac{V}{\pi r^2} = h \quad \text{Simplify.}$$

b. Substitute the given values for V and r.

$$h = \frac{V}{\pi r^2} = \frac{500{,}000 \text{ cm}^3}{\pi(30 \text{ cm})^2} \approx 176.8 \text{ cm}$$

The barrel needs to be about 180 cm high to contain 500,000 cm^3 of liquid.

Solution 2

a. Use the SOLVE ( ) command on a CAS to solve for h. One CAS display is shown below. Entering ",h" after the formula in the command line tells the CAS to solve for h.

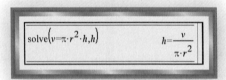

b. Substitute the given values for V and r to find h. We use the SUCH THAT command and the AND connector to enter both variable values at once. The display below shows both an exact answer and an approximate answer.

STOP **See Quiz Yourself 1 at the right.**

The formulas $V = \pi r^2 h$ and $h = \dfrac{V}{\pi r^2}$ are *equivalent formulas* because any V, r, and h that satisfy one of them also satisfies the other. Two formulas are **equivalent formulas** when the values of the variables that satisfy them are the same.

▶QUIZ YOURSELF 1

Find h when
$V = 500{,}000$ cm^3
and $r = 32$ cm.

Example 2

The formula $c = \dfrac{65t}{r - 65}$ computes the number c of hours needed for one car driving r miles per hour to catch up to another car that is t hours ahead and driving at 65 mph. Solve the formula for t in terms of c and r.

Solution Solve the formula for t by hand. Compare these steps to the computations on a CAS.

By hand		On a CAS
$c = \dfrac{65t}{r - 65}$	Write down the formula (at left) and enter it on a CAS (at right).	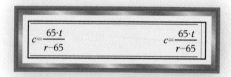
$c \cdot (r - 65) = 65t$	Multiply each side by $(r - 65)$. You may see the word **ANS** on your screen before you push **ENTER**. ANS refers to the last thing displayed on the CAS. In this case, it is the whole formula for c.	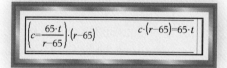
$cr - 65c = 65t$	Use the Distributive Property. The CAS shown here does this by using the **EXPAND()** command.	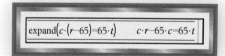
$\dfrac{cr - 65c}{65} = t$	Divide each side by 65.	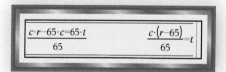

The formula is solved for t in terms of c and r. The result is the same, whether done by hand or with CAS.

🛑 **See Quiz Yourself 2 at the right.**

Using a CAS as an aid when solving equations has some advantages. It can be faster and more efficient than computing by hand. Using a CAS can also simplify the process for solving for a variable in an equation, especially when that variable's exponent is greater than 1, or when there is more than one variable, as Example 3 illustrates. The process can also be simplified by substituting values for some of the variables to *reduce* the problem to a function of a single variable.

> ▶ **QUIZ YOURSELF 2**
>
> **a.** Why is $c = \dfrac{65t}{r - 65}$ undefined when $r = 65$?
>
> **b.** Find t if $r = 70$ and $c = 10$. Explain your result in the context of the problem.

Example 3

a. Solve $V = \pi r^2 h$ for r.

b. Assume a fixed volume of 500,000 cm³. Write the formula for r in terms of h only.

Solution

a. Solve the formula for r on a CAS. The CAS result at the right shows two different values for r.

So, $r = \dfrac{-\sqrt{\frac{V}{h}}}{\sqrt{\pi}}$ or $r = \dfrac{\sqrt{\frac{V}{h}}}{\sqrt{\pi}}$.

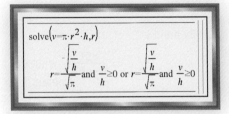

The first expression is negative, which is impossible in this situation, since $r > 0$.

So, $r = \dfrac{\sqrt{\frac{V}{h}}}{\sqrt{\pi}}$. You could also write this as $r = \dfrac{\sqrt{V}}{\sqrt{h\pi}}$.

b. Substitute 500,000 for V. This can be done by hand, or by using the **SUCH THAT** command on a CAS.

Notice that the CAS simplified $\sqrt{500000}$ but left the answer in exact form. So $r = 500 \cdot \dfrac{\sqrt{2}}{\sqrt{h \cdot \pi}}$ is a formula for r in terms of h.

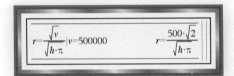

The volume formula solved for r is substantially more complicated than the formula solved for h. The added complexity is not surprising because in the original formula, r was squared, and h was not.

Questions

COVERING THE IDEAS

In 1–2, refer to the formula $V = \pi r^2 h$ in the lesson.

1. What is the volume of a cylindrical barrel of radius 5 cm and height 25 cm?

2. The formula $\dfrac{V}{\pi r^2} = h$ is solved for __?__.

3. Graph the function from Example 3b giving r in terms of h when V is assumed to be 500,000. Use a window that shows x and y values as large as 100.

In 4 and 5, complete the sentence "__?__ is written in terms of __?__" for the given formula.

4. $A = s^2$ (area of a square)

5. $V = \frac{1}{3}Bh$ (volume of a pyramid)

6. Refer to the formula $c = \dfrac{65t}{r - 65}$ from Example 2. Assume the first car has a 2-hour head start over the second car. Find the second car's catch up time for each speed. Explain what your answers mean in the context of the problem.

   a. 75 mph   b. 85 mph   c. 45 mph

7. Refer to Example 2. Solve the formula for r.

APPLYING THE MATHEMATICS

8. Imagine you are riding in the second car in the situation of Example 2. Let t be the number of hours head start the other car had at a speed of $65 \frac{\text{mi}}{\text{hr}}$. Let c be the number of hours it will take you to catch up at a rate $r \frac{\text{mi}}{\text{hr}}$. So, in time c, you will have traveled rc miles, while the other car will have traveled $65(t + c)$ miles. Thus, when you catch up, $rc = 65(t + c)$. Solve for c.

9. **Multiple Choice** Which formula is easiest to use if you want to find W and you are given G and k?

   A $G = 17Wk$ B $W = \dfrac{G}{17k}$ C $k = \dfrac{G}{17W}$

10. **Multiple Choice** In chemistry, a given mass and volume of gas will satisfy the equation $\dfrac{T_1}{P_1} = \dfrac{T_2}{P_2}$, where P_1 and T_1 are the pressure and temperature at one time, and P_2 and T_2 are the pressure and temperature at another time. Which is an equivalent formula solved for P_2?

   A $P_2 = \dfrac{T_2 T_1}{P_1}$ B $P_2 = \dfrac{T_1}{P_1 T_2}$ C $P_1 = \dfrac{P_2 T_1}{T_2}$ D $P_2 = \dfrac{P_1 T_2}{T_1}$

In **11** and **12**, the *pitch P* of a gabled roof is a measure of the steepness of the slant of the roof. Pitch is the ratio of the vertical rise *R* to half the *span S* of the roof: $P = \dfrac{R}{0.5\,S}$. (The span is the longest horizontal distance from one side of the roof to the other.)

11. a. Solve the pitch formula for *S*.

    b. If a builder wants a roof to have a pitch of $\dfrac{6}{15}$ and a rise of 8 feet, what must be the span of the roof?

12. The photograph at the right shows the roof of the Silver Pagoda of the Cambodian Royal Palace in Phnom Penh. Using the interior triangle, measure the rise and span of the roof and estimate its pitch.

13. If $y = kx^2$ and $y = 12$ when $x = 4$, solve for k.

14. The formula $C = \pi d$ gives the circumference of a circle in terms of its diameter d.

   a. Solve this formula for π.

   b. Use your result in Part a to write an English sentence that gives the definition of π.

In 15 and 16, use the formula $A = \frac{1}{2}h(b_1 + b_2)$ for the area A of a trapezoid with bases of length b_1 and b_2 and height h.

15. Solve the formula for b_1.

16. a. In the original formula, how does the value of A change if the values of b_1 and b_2 are switched? To what geometrical situation does this correspond?

   b. How does the original formula change if $b_1 = b_2$? To what geometrical situation does this correspond?

   c. How does the original formula change if $b_2 = 0$? To what geometrical situation does this correspond?

REVIEW

17. Solve the equation. (**Lesson 1-6**)

   a. $\frac{5}{8}x = 10$

   b. $\frac{5}{8}x + 30 = 10$

   c. $\frac{5}{8}x + 30 = 10 + \frac{1}{3}x$

18. Describe how you would use your calculator to generate a table for the function with equation $y = x^2 - 2x$ and with values starting at –5 in increments of 0.1 (**Lesson 1-5**)

19. a. Graph the function f when $f(x) = 3x - \frac{x^2}{9}$ for all real numbers x.

   b. What are its domain and range?

   c. What is its range if its domain is restricted to $\{x \mid 2 \leq x \leq 6\}$? (**Lesson 1-4**)

20. The velocity v of an object that starts from rest and accelerates at rate a over a distance d is given by the formula $v = \sqrt{2ad}$. What is the velocity of an object that started at rest and accelerated at $6 \frac{\text{m}}{\text{s}^2}$ over a distance of 6.75 m? (**Lesson 1-1**)

EXPLORATION

21. Formulas with three or more variables are common in science and financial applications. Find such a formula and describe what its variables represent and what it computes.

Lesson 1-8

Explicit Formulas for Sequences

Vocabulary

sequence

term of a sequence

subscript

index

explicit formula

discrete function

▶ **BIG IDEA** Sequences can be thought of as functions, but they have their own notation different from other functions. If their terms are real numbers, they are real functions and can be graphed.

Recognizing and Representing Sequences

In previous mathematics courses you have seen many sequences. A *sequence* is an ordered list of numbers or objects. Specifically, a **sequence** is defined as a function whose domain is the set of all positive integers, or the set of positive integers from *a* to *b*. Each item in a sequence is called a **term of the sequence**. In the following Activity, you will explore a sequence.

Mental Math

Let $a = -3$ and $b = 3$. Evaluate.

a. $a^2 - b^2$

b. $b^2 - a^2$

c. $(a - b)^2$

d. $(b - a)^2$

e. $(ab)^2$

Activity

The collections of dots below form the first five terms of a sequence of triangular arrays. The numbers of dots in each collection form a sequence of numbers.

Step 1 Complete the table to show the number of dots in each of the terms pictured.

Step 2 Notice that after the first term of the sequence, each subsequent term adds a predictable and increasing number of dots to the previous term. Use this fact to complete the table for the next four terms.

Step 3 This process can be continued for as long as you want. You can even think of it as going on forever. Explain why the set of ordered pairs (term number, number of dots) describes a function.

Term Number	Number of Dots
1	
2	
3	
4	
5	

Term Number	Number of Dots
6	
7	
8	
9	

The sequence you explored in the Activity is the sequence of *triangular numbers*. This sequence defines a function whose domain is the set of all positive integers. If you call this function T, then $T(1) = 1$, $T(2) = 3$, $T(3) = 6$, ... A notation for sequences more common than $f(x)$ notation is to put the argument in a *subscript*. A **subscript** is a label that is set lower and smaller than regular text. Using subscripts, $T_1 = 1$, $T_2 = 3$, $T_3 = 6$, ... The notation $T_3 = 6$ is read "T sub three equals six." The subscript is often called an **index** because it *indicates* the position of the term in the sequence.

 See Quiz Yourself 1 at the right.

▶ QUIZ YOURSELF 1

What are T_4 and T_5?

Writing Explicit Formulas for Sequences

Many sequences can be described by a rule called an **explicit formula** for the nth term of the sequence. Explicit formulas are important because they can be used to calculate any term in the sequence by substituting a particular value for n.

To find an explicit formula for the nth triangular number T_n, you can use the fact that the area of a triangle is half the area of a rectangle.

Notice that each triangular array of dots can be arranged to be half of a rectangular array.

For instance, the number of dots representing the 4th triangular number is half the number of dots in a 4 by 5 rectangular array.

$$T_4 = \frac{1}{2} \cdot 4 \cdot 5 = 10$$

You can generalize this idea to develop a formula for T_n.

Term Number	Value of Term (number of dots)
1	$T_1 = \frac{1}{2} \cdot 1 \cdot 2 = 1$
2	$T_2 = \frac{1}{2} \cdot 2 \cdot 3 = 3$
3	$T_3 = \frac{1}{2} \cdot 3 \cdot 4 = 6$
4	$T_4 = \frac{1}{2} \cdot 4 \cdot 5 = 10$
⋮	⋮
n	$T_n = \frac{1}{2} \cdot n \cdot (n + 1)$

The number of dots in the nth rectangle is $n(n + 1)$. T_n is half that.

$$T_n = \frac{1}{2} \cdot n \cdot (n + 1) = \frac{n(n + 1)}{2}$$

Thus an explicit formula for the number of dots in the nth term is

$$T_n = \frac{n(n + 1)}{2}.$$

 See Quiz Yourself 2 at the right.

Example 1

Suppose you flip a fair coin until it comes up tails. The probability that you will not have had an outcome of tails after n flips is given by the sequence

$$p_n = \left(\frac{1}{2}\right)^n.$$

a. Compute and graph the first four terms of this sequence.

b. Evaluate p_{20}, and explain what it represents.

Solution

a. Substitute 1, 2, 3, and 4 for n in the formula and graph the ordered pairs (n, p_n).

$$p_1 = \left(\frac{1}{2}\right)^1 = \frac{1}{2}$$

$$p_2 = \left(\frac{1}{2}\right)^2 = \frac{1}{4}$$

$$p_3 = \left(\frac{1}{2}\right)^3 = \frac{1}{8}$$

$$p_4 = \left(\frac{1}{2}\right)^4 = \frac{1}{16}$$

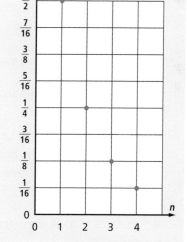

b. Substitute $n = 20$ into the formula.

$$p_{20} = \left(\frac{1}{2}\right)^{20} = \frac{1}{1,048,576}$$

p_{20} is the probability that you will not have had an outcome of tails after 20 flips.

See Quiz Yourself 3 at the right.

Using Explicit Formulas for Sequences

Sequences arise naturally in many situations in science, business, finance, and other areas. Example 2 looks at a sequence in finance.

GUIDED

Example 2

It is common for people to save money in savings accounts such as Certificates of Deposites (CDs) that yield a high interest rate paid once a year. Suppose you deposited $28,700 and expected a 4.1% interest rate to be *compounded* annually. Then the formula $S_n = 28{,}700(1.041)^{n-1}$ gives your total savings at any time during the year leading up to the nth anniversary.

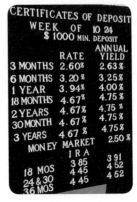

a. Compute the first five terms of the sequence.

b. Compute the hundredth term of the sequence.

c. What does your answer to Part b mean in the context of this problem?

Rates for Certificates of Deposit

Solution 1

a. Define the sequence using function notation on a CAS and compute the first five values.

$S(1) = \underline{\ ?\ }$

$S(2) = \underline{\ ?\ }$

$S(3) = \underline{\ ?\ }$

$S(4) = \underline{\ ?\ }$

$S(5) = \underline{\ ?\ }$

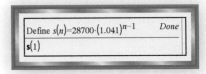

b. Compute $S(100)$ in the same way.

$S(100) = \underline{\ ?\ }$

c. This sequence gives the total savings at the end of the nth year. So, $S(100) = \underline{\ ?\ }$ means that **on the 100th anniversary of the account opening, there will be** $\underline{\ ?\ }$ **in the account.**

Solution 2

a. Enter the formula into a grapher and generate a table to view the first five values.

The table start value is n = $\underline{\ ?\ }$.

The increment is $\underline{\ ?\ }$.

The table end value is n = $\underline{\ ?\ }$.

b. Scroll down to see the value of $S(n)$ when $n = 100$.

$S(100) = \underline{\ ?\ }$

c. After 100 years, there will be $\underline{\ ?\ }$ in the account.

A sequence is an example of a *discrete function*. A **discrete function** is a function whose domain can be put into one-to-one correspondence with a finite or infinite set of integers, with gaps, or intervals, between successive values in the domain. The graphs of discrete functions consist of unconnected points. The gaps in the domain of a sequence are the intervals between the positive integers. The graph of gold prices on page 5 and the graph in Example 1 of this lesson are both examples of graphs of discrete functions.

Questions

COVERING THE IDEAS

1. Consider the increasing sequence 1, 3, 5, 7, ... of positive odd numbers.
   a. 13 is the 7th __?__ of the sequence.
   b. If this sequence is called D, what is D_{11}?

2. Consider the equation $a_{11} = 22.83$.
   a. Which number is the subscript?
   b. What does the *number that is not the subscript* represent?
   c. Which term of the sequence is this?
   d. Rewrite the equation using function notation.
   e. Rewrite the equation in words.

In 3 and 4, consider the sequence T of triangular numbers in the Activity on page 53.

3. Compute the 20th triangular number.
4. If $T_n = 15$, what is the value of n?

5. a. Draw a possible next term in the sequence at the right.
   b. How many dots does it take to draw each of the first 5 terms?
   c. Determine an explicit formula for the sequence S_n if S_n = the number of dots in the nth term.

6. Consider the sequence h whose first six terms are 231, 120, 91, 66, 45, 28.
   a. What number is the 4th term?
   b. How is the sentence "$h_5 = 45$" read?
   c. $h_6 = $ __?__

In 7 and 8, an explicit formula for a sequence is given. Write the first four terms of the sequence.

7. $a_n = 7.3 - 3n$
8. $S_n = \dfrac{n(n+1)(2n+1)}{6}$ (sum of the first n squares)

9. Refer to Solution 2 of Example 2.

   a. Graph the values from the table with the anniversary n on the horizontal axis and the total savings S_n on the vertical axis.

   b. Should the points on the graph be connected? Why or why not? Relate your answer to the domain of the function.

10. a. **Multiple Choice** Which could be a formula for the nth term of the sequence 3, 9, 27, 81, … ?

     A $t_n = 3n$ **B** $t_n = 3n^3$ **C** $t_n = 3^n$

   b. For the choices not used, write the first four terms of the sequence being represented.

APPLYING THE MATHEMATICS

11. a. Generate a table of the 4th through 7th terms of the sequence C defined by $C_n = \dfrac{n^2(n+1)^2}{4}$.

   b. This sequence gives the sum of the first n cubes. What does C_7 represent?

In Questions 12 and 13, consider the story *Anno's Magic Seeds* by Mitsumasa Anno (1992). It is the story of Jack, who receives two magic seeds. A person who eats one of these seeds will be full for one year, and planting a seed yields two seeds.

MITSUMASA ANNO

ANNO'S MAGIC SEEDS

12. Jack begins the first year with two magic seeds. He eats one and plants the other. It grows and produces two seeds at the start of the second year. In this second year, Jack repeats his behavior by eating one seed and planting the other to get two seeds for the start of the third year. Jack continues the trend each year, eating one seed and planting the remaining seed to get two for the following year.

   a. Create a table and a graph illustrating the relationship between the year and the number of seeds Jack has at the beginning of the year for the first five years.

   b. Does this situation determine a function? Why or why not?

   c. Does this situation determine a sequence? Why or why not?

13. Suppose Jack decides to forgo eating a seed in the first year and instead plants both seeds to end up with four seeds at the start of the second year. In the second year, Jack eats one seed and plants the remaining three to end up with six seeds at the start of the third year. The third year, Jack eats one seed and plants the remaining five to end up with ten seeds at the start of the fourth year. Jack continues his behavior of eating one seed and planting what is left.

   a. Write the first six ordered pairs that relate the year to the number of seeds.

   b. Find an explicit formula for the number of seeds at the beginning of the nth year, for all $n > 1$.

14. Some common bacterial cells, such as *E. coli,* can divide and double every 20 minutes. The doubling process takes place when a microbe reproduces by splitting to make 2 cells. Each of these cells then splits in half to make a total of 4 cells. Each of these 4 cells then splits to make a total of 8, and so on. Each splitting is called a *generation*. If a colony begins with 125 microbes, the equation $P_n = 125(2)^{n-1}$ gives the number of microbes in the nth generation (assuming all microbes survive).

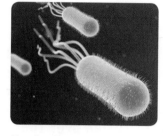

E. coli cells

   a. Calculate the first three terms of the sequence.

   b. Identify the independent variable and dependent variable of the function.

REVIEW

15. Solve for s in the formula $d = 7s - 13$. (**Lesson 1-7**)

16. The measure θ of an exterior angle of a regular polygon is given by $\theta = \frac{360}{n}$, where n is the number of sides of the polygon. Solve for n in terms of θ. (**Lesson 1-7**)

In 17 and 18, an equation is given. (Lesson 1-6)

   a. Solve the equation.

   b. Check your answer.

17. $5r - (2r + 1) = 6$ **18.** $7t - (9 - 4t) = 9$

19. A medium pizza costs $12.50 plus $1.50 for each topping. If C is the total cost of the pizza and t is the number of toppings ordered, then $C(t) = 12.50 + 1.50t$ gives c as a function of t. (**Lesson 1-4**)

   a. Specify the domain of this function.

   b. Write the four smallest numbers in the range of this function.

   c. Graph this function.

20. **Fill in the Blank** Let $h: a \to a^3 + 2$. Then $h: -3 \to \underline{\ ?\ }$.
(**Lesson 1-3**)

EXPLORATION

21. Triangular numbers have many curious properties. For example, a triangular number can never end with the digits 2, 4, 7, or 9. Find one more property of the triangular numbers.

Chapter 1 Projects

A project represents an opportunity for you to extend your knowledge of a topic related to the material of this chapter. You should allow more time for a project than you do for typical homework questions.

1 Value over Time

Recall the first pages of the chapter describing the fluctuation of gold prices over time. Find an object, such as a painting masterpiece, whose value has tended to increase over time. Find another object, such as a new car, whose value usually decreases over time. In tables and graphs, record values of these objects over reasonable domains. Write a paragraph explaining why the values increase in one case while decreasing in the other.

Van Gogh's *L'Arlésienne*, 1890 (oil on canvas), Vincent Van Gogh

2 Functions and Graphs on the Internet

Search for coordinate graphs on the Internet. Find six to eight graphs you think are interesting. Look for a variety of subjects and shapes of graphs. For each graph, tell whether it represents a function. Identify its domain and range. Make up at least one question that can be answered by the graph.

3 Triangular Numbers

Triangular numbers appear in several mathematical applications and have many interesting properties.

a. Research the applications and properties of triangular numbers and prepare a report on your findings.

b. Pentagonal numbers are similar to triangular numbers. Draw the first five geometric representations of the pentagonal sequence using dots and find an explicit formula for the nth term.

4 Cooking a Turkey

The amount of time it takes to cook a turkey depends on several factors, such as the turkey's weight. Use the Internet to find at least three different tables or formulas for cooking times for turkey. Compare the tables and formulas and explain which factors each table or formula takes into account.

5 Height and Area of an Isosceles Triangle

Consider the triangle below, with two sides of length 10, base of length x, and height h.

a. Find a formula for h in terms of x. What is the domain of this function? (*Hint:* Use the Triangle Inequality.)

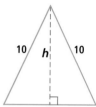

b. Use your graphing utility to graph the function you found in Part a over its domain. How does h change as x changes?

c. Find a formula for the area of the triangle in terms of x only.

d. Use your graphing utility to graph the function you found in Part c. Draw a sketch of the graph.

e. How does the area of the triangle change as x changes? Use the TRACE command on your grapher to find the value of x when the area of the triangle is the largest.

f. Find the value of h when the area of the triangle is the largest. Draw a sketch of the triangle of largest area.

Chapter 1 Summary and Vocabulary

In the columns at the right are the most important terms and phrases for this chapter. You should be able to give a general description and a specific example of each and a precise definition for those marked with an asterisk (*).

◉ The language of algebra is based on numbers and **variables**. These are combined in **expressions**, and two or more expressions connected by a verb make up an **algebraic sentence**.

◉ **Formulas** are equations stating that a single variable is equal to an expression with one or more variables on the other side. Formulas are evaluated and can be rewritten using the rules for the **order of operations** and properties of equality.

◉ A **function** is defined as a set of ordered pairs in which each first coordinate can be paired with exactly one second coordinate. The first coordinate of the ordered pair is the **independent variable**; the second coordinate is the **dependent variable**. Thus, a function pairs two variables in such a way that each value of the independent variable corresponds to exactly one value of the dependent variable. The **domain** of a function is the set of possible values for the independent variable, while the **range** is the set of values obtained for the dependent variable. Many functions are **mathematical models** of real situations or other mathematical properties.

◉ Functions can be used to represent situations where one value of an ordered pair is known and the other unknown. They can also be used to describe situations in which objects are transformed or otherwise mapped onto other objects. Each of these types of situations may arise in everyday contexts and in all branches of mathematics.

◉ Functions are often named with a single letter. Euler's **f(x)** notation $f(x) = x + 20$, and the **mapping notation** $f : x \rightarrow x + 20$, both describe the same function. Graphing utilities use $f(x)$ notation and often name functions y_1, y_2, y_3, and so on. A CAS defines functions with $f(x)$ notation, but also accepts equations and formulas that are not solved for one variable.

Vocabulary

1-1
*variable
algebraic expression
expression
algebraic sentence
evaluating an expression
order of operations
vinculum
*equation
*formula

1-2
relation
*function
*independent variable
*dependent variable
input, output
mathematical model

1-3
*$f(x)$ notation
argument of a function
value of a function
mapping notation

1-4
real function
*domain of a function
*range of a function
set builder notation
standard window
*natural numbers
*counting numbers
*whole numbers
*integers
rational numbers
real numbers
irrational numbers

Along with words and formulas, functions are also represented by tables and graphs.

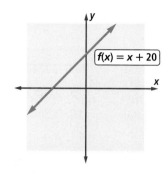

x	y
–2	18
–1	19
0	20
1	21
2	22

$f(x) = x + 20$

A **sequence** is a function whose domain is the set of all positive integers or the set of positive integers from a to b. Sequences may be defined with an **explicit formula** in which the nth **term** can be calculated directly from n. In a sequence S, the nth term is identified as S_n.

Theorems and Properties

Distributive Property (p. 41)
Opposite of a Sum Theorem (p. 44)

Chapter 1
Self-Test

Take this test as you would take a test in class. You will need a calculator. Then use the Selected Answers section in the back of the book to check your work.

1. If $f(x) = x^2 - 4^x$, find $f(-1.5)$ to the nearest hundredth.

2. Suppose that f is a function and a is a real number. Describe the difference in the statements "$f(a) = 7$" and "$f(7) = a$."

3. The table below defines the function g. Find $g(-2)$ and $g(2)$.

r	-1	-2	-1	0	1	2	3
$g(r)$	6	17	15	2	-4	3	-2

In 4 and 5, let f be the function graphed below.

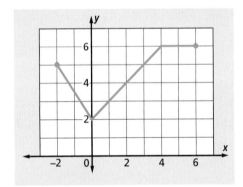

4. Find the domain and range of f.

5. Use the graph to estimate each of the following:
   a. $f(3)$
   b. All values of x such that $f(x) = 4$.

6. Suppose $S: n \rightarrow \frac{n}{2}(3(n - 1))$. Evaluate $S(10)$.

7. Which formula below is solved for c?
   A $A = \frac{1}{2}c_1c_2$
   B $c = 2\pi r$
   C $a^2 + b^2 = c^2$

8. Suppose you travel m miles in 4 hours. Write a formula that gives your speed s in miles per hour.

In 9–11, solve for the variable.

9. $5(7x - 4) = 50$

10. $3.2a = 0.75 + 1.2a$

11. $p + 0.2(1200 - p) = 244.4p$

In 12–14, use the formula $V = \frac{1}{3}\pi r^2 h$ for the volume of a cone.

12. Find the volume of the cone at right to the nearest cubic centimeter.

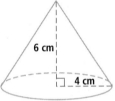

6 cm
4 cm

13. Suppose $h = 6$ and the domain of the volume function is $0 < r \leq 10$. What is the range of the function?

14. a. Solve this formula for h in terms of V and r.
    b. Solve this formula for r in terms of V and h.

15. A function f contains only the points $(0, 7)$, $(4, 8)$, $(8, 9)$ and $(16, 11)$. Give its domain and range.

16. **Multiple Choice** Which of the following are graphs of functions?

A
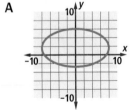

B
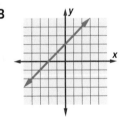

C
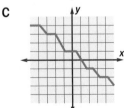

D
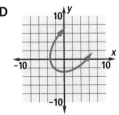

In 17 and 18, determine whether the relation is a function. Justify your answer.

17. $\{(9, 7), (7, 5), (18, 12), (9, 4)\}$

18.

x	0	1	2	4
y	0	2	3	3

(Consider x to be the independent variable.)

19. a. Write the first five terms of the sequence $a_n = -4^n + 2$.

b. What is a_7?

20. Use a graphing utility to graph the function $f(t) = 6 - t^2$ with a domain $\{t | -5 \le t \le 5\}$.

a. What is the range of this function?

b. To the nearest hundredth, estimate where $f(t) = 0$.

21. Create a table for $y = -3 \cdot \left(\frac{1}{2}\right)^x$ with a start value of $x = -6$, a table increment of 2 and, if necessary, an end value of 2.

a. List the first five ordered pairs in the table.

b. For what value of x does $y = -0.75$?

22. The following table of values represents the wind chill index for various temperatures when there is a 10 mph wind. Use the table to answer the following questions.

A = Actual temperature (°F)	30	20	10	0	–10	–20	–30
$W(A)$ = Wind-chill index (°F)	21	9	–4	–16	–28	–41	–53

Source: National Oceanic and Atmospheric Administration

Let A = the actual temperature and let W be the name of the function that maps A onto the wind-chill index.

a. Evaluate $W(10)$ and explain what it means in terms of the wind chill.

b. For what value of A does $W(A) = -41$?

23. Refer to the graph at the right.

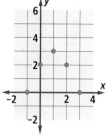

a. List the ordered pairs in the graph.

b. Is this a function? Explain why or why not.

c. State the domain and range.

24. The population, $P(y)$ in millions, of a certain city is modeled by the function with equation $P(y) = 2 \cdot 1.005^y$, where y is the year minus 2000 (for example, in 2006, $y = 6$). How many more people will be living in the town in the year 2020 than in 2006?

25. **Multiple Choice** What CAS command would you use to rewrite the expression $(a^2 + 1)(a + 1)(a - 1)$?

A SOLVE()   B FUNCTION()

C EXPAND()   D DEFINE()

26. The probability of tossing a six-sided die and rolling a 3 n times in a row can be represented by the sequence $p_n = \left(\frac{1}{6}\right)^n$. What is the probability of rolling a 3 four times in a row?

27. On a recent flight between St. Louis and Los Angeles, $\frac{1}{4}$ of the tickets sold were for seats in business class, $\frac{1}{10}$ in first class, and the remaining 195 tickets were in coach class. How many tickets were sold altogether?

28. The Pythagorean Theorem states that if a and b are the legs of a right triangle and c is its hypotenuse, then $a^2 + b^2 = c^2$. What happens to the value of c if you switch the values of a and b?

Chapter 1 Chapter Review

SKILLS
PROPERTIES
USES
REPRESENTATIONS

SPUR stands for Skills, Properties, Uses, and Representations. The Chapter Review Questions are grouped according to the SPUR Objectives in this chapter.

SKILLS Procedures used to get answers

OBJECTIVE A Evaluate expressions and formulas, including correct units in answers. (Lesson 1-1)

1. Evaluate $100 \cdot (1 + r)^t$ when $r = 0.06$ and $t = 3$.

2. If $m = \dfrac{4y(y + 2)}{9}$, find m when $y = 16$.

3. If $d = \frac{1}{2}gt^2$, find d when $g = 9.8\dfrac{m}{sec^2}$ and $t = 3$ sec.

4. Evaluate $\dfrac{n}{2^{2(n-1)}}$ when $n = 4$.

OBJECTIVE B Use mapping and $f(x)$ notation for functions. (Lesson 1-3)

5. If $f(x) = -5x + 10$, what is $f(5)$?

6. **Fill in the Blank** Suppose $g: t \rightarrow t - 2t^3$. Then $g: -1 \rightarrow$ __?__.

In 7 and 8, a function is described by an equation.

a. Rewrite the function in mapping notation.

b. Evaluate the function at $x = 9$.

7. $h(x) = 12x - 2\sqrt{x}$

8. $j(x) = 3 - 4 \cdot 2^x$

In 9 and 10, a function is given in mapping notation.

a. Rewrite the function in $f(x)$ notation.

b. Evaluate the function at $x = 8$.

9. $f: x \rightarrow 4 - 27x$

10. $c: x \rightarrow \dfrac{\sqrt{2x}}{5}$

OBJECTIVE C Solve and check linear equations. (Lesson 1-6)

In 11–18, solve and check.

11. $12x = \frac{4}{5}$

12. $\frac{3}{8}(b + 5) = 9$

13. $-\frac{5}{3} = \frac{4}{3} - v$

14. $L = 2L - (6 - 6L)$

15. $-s + 5s = 4(2s + 3)$

16. $0.02(800z + 50) = -500 - 24z$

17. $\frac{m}{8} + \frac{m}{6} - 2 = -3m$

18. $y - 8(y + 5) = 13y$

OBJECTIVE D Solve formulas for their variables. (Lesson 1-7)

19. Solve for t in the formula $x = 12 - 6t$.

20. Solve for q in the formula $E = \dfrac{kq}{r^2}$.

21. Solve for h in the formula $A = \frac{1}{3}(\pi r_1^2 h - \pi r_2^2 h.)$

22. Solve for n in the formula $t = a + (n - 1)d$.

23. Explain why the equation $s = \dfrac{4\pi}{\sqrt{2bds}}$ is not solved for s.

24. **Multiple Choice** Which of the following is not a formula solved for the volume, radius, or height of a cylindrical barrel?

A $V = hr^2 \cdot \pi$ B $\dfrac{V}{\pi r^2} = h$

C $r^2 = \dfrac{V}{h\pi}$ D $\sqrt{\dfrac{V}{\pi h}} = r$

OBJECTIVE E Find terms of sequences. (Lesson 1-8)

In 25 and 26, write the first five terms of the sequence. These sequences are defined for all integers $n \geq 1$.

25. $a_n = -3 - 6n$

26. $b_n = -5 \cdot 2^n$

27. If $c_n = n^2 - n$, find $c_4 + c_3$.

28. When $d_n = (-0.5)^n$, find $3d_2 + d_1$.

OBJECTIVE F Use a CAS to solve equations or expand expressions. (Lesson 1-7)

29. Use the SOLVE( ) function on your CAS to solve the formula for the volume of a cone, $V = \frac{1}{3}\pi r^2 h$, for r. What two solutions does it give? Which one is the correct formula for r?

In 30 and 31, suppose that while solving an equation for the variable c on a CAS, you encounter the intermediate equation $5 + 5r + ab = 5c$.

30. What should you enter next? Finish solving this equation on your CAS.

31. **Multiple Choice** A student solving the equation obtains the CAS result $c = \frac{1}{5}(ab + 5r + 5)$. What CAS command will yield a simplified solution?

    A ANS*5   B EXPAND(ANS)

    C ANS/5   D SOLVE(ANS, c)

PROPERTIES Principles behind the mathematics

OBJECTIVE G Determine whether a given relation is a function. (Lessons 1-2, 1-5)

In 32–33, determine whether the given table or set of ordered pairs describes y as a function of x.

32.

x	1	3	5	7	9
y	3	5	7	9	1

33. $\{(0, 3), (1, -2), (2, -1), (1, 2), (-1, 5)\}$

In 34–37, determine whether the given graph represents the graph of a function. Justify your answer.

34.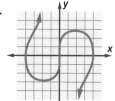

35.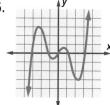

36.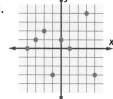

37.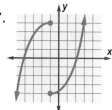

OBJECTIVE H Determine the domain and range of a function. (Lessons 1-4, 1-5)

In 38–43, find the domain and range of the function described by the given table, set of ordered pairs, equation, or graph.

38.

t	2	3	5	7	11
e(t)	1	2	3	4	5

39. $\{(0, -1), (1, -1), (2, -1), (-1, -1), (-2, -1)\}$

40. $f(x) = x^2$ 41. $g(y) = 3y^5$

42. 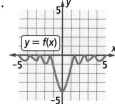   43.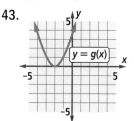

44. From the graph in Question 42, find the approximate values of x for which $f(x) = 0$.

45. What is $g(-3)$ in Question 43?

46. Graph the function $h(x) = (x - 2)^2$ on a grapher using a standard window, and use the graph to determine the function's domain and range.

OBJECTIVE I Describe relationships between variables in a formula. (Lesson 1-7)

47. Isaac Newton's Law of Universal Gravitation states that every object exerts a gravitational force on every other object. The formula $F = \dfrac{Gm_1 m_2}{r^2}$ gives the magnitude of the gravitational force F that a body of mass m_1 exerts on a body of mass m_2 if they are a distance r apart. G is the gravitational constant. What happens to the value of F if m_1 and m_2 are switched? What does this imply about the force that the body of mass m_2 exerts on the body of mass m_1?

48. Rewrite the formula $V = \ell wh$, for the volume of a rectangular box with length ℓ, width w, and height h, in terms of ℓ when $w = \ell = h$. What situation does this formula represent?

49. **Multiple Choice** Given $x \neq 0$, $y \neq 0$, and $k \neq 0$, which of the following formulas is equivalent to the formula $y = \frac{k}{x}$?

    A $k = \frac{y}{x}$ B $x = ky$ C $y = \frac{x}{k}$

    D $x = \frac{k}{y}$ E $k = \frac{x}{y}$

USES Applications of mathematics in real-world situations

OBJECTIVE J Evaluate and interpret values of functions in real-world situations. (Lessons 1-3, 1-4)

50. Michelle's annual salary is $43,000. She gets an increase of 8% at the end of each year. The sequence $a_n = 43{,}000 \cdot 1.08^n$ gives Michelle's salary in the nth year. At this growth rate, what will Michelle's salary be in the 5th, 10th, 15th, and 20th years?

51. The formula $S(x) = x + \dfrac{x^2}{20}$ relates a car's stopping distance $S(x)$ in feet to the car's speed x in miles per hour. Determine how many more feet it takes a car traveling at 85 mph to stop than a car traveling at 55 mph.

52. The distance in feet $d(t)$ a dropped object falls in t seconds is given by the function $d(t) = 16t^2$. Suppose you drop a ball from a height of 60 feet.

    a. About how far will the ball fall in 1.3 seconds?

    b. Find $d(2)$. Explain why this result does not make sense in this situation.

53. Janelle opened a savings account with $200. Each month she adds $35 to her account.

    a. Create a table with the amounts in her account for the first six months.

    b. Write a formula for the sequence giving her savings account balance at the end of each month.

In 54–56, let $M(x)$ and $W(x)$ be the populations of Minnesota and Wisconsin, respectively, in the year x.

Year x	Population of Minnesota $M(x)$	Population of Wisconsin $W(x)$
1970	3,806,103	4,417,821
1980	4,075,970	4,705,642
1990	4,375,099	4,891,769
2000	4,919,479	5,363,675

Source: U.S. Census Bureau

54. What does $M(1970)$ represent?

55. Calculate $W(1980) - M(1980)$. What does this represent?

56. Calculate $W(2000) - W(1990)$. What does this represent?

OBJECTIVE K Use linear equations to solve real-world problems. (Lesson 1-6)

57. Suppose a baby blue whale weighs 4,000 lb at birth and gains 200 lb per day while nursing. A formula that gives its weight W after d days of nursing is $W = 4000 + 200d$.

    a. Write an equation that can be used to find the number of days d a young blue whale has been nursing if it weighs W lb.

    b. Use your answer to Part a to find d for $W = 14,000$.

58. At George Washington High School, all students are in grade 10, 11, or 12. This year $\frac{3}{10}$ of the students are in grade 10, 400 students are in grade 11, and $\frac{1}{5}$ of all students are in grade 12. How many students are at George Washington High this year?

59. Cara wants to park her car in a parking lot that charges $6 for the first hour and $1.50 for each additional hour. So the cost c in dollars to park for h hours is $c = 6 + 1.5(h - 1)$. How long can Cara park if she has $20? (Parking time is always rounded up to the next hour, so if she parks for 70 minutes, she must pay for 2 hours.)

60. Frederick is paying off a $3000 loan in installments of $160 per month. So the amount left to pay in dollars after m months will be $3000 - 160m$. After how many full months will Frederick be out of debt?

REPRESENTATIONS Pictures, graphs, or objects that illustrate concepts

OBJECTIVE L Use a graphing utility to graph functions and generate tables for functions. (Lessons 1-4, 1-5, 1-8)

In 61 and 62, a function is given.

a. Use a graphing utility to graph f in the standard window.

b. Use your graph to find all values of x when $f(x) = 0$.

61. $f(x) = x^3 - 8x^2 + 19x - 12$

62. $f(x) = -2 + 2^{2 - x^2}$

63. Use a graphing utility to graph A when $A(t) = 50 \cdot 1.04^t$, with the domain restricted to $\{t \mid t \geq 0\}$. What is the range of this function?

64. Use a graphing utility to graph B when $B(t) = 50 \cdot 0.96^t$, with the domain restricted to $\{t \mid t \geq 0\}$. What is the range of this function?

In 65 and 66, a function f is given.

a. Create a table for f with a start value of 1 and a table increment of 0.2. List the first five ordered pairs in the table.

b. Give a value of x for which f(x) is within 0.05 of –0.4.

c. **Multiple Choice** What appears to be true of $f(x)$?

    A $f(x)$ is always positive.

    B $f(x)$ is always negative.

    C $f(x)$ starts off positive and becomes negative.

    D none of the above

65. $f(x) = 2^x - 4^x + 3^x$

66. $f: x \rightarrow \dfrac{2x - 1}{6 - 5x}$

Chapter

2

Variation and Graphs

As you ride your bicycle on level ground, you control the speed by moving the pedals. The faster you pedal, the more revolutions R the pedals make each minute, and the faster the bicycle travels. However, an experienced cyclist knows that speed is also dependent upon the *gear ratio* between the numbers of teeth on the front and back gears. Increasing the number F of front-gear teeth increases speed, but increasing the number B of back-gear teeth reduces the speed. Thus, speed varies as R, F, and B change. But how does it vary?

Suppose Mario knows that when he is
pedaling at 80 revolutions per minute (rpm)
with 35 teeth on the front gear and 15 teeth
on the back gear, he travels at 14.5 miles
per hour. How would his speed be affected
if he increases his revolutions to 100 rpm?
You will learn how to answer this question in
this chapter about *variation*, that is, how one
quantity changes as others are changed.

Lesson 2-1

Direct Variation

Vocabulary

varies directly as

direct-variation equation

constant of variation

direct-variation function

directly proportional to

▶ **BIG IDEA**   When two variables x and y satisfy the equation $y = kx^n$ for some constant value of k, we say that y varies directly as x^n.

When possible, Lance puts his bike into its highest gear for maximum speed. In highest gear, the ratio of the number of front-gear teeth to the number of back-gear teeth is $\frac{52 \text{ teeth}}{11 \text{ teeth}} \approx 4.73$. This means that as Lance turns the pedals one complete revolution, the back wheel turns almost 5 times. So if w is the number of back-wheel turns per minute and p is the number of pedal turns per minute, then

$$w = \frac{52}{11}p.$$

If Lance starts pedaling twice as fast, then the back wheel will also turn twice as fast as it previously did. We say that w **varies directly as** p, and we call $w = \frac{52}{11}p$ a **direct-variation equation**. In a direct variation, both quantities increase or decrease together.

Suppose Lance changes to a lower gear with a gear ratio of $\frac{42 \text{ teeth}}{21 \text{ teeth}} = 2$. A direct-variation equation for this situation is $w = 2p$.

Recall the formula $A = \pi r^2$ for the area of a circle. This, too, is a direct-variation equation; as the radius r increases, the area A also increases. In this case, A varies directly as r^2. Often this wording is used: the area A varies directly as the *square* of r.

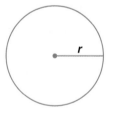

$A = \pi r^2$

Mental Math

a. Which is larger: $\frac{7}{9}$ or $\frac{7}{10}$?

b. Which is closer to $\frac{1}{2}$: $\frac{3}{5}$ or $\frac{4}{7}$?

c. Which is closer to zero: $\frac{1}{2} - \frac{1}{3}$ or $\frac{1}{3} - \frac{1}{4}$?

d. Which is closer to 1: $\frac{1}{2} + \frac{1}{4} + \frac{1}{6}$ or $\frac{1}{3} + \frac{1}{5} + \frac{1}{7}$?

Direct-Variation Functions

The formulas $w = \frac{52}{11}p$, $w = 2p$, and $A = \pi r^2$ are all of the form $y = kx^n$, where k is a nonzero constant, called the **constant of variation**, and n is a positive number. These formulas all describe *direct-variation functions*.

Definition of Direct-Variation Function

A **direct-variation function** is a function that can be described by a formula of the form $y = kx^n$, with $k \neq 0$ and $n > 0$.

When y varies directly as x^n we also say that y is **directly proportional to** x^n. For instance, the formula $A = \pi r^2$ can be read "the area of a circle is directly proportional to the square of its radius." Here $n = 2$ and $k = \pi$, so π is the constant of variation.

 QY1

If you know one point (x, y) of a direct-variation function, you can determine k and, thus, know precisely the function.

The formula $A = \pi r^2$ is one example of a general theorem from geometry: in a set of similar figures (in this case, circles), area is proportional to the square of length. Likewise, in a set of similar figures, volume is proportional to the cube of edge length. Example 1 is an instance of that theorem.

> **QY1**
>
> Identify the constant of variation and the value of n in the formula $w = \frac{52}{11}p$.

Example 1

The volume V of a regular icosahedron is directly proportional to the cube of the length ℓ of an edge.

a. Identify the dependent and independent variables and write an equation relating them.

b. A regular icosahedron with an edge length of 4 cm has a volume of about 140 cm^3. Determine the constant of variation k.

c. Rewrite the variation equation using your result from Part b.

d. Approximate to the nearest cubic centimeter the volume of an icosahedron with an edge length of 5 cm.

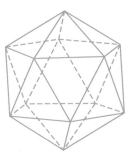

Regular Icosahedron (20 faces)

Solution

a. Because V is directly proportional to ℓ, the dependent variable is V and the independent variable is ℓ. In this problem $n = 3$, so an equation for the direct variation is $V = k\ell^3$, where k is a constant.

b. To determine k, substitute $V = 140$ cm^3 and $\ell = 4$ cm into your direct-variation equation from Part a.

$$140 \text{ cm}^3 = k \cdot (4 \text{ cm})^3$$
$$140 \text{ cm}^3 = (64 \text{ cm}^3) \cdot k$$
$$k = \frac{140 \text{ cm}^3}{64 \text{ cm}^3} = 2.1875$$

(continued on next page)

c. Substitute $k = 2.1875$ to get a formula relating the edge length and volume.

$$V = 2.1875\, \ell^3$$

d. Evaluate your formula when $\ell = 5$ cm.

$$V = 2.1875(5 \text{ cm})^3 \approx 270 \text{ cm}^3$$

Direct-Variation Functions

The four parts of Example 1 illustrate a procedure you can use to solve variation problems. First, write a general equation that describes the variation. Next, substitute the given values into the general equation and solve for k. Then use the k-value to write the variation function. Finally, evaluate the function at the specified point to find the missing value.

Guided Example 2 illustrates how to use this procedure to solve a typical direct-variation problem.

GUIDED

Example 2

Suppose b varies directly as the sixth power of g. If $b = 729$ when $g = 2$, find b when $g = 10$.

Solution Write an equation with variables b, g, and k that describes the variation.

$$\underline{\ ?\ } = k \cdot \underline{\ ?\ }^6$$

Find the constant of variation. You are given that $b = 729$ when $g = 2$. Substitute these values into the variation formula to find k.

$$\underline{\ ?\ } = k \cdot \underline{\ ?\ }^6$$
$$\underline{\ ?\ } = k \cdot \underline{\ ?\ }$$
$$\underline{\ ?\ } = k$$

Now rewrite the variation formula using the value of the constant you found.

$$\underline{\ ?\ } = \underline{\ ?\ } \cdot \underline{\ ?\ }^6$$

Finally, use the formula to find b when $g = 10$.

$$b = \underline{\ ?\ } \cdot \underline{\ ?\ }^6$$
$$b = \underline{\ ?\ }$$

So, when $g = 10$, $b = \underline{\ ?\ }$.

Activity

You can solve direct-variation problems on a CAS by defining two functions. The first function calculates the constant of variation k. The second function calculates values of the direct-variation function using that k and a value of the independent variable.

Step 1 Plan your first function.

    a. Give a meaningful name to the function so you can use it in the future. We call the first function *dirk*, short for *direct-variation k-value*.

    b. Think about the inputs you need to calculate k and give them names. In Part b of Example 1, you used initial values of the independent and dependent variables and the value of the exponent to calculate k. Good names for the initial values of the independent and dependent variables are xi and yi, respectively. We call the exponent n, as in the direct-variation formula $y = kx^n$.

    c. Generalize the result of solving for k in Part b of Example 1. In the Example,
$$xi = 4, yi = 140, n = 3, \text{ and } k = \frac{140}{4^3}.$$
So, in general, $k = \frac{yi}{xi^n}$.

Step 2 Clear the values for $xi, yi, k,$ and n in the CAS memory. Then define the *dirk* function using its three inputs and the general formula for k you found above.

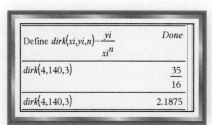

Step 3 Use the *dirk* function to calculate k for the situation in Example 1. The input values are $xi = 4, yi = 140,$ and $n = 3$. The display at the right shows the value of k in both fraction and decimal form.

Step 4 A good name for the second function is *dvar*, for *direct variation*. There are three inputs for this function also: another known value x for the independent variable, the constant of variation k, and the exponent n. Clear x from the CAS memory and define *dvar* using the general form of the direct-variation formula, as shown at the right.

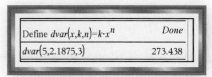

Step 5 Use *dvar* to find the missing function value for $x = 5$, $k = 2.1875,$ and $n = 3$. Compare your answer to Part d of Example 1.

STOP **QY2**

You may use ***dirk*** and ***dvar*** to answer the Questions.

> **QY2**
>
> Use *dirk* and *dvar* to check your answers to Example 2.

Questions

COVERING THE IDEAS

1. Provide an example of direct variation from everyday life.

2. Give an example of a direct-variation function from geometry.

3. **Fill in the Blanks** In the function $y = 5x^2$, __?__ varies directly as __?__, and __?__ is the constant of variation.

4. Suppose pay varies directly as time worked and Paul makes $400 for working 25 hours.

   a. What is the constant of variation?

   b. How much will Paul make for working 32 hours?

5. The lengths of the femur and tibia within a species of mammal are typically directly proportional. If the femur of one household cat is 116.0 mm long and its tibia is 122.8 mm long, how long is the tibia of a cat whose femur is 111.5 mm long?

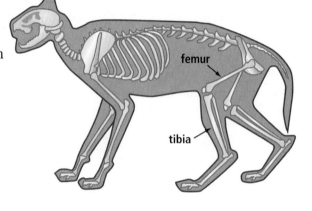

6. Suppose $s = 4.7t$.

   a. Find s when $t = 3.1$.

   b. Is this an equation for a direct-variation function? How can you tell?

7. Assume that y is directly proportional to the square of x.

   a. **Multiple Choice** Which equation represents this situation?

   A $y = 2x$ B $y = kx^2$ C $x = ky^2$ D $y = 2x^k$

   b. **Multiple Choice** It is also true that x is directly proportional to

   A the square of y.   B the square root of y.

   C twice y.   D half of y.

In 8 and 9, refer to Example 1.

8. What is the volume of a regular icosahedron with an edge of length 2 cm?

9. Use the solve command on a CAS to find the length of the edge of a regular icosahedron with a volume of 20 cm³.

10. Suppose W varies directly as the fourth power of z, and $W = 27$ when $z = 3$.

    a. Find the constant of variation.

    b. Find W when $z = 9$.

11. Suppose y varies directly as the cube of x, and $y = 27.6$ when $x = 0.5$. Find y when $x = 3.2$.

12. Suppose f is a function defined by $f(x) = \frac{x^3}{7}$. Is f a direct-variation function? Justify your answer.

13. When lightning strikes in the distance, you do not see the flash and hear the thunder at the same time. You first see the lightning, then you hear the thunder.

    a. Write an equation to represent the following situation: The distance d (in miles) from the observer to the flash varies directly as the time t (in seconds) between seeing the lightning and hearing the thunder.

    b. Suppose that lightning strikes a point 4 miles away and that you hear the thunder 20 seconds later. How far away has lightning struck if 12 seconds pass between seeing a flash and hearing its thunder?

14. The speed of sound in air is about 1,088 feet per second.

    a. Convert the speed of sound to miles per second.

    b. Write the relationship between the number of miles sound travels and the number of seconds it takes sound to travel that distance as a direct variation. Be sure to identify what your variables represent.

    c. Use your answer from Part b to find the time it takes sound to travel four miles in air.

15. Refer to the formula S.A. $= 6s^2$ for the surface area of a cube with an edge of length s.

    a. Complete the table at the right.

    b. How many times as large is the area when $s = 4$ as when $s = 2$?

    c. How many times as large is the area when $s = 6$ as when $s = 3$?

    d. How many times as large is the area when $s = 10$ as when $s = 5$?

    e. **Fill in the Blank** Relate the length of the edge to the surface area of the cube. When the edge length doubles, the surface area ___?___.

    f. **Fill in the Blank** When the edge length triples, the surface area ___?___.

s	S.A.
1	?
2	?
3	?
4	?
5	?
6	?
7	?
8	?
9	?
10	?

16. The table at the right gives a typical thinking distance, braking distance, and overall stopping distance for a car traveling at various speeds.

    a. Show that the thinking distance t is directly proportional to the speed s by finding the constant of variation and writing a direct-variation formula.

    b. Show that the braking distance b is directly proportional to the square of the speed s by finding the constant of variation and writing a direct-variation formula.

    c. Is the overall stopping distance d directly proportional to the speed? Explain how you know.

Speed	Thinking Distance	Braking Distance	Overall Stopping Distance
20 mph	20 ft	20 ft	40 ft
30 mph	30 ft	45 ft	75 ft
40 mph	40 ft	80 ft	120 ft
50 mph	50 ft	125 ft	175 ft
60 mph	60 ft	180 ft	240 ft
70 mph	70 ft	245 ft	315 ft

REVIEW

17. The distance d in feet a body falls in t seconds is given by the formula $d = \frac{1}{2}gt^2$, where $g = 32 \frac{\text{ft}}{\text{sec}^2}$ is the constant of gravitational acceleration. (**Lesson 1-4**)

    a. How far will a body fall in 7 seconds?

    b. How long did a body fall if it traveled 83 feet?

18. Graph the function f with the equation $f(x) = \frac{1}{x^2}$ in a standard window. What is the domain of f? (**Lesson 1-4**)

19. In the table at the right, is y a function of x? Justify your answer. (**Lesson 1-2**)

20. A dime has a diameter of 17.91 mm. (**Lesson 1-1**)

    a. You place 15 dimes in a row. What is the length of your row of dimes?

    b. Write an expression to describe what happens to the length of your row of dimes when you remove x dimes.

x	y
−1	1
−1	−1
−4	2
0	0
−18	−4

21. Suppose $\triangle ABC \sim \triangle DEF$ (that is, the two triangles are similar).

    a. Sketch a possible diagram of this situation.

    b. What can you say about the ratios $\frac{AB}{BC}$ and $\frac{DE}{EF}$? (**Previous Course**)

In Questions 22 and 23, write the expression as a power of 5. (**Previous Course**)

22. $5^2 \cdot 5^3$

23. $(5 \cdot 5)^3$

EXPLORATION

24. Use the Internet to find some common gear ratios that are used for different speed settings on racing bikes. Write and solve a direct-variation problem involving one of these gear ratios.

Lesson 2-2

Inverse Variation

Vocabulary

inverse-variation function

varies inversely as

inversely proportional to

▶ **BIG IDEA**   When two variables x and y satisfy the equation $y = \frac{k}{x^n}$ for some constant value of k, we say that y varies inversely as x^n.

The Condo Care Company has been hired to paint the hallways in a condominium community. A few years ago, it took 8 workers 6 hours (that is, 48 worker-hours) to do this job. If w equals the number of workers and t equals the time (in hours) that each worker paints, then the product wt is the total number of hours worked. Since it takes 48 worker-hours to finish the job,

$$wt = 48, \text{ or } t = \frac{48}{w}.$$

Certain combinations of w and t that could finish the job are given below.

Mental Math

Let $g(x) = 2x^2$. Find:

a. $g(2)$

b. $g(0.4)$

c. $g(3n)$

d. $g(3n) - g(2) + g(1)$

Number of Workers w	1	3	5	6	8	12	15
Time t (hr)	48	16	9.6	8	6	4	3.2

 QY1

Inverse-Variation Functions

The formula $t = \frac{48}{w}$, which determines the values in the table above, has the form $y = \frac{k}{x^n}$ where $k = 48$ and $n = 1$. This is an example of an *inverse-variation function*.

▶ **QY1**

If 20 workers were to divide the painting job equally, how many hours would each one have to paint?

Definition of Inverse-Variation Function

An **inverse-variation function** is a function that can be described by a formula of the form $y = \frac{k}{x^n}$, with $k \neq 0$ and $n > 0$.

For the inverse-variation function with equation $y = \frac{k}{x^n}$, we say y **varies inversely as x^n**, or y is **inversely proportional to x^n**. In an inverse variation, as either quantity increases, the other decreases. In the painting example, as the number of workers increases, the number of hours each must work decreases.

As with direct variation, inverse variation occurs in many kinds of situations.

Example 1

The speed S of a bike varies inversely with the number B of back-gear teeth on the rear wheel. Write an equation that expresses this relationship.

Solution Use the definition of an inverse-variation function. In this case, $n = 1$. So,

$$S = \frac{k}{B}.$$

Solving Inverse-Variation Problems

Many scientific principles involve inverse-variation functions. For example, imagine that a person is sitting on one end of a seesaw. According to the *Law of the Lever*, in order to balance the seesaw another person must sit a certain distance d from the pivot (or fulcrum) of the seesaw, and that distance is inversely proportional to his or her weight w. Algebraically, $d = \frac{k}{w}$. Since d is inversely proportional to w, as d increases, w will decrease. This means a lighter person can balance the seesaw by sitting farther from the pivot, or a heavy person can balance the seesaw by sitting closer to the pivot.

Example 2

Ashlee and Sam are trying to balance on a seesaw. Suppose Sam, who weighs 42 kilograms, is sitting 2 meters from the pivot. Ashlee weighs 32 kilograms. How far away from the pivot must she sit to balance Sam?

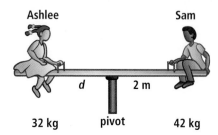

Ashlee   Sam

d 2 m

32 kg   pivot   42 kg

Solution Let d = a person's distance (in meters) from the pivot. Let w = that person's weight (in kilograms).

First write a variation equation relating d and w. From the *Law of the Lever*,

$$d = \frac{k}{w}.$$

To find k, substitute Sam's weight and distance from the pivot into $d = \frac{k}{w}$ and solve for k.

$$2m = \frac{k}{42 \text{ kg}}$$
$$k = 2 \text{ m} \cdot 42 \text{ kg}$$
$$k = 84 \text{ meter-kilograms}$$

Substitute the value found for k into the formula.

$$d = \frac{84}{w}$$

Substitute Ashlee's weight into the formula above to find the distance she must sit from the pivot.

$$d = \frac{84}{32} = 2.625 \text{ m}$$

Ashlee must sit about 2.6 meters away from the pivot to balance Sam.

Check Since $d = \frac{k}{w}$, $k = dw$. So the product of Ashlee's distance from the pivot and her weight should equal the constant of variation. Does 2.625 meters · 32 kilograms = 84 meter-kilograms? Yes, the numbers and the units agree.

 QY2

An Inverse-Square Situation

Just as one variable can vary directly as the square of another, one variable can also vary inversely as the square of another. For example, in the figure on the next page, a spotlight shines onto a wall through a square window that measures 1 foot on each side. Suppose the window is 5 feet from the light and the wall is 10 feet from the light. The light that comes through the window will illuminate a square on the wall that is 2 feet on a side. The same light that comes through the 1-square foot window now covers 4 square feet.

> ▸ **QY2**
>
> If Saul takes Sam's place on the seesaw and Saul weighs 55 kg, what is the new constant k of variation?

Since the same amount of light illuminates four times the area, the intensity of light on the wall is $\frac{1}{4}$ of its intensity at the window. As distance from the light source increases, the area the light illuminates increases, and the intensity of the light decreases. This is an example of inverse variation: the intensity I of light is inversely proportional to the square of the distance d from the light source.

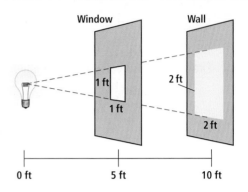

$$I = \frac{k}{d^2}$$

GUIDED

Example 3

Suppose the intensity of the light 4 meters from a light source is 40 lumens. (A lumen is the amount of light that falls on a 1-square foot area that is 1 foot from a candle.) Find the constant of variation and determine the intensity of the same light 6 meters from its source.

Solution Write an equation relating d and I, where $d =$ the distance from the light source in meters and $I =$ the light's intensity in lumens.

$$I = \frac{k}{?}$$

To find k, substitute $d = \underline{\ ?\ }$ and $I = \underline{\ ?\ }$ into your equation and solve for k.

$$\frac{?}{} = \frac{k}{?}$$

$$\underline{\ ?\ } \cdot \underline{\ ?\ } = k$$

$$\underline{\ ?\ } = k$$

Substitute k back into the equation to find the inverse-variation formula for this situation.

$$I = \frac{?}{d^2}$$

Evaluate this formula when $d = 6$ meters.

$$I = \frac{?}{?}$$

$$I = \underline{\ ?\ } \text{ lumens}$$

As you did in Lesson 2-1 for direct-variation problems, you can define functions on your CAS to help solve inverse-variation problems.

Activity

MATERIALS CAS

Step 1 Clear all variable values on your CAS.

Define the function $ink(xi, yi, n) = xi^n \cdot yi$.

This function calculates the constant of variation k from three inputs: an *initial* independent variable value xi, an *initial* dependent variable value yi, and the exponent n.

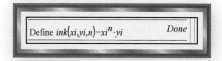

Define $ink(xi,yi,n)=xi^n \cdot yi$ *Done*

Step 2 Define the function $invar(x, k, n) = \dfrac{k}{x^n}$.

This function calculates an inverse-variation value from three inputs: *any* independent variable value x, the constant of variation k calculated by *ink*, and the exponent n.

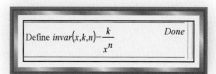

Define $invar(x,k,n)=\dfrac{k}{x^n}$ *Done*

Step 3 Check your solution to Example 3 by using *ink* to find k for $xi = 4$, $yi = 40$ and $n = 2$. Use *invar* with the appropriate inputs to verify the rest of your solution.

Questions

COVERING THE IDEAS

1. **Fill in the Blank** In the Condo Care Company problem at the beginning of this lesson, the time to finish the job varies inversely as the ___?___.

2. **Fill in the Blank** The equation $s = \dfrac{k}{r^2}$ means s varies inversely as ___?___.

3. **Multiple Choice** Assume k is a nonzero constant. Which equation does *not* represent an inverse variation?

    A $y = kx$ B $xy = k$ C $y = \dfrac{k}{x}$ D $y = \dfrac{k}{x^2}$

4. Refer to Example 1. Find the constant of variation if you are pedaling 21 mph and have 11 teeth on the back gear.

5. Refer to Example 2. If Sam sits 2.5 meters from the pivot, how far away from the pivot must Ashlee sit to balance him?

6. Suppose the seesaw at the right is balanced.
    a. Find the missing distance.
    b. If the 80 lb person sits farther from the pivot, which side of the seesaw will go up?

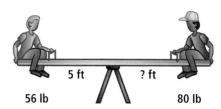

5 ft   ? ft

56 lb   80 lb

7. Refer to Example 3. Find the intensity of the light 9 meters from the source. How does this compare to the intensity of the light 6 meters from the source?

APPLYING THE MATHEMATICS

8. Translate this statement into a variation equation.

   The time t an appliance can run on 1 kilowatt-hour of electricity is inversely proportional to the wattage rating w of the appliance.

9. If y varies inversely as x^3, and $y = 12$ when $x = 5$, find the value of y when $x = 2$.

10. The weight W of a body above the surface of Earth varies inversely as the square of its distance d from the center of Earth. Use 4,000 miles for the radius of Earth.

    a. Write an inverse-variation function to model this situation.

    b. When an astronaut is 300 miles above Earth, what is the value of d?

    c. Suppose an astronaut weighs 170 pounds on the surface of Earth. What will the astronaut weigh in orbit 300 miles above Earth?

    d. What will be the astronaut's weight 2,000 miles above the surface of Earth?

11. Consider again the Condo Care Company situation at the beginning of the lesson.

    a. Complete the table below by filling in the missing values.

w	2	3	4	5	6	7	8	9	10	11	12	15	20
t	?	16	?	9.6	8	?	6	?	4.8	?	?	?	?

    b. **Fill in the Blank** Compare the values of t when $w = 2$ and $w = 4$. Also compare the values of t when $w = 4$ and $w = 8$, and again when $w = 6$ and $w = 12$. Make a conjecture. When the number of people working doubles, the mean time each person needs to work ___?___.

    c. **Fill in the Blank** Follow a similar procedure to complete the following conjecture. When the number of people working triples, the mean time ___?___.

    d. Prove your conjecture from Part b or Part c.

Fill in the Blank In 12–15, complete the sentence with the word *directly* or *inversely*.

12. The surface area of a sphere varies ___?___ as the square of its radius.

13. The number of hours required to drive a certain distance varies ___?___ as the speed of the car.

14. My hunger roughly varies ___?___ as the time since I last ate.

15. My hunger roughly varies ___?___ as the amount of food I have eaten.

REVIEW

16. At Percy's Priceless Pizza, the price of a pepperoni pizza is proportional to the square of its diameter. If Percy charges $11.95 for a 10-inch diameter pizza, how much does Percy charge for a 14-inch pizza? **(Lesson 2-1)**

17. At 7'7" and 303 pounds, Gheorghe Muresan was one of the tallest people ever to play professional basketball. For people with similar body shapes, weight varies directly with the cube of height. How much would you expect someone with Gheorghe's body shape to weigh if that person were 5'10"? **(Lesson 2-1)**

18. If $f(d) = 3d^3$ for all d, find $f(2x)$. **(Lesson 1-3)**

In **19** and **20**, simplify and indicate the general property. **(Previous Course)**

19. $\dfrac{x^{11}}{x^4}$

20. $(2x)^4$

21. At a certain time of day, a 13' tree casts a shadow 7' long.
    a. Draw a picture of this situation and mark a right triangle in your picture.
    b. A nearby tree is 18' tall. How long would its shadow be at the same time of day? **(Previous Course)**

22. In the figure at the right, line ℓ is parallel to line m. Find x. **(Previous Course)**

EXPLORATION

23. The *inverse-square law* in physics governs the way various things happen as distance varies, such as how the light intensity decreases as the distance from the source increases, as discussed in Example 3. Research the inverse-square law, and find three other situations where it applies.

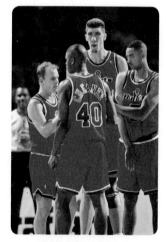

Gheorghe Muresan of the Washington Bullets

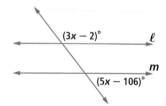

$(3x - 2)°$ ℓ

$(5x - 106)°$ m

QY ANSWERS

1. 2.4 hours

2. 110 meter-kilograms

Lesson 2-3

The Fundamental Theorem of Variation

▶ **BIG IDEA** When y varies directly as x^n, multiplying x by c causes y to be multiplied by c^n; when y varies inversely as x^n, multiplying x by c causes y to be divided by c^n.

In previous lessons, you explored the effects of doubling or tripling the length s of the edge of a cube on the surface area S.A. of the cube. You also explored the effects of doubling or tripling the value of w on the value of t in the equation $t = \frac{48}{w}$. In this lesson you will see how to generalize the findings of these problems.

This Activity explores how changes in the independent variable result in changes in the dependent variable in two different variation functions.

Activity

Fruit Roll Industries makes mini and regular fruit pops and is considering making larger jumbo pops. Assume the production cost p of a fruit pop varies directly with the volume V of candy on the stick. So, $p = k_0 \cdot V$. Also assume that a pop is approximately a sphere, so $V = \frac{4}{3}\pi r^3$, where r is the mean radius of pops of a particular size. Then $p = k_0 \cdot \frac{4}{3}\pi r^3$. Since $\frac{4}{3}$, π, and k_0 are all constants, $\frac{4}{3}\pi k_0$ is a constant and we can write $p = kr^3$. This means that p varies directly with the cube of r.

Step 1 Make a table like the one below based on the estimated radius of different-size pops. Fill in the blank cells in all columns except the one labeled Production Cost Estimate.

Size	Mean Radius r (cm)	r^3 (cm³)	Ratio of Radius to Mini's Radius	Ratio of r^3 to Mini's r^3	Production Cost Estimate
Mini	0.7	0.343	1:1	1:1	$\frac{1}{2}$ cent
Regular	1.4	?	?	?	?
Jumbo	2.1	?	?	?	?

Mental Math

Claudia is building an in-ground pool in her back yard. The pool is 12 feet wide and 20 feet long.

a. If the pool is 5 feet deep, how much water can it hold?

b. Claudia is building a cement walkway 2 feet wide around the pool. If she builds a fence around the walkway, how long will the fence be?

c. If she places a fence post every 4 feet around the walkway, how many posts will she need?

National Lollipop Day is celebrated on July 20.

Step 2 Compare r and r^3 for the mini and regular pops.

    a. The radius of the regular pop is __?__ times as big as the mini pop, but r^3 for the regular pop is __?__ times as big as for the mini pop. Since p varies directly as r^3, p is proportional to r^3. So, the cost of production of a regular pop should be __?__ times that of the mini, pop, or __?__ cents.

    b. Make the same comparisons for the mini and jumbo pops. The jumbo cost should be __?__ times the mini cost, or __?__ cents. Note that when the mini's radius is multiplied by 3 to get the jumbo radius, you can find the cost of the jumbo pop by multiplying the cost of the mini pop by __?__ to the third power.

    c. Suppose there is a super jumbo pop with a radius of 2.8 cm. Fill in the table to find the estimated cost of production? Justify your answer.

Size	Mean Radius r (cm)	r^3 (cm³)	Ratio of Radius to Mini's Radius	Ratio of r^3 to Mini's r^3	Production Cost Estimate
Super Jumbo	2.8	?	?	?	?

Step 3 Complete the following sentence to generalize your findings: If the production cost varies directly as r^3, and r is multiplied by a number c, then the cost is multiplied by __?__.

Step 4 Suppose there is only one size of carton used for fruit pop shipments and each carton contains fruit pops of only one size. As the radius of the fruit pop gets larger, what happens to the number of pops that fit in the carton? The number n of pops that fit in a carton varies ___?___ as the cube of the radius. Algebraically, $n = \dfrac{k}{r^3}$.

    a. Make a table like the one below and fill in the ratio column.

Size	Radius r (cm)	Ratio of r^3 to Mini's r^3	Number of Pops n in a Carton
Mini	0.7	1:1	270
Regular	1.4	?	?
Jumbo	2.1	?	?
Super Jumbo	2.8	?	?

    b. Note that since n varies inversely as r^3, n will __?__ as r^3 increases. Since r^3 for the regular pop is __?__ r^3 for the mini pop, $\dfrac{270}{?} =$ __?__ regular pops will fit in a carton. Write this number in the table.

(continued on next page)

c. Note that when the mini's radius is multiplied by 3 to get the jumbo's radius, you can find the number of jumbo pops that fit in the carton by __?__ the number of mini pops by __?__ to the __?__ power. You can find the number of super jumbo pops by __?__ the number of mini pops by __?__ to the __?__ power. Use this information to complete the rightmost column of the table.

Step 5 Complete the following sentence to generalize your findings: If the number of pops in a carton varies inversely as r^3, and r is multiplied by a number c, then the number of pops is divided by __?__.

 QY1

The generalizations you made in Steps 3 and 5 of the Activity are instances of the *Fundamental Theorem of Variation*. In your generalizations, $n = 3$. In the theorem, n can be any positive number.

> ▶ **QY1**
>
> If an all-day giant pop has a radius that is 5 times the radius of the mini pop, what is its production cost?

The Fundamental Theorem of Variation

1. If $y = kx^n$, that is, y varies *directly* as x^n, and x is multiplied by c, then y is multiplied by c^n.

2. If $y = \dfrac{k}{x^n}$, that is, y varies *inversely* as x^n, and x is multiplied by a non-zero constant c, then y is divided by c^n.

Proof of 1   Let y_1 = original value before multiplying x by c.

$$y_1 = kx^n \qquad \text{definition of direct variation}$$

Let y_2 = value when x is multiplied by c.

To find y_2, x must be multiplied by c.

$$y_2 = k(cx)^n \qquad \text{definition of } y_2$$
$$y_2 = k(c^n x^n) \qquad \text{Power of a Product Postulate}$$
$$y_2 = c^n(kx^n) \qquad \text{Associative and Commutative Properties of Multiplication}$$
$$y_2 = c^n y_1 \qquad \text{substitution of } y_1, \text{ for } kx^n$$

Proof of 2   The proof of this part is left for you to do in Question 17.

 QY2

> ▶ **QY2**
>
> If y varies *inversely* as x, and x is divided by c, what is the effect on y?

Example 1

Fruit Roll Industries needs a cost estimate for the wrappers of the new super jumbo pop. Because the pop is roughly a sphere, the area of the wrapper is a multiple of the surface area of a sphere, where $S.A. = 4\pi r^2$. So, the cost w of a wrapper varies directly with the surface area. The company knows the wrappers for the mini pops cost 1.5¢ each. How much do the wrappers of the super jumbo pop cost?

Solution Because the cost w_1 of a mini pop wrapper varies directly with the surface area, $w_1 = kr^2$ when r is the radius of the mini pop. In the Activity, the radius of the super jumbo pop was 4 times as long as the radius of the mini pop, or $4r$. So, if w_2 is the cost of a super jumbo pop wrapper, $w_2 = k(4r)^2$. Then

$w_2 = k(4^2 r^2)$ Power of a Product Property

$w_2 = 16(kr^2)$ Associative and Commutative Properties of Multiplication

$w_2 = 16w_1$ Substitution of w_1 for kr^2

So, the cost of a super jumbo pop wrapper is 16, or 4^2, times the cost of a mini pop wrapper. Because a mini pop wrapper costs 1.5¢, a super jumbo pop wrapper costs $16(1.5) = 24$¢.

GUIDED

Example 2

Generally, the weight of a land animal of a particular type varies directly with the cube of its femur diameter. Phoberomys, an extinct rodent that lived over 5 million years ago, is an ancestor of the modern guinea pig. Its femur diameter is 18 times that of today's average guinea pig. Estimate how the weight of this ancient rodent compares to the weight of a modern guinea pig.

Solution 1 Let d = the animal's femur diameter and w = the animal's weight. Since weight varies directly as the cube of femur diameter, an equation for the variation function is $w = kd^3$. Now apply the Fundamental Theorem of Variation. When d is multiplied by __?__, w is multiplied by __?__^{3} = __?__. Thus, the ancient rodent Phoberomys weighed about __?__ times as much as a modern guinea pig.

This is what a Phoberomys may have looked like.

Illustration by Carin L. Cain © Science

(continued on next page)

Solution 2 Set the problem up as in Example 1.

An equation for the variation function is $w = kd^3$. Let $w_1 =$ the weight of the guinea pig, $d =$ the femur diameter of the guinea pig, and $w_2 =$ the weight of Phoberomys with femur diameter $18d$.

$$w_1 = kd^3$$
$$w_2 = k(\underline{\ ?\ })^3$$
$$= k(\underline{\ ?\ }d^3)$$
$$= 18^3(\underline{\ ?\ })$$
$$= 18^3 w_1$$
$$= \underline{\ ?\ }w_1$$

So, $w_2 \approx 5800w_1$. Phoberomys weighed about __?__ times as much as the modern guinea pig.

A modern adult male guinea pig weighs about 2 pounds. So, with this model, an estimate of Phoberomys' weight is 11,600 pounds, or almost 6 tons! Other measurements of Phoberomys show that the $w = kd^3$ model overestimates the weight of the rodent. A better model is $w = kd^{2.5}$. You will study noninteger exponents in a later chapter.

Questions

COVERING THE IDEAS

In 1 and 2, refer to the Activity.

1. What is the production-cost estimate of a jumbo pop?

2. How many super jumbo pops fit in a carton?

In 3 and 4, consider the formula $S.A. = 4\pi r^2$ for the surface area of a sphere from Example 1.

3. **Fill in the Blank** The pairs $(2, 16\pi)$ and $(4, 64\pi)$ represent (r, s) for two spheres. They are instances of this pattern: if the radius r is doubled, the surface area is multiplied by __?__.

4. Show that the Fundamental Theorem of Variation is true for the points in Question 3.

5. **Fill in the Blank** If $y = kx^n$, and x is divided by c, then y is __?__.

6. **Fill in the Blank** If $y = \dfrac{k}{x^n}$, and x is multiplied by c $(c \neq 0)$, then y is __?__.

7. Refer to Example 2. Suppose the diameter of the femur bone of another ancient rodent is 3.2 times that of the modern guinea pig. Compare the weight of this ancient rodent to the weight of a modern one.

In 8–10, suppose $y = 3x^4$.

8. Complete the table of values at the right.

9. Describe the change in y when x is doubled. Explain your reasoning.

10. When answering the question, "Describe the change in y when x is divided by 3," Laura's response was "y is multiplied by $\frac{1}{81}$." Do you agree with Laura? Why or why not?

11. The volume of a cube varies directly with the cube of its edge length.
    a. If the edge length is multiplied by 8, what effect does that have on the volume?
    b. If the edge length is divided in half, what effect does that have on the volume?

x	y
1	?
2	?
3	?
4	?
6	?
8	?
9	?

APPLYING THE MATHEMATICS

12. Marta went to the farmer's market to buy oranges. The oranges that are 3 inches in diameter cost 25 cents per dozen. The oranges that are 4 inches in diameter are 50 cents per dozen. Marta chose the 3-inch oranges. Did she make the more economical decision? Explain your answer?

13. The Brobdingnagians in Jonathan Swift's *Gulliver's Travels* are similar to us, but they are 12 times as tall.
    a. How would you expect the weight of these giants to compare to our weight?
    b. How would you expect their surface area to compare to ours?

14. The inverse square law for light intensity, $I = \frac{k}{d^2}$, models the relationship between distance d from a light source and the intensity I of the light.
    a. When Booker is working at his computer, he is twice the distance from the floor lamp in his room as he is when he is working at his desk. Compare the light intensity of the floor lamp at his computer to the intensity at his desk. Justify your answer.
    b. Suppose Booker moves his computer so that it is three times as far from the floor lamp as his desk. Compare the intensity of the lamp light at his computer and his desk. Justify your answer.

In 15 and 16, refer to the illustration of the micro- and standard-size CDs at the right. The radius of the standard CD is 5 inches and the radius of the micro CD is 3 inches.

15. What is the ratio of the circumferences of the larger to the smaller CD?

16. Use the Fundamental Theorem of Variation to calculate the ratio of the surface areas of the larger to the smaller CD, assuming the CDs had no hole in the center.

17. Complete the proof of Part 2 of the Fundamental Theorem of Variation.

A standard and micro CD.

REVIEW

18. Translate this statement into a variation equation: The number of oranges that fit into a crate is inversely proportional to the cube of the radius of each orange. (**Lesson 2-2**)

19. **Multiple Choice** Most of the power of a boat motor results in the generation of a wake (the track left in the water). The engine power P used in generating the wake is directly proportional to the seventh power of the boat's speed s. Which equation models this variation? (**Lesson 2-1**)

   A $P = 7s$ B $s = kP^7$ C $P = ks^7$ D $P = k^7s$

20. Suppose V varies directly as the third power of r. If $V = 32$ when $r = 8$, find V when $r = 5$. (**Lesson 2-1**)

21. a. If one blank CD costs c cents, how much do n blank CDs cost? (**Lesson 1-1**)

   b. If two blank CDs cost d cents, how much do m blank CDs cost? (**Lesson 1-1**)

22. **Skill Sequence** Find all solutions. (**Previous Course**)

   a. $x^2 = 25$ b. $25y^2 = 36$ c. $3z = \frac{25}{3z}$

23. In the diagram at the right, $\triangle DIG \sim \triangle ART$. Find TR. (**Previous Course**)

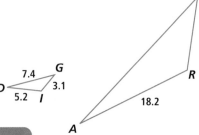

EXPLORATION

24. Go to the store or use the Internet to find a product that comes in two different sizes (for example, a regular and a large-screen television). Compare prices of the smaller and larger versions of the product. Using the Fundamental Theorem of Variation, decide whether the prices of the two products are in the proper ratio. Explain your decision. What factors other than size might affect the price?

QY ANSWERS

1. 62.5 cents

2. y is multiplied by c.

The Graph of $y = kx$

▶ **BIG IDEA** The graph of the set of points (x, y) satisfying $y = kx$, with k constant, is the line containing the origin and having slope k.

Recall from the Questions in Lesson 2-1 that the distance d you are from a lightning strike varies directly with the time t elapsed between seeing the lightning and hearing the thunder. The formula $d = \frac{1}{5}t$ describes this situation for the values given in that lesson. This direct-variation function can also be represented graphically. A table and a graph for the equation $d = \frac{1}{5}t$ are shown here.

Time t (sec)	0	5	10	15	20	25	30
Distance d (mi)	0	1	2	3	4	5	6

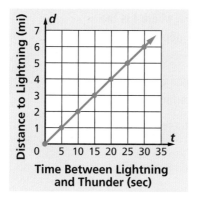

Time Between Lightning and Thunder (sec)

Note that the domain of this function is the set of nonnegative real numbers, and the range is also the set of nonnegative real numbers. When all real-world solutions to the equation $d = \frac{1}{5}t$ are plotted in the coordinate plane, the graph is a ray starting at the origin and passing through the first quadrant. There are no points on the graph of $d = \frac{1}{5}t$ in any other quadrants.

 QY1

▶ **QY1**

Explain why there could be no real-world solutions in Quadrants II, III, or IV.

The Slope of a Line

Recall that the steepness of a line is measured by a number called the *slope*. The slope of a line is the **rate of change** of y with respect to x and can be calculated using the coordinates of two points on the line. Let (x_1, y_1) and (x_2, y_2) be the two points. Then $y_2 - y_1$ is the vertical change (the change in the dependent variable), and $x_2 - x_1$ is the horizontal change (the change in the independent variable). The slope, or rate of change, is the quotient of these changes.

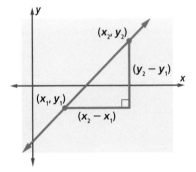

$$\text{slope} = \frac{\text{change in vertical distance}}{\text{change in horizontal distance}}$$

$$= \frac{\text{change in dependent variable}}{\text{change in independent variable}}$$

$$= \frac{y_2 - y_1}{x_2 - x_1}$$

Definition of Slope

The **slope** of the line through the two points (x_1, y_1) and (x_2, y_2) is $\frac{y_2 - y_1}{x_2 - x_1}$.

Example

Determine the slope of the line with equation $d = \frac{1}{5}t$, where t is the independent variable time (in seconds) and d is the dependent variable distance (in miles).

Solution Use the definition of slope. Because d is on the vertical axis and t is on the horizontal axis, the ordered pairs are of the form (t, d).

Find two points on the line; either point can be considered (t_1, d_1). Here we use $(t_1, d_1) = (10, 2)$ and $(t_2, d_2) = (15, 3)$.

$$\text{slope} = \frac{d_2 - d_1}{t_2 - t_1} = \frac{3 \text{ mi} - 2 \text{ mi}}{15 \text{ sec} - 10 \text{ sec}} = \frac{1}{5}\frac{\text{mi}}{\text{sec}}$$

Refer to the graph at the beginning of this lesson. Notice that for every change of 5 units to the right, there is a change of 1 unit up. This is equivalent to saying that for every change of 1 horizontal unit, there is a change of $\frac{1}{5}$ of a vertical unit. Notice that because of the difference in units, you cannot visually see the slope as $\frac{1}{5}$.

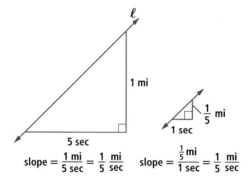

$$\text{slope} = \frac{1 \text{ mi}}{5 \text{ sec}} = \frac{1}{5} \frac{\text{mi}}{\text{sec}} \qquad \text{slope} = \frac{\frac{1}{5}\text{ mi}}{1 \text{ sec}} = \frac{1}{5} \frac{\text{mi}}{\text{sec}}$$

 QY2

▸ **QY2**

What is the slope of the line through (9, 36) and (25, 100)?

The Slope of $y = kx$

Observe from the Example that the graph of $y = \frac{1}{5}x$ has slope $\frac{1}{5}$. This is an instance of the following theorem.

> **Slope of $y = kx$ Theorem**
>
> The graph of the direct-variation function with equation $y = kx$ has constant slope k.

Proof Let (x_1, y_1) and (x_2, y_2) be any two distinct points on $y = kx$, with $k \neq 0$. Since the points are on the line,

$$y_1 = kx_1 \qquad \text{substitution}$$
$$\text{and } y_2 = kx_2$$

Now solve this system of equations for k.

$$y_2 - y_1 = kx_2 - kx_1 \qquad \text{Subtraction Property of Equality}$$
$$y_2 - y_1 = k(x_2 - x_1) \qquad \text{Distributive Property}$$
$$\frac{y_2 - y_1}{x_2 - x_1} = k \qquad \text{Division Property of Equality}$$

So by the definition of slope, k is the slope of the line through these points.

Thus, k is the slope of the line with equation $y = kx$.

All the lines with equation $y = kx$, where k is any real number, make up a family of lines through the origin with different slopes k. They are direct-variation functions. In general, the domain of a function with equation $y = kx$ is the set of real numbers, and the range is the set of real numbers.

To explore how the value of k affects the graph of $y = kx$, you can graph $y = kx$ using several different values of k. This approach lets you view several graphs simultaneously and compare them. The graphs of $y = kx$ for four different values of k are shown on the axes below.

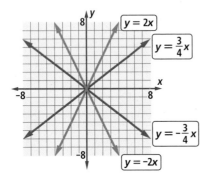

Activity

MATERIALS CAS, graphing calculator, or variation graph application

In this Activity, you will vary k to see how the graph of $y = kx$ changes.

Step 1 Consider the family of curves $y = kx$. Examine the graph for various positive values of k. Describe, in terms of slope, how the graph of the line $y = kx$ behaves for values of $k > 0$. How is the steepness of the graph affected as k increases?

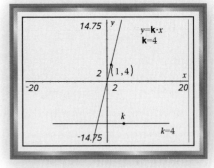

Step 2 Examine the graph for various negative values of k. Describe, in terms of slope, how the graph of the line $y = kx$ behaves for values of $k < 0$. How is the steepness of the graph affected as k decreases?

Step 3 Describe the graph of the line $y = kx$ when $k = 0$.

Step 4 Choose any k and note the coordinates of the point $A = (1, k)$ on the graph. A always has x-coordinate 1, but its y-coordinate changes with the value of k. Use the coordinates of point A and the fact that the line passes through the origin to find the slope of the line.

Compare this slope to the y-coordinate of A and the value of k. What do you notice?

Questions

COVERING THE IDEAS

1. By definition, $\frac{y_2 - y_1}{x_2 - x_1}$ is the slope of the line through which points?

2. Use the expression $\frac{y_2 - y_1}{x_2 - x_1}$ for the slope of a line.

   a. Let $(x_1, y_1) = (2, 4)$ and $(x_2, y_2) = (4.2, -5.3)$. Find the slope of the line through these two points.

   b. Let $(x_1, y_1) = (4.2, -5.3)$ and $(x_2, y_2) = (2, 4)$. Again, find the slope of the line through these points. Compare your answer to your answer to Part a.

   c. Mingmei incorrectly calculated the slope of the line through $(x_1, y_1) = (4.2, -5.3)$ and $(x_2, y_2) = (2, 4)$ as follows: $\frac{4 - (-5.3)}{4.2 - 2} \approx 4.23$. What error did she make? How does her answer compare to the answer you found in Part b?

   d. Given any two points $A = (c, d)$ and $B = (j, k)$, is the slope of the line through A and B the same as the slope of the line through B and A? Explain using the results you found in Parts a and b.

3. **Fill in the Blanks** Slope $= \dfrac{\text{change in the } \underline{\ ?\ } \text{ variable}}{\text{change in the } \underline{\ ?\ } \text{ variable}}$

4. **Fill in the Blanks** A slope of $-\frac{2}{5}$ means that for every change of 5 units to the right there is a change of __?__ units __?__. It also means that for every change of 1 horizontal unit there is a vertical change of __?__ unit.

5. In the graph of $d = \frac{1}{5}t$, the slope of the line is $\frac{1}{5}$. Write a sentence to describe this slope in the context of lightning and thunderstorms, using the appropriate units.

6. In the lesson, a triangle is drawn to show that a line has slope $\frac{1}{5}$. Draw a similar diagram that shows that a line has a slope of $\frac{7}{3}$.

7. **Fill in the Blanks** The graph of $y = kx$ slants up as you read from left to right if k is __?__. It slants down as you read from left to right if k is __?__.

8. When k is negative, in which quadrants is the graph of $y = kx$?

9. **Fill in the Blanks** The graph of every direct-variation function $y = kx$ is a __?__ with slope __?__ and passing through the point __?__.

APPLYING THE MATHEMATICS

In 10–12, compute the rate of change for the given situation. Include units where appropriate.

10. An escalator drops 2 feet for every 7 seconds traveled.

11. A car travels forward 60 miles every hour.

12. A line passes through the points $\left(-2, \frac{1}{3}\right)$ and $\left(\frac{2}{5}, 10\right)$.

13. Refer to Step 4 of the Activity. Find an additional point B on the line $y = kx$ for your chosen value of k. Calculate the slope of the line using points A and B. Compare this to the slope you calculated using A and O. How do the slopes compare to each other and to k?

14. Bicycle manufacturers have found that, for the average person and a given style of bicycle, the proper seat height h varies directly with the inseam measurement i of the rider's pants.
    a. If the seat height is 27 inches for an inseam of 25 inches, determine an equation for the variation function.
    b. Should the domain and range be all real numbers? Why or why not?
    c. What is the slope of the graph of this function?
    d. What does the slope represent in this situation?

15. Plot the points $(3, 1)$, $(9, 3)$, $(12, 6)$ on a coordinate plane. Do these points lie on a line that represents a direct-variation function of the form $y = kx$? If so, compute the rate of change of that function. If not, explain why not.

16. The federal minimum wage was $5.15/hr between 1997 and the summer of 2007.
    a. Determine the direct-variation function that calculates the amount of money m that someone earned (before taxes) by working w hours at this minimum wage.
    b. A typical work week has 40 hours. Determine a reasonable domain and range for the function that maps time worked in a week onto the amount earned.
    c. Sketch a detailed graph of the function.
    d. In July 2009, the minimum wage increased to $7.25 an hour. Describe how the graph of the function would change.

17. The graph at the right shows the four equations $y = 4x$, $y = -4x$, $y = \frac{1}{4}x$, and $y = -\frac{1}{4}x$. Match each graph with its equation.

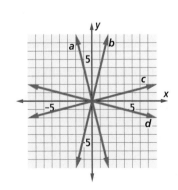

18. Below is a graph of $y = kx$. Explain how similar triangles can be used to show that the slope of the line is the same no matter which points are chosen to find the slope.

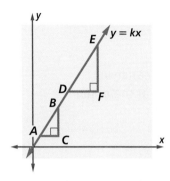

19. To program the formula for slope into a CAS, use the DEFINE command and enter slope $(x1, y1, x2, y2) = \frac{y_2 - y_1}{x_2 - x_1}$. Then use your programmed formula to find the slope of the line through the given points.

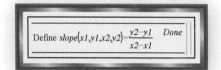

a. (152, 278) and (194, 360)   b. (8.4, –3.6) and (6, 10)

REVIEW

20. In the variation function $W = \frac{k}{d^3}$, what is the effect on W if
    a. d is tripled?
    b. d is halved? (Lesson 2-3)

21. Assume that the cost of a square cake varies directly as the square of the length of a side. What would be the ratio of the cost of a 6-inch square cake to a 10-inch square cake? (Lesson 2-3)

In 22–25, state whether the equation is of a direct-variation function, an inverse-variation function, or neither. (Lessons 2-1, 2-2)

22. $y = -\frac{3}{x}$ 23. $y = -\frac{x}{3}$

24. $y = -\frac{3}{x^2}$ 25. $y = x - 3$

26. Ohm's Law, $I = \frac{V}{R}$, relates current I (in amperes) to voltage V (in volts) and resistance R (in ohms). (Lesson 1-7)
    a. Solve this formula for R.   b. Solve this formula for V.

EXPLORATION

27. Each of the following terms is a synonym for *slope*. Research each term to find out who might use each one.
    a. marginal cost   b. pitch
    c. grade   d. rise over run

QY ANSWERS

1. In Quadrants II, III, and IV, t or d or both would be negative. The distance d cannot be negative, and the thunder cannot occur before the lightning, so the time t cannot be negative.

2. $\frac{100 - 36}{25 - 9} = \frac{64}{16} = 4$

Lesson 2-5

The Graph of $y = kx^2$

Vocabulary

parabola

vertex of a parabola

reflection-symmetric

line of symmetry

▶ **BIG IDEA** The graph of the set of points (x, y) satisfying $y = kx^2$, with k constant, is a parabola with vertex at the origin and containing the point $(1, k)$.

In Lesson 1-4, you studied functions whose values gave the stopping distances of a vehicle at various speeds. Stopping distance includes reaction time, braking distance, and other factors. Braking distance is simply the distance needed to stop a vehicle after applying the brake. Using mathematics you will study in calculus, it can be proved that braking distance varies directly with the square of a vehicle's speed. In Question 16 of Lesson 2-1, you found that the formula $d = \frac{1}{20}s^2$ describes this relation for a typical car. The table and graph below represent this relation.

Mental Math

Suppose that packs of gum cost 39 cents each.

a. What will be your change if you pay for two packs with a $1 bill?

b. What will be your change if you pay for eight packs with a $5 bill?

c. What will be your change if you pay for twenty packs with a $10 bill?

d. Now suppose you want 20 packs and the store is having a buy-three-packs, get-one-free sale. What will be your change if you pay with a $10 bill?

Speed s (mph)	Braking Distance d (ft)
0	0
10	5
20	20
30	45
40	80
50	125
60	180
70	245

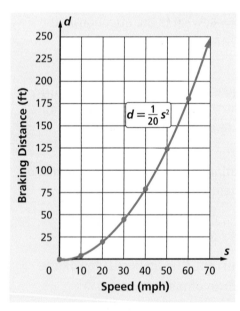

Rates of Change

The points on the graph above do not lie on a straight line. You can verify this by calculating the slopes of the lines through different pairs of points on the graph and seeing that they are not equal.

Example 1

Find the following rates of change and explain what each means in terms of braking distance.

a. r_1, the rate of change from $(20, 20)$ to $(40, 80)$

b. r_2, the rate of change from $(40, 80)$ to $(60, 180)$

 Solution

a. Use the definition of slope.
$$r_1 = \frac{80 \text{ ft} - 20 \text{ ft}}{40 \text{ mph} - 20 \text{ mph}} = \frac{60 \text{ ft}}{20 \text{ mph}} = \frac{3 \text{ ft}}{\text{mph}}$$
This means that, on average, when driving between 20 mph and 40 mph, for every increase of 1 mph in speed, you need 3 more feet to stop your car.

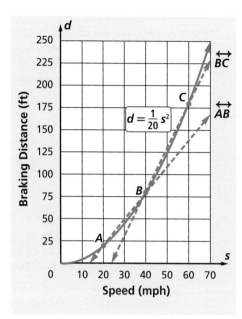

b. Similarly, $r_2 = \frac{180 \text{ ft} - 80 \text{ ft}}{60 \text{ mph} - 40 \text{ mph}} = \frac{100 \text{ ft}}{20 \text{ mph}} = 5 \frac{\text{ft}}{\text{mph}}$.
So on average, between $s = 40$ and $s = 60$, for every change of 1 mph (the horizontal unit), there is a change of 5 feet of braking distance (the vertical unit).

Check Look at the points on the graph. Let $A = (20, 40)$, $B = (40, 80)$, and $C = (60, 180)$. Is $\overleftrightarrow{BC}$ steeper than $\overleftrightarrow{AB}$? Yes, it is.

 QY1

> ▶ **QY1**
>
> Calculate the average rate of change between $A = (20, 40)$ and $C = (60, 180)$ in Example 1 and explain what it means.

The rate of change between different pairs of points on the graph of $d = \frac{1}{20}s^2$ is not constant. Two conclusions can be drawn:

1. The graph of $d = \frac{1}{20}s^2$ is not a line.

2. A single number does not describe the slope of the whole graph.

Notice that the slope is larger where the graph is steeper, meaning the braking distance increases more and more rapidly as the speed increases.

The equation $d = \frac{1}{20}s^2$ represents a direct-variation function of the form $y = kx^2$. All graphs of equations of this form share some properties. You should be able to sketch graphs of any equation of this form.

Activity

MATERIALS CAS, graphing calculator, or variation graph application

In this Activity, you will explore graphs of the family of curves $y = kx^2$.

Step 1 Consider the family of curves $y = kx^2$. The graph has point $A = (1, k)$ on it. Verify that the coordinates for point A are correct by substituting $x = 1$ into $y = kx^2$.

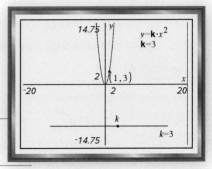

Step 2 Grab and drag point A to vary the shape of the graph. Since point A has y-coordinate k, the value of k varies as you drag A.

Step 3 When $k > 0$, and k increases in value, how does the graph of $y = kx^2$ change?

Step 4 When $k = 0$, describe the appearance of the graph $y = kx^2$. What are the coordinates of A?

Step 5 When $k < 0$, and k decreases in value, how does the graph of $y = kx^2$ change? How is the graph of $y = kx^2$ different when k is negative from when k is positive?

The Graph of $y = kx^2$

All of the graphs in the family of curves with the equation $y = kx^2$ with $k \neq 0$ are **parabolas**. No matter what value is chosen for k, the parabola passes through the point $(0, 0)$. This point is the parabola's **vertex**. Further, each parabola in this family coincides with its reflection image over the y-axis. Thus, each parabola is **reflection-symmetric**, and the y-axis is called the **line of symmetry**.

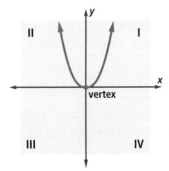

$y = kx^2, k > 0$
parabola opens up
Domain: the set of all reals
Range: the set of nonnegative reals

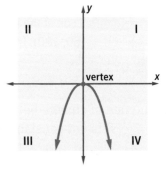

$y = kx^2, k < 0$
parabola opens down
Domain: the set of all reals
Range: the set of nonpositive reals

In general, the domain of the function with equation $y = kx^2$ is the set of all real numbers. As you saw in the Activity, when $k > 0$, the range is the set of nonnegative real numbers, and the graph of the function is a parabola that *opens up*. That is, the vertex of the parabola is its *minimum* point. When $k < 0$, the range is the set of nonpositive real numbers and the corresponding parabola *opens down*. That is, the vertex of the parabola is its *maximum* point.

▶ **READING MATH**

Parabola (a curve) and *parable* (a story) come from the same Greek word *parabolē* meaning "comparison." The *parabola* is named for a comparison of two distances that are equal.

QY2

The equation $y = kx^2$ is not the only equation whose graph is a parabola. You will study other equations leading to parabolas in Chapter 6.

Questions

COVERING THE IDEAS

In 1 and 2, refer to the formula $d = \frac{1}{20}s^2$ relating speed and braking distance.

1. Find the average rate of change between each pair of points and explain what it means.
   a. $(10, 5)$ and $(20, 20)$
   b. $(50, 125)$ and $(60, 180)$
   c. (a, b) and (c, d)

2. Your answers to Questions 1a and 1b should be different numbers. What does that tell you about the graph of $d = \frac{1}{20}s^2$?

3. Name the type of curve that results from graphing $y = kx^2$ $(k \neq 0)$.

4. Explain what it means to say that the graph of $y = kx^2$ is symmetric to the y-axis.

5. Suppose $k < 0$. State the domain and range of the function $f: x \rightarrow kx^2$.

6. Refer to the Activity. Describe how the graph of $y = kx^2$ changes in each situation.
   a. k increases from 1 to 10
   b. k decreases from 1 to 0

7. For what values of k does the graph of $y = kx^2$
   a. open up?
   b. open down?

In 8 and 9, one variable varies directly as another.
   a. Name the variables.
   b. Name the constant of variation.
8. $S.A. = 4\pi r^2$ (surface area of a sphere)
9. $A = \frac{s^2}{4}\sqrt{3}$ (area of an equilateral triangle)

APPLYING THE MATHEMATICS

10. **Matching** Match each equation with the proper graph. Assume each graph has the same scale.

    a. $y = \frac{1}{4}x^2$　　b. $y = -3x$　　c. $y = -2x^2$　　d. $y = 2x^2$

    i. 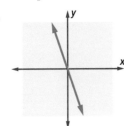　　ii. 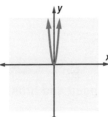　　iii. 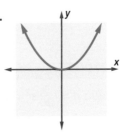　　iv.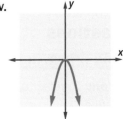

11. a. Explain why the point $(0, 0)$ is on the graph of $y = kx^2$ for all values of k.

    b. Explain how to use the point $(1, f(1))$ to determine the value of k if $f(x) = kx^2$.

12. On some cameras, you can control the diameter of the aperture, or the opening through which light passes to the film, by setting an f-stop. You also control the *exposure*, or the length of time the aperture is open, to ensure that the proper amount of light reaches the film. Let E be the exposure (in seconds) and f be the f-stop of a camera. Suppose you know that E varies directly as either f or f^2, and the constant of variation is $\frac{1}{6000}$.

    a. Assume that E varies directly as f, so $E = \frac{1}{6000}f$. Make a table of ordered pairs for this function. For values of f use 1, 1.4, 2, 2.8, 4, and 5.6.

    b. Using a special camera, you set an f-stop of 4.8. Use your table to find two values close to the required exposure. Average the two values to make an estimate of E for $f = 4.8$.

    c. Graph the ordered pairs from the table on a coordinate grid and connect the points with a smooth curve. Use the graph to estimate the required exposure.

    d. Using the variation equation, determine the actual exposure required for an f-stop of 4.8. How do the three values you found for the exposure compare?

    e. Repeat Parts a–d, but this time assume that E varies directly as f^2, so $E = \frac{1}{6000}f^2$. How do the three values you found for the exposure compare? How does the shape of the graph affect the accuracy of these methods for estimating E?

13. The maximum load a beam can safely support varies directly as its width. A contractor suggests replacing an 8"-wide beam that can hold 6000 pounds with two 4"-wide beams made of the same material as the 8" beam, one next to the other. Can the two 4" beams replace a single 8" beam safely? Justify your answer.

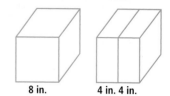

8 in.   4 in. 4 in.

14. Suppose (3, 5) is on the graph of $y = kx^2$. Is this sufficient information to determine k? If so, find k. If not, why not?

REVIEW

15. Consider the following four equations. (**Lesson 2-4, Previous Course**)

$$y = 3x \qquad y = \tfrac{1}{3}x \qquad y = -\tfrac{1}{3}x \qquad y = -3x$$

  a. Graph the equations on one set of axes.

  b. Find the slope of each line.

  c. Give the equations of two lines that appear to be perpendicular.

16. Find the slope of a submarine dive if the submarine drops 1,200 feet while moving forward 4,000 feet. (**Lesson 2-4**)

17. Consider the sequence t defined by $t_n = 3n^2 - 2n + 2$. (**Lessons 1-2, 1-5, 1-8**)

  a. What is the domain of the sequence?

  b. Use your CAS to generate a table containing the first forty terms of this sequence. Write down the last five.

18. Onida's annual salary is $42,000. She gets an increase of 6% at the end of each year. The formula $a_n = 42{,}000(1.06)^n$ gives Onida's salary a_n at the end of the nth year. (**Lesson 1-8**)

  a. Find Onida's salary at the end of each year for the first five years.

  b. What would Onida's salary be after 20 years with this company?

19. a. Write a letter of the alphabet that has exactly one line of symmetry.

  b. Write a letter of the alphabet that has exactly two lines of symmetry. (**Previous Course**)

EXPLORATION

20. The f-stop settings on a camera are related to the diameter of the aperture of the camera. Increasing the f-stop decreases the diameter of the aperture. Research to find out the relationship between these two variables.

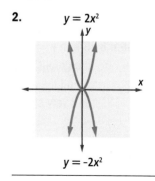

Lesson 2-6

The Graph of $y = \dfrac{k}{x}$ and $y = \dfrac{k}{x^2}$

Vocabulary

hyperbola

branches of a hyperbola

vertical asymptote

horizontal asymptote

discrete set

inverse-square curve

▶ **BIG IDEA** The graph of the set of points (x, y) satisfying $y = \dfrac{k}{x}$, with k constant, is a hyperbola with the x- and y-axes as asymptotes; the graph of the set of points (x, y) satisfying $y = \dfrac{k}{x^2}$ looks somewhat similiar but is closer to the axes.

In previous lessons you explored the graphs of the equations of the direct-variation functions $y = kx$ and $y = kx^2$. Activity 1 explores the quite different graph of the equation of an inverse-variation function, $y = \dfrac{k}{x}$.

Mental Math

$m\angle 1 = 72$.
What is $m\angle 2$ if

a. $\angle 1$ and $\angle 2$ are supplementary?

b. $\angle 1$ and $\angle 2$ are the two acute angles in a right triangle?

c. $\angle 1$ and $\angle 2$ are vertical angles?

d. $\angle 1$ and $\angle 2$ form a linear pair?

Activity 1

MATERIALS graphing utility

Step 1 Graph the function $y = \dfrac{10}{x}$ in a standard window. The resulting graph has points only in Quadrants I and III.

Step 2 Trace along the graph starting from a positive value of x and moving to the right. You can trace beyond the window edge if you want. It may also help if you zoom in on the graph for large values of x. Describe what happens to y as x increases in value.

Step 3 Now trace to the left starting from a negative value of x. Describe what happens to y as x decreases in value.

Step 4 Does the graph of $y = \dfrac{10}{x}$ ever intersect the x-axis? If it does, give the coordinates of any point(s) of intersection. If it does not, explain why not.

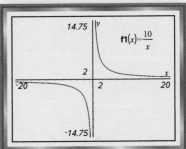

Step 5 Now trace closer and closer to $x = 0$ from both the positive and negative directions. Does the graph of $y = \dfrac{10}{x}$ ever intersect the y-axis? If it does, give the coordinates of any point(s) of intersection. If it does not, explain why not.

Step 6 Graph $y = \dfrac{k}{x}$ for a few nonzero values of k other than 10. Trace along the graphs past the edges of the standard window and near $x = 0$. Do all the graphs have the same behavior as x gets very far from 0 and very close to 0?

The Graph of $y = \frac{k}{x}$

The graph of every function with an equation of the form $y = \frac{k}{x}$, where $k \neq 0$, is a **hyperbola**. In Activity 1 you graphed the hyperbola $y = \frac{10}{x}$. All hyperbolas with equation $y = \frac{k}{x}$ share some properties. For example, the functions they represent all have the same domain and range.

Example 1

What are the domain and range of the function with equation $y = \frac{k}{x}$?

Solution To find the domain, think: What values can x have? You can substitute any number for x except 0 into the equation $y = \frac{k}{x}$.

So, the domain of the function with equation $y = \frac{k}{x}$ is $\{x \mid x \neq 0\}$.

To find the range, think: What values can y have? Recall from the Activity that y can be positive or negative, large or small. But if $y = 0$, you would have $0 = \frac{k}{x}$, or $0 \cdot x = k$, which is impossible because, in the formula $y = \frac{k}{x}$ for a hyperbola, k cannot be zero. So, the range of the function with equation $y = \frac{k}{x}$ is $\{y \mid y \neq 0\}$.

The graphs in the $y = \frac{k}{x}$ family of curves share another property. Each hyperbola in this family consists of two separate parts, called **branches**. When $k > 0$, the branches of $y = \frac{k}{x}$ lie in the first and third quadrants; if $k < 0$, the branches lie in the second and fourth quadrants, as shown below.

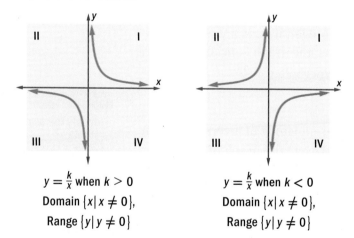

$y = \frac{k}{x}$ when $k > 0$
Domain $\{x \mid x \neq 0\}$,
Range $\{y \mid y \neq 0\}$

$y = \frac{k}{x}$ when $k < 0$
Domain $\{x \mid x \neq 0\}$,
Range $\{y \mid y \neq 0\}$

Asymptotes

When $x = 0$, $y = \frac{k}{x}$ is undefined, so, the curve does not cross the y-axis. However, when x is *near* 0, the function is defined.

You found in Activity 1 that for $y = \frac{k}{x}$, $k > 0$, as x approaches zero from the positive direction, the y-value gets larger and larger. Similarly, as x approaches zero from the negative direction, the y-value gets smaller and smaller (more and more negative). As the values of x get closer and closer to a certain value a, if the values of the function get larger and larger or smaller and smaller, then the vertical line with equation $x = a$ is called a **vertical asymptote** of the graph of the function. The y-axis is a vertical asymptote to the graph of $y = \frac{k}{x}$, for $k \neq 0$.

Similarly, the x-axis is a **horizontal asymptote** of the curve $y = \frac{k}{x}$. As x gets very, very large (or very, very small), the value of y gets closer and closer to zero.

In some situations, only one branch of a hyperbola is relevant. For instance, recall the Condo Care Company example of Lesson 2-2. The time t it takes to complete the job and the number of workers w are related by the equation $t = \frac{48}{w}$. A table of values for this equation is shown below, along with a graph of the points in the table at the right. Because you cannot have a negative number of workers, there are no points in Quadrant III.

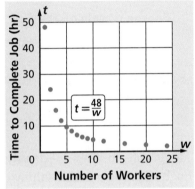

w	1	2	3	4	5	6	7	8	9	10	12	16	20	24
t	48	24	16	12	$9\frac{3}{5}$	8	$6\frac{6}{7}$	6	$5\frac{1}{3}$	$4\frac{4}{5}$	4	3	$2\frac{2}{5}$	2

In this example the function has a *discrete* domain, the set of positive integers. A **discrete set** is one in which there is a positive distance greater than some fixed amount between any two elements of the set. Because the number of workers is an integer, it does not make sense to connect the points of this graph. Thus, the graph consists of a set of discrete points on one branch of the hyperbola $y = \frac{48}{x}$.

 QY

Activity 2 explores the graphs of a different family of inverse-variation functions, those with equation $y = \frac{k}{x^2}$. We call the graph of such an inverse-variation function an **inverse-square curve**.

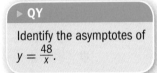

▶ QY

Identify the asymptotes of $y = \frac{48}{x}$.

Activity 2

Step 1 Graph the function $y = \frac{10}{x^2}$ in a standard window. The resulting graph appears in Quadrants I and II. Why do you think this graph does not appear in Quadrants III or IV?

Step 2 Trace along the curve to a positive value of x. Continue tracing to the right. Describe what happens to y as x increases.

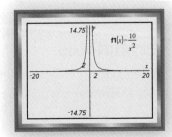

Step 3 Now trace to a negative value of x. Continue tracing to the left. Describe what happens to y as x decreases.

Step 4 Does the graph of $y = \frac{10}{x^2}$ ever intersect the x-axis? If it does, give the coordinates of the point(s) of intersection. If it does not, explain why not.

Step 5 Does the graph of $y = \frac{10}{x^2}$ ever intersect the y-axis? If it does, give the coordinates of the point(s) of intersection. If it does not, explain why not.

Step 6 Trace the graph close to $x = 0$ from both the positive direction and the negative direction. How does the graph behave near $x = 0$? Why does it behave this way?

The Graph of $y = \frac{k}{x^2}$

Example 2 examines the graph of $y = \frac{k}{x^2}$ for two values of k, one positive value and one negative value.

Example 2

a. Graph $y = \frac{24}{x^2}$ in a window with $\{x \mid -5 \le x \le 5\}$ and $\{y \mid -10 \le y \le 10\}$. Sketch the graph on your paper. Then clear the screen and graph $y = -\frac{24}{x^2}$.

b. Describe the symmetry of each graph.

Solution

a. The graphs below were generated using a graphing utility.

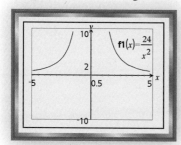 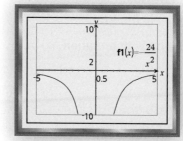

b. If either graph were reflected over the y-axis, the preimage and the image would coincide. So, each graph is symmetric about the y-axis.

Notice that the inverse-square curve, like a hyperbola, has two distinct branches. These two branches, however, do not form a hyperbola because the shape and relative position of the branches differ from a hyperbola. The domain of every inverse square function is $\{x \mid x \ne 0\}$. The range depends on the value of k. When k is positive, the range is $\{y \mid y > 0\}$. When k is negative, the range is $\{y \mid y < 0\}$. Graphs of the two types of inverse-square curves are shown on the next page.

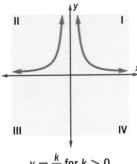

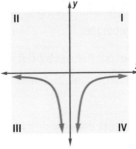

$y = \frac{k}{x^2}$ for $k > 0$
Domain $\{x \mid x \neq 0\}$
Range $\{y \mid y \neq 0\}$

$y = \frac{k}{x^2}$ for $k < 0$
Domain $\{x \mid x \neq 0\}$
Range $\{y \mid y < 0\}$

Notice that the graph of $y = \frac{k}{x^2}$ has the same vertical and horizontal asymptotes as the graph of $y = \frac{k}{x}$.

In general, asymptotes may be vertical, horizontal, or oblique, but not all graphs have asymptotes. For example, the parabola $y = kx^2$ does not have any asymptotes. When a graph has an asymptote, that asymptote is *not* part of the graph.

Questions

COVERING THE IDEAS

1. Why is 0 not in the domain of the functions with equations $y = \frac{k}{x}$ and $y = \frac{k}{x^2}$?

2. For what values of k is the range of the function $y = \frac{k}{x^2}$ the set $\{y \mid y < 0\}$?

3. Suppose $k \neq 0$. What shape is the graph of $y = \frac{k}{x}$?

In 4 and 5, consider the function f with equation $f(x) = \frac{15}{x}$.

4. State the domain and range of f.

5. a. Evaluate $f(-10)$ and $f(15)$.
   b. Graph f in the window $-12 \leq x \leq 12$ and $-12 \leq y \leq 12$ and sketch the graph.

6. Refer to the Condo Care Company situation and the graph of $t = \frac{48}{w}$ in this lesson. Why is the domain discrete?

7. Suppose $k \neq 0$. Is the graph of the equation symmetric to the y-axis?
   a. $y = \frac{k}{x}$
   b. $y = \frac{k}{x^2}$

8. In which quadrants are the branches of $y = \frac{k}{x^2}$
   a. if k is positive?
   b. if k is negative?

9. What happens to the y-coordinates of the graph of $y = \frac{21}{x}$ when x is positive and is getting closer and closer to 0?

10. Identify the asymptotes of the graph of $y = \frac{4}{x}$.

APPLYING THE MATHEMATICS

11. Match each equation with the proper graph.

   a. $y = \frac{5}{x^2}$ **b.** $y = -\frac{10}{x^2}$ **c.** $y = \frac{20}{x}$ **d.** $y = -\frac{5}{x}$

   i. **ii.** **iii.** 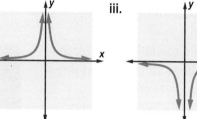 **iv.**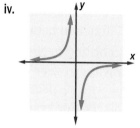

   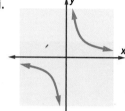

12. Compare and contrast the graphs of $y = -\frac{k}{x^2}$ and $y = \frac{k}{x^2}$ when $k = \frac{1}{10}$.

13. *The Doorbell Rang*, by Pat Hutchins (1986), tells the story of 12 cookies to be divided among an increasing number of people. The story begins with 2 kids who want to share the cookies. Then another kid comes to the door and they must divide the cookies 3 ways, then 4, and so on.

   a. The table at the right shows some values of the function described in the story. Fill in the missing values.

   b. Write an equation for a function that contains the coordinate pairs in the table. Explain the meaning of your variables.

   c. Explain why this function has a discrete domain.

Number of Kids	Number of Cookies Each Kid Gets
1	12
2	6
3	?
4	3
5	$\frac{12}{5} = 2.4$
6	?
7	$\frac{12}{7} \approx 1.7$
8	?
9	$\frac{12}{9} \approx 1.3$
10	?

14. Quentin is on a seesaw. He weighs 120 pounds and is sitting 5 feet from the pivot. Recall that the Law of the Lever is $d = \frac{k}{w}$, where d is distance from the pivot, w is weight, and k is a constant of variation.

   a. Find k and write a variation equation for this situation.

   b. Make a table of 10 different weights and their distances from the pivot that would balance Quentin.

   c. Plot your values from Part b.

   d. Is the domain for the context discrete? Explain why or why not.

15. Consider the functions f and g described by the equations
$f(x) = \frac{36}{x}$ and $g(x) = \frac{36}{x^2}$.

   a. Use a graphing utility to graph both functions on the same axes, with the window $-12 \le x \le 12$ and $-60 \le y \le 60$. Describe some similarities and differences between the graphs.

   b. Find and compare the average rates of change of each function between $x = 1$ and $x = 6$.

   c. Find and compare the average rates of change of each function between $x = 6$ and $x = 12$.

16. According to Newton's Law of Gravitation, the force of gravity between two objects obeys an inverse-square law with respect to the distance between the two objects. For example, the force of gravity between Earth and the Voyager 1 space probe launched in 1977 is approximately $F = \frac{1.56 \cdot 10^{12}}{d^2}$, where d is the distance between Earth and Voyager 1 in kilometers, and F is the force in newtons.

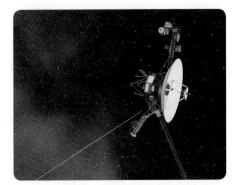

The space probe Voyager 1

   a. In the summer of 2002, Voyager 1 was approximately $1.3 \cdot 10^{10}$ km from Earth. What was the force of gravity between Voyager 1 and Earth at that time?

   b. Will Voyager 1 ever completely escape Earth's gravity? Explain your answer.

REVIEW

17. In the graph at the right, parabolas P_1 and P_2 are reflection images of each other over the x-axis. If parabola P_1 has equation $y = 3x^2$, what is an equation for parabola P_2? (**Lesson 2-5**)

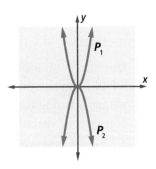

18. Does the line m graphed below at the right illustrate an example of direct variation, inverse variation, or neither? Justify your answer. (**Lesson 2-4**)

19. Dominique, a scuba diver 99 feet below the water's surface, inflates a balloon until it has a diameter of 6 inches. The air in the balloon expands as Dominique ascends. When Dominique reaches the surface, the balloon has a diameter of 9.5 inches. (**Lesson 2-3**)

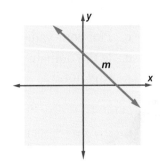

   a. Write the ratio of the radius of the balloon at the surface to the radius 99 feet below the surface.

   b. The volume of the balloon at the surface is how many times the volume of the balloon at 99 feet?

   c. Explain how your answer to Part a can be used to answer Part b.

Multiple Choice In 20 and 21, which could be a formula for the *n*th term of the sequence? **(Lesson 1-8)**

20. 2, 4, 8, 16, 32, …

    A $t_n = 2n$ **B** $t_n = n^2$ **C** $t_n = 2^n$

21. 2, 9, 28, 65, 126, …

    A $t_n = 7n - 5$ **B** $t_n = 7n^2 - 2$ **C** $t_n = n^3 + 1$

22. Solve for x: $y = -\frac{1}{\pi}x$. **(Lesson 1-7)**

23. Bob decided to bake a frozen pizza for dinner. It took 10 minutes to preheat the oven to 425° from room temperature of 75°. The pizza baked for 17 minutes, and then it took another 30 minutes for the oven to return to room temperature after Bob turned it off. **(Lessons 1-4, 1-2)**

    a. Identify the independent and dependent variables in this situation.

    b. Sketch a graph of the situation. Label the axes and explain the meaning of your variables.

    c. Identify the domain and range of the function mapping time onto the temperature of the oven.

24. **Multiple Choice** The relationship between $\triangle ABC$ and $\triangle PRQ$, as shown below, is best described by which of the following? **(Previous Course)**

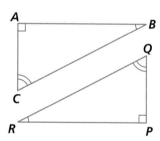

    A congruent

    B similar

    C both congruent and similar

    D neither congruent nor similar

EXPLORATION

25. Use a graphing utility to graph $y = \frac{10}{x^3}$, $y = \frac{10}{x^4}$, and $y = \frac{10}{x^5}$. What pattern do you notice? What generalization can you make about the value of n and the graph of $y = \frac{10}{x^n}$? Is the same generalization true for every function with equation $y = \frac{k}{x^n}$, $k \neq 0$?

Fitting a Model to Data I

▶ **BIG IDEA** If you determine from a particular set of data that *y* varies directly or inversely as *x*, you can graph the data to see what relationship is reasonable. Using that relationship, you can then substitute a known pair of values of *x* and *y* to obtain the constant *k* of variation, and compare what the variation equation predicts to the original data.

Recall from Chapter 1 that a mathematical model for a real situation is a description of that situation using the language and concepts of mathematics. Models can be created from collected data or from mathematical properties. Using data to create a mathematical model that describes those data is called *fitting a model to data*.

A Model of Direct Variation

In 1638, the Italian scientist Galileo Galilei (1564–1642) published *Dialogues Concerning Two New Sciences*, in which he first proposed the Law of Free-Falling Objects. This law states that near the surface of Earth, all heavier-than-air objects dropped from the same height fall to Earth in the same amount of time, assuming no resistance on the objects. (Aristotle, 1900 years earlier, wrote that heavier objects fall faster, and people believed Aristotle.) Equipment for timing free-falling bodies with sufficient precision did not exist, so Galileo tested his theory and developed a model by rolling objects down an inclined plane.

Today, scientists have more precise measuring techniques. For example, scientists can use slow-motion film to determine the distance an object in free fall travels over different periods of time. The table below gives the distance *d* in meters that a ball travels in *t* seconds after it is dropped from the top of a cliff. The ordered pairs in the table are graphed at the right.

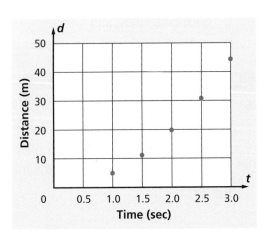

Time in Air *t* (sec)	1	1.5	2	2.5	3
Distance Fallen *d* (m)	4.9	11.0	19.6	30.6	44.1

Because the distance the ball travels depends on the elapsed time, distance is the dependent variable and is placed on the vertical axis. How is the distance traveled related to time? Notice that distance increases as more time elapses; this implies direct variation. However, the points do not all lie on a straight line. The points suggest a direct-variation model of the form $y = kx^2$.

GUIDED

Example 1

Find a variation equation to describe the free-falling object data on the previous page.

Solution The shape of the graph suggests the formula $d = kt^2$. Substitute one of the ordered pairs into this formula to find k. The easiest pair to use is $(1, 4.9)$.

$$\underline{\ ?\ } = k\ \underline{\ ?\ }^2$$

$$k = \underline{\ ?\ }\frac{m}{sec^2}$$

So, a variation equation describing the situation is $d = \underline{\ ?\ }t^2$.

Check See how close the values predicted by the equation are to the values observed.

Time in Air t (sec)	1	1.5	2	2.5	3
Observed Distance Fallen d (m)	4.9	11.0	19.6	30.6	44.1
Predicted Distance Fallen d (m)	?	?	?	?	?

 QY1

After fitting a model to a set of data, you can use the model to predict other data points that were not in the original set. The model $d = 4.9t^2$ predicts that the distance traveled in 6.5 seconds would be

$$d = (4.9)(6.5)^2 = 207.025.$$

That is, the distance would be about 207 meters.

 QY2

A Model of Inverse Variation

Anna Lyzer and her sister Jenna went to a playground to investigate the validity of the Law of the Lever. Anna sat at a fixed point on one side of the seesaw while nine of her friends took turns sitting on the other side. When the seesaw was in balance, Jenna measured each person's distance d from the pivot and recorded their weight w. The results are shown on the next page.

▸ **QY1**

If $k = 4.9$ m/sec^2 and $d = kt^2$, why is the unit for d meters?

▸ **QY2**

According to the model, approximately how far will an object fall in 8.25 seconds?

The shape of the graph shows that distance decreased as weight increased and suggests an inverse variation. You have seen graphs like this from two possible models: $d = \frac{k}{w^2}$ and $d = \frac{k}{w}$. The Law of the Lever says that $d = \frac{k}{w}$ is the more appropriate model, but how can the data tell us that? This Activity shows one way.

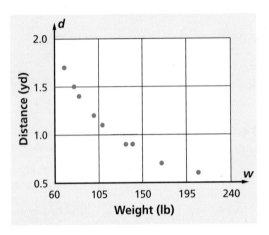

w (lb)	d (yd)
70	1.7
80	1.5
85	1.4
100	1.2
109	1.1
133	0.9
140	0.9
170	0.7
206	0.6

Activity

Step 1 First test $d = \frac{k}{w^2}$. To find k, select the point $(w, d) = (85, 1.4)$. Use the k-value you find to write an equation to model the situation.

Step 2 Use a spreadsheet. Enter the weights from the table above in the first column. Enter your variation equation from Step 1 at the top of the second column to generate a table of values. Compare the generated table to the values that Anna and Jenna observed. Do the values of d predicted by your model fit the observed data?

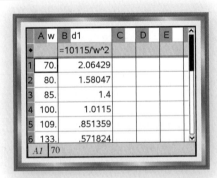

Step 3 Now test $d = \frac{k}{w}$. Use the same point $(w, d) = (85, 1.4)$ that you used in Step 1. Find k and write an equation to model this situation.

Step 4 Enter your variation equation from Step 3 at the top of the third column to generate a table of values. Compare the values predicted by this equation to the observed values that Anna and Jenna found.

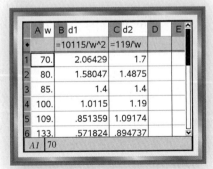

Step 5 Do your findings confirm that $d = \frac{k}{w}$ is the better model? Why or why not?

Another way to approach this problem is to use the Fundamental Theorem of Variation.

Example 2

a. Use the Fundamental Theorem of Variation to determine an appropriate model that relates the weights and distances in Jenna's table of experimental values.

b. Predict the distance in yards that a person weighing 90 pounds must be from the pivot in order to balance Anna's weight.

Solution

a. When d varies inversely with the square of w, then if w is doubled, d is divided by 4. Find a pair of ordered pairs (w_1, d_1) and (w_2, d_2) where the ratio $\frac{w_2}{w_1}$ equals 2. One such pair of points is $(85, 1.4)$ and $(170, 0.7)$. Since $0.7 \neq \frac{1.4}{4}$, as the w-coordinate doubles, the d-coordinate is not divided by 4. Therefore, d does not vary inversely with the square of w. However, as the w-coordinate doubles, the d-coordinate is halved $(0.7 = \frac{1.4}{2})$. Therefore, the more appropriate model for these data is $d = \frac{k}{w}$. To find k, solve the formula for k and substitute an ordered pair into the equation.

$d = \frac{k}{w}$

$k = wd$ Multiply both sides by w.

$k = (85)(1.4) = 119$ Substitute the values of one ordered pair.

Using these two points, the data are modeled by

$d = \frac{119}{w}$.

b. If a person weighs 90 pounds, then $w = 90$. Using this model,

$d = \frac{119}{90} \approx 1.32$ yards.

So, sitting about 4 feet from the pivot will balance Anna.

 QY3

> ▶ **QY3**
>
> If Hector weighs 160 pounds, how far from the pivot should he sit to balance Anna? Is this distance approximately half of the distance from the pivot that the person who weighed 80 pounds sat?

Questions

COVERING THE IDEAS

In 1–4, refer to the data about a free-falling object.

1. Describe in words the variation relationship between distance and time.

2. Use the model $d = 4.9t^2$ to predict the distance that a free-falling ball will fall in 4.5 seconds.

3. If a ball is dropped from a height of 500 meters, how many seconds will it take to reach the ground? (Ignore the effects of air resistance.)

Sky divers are not free-falling due to wind resistance.

4. Suppose a second ball is three times as heavy as the ball in Question 3. Compare the times it will take the balls to hit the ground.

In 5–7 refer to the Activity.

5. Use one of the data points in Jenna's table to show that $d = \dfrac{k}{w^2}$ is not a good model for the data.

6. Use the better model to predict the distance that a 180 lb person must be from the pivot in order to balance the seesaw with Anna.

7. How far from the pivot should a 20 lb baby sit to balance Anna? Does this seem possible in this situation?

APPLYING THE MATHEMATICS

8. Consider the equation $d = 4.9t^2$ where d is measured in meters and t is measured in seconds.
   a. Find the rate of change between the following pairs of points:
      (1, 4.9) and (1.5, 11.0)
      (1.5, 11.0) and (2, 19.6)
      (2, 19.6) and (2.5, 30.6)
   b. Is the rate of change a constant value?
   c. For this model, the rate of change is measured in what units?

9. Malcolm is blowing up a balloon. Refer to the table and graph, which give the number n of breaths he has blown into the balloon and the volume V of the balloon in cubic inches.

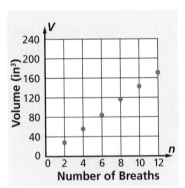

n	2	4	6	8	10	12
V	28.6	57.3	85.9	114.5	143.2	171.8

   a. **Multiple Choice** Which of the following equations is a good model for these data?
      i. $V = kn$ ii. $V = kn^2$
      iii. $V = \dfrac{k}{n}$ iv. $V = \dfrac{k}{n^2}$
   b. Find the constant k for your model.
   c. Use your model to predict the value of V when n is 14.

10. a. **Multiple Choice** Which formula best models the data graphed at the right?
      i. $L = ks$ ii. $L = ks^2$
      iii. $L = \dfrac{k}{s}$ iv. $L = \dfrac{k}{s^2}$
   b. Justify your answer to Part a.
   c. Predict the value of L when $s = 60$. Use the Fundamental Theorem of Variation to explain your prediction.

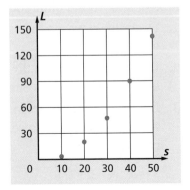

11. Refer to the table below and the graph at the right which show the intensity I of the sound, measured in decibels (dB), emitted from a 150-watt speaker at a distance d meters from the speaker.

d	1	2	3	4	5	6
I	11.9	3.0	1.3	0.75	0.48	0.33

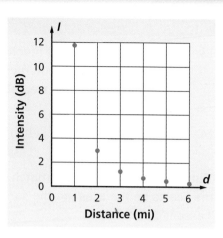

a. **Multiple Choice** Which of the following equations is a good model for these data?

   i. $I = kd$ ii. $I = kd^2$

   iii. $I = \dfrac{k}{d}$ iv. $I = \dfrac{k}{d^2}$

b. Find the constant k for your model.

c. Alex looked at the graph and predicted that when d is 8, I is 0.62. Do you agree with him? Why or why not?

d. Use your model to predict the value of I when d is 10.

12. The deeper a diver goes below sea level, the greater the water pressure on the diver. To model this relationship, pressure data (in pounds per square inch, or psi) was recorded at various depths (in feet).

Depth d (ft)	0	15	20	25	30	35
Pressure p (psi)	0	6.5	8.6	10.8	12.9	15

a. Draw a graph of these data points. Let the depth d in feet be the independent variable and the pressure p in pounds per square inch be the dependent variable.

b. Write a variation function to model these data. Use one data point to calculate k and check the model with other data points.

c. Use your model to predict the value of p when $d = 40$.

d. Depth below sea level is often measured in *atmospheres*. Each atmosphere in fresh water equals 34 feet of vertical distance. For each additional atmosphere below sea level, there is an increase of 14.7 pounds per square inch (psi) of pressure on a diver. Explain whether or not this information is consistent with the model you have found in Parts b and c.

Deep sea divers must wear special suits to combat the enormous underwater pressure.

REVIEW

13. Pat weighs 2.5 times as much as her brother Matt. If they balance a seesaw, how do their distances from the pivot compare? Give sample weights and distances to support your answer. (Lesson 2-6)

14. Find the average rate of change between $x = 1$ and $x = 5$ for the function f with $f(x) = \frac{13}{x}$. (Lesson 2-5)

Multiple Choice In 15–18, match each graph to its equation. The scales on the axes are the same for all four graphs. (Lessons 2-4, 2-5, 2-6)

A $y = 2x$ B $y = \frac{2}{x}$ C $y = \frac{2}{x^2}$ D $y = -\frac{2}{x}$ E $y = -\frac{1}{2}x^2$

15. 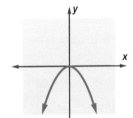 16. 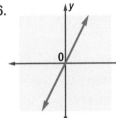 17. 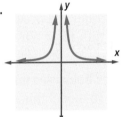 18.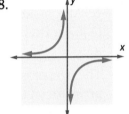

19. Suppose the value of x is tripled. How is the value of y changed if y is directly proportional to x^{23}? (Lesson 2-3)

20. Give the domain and range for each function graphed in Questions 15–18. (Lesson 1-4, 2-1, 2-2)

21. The table at the right provides data on the number of car sales in the U.S. from 1995 to 2004. (Lessons 1-3, 1-5)

   a. Let f be the function that maps the year y onto the number n of cars sold. Find $f(1999)$.

   b. For which year y is $f(y) = 8,104,000$?

22. Suppose V is the volume of a circular cylinder in cubic inches and r is the radius of the cylinder in inches. Let $h = \frac{V}{\pi r^2}$. What is h, and in what unit is h measured? (Previous Course)

EXPLORATION

23. Research Galileo's Law of Free-Falling Objects. Find out how he was able to use the results of rolling objects down an inclined plane as the basis for his conclusions on free-falling objects.

Year y	Car Sales n = f(y) (thousands)
1995	8,635
1996	8,526
1997	8,272
1998	8,142
1999	8,698
2000	8,847
2001	8,423
2002	8,104
2003	7,610
2004	7,506

QY ANSWERS

1. The units multiply.
   $\frac{m}{sec^2} \cdot sec^2 = m$.

2. about 333.5 meters

3. about 0.74 yard; Yes, this distance is about half.

Lesson

2-8

Fitting a Model to Data II

> ▶ **BIG IDEA** When a situation involves more than two variables that vary directly or inversely as each other, by pairing the dependent variable and each independent variable you can see how the variables fit into one *combined variation*.

So far in this chapter you have studied situations in which two quantities vary. In many real-world situations there are more than two variables. Consider the problem of determining the maximum weight that can be supported by a board.

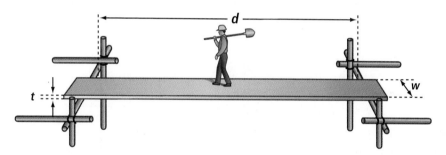

Mental Math

Find the point of intersection of the two lines.

a. $x = 2$ and $y = 7$

b. $x = 4$ and $y = 2x + 1.5$

c. $3x + y = 18$ and $y = 3$

Three quantities that influence the maximum weight M that a board can hold are the board's width w, thickness t, and the distance d between supports. The goal is to find an equation relating w, t, d, and the dependent variable M.

In Lesson 2-7, you used a graph in two dimensions to determine equations relating two variables. You cannot use a single graph to determine the equation representing this situation because there are four variables to be considered. However, by keeping *all but one* independent variable constant, you can investigate separately the relationship between the dependent variable M and that independent variable.

To obtain equations that model these relationships, you can use the converse of the Fundamental Theorem of Variation.

Converse of the Fundamental Theorem of Variation

a. If multiplying every x value of a function by c results in multiplying the corresponding y values by c^n, then y varies directly as the nth power of x; that is, $y = kx^n$.

b. If multiplying every x value of a function by c results in dividing the corresponding y values by c^n, then y varies inversely as the nth power of x; that is, $y = \frac{k}{x^n}$.

Below is an explanation of how to find a model for the board problem. The data are made up, but the idea is not.

Maximum Weight as a Function of Width

Perry worked on this problem and found the model by working with the variables two at a time. First, Perry held the two independent variables d and t constant. He did this by placing supports at the ends of 12-foot long, 1.5-inch-thick boards of different widths. He measured how much weight a board of a given width could support before it broke. Perry obtained the following data.

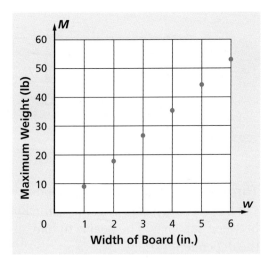

Width of Board w (in.)	1	2	3	4	5	6
Observed Maximum Weight M (lb)	9	18	26	36	44	53

The graph shows how the maximum weight M depends on the width w of the board. Because the points seem to lie on a line through the origin, Perry concluded that M varies directly as w.

GUIDED

Example 1

Verify that M varies directly as w.

Solution Assume $M = kw$. Select a data point (__?__ , __?__) from the table and use it to find k.

__?__ = k __?__

__?__ = k

Use your k to write a variation formula for M as a function of w.

__?__ = __?__ · __?__

Use your formula to fill in values in the table predicted by this model.

Width of Board w (in.)	1	2	3	4	5	6
Predicted Maximum Weight M (lb)	?	?	?	?	?	?

Your results should be almost identical to Perry's data, so this verifies that M varies ___?___ as ___?___.

Maximum Weight as a Function of Thickness

Perry then investigated the relationship between M and board thickness t. He held the distance d between supports constant at 12 feet, and the width w constant at 2 inches. He varied the thickness and measured the maximum weight that could be supported. The table below and graph at the right present his findings. On the graph, the points seem to lie on a parabola through the origin.

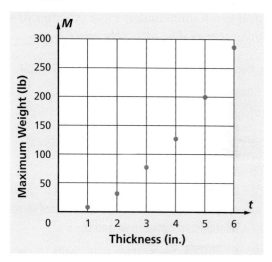

Thickness t (in.)	1	2	3	4	5	6
Predicted Maximum Weight M (lb)	8	32	72	128	200	288

Example 2

Use the Converse of the Fundamental Theorem of Variation to determine how *M* varies with *t* and write a variation equation for the situation.

Solution Select two data points (t_1, M_1) and (t_2, M_2) from the table such that t_2 is *double* the value of t_1.

$(t_1, M_1) = (2, 32)$ and $(t_2, M_2) = (4, 128)$.

$\frac{M_2}{M_1} = \frac{128}{32} = 4$, so, as t doubles, M is multiplied by $4 = 2^2$.

Select two other data points (t_3, M_3) and (t_4, M_4) from the table such that t_4 is *triple* the value of t_3.

$(t_3, M_3) = (1, 8)$ and $(t_4, M_4) = (3, 72)$

$\frac{M_4}{M_3} = \frac{72}{8} = 9$, so, as t triples, M is multiplied by $9 = 3^2$.

So, M varies directly as the square of t.

 QY1

> **▶ QY1**
>
> In Example 2, $M = kt^n$.
> What is the value of *n*?
> What is the value of *k*?

Maximum Weight as a Function of Distance between Supports

Next, Perry investigated the relationship between M and d by holding t and w constant. He chose boards 1.5 inches thick and 2 inches wide and measured the maximum weight that boards of different lengths would hold. Perry obtained the following data. The graph shows how M depends on d. It is not immediately clear whether M varies inversely as d or inversely as d^2.

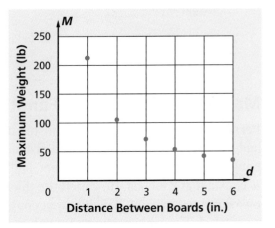

Distance Between Boards (in.)

Distance d (ft)	1	2	3	4	5	6
Observed Maximum Weight M (lb)	212	106	71	53	42	35

GUIDED

Example 3

Use the Converse of the Fundamental Theorem of Variation to determine how M varies with d. Then find the constant of variation k and use it to verify the relationship between M and d.

Solution Select three data points (d_1, M_1), (d_2, M_2), and (d_3, M_3) from the table such that d_2 is double the value of d_1 and d_3 is triple the value of d_1.

$(d_1, M_1) = (\underline{}, \underline{})$, $(d_2, M_2) = (\underline{}, \underline{})$,
and $(d_3, M_3) = (\underline{}, \underline{})$

$\dfrac{M_2}{M_1} = \underline{}$, so, as d doubles, M is divided by $\underline{}$.

$\dfrac{M_3}{M_1} = \underline{}$, so, as d triples, M is divided by $\underline{}$.

So, M varies $\underline{}$ as $\underline{}$, and $M = \dfrac{k}{?}$.

Now select a data point from the table and use it to find k.

$\underline{} = \dfrac{k}{?}$

$\underline{} = k$

Use k to write a variation formula for M as a function of d:

$\underline{}$.

Verify the relationship by using your formula to fill in the values of M in the table below.

Distance d (ft)	1	2	3	4	5	6
Predicted Maximum Weight M (lb)	?	?	?	?	?	?

Compare your results to Perry's data. Based on this comparison, you can confirm that M varies ___?___ as __?__.

 QY2

▶ QY2

Why are there different values for k in Examples 1, 2, and 3?

Summarizing the Results with a Single Model

Perry summarized his findings as follows:

M varies directly as w and the square of t, while M varies inversely as d.

These relationships can be expressed in the single formula

$$M = \frac{kwt^2}{d},$$

where M is in pounds, w and t are in inches, d is in feet, and k is the constant of variation. Notice that the independent variables that have direct-variation relationships with M are in the numerator, and the independent variable that has an inverse-variation relationship with M is in the denominator. The formula tells you that the greater the width and the thickness of the board, and the shorter the distance between supports, the stronger the board will be. This situation is an example of *combined variation*. You will learn more about combined variation in the next lesson.

Questions

COVERING THE IDEAS

1. Which variables in the Examples are the independent variables?

2. Why did it take three different tables to develop the single formula $M = \frac{kwt^2}{d}$?

In 3–9, refer to the tables and graphs in this lesson.

3. What is the maximum weight supported by a board 12 ft long, 2 in. wide, and 2 in. thick?

4. What is the maximum weight supported by a board 12 ft long, 6 in. wide, and 1.5 in. thick?

5. What is the shape of the graph relating M and w when t and d are held constant?

6. Refer to the data for M as a function of t. Find an equation to model the data when $d = 12$ ft and $w = 2$ in.

7. Use the Converse of the Fundamental Theorem of Variation to show that M does not vary inversely as d^2.

8. Suppose you run an experiment similar to the ones in the lesson and find that the maximum weight the board can hold is 50 pounds. Would you increase or decrease each of the following variables to increase the maximum weight the board could hold to 200 pounds?

   a. w **b.** t **c.** d

9. If M varies directly as a variable in the formula $M = \dfrac{kwt^2}{d}$, is that variable in the numerator or denominator of the expression on the right side of the formula?

APPLYING THE MATHEMATICS

10. **Multiple Choice** The two graphs below show the relationships between a dependent variable y and the independent variables x and z.

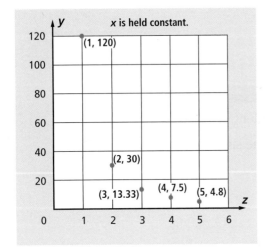

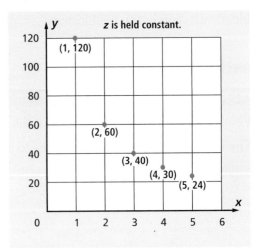

Which equation best models this situation?

A $y = \dfrac{kx^2}{z^2}$ B $y = \dfrac{kx}{z}$ C $y = \dfrac{k}{xz^2}$ D $y = \dfrac{k}{x^2z}$

11. Carrie was trying to determine how the pressure P that is exerted on the floor by the heel of a shoe depends on the heel width h and the weight w of the person wearing the shoe. She started by measuring the pressure (in psi) exerted by several people of different weights wearing a shoe with a heel 3 inches wide.

a. A table and graph of these data points are shown, using P as the dependent variable. How does the pressure exerted on the floor appear to relate to a person's weight? Use the Converse of the Fundamental Theorem of Variation to test your conclusion for two ordered pairs in the table.

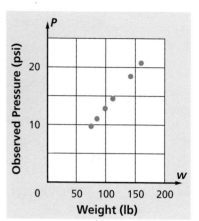

Weight w (lb)	75	85	99	112	142	160
Observed Pressure P (psi)	9.8	11.1	12.9	14.6	18.5	20.8

b. Carrie then had her sister Candice, who weighs 135 pounds, try on shoes with different heel widths h, and she measured the pressure exerted. The data are summarized in the table and graph below, again using P as the dependent variable. Use the Converse of the Fundamental Theorem of Variation to determine if P varies inversely as h or as h^2.

Heel width h (in.)	1.25	1.5	1.75	2	2.25	2.5
Observed Pressure P (psi)	101.1	70.2	51.6	39.5	31.2	25.3

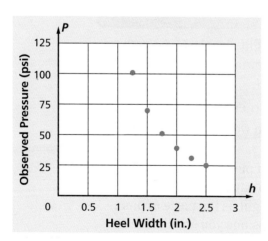

c. What amount of pressure would you expect to be exerted by Candice if she were wearing a 3-inch wide heel? Explain your reasoning.

d. Write an equation modeling the variation relationships between P, w, and h. Do not solve for k.

REVIEW

12. Consider $y = -\dfrac{7}{x^2}$ (**Lesson 2-6**)

    a. What real number is excluded from the domain of x?

    b. **Multiple Choice** Which could be the graph of the equation?

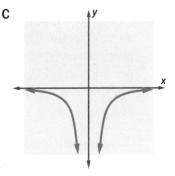

A   B   C

13. Identify all asymptotes of the graph of the equation.
    (**Lessons 2-4, 2-5, 2-6**)

    a. $y = \dfrac{7x}{4}$ b. $y = \pi x$ c. $y = -\dfrac{3}{x}$ d. $y = \dfrac{\frac{1}{2}}{x^2}$

14. **Fill in the Blanks** The graph of $y = \dfrac{5x}{3}$ is a ___?___ with slope ___?___. (**Lesson 2-4**)

15. Suppose r varies inversely as the cube of t. If r is 8 when t is 4, find r when $t = 9$. (**Lesson 2-2**)

16. If you buy x granola bars at y cents per bar, what is the total cost? (**Lesson 1-1**)

EXPLORATION

17. The ability of a board to support a weight also depends on the type of wood. That is, the constant of variation k in the formula $M = \dfrac{kwt^2}{d}$ depends on the type of wood.

    a. For a stronger kind of wood, is k larger or smaller?

    b. Do some research to find out which wood is strongest: oak, balsa, or pine.

    c. Suppose you have a 10-foot piece of pine that is 1 inch thick and 6 inches wide. If you wanted to cut a piece of oak that could hold the same amount of weight as the piece of pine, what do you know about the dimensions you should use for the piece of oak?

QY ANSWERS

1. $n = 2; k = 8$

2. Examples 1, 2, and 3 all describe different relationships, and as such they have different values of k.

Lesson

2-9

Combined and Joint Variation

Vocabulary

combined variation

joint variation

▶ **BIG IDEA** The same methods used to solve variation problems involving two variables can be applied to variation problems involving more than two variables.

Combined Variation

At the beginning of this chapter, you read about how adjusting the number of teeth on the gears of a bicycle changes its speed. The speed S of a bicycle varies directly with the number of revolutions per minute (rpm) R that you turn the pedals and with the number F of teeth on the front gear. The speed also varies inversely with the number B of teeth on the back gear. This situation is modeled by the equation

$$S = \frac{kRF}{B}.$$

This equation is read "S varies directly as R and F and inversely as B." When both direct and inverse variations occur together in a situation, we say the situation is one of **combined variation**.

You saw another example of combined variation in Lesson 2-8, where the maximum weight M of a board varied directly with its width w and the square of its thickness t, and inversely with the distance d between its supports. This relationship was modeled by the equation

$$M = \frac{kwt^2}{d}.$$

 QY1

A combined-variation equation has two or more independent variables, and the independent variables can have any positive exponent. To find k in a combined-variation model, use the same strategy as in a variation problem with one independent variable:

- Find one instance that relates all the variables simultaneously.

- Substitute known values into the general variation equation.

- Solve for k.

Mental Math

Jeff is experimenting with a balance scale. He finds that 8 erasers balance 1 apple.

a. His calculator weighs 2.5 times as much as an apple. How many erasers does he need to balance his calculator?

b. A pair of scissors weighs the same as 2 erasers. How many pairs of scissors will balance the calculator?

c. There are two pairs of scissors on one side of the scale and the calculator on the other. How many erasers should he add to the pan with the scissors to balance the calculator?

▶ **QY1**

Write an equation that represents this statement: y varies directly as the square of x and inversely as z.

Example 1

Mario is pedaling a bike at 80 revolutions per minute using a front gear with 35 teeth and a back gear with 15 teeth. At these settings, he is traveling 14.5 miles per hour. Describe how his speed would change if he increased his pedaling to 100 rpm.

Solution Use the combined variation equation $S = \frac{kRF}{B}$. Substitute $S = 14.5$, $R = 80$, $F = 35$, and $B = 15$ and solve for k.

$$14.5 = \frac{k(80)(35)}{15}$$

$$14.5 \approx k(187)$$

$$0.078 \approx k$$

So the variation formula for this situation is $S = \frac{0.078RF}{B}$.

To find Mario's new speed, substitute $R = 100$, $F = 35$, and $B = 15$ into your formula and solve for S.

$$S = \frac{0.078(100)(35)}{15} \approx 18$$

This means that by increasing his rpm to 100, Mario increases his speed by about 3.5 mph (from 14.5 mph to 18 mph).

Another Example of Combined Variation

Photographers are always looking for ways to make their pictures sharper. One way is to focus the lens at the *hyperfocal distance*. Focusing the lens at this distance will produce a photograph with the maximum number of objects in focus.

Photographers often use the hyperfocal distance when taking pictures of landscapes. The two images at the right were shot with the same camera at the same settings, except the one on the bottom was taken with the lens focused at the hyperfocal distance, giving the extra degree of sharpness.

Example 2 explores how the hyperfocal length H can be calculated using the focal length L of the camera lens in millimeters and the f-stop, or aperture, f. The *aperture* is a setting that tells you how wide the lens opening is on a camera. It is the ratio of the focal length to the diameter of the lens opening and so has no units.

Example 2

The hyperfocal length H in meters varies directly with the square of the focal length L in millimeters, and inversely with the selected aperture f.

a. Write a general variation equation to model this situation.

b. Find the value and unit of k when the hyperfocal length H is 10.42 m when using a 50 mm lens (L) and the aperture f is set at 8.

c. Write a variation formula for the situation using your answer to Part b.

d. Find the hyperfocal length needed if you want to shoot with a 300 mm lens at an aperture setting of 2.8. Include the units in your calculations.

Solution

a. Because the hyperfocal length varies directly with the square of the focal length of the lens, L^2 will be in the numerator of the expression on the right side of the formula. Because the hyperfocal length varies inversely with the aperture setting, f will be in the denominator of the expression. *A general equation is* $H = \frac{kL^2}{f}$.

b. Use your formula from Part a with: $H = 10.42$ m, $L = 50$ mm, and $f = 8$. Include units when making substitutions. Here we show how to do it by hand and with a CAS.

By hand:

$$10.42 \text{ m} = \frac{k(50 \text{ mm})^2}{8}$$

Substitute the given values for H, L, and f.

$$10.42 \text{ m} = \frac{2500 \text{ mm}^2 \cdot k}{8}$$

Square 50 mm.

$$83.36 \text{ m} = 2500 \text{ mm}^2 \cdot k$$

Multiply both sides by 8.

$$0.033 \frac{\text{m}}{\text{mm}^2} \approx k$$

Divide both sides by 2500 mm².

With a CAS:

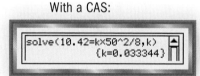

solve(10.42=k×50^2/8,k)
{k=0.033344}

c. Substitute k into the equation from Part a.

$$H = \frac{0.033 \, L^2}{f}$$

d. Use the formula from Part c to calculate H when $L = 300$ mm and $f = 2.8$. Include the units when you substitute.

$$H = \frac{0.033 \frac{\text{m}}{\text{mm}^2}(300 \text{ mm})^2}{2.8}$$

$$H = \frac{0.033 \frac{\text{m}}{\text{mm}^2}(90{,}000 \text{ mm}^2)}{2.8}$$

$$H \approx 1060.7 \text{ m}$$

The hyperfocal length is about 1,060 meters.

Joint Variation

Sometimes one quantity varies directly as powers of two or more independent variables, but not inversely as any variable. This is called **joint variation**. The simplest joint-variation equation is

$$y = kxz,$$

where k is the constant of variation. The equation is read "y varies jointly as x and z" or "y varies directly as the product of x and z." Guided example 3 explores a joint-variation situation in geometry.

GUIDED

Example 3

The volume of a solid with a circular base varies jointly as the height of the solid and the square of the radius of the base.

a. Write a general equation to model this situation.

b. If the volume of the solid is approximately 75.4 cubic centimeters when the radius is 2 centimeters and the height is 6 centimeters, find the value of k.

c. The value of k is approximately equal to what famous mathematical value?

d. Write a variation formula using your answer to Part b.

e. What well-known kind of solid figure could this be?

Solution

a. Let V be the volume of the solid, h be the height of the solid, and r be the radius of the base.

A general equation is $V = k\underline{\ ?\ }$.

b. Use your formula from Part a with $V = 75.4$, $h = 6$, and $r = 2$ to solve for k.

$75.4 = k\ \underline{\ ?\ }\ \underline{\ ?\ }\,^2$

$75.4 = k\ \underline{\ ?\ }$

$\underline{\ ?\ } \approx k$

c. The value of k is approximately equal to $\underline{\ ?\ }$.

d. Substitute k into your equation from Part a. So, $V = \underline{\ ?\ }$.

e. Based on the formula in Part d, this solid could be a(n) $\underline{\ ?\ }$.

Many other geometry formulas can be interpreted as direct- or joint-variation equations.

 QY2

> ▶ QY2
>
> Translate this statement into an equation: The area of a triangle varies jointly with the height and the base of the triangle. What is the constant of variation for this formula?

As in combined variation, a joint-variation equation can have more than two independent variables, and the independent variables can have any positive exponent.

Questions

1. a. What is combined variation?

   b. How is joint variation different from combined variation?

2. Translate into a single formula: M varies directly as t and r^2 and inversely as d.

3. Refer to Example 1. Suppose Mario slowed his pedaling to 75 rpm. What would his speed be?

4. In Example 2, assume that a photographer calculated a hyperfocal length of 6.54 meters. If he used a lens with a focal length of 28 millimeters, at what aperture was the lens set?

5. Sonia calculated the volume of a cylindrical can of cat food with a radius of 1.5 inches and a height of 1.2 inches to be about 9.2 cubic inches.

   a. Use the equation $V = kr^2h$ to calculate the constant of variation k that she used.

   b. Did Sonia use the correct formula? How do you know?

6. Refer to Example 3. If the volume of the solid is 25.13 cm^3 with a radius of 2 cm, what is its height?

7. The formula $F = ma$ gives the force F on an object with mass m and acceleration a.

   a. Rewrite the formula in words using the language of variation.

   b. What is the constant of variation?

8. The volume of a solid with a circular base varies jointly with the square of the base radius and the height. When the volume is 83.78 cm^3, the height is 5 cm and the base radius is 4 cm.

   a. Write a general equation to model this situation and find k.

   b. k is a multiple of π. Write k in terms of π.

   c. Write a variation formula for the volume of this solid in terms of π.

   d. What kind of well-known solid is this?

9. The volume of a certain solid varies jointly as its height, width, and length.

    a. Write a general equation to model this situation.

    b. **Fill in the Blank** The answer to Part a suggests that this is a ___?___ solid.

    c. Based on your answer to Part b, find the value of k.

10. Suppose y varies directly as x and inversely as z. Describe how y changes when x and z are each tripled. Explain your answer.

11. The cost C of polyvinyl chloride (PVC) piping in dollars varies jointly as the length L of the pipe and the difference between the squares of its outer and inner radii, $R_o^2 - R_i^2$. Suppose that a foot of PVC piping with an outer radius of 0.25 foot and an inner radius of 0.20 foot costs $3.72.

    a. Using the given variables, write a joint variation equation.

    b. Find the constant of variation.

    c. Rewrite the variation equation using the constant from Part b.

    d. Find the cost of 10 feet of PVC piping with an outer radius of 0.5 foot and in inner radius of 0.48 foot.

    e. Determine the unit of the constant of variation.

REVIEW

12. The Ideal Gas Law in chemistry relates the pressure P (in atmospheres) exerted by a gas to the temperature T (in Kelvins) of the gas and volume V (in liters) of its container. Chantel obtained the following data using a 5-liter container in the lab. (Lessons 2-7, 2-8)

T (K)	235	260	285	305	500
P (atm)	0.7285	0.8060	0.8835	0.9455	1.55

    a. Graph the data points.

    b. How does P vary with T?

    c. With a temperature of 350 Kelvins, Chantel manipulated the volume of the container and obtained the following data. Graph these points on a different set of axes.

V (L)	1	2	3	4	5	6
P (atm)	5.425	2.713	1.808	1.356	1.085	0.603

    d. How does P vary with V?

    e. Write an equation that relates P, T, and V. You do not need to find the constant of variation.

13. Let $g: x \to 3x^2$. (**Lesson 2-5**)

    a. Graph g over the domain $-2.5 \le x \le 2.5$.

    b. What is the name of the shape of this graph?

    c. Find the average rate of change of g between $x = 1$ and $x = 2$.

    d. Would the answer to Part c be the same for any two values of x that differ by 1? Explain your answer.

14. If y varies directly as w^3, and $y = 25$ when $w = 5$, find the value of y when $w = 2$. (**Lesson 2-1**)

15. One general equation for a combined variation is $y = k\frac{xz}{w}$. Solve for k in terms of the other variables. (**Lesson 1-7**)

16. When Clara tried to solve the equation $\frac{1}{5}x + \frac{1}{7}x + 2 = 5$, her first step led to $7x + 5x + 70 = 175$. (**Lesson 1-6**)

    a. Explain what Clara did.

    b. Finish solving the equation.

17. Given the function $f: x \to 4x^3 - 2x + 1$, find $f(\pi)$. (**Lesson 1-3**)

EXPLORATION

18. There are many instances of mathematics on TV shows. Watch a show in which mathematics plays a role and record any mathematics used on the show. Note whether the mathematics was used accurately and try to explain any mistakes.

QY ANSWERS

1. $y = \frac{kx^2}{z}$

2. $A = kbh$. In this case, $k = \frac{1}{2}$.

Chapter 2 Projects

Photographers use light meters to help create the desired lighting conditions for a photo.

1 The Law of the Lever

Use a rigid ruler as a lever and a triangular object as a pivot. Place one penny on the ruler 6 inches from the pivot.

a. Stack two pennies on top of each other and place them on the ruler where they will balance the one penny on the other end. Record how far from the pivot the pennies are placed. Repeat the process for three pennies and four pennies. Does your experiment confirm the Law of the Lever?

b. Place a penny 4 inches from the pivot and then place a quarter so that it balances the penny. Repeat this process with a dime instead of a quarter. Use the Law of the Lever to determine how many times heavier a quarter is than a penny, and how many times heavier a penny is than a dime.

c. Use your findings in Part b to predict how far from the pivot a quarter should be placed to balance a dime 6 inches from the pivot. Test your prediction. How accurate was your prediction?

Archimedes wrote about the Law of the Lever in his work titled *On the Equilibrium of Planes*.

2 Variation and Light

Use a small, bright, pocket-sized flashlight and a sheet of printer paper. Put the paper on a desk in a dimly lit room and shine the light straight down on the paper from a sequence of increasing distances, for example, 1 inch, 2 inches, 4 inches, 7 inches, and so on, until the image becomes too dim to see.

a. At each distance, measure the diameter of the circular image produced by the beam of light. Create a table with the height h of the flashlight above the paper in the first column and the diameter d of the circular image in the second.

b. Graph your data from Part a.

c. Write a formula showing how the diameter of the image depends on the height of the flashlight. Does the diameter vary directly or inversely as h?

d. Borrow a light meter from a photographer or a science teacher. Repeat the process in Part a, but measure the intensity or brightness of the image at each height rather than the diameter of the image.

e. Theoretically, the intensity I of light varies inversely as the square of the height h of the flashlight above the image. How closely do your data fit this model?

3 The Maximum Load of a Balsa Board

Refer to Perry's experiments in Lesson 2-8. Collect pieces of balsa wood of various lengths, widths, and thicknesses. Reconstruct the experiments regarding the maximum load M a board can hold. Find an equation relating d, w, t, and M for balsa wood.

4 Exploring a Connection between $y = \frac{1}{x}$ and the Altitudes of a Triangle

Use a DGS to graph the function $y = \frac{1}{x}$.

a. On one branch of the hyperbola, construct $\triangle ABC$ with its vertices on the hyperbola.

b. The *orthocenter* of a triangle is the intersection of the lines that contain the altitudes of the triangle. Construct M, the orthocenter of $\triangle ABC$.

c. Vary the location of the vertices of $\triangle ABC$ and observe how the location of M changes.

d. Give the coordinates of M and verify that it is located on the other branch of the hyperbola.

5 Galileo's Law of Free-Falling Objects

The Exploration in Lesson 2-7 asks you to find how Galileo used an inclined plane as the basis for his conclusions about free-falling objects. Recreate Galileo's experiment using a long board, a marble, and a stopwatch, or use a computer-based laboratory device. Write a report about your experiment. How close to the constant $k = 4.9$ m/sec^2 did you come?

6 Average Rates of Change

On a CAS, define a function *arc* that has four inputs and computes the *average rate* of change between two given points.

a. Use *arc* to calculate the average rate of change for $y = 2x^2$ between each set of consecutive ordered pairs in a table of values where $x = -5, -4, -3, \ldots, 4, 5$. Repeat the calculations for $y = -2x^2$. Explain the connections between the average rates of change of each function and the graph of each function.

b. Repeat Part a using $y = 2x^3$ and $y = -2x^3$.

c. Repeat Part a using $y = \frac{1}{x}$.

d. Look up the term *concavity* and explain how it relates to the average rate of change you found in Parts a–c.

Galileo used an apparatus similar to this to test and develop the Law of Free-Falling Objects.

Chapter 2 Summary and Vocabulary

● The functions studied in this chapter are all based on direct and inverse variation. When $k \neq 0$ and $n > 0$, formulas of the form $y = kx^n$ define **direct-variation functions**, and those of the form $y = \frac{k}{x^n}$ define **inverse-variation functions**. Four special cases of direct and inverse variation commonly occur. The equation, graph, name of the curve, domain D, range R, asymptotes, and symmetries of each special case are summarized on the next page.

● In a formula where y is given in terms of x, it is logical to ask how changing x (the independent variable) affects the value of y (the dependent variable). In direct or inverse variation, when x is multiplied by a constant, the change in y is predicted by the **Fundamental Theorem of Variation**. If y varies directly as x^n, then when x is multiplied by c, y is multiplied by c^n. If y varies inversely as x^n, then when x is multiplied by c, y is divided by c^n. The converse of this theorem is also true.

● The rate of change $\frac{y_2 - y_1}{x_2 - x_1}$ between (x_1, y_1) and (x_2, y_2) is the **slope** of the line connecting the points. For graphs of equations of the form $y = kx$, the rate of change between any two points on the graph is the constant k. For nonlinear curves, the rate of change is not constant, but varies depending on which points are used to calculate it.

● Variation formulas may involve three or more variables, one dependent and the others independent. In a **joint variation**, all the independent variables are multiplied. If the independent variables are not all multiplied, the situation is a **combined variation**. Variation formulas can be derived from real data by examining two variables at a time and comparing their graphs with those given on the following page.

● There are many applications of direct, inverse, joint, and combined variation. They include perimeter, area, and volume formulas, the inverse-square laws of sound and gravity, a variety of relationships among physical quantities such as distance, time, force, and pressure, and costs that are related to these other measures.

Vocabulary

Lesson 2-1
varies directly as
directly proportional to
direct-variation equation
constant of variation
*direct-variation function

Lesson 2-2
*inverse-variation function
varies inversely as
inversely proportional to

Lesson 2-4
*rate of change, slope

Lesson 2-5
parabola
vertex of a parabola
reflection-symmetric
line of symmetry

Lesson 2-6
hyperbola
branches of a hyperbola
vertical asymptote
horizontal asymptote
discrete set
inverse-square curve

Lesson 2-9
combined variation
joint variation

▷ Some **direct-variation functions**

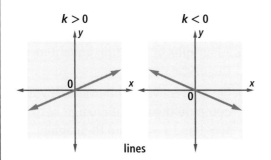

$$y = kx$$
y varies directly as x.

$k > 0$ $k < 0$

lines

$D = R =$ set of all real numbers

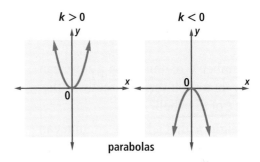

$$y = kx^2$$
y varies directly as the square of x.

$k > 0$ $k < 0$

parabolas

$D =$ set of all reals $D =$ set of all reals
$R = \{y \mid y \geq 0\}$ $R = \{y \mid y \leq 0\}$
reflection-symmetric to the y-axis

▷ Some **inverse-variation functions**

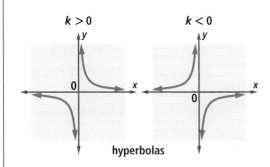

$$y = \frac{k}{x}$$
y varies directly as x.

$k > 0$ $k < 0$

hyperbolas

$D = R = \{x \mid x \neq 0\}$
asymptotes: x-axis, y-axis

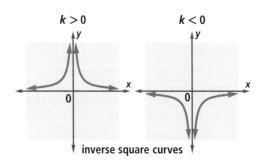

$$y = \frac{k}{x^2}$$
y varies inversely as the square of x.

$k > 0$ $k < 0$

inverse square curves

$D = \{x \mid x \neq 0\}$ $D = \{x \mid x \neq 0\}$
$R = \{y \mid y > 0\}$ $R = \{y \mid y < 0\}$
asymptotes: x-axis, y-axis
reflection-symmetric to the y-axis

Theorems

Fundamental Theorem of Variation (p. 88)
Slope of $y = kx$ Theorem (p. 95)
Converse of the Fundamental Theorem of Variation (p. 122)

Chapter 2 Self-Test

Take this test as you would take a test in class. Then use the Selected Answers section in the back of the book to check your work.

In 1 and 2, translate the statement into a variation equation.

1. The number n of items that a store can display on a shelf varies directly as the length ℓ of the shelf.

2. The weight w that a bridge column can support varies directly as the fourth power of its diameter d and inversely as the square of its length L.

3. Write the variation equation $s = \frac{k}{p^4}$ in words.

4. If T varies directly as the third power of s and the second power of w, and if $T = 10$ when $s = 2$ and $w = 1$, find T when $s = 8$ and $w = \frac{1}{2}$.

5. For the variation equation $y = \frac{5}{x^2}$, how does the y-value change when an x-value is doubled? Give two specific ordered pairs (x, y) that support your conclusion.

6. If y is multiplied by 8 when x is doubled, how is y related to x? Write a general equation to describe the relationship.

7. Find the average rate of change of $y = x^2$ between points A and B shown on the graph.

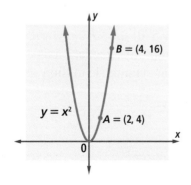

$B = (4, 16)$
$y = x^2$
$A = (2, 4)$

8. **True or False**

   a. All graphs of variation equations contain the origin.

   b. On a direct-variation graph, if you compute the rate of change between $x = 1$ and $x = 2$, and then again between $x = 2$ and $x = 3$, then you will get the same result.

9. a. **Fill in the Blanks** The graph of $y = kx^2$ is called a ___?___ and opens upward if ___?___.

   b. How does the graph of $y = kx^2$ change as k gets closer to 0?

10. Suppose f is a function with $f(d) = \frac{12}{d^2}$. What is the domain of f?

11. **Fill in the Blank** Complete each sentence with *inversely, directly,* or *neither inversely nor directly*. Explain your reasoning.

    a. The surface area of a sphere varies ___?___ as the square of its radius.

    b. If you have exactly $5,000 to invest, the number of shares of a stock you can buy varies ___?___ as the cost of each share.

12. a. Make a table of values for $y = -\frac{1}{2}x$. Include at least five pairs.

    b. Draw a graph of the equation in Part a.

    c. Find the slope of the graph in Part b.

13. **a.** Use a graphing utility to sketch a graph of $y = \frac{4}{x}$.

 b. Identify any asymptotes to the graph in Part a.

14. **a. Multiple Choice** Which equation's graph looks the most like the graph below?

 A $y = 2x$ **B** $y = -\frac{2}{x}$

 C $y = \frac{2}{x^2}$ **D** $y = -\frac{x}{2}$

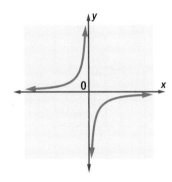

 b. State the domain and range of your answer to Part a.

15. Bashir's family is buying a new car. After looking at several models they became interested in the relationship between a car's fuel economy F in miles per gallon (mpg) and its weight w in pounds. Bashir collected data in the table below.

Weight w (lb)	Fuel Economy F (mpg)
2,655	40
2,991	34
3,263	32
3,737	28
3,962	26
5,577	19

 a. Graph the data points in the table.

 b. Which variation equation is a better model for Bashir's data, $F = \frac{k}{w}$ or $F = \frac{k}{w^2}$? Justify your answer and find the constant of variation.

 c. A family is thinking about buying a 5,900-pound sport-utility vehicle (SUV). Based on the model you found in Part b, what would you predict its fuel economy will be?

 d. In you own words, explain what the model tells you about the relationship between fuel economy and vehicle weight.

16. Suppose that variables V, h, and g are related as illustrated in graphs I and II below. The points on graph I lie on a parabola. The points on graph II lie on a line through the origin. Write a single equation that represents the relationship among V, h, and g.

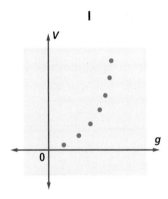

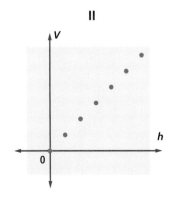

17. For a cylinder of fixed volume, the height is inversely proportional to the square of the radius.

    a. If a cylinder has height 10 cm and radius 5 cm, what is the height of another cylinder of equal volume and with a 10 cm radius?

    b. Generalize Part a. If two cylinders have the same volume, and the radius of one is double the radius of the other, what is the relationship between their heights?

18. Paula thinks the mass of a planet varies directly with the cube of its diameter.

    a. Assume Paula is correct. The diameter of Earth is approximately 12,700 km, and its mass is approximately 6.0×10^{24} kg. Compute the constant of variation, and write an equation to model the situation.

    b. The diameter of Jupiter is approximately 11 times that of Earth. What does the model from Part a predict Jupiter's mass to be?

    c. The mass of Jupiter is actually about 1.9×10^{27} kg. Why do you think this differs from the answer to Part b?

19. A cylindrical tank is being filled with water and the water depth is measured at 30-minute intervals. Below are a table and a graph showing the time passed t and the depth d of the water in the tank.

t (min)	30	60	90	120
d (in.)	41	80	122	162

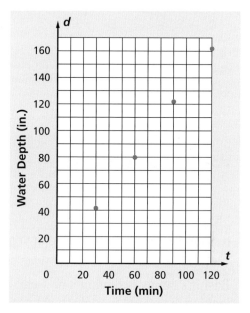

    a. Which variation equation models this data better, $d = kt$ or $d = kt^2$? Justify your answer.

    b. Find the constant of variation and write the variation equation.

    c. Based on your model, what is the depth of the water after 3 hours?

    d. If the tank is 50 feet tall, how long can the water run before the tank is full?

Chapter 2 Chapter Review

SKILLS Procedures used to get answers

OBJECTIVE A Translate variation language into formulas and formulas into variation language. (Lessons 2-1, 2-2, 2-9)

In 1–3, translate the sentence into a variation equation.

1. y varies directly as x.

2. d varies inversely as the square of t.

3. z varies jointly as x and t.

4. **Fill in the Blanks** In the equation $w = kyz$, where k is a constant, __?__ varies __?__ as __?__.

5. **Fill in the Blanks** If $y = \frac{kx}{v^2}$, where k is a constant, then y varies __?__ as __?__, and varies __?__ as __?__.

In 6–8, write each variation equation in words.

6. $y = kw^3$

7. $y = \frac{k}{w^3}$

8. $y = \frac{kz}{w^3}$

OBJECTIVE B Solve variation problems. (Lessons 2-1, 2-2, 2-9)

9. Suppose y varies directly as x. If $x = 3$, then $y = -21$. Find y when $x = -7$.

10. If y varies directly as the cube of x, and $y = \frac{3}{2}$ when $x = 3$, find y when $x = -6$.

11. Suppose y varies inversely as the square of x. When $x = 4$, $y = \frac{3}{4}$. Find y when $x = -6$.

12. z varies directly as x and inversely as the cube of y. When $x = 12$ and $y = 2$, $z = 39$. Find z when $x = 7$ and $y = 3$.

OBJECTIVE C Find slopes and rates of change. (Lessons 2-4, 2-5)

13. Find the slope of the line passing through $(3, -12)$ and $(10, 20)$.

14. What is the slope of the line graphed below?

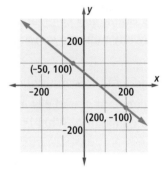

In 15 and 16, let $y = 2x^3$.

15. Find the average rate of change from $x = -2$ to $x = 0$.

16. Find the average rate of change from $x = 2$ to $x = 3$.

In 17 and 18, find the average rate of change from $x = 10$ to $x = 20$.

17. $y = \frac{5}{x}$

18. $y = \frac{5}{x^2}$

PROPERTIES Principles behind the mathematics

OBJECTIVE D Use the Fundamental Theorem of Variation. (Lessons 2-3, 2-8)

In 19–21, suppose x is tripled. Tell how the value of y changes under the given condition.

19. y varies directly as x.

20. y varies inversely as x^2.

21. y varies directly as x^n.

22. Suppose y varies inversely as x^2. How does y change if x is divided by 10?

23. **Fill in the Blank** If $y = \frac{k}{x^n}$, and x is multiplied by any nonzero constant c, then y is ___?___.

24. **Fill in the Blank** If $y = \frac{k}{x^n}$, and x is divided by any nonzero constant c, then y is ___?___.

25. Suppose $y = \frac{kx^n}{z^n}$, x is multiplied by nonzero constant a, and z is multiplied by a nonzero constant b. What is the effect on y?

26. Suppose that when x is divided by 3, y is divided by 27. How is x related to y? Express this relationship in words and in an equation.

OBJECTIVE E Identify the properties of variation functions. (Lessons 2-4, 2-5, 2-6)

27. **Fill in the Blank** The graph of the equation $y = kx^2$ is a ___?___.

28. **Fill in the Blanks** Graphs of all direct variation functions have the point (___?___ , ___?___) in common.

In 29–31, refer to the four equations below.

A $y = kx$ B $y = kx^2$
C $y = \frac{k}{x}$ D $y = \frac{k}{x^2}$

29. Which equations have graphs that are symmetric to the y-axis?

30. The graph of which equation is a hyperbola?

31. **True or False** When $k > 0$, all of the equations have points in Quadrant I.

32. a. Identify the domain and range of $f(x) = \frac{k}{x^2}$ when $k < 0$.
    b. What are the asymptotes of f if $f(x) = \frac{k}{x}$?

33. For what domain values are the inverse and inverse-square variation functions undefined?

34. Write a short paragraph explaining how the functions $y = -2x^2$ and $y = -\frac{2}{x^2}$ are alike and how they are different.

USES Applications of mathematics in real-world situations

OBJECTIVE F Recognize variation situations. (Lessons 2-1, 2-2)

In 35–37, translate the sentence into a variation equation.

35. The number n of hamsters that can safely live in a square hamster cage is proportional to the square of the length ℓ of a side of the cage.

36. The length of time t required for an automobile to travel a given distance is inversely proportional to the velocity v at which it travels.

37. The electromagnetic force F between two bodies with a given charge is inversely proportional to the square of the distance d between them.

38. Translate Einstein's famous equation for the relationship between energy and mass, $E = mc^2$, into variation language. (E = energy, m = mass, c = the speed of light, which is a constant.)

Fill in the Blank In 39–42, complete each sentence with *directly*, *inversely*, or *neither directly nor inversely*.

39. The number of people that can sit comfortably around a circular table varies ___?___ with the radius of the table.

40. The volume of a circular cylinder of a given height varies ___?___ as the square of the radius of its base.

41. The wall area that can be painted with a gallon of paint varies ___?___ as the thickness of the applied paint.

42. Your height above the ground while on a Ferris wheel varies ___?___ as the number of minutes you have been riding it.

OBJECTIVE G Solve real-world variation problems. (Lessons 2-1, 2-2, 2-9)

43. One of Murphy's Laws says that the time t spent debating a budget item is inversely proportional to the number d of dollars involved. According to this law, if a committee spends 45 minutes debating a $5,000 item, how much time will they spend debating a $10,000,000 item?

44. Recall that the weight of an object is inversely proportional to the square of its distance from the center of Earth. If you weigh 115 pounds on the surface of Earth, how much would you weigh in space 50,000 miles from Earth's surface? (The radius of Earth is about 4,000 miles.)

45. Computer programmers are often concerned with the efficiency of the algorithms they create. Suppose you are analyzing large data sets and create an algorithm that requires a number of computations varying directly with n^4, where n is the number of data points to be analyzed. You first apply your algorithm to a test data set with 10,000 data points, and it takes the computer 5 seconds to perform the computations. If you then apply the algorithm to your real data set, which has 1,000,000 data points, how long would you expect the computation to take? (Assume the computer takes the same amount of time to do each computation.)

46. A credit card promotion offers a discount on all purchases made with the card. The discount is directly proportional to the size of the purchase. If you save $1.35 on a $20 purchase, how much will you save on a $173 purchase?

47. The force required to prevent a car from skidding on a flat curve varies directly as the weight of the car and the square of its speed, and inversely as the radius of the curve. It requires 290 lb of force to prevent a 2,200 lb car traveling at 35 mph from skidding on a curve of radius 520 ft. How much force is required to keep a 2,800 lb car traveling at 50 mph from skidding on a curve of radius 415 ft?

48. An object is tied to a string and then twirled in a circular motion. The tension in the string varies directly as the square of the speed of the object and inversely as the length of the string. When the length of the string is 2 ft and the speed is 3 ft/sec, the tension on the string is 130 lb. If the string is shortened to 1.5 ft and the speed is increased to 3.4 ft/sec, find the tension on the string.

OBJECTIVE H Fit an appropriate model to data. (Lessons 2-7, 2-8)

In 49 and 50, a situation and question are given.

a. Graph the given data.

b. Find a general variation equation to represent the situation.

c. Find the value of the constant of variation and use it to rewrite the variation equation.

d. Use your variation equation from Part c to answer the question.

49. Officer Friendly measured the length L of car skid marks when the brakes were applied at different starting speeds S. He obtained the following data.

S (kph)	40	60	80	100	120
L (m)	9	20	36	56	81

How far would a car skid if the brakes were applied at 150 kph?

50. Two protons will repel each other with a force F that depends on the distance d between them. Some values of F are given in the following table. Note that d is measured in 10^{-10} meters, and F is measured in 10^{-10} newtons.

d	1.5	3	4.5	6	7.5
F	103	26	11	6	4

With what force will two protons repel each other if they are 1×10^{-10} m apart?

51. Jade performed an experiment to determine how the pressure P of a liquid on an object is related to the depth d of the object and the density D of the liquid. She obtained the graph on the left by keeping the depth constant and measuring the pressure on an object in solutions with different densities. She obtained the graph on the right by keeping the density constant and measuring the pressure on an object in solutions at various depths.

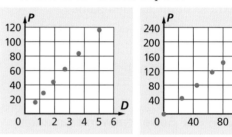

Write a general equation relating P, d, and D. Do not find the constant of variation. Explain your reasoning.

REPRESENTATIONS Pictures, graphs, or objects that illustrate concepts

OBJECTIVE I Graph variation equations. (Lessons 2-4, 2-5, 2-6)

In 52 and 53, an equation is given.

a. Make a table of values.

b. Graph the equation for values of the independent variable from –3 to 3.

52. $r = -\frac{1}{3} x^2$

53. $p = \frac{2}{q^2}$

In 54 and 55, use graphing technology to graph the equation.

54. $y = \frac{216}{x^3}$

55. $z = -\frac{4.85}{m}$

56. Describe how the graph of $y = kx^2$ changes as k decreases from 1 to –1.

57. Describe how the graph of $y = -\dfrac{k}{x^2}$ changes as k increases from 1 to 100.

OBJECTIVE J Identify variation equations from graphs. (Lessons 2-4, 2-5, 2-6)

Multiple Choice In 58 and 59, select the equation whose graph is most like the one shown. Assume the scales on the axes are the same.

58.

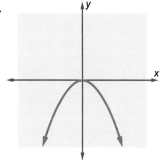

A $y = -x$ **B** $y = -x^2$

C $y = \dfrac{1}{x^2}$ **D** $y = x^2$

59.

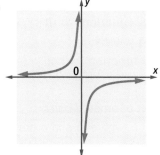

A $y = \dfrac{x^2}{8}$ **B** $y = \dfrac{8}{x}$

C $y = -\dfrac{8}{x}$ **D** $y = -\dfrac{8}{x^2}$

60. In the graph of $y = kx^2$, what type of number must k be?

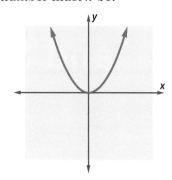

61. In the graph of $y = \dfrac{k}{x}$, what type of number must k be?

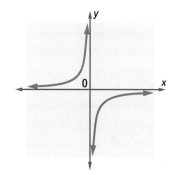

Chapter

3 Linear Functions and Sequences

Contents

In Chapter 2, you studied direct variation situations modeled by functions with equations of the form $y = kx$, or $f(x) = kx$. f is a linear function whose graph contains (0, 0).

A linear function is a set of ordered pairs (x, y) in which $y = mx + b$, where m and b are constants. Many situations can be modeled mathematically using linear functions. For example, your body mass index (BMI) is a measure of body fat based on height and weight. The table on the next page shows some data about the average BMI for males of different ages. The graph shows that the data appear to be linear. In fact, if y is the average BMI of a person of age x, then $y \approx 0.07x + 24.9$.

Age	BMI
25	26.6
35	27.5
45	28.4
55	28.7

Source: Centers for Disease Control and Prevention (CDC)

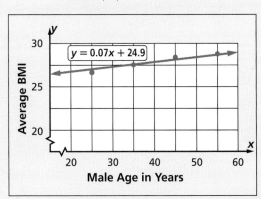

In this chapter, you will create linear models of real-world situations. While you are doing this, you will learn about relationships among lines, geometry, and statistics.

This chapter also extends the study of sequences you began in Chapter 1. Sequences with formulas of the form $a_n = mn + b$, where m and b are constants, model linear situations of constant rates of change and are examples of linear functions.

Lesson 3-1

Constant Change and the Graph of $y = mx + b$

Vocabulary

y-intercept

linear function

slope-intercept form

x-intercept

▶ **BIG IDEA** If *y* changes by a constant amount *m* as *x* increases by 1, then $y = mx + b$ for all (x, y) and the graph of the points (x, y) is a line with slope *m* and *y*-intercept *b*.

Constant-Increase and Constant-Decrease Situations

In many real-world situations, there is an initial condition and a constant increase or decrease applied to that condition. This type of situation can be modeled by a linear equation.

Mental Math

a. Rafael graduated high school in May 2007 at the age of 18. How old will he be in October 2020?

b. Rafael's birthday is August 19. How old was he in November 2005?

c. Rafael's sister Bianca is 2 years and 8 months younger than he is. How old was Bianca in November 2005?

Example 1

Noah usually waits until he has only 1 gallon of gas left in his car before filling up. Suppose the gas pump pumps at a rate of 6 gallons per minute. Write an equation for a function representing the amount of gas in Noah's tank *x* minutes after he starts pumping gas. The tank holds 17.5 gallons when full.

Reighard's gas station, established in 1909, claims to be the oldest gas station in the United States.

Solution Let *A*(*x*) be the amount of gas in Noah's tank *x* minutes after he starts pumping. The function will have the form

$A(x) =$ (amount in tank at start) + (amount added).

The amount in the tank at the start is 1 gallon. Every minute, 6 gallons are added, so after *x* minutes, 6*x* gallons have been added. The function with equation

$$A(x) = 1 + 6x$$

or

$$A(x) = 6x + 1$$

gives the amount of gas in his tank after *x* minutes.

A graph of the function *A* in Example 1 is shown on the next page.

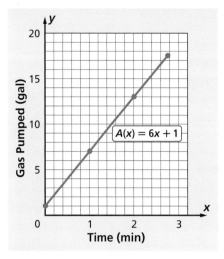

 QY1

Outside the context of Noah's gas tank, $y = 6x + 1$ is an equation describing a function whose domain and range are each the set of real numbers. The graph of the equation is the line containing the segment graphed above.

The y-value of the point where a graph crosses the y-axis is called its **y-intercept**. The graph of $y = 6x + 1$ crosses the y-axis at the point $(0, 1)$, so its y-intercept is 1. The y-intercept is the initial condition, or starting point, of the situation modeled by the graph; in Noah's case it is 1 gallon.

The slope of this line is the rate of change, 6 gallons per minute. You can verify this by finding the slope between two points on the graph, for example $(0, 1)$ and $(2.75, 17.5)$.

$$\frac{y_2 - y_1}{x_2 - x_1} = \frac{(17.5 - 1)\,\text{gallons}}{(2.75 - 0)\,\text{minutes}} = \frac{16.5\,\text{gallons}}{2.75\,\text{minutes}} = 6\,\frac{\text{gallons}}{\text{minute}}$$

Notice that the unit of the slope matches the unit of the constant change.

Slope-Intercept Form

A function whose graph is a line or a part of a line is called a **linear function**. In general, a linear function is a function that can be represented by an equation in the form $y = mx + b$. The form $y = mx + b$, or $f(x) = mx + b$, is called the **slope-intercept form** of a linear equation. Although the letter b is commonly used for the y-intercept, and the letter m commonly represents the slope, any other letters could be used in their place. Example 2 illustrates how the slope and y-intercept can be used to graph a line.

> ▶ **QY1**
>
> In Example 1, how long will it take to fill up Noah's gas tank?

GUIDED

Example 2

A pond with about 26 acre-feet of water needs to be drained for repairs to its levy. Engineers determine that 8 acre-feet of water can be safely drained per day.

a. Write a linear equation to model this situation.

b. In how many days will the pond be empty?

c. What are the *y*-intercept and the slope of the line represented by your equation in Part a?

d. Graph the line and indicate the point where the reservoir is empty.

Solution

a. Let $M(t)$ be the amount of water in the pond after t days.

$M(t) = \underline{\ ?\ } - \underline{\ ?\ }$

b. The pond will be empty when $M(t) = 0$.

Solve $\underline{\ \ ?\ \ } = 0$ to find $t = \underline{\ ?\ }$.

The reservoir will be empty after $\underline{\ ?\ }$ days.

c. The *y*-intercept is the starting point; the slope is the rate of change. So in this situation, the *y*-intercept is $\underline{\ ?\ }$ and the slope is $\underline{\ ?\ }$.

d. To graph this line, first locate the *y*-intercept 26, which corresponds to a full pond when $t = 0$. Use the slope to locate another point. A slope of $-8 = \frac{-8}{1}$ means that every horizontal change of 1 unit to the right corresponds to a vertical change of 8 units down. This gives the new point $(0 + 1, 26 - 8) = (1, 18)$. Plot $(1, 18)$ and draw the line. Label it $M(t) = -8t + 26$. The pond is empty at the point where the line crosses the *x*-axis, at the point $(3.25, 0)$.

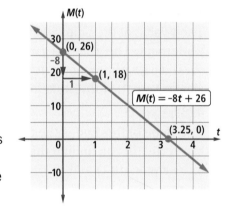

The *x*-value of the point where a line crosses the *x*-axis is called the **x-intercept** of the line. Because the line in Example 2 intersects the *x*-axis at $(3.25, 0)$, 3.25 is the *x*-intercept of the line.

In constant-increase situations, such as Example 1, as the value of *x* increases, so does the value of *y*. The slope of the line is positive. The graph slants up to the right and the function is said to be *increasing*.

In constant-decrease situations, such as Example 2, as the value of *x* increases, the value of *y* decreases. The slope of the line is negative. The graph slants down to the right and the function is said to be *decreasing*.

When the slope $m = 0$, the graph does not rise or fall; the function is *constant*. A line with a slope of 0 is horizontal. Vertical lines do not represent functions and have an undefined slope.

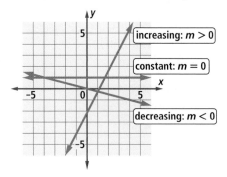

STOP QY2

▶ QY2

Why is a vertical line not the graph of a function?

Parallel Lines and Slope

Consider the graphs of $y = 4x + 7$ and $y = 4x - 2$ at the right. Both lines have slope 4; on each line, as you move 1 unit to the right, the line rises 4 units. In the figure, right triangles $\triangle ABC$ and $\triangle DEF$ are congruent by the SAS Congruence Theorem. So $\angle CAB \cong \angle FDE$, and these lines form congruent corresponding angles with the y-axis at A and D. Consequently, $\overleftrightarrow{AB}$ and $\overleftrightarrow{DE}$ are parallel.

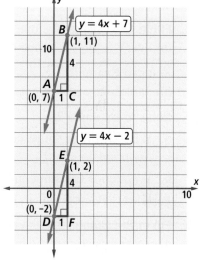

This argument can be repeated with any two lines that have the same slope. We also say that a line is parallel to itself. Thus, we can conclude the following:

If two lines have the same slope, then they are parallel.

The converse of this statement is: If two lines are parallel, then they have the same slope. This can be proved for all nonvertical lines as follows. Suppose lines s and t are parallel and not vertical, as shown at the right. Let m_1 be the slope of s, and m_2 be the slope of t. We want to show that $m_1 = m_2$. Draw line ℓ with equation $x = 1$. Note that ℓ is a transversal to the parallel lines. Draw horizontal segments from the y-intercepts of lines s and t to line ℓ as shown, forming right angles at C and F. Then $m_1 = $ slope of $s = \frac{AC}{BC} = \frac{AC}{1} = AC$, and $m_2 = $ slope of $t = \frac{DF}{EF} = \frac{DF}{1} = DF$.

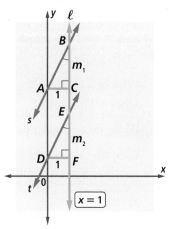

Recall from geometry that corresponding angles formed by parallel lines and a transversal are congruent. So $\angle ABC \cong \angle DEF$. Note also that $AC = DF$, so $\overline{AC} \cong \overline{DF}$. Since $\angle BCA \cong \angle EFD$, $\triangle ABC$ and $\triangle DEF$ are congruent by the AAS Congruence Theorem, and $BC = EF$. Because $m_1 = BC$ and $m_2 = EF$, the slopes are equal.

Consequently, we have shown that the following is true:

If two non-vertical lines are parallel, then they have the same slope.

Because the original statement and its converse above are both true, you can combine them into one biconditional (if and only if) statement that is important enough to be labeled as a theorem.

Parallel Lines and Slope Theorem

Two non-vertical lines are parallel if and only if they have the same slope.

Questions

COVERING THE IDEAS

1. **Fill in the Blanks** In the equation $y = mx + b$, the slope is __?__ and the y-intercept is __?__.

2. **Fill in the Blanks** A slope of 3 means a __?__ change of 3 units for every __?__ change of one unit.

3. Refer to Example 1. Suppose a car has 4 gallons in its tank and the tank holds 14.5 gallons. How long will it take to fill the tank?

4. Refer to Example 2.
   a. What does the y-intercept mean in the context of draining the pond?
   b. What does the slope mean in this context?
   c. How much water is in the pond at the end of the second day?

5. **Fill in the Blanks** In real-world constant-increase or constant-decrease situations, the initial condition can be represented by the __?__, and the constant change can be represented by the __?__.

In 6 and 7, find the slope of the line containing the points.

6. $(7, 2)$ and $(5, -4)$
7. $(32, 14)$ and $(-8, -6)$

8. Why are vertical lines excluded from the Parallel Lines and Slopes Theorem?

9. Write an equation for the line with y-intercept of -3 and slope of $\frac{2}{3}$.

10. Give the slope and y-intercept of $y = 5 - 2x$.

11. Which lines are parallel? (There may be more than one pair.)
    a. $y = 2x - 7$
    b. $y = -3x + 4$
    c. $2x - y = 7$
    d. $y = 12 - 3x$
    e. $-4x + 2y = 12$

APPLYING THE MATHEMATICS

12. A stack of Sunday newspapers sits on a skid 3" off the ground. Each newspaper is $1\frac{3}{4}$" thick.

    a. How high is the stack if there are n newspapers in it?

    b. If the top of the stack is 3' high, about how many newspapers are in the stack?

13. The Clear-Bell Cell phone company has a plan that costs customers $29.99 for a monthly service fee plus 3¢ per minute used.

    a. Write a linear equation to model this situation.

    b. Identify the slope and the y-intercept. Interpret the slope and y-intercept in the context of this problem.

    c. If Polly talks for 136 minutes, how much would her cell phone bill be (excluding taxes)?

14. The function $slope(a,b,c,d) = \frac{(d-b)}{(c-a)}$ calculates the slope of the line connecting two points. To use this function, define slope on a CAS, with the coordinates of the points (a, b) and (c, d) as arguments. Use the function to find the slope of the line through

    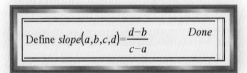

    a. $(2, -3)$ and $(5, 3)$.

    b. $(4.2, 8.3)$ and $(-3.4, 5.01)$.

15. Find the slope of a line with y-intercept 4 that contains the point $(2, 6)$.

In **16** and **17**, use graph paper to draw an accurate graph of these lines.

16. $y = -\frac{2}{3}x + 4$

17. $y = 5$

18. The line with equation $y = \frac{5}{2}x + 4$ is graphed at the right.

    a. What is its slope?

    b. Find the average rate of change between $(0, 4)$ and $(4, 14)$ to verify your answer to Part a.

    c. What is the y-intercept?

    d. Is the point $(-4, 3)$ on the line? Explain how you know.

    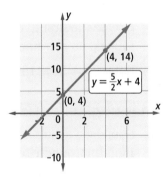

19. Stephanie is a taxi driver. She rents her taxi for $300/week. She feels she can take in $120/day in fares. Let D be the number of days she drives her taxi in a week and T be the total earnings she expects after subtracting the rental charge.

   a. Identify four ordered pairs (D, T).

   b. Write an equation that gives T as a function of D.

   c. What is the most she can earn in a year at this rate?

20. Suppose a varies directly as b, and that a is 10 when b is 3.

   a. Find the constant of variation and write an equation describing the variation.

   b. Make a table of values and graph the equation.

   c. Write the variation equation in slope-intercept form and identify the slope and y-intercept.

   d. Which type of variation does this represent: constant increase or constant decrease?

REVIEW

In 21–24, decide whether the variables in the equation exemplify direct, inverse, combined, or joint variation. (Lessons 2-1, 2-2, 2-9)

21. $ab = 5c$ 22. $x = y$ 23. $m = \dfrac{61g}{hz}$ 24. $2 = pr$

25. Consider the table of data at the right, which relates the number p of snowplows out on city streets to the number c of car crashes during a snowy day. (Lesson 2-7)

   a. Does c vary directly or inversely as a power of p?

   b. Based on your answer to Part a , either $c = kp^n$ or $c = \dfrac{k}{p^n}$. Find n for the appropriate relation.

26. Create the first five rows of a table for $y = \frac{3}{5}x$ with a start value of –10 and an increment of 5. (Lesson 1-5)

27. Solve $\frac{1}{3}y + 2x = 16$ for y. (Lesson 1-7)

Plows p	Crashes c
1	50
2	25
3	16
4	12
5	10
6	7

EXPLORATION

28. Suppose you are going on vacation outside the United States. Find the roundtrip airfare from your local airport to this destination. Estimate your daily expenses for hotel and food. Use the information to write a function that gives the cost of a trip to stay for d days. Is this function a linear function?

QY ANSWERS

1. 2.75 min, or 2 min 45 sec

2. Two points on a vertical line have different second coordinates but the same first coordinate. So, these lines violate the definition of a function.

Lesson 3-2
Linear Combinations and $Ax + By = C$

Vocabulary

linear combination

standard form

▶ **BIG IDEA** When the sum of multiples of x and y is a constant, then $Ax + By = C$ for all (x, y).

The form $y = mx + b$ is convenient for graphing lines because the y-intercept and slope are obvious, but you have also seen many equations for lines in different forms. For example, in the rectangle with length ℓ, width w, and perimeter 20 inches, $20 = 2\ell + 2w$. This is a linear equation, and the expression $2\ell + 2w$ is called a *linear combination* of ℓ and w.

Mental Math

Solve for x.

a. $|x| < 12$

b. $|2x| < 12$

c. $|2x - 4| < 12$

d. $|2x - 4| \geq 12$

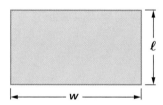

A **linear combination** is an expression in which all variables are raised to the first power and are not multiplied or divided by each other. Linear combinations may have 2, 3, 4, or more variables.

Activity

At Harry's Hamburger Hovel, hamburgers cost $2.50 and hot dogs cost $2. Suppose you have $30 to spend for hamburgers and hot dogs for a family gathering.

Step 1 Write an equation relating the number b of hamburgers and the number d of hot dogs you can purchase to spend all the money.

Step 2 Solve the equation in Step 1 for d.

Step 3 Graph the equation in Step 2 without taking the real-world context into account.

Step 4 Identify all points on the graph that correspond to a possible combination of hamburgers and hot dogs you can buy.

In Step 3 of the Activity you should have found that the graph of the equation is a line. This function is a *continuous* function whose domain and range are the set of real numbers. However, the entire line, as shown at the left below, is not an appropriate graph of this situation because counts of hamburgers and hot dogs are whole numbers. The linear function in this situation is *discrete; x* and *y* must be nonnegative integers. At the right below is a graph of the four ordered pairs of nonnegative integers that satisfy the equation.

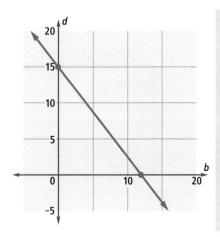
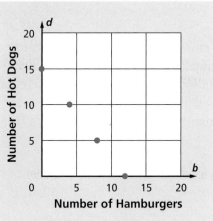

 QY

► **QY**

Give the domain and range of the discrete function graphed above at the right.

GUIDED

Example 1

Give two different situations that could be modeled by the linear combination equation $6x + 3y = 30$. For each situation, state whether the linear function is discrete or continuous, and explain what the point $(3, 4)$ represents.

Solution Each situation should specify what the numbers and variables in the equations represent. Here are two possible solutions.

a. You jog for x hours at 6 miles per hour and walk for y hours at 3 miles per hour, going a total of __?__ miles. The point $(3, 4)$ represents __?__ hours of jogging and __?__ hours of walking.

   You can jog and walk for any part of an hour; it does not make sense to restrict the domain to whole numbers, so the function is __?__.

b. You buy x adult tickets at __?__ each and y child tickets at __?__ each. The total cost is __?__. The point $(3, 4)$ represents __?__.

   You cannot buy part of a ticket. The function is __?__.

The equation $6x + 3y = 30$ is in the form $Ax + By = C$, with $A = 6$, $B = 3$, and $C = 30$. When A and B are not zero, the equation $Ax + By = C$ is called the **standard form** of an equation for a line.

Linear equations commonly occur in chemistry. For example, chemists describe the concentration of a solution in terms of the number of moles of the substance per liter (mol/L). One mole of a substance is approximately 6.02×10^{23} molecules. So, for example, 2 liters of a solution of 5 mol/L hydrochloric acid contains $2 \text{ L} \cdot 5 \text{ mol/L} = 10$ moles, or about 6.02×10^{24} molecules, of hydrochloric acid. Although a well-equipped lab usually stocks solutions in various concentrations, a chemist often has to mix his or her own solution at a particular concentration by combining other stock solutions. Pharmacists, painters, and others who deal with mixtures use the same mathematics.

Example 2

Suppose a chemist wants to mix x liters of a 2.5 mol/L solution of acid with y liters of a 7.25 mol/L solution to obtain a mixture with 10 moles of acid.

a. Write an equation that relates x, y, and the total number of moles of acid.

b. How many liters of the 7.25 mol/L solution must be added to 1.3 liters of the 2.5 mol/L solution to get 10 moles of acid in the final mixture? Answer to the nearest 0.1 liter.

Solution

a. To find the number of moles of acid in each solution, multiply the concentration by the quantity.

(L)(mol//L) = moles, so

x liters of 2.5 mol/L solution contain $2.5x$ moles of acid.

y liters of 7.25 mol/L solution contain $7.25y$ moles of acid.

So, an equation representing the situation is $2.5x + 7.25y = 10$.

b. Think: I want to know y when $x = 1.3$. On one CAS this translates to
solve(2.5x + 7.25y = 10,y)|x = 1.3.

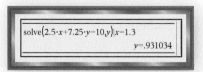

This CAS gives $y \approx 0.9$. So, you need to add approximately 0.9 liter of the 7.25 mol/liter solution.

Check 1 Substitute 1.3 for x and 0.9 for y in the original equation. $2.5(1.3) + 7.25(0.9) = 9.775$, which is close but not exactly equal because of rounding. It checks.

Check 2 Solve by hand. Substitute 1.3 for x. Then $2.5 \cdot 1.3 + 7.25y = 10$, so $7.25y = 6.75$. Then $y = \dfrac{6.75}{7.25} = \dfrac{27}{29} \approx 0.931$. It checks.

Questions

COVERING THE IDEAS

1. a. **Fill in the Blank** The expression $15P + 12L$ is called a ___?___ of P and L.

   b. Give a situation that could be modeled by the equation $15P + 12L = 600$.

2. **Multiple Choice** Which of the following is not a linear combination of x and y?

   A $x + y$ **B** $y - x$ **C** $4 \cdot x + x \cdot y$ **D** $x \cdot 2 + y \cdot -8$

3. At a vegetable market, Farmer Bob sells T tomatoes at $0.50 each, S squash at $0.75 each, and K bunches of kale at $1.25 each. Write a linear combination to express the total amount of money he takes in.

In 4 and 5, a situation is given.
   a. Write a linear combination equation describing the situation.
   b. Determine whether the situation is discrete or continuous.

4. Ayani is planning a trip across Europe. He will travel b km by bus and t km by train over a 5,000 km route.

5. Jewel is planning a trip across the United States. She wants to visit 12 cities along the way, taking a bus to b cities and taking a train to t cities.

6. Refer to Example 2.

   a. **Fill in the Blanks** The point $(0.8, 1.1)$ is an approximate solution to the equation $2.5x + 7.25y = 10$. This solution means that you used about __?__ liters of 2.5 mol/L solution, about __?__ liters of 7.25 mol/L solution, with a total of about __?__ moles of acid.

   b. What is the domain of the relation $2.5x + 7.25y = 10$ in this situation?

   c. Find the y-intercept of the graph. What combination does it represent?

7. Sodium hydroxide, NaOH, is a common compound used in chemistry. Suppose that x liters of a solution that is 5.2 mol/L sodium hydroxide are combined with y liters of a solution that is 7.8 mol/L sodium hydroxide. In Parts a–c, write an expression for the number of moles of NaOH in

   a. the 5.2 mol/L solution.

   b. the 7.8 mol/L solution.

   c. the two solutions combined.

d. If Tyra wants 3.6 moles of NaOH in the final mixture, what equation relates x, y, and the 3.6 total moles of NaOH?

e. How many liters of the 7.8 mol/L solution must be added to 0.4 liter of the 5.2 mol/L solution to get 3.6 moles of NaOH in the final mixture?

APPLYING THE MATHEMATICS

8. Refer to the Activity. Suppose that you have $40 to spend on hamburgers and hot dogs.

   a. In standard form, write a new equation relating the costs of hamburgers and hot dogs.

   b. Graph your equation on your graphing utility. Set the window to $\{x \mid 0 \leq x \leq 16\}$ and $\{y \mid 0 \leq y \leq 20\}$. Sketch the graph.

   c. Make a table showing all the possible combinations of hamburgers and hot dogs that you could purchase for exactly $40.

9. A charity makes and sells piñatas as a fundraiser. Large piñatas cost $25 and small piñatas cost $15. Let L be the number of large piñatas and S be the number of small piñatas that are sold.

   a. What kinds of numbers make sense for S and L in this context?

   b. How much money will the charity take in if 4 large piñatas and 6 small piñatas are sold?

   c. If the charity takes in a total of $225, write an equation relating S, L, and the amount of money taken in.

   d. Graph the equation from Part c. Use L as the independent variable.

   e. Make a discrete graph of the solution set appropriate to this situation.

   f. Give all possible pairs (L, S) of large and small piñatas the charity could have sold to earn $225.

A traditional piñata is usually made of paper and is filled with toys or sweets.

10. Driving from the city to visit a friend in the country, suppose a person drives at an average speed of 15 miles per hour on city streets and 40 miles per hour on the highway. Suppose the person spends C hours on city streets and H hours driving on the highway, and the friend lives 60 miles away.

   a. Write an equation relating C, H, and the total distance.

   b. If the person spent 2 hours driving on city streets, how many minutes did the person spend on the highway?

11. How many pounds of cashews at \$3.29/lb should be mixed with peanuts at \$1.79/lb to create a mixture of 5 pounds of nuts worth \$2.29/lb?

REVIEW

12. **Multiple Choice** Which of the following could be an equation of the graph at the right? (Lesson 3-1)

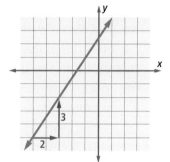

A $y = \frac{2}{3}x + 1$ B $y = \frac{3}{2}x - 1$ C $y = 3x - 2$

D $y = x + \frac{2}{3}$ E $y = \frac{3}{2}x + \frac{5}{2}$

13. If t varies jointly with s and r and r varies directly with m, show that t varies jointly with s and m. (Lesson 2-9)

14. Consider the points (3, 1), (−4, 4.5), and (5, 0). Determine if they lie on a line. If they do, state the slope of the line; if they do not, explain why not. (Lesson 2-4)

15. Michelangelo's *David*, a statue in Florence, Italy, is 17 feet high and made out of marble. Over the years, many replicas have been made of this statue. The weight w of a marble replica varies directly as the cube of the replica's height h. (Lesson 2-1)

    a. Write an equation modeling this situation.

    b. **Fill in the Blank** The weight of a 12-inch marble replica would be __?__ the weight of the original statue.

16. Graph $y = \sqrt{x^2 - 4}$ on a calculator. (Lesson 1-4)

    a. Does this graph appear to be a graph of a function?

    b. What is the domain?

    c. What is the range?

EXPLORATION

17. In many schools, a student's grade point average is calculated using linear combinations. Some schools give 4 points for each A, 3 points for each B, 2 points for each C, and 1 point for each D. Suppose a person gets 7 As, 3 Bs and 2 Cs.

    a. Calculate this person's total number of points.

    b. Divide your answer in Part a by the total number of classes to get the grade point average.

    c. Calculate your grade point average for last year using this method.

Lesson

3-3

The Graph of $Ax + By = C$

> ▶ **BIG IDEA** If a linear combination of two variables x and y is a constant, then the graph of all the points (x, y) is a line.

Recall the equation $2.5x + 2y = 30$ from the previous lesson. This equation represents allowable $30 purchases of x hamburgers at $2.50 each and y hot dogs at $2.00 each from Harry's Hamburger Hovel. Because you do not buy fractions of sandwiches, both x and y are nonnegative integers. So a graph of the solution is a set of discrete points. However, if you allow x and y to be any real numbers, then the graph of $2.5x + 2y = 30$ is shown at the right.

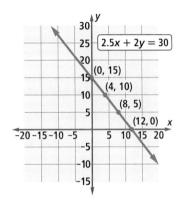

The equation $2.5x + 2y = 30$ is of the form $Ax + By = C$, with $A = 2.5$, $B = 2$, and $C = 30$. When A and B are not both zero, the graph of $Ax + By = C$ is always a line.

Standard Form of an Equation of a Line Theorem

The graph of $Ax + By = C$, where A and B are not both zero, is a line.

Proof There are two cases to consider: (1) if $B = 0$ and (2) if $B \neq 0$.

(1) If $B = 0$, then $A \neq 0$, and the equation is simply $Ax = C$.
Multiply both sides by $\frac{1}{A}$ to obtain the equivalent equation $x = \frac{C}{A}$.
The graph of this equation is a vertical line.

(2) If $B \neq 0$, then solve the given equation for y:

$$Ax + By = C \qquad \text{Given}$$
$$By = -Ax + C \qquad \text{Add } -Ax \text{ to both sides.}$$
$$y = -\frac{A}{B}x + \frac{C}{B} \qquad \text{Divide both sides by } B.$$

Mental Math

Give a general variation equation based on the description.

a. The cost c of painting the interior of a house varies directly as the number n of rooms to be painted.

b. The amount p of paint needed for a wall varies jointly as the length ℓ and height h of the wall.

c. The time t it will take to paint varies inversely as the number n of painters hired.

d. The time d needed for the paint to dry varies directly as the thickness t of the paint applied and inversely as the square of the amount a of air circulation in the room.

QY1

▶ **QY1**

When $B \neq 0$, what are the slope and y-intercept of the line with equation $Ax + By = C$?

Graphing a Line Using Intercepts

Because the form $Ax + By = C$ can describe any line, it is called the *standard form of an equation for a line.* Although, if $B \neq 0$, you could rewrite such an equation in slope-intercept form in order to make a graph, it is often much quicker to graph such equations by hand using x- and y-intercepts. If A, B, and C are all nonzero, the line with equation $Ax + By = C$ has distinct x- and y-intercepts, so the intercepts can be used to graph the line.

Example 1

Graph the equation $-5x + 2y = 10$ using its x- and y-intercepts.

Solution To find the x-intercept, substitute 0 for y, and solve for x.

$$-5x + 2(0) = 10$$
$$x = -2$$

The x-intercept is -2.

To find the y-intercept, substitute 0 for x, and solve for y.

$$-5(0) + 2y = 10$$
$$y = 5$$

The y-intercept is 5.

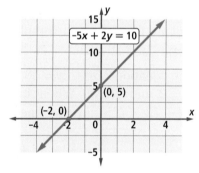

Plot $(-2, 0)$ and $(0, 5)$ and draw the line containing them, as shown at the right.

Check Find a third ordered pair that satisfies $-5x + 2y = 10$. For example, when $x = 2$, $-10 + 2y = 10$, so $2y = 20$ and $y = 10$. Thus, $(2, 10)$ should be on the graph. Is it? Yes. It checks.

This technique does not work when A, B, or C is zero. If $A = 0$, the slope is $-\frac{0}{B} = 0$. The line is horizontal, and so there is no x-intercept. If $B = 0$, the slope of the line is $-\frac{A}{0}$, which is undefined. The line is vertical, and so there is no y-intercept.

▶ **QY2**

Why can you not use the x- and y-intercepts to graph $Ax + By = C$ when $C = 0$?

 QY2

Example 2

Graph $x + 0y = 3$.

a. Is this the graph of a line?

b. Is this the graph of a function?

Solution

a. The equation simplifies to $x = 3$. The value of x is always 3, regardless of the value of y. The graph is a vertical line.

b. Create a table of values.

It is not the graph of a function because more than one ordered pair has the same x-coordinate.

x	y
3	−5
3	0
3	2

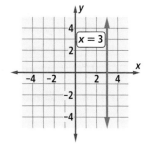

Equivalent Equations for Lines

One drawback of the standard form is that the same line can have many different, but equivalent, equations in standard form. Recall that multiplying both sides of an equation by a nonzero real number yields an equivalent equation. Since lines have a unique equation in slope-intercept form, you can test equations for equivalence by putting them in slope-intercept form.

GUIDED

Example 3

Find which equations below, if any, represent the same line.

(1) $4x + 1.5y = 12$ (2) $8x + 3y = 24$

(3) $8x + 3y = 12$ (4) $16x + 6y = 12$

Solution 1 Rewrite each line in slope-intercept form.

(1) $y =$ __?__   (2) $y =$ __?__

(3) $y =$ __?__   (4) $y =$ __?__

Equations __?__ are equivalent. Equations __?__ are not equivalent to any other given equations.

Solution 2 If I multiply both sides of Equation __?__ by 2, Equation __?__ results. So Equations __?__ and __?__ are equivalent. Since the right side of three of the given equations is 12, no other equations are equivalent.

Equations (2), (3), and (4) from Example 3 represent lines with the same slope but different y-intercepts. This suggests that the graphs of $Ax + By = C$ and $Ax + By = D$ are distinct parallel lines when $C \neq D$.

Questions

COVERING THE IDEAS

1. a. **Fill in the Blank** If A and B are not both 0, the graph of $Ax + By = C$ is a ___?___.
    b. If $A \neq 0$ but $B = 0$, what kind of line is the graph?
    c. If $B \neq 0$ but $A = 0$, what kind of line is the graph?

2. **Fill in the Blank** $Ax + By = C$ is in the ___?___ of an equation of a line.

3. What is true about the slope of a vertical line?

4. **True or False** Every line in standard form can be graphed by drawing the line containing its x- and y-intercepts.

In 5 and 6, an equation for a line is given.
    a. Find its x-intercept.
    b. Find its y-intercept.
    c. Graph the line using your answers to Parts a and b.

5. $4x + 9y = 36$ 6. $4x - 5y = 10$

7. Consider $Ax + By = C$.
    a. Find the x-intercept of the line. What happens when $A = 0$?
    b. Find the y-intercept of the line. What happens when $B = 0$?

APPLYING THE MATHEMATICS

8. Write an equation in standard form for the line graphed at the right.

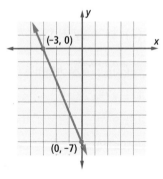

9. Find the value of C such that the point $(4, -1)$ lies on the graph of $10x - 2y = C$.

10. Delaney's Deli makes ham and cheese sandwiches and turkey sandwiches. Each ham and cheese sandwich uses $\frac{1}{8}$ lb of cheese, while each turkey sandwich uses no cheese. Let x be the number of ham and cheese sandwiches the deli prepares. Let y be the number of turkey sandwiches the deli prepares.
    a. Write an equation stating that the total amount of cheese the deli uses is 5 lb.
    b. Graph your equation from Part a.

11. **a.** Graph the line with equation $0x + 4y = 14$.

  b. Find two ordered pairs satisfying the equation.

  c. Compute the slope of the line through the two points.

In 12 and 13, find an equation for a line in standard form with the given properties.

12. y-intercept –3 and slope 2

13. no y-intercept and passes through the point (17, 29.93)

14. Mallory combines n liters of a solution that is 3 mol/L chlorine with y liters of a solution that is 5 mol/L chlorine. She ends up with a mixture that contains 2 moles of chlorine.

  a. Write an equation relating n, y, and the total amount of chlorine in the mixture.

  b. Graph the equation you obtained in Part a by finding the n- and y-intercepts. Consider n to be the independent variable.

  c. Use your graph to find about how many liters of the 5 mol/L solution Mallory must add to 0.4 liter of the 3 mol/L solution to get the final mixture.

15. Consider the graphs of $Ax + By = C$ and $Ax + By = D$ if A and B are not both zero and $C \neq D$.

  a. Rewrite each equation in slope-intercept form.

  b. What is the relationship between the slopes of the lines? What does that tell you about the lines?

  c. What is the relationship between the y-intercepts of the lines? What does that tell you about the lines?

  d. Write the conclusions of Parts b and c as one if-then statement.

  e. Use the if-then statement from Part d to give the equations of several lines parallel to $16x - 13y = 11$.

16. Use a CAS expand command to show that multiplying the equation $Ax + By = C$ by a nonzero number k yields another equation in standard form. Why must this new equation describe the same line as the original?

REVIEW

17. At a library book sale, paperbacks are being sold for $0.50 each and hardcover books are $1. If you want to buy P paperbacks and H hardcover books, write a linear combination that expresses the amount you will have to pay. **(Lesson 3-2)**

18. Suppose your car uses 0.035 gallon of gas to travel 1 mile in the city and 0.027 gallon of gas to travel 1 mile on the highway. (**Lesson 3-2**)

    a. Write a linear combination to express the number of gallons of gas you would use to travel C miles in the city and H miles on the highway.

    b. If your car has 14 gallons of gas, how many city miles can you drive without refilling if you also make a 200-mile highway trip?

19. A car starts out 400 miles from St. Louis and drives directly toward St. Louis at 60 mph. (**Lesson 3-1**)

    a. Find an equation for the distance d from St. Louis as a function of time t in hours from the start of the trip.

    b. Does the equation in Part a describe a constant-increase or a constant-decrease situation?

20. The sum S of the measures of the interior angles of a convex polygon varies directly as $n - 2$, where n is the number of sides of the polygon. (**Lesson 2-1, Previous Course**)

    a. Find the constant of variation.

    b. Graph the function.

21. The independent variable of a function is given. State a reasonable domain for the function. (**Lesson 1-4**)

    a. h = number of hours worked in a day

    b. d = distance traveled away from home while on vacation

    c. t = temperature in Indianapolis, Indiana, in February

Gateway Arch in St. Louis is the tallest national monument in the United States. The shape of the arch is known as a *catenary curve*.

EXPLORATION

22. Consider the lines $Ax + By = C$ and $Bx + Ay = C$. Explore the connections between slopes and intercepts of these lines.

QY ANSWERS

1. The slope is $-\frac{A}{B}$, and the y-intercept is $\frac{C}{B}$.

2. When $C = 0$, the x- and y-intercepts are both zero. The line passes through the origin, and the x- and y-intercepts are not distinct.

Lesson
3-4
Finding an Equation of a Line

Vocabulary

point-slope form

piecewise linear function

▶ **BIG IDEA** Postulates and theorems of geometry about lines tell when exactly one line is determined from given information. An equation of that line can often be determined algebraically.

In geometry you learned that *there is exactly one line through a given point parallel to a given line.* Since nonvertical lines are parallel if and only if they have the same slope, this means that there is exactly one line through a given point with a given slope. Using algebra you can determine the equation of this line.

Mental Math

Suppose you pull a chip from a bag. What is the probability of choosing a blue chip if

a. there are 8 blue chips and 4 red chips?

b. there are 3 blue chips, 7 red chips, and 1 white chip?

c. there are b blue chips, r red chips, and w white chips?

Finding a Linear Equation

Example 1

In a physics experiment, a spring is 11 centimeters long with a 13-gram weight attached. Its length increases 0.5 centimeter with each additional gram of weight. This is a constant-increase situation up to the spring's elastic limit. Write a formula relating spring length L and weight W. Then graph the equation.

Solution The slope is $0.5 \frac{cm}{gram}$. Because the unit for slope is centimeters per gram, length is the dependent variable and weight is the independent variable. The point (13, 11) is on the line. Substitute these values into the slope formula.

$$\frac{L - 11}{W - 13} = 0.5$$

To put this equation in slope-intercept form, multiply both sides by $W - 13$.

$$L - 11 = 0.5(W - 13)$$

Then solve for L.

$$L - 11 = 0.5W - 6.5$$
$$L = 0.5W + 4.5$$

Now you know the y-intercept of this line as well as the point (13, 11). Use this information to graph the equation, as shown at the right.

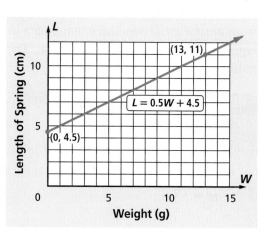

QY1

▶ **QY1**

What does the y-intercept in the equation

$L = 0.5W + 4.5$

represent in Example 1?

Point-Slope Form of a Line

Each of the equations $L = 0.5W + 4.5$ and $L - 11 = 0.5(W - 13)$ describes the situation of Example 1. The slope-intercept form is useful for computing values of L quickly if you know values of W. The form $L - 11 = 0.5(W - 13)$ shows the slope and a specific point on the graph, and it can be used to determine the slope-intercept form.

Point-Slope Theorem

If a line contains the point (x_1, y_1) and has slope m, then it has the equation $y - y_1 = m(x - x_1)$.

Proof Let ℓ be the line with slope m containing (x_1, y_1). If (x, y) is any other point on ℓ, then using the definition of slope,

$$m = \frac{y - y_1}{x - x_1}.$$

Multiplying both sides by $x - x_1$ gives

$$y - y_1 = m(x - x_1).$$

This is the desired equation of the line.

The equation $y - y_1 = m(x - x_1)$ is called a **point-slope form** of an equation for a line. Solving for y in the point-slope form of a line gives a form that combines aspects of the slope-intercept and point-slope forms.

$$y = m(x - x_1) + y_1$$

Finding an Equation of a Line through Two Points

Two points determine a line. You use this idea every time you draw a line through two points with a straightedge. It is another postulate from geometry. If you know two points on a line, you can find its equation by first computing its slope. Then use either point in the point-slope form $y - y_1 = m(x - x_1)$ or $y = m(x - x_1) + y_1$.

GUIDED

Example 2

Find an equation of the line p through $(4, 7)$ and $(6, -2)$.

Solution First, compute the slope of the line p.

$$m = \underline{\quad ? \quad}$$

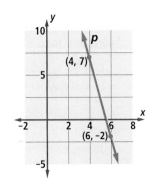

Second, because you know two points on the line and neither of them is the *y*-intercept, use one of the points in the point-slope form.

$$y - \underline{\ ?\ } = \underline{\ ?\ }(x - \underline{\ ?\ })$$

This is one equation for the line *p*. In slope-intercept form, this equation is $\underline{\ ?\ }$.

 QY2

▶ **QY2**

Check your solution to Example 2 by finding an equation of the line using the point you did not use before and putting the equation in slope-intercept form.

Piecewise Linear Functions and Graphs

In Chapter 1 you saw the graph at the right of Frank's bicycle trip to Chuck's house. You focused on the domain and range of the whole trip and on function representations for different parts of the trip. For instance, at time $t = 60$ minutes, Frank was about 6 miles from home, so $d = f(60) = 6$.

Because this graph does not have a constant rate of change over its domain $\{t\mid 0 \le t \le 210\}$, it cannot be represented by a single linear equation. But you can write an equation for each segment of the graph that does have a constant rate of change. Because the graph is described in pieces, the graph and the function are called **piecewise linear**. The graph is the union of several segments. Each segment can be described by a linear function that has a restricted domain indicating where the segment starts and stops.

Writing a function for a piecewise linear graph requires writing the equation of each line segment.

Example 3

Write a piecewise linear function for the first two segments of Frank's bicycle trip as described in the graph above. Estimate the values of segment endpoints where necessary.

Solution **First piece:** This segment appears to begin when $t = 0$ and end when $t = 45$. The ordered pairs represented by the endpoints are $(0, 0)$ and about $(45, 6)$. Write an equation for the line through these two points.

Find the slope: $m = \dfrac{6 - 0}{45 - 0} = \dfrac{6}{45} = \dfrac{2}{15}$.

Because $(0, 0)$ is on the line, 0 is the *d*-intercept. So use the slope-intercept form of a line to get

$$d = \frac{2}{15}t + 0 \ \text{ or } \ d = \frac{2}{15}t.$$

(continued on next page)

Since this equation describes only the part of the situation for t-values from 0 through 45, write

$$d = \frac{2}{15}t \text{ for } 0 \leq t \leq 45.$$

Second piece: This appears to be a constant function where $d = 6$ between $t = 45$ and $t = 60$. So,

$$d = 6 \text{ for } 45 < t \leq 60.$$

In the second piece, we have a choice whether to include the value 45 in the domain. This is because (45, 6) is already included in the first piece and it is not necessary to repeat it.

The piecewise function: Combine the functions for the two pieces and their domains into one formula using a brace.

$$d = f(t) = \begin{cases} \frac{2}{15}t & \text{for } 0 \leq t \leq 45 \\ 6 & \text{for } 45 < t \leq 60 \end{cases}$$

Writing a piecewise formula for the last four segments of the trip is left to you in the Questions.

The piecewise formula can be used to determine Frank's distance from home at any time, even if the distance cannot be easily determined from the graph. For instance, to determine how far Frank was from home after 20 minutes, use the first equation because $t = 20$ is in the domain $0 \leq t \leq 45$. So, when $t = 20$, $d = f(20) = \frac{2}{15}(20) \approx 2.67$ miles.

 QY3

▶ QY3

How far from home was Frank after 55 minutes?

If you know the formula for a piecewise function, a CAS or graphing utility can be used to graph it.

Activity

Step 1 Open a graphing utility and clear any functions that have been entered. Find out how to enter a piecewise function. On the CAS pictured at the right, a template is used to enter a piecewise function.

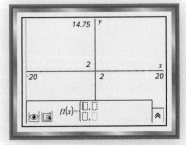

Step 2 Enter the piecewise formula you found in Example 3. Include the domain restrictions for each line of the formula.

Step 3 Graph the function. Adjust your window as necessary to see the whole graph.

Step 4 Compare your graph to the first part of the graph found in this lesson. Do they look the same?

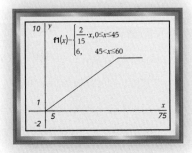

Questions

1. How many points determine a line?

2. Name a point on the line $L = 0.5W + 4.5$ other than the two used in Example 1.

3. **True or False** You can determine the equation of a line knowing only its slope.

4. **Fill in the Blank** The point-slope form of the equation for a line with slope m and passing through point (x_1, y_1) is ___?___.

5. Give a strategy for finding an equation for a line when you know two points on the line.

6. A line passes through the points $\left(\frac{1}{3}, \frac{2}{5}\right)$ and $\left(\frac{7}{3}, \frac{9}{10}\right)$.
    a. Compute the slope of the line.
    b. Use $\left(\frac{1}{3}, \frac{2}{5}\right)$ and the slope to determine an equation of the line.
    c. Check that $\left(\frac{7}{3}, \frac{9}{10}\right)$ satisfies the equation.

In 7 and 8, write an equation for the line with the given information.

7. slope 6 and y-intercept $\sqrt{3}$

8. slope $-\frac{3}{2}$ and passing through $(-4, 1)$

9. Find an equation in point-slope form for the line graphed at the right.

10. Refer to Example 3.
    a. Write the equations describing the last four segments of Frank's bicycle trip.
    b. How far from home was Frank after 132 minutes?
    c. Graph the whole piecewise function on a CAS.

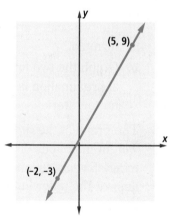

11. A club will be charged $247.50 for printing 150 t-shirts and $501.25 for printing 325 t-shirts. Let c be the cost of printing s shirts.
    a. Write c as a linear function of s.
    b. How much will it cost to print 0 t-shirts? (This is the set-up cost.)
    c. How much will it cost to print 100 t-shirts?

12. Find an equation for the line through $(5, 3)$ and parallel to $y = \frac{4}{7}x - 12$.

13. Refer to the `slope` function defined in Question 14 of Lesson 3-1.

    a. Use `slope` to find the slope of the line through $P = (17.3, 2.4)$ and $Q = (43.9, 22.6)$.

    b. Define a new function `ptslope` that will find the equation of a line with slope m and passing through the point (a, b) in point-slope form. A sample from one CAS is shown at the right.

    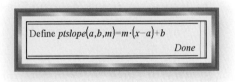

    Define $ptslope(a,b,m)=m \cdot (x-a)+b$

    Done

    c. Use `ptslope` to find an equation of $\overleftrightarrow{PQ}$ from Part a. Use $a = 17.3$, $b = 2.4$, and the slope you found in Part a as m.

    d. You can define one function, call it `line2pt(a,b,c,d)`, that combines the calculations of `slope` and `ptslope`. Define this function on a CAS.

    e. Input the coordinates of points P and Q into `line2pt`. Are your results the same as in Part c?

14. An approximate conversion from degrees C on the Celsius scale to degrees F on the Fahrenheit scale is $F \approx 2C + 32$.

    a. Use the facts that $32°F = 0°C$, $100°C = 212°F$, and that the two scales are related linearly to write a more accurate formula.

    b. Graph the two formulas to determine when the approximation is within $3°$ of the actual Fahrenheit temperature.

15. The graph at the right represents the average weight in pounds of a child from age 3 months to age 5 years (60 months). Use the graph to answer the following questions.

    a. Write a piecewise linear function for this situation.

    b. What is the average weight of a 30-month-old child?

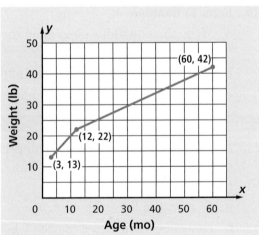

16. In the book, *The Lion, the Witch, and the Wardrobe* (C.S. Lewis, 1961), a group of children find a portal to the world of Narnia where time passes at a different rate than on Earth. The children are in Narnia for 15 years while only 5 minutes pass on Earth. Peter, the oldest child in the group, is 13 when he enters Narnia. Write a linear equation that shows Peter's age n in Narnia for each minute m that passes on Earth.

REVIEW

17. **Fill in the Blanks** The equation $Ax + By = C$ describes a line with an undefined slope whenever A is ___?___ and B is ___?___. (**Lesson 3-3**)

18. Write an equation of the line with x-intercept –2 and y-intercept 4 in standard form. (**Lesson 3-3**)

19. **Multiple Choice** Which of the following is not a linear combination? (**Lesson 3-2**)

 A $4a + (-3)b$ B $12x + 12y + 12z$ C $T - F$
 D $7A + 3C^2$ E $e + e + e + 3f$

20. Makayla finds that the amount she studies varies inversely as the number of phone calls she receives. One day she studied 3.5 chapters with 3 phone calls. How many chapters can she study if she gets 5 phone calls? (**Lesson 2-2**)

21. Consider the sequence r whose first 7 terms are –5, 3, 1700, –65.4, 0.354, 29, and –1327. (**Lesson 1-8**)
 a. What is r_3?
 b. Write the sentence $r_7 = -1327$ in words as you would read it.

22. The following table of values shows the height of Azra's dog for certain numbers of months after she bought him. (**Lesson 1-5**)

Month	0	1	2	3	4	5	6
Height (in.)	10	14	16	19	22	23	24

 a. Graph the values in the table on the coordinate axes, with time on the horizontal axis and height on the vertical axis.
 b. Azra thinks the domain of her function is all positive real numbers between 0 and 6, but her brother Marquis thinks it is only the whole numbers from 0 to 6. Who is right and why?

EXPLORATION

23. In 1912, the Olympic record for the men's 100-meter dash was 10.8 seconds. In 1996 it was 9.84 seconds. Assume a linear relationship between the year and the record time.
 a. According to the information given, determine a linear equation to model the given data.
 b. Using your model, compute the predicted world record in 2006.
 c. A new world record of 9.76 seconds for the 100-meter dash was set in 2006. Compare your predicted record with the actual 2006 value.

QY ANSWERS

1. 4.5 cm, the length of the spring with no weights attached

2. Answers vary. Sample: $y + 2 = -4.5(x - 6)$. In slope-intercept form: $y = -4.5x + 25$.

3. 6 miles

Finding an Equation of a Line **175**

Lesson
3-5
Fitting a Line to Data

Vocabulary

linear regression

line of best fit,
   least-squares line,
   regression line

deviation

▶ **BIG IDEA** When data are almost on a line, it is often helpful to approximate the data by a line that minimizes the sum of the squares of the distances from the data points to the corresponding points on the line.

Mental Math

Find the slope of the line.

a. $y = 2.5x + 7$

b. $y = 14$

c. $y - 0.2 = 0.144(x - 0.4)$

d. $16x + 20y = 45$

Square Grabber, which may be provided by your teacher, is a fun game with simple rules. You control a black square. Your job is to capture the other black squares by touching them while avoiding the red squares.

Play the game once. At the end of the game you are told the number of squares you captured and are given a score.

You can write the numbers as an ordered pair (number of squares, score). If you play many times and generate many ordered pairs, you can graph the ordered pairs and use the graph to find a model for the relationship between the number of squares you capture and your score. Is this a simple linear relationship?

Activity 1

MATERIALS internet connection

Step 1 Play Square Grabber 15 times. Record the number of squares you capture and your score for each game in a table like the one below.

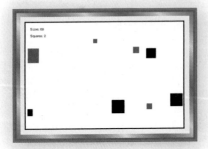

Number of Squares (n)	Score (s)
43	855
23	461
12	235

Step 2 Graph your data. Let $n =$ number of squares and $s =$ score. Use n as the independent variable and s as the dependent variable. Does it appear that a line could be a good model for your data?

Step 3 Eyeball a line that comes close to modeling all of the points in your data set. Draw it on your graph.

Step 4 Estimate the coordinates of any two points on your line. (They do not have to be actual data points.) Find an equation for the line through these points.

Step 5 Your equation from Step 4 can be used to estimate your score based on the number of squares you captured. Use your model to estimate your score for capturing 100 squares.

The Regression Line

In Activity 1, you approximated a *line of best fit* for your data by eye. How can you tell which line fits the data the *best*? If you passed your graph of ordered pairs around the room and asked each of your classmates to find a *line of best fit*, you might get as many different lines as you have classmates.

To solve this problem, statisticians have developed a method called **linear regression** that uses all the data points to find the line. A line found by using regression is what people call the **line of best fit**. It is also called the **least-squares line**, or simply the **regression line**. You will learn how regression works in a later course. For now, to find the line of best fit, use a statistics application. For details on how your application works, check the manual or ask your teacher.

Activity 2

Navy divers who remain underwater for long periods of time cannot come quickly back to the surface due to the high pressure under the water. They must make what are known as *decompression stops* on the way up. If divers skip this procedure they risk a serious medical condition known as *the bends*. The U.S. Navy has created tables that allow divers to know when and for how long they should stop on the way to the surface. The table below gives the decompression time needed (including ascent time) based on how many minutes were spent at a given depth. The points in the table are graphed at the right. Calculate the regression line for the decompression data.

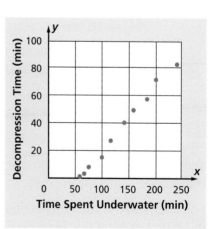

Time Spent at a Maximum Depth of 60 feet (min)	60	70	80	100	120	140	160	180	200	240
Decompression Time Needed (min)	1	3	8	15	27	40	49	57	71	82

(continued on next page)

Step 1 Enter the data into columns in your statistics or spreadsheet application. Name the first column *timespent* and the second column *decomptime* to indicate which is the independent and which is the dependent variable.

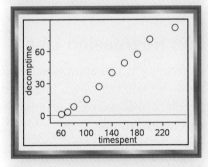

A timespent	B decomptime	C	D
1	60	1	
2	70	3	
3	80	8	
4	100	15	
5	120	27	
6	140	40	

A1 | 60

Step 2 Create a scatterplot of the data. You may be prompted to choose one column to use for the independent variable data and choose one column to use for the dependent variable data.

Step 3 If possible, add a movable line to your scatterplot. Adjust the position of the line to eyeball a line of best fit. Record the equation of your movable line. Also record the predicted values when $x = 80$ and when $x = 220$.

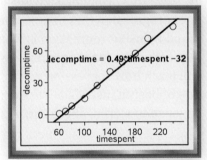

decomptime = 0.49*timespent −32

Step 4 Remove your movable line. Choose to show a linear regression line. The application will graph the regression line on the same axes as the data. Record the equation of the regression line. How does your movable line compare to the regression line?

Step 5 Compare your movable line's predictions to those of the regression line. When $x = 220$, by how much do the predicted values differ from each other? When $x = 80$, by how much does the predicted value differ from the value in the table?

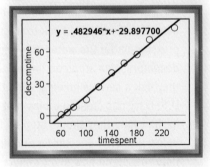

y = .482946*x+-29.897700

You can see from the graphs that the regression line is a reasonable model for the data.

For each value of the independent variable, the difference between the actual value of the dependent variable and the value predicted by the model is called the **deviation**. The line of best fit has the following property: *The sum of the squares of the deviations of its predicted values from the actual values is the least among all possible lines that could model the data.* This is why it is called the least-squares line.

Questions

COVERING THE IDEAS

In 1–3, refer to Activity 1.

1. A student playing Square Grabber recorded a data point of (48, 985). What does this ordered pair mean in this context?

2. a. Find the regression line for the data you collected in Activity 1.
   b. Pick two values of x to compare how well the regression line predicts y-values compared to the line you eyeballed.

3. A person found $y = 19.0x + 35$ to be an equation for the line of best fit for Activity 1.
   a. What is the slope of the line?
   b. What does the slope mean in this situation?
   c. What is the y-intercept of the line?
   d. What does the y-intercept mean in this situation?
   e. Does the x-intercept have a practical meaning in this case?

4. Refer to Activity 2. If a diver spent 130 minutes at a maximum depth of 60 feet, estimate his decompression time using the regression line.

APPLYING THE MATHEMATICS

5. a. Use regression to find an equation of the line through the points (1, 4) and (–2, 8).
   b. Verify your equation in Part a by finding the slope of the line and using the Point-Slope Theorem.

6. The table below gives the total payroll for the Chicago Cubs from 1998 to 2006.

Year	2006	2005	2004	2003	2002	2001	2000	1999	1998
Payroll (millions of dollars)	94.4	87.0	90.6	79.9	75.7	64.5	62.1	55.4	49.4

Source: http://content.usatoday.com/sports/baseball/salaries/teamresults.aspx?team=17

   a. Draw a scatterplot and eyeball a line of best fit to the data.
   b. Which data point has the greatest deviation from your line?
   c. Find an equation for the regression line for the data.
   d. Which data point has the greatest deviation from the regression line?
   e. Use the regression line and your line to predict the 2007 Chicago Cubs payroll.
   f. Which gives the closer prediction to the actual value of 99.7 million dollars, your eyeballed line or the regression line?

7. Recall the data at the right from the beginning of the chapter on the average body mass index b for a male of age a. Verify that $b = 0.07a + 24.9$ is an equation for the line of best fit to the data.

Age	BMI
25	26.6
35	27.5
45	28.4
55	28.7

8. The table below shows nutritional information on various items from a fast-food menu. Included for each item are the number of calories and the fat content. A scatterplot of the data is shown.

Item	Calories	Fat Content (g)
chicken pieces	250	15
Asian salad	300	10
cheeseburger	300	12
fish fillet	380	18
chicken sandwich	360	16
large French fries	570	30
big hamburger	460	24
big cheeseburger	510	28
big breakfast	720	46

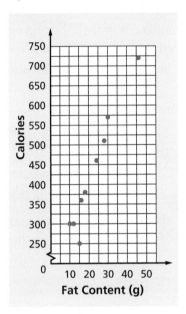

a. Describe the general relationship between calories and fat content.

b. What are the independent and dependent variables?

c. Find an equation of the regression line for this data set.

d. What is the slope of the regression line, and what does it mean in the context of the problem? What is the unit for the slope?

e. What is the y-intercept of the regression line, and what does it mean in context of the problem? Is the value practical in this situation?

f. A salad with chicken is not on this menu. It contains 320 calories. Use the regression line to estimate the number of grams of fat in the salad.

9. Add a large vanilla milkshake with 740 calories and 18 grams of fat to the menu in Question 8. Recalculate the regression line.

REVIEW

10. **a.** Find an equation for the line that passes through the points $(2, 5)$ and $\left(-\frac{4}{3}, 9\right)$.

    b. Graph the line from Part a. **(Lesson 3-4)**

11. **a.** Write an equation of a line with slope $-\frac{5}{2}$ and y-intercept 3.

    b. Write an equation of a line parallel to the line in Part a that passes through the point $(-1, -2)$. **(Lessons 3-4, 3-1)**

12. Consider the formula for the volume of a sphere, $V = \frac{4}{3}\pi r^3$. If the radius of a sphere is divided by three, how many times smaller is the volume of the resulting sphere? **(Lesson 2-3)**

13. Consider the sequence $H_n = (-1)^n(n + 5)$, for integers $n \geq 1$. **(Lesson 1-8)**

    a. Write the first four terms.

    b. Without explicitly calculating, is the 25th term positive or negative?

14. Jayla's cake company charged $25 per cake, and each week she paid $50 for supplies and the upkeep of her equipment. She found that $I = 25c - 50$ is an explicit formula for her income I based on the number of cakes c that she sold. **(Lesson 1-5)**

    a. Use a graphing utility to generate a table showing the number of cakes sold and Jayla's income.

    b. Graph the first six data values from your table on coordinate axes.

15. Given that $b: x \rightarrow \dfrac{3x + 5}{7 - 8x}$ evaluate **(Lesson 1-3)**

    a. $b(2)$. **b.** $b(0)$.

    c. $b(a)$. **d.** $b(-2) + b(-4)$.

EXPLORATION

16. When you calculate a regression equation, some calculators and software include an extra statistic r called *correlation* along with the slope and intercept. Find out what correlation means and what it tells you about the regression line. Go back and examine r for the data sets in Questions 6–8 and see if you can interpret its value in the context of the problems.

Lesson 3-6

Recursive Formulas for Sequences

Vocabulary

recursive formula
recursive definition

▶ **BIG IDEA** Some sequences can be defined by giving their first terms and indicating how later terms are related to each other.

Recall from Chapter 1 that a *sequence* is a function whose domain is the set of natural numbers or a subset of the natural numbers less than a given number. If you know an *explicit formula* for a sequence, then you can write the terms of a sequence rather quickly.

Consider a freight train consisting of one engine and a series of identical boxcars. The engine is 72 feet long and each boxcar is 41 feet long. So the length of a train with n boxcars is defined by the explicit formula $L_n = 72 + 41n$. The first four terms of the sequence are:

$$L_1 = 72 + 41 \cdot 1 = 113$$
$$L_2 = 72 + 41 \cdot 2 = 154$$
$$L_3 = 72 + 41 \cdot 3 = 195$$
$$L_4 = 72 + 41 \cdot 4 = 236$$

We say the formula $L_n = 72 + 41n$ *generates* the sequence 113, 154, 195, 236, ….

 QY1

Mental Math

Consider the sequence
$t_n = \frac{n^2}{2} + 7$.

a. Find t_7.

b. Find t_{100}.

c. Find $t_{12} - t_{10}$.

d. What is the domain of the sequence?

Generating Sequences Using a Calculator

You can use a calculator to generate sequences.

▶ **QY1**

Generate the first five terms of a sequence S when $S_n = 34 - 12(n - 1)$.

Activity

Use a calculator to generate the sequence of freight train lengths.

Step 1 Store the first value of this sequence, 113. When we enter 113 and press ENTER, this calculator stores it in the variable ans.

Step 2 Add 41 to ans, and repeatedly press ENTER. The first five terms of the sequence are shown at the right. To find additional terms, continue pressing ENTER.

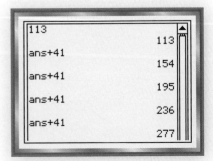

QY2

Any sequence of numbers that is defined by adding, subtracting, multiplying, or dividing by a constant can be generated on your calculator using the approach of the Activity.

▶ **QY2**

In the Activity, if you pressed ENTER 35 times, what term of the sequence would appear?

What Is a Recursive Formula?

Notice that no explicit formula was used to generate the sequence in the Activity. Instead, each term of the sequence was derived from the preceding one. For instance, to find the 6th term of the sequence, you add 41 to the preceding term (the 5th term, 271). This process generates the sequence *recursively*.

> **Definition of Recursive Formula**
>
> A **recursive formula** or **recursive definition** for a sequence is a set of statements that
>
> a. indicates the first term (or first few terms) of the sequence, and
> b. tells how the next term is calculated from the previous term or terms.

Consider the sequence $L = 113, 154, 195, 236, \ldots$ of lengths of a train with n boxcars. You can write this sequence recursively as

$$\begin{cases} L_1 = 113 \\ L_n = \text{previous term} + 41, \text{ for integers } n \geq 2. \end{cases}$$

The brace { indicates that both lines are needed for the recursive definition. You should be able to read and evaluate recursive definitions.

GUIDED

Example 1

Verify that the recursive definition of L above generates the first four terms of the sequence.

Solution The first term is given,

$L_1 = 113$.

According to the second line of the definition,

$L_2 = \text{previous term} + 41 = \underline{\ ?\ } + 41 = \underline{\ ?\ }$.

To find L_3, use the definition again.

$L_3 = \text{previous term} + 41 = \underline{\ ?\ } + 41 = \underline{\ ?\ }$.

(continued on next page)

L_4 = previous term + 41 = __?__ + 41 = __?__.

The first four terms are 113, __?__, __?__, __?__, which checks according to what was calculated before.

Notation for Recursive Formulas

In Example 1, the sequence is described using the symbol L and the symbol L_n denotes term number n. The term that precedes term n is term number $(n - 1)$, and the term that follows term n is term number $(n + 1)$. So the symbol L_{n-1} can be used in place of the words *previous term*. Thus the sequence of Example 1 can be defined as follows:

$$\begin{cases} L_1 = 113 \\ L_n = L_{n-1} + 41, \text{ for integers } n \geq 2 \end{cases}$$

Or, if you prefer, you could let L_n be the previous term and L_{n+1} be the next term. In this case, the definition would be

$$\begin{cases} L_1 = 113 \\ L_{n+1} = L_n + 41, \text{ for integers } n \geq 1. \end{cases}$$

A recursive definition always includes at least two lines. One or more lines define the initial values of the sequence, and another line defines the relationship between consecutive terms.

GUIDED

Example 2

Consider the sequence $A = 10{,}000$; 9600; 9200; ... of dollar amounts left on a car loan if $400 is repaid each month. Write a recursive definition for the sequence A_n.

Solution The first line of the definition gives the initial term. So, the first line is

$A_1 = $ __?__.

The second line relates each term to the previous term. So, the second line is
$A_{n+1} = A_n - $ __?__, for integers n $\geq$ __?__.

If you can describe a sequence in words, then you can use that description to write a recursive formula for the sequence.

Example 3

The NCAA Women's Basketball Tournament originally consists of 64 teams, which are paired in single-elimination contests. At the end of each round of play, the number of teams proceeding to the next round is half the previous number.

The sequence 64, 32, 16, 8, 4, 2 gives the number of teams in the tournament at the beginning of round n.

a. Use words to write a recursive definition of this sequence.

b. Write a recursive formula for this sequence.

Solution

a. Identify the first term and the rule for generating all following terms. **The first term is 64. Each term after the first is found by dividing the previous term by 2.**

b. Choose a name for the sequence. We call it b_n. Then write the recursive formula. Note that there are only six terms.

$$\begin{cases} b_1 = 64 \\ b_n = \tfrac{1}{2}b_{n-1}, \text{ for integers } n = \{2, 3, 4, 5, 6\} \end{cases}$$

CAUTION: The first term of the sequence in Example 3 is 64, the term written in the first line of the recursive definition. Although 32 is the first term given by the rule for b_n in the solution, it is the second term of the sequence. Remember, b_n represents the nth term in this definition, and the formula for b_n is only defined for $n = \{2, 3, 4, 5, 6\}$.

Recursive Definitions and Spreadsheets

A spreadsheet is another efficient tool for generating a sequence from its recursive definition.

Example 4

Use a spreadsheet to generate the first five terms of the sequence

$$\begin{cases} L_1 = 113 \\ L_n = L_{n-1} + 41, \text{ for integers } n \geq 2. \end{cases}$$

Solution Enter 113 in cell A1.

Then in A2, enter =A1+41.

Each cell after A2 can be defined as the value in the previous cell + 41. A shortcut to do this is to copy cell A2 and paste it into cells A3–A5. For example, after the paste, the screen at right shows that the formula in A5 has been updated to refer to cell A4. You could also use the fill down function on the spreadsheet instead of copying and pasting.

So the value in cell A5 $= L_5 = 277$.

The first five terms are 113, 154, 195, 236, and 277.

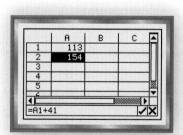

Example 5

Use a spreadsheet to generate the six terms of the sequence in Example 3.

Solution Enter 64 in cell __?__.

Enter ___?___ in cell A2.

To see the six terms of this sequence in your spreadsheet, copy cell __?__ and then paste it into cells __?__ through __?__.

The six terms of the sequence are __?__, __?__, __?__, __?__, __?__, and __?__.

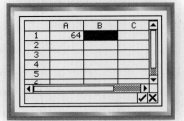

Questions

COVERING THE IDEAS

In 1 and 2, refer to the Activity.

1. What is the meaning of ans on the CAS?

2. How long is the freight train with 12 boxcars?

3. What is a recursive definition for a sequence?

4. Suppose t_n denotes the nth term of a sequence.
   a. What does t_{n-1} denote?   b. What does t_{n-2} denote?
   c. What does t_{n+1} denote?

In 5 and 6, refer to Example 3.

5. Rewrite the recursive formula if b_n is the previous term instead of the next term.

6. Suppose that the NCAA Women's Basketball Tournament is expanded to include 128 teams.
   a. How many terms are now in the sequence describing the tournament?
   b. Write a recursive formula for this sequence using recursive notation.

In 7 and 8, a description of a sequence is given.
   a. Write the first five terms of the sequence.
   b. Write the recursive formula.

7. The first term is –2; each term after the first is 8 more than the previous term.

8. The first term is 4; each term after the first is $-\frac{1}{4}$ times the previous term.

9. Explain in your own words why a recursive formula must have at least two parts.

10. Consider the sequence that begins $-19, -26, -33, -40, \ldots$.
    a. What is the first term?
    b. **Fill in the Blank** From the second term on, each term is ___?___ the previous term.
    c. Write a recursive formula for the sequence.

11. Consider the decreasing sequence of negative even integers, beginning with $-2, -4, -6, \ldots$. Write a recursive definition for this sequence if t_{n+1} is the next term in the second line.

12. Generate the first eight terms of the sequence below by hand or using a spreadsheet.
$$\begin{cases} S_1 = 12 \\ S_n = 2S_{n-1} + 3, \text{ for integers } n \geq 2 \end{cases}$$

APPLYING THE MATHEMATICS

13. In an auditorium with 25 rows, the first row has 8 seats and each succeeding row has 2 more seats than the row in front of it.
    a. Write a recursive formula for a sequence that gives the number S_n of seats in row n.
    b. Find the number of seats in the 14th row.

14. You are given a penny on the first day of September. On each subsequent day you are given twice as many as the previous day.
    a. Write a recursive formula to represent how much you will receive on the nth day of September.
    b. How much money will you receive on September 30?

15. Let S be a sequence. Is the set of ordered pairs of the form (n, S_n) a function? Why or why not?

16. The table at the right gives the cost per book of printing 100 or more books, based on the number of color pages in the book.
    a. Describe in words the sequence that gives the cost per book for books with n color pages.
    b. Write a recursive formula for the sequence that gives the cost per book for books with n color pages.

17. Consider the sequence $6.42, 3.69, 0.96, -1.77, -4.5, \ldots$.
    a. Write a description of this sequence in words.
    b. Write a recursive definition of this sequence.
    c. Find the eighth term of the sequence.

Number of Color Pages	Cost ($)
1	5.55
2	5.70
3	5.85
4	6.00
5	6.15
6	6.30
7	6.45
8	6.60

18. The first term of a sequence is 1. Suppose each term after that is the sum of its term number and the previous term.

   a. Write a recursive definition for this sequence.

   b. Explain how you would use a spreadsheet to generate this sequence through term 12.

   c. What is the 12th term?

REVIEW

19. The table at the right gives the total attendance and payroll of some of the Major League Baseball teams in 2006. Make a scatterplot of the data and find the line of best fit. Does there appear to be a linear relationship between attendance and payroll? **(Lesson 3-5)**

20. **True or False** The line of best fit for a set of collinear points is the line through those points. **(Lesson 3-5)**

21. Suppose a store sells custom-made baseball caps with a graduated pricing scheme. The first 10 caps cost $14 each, the next 10 cost $12 each, and each cap after the 20th costs $10. **(Lesson 3-4)**

   a. Draw a graph showing how the total cost C in dollars relates to the number of caps n that you order.

   b. Express the relation between C and n as a piecewise function with linear pieces.

Team	Attendance (thousands)	Payroll (millions of dollars)
Yankees	4200	198.6
Angels	3406	103.6
White Sox	2957	102.8
Red Sox	2930	120.1
Rockies	2105	41.1
Indians	1998	56.8
Pirates	1861	46.8
Royals	1372	47.3

22. A formula for the area of a trapezoid with height h and bases b_1 and b_2 is $A = \frac{1}{2}h(b_1 + b_2)$. Solve this formula for h. **(Lesson 1-7)**

23. When you drop an object, its distance fallen in feet is modeled by $d(t) = 4.9t^2$, where t is time in seconds. **(Lesson 1-3)**

   a. What does the sentence $d(3) = 44.1$ mean in English?

   b. Does the sentence $d(-1) = 4.9$ have meaning? Why or why not?

EXPLORATION

24. There is an expression *two steps forward, one step back.*

   a. Write a recursive definition for the sequence of stopping points on a path following these directions. Let the nth term of the sequence be your position after you have followed these directions n times.

   b. What does the phrase mean?

3-7 Graphs of Sequences

Vocabulary

Fibonacci sequence

> ▶ **BIG IDEA** Sequences are graphed like other functions. The major differences are that the graph of a sequence is discrete and you can obtain some values of sequences using a recursive definition.

As you saw in Lesson 1-8 and in the last lesson, sequences can be described in two ways:

- An *explicit formula* gives an expression for the nth term of a sequence in terms of n. An example is $a_n = 4n - 6$.

- A *recursive formula* gives a first term or first few terms and an expression for the nth term of a sequence in terms of previous terms. An example is

$$\begin{cases} a_1 = -2 \\ a_n = a_{n-1} + 4, \text{ for } n > 1. \end{cases}$$

To graph a sequence, plot each ordered pair (n, a_n). You can generate the ordered pairs using a written description of a sequence, an explicit formula, or a recursive formula. The next three examples explore these possibilities.

Mental Math

a. What is 30% of 70x?

b. What is 70% of 30x?

c. What is 75% of $8x - 24y$?

GUIDED

Example 1

Consider the sequence with recursive formula $\begin{cases} a_1 = -2 \\ a_n = a_{n-1} + 4, \text{ for } n > 1. \end{cases}$

a. Make a table of values of the first six terms of this sequence.

b. Graph the first six terms of the sequence.

Solution

a. Make a table with n in one column and a_n in another column, as shown at the right. From the recursive definition, $a_1 = -2$ and each succeeding term is 4 larger than the previous term.

(continued on next page)

n	a_n
1	-2
2	?
3	?
4	?
5	?
6	?

b. Plot the points from your table on a coordinate grid, with n as the independent variable. The points should be collinear. (The graph at the right is not complete.)

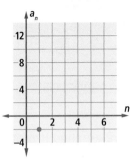

In Example 1, you could calculate each term of the sequence by hand. Using a spreadsheet, you can automatically generate the terms of a sequence if you know an explicit or recursive formula for it. Example 2 shows how to do this, and also points out an important difference between the graph of a line and the graph of a sequence.

Example 2

Recall the explicit formula for the sequence of lengths L_n of a train with n boxcars from Lesson 3-6, $L_n = 72 + 41n$.

a. Graph the first six terms of L_n using the explicit formula.
b. Graph the function $L(n) = 72 + 41n$ using a domain of the set of all reals on the same axes as the graph of the sequence.
c. Compare and contrast the graphs.

Solution

a. Enter the index values 1 through 6 into column A of a spreadsheet. Then use the explicit formula to generate values for the train-length sequence. On one calculator, we do this by entering the formula at the top of column B, as shown at the right. On other machines, you might do this by using the fill function.

Create a scatterplot of the sequence, as shown below.

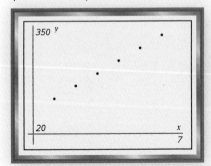

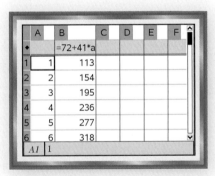

b. Graph $L(n) = 72 + 41n$ on your plot from Part a, as shown at the right.

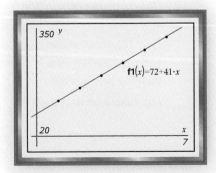

c. Notice the similarities and differences between the two graphs.

Both graphs have the same constant rate of change of 41 feet per car. The difference is that the line graph is continuous, while the sequence graph is discrete.

Graphing a Sequence Using a Recursive Formula

Sometimes, it is easier to write a recursive formula for a sequence than an explicit formula. Example 3 illustrates such a situation. It also shows how spreadsheets can be used to generate terms and graph a recursively defined sequence.

Example 3

The sneezewort plant (also called sneezeweed or *Achillea ptarmica*) starts as a single stem. After two months of growth, the stem sends off a shoot that becomes a new stem, and produces a new shoot every month thereafter. The new shoots must mature for two months before they are strong enough to produce shoots of their own.

Let F_n be the number of stems in month n. The first two terms are equal to 1. Beginning with the third term, each term is found by adding the previous two terms.

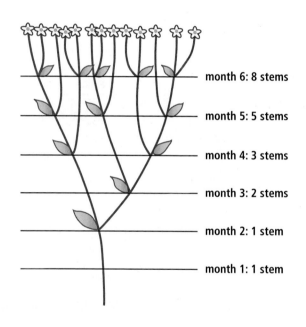

month 6: 8 stems

month 5: 5 stems

month 4: 3 stems

month 3: 2 stems

month 2: 1 stem

month 1: 1 stem

a. Using F_n to represent the number of stems and shoots in month n, write a recursive formula for the sequence.

b. Graph the first ten terms of the sequence.

Solution

a. After the first two terms, each term of the sequence is calculated by adding the previous two terms. As you know, when F_n is the nth term, then the previous $(n-1)$st term is F_{n-1}. Similarly, the term preceding F_{n-1} is the $(n-2)$nd term, or F_{n-2}. So a recursive definition is

$$\begin{cases} F_1 = 1 \\ F_2 = 1 \\ F_n = F_{n-1} + F_{n-2}, \text{ for } n \geq 3. \end{cases}$$

(continued on next page)

Alternatively, you could write

$$\begin{cases} F_1 = 1 \\ F_2 = 1 \\ F_{n+1} = F_n + F_{n-1}, \text{ for } n \geq 1. \end{cases}$$

b. Use a spreadsheet. Enter the index numbers 1 through 10 into column A. The terms of the sequence will be in column B.

The recursive definition from Part a says that $F_1 = F_2 = 1$. So enter 1 in each cell B1 and B2.

Each of the terms F_3 through F_{10} is defined as the sum of the two previous terms. So, each cell from B3 through B10 needs to be defined as the sum of the previous two cells. Enter =B1+B2 in cell B3, as shown at the right. Then copy and paste this formula into cells B4 through B10 to generate the rest of the desired terms of the sequence. When this is done, 55 should appear in cell B10, as shown below at the left.

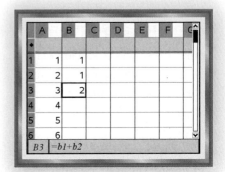

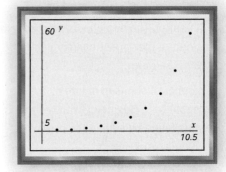

Create a scatter plot of the sequence, as shown above at the right.

The sequence in Example 3 is called the **Fibonacci** (pronounced "Fee-boh-NOTCH-ee") **sequence**. It is named after Leonardo of Pisa, a 12th century mathematician who wrote under the name Fibonacci. The Fibonacci numbers arise in a wide variety of contexts, and are so mathematically rich that there is an entire publication, the *Fibonacci Quarterly*, devoted to the mathematics arising from them.

 QY

You may also have noticed that the points on the graphs in Examples 1 and 2 are collinear, but the points on the graph in Example 3 are not. This is because in the sequences in Examples 1 and 2, there is a constant difference between terms, but in the Fibonacci sequence, the difference between terms is not constant. The sequences in Examples 1 and 2 are examples of *arithmetic sequences*. You will learn more about arithmetic sequences in the next lesson.

▶ QY

In the sequence F of Example 3, find F_{11}.

Questions

COVERING THE IDEAS

1. Refer to Example 1.
   a. What is a_{10}?
   b. Write a recursive rule for a_{n+1}.

2. Refer to Example 2.
   a. Write the recursive formula for L_n.
   b. Explain why it does not make sense to let n be any real number.

3. **Multiple Choice** Which formula below gives the sequence graphed at the right?

   A $s_n = n^2$

   B $\begin{cases} s_1 = 1 \\ s_n = s_{n-1} + 1, n > 1 \end{cases}$

   C $\begin{cases} s_1 = 40 \\ s_n = s_{n-1} - 3, n > 1 \end{cases}$

   D $\begin{cases} s_1 = 4 \\ s_n = s_{n-1} + 3, n > 1 \end{cases}$

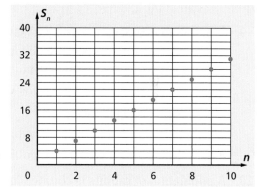

4. Give one reason why you might choose to model a situation with a recursively defined sequence rather than an explicitly defined one.

5. a. What is the tenth Fibonacci number?
   b. How many 3-digit Fibonacci numbers are there?
   c. How many 5-digit Fibonacci numbers are there?

APPLYING THE MATHEMATICS

6. a. Graph the first ten terms of the sequence whose explicit formula is $s_n = n^2$.
   b. On the same axes, graph the first ten terms of the sequence whose recursive formula is
   $\begin{cases} r_1 = 1 \\ r_n = r_{n-1} + 2n - 1, n > 1. \end{cases}$
   c. What do the graphs suggest about the sequence of numbers 1, 4, 9, 16, ... ?

7. Forrest, a crime scene investigator, is called to investigate a missing person's case. On entering the person's apartment, he discovers hundreds of bugs crawling around. Suppose that the number of bugs increases by 20% every day the apartment is left unoccupied, and that on the first day the apartment was left unoccupied, there were about 10 bugs.
   a. Write a recursive definition that describes the sequence giving the number b_n of bugs on day n.
   b. Graph the first 20 terms of the sequence you found in Part a.
   c. A detailed census determines that there are 450 bugs in the apartment. For how long has the apartment been unoccupied?

8. The *tribonacci* numbers R_n are defined as follows:

   (1) The first three tribonacci numbers are 1, 1, and 2.

   (2) Each later tribonacci number is the sum of the *three* preceding numbers.

   a. Using the rule, the fourth tribonacci number is $1 + 1 + 2 = 4$. Compute the fifth, sixth, and seventh tribonacci numbers.

   b. Give a recursive formula for the tribonacci numbers.

   c. Use a spreadsheet to find the 17th tribonacci number greater than 1000.

9. Kamilah gets a part-time job and saves $75 each week. At the start of the summer, her bank account has a balance of $500. Let b_n be the balance at the start of week n (so $b_1 = \$500$).

   a. Make a table of values of b_n for $n = 1, 2, \dots , 6$.

   b. Write a recursive formula for b_n.

   c. Make a graph of the first 20 values of b_n.

   d. Are the points on the graph collinear? Justify your answer.

10. A population of rabbits is counted annually; the number of rabbits in year n is r_n. The values of r_n can be modeled by a recursively defined sequence:

$$\begin{cases} r_1 = 150 \\ r_{n+1} = 0.008 r_n \left(200 - r_n\right), \text{ for } n \geq 1. \end{cases}$$

The largest litter of rabbits on record is 24. This occured in 1978 and 1999.

   a. Graph the first 20 terms of this sequence.

   b. In the long term, what happens to the population of rabbits? Does it increase, decrease, stabilize at a particular value, or follow some other pattern you can describe?

   c. Try graphing the sequence with several different values of r_1, for $2 \leq r_1 < 200$. How does the starting value affect the long-term population?

11. a. Graph the first ten terms of the sequence given by the recursive formula $a_n = \frac{1}{2}\left(a_{n-1} + \frac{9}{a_{n-1}}\right)$, $n > 1$, with $a_1 = 20$.

   b. Describe any patterns you see in the graph in words. For example, do the terms get larger or smaller? Do they appear to get closer to a particular number?

   c. Repeat Parts a and b using the recursive formula $a_1 = 20$, $a_n = \frac{1}{2}\left(a_{n-1} + \frac{25}{a_{n-1}}\right)$. How do your answers change?

REVIEW

12. Write the first five terms of the sequence defined by

$$\begin{cases} b_1 = 2 \\ b_{n+1} = 2b_n + 1, \text{ for integers } n \geq 1. \end{cases}$$ **(Lesson 3-6)**

13. Consider the sequence that begins 81, 27, 9, 3, ... **(Lesson 3-6)**
    a. **Fill in the Blank** From the second term on, each term is
    ___?___ the previous term.
    b. Write a recursive formula for this sequence.
    c. Compute the next four terms in the sequence.

14. The table at the right lists body lengths and weights for several
    humpback whales. Compute the regression line for these data
    and use it to estimate the weight of a humpback whale that is
    12 meters long. **(Lesson 3-5)**

Length (meters)	Weight (metric tons)
12.5	23
13.4	30
13.5	29
14.5	36
13.0	27
14.0	32

15. **Multiple Choice** Which of the following equations describes a
    line that does not have an x-intercept? **(Lesson 3-3)**
    A $4x + 3y = 0$ B $2x - 2y = 1$ C $0x + 0y = 3$
    D $-3x + 0y = 0$ E $0x + 2y = -6$

16. Two perpendicular lines
    $\overleftrightarrow{AB}$ and $\overleftrightarrow{AC}$ have been
    drawn and their slopes
    measured. If the slope of
    $\overleftrightarrow{AB}$ is r, then the slope s
    of $\overleftrightarrow{AC}$ is a function of r.
    The point (r, s) is plotted
    as the lines rotate.
    (Lesson 2-7)

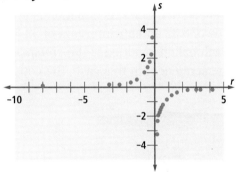

Slope $\overleftrightarrow{AB}$; r	Slope $\overleftrightarrow{AC}$; s
4.11	-0.24
3.30	-0.30
3.05	-0.33
2.35	-0.42
1.76	-0.57
1.40	-0.71
1.17	-0.86
0.99	-1.01
0.86	-1.16
0.77	-1.29
0.70	-1.44
0.63	-1.58
0.58	-1.72
0.53	-1.87
-18.08	0.06

    a. Use the graph to determine whether there is a direct or
    inverse variation relationship between r and s.
    b. Ana used the table function of her calculator to generate
    the table of s and r values at the right. Based on her data,
    determine an equation relating r and s.

EXPLORATION

17. The Fibonacci numbers have many arithmetic properties. Label
    the Fibonacci numbers $F_1 = 1$, $F_2 = 1$, $F_3 = 2$, and so on. Use
    a list of the first 15 Fibonacci numbers to explore each of the
    following.
    a. Let s_n be the sum of the first n Fibonacci numbers. That is,
    $s_1 = F_1$, $s_2 = F_1 + F_2$, $s_3 = F_1 + F_2 + F_3$, etc. Make a table of
    values of s_n for $n = 1, 2, ... , 14$.
    b. Find a pattern in your table that allows you to quickly
    compute s_n from a list of Fibonacci numbers.
    c. Do some research and find some other arithmetical
    properties of the Fibonacci numbers.

QY ANSWER

$F_{11} = 89$

Lesson

3-8

Formulas for Linear (Arithmetic) Sequences

Vocabulary

linear sequence,
  arithmetic sequence

▶ **BIG IDEA**   As with other linear functions, two points on the graph of a *sequence*, or one point on the graph and its constant difference, are enough to determine an explicit formula.

Demetrius is adding water to a fish tank one gallon at a time. The tank weighs 32 pounds when empty and each gallon of water weighs 8.33 pounds. Let $W(g) = 32 + 8.33g$ represent the weight of the tank after g gallons of water are added. Notice that this equation is in slope-intercept form with a slope of 8.33 and a y-intercept of 32. The increase of 8.33 pounds after each gallon is added is the *constant difference* between terms of the sequence. In this case, the constant difference is a constant increase.

Mental Math

Tell whether the graph of the function is a line, a parabola, a hyperbola, or none of these.

a. $f(x) = 6x^2$

b. $g(x) = -2x$

c. $h(x) = -\dfrac{3}{x^2}$

d. $j(x) = \dfrac{17}{x}$

A sequence with a *constant difference* between successive terms is called a **linear** or **arithmetic sequence**. (Here the word *arithmetic* is used as an adjective; it is pronounced a-rith-MEH-tik.) Because there is a constant difference between successive terms, a linear sequence is a linear function with the domain of a sequence.

Developing an Explicit Formula for an Arithmetic Sequence

As you learned in Lesson 1-8, an explicit formula for a sequence can be developed by examining the pattern in a table.

Example 1

a. Write an explicit formula for the arithmetic sequence 4, 7, 10, 13, ...

b. Compute the 30th term of the sequence.

Solution

a. To develop an explicit formula, use the constant difference to write each term after the first. Consider the pattern in the table at the right.

Notice that in term n, the number of 3s added to the initial term is 1 less than n. So, an explicit formula for the sequence is

$a_n = 4 + (n - 1) \cdot 3$.

$a_n = 3n + 1$

b. a_{30} is the 30th term of the sequence. Substitute 30 for n.

$a_{30} = 3(30) + 1$

$a_{30} = 91$

n = number of term	term = a_n
1	4
2	$4 + 1 \cdot 3 = 7$
3	$4 + 2 \cdot 3 = 10$
4	$4 + 3 \cdot 3 = 13$
5	$4 + 4 \cdot 3 = 16$
⋮	⋮
n	$4 + (n - 1) \cdot 3 = a_n$

Because the sequence in Example 1 is a function, it can be written as a set of ordered pairs: $\{(1, 4), (2, 7), (3, 10), \ldots\}$. These points lie on a line with slope, or constant increase, of 3. Substitute this slope and the point $(1, 4)$ in the point-slope form of the equation of a line.

$y - 4 = 3(x - 1)$

$a_n - 4 = 3(n - 1)$ Substitute.

$a_n = 4 + 3(n - 1) = 3n + 1$ Solve for a_n.

This is the formula found in Part a of Example 1. This suggests the following theorem.

*n*th Term of an Arithmetic Sequence Theorem

The nth term a_n of an arithmetic (linear) sequence with first term a_1 and constant difference d is given by the explicit formula $a_n = a_1 + (n - 1)d$.

Proof Each ordered pair of the arithmetic sequence is of the form $(x, y) = (n, a_n)$. The first ordered pair is $(x_1, y_1) = (1, a_1)$. Because arithmetic sequences represent constant-increase or constant-decrease situations, a graph of the sequence consists of points that lie on a line. The slope of the line is the constant difference d. Substitute these values into the point-slope form of a linear equation.

$y - y_1 = m(x - x_1)$ Point-slope form

$a_n - a_1 = d(n - 1)$ Substitute.

$a_n = a_1 + (n - 1)d$ Solve for a_n.

 QY

▶ QY

Use the theorem to find an explicit formula for the sequence 6, 10, 14, 18, ...

Recursive Notation for Arithmetic Sequences

A recursive formula for the arithmetic sequence in Example 1 is

$$\begin{cases} a_1 = 4 \\ a_n = a_{n-1} + 3, \text{ for integers } n \geq 2. \end{cases}$$

The second line of this formula can be rewritten as

$$a_n - a_{n-1} = 3.$$

This shows the constant difference between term n and term $(n - 1)$. The constant difference 3 is the slope between the points $(n - 1, a_{n-1})$ and (n, a_n).

More generally, suppose a sequence is defined recursively as

$$\begin{cases} a_1 \\ a_n = a_{n-1} + d, \text{ for integers } n \geq 2. \end{cases}$$

When $n \geq 2$, we can rewrite the second line as $a_n - a_{n-1} = d$. This means that the difference between consecutive terms is the constant d. By definition, the sequence is arithmetic. This proves the following theorem.

Constant-Difference Sequence Theorem

The sequence defined by the recursive formula

$$\begin{cases} a_1 \\ a_n = a_{n-1} + d, \text{ for integers } n \geq 2 \end{cases}$$

is the arithmetic sequence with first term a_1 and constant difference d.

Example 2

A cell-phone company charges 25¢ per minute for overseas calls along with a 30¢ service charge.

a. Write a recursive formula for the arithmetic sequence that represents the cost of a call lasting n minutes.

b. Graph the sequence.

Solution

a. When the call begins you are immediately charged 55¢ (25¢ for your first minute plus the 30¢ service charge). Then you pay 25¢ for each additional minute that the call lasts, so the difference between successive terms is $d = 25$. A recursive formula for this arithmetic sequence is

$$\begin{cases} a_1 = 55 \\ a_n = a_{n-1} + 25, \text{ for integers } n \geq 2. \end{cases}$$

b. You can graph arithmetic sequences using the methods you used in Lesson 3-7. The graph of the first ten values of the sequence is shown at the right.

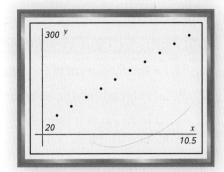

You should be able to translate from a recursive formula to an explicit formula for an arithmetic sequence, and vice versa. To go from recursive to explicit form you can use the theorems on the previous two pages. Use the second theorem to find a_1 and d and then substitute the values into the first theorem. To go from explicit to recursive form, you can use the explicit formula to find a_1, and substitute the known values of a_1 and d into the recursive pattern for an arithmetic sequence.

Example 3

Some car dealers offer interest-free loans, provided that you pay back the amount borrowed by a certain date. Suppose a car that costs $14,736 requires monthly payments of $245.60. Let a_n be the amount you still owe after n months.

a. Write a recursive formula for this sequence.

b. Write an explicit formula for this sequence.

Solution

a. After 1 month you owe $14,736 − $245.60 = $14,490.40, so a_1 = 14,490.40. Each month you pay $245.60, so the difference between successive terms is d = −245.60. A recursive formula for this sequence is

$$\begin{cases} a_1 = 14{,}490.40 \\ a_n = a_{n-1} - 245.60, \text{ for integers } n \geq 2. \end{cases}$$

b. You know that a_1 = 14,490.40 and d = −245.60.

So, a_n = 14,490.40 + $(n − 1)(−245.60)$

a_n = −245.60n + 14,736

where n = the number of payments and a_n = the current balance owed.

Many calculators have a sequence command seq that lets you generate the terms of a sequence if you know its explicit form. At the right is how one calculator generates the first four terms of the sequence in Example 3. The "x,1,4" tells the machine to start at x = 1 and end at x = 4.

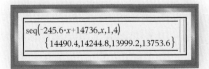

Questions

COVERING THE IDEAS

1. Write the 50th term of the sequence from Example 1.

2. What is an arithmetic sequence?

3. What is an explicit formula for the nth term of an arithmetic sequence with first term a_1 and constant difference d?

4. What is the connection between the slope of a linear function and the constant difference of an arithmetic sequence?

5. Consider the arithmetic sequence 12, 16, 20, 24, 28, … .
   a. Write an explicit formula for the nth term of the sequence.
   b. Write a recursive definition of the sequence.
   c. Calculate the 47th term of this sequence.

In 6 and 7, refer to Example 2.

6. What term of the sequence gives the cost of a 20-minute overseas phone call?

7. Rewrite the formula if the service charge is $1.00 and the cost per minute is 30¢.

8. An arithmetic sequence has an initial term of –127 and a constant difference of 42.
   a. Write an explicit formula for the sequence.
   b. Use the seq command on a calculator to find the first seven terms.

9. Suppose $t_n = 12 + 7(n - 1)$.
   a. Write a recursive formula for the arithmetic sequence t.
   b. Compute t_{77}.

10. Consider the sequence $\begin{cases} a_1 = 10.5 \\ a_n = a_{n-1} + 4.3, \text{ for integers } n \geq 2. \end{cases}$
   a. Write its first three terms.
   b. Write an explicit formula for the sequence.

11. Write a recursive formula for the linear sequence –70, –47.5, –25, –2.5, … .

APPLYING THE MATHEMATICS

12. In Chapter 1, the sequence of triangular numbers was described by the explicit formula $t(n) = \frac{n(n+1)}{2}$. Is this an example of an arithmetic sequence? Why or why not?

13. At the right is a graph of the first five terms of an arithmetic sequence. Write an explicit formula for the sequence.

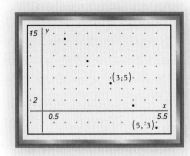

14. A formula for the sum of the measures of the interior angles of a convex polygon is $S_n = 180(n - 2)$ for $n \geq 3$, where n is the number of sides of the polygon.

    a. Evaluate S_n for $n = 3, 4, 5, 6, 7$.

    b. Explain why the results of Part a represent the terms of an arithmetic sequence.

    c. Find a recursive formula for S_n.

In 15 and 16, a local radio station holds a weekly contest to give away a cash prize. The announcer calls a number and if the person who answers guesses the correct amount of money in the pot, he or she wins the money. If the person misses, $25 is added to the money pot.

15. On the 12th call, a contestant won $675. How much was in the pot at the beginning?

16. Suppose the pot starts with $140. On what call could the winner receive $1,115?

17. A 16 ounce jar candle is advertised to burn an average of 110 hours. Suppose that the candle burns at a constant rate.

    a. Write the first three terms of the sequence that shows how many ounces of the candle remain after each hour it burns.

    b. Write the explicit formula for r_n, the number of ounces of the candle left after n hours of burning.

REVIEW

18. Graph the first eight terms of the sequence $r_n = \dfrac{n^3}{n!}$. Describe how the value of r_n changes as n increases. Recall that $n! = n \cdot (n - 1) \cdot (n - 2) \cdot \ldots \cdot 3 \cdot 2 \cdot 1$. **(Lesson 3-7)**

19. Verify that the explicit formula for the nth triangular number $T_n = \dfrac{n(n+1)}{2}$ satisfies the recursive formula
$$\begin{cases} T_1 = 1 \\ T_n = T_{n-1} + n. \end{cases}$$ **(Lessons 3-6, 1-8)**

20. Write the first twelve terms of the sequence defined by
$$\begin{cases} p_1 = 1 \\ p_2 = -1 \\ p_n = p_{n-1} \cdot p_{n-2}, \text{ for integers } n \geq 3. \end{cases}$$ **(Lesson 3-6)**

21. **Multiple Choice** Which of the following are *not* sufficient criteria to determine a (unique) line? **(Lesson 3-4)**

    A   two distinct points

    B   a point and a slope

    C   a slope and a y-intercept

    D   a slope and an x-intercept

    E   All of the above uniquely determine a line.

22. Suppose you have two solutions of oxalic acid, one at 0.1 mol/L, the other 0.5 mol/L. **(Lessons 3-2)**

    a.   If you mix A liters of the first solution with B liters of the second, what will be the concentration of the resulting solution?

    b.   How many liters of the 0.5 mol/L solution must be added to 4 liters of the 0.1 mol/L solution to get 1 mole of acid in the final mixture?

23. Suppose y varies directly as x, and $y = -24$ when $x = 48$. **(Lesson 2-4, 2-1)**

    a.   Find the constant of variation k.

    b.   How is k represented on the graph of the function?

EXPLORATION

24. The following recursively defined sequence generates numbers known as *hailstone numbers*.

$$\begin{cases} a_1 \text{ is any positive integer.} \\ \text{For } n > 1, \text{ If } a_{n-1} \text{ is even, } a_n = \frac{a_{n-1}}{2}. \\ \qquad \text{If } a_{n-1} \text{ is odd, } a_n = 3a_{n-1} + 1. \end{cases}$$

For example, if $a_1 = 46$, then $a_2 = 23$, $a_3 = 70$, $a_4 = 35$, $a_5 = 106\ldots$.

Hailstones rise and fall within clouds, growing larger and larger until they are so heavy they fall out of the cloud.

    a.   Explore this sequence rule for at least five different values of a_1. Continue generating terms until you can predict what will happen in the long run.

    b.   Look up hailstone numbers on the Internet to find out how they got their name and what is known about them. Briefly describe what you find.

QY ANSWER

$a_n = 6 + (n - 1) \cdot 4 = 4n + 2$

Lesson

3-9

Step Functions

Vocabulary

step function

floor symbol ⌊ ⌋

ceiling symbol ⌈ ⌉

floor function, greatest-integer function, rounding-down function, int function

ceiling function, rounding-up function

▶ **BIG IDEA** *Step functions* have applications in many situations that involve rounding.

In 2007, the U.S. postage rate for first class flats (certain large envelopes) was $0.70 for the first ounce plus $0.17 for each additional ounce or part of an ounce. First-class mail rates for flats up to 13 ounces are given in the table below. Notice that the phrase "up to and including the given weight" means that the weight is *rounded up* to the nearest ounce. For instance, an envelope weighing 4.4 ounces is charged at the 5-ounce rate.

Mental Math

Each sequence below is either arithmetic or consists of consecutive powers of a number. Give the next two terms in the sequence.

a. 2, 4, 6, 8, ...

b. 23, 17, 11, 5, ...

c. 1, –1, 1, –1, ...

2007 First-Class Mail Rates for Flats*			
Weight (oz)	Rate (dollars)	Weight (oz)	Rate (dollars)
1	0.70	8	1.89
2	0.87	9	2.06
3	1.04	10	2.23
4	1.21	11	2.40
5	1.38	12	2.57
6	1.55	13	2.74
7	1.72		

*Rate is for a flat up to and including the given weight.

The graph at the right shows the cost of mailing a first-class flat for weights up to 13 ounces. Because the cost is rounded up, the left end of each segment is not included on the graph and the right end of each segment is included. Because no single weight has two costs, this graph pictures a function.

The domain is the set of possible weights of a flat in ounces between 0 and 13 ounces, and the range is the set of costs {$0.70, $0.87, $1.04, ... , $2.74}.

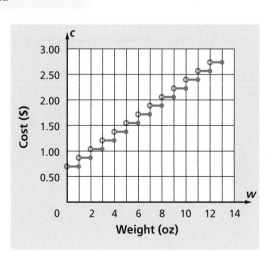

The mail-rates function is not a linear function, but it is a piecewise linear function. Because its graph looks like a series of steps, it is called a **step function**. Each step is part of a horizontal line. Two step functions commonly used are the *floor function* and the *ceiling function*.

The Floor and Ceiling Functions

The **floor symbol** $\lfloor\ \rfloor$ and the **ceiling symbol** $\lceil\ \rceil$ are defined as follows.

> ### Definition of Greatest Integer/Least Integer
>
> $\lfloor x \rfloor$ = the greatest integer less than or equal to x, and
>
> $\lceil x \rceil$ = the least integer greater than or equal to x.

The **floor function** is the function f with $f(x) = \lfloor x \rfloor$, for all real numbers x. It is also called the **greatest-integer function**, or the **rounding-down function**. On some calculators and in some computer languages it is called the **int function**. Another notation you may see for the floor function is $f(x) = [\![x]\!]$.

The **ceiling function** is the function f with $f(x) = \lceil x \rceil$, for all real numbers x. It is also called the **rounding-up function**.

 QY1

> ▶ **QY1**
>
> What names does your calculator use for the floor and ceiling functions?

GUIDED

Example 1

Evaluate each of the following.

a. $\left\lfloor 5\frac{7}{8} \right\rfloor$ b. $\lfloor -4.2 \rfloor$ c. $\lceil \pi \rceil$ d. $\lceil 13 \rceil$

Solution

a. $\left\lfloor 5\frac{7}{8} \right\rfloor$ is the greatest integer less than or equal to $5\frac{7}{8}$. So, $\left\lfloor 5\frac{7}{8} \right\rfloor = \underline{\ ?\ }$.

b. $\lfloor -4.2 \rfloor$ is the $\underline{\ ?\ }$ less than or equal to -4.2. So, $\lfloor -4.2 \rfloor = \underline{\ ?\ }$.

c. $\lceil \pi \rceil$ is the least integer greater than or equal to $\pi \approx \underline{\ ?\ }$. So, $\lceil \underline{\ ?\ } \rceil = \underline{\ ?\ }$

d. $\lceil 13 \rceil$ is the $\underline{\ ?\ }$ greater than or equal to 13. So, $\lceil \underline{\ ?\ } \rceil = \underline{\ ?\ }$.

The Graph of the Floor Function

One way to sketch the graph of a step function is to make a table of values so you can see the pattern.

Example 2

Graph the function f defined by $f(x) = \lfloor x \rfloor$.

Solution Make a table of values. For all x greater than or equal to 0 but less than 1, the greatest integer less than or equal to x is 0. For all x greater than or equal to 1 but less than 2, the greatest integer less than or equal to x is 1. In a similar manner, you can get the other values in the table below. The graph is at the right below.

x	$f(x) = \lfloor x \rfloor$
$-3 \le x < -2$	-3
$-2 \le x < -1$	-2
$-1 \le x < 0$	-1
$0 \le x < 1$	0
$1 \le x < 2$	1
$2 \le x < 3$	2
$3 \le x < 4$	3

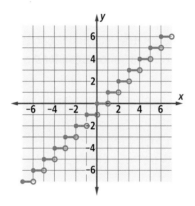

In the graph in Example 2, the open circles at (1, 0), (2, 1), (3, 2), and so on, indicate that these points do not lie on the graph of $f(x) = \lfloor x \rfloor$. At these points, the function value jumps to the next step. The solid circles indicate that the points (1, 1), (2, 2), (3, 3), and so forth, do lie on the graph. Notice that the domain of the greatest-integer function is the set of real numbers, but the range is the set of integers.

If your graphing utility has the int, or floor function, it will graph the greatest-integer function for you. The graph from one graphing utility is shown at the right. By default, some graphing utilities connect successive pixels, so they may join successive steps. This makes it appear as if the graph

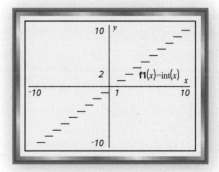

does not represent a function. On these graphing utilities, you can get the correct graph by switching from connected mode to dot mode.

Applications of Step Functions

The floor or ceiling function is appropriate when function values must be integers and other formulas would give noninteger values.

Example 3

In March 2008, New York City taxi rates were an initial fee of $2.50 plus $0.40 for each full $\frac{1}{5}$-mile traveled.

a. Write a formula for $T(m)$, the charge for a trip of m miles.

b. What is the charge for an 8.75-mile trip in a New York City taxi?

Solution

a. Because there are 5 one-fifths of a mile in each mile, multiply the miles by 5 to determine the number of $\frac{1}{5}$-miles traveled. This number, $5m$, may not be a whole number, so use the greatest-integer function to change it to an integer before multiplying by $0.40. An equation for this function is
$$T(m) = 2.50 + 0.40 \lfloor 5m \rfloor.$$

b. The charge for a trip of 8.75 miles can be computed by substituting $m = 8.75$ into the formula for $T(m)$.
$$T(m) = \$2.50 + \$0.40 \lfloor 5 \cdot 8.75 \rfloor$$
$$= \$2.50 + \$0.40 \lfloor 43.75 \rfloor$$
$$= \$2.50 + \$0.40 \cdot 43$$
$$= \$19.70$$

The taxi-to-resident ratio in New York City is 1:149.

 QY2

> ▶ QY2
>
> What is the charge for a 15.3-mile trip in a NYC taxi?

GUIDED

Example 4

Users of pre-paid calling cards are billed in 1-minute increments. This means that customers are billed for a full minute when any part of a minute is used. If the Call-Me-Often Phone Card Company charges $0.03 per minute with a 1-minute billing increment, what is the charge for a 5-minute, 40-second phone call?

Solution Call-Me-Often's charge is rounded up to the nearest minute, so use a ceiling function.

$$0.03 \left\lceil 5\tfrac{40}{60} \right\rceil = 0.03(\underline{\ ?\ }) = \underline{\ ?\ }$$

Call-Me-Often charges __?__ cents for the call.

Questions

COVERING THE IDEAS

In **1** and **2**, refer to the postage example at the beginning of this lesson.

1. What is the cost to mail a letter weighing 4.3 ounces?

2. What is the domain of the function?

3. In your own words, write the meaning of $\lceil x \rceil$. Why do you think it is also called the ceiling function?

In **4–7**, evaluate.

4. $\left\lfloor 4\frac{3}{4} \right\rfloor$
5. $\lfloor 4\pi \rfloor$
6. $\lfloor -5.87 \rfloor$
7. $\lceil 7 - 0.5 \rceil$

8. a. **Fill in the Blanks** The function f defined by $f(x) = \lfloor x \rfloor$ is called the ___?___ or ___?___ function.
   b. The range of $f: x \rightarrow \lfloor x \rfloor$ is ___?___.
   c. Why are there open circles at $(1, 0)$, $(2, 1)$, $(3, 2)$, and so on in the graph of f?

9. Give the domain and range of the function.
   a. $f(x) = \lceil x \rceil$
   b. the function in Example 3

10. Refer to Example 4. A 2-minute billing increment charges for parts of minutes as if they were the next even minute (for example, a 3-minute call is billed for 4 minutes). If Call-Me-Often Phone Card Company charges $0.03 per minute with a 2-minute billing increment, what does an 18-minute, 10-second phone call cost?

APPLYING THE MATHEMATICS

11. Let $r(x) = \lfloor x + 0.5 \rfloor$.
    a. Find $r(1.2)$.
    b. Find $r(1.7)$.
    c. What kind of rounding does r do?

In **12** and **13**, an auditorium used for a high school graduation has 750 seats available for its g graduates.

12. **Multiple Choice** If the tickets are divided evenly among the graduates, which of the following represents the number of tickets each graduate may have?

A $\left\lceil \dfrac{750}{g} \right\rceil$ B $\left\lfloor \dfrac{750}{g} \right\rfloor$ C $\left\lceil \dfrac{g}{750} \right\rceil$ D $\left\lfloor \dfrac{g}{750} \right\rfloor$

13. Write an expression for the number of tickets left over, if any, after each graduate gets his or her tickets.

14. A used-car salesperson is paid $350 per week plus a commission of $100 for each $1500 in sales during the week.

    a. Find the salesperson's salary during a week in which he or she had $3500 in sales.

    b. When the person has d dollars in sales, write an equation that gives the weekly earnings E.

    c. Is it possible for the salesperson to earn exactly $1000 a week? Why or why not?

15. The table at the right shows the typical fees charged by the postal service for its COD (collect on delivery) service as a function of the amount of money to be collected from the recipient (as of 2008).

    a. Can these data be modeled by a step function?

    b. **Fill in the Blank** Complete the following piecewise definition of a function that gives the COD fee $F(a)$ (in dollars) as a function of the amount a (in dollars) to be collected.

    $$F(a) = \begin{cases} 5.10, & \text{if } a \le 50 \\ \underline{\quad?\quad}, & \text{if } 50 < a \le 1000 \end{cases}$$

Amount Collected from Recipient (dollars)	COD Fee (dollars)
0.01 to $50.00	5.10
50.01 to 100.00	6.25
100.01 to 200.00	7.40
200.01 to 300.00	8.55
300.01 to 400.00	9.70
400.01 to 500.00	10.85
500.01 to 600.00	12.00
600.01 to 700.00	13.15
700.01 to 800.00	14.30
800.01 to 900.00	15.45
900.01 to 1000.00	16.60

16. The Fine Furniture Factory pays employees a bonus based on their monthly sales. For sales of $5,000 up to $25,000 the bonus is $500. For sales of $25,000 up to $40,000, the bonus is $1,000. For sales of $40,000 or more, the bonus is $2,000. Write a piecewise linear function to give the bonus b for monthly sales m.

17. The formula $W = d + 2m + \left\lfloor \frac{3(m+1)}{5} \right\rfloor + y + \left\lfloor \frac{y}{4} \right\rfloor - \left\lfloor \frac{y}{100} \right\rfloor + \left\lfloor \frac{y}{400} \right\rfloor + 2$ gives the day of the week based on our current calendar where d = the day of the month of the given date; m = the number of the month in the year with January and February regarded as the 13th and 14th months of the previous year; that is, 2/22/90 is 14/22/89. The other months are numbered 3 to 12 as usual; and y = the year as a 4-digit number. Once W is computed, divide by 7 and the remainder is the day of the week, with Saturday = 0, Sunday = 1, ..., Friday = 6. Enter the formula into a spreadsheet to answer the questions.

    a. On what day of the week were you born?

    b. On what day of the week was the Declaration of Independence adopted?

    c. On what day of the week was January 1, 2001, the first day of the current millennium?

REVIEW

18. **Multiple Choice** Which of the following is *not* an arithmetic sequence? (**Lesson 3-8**)

    A $a_n = 3 - 7n$

    B $b_n = n + n$

    C $\begin{cases} c_1 = 1 \\ c_2 = 5 \\ c_n = c_{n-1} + c_{n-2}, \text{ for } n \geq 3 \end{cases}$

    D $\begin{cases} d_1 = 1 \\ d_2 = 5 \\ d_n = d_{n-1} + 4, \text{ for } n \geq 3 \end{cases}$

    E $\begin{cases} e_1 = -6 \\ e_n = e_{n-1} + 4, \text{ for } n \geq 2 \end{cases}$

19. Consider the arithmetic sequence
    $\sqrt{2}, \sqrt{2} + 2\sqrt{3}, \sqrt{2} + 4\sqrt{3}, \sqrt{2} + 6\sqrt{3}, \ldots$. (**Lesson 3-8**)
    a. Write a recursive definition of the sequence.
    b. Write an explicit formula for the nth term of the sequence.
    c. Find the 101st term of the sequence.

20. The table at the right shows the number of voters (in thousands) that voted in each of the presidential elections in the United States from 1980 to 2004. (**Lesson 3-5**)
    a. Find an equation for the regression line for these data.
    b. According to the answer in Part a, what would be the predicted voter turnout in 2008?
    c. What is the slope of the line you found in Part a? Name a real-life factor that may influence this slope.

Year	Voters (thousands)
1980	86,515
1984	92,652
1988	91,594
1992	104,405
1996	96,456
2000	105,586
2004	122,294

21. A line passes through the points $(2, 2)$ and $(0, -3)$. (**Lesson 3-4**)
    a. Find an equation for this line in point-slope form using the point $(2, 2)$.
    b. Find an equation for this line in point-slope form using the point $(0, -3)$.
    c. Verify that your equations from Parts a and b are equivalent.

EXPLORATION

22. a. Solve the equation $\left\lfloor \frac{x}{2} \right\rfloor = \frac{x}{2}$.
    b. Generalize Part a to solve $\left\lfloor \frac{x}{n} \right\rfloor = \frac{x}{n}$.

Chapter 3 Projects

1 Residual Squares

A CAS can be used to illustrate why the line of best fit is sometimes called the *least squares line*.

a. The decompression-time data from Lesson 3-5 are given in the table below. Use a graphing utility to create a scatterplot of the data. Add a movable line and adjust it until you think it is the line of best fit. Record the equation of your line.

Time Spent at a Maximum Depth of 60 Feet (min)	Decompression Time Needed (min)
60	1
70	3
80	8
100	15
120	27
140	40
160	49
180	57
200	71
240	82

b. Show the residual squares on your plot. Some graphing utilities will show the sum of the squares on the screen. Continue to adjust your line until you think you have minimized the sum of the squares. Record the equation of this adjusted line.

c. Remove your movable line and add a regression line to the plot. Record the equation of the regression line. Compare it to your equations from Parts a and b. Were you able to make a closer estimate to the regression line by using the residual squares?

2 Penalties for Speeding

States in the United States have speed limits for motor vehicles traveling on public roadways. States use point systems to penalize drivers who speed and to identify repeat offenders. In general, the number of points you receive is a function of the amount that your speed exceeds the legal speed limit. In some states these functions are linear functions; in other states they are piecewise linear.

a. Find out what the point penalties are for speeding (in a car or a truck) in your state. Describe the point-penalty function using a table and a graph. What are the domain and range of this function? Does this function belong to any of the categories of functions you have studied in this course? If so, what kind of function is it? If possible, find an equation for the point-penalty function.

b. Find a state that has a different set of point penalties for speeding than your state. Describe those fines with a table, graph, and equation. Describe some ways the two point-penalty functions are alike and the ways they are different.

3 Linear Combinations

In the sport of Rugby, different point values are awarded for various actions. A *try* is worth 5 points and allows for a follow-up kick through two uprights worth an additional 2 points (called *converting a try*, or a *conversion*). Teams may also attempt to drop kick the ball through the uprights without having scored a try. A successful drop kick (a *dropped goal*) is worth 3 points. So a Rugby team may score 3, 5, 6, etc., points, but cannot score 1, 2, 4, etc., points.

a. What is the largest number of points that a Rugby team cannot score? Explain how you got your answer. (*Hint:* The number of conversions must always be equal to or less than the number of tries.)

b. Imagine you are creating your own sport with its own scoring system. A team achieves x points for scoring with their hands, and y points for scoring with their feet. If x and y have no common factors, what is the greatest number of points a team cannot score? Explain how you got your answer.

4 A Graphical Investigation

Graphs of equations of the form $Ax + By = C$, where A, B, and C are consecutive terms in an arithmetic sequence, have something in common.

a. Graph five equations of this type, such as the following:
$$x + 2y = 3$$
$$3x + 5y = 7$$
$$8x + 6y = 4$$
$$-2x - 3y = -4$$
$$3x - y = -5.$$

b. Make a conjecture based on these five graphs.

c. Test your conjecture with a few more graphs.

d. Use the definitions or theorems about arithmetic sequences to verify your conjecture.

5 Time Series Data

When the value of a dependent variable changes over time, like daily temperatures and world population, the data are called *time-series data*.

a. Find an example of time-series data in which the dependent variable appears to vary linearly with time. Make a scatterplot of the data.

b. Use a computer or calculator to find a line of best fit and draw it on the graph with the scatterplot.

c. According to your model, what will the value of the variable be in the year 2025, 2075, 3000, and 3050? Do your predictions seem reasonable? Why or why not?

Chapter 3 Summary and Vocabulary

A **linear equation** is one that is equivalent to an equation of the form $Ax + By = C$, where A and B are not both zero. The graph of every linear equation is a line. If the line is not vertical, then its equation represents a linear function and can be put into the **slope-intercept form** $y = mx + b$, where m is its slope and b is its **y-intercept**. Horizontal lines have slope 0 and equations of the form $y = b$. Slope is not defined for vertical lines, which have equations of the form $x = a$. In these last three forms, m, b, and a can be any real numbers.

If the slope m and one point (x_1, y_1) on a line are known, then an equation for the line is $y - y_1 = m(x - x_1)$, or $y = m(x - x_1) + y_1$.

In many real-world situations, a set of data points is roughly linear. In such cases, a regression or least squares line, **the line of best fit**, can be used to describe the data and make predictions.

Linear equations can model two basic kinds of real-world situations: constant increase or decrease, and **linear combinations**. The graph of a constant-increase or constant-decrease situation is a line, with the slope of the line representing the constant change and the y-intercept representing the initial condition.

Sequences with a constant difference between terms are called **linear** or **arithmetic sequences**. If a_n is the nth term of an arithmetic sequence with constant difference d, then the sequence can be described explicitly as $a_n = a_1 + (n - 1)d$, for integers $n \geq 1$, or recursively as $\begin{cases} a_1 \\ a_n = a_{n-1} + d \end{cases}$, for integers $n \geq 2$.

A function whose graph is the union of segments or rays is called **piecewise linear**. **Step functions** are instances of piecewise linear functions. Step functions represent situations in which rates are constant for a while but change to a different constant rate at known points.

Theorems

Parallel Lines and Slope Theorem (p. 154)
Standard Form of an Equation of a Line Theorem (p. 163)
Point-Slope Theorem (p. 170)
nth Term of an Arithmetic Sequence Theorem (p. 197)
Constant-Difference Sequence Theorem (p. 198)

Vocabulary

Lesson 3-1
y-intercept
*linear function
*slope-intercept form
x-intercept

Lesson 3-2
*linear combination
*standard form

Lesson 3-4
*point-slope form
piecewise linear

Lesson 3-5
linear regression
least-squares line,
    regression line,
    line of best fit
deviation

Lesson 3-6
*recursive formula
recursive definition

Lesson 3-7
Fibonacci sequence

Lesson 3-8
linear sequence,
    *arithmetic sequence

Lesson 3-9
*step function
floor symbol $\lfloor \ \rfloor$
ceiling symbol $\lceil \ \rceil$
*floor function,
    greatest-integer function,
    rounding-down function,
    int function
ceiling function,
    rounding-up function

Chapter 3 Self-Test

Take this test as you would take a test in class. Then use the Selected Answers section in the back of the book to check your work.

1. Graph the line with equation $y = \frac{1}{2}x - 3$.

2. Consider the line with equation $5x - 3y = 10$.
   a. What is its slope?
   b. What are its x- and y-intercepts?

3. If $m < 0$, does the equation $y = mx + b$ model a constant-increase or a constant-decrease situation?

4. Write an equation for the line graphed below.

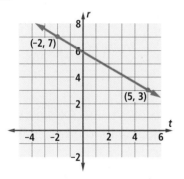

5. Write an equation for the line parallel to $y = -\frac{2}{3}x - 7$ that contains $(-6, 3)$.

6. a. Slope is undefined for which type of line?
   b. Which type of line has a slope of zero?

7. A restaurant sells hot dogs for $2.25 and tacos for $3.75. Write an expression that gives the amount of money the restaurant takes in by selling H hot dogs and T tacos.

8. A crane lowers a 500-pound load from the top of a 310-foot tall building at the rate of 20 feet per minute. Write a formula for the height h of the load above the ground after t seconds.

9. Consider the sequence defined by $a_n = 50 - 3(n - 1)$.
   a. Is this an explicit or recursive formula for the sequence? Explain your answer.
   b. Write the sequence's first five terms.
   c. Is the sequence arithmetic? Explain your answer.

10. A cell-phone plan costs $23.50 a month for the first 200 minutes of calls, plus 8 cents per minute for all calls after that.
    a. In July, Michelle's calls total 453 minutes. How much will she be charged?
    b. Write an equation for the piecewise linear function that gives the monthly cost C in terms of the minutes m spent on the phone.

11. The Lucas sequence L begins with the terms 1 and 3. After that, each term is the sum of the two preceding terms. The first six Lucas numbers are 1, 3, 4, 7, 11, and 18.
    a. Write the next four Lucas numbers.
    b. Let L_n be the nth Lucas number. Write a recursive formula for the Lucas sequence L.

12. Temperature F in degrees Fahrenheit and temperature K in kelvins are related by a linear equation. Two pairs of corresponding temperatures are $32°F = 273.2$ kelvins and $90°F = 305.4$ kelvins. Write a linear equation relating F and K, and solve for K.

13. Write an explicit formula for the arithmetic sequence $-25, -45, -65, -85, \ldots$.

14. **Multiple Choice** Which graph most closely describes a piecewise function where each piece has a slope greater than the previous piece? Explain your response. (The scales on all four sets of axes are the same.)

A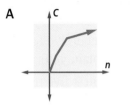
B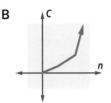

C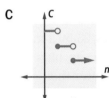
D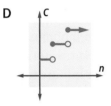

15. Suppose the U.S. Postal Service sought to set the price of mailing a letter weighing 1 ounce or less at 50¢, with each additional ounce or part of an ounce costing an extra 30¢. Graph the function that maps weight w (in ounces) of a letter onto the cost c (in cents) to mail the letter.

16. a. **Multiple Choice** Which equation describes the function in Question 15?

   A $c = 50 - 30\lceil 1 - w \rceil$
   B $c = 50 + 30\lceil 1 - w \rceil$
   C $c = 50 - 30\lceil w - 1 \rceil$
   D $c = 50 + 30\lceil w - 1 \rceil$

   b. Find the cost of mailing a 3.2-ounce letter.

17. Sketch a graph of $y = \lfloor x \rfloor + 1$ for $-2 \le x \le 2$.

18. The table and the scatterplot below give the life expectancy in the U.S. at selected ages in 2003.

Age	Expected Years of Life
0	77.4
10	68.1
20	58.4
30	48.9
40	39.5
50	30.5
60	22.2
70	14.8
80	8.9
90	4.8
100	2.5

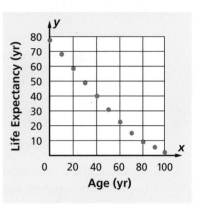

Source: Centers for Disease Control and Prevention

a. Using the data points for ages 0 to 80, find an equation of the regression line.

b. Using your answer to Part a, estimate the life expectancy of someone who is currently 42.

c. Which data point is farthest vertically from the regression line? What does this mean in context?

d. Why is it unreasonable to use a linear model for ages over 80?

SKILLS
PROPERTIES
USES
REPRESENTATIONS

SKILLS Procedures used to get answers

OBJECTIVE A Determine the slope and intercepts of a line given its equation. (Lessons 3-1, 3-3)

In 1–6, an equation for a line is given.

a. Give its slope.

b. Give its x-intercept.

c. Give its y-intercept.

1. $y = 3x - 12$
2. $4y = 6 + 5x$
3. $y = -17$
4. $x = 8$
5. $300x - 250y = -100$
6. $x + y = 1.46$

OBJECTIVE B Find an equation of a line given two points on it or given a point on it and its slope. (Lesson 3-4)

In 7–10, find an equation of the line satisfying the given conditions.

7. The line has a slope of $\frac{2}{5}$ and contains (-5, 10).

8. The line contains the point (8, 2) and goes through the origin.

9. The line contains the points (-3, 4) and (5, -4).

10. The line is parallel to $y = 6x - 1$ and contains the point (7, 1).

OBJECTIVE C Evaluate expressions based on step functions. (Lesson 3-9)

In 11–13, evaluate the expression.

11. a. $\lfloor 13.5 \rfloor$ b. $\lceil -13.5 \rceil$
12. a. $\lfloor x - 0.4 \rfloor$ when $x = 3.6$
    b. $\lceil x + 0.4 \rceil$ when $x = 3.6$
13. $4\lfloor 2n + 1.4 \rfloor$ when $n = 0.4$

OBJECTIVE D Evaluate or find explicit and recursive formulas for sequences. (Lessons 3-6, 3-8)

In 14 and 15, an arithmetic (linear) sequence is given. For the sequence:

a. find an explicit formula.

b. write a recursive formula.

c. find the fourth through the eighth terms.

14. -2, 1, 4, …   15. 37, 16, -5, …

16. Write a recursive definition of the sequence whose nth term is $a_n = -4n + 15$.

17. Use this recursively-defined sequence.
$$\begin{cases} a_1 = -\frac{2}{3} \\ a_n = a_{n-1} + \frac{1}{4} \text{ for integers } n \geq 2 \end{cases}$$

a. Write an explicit formula for the nth term of this sequence.

b. Generate the first five terms.

PROPERTIES Principles behind the mathematics

OBJECTIVE E Recognize properties of linear functions. (Lessons 3-1, 3-2, 3-3)

18. **Multiple Choice** Which of the following does *not* mean a line has a slope of $-\frac{2}{3}$?

A It has a vertical change of $\frac{2}{3}$ unit for a horizontal change of -1 unit.

B It has the equation $y = -\frac{2}{3}x + 5$.

C It has the equation $2x - 3y = 7$.

D It is parallel to the line with equation $2x + 3y = 7$.

19. **True or False** Two lines in a plane are parallel if they have equal slopes.

20. Consider $Ax + By = C$. For what values of A and B does this equation not represent a function? Explain your response.

21. What is the x-intercept of $y = mx + b$?

22. **Multiple Choice** Three points A, B, and C are on a line with B between A and C and $AB = 3(BC)$. The slope determined by B and C

   A is 3 times the slope determined by A and B.

   B is $\frac{1}{3}$ the slope determined by A and B.

   C equals the slope determined by A and B.

   D None of the above is true.

23. Suppose a line has slope $-\frac{3}{4}$. Then if (x, y) is a point on the line, name another point on the line.

OBJECTIVE F Recognize properties of linear or arithmetic sequences. (Lesson 3-8)

24. What is a linear sequence?

25. Describe the graph of an arithmetic sequence.

In 26 and 27, tell whether the numbers could be the first four terms of an arithmetic sequence.

26. $-4, -6, -8, -10$

27. $3, 4, 6, 9$

In 28 and 29, does the formula generate an arithmetic sequence?

28. $\begin{cases} a_1 = 3 \\ a_n = \frac{1}{2}a_{n-1} - 1, \text{ for integers } n \geq 2 \end{cases}$

29. $\begin{cases} a_1 = 3 \\ a_{n+1} = a_n - 5, \text{ for integers } n \geq 1 \end{cases}$

30. Find the nth term of a linear sequence whose 1st term is 11 and whose constant difference is -4.

31. If the 10th term of an arithmetic sequence is 8 and the 20th term is 16, what is the first term?

USES Applications of mathematics in real-world situations

OBJECTIVE G Model constant-increase or constant-decrease situations or situations involving arithmetic sequences.
(Lessons 3-1, 3-8)

32. A truck weighs 2000 kg when empty. It is loaded with crates of oranges weighing 17 kg each.

   a. Write an equation relating the total weight w and the number c of crates.

   b. Find the weight when there are 112 crates in the truck.

33. On his way to work, Rusty drives over a nail that punctures his tire. The tire begins to lose pressure at about 2 pounds per square inch (psi) per hour.

   a. If Rusty's tire had 44 psi of pressure before the puncture, write an equation to show how much pressure p the tire has after t hours.

   b. In order to drive safely, the tire must have at least 26 psi of pressure. How long can Rusty wait to replace his tire?

OBJECTIVE H Model linear combination situations. (Lesson 3-2)

34. A crate contains grapefruits and oranges. On average, an orange weighs 0.3 pound and a grapefruit weighs 1.1 pounds. The contents of the crate weigh a total of 30 pounds. Let x be the number of oranges and let y be the number of grapefruits.

   a. Write an equation to model this situation.

   b. If there are 15 grapefruits in the crate, how many oranges are there?

35. A chemist combines A liters of a solution that is 2.5 moles/liter bromic acid with B liters of a solution that is 6.25 moles/liter bromic acid.

    a. Write an expression for how many liters of solution there are altogether.

    b. How many moles of bromic acid are there altogether?

    c. The chemist preparing the solution needs a total of 0.75 mole of bromic acid in a solution. Write an equation that describes this situation.

    d. List three ordered pairs that are realistic solutions to the equation in Part c.

OBJECTIVE I In a real-world context, find an equation for a line containing two points. (Lesson 3-4)

36. On a trip abroad, Salena buys 7000 Indian rupees for 150 U.S. dollars, and then buys 11,200 Indian rupees for 250 U.S. dollars.

    a. Assume a linear relationship exists between the number of Indian rupees and the cost in U.S. dollars. Write an equation representing the relationship.

    b. How much will it cost to buy 20,000 Indian rupees?

37. Gerald finds that it takes 30 mL of a standard solution to neutralize 12 mL of a solution of unknown concentration. If he starts with 20 mL of the unknown solution, it takes 50 mL of standard solution to neutralize it. Assuming a linear relationship exists between the amount of unknown solution and the amount of standard solution required to neutralize it, how much unknown solution can be neutralized with 175 mL of standard solution?

OBJECTIVE J Fit lines to data. (Lesson 3-5)

38. The display below shows the number of tons of sulfur dioxide, a major form of air pollution, in the United States from 1990 to 2002. An equation of the regression line is $y = -652.51x + 23{,}846$, with $x = 1$ in 1990.

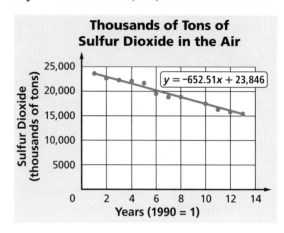

a. What value does the regression equation predict for 1998 (year 9)?

b. The actual value for 1998 is 18,944. What is the difference between the value predicted by the regression equation and the actual value?

c. What is the percent decrease from 1990 (23,760,000 tons) to 2002 (15,353,000 tons)?

d. If this linear trend continued, what would have been the approximate number of tons of sulfur dioxide in the air in 2006?

39. The table below gives a measure of the purchasing power of the U.S. dollar from 1991 to 2004. Here year 1 = 1991 and year 14 = 2004.

Year	Purchasing Power	Year	Purchasing Power
1	0.734	8	0.613
2	0.713	9	0.600
3	0.692	10	0.581
4	0.675	11	0.565
5	0.656	12	0.556
6	0.638	13	0.543
7	0.623	14	0.529

a. Make a scatterplot of these data.

b. Find an equation for the regression line of these data.

c. Does it appear that a linear equation would be a good model for these data? Explain why or why not.

OBJECTIVE K Model situations leading to piecewise linear functions or to step functions. **(Lessons 3-4, 3-9)**

In 40 and 41, an online music service charges 99¢ for each of the first 20 music downloads and then 89¢ for each additional download.

40. Find the total cost for a user who completes

a. 15 downloads.   b. 27 downloads.

41. Describe the situation with a function mapping the number d of downloads onto the total cost C.

42. Multiple Choice Suppose a salesperson earns a $50 bonus for each $1000 in sales that he or she makes. What is a rule for the function that relates sales s to the amount b of bonus?

A $b = 50 \cdot \left\lfloor \frac{s}{1000} \right\rfloor$ B $b = \left\lfloor \frac{50s}{1000} \right\rfloor$

C $b = 50 + \left\lfloor \frac{s}{1000} \right\rfloor$ D $b = 50 + \left\lfloor \frac{1000}{s} \right\rfloor$

43. The graph below shows Imani's trip directly from home to school. After attending school all day she went to her friend's house where they did their math homework. Then she returned to her house.

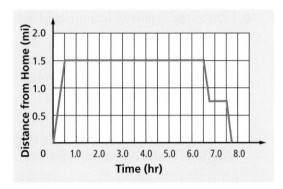

a. At what rate did she walk to school?

b. At what rate did she walk to her friend's house?

c. Write a piecewise linear function for this graph.

d. Was her speed walking home the same, faster, or slower than her speed walking to school? Explain how you know.

44. A cell-phone plan charges $39.99 for the first 450 minutes plus $0.45 for each additional minute or part of a minute. Write an equation to model this situation.

OBJECTIVE L Model situations with recursive formulas. **(Lessons 3-6, 3-7)**

45. On the first day of winter, Charo Chipmunk has a pile of 2300 nuts. Each day, she eats 26 nuts. Let a_n be the number of nuts in her pile on day n.

a. Write a recursive formula for a_n.

b. Write an explicit formula for a_n.

c. The winter typically lasts 80 days. Will Charo's supply of nuts last her through the winter?

46. Martin raises emus. The number a_n of emus on Martin's farm in year n is given by the table below.

n	1	2	3	4
a_n	10	20	40	80

a. Assume the doubling pattern in the table continues. Write a recursive formula for a_n.

b. How many emus should he expect to have in year 10?

REPRESENTATIONS Pictures, graphs, or objects that illustrate concepts

OBJECTIVE M Graph or interpret graphs of linear equations. (Lessons 3-1, 3-3)

47. Graph the line with slope 4 and y-intercept 8.

48. Graph the line $-2x + 8y = 12$ using its intercepts.

49. Graph $x = -5$ in the coordinate plane.

50. Graph $y = 2$ in the coordinate plane.

In 51 and 52, tell whether the slope of the line is positive, negative, zero, or undefined.

51.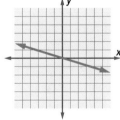

52.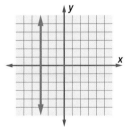

53. What is an equation of the line graphed below?

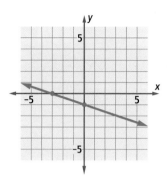

OBJECTIVE N Graph or interpret graphs of piecewise linear functions, step functions, or sequences. (Lessons 3-4, 3-7, 3-9)

54. Consider the function f, where
$$f(x) = \begin{cases} x - 2, & \text{for } x < 0 \\ 2, & \text{for } x \geq 0 \end{cases}.$$

a. Draw a graph of $y = f(x)$.

b. Find the domain and range of f.

55. a. Graph the function h, where $h(x) = 3 + \lfloor x \rfloor$.

b. Give the domain and range of h.

56. A personal trainer earns a bonus of $150 for every 2 pounds of weight a client loses. Draw a graph of the bonuses as a function of the number of pounds lost.

57. A graph of the first five terms of an arithmetic sequence is shown at the right. Write an explicit formula for the sequence.

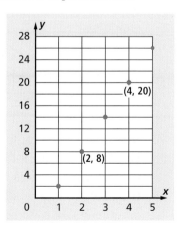

58. Graph this sequence.
$$\begin{cases} a_1 = -5 \\ a_n = a_{n-1} + 3.5, & \text{for integers } n \geq 2 \end{cases}$$

Chapter

4 Matrices

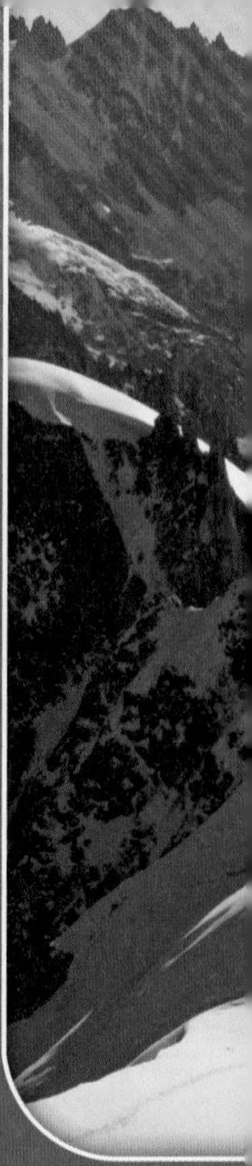

A *matrix* is a rectangular arrangement of objects or numbers, often used to store data. We use brackets [ ] to identify a matrix. In the matrix below, the numbers give information about the largest seven countries in land area. The titles of the rows and columns are not part of the matrix.

Country	Land Area (km^2)	Water Area (km^2)	Highest Point (m)	Lowest Point (m)
Russia	16,995,800	79,400	5633	−28
China	9,326,410	270,550	8850	154
USA	9,161,923	664,707	6198	−86
Canada	9,093,507	891,163	5959	0
Brazil	8,456,510	55,455	3014	0
Australia	7,617,930	68,920	2229	−15
India	2,973,190	314,400	8598	0

Source: www.cia.gov/cia/publications/factbook

Matrices can also describe geometric transformations. In the graph at the right, *PENTA* has been reflected over the *x*-axis, and then translated left by 3 units and down by 5 units. The vertices of *PENTA* may be described by the matrix below the graph. You will learn that the coordinates of the image *P'E'N'T'A'* can be found through matrix operations.

In this chapter you will study various matrix operations, and learn how those operations can be applied to geometric transformations and real-life situations.

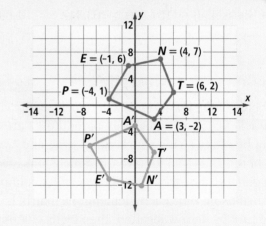

$$\begin{array}{ccccc} P & E & N & T & A \\ \begin{bmatrix} -4 & -1 & 4 & 6 & 3 \\ 1 & 6 & 7 & 2 & -2 \end{bmatrix} \end{array}$$

Lesson

4-1 Storing Data in Matrices

Vocabulary

matrix

element

dimensions

equal matrices

point matrix

▶ **BIG IDEA** A variety of types of data, from numerical information to coordinates of points, can be stored in *matrices*.

A rectangular arrangement of objects or numbers is called a **matrix**. The plural of *matrix* is *matrices*. Each object in a matrix is called an **element** of the matrix. Matrices are useful for storing data of all kinds.

For example, the median salaries of collegiate head coaches for three different sports, based on the highest degree an institution grants, are shown in the matrix below. Entries are in dollars.

	column 1 ↓ Doctoral	column 2 ↓ Master's	column 3 ↓ Bachelor's	column 4 ↓ Associate's
row 1 → Football	185,000	72,070	62,499	31,314
row 2 → Baseball	72,975	46,654	41,105	44,799
row 3 → Basketball	157,500	63,347	51,092	45,982

Source: *Chronicle of Higher Education*, March 2006

Dimensions of a Matrix

The elements of the above matrix are enclosed by large square brackets. (Sometimes large parentheses are used in place of brackets.) This matrix has 3 *rows* and 4 *columns*. Because of this, it is said to have the **dimensions** 3 by 4, written 3×4. In general, a matrix with m rows and n columns has dimensions $m \times n$. Each element of a matrix is identified first by its row location, then by its column location. For example, the element in the 3rd row and 2nd column of this matrix is 63,347. Headings are placed outside the matrix, like the sports and degrees above.

A rectangular block of cells in a spreadsheet also constitutes a matrix. Spreadsheets use the reverse order for identifying an element—column first (a letter) and row second (a number). Like matrices, spreadsheets can have headings to identify their row(s) and column(s).

Mental Math

Find an equation for a line satisfying the conditions.

a. slope 4 and y-intercept 2.5

b. undefined slope and passing through (−7, 2)

c. slope $\frac{1}{3}$ and passing through $(0, \frac{9}{10})$

d. passing through (17, 12) and (0.4, 12)

Many calculators let you enter and manipulate matrices. Use the Guided Example to see how to enter a matrix into a CAS and to store a matrix as a variable.

GUIDED

Example

According to the *Statistical Abstract of the United States*, in 1980, approximately 3.5 million males and 1.9 million females participated in high school athletic programs. Ten years later, 3.4 million males and 1.9 million females participated. In 2000, 3.9 million males and 2.8 million females participated.

a. Store the high school athletic participation information in a matrix *M*.

b. What are the dimensions of the matrix?

c. Enter the matrix from Part a into a CAS and store it as a variable.

Solution

a. You can write either of the two matrices below. Matrix *M*1 has the years as rows, and matrix *M*2 has the years as columns. Either matrix is an acceptable way to store the data.

$$\text{Matrix } M1: \begin{array}{r} \\ 1980 \\ 1990 \\ 2000 \end{array} \begin{array}{cc} \text{Males} & \text{Females} \\ \left[\begin{array}{cc} ?\,3.5 & ?\,1.9 \\ ?\,3.4 & ?\,1.9 \\ ?\,3.9 & ?\,2.8 \end{array}\right] \end{array}$$

$$\text{Matrix } M2: \begin{array}{r} \\ \text{Males} \\ \text{Females} \end{array} \begin{array}{ccc} 1980 & 1990 & 2000 \\ \left[\begin{array}{ccc} ? & ? & ? \\ ? & ? & ? \end{array}\right] \end{array}$$

b. Matrix *M*1 has __?__ rows and __?__ columns.
The dimensions of M1 are __?__.
Matrix *M*2 has __?__ rows and __?__ columns.
The dimensions of M2 are __?__.

c. Use a CAS. Clear *M*1 or *M*2 before storing your matrix.

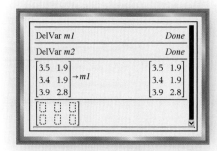

Although matrices *M*1 and *M*2 in the Example are both acceptable ways to store and represent the data, the two matrices are not considered equal. Matrices are **equal matrices** if and only if they have the same dimensions *and* their corresponding elements are equal.

Matrices and Geometry

Points and polygons can also be represented by matrices.

The ordered pair (x, y) is generally represented by the matrix $\begin{bmatrix} x \\ y \end{bmatrix}$.

This 2×1 matrix is called a **point matrix**. Notice that the element in the first row is the x-coordinate and the element in the second row is the y-coordinate. For instance, the point $(5, -1)$ is represented by the matrix $\begin{bmatrix} 5 \\ -1 \end{bmatrix}$.

Similarly, polygons can be written as matrices. Each column of the matrix contains the coordinates of a vertex of the polygon in the order in which the polygon is named. The Activity illustrates this.

Activity

MATERIALS matrix polygon application

Step 1 Pentagon $ABCDE$ with vertices $A = (3, -5)$, $B = (6, -1)$, $C = (4, 5)$, $D = (-2.5, 4)$, and $E = (-5, -0.75)$, is shown at the right. Write a matrix representing the coordinates of the vertices of the pentagon, starting with point A.

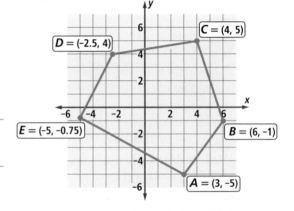

Step 2 Write two other matrices representing the pentagon. (*Hint:* Start with a different vertex.)

Step 3 Verify that your matrices from Step 2 are correct by plotting the pentagon each matrix describes. Each picture should be the same as pentagon $ABCDE$. You can do this by using a matrix polygon application supplied by your teacher.

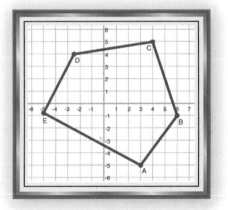

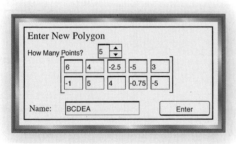

Step 4 Plot
$$\begin{array}{ccccc} B & A & C & D & E \\ \begin{bmatrix} 6 & 3 & 4 & -2.5 & -5 \\ -1 & -5 & 5 & 4 & -0.75 \end{bmatrix} \end{array}.$$

Explain why $BACDE$ does not describe a pentagon.

STOP QY

> **QY**
>
> Are the four matrices in the Activity equal? Explain your answer.

Questions

COVERING THE IDEAS

1. What is a matrix?

In 2–4, refer to the matrix regarding coaching salaries at the beginning of the lesson.

2. a. What is the element in row 2, column 3?

   b. What does this element represent?

3. Write instructions that someone could use to enter the matrix on your CAS.

4. What would be the dimensions of the matrix if *highest degrees* were rows, and *sports* were columns?

5. Refer to the Example. In 1985, there were 3,344,275 males and 1,807,121 females participating in high school athletic programs. Construct a 2 × 4 matrix that incorporates this new information with the old.

6. In the fall of 2008, mathematics classes at a local community college had a total enrollment of 2850, compared to 2241 in the fall of 2007. Additionally, English classes had total enrollments of 2620 and 2051, biology classes had enrollments of 1160 and 1572, and psychology classes had enrollments of 740 and 784, all respectively.

   a. Arrange the data into a matrix on a CAS or graphing calculator, representing years as rows.

   b. Now arrange the data into a matrix representing years as columns.

   Many college campuses have inner courtyards called quadrangles.

7. a. Write the ordered pair (x, y) as a matrix.

   b. What is this matrix called?

8. **Multiple Choice** Which matrix represents the point $(15\sqrt{2}, -7.3)$?

   A $\begin{bmatrix} 15\sqrt{2} & -7.3 \end{bmatrix}$

   B $\begin{bmatrix} -7.3 & 15\sqrt{2} \end{bmatrix}$

   C $\begin{bmatrix} -7.3 \\ 15\sqrt{2} \end{bmatrix}$

   D $\begin{bmatrix} 15\sqrt{2} \\ -7.3 \end{bmatrix}$

9. Write $\triangle ABC$ at the right as a matrix.

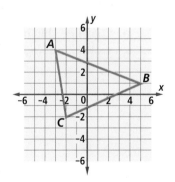

10. Refer to the Activity.

a. The matrix at the right uses the same points as the Activity. Use a matrix polygon application to draw (or plot by hand) and connect the points in the matrix. Use the same window as in the Activity. Why is the picture different from pentagon *ABCDE*?

$$\begin{array}{ccccc} A & B & C & E & D \\ \begin{bmatrix} 3 & 6 & 4 & -5 & -2.5 \\ -5 & -1 & 5 & -0.75 & 4 \end{bmatrix} \end{array}$$

b. How many matrices can represent the pentagon *ABCDE*?

11. The matrix at the right gives the numbers of professional degrees earned in 2000 in four professions, separated by gender.

a. What are the dimensions of this matrix?

b. What does the sum of the elements in row 3 represent?

c. What does the sum of the elements in column 2 represent?

	Males	Females
Medicine	8,759	6,527
Dentistry	2,546	1,704
Law	20,640	17,512
Theology	4,339	1,790

APPLYING THE MATHEMATICS

12. **Fill in the Blanks** If $\begin{bmatrix} -6 & 4.3 \\ \frac{1}{2} & w \end{bmatrix} = \begin{bmatrix} -6 & r \\ \frac{1}{2} & 0.9 \end{bmatrix}$, then $w =$ __?__ and $r =$ __?__.

Many college graduations use different colored tassels or sashes to denote individual majors.

13. **Fill in the Blanks** If $\begin{bmatrix} 2a - 3 \\ h + 0.4 \end{bmatrix} = \begin{bmatrix} -9 \\ \frac{1}{2} \end{bmatrix}$, then $a =$ __?__ and $h =$ __?__.

14. Recall on your CAS the matrix $M1$ from the Guided Example.

If $M1 = \begin{bmatrix} x & y \\ z - 1 & y \\ 3w & 2.8 \end{bmatrix}$, find w, x, y, and z.

15. In the English language, the vowels A, E, I, O, and U show up with frequencies among all letters of about 8%, 13%, 7%, 8%, and 3%, respectively. In the board game SCRABBLE®, these letters show up with frequencies 9%, 12%, 9%, 8%, and 4%, respectively.

a. Arrange this information into a 2×5 matrix.

b. Explain how to enter this matrix into a CAS.

16. The endpoints of $\overline{PA}$ on line m are given by the matrix $\begin{bmatrix} \frac{1}{2} & 2 \\ \frac{17}{2} & 13 \end{bmatrix}$.

    The endpoints of $\overline{LN}$ on line n are defined by the matrix $\begin{bmatrix} 4 & -1 \\ 8 & -7 \end{bmatrix}$.

    Prove that lines m and n are parallel.

17. Use a matrix polygon application to draw (or plot by hand) the

    octagon $\begin{bmatrix} 0 & 1 & 5 & 1 & 0 & -1 & -5 & -1 \\ 5 & 1 & 0 & -1 & -5 & -1 & 0 & 1 \end{bmatrix}$. Sketch a picture of the output,

    and explain if the polygon is convex or nonconvex.

REVIEW

18. Evaluate the following expressions. **(Lesson 3-9)**
    a. $r\lceil \pi \rceil - r\lfloor \pi \rfloor$
    b. $n\lceil 10 \rceil - n\lfloor 10 \rfloor$
    c. $\lceil -\frac{1}{2} \rceil$
    d. $\lfloor -3.6 \rfloor - \lfloor -3 \rfloor$
    e. $\lfloor 56.63 \rfloor - \lfloor -56.9 \rfloor$
    f. $\lceil 5 \rceil - \lceil 5.02 \rceil$
    g. $\lfloor -4.5 \rfloor + \lfloor 9.7 \rfloor$

19. Shelby put $50 on her public transportation card. For every bus or train ride she takes, $2 is deducted from her total. Express the total left on her card as an explicit formula of an arithmetic sequence dependent on the number of train or bus rides taken. **(Lesson 3-8)**

20. **Fill in the Blank** $Ax + By = C$ is the ___?___ form of an equation for a line. **(Lesson 3-2)**

21. If the volume of cube A is 27 times the volume of cube B, how do the lengths of their edges compare? **(Lesson 2-3)**

22. a. What does the Commutative Property of Addition say?
    b. Is subtraction commutative? If so, explain. If not, give a counterexample. **(Previous Course)**

EXPLORATION

23. What is a dot-matrix printer? How is it related to the matrices discussed in this lesson?

QY ANSWER

No, corresponding elements are not equal.

4-2 Matrix Addition

Vocabulary

matrix addition,
   sum of two matrices
scalar multiplication,
   scalar product
difference of two matrices

▶ **BIG IDEA** Matrices with the same dimensions can be added in a very natural way.

How Are Matrices Added?

There are many situations which require adding the information stored in matrices. For instance, suppose matrix C represents the current inventory of cars at Rusty's Car Dealership.

Current Inventory

$$\begin{array}{c} \\ \text{Turbo} \\ \text{Cruiser} \\ \text{Clunker} \\ \text{Vacationer} \end{array} \begin{array}{ccccc} \text{Red} & \text{Blue} & \text{White} & \text{Silver} & \text{Tan} \\ \left[\begin{array}{ccccc} 12 & 10 & 7 & 8 & 2 \\ 14 & 12 & 7 & 12 & 7 \\ 17 & 8 & 2 & (12) & 5 \\ 15 & 4 & 14 & 13 & 3 \end{array}\right] \end{array} = C$$

A new shipment of cars arrives and the numbers stored in matrix D are the quantities of the new cars received by Rusty.

Deliveries

$$\begin{array}{c} \\ \text{Turbo} \\ \text{Cruiser} \\ \text{Clunker} \\ \text{Vacationer} \end{array} \begin{array}{ccccc} \text{Red} & \text{Blue} & \text{White} & \text{Silver} & \text{Tan} \\ \left[\begin{array}{ccccc} 6 & 2 & 3 & 5 & 1 \\ 7 & 4 & 2 & 10 & 2 \\ 4 & 7 & 1 & (8) & 1 \\ 3 & 3 & 5 & 9 & 1 \end{array}\right] \end{array} = D$$

The current total inventory is found by adding matrices C and D. This **matrix addition** is performed according to the following rule.

Definition of Matrix Addition

If two matrices A and B have the same dimensions, their **sum** $A + B$ is the matrix in which each element is the sum of the corresponding elements in A and B.

Add corresponding elements of C and D to find the elements of $C + D$. One set of corresponding elements is circled in the two matrices on this page and the matrix at the top of the following page: $12 + 8 = 20$.

Mental Math

At the movies, a bag of popcorn costs $4.50, a soda costs $3.25, and a package of candy costs $3.00.

a. How much do one bag of popcorn and two sodas cost?

b. How much do two sodas and two packages of candy cost?

c. How much do one bag of popcorn, one soda, and one package of candy cost?

d. How much do p bags of popcorn, s sodas, and c packages of candy cost?

Custom paint jobs such as the one above can give cars a unique look.

New Inventory

	Red	Blue	White	Silver	Tan
Turbo	18	12	10	13	3
Cruiser	21	16	9	22	9
Clunker	21	15	3	⑳	6
Vacationer	18	7	19	22	4

$= C + D$

Since matrix addition involves adding corresponding elements, only matrices with the same dimensions can be added. As the car dealership example shows, the sum matrix will have the same dimensions as the matrices that were added.

Because addition of real numbers is commutative, *addition of matrices is commutative*. Thus, if two matrices A and B can be added, then $A + B = B + A$. Also, *addition of matrices is associative*, meaning that for all matrices A, B, and C with the same dimensions, $(A + B) + C = A + (B + C)$.

Scalar Multiplication

Matrix addition is related to a special multiplication involving matrices called *scalar multiplication*.

Activity

MATERIALS CAS

Step 1 Enter the matrix $\begin{bmatrix} a & b & c \\ d & e & f \end{bmatrix}$ into your CAS and store it as $M1$.

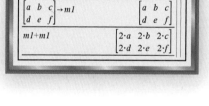

Step 2 Add the matrix from Step 1 to itself and write down the result. You should see a display similar to the one at the right.

Step 3 Add the result from Step 2 to the original matrix in Step 1 and write down the result.

Step 4 If you add the result from Step 3 to the matrix in Step 1, what result do you expect? Why?

Step 5 Clear your screen and enter $2 * \begin{bmatrix} a & b & c \\ d & e & f \end{bmatrix}$ and write down the result.

Step 6 Enter $3 * \begin{bmatrix} a & b & c \\ d & e & f \end{bmatrix}$ and write down the result.

Step 7 If you enter $4 * \begin{bmatrix} a & b & c \\ d & e & f \end{bmatrix}$, what result do you expect? Why?

Step 8: Is there a connection between your responses to Step 4 and Step 7? If so, what is it?

With real numbers, you can use multiplication as shorthand for repeated addition. For instance, $a + a + a$ can be written as $3 \cdot a$ or $3a$. The Activity gives evidence that this rule also holds true for matrices. Repeated matrix addition gives rise to an operation called **scalar multiplication**.

Definition of Scalar Multiplication

The **scalar product** of a real number k and a matrix A is the matrix kA in which each element is k times the corresponding element in A.

Example 1

Find the scalar product $6\begin{bmatrix} -2.3 & 8 \\ 7.1 & \frac{1}{2} \end{bmatrix}$.

Solution Multiply each element in the matrix by 6.

$$6\begin{bmatrix} -2.3 & 8 \\ 7.1 & \frac{1}{2} \end{bmatrix} = \begin{bmatrix} 6 \cdot -2.3 & 6 \cdot 8 \\ 6 \cdot 7.1 & 6 \cdot \frac{1}{2} \end{bmatrix} = \begin{bmatrix} -13.8 & 48 \\ 42.6 & 3 \end{bmatrix}$$

How Are Matrices Subtracted?

In the set of real numbers, subtraction has the following property: $a - b = a + -b = a + -1 \cdot b$. Matrices are subtracted in a similar manner:

$$A - B = A + (-1)B = A + (-B).$$

Definition of Matrix Subtraction

Given two matrices A and B with the same dimensions, their **difference** $A - B$ is the matrix in which each element is the difference of the corresponding elements in A and B.

Example 2

The matrices P1 and P2 below represent the degrees earned by men and women in four professions during 2003 and 2004.

$$P1 = \begin{array}{c} \\ \text{Medicine} \\ \text{Dentistry} \\ \text{Law} \\ \text{Theology} \end{array} \overset{\begin{array}{cc} \text{Male} & \text{Female} \end{array}}{\left[\begin{array}{cc} 8{,}221 & 6{,}813 \\ 2{,}653 & 1{,}691 \\ 19{,}916 & 19{,}151 \\ 3{,}499 & 1{,}852 \end{array} \right]}$$

2003

$$P2 = \begin{array}{c} \\ \text{Medicine} \\ \text{Dentistry} \\ \text{Law} \\ \text{Theology} \end{array} \overset{\begin{array}{cc} \text{Male} & \text{Female} \end{array}}{\left[\begin{array}{cc} 8{,}273 & 7{,}169 \\ 2{,}532 & 1{,}803 \\ 20{,}332 & 19{,}877 \\ 3{,}511 & 1{,}821 \end{array} \right]}$$

2004

Source: U.S. Census Bureau

a. Create a new matrix that shows the change in the number of degrees awarded from 2003 to 2004, separated by gender.

b. Which group, male or female, had the greater total increase in degrees awarded?

Solution

a. The change by gender of degrees awarded can be found by creating $P2 - P1$. Store P1 and P2 on a CAS. Then subtract the matrices as shown at the right.

b. The sum of the elements in the first column in matrix $P2 - P1$ gives the overall change for males, and the sum of the elements in the second column gives the overall change for females.

Overall change for males: $52 + {-}121 + 416 + 12 = 359$

Overall change for females: $356 + 112 + 726 + {-}31 = 1163$

Females had the greater total increase in degrees awarded with 1163 more degrees in 2004 than 2003.

 QY

When more than one matrix operation appears in an expression, you should follow the same order of operations as with expressions involving real numbers. That is, you should perform scalar multiplication before addition or subtraction. You are asked to evaluate expressions with more than one matrix operation in Questions 4 and 5.

> **QY**
>
> Check your answer to Part a of Example 2 by performing the subtraction of the first row by hand.

Questions

COVERING THE IDEAS

1. Can you add $\begin{bmatrix} 6 & -8 \end{bmatrix}$ to $\begin{bmatrix} 4 \\ 2 \end{bmatrix}$? Explain why or why not.

In 2–5, perform the indicated operations.

2. $\begin{bmatrix} -2 & 5 \\ 0 & -7 \end{bmatrix} + \begin{bmatrix} 1 & -3 \\ 8 & -2 \end{bmatrix}$

3. $\begin{bmatrix} 13 & -9 \\ -21 & 6 \end{bmatrix} - \begin{bmatrix} 8 & 6 \\ -15 & 3 \end{bmatrix}$

4. $2\begin{bmatrix} -1 & -3 \\ 1 & \frac{5}{2} \\ \frac{1}{2} & \end{bmatrix} + \begin{bmatrix} 6 & -4 \\ 2 & 0 \end{bmatrix}$

5. $-3\begin{bmatrix} 0 & 1 \\ \frac{1}{2} & -2 \end{bmatrix} - 4\begin{bmatrix} \frac{1}{4} & 0 \\ \frac{3}{4} & -1 \end{bmatrix}$

In 6 and 7, refer to Rusty's car inventory matrices C and D at the beginning of this lesson.

6. Does $C + D = D + C$? Explain why or why not.

7. Suppose matrix D represents cars in the current inventory that were damaged in a hailstorm. Find $C - D$, and explain what it represents.

8. Refer to the matrix $P2 - P1$ from Example 2.
   a. For which profession was there the greatest change in the number of degrees awarded?
   b. Is the change in Part a an increase or decrease?

9. a. **True or False** Subtraction of matrices is commutative.
   b. Let $B = \begin{bmatrix} 1 & 4 \\ 7 & 10 \end{bmatrix}$ and $C = \begin{bmatrix} 2 & 5 \\ 8 & 11 \end{bmatrix}$. Support your answer to Part a by evaluating $B - C$ and $C - B$.

APPLYING THE MATHEMATICS

10. The matrices $N1$, $N2$, and $N3$ give the enrollments by gender and grade at North High School over a 3-year period, beginning with $N1$ and ending with $N3$. In each matrix, row 1 gives the number of boys and row 2 the number of girls. Columns 1 to 4 give the number of students in grades 9 through 12, respectively.

Grades

$$N1 = \begin{bmatrix} 289 & 282 & 276 & 270 \\ 299 & 288 & 276 & 264 \end{bmatrix} \begin{matrix} \text{boys} \\ \text{girls} \end{matrix}$$

$$N2 = \begin{bmatrix} 240 & 228 & 216 & 204 \\ 240 & 234 & 228 & 222 \end{bmatrix} \begin{matrix} \text{boys} \\ \text{girls} \end{matrix}$$

$$N3 = \begin{bmatrix} 154 & 149 & 143 & 138 \\ 143 & 143 & 138 & 132 \end{bmatrix} \begin{matrix} \text{boys} \\ \text{girls} \end{matrix}$$

(columns headed 9, 10, 11, 12)

   a. Find matrix T that shows the total enrollment in each grade by gender over the 3-year time span.
   b. What was the change in enrollment for boys and girls in each grade from the first year to the last year?

11. Some key results for four teams from the National Football League for the 2005 and 2006 regular season are given in the matrices below. W = number of wins, L = number of losses, PF = points scored for the team, and PA = points scored against the team.

2006

	W	L	PF	PA
New England	12	4	385	237
Pittsburgh	8	8	353	315
Indianapolis	12	4	427	360
Denver	9	7	319	305

2005

	W	L	PF	PA
New England	10	6	379	338
Pittsburgh	11	5	389	258
Indianapolis	14	2	439	247
Denver	13	3	395	258

Source: www.NFL.com

a. Subtract the right matrix from the left matrix. Call the difference M.

b. What is the meaning of the 4th column of M?

c. Interpret each number in the 1st row of M.

12. Pearl, a puzzle maker, constructs three types of animal puzzles for children in two different styles. Pearl's output last year is given in the matrix P below.

	Cat	Dog	Mouse
Wood	12	20	8
Cardboard	18	14	11

$= P$

The Pittsburgh Steelers celebrating their 2006 Super Bowl win at the White House

a. Suppose Pearl wants to increase output by 25%. Write the matrix that represents the needed output. Round all decimals to the nearest whole number.

b. If your answer to Part a is the matrix kP, what is k?

13. Let $F = \begin{bmatrix} \frac{1}{3} & \frac{1}{6} \\ \frac{1}{9} & \frac{1}{12} \end{bmatrix}$. Let $G = \begin{bmatrix} 1 & \\ \frac{1}{3} & \frac{1}{4} \end{bmatrix}$.

a. Solve $F = kG$ for k.   b. Solve $G = \ell F$ for ℓ.

14. Is $D = \begin{bmatrix} 20 & 12 \\ 50 & 40 \end{bmatrix}$ a scalar multiple of $E = \begin{bmatrix} \frac{1}{3} & \frac{1}{5} \\ \frac{5}{6} & \frac{3}{2} \end{bmatrix}$? If so, identify the scalar k. If not, explain why not.

In 15 and 16, solve for a, b, c and d.

15. $\begin{bmatrix} a & 18 \\ 54 & 35 \end{bmatrix} + \begin{bmatrix} 17 & b \\ c & d \end{bmatrix} = \begin{bmatrix} 21 & 81 \\ 25 & 30 \end{bmatrix}$

16. $2\begin{bmatrix} a & b \\ 14 & 4 \end{bmatrix} - 7\begin{bmatrix} 3 & -1 \\ c & -2.5 \end{bmatrix} = \begin{bmatrix} \frac{1}{2} & 0 \\ 4 & d \end{bmatrix}$

REVIEW

17. Write a matrix to represent the triangle with vertices at the origin, (0, 4), and (–1, 0). **(Lesson 4-1)**

18. According to the 2000 United States Census, in 2000 there were 7,229,068 foreign-born and 46,365,310 native-born people living in the Northeast; 3,509,937 foreign-born and 60,882,839 native-born people living in the Midwest; 8,608,441 foreign-born and 91,628,379 native-born people living in the South; and 11,760,443 foreign-born and 51,437,489 native-born people living in the West. Write a matrix to store this information in millions, rounding each population to the nearest million. **(Lesson 4-1)**

19. Graph $f(x) = -\lfloor x \rfloor$ and $g(x) = \lfloor -x \rfloor$ separately. Explain the difference in the two graphs. **(Lesson 3-9)**

20. A plumber charges a fixed fee to come to your house, plus an hourly fee for the time spent on the repair job. Suppose the plumber comes to your house, works for an hour, and charges $120. The next week the plumber has to return to finish the job and spends one and a half hours working, and charges $160. Write a possible equation that describes how much the plumber charges based on hours of work. Use it to figure out the fixed fee. **(Lesson 3-4)**

21. A rectangle has sides of length $4x$ and $0.5x$. **(Lesson 2-5)**
    a. Write an equation for the area of the rectangle as a function of the lengths of its sides.
    b. Graph this equation in an appropriate window.
    c. What is the domain of the function you graphed?
    d. What is the domain of the function if the context is ignored?
    e. Graph the function again over the domain in Part d.

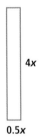

4x

0.5x

EXPLORATION

22. The matrix $\begin{bmatrix} 1 & 3 \\ 2 & 4 \end{bmatrix}$ gives the endpoints of a line segment.
    a. Find the slope of the line containing the segment and graph it.
    b. Find the slope of the line through points $\begin{bmatrix} -2 & -4 \\ 1 & 3 \end{bmatrix}$ and graph it.
    c. Compare the two matrices. Then generalize Parts a and b.

Lesson 4-3

Matrix Multiplication

Vocabulary

row-by-column multiplication

matrix multiplication

matrix product

▶ **BIG IDEA** Matrices with compatible dimensions can be multiplied using row-by-column multiplication.

If you want to buy four $12 pizzas, you find the total cost by multiplying: $12 \frac{\text{dollars}}{\text{pizza}} \cdot 4$ pizzas = $48. But consider a more complicated situation: the Forensics Club (F), the Jazz Band (J), and the Volleyball team (V) are each holding pizza parties, are each ordering different quantities of cheese (C), mushroom (M), and "garbage" (G) pizzas (which have everything on them), and want to compare prices at two different pizzerias, Lorenzo's and Billy's.

These data have been organized into the two matrices below. One matrix contains the prices in dollars for each type of pizza at the two pizzerias, and the second matrix contains the number of each type of pizza ordered by each club.

Mental Math

Estimate the tip to the nearest dime.

a. a 10% tip on a bill of $24.95

b. a 15% tip on a bill of $32.20

c. a 20% tip on a bill of $73.83

Price per Pizza

$$
\begin{array}{c}
\text{C}\text{M}\text{G} \\
\begin{array}{l}\text{Lorenzo's Prices}\\ \text{Billy's Prices}\end{array}
\left[\begin{array}{ccc} 12 & 15 & 20 \\ 11 & 17 & 18 \end{array}\right]
\end{array}
$$

Pizzas per Club

$$
\begin{array}{c}
\text{F}\text{J}\text{V} \\
\begin{array}{l}\text{Cheese}\\ \text{Mushroom}\\ \text{Garbage}\end{array}
\left[\begin{array}{ccc} 4 & 1 & 3 \\ 3 & 5 & 3 \\ 2 & 3 & 3 \end{array}\right]
\end{array}
$$

 A "garbage" pizza of vegetables

🛑 **QY1**

▶ **QY1**

a. What is the price of a cheese pizza at Lorenzo's?

b. How many mushroom pizzas has the Forensics Club ordered?

Example 1

Write a single expression giving the total price for pizzas at Lorenzo's for the Forensics Club and evaluate it.

Solution The prices at Lorenzo's are in the first row of the Price-per-Pizza matrix. The numbers of each kind of pizza ordered by the Forensics Club are in the first column of the Pizza-per-Club matrix.

Lorenzo's Prices per Pizza

$$
\begin{array}{c}
\text{C}\text{M}\text{G} \\
\left[\begin{array}{ccc} 12 & 15 & 20 \end{array}\right]
\end{array}
$$

Pizzas Ordered by the Forensics Club

$$
\begin{array}{c}
\text{C}\left[\begin{array}{c} 4 \\ \end{array}\right. \\
\text{M}3 \\
\text{G}\left.\begin{array}{c} 2 \end{array}\right]
\end{array}
$$

Total Price for the Forensics Club Order at Lorenzo's

$12 \cdot 4 + \$15 \cdot 3 + \$20 \cdot 2 = \$133$

Multiplying Two Matrices

The calculation $\$12 \cdot 4 + \$15 \cdot 3 + \$20 \cdot 2 = \$133$ in Example 1 illustrates an idea that is used to calculate the product of these matrices. The idea is called **row-by-column multiplication**. The solution to Example 1 is shown in matrix notation below. A row matrix is multiplied by a column matrix to get a 1×1 product matrix.

$$\begin{bmatrix} 12 & 15 & 20 \end{bmatrix} \cdot \begin{bmatrix} 4 \\ 3 \\ 2 \end{bmatrix} = [12 \cdot 4 + 15 \cdot 3 + 20 \cdot 2] = [133]$$

This process can be generalized into the algorithm below.

Algorithm for Row-by-Column Multiplication

Step 1 Multiply the first element in the row matrix by the first element in the column matrix. Multiply the second element in the row matrix by the second element in the column matrix. Continue multiplying the nth element in the row by the nth element in the column until you reach the end of the row and the column.

Step 2 Add the products in Step 1.

Notice that the row and the column have to have the same number of elements in order for this multiplication to work. Also, the row is on the left and the column is on the right.

The product $A \cdot B$ or AB of two matrices A and B is found by using the above algorithm to multiply each row of A times each column of B. For example, if matrix A has 2 rows and matrix B has 3 columns, then there are 6 ways to multiply a row by a column. The 6 products of these rows and columns are the 6 elements of the product matrix AB. The entry in row 2, column 3 of the product comes from multiplying row 2 of the first matrix by column 3 of the second matrix, as shown below.

$$\begin{bmatrix} 12 & 15 & 20 \\ 11 & 17 & 18 \end{bmatrix} \cdot \begin{bmatrix} 4 & 1 & 3 \\ 3 & 5 & 3 \\ 2 & 3 & 3 \end{bmatrix} = \begin{bmatrix} 133 & 147 & 141 \\ 131 & 150 & 138 \end{bmatrix}$$

If you multiply the Price-per-Pizza matrix by the Pizzas-per-Club matrix, you will get a product matrix for price per club. The calculator display at the right shows the product. This is not scalar multiplication; it is **matrix multiplication**.

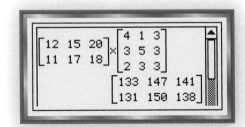

QY2

In the 2×3 product matrix, the rows represent the 2 pizzerias and the columns represent the 3 clubs as shown at the right.

$$\begin{array}{c} \\ \text{Lorenzo's Prices} \\ \text{Billy's Prices} \end{array} \begin{array}{ccc} F & J & V \\ \left[\begin{array}{ccc} 133 & 147 & 141 \\ 131 & 150 & 138 \end{array}\right] \end{array}$$

▶ QY2

Use your calculator to verify the matrix product on the previous page.

Each element in this product matrix is the total price of a pizza order. For example, the Forensics Club would pay a total of $133 at Lorenzo's, and the Jazz Band would pay $150 at Billy's.

GUIDED

Example 2

Let $A = \begin{bmatrix} 1 & 3 \\ -2 & 4 \end{bmatrix}$ and $B = \begin{bmatrix} 0 & 5 & 2 \\ 3 & -1 & 4 \end{bmatrix}$. Find their product AB by hand.

Solution A has __?__ rows and B has __?__ columns. So the matrix AB will have __?__ entries.

Use the algorithm for row-by-column multiplication to fill in the missing entries below.

$$AB = \begin{bmatrix} 1 \cdot 0 + 3 \cdot 3 & 1 \cdot 5 + \underline{} & \underline{} \\ -2 \cdot 0 + 4 \cdot 3 & \underline{} + 4 \cdot -1 & \underline{} \end{bmatrix} = \begin{bmatrix} 9 & \underline{} & \underline{} \\ 12 & \underline{} & \underline{} \end{bmatrix}$$

In general, to multiply matrices A and B, find all possible products using rows from matrix A and columns from matrix B. This leads to the following definition.

Definition of Matrix Multiplication

If A is an $m \times n$ matrix and B is a $n \times q$ matrix, then the **matrix product** $A \cdot B$ (or AB) is an $m \times q$ matrix whose element in row r, column c is the product of row r of A and column c of B.

Caution! The product of two matrices A and B exists only when the *number of columns of A equals the number of rows of B.* So if A is $m \times n$, B must be $n \times q$ for AB to exist.

These matrices can be multiplied.

$$\underset{2 \times 2}{\begin{bmatrix} 4 & -1 \\ 6 & 2 \end{bmatrix}} \underset{2 \times 3}{\begin{bmatrix} 3 & -2 & 7 \\ 5 & 3 & 0 \end{bmatrix}} = \underset{2 \times 3}{\begin{bmatrix} 7 & -11 & 28 \\ 28 & -6 & 42 \end{bmatrix}}$$

equal

dimensions of product

These matrices cannot be multiplied.

$$\underset{2 \times 3}{\begin{bmatrix} 3 & -2 & 7 \\ 5 & 3 & 0 \end{bmatrix}} \underset{2 \times 2}{\begin{bmatrix} 4 & -1 \\ 6 & 2 \end{bmatrix}}$$

not equal

no product

These two cases indicate that, in general, *multiplication of matrices is not commutative*. When both *AB* and *BA* exist, we can avoid confusion by saying "Multiply *A* on the left by *B*" to mean "find *BA*."

 QY3

When matrices arise from real situations, you often have several choices for ways of arranging your data. However, when you are multiplying two matrices, the number of columns of the first matrix must match the number of rows of the second matrix. In order to set up matrices for multiplication, think about the units involved: the *headings of the columns* of the left matrix must match the *headings of the rows* of the right matrix. That is how the matrices were set up in the opening example.

▶ **QY3**

Can you multiply a 3 × 4 matrix on the right by a 4 × 2 matrix? If so, what would the dimensions of the product be? If not, why not?

Multiplying More Than Two Matrices

The Activity below suggests that an important property of real-number multiplication extends to matrix multiplication.

Activity

MATERIALS CAS or graphing calculator

Step 1 Store these matrices in your calculator:

$$M1 = \begin{bmatrix} 4 & 3 \end{bmatrix}, M2 = \begin{bmatrix} -2 & 4 & 1 \\ 3 & 6 & -5 \end{bmatrix}, \text{ and } M3 = \begin{bmatrix} 6 \\ 11 \\ 2 \end{bmatrix}.$$

Step 2 Calculate $(M1 \cdot M2) \cdot M3$.

Step 3 Calculate $M1 \cdot (M2 \cdot M3)$.

Step 4 Compare the results of Steps 2 and 3. What property holds true for matrices *M1*, *M2*, and *M3*?

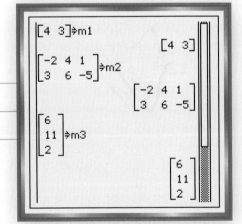

This property holds for all matrices. For any matrices *A*, *B*, and *C*, if the applicable products exist, $(A \cdot B) \cdot C = A \cdot (B \cdot C)$.

Matrix multiplication has many applications. In this course, you will use it to model linear combinations, to perform geometric transformations, and to solve systems of equations.

Questions

COVERING THE IDEAS

In 1–3, refer to the Price-per-Pizza, Pizzas-per-Club, and Price-per-Club matrices at the beginning of this lesson.

1. How many mushroom pizzas did the Volleyball team order?

2. How many garbage pizzas were ordered by all the clubs together?

3. Write an expression to calculate the total cost for the Jazz Band's order from Billy's pizzeria.

In 4 and 5, perform the multiplication.

4. $\begin{bmatrix} 3 & 5 \end{bmatrix} \cdot \begin{bmatrix} 1 \\ 0 \end{bmatrix}$

5. $\begin{bmatrix} 1 & 4 & -5 & 6 \end{bmatrix} \cdot \begin{bmatrix} 2 \\ -3 \\ 11 \\ 7 \end{bmatrix}$

In 6 and 7, use the matrices $A = \begin{bmatrix} 5 & 1 \\ 6 & 2 \\ 3 & -1 \end{bmatrix}$ and $B = \begin{bmatrix} -4 \\ 4 \end{bmatrix}$.

6. a. **Fill in the Blanks** The product AB has __?__ row(s) and __?__ column(s).

   b. **Fill in the Blanks** The element in the second row and first column of AB is __?__ · __?__ + __?__ · __?__.

   c. Compute the product AB.

7. a. Is it possible to compute the product BA? Explain why or why not.

   b. What property of matrix multiplication does the answer to Part a illustrate?

8. Multiply the matrices at the right in two different ways to show that matrix multiplication is associative.

9. Multiply the matrices at the right in two different orders to show that matrix multiplication is not commutative.

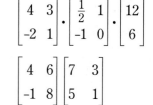

$$\begin{bmatrix} 4 & 3 \\ -2 & 1 \end{bmatrix} \cdot \begin{bmatrix} \frac{1}{2} & 1 \\ -1 & 0 \end{bmatrix} \cdot \begin{bmatrix} 12 \\ 6 \end{bmatrix}$$

$$\begin{bmatrix} 4 & 6 \\ -1 & 8 \end{bmatrix} \begin{bmatrix} 7 & 3 \\ 5 & 1 \end{bmatrix}$$

10. Consider the matrices M and N below.

$$M = \begin{bmatrix} 11 & 12 & 13 & 14 & 15 \\ 21 & 22 & 23 & 24 & 25 \\ 31 & 32 & 33 & 34 & 35 \\ 41 & 42 & 43 & 44 & 45 \end{bmatrix}, N = \begin{bmatrix} 11 & 12 & 13 & 14 \\ 21 & 22 & 23 & 24 \\ 31 & 32 & 33 & 34 \\ 41 & 42 & 43 & 44 \end{bmatrix}$$

   a. Only one of the products MN and NM exists. Which is it?

   b. What are the dimensions of the product?

   c. Write an expression for the element in the third row, second column of the product.

APPLYING THE MATHEMATICS

11. Lenora runs a small shipping company. One afternoon, she gets requests from two sporting-goods stores to transport shipments of sports equipment. Store 1 wants to transport 20 cases of table tennis balls (TTB), 50 cases of tennis balls (TB), and 12 bowling balls (BB). Store 2 wants to transport 60 cases of table tennis balls, 10 cases of tennis balls, and 15 bowling balls. One case of table tennis balls weighs 1.5 pounds, one case of tennis balls weighs 5 pounds, and one bowling ball weighs 12 pounds. In addition, each case of table tennis balls or tennis balls takes up 1 cubic foot of cargo space, but each bowling ball takes 0.6 cubic foot of space. Use matrix multiplication to find the total weight and cargo space required for each order.

Bowling balls are available in a wide variety of weights, patterns, and colors.

12. Refer to the pizza story in this lesson. According to the product matrix, is there any group that will save money by ordering their pizza at Lorenzo's? How much would the group save?

13. Consider the point matrix $A = \begin{bmatrix} x \\ y \end{bmatrix}$. Perform the indicated multiplication. Call the product B. Then describe the relationship between points A and B.

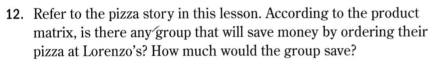

a. $\begin{bmatrix} 1 & 1 \\ 1 & 1 \end{bmatrix} \cdot A$ b. $\begin{bmatrix} 1 & 0 \\ 0 & 1 \end{bmatrix} \cdot A$ c. $\begin{bmatrix} 0 & 1 \\ 1 & 0 \end{bmatrix} \cdot A$

d. $\begin{bmatrix} 0 & -1 \\ 1 & 0 \end{bmatrix} \cdot A$ e. $\begin{bmatrix} 2 & 0 \\ 0 & 2 \end{bmatrix} \cdot A$

14. You own the Sweet Cakes Bakery, which produces three sizes of muffins: mini, regular, and large. You are concerned with three main ingredients: flour, butter, and sugar. Matrix R shows the amount of each ingredient (in pounds) required per dozen of each muffin size. Matrix Q shows the average number of dozens of each muffin size sold on Mondays and Fridays.

	Mini	Regular	Large
Flour	0.3	0.5	0.8
Butter	0.1	0.2	0.3
Sugar	0.1	0.2	0.3

$= R$

	Mon	Fri
Mini	10	20
Regular	15	35
Large	20	50

$= Q$

a. What does the number in the third row, first column of Q represent in this situation?

b. Compute RQ.

c. What does the number in the second row, second column of RQ represent?

d. One Saturday, the weekly sugar shipment does not come in. There are 8 pounds of sugar on hand. How will that impact your preparations for Monday?

15. Suppose $\begin{bmatrix} 4 & 3 \\ 2 & x \end{bmatrix} \cdot \begin{bmatrix} y \\ 1 \end{bmatrix} = \begin{bmatrix} 23 \\ 13 \end{bmatrix}$. Find the value of x.

16. **a.** The matrix $I = \begin{bmatrix} 1 & 0 \\ 0 & 1 \end{bmatrix}$ is called the *2 × 2 identity matrix*.

    If $A = \begin{bmatrix} 4 & 2 \\ -3 & 6 \end{bmatrix}$, compute IA and AI.

    b. Why is I called the identity matrix?

REVIEW

17. Compute $\begin{bmatrix} 6 & -4 \\ -1 & 20 \end{bmatrix} + 2\begin{bmatrix} -3 & 2 \\ \frac{1}{2} & -10 \end{bmatrix}$. **(Lesson 4-2)**

18. If $\begin{bmatrix} a & 3.3 & 5c \\ 7 & -2e & -9 \end{bmatrix} = k\begin{bmatrix} 9 & -11 & -8c \\ d & \frac{4}{9} & 5f \end{bmatrix}$, find a, c, d, e, f, and k. **(Lesson 4-2)**

19. What are the dimensions of the matrix at the right? **(Lesson 4-1)**

$\begin{bmatrix} 4 & 0 & 0 \\ 0 & 2 & 0 \\ 0 & 0 & -1 \\ 1 & 0 & \pi \end{bmatrix}$

20. **Multiple Choice** Which formula best models the data in the table below? **(Lesson 2-7)**

    A $P = km$ **B** $P = km^2$ **C** $P = \frac{k}{m}$ **D** $P = \frac{k}{m^2}$

m	1	2	3	4	5	6
P	36	9	4	2.25	1.44	1

21. Let $K = (1, 3)$, $L = (-4, 3)$, $M = (-4, 8)$, and $N = (1, 8)$. **(Previous Course)**
    a. What kind of figure is *KLMN*?
    b. Divide each coordinate of the vertices of *KLMN* by 2 to find the coordinates of *K'L'M'N'*.
    c. What kind of figure is *K'L'M'N'*? How do you know?

22. Find the distance between each pair of points. **(Previous Course)**
    a. (2, 4) and (8, 16) b. (1.2, 0.7) and (6.3, −3.75)
    c. $\left(-\frac{3}{4}, \frac{2}{3}\right)$ and $\left(-\frac{4}{9}, \frac{1}{2}\right)$

EXPLORATION

23. Suppose the product of two 2 × 2 matrices A and B is $\begin{bmatrix} 0 & 0 \\ 0 & 0 \end{bmatrix}$, but neither matrix is $\begin{bmatrix} 0 & 0 \\ 0 & 0 \end{bmatrix}$. What is the largest number of nonzero elements possible in A and B in order for this to happen?

QY ANSWERS

1a. $12

1b. 3

2. Calculator display should be similar to the one shown on page 236.

3. Yes; the product would be a 3 × 2 matrix.

Lesson
4-4

Matrices for Size Changes

▶ **BIG IDEA**   2 × 2 matrices can represent geometric transformations.

Vocabulary

transformation

preimage

image

size change

center of a size change

magnitude of a size change

identity matrix

identity transformation

similar

In the next few lessons, you will see how matrices can represent a variety of geometric transformations. Recall from geometry that a **transformation** is a one-to-one correspondence between the points of a **preimage** and the points of an **image**. Consider the transformation S_k that maps (x, y) onto (kx, ky). If $k = 5$, the transformation images of a few points under S_5 are written as follows:

$$S_5(2, -3) = (10, -15);$$
$$S_5(-3, 0) = (-15, 0);$$
$$S_5(-2, -1.5) = (-10, -7.5);$$

and, in general, $S_5(x, y) = (5x, 5y)$.

Definition of Size Change

For any $k \neq 0$, the transformation that maps (x, y) onto (kx, ky) is called the **size change** with **center** $(0, 0)$ and **magnitude** k, and is denoted S_k.

From the definition, $S_k(x, y) = (ky, ky)$,

or $S_k: (x, y) \rightarrow (kx, ky)$.

These are read "Under the size change of magnitude k and center $(0, 0)$, the image of (x, y) is (kx, ky)," or "The size change of magnitude k and center $(0, 0)$ maps (x, y) onto (kx, ky)."

Example 1

a. What is the magnitude of the size change that maps the preimage $\triangle TRI$ onto its image $\triangle T'R'I'$ as shown at the right?

b. Write the size change using mapping notation.

Solution

a. Pick one coordinate pair on the preimage and compare it to the corresponding coordinate pair of the image.

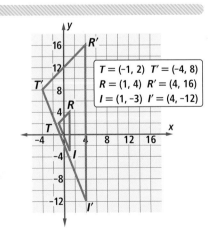

$T = (-1, 2)$ $T' = (-4, 8)$
$R = (1, 4)$ $R' = (4, 16)$
$I = (1, -3)$ $I' = (4, -12)$

T = (–1, 2) and T′ = (–4, 8). The coordinates of T have been multiplied by 4. So the magnitude of the size change is 4.

b. This size change maps (x, y) onto (4x, 4y).

$$S_4: (x, y) \rightarrow (4x, 4y)$$

Activity

MATERIALS graph paper, ruler

Work with a partner.

Step 1 You and your partner should each copy polygon *NUMER* shown at the right, then together write a matrix to represent it.

Step 2 One of you should multiply *NUMER* on the left by $\begin{bmatrix} 2 & 0 \\ 0 & 2 \end{bmatrix}$.

The other partner should multiply *NUMER* on the left by $\begin{bmatrix} \frac{1}{3} & 0 \\ 0 & \frac{1}{3} \end{bmatrix}$.

Each of you should draw the image polygon on your graph of *NUMER* and label it *N′U′M′E′R′*. Complete Steps 3–6 for your own graph.

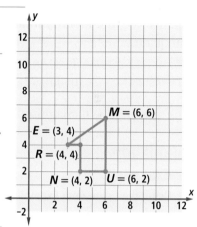

Step 3 How do the coordinates of *NUMER* compare to the coordinates of *N′U′M′E′R′*?

Step 4 Draw a ray from the origin (point O) through vertex E of *NUMER*. Then draw a ray from the origin through E′.

Step 5 Measure the distances *OE* and *OE′* and compare them. What is $\frac{OE'}{OE}$?

Step 6 Repeat Steps 4 and 5 for the other vertices of *NUMER* and *N′U′M′E′R′*.

Step 7 Discuss each other's results and generalize. If the matrix for *NUMER* is multiplied on the left by $\begin{bmatrix} k & 0 \\ 0 & k \end{bmatrix}$ to get polygon *N′U′M′E′R′*, where k is a real number, how do the vertices of the polygon *NUMER* compare to those of the polygon *N′U′M′E′R′*?

 QY

▶ **QY**

How do the lengths of sides of the polygons *NUMER* and *N′U′M′E′R′* from Step 2 compare?

Using Matrices to Perform Size Changes

As the Activity illustrates, size change images can be found by multiplying matrices. When the matrix for the point (x, y) is multiplied on the left by $\begin{bmatrix} k & 0 \\ 0 & k \end{bmatrix}$, the matrix for the point (kx, ky) results.

$$\begin{bmatrix} k & 0 \\ 0 & k \end{bmatrix} \begin{bmatrix} x \\ y \end{bmatrix} = \begin{bmatrix} kx + 0y \\ 0x + ky \end{bmatrix} = \begin{bmatrix} kx \\ ky \end{bmatrix}$$

This proves the following theorem.

Size Change Theorem

$\begin{bmatrix} k & 0 \\ 0 & k \end{bmatrix}$ is the matrix for S_k.

When $k = 1$, the **identity matrix** $\begin{bmatrix} 1 & 0 \\ 0 & 1 \end{bmatrix}$ maps each point $\begin{bmatrix} x \\ y \end{bmatrix}$ of a figure onto itself.

$$\begin{bmatrix} 1 & 0 \\ 0 & 1 \end{bmatrix} \begin{bmatrix} x \\ y \end{bmatrix} = \begin{bmatrix} 1 \cdot x + 0 \cdot y \\ 0 \cdot x + 1 \cdot y \end{bmatrix} = \begin{bmatrix} x \\ y \end{bmatrix}$$

This size change of magnitude 1 is called the **identity transformation**.

GUIDED

Example 2

Using the figure from Example 1, perform the appropriate matrix multiplication to find the vertices of the image △T'R'I' from the vertices of the preimage △TRI.

Solution Write △TRI and S_4 in matrix form and multiply.

$$\underset{S_4}{\begin{bmatrix} \underline{\;?\;} & 0 \\ \underline{\;?\;} & 4 \end{bmatrix}} \underset{\begin{matrix} T & R & I \end{matrix}}{\begin{bmatrix} \underline{\;?\;} & 1 & \underline{\;?\;} \\ 2 & \underline{\;?\;} & -3 \end{bmatrix}} = \underset{\begin{matrix} T' & R' & I' \end{matrix}}{\begin{bmatrix} -4 & \underline{\;?\;} & 4 \\ \underline{\;?\;} & 16 & \underline{\;?\;} \end{bmatrix}}$$

So △T'R'I' has vertices T' = (-4, __?__), R' = (__?__, 16), and I' = (4, -12).

Check Compare the vertices from the matrix with the graph of △T'R'I'. The coordinates are the same.

Size Changes and the Multiplication of Distance

Recall that the Pythagorean Theorem can be applied to calculate the distance between two points given their coordinates.

Suppose you wish to find the distance AB when $A = (x_1, y_1)$ and $B = (x_2, y_2)$. If you don't remember the general formula, let $C = (x_2, y_1)$ to create right triangle △ABC. In that triangle,

$$AB^2 = AC^2 + BC^2.$$

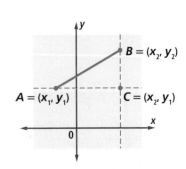

$\overline{AC}$ and $\overline{BC}$ lie on horizontal and vertical lines. So their lengths are like distances on a number line.

$$AB^2 = |x_2 - x_1|^2 + |y_2 - y_1|^2$$

Take the square root of each side and you have the Pythagorean Distance Formula.

> **Pythagorean Distance Formula**
>
> If $A = (x_1, y_1)$ and $B = (x_2, y_2)$, then
> $$AB = \sqrt{|x_2 - x_1|^2 + |y_2 - y_1|^2}.$$

So, for instance, in Example 1, where $T = (-1, 2)$ and $I = (1, -3)$, then

$$TI = \sqrt{|1 - (-1)|^2 + |-3 - 2|^2} \quad \text{and} \quad T'I' = \sqrt{|4 - (-4)|^2 + |-12 - 8|^2}$$
$$= \sqrt{2^2 + 5^2} \qquad\qquad\qquad = \sqrt{8^2 + 20^2}$$
$$= \sqrt{29} \qquad\qquad\qquad\qquad = \sqrt{464}$$
$$\qquad\qquad\qquad\qquad\qquad = 4\sqrt{29}.$$

This example illustrates that, in the size change S_k, the distance between images is $|k|$ times the distance between their preimages.

Similarity and Size Changes

Size changes are fundamental in the study of similarity. Recall from geometry that two figures are **similar** if and only if one is the image of the other under a composite of reflections and size changes. Composites of reflections, such as rotations, translations, and reflections themselves, preserve distance. So, if two figures are similar, as are $\triangle TRI$ and $\triangle T'R'I'$ in Example 1, distances in one are a constant multiple of distances in the other.

Example 3

Refer to triangles $\triangle TRI$ and $\triangle T'R'I'$ in Example 1. Calculate the following to verify that they are similar.

a. The ratios of each pair of corresponding sides of the triangles

b. Measures of corresponding angles of the triangles

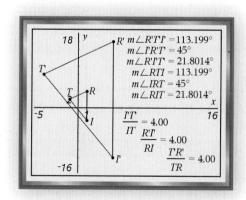

Solution Plot both triangles in your dynamic geometry system (DGS). Use the DGS to show measures of the angles of the triangles and the ratios of the sides. Answers to Parts a and b are shown in the display at the right.

Based on these measures, the triangles appear to be similar. The ratios of sides are all equal to the magnitude of the size change, 4. The corresponding angles have equal measure.

Questions

COVERING THE IDEAS

1. Refer to the Activity. How does $N'U'M'E'R'$ compare to $NUMER$ if the matrix $NUMER$ is multiplied by $\begin{bmatrix} k & 0 \\ 0 & k \end{bmatrix}$ and k is:

   a. 4? b. $\frac{1}{2}$? c. 1?

In 2 and 3, how is the expression read?

2. $S_6(3, -2) = (18, -12)$

3. $S_{1.2}: (4, 3.4) \rightarrow (4.8, 4.08)$

4. What matrix would you use to change the size of a polygon by a magnitude of $\frac{1}{5}$?

5. **Multiple Choice** If $S_k: (2, 5) \rightarrow \left(1, \frac{5}{2}\right)$, what is the value of k?

   A 2 B 4 C $\frac{1}{2}$ D $\frac{1}{4}$

6. a. Write a matrix to describe the vertices of the quadrilateral $QUAD$ with coordinates $Q = (3, 7)$, $U = (5, -9)$, $A = (-2, -8)$, and $D = (0, 4)$.

   b. Give the coordinates of the vertices of $Q'U'A'D'$, the image of the quadrilateral $QUAD$ in Part a, under S_6.

   c. Verify that $Q'U' = 6 \cdot QU$.

7. Refer to Example 2. Write a size change matrix that transforms $\triangle T'R'I'$ back to $\triangle TRI$.

8. **True or False** To map a point onto itself, multiply the point matrix on the left by $\begin{bmatrix} 0 & 1 \\ 1 & 0 \end{bmatrix}$.

In 9–11, answer *always, sometimes but not always,* or *never.*

9. Under a size change, an angle and its image are congruent.

10. Under a size change, a segment and its image are congruent.

11. Under a size change, a figure and its image are similar.

12. $\triangle ABC$ has matrix $\begin{bmatrix} 3 & -9 & 6 \\ 8 & 12 & 4 \end{bmatrix}$.

    a. Graph $\triangle ABC$ and its image $\triangle A'B'C'$ under $S_{\frac{1}{3}}$.

    b. Explain why $\triangle ABC$ and $\triangle A'B'C'$ are similar.

APPLYING THE MATHEMATICS

13. A Chicago souvenir store sells an exact scale replica of the Sears Tower that is 7.5 inches tall. The actual Sears Tower is 1730 feet tall. What transformation matrix could be used to change the size of the scale model to the size of the actual tower?

The Sears Tower, renamed Willis Tower in 2009, was built in 1970–73 at a cost of $250 million.

14. The Japanese fairy tale *Issunbōshi* (Suyeoka, Goodman, & Spicer, 1974), tells the story of a 2.5 cm tall boy named Issunbōshi who goes on a journey and eventually becomes a full-sized man.

    a. If the average height of a man is 170.2 cm, what size-change magnitude k is needed to transform Issunbōshi's height?

    b. Issunbōshi has a cricket for a pet. The average cricket length is 3.8 cm. If the length of the cricket were transformed by the same size change as Issunbōshi, how long would the cricket be?

15. Refer to the drawing at the right.

    a. Find the matrix of the transformation mapping *ABCD* onto *A′B′C′D′*.

    b. Find the slope of $\overleftrightarrow{BC}$.

    c. Find the slope of $\overleftrightarrow{B′C′}$.

    d. Is $\overleftrightarrow{BC}$ parallel to $\overleftrightarrow{B′C′}$? Why or why not?

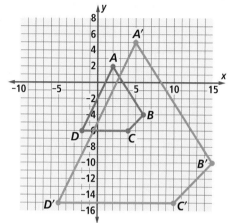

16. Define a function pdf (Pythagorean distance function) on a CAS with inputs xa, ya, xb, and yb that finds the Pythagorean Distance between any two ordered pairs (xa, ya) and (xb, yb). (We use these names because x1, y1, x2 and y2 are reserved names on many calculators.) Use pdf to check the lengths of *TI* and *T′I′* in the lesson.

17. a. Refer to Example 1. Find the image $\triangle T^*R^*I^*$ of $\triangle TRI$ under a size change of magnitude –4.

    b. What are some differences between the image $\triangle T^*R^*I^*$ and the image $\triangle T′R′I′$?

REVIEW

18. **Multiple Choice** Refer to the matrices at the right. Which of the following matrix multiplications are defined? (There may be more than one correct answer.) **(Lesson 4-3)**

    A *AB*   B *AC*   C *BA*

    D *BC*   E *CA*   F *CB*

$$A = \begin{bmatrix} 1 & 0 & 0 \\ 0 & 2 & 0 \\ 3 & 3 & -2 \end{bmatrix} \; {\scriptstyle 3 \times 3}$$

$$B = \begin{bmatrix} a & b & c \\ 8 & 8 & 8 \end{bmatrix} \; {\scriptstyle 2 \times 3}$$

$$C = \begin{bmatrix} 2 & 1 \\ 1 & 2 \end{bmatrix} \; {\scriptstyle 2 \times 2}$$

19. Let $Q = \begin{bmatrix} 1 \\ 1 \end{bmatrix}$. **(Lessons 4-3, 4-1)**

    a. Compute $\begin{bmatrix} 2 & 0 \\ 0 & 1 \end{bmatrix} Q, \begin{bmatrix} 4 & 0 \\ 0 & 1 \end{bmatrix} Q, \begin{bmatrix} 1 & 0 \\ 0 & 2 \end{bmatrix} Q$, and $\begin{bmatrix} 1 & 0 \\ 0 & 4 \end{bmatrix} Q$.

    b. Plot Q and the four answers to Part a as points in the plane.

20. Compute $\begin{bmatrix} a & b \\ c & d \end{bmatrix} + \begin{bmatrix} b & d \\ a & c \end{bmatrix} + \begin{bmatrix} d & c \\ b & a \end{bmatrix} + \begin{bmatrix} c & a \\ d & b \end{bmatrix}$. **(Lesson 4-2)**

21. You know the volume of a rectangular solid is given by the formula $V = \ell wh$. (**Lesson 2-9**)

    a. Solve this equation for ℓ.

    b. **Fill in the Blanks** From Part a, ℓ varies directly as __?__ and inversely as __?__ and __?__.

    c. If w is multiplied by 8, h is multiplied by 9, and V is multiplied by 20, by what is ℓ multiplied?

EXPLORATION

22. Suppose $\triangle MIA$ is represented by the matrix
$\begin{bmatrix} 4 & 8 & 10 \\ -2 & 5 & -3 \end{bmatrix}$.

    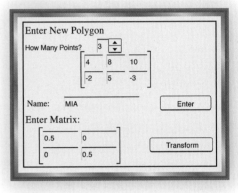

    a. Find the product $\begin{bmatrix} \frac{1}{2} & 0 \\ 0 & \frac{1}{2} \end{bmatrix} \begin{bmatrix} 4 & 8 & 10 \\ -2 & 5 & -3 \end{bmatrix}$.

    b. The product matrix in Part b represents $\triangle M'I'A'$, the image of $\triangle MIA$ under a size change of what magnitude? Draw $\triangle MIA$ and $\triangle M'I'A'$.

    c. You can do Parts a and b by using a matrix polygon application. First input a matrix for $\triangle MIA$ and a matrix for a size change of a magnitude of your choosing. Then run the program for the product of those matrices.

    d. How are the lengths of the sides of $\triangle M'I'A'$ related to the lengths of the sides of $\triangle MIA$?

    e. How are the areas of $\triangle M'I'A'$ and $\triangle MIA$ related?

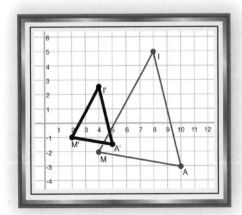

QY ANSWER

The lengths of sides of $N'U'M'E'R'$ are two times (or $\frac{1}{3}$ of) the lengths of corresponding sides of $NUMER$.

Lesson
4-5
Matrices for Scale Changes

Vocabulary

scale change

horizontal magnitude

vertical magnitude

stretch

shrink

▶ **BIG IDEA** Matrices can represent scale changes.

What Is a Scale Change?

In the previous lesson you studied size changes, which are transformations in which the changes to the preimage have the same magnitude in both the horizontal and vertical directions. Sometimes it is more useful to apply a transformation in which the horizontal and vertical changes have *different* magnitudes.

For example, imagine that a friend hands you a rubber band and says there is a secret message written on it. The writing on the rubber band looks like the picture at the left below. If you apply a size change, and stretch the rubber band by the same amount in both directions, it looks like the picture in the center. It is still unreadable. However, if you stretch the rubber band more in the horizontal direction than the vertical direction, it looks like the picture at the right, and you can read the message.

  I love Advanced Algebra!

Transformations that multiply coordinates by constants are called *scale changes*.

Definition of Scale Change

For any nonzero numbers a and b, the transformation that maps (x, y) onto (ax, by) is called the **scale change** with **horizontal magnitude** a and **vertical magnitude** b, and is denoted $S_{a,b}$.

$$S_{a,b}(x, y) = (ax, by) \text{ or}$$
$$S_{a,b} : (x, y) \rightarrow (ax, by)$$

When $|a| > 1$ (or $|b| > 1$), the scale change is a **stretch** in the horizontal (or vertical) direction. When $|a| < 1$ (or $|b| < 1$), the scale change is a **shrink** in the horizontal (or vertical) direction.

Souvenirs are often scale models of the actual things they represent.

Like size changes, scale changes are functions because each point in the preimage maps to one and only one point in the image.

Example 1

Consider the triangle at the right. Find its image under $S_{4,0.5}$.

Solution $S_{4,0.5}(x, y) = (\underline{\;?\;}x, \underline{\;?\;}y)$, so multiply all x-coordinates of the preimage points by $\underline{\;?\;}$ and all y-coordinates of the preimage points by $\underline{\;?\;}$. Images of the vertices determine the image polygon.

$A': S_{4,0.5}(0, 4) = (0, 2)$

$B': S_{4,0.5}(5, 6) = (\underline{\;?\;}, \underline{\;?\;})$

$C': S_{4,0.5}(6, 0) = (\underline{\;?\;}, \underline{\;?\;})$

The image of $\triangle ABC$ is $\triangle A'B'C'$ with the vertex coordinates above.

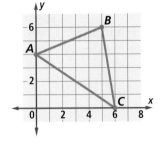

Check Check using the graph at the right. The image of every point should be 4 times as far from the y-axis and half as far from the x-axis as its preimage. For instance, the image $B' = (20, 3)$ is 4 times as far from the y-axis and half as far from the x-axis as its preimage $B = (5, 6)$.

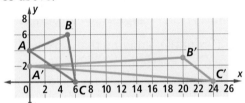

Example 1 demonstrates that scale changes might not multiply both vertical and horizontal distances by a constant amount, and so images are not necessarily similar to their preimages.

Matrices for Scale Changes

Because a size change is represented by a matrix, it is reasonable to expect that a scale change has a matrix.

MATERIALS CAS or graphing calculator

Step 1 Sketch a graph of $\triangle ABC$ from Example 1. Write a matrix to represent $\triangle ABC$ and store the matrix as M on a CAS.

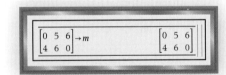

Step 2 Define a second matrix S1 on your CAS, with $S1 = \begin{bmatrix} 4 & 0 \\ 0 & 0.5 \end{bmatrix}$. Calculate $S1 \cdot M$, the matrix for $\triangle A'B'C'$.

Step 3 Sketch $\triangle A'B'C'$ on the same graph. Compare your graph to the graphs in Example 1. What transformation do you think matrix S1 represents?

Step 4 Define a third matrix $S2 = \begin{bmatrix} 1.2 & 0 \\ 0 & 3 \end{bmatrix}$. Calculate $S2 \cdot M$, the matrix for $\triangle A''B''C''$. Sketch $\triangle A''B''C''$ on the same graph.

Step 5 Compare the coordinates and graph of △ABC to the coordinates and graph of △A″B″C″. What transformation do you think S2 represents?

Step 6 Generalize your findings. Make a conjecture about the matrix for the scale change $S_{a,b}$.

Algebra easily proves the results of the Activity. Suppose that $S_{a,b}$ has the matrix

$$\begin{bmatrix} e & f \\ g & h \end{bmatrix}$$

where e, f, g, and h are real numbers. Because $S_{a,b}$ maps (x, y) onto (ax, by), we want to find e, f, g, and h so that for all x and y,

$$\begin{bmatrix} e & f \\ g & h \end{bmatrix}\begin{bmatrix} x \\ y \end{bmatrix} = \begin{bmatrix} ax \\ by \end{bmatrix}.$$

By matrix multiplication, $\begin{bmatrix} ex + fy \\ gx + hy \end{bmatrix} = \begin{bmatrix} ax \\ by \end{bmatrix}.$

Thus for all x and y, $ex + fy = ax$, so $e = a$ and $f = 0$, and $gx + hy = by$, so $g = 0$ and $h = b$. We have proved the following theorem.

Scale Change Theorem

$\begin{bmatrix} a & 0 \\ 0 & b \end{bmatrix}$ is the matrix for $S_{a,b}$.

 QY

▶ **QY**

What is the matrix for $S_{7,4}$?

Example 2

Consider the quadrilateral ABCD with A = (−2, 5), B = (0, 7), C = (4, 1), and D = (−4, −1). Find its image A′B′C′D′ under $S_{3,2}$.

Solution 1 Write $S_{3,2}$ and ABCD in matrix form. Calculate the product by hand.

$$\begin{array}{cccc} & A & B & C & D \\ S_{3,2} \end{array}$$

$$\begin{bmatrix} 3 & 0 \\ 0 & 2 \end{bmatrix}\begin{bmatrix} -2 & 0 & 4 & -4 \\ 5 & 7 & 1 & -1 \end{bmatrix} = \begin{matrix} A' & B' & C' & D' \\ \begin{bmatrix} -6 & 0 & 12 & -12 \\ 10 & 14 & 2 & -2 \end{bmatrix} \end{matrix}$$

Solution 2 Use technology to do the calculations. Let m represent the transformation matrix and n represent quadrilateral ABCD.

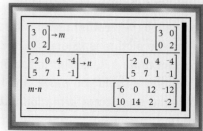

Notice that a scale change may or may not stretch or shrink by different factors in the horizontal and vertical directions. If the factors are the same in both directions, then the scale-change matrix has the form $\begin{bmatrix} k & 0 \\ 0 & k \end{bmatrix}$, and the transformation is just a size change. Conversely, a size change with magnitude k is a scale change with horizontal magnitude k and vertical magnitude k. Thus *size changes are special types of scale changes*. In symbols, $S_k = S_{k,k}$.

Questions

COVERING THE IDEAS

1. **Fill in the Blank** $S_{0.8,\frac{4}{7}} : (x, y) \rightarrow$ ___?___

2. **Multiple Choice** Which of the following mappings gives a horizontal shrink and a vertical stretch?

   A $S_{\frac{1}{5},\frac{2}{3}}$ **B** $S_{9,0.75}$ **C** $S_{0.36,1.03}$ **D** $S_{5,5}$

3. **a.** What is the image of $(3, 7.5)$ under $S_{4,2}$?
   b. Describe $S_{4,2}$ in words.

4. **a.** Draw $\triangle FLY$ with $F = (-2, 3)$, $L = (3, 1)$, and $Y = (1, 0)$.
   b. Draw its image $\triangle F'L'Y'$ under $S_{3,1}$.
   c. Which component changed, the horizontal or vertical?

In 5 and 6, refer to Example 2.

5. What are the coordinates of the image quadrilateral?

6. Find the image matrix of $ABCD$ under $S_{1,4}$.

7. **True or False** All scale changes produce images that are not similar to their preimages.

8. Describe the scale change with a horizontal shrink of magnitude $\frac{1}{8}$ and a vertical stretch of magnitude 2
   a. in $f(x)$ notation.   **b.** in mapping notation.   **c.** as a matrix.

9. **Fill in the Blanks** The scale change $S_{8,8}$ can be thought of as the ___?___ change identified as ___?___.

APPLYING THE MATHEMATICS

In 10 and 11, consider this information. Pictures in word processors can usually be resized under both scale changes and size changes. When the picture is selected, a rectangle with small boxes (handles) appears around it, similar to the one at the right. Which of the handles A through H could you move, and how would you need to move it, to apply the given transformation to the preimage at the right?

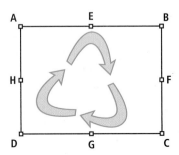

10. $S_{3,\frac{3}{4}}$

11. $S_{1.6}$

12. The transformation $S_{2,1.5}$ is applied to the rectangular fenced-in plot at the right. Coordinates are given for each vertex of the plot.

    a. Find the perimeter P and area A of the pictured fenced-in plot.

    b. Find the coordinates of the vertices of the image of the plot.

    c. Find the perimeter P and area A of the new image.

    (3, 18)   37   (40, 18)

    18   18

    (3, 0)   37   (40, 0)

13. $\triangle BLT$ is represented by the matrix $\begin{bmatrix} 0 & 0 & 10 \\ 0 & 10 & 0 \end{bmatrix}$.

    a. Graph the triangle. Classify $\triangle BLT$ as isosceles or scalene, and as acute, right, or obtuse.

    b. Find the matrix for $\triangle B'L'T'$, the image of $\triangle BLT$ under the scale change represented by $\begin{bmatrix} 4 & 0 \\ 0 & 1.2 \end{bmatrix}$.

    c. Graph $\triangle B'L'T'$. Would you classify $\triangle B'L'T'$ the same way as $\triangle BLT$? If not, what is different?

14. Prove or disprove the following: If $\overline{A'B'}$ is the image of $\overline{AB}$ under a scale change, then $\overline{AB} \parallel \overline{A'B'}$.

15. Consider the matrix equation below that maps $\triangle FOR$ onto $\triangle F'O'R'$.

    $$\begin{bmatrix} 5 & 0 \\ 0 & 7 \end{bmatrix} \begin{bmatrix} a & b & c \\ d & e & f \end{bmatrix} = \begin{bmatrix} 25 & 5 & 20 \\ 14 & 28 & -7 \end{bmatrix}$$

    a. What transformation is applied to $\triangle FOR$?

    b. Find the coordinates of the vertices of $\triangle FOR$.

    c. Find $\dfrac{F'O'}{FO}$ and $\dfrac{O'R'}{OR}$. (*Hint:* Use the formula for distance between two points or a DGS.)

    d. Should the ratios in Part c be the same? Why or why not?

REVIEW

16. Let $R = \begin{bmatrix} 5 & -1 & -1 & 5 \\ 3 & 3 & 6 & 5 \end{bmatrix}$. (**Lesson 4-4**)

    a. What matrix will produce a size change of magnitude $\frac{2}{3}$ on R?

    b. Multiply your matrix in Part a by R. Call your answer R'.

    c. Is the rectangle whose vertices are given by R similar to the rectangle whose vertices are given by R'? Justify your answer.

17. **Fill in the Blanks** Let $T = \begin{bmatrix} 4 & 0 \\ 0 & 4 \end{bmatrix}$, and $V = \begin{bmatrix} \frac{1}{2} & 0 \\ 0 & \frac{1}{2} \end{bmatrix}$.

    (Lessons 4-4, 4-3)

    a. T is the matrix for a size change of magnitude __?__.

    b. V is the matrix for a size change of magnitude __?__.

    c. Compute TV. TV is the matrix for a size change of magnitude __?__.

18. Let $P = \begin{bmatrix} 2 \\ 4 \end{bmatrix}$. (Lessons 4-3, 4-1, 2-7)

    a. Compute $\begin{bmatrix} \frac{1}{2} & 0 \\ 0 & \frac{1}{2} \end{bmatrix} P$, $\begin{bmatrix} 1 & 0 \\ 0 & 1 \end{bmatrix} P$, $\begin{bmatrix} 2 & 0 \\ 0 & 2 \end{bmatrix} P$, and $\begin{bmatrix} 4 & 0 \\ 0 & 4 \end{bmatrix} P$.

    b. If you consider all your answers to Part a as representing points (x, y), what single direct-variation equation do x and y satisfy?

19. **Fill in the Blank** The explicit formula for the nth term of an arithmetic sequence with first term a_1 and constant difference d is __?__. (Lesson 3-8)

20. Suppose a line goes through the points $(-4, -4)$ and $(2, 1)$.

    (Lesson 3-4)

    a. Write an equation for this line in point-slope form.

    b. This line goes through a point $(7, a)$. What is the value of a?

21. a. Name three properties that are preserved under reflections.

    b. A reflection is a type of isometry. What is an isometry? Name two other types of isometries. (**Previous Course**)

QY ANSWER

$\begin{bmatrix} 7 & 0 \\ 0 & 4 \end{bmatrix}$

EXPLORATION

22. Let $ABCD$ be the square defined by the matrix $\begin{bmatrix} 0 & 2 & 2 & 0 \\ 0 & 0 & 2 & 2 \end{bmatrix}$.

    a. Transform $ABCD$ by multiplying its matrix on the left by each of the following matrices (and by some others of your own choice).

    i. $\begin{bmatrix} 3 & 0 \\ 0 & 4 \end{bmatrix}$ ii. $\begin{bmatrix} 3 & 0 \\ 0 & 1 \end{bmatrix}$

    iii. $\begin{bmatrix} 3 & 0 \\ 0 & 2 \end{bmatrix}$ iv. $\begin{bmatrix} 5 & 0 \\ 0 & 5 \end{bmatrix}$

    b. Find the area of each image. Enter your results in a table like the one shown at the right.

    c. What is the connection between the elements a and b of the scale-change matrix $\begin{bmatrix} a & 0 \\ 0 & b \end{bmatrix}$ and the effect the scale change has on area?

Transformation Matrix	Preimage Area	Image Area
$\begin{bmatrix} 3 & 0 \\ 0 & 4 \end{bmatrix}$	4 sq. units	? sq. units
$\begin{bmatrix} 3 & 0 \\ 0 & 1 \end{bmatrix}$	4 sq. units	? sq. units
$\begin{bmatrix} 3 & 0 \\ 0 & 2 \end{bmatrix}$	?	?
$\begin{bmatrix} 5 & 0 \\ 0 & 5 \end{bmatrix}$	?	?

Lesson 4-6

Matrices for Reflections

> ▶ **BIG IDEA** Matrices can represent reflections.

Vocabulary

reflection image of a point over a line

reflecting line, line of reflection

reflection

What Is a Reflection?

Recall from geometry that the **reflection image of a point A over a line m** is:

1. the point A, if A is on m;
2. the point A' such that m is the perpendicular bisector of $\overline{AA'}$, if A is not on m.

The line m is called the **reflecting line** or **line of reflection**.

A **reflection** is a transformation that maps a figure to its reflection image. The figure on the right is the *reflection image* of a drawing and the point A over the line m. This transformation is called r_m, and we write $A' = r_m(A)$.

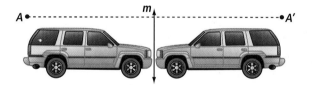

Mental Math

Use matrices A, B, and C below. Tell whether it is possible to calculate each of the following.

$$A = \begin{bmatrix} 2 & 3 \\ -1 & 7 \end{bmatrix}$$

$$B = \begin{bmatrix} 4 & 0.5 \end{bmatrix}$$

$$C = \begin{bmatrix} 6 & 3 \\ 0 & -12 \end{bmatrix}$$

a. $A + B$

b. AB

c. BA

d. BC

Reflection over the y-axis

Suppose that $A = (x, y)$ and $B = (-x, y)$, as shown at the right. Notice that the slope of $\overline{AB}$, like the slope of the x-axis, is zero, which means that $\overline{AB}$ is parallel to the x-axis, and perpendicular to the y-axis. Also, because the y-coordinates are the same and the x-coordinates are opposites, the points are equidistant from the y-axis. So the y-axis is the perpendicular bisector of $\overline{AB}$. This means that $B = (-x, y)$ is the reflection image of $A = (x, y)$ over the y-axis. Reflection over the y-axis can be denoted $r_{y\text{-axis}}$ or r_y. In this book we use r_y.

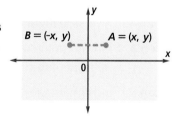

You can write

$$r_y: (x, y) \rightarrow (-x, y) \text{ or } r_y(x, y) = (-x, y).$$

Both are read "the reflection over the y-axis maps point (x, y) onto point $(-x, y)$."

Notice that $\begin{bmatrix} -1 & 0 \\ 0 & 1 \end{bmatrix}\begin{bmatrix} x \\ y \end{bmatrix} = \begin{bmatrix} -1 \cdot x + 0 \cdot y \\ 0 \cdot x + 1 \cdot y \end{bmatrix} = \begin{bmatrix} -x \\ y \end{bmatrix}$.

This means that there is a matrix associated with r_y and proves the following theorem.

Matrix for r_y Theorem

$\begin{bmatrix} -1 & 0 \\ 0 & 1 \end{bmatrix}$ is the matrix for r_y.

Example 1

If $J = (1, 4)$, $K = (2, 4)$, and $L = (1, 7)$, find the image of $\triangle JKL$ under r_y.

Solution Represent r_y and $\triangle JKL$ as matrices and multiply.

$$\overset{r_y}{\begin{bmatrix} -1 & 0 \\ 0 & 1 \end{bmatrix}} \overset{\triangle JKL}{\begin{bmatrix} 1 & 2 & 1 \\ 4 & 4 & 7 \end{bmatrix}} = \overset{\triangle J'K'L'}{\begin{bmatrix} -1 & -2 & -1 \\ 4 & 4 & 7 \end{bmatrix}}$$

The image $\triangle J'K'L'$ has vertices $J' = (-1, 4)$, $K' = (-2, 4)$, and $L' = (-1, 7)$.

Check Use a DGS to plot $\triangle JKL$ and its image $\triangle J'K'L'$. The preimage and image are graphed at the right. It checks.

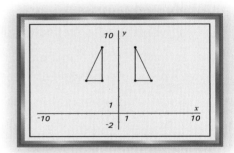

Remembering Transformation Matrices

You have now seen matrices for size changes, scale changes, and one reflection, r_y. You may wonder: Is there a way to generate a matrix for any transformation A so that I do not have to just memorize them? One method of generating transformation matrices that works for reflections, rotations, and scale changes is to use the following two-step algorithm.

Step 1 Find the image of $(1, 0)$ under A and write the coordinates in the first column of the transformation matrix.

Step 2 Find the image of $(0, 1)$ under A and write these coordinates in the second column.

For example, here is a way to remember the matrix for r_y, the reflection over the y-axis.

The image of $(1, 0)$ under r_y is $(-1, 0)$. $\begin{bmatrix} -1 & 0 \\ 0 & 1 \end{bmatrix}$

The image of $(0, 1)$ under r_y is $(0, 1)$.

This general property is called the *Matrix Basis Theorem*.

> ### Matrix Basis Theorem
>
> Suppose A is a transformation represented by a 2×2 matrix. If $A : (1, 0) \rightarrow (x_1, y_1)$ and $A : (0, 1) \rightarrow (x_2, y_2)$, then A has the matrix $\begin{bmatrix} x_1 & x_2 \\ y_1 & y_2 \end{bmatrix}$.

Proof Let the 2×2 transformation matrix for A be $\begin{bmatrix} a & b \\ c & d \end{bmatrix}$, and suppose

$A : (1, 0) \rightarrow (x_1, y_1)$ and $A : (0, 1) \rightarrow (x_2, y_2)$. Then

$$\begin{bmatrix} a & b \\ c & d \end{bmatrix} \begin{bmatrix} 1 & 0 \\ 0 & 1 \end{bmatrix} = \begin{bmatrix} x_1 & x_2 \\ y_1 & y_2 \end{bmatrix}.$$

Multiply the 2×2 matrices on the left side of the equation.

$$\begin{bmatrix} a \cdot 1 + b \cdot 0 & a \cdot 0 + b \cdot 1 \\ c \cdot 1 + d \cdot 0 & c \cdot 0 + d \cdot 1 \end{bmatrix} = \begin{bmatrix} x_1 & x_2 \\ y_1 & y_2 \end{bmatrix}$$

$$\begin{bmatrix} a & b \\ c & d \end{bmatrix} = \begin{bmatrix} x_1 & x_2 \\ y_1 & y_2 \end{bmatrix}$$

Thus, the matrix for the transformation A is $\begin{bmatrix} x_1 & x_2 \\ y_1 & y_2 \end{bmatrix}$.

Reflections over Other Lines

Other transformation matrices let you easily reflect polygons over lines other than the *y*-axis.

> **GUIDED**
>
> ### Example 2
>
> Use the Matrix Basis Theorem to find the transformation matrix for $r_{y=x}$.
>
> **Solution** Find the image of $(1, 0)$ under $r_{y=x}$ and write its coordinates in the first column of the matrix for $r_{y=x}$.
>
> $r_{y=x} (1, 0) = (\underline{\quad?\quad}, \underline{\quad?\quad})$
>
> $\begin{bmatrix} \underline{?} & \underline{?} \\ \underline{?} & \underline{?} \end{bmatrix}$
>
> Find the image of $(0, 1)$ under $r_{y=x}$ and write its coordinates in the second column of the matrix for $r_{y=x}$.
>
> $r_{y=x} (0, 1) = (\underline{\quad?\quad}, \underline{\quad?\quad})$
>
> $\begin{bmatrix} \underline{?} & \underline{?} \\ \underline{?} & \underline{?} \end{bmatrix}$ is the matrix for $r_{y=x}$.

Activity

MATERIALS matrix polygon application

Step 1 Familiarize yourself with how a matrix polygon application lets you draw and transform polygons.

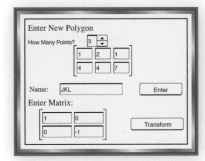

Step 2 Enter the matrix $\begin{bmatrix} 1 & 2 & 1 \\ 4 & 4 & 7 \end{bmatrix}$, which represents $\triangle JKL$.

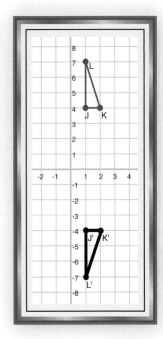

Step 3 Enter the transformation matrix $T1 = \begin{bmatrix} 1 & 0 \\ 0 & -1 \end{bmatrix}$.

Step 4 Graph $\triangle J'K'L'$, the image of $\triangle JKL$ under the transformation represented by $T1$. How do the coordinates of corresponding points on $\triangle J'K'L'$ and $\triangle JKL$ compare? What transformation maps $\triangle JKL$ onto $\triangle J'K'L'$? If you are not sure, enter another polygon and transform it.

Step 5 Enter two more transformation matrices,

$$T2 = \begin{bmatrix} 0 & 1 \\ 1 & 0 \end{bmatrix} \text{ and } T3 = \begin{bmatrix} 0 & -1 \\ -1 & 0 \end{bmatrix}.$$

Repeat Step 4, answering the same questions for each of these transformations.

The Activity verifies that matrix multiplication can be used to reflect $\triangle JKL$ over three lines: the x-axis, the line $y = x$, and the line $y = -x$. Graphs of the triangle and its images are shown below.

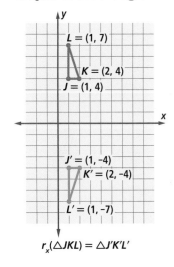

$r_x(\triangle JKL) = \triangle J'K'L'$

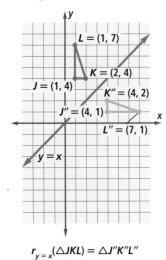

$r_{y=x}(\triangle JKL) = \triangle J''K''L''$

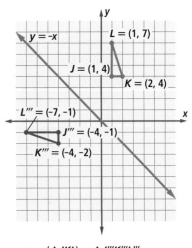

$r_{y=-x}(\triangle JKL) = \triangle J'''K'''L'''$

In general, in mapping notation,

r_x: $(x, y) \rightarrow (x, -y)$, $r_{y=x}$: $(x, y) \rightarrow (y, x)$, and $r_{y=-x}$: $(x, y) \rightarrow (-y, -x)$.

It is easy to prove that the matrices for r_x, $r_{y=x}$, and $r_{y=-x}$ are as stated in the next theorem.

Matrices for r_x, $r_{y=x}$, and $r_{y=-x}$ Theorem

1. $\begin{bmatrix} 1 & 0 \\ 0 & -1 \end{bmatrix}$ is the matrix for r_x.

2. $\begin{bmatrix} 0 & 1 \\ 1 & 0 \end{bmatrix}$ is the matrix for $r_{y=x}$.

3. $\begin{bmatrix} 0 & -1 \\ -1 & 0 \end{bmatrix}$ is the matrix for $r_{y=-x}$.

Proof of 1

$$\begin{bmatrix} 1 & 0 \\ 0 & -1 \end{bmatrix} \begin{bmatrix} x \\ y \end{bmatrix} = \begin{bmatrix} 1 \cdot x + 0 \cdot y \\ 0 \cdot x + -1 \cdot y \end{bmatrix} = \begin{bmatrix} x \\ -y \end{bmatrix}$$

You are asked to prove 2 and 3 in Questions 14 and 15.

STOP QY

You have seen that matrix multiplication can be used to perform geometric transformations such as size changes, scale changes, and reflections. It is important to note how reflections, size changes, and scale changes differ. All reflections preserve shape and size, so reflection images are always congruent to their preimages. All size-change images are similar to their preimages, but only S_1 and S_{-1} yield congruent images. In general, scale-change images are neither congruent nor similar to their preimages.

> ▶ **QY**
>
> Let $A = (1, 3)$, $B = (4, 5)$, and $C = (-2, 6)$. Use matrix multiplication to find the image of $\triangle ABC$ under r_x.

Questions

COVERING THE IDEAS

1. Consider the photograph of a duck and its reflection image shown at the right. For the preimage, assume that the coordinates of the pixel at the tip of the bill are $(-30, 150)$. What are the coordinates of the pixel at the tip of the bill of the image if the reflecting line is the x-axis?

2. Sketch $\triangle DEF$ with $D = (0, 0)$, $E = (0, 3)$ and $F = (4, 0)$. Sketch its reflection image $\triangle D'E'F'$ over the y-axis on the same grid.

3. **Fill in the Blank** Suppose that A' is the reflection image of A over m, and A' is not on line m. Then m is the perpendicular bisector of ___?___.

4. What is the reflection image of a point C over a line m if C is on m?

5. Write in symbols in two ways: "The reflection over the x-axis maps point (x, y) onto point $(x, -y)$."

6. Translate "$r_{y=x}(x, y) = (y, x)$" into words.

Multiple Choice In 7–9, choose the matrix that corresponds to the given reflection.

A $\begin{bmatrix} 1 & 0 \\ 0 & -1 \end{bmatrix}$ B $\begin{bmatrix} -1 & 0 \\ 0 & 1 \end{bmatrix}$ C $\begin{bmatrix} -1 & 0 \\ 0 & -1 \end{bmatrix}$ D $\begin{bmatrix} 0 & 1 \\ 1 & 0 \end{bmatrix}$ E $\begin{bmatrix} 0 & -1 \\ -1 & 0 \end{bmatrix}$

7. $r_{y=x}$ 8. r_x 9. r_y

10. a. Write a matrix for quadrilateral $RUTH$ shown at the right.

    b. Use matrix multiplication to draw $R'U'T'H'$, its reflection image over the line with equation $y = -x$.

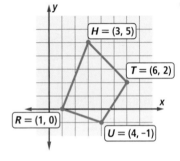

$H = (3, 5)$
$T = (6, 2)$
$R = (1, 0)$
$U = (4, -1)$

True or False In 11 and 12, if the statement is false, provide a counterexample.

11. Reflection images are always congruent to their preimages.

12. Reflection images are always similar to their preimages.

13. **Fill in the Blanks** The matrix equation $\begin{bmatrix} 0 & 1 \\ 1 & 0 \end{bmatrix}\begin{bmatrix} 4 \\ 6 \end{bmatrix} = \begin{bmatrix} 6 \\ 4 \end{bmatrix}$ shows that the reflection image of the point ___?___ over the line with equation ___?___ is the point ___?___.

APPLYING THE MATHEMATICS

14. Prove that $\begin{bmatrix} 0 & 1 \\ 1 & 0 \end{bmatrix}$ is a matrix for the reflection over the line with equation $y = x$.

15. Prove that $\begin{bmatrix} 0 & -1 \\ -1 & 0 \end{bmatrix}$ is a matrix for the reflection over the line with equation $y = -x$.

16. Suppose $P = (x, y)$ and $Q = (y, x)$. Let $R = (a, a)$ be any point on the line with equation $y = x$.

    a. Verify that $PR = QR$. (*Hint:* Use the Pythagorean Distance Formula.)

    b. **Fill in the Blank** Therefore, the line with equation $y = x$ is the ___?___ of $\overline{PQ}$.

    c. What does your answer to Part b mean in terms of reflections?

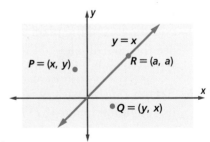

$y = x$
$P = (x, y)$
$R = (a, a)$
$Q = (y, x)$

REVIEW

17. **a.** What is the image of $(2, -3)$ under $S_{0.5, 2}$?
    b. Write a matrix for $S_{0.5, 2}$.
    c. Describe $S_{0.5, 2}$ in words. **(Lesson 4-5)**

18. When does the scale change matrix $\begin{bmatrix} a & 0 \\ 0 & b \end{bmatrix}$ represent a size change? **(Lessons 4-5, 4-4)**

19. Let $P = \begin{bmatrix} -2 \\ 3 \end{bmatrix}$, and $Q = \begin{bmatrix} 4 \\ 4 \end{bmatrix}$. **(Lesson 4-4)**

    a. Compute the distance between P and Q.
    b. Compute the distance between P' and Q' if
    $$P' = \begin{bmatrix} \frac{1}{2} & 0 \\ 0 & \frac{1}{2} \end{bmatrix} P, \text{ and } Q' = \begin{bmatrix} \frac{1}{2} & 0 \\ 0 & \frac{1}{2} \end{bmatrix} Q.$$
    c. Does the transformation given by the matrix $\begin{bmatrix} \frac{1}{2} & 0 \\ 0 & \frac{1}{2} \end{bmatrix}$ preserve distance?

20. **a.** Write the first 6 terms of the sequence defined by
    $$\begin{cases} c_1 = 1 \\ c_n = (-1)^n c_{n-1} - 3, \text{ for integers } n \geq 2 \end{cases} \cdot \text{ (Lesson 3-6)}$$
    b. Rewrite the second line of this recursive definition if the previous term is called c_n.

21. The surface area A of a box with length ℓ, height h, and width w is given by the equation $A = 2\ell h + 2\ell w + 2hw$. **(Lesson 1-7)**
    a. Rewrite this equation if all edges are the same length, x.
    b. Solve the equation in Part a for x.

22. On the television show *The Price is Right*, contestants spinning the Big Wheel have to make sure it turns at least one full revolution. Describe the range in degrees d that the Wheel must turn in order for the spin to count. **(Previous course)**

EXPLORATION

23. **a.** Either by graphing manually, or by using a DGS, find the matrix for the reflection $r_{y=2x}$.
    b. To check this matrix, test at least five points. Make sure that at least two of the points are on the line of reflection.

QY ANSWER

$$\begin{bmatrix} 1 & 0 \\ 0 & -1 \end{bmatrix} \begin{bmatrix} 1 & 4 & -2 \\ 3 & 5 & 6 \end{bmatrix} = $$
$$\begin{bmatrix} 1 & 4 & -2 \\ -3 & -5 & -6 \end{bmatrix};$$
$A' = (1, -3), B' = (4, -5),$ and $C' = (-2, -6)$

Lesson

4-7

Transformations and Matrices

Vocabulary

composite of two
 transformations

▶ **BIG IDEA** The product of two matrices corresponds to the composite of the transformations that the matrices represent.

Properties of 2 × 2 Matrix Multiplication

Multiplication of 2×2 matrices has some of the same properties as multiplication of real numbers, such as closure, associativity, and the existence of an identity.

1. Closure: *The set of 2 × 2 matrices is closed under multiplication.*
   Closure means that when an operation is applied to two elements in a set, the result is an element of the set. From the definition of multiplication of matrices, if you multiply two 2×2 matrices, the result is a 2×2 matrix.

2. Associativity: *Multiplication of 2 × 2 matrices is associative.* For any 2×2 matrices A, B, and C, it can be shown that $(AB)C = A(BC)$. You are asked to prove this in the Questions.

3. Identity: *The matrix* $I = \begin{bmatrix} 1 & 0 \\ 0 & 1 \end{bmatrix}$ *is the identity for multiplication of 2 × 2 matrices.*

   Here is a proof of the Identity Property. Recall that for the real numbers, 1 is the identity for multiplication because for all a, $1 \cdot a = a \cdot 1 = a$. In the matrix version of this property, we need to show that for all 2×2 matrices M, $\begin{bmatrix} 1 & 0 \\ 0 & 1 \end{bmatrix} \cdot M = M$, and $M \cdot \begin{bmatrix} 1 & 0 \\ 0 & 1 \end{bmatrix} = M$.

   Let $M = \begin{bmatrix} a & b \\ c & d \end{bmatrix}$.

   Then $\begin{bmatrix} 1 & 0 \\ 0 & 1 \end{bmatrix}\begin{bmatrix} a & b \\ c & d \end{bmatrix} = \begin{bmatrix} a & b \\ c & d \end{bmatrix}$ and $\begin{bmatrix} a & b \\ c & d \end{bmatrix}\begin{bmatrix} 1 & 0 \\ 0 & 1 \end{bmatrix} = \begin{bmatrix} a & b \\ c & d \end{bmatrix}$.

   Thus, $I = \begin{bmatrix} 1 & 0 \\ 0 & 1 \end{bmatrix}$ is the identity for multiplication of 2×2 matrices.

STOP QY1

Mental Math

Azra has a collection of 35 china elephants. All the elephants are gray, white, or pink. She has twice as many pink elephants as white elephants and half as many gray elephants as white elephants.

a. How many of Azra's elephants are white?

b. How many are gray?

c. How many are pink?

▶ **QY1**

Verify that both matrix products $\begin{bmatrix} 1 & 0 \\ 0 & 1 \end{bmatrix}\begin{bmatrix} a & b \\ c & d \end{bmatrix}$ and $\begin{bmatrix} a & b \\ c & d \end{bmatrix}\begin{bmatrix} 1 & 0 \\ 0 & 1 \end{bmatrix}$ equal $\begin{bmatrix} a & b \\ c & d \end{bmatrix}$.

Notice that both multiplications were needed in the proof because multiplication of 2×2 matrices is not always commutative.

Multiplying the Matrices of Two Transformations

The product of two 2×2 transformation matrices yields a third 2×2 matrix that also represents a transformation.

Activity

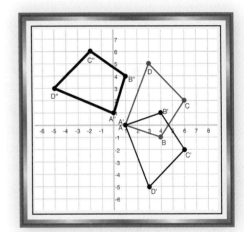

MATERIALS matrix polygon application

Step 1 Use a matrix polygon application to draw quadrilateral $ABCD$ with $A = (1, 0)$, $B = (4, -1)$, $C = (6, 2)$, and $D = (3, 5)$.

Step 2 Reflect $ABCD$ over the x-axis and call its image $A'B'C'D'$.

Step 3 Reflect $A'B'C'D'$ over the line with equation $y = x$ and call its image $A''B''C''D''$.

Step 4 Compare the coordinates of the vertices of the final image $A''B''C''D''$ to the coordinates of the vertices of your original preimage $ABCD$. What single transformation could you apply to $ABCD$ to get the image $A''B''C''D''$?

Step 5 Multiply the matrix for r_x on the left by the matrix for $r_{y=x}$. Call the product M.

Step 6 Clear the screen and redraw $ABCD$. Then draw the image of $ABCD$ after applying the transformation defined by M. Compare your results to the results of Step 3. What do you notice?

Composites of Transformations

We call the final quadrilateral $A''B''C''D''$ from the Activity the image of $ABCD$ under the *composite* of the reflections r_x and $r_{y=x}$. In general, any two transformations can be composed.

Definition of Composite of Transformations

Suppose transformation T_1 maps figure G onto figure G', and transformation T_2 maps figure G' onto figure G''. The transformation that maps G onto G'' is called the **composite** of T_1 and T_2, written $T_2 \circ T_1$.

The symbol $\circ$ is read "following." In the Activity, r_x came first and then $r_{y=x}$, so we write $r_{y=x} \circ r_x$ and say "$r_{y=x}$ following r_x" or "the composite of r_x and $r_{y=x}$."

The Composite of $r_{y=x}$ and r_x

To describe $r_{y=x} \circ r_x$ as one transformation, consider the results of the previous Activity. Quadrilateral $ABCD$ is the preimage, $A'B'C'D'$ is the first image, and $A''B''C''D''$ is the final image. In Step 6 of the Activity, you plotted only the preimage $ABCD$ and the final image $A''B''C''D''$ to get a graph like the one at the right below.

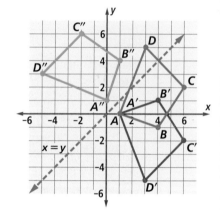

The graph shows that the composite is a *rotation* $90°$ counterclockwise about $(0, 0)$. We denote this rotation by R_{90}. This rotation is the composite of the two reflections: $R_{90} = r_{y=x} \circ r_x$.

In geometry, you should have encountered the Two-Reflection Theorem for Rotations: The composite of two reflections $r_m \circ r_\ell$ over intersecting lines ℓ and m is a rotation whose center is the intersection of ℓ and m. The magnitude of rotation is twice the measure of an angle formed by ℓ and m, measured from ℓ to m. Because the x-axis and the line with equation $y = x$ intersect at the origin at a $45°$ angle, this theorem explains why the composite $r_{y=x} \circ r_x$ is a rotation about the origin with magnitude $90°$.

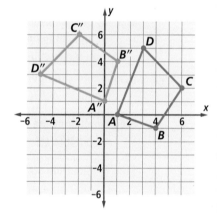

Matrices and Composites of Transformations

How can you find the matrix associated with R_{90}? In Steps 5 and 6 of the Activity, you multiplied the matrices for the two reflections, then multiplied the matrix for $ABCD$ on the left by the product. The result was the matrix for $A''B''C''D''$. In symbols, we can show this as follows.

$$\overset{r_{y=x} \ \circ \ r_x}{\left(\begin{bmatrix} 0 & 1 \\ 1 & 0 \end{bmatrix} \begin{bmatrix} 1 & 0 \\ 0 & -1 \end{bmatrix} \right)} \overset{(A \ B \ C \ D)}{\begin{bmatrix} 1 & 4 & 6 & 3 \\ 0 & -1 & 2 & 5 \end{bmatrix}} =$$

$$\overset{R_{90}}{\begin{bmatrix} 0 & -1 \\ 1 & 0 \end{bmatrix}} \cdot \overset{(A \ B \ C \ D)}{\begin{bmatrix} 1 & 4 & 6 & 3 \\ 0 & -1 & 2 & 5 \end{bmatrix}} = \overset{A'' \ B'' \ C'' \ D''}{\begin{bmatrix} 0 & 1 & -2 & -5 \\ 1 & 4 & 6 & 3 \end{bmatrix}}$$

This tells us that $\begin{bmatrix} 0 & -1 \\ 1 & 0 \end{bmatrix}$ is a matrix for R_{90}. The general idea is summarized in the following theorem.

Matrices and Composites Theorem

If $M1$ is the matrix for transformation T_1, and $M2$ is the matrix for transformation T_2, then $M2M1$ is the matrix for $T_2 \circ T_1$.

 QY2

▶ **QY2**

If $M1$ represents $r_{y=x}$ and $M2$ represents r_x, does $M1M2 = M2M1$? Why or why not?

Example

Golfers strive to achieve the elusive perfect swing. Assume that when looking at a right-handed golfer from the front, as a golfer begins the swing, the golf club rotates clockwise about a point near the tip of the shaft. The club starts from a vertical position at setup, then moves along a 90° arc until it is horizontal at some point during the backswing. The rotation is R_{-90}. For the left-handed golfer, the rotation will be counterclockwise, or R_{90}. The preimage golf club and the image golf club are pictured in the graph below at the right. Three points on the original golf club are $P = (1, 0)$, $G = (1, -6)$, and $A = (2, -6)$. Use matrix multiplication to find the image of PGA under R_{-90}.

Solution A counterclockwise rotation of 90° is a composite of two reflections given by $R_{90} = r_{y=x} \circ r_x$.

To find R_{-90}, a rotation 90° clockwise about $(0, 0)$, you need to measure the angle formed in the opposite direction, that is, the angle formed by the line with equation $y = x$ and the x-axis, measured from $y = x$ to the x-axis. So reverse the order of composition: $R_{-90} = r_x \circ r_{y=x}$.

So the matrix for R_{-90} is
$$\overset{r_x}{\begin{bmatrix} 1 & 0 \\ 0 & -1 \end{bmatrix}} \overset{r_{y=x}}{\begin{bmatrix} 0 & 1 \\ 1 & 0 \end{bmatrix}} = \overset{R_{-90}}{\begin{bmatrix} 0 & 1 \\ -1 & 0 \end{bmatrix}}.$$

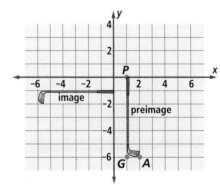

To find the image of golf club PGA, multiply its matrix by the matrix for R_{-90}.

$$R_{-90}(PGA) = \overset{R_{-90}}{\begin{bmatrix} 0 & 1 \\ -1 & 0 \end{bmatrix}} \overset{(P\ G\ A)}{\begin{bmatrix} 1 & 1 & 2 \\ 0 & -6 & -6 \end{bmatrix}} = \overset{P'\ G'\ A'}{\begin{bmatrix} 0 & -6 & -6 \\ -1 & -1 & -2 \end{bmatrix}}$$

Questions

COVERING THE IDEAS

1. **a.** When a 2×2 matrix is multiplied by a 2×2 matrix, what are the dimensions of the product matrix?

   b. What property of 2×2 matrix multiplication does this demonstrate?

2. What is the *multiplicative identity* for 2×2 matrices?

3. Write this product as a single matrix: $\begin{bmatrix} 1 & 0 \\ 0 & 1 \end{bmatrix} \cdot \begin{bmatrix} 5 & 2 \\ \sqrt{8} & \pi \end{bmatrix}$.

4. **Fill in the Blank** The identity transformation maps each point of a preimage onto ___?___.

5. In general, multiplication of 2×2 matrices is not commutative, but sometimes it is.

   a. Suppose I is the 2×2 identity matrix and A is any other 2×2 matrix of your choosing. Does $IA = AI$?

   b. Pick any 2×2 size change matrix A and any 2×2 matrix B other than the identity matrix. Does $AB = BA$?

   c. Give an example of two 2×2 matrices A and B such that $AB \ne BA$.

Multiple Choice In 6–10, identify which property is illustrated by the statement.

A closure   B associativity   C existence of an identity

D commutativity in some cases   E noncommutativity

6. $\begin{bmatrix} 2 & 3 \\ 7 & 4 \end{bmatrix}\begin{bmatrix} 1 & 0 \\ 0 & 1 \end{bmatrix} = \begin{bmatrix} 2 & 3 \\ 7 & 4 \end{bmatrix}$

7. $\begin{bmatrix} 2 & 3 \\ 7 & 4 \end{bmatrix}\left(\begin{bmatrix} 0 & 3 \\ 8 & 6 \end{bmatrix}\begin{bmatrix} 5 & 9 \\ 1 & 2 \end{bmatrix}\right) = \left(\begin{bmatrix} 2 & 3 \\ 7 & 4 \end{bmatrix}\begin{bmatrix} 0 & 3 \\ 8 & 6 \end{bmatrix}\right)\begin{bmatrix} 5 & 9 \\ 1 & 2 \end{bmatrix}$

8. $\begin{bmatrix} 2 & 3 \\ 7 & 4 \end{bmatrix}\begin{bmatrix} 0 & 3 \\ 8 & 6 \end{bmatrix} \ne \begin{bmatrix} 0 & 3 \\ 8 & 6 \end{bmatrix}\begin{bmatrix} 2 & 3 \\ 7 & 4 \end{bmatrix}$

9. The product of any 2×2 matrix multiplication is a 2×2 matrix.

10. $\begin{bmatrix} 3 & 0 \\ 0 & 3 \end{bmatrix}\begin{bmatrix} 3 & 5 \\ 7 & 8 \end{bmatrix} = \begin{bmatrix} 3 & 5 \\ 7 & 8 \end{bmatrix}\begin{bmatrix} 3 & 0 \\ 0 & 3 \end{bmatrix}$

11. Noel thought that by applying the Associative Property, the following matrix equation would be true.

$$\begin{bmatrix} 0 & 1 \\ 1 & 0 \end{bmatrix}\left(\begin{bmatrix} 1 & 0 \\ 0 & -1 \end{bmatrix}\begin{bmatrix} 1 & 1 \\ 2 & 6 \end{bmatrix}\right) = \left(\begin{bmatrix} 0 & 1 \\ 1 & 0 \end{bmatrix}\begin{bmatrix} 1 & 1 \\ 2 & 6 \end{bmatrix}\right)\begin{bmatrix} 1 & 0 \\ 0 & -1 \end{bmatrix}$$

However, when checking the matrix multiplication with a calculator, the left side of the equation did not equal the right side of the equation. Explain the mistake in Noel's logic.

12. If T_1 and T_2 are two transformations, what does $T_1 \circ T_2$ mean?

13. Does $r_{y=x} \circ r_x = r_x \circ r_{y=x}$? Explain why or why not.

14. **Fill in the Blanks** R_{-90} represents a rotation of ___?___ degrees about ___?___ in a ___?___ direction.

15. Refer to the Example. As a right-handed golfer takes the club back the golfer cocks his or her wrist at the top of the backswing as shown in the picture. When the golfer is viewed from the back, the club appears to have rotated 270 degrees (R_{270}) from vertical.

   a. Use the Matrix Basis Theorem to find the matrix for R_{270}.
   b. Find the image of the golf club under a $270°$ counterclockwise rotation.
   c. Compare the results of Part b to the results of the Example. Explain your findings.

APPLYING THE MATHEMATICS

16. Let $G = (-1, 4)$, $R = (-3, 6)$, $E = (2, 6)$, $A = (2, -1)$, and $T = (-1, -4)$.
   a. Write the matrix for *GREAT*.
   b. Write the matrix for $R_{90}(GREAT)$.
   c. Write the matrix for $r_{y=x} \circ R_{90}(GREAT)$.
   d. What single transformation is equal to $r_{y=x} \circ R_{90}$?

17. Refer to the Activity and quadrilateral *ABCD* with $A = (1, 0)$, $B = (4, -1)$, $C = (6, 2)$, and $D = (3, 5)$.
   a. Find the image of *ABCD* under the transformation $S_{\frac{1}{2}} \circ r_y$.
   b. Is the image congruent to *ABCD*?
   c. Is the image similar to *ABCD*?

18. a. To what single transformation is each of the following equivalent?
   i. $r_x \circ r_x$ ii. $r_y \circ r_y$ iii. $r_{y=x} \circ r_{y=x}$
   b. Explain the geometric meaning of the results of Part a.

19. Consider the matrix multiplications shown at the right. Laura hypothesizes that reversing the order of a 2×2 matrix multiplication always results in switching the diagonal elements of the original 2×2 product matrix. Let $A = \begin{bmatrix} a & b \\ c & d \end{bmatrix}$ and $B = \begin{bmatrix} e & f \\ g & h \end{bmatrix}$.

$$\begin{bmatrix} 5 & 2 \\ 8 & 3 \end{bmatrix}\begin{bmatrix} 3 & 4 \\ -2 & 5 \end{bmatrix} = \begin{bmatrix} 11 & 30 \\ 18 & 47 \end{bmatrix}$$

$$\begin{bmatrix} 3 & 4 \\ -2 & 5 \end{bmatrix}\begin{bmatrix} 5 & 2 \\ 8 & 3 \end{bmatrix} = \begin{bmatrix} 47 & 18 \\ 30 & 11 \end{bmatrix}$$

   a. Find *AB* and *BA*.
   b. Is Laura's conjecture correct? Why or why not?

20. a. Use a CAS to show that
$$\left(\begin{bmatrix} a & b \\ c & d \end{bmatrix}\begin{bmatrix} e & f \\ g & h \end{bmatrix}\right)\begin{bmatrix} i & j \\ k & \ell \end{bmatrix} = \begin{bmatrix} a & b \\ c & d \end{bmatrix}\left(\begin{bmatrix} e & f \\ g & h \end{bmatrix}\begin{bmatrix} i & j \\ k & \ell \end{bmatrix}\right).$$
   b. What property does Part a demonstrate?

REVIEW

21. Name three points that are their own images under the transformation represented by $\begin{bmatrix} 0 & 1 \\ 1 & 0 \end{bmatrix}$. (**Lesson 4-6**)

22. Suppose T is a transformation with T: $(1, 0) \rightarrow (5, 0)$, and T: $(0, 1) \rightarrow \left(0, \frac{4}{5}\right)$. (**Lessons 4-6, 4-5**)

    a. Find a matrix that represents T.

    b. What kind of transformation is T?

23. Find a matrix for the size change that maps the triangle $\begin{bmatrix} -6 & 4 & 7 \\ 2 & 2 & 8 \end{bmatrix}$ onto the triangle $\begin{bmatrix} -9 & 6 & 10.5 \\ 3 & 3 & 12 \end{bmatrix}$. (**Lesson 4-4**)

24. a. Suppose a is a real number. For what values of a is there a unique line containing $(a, 0)$ and $(0, a)$?

    b. If a is such that there is a unique line, find an equation satisfied by every point (x, y) on it. (**Lesson 3-4**)

25. There are two prices for used movies at the Lights, Camera, Action! video store. The regular price is $15, but some are on sale for $9. Let s represent the number of movies Bob bought on sale, and r represent the number of movies he bought at regular price. (**Lesson 3-2**)

    a. Suppose Bob spent $78 on movies. Write an equation relating s, r, and the money Bob spent.

    b. Graph the equation from Part a and list all possible combinations of regular and sale movies Bob could have bought.

EXPLORATION

26. Test whether the properties of 2×2 matrix multiplication extend to 3×3 matrices.

    a. Explain whether you think that the closure property holds for 3×3 matrix multiplication.

    b. There is also an identity matrix M for 3×3 matrix multiplication. Use your knowledge of matrices to guess the values of the elements of matrix M. Test your identity matrix M by multiplying it by a general 3×3 matrix.

    c. Test the commutative property for 3×3 matrix multiplication with a specific example.

    d. Test the associative property for 3×3 matrix multiplication with a specific example.

QY ANSWERS

1. $\begin{bmatrix} 1 \cdot a + 0 \cdot c & 1 \cdot b + 0 \cdot d \\ 0 \cdot a + 1 \cdot c & 0 \cdot b + 1 \cdot d \end{bmatrix}$

$= \begin{bmatrix} a & b \\ c & d \end{bmatrix}$, and

$\begin{bmatrix} a \cdot 1 + b \cdot 0 & a \cdot 0 + b \cdot 1 \\ c \cdot 1 + d \cdot 0 & c \cdot 0 + d \cdot 1 \end{bmatrix}$

$= \begin{bmatrix} a & b \\ c & d \end{bmatrix}$

2. No; $M1M2 =$

$\begin{bmatrix} 0 & 1 \\ 1 & 0 \end{bmatrix}\begin{bmatrix} 1 & 0 \\ 0 & -1 \end{bmatrix} = \begin{bmatrix} 0 & -1 \\ 1 & 0 \end{bmatrix}$

and $M2M1 =$

$\begin{bmatrix} 1 & 0 \\ 0 & -1 \end{bmatrix}\begin{bmatrix} 0 & 1 \\ 1 & 0 \end{bmatrix} = \begin{bmatrix} 0 & 1 \\ -1 & 0 \end{bmatrix}$.

The matrices are not equal.

Lesson

4-8

Matrices for Rotations

Vocabulary

rotation

center of a rotation

magnitude of a
    rotation

▶ **BIG IDEA** Matrices can represent rotations about the origin.

Recall from geometry and Lesson 4-7 that a turn, or **rotation**, is described by its *center* and *magnitude*. The **center of a rotation** is a point that coincides with its image. The **magnitude of a rotation** is positive if the rotation is counterclockwise, while the magnitude is negative if the rotation is clockwise. The rotation of magnitude $x°$ around the origin is denoted R_x.

Note that we denote a rotation with a capital R, while we identify a reflection with a lowercase r.

Mental Math

Solve:

a. $V = \frac{1}{3}\pi r^2 h$ for h.

b. $h = 4.9t^2$ for t.

c. $M = \frac{kwt^2}{d}$ for k.

d. $y = 4x + 12$ for x.

The Composite of Two Rotations

Rotations often occur one after the other, as when a spinner is spun twice.

Activity

MATERIALS tracing paper, protractor
**At the right, figures I, II, and III are images of each other under rotations
with center O.**

Step 1 Trace the figures at the right.

Step 2 Measure ∠AOE to determine the magnitude of the rotation
that maps figure I onto figure II.

Step 3 Measure ∠EOI to determine the magnitude of the rotation
that maps figure II onto figure III.

Step 4 What is the magnitude of the rotation that maps figure I onto
figure III?

Step 5 What is the magnitude of the rotation that maps figure III onto
figure I?

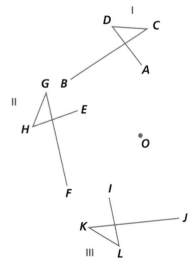

The relationships in the Activity result from a fundamental property of rotations, which itself is a consequence of the Angle Addition Postulate in geometry.

> ### Composite of Rotations Theorem
>
> A rotation of $b°$ following a rotation of $a°$ with the same center results in a rotation of $(a + b)°$. In symbols, $R_b \circ R_a = R_{a+b}$.

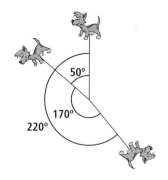

Matrices for Rotations

Rotations centered at the origin are the only rotations that can be represented by 2×2 matrices. This is because any transformation represented by a 2×2 matrix maps $(0, 0)$ onto itself: $\begin{bmatrix} a & b \\ c & d \end{bmatrix}\begin{bmatrix} 0 \\ 0 \end{bmatrix} = \begin{bmatrix} 0 \\ 0 \end{bmatrix}$.

As you will see later in this course, you can use trigonometry to develop a rotation matrix of any magnitude. However, some rotations have matrices whose elements can be derived without trigonometry. For example, in the previous lesson you derived a matrix for R_{90}.

> ### Matrix for R_{90} Theorem
>
> $\begin{bmatrix} 0 & -1 \\ 1 & 0 \end{bmatrix}$ is the matrix for R_{90}.

You can verify this theorem using the Matrix Basis Theorem.

The image of $(1, 0)$ under R_{90} is $(0, 1)$.

The image of $(0, 1)$ under R_{90} is $(-1, 0)$.

So, the matrix for R_{90} is $\begin{bmatrix} 0 & -1 \\ 1 & 0 \end{bmatrix}$.

Using matrices to represent rotations lets you describe rotation images algebraically. For example, for R_{90},

$$\begin{bmatrix} 0 & -1 \\ 1 & 0 \end{bmatrix}\begin{bmatrix} x \\ y \end{bmatrix} = \begin{bmatrix} 0 \cdot x + -1 \cdot y \\ 1 \cdot x + 0 \cdot y \end{bmatrix} = \begin{bmatrix} -y \\ x \end{bmatrix}.$$

Thus, $R_{90}(x, y) = (-y, x)$. This is an important result you should memorize.

STOP QY1

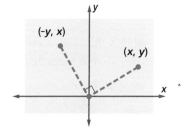

> ▶ QY1
>
> What is the image of $(3, 2)$ under R_{90}?

Matrices for Rotations of Multiples of 90°

The rotations R_{180} and R_{270} are especially important, and their matrices can be computed straightforwardly from the matrix for R_{90}.

Example

Find the matrix for

a. R_{180}. **b.** R_{270}.

Solution

a. Because $90° + 90° = 180°$, a rotation of $180°$ can be considered as the composite of a $90°$ rotation following a first $90°$ rotation. That is,

$R_{180} = R_{90} \circ R_{90}$.

Use matrix multiplication:

The matrix for $R_{180} = R_{90} \circ R_{90}$ is $\begin{bmatrix} 0 & -1 \\ 1 & 0 \end{bmatrix}\begin{bmatrix} 0 & -1 \\ 1 & 0 \end{bmatrix} = \begin{bmatrix} \underline{?} & \underline{?} \\ \underline{?} & \underline{?} \end{bmatrix}$.

Another way to write the matrix for R_{180} is $\begin{bmatrix} 0 & -1 \\ 1 & 0 \end{bmatrix}\begin{bmatrix} 0 & -1 \\ 1 & 0 \end{bmatrix} = \begin{bmatrix} 0 & -1 \\ 1 & 0 \end{bmatrix}^2$.

b. $R_{270} = R_{90} \circ R_{\underline{?}}$.

So, the matrix for R_{270} is $\begin{bmatrix} 0 & -1 \\ 1 & 0 \end{bmatrix}\begin{bmatrix} \underline{?} & \underline{?} \\ \underline{?} & \underline{?} \end{bmatrix} = \begin{bmatrix} \underline{?} & \underline{?} \\ \underline{?} & \underline{?} \end{bmatrix}$.

 QY2

> ▶ **QY2**
>
> What is the matrix for R_{-90}?

Questions

COVERING THE IDEAS

1. A rotation with negative magnitude represents a turn in which direction?

2. **a.** Verify that $\begin{bmatrix} a & b \\ c & d \end{bmatrix}\begin{bmatrix} 0 \\ 0 \end{bmatrix} = \begin{bmatrix} 0 \\ 0 \end{bmatrix}$.

   b. What does this computation prove about representing rotations with 2×2 matrices?

3. **a. Fill in the Blank** A rotation of 45° followed by a rotation of 90° is equivalent to a rotation of ___?___.

   b. Write Part a using R notation.

4. A rotation of –120° is the same as a rotation of what positive magnitude?

5. Write the matrices for R_{90}, R_{180}, R_{270}, and R_{360}.

6. Find the image of (x, y) under R_{270}.

7. Find the image of the square $\begin{bmatrix} 2 & 8 & 8 & 2 \\ 3 & 3 & 9 & 9 \end{bmatrix}$ under R_{180}.

APPLYING THE MATHEMATICS

8. Let $A = (6, 2)$, $B = R_{90}(A)$, $C = R_{180}(A)$, and $D = R_{270}(A)$. Prove that $ABCD$ is a square by showing that $AB = BC = CD = AD$ and $AO = BO = CO = DO$, where $O = (0, 0)$.

9. **a.** Calculate a matrix for $R_{90} \circ r_x$.

   b. Calculate a matrix for $r_x \circ R_{90}$.

   c. Your answers to Parts a and b should be different. What property of matrix multiplication does this illustrate? What property of transformations does this illustrate?

10. The matrix for R_{60} is approximately $\begin{bmatrix} 0.5 & -0.866 \\ 0.866 & 0.5 \end{bmatrix}$. Use the fact that a 150° rotation is the result of a 60° rotation followed by a 90° rotation to find an approximate matrix for R_{150}.

11. **a.** Find a matrix for $R_{90} \circ S_{0.5,2} \circ R_{-90} \circ S_{2,0.5}$.

    b. What does this transformation represent geometrically?

REVIEW

12. **a.** By what matrix can you multiply the size change matrix $\begin{bmatrix} 8 & 0 \\ 0 & 8 \end{bmatrix}$ to get the identity matrix?

    b. What does your answer to Part a represent geometrically? **(Lessons 4-7, 4-4)**

13. **True or False** The matrix for a transformation can be determined by finding the image of the point $(1, 1)$ under the transformation. **(Lesson 4-6)**

14. Consider the triangle $\triangle ELK$ represented by the matrix
$\begin{bmatrix} 0 & 8 & 9 \\ -2 & -4 & -6 \end{bmatrix}$, and the triangle $\triangle RAM$ represented by the matrix
$\begin{bmatrix} 0 & 4 & 4.5 \\ -3 & -6 & -9 \end{bmatrix}$. (Lesson 4-5)

a. Find a matrix for the scale change that maps $\triangle ELK$ onto $\triangle RAM$.

b. Find a matrix for the scale change that maps $\triangle RAM$ onto $\triangle ELK$.

15. In 2006, about 45 million people in the U.S. did not have health insurance. The four states with the highest percent of uninsured residents were Texas (24.1%), New Mexico (21.0%), Florida (20.3%), and Arizona (19.0%). Their populations in 2006 were as follows: Texas 23,508,000; New Mexico 1,955,000; Florida 18,090,000; and Arizona 6,166,000. Use matrix multiplication to determine how many people in these four states had no health insurance. (Lessons 4-3, 4-1)

The population of New Mexico increased by 20.1% between 1990 and 2000.

EXPLORATION

16. You can find a matrix for R_{45} in the following way. Suppose the matrix is $\begin{bmatrix} a & b \\ c & d \end{bmatrix}$. Then, since $R_{45} \circ R_{45} = R_{90}$,

$$\begin{bmatrix} a & b \\ c & d \end{bmatrix}\begin{bmatrix} a & b \\ c & d \end{bmatrix} = \begin{bmatrix} 0 & -1 \\ 1 & 0 \end{bmatrix}.$$

a. Multiply the matrices on the left side of the equation to obtain one matrix.

b. Equate the elements of the product matrix and the matrix for R_{90}. You will have four equations in a, b, c, and d.

c. From the equations, explain why $a^2 = d^2$, but $a \neq -d$.

d. From the equations, explain why $b = -c$.

e. Either by hand or with a CAS, find the two possible matrices $\begin{bmatrix} a & b \\ c & d \end{bmatrix}$.

f. One solution is the matrix for R_{45}. The other solution is the matrix for what transformation?

QY ANSWERS

1. $(-2, 3)$

2. $\begin{bmatrix} 0 & 1 \\ -1 & 0 \end{bmatrix}$

4-9 Rotations and Perpendicular Lines

▶ **BIG IDEA** If a line has slope m, any line perpendicular to it has slope $-\frac{1}{m}$.

If a rotation of magnitude 90° is applied to a line, then the image line is perpendicular to the preimage line. This is true regardless of the center of the rotation, and it explains a very nice relationship between the slopes of perpendicular lines.

Mental Math

Pete has $9.50 in nickels and dimes. How many dimes does he have if he has

a. 50 nickels?

b. 150 nickels?

c. n nickels?

Activity

MATERIALS matrix polygon application

Consider $\overline{AB}$ with endpoints $A = (-8, 3)$ and $B = (5, -1)$.

Step 1 Use a matrix polygon application to enter W, the 2 × 2 matrix of endpoints of $\overline{AB}$. Graph $\overline{AB}$.

Step 2 Enter the R_{90} transformation matrix as R.

Step 3 Calculate $R \cdot W$ to graph $\overline{A'B'}$, the image of $\overline{AB}$ under R_{90}. Describe any relationships you notice between these two line segments.

Step 4 Calculate the slope of $\overline{AB}$ and the slope of $\overline{A'B'}$. How is one slope related to the other?

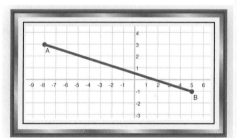

The results of the Activity can be generalized to prove the following theorem.

Perpendicular Lines and Slopes Theorem (Part 1)

If two lines with slopes m_1 and m_2 are perpendicular, then $m_1 \cdot m_2 = -1$.

Proof You are given two perpendicular lines with slopes m_1 and m_2 and need to show that the product $m_1 m_2$ is –1. The given lines either contain the origin or they are parallel to lines that contain the origin. Here we prove the theorem for two lines through the origin. In the next lesson you will show that this property holds true for perpendicular lines elsewhere.

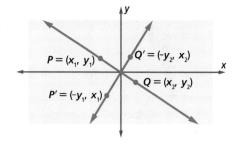

Let $P = (x_1, y_1)$ and $Q = (x_2, y_2)$ be two points on a line $\overleftrightarrow{PQ}$ that contains the origin.

From Lesson 4-8, $R_{90}(P) = (-y_1, x_1) = P'$, and $R_{90}(Q) = (-y_2, x_2) = Q'$. $\overleftrightarrow{P'Q'}$ is perpendicular to $\overleftrightarrow{PQ}$ since $\overleftrightarrow{P'Q'}$ is the image of $\overleftrightarrow{PQ}$ under a rotation of magnitude $90°$.

Now, let the slope of the preimage be m_1 and the slope of the image be m_2.

$$m_1 = \text{slope of } \overleftrightarrow{PQ} = \frac{y_2 - y_1}{x_2 - x_1}$$

$$m_2 = \text{slope of } \overleftrightarrow{P'Q'} = \frac{x_2 - x_1}{-y_2 - (-y_1)} = \frac{x_2 - x_1}{-(y_2 - y_1)} = -\frac{x_2 - x_1}{y_2 - y_1}$$

The product of the slopes is

$$m_1 \cdot m_2 = \frac{y_2 - y_1}{x_2 - x_1} \cdot \left(-\frac{x_2 - x_1}{y_2 - y_1}\right) = -1.$$

This proves the theorem.

STOP **QY1**

Do you see why the product of the slopes is –1? The rotation of $90°$ switches x- and y-coordinates and changes the first coordinate to its opposite. In the slope formula, this switches numerator and denominator and multiplies the denominator by –1. So the slope $\frac{a}{b}$ becomes $\frac{-b}{a}$, and the product is –1.

> ▶ **QY1**
>
> Two perpendicular lines have slopes $\frac{1}{2}$ and s. What is s?

Example

Line n contains $(4, -3)$ and is perpendicular to line ℓ, whose equation is $y = \frac{2}{5}x + 3$. Find an equation for line n.

Solution The slope of line ℓ is ___?___. So by the theorem on the previous page, the slope of line n is ___?___.

Since line n contains $(4, -3)$, an equation for line n in point-slope form is ___?___. $y + 3 = -\frac{5}{2}(x - 4)$

Consider the converse of the preceding theorem. Suppose line ℓ_1 has slope m_1 and line ℓ_2 has slope m_2, and $m_1 m_2 = -1$. Are ℓ_1 and ℓ_2 always perpendicular? The answer is yes.

Proof We want to show that ℓ_1 and ℓ_2 are perpendicular. Think of a third line ℓ_3 with slope m_3 in the same plane as ℓ_1 and ℓ_2, and with $\ell_3 \perp \ell_1$.

Then $m_1 m_3 = -1$. So, $m_1 m_3 = m_1 m_2$. Thus, $m_3 = m_2$, so ℓ_3 and ℓ_2 have the same slope. So, $\ell_3 \parallel \ell_2$. But $\ell_1 \perp \ell_3$. Now use the theorem from geometry that if a line is perpendicular to one of two parallel lines, it must be perpendicular to the other. So, $\ell_1 \perp \ell_2$. This proves the converse of the previous theorem.

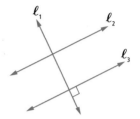

Perpendicular Lines and Slopes Theorem (Part 2)

If two lines have slopes m_1 and m_2 and $m_1 m_2 = -1$, then the lines are perpendicular.

Because the original theorem and its converse are both true, we can conclude the following biconditional.

Perpendicular Lines and Slopes Theorem

Two lines with slopes m_1 and m_2 are perpendicular if and only if $m_1 m_2 = -1$.

STOP **QY2**

> ▶ QY2
>
> Are the lines with slopes 5 and –0.2 perpendicular? Justify your answer.

Questions

COVERING THE IDEAS

1. Write a 2 × 2 matrix that represents the line containing the points (x_1, y_1) and (x_2, y_2).

2. Let $\overleftrightarrow{AB}$ contain points $A = (4, 7)$ and $B = (-3, 5)$.
   a. Find two points on the image of $\overleftrightarrow{AB}$ under R_{90}.
   b. Graph $\overleftrightarrow{AB}$ and its image $\overleftrightarrow{A'B'}$.
   c. Find the slopes of $\overleftrightarrow{AB}$ and of $\overleftrightarrow{A'B'}$.
   d. What is the product of the slopes?

In 3–5, indicate whether each statement is *true* or *false*.

3. If two lines have slopes m_1 and m_2, and $m_1 m_2 = -1$, then the lines are perpendicular.

4. If two lines are perpendicular, and they have slopes m_1 and m_2, then $m_1 m_2 = -1$.

5. Suppose m_1 and m_2 are the slopes of two perpendicular lines. Then m_1 is the reciprocal of m_2.

6. A line has slope 0.25. What is the slope of a line
   a. parallel to this line?
   b. perpendicular to this line?

7. Find an equation of the line through (6, 1) and perpendicular to the line with equation $y = \frac{3}{5}x - 2$.

8. Find an equation of the line perpendicular to the line with equation $y - 2 = -5(x - 1)$ passing through the point (-10, 7).

APPLYING THE MATHEMATICS

9. Use the graph at the right. Find an equation for the line perpendicular to $\overline{BC}$ through point A.

10. A line ℓ_1 contains the points (5, -3) and (9, -1).
    a. Graph ℓ_1 and write a matrix $M1$ representing this line.
    b. Find $M2 = R_{90}(M1)$ and graph the line ℓ_2 that $M2$ represents. Are ℓ_1 and ℓ_2 perpendicular?
    c. Let matrix $M3 = R_{90}(M2)$ represent line ℓ_3. What is the relationship between ℓ_1 and ℓ_3? Explain your answer.

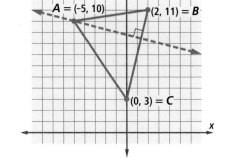

11. Why do the statements of the theorems in this lesson apply only to lines with nonzero slopes?

12. **Multiple Choice** What is the slope of a line perpendicular to the line with equation $x = 5$?

    A 0   B The slope is not defined.
    C $-\frac{1}{5}$ D 5

13. Find an equation for the line through (6, 3) and perpendicular to the line with equation $y = 8$.

14. Refer to the graph at the right. Find an equation for the perpendicular bisector of $\overline{PQ}$. (*Hint:* First find the midpoint of $\overline{PQ}$.)

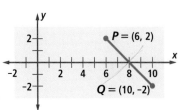

15. Let $A = (6, 2)$ and $B = (-5, 0)$.

    a. **Fill in the Blank** A counterclockwise rotation of $270°$ is the same as a clockwise rotation of ___?___.

    b. Find the coordinates of A' and B', the images of A and B under R_{270}.

    c. Find the slopes of $\overleftrightarrow{AB}$ and $\overleftrightarrow{A'B'}$.

    d. What relationship exists between the slopes? What does this tell you about the lines?

16. Consider $ABCD$, where $A = (0, 5)$, $B = (4, 3)$, $C = (1, -2)$, and $D = (-4, -1)$. Use the Perpendicular Lines and Slopes Theorem to determine if $ABCD$ is a rectangle.

Fill in the Blanks In 17–20, assume all lines lie in the same plane.

    a. Fill each blank with $\parallel$ or $\perp$.

    b. Draw a picture to illustrate each situation.

17. If $\ell \parallel m$ and $m \parallel n$, then ℓ ___?___ n.

18. If $\ell \parallel m$ and $m \perp n$, then ℓ ___?___ n.

19. If $\ell \perp m$ and $m \parallel n$, then ℓ ___?___ n.

20. If $\ell \perp m$ and $m \perp n$, then ℓ ___?___ n.

21. If M is a matrix representing a line that does not contain the origin, tell if each operation on M gives a matrix representing a line that is parallel to M, perpendicular to M, or neither perpendicular nor parallel to M.

    a. $\begin{bmatrix} 0 & -1 \\ 1 & 0 \end{bmatrix} M$ b. $\begin{bmatrix} 0 & 1 \\ 1 & 0 \end{bmatrix} M$

    c. $3M$ d. $\begin{bmatrix} -1 & 0 \\ 0 & -1 \end{bmatrix} M$

    e. $\begin{bmatrix} 3 & 3 \\ -5 & -5 \end{bmatrix} + M$ f. $\begin{bmatrix} 3 & 0 \\ 0 & 5 \end{bmatrix} M$

22. Without doing any computation, find $\begin{bmatrix} 0 & -1 \\ 1 & 0 \end{bmatrix}^4$. Explain how you know your answer is true.

Figures and their rotation images can be found in Hawaiian quilts, batik designs, and Celtic knots.

REVIEW

23. Find a matrix for $R_{80} \circ R_{80} \circ R_{200}$. **(Lesson 4-8)**

24. **a.** Use a DGS to rotate the points $(1, 0)$ and $(0, 1)$ $80°$ about the origin. Find the coordinates of their image points.

    b. Use the results of Part a to find an approximate matrix for R_{80}. **(Lessons 4-8, 4-6)**

25. **Multiple Choice** Which of the following is *not* a property of multiplication of 2×2 matrices? **(Lesson 4-7)**

    A associativity   **B** commutativity

    C existence of an identity   **D** closure

26. Architects designing auditoriums use the fact that sound intensity I is inversely proportional to the square of the distance d from the sound source. **(Lessons 2-3, 2-2)**

    a. Write a variation equation that represents this situation.

    b. A person moves to a seat that is 4 times as far from the sound source. How will the intensity of the sound be affected?

Sydney Opera House

EXPLORATION

27. **a.** Consider the lines with equations $\pi x + \sqrt{10}y = \sqrt[3]{6}$ and $\sqrt{10}x + \pi y = \sqrt[3]{6}$. How is the graph of one related to the graph of the other?

    b. Generalize Part a.

QY ANSWERS

1. $s = -2$

2. Yes; because $5(-0.2) = -1$, the lines are perpendicular.

Rotations and Perpendicular Lines   **279**

Lesson

4-10

Translations and Parallel Lines

▶ **BIG IDEA** By adding matrices, translation images of figures in the plane can be described.

In this chapter you have found images for many transformations by multiplying 2×2 matrices. There is one transformation for which images can be found by *adding* matrices.

Translations

Consider $\triangle ABC$ and its image $\triangle A'B'C'$ at the left below. Matrices M and M' representing the vertices of these triangles are given at the right below.

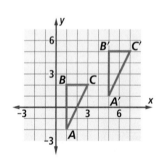

$\triangle ABC$

$$M = \begin{bmatrix} 1 & 1 & 3 \\ -2 & 2 & 2 \end{bmatrix}$$

$\triangle A'B'C'$

$$M' = \begin{bmatrix} 5 & 5 & 7 \\ 1 & 5 & 5 \end{bmatrix}$$

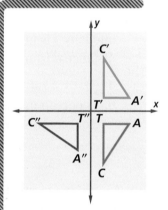

Use the graph above to identify the ransformation that maps

a. $\triangle CAT$ onto $\triangle C'A'T'$.

b. $\triangle C'A'T'$ onto $\triangle C''A''T''$.

c. $\triangle CAT$ onto $\triangle C''A''T''$.

Activity

Use the matrices M and M' above.

Step 1 Calculate $D = M' - M$.

Step 2 What does the first row of D represent?

Step 3 What does the second row of D represent?

Step 4 Calculate $D + M$. What does $D + M$ represent?

Step 5 Complete this sentence: If (x, y) is any point in the preimage, then its image is ___?___.

The transformation in the Activity is an example of a *translation* or *slide*.

Definition of Translation

The transformation that maps (x, y) onto $(x + h, y + k)$ is a **translation** of h units horizontally and k units vertically, and is denoted by $T_{h,k}$.

Reading Math

The Latin prefix *trans* means "across." When we translate a geometric figure, we move it across the plane. Another term in this chapter with the same prefix is *transformation*.

Using mapping notation, $T_{h,k}: (x, y) \rightarrow (x + h, y + k)$.

Using $f(x)$ notation, $T_{h,k}(x, y) = (x + h, y + k)$.

In the figure on the previous page, $\triangle A'B'C'$ is the image of $\triangle ABC$ under the translation $T_{4,3}$.

Matrices for Translations

There is no single matrix representing a specific translation because the dimensions of the translation matrix depend on the figure being translated. Example 1 shows you how to find image coordinates with and without matrices.

GUIDED

Example 1

A quadrilateral has vertices $Q = (1, 1)$, $U = (6, 2)$, $A = (8, 4)$, and $D = (5, 5)$.

a. Find its image under the translation $T_{-8,2}$.

b. Graph the image and preimage on the same set of axes.

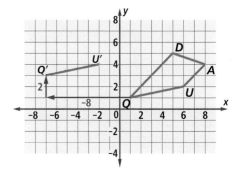

Solution

a. Using $f(x)$ notation:

$T_{-8,2}(x, y) = (x - 8, y + 2)$

$Q' = T_{-8,2}(1, 1) = (1 - 8, 1 + 2) = (-7, 3)$

$U' = T_{-8,2}(6, 2) = (\underline{?}, \underline{?}) = (-2, 4)$

$A' = T_{-8,2}(8, 4) = (\underline{?}, \underline{?}) = (\underline{?}, \underline{?})$

$D' = T_{-8,2}(5, 5) = (\underline{?}, \underline{?}) = (\underline{?}, \underline{?})$

To use matrices, construct the translation matrix by showing the point $(-8, 2)$ in each of four columns.

$$
\begin{array}{ccccccc}
 & T_{-8,2} & & + & Q & U & A & D & = & Q' & U' & A' & D' \\
\end{array}
$$

$$
\begin{bmatrix} -8 & -8 & \underline{?} & \underline{?} \\ 2 & 2 & \underline{?} & \underline{?} \end{bmatrix} + \begin{bmatrix} 1 & 6 & \underline{?} & \underline{?} \\ 1 & 2 & \underline{?} & \underline{?} \end{bmatrix} = \begin{bmatrix} -7 & -2 & \underline{?} & \underline{?} \\ 3 & 4 & \underline{?} & \underline{?} \end{bmatrix}
$$

b. $Q'U'A'D'$ is the image of $QUAD$ under a translation 8 units to the left and 2 units up. Copy and complete the graph shown above.

 QY

▶ **QY**

Calculate the slopes of $\overleftrightarrow{QU}$ and $\overleftrightarrow{Q'U'}$. What is the relationship between $\overleftrightarrow{QU}$ and $\overleftrightarrow{Q'U'}$?

Properties of Translation Images

The QY illustrates a special case of the following more general result.

> ## Parallel Lines and Translations Theorem
>
> Under a translation, a preimage line is parallel to its image.

Proof Let $P = (x_1, y_1)$ and $Q = (x_2, y_2)$ be two different points
on the line $\overleftrightarrow{PQ}$. The image of the line under $T_{h,k}$ contains
the points P' and Q' such that $P' = (x_1 + h, y_1 + k)$ and
$Q' = (x_2 + h, y_2 + k)$.

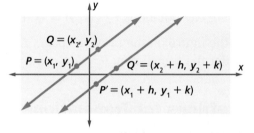

Case 1: $\overleftrightarrow{PQ}$ is a *vertical* line. Then $x_1 = x_2$, and so
$x_1 + h = x_2 + h$.
From this, $\overleftrightarrow{P'Q'}$ is also a vertical line. Thus, in this
case $\overleftrightarrow{PQ} \parallel \overleftrightarrow{P'Q'}$.

Case 2: $\overleftrightarrow{PQ}$ is *not vertical*. Then $x_1 \neq x_2$, and both $\overleftrightarrow{PQ}$ and $\overleftrightarrow{P'Q'}$ have slopes.

Let $m_1 = $ slope of $\overleftrightarrow{PQ} = \frac{y_2 - y_1}{x_2 - x_1}$.

Let $m_2 = $ slope of $\overleftrightarrow{P'Q'} = \frac{(y_2 + k)-(y_1 + k)}{(x_2 + h)-(x_1 + h)} = \frac{y_2 - y_1}{x_2 - x_1}$.

The slopes are equal. So, $\overleftrightarrow{PQ} \parallel \overleftrightarrow{P'Q'}$.

This theorem lets you easily find an equation for a translation image
of a line.

Architect Frank Lloyd Wright used many parallel lines in designing his houses as seen in his Robie House above.

> **GUIDED**
>
> ## Example 2
> Find the image of the line with equation $y = 3x - 5$ under the
> translation $T_{2, -1}$.
>
> **Solution** Because the image is parallel to the original line, the slopes are
> equal. So, the slope of the image is __?__.
>
> Pick a point on the original line. An easy one is $(0, -5)$. Its translation image
> is $T_{2, -1} (0, -5) = (\underline{?}, \underline{?})$.
>
> Using the Point-Slope Theorem, an equation for the line is
> $y = \underline{?}(x - \underline{?}) + \underline{?}$.
>
> So an equation of the line is $y = \underline{?}x + \underline{?}$.

Questions

COVERING THE IDEAS

1. Refer to $\triangle ABC$ and $\triangle A'B'C'$ at the beginning of the lesson. Using $f(x)$ notation, describe the translation that maps
   a. $\triangle ABC$ onto $\triangle A'B'C'$.
   b. $\triangle A'B'C'$ onto $\triangle ABC$.

2. **Fill in the Blank** A translation $T_{h,k}$ is a transformation mapping (x, y) onto ___?___.

3. **Fill in the Blanks** $T_{h,k}$ is a translation of __?__ units horizontally and __?__ units vertically.

4. Find the image of the point under $T_{5,-7}$.
   a. $(0, 0)$
   b. $(-50, 83)$
   c. (a, b)

5. Consider $\triangle PQR$ represented by the matrix $\begin{bmatrix} 4 & 12 & -5 \\ -3 & 0 & 7 \end{bmatrix}$. Use matrix addition to find a matrix for $\triangle P'Q'R'$, the image of $\triangle PQR$ under a translation 7 units to the left and 2 units down.

6. The matrix $\begin{bmatrix} 7 & 2 & 1 & -1 & 4 \\ 2 & 13 & 8 & 4 & -5 \end{bmatrix}$ represents pentagon $AHMED$.
   a. Apply the translation $T_{-3,8}$ to the pentagon. Call the image $A'H'M'E'D'$.
   b. Graph the preimage and the image on the same set of axes.
   c. Verify that $AA' = HH'$.
   d. Why is the result in Part c not a surprise?

7. Refer to Example 1.
   a. What is the slope of $\overleftrightarrow{QA}$?
   b. What is the slope of $\overleftrightarrow{Q'A'}$?
   c. Is $\overleftrightarrow{QA} \parallel \overleftrightarrow{Q'A'}$? Justify your answer.

8. Suppose lines ℓ_1 and ℓ_2 are not parallel. Can they be translation images of each other? Explain your reasoning.

9. Refer to the graph at the right.
   a. What translation maps $ABCDE$ onto $A'B'C'D'E'$?
   b. Verify that $\overline{BC} \parallel \overline{B'C'}$.
   c. Verify that $BC = B'C'$.
   d. $BB'C'C$ is what kind of quadrilateral?

10. Consider the line with equation $y = -2x + 7$. Find an equation for the image of this line under $T_{-4,5}$.

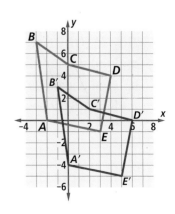

APPLYING THE MATHEMATICS

11. Under $T_{-2,5}$ the image of $\triangle CUB$ is $\triangle C'U'B'$. $\triangle C'U'B'$ is then translated under $T_{7,3}$ to get $\triangle C''U''B''$.

    a. What single translation will give the same result as $T_{7,3} \circ T_{-2,5}$?

    b. Is $T_{7,3} \circ T_{-2,5} = T_{-2,5} \circ T_{7,3}$?

    c. In general, is the composition of two translations commutative? Why or why not?

12. Line ℓ has the equation $y = 3x - 7$. Line ℓ' is the image of ℓ under a translation.

    a. If ℓ' contains the point $(0, 5)$, find an equation for ℓ'.

    b. Give an example of a translation that maps line ℓ onto line ℓ'.

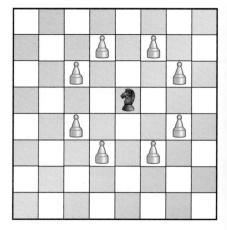

13. In chess, each knight can move 2 squares vertically or horizontally and then one square at a right angle to the first part of its move. In the figure at the right, the black knight can move to any of the places occupied by a white pawn. As a translation, two of the knight's possible moves can be written as $T_{2,1}$ (two squares right and one square up) and $T_{-1,2}$ (one square left and two squares up). The knight has 6 other possible moves; write them as translations.

14. Consider the segments $\overline{QU}$ and $\overline{Q'U'}$ from Example 1.

    a. Find the lengths of both segments. How do they compare?

    b. Do you think it is generally true that a segment and its image under a translation are congruent? If so, try to prove it.

    c. Are there any other transformations under which a segment and its image are congruent? If so, which ones?

15. a. Find the image of the ordered pair (x, y) under $T_{h,k} \circ R_{90}$ and $R_{90} \circ T_{h,k}$.

    b. Is translating a rotation image of a point the same as rotating its translation image? Why or why not?

REVIEW

16. Let $A = (2, 1)$ and $B = (-4, 4)$. Let $A' = R_{90}(A)$ and $B' = R_{90}(B)$. (Lessons 4-9, 4-8)

    a. Find the coordinates of A' and B'.

    b. Find equations for $\overleftrightarrow{AB}$ and $\overleftrightarrow{A'B'}$.

    c. Find the slopes of the two lines and verify that their product is -1.

17. A rotation of 180° can be considered as the composite of two rotations of 90° or as the composite of a reflection over the x-axis followed by a reflection over the y-axis. (**Lessons 4-7, 4-6, 4-3**)

   a. Compute $N \cdot N$, where $N = \begin{bmatrix} 0 & -1 \\ 1 & 0 \end{bmatrix}$ is the matrix for R_{90}.

   b. Compute $\begin{bmatrix} 1 & 0 \\ 0 & -1 \end{bmatrix} \cdot \begin{bmatrix} -1 & 0 \\ 0 & 1 \end{bmatrix}$. Do you get the same answer as in Part a?

18. Consider the line with equation $2x - 3y = 12$. (**Lessons 4-6, 3-3**)

   a. Find the x- and y-intercepts of this line.

   b. Find the images of the x- and y-intercept points under R_{90}.

   c. Write an equation in standard form for the line that contains the two points from Part b.

19. The Central High School debate team is selling T-shirts to raise money. They are selling small, medium, and large shirts in blue, grey, and tan. (**Lessons 4-2, 4-1**)

   a. The debate team ordered 8 of each color in small, 12 of each in medium, and 8 of each in large. Write a matrix to represent their T-shirt inventory.

   b. The first five customers bought 2 medium blue T-shirts, 1 medium grey T-shirt, 1 large grey T-shirt, and 1 small tan T-shirt. Write a matrix to represent these T-shirt purchases.

   c. Use matrix subtraction to write a matrix that represents the T-shirt inventory after the first five purchases.

This T-shirt has a logo that can be read both as printed and when it is rotated 180°.

20. At Mimi's Pizzeria, a large pizza costs $8, and each additional topping costs $0.90. Write an equation describing the total cost y of a large pizza with x toppings. Graph the resulting line, and label the y-intercept. (**Lesson 3-1**)

EXPLORATION

21. Using translations you can find a rule for finding image points for a rotation of 90° with any point as center. Consider the rotation of 90° with center $(-3, -1)$. Let (x, y) be any point.

   a. What is the image of (x, y) under the translation that maps $(-3, -1)$ onto $(0, 0)$?

   b. What is the image of your answer to Part a under a rotation of 90° about the origin?

   c. What is the image of your answer to Part b under the translation that maps $(0, 0)$ onto $(-3, -1)$?

   d. Check that your answer to Part c is the image of (x, y) under a rotation of 90° about $(-3, -1)$.

QY ANSWER

The slope of each line is $\frac{1}{5}$. They are parallel.

Chapter 4 Projects

1 History of Matrices

Investigate the development of matrices and the early work of Arthur Cayley and James Sylvester in the mid 1800s.

a. For what purposes were matrices used before they were given that name?

b. When were matrices first used to describe transformations?

c. What other mathematicians and terms are associated with the history and use of matrices?

Arthur Cayley

James Sylvester

2 Matrices and the Fibonacci Sequence

Let $A = \begin{bmatrix} 0 & 1 \\ 1 & 1 \end{bmatrix}$ and $P = \begin{bmatrix} 1 \\ 1 \end{bmatrix}$.

a. Calculate $A \cdot P, A \cdot A \cdot P, A \cdot A \cdot A \cdot P$, and $A \cdot A \cdot A \cdot A \cdot P$.

b. What do the calculations you did in Part a have to do with the Fibonacci sequence? Explain why this happens.

c. Use matrices to find the first 10 terms of the sequence a_n such that $a_1 = 3, a_2 = 4$, and $a_n = 3a_{n-1} + 2a_{n-2}$ for $n \geq 3$.

3 Pitch, Yaw, and Roll

Rotations are not limited to the two-dimensional coordinate plane. In order to keep their aircrafts properly oriented in space, pilots need to pay attention to *attitude, pitch, yaw,* and *roll*. Research these terms and explain how they relate to rotations in three dimensions.

4 Transpose of a Matrix

The *transpose* of a matrix M is a matrix M^T in which the rows and columns of M are switched. For example, the transpose of matrix $A = \begin{bmatrix} 1 & 2 \\ 3 & 4 \end{bmatrix}$ is $A^T = \begin{bmatrix} 1 & 2 \\ 3 & 4 \end{bmatrix}^T = \begin{bmatrix} 1 & 3 \\ 2 & 4 \end{bmatrix}$.
Use a CAS to explore the following.

a. Find $(A^T)^T$. What does this suggest about the transpose of a transpose matrix?

b. Define a 2×2 matrix B and explore the connection between $A^T + B^T$ and $(A + B)^T$. What do you notice?

c. Using the same matrix B, explore the products $(AB)^T, A^TB^T$, and B^TA^T. What do you notice?

d. Is there a 2×2 matrix C such that $C^T = C$? Explain your answer.

e. Does a matrix have to be square to have a transpose? Give an example to support your answer. Do the properties in Parts a–d apply to nonsquare matrices?

5 Computer Graphics

One application of matrices is in computer graphics. A graphic image is divided into a matrix of pixels, each of which is typically assigned three values: hue (or color), saturation (purity of the color), and intensity (brightness). For example, a 5 × 5 matrix of pixels represents an image with 25 pixels. If you use a 10-point gray scale (10 hues), you could assign each pixel a value from 1 to 10 that represents the picture's level of grayness, where 1 = white and 10 = black.

a. Find a black-and-white picture and overlay a 5-by-5 grid on it. Use a 10-point scale to assign each entry in the resulting matrix a value based on an average level of grayness in the grid square.

b. Using only your matrix and a blank 5-by-5 grid, recreate the picture using the levels of grayness you assigned. You will notice that the resolution of your image is not very good. How could you improve your resolution?

c. Computers use a grid that is much finer (has more pixels) and are able to assign values based on color rather than just shades of gray. Research what size grids are used and what properties are numerically assigned to pixels of a graphic.

6 Predicting the Weather

Matrix multiplication can be used to forecast the weather. Suppose the probability of rain or sunshine on a given day depends on the weather the previous day as shown in matrix Q below.

$$\text{Next Day} \quad \begin{matrix} & \text{Previous Day} \\ & \text{Rain} \quad \text{Shine} \\ \text{Rain} \\ \text{Shine} \end{matrix} \begin{bmatrix} 0.6 & 0.2 \\ 0.4 & 0.8 \end{bmatrix} = Q$$

a. A 30% chance of rain, 70% chance of sunshine on Monday is given by the matrix $M = \begin{bmatrix} 0.3 \\ 0.7 \end{bmatrix}$. Calculate QM to find the probabilities for Tuesday. Multiply by Q again (that is, compute QQM) to find the probabilities for Wednesday. Multiply by Q again ($QQQM$) to find the probabilities for Thursday.

b. Find the probability of rain for the remaining days of the week. What happens to the probabilities as the week goes on?

c. This matrix application is called a *Markov chain*. Find out what you can about Markov chains and report your results.

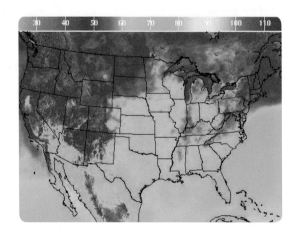

Chapter 4 Summary and Vocabulary

○ A **matrix** is a rectangular array of objects. Matrices are frequently used to store data and to represent **transformations**. Matrices can be added or subtracted if they have the same **dimensions**. Addition of matrices can be used to obtain translation images of figures.

○ Any matrix can be multiplied by a real number, called a *scalar*. Multiplying each element in the matrix by the scalar yields the **scalar product**. However, not all matrices can be multiplied by other matrices. The product of two matrices exists only if the number of columns of the left matrix equals the number of rows of the right matrix. The element in row r and column c of AB is the product of row r of A and column c of B. Matrix multiplication is associative but not commutative.

○ Matrices with 2 rows can represent points, segments, lines, polygons, and other figures in the coordinate plane. Multiplying such a matrix by a 2×2 matrix on its left may yield a transformation image of the figure. Transformations for which 2×2 matrices are given in this chapter include **reflections**, **rotations**, **size changes**, and **scale changes**. They are summarized on the next two pages. The rotation of 90° about the origin is a particularly important transformation. Based on that transformation, it can be proved that two nonvertical lines are perpendicular if and only if the product of their slopes is –1.

○ The set of 2×2 matrices is closed under multiplication. The **identity matrix** for multiplying 2×2 matrices is $\begin{bmatrix} 1 & 0 \\ 0 & 1 \end{bmatrix}$. The **identity transformation** maps any figure onto itself.

The **Matrix Basis Theorem** provides a way to generate and remember matrices for transformations. When a transformation A is represented by a 2×2 matrix, if $A(1, 0) = (x_1, y_1)$ and $A(0, 1) = (x_2, y_2)$, then A has the matrix $\begin{bmatrix} x_1 & x_2 \\ y_1 & y_2 \end{bmatrix}$.

Vocabulary

4-1
*matrix
element
*dimensions
equal matrices
point matrix

4-2
*matrix addition,
    sum of two matrices
scalar multiplication,
    scalar product
difference of two matrices

4-3
row-by-column
    multiplication
*matrix multiplication
matrix product

4-4
*transformation
preimage
image
*size change
center of a size change
magnitude of a size change
*identity matrix
*identity transformation
similar

4-5
*scale change
horizontal magnitude
vertical magnitude
stretch
shrink

4-6
reflection image of a point
    over a line
reflecting line, line of
    reflection
*reflection

○ **Reflections, rotations,** and **translations** preserve distance. A size change S_k multiplies distances by k. These properties can be proved using the Pythagorean Distance Formula for the distance d between two points (x_1, y_1) and (x_2, y_2):

$$d = \sqrt{|x_2 - x_1|^2 + |y_2 - y_1|^2}.$$

Matrices for many specific transformations were discussed in this chapter.

○ **Transformations Yielding Images Congruent to Preimages**

Reflections:

over x-axis	over y-axis	over the line $y = x$	over the line $y = -x$
$\begin{bmatrix} 1 & 0 \\ 0 & -1 \end{bmatrix}$	$\begin{bmatrix} -1 & 0 \\ 0 & 1 \end{bmatrix}$	$\begin{bmatrix} 0 & 1 \\ 1 & 0 \end{bmatrix}$	$\begin{bmatrix} 0 & -1 \\ -1 & 0 \end{bmatrix}$
$r_x: (x, y) \rightarrow (x, -y)$	$r_y: (x, y) \rightarrow (-x, y)$	$r_{y=x}: (x, y) \rightarrow (y, x)$	$r_{y=-x}: (x, y) \rightarrow (-y, -x)$

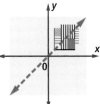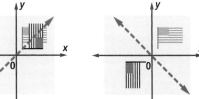

Rotations with center (0, 0):

magnitude 90°	magnitude 180°	magnitude 270°
$\begin{bmatrix} 0 & -1 \\ 1 & 0 \end{bmatrix}$	$\begin{bmatrix} -1 & 0 \\ 0 & -1 \end{bmatrix}$	$\begin{bmatrix} 0 & 1 \\ -1 & 0 \end{bmatrix}$
$R_{90}: (x, y) \rightarrow (-y, x)$	$R_{180}: (x, y) \rightarrow (-x, -y)$	$R_{270}: (x, y) \rightarrow (y, -x)$

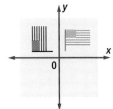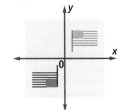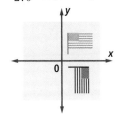

Translations:

No general matrix
$T_{h,k}: (x, y) \rightarrow (x + h, y + k)$

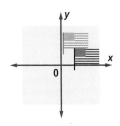

Transformations Yielding Images Similar to Preimages

Size changes with center (0, 0) and magnitude k:

$$\begin{bmatrix} k & 0 \\ 0 & k \end{bmatrix}$$

$S_k: (x, y) \rightarrow (kx, ky)$

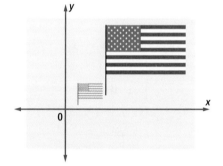

Other Transformations

Scale changes with horizontal magnitude a *and vertical magnitude* b:

$$\begin{bmatrix} a & 0 \\ 0 & b \end{bmatrix}$$

$S_{a,b}: (x, y) \rightarrow (ax, by)$

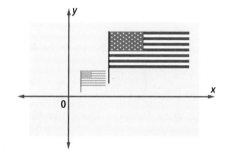

Theorems

Chapter

4   Self-Test

Take this test as you would take a test in class. You will need a calculator. Then use the Selected Answers section in the back of the book to check your work.

1. Write a matrix to represent polygon *HULK* if $H = (-2, 4)$, $U = (5, 1)$, $L = (-2, -2)$, and $K = (-4, 2)$.

2. One day on a Veggie-Air flight from Iceburg, there were 16 first-class and 107 economy passengers going to Jicamaport, 4 first-class and 180 economy passengers bound for Okraville, and 2 first-class and 321 economy passengers flying to Potatotown.

   a. Write a 2 × 3 matrix to store this information. Include appropriate column and row headings.

   b. Write a 3 × 2 matrix to store this information. Include appropriate column and row headings.

In 3–6, use matrices A, B, and C below.

$$A = \begin{bmatrix} 2 & 0 \\ 4 & -2 \\ -\frac{1}{2} & \frac{1}{2} \end{bmatrix} \quad B = \begin{bmatrix} 5 & -2 & -1 \\ \frac{1}{8} & 6 & 5 \end{bmatrix} \quad C = \begin{bmatrix} 17 & 4 \\ \frac{1}{3} & \sqrt{2} \end{bmatrix}$$

3. Determine which of the following products exist: *AB*, *BA*, *AC*, *CA*, *BC*, and *CB*.

4. If possible, find *BA*. If it is not possible, explain why.

5. If possible, find $A - C$. If it is not possible, explain why.

6. Calculate $\frac{1}{3}B$.

7. Why is $\begin{bmatrix} 1 & 0 \\ 0 & 1 \end{bmatrix}$ called the identity matrix?

8. Are the two lines with equations $y = 5x - 3$ and $y = \frac{1}{5}x + 2$ perpendicular? Explain your answer.

9. Find an equation for the line through $\left(\frac{1}{4}, -1\right)$ that is perpendicular to $y = \frac{1}{7}x + 4$.

10. Calculate the matrix for $r_y \circ R_{270}$.

In 11–12, refer to the graph below.

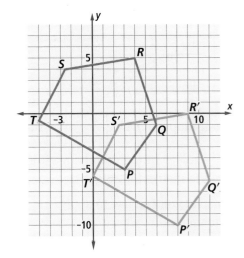

11. What translation maps *PQRST* onto *P'Q'R'S'T'*?

12. Graph the image of *PQRST* under the transformation r_x.

13. The 4-Star Movie Theater has four screens. The number and type of theater attendees is summarized in the following matrix for one show time of four different movies.

	Children	Adults	Students
Movie 1	38	135	169
Movie 2	84	101	152
Movie 3	84	118	135
Movie 4	67	236	34

Ticket prices for children, adults, and students are $4.00, $6.50, and $5.00, respectively. Use matrix multiplication to determine the total ticket sales revenue for each movie.

14. Savannah Reed and Denise Wright have decided to merge their book inventories and hold a book sale. If the matrices below represent each person's inventory, write a matrix for the inventory of books after merging.

	Savannah		Denise	
	Fiction	Nonfiction	Fiction	Nonfiction
Paperbacks	42	15	5	2
Hardbacks	10	2	3	1
Audiotapes	7	1	0	3
Audio CDs	4	0	2	0

15. Show that multiplying $A = \begin{bmatrix} 2 & -9 \\ 7 & 5 \end{bmatrix}$ by the scalar 2 is equivalent to multiplying A on the left by the matrix for S_2.

16. Write a matrix you can use to apply the transformation T, where $T: (x, y) \rightarrow (-y, x)$, to a figure in the coordinate plane.

In 17 and 18, $\triangle XYZ$ has vertices $X = (-4, 5)$, $Y = (2, 6)$, and $Z = (-3, 1)$.

17. Write a sentence that describes this matrix multiplication geometrically:
$$\begin{bmatrix} -1 & 0 \\ 0 & 1 \end{bmatrix} \begin{bmatrix} -4 & 2 & -3 \\ 5 & 6 & 1 \end{bmatrix} = \begin{bmatrix} 4 & -2 & 3 \\ 5 & 6 & 1 \end{bmatrix}.$$

18. Graph $\triangle XYZ$ and $R_{270}(\triangle XYZ)$.

19. Find the matrix of the image of $\begin{bmatrix} 2.5 & 1 & -6 \\ -5 & 4 & -12 \end{bmatrix}$ under each transformation.

    a. $S_{4, 3}$ b. $T_{2.5,-1}$

True or False In 20 and 21, if the statement is false, give an example to show that it is false.

20. A line and its translation image are always parallel.

21. A line and its rotation image are always perpendicular.

22. Consider the transformation S_7.

    a. What is the image of (a, b) under S_7?

    b. If $P = (2, 5)$ and $Q = (3, 9)$, show that the distance between $S_7(P)$ and $S_7(Q)$ is 7 times the distance between P and Q.

23. Find an equation for the perpendicular bisector of $\overline{PQ}$ in Question 22.

Chapter 4 Chapter Review

SKILLS Procedures used to get answers

OBJECTIVE A Write matrices for points and polygons. (Lesson 4-1)

1. Write a matrix to represent each point.
   a. $U = (4, 5)$
   b. $E = (3, -4)$
   c. $D = (-4, 1)$
   d. $G = (-2, -4)$
   e. $F = (3, 1)$

2. Write a matrix to represent polygon *FUDGE*, if its vertices are those defined in Question 1.

OBJECTIVE B Add, subtract, and find scalar multiples of matrices. (Lesson 4-2)

In 3–6, let $A = \begin{bmatrix} 2 & 2 & 5 \\ 6 & 4 & -2 \\ 0 & -3 & -3 \end{bmatrix}$ and

$B = \begin{bmatrix} -2 & 4 & 7 \\ 3 & 3 & -5 \\ 4 & 1 & -10 \end{bmatrix}$. Calculate.

3. $A - B$
4. $A - 3B$
5. $B - 2A$
6. $B + A$

In 7 and 8, find p and q.

7. $2\begin{bmatrix} 4 & p \\ q & -3 \end{bmatrix} + \begin{bmatrix} 2 & 3 \\ 5q & 7 \end{bmatrix} = \begin{bmatrix} 10 & 9 \\ 21 & 1 \end{bmatrix}$

8. $\begin{bmatrix} 8 & -2 \\ -3 & p \end{bmatrix} - 2\begin{bmatrix} q & 4 \\ -7 & 3 \end{bmatrix} = \begin{bmatrix} 5 & -10 \\ 11 & 7 \end{bmatrix}$

OBJECTIVE C Multiply matrices. (Lesson 4-3)

In 9–12, multiply the matrices if possible.

9. $\begin{bmatrix} 4 & 1 \\ 6 & -3 \end{bmatrix}\begin{bmatrix} 1 & 2 \\ -3 & 4 \end{bmatrix}$

10. $\begin{bmatrix} 4 \\ 3 \end{bmatrix}\begin{bmatrix} 1 & -2 & 0 \end{bmatrix}$

11. $\begin{bmatrix} 2 & 2 & 5 \\ 6 & 4 & -2 \end{bmatrix}\begin{bmatrix} -2 & 4 \\ 3 & 3 \\ 4 & 1 \end{bmatrix}$

12. $\begin{bmatrix} 1 & 2 \\ -1 & 0 \end{bmatrix}\begin{bmatrix} 4 & 3 & -1 \\ -1 & 1 & -1 \end{bmatrix}\begin{bmatrix} 2 \\ 2 \\ 2 \end{bmatrix}$

In 13 and 14, find p and q.

13. $\begin{bmatrix} p & 0 \\ 0 & q \end{bmatrix}\begin{bmatrix} 4 \\ -2 \end{bmatrix} = \begin{bmatrix} 16 \\ -4 \end{bmatrix}$

14. $\begin{bmatrix} 0 & -2 \\ 1 & 0 \end{bmatrix}\begin{bmatrix} p \\ q \end{bmatrix} = \begin{bmatrix} 12 \\ 5 \end{bmatrix}$

OBJECTIVE D Determine equations of lines perpendicular to given lines. (Lesson 4-9)

15. Find an equation of the line through $(2, -5)$ perpendicular to the line $y = \frac{1}{3}x + 1$.

16. Find an equation of the line through $(4, -6)$ perpendicular to the line $y = 5$.

17. Given $A = (4, 7)$ and $B = (-6, 1)$, find an equation for the perpendicular bisector of $\overline{AB}$.

18. Consider two lines. One is the image of the other under R_{90}. The slope of one of the lines is $\frac{1}{8}$. What is the slope of the other line?

PROPERTIES Principles behind the mathematics

OBJECTIVE E Recognize properties of matrix operations. (Lessons 4-2, 4-3, 4-7)

In 19–22, a statement is given.
   a. Is the statement true or false?
   b. Give an example to support your answer.

19. Matrix multiplication is associative.

20. Matrix multiplication is commutative.

21. Matrix subtraction is commutative.

22. Scalar multiplication of matrices is commutative.

In 23 and 24, suppose Y and P are matrices. Y has dimensions 1×7 and P has dimensions $m \times n$.

23. If the sum $Y + P$ exists, what are the values of m and n?

24. If the product PY exists, what is the value of n?

25. What matrix is the identity for multiplication of 2×2 matrices?

OBJECTIVE F Recognize relationships between figures and their transformation images. (Lessons 4-4, 4-5, 4-6, 4-8, 4-9, 4-10)

In 26 and 27, fill in the blank with A, B, or C to make a true statement.

   A not necessarily similar or congruent
   B similar, but not necessarily congruent
   C congruent

26. A figure and its size change image are __?__.

27. A figure and its reflection image are __?__.

28. Give an example to show that a figure and its image under $S_{3,\frac{2}{3}}$ are not similar.

29. Use the Pythagorean Distance Formula to show that S_2 multiplies distances by 2.

30. Find an equation for the image of the line $2x + 3y = 60$ under R_{90}.

31. Repeat Question 30 if the transformation is the translation $T_{-1, 3}$.

OBJECTIVE G Relate transformations to matrices, and vice versa. (Lessons 4-4, 4-5, 4-6, 4-7, 4-8, 4-10)

32. Translate the matrix equation
$$\begin{bmatrix} -1 & 0 \\ 0 & -1 \end{bmatrix} \begin{bmatrix} 4 \\ 3 \end{bmatrix} = \begin{bmatrix} -4 \\ -3 \end{bmatrix}$$ into English by filling in the blanks.

The image of the point __?__ under a rotation with center __?__ and magnitude __?__ is the point __?__.

33. Multiply the matrix for $r_{y=x}$ by itself, and tell what transformation the product represents.

34. Write a matrix for a scale change with horizontal magnitude 5 and vertical magnitude 3.5.

35. a. Calculate a matrix for $R_{180} \circ r_x$.
    b. What single transformation corresponds to your answer?

36. a. Find two reflections whose composite is R_{180}.
    b. Use matrix multiplication and your answer to Part a to generate the matrix for R_{180}.

37. a. What size change maps *FIG* onto *F'I'G'* as shown below?

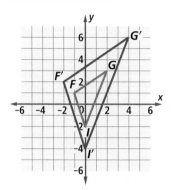

b. Explain how to use a matrix operation to transform *FIG* to *F'I'G'*.

Multiple Choice In 38–40, choose the matrix for each transformation.

A $\begin{bmatrix} 0 & -1 \\ 1 & 0 \end{bmatrix}$ B $\begin{bmatrix} 0 & 1 \\ 1 & 0 \end{bmatrix}$ C $\begin{bmatrix} 0 & 3 \\ 3 & 0 \end{bmatrix}$

D $\begin{bmatrix} -1 & 0 \\ 0 & 1 \end{bmatrix}$ E $\begin{bmatrix} 1 & 0 \\ 0 & -1 \end{bmatrix}$ F $\begin{bmatrix} 3 & 0 \\ 0 & 3 \end{bmatrix}$

38. $r_{y=x}$ **39.** S_3 **40.** R_{90}

41. Find the image of $\begin{bmatrix} 2 & -1 & 4 & 3 \\ 0.7 & 0 & -1 & 5 \end{bmatrix}$ under r_x.

42. *HARP* has coordinates $H = (-2, 2)$, $A = (-1, -2)$, $R = (0, 0)$, and $P = (1, 4)$. Find the matrix of *HARP* under R_{270}.

43. Find the matrix of $\begin{bmatrix} 0 & -8 & 6 \\ 4 & 0 & 4 \end{bmatrix}$ under $S_{0.5}$.

OBJECTIVE H Given their slopes, determine whether lines are parallel or perpendicular to each other, and vice versa. (Lessons 4-9, 4-10)

44. Line ℓ has equation $y = 3x$, and line m has equation $y = kx$. If $\ell \perp m$, find the value of k.

45. Suppose $A = (0, 3)$, $B = (2, 4)$, $C = (7, 8)$, and $D = (2, 9)$. Is $\overleftrightarrow{AB} \parallel \overleftrightarrow{CD}$? Explain.

46. Multiple Choice Line a has slope -4. Which of the following is the slope of a line perpendicular to a?

A 4 B $\frac{1}{4}$ C $-\frac{1}{4}$ D -4

47. Lines j and k are parallel. j has slope 8 and k passes through the point $(0, 0)$. Find another point on k.

48. Let $\triangle NSA$ be represented by the matrix $\begin{bmatrix} 152 & -12 & 87 \\ 16 & 113 & -23 \end{bmatrix}$. Let $\triangle N'S'A' = R_{270}(\triangle NSA)$.

a. What is the slope of $\overleftrightarrow{NA}$?

b. What is the product of the slopes of $\overleftrightarrow{NA}$ and $\overleftrightarrow{N'A'}$?

c. Use your answers to Parts a and b to find the slope of $\overleftrightarrow{N'A'}$.

USES Applications of mathematics in real-world situations

OBJECTIVE I Use matrices to store data. (Lesson 4-1)

49. In 2005, the average loss for people who were victims of three common online scams was $240 for credit or debit card fraud, $410 for nondelivery of merchandise, and $2000 for investment fraud. In 2006, the amounts lost to these scams averaged $427.50, $585, and $2694.99, respectively. Store these data in a 2 × 3 matrix.

50. The recommended daily allowance (RDA) of vitamin K is 60 µg for a 9- to 13-year-old male, 75 µg for a 14- to 18-year-old male, and 120 µg for a 19- to 30-year-old male. (Note: µg means micrograms, or one-millionth of a gram.) The RDA of thiamin is 0.9 mg for a 9- to 13-year-old male, 1.2 mg for a 14- to 18-year-old male, and 1.2 mg for a 19- to 30-year-old male. The RDAs of vitamin C for these age categories are 45 mg, 75 mg, and 90 mg, respectively. The RDAs of niacin for these age categories are 12 mg, 16 mg, and 16 mg, respectively. Store these data in a 3 × 4 matrix.

In 51 and 52, the matrix gives the time (in minutes and seconds) behind Lance Armstrong's time each competitor in the Tour de France finished in each of three years.

	2003	2004	2005
Lance Armstrong	0:00	0:00	0:00
Jan Ulrich	1:01	8:50	6:21
Francisco Mancebo	19:15	18:01	9:59

51. Which element represents the time behind Armstrong's time that Jan Ullrich finished in 2003?

52. How much time behind Jan Ullrich's time did Francisco Mancebo finish in 2005?

OBJECTIVE J Use matrix addition, matrix multiplication, and scalar multiplication to solve real-world problems. (Lessons 4-2, 4-3)

53. The matrices below contain box office data for three popular movie series. One matrix contains the movies' domestic gross earnings, in millions of dollars, for both the first and second movie in the series, while the other matrix contains the movies' foreign gross earnings.

Domestic Gross (10^6 dollars)

	Spiderman	The Matrix	Harry Potter
Original	404	171	318
Sequel	374	282	262

Foreign Gross (10^6 dollars)

	Spiderman	The Matrix	Harry Potter
Original	418	285	659
Sequel	410	457	617

a. Calculate the matrix that stores the worldwide total amount of money each movie made (in millions of dollars).

b. Of these movies, which made the most money worldwide?

c. Which sequel made more money worldwide than its original movie?

54. Suppose the New York Yankees and Seattle Mariners both decided to raise their ticket prices by 5%. Prices of tickets in dollars in 2008 for three tiers are given in the matrix below.

	Bleachers	Premium Box	Upper Deck
Yankees	14	95	80
Mariners	7	60	20

a. What scalar multiplication will yield the new ticket prices?

b. Find a matrix that stores the new ticket prices for each team.

55. In basketball, a free throw is worth 1 point, a shot made from inside the three-point arc is worth 2 points, and a shot made from behind the three-point arc is worth 3 points. Suppose that in one game, Brenda made 9 free throws, 11 shots from inside the three-point arc, and 2 shots from behind the three-point arc. In the same game, Marisa made 5 free throws, 7 shots from inside the three-point arc, and 5 shots from behind the three-point arc. Write a matrix B for the number of each type of basket each player made and a matrix P for the number of points the baskets are worth, then calculate BP to find the total points each player scored.

56. An office supply store sells three different models of graphing calculators. Model X sells for $150, model Y sells for $130, and model Z sells for $100. The following matrix multiplication represents a teacher's order at the store.

$$\begin{bmatrix} 5 & 11 & 7 \end{bmatrix} \begin{bmatrix} 150 \\ 130 \\ 100 \end{bmatrix}$$

a. How many of each model did the teacher order?

b. What is the total cost of the order?

REPRESENTATIONS Pictures, graphs, or objects that illustrate concepts

| OBJECTIVE K Graph figures and their transformation images. (Lessons 4-4, 4-5, 4-6, 4-7, 4-8, 4-10)

57. a. Graph the polygon *HELP* described by the matrix $\begin{bmatrix} -3 & 1 & 1 & -3 \\ 2 & 1 & 5 & 7 \end{bmatrix}$.

 b. Use matrix multiplication to find the image of *HELP* under $S_{2,3}$.

 c. Graph the image of *HELP*.

 d. Is the image similar to the preimage? Explain why or why not.

58. A popular road atlas company uses a 1:7,500,000 scale to represent roads on a map. Consider the actual road to be the preimage. Write a matrix that could be used to transform the road to its map representation.

59. Consider $\triangle ABC$ defined by $\begin{bmatrix} -5 & 0 & -3 \\ 1 & 0 & -3 \end{bmatrix}$.

 a. Graph $\triangle ABC$ and $\triangle A'B'C'$, the image of $\triangle ABC$ under $r_{y=x} \circ r_x$.

 b. What single transformation maps $\triangle ABC$ to $\triangle A'B'C'$?

In 60 and 61, refer to polygon BRUCE and its image $B'R'U'C'E'$ shown below.

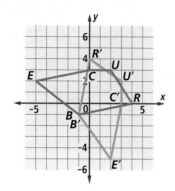

60. a. What transformation maps *BRUCE* onto $B'R'U'C'E'$?

 b. Show how the coordinates of $B'R'U'C'E'$ can be derived by matrix multiplication.

61. a. Write the image $B''R''U''C''E''$ of $B'R'U'C'E'$ under r_x as a matrix and draw the image.

 b. Is $B''R''U''C''E''$ congruent to $B'R'U'C'E'$? Explain.

 c. Are the two images similar? Explain.

Chapter

5 Systems

Contents

The Fédération Internationale de Football Association (FIFA) is the international body that governs the game known in the United States as soccer. FIFA makes the rules for the soccer balls allowed in tournament play. Some requirements are:

· The circumference of the ball must be from 68 to 70 cm.
· The weight of the ball must be from 410 to 450 grams at the start of a match.
· The inflated pressure of the ball should be from 8.5 to 15.6 psi (pounds per square inch) at sea level.

Suppose C represents the circumference of the ball, w represents its weight, and p represents its pressure. The requirements on the previous page imply that a soccer ball must satisfy the following constraints:

$$\begin{cases} 68 \text{ cm} \leq C \leq 70 \text{ cm} \\ 410 \text{ grams} \leq w \leq 450 \text{ grams} \\ 8.5 \text{ psi} \leq p \leq 15.6 \text{ psi} \end{cases}$$

Notice that each *constraint* above involves just one variable. In other situations, a constraint may involve more than one variable.

When mathematical conditions are joined by the word *and*, the set of conditions or sentences is called a *system*. Systems have many applications. For instance, in the early 19th century, the German mathematician Carl Friedrich Gauss used systems of equations to calculate the orbits of asteroids from sightings made by a few astronomers. During World War II, the American mathematician George Dantzig was charged with the task of optimizing the distribution of forces and equipment for the U.S. Air Force. His work in this area led to advances in the field of *linear programming*. In 1945, the American economist George Stigler used linear programming to determine a best diet for the least cost. For his work, Stigler received a Nobel Prize in 1982. (This prize was in economics; there is no Nobel Prize in mathematics.) In this chapter you will study systems of equations and inequalities and their applications in the field of linear programming.

Lesson

5-1

Inequalities and Compound Sentences

Vocabulary

compound sentence

intersection of two sets

double inequality

union of two sets

▶ **BIG IDEA** The graph of a linear inequality in one variable is a ray either with or without its endpoint.

Compound Sentences

In English, you can use the words *and* and *or* as conjunctions to join two or more clauses. In mathematics, these two words are used in a similar way. A sentence in which two clauses are connected by the word *and* or by the word *or* is called a **compound sentence**. For example, the requirement that a soccer ball's circumference C satisfy the single sentence 68 cm $\leq C \leq$ 70 cm is mathematical shorthand for the compound sentence "68 cm $\leq C$ *and* $C \leq$ 70 cm."

You can enter compound sentences on a CAS.

Mental Math

Find the union or intersection.

a. $\{1, 4, 7, 11\} \cap \{3, 5, 7, 9, 11\}$

b. the set of all odd integers $\cup$ the set of all even integers

c. the set of all odd integers $\cap$ the set of all even integers

d. the set of all real numbers $\cup$ the set of all integers

Activity

MATERIALS CAS

Step 1 Make a table like the one below. Clear all variables in your CAS memory, then enter each compound sentence into your CAS and record the output.

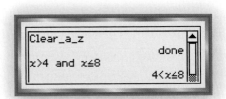

```
Clear_a_z
                              done
x>4 and x≤8
                          4<x≤8
```

Compound Sentence Entry	$x > 4$ and $x \leq 8$	$x > 4$ or $x \leq 8$	$a < 5$ and $a < 20$	$a < 5$ or $a < 20$	$n < 0$ and $n > 5$	$n < 0$ or $n > 5$
CAS Output	?	?	?	?	?	?

Step 2 In some cases, the output is only one word, either true or false. What do you notice about situations that lead to these results?

Step 3 Examine each pair of compound sentences that only differ by the connecting word *and* versus *or* (such as "$x > 4$ and $x \leq 8$" and "$x > 4$ or $x \leq 8$"). In each pair, which output includes more values in its solution? Explain why you think this is the case.

Step 4 Store values for variables x, a, and n, as indicated in the table below. Then enter each sentence and record the output.

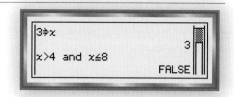

Stored Variable	$3 \to x$	$3 \to x$	$1 \to a$	$1 \to a$	$2 \to n$	$2 \to n$
Compound Sentence Entry	$x > 4$ and $x \leq 8$	$x > 4$ or $x \leq 8$	$a < 5$ and $a < 20$	$a < 5$ or $a < 20$	$n < 0$ and $n > 5$	$n < 0$ or $n > 5$
CAS Output	?	?	?	?	?	?

Step 5 Summarize the results of Step 4. In general, when are outputs of compound sentences with assigned variables true and when are they false?

 QY1

▶ **QY1**

If 8 is stored as x, what output do you expect from your CAS if you enter "$x = 8$ and $x = 5$"? If you enter "$x = 8$ or $x = 5$"?

Compound Sentences Using the Word *and*

The solution set for a compound sentence using *and* consists of the *intersection* of the solution sets of the individual sentences. Recall that the **intersection of two sets** is the set consisting of those values common to both sets. The graph of the intersection consists of the points common to the graphs of the individual sets.

For instance, roller coasters are classified according to maximum height achieved during the ride. A gigacoaster is a roller coaster that reaches a height of at least 300 feet, but less than 400 feet. In symbols, $H \geq 300$ and $H < 400$. The solution sets of the individual sentences, $\{H \mid H \geq 300\}$ and $\{H \mid H < 400\}$, are *intervals*.

The graph of the interval $\{H \mid H \geq 300\}$ is a ray. The graph of $\{H \mid H < 400\}$ is a ray without its endpoint.

$\{H \mid H \geq 300\}$

$\{H \mid H < 400\}$

The intersection of these two graphs is the graph of required height. The intersection can be described by the single compound sentence $300 \leq H < 400$, which is called a **double inequality** because it has two inequality symbols.

$300 \leq H < 400$

Recall that the symbol used for intersection is $\cap$. So the intersection of sets A and B is written $A \cap B$. In set notation,

$$\{H \mid 300 \leq H < 400\} = \{H \mid H \geq 300\} \cap \{H \mid H < 400\}.$$

This can be read "the set of numbers from 300 up to but not including 400 equals the intersection of the set of numbers greater than or equal to 300 and the set of numbers less than 400."

When describing an interval in English, it is sometimes difficult to know whether endpoints are included. In this book, we use the following language:

"x is *from* 3 to 4" means $3 \leq x \leq 4$. The endpoints are included.

"x is *between* 3 and 4" means $3 < x < 4$. The endpoints are not included.

This is consistent with the use of the word "between" in geometry. When just one endpoint is included, as in "$3 \leq x < 4$," we say "x is 3 or between 3 and 4," or "x is at least 3 and less than 4."

 QY2

Compound Sentences Using the Word *or*

Recall that the **union of two sets** is the set consisting of those values in *either one or both* sets. This meaning of *or* is somewhat different from the everyday meaning *either, but not both*. The symbol often used for union is ∪. The union of sets A and B is written $A \cup B$.

The solution set for a compound sentence using *or* consists of the union of the solution sets to the individual sentences. For example, suppose you are the quality-control examiner in a factory that produces cell-phone batteries. For one type of battery, the specification states that battery length L must be 3 cm ± 0.06 cm. (Recall that the symbol ± means *plus or minus*.) Batteries produced with a length outside the acceptable range are rejected. L must lie in the interval

$$3 \text{ cm} - 0.06 \text{ cm} \leq L \leq 3 \text{ cm} + 0.06 \text{ cm, or}$$

$$2.94 \text{ cm} \leq L \leq 3.06 \text{ cm.}$$

So, the battery will be rejected if $L < 2.94$ or $L > 3.06$.

In set notation, $\{L \mid L < 2.94 \text{ or } L > 3.06\} =$
$\{L \mid L < 2.94\} \cup \{L \mid L > 3.06\}.$

These sets are graphed below.

 QY3

▶ QY2

Graph

$\{x \mid 4 < x \text{ and } x \leq 8\}.$

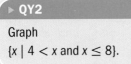

▶ QY3

Graph

$\{A \mid A < 12 \text{ or } A > 65\}.$

Solving Linear Inequalities

Solving a linear inequality is very much like solving a linear equation. The only difference is that when you multiply or divide each side of an inequality by a negative number, you must *reverse* the inequality sign.

Properties of Inequality

For all real numbers a, b, and c:

Addition Property of Inequality
If $a < b$, then $a + c < b + c$.

Multiplication Property of Inequality
If $a < b$ and $c > 0$, then $ac < bc$.
If $a < b$ and $c < 0$, then $ac > bc$.

This Example illustrates an application of linear inequalities.

Example

Penny Nichols has $500 to buy stock options at $17.50 per option. She wants to stop buying options as soon as she has less than $100. (Stock options from a company give an employee the right to buy shares of the company's stock at a designated price.)

a. How many options should she buy before stopping?
b. How much money will she have left?

Solution

a. Let s = the number of options bought. After buying s options Penny will have $500 - 17.50s$ dollars left. She will stop buying stock as soon as $500 - 17.50s < 100$. Solve the inequality.

$$500 - 17.50s < 100$$
$$-17.50s < -400 \quad \text{Subtract 500 from both sides.}$$
$$s > \frac{-400}{-17.5} \quad \text{Divide both sides by } -17.5. \text{ (Notice that the inequality sign is reversed in this step.)}$$
$$s > 22.86$$

So, Penny will have less than $100 left when $s > 22.86$. Penny should purchase 23 options before stopping.

b. To find how much money Penny has left, we substitute 23 into the expression in Part a. Penny has
$500 - $17.50(23) = $97.50 left.

Check To check the solution, use the solve command on a CAS. It gives the same result.

```
solve(500-17.5s<100,s)
          {s>22.85714286}
```

Questions

COVERING THE IDEAS

In 1–3, translate the statement into a mathematical inequality and graph the inequality.

1. To ride the T-Bar in Blizzard Beach at Disney World, you must be under 4' tall.

2. In order to fit snugly into the box, the paperweight has to be between 7.5 and 7.6 cm in diameter.

3. In Louisville, Kentucky, the number of hours of daylight in a day ranges over a year from about 14 hr 48 min to about 9 hr 29 min.

4. A new soccer ball was manufactured for the 2006 World Cup Soccer Tournament. The ball had a circumference between 69 and 69.25 cm, a weight between 441 and 444 g, and a possible weight gain due to water absorption less than 0.1%.

   a. Represent circumference C, weight w, and percent weight gain g due to water absorption as inequalities.

   b. Refer to the first page of the chapter. Does this ball meet the FIFA requirements for circumference and weight?

5. **Multiple Choice** Which inequality represents the statement "The weight L of airplane carry-on luggage may be no more than 25 pounds"?

   A $25 < L$ **B** $L \le 25$

   C $25 > L$ **D** $L \ge 25$

6. Write an inequality for the set of numbers graphed at the right.

7. a. The solution set for a compound sentence using *and* is the _____?_____ of the solution sets to the individual sentences.

   b. The solution set for a compound sentence using *or* is the _____?_____ of solution sets to the individual sentences.

8. Assume all variables are cleared in a CAS memory.

   a. **Matching** Match each CAS entry at the left with its output at the right.

       i. $x > 1$ and $x < 4$ **A** `x < 1 or x > 4`

       ii. $x > 1$ or $x < 4$ **B** `false`

       iii. $x < 1$ or $x > 4$ **C** `1 < x < 4`

       iv. $x < 1$ and $x > 4$ **D** `true`

   b. Graph the solution set for each compound sentence.

In 9 and 10, graph the solution set on a number line.

9. $x \le -7$ or $x > -2$ 10. $t \le -7$ and $t > -2$

11. Three roller coasters and their maximum heights are listed below. Which ones are gigacoasters?

Rollercoaster	Location	Maximum Height (ft)
Kingda Ka	Jackson, NJ	456
Steel Dragon 2000	Nagashima, Japan	318
Superman El Último Escape	Mexico City, Mexico	220

Source: Ultimate Rollercoaster

12. In the Example, suppose Penny had $750 to spend and stock options cost $19.00 per share. If she stops buying stock when she first has $50 or less left, how many shares of stock will she buy?

13. **Fill in the Blanks** When you ___?___ or ___?___ both sides of an inequality by a negative number, you must ___?___ the inequality sign.

In **14 and 15**, solve by hand or with a CAS. Graph all solutions.

14. $\frac{2}{3}x \le \frac{1}{2}$

15. $-5m - 0.4 > 1$

APPLYING THE MATHEMATICS

16. Suppose a small plane weighs 1615 pounds and an average passenger weighs 146 pounds.
   a. Write an expression representing the total weight of the airplane with p people on board.
   b. How many people can the airplane hold if the total weight limit is 2300 pounds?

17. *From The Mixed-up Files of Mrs. Basil E. Frankweiler* (Konigsburg, 1967) tells the story of Claudia and Jamie, who run away from home and live in the Metropolitan Museum of Art in New York City for a short while. Short on money, Jamie argues with Claudia about spending money to ride the bus. He wants to save their $24.43 for food and other necessities.

Six New York City bus routes have stops for the Metropolitan Museum of Art.

   a. In 1967, it cost 20 cents per person to ride the bus in New York City. If Jamie and Claudia both ride the bus, how many rides r could they take together? Write your answer as a double inequality that includes all possible values.
   b. In 2008, it cost $2.00 to ride the bus in New York City. Write a double inequality that includes all possible values for r in 2008.

18. Omar solved $x^2 = 5$ and wrote "$x = \sqrt{5}$ and $x = -\sqrt{5}$." What is wrong with Omar's answer?

19. Solve the compound sentence $7m + 2 < 23$ and $8 - 3m \leq 9$ for m. Write the final answer using set-builder notation and provide a graph of the solution set.

REVIEW

In 20 and 21, line m has equation $y = -2x + 8$. Line m' is the image of line m under a translation. Suppose line m' contains the point (4, 8).

20. Find an equation for m'. (Lessons 4-10, 3-4)

21. Find a translation that maps m onto m'. (Lesson 4-10)

22. Find an equation for the line that passes through the origin and is perpendicular to the line with equation $2x - 3y = -6$. (Lessons 4-9, 3-4)

23. **Multiple Choice** Which of the equations below is equivalent to the equation $4x - 3y = 12$? (Lesson 3-3)

A $4x + 3y = 12$ 　　　　　B $-3x + 4y = 12$

C $8x - 6y = 24$ 　　　　　D $-8x + 6y = 2$

24. A *duathlon* is a sporting event involving running and cycling. While training for a duathlon, Gustavo ran for R hours at 10 kilometers per hour and cycled for C hours at 28 kilometers per hour. He went a total of 50 kilometers. (Lessons 3-3, 3-2)

　a. Write a linear combination equation describing this situation.

　b. Graph your equation from Part a with R as the independent variable.

　c. What does the point (1.5, 1.25) represent on the graph?

EXPLORATION

25. Data on average daily temperature highs and lows are collected for many cities by month, and published on the Internet. Find January temperature data for three different cities. For each city, write the temperature range of average daily lows and highs as a compound inequality.

26. In ordinary usage, replacing *or* by *and* can dramatically change the meaning of a sentence. For instance, "Give me liberty and give me death" differs from Patrick Henry's famous saying only by that one word. Find examples of other sayings that have a change in meaning when "and" is replaced with "or," or vice versa.

 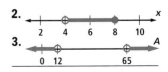

Lesson

5-2

Solving Systems Using Tables, Graphs, or a CAS

Vocabulary

system

solution set of a system

▶ **BIG IDEA** The solution(s) to a system of equations in two variables can be estimated by examining a table or graph and often solved exactly by using a CAS.

Minnie Strikes and Noah Spares are in a bowling league in which players are ranked using a handicap system. The maximum handicap is 50 pins per game and depends on the player's average. Minnie has a high average, so she has a low 5-pin handicap. Noah has a low average; his handicap is 45 pins. Noah practices, and his handicap decreases by 5 pins each month. Minnie does not practice, and her handicap increases by 3 pins per month. So after x months; Noah's handicap $N(x)$ is $45 - 5x$ and Minnie's handicap $M(x)$ is $5 + 3x$.

If the situation continues, then at some point Minnie's and Noah's handicaps will be the same. To determine when this will happen, you can solve a *system of equations*. Remember that a **system** is a set of conditions joined by the word *and*. Thus, if we call the handicaps y, the following compound sentence models the situation where the handicaps are equal: $y = 45 - 5x$ *and* $y = 5 + 3x$.

A system is often denoted by a brace: $\begin{cases} y = 45 - 5x \\ y = 5 + 3x \end{cases}$.

The **solution set of a system** is the intersection of the solution sets of the individual sentences in the system.

Mental Math

Suppose that the amount of hair a cat sheds varies directly as the square of the cat's height.

a. Fraidy is 3 times as tall as she was when she was a kitten. How much hair does she shed now compared to when she was a kitten?

b. Mittens is 1.2 times as tall as Fluffy. How much hair does Mittens shed compared to Fluffy?

c. Suppose that a Burmese sheds four times as much as a Siamese. How many times as tall as the Siamese is the Burmese?

Finding Solutions Using Tables and Graphs

To solve a system, you can create a table or a graph. To speed up graphing, use the equations in slope-intercept form.

Noah's handicap $N(x) = -5x + 45$
Minnie's handicap $M(x) = 3x + 5$

The table and graph show that the handicaps $M(x)$ and $N(x)$ both equal 20 when $x = 5$.
They also show that when $x < 5$, Minnie has a lower handicap than Noah. When $x > 5$, Noah's handicap is lower.

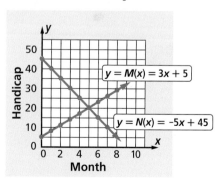

Month	$N(x)$	$M(x)$
0	45	5
1	40	8
2	35	11
3	30	14
4	25	17
5	20	20
6	15	23
7	10	26

Solutions to systems of equations can be written in several ways. For example, the solution to Minnie and Noah's system can be expressed by

   (1) listing the solution: $(5, 20)$.

   (2) writing the solution set: $\{(5, 20)\}$.

   (3) writing a simplified equivalent system: $\begin{cases} x = 5 \\ y = 20 \end{cases}$.

 QY1

> ▶ **QY1**
>
> If Minnie's handicap increased by 5 pins each month, when would her handicap be equal to Noah's?

Solving Systems with a CAS

You can also use a CAS to solve a system of equations. This is a good method if solutions are not easily found in tables or on graphs.

> **Example 1**
>
> Solve the system $\begin{cases} y = \frac{1}{2}x - 5 \\ y = 2x - 1 \end{cases}$.
>
> **Solution** Most CAS have more than one way to enter and solve systems. Two different approaches on one machine are shown at the right. The first approach enters the system as a compound sentence. The second enters the system using a brace. You should find out how to do this on your CAS.
>
> So the solution is $\left(-\frac{8}{3}, -\frac{19}{3}\right)$.
>
> $\mathrm{solve}\left(y = \frac{1}{2} \cdot x - 5 \text{ and } y = 2 \cdot x - 1, \{x, y\}\right)$
>
> $\qquad x = \dfrac{-8}{3} \text{ and } y = \dfrac{-19}{3}$
>
> $\mathrm{solve}\left(\begin{cases} y = \dfrac{1}{2} \cdot x - 5 \\ y = 2 \cdot x - 1 \end{cases}, \{x, y\}\right)$
>
> $\qquad x = \dfrac{-8}{3} \text{ and } y = \dfrac{-19}{3}$

STOP QY2

> ▶ **QY2**
>
> Verify the solution to Example 1 on your CAS.

Solving Nonlinear Systems

Systems can involve nonlinear equations.

Example 2

Kaila is a packaging engineer at Healthy Soup Company. The marketing department has requested that she design a new soup can. The new can must hold 360 mL of soup and be in the shape of a cylinder. An eye-catching new label must be large enough to show the product name and all the nutrition content data. Marketing wants the area of the label to be about 225 cm². What are the dimensions of this new can?

Solution Define variables. Let h = the height of the can and r = the radius of the can.

Draw a picture, as shown at the right.

Write a system of equations. The volume of the can is given by the formula $V = \pi r^2 h$. The can must hold 360 cm³ of soup (1 mL = 1 cm³). So, $\pi r^2 h = 360$.

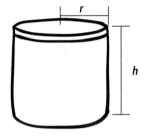

The area of the rectangular label that wraps around the can is the product of the can's circumference $2\pi r$ and its height h. Marketing wants the area to be about 225 cm². So, $2\pi rh = 225$.

So, a **system describing the situation is** $\begin{cases} \pi r^2 h = 360 \\ 2\pi rh = 225 \end{cases}$.

Solve the system. One way to solve this system is by graphing. First, solve each equation for h so that you can enter the functions into a graphing utility.

$$\begin{cases} h = \dfrac{360}{\pi r^2} \\ h = \dfrac{225}{2\pi r} \end{cases}$$

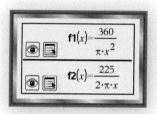

Think of r as the independent variable and h as the dependent variable. Enter the equations into a graphing utility as shown at the right. r becomes x and h becomes $f(x)$.

Both r and h must be positive, so only consider the branches of these inverse variation and inverse square variation graphs in the first quadrant. These branches intersect at only one point. Use the `intersect` command on your graphing utility to find the coordinates of the point of intersection.

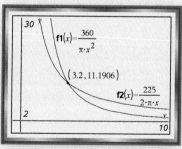

The display shows the point of intersection is about (3.2, 11.2), so the radius of the can is approximately 3.2 cm and the height is approximately 11.2 cm.

This is the only solution that meets the criteria of a 360 mL volume and a 225 cm² label area.

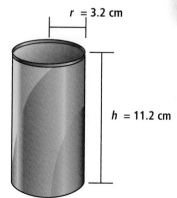

r = 3.2 cm

h = 11.2 cm

When a system is nonlinear, it is possible for it to have more than one solution, as in Example 3.

GUIDED

Example 3

Solve the system $\begin{cases} y = 4x^2 \\ 2y + 6x = 14 \end{cases}$.

Solution 1 First, solve both equations for y.

The first equation is already solved for y. The second equation is equivalent to $y = \underline{\ ?\ } + 7$.

Graph both equations in a window that shows all the points of intersection.

Window dimensions: $\underline{\ ?\ } \leq x \leq \underline{\ ?\ }$ and $\underline{\ ?\ } \leq y \leq \underline{\ ?\ }$.

Use the intersection feature of the graphing utility to identify both intersection points.

The graphs intersect at $(\underline{\ ?\ }, \underline{\ ?\ })$ and $(\underline{\ ?\ }, \underline{\ ?\ })$.

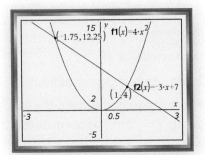

Solution 2 Solve on a CAS.

So the two possible solutions are $(\underline{\ ?\ }, \underline{\ ?\ })$ and $(\underline{\ ?\ }, \underline{\ ?\ })$.

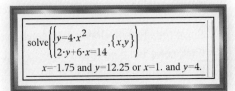

You will learn an algebraic method of solution later in this book.

Questions

COVERING THE IDEAS

1. What is a system?

In 2 and 3, refer to Minnie and Noah's situation.

2. After how many months is Noah's handicap less than Minnie's?

3. Suppose Noah's handicap has decreased to 40 pins, while Minnie's handicap has increased to 20 pins. If Noah's handicap continues decreasing by 3 pins each month and Minnie's declines by 1 pin each month, when will their handicaps be equal?

   a. Solve this new system of equations by using the solve command.

   b. Solve the system by graphing.

   c. Which method do you prefer and why?

4. **a.** Use brace notation to write the system $y = 7x - 4$ and $y = 10$.

   b. Solve the system in Part a using any of the four methods below. Check your solution with any of the other methods.

   i. by hand

   ii. using the `solve` command

   iii. using a graphing utililty to find the intersection points

   iv. using a table of values

   c. Explain why you chose the methods that you used in Part b.

5. **a.** Solve the system $\begin{cases} y = \frac{1}{2}x - 1.5 \\ y = 2x + 3 \end{cases}$ by graphing.

   b. What do you notice about the relationship between the two lines graphed in Part a?

   c. Check your answer to Part a on a CAS using the `solve` command.

In 6 and 7, refer to Example 2.

6. We only looked for intersection points in the first quadrant. Why?

7. **a.** How can you tell that (3.2, 11.2) is an approximate and not an exact solution?

   b. A CAS gives the solution (3.2, 11.190582). Is it an exact solution? Justify your answer.

8. Refer to Example 3. Would a window with $-2 \le x \le 4$ and $1 \le y \le 13$ show both intersections of the graphs? Why or why not?

APPLYING THE MATHEMATICS

In 9–11, use the given systems and their graphs.

   a. Tell how many solutions the system has.

   b. Estimate the solutions, if there are any, to the nearest hundredth.

   c. Verify that your solutions satisfy all equations of the system.

9. $\begin{cases} y = 2x^2 \\ y = -3x + 4 \end{cases}$ 10. $\begin{cases} y = -x \\ xy = 4 \end{cases}$ 11. $\begin{cases} y = 7 \\ x^2y = 4 \end{cases}$

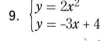

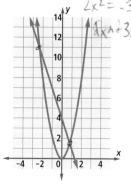 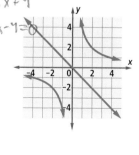 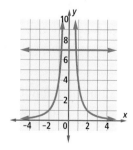

In **12** and **13**, a system is given.

a. **Graph the system.**

b. **Tell how many solutions the system has.**

c. **Estimate any solutions to the nearest tenth.**

12. $\begin{cases} y = -3x^2 \\ x - y = 1 \end{cases}$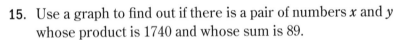

13. $\begin{cases} y = \frac{-6}{t^2} \\ y = 12 \end{cases}$

14. The Aguilar family needs to hire a moving company. They received two quotes.

    Company 1: $1000 plus $0.10 per pound
    Company 2: $750 plus $0.20 per pound

The Aguilars would like to find the value of the *break-even* point, where the cost is the same with both movers.

a. Write a system describing this situation.

b. Find the break-even point on a CAS by making a table of values starting at 0 and with a step size of 500.

c. What is an appropriate domain for the functions of this system? Explain.

15. Use a graph to find out if there is a pair of numbers x and y whose product is 1740 and whose sum is 89.

16. Tomi is planting a vegetable garden that will be rectangular in shape. He has purchased 72 linear feet of fencing material to enclose the garden. He has enough fertilizer to take care of 320 square feet of garden.

a. Write a system of equations relating the length L and width W of the rectangle to represent this situation.

b. Assume that L is the independent variable. Rewrite each equation of the system to give W in terms of L.

c. Solve the system using any method.

REVIEW

17. Solve $2 - 4x < 5$ and graph your solution on a number line. (**Lesson 5-1**)

18. Alkas is throwing a party. She tells Felix to bring a friend, but no more than 5 friends. Write a double inequality describing the number F of friends Felix can bring to the party without upsetting Alkas. (**Lesson 5-1**)

19. Find the 2×2 matrix for a transformation that maps the point (1, 0) onto (–4, 1) and the point (0, 1) onto (–1, –4). (**Lesson 4-7**)

20. **True or False** If $Ax + By$ is a linear combination of x and y, and $Cx + Dy$ is another linear combination of x and y, then the sum of these two expressions is a linear combination of x and y. **(Lesson 3-2)**

21. In basketball, a player's effective shooting percentage e is given by the formula $e = \frac{f + 1.5t}{a}$, where f is the total number of two-point shots made, t is the number of three-point shots made, and a is the total number of shots attempted. Write a formula for the number t of three-point shots a player has made in terms of e, f, and a. **(Lesson 1-7)**

EXPLORATION

22. The break-even point is used often when making business decisions. Do some research about how the break-even point is applied in different types of financial analyses. Write a summary of your findings.

Lesson

5-3

Solving Systems Using Substitution

▶ **BIG IDEA**   Some systems of equations in two (or more) variables can be solved by solving one equation for one variable, substituting the expression for that variable into the other equation, and solving the resulting equation.

Tables and graphs can be used to solve systems, but they do not always give exact solutions. You can find exact solutions with paper and pencil or on a CAS by using the **Substitution Property of Equality**: if $a = b$, then a may be substituted for b in any arithmetic or algebraic expression. For instance, if $H = 4V$, you can substitute $4V$ for H in any other expression. The following Activity illustrates how to use the Substitution Property of Equality as part of a **substitution method** to solve a system of two equations.

Mental Math

Find the perimeter of the polygon.

a. a square with side length $x + 1$

b. a right triangle with leg lengths 3 and 5

c. a parallelogram with one side of length 5 and one side of length L

d. a regular n-gon with side length $\dfrac{\sqrt{3}}{2}$

Activity

MATERIALS CAS

The Policeville Bandits sports stadium seats 60,000 people. Suppose 600 tickets are reserved for the two teams. The home team gets 4 times as many tickets as the visiting team. Let H be the number of tickets for the Bandits and V be the number of tickets for the visiting team. How many tickets does each team receive?

Step 1   Work with a partner. Write a system of equations describing this situation.

Step 2   Have one partner solve one equation in the system for H and use the Substitution Property of Equality to rewrite the other equation in terms of V only. Have the other partner solve one equation in the system for V and then rewrite the other equation in terms of H only.

Step 3   Solve your last equation in Step 2 for the remaining variable. Then substitute your solution in one of the equations in the original system and solve for the other variable. How many tickets does each team receive?

Step 4   Check your answers from Step 3 by comparing them to your partner's answers. Was one substitution easier than the other? Why or why not?

Step 5 A CAS applies the Substitution Property of Equality when the `such that` command is used. Enter the equations into a CAS using the `such that` command, and explain the result.

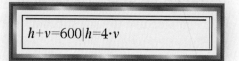

$$h+v=600 \mid h=4 \cdot v$$

Step 6 Solve the result of Step 5 for V.

Step 7 Use the second equation in the system and the such that command to find H for your value of V from Step 6.

Solving Systems with Three or More Linear Equations

You can also use the substitution method when there are more than two variables and two equations.

GUIDED

Example 1

Able Baker makes a total of 160 dozen regular muffins, mini-muffins, and jumbo muffins for his stores each day. He makes twice as many regular muffins as mini-muffins and 5 times as many jumbo muffins as mini-muffins. How many of each does he make?

Solution Solve by hand using substitution.

Let R = the number of dozens of regular muffins,

   M = the number of dozens of mini-muffins, and

   J = the number of dozens of jumbo muffins.

Then the system to be solved is
$$\begin{cases} R + M + J = 160 \\ R = \underline{\ ?\ } . \\ J = \underline{\ ?\ } \end{cases}$$

Substitute your expressions for R and J in the last two equations into the first equation and solve for M.

$$\underline{\ ?\ } + M + \underline{\ ?\ } = 160$$
$$8M = 160$$
$$M = \underline{\ ?\ } \text{ dozen}$$

Substitute to find R. $R = \underline{\ ?\ }$ dozen

Substitute to find J. $J = \underline{\ ?\ }$ dozen

Carl makes __?__ dozen regular muffins, __?__ dozen mini-muffins, and __?__ dozen jumbo muffins.

Check Does R = 2M? __?__ Does J = 5M? __?__ Does R + M + J = 160? __?__ If the answers are yes, then your solution to the system is correct.

Solving Nonlinear Systems

You can also use the substitution method to solve some systems with nonlinear equations. Write one equation in terms of a single variable, and substitute the expression into the other equation.

Example 2

Solve the system $\begin{cases} y = 4x \\ xy = 36 \end{cases}$.

Solution Substitute $4x$ for y in the second equation and solve for x.

$$x(4x) = 36 \quad \text{Substitute.}$$
$$4x^2 = 36 \quad \text{Simplify.}$$
$$x^2 = 9 \quad \text{Divide both sides by 4.}$$
$$x = 3 \text{ or } x = -3 \quad \text{Take the square root of both sides.}$$

The word *or* means that the solution set is the union of all possible answers. So substitute each value of x into either of the original equations to get two corresponding values of y. We substitute into $y = 4x$.

If $x = 3$, then $y = 4(3) = 12$. If $x = -3$, then $y = 4(-3) = -12$.

The solution set is $\{(3, 12), (-3, -12)\}$.

Check Graph the equations. This calculator display shows the two solutions (3, 12) and (−3, −12). It checks.

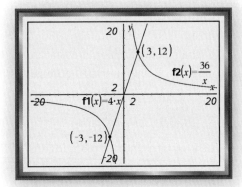

Consistent and Inconsistent Systems

A system that has *one or more solutions* is called a **consistent system**. Examples 1 and 2 involve consistent systems. A system that has *no solutions* is called an **inconsistent system**. Example 3 illustrates an inconsistent system.

Example 3

Solve the system $\begin{cases} x = 4 - 3y \\ 2x + 6y = -2 \end{cases}$.

Solution 1 Substitute $4 - 3y$ for x in the second equation.

$$2(4 - 3y) + 6y = -2$$
$$8 - 6y + 6y = -2 \quad \text{Use the Distributive Property.}$$
$$8 = -2 \quad \text{Add like terms.}$$

We came up with the statement $8 = -2$, which is never true! This false conclusion indicates that what we started with is impossible. That is, **the system has no solutions. In other words, the solution is Ø, the empty set.** A graph of the system shows parallel lines, which supports this conclusion.

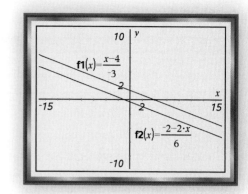

Solution 2 Solving the system on a CAS returns `false`, as shown below. This indicates that **the system is inconsistent and has no solution.**

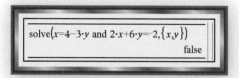
$$\text{solve}(x=4-3{\cdot}y \text{ and } 2{\cdot}x+6{\cdot}y=-2,\{x,y\})$$
$$\text{false}$$

Example 4 illustrates a consistent system with infinitely many solutions.

Example 4

Solve the system $\begin{cases} y = -3x + 2 \\ 15x + 5y = 10 \end{cases}$.

Solution 1 Substitute $-3x + 2$ for y in the second equation.

$15x + 5(-3x+2) = 10$ Substitute.

$15x - 15x + 10 = 10$ Use the Distributive Property.

$10 = 10$ Add like terms.

The statement $10 = 10$ is *always* true. **The solutions to the system are all ordered pairs satisfying either equation:** $\{(x, y)|\ y = -3x + 2\}$. The graph of each equation is the same line. So, this system has an *infinite number of solutions.*

Solution 2 Substituting the first equation into the second on a CAS returns `true`. This means that every solution to the first equation is a solution to the second equation.

$$15{\cdot}x+5{\cdot}y=10|y=-3{\cdot}x+2 \qquad \text{true}$$

 QY

> ▶ **QY**
>
> Show that the two equations in Example 4 are equations describing the same line by putting the second one in slope-intercept form.

Questions

COVERING THE IDEAS

1. State the Substitution Property of Equality.

2. Solve $\begin{cases} x + y = 20 \\ x = 2y \end{cases}$ using the such that comand on a CAS.

In 3 and 4, refer to Example 1.

3. After the expressions in the second and third equations are substituted into the first equation, how many variables are in this new equation?

4. For a holiday when business was slow, Able Baker cut back the total number of muffins he made to 120 dozen. How many jumbo muffins did he make?

5. Verify that $(-3, -12)$ is a solution to the system of Example 2.

In 6 and 7, solve the system by hand using the substitution method.

6. $\begin{cases} y = 4 - 2x \\ 3x + 4y = 11 \end{cases}$

7. $\begin{cases} x - 2y + 4z = 9 \\ x = 2z + 2 \\ y = -4z \end{cases}$

In 8–10, a system is given.

    a. Solve the system using substitution.

    b. Tell whether the system is consistent or inconsistent.

    c. Graph the system to verify your answers to Parts a and b.

8. $\begin{cases} y = 5x \\ xy = 500 \end{cases}$

9. $\begin{cases} y = \frac{2}{3}x + 3 \\ x = \frac{3}{2}y - 4 \end{cases}$

10. $\begin{cases} y = 2.75x - 1 \\ 11x - 4y = 4 \end{cases}$

11. Two lines in a plane are either the same line, parallel non-intersecting lines, or lines that intersect in a single point. What does this tell you about the possible number of points in the solution set of a system of linear equations in two variables?

APPLYING THE MATHEMATICS

12. Cheap Airways offers to fly the Underdog team to the playoffs for $250 per person. Comfort Flights offers to take the team for $1500 plus $100 per person. Let x = the number of team members on the flight and y = the total cost of the flight. The Underdog coach, who is also a math teacher, wants to analyze the rates to get the best deal.
    a. Write a system of equations to describe the relationship between x and y for the two airlines.
    b. For what number of passengers will the cost be the same for both airlines?
    c. What is this cost?
    d. The Underdog team and staff consists of 23 people. Which airline will cost the least? How much is saved over the other airline?

13. Consider the system $\begin{cases} y = x^2 \\ y = x - 3 \end{cases}$.
    a. Solve the system by substitution.
    b. Tell whether the system is consistent or inconsistent.
    c. Graph the system to verify your answer to Parts a and b.

14. For the system $\begin{cases} xy = 4 \\ y = x \end{cases}$, a student wrote "The solution is $x = -2$ or 2 and $y = -2$ or 2." Explain why this is *not* a correct answer.

15. Sand, gravel, and cement are mixed with water to produce concrete. One mixture has these three components in the extended ratio 2:4:1. For a total of 50 cubic yards of concrete, how much sand, gravel, and cement should be mixed?
    a. Write a system of three equations to describe this situation.
    b. Solve the system to determine how much of each ingredient should be used.

16. Suppose the circumference C of a circle is 1 meter longer than the diameter d of the circle.
    a. Write a system of equations relating C and d.
    b. Solve the system.
    c. Estimate the radius of the circle to the nearest millimeter.

To test the consistency of concrete, inspectors perform a slump test by forming a cone of concrete and measuring how much it slumps due to gravity.

REVIEW

17. The Outdoors Club at Larkchester High School budgeted $20 for the year to print color flyers for their trips. Printing costs 9 cents per page. **(Lesson 5-1)**

    a. Write an inequality relating the number x of flyers the Outdoors Club can print this year.

    b. Solve your inequality from Part a.

    c. How many flyers can the club print?

18. a. Write a matrix for $\triangle HAW$ at the right.

    b. Use matrix multiplication to determine the coordinates of $\triangle H'A'W'$, the reflection image of $\triangle HAW$ over the line $y = x$. **(Lessons 4-6, 4-1)**

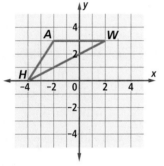

19. Give the domain and range of each variation function. **(Lessons 2-6, 2-5, 2-4)**

    a. $d = \dfrac{10}{t}$

    b. $y = -2.5x^2$

    c. $I = \dfrac{300}{d^2}$

    d. $C = 10.09n$

20. Tell whether each relation is a function. Justify your answer. **(Lesson 1-2)**

    a. $g: x \to 3x^2$

    b. $\{(-2, 0), (1, 4), (2, 14), (1, 3), (-2, -5)\}$

    c.

s	-1	-0.5	0	0.5	1	1.5	2
r	-12	-6	0	-6	-12	-18	-24

    (Treat s as the independent variable.)

    d. The relation that maps Blake's age onto his height at that age.

EXPLORATION

21. Consider the system $\begin{cases} y = x^2 \\ y = x + k \end{cases}$. Explore these equations and find the values of k for which this system has

    a. two solutions.

    b. exactly one solution.

    c. no solution.

Lesson 5-4

Solving Systems Using Linear Combinations

Vocabulary

linear-combination method

▶ **BIG IDEA** Some systems of equations can be solved by creating a new equation that is a linear combination of the original equations.

Mental Math

Write a linear combination to answer the question. How much did I spend if I bought

a. p pairs of shoes at $39 each and s pairs of socks at $4 each?

b. 3 pairs of shoes at $$d_1$ each and 6 pairs of socks at $$d_2$ each?

c. x pairs of shoes at $$d_1$ each and y pairs of socks at $$d_2$ each?

In the last lesson, you solved systems using the Substitution Property, which may be faster than using a CAS in cases when one equation is already solved for one variable in terms of the other variables. When linear equations are written in standard form, it may be more efficient to solve the system using the Addition and Multiplication Properties of Equality.

Recall that an expression of the form $Am + Bn$ is called a *linear combination* of m and n. In this lesson we use what we call the **linear-combination method** of solving systems because it involves adding multiples of the given equations.

Example 1

Solve the system $\begin{cases} x + 2y = 11 \\ 3x - 9y = 23 \end{cases}$ by the linear-combination method.

Solution The equations in this system represent the lines graphed at the right. The solution to the system is the point of intersection of the lines.

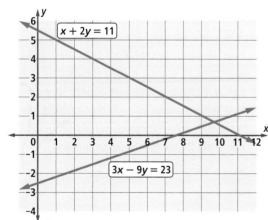

To solve the system using linear combinations, multiply the first equation by –3.

$$-3(x + 2y) = -3 \cdot 11 \quad \text{Multiplication Property of Equality}$$

$$-3x - 6y = -33 \quad \text{Distributive Property, arithmetic}$$

This equation is equivalent to the first equation of the original system. So, rewrite the system using this equation.

$$\begin{cases} -3x - 6y = -33 \\ 3x - 9y = 23 \end{cases}$$

Add these two equations. The result has no term in x.

$$-15y = -10 \quad \text{Addition Property of Equality}$$

(continued on next page)

Solve this equation for y.

$$y = \frac{-10}{-15} = \frac{2}{3}$$

Thus, $\frac{2}{3}$ is the y-coordinate of the point of intersection.

Substitute $\frac{2}{3}$ for y in any equation above to find x. We choose the first equation.

$$x + 2\left(\frac{2}{3}\right) = 11$$

$$x = 11 - \frac{4}{3} = \frac{29}{3}$$

So the solution to the system is $\begin{cases} x = \frac{29}{3} \\ y = \frac{2}{3} \end{cases}$.

You can write $(x, y) = \left(\frac{29}{3}, \frac{2}{3}\right)$.

Check 1 Substitute $\frac{29}{3}$ for x and $\frac{2}{3}$ for y in the second equation. Does $3 \cdot \frac{29}{3} - 9 \cdot \frac{2}{3} = 23$? Yes. It checks.

Check 2 The graphs of the lines with the equations $x = \frac{29}{3}$ and $y = \frac{2}{3}$ intersect at the same point as the lines in the other systems. The last system is equivalent to the other two systems on the previous page.

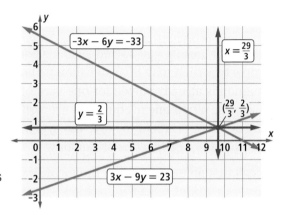

Systems arise in important practical endeavors. For example, we all need to eat foods that meet minimum requirements of protein, vitamins, minerals, and calories.

Activity

MATERIALS CAS (optional)

Long-distance runner Mary Thawn wants to get 800 calories and 50 grams of protein from a dinner course of chicken and pasta with sauce. Protein and calorie counts for each item are given below. How many ounces of each type of food does Mary need to eat? Work with a partner to solve this problem in two different ways using the linear-combination method.

Scientific studies suggest that marathoners eat 0.8–1.7 grams of protein per kilogram of body weight per day.

Step 1 Organize the given information into a table.

	Chicken	Pasta with Sauce	Total Needed
Protein	9 g/oz	2 g/oz	50 g
Calories	65 cal/oz	45 cal/oz	800 cal

Step 2 Together, write a system to model this situation if c = the number of ounces of chicken Mary should eat, and p = the number of ounces of pasta with sauce that Mary should eat.

Step 3 In the linear-combination method, you want to eliminate a variable by making the coefficients for that variable into opposites. Have one partner work to eliminate c and the other to eliminate p. By what numbers can you multiply the equations in the system to make the coefficients of your variable opposites?

Step 4 Carry out your multiplications from Step 3 and add to eliminate your variable. You can do this on a CAS or by hand. If you use a CAS, you might want to assign the two original equations to variables such as eq1 and eq2. Then you can multiply eq1 and eq2 by constants and add the results without retyping whole equations.

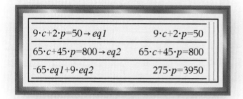

$9 \cdot c + 2 \cdot p = 50 \rightarrow eq1$	$9 \cdot c + 2 \cdot p = 50$
$65 \cdot c + 45 \cdot p = 800 \rightarrow eq2$	$65 \cdot c + 45 \cdot p = 800$
$^{-}65 \cdot eq1 + 9 \cdot eq2$	$275 \cdot p = 3950$

Step 5 Solve for the remaining variable and substitute into one of the original equations to find the value of the other variable. Then answer the question in the problem.

Step 6 Compare your results with those of your partner. Did you get the same solution? Was eliminating p easier than eliminating c, or vice versa? Was using a CAS easier or harder than doing it by hand?

 QY

Linear Combinations with Systems of Three Equations

The linear-combination method can also be used to solve a system with three linear equations. First, use a linear combination of any pair of equations to eliminate one of the variables. Then eliminate the same variable from another pair of equations by using another linear combination. The result is a system of two equations with two variables. Then you can solve the simpler system using the methods of Example 1 and the Activity.

> ▶ **QY**
>
> If m is one of your equations in the Activity, and n is the other equation, then $Am + Bn$ is the linear combination you used to eliminate your variable. What were your A and B? What were your partner's A and B?

GUIDED

Example 2

Solve the system $\begin{cases} 2x + y + 4z = 20 \\ 3x - 3y + 2z = 27 \\ 4x + 5y - 2z = 4 \end{cases}$.

Solution We choose to eliminate z first because its coefficients in the last two equations are already opposites. Add the last two equations to get an equation in terms of x and y.

(continued on next page)

$$3x - 3y + 2z = 27$$
$$4x + 5y - 2z = 4$$
$$\underline{?} + \underline{?} = 31 \quad \text{Add.}$$

Now consider the first two equations. The coefficients of z are 4 and 2, so you can multiply the second equation by -2, then add the result to the first equation to get an equation in terms of x and y.

$$2x + y + 4z = 20 \rightarrow \qquad 2x + y + 4z = 20$$
$$3x - 3y + 2z = 27 \rightarrow \underline{?} + \underline{?} - 4z = \underline{?} \qquad \text{Multiply by } -2.$$
$$\underline{?} + \underline{?} = -34 \qquad \text{Add.}$$

The result is the system below.

$$\begin{cases} -4x + 7y = -34 \\ \underline{?} + \underline{?} = \underline{?} \end{cases}$$

Continue using the linear-combination method on this system of two equations in two variables. We choose to eliminate x.

$$-4x + 7y = -34 \rightarrow \qquad -28x + \underline{?} = \underline{?} \qquad \text{Multiply by 7.}$$
$$7x + 2y = 31 \rightarrow \qquad \underline{?} + 8y = \underline{?} \qquad \text{Multiply by 4.}$$
$$\underline{?} = -114 \qquad \text{Add.}$$
$$y = \underline{?} \qquad \text{Solve for } y.$$

Substitute $y = -2$ into one of the two equations involving just two variables. We substitute into $-4x + 7y = -34$.

$$-4x + 7(-2) = -34$$
$$-4x = \underline{?}$$
$$x = \underline{?}$$

Now substitute $x = 5$ and $y = -2$ into one of the original equations of the system. We use $2x + y + 4z = 20$.

$$2(5) + (-2) + 4z = 20$$
$$4z = \underline{?}$$
$$z = \underline{?}$$

So the solution is $x = \underline{?}$, $y = \underline{?}$, and $z = \underline{?}$. This can be written as the ordered triple $(x, y, z) = (\underline{?}, \underline{?}, \underline{?})$.

Check 1 Substitute the values for x, y, and z in each of the original equations to make sure they check.

Check 2 Solve using a CAS. One CAS solution is shown at the right.

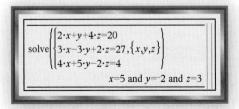

The systems in Examples 1 and 2 are consistent, and each has a unique solution. However, suppose you try to solve systems such as

$$\begin{cases} x + 2y = 11 \\ x + 2y = 12 \end{cases} \quad \text{or} \quad \begin{cases} x + 2y = 11 \\ 2x + 4y = 22 \end{cases}.$$

If you use the linear-combination method with the first system, the result is a false statement, such as $0 = 1$, so the system is inconsistent and has no solutions. However, with the second system, the linear-combination method yields a result that is always true, such as $0 = 0$. This means the system is consistent, and there are infinitely many solutions.

As Example 2 shows, the linear-combination method takes some work with a 3×3 system. This method is more complicated and impractical when there are large numbers of variables and equations in the system. In Lesson 5-6, you will see how matrices provide an efficient way to solve large systems.

Questions

COVERING THE IDEAS

1. Refer to Example 1. What properties are used to obtain the equation $-15y = -10$ from the two original equations?

2. What equation results when $5x - 4y = 12$ is multiplied by -2? What property is being applied?

3. Refer to the Activity. One student graphed the equations she obtained in Steps 4 and 5. Her graph is at the right.
   a. Which variable was eliminated in Step 4?
   b. What is an equation for the horizontal line? What is the equation for the vertical line?
   c. What is the solution to the system?

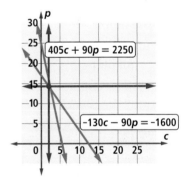

4. Refer to the system involving only x and y in Example 2. Multiply the first equation by $\frac{7}{4}$ and solve the resulting system.

5. The table at the right gives the number of grams of protein and the number of calories in one ounce of each of two foods. Mary Thawn from the Activity still wants to get 800 calories from her meal and obtain 50 grams of protein.

	Chicken Strips	French Fries
Protein	5 g/oz	1 g/oz
Calories	80 cal/oz	95 cal/oz

   a. Let h be the number of ounces of chicken strips and r be the number of ounces of french fries she eats. Write a system of equations that describes these conditions.
   b. How many ounces of each food should Mary eat?

In 6–9, solve the system using the linear-combination method.

6. $\begin{cases} a + b = \frac{1}{2} \\ a - b = \frac{1}{3} \end{cases}$

7. $\begin{cases} 4u - 2v = 24 \\ 5u + 6v = 13 \end{cases}$

8. $\begin{cases} 10g + 15h = 60 \\ 0.02g + 0.12h = 0.3 \end{cases}$

9. $\begin{cases} 2x - y + 2z = 4 \\ 5x + 2y - 3z = 43 \\ x + y - z = 11 \end{cases}$

10. In the process of solving a system using the linear-combination method, a student obtained the result $0 = 0$. How should the result be interpreted?

In 11 and 12, use the linear-combination method to determine whether the system is inconsistent or consistent.

11. $\begin{cases} 5x + y = 16 \\ 10x + 2y = 20 \end{cases}$

12. $\begin{cases} 4a + 10b = 12 \\ 6a + 15b = 18 \end{cases}$

APPLYING THE MATHEMATICS

13. Suppose 3 turkey wraps and 2 juices cost $26.50, while 2 turkey wraps and 3 juices cost $24.75. What is the cost of one juice?

14. N liters of a 1.3 moles per liter (mol/L) nitric acid solution are mixed with A liters of a 6.8 mol/L nitric acid solution. The result is 2 liters of a solution that is 3.5 mol/L.

    a. Write an equation relating N, A, and the total number of liters.

    b. The amount of nitric acid in the resulting solution is 2 liters · 3.5 mol/L, or 7 mols. Write an equation relating the amount of nitric acid in the two initial solutions and the resulting solution.

    c. Solve the system represented by your answers in Parts a and b. How many liters of each solution are needed?

Nitric acid is sometimes used in etching as a *mordant*, a solution that eats away unwanted metal.

15. Refer to Example 2.

    a. Suppose you use the first and second equations to eliminate x. What can you multiply each equation by so that when you add the results, you get an equation in terms of y and z only?

    b. Suppose you use the first and third equations to eliminate x. What can you multiply each equation by so that when you add the results, you get an equation in terms of y and z only?

    c. Do the multiplications in Parts a and b and solve the resulting system of 2 equations in y and z. Then substitute to find x.

    d. **True or False** The method used in Parts a–c gives the same solution as in Example 2.

16. Solve the system $\begin{cases} 3x^2 - 2y^2 = 40 \\ 2x^2 + 4y^2 = 48 \end{cases}$ using the linear-combination method.

REVIEW

17. Solve by substitution and check: $\begin{cases} x - 3y + 2z = 18 \\ y + z = -9 \\ z = 1 \end{cases}$ (Lesson 5-3)

18. Consider the scale change transformations with matrices $\begin{bmatrix} 2 & 0 \\ 0 & 1 \end{bmatrix}$ and $\begin{bmatrix} 1 & 0 \\ 0 & 2 \end{bmatrix}$. (Lessons 4-7, 4-5, 4-4)

   a. What is the matrix for the composite of these two transformations?

   b. What type of transformation is represented by the matrix you found in Part a?

In 19 and 20, find the product. (Lesson 4-3)

19. $\begin{bmatrix} 4 & 3 \\ -2 & 6 \end{bmatrix} \begin{bmatrix} x \\ y \end{bmatrix}$

20. $\begin{bmatrix} 2 & -4 \\ 1 & 8 \end{bmatrix} \begin{bmatrix} 0.4 & 0.2 \\ -0.05 & 0.1 \end{bmatrix}$

21. a. Rewrite the following using scalar multiples.

$$\begin{bmatrix} 4 & -6 & 5.5 \\ 0 & 3 & 0 \end{bmatrix} + \begin{bmatrix} 4 & -6 & 5.5 \\ 0 & 3 & 0 \end{bmatrix} + \begin{bmatrix} 4 & -6 & 5.5 \\ 0 & 3 & 0 \end{bmatrix} - \begin{bmatrix} 7 & 2 & 0 \\ 0 & 0 & -1 \end{bmatrix}$$
$$- \begin{bmatrix} 7 & 2 & 0 \\ 0 & 0 & -1 \end{bmatrix}$$

   b. Find the sum in Part a. (Lesson 4-2)

EXPLORATION

22. In a talk to students, mathematician Raymond Smullyan gave this problem as an example of one that can be solved without algebra:

   A family has 12 pets, a combination of cats and dogs. Every night, they give each cat two treats and each dog three treats. If they give 27 treats a night, how many of each kind of pet do they have?

   a. Show how to solve this problem without any algebra and without using trial-and-error.

   b. Show how to answer Question 13 without algebra or trial-and-error.

Lesson
5-5

Inverses of Matrices

▸ **BIG IDEA** Most 2 × 2 and most 3 × 3 matrices *A* have an inverse, A^{-1}, of the same dimensions, that satisfies $A^{-1}A = AA^{-1} = I$, where *I* is the identity matrix for their dimensions.

Are there ever times when you want to send a friend a note, but you want only your friend to be able to read it? To do this, you could use cryptography. Cryptography, which comes from the Greek words *kryptos* (hidden) and *graphia* (writing), is the study of encoding and decoding messages. In this lesson you will see how to use the *inverses of matrices* in cryptography.

Mental Math

Sheila tosses a fair 6-sided die. What is the probability that the die shows

a. a 4?

b. an even number?

c. a prime number?

d. a rational number?

What Are Inverse Matrices?

The idea of an inverse of an operation runs throughout mathematics. Recall that two real numbers *a* and *b* are *additive inverses* (or *opposites*) if and only if $a + b = 0$. For example, 7.3 and –7.3 are additive inverses. The sum of two additive inverses is 0, the additive identity. Two real numbers *a* and *b* are *multiplicative inverses* (or *reciprocals*) if and only if $ab = 1$. For example, π and $\frac{1}{\pi}$ are multiplicative inverses. The product of two multiplicative inverses is 1, the multiplicative identity.

The definition of *inverse matrices* under multiplication follows the same idea as the inverses mentioned above. The 2 × 2 matrices *A* and *B* are called **inverse matrices** if and only if their product is the 2 × 2 identity matrix for multiplication, that is, if and only if

$$AB = BA = \begin{bmatrix} 1 & 0 \\ 0 & 1 \end{bmatrix}.$$

Because matrix multiplication is not commutative, the definition of *inverse matrices* requires that the product in both orders of multiplication be the identity. To be multiplied in both directions, the matrices must be **square matrices**, those with the same number of rows and columns.

So there can be inverse 3 × 3 matrices, inverse 4 × 4 matrices, and so on. But there cannot be inverse 2 × 3 matrices. Furthermore, as you will see, not all square matrices are *invertible* (have inverses).

Notation for Inverse Matrices

The real number x^{-1} is equal to $\frac{1}{x}$, the multiplicative inverse of x, for all $x \neq 0$. Similarly, the symbol M^{-1} stands for the inverse of the square matrix M, if M has an inverse. Many graphing calculators and all CAS can display matrices and their inverses. To verify that two matrices are inverses, multiply.

Activity 1

MATERIALS CAS or graphing calculator

Step 1 Clear the variable a on your calculator. Store $\begin{bmatrix} -3 & 5 \\ 7 & -11 \end{bmatrix}$ in your calculator as variable a.

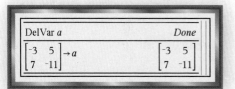

Step 2 Calculate a^{-1} by using the inverse key on your calculator, or by entering a^-1.

Step 3 Find the product aa^{-1}.

Step 4 Find the product $a^{-1}a$.

A Formula for the Inverse of a 2 × 2 Matrix

You do not need a calculator to obtain the inverse of a 2 × 2 matrix. There is a formula.

Inverse Matrix Theorem

If $ad - bc \neq 0$ and $M = \begin{bmatrix} a & b \\ c & d \end{bmatrix}$, then $M^{-1} = \begin{bmatrix} \dfrac{d}{ad-bc} & \dfrac{-b}{ad-bc} \\ \dfrac{-c}{ad-bc} & \dfrac{a}{ad-bc} \end{bmatrix}$.

Proof We need to show that the product of the two matrices in either order is the identity matrix. Below, we show one order. In Question 9, you are asked to verify the multiplication in the reverse order.

$$\begin{bmatrix} a & b \\ c & d \end{bmatrix}\begin{bmatrix} \dfrac{d}{ad-bc} & \dfrac{-b}{ad-bc} \\ \dfrac{-c}{ad-bc} & \dfrac{a}{ad-bc} \end{bmatrix}$$

$$= \begin{bmatrix} \dfrac{ad}{ad-bc} + \dfrac{-bc}{ad-bc} & \dfrac{-ab}{ad-bc} + \dfrac{ab}{ad-bc} \\ \dfrac{cd}{ad-bc} + \dfrac{-cd}{ad-bc} & \dfrac{-bc}{ad-bc} + \dfrac{ad}{ad-bc} \end{bmatrix} \quad \text{Matrix multiplication}$$

$$= \begin{bmatrix} \dfrac{ad-bc}{ad-bc} & \dfrac{0}{ad-bc} \\ \dfrac{0}{ad-bc} & \dfrac{ad-bc}{ad-bc} \end{bmatrix} = \begin{bmatrix} 1 & 0 \\ 0 & 1 \end{bmatrix} \quad \begin{array}{l}\text{Addition of fractions,} \\ \text{definition of subtraction}\end{array}$$

Not only does the Inverse Matrix Theorem tell you how to find an inverse without a calculator, it gives you a quick test for whether an inverse exists. Note that $ad - bc$ cannot be zero because it is in the denominator of all the fractions. The inverse of a 2×2 matrix $\begin{bmatrix} a & b \\ c & d \end{bmatrix}$ exists only if $ad - bc \neq 0$.

GUIDED

Example

Use the Inverse Matrix Theorem to find the inverse of each matrix, if it exists.

a. $\begin{bmatrix} -2 & 4 \\ -1 & 6 \end{bmatrix}$ b. $\begin{bmatrix} 5 & 10 \\ 3 & 6 \end{bmatrix}$

Solution

a. In $\begin{bmatrix} -2 & 4 \\ -1 & 6 \end{bmatrix}$, $a = \underline{\ ?\ }$, $b, = \underline{\ ?\ }$, $c = \underline{\ ?\ }$, and $d = \underline{\ ?\ }$.

So, $ad - bc = (\underline{\ ?\ } \cdot \underline{\ ?\ } - \underline{\ ?\ } \cdot \underline{\ ?\ }) = \underline{\ ?\ }$, and the matrix has an inverse.

Now substitute into the formula. The inverse is $\begin{bmatrix} \underline{?} & \underline{?} \\ \underline{?} & \dfrac{-2}{-8} \end{bmatrix}$, which

simplifies to $\begin{bmatrix} \underline{?} & \underline{?} \\ \underline{?} & \underline{?} \end{bmatrix}$.

b. In $\begin{bmatrix} 5 & 10 \\ 3 & 6 \end{bmatrix}$, $a = \underline{\ ?\ }$, $b, = \underline{\ ?\ }$, $c = \underline{\ ?\ }$, and $d = \underline{\ ?\ }$.

So $ad - bc = (\underline{\ ?\ } \cdot \underline{\ ?\ } - \underline{\ ?\ } \cdot \underline{\ ?\ }) = \underline{\ ?\ }$.

Because $ad - bc = \underline{\ ?\ }$, $\begin{bmatrix} 5 & 10 \\ 3 & 6 \end{bmatrix}$ has $\underline{\quad ?\quad }$.

If a matrix has no inverse, trying to find it on a calculator leads to an error message. Some calculators say the inverse is undefined, while others say the original matrix is a **singular matrix**, meaning that its inverse does not exist.

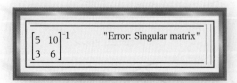

330   Systems

Determinants and Inverses

The formula in the Inverse Matrix Theorem can be simplified using scalar multiplication. When $ad - bc \neq 0$,

$$\begin{bmatrix} \frac{d}{ad-bc} & \frac{-b}{ad-bc} \\ \frac{-c}{ad-bc} & \frac{a}{ad-bc} \end{bmatrix} = \frac{1}{ad-bc}\begin{bmatrix} d & -b \\ -c & a \end{bmatrix}.$$

Because the number $ad - bc$ determines whether or not a matrix has an inverse, it is called the **determinant** of the matrix. We abbreviate the word *determinant* as *det*. Thus, the Inverse Matrix Theorem can be written:

If $M = \begin{bmatrix} a & b \\ c & d \end{bmatrix}$, then $M^{-1} = \frac{1}{\det M}\begin{bmatrix} d & -b \\ -c & a \end{bmatrix}$ if and only if det $M \neq 0$.

Calculators that handle matrices have a determinant function, usually called det.

$$\det\left(\begin{bmatrix} a & b \\ c & d \end{bmatrix}\right) \qquad a \cdot d - b \cdot c$$

 QY

Using Matrices to Encode and Decode

How can matrices and their inverses help in cryptography? At the right is a picture of a decoder ring, created as part of a promotional offer in May, 2000. On the outside circle are the letters of the alphabet and on the inside circle are the numbers 1 through 26. You are able to turn the inside dial so the number 1 can correspond to any of the letters in the alphabet. In the configuration in the photo at the right, 1 corresponds to F. This is the *key* to the code. A message gets encoded as a set of numbers. To decode the message, the recipient sets the dial according to the key and reads the letters corresponding to the numbers. For example, the code 26, 7, 21, 21, 6 in this configuration decodes as HELLO.

While encoding phrases using the ring is easy to do, it is also relatively easy to decode them. For one thing, each number always stands for the same letter. For example, each E is encoded as 7. Also, it is easy to tell the length of a word. For these reasons it would not be too difficult to break the code. Businesses and governments need a more powerful method of encryption. One such method is to use matrices.

 QY

a. Use a calculator to find the determinant of $\begin{bmatrix} -3 & 5 \\ 7 & -11 \end{bmatrix}$.

b. Check by hand.

Activity 2

MATERIALS CAS or graphing calculator

Step 1 To encode a phrase using matrices, first make a *key* by assigning a number to each letter like the ring does on the previous page. For this Activity, let $A = 1, B = 2$, and so on, and assign the number 27 to spaces between words. Use this method to make a key for MEET ME IN THE PARK in the table below. It has been started for you.

M	E	E	T		M	E		I	N		T	H	E		P	A	R	K
13	5	5	20	27	?	?	?	?	?	?	?	?	?	?	?	?	?	?

Step 2 Create a 2 × 2 *encoder matrix* e that has an inverse,

for example, $e = \begin{bmatrix} -3 & 8 \\ 2 & 5 \end{bmatrix}$.

Store this matrix in variable e in a calculator. Although we are using a 2 × 2 matrix for an encoder, any size square matrix with an inverse could be used.

Step 3 Turn the message into a matrix. Because you are using a 2 × 2 matrix as an encoder, use a matrix with 2 rows. Enter your numerical message from Step 2 starting at the left of the matrix and filling the columns from top to bottom, left to right. Fill in the empty element at the end with a 27, like the other spaces.

$$\begin{bmatrix} 13 & 5 & 27 & ? & ? & ? & ? & ? & ? & ? \\ 5 & 20 & ? & ? & ? & ? & ? & ? & ? & ? \end{bmatrix}$$

Step 4 Store your matrix from Step 3 as *m*. Multiply the encoder matrix e by m and store the product as v.

Step 5 To send the encoded message, record it as 1, 51, 145, 110, and so on, reading down the columns of v. Notice that with this encoding method, each E (or any other letter) does not get encoded to the same number, making it a more difficult code to break.

$e \cdot m \to v$
$$\begin{bmatrix} 1 & 145 & 23 & 201 & 85 & 79 & 16 & 47 \\ 51 & 110 & 119 & 145 & 88 & 154 & 41 & 134 \end{bmatrix} \blacktriangleright$$

Step 6 Imagine that you gave your encoded message to a friend. The friend is a *receiver*. The receiver first needs to rewrite the message as a 2 × 10 matrix (the one you named v). Your friend then needs a *decoder matrix* to undo the multiplication by your encoder matrix e. This matrix is e^{-1}. Multiply the inverse matrix by matrix v. This is how the receiver would get your keyed message. He or she could then use the key to read the message.

$e^{-1} \cdot v$

Step 7 If you want to send encoded messages this way, the receiver needs to know only the encoding matrix and the key. The receiver can then find the inverse matrix and apply it.

Use the encoding matrix $\begin{bmatrix} -89 & 120 \\ 35 & 6 \end{bmatrix}$ and the key from this Activity

(A = 1, B = 2, . . .) to decode the following message: –1304, 566, 589, 779, –843, 1023, 2311, 155, 2528, 442

Questions

COVERING THE IDEAS

1. Identify the identity for each operation.
   a. real-number multiplication
   b. real-number addition
   c. 2×2 matrix multiplication

2. Explain why only square matrices can have inverses.

3. If N is a 2×2 matrix and N^{-1} exists, find each product.
   a. NN^{-1}
   b. $N^{-1}N$

4. Verify that $\begin{bmatrix} -\frac{3}{4} & \frac{1}{2} \\ -\frac{1}{8} & \frac{1}{4} \end{bmatrix}$ is the inverse of $\begin{bmatrix} -2 & 4 \\ -1 & 6 \end{bmatrix}$ using matrix multiplication.

5. a. When does the matrix $\begin{bmatrix} a & b \\ c & d \end{bmatrix}$ have an inverse?
   b. What is that inverse?

6. Determine whether each matrix has an inverse.
   a. $\begin{bmatrix} 3 & 2 \\ 3 & 2 \end{bmatrix}$
   b. $\begin{bmatrix} 3 & -2 \\ -3 & 2 \end{bmatrix}$
   c. $\begin{bmatrix} 3 & -2 \\ 3 & 2 \end{bmatrix}$

In 7 and 8, give the determinant of the matrix. Give the inverse of the matrix if it exists.

7. $\begin{bmatrix} 2 & 9 \\ -7 & 6 \end{bmatrix}$

8. $\begin{bmatrix} 12 & -2 \\ 25 & 5 \end{bmatrix}$

9. Complete the second part of the proof of the Inverse Matrix Theorem. That is, show that the identity matrix is the product of the two matrices in the reverse order.

10. Refer to Activity 2. A message has the encoder matrix $\begin{bmatrix} 4 & -8 \\ 10 & -10 \end{bmatrix}$. Find its decoder matrix.

In 11 and 12, use the following key.

A	B	C	D	E	F	G	H	I	J	K	L	M
2	4	6	8	10	12	14	16	18	20	22	24	26
N	O	P	Q	R	S	T	U	V	W	X	Y	Z
1	3	5	7	9	11	13	15	17	19	21	23	25

11. Encode the message MATRICES DO CODES using the matrix $\begin{bmatrix} 2 & 1 \\ 7 & 4 \end{bmatrix}$. Use 27 to represent a space.

12. Kendra got the message 378, –202, 294, –56, 504, –162, 343, –7, 273, –51, 203, 73, the encoder matrix $\begin{bmatrix} 14 & 7 \\ -8 & 3 \end{bmatrix}$, and the key above.

    a. What is the decoding matrix?

    b. Decode the message.

APPLYING THE MATHEMATICS

13. Give an example of a 2 × 2 matrix not mentioned in this lesson that does not have an inverse.

14. If it has one, the inverse of a 2 × 2 matrix can be found by solving a pair of systems of linear equations. For example, if the inverse of $\begin{bmatrix} 1 & 4 \\ 2 & -2 \end{bmatrix}$ is $\begin{bmatrix} a & b \\ c & d \end{bmatrix}$, then $\begin{bmatrix} 1 & 4 \\ 2 & -2 \end{bmatrix}\begin{bmatrix} a & b \\ c & d \end{bmatrix} = \begin{bmatrix} 1 & 0 \\ 0 & 1 \end{bmatrix}$.

    This yields the systems $\begin{cases} a + 4c = 1 \\ 2a - 2c = 0 \end{cases}$ and $\begin{cases} b + 4d = 0 \\ 2b - 2d = 1 \end{cases}$.

    a. Solve these two systems and determine the inverse matrix.

    b. Check your answer to Part a using matrix multiplication.

15. a. Multiply the matrix for R_{90} by the matrix for R_{270} in both orders.

    b. Based on your answer to Part a, what can you say about the two matrices? Why do you think this is so?

    c. Based on your answer to Part b, find the inverse of the matrix for R_{180}. Try to get the answer *without* doing any calculations.

REVIEW

In 16 and 17, a situation is presented and a question is asked.

a. Set up a system of equations to represent the situation.

b. Solve the system using any method and answer the question.

c. Explain why you chose the method you did in Part b.
  (Lessons 5-4, 5-3, 5-2)

16. Javier is making mango-and-banana fruit salad to take to a party. Mangos cost $2.19/pound and bananas cost $0.89/pound. Javier decides to buy 5 pounds of fruit for $7.50. How many pounds of each fruit does he buy?

17. Trevon is going on a long mountain bike ride. During the ride, he wants to make sure he consumes 160 g of carbohydrates, 20 g of protein, and 3 liters of fluids. An energy bar contains 50 g of carbohydrates, 10 g of protein, and no fluids. An energy drink bottle contains 30 g of carbohydrates, no protein, and 750 mL of fluid. A water bottle contains no carbohydrates, no protein, and 750 mL of fluid. How many energy bars, energy drink bottles, and water bottles should Trevon pack for his ride?

18. Translate into a single formula: P varies directly as w and inversely as m and r^2. (Lesson 2-9)

19. A car is traveling at 60 miles per hour. (Lessons 2-4, 2-1)

    a. Write a variation equation to describe the distance d in miles the car travels in t hours.

    b. Graph your equation from Part a.

Banana plants commonly grow to full size in just a few weeks.

EXPLORATION

20. Let $A = \begin{bmatrix} 2 & 1 \\ 1 & 2 \end{bmatrix}$ and $B = \begin{bmatrix} 0 & -1 \\ 3 & 0 \end{bmatrix}$.

    a. Calculate det A, det B, and det AB. What do you notice about these values?

    b. Make a conjecture about the determinant of the product of two 2×2 matrices. Check it for two pairs of 2×2 matrices of your choosing.

QY ANSWER

a. –2

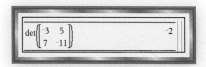

$\det\begin{bmatrix} -3 & 5 \\ 7 & -11 \end{bmatrix}$ –2

b. $-3 \cdot -11 - 5 \cdot 7 = 33 - 35 = -2$

Lesson
5-6
Solving Systems Using Matrices

Vocabulary

matrix form of a system

coefficient matrix

constant matrix

▶ **BIG IDEA** A system of linear equations in standard form can be written as a matrix equation that can often be solved by multiplying both sides of the equation by the inverse of the *coefficient matrix*.

The matrix equation $\begin{bmatrix} 3 & -1 \\ 2 & 4 \end{bmatrix} \begin{bmatrix} x \\ y \end{bmatrix} = \begin{bmatrix} 14 \\ 0 \end{bmatrix}$ is called the **matrix form of the system** $\begin{cases} 3x - y = 14 \\ 2x + 4y = 0 \end{cases}$. The matrix $\begin{bmatrix} 3 & -1 \\ 2 & 4 \end{bmatrix}$ is called the **coefficient matrix** because it contains the coefficients of the variables in the system. The matrix $\begin{bmatrix} 14 \\ 0 \end{bmatrix}$ is called the **constant matrix** because it contains the constants on the right side of the equations in the system.

 QY1

How do you solve a system in matrix form? Think of how you solve the equation $3x = 6$. You might divide both sides of the equation by 3, or multiply both sides of the equation by $\frac{1}{3}$, the multiplicative inverse of 3. This idea is employed to solve linear systems in matrix form.

Mental Math

Calculate the determinant of the matrix.

a. $\begin{bmatrix} 2 & 0 \\ 0 & 2 \end{bmatrix}$

b. $\begin{bmatrix} 7 & 3 \\ 4 & 2 \end{bmatrix}$

c. $\begin{bmatrix} -1 & 5 \\ -2 & -1 \end{bmatrix}$

d. $\begin{bmatrix} a & b \\ c & d \end{bmatrix}$

▶ **QY1**

Expand

$\begin{bmatrix} 3 & -1 \\ 2 & 4 \end{bmatrix} \begin{bmatrix} x \\ y \end{bmatrix} = \begin{bmatrix} 14 \\ 0 \end{bmatrix}$

to verify that it represents

$\begin{cases} 3x - y = 14 \\ 2x + 4y = 0 \end{cases}$.

GUIDED

Example 1

Solve the system $\begin{bmatrix} 3 & -1 \\ 2 & 4 \end{bmatrix} \begin{bmatrix} x \\ y \end{bmatrix} = \begin{bmatrix} 14 \\ 0 \end{bmatrix}$.

Solution First, find the inverse of the coefficient matrix.

$\begin{bmatrix} 3 & -1 \\ 2 & 4 \end{bmatrix}^{-1} = \begin{bmatrix} ? & ? \\ ? & ? \end{bmatrix}$

Next, multiply both sides of the equation on the left by the inverse.

$\begin{bmatrix} ? & ? \\ ? & ? \end{bmatrix} \cdot \begin{bmatrix} 3 & -1 \\ 2 & 4 \end{bmatrix} \cdot \begin{bmatrix} x \\ y \end{bmatrix} = \begin{bmatrix} ? & ? \\ ? & ? \end{bmatrix} \cdot \begin{bmatrix} 14 \\ 0 \end{bmatrix}$

$\begin{bmatrix} ? & ? \\ ? & ? \end{bmatrix} \cdot \begin{bmatrix} x \\ y \end{bmatrix} = \begin{bmatrix} ? \\ ? \end{bmatrix}$

Because you multiplied by the inverse, the left side of the equation should

now be $\begin{bmatrix} 1 & 0 \\ 0 & 1 \end{bmatrix} \cdot \begin{bmatrix} x \\ y \end{bmatrix} = \begin{bmatrix} x \\ y \end{bmatrix}$. Thus $\begin{bmatrix} x \\ y \end{bmatrix} = \begin{bmatrix} ? \\ ? \end{bmatrix}$, so x = __?__ and

y = __?__.

Check Check your answer by substituting your values for x and y into the original matrix equation.

$\begin{bmatrix} 3 & -1 \\ 2 & 4 \end{bmatrix} \begin{bmatrix} 4 \\ -2 \end{bmatrix} = \begin{bmatrix} 14 \\ 0 \end{bmatrix}$ It checks.

This method for solving a system of linear equations with matrices was developed in the middle of the nineteenth century by the British mathematician Arthur Cayley and can be generalized. For any invertible coefficient matrix A and constant matrix B, with $X = \begin{bmatrix} x \\ y \end{bmatrix}$, the solution to the matrix equation $AX = B$ is $X = A^{-1}B$. Example 2 shows how this method works on systems with three linear equations in three variables.

Example 2

Use matrices to solve $\begin{cases} 4x + 2y - 2z = 2 \\ 2x + 4z = 28 \\ 3y - 2z = -16 \end{cases}$.

Solution Write the matrix form of the system.

$$\begin{matrix} A & \cdot & X & = & B \end{matrix}$$
$$\begin{bmatrix} 4 & 2 & -2 \\ 2 & 0 & 4 \\ 0 & 3 & -2 \end{bmatrix} \cdot \begin{bmatrix} x \\ y \\ z \end{bmatrix} = \begin{bmatrix} 2 \\ 28 \\ -16 \end{bmatrix}$$

For each side of the equation use a CAS to multiply on the left by the inverse of the coefficient matrix.

$$\begin{matrix} A^{-1} & \cdot & A & \cdot X & & = & A^{-1} & \cdot & B \end{matrix}$$

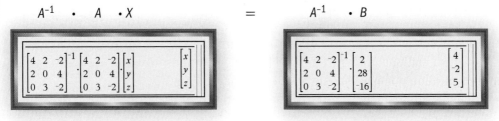

(continued on next page)

Write an equation with the results from the multiplications.

$X = A^{-1} \cdot B$

$$\begin{bmatrix} x \\ y \\ z \end{bmatrix} = \begin{bmatrix} 4 \\ -2 \\ 5 \end{bmatrix}. \text{ So, } x = 4, y = -2, \text{ and } z = 5.$$

Check Substitute for x, y, and z in each of the three given equations.

Does $4(4) + 2(-2) - 2(5) = 2$? Yes, $16 + -4 - 10 = 2$.

Does $2(4) + 4(5) = 28$? Yes, $8 + 20 = 28$.

Does $3(-2) + -2(5) = -16$? Yes, $-6 + -10 = -16$.

The Number of Solutions to Linear Systems

Matrices can be used to determine the number of solutions to a linear system. For example, consider the system $\begin{cases} ax + by = e \\ cx + dy = f \end{cases}$.

If the determinant $ad - bc$ of the coefficient matrix $\begin{bmatrix} a & b \\ c & d \end{bmatrix}$ is not 0, the matrix has an inverse and the system has exactly one solution. When $ad - bc = 0$, the coefficient matrix has no inverse. In that case, the system has either infinitely many solutions or none at all.

> ### System-Determinant Theorem
>
> An $n \times n$ system of linear equations has exactly one solution if and only if the determinant of the coefficient matrix is *not* zero.

In the previous lesson you calculated the determinant of a 2×2 matrix. While the calculation of a determinant of a larger square matrix is tedious by hand, a calculator can find the determinant automatically.

 QY2

When the determinant of the coefficient matrix is zero, to determine whether there are no solutions or infinitely many solutions, find a solution to one of the equations and test it in the other equations. If it satisfies all the other equations, there are infinitely many solutions to the system. If the solution does not satisfy all the other equations, there are no solutions to the system.

▶ **QY2**

Use a calculator to find the determinant of the coefficient matrix in Example 2. Does its value support the solution to that Example?

Example 3

Consider the system $\begin{cases} 3x + 2y = 14 \\ 6x + 4y = 10 \end{cases}$.

a. Show that this system does not have exactly one solution.

b. Determine how many solutions this system has.

Solution 1

a. Let A be the coefficient matrix $\begin{bmatrix} 3 & 2 \\ 6 & 4 \end{bmatrix}$.

Then det $A = (3)(4) - (2)(6) = 0$.

So, by the System-Determinant Theorem, the system does not have exactly one solution.

b. Either the system has infinitely many solutions or no solutions. To decide, find an ordered pair that satisfies one equation.

(4, 1) satisfies the first equation.

Next, does it satisfy the other equation?

Does $6(4) + 4(1) = 10$? No. Thus, because (4, 1) does not satisfy all the equations, the system has no solutions.

Solution 2

a. Use a calculator to solve the system with matrices. One calculator gives the result at the right. The error message means that the coefficient matrix does not have an inverse.

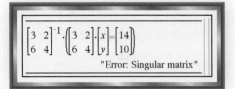
"Error: Singular matrix"

b. Proceed as in Part b of Solution 1.

STOP QY3

> **QY3**
>
> Is the system in Example 3 consistent or inconsistent?

Questions

COVERING THE IDEAS

1. a. Write the matrix form of the system $\begin{cases} 0.5x - y = 1.75 \\ 3x + 8y = 5 \end{cases}$.

   b. Solve the system.

2. a. Write a system of equations whose matrix form is

   $$\begin{bmatrix} 2 & 0 & -1 \\ 5 & 5 & 2 \\ 3 & 1 & 0 \end{bmatrix} \cdot \begin{bmatrix} x \\ y \\ z \end{bmatrix} = \begin{bmatrix} 2 \\ -3 \\ 10 \end{bmatrix}.$$

   b. Which matrix in Part a is the coefficient matrix? Which is the constant matrix?

   c. Use technology to solve this system.

In 3–5, determine how many solutions the system has. Justify your answer.

3. $\begin{cases} 8x + 12y = 40 \\ 4x + 6y = 25 \end{cases}$

4. $\begin{cases} 8x + 12y = 40 \\ 6x + 20y = 52 \end{cases}$

5. $\begin{cases} 8x + 12y = 40 \\ 6x + 9y = 30 \end{cases}$

In 6 and 7, solve each system using matrices.

6. $\begin{cases} 18 = 2d - 3h \\ 7 = 5d + 2h \end{cases}$

7. $\begin{cases} 4x - 3y + z = 4 \\ \quad\quad 2x = -2 \\ \quad 3y + 5z = 40 \end{cases}$

8. Solve the system in Example 3 using the linear-combination method. Explain how that method shows there is no solution.

APPLYING THE MATHEMATICS

9. Set up a system and solve it using matrices to answer this question: Two types of tickets are available for Lincoln High School plays. Student tickets cost $3 each, and nonstudent tickets cost $5 each. On opening night of the play *Proof* by David Auburn, 937 total tickets were sold for $3943. How many of each type of ticket were sold?

10. Forensic scientists and anthropologists can estimate the height h of a person based on the length b of the person's femur using the following equations. All lengths are measured in inches.

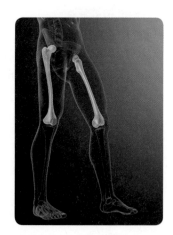

$$\text{Male:} \quad h = 32.010 + 1.880b$$
$$\text{Female:} \quad h = 28.679 + 1.945b$$

a. Even though the equation for male height has a larger h-intercept, there is a point where males and females are expected to have the same height and length of femur. How do you know this from the equations?

b. Use matrices to find the point of intersection of the lines represented by the equations. (Hint: Remember to put the system in standard form before setting up each matrix.)

c. **Fill in the Blanks** According to these linear models, a man and a woman with the same height and femur length are __?__ inches tall and have a femur length of __?__ inches.

11. A hotel has standard, special, and deluxe rooms. A meeting planner needs 15 rooms. If 6 standard, 6 special, and 3 deluxe rooms are booked, the cost will be $2835 a night. Is this enough information to determine the cost of each kind of room? If so, determine the costs. If not, explain why not.

12. a. Write the matrix form of the system $\begin{cases} 4x = 12 \\ 8y = 16 \end{cases}$ and solve.

  b. Give a geometric interpretation of the solution to the system.

In 13 and 14, determine all values of n that satisfy the condition.

13. $\begin{cases} 5x + 3y = 4 \\ 20x - ny = 16 \end{cases}$ has infinitely many solutions.

14. $\begin{cases} 2x + 7y = 1 \\ 4x + 14y = n \end{cases}$ has no solution.

15. Solve the system $\begin{cases} 2w - x - y + z = -1 \\ 3.2w + 1.4z = 2.4 \\ 1.8x - 3y + 7z = 23.8 \\ 5w - 3x + 7y + z = 10 \end{cases}$.

REVIEW

16. A 2×2 *diagonal matrix* is of the form $\begin{bmatrix} a & 0 \\ 0 & b \end{bmatrix}$. Show that the inverse (if it exists) of a diagonal matrix is another diagonal matrix. **(Lesson 5-5)**

17. a. Find the inverse of $\begin{bmatrix} 1 & 1 \\ 0 & 1 \end{bmatrix}$.

  b. Check your answer using matrix multiplication.
    (Lessons 5-5, 4-3)

18. Solve and check: $\begin{cases} A = 19 - 3D \\ 2A - 4D = -12 \end{cases}$. **(Lesson 5-3)**

19. Find the equation in slope-intercept form for the line through the origin and perpendicular to the line with equation $8x - 5y = 20$. **(Lesson 4-9)**

20. A sequence L has the explicit formula $L_n = 4 - 2.5n$. Write a recursive formula for the sequence. **(Lesson 3-8)**

EXPLORATION

21. From a reference, find a formula for the determinant and inverse of a 3×3 matrix. Check the formula for the inverse for the matrix $A = \begin{bmatrix} 2 & -2 & 5 \\ 4 & 6 & 12 \\ 3 & 1 & 7 \end{bmatrix}$.

QY ANSWERS

1. $\begin{bmatrix} 3 & -1 \\ 2 & 4 \end{bmatrix}\begin{bmatrix} x \\ y \end{bmatrix} =$
$\begin{bmatrix} 3x - y \\ 2x + 4y \end{bmatrix}$, so,
$\begin{bmatrix} 3x - y \\ 2x + 4y \end{bmatrix} = \begin{bmatrix} 14 \\ 0 \end{bmatrix}$,
which represents
$\begin{cases} 3x - y = 14 \\ 2x + 4y = 0 \end{cases}$.

2. -52; yes; it does.

3. inconsistent

Lesson 5-7

Graphing Inequalities in the Coordinate Plane

> ▶ **BIG IDEA**   The graph of a linear inequality in two variables consists of the points on one side of a line, either with or without the line.

As you learned in Lesson 5-1, solutions to compound sentences with inequalities involving only one variable can be graphed on a number line. In this lesson, we review the graphs of inequalities in two variables.

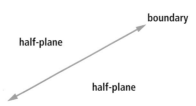

When a line is drawn in a plane, the line separates the plane into three distinct sets of points: two regions called **half-planes** and the line itself. The line is called the **boundary of the half-planes**. The boundary does *not* belong to either half-plane.

Mental Math

Kelley's hair is 8 inches long and grows at a rate of 0.5 inch per month.

a. How long will Kelley's hair be m months from now?

b. How long will Kelley's hair be 3 months from now?

c. How long was Kelley's hair 3 months ago?

d. Four months ago, Kelley got a hair cut. She cut 3 inches off her hair. How long was her hair before her hair cut?

Inequalities with Horizontal or Vertical Boundaries

Inequalities with one variable can be graphed on either a number line or in the coordinate plane. In the plane, think of $x < 3$ as $x + 0 \cdot y < 3$. The solution set is the set of all points (x, y) with x-coordinates less than 3. Graph $x = 3$ with a dashed line because the points on this line are *not* part of the solution set. Shade the half-plane to the left of the line $x = 3$. Shaded is $\{(x, y) \mid x < 3\}$.

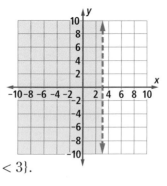

🛑 **QY1**

> ▶ **QY1**
>
> Describe the graph of $\{(x, y) \mid y \geq -100\}$.

Half-Planes with Oblique Boundaries

The line $y = mx + b$ is oblique when $m \neq 0$. The half-planes of this line are described by the inequalities $y > mx + b$ (the half-plane above the line) and $y < mx + b$ (the half-plane below the line). Read $y > mx + b$ as "the set of all points where the y-coordinate is greater than $mx + b$."

Example 1

Graph the linear inequality $y > -\frac{1}{2}x + 2$.

Solution 1 First, graph the boundary $y = -\frac{1}{2}x + 2$. Use a dashed line because the boundary points do not satisfy the inequality. For the points on the boundary, the y-coordinates are equal to $-\frac{1}{2}x + 2$. The y-coordinates are greater than $-\frac{1}{2}x + 2$ for all points above the line. So the solution is the set of all points above the line. Shade the half-plane above the line to show this.

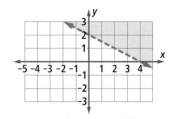

Solution 2 Use a graphing utility to graph the inequality. On the grapher shown, you can enter the inequality symbol directly. On other graphers, you may have to choose "$y >$" from a menu, then enter the rest of the inequality.

Graphers may also differ in the appearance of the boundary line itself. Although this inequality calls for a dotted line, many graphing utilities only show a solid line. The display at the right shows a dotted line, but it is difficult to see.

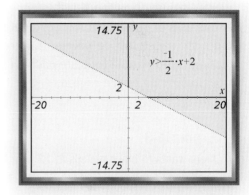

Check Pick a point in the shaded region. We pick (4, 6). Do the coordinates satisfy $y > -\frac{1}{2}x + 2$? Is $6 > -\frac{1}{2}(4) + 2$? Yes, it checks.

 QY2

> **QY2**
>
> How do you enter $y > -\frac{1}{2}x + 2$ on your grapher? Does the boundary line show as dotted or solid?

If the equation for a line is not in slope-intercept form, you can still describe its half-planes rather easily. The half-planes of the line with equation $Ax + By = C$ are described by

$$Ax + By < C \text{ and } Ax + By > C.$$

To decide which inequality describes which side of the line, pick a point not on the line and test it in the inequality.

GUIDED

Example 2

Is the set of points satisfying $3x - 2y > 6$ above or below the line with equation $3x - 2y = 6$?

Solution Refer to the graph of $3x - 2y = 6$ at the right. Pick a point not on the line:

(__?__ , __?__) is a point __?__ (above/below) the line.

Test your point.

Is $3 \cdot$ __?__ $- 2 \cdot$ __?__ > 6?

(continued on next page)

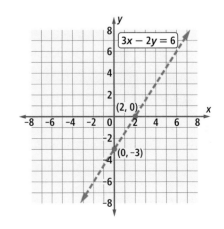

This point ___?___ (does/does not) satisfy the inequality.

Based on your answer, the points satisfying $3x - 2y > 6$ are ___?___ (above/below) the line with equation $3x - 2y = 6$.

Lattice Points

In the examples you have seen, the domain of all variables is the set of real numbers or a subset of the real numbers. In these cases, the graph of an inequality consists of all points in a region, indicated by shading. However, in some situations the domain of each variable is a discrete set, such as the set of integers. In these cases, the solution set consists of points whose coordinates are both integers. These points are called **lattice points**. If there are not too many lattice-point solutions, you should indicate each with a dot on the plane, rather than shading an entire region.

> ▶ READING MATH
>
> A *lattice* is an open framework of interwoven strips forming regular patterned spaces. The grid lines on an *x-y* graph form a lattice. You may see a lattice on a window or gate, or as the top crust of a pie.

Example 3

Aleta wants to buy p pencils at 12 cents each and r erasers at 15 cents each and spend no more than $1. What combinations (p, r) are possible?

Solution The domains of the variables p and r are nonnegative integers, so every solution is a lattice point, $p \geq 0$, and $r \geq 0$.

Write an inequality to represent the situation.
The cost of the pencils is 12p cents.
The cost of the erasers is 15r cents.
So $12p + 15r \leq 100$.

Now graph the corresponding equation $12p + 15r = 100$.
The p-intercept is $\frac{100}{12} = 8\frac{1}{3}$ and the r-intercept is $\frac{100}{15} = 6\frac{2}{3}$.
A graph is shown at the right. Notice that the line is dashed.
This is because the restricted domains of p and r mean that a point on the boundary is included in the solution set only if both coordinates of the point are integers.

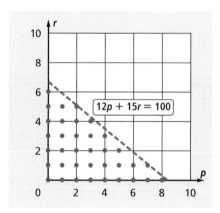

Identify all lattice points satisfying the three inequalities $12p + 15r \leq 100$, $p \geq 0$, and $r \geq 0$ and mark them on the graph. There are 36 possible points.

When there are too many lattice points to graph them all, you should shade the region they are in and make a note that only the appropriate discrete values are solutions.

Questions

COVERING THE IDEAS

1. **Fill in the Blanks** A line separates a plane into two distinct regions called ___?___. The line itself is called the ___?___ of these regions.

2. Graph the solutions to $x \le 4.93$
   a. on a number line.
   b. in the coordinate plane.

3. Does the graph in the coordinate plane of all solutions to $y < 3$ consist of the points above or below the line with equation $y = 3$?

4. To graph the inequality $y > 5 - 3x$, should you shade above or below the line with equation $y = 5 - 3x$?

5. **Matching** Tell which sentence each point satisfies.
   a. $(-4, 4)$
   b. $(11, 2)$
   c. $(0, 0)$

   i. $y > -\frac{1}{2}x + 2$
   ii. $y = -\frac{1}{2}x + 2$
   iii. $y < -\frac{1}{2}x + 2$

In 6 and 7, graph the inequality.

6. $100x - 80y > 200$

7. $y - \frac{x}{3} \le 6$

8. What name is given to a point with integer coordinates?

In 9–11, refer to Example 3.

9. What is the greatest number of pencils Aleta can purchase?

10. What is the greatest number of erasers Aleta can purchase?

11. If Aleta wants an equal number of pencils and erasers, what is the greatest number of each she can purchase?

12. Norma Lee Lucid wants to buy d DVDs at \$19.95 each and m music CDs at \$14.95 each. She wants to spend less than \$100.00.
    a. Write an inequality in d and m describing this situation.
    b. Graph all solutions.
    c. What pairs (d, m) are possible?

APPLYING THE MATHEMATICS

In **13** and **14**, write an inequality that describes the graph.

13.

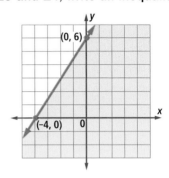

14.

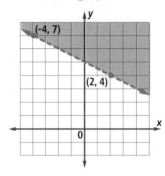

In **15** and **16**, a group of 4 adults and 8 children are planning to attend a family reunion at an amusement park. Children's tickets cost $24.95 each and adult tickets cost $32.50 each. The group budgeted $300.00 for the trip, which is not quite enough money for everyone to attend. Some children and adults will have to stay home.

15. State whether each of these combinations will be able to attend.

    a. 4 children and 4 adults

    b. 6 children and 3 adults

    c. 7 children and 4 adults

16. a. Graph all pairs of numbers of children and adults who can attend.

    b. How many pairs are there?

17. Can the graph of a linear inequality in two variables also be the graph of a function? Explain why or why not.

REVIEW

18. a. Write the matrix form of the system $\begin{cases} 3x + 6y = 18 \\ -4x - 5y = 20 \end{cases}$.

    b. Solve the system. **(Lesson 5-6)**

19. Use the System-Determinant Theorem to determine whether

the system $\begin{cases} A + 3B - 2C = 5.5 \\ 7A - 5B + C = 1 \\ B - C = -1 \end{cases}$ has exactly one solution.

(Use a calculator to find the determinant.) **(Lesson 5-6)**

20. Antonio claims to have a shortcut for determining if the matrix $\begin{bmatrix} a & b \\ c & d \end{bmatrix}$ has an inverse whenever c and d are nonzero. If the fractions $\frac{a}{c}$ and $\frac{b}{d}$ are equal, Antonio argues, then the matrix has no inverse. Otherwise, it does. Does Antonio's method work? Explain your answer. (**Lesson 5-5**)

21. Solve $37w + 3 > 77$ and $8 - 4w \geq -10$. Write the solution in set-builder notation and graph it on a number line. (**Lesson 5-1**)

22. Meteor Crater in Arizona is approximately in the shape of a cylinder 1500 meters in diameter and 180 meters deep. Scientists estimate that the meteor that created it weighed about 300,000 tons at entry. Assuming the volume of the crater varies directly with the weight of the meteor, what volume of crater would you expect for a meteor weighing 500,000 tons? (**Lesson 2-1**)

When the crater was first discovered, it was surrounded by about 30 tons of meteoritic iron chunks, scattered over an area 12 to 15 kilometers in diameter.

EXPLORATION

23. This problem was made up by the Indian mathematician Mahavira and dates from about 850 CE. "The price of nine citrons and seven fragrant wood apples is 107; again, the mixed price of seven citrons and nine fragrant wood apples is 101. Oh you arithmetician, tell me quickly the price of a citron and a wood apple here, having distinctly separated these prices well." At this time algebra had not been developed yet. How could this question be answered by someone without using algebra?

24. Write a system of three inequalities that describes the shaded region below.

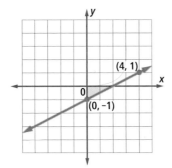

Lesson

5-8

Systems of Linear Inequalities

Vocabulary

feasible set, feasible region

▶ **BIG IDEA** The solution to a system of linear inequalities in two variables is either the empty set, the interior of a polygon, or a region bounded by line segments and rays.

A linear inequality in two or more variables may be used to model a situation in which one or more resources limit, or *constrain*, the values of the variables. In the previous lesson, Aleta's money limited the number of pencils and erasers she could buy. In this lesson we show how systems of linear inequalities can model even more complicated situations.

Mental Math

What inequality is represented by the graph?

a.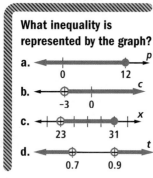

b.

c.

d.

Activity

Suppose you have a collection of 16 square (2 × 2) and 12 long (2 × 4) interlocking blocks to form into tables and chairs. It takes 2 long blocks and 2 square blocks to make a table. It takes 1 long block and 2 square blocks to make a chair. What combinations of tables and chairs can you make with your collection?

	Table	Chair
Long Blocks Needed	2	1
Square Blocks Needed	2	2

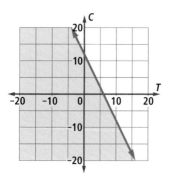

Step 1 Let T = the number of tables and C = the number of chairs. Notice that T and C are nonnegative integers. Write an inequality using T and C that relates the total number of long blocks needed to the number of long blocks available. Graph this inequality.

Step 2 Write an inequality using T and C that relates the total number of square blocks needed to the number of square blocks available. Graph this inequality on the same axes as the inequality from Step 1.

Step 3 Determine whether each ordered pair (T, C) in the table on the next page satisfies the inequalities from Steps 1 and 2. Then plot the four ordered pairs on your graph.

(T, C)	Satisfies Long Block Inequality	Satisfies Square Block Inequality
(2, 4)	?	?
(6, 2)	?	?
(2, 7)	?	?
(5, 4)	?	?

Step 4 Which point(s) from Step 3 satisfy both the long-block inequality and the square-block inequality? Where do these points appear on the graph?

In the Activity, the limited numbers of square and long blocks are constraints to solving the problem. These constraints are represented by the inequalities $2T + C \le 12$, $2T + 2C \le 16$, $T \ge 0$, and $C \ge 0$. The set of possible combinations of tables and chairs is the intersection of the solution sets of all these inequalities, as shown in the graph at the right.

Because the graph of a linear inequality in two variables is a half-plane, the graph of the solution to a system of linear inequalities is the intersection of half-planes. Points in this intersection are often called the **feasible set** or **feasible region** for the system.

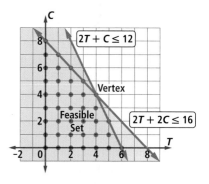

 QY

A feasible region is always bounded by either segments or rays. The intersections of boundary segments or rays are *vertices* of the feasible region. One vertex of the feasible region from the Activity is labeled in the graph. Because the boundary may not be included in the solution set of an inequality, the vertices and boundary segments or rays may not be part of the feasible region.

▶ **QY**

Multiple Choice
Which of these combinations of tables and chairs are in the feasible region from the Activity?

A (5, 2)   B (6, 1)
C (0, 12)   D (0, 0)
E (4, 4)

GUIDED

Example 1

An electronics firm makes two kinds of televisions: plasma and projection. The firm assumes it will sell all the sets it makes. It profits $2000 on each plasma set it sells and $1500 on each projection TV. If it wants a profit of over $100,000, how many plasma sets and how many projection TVs should it make?

Solution Let $L =$ the number of plasma sets made. Then the profit the firm makes on these sets is __?__ $\frac{\text{dollars}}{\text{set}} \cdot$ __?__ sets $=$ __?__ dollars. Similarly, let $R =$ the number of projection sets made. The profit on these sets is __?__ dollars.

(continued on next page)

So the total profit is __?__ + __?__ dollars, and the pair (L, R) of numbers of televisions the firm should make needs to satisfy

$$\underline{\quad?\quad} + \underline{\quad?\quad} > 100{,}000.$$

Divide both sides by 100 to make the numbers easier to manage.

$$\underline{\quad?\quad} + \underline{\quad?\quad} > 1000$$

Because L and R are numbers of televisions, they are nonnegative integers. So $L \geq 0$ and $R \geq 0$. These three inequalities form the boundary of the feasible region.

Graph the three boundary lines. Shade the feasible region as shown at the right.

The firm can make any pair (L, R) of numbers of televisions that is a lattice point in the feasible region.

In this situation, there are too many lattice points to plot distinctly, so the feasible region is shaded and two discrete solutions, (__?__, __?__) and (__?__, __?__), are noted on the graph. These points mean that the firm could sell __?__ plasma sets and __?__ projection sets or __?__ plasma sets and __?__ projection sets and meet their goal.

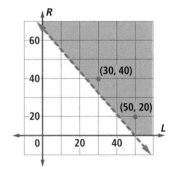

In real life, there may be other constraints on the situation in Example 1. It takes time to make each set. It takes different amounts of various kinds of metals, plastics, and electronics to make each set. Each constraint may add a line to the boundary and make the problem a little more complicated.

Example 2

Consider the system $\begin{cases} y > -\frac{1}{2}x \\ y \leq \frac{1}{2}x + 4. \\ y \leq 6 \end{cases}$

a. Graph the solution to the system by hand and check your solution.

b. Find the vertices of the feasible region.

Solution

a. The boundary lines for all the inequalities are easily sketched because the inequalities are in slope-intercept form.

The first inequality has a dotted boundary, and the half-plane above is shaded. The second has a solid boundary, and the half-plane below it is shaded. The third inequality has a solid boundary, and the half-plane below it is shaded. The three inequalities are graphed on the same grid shown here.

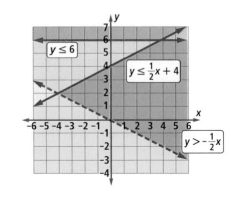

The solution to the system is all the points in the intersection of the three half-planes. A graph of the feasible region is shown at the right. The region extends forever to the right, below and on $y = 6$ and $y = -\frac{1}{2}x + 4$ and above $y = -\frac{1}{2}x$.

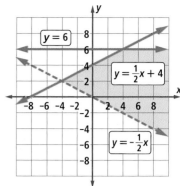

Choose an ordered pair in the feasible region to see if it satisfies all three inequalities. We choose $(1, 0)$.

Is $0 > -\frac{1}{2}(1)$? Yes.
Is $0 \le \frac{1}{2}(1) + 4$? Yes.
Is $0 \le 6$? Yes, so the solution checks.

b. The coordinates of each vertex of the feasible region can be found by reading the graph or by solving pairs of equations using the substitution method.

$$\begin{cases} y = \frac{1}{2}x + 4 \\ y = 6 \end{cases}$$

Substitute 6 for y in $y = \frac{1}{2}x + 4$.

$6 = \frac{1}{2}x + 4$
$x = 4$

$(4, 6)$ is one vertex.

$$\begin{cases} y = \frac{1}{2}x + 4 \\ y = -\frac{1}{2}x \end{cases}$$

Substitute $-\frac{1}{2}x$ for y in $y = \frac{1}{2}x + 4$.

$-\frac{1}{2}x = \frac{1}{2}x + 4$
$x = -4$

Substitute –4 for x in either equation to find that $y = 2$.
$(-4, 2)$ is the other vertex.

A graphing utility can graph systems of inequalities. However, the feasible region may not be clear on all machines. A graph of the system from Example 2 on one grapher is shown below. The darkest region is the feasible region.

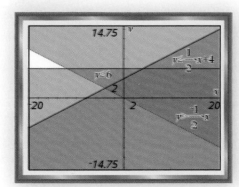

Questions

COVERING THE IDEAS

1. **Fill in the Blank** The solution to a system of linear inequalities can be represented by the ___?___ of half-planes.

2. The solution set to a system of linear inequalities is often called the ___?___.

In 3–5, refer to the Activity.

3. Use your graph to determine whether each combination of tables and chairs is possible. If it is not, describe which block, long or square, is in short supply.
   a. 7 tables and 2 chairs
   b. 5 tables and 2 chairs
   c. 2 tables and 5 chairs
   d. 2 tables and 7 chairs

4. The point (3, 2.5) satisfies all the constraints, but it is not a solution in this situation. Explain why not.

5. a. What system of equations can you solve to find the vertex of the feasible region located in the first quadrant?
   b. Solve the system from Part a to find the coordinates of the vertex.

6. The graph at the right represents a system of inequalities. In what region(s) are the solutions to each system, with $x > 0$ and $y > 0$?
   a. $y + 2x < 600$ and $x + 2y < 600$
   b. $y + 2x > 600$ and $x + 2y < 600$
   c. $y + 2x < 600$ and $x + 2y > 600$
   d. $y + 2x > 600$ and $x + 2y > 600$

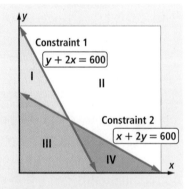

In 7 and 8, refer to Example 1.

7. How much profit will the company make if it sells
   a. 50 plasma sets and 40 projection TVs?
   b. 20 plasma sets and 40 projection TVs?
   c. 50 plasma sets and 5 projection TVs?

8. Will the firm meet its goal of surpassing a $100,000 profit if it sells 11 plasma sets and 45 projection TVs?

Most projection TVs are rear-projection systems; the image is displayed on the back of the screen and the projector is contained in the TV.

In 9 and 10, a system of inequalities is given.

    a. Graph the solution set.

    b. Find the coordinates of each vertex of the feasible region.

9. $\begin{cases} y \le 2x + 3 \\ y > -3x - 4 \end{cases}$

10. $\begin{cases} x + 2y \le 16 \\ 3x + y < 18 \\ y \le 7 \\ x \ge 0 \end{cases}$

APPLYING THE MATHEMATICS

11. A school's film club shows movies after school and sells popcorn in small and large sizes. The bags cost $0.15 for a small and $0.20 for a large. A small bag holds 1.25 ounces of popcorn; a large bag holds 2.5 ounces of popcorn. The club has 400 ounces of popcorn and a budget of $40 for popcorn bags.

    a. **Fill in the Blanks** Let S be the number of small bags and L be the number of large bags. Complete the translation of this situation into a system of inequalities.

$$\begin{cases} \times S \ge \underline{\ ?\ } \\ y L \ge \underline{\ ?\ } \\ 0.15S + 0.2L \le \underline{\ ?\ } \\ \underline{\ ?\ } \le 400 \end{cases}$$

    b. Graph the feasible region of points (S, L) for this system and label the vertices.

Americans consume 17.3 billion quarts of popcorn in an average year.

12. Graph the solutions to the system $\begin{cases} y < 2x + 4 \\ 4x - 2y \le 6 \end{cases}$.

13. In his shop, Hammond Wrye makes two kinds of sandwiches: plain turkey and turkey-and-cheese. Each sandwich uses 2 pieces of bread. The plain turkey sandwiches use 3.5 ounces of turkey, while the turkey-and-cheese sandwiches use 2.5 ounces of turkey and 1 slice of cheese. Hammond has 100 slices of bread, 40 slices of cheese, and 150 ounces of turkey. Let x be the number of turkey-and-cheese sandwiches and y be the number of plain turkey sandwiches he makes.

    a. Translate the situation into a system of inequalities.

    b. Graph the feasible set for this system, and label the vertices.

14. Refer to Example 1. Suppose it takes 30 worker-hours to make a plasma TV and 25 worker-hours to make a projection TV. If the company has 15,000 worker-hours available, can it make its desired profit of $100,000? Explain your answer.

REVIEW

In 15 and 16, graph the inequality. (Lesson 5-7)

15. $y \leq -4x + 3$

16. $3x - 2y > 14$

17. **True or False** If the coefficient matrix of a system has determinant zero, then the system has no solutions. (Lesson 5-6)

18. Recall from geometry that a region of the plane is said to be *convex* if and only if any two points of the region can be connected by a line segment which is itself entirely within the region. The pentagonal region at the right is convex, but the quadrilateral region is not.

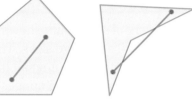

Tell whether the shaded region is convex. (Previous Course)

a. b. c. d.

19. Cesar runs for x to $x + 5$ minutes every Tuesday. Every year has either 365 or 366 days in it (depending on whether or not it is a leap year). What are the minimum and maximum amounts of time Cesar could spend on his weekly runs in any given year? (Previous Course)

EXPLORATION

20. Consider the system $\begin{cases} x > 0 \\ y > 0 \\ y < mx + b \end{cases}$.

    a. For what values of m and b is the graph of the solution set to this system the interior of a triangle?

    b. Find the area of the triangle in terms of m and b.

Linear Programming

Vocabulary

linear-programming problem

▶ **BIG IDEA** A *linear-programming problem* is one in which you wish to find a solution to a system of inequalities that minimizes or maximizes a linear combination of the variables.

In Lesson 5-8, you graphed a system of linear inequalities to see the numbers of tables and chairs that could be made from a certain set of blocks. In this lesson, we consider combinations of actual tables and chairs that a furniture maker, Tim Burr, can build under the same constraints:

$$\begin{cases} 2T + C \le 12 \\ 2T + 2C \le 16 \\ T \ge 0 \\ C \ge 0 \end{cases}$$

Suppose that Tim earns $900 for each table and $600 for each chair he makes and sells. Under these constraints and assuming that Tim sells all the tables and chairs that he makes, how many of each should he produce to maximize revenue?

If T tables and C chairs are sold, the revenue R in dollars is given by the formula

$$900T + 600C = R.$$

For instance, suppose 3000 is substituted for R in the formula. The solutions (T, C) to $900T + 600C = 3000$ are ordered pairs that represent combinations of tables and chairs that will yield $3000 in earnings. Two such solutions are $(2, 2)$ and $(0, 5)$. This means that Tim could make 2 tables and 2 chairs, or 0 tables and 5 chairs, and he would earn $3000.

The feasible set for Tim's system of inequalities is the set of lattice points in the shaded region of the graph on the next page. The graph also includes the line for a $3000 revenue and five other revenue lines that result from substituting different values of R into the revenue formula. Notice that all lines with equations of the form $900T + 600C = R$ are parallel because each has a slope of -1.5.

Mental Math

Suppose a cleaning service charges a flat rate of $25 for the first hour plus $10 for each additional half hour or part of a half hour. What would the cleaning service charge if the time spent cleaning were

a. 2 hours?

b. 1 hour and 35 minutes?

c. 3 hours and 15 minutes?

d. 4 hours and 5 minutes?

Some of these lines intersect the feasible region and some do not. Lines such as L_3 and L_4 that do intersect the feasible region indicate possible revenues. The highest revenue line that intersects the feasible region represents the greatest possible revenue. This is the line L_3, where the revenue line passes through vertex $(4, 4)$. So, to maximize revenue, Tim should make 4 tables and 4 chairs each week. The maximum revenue under these conditions is $900 \cdot 4 + 600 \cdot 4 = \$6000$.

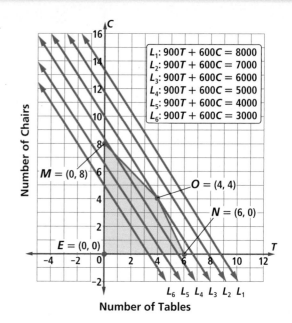

L_1: $900T + 600C = 8000$
L_2: $900T + 600C = 7000$
L_3: $900T + 600C = 6000$
L_4: $900T + 600C = 5000$
L_5: $900T + 600C = 4000$
L_6: $900T + 600C = 3000$

$M = (0, 8)$
$O = (4, 4)$
$N = (6, 0)$
$E = (0, 0)$

Number of Chairs

L_6 L_5 L_4 L_3 L_2 L_1

Number of Tables

Problems such as this one, that involve wanting to maximize or minimize a quantity based on solutions to a system of linear inequalities, are called **linear-programming problems**. The word *programming* does not refer to a computer; it means that the solution gives a program, or course of action, to follow.

In 1826, the French mathematician Jean-Baptiste Joseph Fourier proved the following theorem. It tells you where to look for the greatest or least value of a linear-combination expression in a linear-programming situation, without having to draw any lines through the feasible region.

Linear-Programming Theorem

The feasible region of every linear-programming problem is convex, and the maximum or minimum quantity is determined at one of the vertices of this feasible region.

Linear programming is often used in industries in which all the competitors make the same product (such as gasoline, paper, appliances, clothing, and so on). Their efficiency in the use of labor and materials greatly affects their profits. These situations can involve as many as 5000 variables and 10,000 inequalities.

Testing vertices is a good solution method for problems involving a few constraints and only two variables. There are procedures for solving large linear-programming problems such as the *simplex algorithm* invented in 1947 by the mathematician George Dantzig, who worked on it with the econometrician Leonid Hurwicz, and the mathematician T.C. Koopmans, all from the United States. It is for this work, as well as other contributions, that Koopmans shared the Nobel Prize in 1975.

The Nobel Prize medal is made of 18-karat green gold plated with 24-karat gold.

Example

For a certain company, a bed sheet requires 2 pounds of cotton, 9 minutes of dyeing time, and 2 minutes of packaging time. A set of pillowcases requires 1 pound of cotton, 1.5 minutes of dyeing time, and 7.25 minutes of packaging time. Each day, the company has available 110 pounds of cotton, 405 minutes of dyeing time, and 435 minutes of packaging time.

a. Let $b =$ the number of bed sheets made per day and $p =$ the number of sets of pillowcases made per day. Find the vertices of the region of pairs (b, p) the company can make.

b. The company's daily revenue will be $12 for each sheet and $8 for each set of pillowcases. Assuming they sell everything they make, how many of each product should the company produce per day to maximize revenue?

Solution

a. From the given information, write the constraints.

Available cotton: $2b + p \leq 110$

Dyeing time: $9b + 1.5p \leq 405$

Packaging time: $2b + 7.25p \leq 435$

$b \geq 0, p \geq 0$

Graph the constraints. The graph at the right shows the feasible region. Because of the Linear-Programming Theorem, the revenue will be maximized at one of the vertices of this region.

There are 5 vertices to consider for the solution of this linear-programming problem. Each of these vertices is the intersection of two lines and is the solution to one of the five systems shown below.

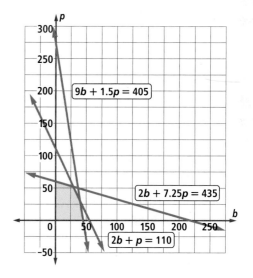

$$\begin{cases} p = 0 \\ 9b + 1.5p = 405 \end{cases} \quad \begin{cases} 2b + p = 110 \\ 9b + 1.5p = 405 \end{cases} \quad \begin{cases} 2b + p = 110 \\ 2b + 7.25p = 435 \end{cases}$$

$$\begin{cases} b = 0 \\ 2b + 7.25p = 435 \end{cases} \quad \begin{cases} b = 0 \\ p = 0 \end{cases}$$

Use solve to find each vertex (b, p).

The vertices are (45, 0), (40, 30), (29, 52), (0, 60), and (0, 0).

(continued on next page)

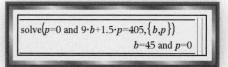

$\mathrm{solve}(p{=}0 \text{ and } 9{\cdot}b{+}1.5{\cdot}p{=}405, \{b,p\})$

$b{=}45 \text{ and } p{=}0$

b. The company wants to maximize daily revenue R. Because the company sells bed sheets for $12 and sets of pillowcases for $8, a revenue formula is $R = 12b + 8p$.

Use the Linear-Programming Theorem. The maximum value of R occurs at a vertex of the feasible region. Evaluate R at each vertex to see which combination of p and b gives the maximum revenue.

The maximum daily revenue of $764 is obtained when $b = 29$ and $p = 52$, that is, when the company produces 29 bed sheets and 52 sets of pillowcases.

Define $f(b,p)=12 \cdot b + 8 \cdot p$	Done
$f(45,0)$	540
$f(40,30)$	720
$f(29,52)$	764
$f(0,60)$	480
$f(0,0)$	0

Steps for Solving Linear-Programming Problems

The steps at the right are a good way to organize your work when solving a linear-programming problem. Use the steps in the following Activity.

Activity

MATERIALS CAS (optional)

A Prom Committee is responsible for prom decorations. There are two types of decorations needed; centerpieces for tables and topiaries for the dance floor. The committee decides to make its own decorations.

Each centerpiece requires 5 packages of artificial flowers and 2 twinkle lights; each topiary requires 3 packages of artificial flowers and 6 twinkle lights. A local florist helps out by donating 115 packages of artificial flowers and 130 twinkle lights.

Twenty prom committee members will meet for 6 hours to assemble the decorations, so they must complete the assembly job in 120 person-hours. Each centerpiece requires 2.5 hours, while each topiary requires 4 hours of labor. At least 10 centerpieces and at least 1 topiary will be needed.

After the prom, the committee plans on selling the decorations. Centerpieces will sell for $20 each and topiaries will sell for $15 each.

Step 1 *Identify the variables and write a system of constraints.*
Let x = number of centerpieces and y = number of topiaries. Complete a table like the one at the top of the next page to organize the information. Then write three constraints that summarize the table, and two more that represent the minimum values for x and y.

> **LINEAR-PROGRAMMING PROBLEMS**
>
> 1. Identify the variables and write a system of constraints.
> 2. Determine which intersections of lines define the vertices of the feasible region. (A sketch of the inequalities found in Step 1 may help.)
> 3. Find the vertices of the feasible region.
> 4. Write a formula or an expression to be maximized or minimized.
> 5. Apply the Linear-Programming Theorem.
> 6. Interpret the results.

	Number of Centerpieces (x)	Number of Topiaries (y)	Constraint Value
No. of packages of flowers	?	?	115
No. of packages of twinkle lights	?	?	?
No. of hours to make	?	?	?

Step 2 *Determine which intersections of lines define the vertices of the feasible region.* Make a sketch of the feasible region. There are four vertices of the feasible region, defined by the solutions of four systems. Fill in the blanks below to record the systems. Which constraint, if any, does not help define the feasible region?

$$\begin{cases} x = 10 \\ y = 1 \end{cases} \qquad \begin{cases} x = 10 \\ \underline{\quad ? \quad} \end{cases} \qquad \begin{cases} y = 1 \\ \underline{\quad ? \quad} \end{cases} \qquad \begin{cases} \underline{\quad ? \quad} \\ \underline{\quad ? \quad} \end{cases}$$

Step 3 *Find the vertices of the feasible region.* Use a CAS to solve the four systems from Step 2.

The four vertices of the region are __?__, __?__, __?__, and __?__.

> solve(x=10 and 2·x+6·y=130,x,y)
>
> x=10. and y=18.3333

Step 4 *Write a formula or an expression to be maximized or minimized.* Write an equation for the revenue R if all decorations made are sold.

Step 5 *Apply the Linear-Programming Theorem.* Using your results from Steps 3 and 4, find the revenue for each vertex. Round vertex coordinates down because the committee cannot sell partial centerpieces or topiaries.

Step 6 *Interpret the results.* To maximize profits, how many centerpieces and topiaries should be made? What would be the maximum revenue from the sales?

Questions

COVERING THE IDEAS

1. Refer to the discussion of Tim Burr at the beginning of this lesson.

   a. What is Tim trying to maximize? How is it represented?

   b. Find the revenue for making 5 tables and 1 chair.

   c. What do the coordinates of points with integer coordinates inside the feasible region represent?

   d. **True or False** The line $900T + 600C = 950$ intersects the feasible region.

2. Refer to the Example.

   a. What does the linear combination R represent?

   b. Why is the vertex (52, 29) a solution to the problem?

   c. Why is $p \geq 0$ and $b \geq 0$?

   d. Assume that the available cotton has changed from 110 pounds to 120 pounds. Write a new constraint for cotton.

3. What does Jean-Baptiste Fourier have to do with the content of this lesson?

4. Why do industries test vertices of the feasible region of a system of inequalities?

In 5 and 6, refer to the Activity.

5. Write the coordinates of the vertex that maximizes revenue. What does this vertex mean for the committee?

6. Why should you not round the vertex values up to the next integer value?

7. What is the simplex algorithm?

8. Who developed the simplex algorithm, and when?

9. Use the feasible set graphed at the right.

   a. Which vertex maximizes the profit equation $P = 30x + 18y$?

   b. Which vertex minimizes the cost equation $C = 25x + 13y$?

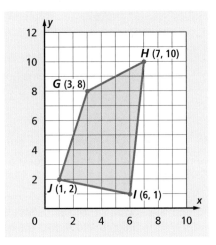

APPLYING THE MATHEMATICS

10. Apples and oranges can meet certain nutritional needs. Suppose a person needs to consume at least 1000 mg of calcium and at least 1000 IU (international units) of vitamin A each day. An apple has about 73 IU of vitamin A and 10 mg calcium and an orange has about 20 IU of vitamin A and 60 mg calcium. Suppose an apple costs 30¢ and an orange costs 45¢. You want to minimize costs, yet meet the nutritional needs.

   a. Identify the variables and constraints. Translate them into a system of inequalities.

   b. Use a CAS and find the vertices of the feasible region.

   c. Write the expression to be minimized.

   d. Find the number of apples and oranges that will minimize the cost.

11. A muffin recipe uses $\frac{9}{16}$ pound of flour, $\frac{3}{4}$ pound of sugar, and $\frac{3}{4}$ pound of nuts to make 1 tray of muffins. A nut-bread recipe uses $\frac{1}{2}$ pound of flour, $\frac{3}{4}$ pound of sugar, and $\frac{1}{6}$ pound of nuts to yield 1 loaf of nut bread. Able Baker Carl has 50 pounds of flour, 40 pounds of sugar, and 15 pounds of nuts available.

   a. Identify the variables and constraints. Translate them into a system of inequalities.

   b. Sketch a graph of the feasible region. Which constraint from Part a is not a boundary of the feasible region?

   c. How many of each item should Carl make to maximize his revenue, if each tray of muffins sells for $8.00 and each loaf of nut bread sells for $12.00?

12. Landscaping contractor Pete Moss uses a combination of two brands of fertilizers, each containing different amounts of phosphates and nitrates as shown in the table. A certain lawn requires a mixture of at least 24 lb of phosphates and at least 16 lb of nitrates. If Pete uses a packages of Brand A and b packages of Brand B, then the constraints of the problem are given by the system of inequalities at the right.

	Brand A	Brand B
Phosphates (lb)	4	6
Nitrates (lb)	2	5

$$\begin{cases} a \geq 0 \\ b \geq 0 \\ 4a + 6b \geq 24 \\ 2a + 5b \geq 16 \end{cases}$$

   a. Graph the feasible region for this situation.

   b. If a package of Brand A costs $6.99 and a package of Brand B costs $17.99, which acceptable combination of packages (a, b) will cost Pete the least?

REVIEW

In 13 and 14, graph the solution set to the system. (Lesson 5-8)

13. $\begin{cases} y > 3 - x \\ y < x - 3 \end{cases}$

14. $\begin{cases} x + y \geq 10 \\ x - y \leq 10 \end{cases}$

15. Write an inequality to describe the shaded region at the right. (Lesson 5-7)

In 16–18, consider the system $\begin{cases} A = s^2 \\ A + 2s = 8 \end{cases}$.

16. Solve the system by substitution. (Lesson 5-3)

17. Solve the system using a CAS. (Lesson 5-2)

18. Solve the system by graphing. (Lesson 5-1)

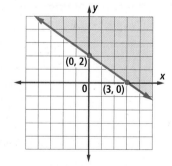

EXPLORATION

19. Suppose $a, b, c, d, e,$ and f are positive integers and $a + b + c + d + e + f = 1000$. What integer values of a, b, c, d, e and f will make the product $abcdef$

   a. as large as possible?   b. as small as possible?

Chapter 5 Projects

1 Nutrition: Low Cost vs. Low Cal

a. Look up the Dietary Reference Intake (DRI) for protein, fiber, calcium, vitamin A, and vitamin C for a person of your age and gender.

b. Name two foods that contain these nutrients and find the cost and number of calories per serving for each food.

c. Use linear programming to find out how many servings of these foods you would need to eat each day in order to get the DRI of these nutrients at the lowest possible cost.

d. Use linear programming to find out how many servings of these foods you would need to eat each day in order to get the DRI of these nutrients with the fewest number of calories.

e. Compare and contrast the low-cost diet to the low-calorie diet.

2 Simplexes

What is a *simplex*? Describe how the simplex algorithm works.

3 The Enigma Machine

One of the biggest decoding breakthroughs of World War II was deciphering messages encoded by the German Enigma machine. Research the Enigma machine and the mathematicians associated with deciphering this code algorithm. Write a summary of your findings.

4 Using Matrices to Code and Decode Messages

Between 1929 and 1931, the mathematician Lester Hill devised a method of encoding messages using matrices. Every positive integer is assigned a letter according to the scheme:

$1 \rightarrow A, 2 \rightarrow B, 3 \rightarrow C, \ldots, 25 \rightarrow Y, 26 \rightarrow Z, 27 \rightarrow A, 28 \rightarrow B, \ldots$.

To encode or encipher a message, follow these steps:

Step 1 Put the numbers for the letters into

2×2 matrices: $\begin{bmatrix} \text{1st letter} & \text{2nd letter} \\ \text{3rd letter} & \text{4th letter} \end{bmatrix}$.

For instance, to encode the word FOUR, with $6 = F$, $15 = O$, $21 = U$, $18 = R$, use the

matrix $\begin{bmatrix} 6 & 15 \\ 21 & 18 \end{bmatrix}$.

Step 2 Choose a 2 × 2 matrix as a coding matrix, such as $\begin{bmatrix} 0 & 1 \\ 1 & 2 \end{bmatrix}$. Multiply each 2 × 2 matrix from Step 1 on the left by the coding matrix. For instance, $\begin{bmatrix} 0 & 1 \\ 1 & 2 \end{bmatrix}\begin{bmatrix} 6 & 15 \\ 21 & 18 \end{bmatrix} = \begin{bmatrix} 21 & 18 \\ 48 & 51 \end{bmatrix}$.

Step 3 Change the resulting matrices back to letters to write the coded message. In our example, since 21 = U, 18 = R, 48 = V, and 51 = Y, FOUR is encoded as URVY.

To decode or decipher a message, follow these steps:

Step 1 Break the message into groups of four letters and write as matrices using the corresponding numbers: $\begin{bmatrix} \text{1st letter} & \text{2nd letter} \\ \text{3rd letter} & \text{4th letter} \end{bmatrix}$.

Step 2 Find the inverse of the coding matrix and multiply each letter-group matrix by the inverse.

Step 3 Translate the resulting matrices back into letters to read the message. (Note that 0 → Z, –1 → Y, –2 → X, and so on.)

a. Code MEET ME AT NOON using the coding matrix $\begin{bmatrix} 0 & 1 \\ 1 & 2 \end{bmatrix}$.

b. A message was encoded as YTKOFOTISBGVITWKOULO using the coding matrix $\begin{bmatrix} 0 & 1 \\ 1 & 2 \end{bmatrix}$. What was the original message?

c. Make up a coding matrix and a coded message of your own. (Note: In order for Hill's method to work, the determinant of your coding matrix must be 1 or –1.)

5 **Programming and CAS**

Many CAS programs use the words *and* and *or*. In the program below, X is assigned each value of 1 through 50 incrementally. The IF/THEN statements test if the value Y is in the appropriate interval for display. Research how your CAS can be used to run the following program.

```
For (X, 1, 50)
2X → Y
If Y > 50 and Y < 84
Then: Disp X
EndIf
EndFor
```

a. What numbers will be displayed when this program is run?

b. Explain why this program will display an arithmetic sequence.

c. What numbers will be displayed if the word AND in the program is changed to OR?

d. If the word AND in the third line is changed to XOR, the numbers displayed are (1, 2, 3, 4, …, 25, 42, 43, 44, …, 50). What do you think XOR means?

e. Change the line "If Y > 50 XOR Y < 84" so that the numbers displayed will be (1, 2, 3, 4, …, 12, 45, 46, 47, …, 50).

Chapter 5

Summary and Vocabulary

○ A **compound sentence** is the result of joining two or more sentences with the word *and* or *or*. The solution set for *A or B* is the **union** of the solution sets of *A* and *B*. The compound sentence *A and B* is called a **system.** The solution set to the system *A and B* is the **intersection** of the solution sets of *A* and *B*.

○ Systems may contain any number of variables. If the system contains one variable, then its solutions may be graphed on a number line. If the system contains two variables, then its solutions may be graphed in the plane. The graph often tells you the number of solutions the system has, but may not yield the exact solutions.

○ Systems of equations can be solved by hand or with technology. Some systems can be solved with tables and graphs. Algebraic methods use **linear combinations, substitution,** and matrices. The matrix method converts a system of *n* equations in *n* unknowns to a single matrix equation. To find the solution to a system in **matrix form,** multiply both sides of the equation by the **inverse** of the **coefficient matrix.**

○ The graph of a linear inequality in two variables is a **half-plane** or a half-plane with its **boundary.** For a system of two linear inequalities, if the boundary lines intersect, then the **feasible region** is the interior of an angle plus perhaps one or both of its sides.

○ Systems with two variables but more than two inequalities arise in **linear-programming problems.** In such a problem, you look for a solution to the system that maximizes or minimizes the value of a particular expression or formula.

○ The Linear-Programming Theorem states that the feasible region of every linear-programming problem is always convex and that the solution that will maximize or minimize the pertinent expression must be a vertex of the feasible region.

Theorems and Properties

Addition Property of Inequality (p. 303)	Inverse Matrix Theorem (p. 329)
Multiplication Property of Inequality (p. 303)	System-Determinant Theorem (p. 338)
	Linear-Programming Theorem (p. 356)
Substitution Property of Equality (p. 314)	

Vocabulary

Lesson 5-1
compound sentence
double inequality
*union of two sets
*intersection of two sets

Lesson 5-2
system
*solution set of a system

Lesson 5-3
substitution method
consistent system
inconsistent system

Lesson 5-4
linear-combination method

Lesson 5-5
*inverse matrices
square matrix
singular matrix
*determinant

Lesson 5-6
matrix form of a system
coefficient matrix
constant matrix

Lesson 5-7
half-planes
boundary of the half-planes
lattice point

Lesson 5-8
feasible set, feasible region

Lesson 5-9
linear-programming
  problem

Chapter

5 Self-Test

Take this test as you would take a test in class. You will need a calculator. Then use the Selected Answers section in the back of the book to check your work.

1. Solve $-4n + 18 < 30$ and graph the solution set on a number line.

2. On a number line, graph $\{x \mid x \leq -3 \text{ or } x > 4\}$.

3. A graph of the system $\begin{cases} y = 0.5x - 2 \\ y = -x^2 \end{cases}$ is shown below. Approximate the solutions to the system to the nearest tenth.

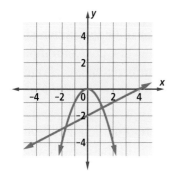

4. Consider the system $\begin{cases} x + 4y = 16 \\ 2x + 8y = 30 \end{cases}$.

   Is this system consistent or inconsistent? How do you know?

In 5 and 6, solve each system using an algebraic method and show how to check your answer.

5. $\begin{cases} p = 3q \\ r = q + 4 \\ 3r - 2p = 3 \end{cases}$

6. $\begin{cases} 7x + 5y = 12 \\ 2x + 9y = 14 \end{cases}$

7. The Natural Nut Company sells organic nuts. How many pounds of organic hazelnuts priced at $7.83 per pound should be mixed with organic pecans priced at $11.50 per pound to obtain 20 pounds of mixed nuts priced at $10 per pound?

   a. Write a system of equations representing this situation. Tell what each variable stands for in this problem.

   b. Solve the system and answer the question.

8. Consider the system $\begin{cases} 3x + 2y = 24 \\ -2x + 7y = 39 \end{cases}$.

   a. What is the coefficient matrix?

   b. Find the inverse of the coefficient matrix.

   c. Use a matrix equation to solve the system.

9. a. Give an example of a 2×2 matrix that does not have an inverse.

   b. How can you tell that the inverse does not exist?

10. Graph the solution set of $y > -\frac{5}{2}x - 2$.

11. The graph below shows the feasible region for what system?

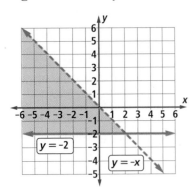

12. Tell whether the following points are solutions to the system from Question 11.

a. (2.5, 3.2)

b. (–4, 0.5)

c. (–2, 2)

13. Surehold Shelving Company produces two types of decorative shelves. The Olde English style takes 20 minutes to assemble and 10 minutes to finish. The Cool Contemporary style takes 10 minutes to assemble and 20 minutes to finish. Each day there are at most 48 worker-hours available in the assembly department and at most 64 worker-hours available in the finishing department. If Surehold Shelving makes a $15 profit on each Olde English shelf and a $12 profit on each Cool Contemporary shelf, how many of each shelf should the company make to maximize its profit?

a. Let e = number of Olde English shelves and c = number of Cool Contemporary shelves the company makes. Translate the constraints into a system of linear inequalities.

b. Graph the system of inequalities and find the vertices of the feasible region.

c. Apply the Linear-Programming Theorem and interpret the results.

Chapter 5 Chapter Review

SKILLS Procedures used to get answers

OBJECTIVE A Solve 2 × 2 and 3 × 3 systems using the linear-combination method or substitution. (Lessons 5-3, 5-4)

1. **Multiple Choice** After which of the following operations does the system $\begin{cases} -2x + 3y = 13 \\ 6x + y = 5 \end{cases}$ yield $20x = 2$?

   A Multiply the first equation by 3 and add.

   B Multiply the second equation by 3 and add.

   C Multiply the first equation by –3 and add the second equation.

   D Multiply the second equation by –3 and add the first equation.

In 2–7, solve and check.

2. $\begin{cases} 3z - 6w = 15 \\ 0.5z - w = 22 \end{cases}$

3. $\begin{cases} 2x + 10y = 16 \\ x = -3y \end{cases}$

4. $\begin{cases} r = s - 5 \\ 2r - s = -3 \end{cases}$

5. $\begin{cases} -7 = 2x + 6y \\ 0 = -4x - 8y + 22 \end{cases}$

6. $\begin{cases} 3r + 12t = 6 \\ r = 2s \\ t = \frac{7}{12}s \end{cases}$

7. $\begin{cases} a = 2b - 4 \\ b = 2c + 2 \\ c = 4a + 6 \end{cases}$

8. Consider the system $\begin{cases} y = -3x \\ -5x + 3y = -28 \end{cases}$.

   a. Which method do you prefer to use to solve this system, substitution or the linear-combination method?

   b. Solve and check the system using your method from Part a.

OBJECTIVE B Find the determinant and inverse of a square matrix. (Lesson 5-5)

In 9–12, a matrix is given.

a. Calculate its determinant.

b. Find the inverse, if it exists.

9. $\begin{bmatrix} 4 & 0 \\ 0 & 1 \end{bmatrix}$

10. $\begin{bmatrix} 5 & 2 \\ -3 & 2 \end{bmatrix}$

11. $\begin{bmatrix} 2 & -2 \\ 5 & -5 \end{bmatrix}$

12. $\begin{bmatrix} w & x \\ y & z \end{bmatrix}$

13. Suppose $M = \begin{bmatrix} 1 & 2 \\ -5 & 3 \end{bmatrix}$. Find M^{-1}.

14. If the inverse of $\begin{bmatrix} a & b \\ c & d \end{bmatrix}$ does not exist, what must be true about its determinant?

15. Explain why the matrix $\begin{bmatrix} 3 & 1 \\ 6 & 2 \end{bmatrix}$ does not have an inverse.

16. a. Find $\begin{bmatrix} -9 & 8 & 7 \\ 6 & 5 & 4 \\ 3 & 2 & 1 \end{bmatrix}^{-1}$ using a calculator.

    b. Check your answer using matrix multiplication.

OBJECTIVE C Use matrices to solve systems of two or three linear equations. (Lesson 5-6)

In 17–20, solve each system using matrices.

17. $\begin{cases} 3x - 4y = 12 \\ 4x - 6y = 48 \end{cases}$ 18. $\begin{cases} 4a - 2b = -3 \\ 3a + 5b = 15 \end{cases}$

19. $\begin{cases} 2m = 5n + 4 \\ 3m = 6n - 3 \end{cases}$ 20. $\begin{cases} 24 = 3x - 4y + 2z \\ 6 = x + 9y \\ 12 = 2x - 3y + 6z \end{cases}$

PROPERTIES Principles behind the mathematics

OBJECTIVE D Recognize properties of systems of equations. (Lessons 5-2, 5-3, 5-4, 5-6)

21. Are the systems $\begin{cases} 7x - 3y = 7 \\ 3x + 2y = 19 \end{cases}$ and $\begin{cases} y = 5 \\ x + y = 8 \end{cases}$ equivalent? Why or why not?

22. Give the simplest system equivalent to $4x = 8$ and $x + y = 6$.

23. What is a system with no solutions called?

In 24–27, a system is given.

a. Identify the system as inconsistent or consistent.

b. Determine the number of solutions.

24. $\begin{cases} 2x + 7y = 14 \\ 2x + 7y = 28 \end{cases}$ 25. $\begin{cases} 6m - 2n = 7 \\ -3m = -2n - \frac{7}{2} \end{cases}$

26. $\begin{cases} 4a - 5b = 20 \\ 2a + 3b = -6 \end{cases}$ 27. $\begin{cases} r = -t^2 \\ r = t - 3 \end{cases}$

28. For what value of k does $\begin{cases} 3x + ky = 6 \\ 15x + 5y = 30 \end{cases}$ have infinitely many solutions?

29. Find a value of t for which $\begin{cases} 2x + 8y = t \\ 3x + 12y = 7 \end{cases}$ has no solutions.

30. Suppose the determinant of the coefficient matrix of a system of equations is not zero. What can you conclude about the system?

OBJECTIVE E Recognize properties of systems of inequalities. (Lessons 5-8, 5-9)

31. **True or False** The boundaries are included in the graph of the solution set of $\begin{cases} y > 5 \\ y < 6 - x \end{cases}$.

32. A system of inequalities is graphed below. Tell whether the point is a solution to the system. Justify your answer.

    a. $(2, 4)$ b. $(8, 2)$

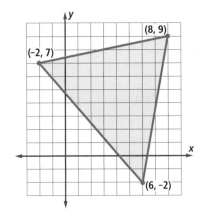

33. Tell whether the shaded region could be a feasible set in a linear-programming situation. Justify your answer.

a.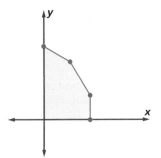

b.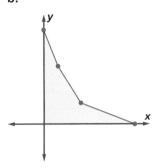

c.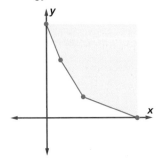

d.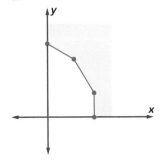

34. Where in a feasible set are the possible solutions to a linear-programming problem?

35. Does the point *M* in the region below represent a possible solution to a linear programming problem? Why or why not?

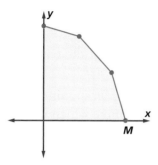

USES Applications of mathematics in real-world situations

OBJECTIVE F Use systems of two or three linear equations to solve real-world problems. (Lessons 5-2, 5-3, 5-4, 5-6)

36. A painter is creating a shade of pink by mixing 5 parts white for every 2 parts red. If 3 gallons of this shade are needed, how much white and how much red is needed?

37. To make longer dives, scuba divers breathe oxygen-enriched air. A diver wants to create a tank of air that contains 36% oxygen by combining two sources. The first source is standard compressed air that contains 20% oxygen. The second source contains 60% oxygen. What percent of each source will the diver need to put into the tank?

38. After three tests, Rachelle's average in math was 76. After four tests, her average was 81. If the teacher drops Rachelle's lowest test score, Rachelle's average will be 85. What was Rachelle's lowest test score?

39. A nutritionist plans a dinner menu that provides 5 grams of protein and 6 grams of carbohydrates per serving. The meat she is serving contains 0.6 gram of protein and 0.4 gram of carbohydrates per ounce. The vegetables have 0.2 gram of protein and 0.8 gram of carbohydrates per ounce. How many ounces of meat and how many ounces of vegetables will be in each serving?

40. Molly Millions makes $6000 per month, while Mike Myzer makes $2800 per month. The following table shows average daily expenses for Molly and Mike. Both are paid on the first day of the month.

	Molly	Mike
Housing	$127	$28
Transportation	$22	$3
Food	$21	$6
Other	$15	$2

a. Who has more money left from their paycheck at the end of the month?

b. On which day of the month do Mike and Molly have about the same amount left from their pay?

OBJECTIVE G Solve problems using linear programming. (Lesson 5-9)

41. A real estate developer is planning a building that will combine commercial space (offices and stores) with residential space. The total development will have 40,000 square feet. Construction costs are $225 per square foot for residential space and $185 per square foot for commercial space, and the developer has a construction budget of $8,500,000. Zoning laws dictate that there be no more than 30,000 square feet of commercial space in the development. The developer plans to sell the property and anticipates a profit of $70 per square foot for residential space and $60 per square foot for commercial space.

    a. Let c = number of square feet of commercial space and r = number of square feet of residential space. Translate the constraints into a system of linear inequalities.

    b. Graph the system of inequalities and find the vertices of the feasible region.

    c. Find the solution (c, r) that maximizes profit.

42. Some parents shopping for their family want to know how much hamburger and how many potatoes to buy. From a nutrition table, they find that one ounce of hamburger has 0.7 mg of iron, 0 IU of vitamin A, and 7.4 grams of protein. One medium potato has 1.9 mg of iron, 17.3 IU of vitamin A, and 4.3 grams of protein. For this meal the parents want each member of the family to have at least 5 mg of iron, 15 IU of vitamin A, and 35 grams of protein. One potato costs $0.10 and 1 ounce of hamburger costs $0.15. The parents want to minimize their costs, yet meet daily requirements. What quantities of hamburger and potatoes should they buy for the family?

a. Identify the variables for this problem.

b. Translate the constraints of the problem into a system of inequalities. (You should have five inequalities; a table may help.)

c. Graph the system of inequalities from Part b and find the vertices of the feasible region.

d. Write an expression for the cost to be minimized.

e. Apply the Linear-Programming Theorem to determine which vertex minimizes the cost expression of Part d.

f. Interpret your answer to Part e. What is the best buy for this family?

REPRESENTATIONS Pictures, graphs, or objects that illustrate concepts

OBJECTIVE H Solve and graph linear inequalities in one variable. (Lesson 5-1)

43. Other weather conditions aside, the Space Shuttle will not delay or cancel a launch as long as the temperature is greater than 48°F and no more than 99°F.

    a. Graph the allowable launch temperatures.

    b. Write the set of allowable launch temperatures in set-builder notation.

44. **Multiple Choice** Which inequality is graphed below?

A $x > -\dfrac{3}{2}$　　　　B $x < -\dfrac{3}{2}$

C $x \geq -\dfrac{3}{2}$　　　　D $x \leq -\dfrac{3}{2}$

In 45 and 46, solve the inequality and graph its solution set.

45. $-3x + 9 > 21$　　　46. $2n + 3(n - 12) \geq 9$

47. Write an inequality that describes the graph below.

In 48–51, graph on a number line.

48. $\{x \mid x > 7 \text{ and } x < 11\}$

49. $\{t \mid -3 \le t < 5\} \cap \{t \mid t \ge 1\}$

50. $\{n \mid n > 9\} \cup \{n \mid n > 4\}$

51. $\{y \mid y \le 7 \text{ or } 8 \le y \le 10\}$

52. Write the compound sentence that is graphed below.

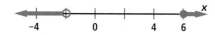

OBJECTIVE I Estimate solutions to systems by graphing. (Lesson 5-2)

In 53–55, estimate all solutions to the system by graphing.

53. $\begin{cases} y = 4x - 5 \\ y = 0.5x + 3 \end{cases}$

54. $\begin{cases} 3x - 2y = -6 \\ \qquad\quad y = x^2 \end{cases}$

55. $\begin{cases} 3x + 5y = -19 \\ \qquad\quad xy = 7 \end{cases}$

OBJECTIVE J Graph linear inequalities in two variables. (Lesson 5-7)

In 56–59, graph on a coordinate plane.

56. $x < -3$ or $y \ge 1$

57. $x \ge 7$ and $y \ge 13$

58. $y \ge -2x + 2$

59. $2x - 5y < 8$

60. Write an inequality to describe the shaded region below.

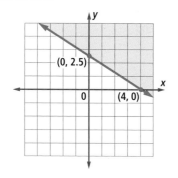

OBJECTIVE K Solve systems of inequalities by graphing. (Lesson 5-8)

In 61–63, graph the solution set.

61. $\begin{cases} x \ge 3 \\ y \le -2 \end{cases}$

62. $\begin{cases} 5c + 7d < 35 \\ 5c - 7d > -1 \end{cases}$

63. $\begin{cases} \qquad\ \ 3x \ge -6 \\ 2(x + y) \le 6 \\ \qquad 6 > y - 2x \end{cases}$

64. Use a compound sentence to describe the shaded region below.

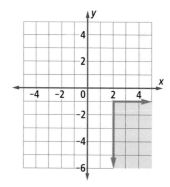

65. Multiple Choice Which of the following systems describes the shaded region below?

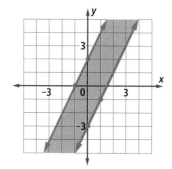

A $\begin{cases} y < \frac{1}{2}x + 2 \\ y < \frac{1}{2}x - 3 \end{cases}$

B $\begin{cases} y > 2x + 2 \\ y < x - 3 \end{cases}$

C $\begin{cases} y \le 2x + 2 \\ y \ge 2x - 3 \end{cases}$

D $\begin{cases} y \le x - 1 \\ y \ge 2x + 1.5 \end{cases}$

Chapter

6 Quadratic Functions

Contents

The word *quadratic* comes from the Latin word *quadratus*, which means "to make square." Many situations lead to quadratic functions. The area formulas $A = s^2$ (for a square) and $A = \pi r^2$ (for a circle) are equations for quadratic functions. You studied direct-variation quadratic functions of the form $f(x) = kx^2$ in Chapter 2. The area of a rectangle with length x and width y is xy, a quadratic expression in the two variables x and y.

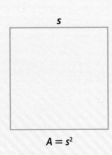

$A = s^2$

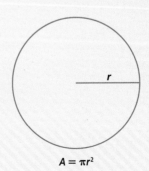

$A = \pi r^2$

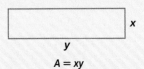

$A = xy$

Quadratic functions also arise from studying the paths of objects. Consider the picture at the right of water spouting from a water fountain. Let y be the height (in inches) of the water above the spout and let x be the horizontal distance (in inches) from the spout. Then quadratic regression can be used to find the equation $y = -0.58x^2 + 2.7x$ to represent the path of the water.

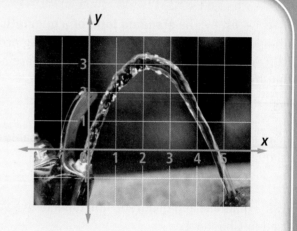

The water's path closely follows a parabola, as do the paths traveled by a batted baseball or thrown basketball. All parabolas can be described by quadratic equations.

In this chapter, you will study many uses of quadratic expressions and functions. You also will learn how to solve all quadratic equations, including those whose solutions involve square roots of negative numbers.

Lesson
6-1

Quadratic Expressions, Rectangles, and Squares

Vocabulary

quadratic expression

quadratic equation

quadratic function

standard form of a quadratic

binomial

▶ **BIG IDEA** Some quadratic expressions arise from problems involving the areas of rectangles.

Recall that the degree of a monomial is the sum of the exponents of the variables in the monomial, and the degree of a polynomial is the highest degree among its monomial terms. For instance, the monomials $20x^5$ and $\frac{1}{3}ab^4$ have degree 5, while the polynomial $x^5 - x^6$ has degree 6. The word *quadratic* in today's mathematics refers to expressions, equations, and functions that involve sums of constants and first and second powers of variables and no higher powers. That is, they are of degree 2. Specifically, if a, b, and c are real numbers, $a \neq 0$, and x is a variable,

$ax^2 + bx + c$ is the general **quadratic expression** in x,

$ax^2 + bx + c = 0$ is the general **quadratic equation** in x, and

$f: x \rightarrow ax^2 + bx + c$ is the general **quadratic function** in x.

We call $ax^2 + bx + c$ the **standard form of a quadratic**. In standard form, the powers of the variable are in decreasing order. Some quadratic expressions, equations, and functions are not in standard form, but they can be rewritten in standard form.

 QY1

There can also be quadratics in two or more variables. The general quadratic expression in two variables is

$$Ax^2 + Bxy + Cy^2 + Dx + Ey + F.$$

Quadratic equations and quadratic functions in two variables are discussed in Chapter 12.

The simplest quadratic expression x^2 is the product of the simplest linear expressions x and x. More generally, the product of any two linear expressions $ax + b$ and $cx + d$ is a quadratic expression for any real numbers a, b, c, and d, provided a and c are not zero. Because all area formulas involve the product of two lengths, they all involve quadratic expressions.

Mental Math

A 6' by 3.5' mural is surrounded by a 1.5'-wide border.

a. What is the area of the mural?

b. What is the area of the mural plus the border?

c. What is the area of just the border?

d. The border is narrowed to 1 foot wide. Now what is the area of the mural and border?

▶ **QY1**

Write $3x - (5 - 2x^2)$ in standard form.

Quadratic Expressions from Rectangles

Example 1

Hector and Francisca are remodeling their kitchen. They purchase a 6-foot by 2-foot pantry door, and are looking at different widths of molding to trim the door frame. If the trim is w inches wide, write the total area of the door and trim in standard form.

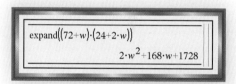

Solution Draw a picture. The door is surrounded by trim on 3 sides. The door with trim occupies a rectangle with length $72 + w$ inches and width $24 + 2w$ inches. The area of this rectangle is $(72 + w)(24 + 2w)$ square inches.

Use the Distributive Property to multiply $(72 + w)(24 + 2w)$. Think of $(72 + w)$ as a single number.

$$(72 + w)(24 + 2w) = (72 + w) \cdot 24 + (72 + w) \cdot 2w$$

Now apply the Distributive Property twice more.

$$
\begin{aligned}
&= 72 \cdot 24 + w \cdot 24 + 72 \cdot 2w + w \cdot 2w \\
&= 1728 + 24w + 144w + 2w^2 \qquad \text{Arithmetic} \\
&= 1728 + 168w + 2w^2 \qquad \text{Combine like terms.}
\end{aligned}
$$

In standard form, the total area of the door and the trim is $2w^2 + 168w + 1728$ square inches.

Check Use a CAS to expand the expression.

$$\text{expand}\big((72+w)\cdot(24+2\cdot w)\big)$$
$$2 \cdot w^2 + 168 \cdot w + 1728$$

 QY2

> **QY2**
>
> The expression
> $2w^2 + 168w + 1728$ is
> in the form $ax^2 + bx + c$.
> What are a, b, c, and x?

Quadratic Expressions from Squares

The expression $x + y$ is an example of a *binomial*. In general, a **binomial** is an expression with two terms. The square of a binomial can be thought of as the area of a square whose side length is the binomial.

Example 2

Write the area of the square with sides of length $x + y$ in standard form.

Solution 1 Draw a picture of the square. Notice that its area is the sum of four smaller areas: a square of area x^2, two rectangles, each with area xy, and a square with area y^2. So, the area of the original square is $x^2 + 2xy + y^2$.

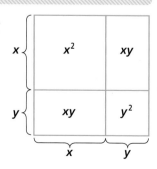

(continued on next page)

Solution 2 The area of a square with side $x + y$ is $(x + y)^2$.
Rewrite $(x + y)^2$.

$$(x + y)^2 = (x + y)(x + y) \quad \text{Definition of second power}$$
$$= (x + y)x + (x + y)y \quad \text{Distributive Property}$$
$$= x^2 + yx + xy + y^2 \quad \text{Distributive Property}$$
$$= x^2 + 2xy + y^2 \quad \text{Commutative Property of Multiplication and Distributive Property}$$

The area of the square is $x^2 + 2xy + y^2$.

When a linear expression is multiplied by itself, or squared, the result is a quadratic expression. In Example 2, the linear expression $x + y$ is squared. You can also say it is "taken to the 2nd power." Writing this power as a quadratic expression is called *expanding* the power. Squares of binomials occur so often that their expansions are identified as a theorem.

Binomial Square Theorem

For all real numbers x and y,
$$(x + y)^2 = x^2 + 2xy + y^2 \text{ and}$$
$$(x - y)^2 = x^2 - 2xy + y^2.$$

Solution 2 of Example 2 provides the proof of the first part of this theorem. You are asked to complete the second part of this proof in Question 13. The Binomial Square Theorem is so useful that you will want to be able to apply it automatically.

STOP QY3

▶ **QY3**

Use the Binomial Square Theorem to expand $(3x - y)^2$.

GUIDED

Example 3

A city wants to cover the seating area around a circular fountain with mosaic tiles. The radius of the fountain, including s feet for seating, is 15 feet.

a. Write a quadratic expression in standard form for the area of the fountain, not including the seating area.

b. How many square feet are in the seating area in terms of s?

Solution

a. Draw a picture. The radius of the fountain without seating is ___?___ ft.
So, the fountain area without seating is π (___?___)^2 ft^2.
To expand (___?___ − ___?___)^2, use the Binomial Square Theorem with $x = $ ___?___ and $y = s$.

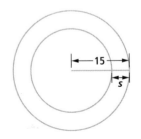

$$(\underline{\ ?\ } - s)^2 = \underline{\ ?\ }^2 - 2 \cdot \underline{\ ?\ } \cdot s + s^2$$
$$= s^2 - \underline{\ ?\ } s + \underline{\ ?\ }$$

So the area of the fountain not including seating is
$\pi(\underline{\ ?\ })^2$ or $\pi \underline{\ ?\ } - 30\pi s + 225\pi$ ft².

b. The seating area is the total area minus the fountain area without seating.

$$15^2\pi - (\underline{\ ?\ })^2\pi = 225\pi - (\underline{\ ?\ } - \underline{\ ?\ } + \underline{\ ?\ })\pi$$
$$= \underline{\ ?\ } - \underline{\ ?\ }$$

So the area of the seating area is $\underline{\ ?\ }$ ft².

Questions

COVERING THE IDEAS

1. **Multiple Choice** Which is not a quadratic equation? Explain your answer.

   A $y = \dfrac{x^2}{6}$

   B $y = 2(x - 4)$

   C $\dfrac{x^2}{4} - \dfrac{y^2}{9} = 25$

   D $y = (x + 3)(2x - 5)$

2. Is $x^2 + \sqrt{7}$ a quadratic expression? Explain your answer.

3. Name two geometric figures for which quadratic expressions describe their area.

4. A door is 7 feet high and 30 inches wide, with trim of w inches wide on three sides of the door frame. Write the total area (in square inches) of the door and trim in standard form.

In 5–8, the product of two linear expressions is given.
   a. Rewrite the product as a single polynomial.
   b. Check your results using the **expand command** on a **CAS**.

5. $(3x + 5y)(2x + 7y)$

6. $(x - 3)(x + 2)$

7. $(1 - 2y)(2 - 3y)$

8. $(6 + b)(2 - b)$

In 9–11, expand the square of the binomial.

9. $(10 + 3)^2$

10. $(d - 6)^2$

11. $(p + w)^2$

12. Draw a geometric diagram of the expansion of $(x + 5)^2$.

13. Prove the second part of the Binomial Square Theorem.

In 14 and 15, rewrite the expression in the form $ax^2 + bx + c$.

14. $\left(3t - \dfrac{1}{3}\right)^2$

15. $(1 - p)^2$

This doorway in Dublin, Ireland, has very ornate trim.

16. Refer to the quadratic expression in Example 1. Graph
$y = (72 + x)(24 + 2x)$ and $y = 2x^2 + 168x + 1728$ in the window
$\{x \mid -160 \leq x \leq 160\}, \{y \mid -2250 \leq y \leq 5750\}$.

   a. Trace the graph and toggle between the two graphs for at
   least three different values of x. What do you notice about the
   ordered pairs when you switch between the graphs?

   b. Look at a table of values for the two functions. Does the table
   support your observation from Part a?

   c. Based on your results in Parts a and b, what do you conclude
   about the two equations you graphed?

17. Suppose a rectangular swimming pool with dimensions of
100 feet by 12 feet is surrounded by a walkway of width w.

   a. Write a quadratic expression in standard form that
   gives the area of the pool and walkway together.

   b. Write an expression that gives the area of the
   walkway only.

APPLYING THE MATHEMATICS

In **18** and **19**, rewrite the expression in the form $ax^2 + bxy + cy^2$.

18. $(2a + 3b)^2$

19. $\left(2t - \frac{k}{3}\right)^2$

20. Refer to Example 3. If $s = 3$ feet and 12 mosaic tiles cover one
square foot, how many tiles would be needed to cover the
seating area around the fountain?

21. Certain ceramic tiles are 4 inches by 8 inches and
are separated by grout seams that are x inches wide.

   a. Write a quadratic expression in standard form for
   the area covered by each tile and its share of the
   grout. (The grout in each seam is shared by two
   tiles, so each tile's share is only half the grout in
   each seam.)

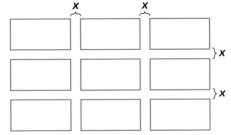

   b. If the grout seams are $\frac{1}{2}$ inch wide, approximately
   how many tiles will it take to cover a 5-foot by 10-foot wall?

   c. What percent of the wall in Part b is grout?

   d. If the grout seams are $\frac{3}{4}$-inch wide, what percent of the wall
   will be grout?

22. a. Expand $(x - y)^2 - (x + y)^2$ on a CAS.

   b. Verify the CAS solution by using the
   Binomial Square Theorem.

In **23** and **24**, find h so that the given equation is true.

23. $x^2 + 20x + 100 = (x + h)^2$

24. $x^2 - hx + h = (x - \sqrt{h})^2$

REVIEW

In 25 and 26, use the following data about the United States National Parks. (Lessons 3–5, 2–4)

Year	Number of Recreation Visits (millions)	Federal Appropriations (billions of $)
1999	287.1	2.030
2000	285.9	2.112
2001	279.9	2.568
2002	277.3	2.654
2003	266.1	2.546

Source: National Park Service

25. Find the rate of change from 1999 to 2003 for each of the following:
    a. number of recreation visits   b. federal appropriations

26. a. Find an equation for the line of best fit describing the number of recreation visits as a function of the year using only the years 1999, 2001, and 2003. Let x be the number of years after 1999.
    b. How well does the line of best fit predict the value for 2002?

27. a. Draw $\triangle ABC$ with vertices $A = (0, 0)$, $B = (1, 1)$, and $C = (2, 4)$.
    b. Draw $\triangle A'B'C'$, its image under the transformation $(x, y) \rightarrow (x - 5, y + 2)$.
    c. Describe the effect of this transformation on $\triangle ABC$. (Lesson 4-10)

28. If m varies inversely as t^2, and $m = 14$ when $t = 2.5$, find the value of m when $t = 7$. (Lesson 2-2)

EXPLORATION

29. Doorway trim comes in various widths.
    a. Find out the prices of at least three different widths of doorway trim.
    b. Suppose you want to frame a 7-foot by 3-foot door. Find the area of the door moldings for each of the three different sizes you have found.
    c. How much will it cost to frame the door with each size of molding?
    d. Does the cost in Part b vary directly as the area of the molding? If so, how?

QY ANSWERS

1. $2x^2 + 3x - 5$

2. $a = 2, b = 168$, $c = 1728$, and $x = w$

3. $(3x)^2 - 2(3x)y + y^2 = 9x^2 - 6xy + y^2$

Lesson 6-2

Absolute Value, Square Roots, and Quadratic Equations

Vocabulary

absolute value
absolute-value function
square root
rational number
irrational number

▶ **BIG IDEA** Geometrically, the *absolute value* of a number is its distance on a number line from 0. Algebraically, the absolute value of a number equals the nonnegative square root of its square.

The **absolute value** of a number n, written $|n|$, can be described geometrically as the distance of n from 0 on the number line. For instance, $|42| = 42$ and $|-42| = 42$. Both 42 and –42 are 42 units from zero.

Mental Math

A company makes $6 dollars in revenue for every teacup it sells and $5 in revenue for every saucer it sells. How much revenue will the company make if they sell

a. 500 teacups and no saucers?

b. 400 teacups and 200 saucers?

c. 500 saucers and no teacups?

Algebraically, the absolute value of a number can be defined piecewise as follows.

$$|x| = \begin{cases} x, \text{ for } x \geq 0 \\ -x, \text{ for } x < 0 \end{cases}$$

Examine the definition carefully. Because $-x$ is the opposite of x, $-x$ is positive when x is negative. For instance, $|-7.4| = -(-7.4) = 7.4$. Thus $|x|$ are $|-x|$ are never negative, and, in fact, $|x| = |-x|$.

On many graphing utilities, spreadsheets, and CAS, the absolute-value function is denoted abs. For example, abs(x–3) = |x–3|.

Example 1

Solve for x: $|x - 4| = 8.1$.

Solution Use the algebraic definition of absolute value.

Either $x - 4 = 8.1$ or $x - 4 = -8.1$.

So, $x = 12.1$ or $x = -4.1$.

Check Use a CAS.

 QY1

▶ **QY1**

Suppose $f(x) = |x - 1|$. Write a piecewise definition for f.

The Absolute-Value Function

Because every real number has exactly one absolute value, $f: x \rightarrow |x|$ is a function. The graph of $f(x) = |x|$ is shown at the right. When $x \geq 0$, $f(x) = x$ and the graph is a ray with slope 1 and endpoint $(0, 0)$. This is the ray in the first quadrant. When $x \leq 0$, $f(x) = -x$, and the graph is the ray with slope -1 and endpoint $(0, 0)$. This is the ray in the second quadrant. The graph of $f(x) = |x|$ is the union of two rays, so the graph of $f(x) = |x|$ is an angle.

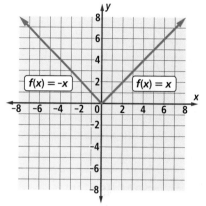

This function is called the **absolute-value function**. Its domain is the set of real numbers, and its range is the set of nonnegative real numbers.

Absolute Value and Square Roots

The simplest quadratic equations are of the form $x^2 = k$. When $k \geq 0$, the solutions to $x^2 = k$ are the positive and negative **square roots** of k, namely $\sqrt{k}$ and $-\sqrt{k}$. Square roots are intimately connected to absolute value.

Activity

Consider the functions f and g with equations
$f(x) = \sqrt{x^2}$ and $g(x) = |x|$.

Step 1 In each row of the table, choose a value of x satisfying the constraint. Then evaluate $f(x)$ and $g(x)$. One row is completed for you.

Step 2 Make a conjecture about the relationship between f and g.

Step 3 Graph f and g on the same axes.

Step 4 Trace and toggle between the graphs to compare $f(x)$ and $g(x)$ for several values of x. Explain the apparent relationship between the graphs of f and g.

Constraint	x	f(x)	g(x)
$x < -10$	?	?	?
$-10 \leq x \leq -1$	?	?	?
$-1 < x < 0$	?	?	?
$x = 0$	0	0	0
$0 < x < 1$	?	?	?
$1 \leq x \leq 10$	?	?	?
$x > 10$	?	?	?

The Activity suggests that, for all real numbers x, $\sqrt{x^2}$ is equal to $|x|$.

> ### Absolute Value–Square Root Theorem
> For all real numbers x, $\sqrt{x^2} = |x|$.

Proof Either $x > 0$, $x = 0$, or $x < 0$.

If $x > 0$, then $\sqrt{x^2} = x$, and also $|x| = x$, so $\sqrt{x^2} = |x|$.

If $x = 0$, then $\sqrt{x^2} = 0$, and also $|x| = |0| = 0$, so $\sqrt{x^2} = |x|$.

If $x < 0$, then $\sqrt{x^2} = -x$, and also $|x| = -x$, so $\sqrt{x^2} = |x|$.

Solving $ax^2 = b$

The Absolute Value–Square Root Theorem can be used to solve quadratic equations of the form $ax^2 = b$.

Example 2

Solve $x^2 = 12$.

Solution 1 Take the positive square root of each side.

$\sqrt{x^2} = \sqrt{12}$

Use the Absolute Value–Square Root Theorem.

$|x| = \sqrt{12}$

So, either $x = \sqrt{12}$ or $x = -\sqrt{12}$.

Check Use your calculator to evaluate $\left(\sqrt{12}\right)^2$ and $\left(-\sqrt{12}\right)^2$. Each equals 12. It checks.

Solution 2 Use a CAS.

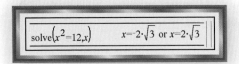

Check The solutions are shown as $-2\sqrt{3}$ and $2\sqrt{3}$, so multiply to show that $(-2\sqrt{3})^2$ and $(2\sqrt{3})^2$ both equal 12.

When $x = a$ or $x = -a$, you can write $x = \pm a$. In Example 2, $x = \pm\sqrt{12} = \pm 2\sqrt{3}$.

Example 3

A square and circle have the same area. The square has side length 15 units. Which is longer, a side of the square or the diameter of the circle?

Solution The area of the square is $15 \cdot 15 = 225$ square units.

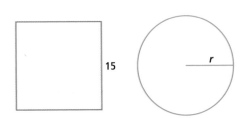

Since we know a formula for the area of a circle in terms of its radius, let r be the radius of the circle.

$$\pi r^2 = 225$$

$$r^2 = \frac{225}{\pi} \qquad \text{Divide by } \pi.$$

$$|r| = \sqrt{\frac{225}{\pi}} \qquad \begin{array}{l}\text{Take the square root of each side and use the} \\ \text{Absolute Value-Square Root Theorem.}\end{array}$$

$$r = \pm\sqrt{\frac{225}{\pi}} \qquad \text{Definition of absolute value}$$

$$\approx \pm 8.46 \text{ units}$$

You can ignore the negative solution because a radius cannot be negative. The radius of the circle is approximately 8.5 units. So the diameter is about 17 units and is longer than a side of the square.

 STOP QY2

> ▶ **QY2**
>
> Check the solution to Example 3.
> Is $\pi(8.5)^2 \approx 225$?

Rational and Irrational Numbers

Recall from earlier courses that a *simple fraction* is a fraction of the form $\frac{a}{b}$, where a and b are integers and $b \neq 0$. And recall from Chapter 1 that a number that can be written as a simple fraction is called a **rational number**. Around 430 BCE, the Greeks proved that unless an integer is a perfect square (like 49, 625, or 10,000), its square root is an *irrational number*. An **irrational number** is a real number that cannot be written as a simple fraction. Irrational numbers, including most square roots, have infinite nonrepeating decimal expansions. The exact answers to Examples 2 and 3 are irrational numbers.

> ▶ **READING MATH**
>
> As used in algebra, the word *rational* comes from the word *ratio*. A *rational number* is a number that can be written as the *ratio* of two integers.

Questions

COVERING THE IDEAS

1. Evaluate without a calculator.
   a. $|17.8|$ b. $|{-}17.8|$ c. $-|17.8|$ d. $-|{-}17.8|$

2. A classmate believes $|{-}t| = t$ for all real numbers t. Is this correct? Explain your answer.

3. A classmate believes $\text{abs}(x) = -\text{abs}(x)$ for all real numbers x. Is this correct? Why or why not?

4. Sketch a graph of f and g with equations $f(x) = |x - 4|$ and $g(x) = 8.1$, and label the coordinates of the points of intersection to verify the answer to Example 1.

5. The two numbers at a distance 90 from 0 on a number line are the solutions to what equation?

In 6 and 7, solve.

6. $|3.4 - y| = 6.5$

7. $|2n + 7| = 5$

8. Consider the function f with equation $f(x) = -|x|$.

a. State its domain and its range.

b. **True or False** The graph of f is piecewise linear. Justify your answer.

9. **Multiple Choice** What is the solution set to $\sqrt{x^2} = |x|$?

A the set of all real numbers

B the set of all nonnegative real numbers

C the set of all positive real numbers

In 10 and 11, find all real–number solutions to the nearest thousandth.

10. $k^2 = 261$

11. $3x^2 = 2187$

12. a. Find the exact radius of a circle whose area is 150 square meters.

b. Estimate the answer to Part a to the nearest thousandth.

13. A circle has the same area as a square with side length 8. What is the radius of the circle to the nearest hundredth?

14. A square has the same area as a circle with radius 9. What is the length of a side of the square to the nearest hundredth?

In 15–20, tell whether the number is rational or irrational. If it is rational, write the number as an integer or a simple fraction.

15. $\sqrt{8}$

16. $\sqrt{100} - 2$

17. $\sqrt{36}$

18. $\frac{0.13}{713}$

19. $\frac{2}{\sqrt{2}}$

20. π

APPLYING THE MATHEMATICS

21. The formula $e = |p - I|$ gives the allowable margin of error e for a given measurement p when I is the ideal measurement. A certain soccer ball manufacturer aims for a weight of 442.5 g with an acceptable value of e being no more than 1.5 g.

a. Use absolute value to write a mathematical sentence for the allowable margin of error for soccer ball weights p.

b. What is the most a soccer ball from this manufacturer should weigh?

22. a. Graph $f(x) = -2\sqrt{(x + 3)^2}$ and $g(x) = -2|x + 3|$ on the same set of axes in a standard window.

b. How do the two graphs appear to be related?

23. The directions on a brand-name pizza box read, "Spread dough to edges of a round pizza pan or onto a 10" by 14" rectangular baking sheet." How big a circular pizza could you make with this dough, assuming it is spread the same thickness as for the rectangular pizza?

24. Graph $f(x) = |x + 2|$ and $h(x) = |x| + 2$ on the same set of axes in a standard window.

    a. According to the graph, for which values of x does $f(x) = g(x)$?

    b. Describe the set of numbers for which $f(x) \neq g(x)$.

25. Use the drawing at the right to explain why $2\sqrt{3} = \sqrt{12}$.

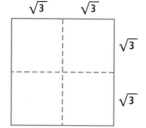

REVIEW

In 26 and 27, multiply and simplify. (Lesson 6-1)

26. $(x + 3y)(x - 2y)$

27. $(8 + x)(8 - x)$

28. Consider the line with equation $y = \frac{4}{3}x + 3$. Find an equation for the image of this line under the translation $T_{3,1}$. (Lesson 4-10)

29. a. Graph the first eight terms of the sequence defined recursively by $\begin{cases} v_1 = 1 \\ v_n = v_{n-1} + n, \text{ for integers } n \geq 2 \end{cases}$.

    b. Rewrite the second line of the formula in Part a if v_n represents the previous term of the sequence. (Lessons 3-7, 3-6)

30. A graph of $y = kx^2$ is shown at the right. Find the value of k. (Lesson 2-5)

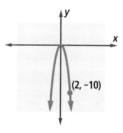

$(2, -10)$

EXPLORATION

31. One way to estimate $\sqrt{k}$ without using a square root command on a calculator or computer uses the following sequence:
$\begin{cases} a_1 = \text{initial guess at the root} \\ a_n = \frac{1}{2}\left(a_{n-1} + \frac{k}{a_{n-1}}\right), \text{ for integers } n \geq 2. \end{cases}$

    a. Let $k = 5$. Give a rational number approximation for $\sqrt{5}$ and use that number as a_1. Then find $a_2, a_3, a_4,$ and a_5. Use a calculator to check the difference between a_5 and $\sqrt{5}$.

    b. Continue to generate terms of the sequence until you are within 0.0001 of $\sqrt{5}$.

    c. Use the sequence to estimate the positive square root of 40 to the nearest millionth.

QY ANSWERS

1. $f(x) = |x-1| = \begin{cases} x - 1, \text{ for } x \geq 1 \\ -x + 1, \text{ for } x < 1 \end{cases}$

2. $\pi(8.5)^2 \approx 226.98$, close enough given that 8.5 is an approximation. (In fact, $\pi(8.46)^2 \approx 224.85$, much closer to 225.)

Lesson

6-3

The Graph-Translation Theorem

Vocabulary

corollary

vertex form

axis of symmetry

minimum, maximum

▶ **BIG IDEA** If you know an equation for a graph, then you can easily find an equation for any translation image of the graph.

You can quickly graph functions whose graphs are translation images of functions with which you are already familiar.

Mental Math

Let $x = 3$. Find

a. $x^2 + 7$.

b. $(x + 7)^2$.

c. $7x^2$.

d. $(7x)^2$.

Activity

MATERIALS graphing utility

Work with a partner.

Step 1 Graph each group of equations below on the same axes. Print or sketch each group and label the individual parabolas.

Group A	Group B	Group C
$f1(x) = x^2$	$f1(x) = x^2$	$f1(x) = x^2$
$f2(x) = (x - 4)^2$	$f2(x) = x^2 - 5$	$f2(x) = (x - 3)^2 + 2$
$f3(x) = (x + 2)^2$	$f3(x) = x^2 + 3$	$f3(x) = (x + 4)^2 - 1$

Step 2 For each group, describe the translations that map the graph of $f(x) = x^2$ onto the graphs of the other two equations.

Step 3 Without graphing, describe the graph of each equation below as a translation image of the graph of $y = x^2$.

a. $y = (x + 5)^2$

b. $y = x^2 - 1$

c. $y = (x + 2)^2 - 5$

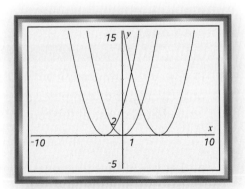

Step 4 Make some conjectures. For any real numbers h and k, what translation maps $y = x^2$ onto the graph of

a. $y = (x - h)^2$? **b.** $y = x^2 + k$? **c.** $y = (x - h)^2 + k$?

Step 5 Test your conjectures from Step 4 for some other positive values of h and k.

The Graph-Translation Theorem

Your sketch of Group A in the Activity should show that when x is replaced by $(x - 4)$, the preimage is translated 4 units to the right, and when x is replaced by $(x + 2)$, the preimage is translated 2 units to the left. In general, replacing x with $x - h$ in a mathematical sentence translates its graph h units horizontally.

Similarly, replacing y with $y - k$ in a sentence translates its graph k units vertically. For example, your sketch of Group B should show that the graph of $y = x^2 + 3$ is 3 units above the graph of $y = x^2$. Note that you can rewrite this equation as $y - 3 = x^2$, so replacing y with $y - 3$ in the equation for a function translates its graph 3 units up.

Recall that the translation $T_{h, k}$ creates an image of a figure h units to the right and k units up from its preimage. The graph of $y - 2 = (x - 3)^2$ is the translation image of the graph $y = x^2$ under $T_{3, 2}$. The results of the Activity are summarized in the Graph-Translation Theorem.

Graph-Translation Theorem

In a relation described by a sentence in x and y, the following two processes yield the same graph:

1. replacing x by $x - h$ and y by $y - k$;
2. applying the translation $T_{h, k}$ to the graph of the original relation.

The Graph-Translation Theorem applies to all relations that can be described by a sentence in x and y.

Example 1

Find an equation for the image of the graph of $y = |x|$ under the translation $T_{-2, 4}$.

Solution Applying $T_{-2, 4}$ is equivalent to replacing x with $x - (-2)$, or $x + 2$, and y with $y - 4$ in the equation for the preimage. An equation for the image is $y - 4 = |x - (-2)|$, or $y = |x + 2| + 4$.

Check $T_{-2, 4}$ is the translation that slides a figure 2 units left and 4 units up. Graph $y = |x|$ and $y = |x + 2| + 4$ on the same set of axes. As shown at the right, the graph of the second equation is the image of the graph of the first equation under a translation 2 units to the left and 4 units up. It checks.

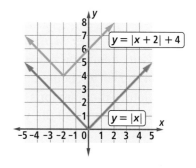

Using the Graph-Translation Theorem to Graph Parabolas

Recall from Chapter 2 that the graph of $y = ax^2$ is a parabola. If we replace x with $x - h$ and y with $y - k$ in the equation $y = ax^2$, we obtain $y - k = a(x - h)^2$. Because a figure is congruent to its translation image, the graph of this equation is also a parabola.

This argument proves the following *corollary* to the Graph-Translation Theorem. A **corollary** is a theorem that follows immediately from another theorem.

Parabola-Translation Theorem

The image of the parabola with equation $y = ax^2$ under the translation $T_{h,k}$ is the parabola with the equation

$$y - k = a(x - h)^2 \text{ or}$$
$$y = a(x - h)^2 + k.$$

The Graph-Translation Theorem and the Parabola-Translation Theorem can help you identify characteristics of a parabola by looking at its equation. As you read about these characteristics below, look at the graphs of $y = ax^2$ and $y - k = a(x - h)^2$ at the right.

- *Vertex* You know that $(0, 0)$ is the vertex of the parabola $y = ax^2$. Under $T_{h,k}$, the translation image of $(0, 0)$ is $T_{h,k}(0, 0) = (0 + h, 0 + k) = (h, k)$. So, the vertex of the parabola with equation $y - k = a(x - h)^2$ is (h, k). For this reason, the equation $y - k = a(x - h)^2$ is called the **vertex form** of an equation of a parabola.

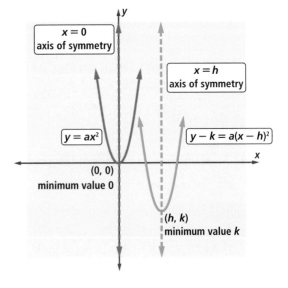

- *Axis of symmetry* The parabola with equation $y = ax^2$ is reflection-symmetric to the y-axis, which has equation $x = 0$. Since the parabola is translated h units to the right under $T_{h,k}$, the line with equation $x = h$ is the symmetry line or **axis of symmetry** of the parabola with equation $y - k = a(x - h)^2$.

- *Maximum or minimum y-value* If $a > 0$, then the parabola with equation $y - k = a(x - h)^2$ opens up and the y-coordinate of the vertex is the **minimum** y-value. The graphs at the right picture the functions when $a > 0$. If $a < 0$, then the parabola opens down and the y-coordinate of the vertex is the **maximum** y-value.

Knowing these facts helps you to quickly sketch parabolas by hand and to better understand what you see when you use a graphing utility.

GUIDED

Example 2

a. State the coordinates of the vertex of the parabola with equation $y - 5 = -2(x + 3)^2$.

b. Write an equation for the axis of symmetry of the parabola.

c. Sketch a graph of the equation by hand.

Solution

a. The equation $y - 5 = -2(x + 3)^2$ results from replacing x with ___?___ and y with ___?___ in $y = -2x^2$. So its graph is the image of $y = -2x^2$ under the translation __?__. The graph is a parabola with vertex (__?__ , __?__).

b. Since the axis of symmetry of $y = -2x^2$ has equation ___?___, the axis of symmetry of this parabola is the line with equation $x = $ __?__.

c. Because the graph of $y = -2x^2$ opens down, so does the graph of $y - 5 = -2(x + 3)^2$. Find a point on the graph other than the vertex. For example, let $x = -2$. Then $y - 5 = -2($ __?__ $+ 3)^2$, so $y = $ __?__ $+ 5$ $= $ __?__, and $(-2, $ __?__ $)$ is a point on the graph. Sketch a parabola with vertex ___?___, opening downward through the point $(-2, $ __?__ $)$ and symmetric to the line $x = $ __?__.

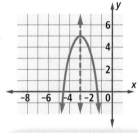

STOP QY

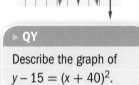

▶ **QY**

Describe the graph of $y - 15 = (x + 40)^2$.

Finding Equations for Parabolas

You can apply the Graph–Translation Theorem to a known parabola to find an equation for its image under a given translation.

Example 3

Consider the parabolas at the right. The one that passes through the origin has equation $y = \frac{1}{5}x^2$. The other is its image under a translation. Find an equation for the image.

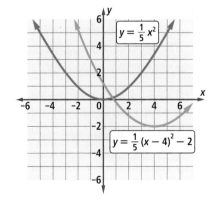

Solution The translation image appears to be 4 units to the right and 2 units down from the preimage. So the translation is $T_{4,-2}$. Applying $T_{4,-2}$ is equivalent to replacing x with $x - 4$ and y with $y - (-2) = y + 2$ in the equation for the preimage. **An equation for the image is $y + 2 = \frac{1}{5}(x - 4)^2$.**

(continued on next page)

> **Check** Use a graphing utility. Because $y + 2 = \frac{1}{5}(x - 4)^2$ is equivalent to $y = \frac{1}{5}(x - 4)^2 - 2$, plot $y = \frac{1}{5}x^2$ and $y = \frac{1}{5}(x - 4)^2 - 2$ in the same window. You should see that the graph of the second equation is the image of the graph of the first under $T_{4,-2}$.

You will see more applications of the Graph-Translation Theorem in later chapters.

Questions

COVERING THE IDEAS

In 1 and 2, tell how the graphs of the two equations are related.

1. $y_1 = x^2$ and $y_2 = (x - 5)^2$

2. $y_1 = x^2$ and $y_2 + 6 = x^2$

3. a. What is the image of (x, y) under $T_{7,0}$?

   b. Under $T_{7,0}$, what is an equation for the image of the graph of $y = x^2$?

4. On the same axes, sketch $y = |x|$ and $y + 1 = |x - 4|$.

   a. Describe how the two graphs are related.

   b. What translation maps the first onto the second?

5. Suppose the translation $T_{4,-7}$ is applied to the parabola with equation $y = \frac{7}{5}x^2$. Find an equation for the image.

6. **Fill in the Blanks** The graph of $y - k = a(x - h)^2$ is __?__ units above and __?__ units to the right of the graph of $y = ax^2$.

7. **True or False** For all values of h and k, the graphs of $y = ax^2$ and $y = a(x - h)^2 + k$ are congruent.

8. a. What is the vertex of the parabola with equation $y - 7 = -3(x + 5)^2$?

   b. What is the vertex of the parabola with equation $y - k = a(x - h)^2$?

9. a. What is an equation of the axis of symmetry of the parabola with equation $y - 7 = -3(x + 5)^2$

   b. What is an equation of the axis of symmetry of the parabola with equation $y - k = a(x - h)^2$?

In 10 and 11, an equation for a parabola is given.

   a. Give the coordinates of the vertex of the parabola.

   b. Give an equation for the axis of symmetry.

   c. Tell whether the parabola opens up or opens down.

   d. Sketch a graph of the equation.

10. $y + 1 = 5(x + 10)^2$ 11. $y = 5 - (x - 4)^2$

Parabolic arcs can often be found in art and architecture, as in Casa Batlló, located in Barcelona, Spain, shown here.

12. Find an equation for the translation image of $y = |x|$ graphed at the left below.

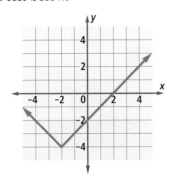 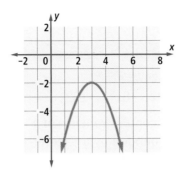

13. The parabola graphed at the right above is a translation image of $y = -x^2$. What is an equation for this parabola?

APPLYING THE MATHEMATICS

14. On the first page of this chapter, an equation for the path of water from a drinking fountain is given as $y = -0.58x^2 + 2.7x$, where x and y are measured in inches. This equation is roughly equivalent to $y - 3.14 = -0.58(x - 2.33)^2$. From the second equation, how high does the water reach?

Centennial Fountain was built in Chicago in 1989 and goes off every hour during the summer months.

15. Consider the graph of $y = -4x^2$. Write an equation for a translation image of the graph

    a. with vertex $(0, 2)$. b. with vertex $(2, 0)$.

16. a. Solve $x^2 = 81$.

    b. Solve $(x - 3)^2 = 81$.

    c. How are the solutions in Parts a and b related to the Graph-Translation Theorem?

17. One solution to $x^2 + 8x + 9 = 57$ is 4. Use this information to find a solution to $(x - 5)^2 + 8(x - 5) + 9 = 57$.

18. Find the x-intercepts and y-intercept of the graph of $y = |x + 3| - 5$.

19. **Fill in the Blanks** The point–slope form of a line, $y - y_1 = m(x - x_1)$, can be thought of as the image of the line with equation ___?___ under the translation $T_{h, k}$, where $h = $ ___?___ and $k = $ ___?___.

20. The parabola $y = 2(x - 4)^2 - 2$ is the image of the parabola $y = 2(x - 3)^2 + 5$ under the translation $T_{h,k}$. What are the values of h and k?

REVIEW

In 21 and 22, solve and check. (Lesson 6–2)

21. $3 \cdot |2d + 3| = 21$

22. $-8x^2 = -162$

23. The competition area for a judo contest consists of a d-meter by d-meter square surrounded by a 3-meter-wide border called the *safety area*. (Lesson 6–1)

   a. Write an expression in standard form for the total area of the competition area.

   b. The rules of judo require $8 \le d \le 10$. What are the minimum and maximum areas a judo competition area can have?

Judo emphasizes flexibility, energy, and balance rather than brute strength.

24. A company makes two kinds of tires: model R (regular) and model S (snow). Each tire is processed on three machines, A, B, and C. To make one model R tire requires $\frac{1}{2}$ hour on machine A, 2 hours on B, and 1 hour on C. To make one model S tire requires 1 hour on A, 1 hour on B, and 4 hours on C. During the upcoming week, machine A will be available for at most 20 hours, machine B for at most 60 hours, and machine C for at most 60 hours. If the company makes a $10 profit on each model R tire and a $15 profit on each model S tire, how many of each tire should be made to maximize the company's profit? (Lesson 5–9)

25. Simplify the expression $\dfrac{(x^2 y)^2}{y^3}$. (Previous Course)

EXPLORATION

26. Investigate how the Graph-Translation Theorem works with other functions. Graph $y = x^3 - 4x$ on your graphing utility. Pick values for h and k, write an equation for its image under each translation below, and graph the image. Verify that the image appears to be a translation image of $y = x^3 - 4x$. Try other values of h and k.

   a. $T_{0,k}$ b. $T_{h,0}$ c. $T_{h,k}$

Lesson
6-4

The Graph of
$y = ax^2 + bx + c$

▶ **BIG IDEA**   The graph of $y = ax^2 + bx + c$, $a \neq 0$, is a parabola that opens upward if $a > 0$ and downward if $a < 0$.

Standard Form for the Equation of a Parabola

Homer King hits a high–fly ball to deep center field. Ignoring air currents, which curve below most closely resembles the flight path of the ball?

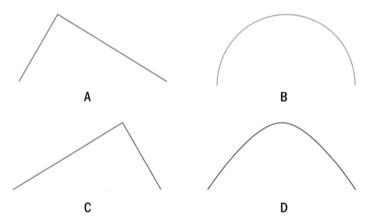

A   B

C   D

Mental Math

Give an example of an equation whose graph contains (1, 3) and is

a. a line.

b. a hyperbola.

c. a parabola.

d. not a line, hyperbola, or parabola.

The answer is D, because high–fly balls and many other projectiles travel in parabolic paths. These paths have equations that can be put into the standard form of a quadratic function, $y = ax^2 + bx + c$. In general, any equation for a parabola that can be written in the vertex form $y - k = a(x - h)^2$ can be rewritten in the standard form $y = ax^2 + bx + c$.

Example 1
Show that the equation $y - 16 = 3(x - 5)^2$ can be rewritten in the form $y = ax^2 + bx + c$, and give the values of a, b, and c.

Solution Solve for y, then expand the binomial, distribute, and simplify.

(continued on next page)

$$y - 16 = 3(x - 5)^2$$

$y = 3(x - 5)^2 + 16$	Add 16 to both sides.
$y = 3(x^2 - 10x + 25) + 16$	Expand the binomial square.
$y = 3x^2 - 30x + 75 + 16$	Distribute the 3.
$y = 3x^2 - 30x + 91$	Arithmetic

So the original equation is equivalent to one in standard form with $a = 3$, $b = -30$, and $c = 91$.

Check 1 Graph both $y = 3(x - 5)^2 + 16$ and $y = 3x^2 - 30x + 91$ on your graphng utility. Use the `trace` feature and toggle between graphs to see if the coordinates match. **The graphs seem to be identical.**

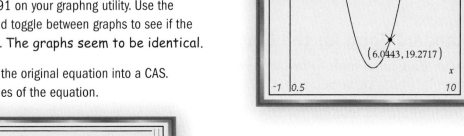

Check 2 Enter the original equation into a CAS. Add 16 to both sides of the equation.

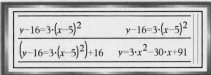

This CAS expands the right side automatically. **It checks.**

 QY1

> **QY1**
>
> In Example 1, subtract the final expression for *y* from the original expression for *y*. What do you get?

In general, to change vertex form to standard form, solve for y and expand.

$$y - k = a(x - h)^2$$

$y = a(x - h)^2 + k$	Add k to each side.
$y = a(x^2 - 2hx + h^2) + k$	Square the binomial.
$y = ax^2 - 2ahx + ah^2 + k$	Use the Distributive Property.

This is in standard form, with $b = -2ah$ and $c = ah^2 + k$. With these substitutions, the equation becomes

$$y = ax^2 + bx + c.$$

Congruent Parabolas

Because the parabola determined by the equation $y - k = a(x - h)^2$ is a translation image of the parabola determined by the equation $y = ax^2$, the two parabolas are congruent. For all h and k, $y - k = a(x - h)^2$ can be written in standard form, so we have the following theorem.

> ### Parabola Congruence Theorem
>
> The graph of the equation $y = ax^2 + bx + c$ is a parabola congruent to the graph of $y = ax^2$.

Recall that a *quadratic function* is any function f whose equation can be put in the form $f(x) = ax^2 + bx + c$, where $a \neq 0$. Thus, the graph of every quadratic function is a parabola, with y-intercept $f(0) = c$. Unless otherwise specified, the domain of a quadratic function is the set of real numbers. When $a > 0$, the range is the set of real numbers greater than or equal to its minimum value. When $a < 0$, the range is the set of real numbers less than or equal to its maximum value.

Applications of Quadratic Functions

Some applications of quadratic functions have been known for centuries. In the early 17th century, Galileo described the height of an object in free fall. Later that century, Isaac Newton derived his laws of motion and the law of universal gravitation. In developing his mathematical equations for the height of an object, Newton reasoned as follows:

Sir Isaac Newton

- Gravity is a force that pulls objects near Earth downward. Without gravity, a ball thrown upward would continue traveling at a constant rate. Then its height would be (initial height) + (upward velocity) · (time). So, if it were thrown at 59 feet per second from an initial height of 4 feet, it would continue traveling at 59 feet per second, and its height after t seconds would be $4 + 59t$.

- Galileo had shown that gravity pulls the ball downward a total of $16t^2$ feet after t seconds. This effect can be subtracted from the upward motion without gravity. Therefore, after t seconds, its height in feet would be $4 + 59t - 16t^2$ feet. The number 16 in the expression is a constant for all objects falling at or near Earth's surface when the distances are measured in feet. When measured in meters, this number is 4.9.

> ### Example 2
> A thrown ball has height $h = -16t^2 + 59t + 4$ after t seconds.
>
> a. Find h when $t = 0, 1, 2, 3,$ and 4.
>
> b. Explain what the pairs (t, h) tell you about the height of the ball for $t = 0, 2,$ and 4.
>
> *(continued on next page)*

c. Graph the pairs (t, h) over the domain of the function.

d. Is the ball moving at the same average rate (speed) between $t = 0$ and $t = 1$ as between $t = 2$ and $t = 3$? Justify your answer.

Solution

a. Use the table feature on your graphing utility or substitute by hand.

b. Each pair (t, h) gives the height h of the ball after t seconds. The pair $(0, 4)$ means that at 0 seconds, the time of release, the ball is 4 feet above the ground. The pair $(2, 58)$ means the ball is 58 feet high after 2 seconds. The pair $(4, -16)$ means that after 4 seconds, the ball is 16 feet below ground level. Unless the ground is not level, it has already hit the ground.

c. The points in Part a are plotted at left below. The points do not tell much about the shape of the graph. More points are needed to show the parabola. By calculating h for other values of t, or by using a graphing utility, you can obtain a graph similar to the one at the right below. The graph is not a complete parabola because the domain of the function is $\{t | t \geq 0\}$.

t (sec)	h (ft)
0	4
1	47
2	58
3	37
4	-16

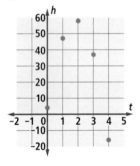
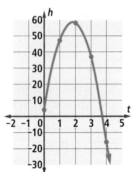

d. The average rate of change between two times is the change in height divided by the change in time. This is the slope of the line through the corresponding points on the graph.

The average rate of change between $t = 0$ and $t = 1$ is $\frac{47-4}{1-0} = 43 \frac{\text{ft}}{\text{second}}$. The average rate of change between $t = 2$ and $t = 3$ is $\frac{37-58}{3-2} = -21 \frac{\text{ft}}{\text{second}}$. (The ball is moving downward on this interval.) The rates are different, meaning the ball travels at different speeds during its flight.

By the Parabola Congruence Theorem, you know that the graph of $h = -16t^2 + 59t + 4$ is a translation image of the graph of $y = -16t^2$.

The equation in Example 2 is a special case of the following general formula that Newton developed for the height h of an object at time t seconds with an initial upward velocity v_0 and initial height h_0.

$$h = -\frac{1}{2}gt^2 + v_0 t + h_0$$

In Example 2, $v_0 = 59\frac{\text{ft}}{\text{sec}}$, the height $h_0 = 4$ ft, and g is a constant denoting *acceleration due to gravity*. Recall that *velocity* involves units like miles per hour, feet per second, or meters per second. Acceleration measures how fast the velocity changes. This "rate of change of a rate of change" involves units like feet per second per second (which is usually abbreviated $\frac{\text{ft}}{\text{sec}^2}$). The acceleration due to gravity varies depending on how close the object is to the center of a massive object. Ignoring the effects of air resistance, near the surface of Earth,

$$g \approx 32\frac{\text{ft}}{\text{sec}^2}, \qquad \text{or} \qquad g \approx 9.8\frac{\text{m}}{\text{sec}^2}.$$

Two common situations are important to note. First, if an object is dropped, not thrown or pushed, its initial velocity $v_0 = 0$. Second, if an object starts at ground level, its initial height $h_0 = 0$.

 QY2

> ▶ **QY2**
>
> An object's height is modeled using the equation
> $h = -16t^2 + 24t + 4.$
> What is the initial velocity? (Do not forget the units.) From what height is it thrown?

Activity

MATERIALS stopwatch, meter stick, tape, rubber ball

Work with a partner to apply Newton's formula for free-falling objects.

Step 1 Copy the table below to record your data.

	Initial height h_0 (m)	Elapsed Time Trial 1 (sec)	Elapsed Time Trial 2 (sec)	Elapsed Time Trial 3 (sec)	Elapsed Time Average t (sec)
Partner 1	?	?	?	?	?
Partner 2	?	?	?	?	?

Step 2 Choose one partner to be the tosser and the other to be the measurer. The tosser chooses a comfortable height from which to toss the ball upward. The measurer records this height and marks it on the meter stick with tape so the tosser can try to consistently release the ball at the same height.

Step 3 The tosser throws the ball upward three times in succession from the height determined in Step 2. With the stopwatch, the measurer records the elapsed time, in seconds, from the initial release of the ball to when it first hits the ground.

Step 4 Reverse roles with your partner and repeat Steps 2 and 3.

Step 5 Calculate and record average times for each partner's tosses.

(continued on next page)

Step 6 Use Newton's formula, $h = -\frac{1}{2}gt^2 + v_0t + h_0$ to calculate the initial upward velocity v_0 for each partner's average toss. (Hint: When did $h = 0$?) Then write an equation to describe each partner's average toss.

Step 7 The ball reaches it maximum height in a little less than half the time it takes the ball to hit the ground. Use your formula to estimate the maximum height of your average toss.

Caution! The equation $h = -\frac{1}{2}gt^2 + v_0t + h_0$ models the height h of the object off the ground at time t. It *does not* describe the path of the object. However, Galileo showed that the actual path of an object thrown at any angle except straight up or straight down is almost parabolic, like the path of water on the second page of the chapter, and an equation for its path is a quadratic equation.

Questions

COVERING THE IDEAS

1. Write the standard form for the equation of a parabola with a vertical line of symmetry.

In 2 and 3, rewrite the equation in standard form.

2. $y = (x - 3)^2$

3. $y = -3(x + 4)^2 - 5$

4. **True or False** For any values of a, b, and c, the graph of $y = ax^2 + bx + c$ is congruent to the graph of $y = ax^2$.

In 5–7, use the equation $h = -\frac{1}{2}gt^2 + v_0t + h_0$ for the height of a body in free fall.

5. Give the meaning of each variable.

    a. h **b.** g **c.** t **d.** v_0 **e.** h_0

6. What value of g should you use if v_0 is measured in $\frac{ft}{sec}$?

7. What is the value of v_0 when an object is dropped?

In 8–11, refer to the graph in Example 2.

8. About how high is the ball after 1.5 seconds?

9. When the ball hits the ground, what is the value of h?

10. At what times will the ball be 20 feet above the ground?

11. What is the average rate of change of the ball's height between 1 second and 3 seconds?

12. Suppose a person throws a ball upward at a velocity of 16 $\frac{m}{sec}$ from the top of a 20-meter tall building.

    a. Write an equation to describe the height of the ball above the ground after t seconds.

    b. How high is the ball after 0.75 second?

    c. Use a graph to estimate the ball's maximum height.

    d. After 6 seconds, is the ball above or below ground level? Justify your answer.

APPLYING THE MATHEMATICS

13. Sketch $y = -x^2 + 4x + 6$ for $-2 \leq x \leq 6$. On your sketch of the graph, label the vertex and the x- and y-intercepts with approximate values.

14. Consider the function f defined by the equation $f(x) = x^2 + 3x - 10$.

    a. Sketch a graph of the function.

    b. Write an equation for the line of symmetry of the parabola.

    c. Estimate the coordinates of the lowest point on the parabola.

In 15 and 16, because the object is dropped, not thrown, its initial velocity is 0.

15. Suppose a penny is dropped from the top of Taipei 101, which in 2004 surpassed the Twin Petronas Towers in Malaysia as the world's tallest building. The roof of Taipei 101 is 1,474 feet above ground.

    a. Write an equation for the penny's height as a function of time.

    b. Graph your equation from Part a over an appropriate domain.

    c. Estimate how much time it would take the penny to fall to the ground.

    d. When the penny falls through the atmosphere, air resistance actually limits its velocity to a maximum of about 94 feet per second. If the penny traveled at a constant rate of 94 feet per second after 2.9 seconds, how much longer would it take to reach the ground?

Taipei 101 in Taipei, Taiwan

16. In an article about education now often circulated as a joke, the late Dr. Alexander Calandra suggested one way to measure the height of a building with a barometer: drop the barometer from the top of the building and time its fall.

    a. Set up an equation for the barometer's height as a function of time, using h_0 for the initial height of the building.

    b. Suppose it takes 3.9 seconds for the barometer to hit the ground. Substitute values into the equation you wrote in Part a and solve for h_0.

The Graph of $y = ax^2 + bx + c$ **399**

17. Find an equation in standard form for the image of the graph of $y = -\frac{1}{4}x^2$ under the translation $T_{4,2}$.

REVIEW

In 18 and 19, two equations are given.
   a. Graph both equations on the same set of axes.
   b. Describe how the graphs of the two equations are related. (Lesson 6–3)

18. $y = x^2$ and $y = (x + 3)^2 + 4$ 19. $y = |x|$ and $y - 5 = |x - 2|$

20. A gallon of paint can cover an area of 450 square feet. Find the diameter of the largest circle that can be covered with a gallon of paint. **(Lesson 6–2)**

21. Write an inequality to describe the shaded region of the graph at the right. **(Lessons 5–7, 3–4)**

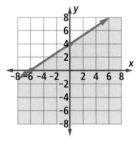

22. Solve the system $\begin{cases} A + B + C = 12 \\ 4A - 4B + 2C = -16 \\ 3A + 3B - C = 4 \end{cases}$ **(Lesson 5–4)**

In 23–25, find n. (Previous Course)

23. $x^2 \cdot x^3 = x^n$ 24. $a^n \cdot a^{16} = a^{64}$ 25. $\dfrac{p^8}{p^2} = p^n$

EXPLORATION

26. How do the values of a, b, and c affect the graph of $y = ax^2 + bx + c$? Here are two suggested methods for investigating:

   Method 1   Use sliders on a DGS or CAS to adjust one coefficient, a, b, or c, at a time.

   Method 2   **a.** Start with $a = 1$ and $b = 6$. Then adjust c and record how the graph changes.

   b. Set $a = 1$ and $c = 4$, then adjust b and note the changes in the graph.

   c. Set $b = 6$ and $c = 4$, then adjust a and note the changes.

   Is the transformation (motion) of the graph simple (like a translation or rotation) for each change of a, b, and c, or is it a compound motion? Which coefficients, if any, affect the graph's size as well as its position?

Completing the Square

Vocabulary

completing the square

perfect-square trinomial

▶ **BIG IDEA** By adding a number to an expression of the form $x^2 + bx$, you can create a new expression that is the square of a binomial.

You have now seen two forms for an equation of a parabola.

Standard form $\quad y = ax^2 + bx + c$

Vertex form $\quad y - k = a(x - h)^2$ or $y = a(x - h)^2 + k$

Because each form is useful, being able to convert between forms is helpful. In Lesson 6-4 you saw how to convert vertex form to standard form. In this lesson, you will see how to convert standard form to vertex form.

What Is Completing the Square?

One method for converting standard form to vertex form is called **completing the square**. Remember that $(x + h)^2 = x^2 + 2hx + h^2$. The trinomial $x^2 + 2hx + h^2$ is called a **perfect-square trinomial** because it is the square of a binomial. At the right you can see that $x^2 + 2hx + h^2$ is the area of a square with side length $x + h$.

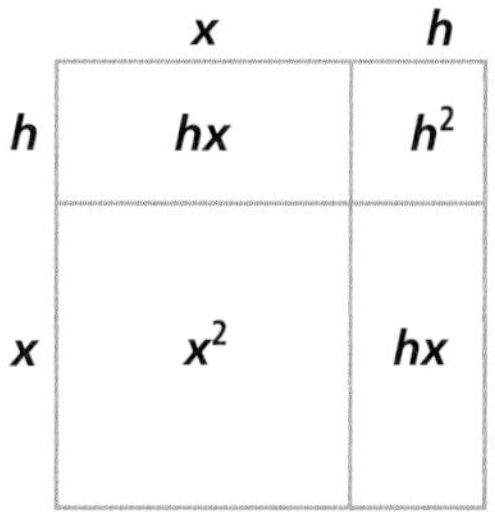

	x	h
h	hx	h^2
x	x^2	hx

Area $= (x + h)^2 = x^2 + 2hx + h^2$

Mental Math

How many solutions does a 2 × 2 system of linear equations have if the two equations in the system represent

a. oblique lines with different slopes?

b. one vertical line and one nonvertical line?

c. the same line?

d. parallel lines with different y-intercepts?

Example 1

a. What number should be added to $x^2 + 6x$ to make a perfect-square trinomial?

b. Write the perfect-square trinomial as the square of a binomial.

(continued on next page)

Solution 1 Use geometry.

a. Draw a picture to represent $x^2 + 6x +$ __?__. Since the sum of the areas of the two rectangles that are not squares must be $6x$, the area of each rectangle is $3x$. Think: What is the area of the missing square in the upper right corner that allows you to complete the larger square? (This is the reason this process is called "completing the square".) A square with area 9 would complete the larger square.

So, 9 must be added to $x^2 + 6x$ to make a perfect-square trinomial.

b. In the picture at the right, the length of the side of the square is $x + 3$. So, $x^2 + 6x + 9 = (x + 3)^2$.

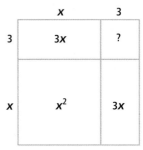

$$\text{Area} = x^2 + 6x + \underline{\quad ? \quad}$$

Solution 2 Use algebra.

a. Compare $x^2 + 6x +$ __?__ with the perfect-square trinomial $x^2 + 2hx + h^2$. The first terms, x^2, are identical. To make the second terms equal, set

$$6x = 2hx.$$

So, $\qquad h = 3.$

The term added to make a perfect-square trinomial should be h^2, or 9.

b. Since $x^2 + 2xh + h^2 = (x + h)^2$, and you found in Part a that $h = 3$,

$$x^2 + 6x + 9 = (x + 3)^2.$$

Check Apply the Binomial Square Theorem to expand $(x + 3)^2$.

$(x + 3)^2 = x^2 + 6x + 9$. It checks.

To generalize Example 1, consider the expression $x^2 + bx +$ __?__. What goes in the blank so that the result is a perfect-square trinomial?

$$x^2 + bx + \underline{\quad ? \quad} = x^2 + 2hx + h^2$$

Because $b = 2h$, $h = \frac{b}{2}$. Then $h^2 = \left(\frac{b}{2}\right)^2$. This illustrates the following theorem.

Completing the Square Theorem

To complete the square on $x^2 + bx$, add $\left(\frac{b}{2}\right)^2$.

Proof $x^2 + bx + \left(\frac{b}{2}\right)^2 = x^2 + bx + \frac{b^2}{4} = \left(x + \frac{b}{2}\right)^2$

 QY

> **QY**
>
> What number should be added to $x^2 - 24x$ to make a perfect square trinomial?

Completing the Square to Find the Vertex of a Parabola

The Completing the Square Theorem can be used to transform an equation of a parabola from standard form into vertex form.

Example 2

a. Rewrite the equation $y = x^2 + 12x + 3$ in vertex form.

b. Find the vertex of the parabola.

Solution

a. Rewrite the equation so that only terms with x are on one side.
$$y - 3 = x^2 + 12x$$

Use the Completing the Square Theorem. Here,
$$b = \underline{12}, \text{ so } \left(\tfrac{b}{2}\right)^2 = \underline{36}.$$

Complete the square on $x^2 + 12x$.

$$y - 3 + \underline{36} = x^2 + 12x + \underline{36} \qquad \text{Add } \left(\tfrac{b}{2}\right)^2 \text{ to both sides.}$$
$$y + \underline{33?} = x^2 + 12x + \underline{36} \qquad \text{Simplify the left side.}$$
$$y + \underline{36?} = (x + \underline{6})^2 \qquad \text{Apply the Binomial Square Theorem.}$$

b. We can read the vertex from the vertex form of the equation. The vertex of the parabola is (_?_ , _?_).

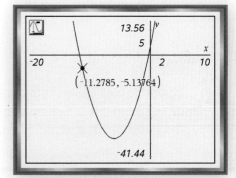

$(-11.2785, -5.13764)$

Check

a. Graph $y = x^2 + 12x + 3$ and the vertex form from Part a on the same set of axes in an appropriate window. Trace and toggle between the graphs for several values of x. They should be identical.

b. Trace to estimate the vertex on the graph.

Equating Expressions to Find the Vertex of a Parabola

Example 2 involves a parabola in which the coefficient of x^2 is 1. Example 3 shows how to find the vertex of a parabola if the coefficient of x^2 is not 1. This kind of expression occurs in describing heights of thrown objects and paths of projectiles.

Example 3

Suppose a ball is thrown straight up from a height of 12 feet with an initial velocity of 32 $\frac{ft}{sec}$. Then the ball's height y after x seconds is given by the formula $y = -16x^2 + 32x + 12$.

a. Rewrite the formula in the vertex form of the equation for a parabola.

b. Find the maximum height of the ball and the time it takes for the ball to reach that point.

Solution

a. Equate the given expression for y with the general vertex form.

$$-16x^2 + 32x + 12 = a(x - h)^2 + k$$

Enter this equation into a CAS. The CAS will automatically expand the square of the binomial, as shown here.

The coefficients of x^2 on the two sides must be equal, so $a = -16$. Substitute this value for a.

$$-16 \cdot x^2 + 32 \cdot x + 12 = a \cdot (x - h)^2 + k$$
$$-16 \cdot x^2 + 32 \cdot x + 12 = a \cdot x^2 - 2 \cdot a \cdot h \cdot x + a \cdot h^2 + k$$

$$+32 \cdot x + 12 = a \cdot x^2 - 2 \cdot a \cdot h \cdot x + a \cdot h^2 + k | a = -16$$
$$-16 \cdot x^2 + 32 \cdot x + 12 = -16 \cdot x^2 + 32 \cdot h \cdot x - 16 \cdot h^2$$

Add $16x^2$ to both sides.

$$\cdot x + 12 = -16 \cdot x^2 + 32 \cdot h \cdot x - 16 \cdot h^2 + k\big) + 16 \cdot x^2$$
$$32 \cdot x + 12 = 32 \cdot h \cdot x - 16 \cdot h^2 + k$$

Equate the coefficients of x to see that $32 = 32h$, so $h = 1$. Substitute this value for h.

$$32 \cdot x + 12 = 32 \cdot h \cdot x - 16 \cdot h^2 + k | h = 1$$
$$32 \cdot x + 12 = 32 \cdot x + k - 16$$

Add $-32x$ to both sides. Then add 16 to both sides. The display at the right show that $k = 28$.

Substitute back for a, h, and k in the square form.

$$(32 \cdot x + 12 = 32 \cdot x + k - 16) + -32 \cdot x \quad\quad 12 = k - 16$$
$$(12 = k - 16) + 16 \quad\quad 28 = k$$

$$y = -16(x - 1)^2 + 28, \text{ or}$$
$$y - 28 = -16(x - 1)^2$$

b. The vertex of this parabola is $(1, 28)$.
So the maximum height of the ball is 28 feet, 1 second after it is thrown.

Check Graph $y = -16x^2 + 32x + 12$. Trace to estimate the maximum value of the parabola. It checks.

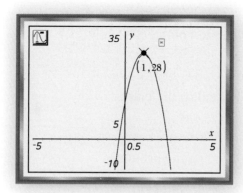

Questions

COVERING THE IDEAS

1. **a.** Give the sum of the areas of the three rectangles at right.
   b. What number must be added to this sum to complete the square?
   c. Fill in the Blanks Complete the equation
   $x^2 + \underline{\quad?\quad} x + \underline{\quad?\quad} = (x + \underline{\quad?\quad})^2$.

Fill in the Blanks In 2–5, find a number to make the expression a perfect-square trinomial.

2. $x^2 + 14x + \underline{\quad?\quad}$
3. $n^2 - n + \underline{\quad?\quad}$
4. $z^2 - 30z + \underline{\quad?\quad}$
5. $t^2 + \frac{2}{3}t + \underline{\quad?\quad}$

In 6 and 7, an equation in standard form is given.
   a. Rewrite the equation in vertex form.
   b. Find the vertex of the parabola represented by each equation.

6. $y = x^2 + 4x + 11$
7. $y = -2x^2 - 6x - 14$

8. Refer to Example 3. Generate a table of values for the equation $h = -16t^2 + 32t + 12$, or fill in a table like the one below by hand.

t	0	0.25	0.5	0.75	1.0	1.25	1.5	1.75	2
h	?	?	?	?	?	?	?	?	?

   a. The points in the table are symmetric to a vertical line through a particular point in the table. Which point?
   b. How does your answer to Part a compare to the vertex found in Example 3?

APPLYING THE MATHEMATICS

9. Suppose a ball is thrown straight up from a height of 8 feet with an initial upward velocity of 64 $\frac{\text{ft}}{\text{sec}}$.

   a. Write an equation to describe the height h of the ball after t seconds.

   b. How high is the ball after 1 second?

   c. Determine the maximum height attained by the ball by completing the square.

   d. Sketch a graph of your equation from Part a.

   e. How long will it take for the ball to land on the ground?

10. **Fill in the Blank** What is the missing term in the expression $x^2 + \frac{b}{a}x + \underline{\ \ ?\ \ }$ if the expression is a perfect-square trinomial?

11. Find an equation in vertex form equivalent to $y = 5x^2 - 2x + 15$.

In **12** and **13**, consider the following. When a quadratic function is graphed, the second coordinate of the vertex of the parabola is always the minimum or maximum value of the function. Commands on some calculators may help you find those values. On one calculator these commands are `fMin` and `fMax`.

12. In the display at the right, `fMin` and `fMax` have been calculated. You are given the x-coordinate of the minimum point of the graph of $y = 3x^2 - 12x + 14$ and the x-coordinate of the maximum point of the graph of $y = -10x^2 + 60x$. Find the coordinates of the vertex of each parabola.

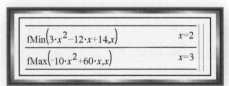

13. At the right, why is the x-coordinate of the minimum value stated as positive or negative infinity?

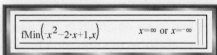

14. You run Twin Wheels bike-rental shop. You currently charge $10 per day and average 56 rentals a day. In researching a price increase, you believe that for every fifty-cent increase in rental price you can expect to lose two rentals a day. Let n = the number of fifty-cent increases.

   a. Write an expression for the new price after n increases.

   b. Write an expression for the expected number of rentals after n increases.

   c. The total income for the day is equal to the price times the number of rentals. Multiply the expressions in Parts a and b to get an expression for the total daily income.

   d. Find the rental price that will maximize the total daily income.

REVIEW

15. Jailah tosses a ball upward from an initial height of 1.6 meters. The ball lands on the ground 4.8 seconds later. What was the upward velocity of Jailah's throw? (**Lesson 6-4**)

16. Rewrite the equation $y = 4(3 - x)^2 - 8$ in standard form. (**Lesson 6-4**)

17. On the coordinate grid at the right are the parabola with equation $y = -x^2$ and its image under a translation. (**Lessons 6-3, 4-10**)

    a. What translation maps the parabola with equation $y = -x^2$ onto the other parabola?

    b. Write an equation for the image.

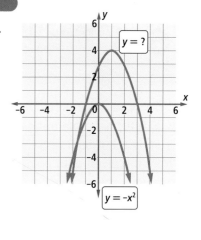

18. Solve the system $\begin{bmatrix} 1 & 0 & -1 \\ 2 & 1 & 3 \\ -1 & 4 & 0 \end{bmatrix} \begin{bmatrix} x \\ y \\ z \end{bmatrix} = \begin{bmatrix} 3 \\ 3 \\ -9 \end{bmatrix}$. (**Lesson 5-6**)

19. The table at the right shows the growth of the European Union (EU) since its inception in 1958. (**Lesson 3-5**)

    a. Draw a scatterplot of this data.

    b. Find an equation for the regression line, and draw the line on your scatterplot.

    c. How many member states does your regression line predict for the year 2058, the 100th anniversary of the founding of the EU?

Years since 1958	Number of Member States
0	6
15	9
23	10
28	12
37	15
46	25
49	27

☐ European Union member state
■ Non-member states

In 20 and 21, find x. (**Previous Course**)

20. $3^{2x} \cdot 3^4 = 3^{24}$

21. $(2^x)^5 = 2^{25}$

EXPLORATION

22. *MNOP* is a square with sides of length $a + b + c$. The areas of three regions inside *MNOP* are shown in the diagram.

    a. Find the areas of the other six regions.

    b. Use the drawing to help you expand $(a + b + c)^2$.

    c. Make a drawing to illustrate the expansion of $(a + b + c + d)^2$.

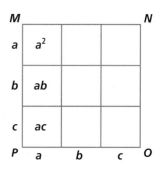

Lesson

6-6

Fitting a Quadratic Model to Data

Vocabulary

quadratic regression

▶ **BIG IDEA** *Quadratic regression* is like linear regression in that it finds the model with the least sum of squares of differences from the given data points to the values predicted by the model.

In Chapter 3, you learned how to find an equation for the line through two points, and how to find a linear model for data that lie approximately on a straight line. You can also fit a *quadratic model* to data that lie approximately on a parabola. **Quadratic regression** fits a model of the form $y = ax^2 + bx + c$ to data.

Mental Math

Find the slope of the image of $3x + 2y = 7$ under the transformation.

a. $T_{6, -1}$

b. R_{90}

c. R_{270}

d. $r_{y = x}$

Activity

MATERIALS old CD that can get scratched up, two $8\frac{1}{2}$-inch by 11-inch pieces of paper, quarter, CAS or graphing calculator

You are going to investigate how the radius x of a circular object affects the probability y of the object landing completely in a fixed region when dropped.

Step 1 A CD like the one pictured at the right has 3 circles on it. Circle *I* represents the hole in the CD. Circle *M* (for middle) represents the start of the silver writing surface. Circle *O* represents the outer edge of the CD. Draw a line across the middle of your $8\frac{1}{2}$-by-11 piece of paper, dividing it into two $5\frac{1}{2}$-by-$8\frac{1}{2}$ rectangular targets.

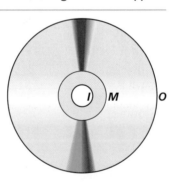

Step 2 On a separate sheet of paper, make four columns labeled Drop Number, *I*, *M*, and *O*. In the Drop Number column, write the integers 1 through 25. Place your divided sheet of $8\frac{1}{2}$-inch-by-11-inch paper on the floor and stand above it holding the CD waist high, parallel to the floor. Drop the CD.

If no part of the CD is touching the paper, drop it again. If any part of the CD is touching the paper, determine the score for each circle, I, M, and O as follows: If a circle on the CD lands completely inside one of the two rectangular targets, give it 1 point for that drop. Pictured below are the four possible situations and scores after a drop. The horizontal line represents either the edge of the paper or the line you drew on the paper.

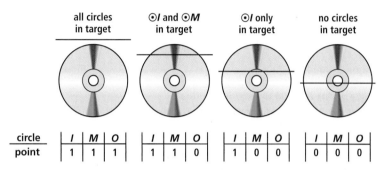

all circles in target			⊙I and ⊙M in target			⊙I only in target			no circles in target		
circle	I	M	O	I	M	O	I	M	O	I	M	O
point	1	1	1	1	1	0	1	0	0	0	0	0

Drop the CD 25 times and fill in your table.

Step 3 Calculate the relative frequency of a circle landing completely in a target. For example, if M has 16 points, then its relative frequency of points per drop is $\frac{16}{25}$. Record the frequencies in a table like the one at the right. The radii of circles on a standard CD have been filled in for you, but you should measure your CD to check.

Circle	Radius (in.)	Relative Frequency
I	$\frac{5}{16}$	?
M	$\frac{7}{8}$	?
O	2	?

Step 4 Create a scatterplot with three data points (radius, frequency) for I, M, and O on your calculator.

Choose the quadratic regression option from the appropriate menu. A sample is shown at the right. On this calculator, the graph of a quadratic model for the data is added to the scatterplot.

The calculator displays the equation for the quadratic model.

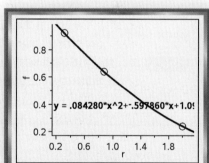

$y = .084280*x^2 + -.597860*x + 1.09$

Step 5 Examine the scatterplot with the graph of the regression equation on it. How well does your model fit your data?

Step 6 Measure the diameter of a quarter and use your regression equation to predict the relative frequency of the quarter landing inside a target rectangle. Drop the quarter 25 times and see if the relative frequency is close to your prediction. Combine your results with other classmates. Compare the combined data with the prediction. Which value is closer to the predicted value—your own data or the combined data?

Finding the Equation of a Given Parabola

You can apply the techniques of solving systems of equations you learned in Chapter 5 and the regression technique in the Activity to find an equation for any parabola on which you know three points.

Example

The parabola at the right contains the points $(-1, 7)$, $(1, -3)$ and $(5, 1)$. Find its equation.

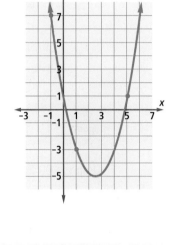

Solution 1 Use a system of equations.

Because the ordered pairs (x, y) are solutions of the equation $y = ax^2 + bx + c$, substitute to get 3 linear equations in a, b, and c.

When $x = -1$, $y = 7$: $7 = a(-1)^2 + b(-1) + c$
When $x = 1$, $y = -3$: $-3 = a(1)^2 + b(1) + c$
When $x = 5$, $y = 1$: $1 = a(5)^2 + b(5) + c$

So a, b, and c are solutions to the system $\begin{cases} 7 = a - b + c \\ -3 = a + b + c \\ 1 = 25a + 5b + c \end{cases}$.

You are asked to solve this system in Question 2.

Solution 2 Use quadratic regression.

Enter the x–coordinates and y–coordinates into lists in your calculator and apply quadratic regression. One calculator gives the solution at the right, so the parabola has equation $y = x^2 - 5x + 1$.

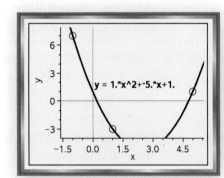

Check Substitute the points into the equation.

Does $(-1)^2 - 5(-1) + 1 = 7$?   Yes.
Does $(1)^2 - 5(1) + 1 = -3$?   Yes.
Does $(5)^2 - 5(5) + 1 = 1$?   Yes; it checks.

Questions

COVERING THE IDEAS

1. Refer to the Activity.
   a. Write an equation for the function your quadratic regression describes.
   b. Use the model from Part a to predict the probability of a mini CD with a radius of 4 centimeters landing completely inside one of the two rectangular targets.

2. Solve the system of the Example to find an equation for the parabola that contains the given points.

In 3 and 4, solve a system of equations to write an equation for the parabola that contains the given points.

3. (4, –6), (–2, 30), (0, 10)   4. (–1, –3), (5, 81), (2, 12)

In 5 and 6, use quadratic regression to find an equation for the parabola that contains the given points.

5. (3, 2), (–1, 5), (8, 7)   6. (–6, 5), (–1, 10), (3, 4)

APPLYING THE MATHEMATICS

7. A quarterback threw a ball from 5 yards behind the line of scrimmage and a height of 6 feet. The ball was 10 feet high as it crossed the line of scrimmage. It was caught 20 yards past the line of scrimmage at a height of 5 feet off the ground.

   a. Find an equation that gives the height y of the ball (in feet) when it was x yards beyond the line of scrimmage.

   b. Graph the equation.

8. In *The Greedy Triangle* by Marilyn Burns, an equilateral triangle keeps changing shape into a regular polygon with more and more sides. As the number of sides of a polygon increase, so do the number of diagonals of the polygon. Recall the formula for the number of diagonals in a polygon. If you have forgotten, you can derive it using quadratic regression.

Peyton Manning throwing for the Indianapolis Colts

   a. Count the number of diagonals in each polygon below and complete the table to the right.

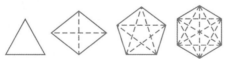

Shape	Number of Sides	Number of Diagonals
Triangle	?	?
Quadrilateral	?	?
Pentagon	?	?
Hexagon	?	?

   b. Use quadratic regression to find a model for the number of diagonals in a polygon.

   c. Is this model exact or approximate?

   d. Use the model to determine the number of diagonals in a 50-sided polygon.

9. Shown below is a table of data on the eight major planets in our solar system. For each planet, the period of the orbit is given in Earth years and the average distance from the Sun is in astronomical units (AU). One AU is approximately the mean distance between Earth and the Sun. Each planet's distance from the Sun is not constant due to its elliptical orbit, so the average distance is given.

Planet	d = Avg. Distance From the Sun (AU)	P = Period of Orbit (in Earth years)
Mercury	0.39	0.24
Venus	0.72	0.62
Earth	1.00	1.00
Mars	1.52	1.88
Jupiter	5.20	11.86
Saturn	9.58	29.46
Uranus	19.20	84.01
Neptune	30.05	164.79

a. Create a scatterplot of the data. Use average distance d as the independent variable and period P as the dependent variable.

b. Fit a quadratic model to these data. Graph the model on the same axes as your scatterplot. Does the model seem like it fits the data? Why or why not?

c. Pluto, a dwarf planet, is an average 39.48 AU from the Sun. It takes Pluto 248.54 Earth years to orbit the Sun. Does Pluto fit your model?

d. Another dwarf planet, Sedna, or 2003 VB12, as it was originally designated, orbits the sun at an average distance of 90 AU. Use your model from Part b to predict how long it takes Sedna to orbit the Sun.

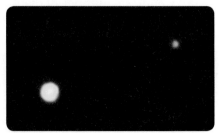

Pluto and one of its moons, Charon, are shown above. Sunlight takes about $5\frac{1}{2}$ hours to reach Pluto. Sunlight reaches Earth in about 8 minutes.

REVIEW

In 10–12, rewrite the equation in vertex form. (Lesson 6-5)

10. $y = x^2 - 14x + 53$

11. $y = 3x^2 - 9x + 9$

12. $y = x^2 - 4gx + 3g^2$

13. Find an equation in standard form for the image of the parabola with equation $y = -2x^2$ under the translation $T_{-4, -0.5}$. (Lessons 6-3, 6-1)

14. a. Find an equation for the line through points (a, b) and (b, a).

   b. Use a rotation matrix to find an equation for the image of the line from Part a under a rotation of 90° about the origin. **(Lessons 4–9, 3-4)**

15. The Moon is much less massive than Earth, and exerts less gravity. Near the surface of the Moon, the acceleration due to gravity is about $5.31 \frac{ft}{sec^2}$. Suppose an astronaut drops an object on the Moon from a height of 5 feet.

   a. How long will it take the object to fall?

   b. How long would it have taken the object to fall if it was dropped from the same height on Earth? **(Lesson 6–4)**

EXPLORATION

16. Follow these steps to determine the theoretical probabilities that the circles of the CD in the Activity will be in one of the rectangular target areas.

   a. Find the area of the region in which the center point of the CD can land so that the CD is touching the paper.

   b. Find the area of the region in which the center point of the CD can land so that circle I lands completely in one of the target areas.

   c. Let a be your answer to Part a, and b your answer to Part b. Calculate $\frac{b}{a}$. This number is the probability that the smallest circle is in the target area.

   d. Repeat Parts b and c for the other two circles.

   e. Do the probabilities seem to agree with the relative frequencies your class found?

Lesson
6-7

The Quadratic Formula

Vocabulary

Quadratic Formula

▶ **BIG IDEA** The *Quadratic Formula* gives the solutions to any quadratic equation in standard form whose coefficients are known.

To help train outfielders, the coach of a baseball team (who is also a math teacher) uses a hitting machine. The machine hits a 4–foot–high pitch, and the ball travels toward the outfield along a nearly parabolic path. Let x be the distance along the ground (in feet) of the ball from home plate, and h be the height (in feet) of the ball at that distance. Using estimated heights of the ball at a various points along its path, he found the following regression equation to model the flight of the ball.

$$h = -0.00132x^2 + 0.545x + 4$$

If an outfielder leaps and catches a ball 10 feet off the ground, how far is he from home plate? To answer this, you can solve the flight equation when $h = 10$.

$$10 = -0.00132x^2 + 0.545x + 4$$

Subtract 10 from each side to rewrite the equation in standard form.

$$0 = -0.00132x^2 + 0.545x - 6$$

You could solve this equation by rewriting it in vertex form, but in previous courses you have probably used a formula that gives the solutions. The **Quadratic Formula** is a theorem that can be proved by completing the square, starting with a general quadratic equation in standard form.

Mental Math

Expand the binomial.

a. $(a - b)^2$

b. $(a - 3b)^2$

c. $(5a - 3b)^2$

d. $(5a^4 - 3b^6)^2$

10 ft   4 ft

Quadratic Formula Theorem

If $ax^2 + bx + c = 0$ and $a \neq 0$, then $x = \dfrac{-b \pm \sqrt{b^2 - 4ac}}{2a}$.

The proof is given on the next page.

Given the equation $ax^2 + bx + c = 0$, where $a \neq 0$.

Proof

1. $x^2 + \frac{b}{a}x + \frac{c}{a} = \frac{0}{a}$ Divide both sides by a so the coefficient of x^2 is 1.

2. $x^2 + \frac{b}{a}x = -\frac{c}{a}$ Add $-\frac{c}{a}$ to each side.

3. $x^2 + \frac{b}{a}x + \frac{b^2}{4a^2} = \frac{b^2}{4a^2} - \frac{c}{a}$ Complete the square by adding $\left(\frac{1}{2} \cdot \frac{b}{a}\right)^2$ to both sides.

4. $\left(x + \frac{b}{2a}\right)^2 = \frac{b^2}{4a^2} - \frac{c}{a}$ Write the left side as a binomial squared.

5. $\left(x + \frac{b}{2a}\right)^2 = \frac{b^2 - 4ac}{4a^2}$ Add the fractions on the right side.

6. $\sqrt{\left(x + \frac{b}{2a}\right)^2} = \sqrt{\frac{b^2 - 4ac}{4a^2}}$ Take the square roots of both sides.

7. $\left|x + \frac{b}{2a}\right| = \sqrt{\frac{b^2 - 4ac}{4a^2}}$ Use the Absolute Value–Square Root Theorem.

8. $x + \frac{b}{2a} = \pm\frac{\sqrt{b^2 - 4ac}}{2a}$ Use the definition of absolute value.

9. $x = \frac{-b \pm \sqrt{b^2 - 4ac}}{2a}$ Add $-\frac{b}{2a}$ to both sides.

If you solve $ax^2 + bx + c = 0$ on a CAS, it is likely to display the solutions in a compound sentence.

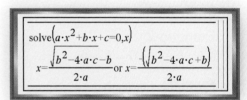

Using the Quadratic Formula

Example 1

To the nearest foot, how far from home plate was the outfielder when he leaped to catch the 10-ft high ball?

Solution Use the Quadratic Formula on the equation $0 = -0.00132x^2 + 0.545x - 6$.

Here $a = -0.00132$, $b = 0.545$, and $c = -6$.

$$x = \frac{-0.545 \pm \sqrt{(0.545)^2 - 4(-0.00132)(-6)}}{2(-0.00132)}$$

(continued on next page)

A calculator can approximate the two solutions in one step. Here are the intermediate steps

$$\approx \frac{-0.545 \pm \sqrt{0.265}}{-0.00264}$$

Estimate the square root and separate the solutions.

$$x \approx \frac{-0.545 + 0.515}{-0.00264} \quad \text{or} \quad x \approx \frac{-0.545 - 0.515}{-0.00264}$$

$$x \approx 11 \text{ ft} \quad \text{or} \quad x \approx 402 \text{ ft}$$

The ball reaches a height of 10 feet in two places. The first is when the ball is on the way up and about 11 feet away from home plate. The second is when the ball is on the way down and about 402 feet from home plate. Between these distances the ball is over 10 feet high. An outfielder is unlikely to be 11 feet from home plate, so he was about 402 feet away.

Check Use the solve command on a CAS.

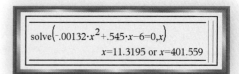

$$\text{solve}(-.00132 \cdot x^2 + .545 \cdot x - 6 = 0, x)$$
$$x = 11.3195 \text{ or } x = 401.559$$

In Example 1, the number $b^2 - 4ac$ is not a perfect square. When this is the case, there are no rational number solutions to the equation. When $b^2 - 4ac$ is a perfect square, as it is in Example 2, the solutions are always rational.

GUIDED

Example 2

Solve $5x^2 + 13x - 6 = 0$.

Solution Use the Quadratic Formula.

$a = \underline{\quad ? \quad} \qquad b = \underline{\quad ? \quad} \qquad c = \underline{\quad ? \quad}$

$$x = \frac{-b \pm \sqrt{b^2 - 4ac}}{2a}$$

$$x = \frac{-(?) \pm \sqrt{(?)^2 - 4(?)(?)}}{2(?)}$$

$$x = \frac{? \pm \sqrt{?}}{?}$$

So, $x = \dfrac{? + ?}{?} \qquad \text{or} \qquad x = \dfrac{? - ?}{?}$

$x = \underline{\quad ? \quad} \qquad \text{or} \qquad x = \underline{\quad ? \quad}$

A quadratic equation must be in standard form before the Quadratic Formula can be applied.

Example 3

Recall that the explicit formula for the sequence t_n of triangular numbers is $t_n = \frac{n(n+1)}{2}$. Is 101,475 a triangular number? If it is, which term of the sequence is it?

Solution Set $t_n = 101,475$ and solve for n.

$$t_n = \frac{n(n+1)}{2} = 101,475$$

Put the equation in standard form.

$$n(n+1) = 202,950 \qquad \text{Multiply both sides by 2.}$$
$$n^2 + n = 202,950 \qquad \text{Expand.}$$
$$n^2 + n - 202,950 = 0 \qquad \text{Add } -202,950 \text{ to both sides.}$$

Use the Quadratic Formula with $a = 1$, $b = 1$, and $c = -202,950$.

$$n = \frac{-1 \pm \sqrt{1^2 - 4(1)(-202,950)}}{2(1)}$$

$$n = \frac{-1 \pm \sqrt{811,801}}{2}$$

$$n = \frac{-1 \pm 901}{2}$$

$$n = -451 \text{ or } n = 450$$

Because 450 is a positive integer, 101,475 is the 450th triangular number.

 QY

▶ **QY**

Use the explicit formula for the triangular numbers to check the solution to Example 3.

Questions

COVERING THE IDEAS

1. If $ax^2 + bx + c = 0$, and $a \neq 0$, write the two values of x in terms of a, b, and c as a compound sentence.

In 2–4 refer to the proof of the Quadratic Formula.

2. Why must a be nonzero in the Quadratic Formula?

3. Why is it necessary to divide both sides by a in the first step?

4. Write $x^2 + \frac{b}{a}x + \frac{b^2}{4a^2}$ as the square of a binomial.

In 5 and 6, a quadratic equation is given. Solve each using the Quadratic Formula.

5. $10z^2 + 13z + 3 = 0$

6. $2n^2 - 11n + 12 = 0$

In 7–10, consider the equation $h = -0.00132x^2 + 0.545x + 4$ from the beginning of the lesson. The left field wall at Fenway Park in Boston is 37 ft tall and 310 ft from home plate.

7. What do x and h represent?

8. Would a ball on the path of this example be a home run? (That is, would it have gone over the left field wall?)

9. Trace a graph of the equation to find the maximum height reached by the ball.

10. How far from home plate would the ball hit the ground, if the outfielder missed it?

11. Refer to Example 2. How do your solutions change when you solve $5x^2 - 13x - 6 = 0$ instead?

Fenway Park, home to the Boston Red Sox

In 12 and 13, refer to Example 3.

12. Use a CAS to check the solution for Example 3.

13. Show that 10,608 is not a triangular number.

APPLYING THE MATHEMATICS

14. What is another way to solve $ax^2 + bx + c = 0$, $a \neq 0$, besides using the Quadratic Formula or a CAS?

15. Consider this sequence of quadratic equations.
    $Q_1: x^2 + 5x + 6 = 0$
    $Q_2: x^2 + 7x + 12 = 0$
    $Q_3: x^2 + 9x + 20 = 0$
    $Q_4: x^2 + 11x + 30 = 0$
    $Q_5: x^2 + 13x + 42 = 0$
    a. Solve each quadratic equation.
    b. Find the product and the sum of each set of answers in Part a.
    c. What is the connection between the solutions to the quadratic equations and the coefficients of the quadratic equations?

16. Consider the parabola with equation $y = m^2 - 2m - 7$.
    a. Find the values of m for which $y = 0$. What are these values called?
    b. Find the vertex of this parabola.
    c. Give an equation for its axis of symmetry.
    d. Sketch a graph of the parabola.

17. The graphs of $y = 6x^2 + x + 2$ and $y = 4$ are shown at the right. Use the Quadratic Formula to find the points of intersection.

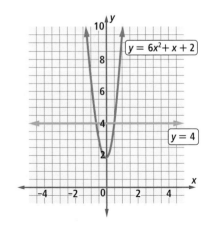

18. **Fill in the Blanks** Given $ax^2 + bx + c = 0$, where $a \neq 0$, provide the missing algebra in this alternate proof of the Quadratic Formula that does not use completing the square.

   a. Multiply both sides by $4a$. _____?_____

   Recognize that the terms $4a^2x^2 + 4abx$ are initial terms of the binomial square $(2ax + b)^2$.

   b. Replace $4a^2x^2 + 4abx$ with $(2ax + b)^2 - b^2$. (Explain why $-b^2$ is added in this step.) _____?_____

   c. Add $b^2 - 4ac$ to both sides. $(2ax + b)^2 = b^2 - 4ac$

   d. Take the square roots of both sides. _____?_____

   e. Use the Absolute Value–Square Root Root Theorem. $|2ax + b| = \sqrt{b^2 - 4ac}$

   f. Use the definition of absolute value. _____?_____

   g. Add $-b$ to both sides. $2ax = -b \pm \sqrt{b^2 - 4ac}$

   h. Divide both sides by $2a$. _____?_____

REVIEW

19. Solve a system of equations to write an equation for the parabola that contains the points $(-1, -7)$, $(0, 3)$, and $(3, 9)$. **(Lesson 6-6)**

20. A ball is thrown straight up from a height of 1.5 meters with initial upward velocity of $6\frac{\text{m}}{\text{s}}$. Find the maximum height of the ball and the time the ball takes to reach this height.
   (Lessons 6-5, 6-4)

21. Let $M = \begin{bmatrix} 4 & 12 \\ -8 & x \end{bmatrix}$. For what value or values of x does M^{-1} not exist? **(Lesson 5-5)**

22. State the domain and range of the function f when $f(x) = -\frac{\sqrt{2}}{x}$. **(Lesson 2-6)**

23. a. Let $g_n = 1 + (-1)^n$. Calculate $g_1, g_2, g_3,$ and g_4.
   b. Describe how to quickly know g_n for any positive integer n.
   c. What is g_{7532}? **(Lesson 1-8)**

EXPLORATION

24. Find out what you can about the formula used to solve the general cubic equation $ax^3 + bx^2 + cx + d = 0$.

Pure Imaginary Numbers

Vocabulary

imaginary number

$\sqrt{k}$

$\sqrt{-1}, i$

pure imaginary number

▶ **BIG IDEA** The square roots of negative numbers are *pure imaginary numbers* and are all multiples of $\sqrt{-1}$, defined as the number i.

Square Roots of Negative Numbers

Consider the quadratic equation $x^2 = 900$. You can solve it for x as follows.

$$x^2 = 900$$
$$\sqrt{x^2} = \sqrt{900} \qquad \text{Take square roots.}$$
$$|x| = 30 \qquad \text{Use the Absolute Value-Square Root Theorem.}$$
$$x = \pm 30 \qquad \text{Solve the absolute value equation.}$$

Mental Math

Solve for x.

a. $|x| = 4$

b. $|x - 3| = 4$

c. $|2x - 3| = 4$

Now consider the quadratic equation $x^2 = -900$. You know that the solution cannot be a real number because the square of a real number is never negative. However, if you followed the solution above you might write:

$$x^2 = -900$$
$$\sqrt{x^2} = \sqrt{-900}$$
$$|x| = ?$$

So far, $\sqrt{x}$ has only been defined for $x \geq 0$. If you try to evaluate $\sqrt{-900}$ on a calculator you may see an error message, or you may see $30 \cdot i$, as shown below. The difference is whether or not your calculator is in complex number mode.

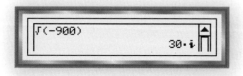

A Brief History of *i*

Why does a calculator display $30 \cdot i$ when it is in complex number mode? The symbol *i* was first used in the 18th century, but work with square roots of negative numbers started sooner. Until the 1500s, mathematicians were puzzled by square roots of negative numbers. They knew that if they solved certain quadratics, they would get negative numbers under the radical sign. However, they did not know what to do with them!

One of the first to work with these numbers was the Italian mathematician Girolamo Cardano. In a book called *Ars Magna* ("Great Art") published in 1545, Cardano reasoned as follows: When *k* is positive, the equation $x^2 = k$ has two solutions, $\sqrt{k}$ and $-\sqrt{k}$. If we solve the equation $x^2 = -k$ in the same way, then the two solutions are $\sqrt{-k}$ and $-\sqrt{-k}$. In this way, he defined symbols for the square roots of negative numbers. Cardano called these square roots of negatives "fictitious numbers."

Working in the 1600s, the French mathematician and philosopher René Descartes called them **imaginary numbers** in contrast to the numbers everyone understood, which he called "real numbers."

Girolamo Cardano

In his book *De Formulis Differentialibus Angularibus,* written in 1777, the Swiss mathematician Leonhard Euler wrote, "in the following I shall denote the expression $\sqrt{-1}$ by the letter *i* so that $i \cdot i = -1$."

Today, people around the world build on the work of Cardano, Descartes, and Euler and use the following definitions.

> **Definition of $\sqrt{k}$ when *k* Is Negative**
>
> When $k < 0$, the two solutions to $x^2 = k$ are denoted $\sqrt{k}$ and $-\sqrt{k}$.

By the definition, when $k < 0$, $\left(\sqrt{k}\right)^2 = k$. This means that we can say, for *all* real numbers *r*,

$$\sqrt{r} \cdot \sqrt{r} = r.$$

Suppose $k = -1$. Then we define the number *i* to be one of the two square roots of -1. That is, *i* is defined as follows.

> **Definition of *i***
>
> $i = \sqrt{-1}$

Thus, *i* is a solution to $x^2 = -1$. The other solution is $-\sqrt{-1}$, which we call $-i$. That is, $i^2 = -1$ and $(-i)^2 = -1$.

Multiples of i, such as $7i$, are called **pure imaginary numbers**. By the definition of i, $7i = 7\sqrt{-1}$. If we assume that multiplication of imaginary numbers is commutative and associative, then

$$(7i)^2 = 7i \cdot 7i$$
$$= 7^2 \cdot i^2$$
$$= 49 \cdot (-1)$$
$$= -49.$$

So $7i$ is a square root of -49. We write $7i = \sqrt{-49}$ and $-7i = \sqrt{-49}$. The following theorem generalizes this result.

> **Square Root of a Negative Number Theorem**
>
> If $k < 0$, $\sqrt{k} = i\sqrt{-k}$.

Thus all square roots of negative numbers are multiples of i.

 QY1

▶ **QY1**

Write $\sqrt{-36}$ as a multiple of i.

Example 1

Solve $x^2 = -900$.

Solution Apply the definition of $\sqrt{k}$ when k is negative.

$$x = \sqrt{-900} \quad \text{or} \quad x = -\sqrt{-900}$$

Now use the Square Root of a Negative Number Theorem.

$$x = i\sqrt{900} \quad \text{or} \quad x = -i\sqrt{900}$$

Simplify.

$$x = 30i \quad \text{or} \quad x = -30i$$

Check Use a CAS in complex mode to solve the equation. On some machines, you use the csolve command to display complex solutions to equations. This CAS uses the solve command and gives the solution at the right. It checks.

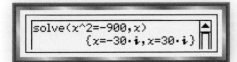

```
solve(x^2=-900,x)
        {x=-30·i,x=30·i}
```

Example 2

a. Show that $i\sqrt{5}$ is a square root of -5.

b. What is the other square root of -5?

Solution

a. Multiply $i\sqrt{5}$ by itself. Assume the Commutative and Associative Properties.

$$i\sqrt{5} \cdot i\sqrt{5} = i \cdot i \cdot \sqrt{5} \cdot \sqrt{5}$$
$$= i^2 \cdot 5$$
$$= -1 \cdot 5$$
$$= -5$$

b. Take the opposite of the square root from Part a. The other square root of –5 is $-i\sqrt{5}$.

 QY2

▶ **QY2**

Use a CAS to show that $-i\sqrt{5}$ is a square root of –5.

Due to the long history of quadratic equations, solutions to them are described in different ways. The following all refer to the same numbers.

the solutions to $x^2 = -5$

the square roots of –5

$\sqrt{-5}$ and $-\sqrt{-5}$

$i\sqrt{5}$ and $-i\sqrt{5}$

The last two forms could also be written $\sqrt{5}i$ and $-\sqrt{5}i$ as a CAS displays them. On handwritten materials and in textbooks you commonly see $i\sqrt{5}$ to clearly show that the i is not underneath the radical sign.

Operations with Pure Imaginary Numbers

The Commutative, Associative, and Distributive Properties of Addition and Multiplication are true for all imaginary numbers, as are all theorems based on these postulates. Consequently, you can use them when working with multiples of i, just as you would when working with multiples of any real numbers.

Example 3

Simplify the following.

a. $(5i)(3i)$ b. $\sqrt{-16} - \sqrt{-64}$ c. $\sqrt{-2} + \sqrt{-2}$ d. $\dfrac{\sqrt{-100}}{\sqrt{-81}}$

Solution

a. $(5i)(3i) = 15i^2 = 15 \cdot -1 = -15$

b. $\sqrt{-16} - \sqrt{-64} = 4i - 8i = -4i$

c. $\sqrt{-2} + \sqrt{-2} = i\sqrt{2} + i\sqrt{2} = 2i\sqrt{2}$

d. $\dfrac{\sqrt{-100}}{\sqrt{-81}} = \dfrac{10i}{9i} = \dfrac{10}{9}$

In operating with imaginary numbers in radical form, be sure to follow the order of operations, treating the square root as a grouping symbol.

Activity

MATERIALS calculator

Work with a partner to calculate with imaginary numbers.

Step 1 Make sure that your calculator is in complex mode. Use your calculator to write each answer in $a + bi$ form.

$(4i)(25i)$	$4i - 5i$	$\sqrt{-4} - \sqrt{-25}$	$\dfrac{\sqrt{-4}}{\sqrt{25}}$	$\sqrt{-4} \cdot \sqrt{25}$	$\sqrt{-4} \cdot \sqrt{-25}$
?	?	?	?	?	?

Step 2 Look for patterns in your results from Step 1. When are the results real numbers and when are they imaginary?

Step 3 Calculate $\sqrt{4} \cdot \sqrt{25}$ and $\sqrt{4 \cdot 25}$. Do your results support the property $\sqrt{a}\,\sqrt{b} = \sqrt{ab}$ for nonnegative real numbers? Why or why not?

Step 4 Calculate $\sqrt{-4 \cdot 25}$ and $\sqrt{-4 \cdot -25}$. Now look at your Step 1 results for $\sqrt{-4} \cdot \sqrt{25}$ and $\sqrt{-4} \cdot \sqrt{-25}$. Does $\sqrt{a}\,\sqrt{b} = \sqrt{ab}$ if either a or b, but not both, is a negative real number? Does $\sqrt{a}\,\sqrt{b} = \sqrt{ab}$ if both a and b are both negative real numbers? Why or why not?

In the Activity, $\sqrt{-4 \cdot -25} = \sqrt{100} = 10$. You can verify that $\sqrt{-4} \cdot \sqrt{-25} = 10$ by hand as follows.

$$\sqrt{-4} \cdot \sqrt{-25} = i\sqrt{4} \cdot i\sqrt{25}$$
$$= 2 \cdot 5 \cdot i \cdot i$$
$$= 10i^2$$
$$= -10$$

So $\sqrt{-4 \cdot -25} \neq \sqrt{-4} \cdot \sqrt{-25}$ is a counterexample showing that the property $\sqrt{a}\,\sqrt{b} = \sqrt{ab}$ *does not hold* for *nonnegative* real numbers a and b.

Questions

COVERING THE IDEAS

1. **True or False** All real numbers have square roots.

2. Write the solutions to $x^2 = -4$.

3. **Multiple Choice** About when did mathematicians begin to use roots of negative numbers as solutions to equations?

A sixth century   B twelfth century

C sixteenth century   D twenty-first century

4. Who first used the term *imaginary number*?

5. Who was the first person to suggest using i for $\sqrt{-1}$?

In 6 and 7, True or False.

6. For all real numbers x, $\sqrt{-x} < 0$.

7. If $b > 0$, $\sqrt{-b} = i\sqrt{b}$.

In 8 and 9, solve for x.
   a. **Write the solutions to each equation with a radical sign.**
   b. **Write the solutions without a radical sign.**

8. $x^2 + 25 = 0$ 9. $x^2 + 16 = 0$

10. **True or False** $i\sqrt{7}$ is a square root of –7. Justify your answer.

11. Show that $-6i$ is a square root of –36.

12. Show that $\sqrt{-3}\,\sqrt{-27} \neq \sqrt{81}$.

In 13–15, simplify.

13. $\sqrt{-53}$ 14. $\sqrt{-121}$ 15. $-\sqrt{72}$

In 16–21, perform the indicated operations. Give answers as real numbers or multiples of i.

16. $-2i + 7i$ 17. $10i - 3i$ 18. $(4i)(17i)$

19. $3\sqrt{-16} + \sqrt{-64}$ 20. $\dfrac{\sqrt{-81}}{\sqrt{-9}}$ 21. $-\sqrt{-49} + \sqrt{-49}$

In 22–24, simplify the product.

22. $\sqrt{-5} \cdot \sqrt{-5}$ 23. $\sqrt{-10} \cdot \sqrt{-30}$ 24. $\sqrt{3} \cdot \sqrt{-3}$

25. a. Does $\sqrt{-3 \cdot 27} = \sqrt{-3} \cdot \sqrt{27}$?

   b. Does your answer in Part a provide a counterexample to $\sqrt{a}\,\sqrt{b} = \sqrt{ab}$ for $a < 0$ and $b > 0$?

26. For what real number values of x and y does $\sqrt{xy} \neq \sqrt{x}\,\sqrt{y}$?

APPLYING THE MATHEMATICS

In 27 and 28, simplify.

27. $\sqrt{-477{,}481}$ 28. $\dfrac{5i + 9i}{2i}$

In 29 and 30, True or False. If false, give a counterexample.

29. The sum of any two imaginary numbers is imaginary.

30. The product of any two imaginary numbers is imaginary.

31. Verify your solutions to $x^2 = -4$ in Question 2 by using the Quadratic Formula to solve $x^2 + 0x + 4 = 0$.

32. Solve $9y^2 + 49 = 0$ using the Quadratic Formula.

In 33 and 34, solve the equation.

33. $a^2 + 12 = 5$ 34. $(b - 5)^2 + 13 = 9$

REVIEW

In 35 and 36, solve for x using the Quadratic Formula. (Lesson 6–7)

35. $8x^2 - 2x - 15 = 0$ 36. $x^2 + dx - 2d^2 = 0$

37. Use quadratic regression to find an equation of the parabola that contains the points (2, 523.3), (4, 1126.3), and (8, 2338.3). (Lesson 6–6)

38. A chef reports that with 5 kilograms of flour, 12 loaves of bread and 6 pizza crusts can be made. With 2 kilograms of flour, 1 loaf of bread and 10 pizza crusts can be made. How much flour is needed to make one loaf of bread? How much is needed for one pizza crust? (Lesson 5–4)

39. Consider the function graphed below. Give its domain and range. (Lesson 1–4)

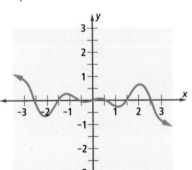

EXPLORATION

40. By definition, $i^2 = -1$. So $i^3 = i^2 \cdot i = -1 \cdot i = -i$ and $i^4 = i^3 \cdot i = -i \cdot i = -i^2 = -(-1) = 1$. Continue this pattern to evaluate and simplify each of i^5, i^6, i^7, and i^8. Generalize your result to predict the value of i^{2009}, i^{2010}, and i^{2020}. Explain how to simplify any positive power of i.

QY ANSWERS

1. $6i$

2.

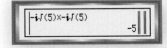

Lesson 6-9

Complex Numbers

▶ **BIG IDEA**   *Complex numbers are numbers of the form $a + bi$, where $i = \sqrt{-1}$, and are operated with as if they are polynomials in i.*

Many aspects of an electrical charge, such as voltage (electric potential) and current (movement of an electric charge), affect the performance and safety of the charge. When working with these two quantities, electricians find it easier to combine them into one number Z, called *impedance*. Impedance in an alternating-current (AC) circuit is the amount, usually measured in ohms, by which the circuit resists the flow of electricity. The two-part number Z is the sum of a real number and an imaginary number, and is called a *complex number*.

What Are Complex Numbers?

Recall from the previous lesson that the set of numbers of the form bi, where b is a real number, are called *pure imaginary numbers*. When a real number and a pure imaginary number are added, the sum is called a *complex number*.

Definition of Complex Number

A **complex number** is a number of the form $a + bi$, where a and b are real numbers and $i = \sqrt{-1}$.

In the complex number $a + bi$, a is the **real part** and b is the **imaginary part**. For example, $-8.5 - 4i$ is a complex number in which the real part is -8.5 and the imaginary part is -4 (not $4i$ or $-4i$).

We say that $a + bi$ and $c + di$ are **equal complex numbers** if and only if their real parts are equal and their imaginary parts are equal. That is, $a + bi = c + di$ if and only if $a = c$ and $b = d$. For example, if $x + yi = 2i - 3$, then $x = -3$ and $y = 2$.

 QY1

▶ **QY1**

If $a + bi = -4 + \sqrt{-3}$, what is a and what is b?

Operations with Complex Numbers

All of the assumed properties of addition, subtraction, multiplication, and division of real numbers hold for complex numbers.

Properties of Complex Numbers Postulate

In the set of complex numbers:

1. Addition and multiplication are commutative and associative.

2. Multiplication distributes over addition and subtraction.

3. $0 = 0i = 0 + 0i$ is the additive identity; $1 = 1 + 0i$ is the multiplicative identity.

4. Every complex number $a + bi$ has an additive inverse $-a + -bi$ and a multiplicative inverse $\frac{1}{a + bi}$ provided $a + bi \neq 0$.

5. The addition and multiplication properties of equality hold.

You can use the properties to operate with complex numbers in a manner consistent with the way you operate with real numbers. You can also operate with complex numbers on a CAS.

Activity

Step 1 Add the complex numbers.

$$(2 + 3i) + (6 + 9i) = \underline{\quad ? \quad}$$

$$(4 - 3i) + (7 + 5i) = \underline{\quad ? \quad}$$

$$(-16 + 5i) + (4 - 8i) = \underline{\quad ? \quad}$$

$(2{+}3i){+}(6{+}9i)$

Step 2 Subtract the complex numbers.

$$(4 - 3i) - (6 + 5i) = \underline{\quad ? \quad}$$

$$(-2 + i) - (7 + 9i) = \underline{\quad ? \quad}$$

$$(8 - 4i) - (1 - i) = \underline{\quad ? \quad}$$

Step 3 Describe, in words and using algebra, how to add and subtract two complex numbers.

Step 4 Check your answer to Step 3 by calculating $(a + bi) + (c + di)$ and $(a + bi) - (c + di)$ on a CAS.

In the Activity, you should have seen that the sum or difference of two complex numbers is a complex number whose real part is the sum or difference of the real parts and whose imaginary part is the sum or difference of the imaginary parts.

The Distributive Property can also be used to multiply a complex number by a real number or by a pure imaginary number.

Example 1

Put $4i(8 + 5i)$ in $a + bi$ form.

Solution

$$4i(8 + 5i) = 4i(8) + 4i(5i) \quad \text{Distributive Property}$$
$$= 32i + 20(i^2) \quad \text{Associative and Commutative Properties of Multiplication}$$
$$= 32i + 20(-1) \quad \text{Definition of } i$$
$$= -20 + 32i \quad \text{Commutative Property of Addition}$$

Check Multiply on a CAS. It checks.

In Example 1, notice that i^2 was simplified using the fact that $i^2 = -1$. Generally, you should write answers to complex number operations in $a + bi$ form. Most calculators use this form as well.

To multiply complex numbers, think of them as linear expressions in i and multiply using the Distributive Property. Then use $i^2 = -1$ to simplify your answer.

Example 2

Multiply and simplify $(6 - 2i)(4 + 3i)$.

Solution

$$(6 - 2i)(4 + 3i) = 24 + 18i - 8i - 6i^2 \quad \text{Distributive Property (Expand.)}$$
$$= 24 + 10i - 6i^2 \quad \text{Distributive Property (Combine like terms.)}$$
$$= 24 + 10i - 6(-1) \quad \text{Definition of } i$$
$$= 30 + 10i \quad \text{Arithmetic}$$

 QY2

> ▶ **QY2**
>
> Multiply $(4 + 3i)(4 - 3i)$.

Conjugate Complex Numbers

The complex numbers $4 + 3i$ and $4 - 3i$ in QY 2 are *complex conjugates* of each other. In general, the **complex conjugate** of $a + bi$ is $a - bi$. Notice that the product $(4 + 3i)(4 - 3i)$ is a real number. In Question 23, you are asked to prove that the product of any two complex conjugates is a real number.

Complex conjugates are useful when dividing complex numbers. To divide two complex numbers, multiply both numerator and denominator by the conjugate of the denominator. This gives a real number in the denominator that you can then divide into each part of the numerator.

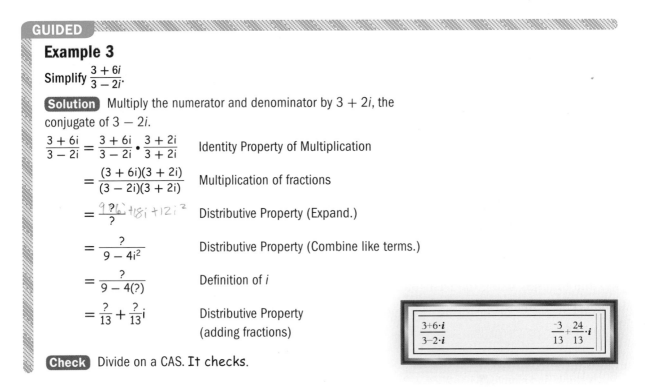

GUIDED

Example 3

Simplify $\dfrac{3 + 6i}{3 - 2i}$.

Solution Multiply the numerator and denominator by $3 + 2i$, the conjugate of $3 - 2i$.

$\dfrac{3 + 6i}{3 - 2i} = \dfrac{3 + 6i}{3 - 2i} \cdot \dfrac{3 + 2i}{3 + 2i}$ Identity Property of Multiplication

$= \dfrac{(3 + 6i)(3 + 2i)}{(3 - 2i)(3 + 2i)}$ Multiplication of fractions

$= \dfrac{9 \; ?\; 6 + 18i + 12i^2}{?}$ Distributive Property (Expand.)

$= \dfrac{?}{9 - 4i^2}$ Distributive Property (Combine like terms.)

$= \dfrac{?}{9 - 4(?)}$ Definition of i

$= \dfrac{?}{13} + \dfrac{?}{13}i$ Distributive Property (adding fractions)

Check Divide on a CAS. **It checks.**

$\dfrac{3+6\cdot i}{3-2\cdot i}$ $\dfrac{-3}{13} + \dfrac{24}{13} \cdot i$

The Various Kinds of Complex Numbers

Because $a + 0i = a$, every real number a is a complex number. Thus, the set of real numbers is a subset of the set of complex numbers. Likewise, every pure imaginary number bi equals $0 + bi$, so the set of pure imaginary numbers is also a subset of the set of complex numbers.

The diagram at the right is a *hierarchy of number sets*. It shows how the set of complex numbers includes some other number sets.

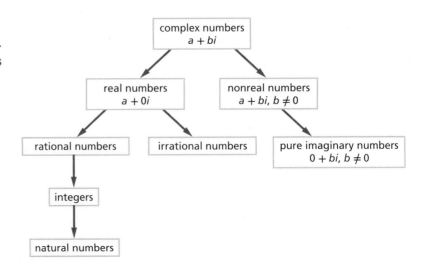

Applications of Complex Numbers

The first use of the term *complex number* is generally credited to Carl Friedrich Gauss. Gauss applied complex numbers to the study of electricity. Later in the 19th century, applications using complex numbers were found in geometry and acoustics. In the 1970s, complex numbers were used in a new field called *dynamical systems*.

Recall that electrical impedance Z is defined as a complex number involving voltage V and current I. A complex number representing impedance is of the form $Z = V + Ii$.

The total impedance Z_T of a circuit made from two connected circuits is a function of the impedances Z_1 and Z_2 of the individual circuits. Two electrical circuits may be connected *in series* or *in parallel*.

In a series circuit, $Z_T = Z_1 + Z_2$. In a parallel circuit, $Z_T = \dfrac{Z_1 Z_2}{Z_1 + Z_2}$.

Thus, to find the total impedance in a parallel circuit, you need to multiply and divide complex numbers.

GUIDED

Example 4

Find the total impedance in a parallel circuit if $Z_1 = 3 + 2i$ ohms and $Z_2 = 5 - 4i$ ohms.

Solution Substitute the values of Z_1 and Z_2 into the impedance formula for parallel circuits.

$$Z_T = \frac{Z_1 Z_2}{Z_1 + Z_2}$$

$$Z_T = \frac{(3 + 2i)(5 - 4i)}{?+?} \quad \text{Substitution}$$

$$= \frac{?+?i}{?-?i}$$

$$= \frac{?+?i}{?-?i} \cdot \frac{?+?i}{?+?i} \quad \text{Multiply numerator and denominator by the conjugate of the denominator.}$$

$$Z_T = \frac{?}{68} \quad \text{Definition of } i \text{ and arithmetic}$$

The total impedance is $\dfrac{?}{68}$ ohms.

parallel circuit

series circuit

The basic properties of inequality that hold for real numbers do not hold for nonreal complex numbers. For instance, if you were to assume $i > 0$, then multiplying both sides of the inequality by i, you would get $i \cdot i > 0 \cdot i$, or $-1 > 0$, which is not true. If you assume $i < 0$, then multiply both sides by i, you get (changing the direction) $i \cdot i > 0 \cdot i$, or again $-1 > 0$. Except for those complex numbers that are also real numbers, there are no positive or negative complex numbers.

Questions

COVERING THE IDEAS

1. **Fill in the Blank** A complex number is a number of the form $a + bi$ where a and b are ___?___ numbers.

In 2–4, give the real and imaginary parts of each complex number.

2. $5 + 17i$
3. $-4 + i\sqrt{5}$
4. i

In 5–8, rewrite the expression as a single complex number in $a + bi$ form.

5. $(7 + 3i) - (4 - 2i)$
6. $(5 + 2i)(6 - i)$
7. $(5 + 2i)(5 - 2i)$
8. $(9 + i\sqrt{2}) + (12 - 3i\sqrt{2})$

9. What is the complex conjugate of $a + bi$?

10. Provide reasons for each step.

$$(9 + 5i)(7 + 2i) = 63 + 18i + 35i + 10i^2 \qquad \text{a.} \ \underline{\ \ ?\ \ }$$
$$= 63 + 53i + 10i^2 \qquad \text{b.} \ \underline{\ \ ?\ \ }$$
$$= 63 + 53i + 10(-1) \qquad \text{c.} \ \underline{\ \ ?\ \ }$$
$$= 53 + 53i \qquad \text{d.} \ \underline{\ \ ?\ \ }$$

11. Find the complex conjugate of each number.
    a. $5 + 2i$
    b. $3i$
    c. $-2 - 3i$
    d. 4

In 12 and 13, write in $a + bi$ form.

12. $\dfrac{5 + 2i}{4 - i}$
13. $\dfrac{13}{2 + 3i}$

14. Two electrical circuits have impedances $Z_1 = 8 + 4i$ ohms and $Z_2 = 7 - 4i$ ohms. Find the total impedance if these two circuits are connected
    a. in series.
    b. in parallel.

15. **True or False** Every real number is also a complex number.

16. Name two fields in which complex numbers are applied.

APPLYING THE MATHEMATICS

17. Write $\sqrt{-25}$ in $a + bi$ form.

18. If $Z_1 = -4 + i$ and $Z_2 = 1 - 2i$, write each expression in $a + bi$ form.
    a. $2Z_1 - Z_2$
    b. $3Z_1Z_2$
    c. $\dfrac{Z_1}{Z_2}$

19. Find two nonreal complex numbers that are not complex conjugates and whose
    a. sum is a real number.
    b. product is a real number.

20. A complex number $a + bi$ is graphed as the point (a, b) with the x-axis as the real axis and the y-axis as the imaginary axis.

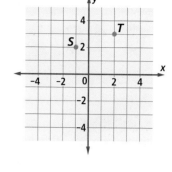

  a. Refer to the graph at the right. S and T are the graphs of which complex numbers?

  b. Graph $S + T$.

  c. Connect $(0, 0)$, S, $S + T$, and T to form a quadrilateral. What type of quadrilateral is formed?

21. a. Solve the equation $x^2 - 6x + 13 = 0$ using the Quadratic Formula. Write the solutions in $a + bi$ form.

  b. How are the solutions to the equation $x^2 - 6x + 13 = 0$ related to each other?

In 22 and 23, consider the complex numbers $a + bi$ and $a - bi$.

22. Find their sum and explain why it is a real number.

23. Find their product and explain why it is a real number.

24. Prove that, if two circuits connected in parallel have impedances Z_1 and Z_2 and the total impedance is Z_T, then $\frac{1}{Z_T} = \frac{1}{Z_1} + \frac{1}{Z_2}$.

REVIEW

In 25 and 26, solve. Write the solutions as real numbers or multiples of i. (Lesson 6-8)

25. $a^2 - 3 = -8$

26. $-r^2 = 196$

27. Explain where the $\pm$ comes from in Step 8 of the derivation of the Quadratic Formula in Lesson 6-7. (Lesson 6-2)

28. Write a piecewise definition for the function g with equation $g(x) = |x + 2|$. (Lessons 6-2, 3-4)

29. The two graphs at the right show the relationships between a dependent variable R and independent variables m and p. Find an equation for R in terms of m and p. (You may leave the constant of variation as k.) (Lesson 2-8)

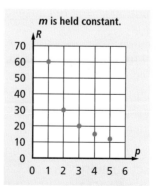
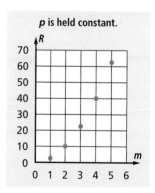

EXPLORATION

30. Refer to Question 20.

  a. Graph $z = 1 + i$ as the point $(1, 1)$.

  b. Compute and graph z^2, z^3, and z^4.

  c. What pattern emerges? Predict where z^5, z^6, z^7, and z^8 will be.

Lesson

6-10

Analyzing Solutions to Quadratic Equations

Vocabulary

discriminant

roots of an equation

zeros of a function

▶ **BIG IDEA**   You can determine whether the solutions to a quadratic equation with real coefficients are real or not real by calculating a value called the *discriminant* of the quadratic.

A Brief History of Quadratics

As early as 1700 BCE, ancient mathematicians considered problems that today would be solved using quadratic equations. The Babylonians described solutions to these problems using words that indicate they had general procedures for solving them similar to the Quadratic Formula. However, the ancients had neither our modern notation nor the notion of complex numbers. The history of the solving of quadratic equations helped lead to the acceptance of irrational numbers, negative numbers, and complex numbers.

The Pythagoreans in the 5th century BCE thought of x^2 as the area of a square with side x. So if $x^2 = 2$, as in the square pictured here, then $x = \sqrt{2}$. The Greeks proved that $\sqrt{2}$ was an irrational number, so a long time ago people realized that irrational numbers have meaning. But they never considered the negative solution to the equation $x^2 = 2$ because lengths could not be negative.

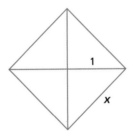

Writings of Indian and Arab mathematicians from 800 to 1200 CE indicate that they could solve quadratic equations. The Arab mathematician Al-Khowarizmi, in 825 CE in his book *Hisab al-jabr w'al muqabala* (from which we get the word "algebra"), solved quadratics like the Babylonians. His contribution is that he did not think of the unknown as having to stand for a length. Thus, the unknown became an abstract quantity. Around 1200, Al-Khowarizmi's book was translated into Latin by Fibonacci, and European mathematicians had a method for solving quadratics.

Mental Math

a. What size change is represented by the matrix
$$\begin{bmatrix} 3 & 0 \\ 0 & 3 \end{bmatrix}?$$

b. Give a matrix for $S_{4.2}$.

c. What scale change is represented by the matrix
$$\begin{bmatrix} 0.5 & 0 \\ 0 & 6 \end{bmatrix}?$$

d. Give a matrix for $S_{7,2}$.

Mathematicians began using complex numbers in the 16th century because these numbers arose as solutions to quadratic and higher-degree equations. In the 19th century, Gauss brought both geometric and physical meaning to complex numbers. The geometric meaning built on Descartes' coordinate plane. Physical meanings of complex numbers occur in a variety of engineering and physics applications. In 1848, Gauss was the first to allow the *coefficients* in his equations to be complex numbers. Today, complex numbers are used in virtually all areas of mathematics.

Predicting the Number of Real Solutions to a Quadratic Equation

Activity

Work in groups of three. Record all the results for Steps 1, 2 and 4 in a single table like the one below. For Steps 1–4, divide the six equations in the table equally among the group members.

$y = ax^2 + bx + c$	Number of x-intercepts of Graph	Solutions to $ax^2 + bx + c = 0$	Number of Real Solutions to $ax^2 + bx + c = 0$	Value of $b^2 - 4ac$
a. $y = 4x^2 - 24x + 27$	?	?	?	?
b. $y = 4x^2 - 24x + 36$	?	?	?	?
c. $y = 4x^2 - 24x + 45$	?	?	?	?
d. $y = -6x^2 + 36x - 54$	?	?	?	?
e. $y = -6x^2 + 36x - 48$	?	?	?	?
f. $y = -6x^2 + 36x - 60$	?	?	?	?

Step 1 Graph each quadratic equation and record the number of x-intercepts of its graph.

Step 2 Solve each equation using the Quadratic Formula and record how many of the solutions are real.

Step 3 Describe any patterns you see in the table so far.

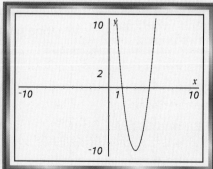

Step 4 When $ax^2 + bx + c = 0$, the value of the expression $b^2 - 4ac$ can be used to predict the number of real solutions to the quadratic equation. This value is the **discriminant** of the quadratic equation. Calculate and record the discriminant for each equation in the table. Make a conjecture about the relationship between the number of real solutions and the value of the discriminant.

(continued on next page)

Step 5 a. Have each person make three new quadratic equations of the form $y = ax^2 + bx + c$, where a, b, and c are real numbers and $a \neq 0$. One equation should have two real solutions, one should have one real solution, and the third should have no real solutions.

b. Add each of the equations in Part a to the table, then graph them on your graphing utility. Solve each equation, then record the number of x–intercepts, the number of real solutions, the values of the real solutions, and the value of the discriminant in the table.

c. Do the nine additions to the table support your conjecture from Step 4?

How Many Real Solutions Does a Quadratic Equation Have?

Now consider the general quadratic equation. When the coefficients a, b, and c are real numbers and $a \neq 0$, the Quadratic Formula gives the following two solutions to $ax^2 + bx + c = 0$: $x = \dfrac{-b \pm \sqrt{b^2 - 4ac}}{2a}$.

 STOP QY

> ▶ **QY**
>
> Write the two solutions to the Quadratic Formula as a compound sentence.

Because a and b are real numbers, the numbers $-b$ and $2a$ are real, so only $\sqrt{b^2 - 4ac}$ could possibly not be real. It is because of this property that the number $b^2 - 4ac$ is called the **discriminant**. It allows you to discriminate the *nature of the solutions* to the equation, as shown below.

If $b^2 - 4ac$ is positive, then $\sqrt{b^2 - 4ac}$ is a positive number. There are two real solutions. The graph of $y = ax^2 + bx + c$ intersects the x-axis in two points.	If $b^2 - 4ac$ is zero, then $\sqrt{b^2 - 4ac} = \sqrt{0} = 0$. Then $x = \dfrac{-b \pm 0}{2a} = \dfrac{-b}{2a}$, and there is only one real solution. The graph of $y = ax^2 + bx + c$ intersects the x-axis in one point.	If $b^2 - 4ac$ is negative, then $\sqrt{b^2 - 4ac}$ is an imaginary number. There will then be two nonreal solutions. Furthermore, because these solutions are of the form $m + ni$ and $m - ni$, they are complex conjugates. The graph of $y = ax^2 + bx + c$ does not intersect the x-axis.

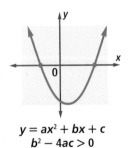

$y = ax^2 + bx + c$
$b^2 - 4ac > 0$
two real solutions

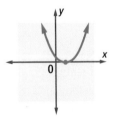

$y = ax^2 + bx + c$
$b^2 - 4ac = 0$
one real solution

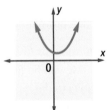

$y = ax^2 + bx + c$
$b^2 - 4ac < 0$
two nonreal solutions

The graphs on the previous page are drawn for positive a, so the parabolas open up. Solutions to quadratic (and other) equations are also called **roots of the equation** $ax^2 + bx + c = 0$, or **zeros of the function** represented by the equation $y = ax^2 + bx + c$. The number i allows you to write square roots of negative numbers as complex solutions. You saw in the Activity that the number of real roots of each quadratic equation equals the number of x-intercepts. To summarize, the results of the Activity should be consistent with the following theorem.

Discriminant Theorem

Suppose a, b, and c are real numbers with $a \neq 0$. Then the equation $ax^2 + bx + c = 0$ has:

(i) two real solutions, if $b^2 - 4ac > 0$.

(ii) one real solution, if $b^2 - 4ac = 0$.

(iii) two complex conjugate solutions, if $b^2 - 4ac < 0$.

GUIDED

Example 1

Determine the nature of the roots of each equation.

a. $4x^2 - 4x + 1 = 0$

b. $25x^2 + 6x + 4 = 0$

c. $3x^2 + 5x - 14 = 0$

Solution Use the Discriminant Theorem. Let $D = b^2 - 4ac$.

a. Here $a = 4$, $b = -4$, and $c = 1$. So $D = (-4)^2 - 4(4)(1) = 0$. Thus, the equation has __?__ real root.

b. Here $a = 25$, $b = 6$, and $c = 4$, so $D = $ __?__. Because D is negative, the equation has __?__ real roots.

c. Here $a = $ __?__, $b = $ __?__, and $c = $ __?__; so $D = $ __?__ > 0. So, this equation has __?__ real roots. Because D is not a perfect square, the roots are irrational.

Check Use a graphing utility. Let $f1(x) = 4x^2 - 4x + 1$, $f2(x) = 25x^2 + 6x + 4$, and $f3(x) = 3x^2 + 5x - 14$. The number of real solutions should equal the number of x-intercepts of the graph.

a. The graph of $f1$ has __?__ x-intercept. So, the equation has __?__ real root. It checks.

b. The graph of $f2$ has __?__ x-intercepts. So, the equation has __?__ real roots. It checks.

c. The graph of $f3$ has __?__ x-intercepts. So, the equation has __?__ real roots. It checks.

Applying the Discriminant Theorem

The number of real solutions to a quadratic equation can tell you something about the situation that led to the equation. The following example shows how.

Example 2

Eight–year old Allie Oop, an aspiring basketball player, shoots five feet from the hoop. She is trying to get the ball above the rim, which is set at a height of 10 feet. She releases the ball from an initial height of 4 feet. The following equation models the path of the ball, where x is the horizontal distance in feet that the ball has traveled and h is the ball's height in feet above the ground.

$$h = -0.4x^2 + 3x + 4$$

Use the Discriminant Theorem to determine whether Allie's ball ever reaches the height of the rim.

Solution The ball will reach the rim if there are real values of x for which

$$10 = -0.4x^2 + 3x + 4.$$

First, rewrite the equation in standard form. $0 = -0.4x^2 + 3x - 6$

Then calculate the value of the discriminant. $3^2 - 4(-0.4)(-6) = -0.6$

The discriminant $D = -0.6$. Because D is negative, there are no real solutions to this equation. This means that the ball will not reach the height of the rim.

Questions

COVERING THE IDEAS

1. What is the relationship between the number of x-intercepts of the graph of a quadratic function and the number of real solutions to the corresponding quadratic equation?

2. Why did the Pythagoreans think there was only one solution to $x^2 = 2$?

3. Consider the equation $ax^2 + bx + c = 0$, where a, b, and c are real numbers.
   a. What is its discriminant?
   b. What are its roots?

4. The discriminant of an equation $ax^2 + bx + c = 0$ is 0. What does this indicate about the graph of $y = ax^2 + bx + c$?

5. **Matching** Match the idea about quadratics at the left with the estimated length of time that idea has been understood.

   a. geometric and physical meanings to complex numbers

   b. problems in which the unknown could be an abstract quantity

   i. about 3700 years

   ii. about 1400 years

   iii. about 1750 years

   iv. about 1150 years

   v. about 150 years

In 6 and 7, a quadratic expression is given.

   a. Set the expression equal to 0. Use the discriminant to determine the nature of the roots to the equation.

   b. Set the expression equal to y. How many x-intercepts does the graph of the equation have?

6. $2x^2 - 3x + 7$

7. $x^2 - 10x + 9$

8. a. Solve $2s^2 - 3s + 7 = 0$, and write the solutions in $a + bi$ form.

   b. **True or False** The roots of this equation are complex conjugates.

9. Sketch a graph of a quadratic function $y = ax^2 + bx + c$, with $a > 0$ and a negative discriminant similar to one of the three graphs on page 436.

In 10 and 11, a graph of a quadratic function $f : x \rightarrow ax^2 + bx + c$ is given.

   a. Tell whether the value of $b^2 - 4ac$ is positive, negative, or zero.

   b. Tell how many real roots the equation $f(x) = 0$ has.

10.

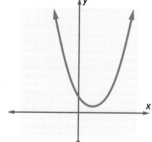

11.

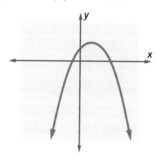

12. Without drawing a graph, tell whether the x-intercepts of $y = 2.5x^2 - 10x + 6.4$ are rational or irrational. Justify your answer.

13. Refer to Example 2. Allie practices her shot, but it never reaches the rim, so her father lowers the hoop to 9 feet. Now will the ball reach the height of the rim? Justify your answer.

APPLYING THE MATHEMATICS

In 14 and 15, True or False. If true, explain why; if false, give a counterexample.

14. The y-intercept of every parabola that has an equation of the form $y = ax^2 + bx + c$ is $(0, b)$.

15. Whenever a parabola opens down and its vertex is above the x-axis, its discriminant must be positive.

16. a. Find the value(s) of k for which the graph of the equation $y = x^2 + kx + 6.25$ will have exactly one x-intercept.

    b. Check your answer to Part a by graphing.

17. Can a quadratic equation have two real roots, one rational and the other irrational? Why or why not?

REVIEW

18. Write $\frac{49 + 28i}{2 + 3i}$ in $a + bi$ form by hand. (**Lesson 6–9**)

19. Solve the equation $x^2 - 8x + 25 = 0$. Write the solutions in $a + bi$ form. (**Lesson 6–9**)

20. Find all the *fourth* roots of 16, that is, the square roots of the square roots of 16. (**Lesson 6–8**)

21. Penny drops a penny from the top of a 1200–foot building. At the exact moment Penny drops the penny, her sister Ellie gets on an elevator at a height of 300 feet which travels upward at a constant rate of $30 \frac{\text{feet}}{\text{second}}$. Ignore air resistance. (**Lessons 6-4, 5-2, 3-1**)

    a. Write an equation for the height p in feet of Penny's penny as a function of the time t in seconds after it is dropped.

    b. Write an equation for Ellie's height m as a function of t.

    c. Graph your two equations from Parts a and b and use the graph to determine at what time are Ellie and the penny at the same height above the ground.

22. The trim around a square window is 2 inches wide. The total length of one side of the window including trim is x inches. Find an expression in standard form for the area of just the window without trim. (**Lesson 6–1**)

EXPLORATION

23. The word "algebra" comes from the Arabic "al-jabr." What does "al-jabr" mean and what does that meaning have to do with algebra?

QY ANSWER

$x = \dfrac{-b + \sqrt{b^2 - 4ac}}{2a}$ or

$x = \dfrac{-b - \sqrt{b^2 - 4ac}}{2a}$

Chapter 6 Projects

1 Sum and Product of Roots

Checking solutions to quadratic equations can be tedious, but there is an easier way to check than by substitution. The method checks the sum and the product of the roots. Look back over some quadratic equations you have solved. Let r_1 and r_2 be the roots of the equation $ax^2 + bx + c = 0$. For each equation, calculate $r_1 + r_2$ and $r_1 \cdot r_1$. What patterns do you notice? Can you prove that your generalizations hold for all quadratics? Use your results to check some of the solutions to quadratic equations in this chapter.

2 Satellite Dishes

Satellite dishes are in the shape of three-dimensional parabolas called *paraboloids*. Find out how satellite dishes work. Why are they in the shape of paraboloids? What determines their dimensions? How are light, sound, or radio waves reflected from the satellite dish? Write a report summarizing your findings.

3 Representing Complex Numbers as Points

The use of the coordinate plane to represent complex numbers is often attributed to Gauss, but the plane defined by one real axis and one imaginary axis is called an *Argand plane* after Jean-Robert Argand (b. 1768). Yet record has it that the Argand plane was first proposed by an amateur mathematician named Caspar Wessel (b. 1745). Research Argand or Wessel and describe their careers and accomplishments.

4 Minimum Distance

Use a DGS.

a. Graph the line $y = 3x$ and plot the dynamic point $(x, 3x)$ on the line.

b. Plot the point (a, b) not on the line in Part a. Construct a line segment from (a, b) to $(x, 3x)$.

c. Calculate the distance from (a, b) to $(x, 3x)$. Call it d. Plot the result as the ordered pair (x, d) for various values of x.

d. Make a conjecture about what an equation describing the graph of all possible ordered pairs (x, d) in Part c might be.

e. When is the distance between $(x, 3x)$ and (a, b) minimized?

5 How High Does Your Parabola Go?

You can make a parabolic shape from water flowing out of a hose or fountain. The curvature and maximum height of the parabola affect the distance at which the parabolic path of the water hits the ground. Work with a partner. Use a hose with a strong, consistent flow of water.

a. Make the water flow from the hose in a parabolic path and try to get the water to reach as high in the air as possible while maintaining the parabolic shape. Have your partner take a picture of the "parabolic" water flow and measure height of the hose opening and the horizontal distance from the hose opening to the point the water flow hits the ground.

b. Place a coordinate system on your photo. Make sure to fit the coordinate system so it reflects the height and distance you measured. Estimate some other points on the water's path and use them to determine the equation of your parabola. How high was the water to the nearest inch?

c. Switch roles and try the experiment again.

d. Compare the horizontal distances and heights to which the water traveled. How do these measures seem to be related?

6 Finding Complex Roots Graphically

Real roots of quadratic equations can be found graphically by finding the x-intercepts of the graph of the equation. Complex roots can also be found graphically.

a. Graph $y = 2(x - 2)^2 + 4.5$ and $y = -2(x - 2)^2 + 4.5$ on the same set of axes. How are the graphs of these two equations related?

b. Find the x-intercepts of the graph of $y = -2(x - 2)^2 + 4.5$. These are the solutions to $0 = -2(x - 2)^2 + 4.5$.

c. The solutions you found in Part b can be expressed in $a \pm b$ form, when a is the x-coordinate of the vertex of the graph of the equation. Find the vertex of the parabola and use it to write your answer to Part b in $a \pm b$ form.

d. Solve $0 = 2(x - 2)^2 + 4.5$. Express the solutions in $a \pm bi$ form. How are the solutions related to your answer to Part c?

e. Generalize Parts a–d to explain how you can find complex solutions to quadratic equations graphically.

Chapter 6

Summary and Vocabulary

- The simplest **quadratic equation** is of the form $x^2 = k$. If $k > 0$, there are two real solutions, $\sqrt{k}$ and $-\sqrt{k}$. When $k < 0$, the solutions are the imaginary numbers $i\sqrt{-k}$ and $-i\sqrt{-k}$, where, by definition, $\sqrt{-1} = i$. Any number of the form $a + bi$, where a and b are real numbers, is a **complex number**. Complex numbers are added, subtracted, and multiplied using the properties that apply to operations with real numbers and polynomials.

- Areas, paths of objects, and relations between the initial velocity of an object and its height over time lead to problems involving quadratic equations and functions. A projectile's height h above the ground on a planet with gravity g at time t after being launched with initial velocity v_0 from an initial height h_0 satisfies
$$h = \tfrac{1}{2}gt^2 + v_0 t + h_0.$$

- When a, b, and c are real numbers and $a \neq 0$, the graph of the general quadratic equation $y = ax^2 + bx + c$ is a parabola. Using a process known as **completing the square,** this equation can be rewritten in **vertex form** $y - k = a(x - h)^2$. This parabola is a translation image of the parabola $y = ax^2$ you studied in Chapter 2. Its vertex is (h, k), its line of symmetry is $x = h$, and it opens up if $a > 0$ and opens down if $a < 0$. If data involving two variables are graphed in a scatterplot that appears to be part of a parabola, you can use three points on the graph to set up a system of equations that will allow you to find a, b, and c in the equation $y = ax^2 + bx + c$.

- The values of x for which $ax^2 + bx + c = 0$ can be found by using the **Quadratic Formula**:
$$x = \frac{-b \pm \sqrt{b^2 - 4ac}}{2a}.$$

- The expression $b^2 - 4ac$ in the formula is the **discriminant** of the quadratic equation, and reveals the nature of its roots.

Vocabulary

6-1
quadratic expression
quadratic equation
quadratic function
standard form of a quadratic
binomial

6-2
absolute value
absolute-value function
square root
rational number
irrational number

6-3
corollary
vertex form
axis of symmetry
minimum, maximum

6-5
completing the square
perfect-square trinomial

6-6
quadratic regression

6-7
Quadratic Formula

6-8
imaginary number
*$\sqrt{k}$
*$\sqrt{-1}$, i
*pure imaginary number

6-9
*complex number
*real part, imaginary part
*equal complex numbers
*complex conjugate

6-10
*discriminant
*root of an equation
*zeros of a function

- If $b^2 - 4ac > 0$, there are two real solutions, as shown below.

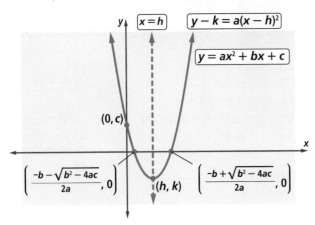

- If $b^2 - 4ac = 0$, there is exactly one solution and the vertex of the parabola is on the x-axis. If a, b, and c are rational numbers and the discriminant is a perfect square, then the solutions are rational numbers.

- If $b^2 - 4ac < 0$, there are no real solutions and the parabola does not intersect the x-axis. The nonreal solutions are **complex conjugates**.

Postulates, Theorems, and Properties

Binomial Square Theorem (p. 376)
Absolute Value–Square Root Theorem (p. 381)
Graph-Translation Theorem (p. 387)
Parabola-Translation Theorem (Graph-Translation Corollary) (p. 388)
Parabola Congruence Theorem (p. 395)

Completing the Square Theorem (p. 402)
Quadratic Formula Theorem (p. 414)
Square Root of a Negative Number Theorem (p. 422)
Properties of Complex Numbers Postulate (p. 428)
Discriminant Theorem (p. 437)

Chapter

6 Self-Test

Take this test as you would take a test in class. You will need a calculator. Then use the Selected Answers section in the back of the book to check your work.

In 1–3, consider the parabola with equation $y = x^2 - 10x + 21$.

1. Rewrite the equation in vertex form.

2. What is the vertex of this parabola?

3. What are the x-intercepts of this parabola?

In 4–7, perform the operations and put the answer in $a + bi$ form.

4. $5i \cdot i$

5. $\sqrt{-12} \cdot \sqrt{-3}$

6. $\dfrac{6 + \sqrt{-64}}{3}$

7. $(1 + 5i)(-4 + 3i)$

8. If $z = 6 - 5i$ and $w = 2 + 5i$, write $z - w$ in $a + bi$ form.

9. **Multiple Choice** How does the graph of $y - 3 = -(x + 4)^2$ compare to the graph of $y = -x^2$? It is translated:

   A 4 units to the right and 3 units down.

   B 4 units to the right and 3 units up.

   C 4 units to the left and 3 units up.

   D 4 units to the left and 3 units down.

10. Graph $y - 3 = -(x + 4)^2$.

11. If $f(x) = -(x + 4)^2 + 3$, find the domain and range of f.

In 12–15, solve the equation. Show your work.

12. $|t + 12| = 21$

13. $\sqrt{(6s + 5)^2} = 9$

14. $17 = (y + 18)^2$

15. $3x^2 + 18x = 4x + 5$

16. This statement is false: $\sqrt{x} = |x|$ for all real values of x. Correct it.

17. Expand: $(2a - 5)^2 + (2a + 5)^2$.

18. The discriminant of an equation $ax^2 + bx + c = 0$ is negative. What does this indicate about the solutions of the equation?

19. Consider the graph of $y = ax^2 + bx + c$. How many x-intercepts does the graph have if its discriminant is

   a. –2?   b. 0?   c. 17?

In 20 and 21, suppose the height h in feet of a ball at time t seconds is given by $h = -16t^2 + 28t + 6$.

20. How high is the ball after 1.5 seconds?

21. When does the ball hit the ground?

22. A 12-foot by 16-foot rectangular garden has an x-foot wide walkway that surrounds it on all sides. Write an expression for the area of the walkway.

23. **Multiple Choice** The graph of each equation is a parabola. Which parabola is not congruent to the others?

   A $y = (x + 4)^2$ B $y = 4x^2$

   C $y + 4 = x^2$ D $y + 2 = x^2$

24. If circles are drawn so every circle intersects every other circle, then there is a pattern to the maximum number of intersection points.

   a. Draw a scatter plot of these data.

   b. Fit a quadratic model to these data.

   c. Predict in how many points 17 circles will intersect if every circle intersects every other circle.

Number of Circles n	Maximum Number of Intersections $f(n)$
1	0
2	2
3	6
4	12
5	20

Chapter 6 Chapter Review

SKILLS
PROPERTIES
USES
REPRESENTATIONS

SKILLS Procedures used to get answers

OBJECTIVE A Expand products and squares of binomials. (Lesson 6-1)

In 1–8, expand.

1. $(7p + 2)(p - 3)$
2. $(r + 2q)(r - q)$
3. $6(3w - 2)(2w + 1)$
4. $(3x + \sqrt{2})(3x - \sqrt{2})$
5. $(a + b)^2$
6. $(2x + 1)^2$
7. $13(y - 3)^2$
8. $7(s + t)^2 - 3(s - t)^2$

OBJECTIVE B Convert quadratic equations from vertex to standard form, and vice versa. (Lessons 6-4, 6-5)

In 9 and 10, rewrite in standard form.

9. $y = 5(x + 1)^2$
10. $y + 3 = 0.25(x - 4)^2$

In 11 and 12, write each equation in vertex form.

11. $y = x^2 + 9x - 5$
12. $3y = 7x^2 - 6x + 5$

13. **Multiple Choice** Which equation is equivalent to $y - 1 = 2(x - 1)^2$?

   A $y = 2x^2 + 4x - 3$
   B $y = 2x^2 + 4x + 3$
   C $y = 2x^2 - 4x + 3$
   D $y = 2x^2 - 4x - 3$

OBJECTIVE C Solve quadratic equations. (Lessons 6-2, 6-7, 6-8, 6-10)

In 14–22, solve.

14. $(x - 3)^2 = 0$
15. $r^2 - 26 = 0$
16. $d^2 = -24$
17. $-25 = y^2$
18. $2x^2 + x - 1 = 0$
19. $3 - s^2 - 5s = 6$
20. $z^2 + 2z - 8 = 7$
21. $k^2 = 6k - 9$
22. $2x(x - 2) = -1 + x^2 - 2x$

OBJECTIVE D Solve absolute-value equations. (Lesson 6-2)

In 23–28, solve.

23. $|11 - y| = 15$
24. $|s + 3| = 5$
25. $-\sqrt{(3x + 2)^2} = -5$
26. $\sqrt{(x - 2)^2} = 3$
27. $|4t - 20| = 0$
28. $|96 - A| = -4$

OBJECTIVE E Perform operations with complex numbers. (Lesson 6-8, 6-9)

In 29–34, simplify.

29. $-i^2$
30. $\sqrt{-49}$
31. $\sqrt{-9} \cdot \sqrt{-25}$
32. $\sqrt{-7} \cdot \sqrt{7}$
33. $2i \cdot 3i$
34. $3\sqrt{-4} + \sqrt{-9}$

In 35 and 36, write the conjugate.

35. $8 - 3i$
36. $-7i$

In 37–42, suppose $r = 7 - i$ and $s = 3i + 2$.
Evaluate and simplify.

37. rs

38. s^2

39. $2r - s$

40. $ir + 3s - i$

41. $\frac{r}{s}$

42. $\frac{is}{r}$

PROPERTIES Principles behind the mathematics

OBJECTIVE F Apply the definition of absolute value and the Absolute Value-Square Root Theorem. (Lesson 6–2)

43. For which real numbers x is $|x| - x > 0$?

44. For which real numbers x is $|x| = -\pi$?

In 45–47, use the Absolute Value-Square Root Theorem to simplify.

45. $\sqrt{(11 + 5)^2}$

46. $-\sqrt{t^2}$

47. $-\sqrt{(-3)^2} + \sqrt{5^2}$

OBJECTIVE G Use the Graph-Translation Theorem to interpret equations and graphs. (Lessons 6-3, 6-4)

48. The preimage graph of $y = x^2$ is translated 50 units to the right and 300 units up. What is an equation for its image?

49. Describe how the graphs of $y = |x|$ and $y = |x + 2|$ are related.

50. **Multiple Choice** Which of the following is not true for the graph of the equation $y - 3 = -\frac{1}{2}(x + 1)^2$?

   A The vertex is (-1, 3).

   B The maximum point is (-1, 3).

   C The equation of the axis of symmetry is $x = -1$.

   D The graph opens up.

51. Assume parabola A is congruent to parabola B in the graph below.

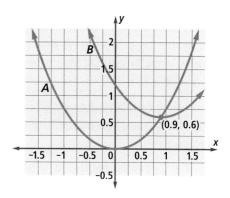

   a. What translation maps parabola A onto parabola B?

   b. What is an equation for parabola B if parabola A has equation $y = \frac{3}{4}x^2$?

52. Compare the solutions to $4 = (x - 1)^2$ with the solutions to $x^2 = 4$.

53. Compare the solutions to $(k + 2)^2 = 25$ with the solutions to $k^2 = 25$.

54. The graph of $y = (5x + 3)^2$ is congruent to the graph of $y = ax^2$. What is a?

OBJECTIVE H Use the discriminant of a quadratic equation to determine the nature of the solutions to the equation. (Lesson 6-10)

In 55–57, an equation is given.

a. Evaluate its discriminant.

b. Give the number of real solutions.

c. Tell whether the solutions are rational, irrational, or non-real.

55. $6x^2 + 9x + 1 = 0$

56. $z^2 = 81z + 81$

57. $5 + k = k^2 + 9$

USES Applications of mathematics in real-world situations

OBJECTIVE I Use quadratic equations to solve area problems or problems dealing with distance, velocity, and acceleration.
(Lessons 6-1, 6-4, 6-5, 6-7)

58. A framed mirror is 28 inches by 34 inches, including the frame. The frame is w inches wide.

   a. Write an expression for the area of the mirror without the frame.

   b. If the area of the mirror is 85% of the total area, how wide is the frame?

In 59 and 60, use the equation, $h = -16t^2 + v_0 t + h_0$ for the height h in feet of an object after t seconds. Ignore wind resistance.

59. A ball is thrown directly upward from an initial height of 5 feet at an initial velocity of 37 feet per second.

   a. Estimate to the nearest hundredth the highest point reached by the ball.

   b. To the nearest hundredth of a second, when does the ball reach its highest point?

   c. How far from the ground will the ball be after 2 seconds?

60. A ball is dropped from the top of a building that is 1000 feet tall. To the nearest tenth of a second, how long after it is dropped will the ball reach the ground?

REPRESENTATIONS Pictures, graphs, or objects that illustrate concepts

OBJECTIVE J Fit a quadratic model to data.
(Lesson 6-6)

61. A researcher marked four evenly-spaced concentric circles on a field. She then released 1000 frogs and returned an hour later to count how many frogs were in each circle. She found 200 frogs: 11 frogs in the inner circle, 38 frogs in the first ring, 60 frogs in the second ring, and 91 frogs in the third ring. The researcher then created another evenly spaced concentric circle and repeated the experiment.

   a. Fit a model to the data that can predict the number of frogs in the fourth ring.

   b. How many frogs should the researcher expect to find in the fourth ring?

   c. Why will the model overestimate the number of frogs in the 20th concentric circle?

62. Consider the sequence $\{1, 5, 12, 22, 35,\ldots\}$, where differences between consecutive terms increase by 3.

   a. Use quadratic regression to write an explicit formula for the terms of this sequence.

   b. Find the 23rd term in this sequence.

63. In 1980, Omaha, Nebraska, had a population of 314,255. By 1990, the population had increased to 344,463. The population continued to increase and reached 390,007 by 2000.

   a. Write a quadratic equation that models Omaha's population growth starting in 1980.

   b. Use your equation to predict Omaha's population in 2010.

   c. Based on your model, in which year will Omaha's population first exceed 1 million?

OBJECTIVE K Graph quadratic functions and absolute-value functions and interpret them. (Lessons 6-2, 6-3, 6-4, 6-10)

In 64–67, sketch a graph of the function, and label the vertex and x-intercepts with their coordinates.

64. $y = 3x^2 + 18x$

65. $y - 6 = -\frac{1}{3}(x + 3)^2$

66. $y - 7 = |x - 2|$

67. $y + 1 = 2x^2$

68. The height of a ball thrown upward at time t is shown on the graph below.

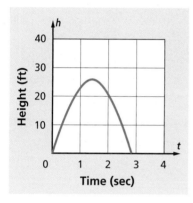

a. About when did the ball get to its maximum height?

b. About how high did the ball get?

c. About when was the ball 10 feet high?

In 69 and 70, refer to the parabolas shown below.

A
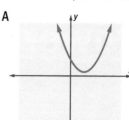

B
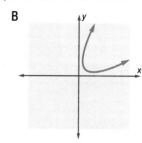

C
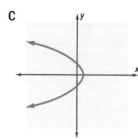

D
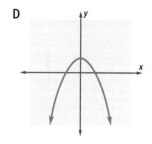

69. Which graph is of the form
$y - k = a(x - h)^2$ with h positive?

70. Which graph is of the form
$y - k = a(x - h)^2$ with a negative?

OBJECTIVE L Use the discriminant of a quadratic equation to determine the number of x-intercepts of a graph of the associated quadratic function. (Lesson 6-10)

In 71 and 72, give the number of x-intercepts of the parabola.

71. $y = 7x^2 + 5x - 13$

72. $y = \frac{1}{7}(x + 3)^2 - 6$

73. Does the parabola $y = 3x^2 - 9x$ ever intersect the line $y = -1$? Explain your reasoning.

74. How many x-intercepts does the graph of $y = -\frac{1}{5}(x - t)^2$ have when $t \neq 0$?

Chapter

7 Powers

The expression x^n stands for the *nth power of x*. In this chapter you will study powers and related functions. You will see how to interpret powers when the exponent n is not an integer, and how to solve equations involving powers.

Powers have many applications. You are familiar with applications involving x^2, the 2nd power of x, in area problems, counting problems, and a variety of physics problems including those that involve projectiles. Expressions involving x^3, the 3rd power of x, arise from problems involving volume. Both positive and negative exponents are used in writing numbers in scientific notation.

Applications involving powers also arise in money matters dealing with interest rates, amounts owed, or amounts paid, as well as in counting problems and probability. Solving equations in any of these situations can lead to powers of the form $x^{\frac{1}{n}}$ or $x^{\frac{m}{n}}$, where m and n are integers.

Sequences determined by the successive powers of a number also have many applications in the physical world, including music and exponential growth and decay. One of these sequences is shown above in the cross-section of a shell of a *chambered nautilus*, a member of the squid family. The full shell is shown at the right.

The chambers in the shell may look like they were made by a machine, but it is a natural shell. As the nautilus matures, it builds successively larger chambers to accommodate its growth. Let L represent the length of some part of the smallest chamber. The lengths of corresponding parts in adjacent chambers are all related by a common ratio, x. The sequence of chamber lengths can, thus, be represented as L, Lx^1, Lx^2, and so on. In this chapter you will determine that in this case $x \approx 2^{0.1} \approx 1.07$.

Lesson

7-1

Power Functions

Vocabulary

powering, exponentiation

base

exponent

power

*n*th-power function

identity function

squaring function

cubing function

▶ **BIG IDEA**   When *n* is a positive integer greater than 1, x^n can be interpreted as repeated multiplication.

Recall that the expression x^n, read "*x* to the *n*th power" or "the *n*th power of *x*", is the result of an operation called **powering** or **exponentiation**. The variable *x* is called the **base**, *n* is called the **exponent**, and the expression x^n is called a **power**.

Defining b^n When *n* Is a Positive Integer

When *n* is a positive integer and *b* is any real number, one way to think of b^n is as the *n*th term of the sequence

$$b^1, b^2, b^3, \dots b^n, \dots .$$

This sequence can be defined recursively as

$$\begin{cases} b^1 = b \\ b^n = b \cdot b^{n-1}, \text{ for } n > 1. \end{cases}$$

For example, $14^1 = 14$, $14^2 = 14 \cdot 14^1 = 196$, $14^3 = 14 \cdot 14^2 = 14 \cdot 196 = 2744$, and so on.

 QY1

As a result of the recursive definition of b^n for any real number *b*, when *n* is a positive integer ≥ 2,

$$b^n = \underbrace{b \cdot b \cdot b \cdot \dots \cdot b}_{n \text{ factors}}.$$

This is the *repeated multiplication* definition of a power. For instance, $x^6 = x \cdot x \cdot x \cdot x \cdot x \cdot x$. The definition enables you to use multiplication to calculate positive integer powers of any number without having to calculate each preceding power. For example,

$$-\left(\frac{2}{3}\right)^5 = -\frac{2}{3} \cdot -\frac{2}{3} \cdot -\frac{2}{3} \cdot -\frac{2}{3} \cdot -\frac{2}{3}$$
$$= -\frac{32}{243}.$$

Mental Math

Give a number that fits the description or tell if no such number exists.

a. an integer that is not a natural number

b. an irrational number that is not a real number

c. a complex number that is not a pure imaginary number

d. a real number that is not a complex number

▶ **QY1**

Calculate 14^4 without a calculator given that $14^3 = 2744$.

An Example of a Power Function

Powers often arise in counting and probability situations. For example, in the play *Rosencrantz & Guildenstern are Dead,* by Tom Stoppard, the character Rosencrantz spends a lot of time flipping a fair coin. It always lands heads up! This, of course, is very unlikely. The probability that a fair coin lands heads up is $\frac{1}{2}$. Because each flip is independent, the probability of as few as four heads in a row is very small:

$$\frac{1}{2} \cdot \frac{1}{2} \cdot \frac{1}{2} \cdot \frac{1}{2} = \left(\frac{1}{2}\right)^4 = 0.0625.$$

Had Rosencrantz been tossing a 6-sided die, his probability of tossing the same number four times in a row would have been even smaller:

$$\frac{1}{6} \cdot \frac{1}{6} \cdot \frac{1}{6} \cdot \frac{1}{6} = \left(\frac{1}{6}\right)^4 \approx 0.0007716.$$

Example 1 generalizes Rosencrantz's situation.

Example 1

Suppose that the probability of an event happening is p. Let A be the probability of the event happening four times in a row.

a. Find a formula for A in terms of p.

b. Make a table of values and a graph for typical values of p.

Solution

a. The probability of the event happening four times in a row is $p \cdot p \cdot p \cdot p$, so $A = p^4$.

b. The probability p must be a number from 0 to 1. A table and graph are shown below.

p	$A = p^4$
0	0
0.1	0.0001
0.2	0.0016
0.25	0.0039
0.3	0.0081
0.4	0.0256
0.5	0.0625
0.6	0.1296
0.7	0.2401
0.8	0.4096
0.9	0.6561
0.95	0.8145
1	1

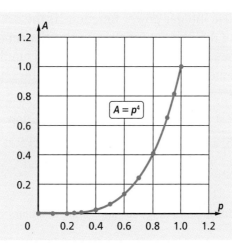

Notice from Example 1 that an event with as large as a 90% chance of happening has only a $(0.90)^4 \approx 0.6561$ probability of happening four times in a row. That's less than a 66% chance.

The general formula is a simple application of nth powers.

> ## Probability of Repeated Independent Events
>
> If an event has probability p, and if each occurrence of the event is independent of all other occurrences, then the probability that the event occurs n times in a row is p^n.

Caution: The events must be independent. For instance, if there is a 40% probability of rain on a typical day in a particular location, there is not a $(40\%)^7$ chance that there will be rain all 7 days because rain on one day can affect the chances of rain on the next day.

 QY2

> ▶ QY2
>
> If the probability of an independent event happening is 0.6, what is the probability that the event will occur n times in a row?

Some Simple Power Functions

In general, the function f defined by $f(x) = x^n$, where n is a positive integer, is called the **nth-power function**. The function with equation $y = x^5$ is the 5th-power function. The graph in Example 1 is a part of the graph of the 4th-power function.

The simplest power function has the equation $f(x) = x^1$ and is called the **identity function** because the output is identical to the input. The quadratic function f with $f(x) = x^2$ is the 2nd-power, or **squaring function**.

Any real number can be raised to the first or second power. So the domain of each of these positive integer power functions is the set of all real numbers. The range of the identity function is also the set of real numbers. However, because the result of squaring a real number is always nonnegative, the range of the squaring function is the set of all nonnegative real numbers.

The function with equation $f(x) = x^3$ is called the **cubing function**. The nth-power functions where $n > 3$ do not have special names.

> ### Example 2
> Draw a graph and state the domain and range for the cubing function.
>
> **Solution** Make a table of values and plot points. Connect the points with a smooth curve.

x	$f(x) = x^3$
-5	$(-5)^3 = -125$
-4	$(-4)^3 = -64$
-3	$(-3)^3 = -27$
-2	$(-2)^3 = -8$
-1	$(-1)^3 = -1$
0	$0^3 = 0$
1	$1^3 = 1$
2	$2^3 = 8$
3	$3^3 = 27$
4	$4^3 = 64$
5	$5^3 = 125$

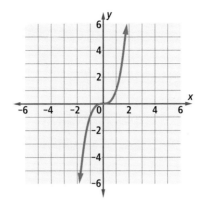

Because x can be any real number, the domain is the set of real numbers.

Because x^3 can be as large or as small as we want, positive or negative, the range is the set of real numbers.

Properties of Power Functions

MATERIALS CAS or graphing calculator

Work with a partner.

Step 1 Graph several functions of the form $y = x^n$ in a window where $\{x|\ -5 \le x \le 5\}$ and $\{y|\ -5 \le y \le 5\}$. One of you should graph functions for even values of n and the other should graph functions for odd values of n. Graph at least four functions each in the same window. Sketch your results on a separate sheet of paper and label each function.

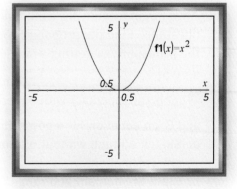

Step 2 State the domain and range for each of the functions you graphed.

Step 3 Compare and contrast your graphs, then discuss the results with your partner. What other properties do these functions seem to have?

The table on the next page summarizes some properties of the *n*th power functions *f* with $f(x) = x^n$, where *n* is a positive integer. As in the Activity, these properties depend on whether *n* is even or odd.

Properties of Power Functions with Equations $y = x^n$, for integers $n > 0$

	n is even.	n is odd.
Some ordered pairs on the graph	(0, 0) (1, 1) (–1, 1)	(0, 0) (1, 1) (–1, –1)
Domain	set of all real numbers	set of all real numbers
Range	$\{y \mid y \geq 0\}$	set of all real numbers
Quadrants	I and II	I and III
Symmetry	Reflection symmetry over the y-axis For all x, $f(x) = f(-x)$.	Rotation symmetry of $180°$ about the origin For all x, $f(-x) = -f(x)$.

 QY3

▶ **QY3**

What type of symmetry does the graph of $y = x^{19}$ have?

Questions

COVERING THE IDEAS

1. Given that $2^{10} = 1024$, calculate 2^{11} and 2^{12} without a calculator.

2. What is the first term of the sequence s with $s_n = (0.9)^n$ in which $s_n < 0.01$?

3. Refer to Example 1. What is the meaning of the number 0.0081 in the table?

4. A spinner has 5 congruent sectors, as pictured at the right. If the spinner is fair, what is the probability of landing in the blue sector 6 times in 6 spins?

5. Alvin forgot to study for a multiple-choice test with 10 independent questions. Find the probability of getting all questions correct if the probability of getting a single question correct is
   a. 0.5.
   b. 0.25.
   c. p.

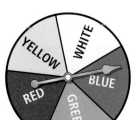

6. What is the function f with equation $f(x) = x^3$ called?

In 7 and 8, an equation for a power function f is given.
   a. Sketch a graph without plotting points or doing any calculations.
   b. State the domain and range of the function.
   c. Describe any symmetry the graph may have.

7. $f(x) = x^5$
8. $f(x) = x^{10}$

9. **Fill in the Blanks** If n is even, the range of $y = x^n$ is __?__ and the graph is in Quadrants __?__ and __?__.

10. **Fill in the Blanks** If n is odd, the range of $y = x^n$ is __?__ and the graph is in Quadrants __?__ and __?__.

11. **Multiple Choice** Which of the following graphs could represent the function with equation $y = x^8$? Justify your answer.

A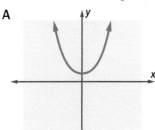

B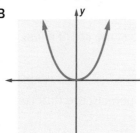

C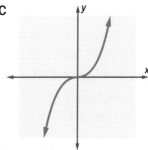

APPLYING THE MATHEMATICS

12. Refer to Example 1. Suppose Rosencrantz gets a new fair coin to flip. This coin always seems to alternate heads and tails with each flip: H T H T H T H T H … . Is the pattern H T H T more likely, less likely, or equally as likely as four heads in a row? Justify your answer.

13. Izzie Wright forgets to study for a history quiz and makes random guesses at each answer. The quiz has 5 true/false questions and 10 multiple-choice questions with 4 choices each.

   a. What is the probability Izzie will guess the correct answer on a true/false question?

   b. What is the probability Izzie will guess the correct answer on a multiple-choice question?

   c. Calculate the probability that Izzie will answer all 15 questions correctly.

14. **True or False** The graphs of the odd power functions have no minimum or maximum values.

15. Consider the graph of each function. For what values of x is the graph above the x-axis, and for what values of x is the graph below the x-axis?

   a. $f : x \rightarrow x^{33}$

   b. $g : x \rightarrow x^n$, where n is even

In **16** and **17**, write an equation for the nth-power function that is graphed.

16.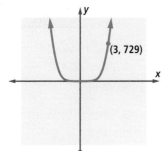

(3, 729)

17.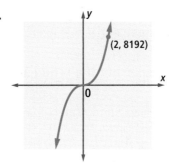

(2, 8192)

18. The point $(-7, 2401)$ is on the graph of an even nth power function. What point (other than the origin, $(1, 1)$, and $(-1, 1)$) must also be on the graph of this function? Why?

19. a. Graph $f(x) = x^2$ and $g(x) = x^6$ on the same set of axes.
    b. For what value(s) of x is $f(x) = g(x)$?
    c. For what values of x is $f(x) > g(x)$?
    d. As x increases from 0 to 1, what happens to the difference between $f(x)$ and $g(x)$?

REVIEW

20. **Fill in the Blanks** The graph of the quadratic function Q where $Q(x) = x^2 + 5x + 2$ intersects the x-axis in __?__ point(s) and the y-axis in __?__ point(s). (**Lesson 6-10**)

21. a. Write $(1 + i)(1 - i)$ in $a + bi$ form.
    b. Calculate $(1 + i)(2 + 3i)(1 - i)$. (**Lesson 6-9**)

22. Give an equation for the quadratic function graphed at the right. (**Lesson 6-3**)

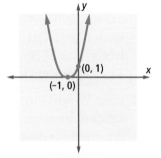

23. Solve $\begin{cases} x + 2y - 4z = -8 \\ 2x + z = 2 \\ 4x - 4y - z = 16 \end{cases}$. (**Lesson 5-6**)

24. In 2006, the St. Louis Cardinals defeated the Detroit Tigers in baseball's World Series. The win-loss records for the two teams in the three regular seasons leading up to this World Series are given in the matrices below. Find a matrix for the total win-loss record of each team over all three seasons. (**Lesson 4-2**)

	2004			2005			2006	
	Wins	Losses		W	L		W	L
Detroit	72	90	D	71	91	D	95	67
St. Louis	105	57	SL	100	62	SL	83	78

25. **Fill in the Blank** $2x^{15} \cdot$ __?__ $= 3x^{18}$ (**Previous Course**)

EXPLORATION

26. Consider nth power functions with equations of the form $f(x) = ax^n$, where a is any real number.
    a. What are the domain and range of f if n is even and a is negative?
    b. What are the domain and range of f if n is odd and a is negative?
    c. How do the graphs of $f(x) = ax^n$ and $f(x) = -ax^n$ compare?

QY ANSWERS

1. $14^4 = 14 \cdot 14^3 = 14 \cdot 2744 = 38{,}416$

2. $(0.6)^n$

3. rotation symmetry about the origin

Lesson

7-2 Properties of Powers

▶ **BIG IDEA** There are single powers equivalent to $x^m \cdot x^n$, $\frac{x^m}{x^n}$, $x^m \cdot y^m$, $\frac{x^m}{y^m}$, and $(x^m)^n$ for all x and y for which these powers are defined.

This lesson reviews the properties of powers for positive integer exponents that you have learned in previous courses. In Lessons 7-3 and 7-7, you will see that these properties also apply to exponents that are negative integers or nonzero rational numbers.

Activity 1

Set a CAS to real-number mode and clear all variables.

Step 1 What does the CAS display when you enter each product of powers?

a. $x^1 x^1$ b. $x^6 x^4$ c. $x^3 x^{12}$

Step 2 Based on your results in Step 1, write the general property: For all positive integers m and n, $x^m x^n = \underline{\ ?\ }$.

Step 3 What does the CAS display when you enter each power of products?

a. $(x \cdot y)^5$ b. $(x \cdot y)^2$ c. $(x \cdot y)^{15}$

Step 4 Based on your results in Step 3, make a conjecture: $(xy)^m = \underline{\ ?\ }$.

Products and Quotients of Powers with the Same Base

The general pattern in Step 2 of Activity 1 is summarized in the following postulate.

Product of Powers Postulate

For any nonnegative base b and nonzero real exponents m and n, or any nonzero base b and integer exponents m and n, $b^m \cdot b^n = b^{m+n}$.

Mental Math

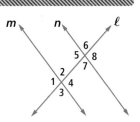

Refer to the diagram above. Assume m and n are parallel. True or false?

a. $\angle 1$ and $\angle 7$ form a linear pair.

b. $\angle 1$ and $\angle 7$ are supplementary.

c. $m\angle 3 = m\angle 6$

d. $m\angle 1 + m\angle 8 = m\angle 5 + m\angle 4$

Consider $10^4 \cdot 10^8 = 10^{12}$. The Product of Powers Postulate shows a correspondence between a product of powers and a sum of exponents.

$$10^4 \cdot 10^8 = 10^{12} \quad \text{and} \quad 4 + 8 = 12.$$

Related facts yield two other correspondences, between quotients of powers and differences of exponents.

$$\frac{10^{12}}{10^4} = 10^8 \quad \text{and} \quad 12 - 4 = 8;$$

$$\frac{10^{12}}{10^8} = 10^4 \quad \text{and} \quad 12 - 8 = 4.$$

In general, from the Product of Powers Postulate, we can prove a theorem about quotients of powers.

Quotient of Powers Theorem

For any positive base b and real exponents m and n, or any nonzero base b and integer exponents m and n, $\frac{b^m}{b^n} = b^{m-n}$.

GUIDED

Example 1
Solve $14^3 \cdot 14^x = 14^{10}$.

Solution Write a correspondence like the one above:

$14^3 \cdot 14^x = 14^{10}$ and $\underline{\ ?\ } + x = \underline{\ ?\ }$.

Now solve for x: $x = \underline{\ ?\ }$.

Power of a Product or a Quotient

The general pattern in Step 4 of Activity 1 is summarized in the following postulate.

Power of a Product Postulate

For any nonnegative bases a and b and nonzero real exponent m, or any nonzero bases a and b and integer exponent m, $(ab)^m = a^m b^m$.

You can think about the Power of a Product Postulate as meaning that *powering distributes over multiplication*.

STOP QY1

> ▶ QY1
>
> Give an example to show that powering does not distribute over addition.

Example 2

How many zeros are at the end of the number $N = 2^6 \cdot 5^6 \cdot 17$ when written in base 10?

Solution Use the Power of a Product Postulate.

$2^6 \cdot 5^6 = (2 \cdot 5)^6 = 10^6$. So, $N = 17 \cdot 10^6$, which is 17 followed by six zeros. There are six zeros at the end of the number.

Check Multiply: $2^6 \cdot 5^6 \cdot 17 = 17{,}000{,}000$. It checks.

 QY2

▶ **QY2**

How many zeros are at the end of $5^3 \cdot 13 \cdot 2^3$ when written in base 10?

From the Power of a Product Postulate, a theorem about the power of a quotient can be deduced.

> ### Power of a Quotient Theorem
>
> For any positive bases a and b and real exponent n, or any nonzero bases a and b and integer exponent n, $\left(\dfrac{a}{b}\right)^n = \dfrac{a^n}{b^n}$.

Zero as an Exponent

Suppose you divide any power by itself. For instance, consider $\dfrac{3^7}{3^7}$. By the Quotient of Powers Theorem, $\dfrac{3^7}{3^7} = 3^{7-7} = 3^0$. But it is also true that $\dfrac{3^7}{3^7} = 1$. These statements and the Transitive Property of Equality prove that $3^0 = 1$.

In general, for any nonzero real number b,

$$\frac{b^n}{b^n} = b^{n-n} \quad \text{Quotient of Powers Theorem}$$
$$= b^0 \quad \text{Arithmetic}$$

Also, $\quad \dfrac{b^n}{b^n} = 1 \qquad$ A number divided by itself is 1.

So, $\quad b^0 = 1 \qquad$ Transitive Property of Equality

This proves the following theorem.

> ### Zero Exponent Theorem
>
> If b is a nonzero real number, $b^0 = 1$.

Notice that the argument above does not work when $b = 0$. 0^0 is not defined because it would involve dividing by 0.

Powers of Powers

Activity 2

Set a CAS to real-number mode and clear all variables.

Step 1 What does the CAS display when you enter each power of powers?

a. $(x^3)^2$ b. $(y^5)^4$ c. $(z^6)^3$ d. $(m^0)^5$

Step 2 Based on your results in Step 1, make a conjecture: $(x^m)^n = \underline{\ ?\ }$.

The general pattern in Activity 2 is the last assumed property of powers.

Power of a Power Postulate

For any nonnegative base b and nonzero real exponents m and n, or any nonzero base b and integer exponents m and n, $(b^m)^n = b^{mn}$.

For example, using the Power of a Power Postulate,

$$(a^7)^4 = a^{7 \cdot 4} = a^{28}.$$

You can check this with the Product of Powers Postulate.

$$(a^7)^4 = a^7 \cdot a^7 \cdot a^7 \cdot a^7 = a^{7+7+7+7} = a^{28}$$

Example 3

A multiple-choice test is 3 pages long. Each page has 5 questions. Each question has c possible choices. How many different ways can you complete the test if you leave no answer blank?

Solution 1 If each of the 5 questions on one page has c possible choices, there are c^5 ways to answer the questions on that page. Because there are 3 pages, there are $(c^5)^3 = c^{5 \cdot 3} = c^{15}$ ways to complete the test.

Solution 2 The whole test has $5 \cdot 3 = 15$ questions, each with c choices. So, there are c^{15} ways to answer the questions on the test.

Using the Properties of Powers

Properties of powers are often used when working with numbers expressed in scientific notation.

Example 4

In 2005, astronomers using the Hubble Space Telescope discovered two tiny moons, named Hydra and Nix, orbiting Pluto. Hydra's mass is believed to be between 1×10^{17} kg and 9×10^{18} kg. Assume that Hydra's mass is actually 2.4×10^{18} kg. The mass of Earth's moon is 7.35×10^{22} kg. About how many times as massive is Earth's moon as Hydra?

Solution Divide the two numbers:

$$\frac{\text{mass of Earth's moon}}{\text{mass of Hydra}} \approx \frac{7.35 \cdot 10^{22} \text{ kg}}{2.4 \cdot 10^{18} \text{ kg}} = \frac{7.35}{2.4} \cdot \frac{10^{22}}{10^{18}}$$

$$\approx 3 \cdot 10^{22-18} = 3 \cdot 10^4$$

Earth's moon is about 30,000 times as massive as Hydra.

Properties of powers can also be used to simplify quotients or products of algebraic expressions containing exponents.

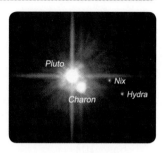

All four of these names have their base in Greek mythology. Nix was named for Nyx, the mother of Charon, and Hydra was named for the nine-headed serpent that guarded Pluto's realm.

GUIDED

Example 5

Farmers often use circular irrigators on square plots of land, leaving the regions at the corners unirrigated. How does the percentage of irrigated land depend on the radius of the circle?

Solution The diameter of the circle and the side of the square are each $2r$. The area of the circle is πr^2 and the area of the square is $(\underline{\ ?\ })^2 = \underline{\ ?\ }$ by the Power of a Product Postulate. To find the percentage of irrigated land, divide:

$$\frac{\text{Area of circle}}{\text{Area of square}} = \frac{\pi r^2}{?} = \frac{\pi}{?} \approx \underline{\ ?\ }.$$

So, regardless of the radius of the circle, $\underline{\ ?\ }$ % of the land is irrigated.

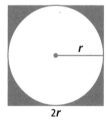

Questions

COVERING THE IDEAS

In 1–5, give an example to illustrate each postulate or theorem.
1. Product of Powers Postulate
2. Power of a Product Postulate
3. Power of a Power Postulate
4. Quotient of Powers Theorem
5. Power of a Quotient Theorem

In 6–8, write the expression as a single power using a postulate or theorem from this lesson. Then, check your answer using the repeated multiplication model of powering.
6. $17^3 \cdot 17^4$
7. $(5^2)^3$
8. $\frac{10^5}{10^2}$

9. What postulate or theorem justifies $\left(\frac{2}{y}\right)^4 = \frac{16}{y^4}$?

10. Solve $6^{2x} \cdot 6^3 = 6^7$ for x.

In 11–16, simplify by hand and check using a CAS.

11. $\frac{n^{77}}{n^7}$

12. $60m^6 \cdot \frac{m^3}{6}$

13. $(3x^4)^3$

14. $\frac{w^8}{(w^4)^2}$

15. $v^2 \cdot v^{a-2}$

16. $j^{289} \cdot j^0$

17. How many zeros are at the end of $2^8 \cdot 3^{10} \cdot 5^6$ when written in base 10?

18. A multiple-choice test has 6 pages and 5 questions on each page. If each question can be answered in a different ways, how many ways are there to complete the test?

19. Refer to Example 4. Pluto itself has a mass of approximately 1.3×10^{22} kg. How many times as massive is Pluto than its moon Hydra?

APPLYING THE MATHEMATICS

20. Refer to Example 5. Suppose the farmer decides to split the field into four smaller squares of side length r and irrigate each one separately.
    a. Compute the total area of the four circles.
    b. Compute the percent of the field that is irrigated in the new arrangement.

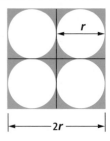

Many farmers irrigate their crops using a center pivot irrigation system.

21. a. Graph $y = (x^3)^2$.
    b. Find a number n such that the graph of $y = x^n$ coincides with the graph in Part a.

22. a. Check that $\left(\frac{x^{12}}{x^{10}}\right)\left(\frac{x}{2}\right)^2 = \frac{x^4}{4}$ by graphing the functions

    $f(x) = \left(\frac{x^{12}}{x^{10}}\right)\left(\frac{x}{2}\right)^2$ and $g(x) = \frac{x^4}{4}$ separately.

    b. Is this check more or less reliable than testing whether the equation is true for a single value of x?

23. The prime factorization of 875 is $5^3 \cdot 7$. The prime factorization of 8575 is $5^2 \cdot 7^3$. Use this information to find the prime factorization of $875 \cdot 8575$.

24. Waneta computed $(40)^3$ and obtained the answer 640. Does her answer end with enough zeros to be correct? How do you know? What is the correct answer?

25. x^3 and x^4 are powers of x whose product is x^7. Find three more pairs of powers of x whose product is x^7.

26. Below is a proof of the Quotient of Powers Theorem. Suppose b is a nonzero real number, and n and m are integers. Fill in each justification.

$$b^{m-n} \cdot b^n = b^{(m-n)+n}$$ 　　a. _____?_____

$$= b^{m+(-n+n)}$$ 　　b. _____?_____

$$= b^{m+0}$$ 　　c. _____?_____

$$b^{m-n} \cdot b^n = b^m$$ 　　d. _____?_____

$$b^{m-n} = \frac{b^m}{b^n}$$ 　　e. _____?_____

In 27–29, write an equivalent expression in which each variable appears once.

27. $\dfrac{18x^4}{(-3x^2)^3}$

28. $\dfrac{8a^2b^6c^8}{(4a)^2b^6c^6}$

29. $\left(\dfrac{8}{w^3}\right)^2 \left(\dfrac{w^2}{4}\right)^3$

REVIEW

30. **Fill in the Blank** The graph of $f(x) = x^n$ has rotation symmetry whenever n is _____?_____. (**Lesson 7-1**)

31. **Fill in the Blank** The graph of $f(x) = x^n$ has reflection symmetry whenever n is _____?_____. (**Lesson 7-1**)

32. How many real roots does the quadratic equation $2p^2 - 4p + 2 = 0$ have? (**Lesson 6-10**)

33. Calculate $\sqrt{-3} \cdot \sqrt{-5} \cdot \sqrt{-15}$. (**Lesson 6-8**)

EXPLORATION

34. a. Copy and fill in the table at the right. What do you notice about the last digits of your answers?

n	1	2	3	4	5	6	7	8	9	10
n^5	?	?	?	?	?	?	?	?	?	?

b. Copy and fill in the table at the right using scientific notation. Round the constant to the nearest tenth.

n	10	20	30	40	50	60	70	80	90	100
n^5	?	?	?	?	?	?	?	?	?	?

c. An old magic trick has a member of the audience pick a number n between 1 and 99, compute n^5, and tell the magician the result. The magician then instantly tells the audience member the original number by using the observation in Part a to determine the number's last digit, and by comparing the size of the number to the values in the second table (which the magician memorizes) to determine the number's first digit. For example, the audience member might say the number 1,419,857. Use the tables above (but no calculator) to determine the original number.

QY ANSWERS

1. Answers vary. Sample: $(a + b)^3 \neq a^3 + b^3$. Let $a = 1, b = 1$. The left side is equal to 8, the right side is equal to 2.

2. 3 zeros

Lesson
7-3
Negative Integer Exponents

▶ **BIG IDEA** The powers x^m and x^{-m} represent reciprocals when they are both defined.

Because of the Quotient of Powers Theorem, for any positive base b and real exponents m and n, or any nonzero base b and integer exponents m and n,

$$\frac{b^m}{b^n} = b^{m-n}.$$

Let us look at this theorem in terms of the values of m and n. If $m > n$, then $m - n$ is positive and b^{m-n} is the product of $(m - n)$ b's. If $m = n$, then $m - n = 0$, and $b^{m-n} = b^0 = 1$. What happens when $m < n$? Then $m - n < 0$, and b^{m-n} has a negative exponent. You have seen negative exponents with base 10 before when writing numbers in scientific notation. For example,

$$10^6 = 1,000,000 = \text{one million}$$
$$10^{-6} = 0.000001 = \text{one millionth.}$$

Note that $1,000,000 \cdot 0.000001 = 1$. This means 10^6 and 10^{-6} are reciprocals. The same kind of relationship holds for bases other than 10.

Activity

MATERIALS CAS

Step 1 What does a CAS display when you enter the following powers of x?
   a. x^{-3} **b.** x^{-8} **c.** x^{-11}

Step 2 Based on the results of Step 1, make a conjecture: $x^{-n} = \underline{\ ?\ }$. Test your conjecture by entering x^{-n} into a CAS. If your calculator does not rewrite x^{-n}, use $\texttt{expand(x\^{}-n)}$ to create an equivalent expression with only positive exponents.

The results of the Activity should be instances of the theorem at the top of the next page.

Negative Exponent Theorem

For any positive base b and real exponent n, or any nonzero base b and integer exponent n, $b^{-n} = \frac{1}{b^n}$.

Proof

1. Suppose $b^{-n} = x$. We want to determine x.

2. $b^n \cdot b^{-n} = b^n \cdot x$ Multiplication Property of Equality
   (Multiply both sides by b^n.)

3. $b^0 = b^n \cdot x$ Product of Powers Postulate

4. $1 = b^n \cdot x$ Zero Exponent Theorem

5. $\frac{1}{b^n} = x$ Divide both sides by b^n
   (which can always be done because $b \neq 0$).

6. Thus $b^{-n} = \frac{1}{b^n}$. Transitive Property of Equality (Steps 1 and 5)

It helps to think of the Negative Exponent Theorem as stating that b^n and b^{-n} are reciprocals. In particular, $b^{-1} = \frac{1}{b}$, so b^{-1} is the reciprocal of b.

GUIDED

Example 1

Write 2^{-5} as a decimal.

Solution By the Negative Exponent Theorem, $2^{-5} = \frac{1}{?}$ as a fraction and __?__ as a decimal.

The Negative Exponent Theorem allows expressions with fractions to be rewritten without them.

Example 2

Critical buckling load is the minimum weight that would cause a column to buckle. Rewrite Euler's formula for critical buckling load, $P = \frac{\pi^2 EI}{L^2}$, using negative exponents and no fractions. (In this formula, E is a constant related to the material used to construct the column, I is the moment of inertia of a cross-section of the column, and L is the length of the column.)

Solution $P = \frac{\pi^2 EI}{L^2}$

$P = \pi^2 EI \frac{1}{L^2}$ Algebraic definition of division

$P = \pi^2 EIL^{-2}$ Negative Exponent Theorem

▶ **READING MATH**

In many math-related fields the term *critical* refers to a point or measurement at which some quality undergoes a drastic change. In the case of critical buckling load, this change is from stable to collapsed.

Properties of Negative Integer Exponents

All the postulates and theorems involving powers stated in Lesson 7-2 hold when the exponents are negative integers.

Example 3

Rewrite each expression as a single power or a single number in scientific notation.

a. $\dfrac{r^5}{r^{12}}$ (Assume $r > 0$.)   b. $4^{-3} \cdot 4^5$ c. $(2 \cdot 10^3)^2 \cdot (1.5 \cdot 10^{-2})$

Solution

a. Use the Quotient of Powers Theorem.

$$\dfrac{r^5}{r^{12}} = r^{\underline{\quad?\quad}-\underline{\quad?\quad}} = r^{\underline{\quad?\quad}}$$

b. Use the Product of Powers Postulate.

$$4^{-3} \cdot 4^5 = 4^{\underline{\quad?\quad}+\underline{\quad?\quad}} = 4^{\underline{\quad?\quad}}$$

c. Use the Power of a Product and Power of a Power Postulates.

$$(2 \cdot 10^3)^2 \cdot (1.5 \cdot 10^{-2}) = (\underline{\quad2\quad} \cdot 10^{\underline{\quad6\quad}}) \cdot (1.5 \cdot 10^{-2})$$

Use the Commutative Property of Multiplication and the Product of Powers Postulate.

$$\underline{\quad?\quad} \cdot 10^{\underline{\quad?\quad}}(10^{-2}) = \underline{\quad?\quad} \cdot 10^{\underline{\quad?\quad}}$$

Check Use a CAS to check whether the original expression and your answer are equivalent.

Caution: A negative number in an exponent does not make the value of an expression negative. All powers of positive numbers are positive.

Example 4

Rewrite $(3c)^{-3}(2c)^4$ as a fraction with the variable c appearing only once.

Solution 1

$$
\begin{aligned}
(3c)^{-3}(2c)^4 &= 3^{-3} \cdot c^{-3} \cdot 2^4 \cdot c^4 && \text{Power of a Product Postulate} \\
&= 3^{-3} \cdot 2^4 \cdot c^{-3} \cdot c^4 && \text{Commutative Property of Multiplication} \\
&= 3^{-3} \cdot 2^4 \cdot c && \text{Product of Powers Postulate} \\
&= \tfrac{1}{3^3} \cdot 2^4 \cdot c && \text{Negative Exponent Theorem} \\
&= \tfrac{16c}{27}
\end{aligned}
$$

Solution 2

$$(3c)^{-3}(2c)^4 = \frac{1}{(3c)^3} \cdot (2c)^4 \qquad \text{Negative Exponent Theorem}$$

$$= \frac{(2c)^4}{(3c)^3} \qquad \text{Multiply.}$$

$$= \frac{2^4 \cdot c^4}{3^3 \cdot c^3} \qquad \text{Power of a Product Postulate}$$

$$= \frac{2^4 c}{3^3} \qquad \text{Quotient of Powers Postulate}$$

$$= \frac{16c}{27}$$

Questions

COVERING THE IDEAS

In 1 and 2, write as a single power.

1. $\frac{t^3}{t^{10}}$

2. $\frac{6^6}{6^7}$

In 3 and 4, write an equivalent expression without negative exponents.

3. x^{-4}

4. 3^{-7}

5. Solve for n: $8^{-7} = \frac{1}{8^n}$.

6. If $p^{-1} = \frac{4}{13}$, find p.

7. Write as a whole number or a fraction without an exponent.
   a. 1.8^0
   b. 1.8^{-1}
   c. 1.8^{-2}

8. Write as a fraction without an exponent.
   a. 9^{-1}
   b. 9^{-3}
   c. 9^{-5}

9. Write 8^{-3}
   a. as a decimal.
   b. in scientific notation.

10. The time t it takes you to read a book is inversely proportional to the number p of pages in the book. Let k be the constant of variation.
    a. Write an inverse variation equation to represent this situation using positive exponents.
    b. Rewrite your inverse variation equation using negative exponents.

In 11–13, write an equivalent expression without negative exponents.

11. $\frac{10y}{15y^{-2}}$

12. $(3e^2)^{-3}(5e)^4$

13. $\frac{2a^{-3}}{b^4}$

In 14–16, write an equivalent expression without an exponent.

14. $10^4 \cdot 10^{-5}$

15. $1.982^0 \cdot 1^{47}$

16. $3^{-3} \cdot \left(\frac{1}{3}\right)^{-1}$

17. **Multiple Choice** If $b > 0$, for what integer values of n is $b^n < 0$?

    A $n < 0$ **B** $0 < n < 1$

    C all values of n **D** no values of n

18. If $y^{-6} = 9$, what is the value of y^6?

APPLYING THE MATHEMATICS

In 19 and 20, rewrite the right side of the formula using negative exponents and no fractions.

19. $A = \dfrac{kh}{g^5}$ 20. $T = \dfrac{k}{m^2 n^5}$

21. Write an expression equivalent to $\left(\dfrac{3x^{-3}y}{2z^{-5}}\right)^{-2}\left(\dfrac{x^4 y^{-3}}{z^3}\right)^8$ without any negative exponents.

22. Write each expression in standard scientific notation $x \cdot 10^n$, where $1 \le x < 10$ and n is an integer.

    a. $(3 \cdot 10^{-4})^2$ **b.** $(6.2 \cdot 10^{-6})(4.6 \cdot 10^9)$

23. Carbon dioxide is a colorless and odorless gas that absorbs radiation from the sun and contributes to global warming. Before the industrial age, carbon dioxide made up about 280 parts per million (ppm) of the atmosphere. Carbon dioxide is released when coal, gasoline, and other fossil fuels are burned, so the level of the gas in the atmosphere has been increasing. Some scientific models have predicted that by 2100, the concentration will range from 490 ppm to 1260 ppm.

    a. Write all three rates as fractions using positive powers of 10.

    b. Write each rate in scientific notation.

The molecular structure of carbon dioxide shows that one carbon atom in the center is flanked by two oxygen atoms.

24. Benjamin Franklin was one of the most famous scientists of his day. In one experiment he noticed that oil dropped on the surface of a lake would not spread out beyond a certain area. In modern units, he found that 0.1 cm^3 of oil spread to cover about 40 m^2 of the lake. About how thick is such a layer of oil? Express your answer in scientific notation. (We now know that the layer of oil stops spreading when it is one molecule thick. Although in Franklin's time no one knew about molecules, Franklin's experiment resulted in the first estimate of a molecule's size.)

In 25 and 26, write an equivalent expression without a fraction.

25. $\dfrac{4x^2}{y^3}$ 26. $\dfrac{12a^4}{19b^5 c^2}$

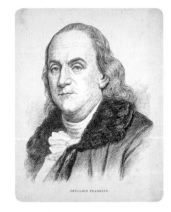

Benjamin Franklin

REVIEW

In 27 and 28, write without a fraction. (Lesson 7-2)

27. $\left(\dfrac{a^3}{a^2}\right)^3 \cdot 20a$ 28. $\dfrac{(xy)^m y^n}{x^2}$

29. **Multiple Choice** Which of the following could be the graph of $y = x^4 + 1$? Justify your answer. (**Lessons 7-1, 6-3**)

A

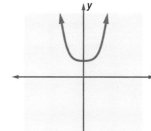

B

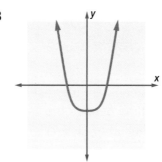

C

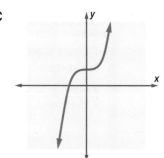

D

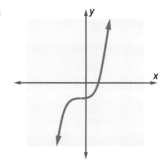

30. A bowler needs 12 strikes in a row to bowl a perfect game of 300.
    a. If the probability that Khadijah gets a strike is $\frac{1}{8}$ and strikes are independent of each other, what is the probability that Khadijah's next game will be a perfect game?
    b. Is your answer to Part a greater than or less than one billionth? (**Lesson 7-1**)

31. a. What are the domain and range of the function with equation $y = -\frac{4}{x^2}$?
    b. What are the domain and range of the function with equation $y = -\frac{4}{x}$? (**Lesson 2-6**)

EXPLORATION

32. a. Examine the table at the right closely. Describe two patterns relating the powers of 5 on the left to the powers of 2 on the right.
    b. Make a chart similar to the one in Part a using the powers of 4 and of 2.5.
    c. Describe how the patterns in the chart from Part b are similar to the patterns in Part a.
    d. Find another pair of numbers with the same properties.

Powers of 5	Powers of 2
$5^6 = 15{,}625$	$2^6 = 64$
$5^5 = 3125$	$2^5 = 32$
$5^4 = 625$	$2^4 = 16$
$5^3 = 125$	$2^3 = 8$
$5^2 = 25$	$2^2 = 4$
$5^1 = 5$	$2^1 = 2$
$5^0 = 1$	$2^0 = 1$
$5^{-1} = 0.2$	$2^{-1} = 0.5$
$5^{-2} = 0.04$	$2^{-2} = 0.25$
$5^{-3} = 0.008$	$2^{-3} = 0.125$
$5^{-4} = 0.0016$	$2^{-4} = 0.0625$
$5^{-5} = 0.00032$	$2^{-5} = 0.03125$
$5^{-6} = 0.000064$	$2^{-6} = 0.015625$

Lesson

7-4

Compound Interest

▶ **BIG IDEA** If money grows at a constant interest rate r in a single time period, then after n time periods the value of the original investment has been multiplied by $(1 + r)^n$.

Interest Compounded Annually

Penny Wise, a high school junior, works part-time during the school year and full-time during the summer. Suppose that Penny decides to deposit $3000 this year in a 3-year certificate of deposit (CD). The investment is guaranteed to earn interest at a yearly rate of 4.5%. The interest is added to the account at the end of each year. If no money is added or withdrawn, then after one year the CD will have the original amount invested, plus 4.5% interest.

amount after 1 year:
$$3000 + 0.045(3000) = 3000(1 + 0.045)^1$$
$$= 3000(1.045)$$
$$= 3135$$

Penny's CD is worth $3135 after one year.

Notice that to find the amount after 1 year, you do not have to add the interest separately; you can just multiply the original amount by 1.045. Similarly, at the end of the second year, there will be 1.045 times the *balance* (the ending amount in the account) from the first year.

amount after 2 years:
$$3000(1.045)(1.045) = 3000(1.045)^2$$
$$\approx 3276.075$$

Since banks round down, Penny's CD is worth $3276.07 after two years.

amount after 3 years:
$$3000(1.045)^2(1.045) = 3000(1.045)^3$$
$$\approx 3423.49$$

The value of Penny's CD has grown to $3423.49 after three years. Notice the general pattern.

amount after t years: $\qquad 3000(1.045)^t$

When interest is earned at the end of each year, it is called *annual compound interest*. To find a more general formula for interest, replace 4.5% with *r*, the annual interest rate, and 3000 with *P*, the **principal** or original amount invested.

> **Annual Compound Interest Formula**
>
> Let *P* be the amount of money invested at an annual interest rate *r* compounded annually. Let *A* be the total amount after *t* years. Then
> $$A = P(1 + r)^t.$$

In the Annual Compound Interest Formula, notice that *A* varies directly as *P*. For example, doubling the principal doubles the amount at the end. However, *A* does not vary directly as *r*; doubling the rate does not necessarily double the amount earned.

 QY

Interest Compounded More Than Once a Year

In most savings accounts, interest is compounded more than once a year. If money is compounded **semi-annually**, the interest rate at each compounding is *half of the annual interest rate* but there are *two compoundings each year* instead of just one. So if your account pays 4.5% compounded semi-annually, you earn 2.25% on the balance every six months. At the end of *t* years, interest paid semi-annually will have been paid 2*t* times. Therefore, the compound interest formula becomes
$$A = P\left(1 + \frac{r}{2}\right)^{2t}.$$

If money is compounded *quarterly*, the compound interest formula becomes
$$A = P\left(1 + \frac{r}{4}\right)^{4t}.$$

This pattern leads to a general compound interest formula.

> **General Compound Interest Formula**
>
> Let *P* be the amount invested at an annual interest rate *r* compounded *n* times per year. Let *A* be the amount after *t* years. Then
> $$A = P\left(1 + \frac{r}{n}\right)^{nt}.$$

The number of times that the interest is compounded makes a difference in the amount of interest earned.

> ▶ **QY**
>
> How much more money would Penny earn if she were able to earn 9% rather than the 4.5% for 3 years?

Example 1

Suppose $10,000 is placed into an account that pays interest at a rate of 5%. How much will be earned in the account in the first year if the interest is compounded as indicated?

a. annually   b. semi-annually   c. quarterly

Solution

a. Since interest is compounded only once, the interest is simply $0.05 \cdot \$10{,}000 = \$500$. The account will earn $500.

b. and c. Substitute into the General Compound Interest Formula to determine the account's value.

For Part b, $P = \$10{,}000$;
$r = 5\%$, $n = 2$, and $t = 1$ year.

$$A = P\left(1 + \frac{r}{n}\right)^{nt}$$
$$= \$10{,}000\left(1 + \frac{0.05}{2}\right)^{2 \cdot 1}$$
$$= \$10{,}000(1.025)^2$$
$$= \$10{,}000(1.050625)$$
$$= \$10{,}506.25$$

For Part c, $P = \$10{,}000$;
$r = 5\%$, $n = 4$, and $t = 1$ year.

$$A = P\left(1 + \frac{r}{n}\right)^{nt}$$
$$= \$10{,}000\left(1 + \frac{0.05}{4}\right)^{4 \cdot 1}$$
$$= \$10{,}000(1.0125)^4$$
$$\approx \$10{,}000(1.050945)$$
$$\approx \$10{,}509.45$$

Now subtract the $10,000 principal to find the amount of interest that was earned.

The account will earn $506.25.

The account will earn $509.45.

In Example 1, the difference after one year between compounding semi-annually and compounding quarterly is only $3.20. However, if you withdraw your money before a year is up, you may have received interest in the account that pays quarterly while you may not have received interest in the account that pays semi-annually. For instance, if interest is compounded quarterly and you withdraw your money after 10 months, you will have received 3 of the 4 quarterly compound interest payments and have a total of $10{,}000\left(1 + \frac{0.05}{4}\right)^{3 \cdot 1} \approx \$10{,}379.70$.

However, if interest is compounded semi-annually, then after 10 months you will have received 1 of 2 semi-annual compound interest payments and have only $10{,}000\left(1 + \frac{0.05}{2}\right)^{1 \cdot 1} = \$10{,}250$, a difference of over $125!

To avoid angering their customers, most savings institutions guarantee that accounts will earn interest "from the date of deposit until the date of withdrawal." They can do this by **compounding daily**. Daily compounding uses either 360 or 365 as the number of days in a year.

Annual Percentage Yield

Because of the many different ways of calculating interest, savings institutions are required by federal law to disclose the **annual percentage yield**, or **APY**, of an account after all the compoundings for a year have taken place. This allows consumers to compare savings plans. For instance, to determine the APY of an account paying 5% compounded quarterly (as in Example 1), find the interest $1 would earn in the account in one year.

$$1 \cdot \left(1 + \frac{0.05}{4}\right)^{4 \cdot 1} \approx 1.0509$$

So the interest earned is $1.0509 − $1 = $0.0509. This means that the APY on an account paying 5% compounded quarterly is 5.09%.

> **GUIDED**
>
> ### Example 2
> What is the APY for a 5.5% interest rate compounded
>
> a. quarterly?
> b. daily, for 365 days per year?
>
> **Solution** To find the APY on an account, use $1 as the principal amount to keep the computations simple.
>
> a. $1 \cdot \left(1 + \frac{?}{4}\right)^{4 \cdot 1} \approx$ __?__
> So, the interest earned is $__?__ − $1 = $__?__.
> This is an APY of __?__%.
>
> b. $1 \cdot \left(1 + \frac{?}{?}\right)^{?} \approx$ __?__
> So, the interest earned is $__?__ − $1 = $__?__.
> This is an APY of __?__%.

Going Back in Time

In both compound interest formulas, you can think of P either as the principal or as the *present amount*. In each of the previous examples, A is an amount that is determined after compounding. Then, because A comes after P, the time t is represented by a positive number. But it is also possible to think of A as an amount some years ago that was compounded to get the present amount P. Then the time t is represented by a negative number.

Example 3

Zero-coupon bonds do not pay interest during their lifetime, typically 20 or 30 years. They are bought for much less than their final value and earn a fixed rate of interest over their life. When a bond *matures*, its value is equal to the initial investment plus all the interest earned over its lifetime. Suppose a 30-year zero-coupon bond has a value at maturity of $20,000 and is offered at 5.5% interest compounded semi-annually. How much do you need to invest to buy this bond?

Series EE savings bonds are sold at face value and earn a fixed interest determined at purchase for up to 30 years. Benjamin Franklin appears on the $1,000 Series EE bond.

Solution 1 Think of how much you would need to have invested 30 years ago to have $20,000 now. Use the General Compound Interest Formula with a present value of $P = 20,000$, $r = 0.055$, $n = 2$, and $t = -30$.

$$A = P\left(1 + \frac{r}{n}\right)^{nt}$$ General Compound Interest Formula

$$A = 20,000\left(1 + \frac{0.055}{2}\right)^{2 \cdot -30}$$ Substitution

$$A = \frac{20,000}{1.0275^{60}}$$ Arithmetic and Negative Exponent Theorem

$$A \approx 3927.54$$ Arithmetic

You need to invest $3927.54 to buy this bond.

Solution 2 Use the General Compound Interest Formula. You know $A = 20,000$, $r = 0.055$, $n = 2$, and $t = 30$. Solve for P.

$$A = P\left(1 + \frac{r}{n}\right)^{nt}$$ General Compound Interest Formula

$$20,000 = P\left(1 + \frac{0.055}{2}\right)^{2 \cdot 30}$$ Substitution

$$20,000 \approx P(1.0275)^{60}$$ Arithmetic

$$P \approx 3927.54$$ Divide both sides by 1.0275^{60}

You need to invest $3927.54 to buy this bond.

Questions

COVERING THE IDEAS

1. Suppose Penny Wise buys $2500 worth of government bonds that pay 3.7% interest compounded quarterly. If no money is added or withdrawn, find out how much the bonds will be worth after 1, 2, 3, 4, and 5 years.

2. Find the interest earned in the fourth year for Penny's 4.5% CD described in this lesson.

3. Write the compound interest formula for an account that earns interest compounded
   a. monthly.
   b. daily in a leap year.

4. To what amount will $8000 grow if it is invested for 12 years at 6% compounded quarterly?

5. Find the APY of a savings account earning 4% interest compounded daily. Use 360 for the number of days in a year.

6. Suppose a zero-coupon bond matures and pays the owner $30,000 after 10 years, paying 4.5% interest annually. How much was invested 10 years ago?

7. **True or False** An account earning 8% compounded annually earns exactly twice as much interest in 6 years as an account earning 8% compounded annually earns in 3 years. Explain your answer.

8. **True or False** Justify your answer. In the General Compound Interest Formula,
   a. A varies directly as t.
   b. A varies directly as P.
   c. A varies inversely as n.
   d. A varies directly as r.

APPLYING THE MATHEMATICS

9. Refer to Penny's certificate of deposit.
   a. Calculate the value of Penny's CD at the end of 3 years using the **simple interest formula** $I = Prt$, where I is the amount of interest earned, P is the principal, r is the annual percentage rate, and t is time in years.
   b. How much money would Penny have if she earned annual compound interest over the same 3 years?
   c. Should Penny prefer simple interest to annual compound interest? Explain why or why not.

10. On a CAS, define a function `gencompint(p,r,n,t)` to calculate the value of an investment using the General Compound Interest Formula. Use the function to verify the answers to Example 1.

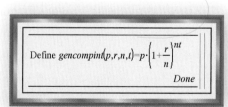

Define $gencompint(p,r,n,t) = p \cdot \left(1 + \dfrac{r}{n}\right)^{nt}$

Done

11. Rich takes a $2000 cash advance against his credit card to fund an investment opportunity he saw on the Internet. The credit card charges an annual rate of 18% and compounds the interest monthly on all cash advances. How much interest does Rich owe if he does not make any payments for 3 months?

12. Stores often advertise a "90-day same as cash" method for making purchases. This means that interest is compounded starting the day of purchase (the interest *accrues*), but no interest is charged if the bill is paid in full before 90 days go by. However, on the 91st day, the accrued interest is added to the purchase price. Suppose Manny purchased a sofa for $3000 under this plan with an annual interest rate of 20.5% compounded monthly. If Manny forgets to pay the purchase in full, how much will he owe on the 91st day?

Many stores offer 90 days same as cash during sales.

REVIEW

13. Rewrite $\left(\frac{1}{3^{-3}}\right)^{-2}$ without exponents. (**Lesson 7-3**)

14. The Stefan-Boltzmann constant in physics is $\frac{2\pi^5 k^4}{15h^3 c^2}$. Rewrite this expression in the form $\frac{2}{15}M$, where M is an expression that does not involve fractions. (**Lesson 7-3**)

15. **True or False** For all positive integers a, b, and c, $(a^b)^c = a^{(b^c)}$. Justify your answer. (**Lesson 7-2**)

16. Solve $x^2 - 3x + 7 = 2x^2 + 5x - 9$ by
    a. graphing.
    b. using the Quadratic Formula. (**Lessons 6-7, 6-4**)

17. What translation maps the graph of $y = \sqrt{x - 3} + 5$ onto the graph of $y = \sqrt{x} - 3$? (**Lesson 6-3**)

18. The graph at the right appears to have a certain symmetry. Give a matrix describing a transformation that would map this graph onto itself. (**Lesson 4-8**)

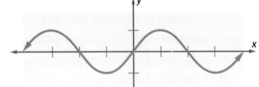

19. Consider the sequence t_n defined recursively as follows. (**Lessons 3-6, 1-8**)
$$\begin{cases} t_1 = 4 \\ t_n = 3t_{n-1} \text{ for } n \geq 2 \end{cases}$$
    a. Find the first five terms of the sequence.
    b. Is t an arithmetic sequence? Why or why not?

EXPLORATION

20. Find an interest rate and an APY for a 3-year CD at a savings institution in your area. Show how to calculate the APY from the interest rate.

QY ANSWER

$461.59

Lesson
7-5
Geometric Sequences

Vocabulary

geometric sequence,
  exponential sequence

constant multiplier

constant ratio

▶ **BIG IDEA** Geometric sequences are generated by multiplying each term by a constant to get the next term, just as arithmetic sequences are generated by adding a constant to each term to get the next.

In 2005, a major television manufacturer filmed an advertisement featuring 250,000 balls bouncing down the hills of San Francisco. The maximum height of a ball after each bounce can be modeled by a *geometric sequence*.

Recursive Formulas for Geometric Sequences

Recall that, in an arithmetic (linear) sequence, each term after the first is found by adding a constant difference to the previous term. If, instead, each term after the first is found by *multiplying* the previous term by a constant, then a **geometric** (or **exponential**) **sequence** is formed.

The constant in a geometric sequence is called a **constant multiplier**. For instance, the geometric sequence with first term 30 and constant multiplier 2 is

$$30, 60, 120, 240, 480, 960, \ldots .$$

Replace 30 by g_1 and 2 by r and you have the general form for a geometric sequence.

Mental Math

Tell whether the sequence could be arithmetic. If it could be, give the constant difference.

a. $1\frac{5}{6}, \frac{5}{6}, -\frac{5}{6}, -1\frac{5}{6}, -2\frac{5}{6}, \ldots$

b. $17, 28, 39, 50, \ldots$

c. $1.017, 0.917, 0.817, 0.717, \ldots$

d. $2, -4, 6, -8, \ldots$

Recursive Formula for a Geometric Sequence

Let r be a nonzero constant. The sequence g defined by the recursive formula

$$\begin{cases} g_1 = x \\ g_n = rg_{n-1}, \text{ for integers } n \geq 2 \end{cases}$$

is the geometric, or exponential, sequence with first term x and constant multiplier r.

Solving the sentence $g_n = rg_{n-1}$ for r yields $\frac{g_n}{g_{n-1}} = r$. This indicates that in a geometric sequence, the ratio of successive terms is constant. For this reason, the constant multiplier r is also called the **constant ratio**.

Alternatively, you can write a recursive formula for a geometric sequence g using the $(n + 1)$st term as

$$\begin{cases} g_1 = x \\ g_{n+1} = rg_n, \text{ for integers } n \geq 1. \end{cases}$$

Example 1

Give the first six terms and the constant multiplier of the geometric sequence g where

$$\begin{cases} g_1 = 6 \\ g_n = 3g_{n-1}, \text{ for integers } n \geq 2. \end{cases}$$

Solution 1 The value $g_1 = 6$ is given. The rule for g_n tells you that each term after the first is found by multiplying the previous term by 3. The constant multiplier is 3.

$$g_2 = 3g_1 = 3 \cdot 6 = 18$$
$$g_3 = 3g_2 = 3 \cdot 18 = 54$$
$$g_4 = 3g_3 = 3 \cdot 54 = 162$$
$$g_5 = 3g_4 = 3 \cdot 162 = 486$$
$$g_6 = 3g_5 = 3 \cdot 486 = 1458$$

The first six terms of the sequence are 6, 18, 54, 162, 486, 1458.

Solution 2 Use a spreadsheet.

Enter 6 in cell A1. Then enter $= 3*A1$ in cell A2.

Copy and paste cell A2 into cells A3–A6.

	A	B	C	D	E	F
1	6					
2	18					
3	54					
4	162					
5	486					
6	1458					

A6 $= 3 \cdot a5$

 QY

Explicit Formulas for Geometric Sequences

You may have received a letter or e-mail that promises good luck as long as you send the letter to five friends asking each to forward it to five of their friends, and so on. Such chain letters are illegal in the U.S. if the mailer asks for money.

> ▶ **QY**
>
> Write a recursive formula for the geometric sequence in Example 1 using g_{n+1}.

Part of the appeal of chain letters is that a very large number of people can receive them quickly. This is because the number of letters sent by each *generation* of mailers forms a geometric sequence. The data at the right represent the first five generations of a chain letter in which a person sends an e-mail to 12 people and asks each person receiving the e-mail to forward it to 4 other friends, and no person receives two letters.

Generation	Number of Letters Sent
1	12
2	$12 \cdot 4 = 48$
3	$12 \cdot 4^2 = 192$
4	$12 \cdot 4^3 = 768$
5	$12 \cdot 4^4 = 3072$

In this sequence, the constant multiplier is $r = 4$. If $g_1 = 12$, the number of letters the first person sends out, then g_2 is the number of letters sent out by everyone in generation 2, and $g_n = 12(4)^{n-1}$ is the number of letters sent out in generation n. This pattern can be generalized to find an explicit formula for the nth term of any geometric sequence.

Explicit Formula for a Geometric Sequence

In the geometric sequence g with first term g_1 and constant ratio r,
$$g_n = g_1(r)^{n-1}, \text{ for integers } n \geq 1.$$

Notice that in the explicit formula, the exponent of the nth term is $n - 1$. When you substitute 1 for n to find the first term, the constant multiplier has an exponent of zero.

$$g_1 = g_1(r)^{1-1}$$
$$= g_1 r^0$$

This is consistent with the Zero Exponent Theorem which states that for all $r \neq 0$, $r^0 = 1$.

Constant multipliers in a geometric sequence can be negative. Then the terms of the sequence alternate between positive and negative values.

GUIDED

Example 2

Write the 1st, 5th, 10th, 35th, and 50th terms of the sequence a defined by $a_n = 4(-3)^{n-1}$.

Solution Substitute $n = 1, 5, 10, 35,$ and 50 into the formula for the sequence.

$$a_1 = 4(-3)^{1-1} = 4 \cdot \underline{} = \underline{}$$
$$a_5 = 4(-3)^{\underline{?}} = 4 \cdot \underline{} = \underline{}$$
$$a_{10} = \underline{} = \underline{} = -78{,}732$$

A calculator display of a_{35} and a_{50} is shown at the right.

$4 \cdot (-3)^{34}$	66708726798666276
$4 \cdot (-3)^{49}$	-957197316922470118360332

Constant multipliers in a geometric sequence can also be between 0 and 1. This is the case for the sequence that models the height of a bouncing ball.

Example 3

Suppose a ball is dropped from a height of 10 meters, and it bounces up to 80% of its previous height after each bounce. (A bounce is counted when the ball hits the ground.) Let h_n be the maximum height of the ball after the nth bounce.

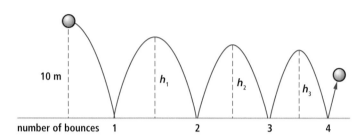

a. Find an explicit formula for h_n.
b. Find the maximum height of the ball after the seventh bounce.

Solution

a. Because each term is 0.8 times the previous term, the sequence is geometric. On the first bounce, the ball bounces up to 10(0.8) meters, or 8 meters. So, $h_1 = 8$. Also, $r = 0.8$. So,
$$h_n = 8(0.8)^{n-1}.$$

b. On the seventh bounce, $n = 7$, so
$$h_7 = 8(0.8)^{7-1} \approx 2.10.$$
So, after the seventh bounce, the ball will rise to a height of about 2.10 meters.

Look back at the sequences generated in Examples 1 to 3. Notice that in Example 1, $r > 1$ and g_n increases as n increases. In Example 3, $0 < r < 1$ and g_n decreases as n increases. In Example 2, $r < 0$ and as n increases, g_n alternates between positive and negative values. These properties are true for all geometric sequences.

Questions

COVERING THE IDEAS

1. **Fill in the Blanks** In an arithmetic sequence, each term after the first is found by ___?___ a constant to the previous term. In a geometric sequence, each term after the first is found by ___?___ the previous term by a constant.

In 2–4, state whether the numbers can be consecutive terms of a geometric sequence. If they can, find the constant ratio and write an explicit formula.

2. 12, 48, 192, ...
3. 5, 10, 20, 25, ...
4. $12, \frac{4}{3}, \frac{1}{6}, \frac{1}{48}$ ...

5. Let g be a sequence with $g_n = 200 \cdot (0.10)^{n-1}$.

   a. Find terms 1, 5, 20, and 50.

   b. Use a spreadsheet to generate the first 15 terms of g.

6. Find the first five terms of the sequence
$$\begin{cases} t_1 = 1.2 \\ t_n = 5 \cdot t_{n-1}, \text{ for integers } n \ge 2. \end{cases}$$

7. Consider $\begin{cases} g_1 = x \\ g_{n+1} = rg_n, \text{ for integers } n \ge 1 \end{cases}$.

   a. Write r in terms of g_{n+1} and g_n.

   b. Write an explicit formula for g_{n+1}.

8. a. Write the first six terms of the geometric sequence whose first term is –3 and whose constant ratio is –2.

   b. Give a recursive formula for the sequence in Part a.

   c. Give an explicit formula for the sequence in Part a.

9. Suppose $g_n = 1.85 \cdot 0.38^{n-1}$ for integers $n \ge 1$.

   a. Find each ratio.

   i. $\dfrac{g_2}{g_1}$ ii. $\dfrac{g_3}{g_2}$ iii. $\dfrac{g_4}{g_3}$ iv. $\dfrac{g_{20}}{g_{19}}$

   b. What is true about the values in Part a?

10. **Matching** Each graph below is a graph of a geometric sequence. Match each graph with its range of possible common ratios r.

   i. $r > 1$ ii. $r < 0$ iii. $0 < r < 1$

   a.

   b.
   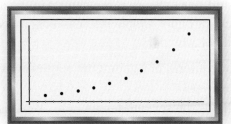

   c.
   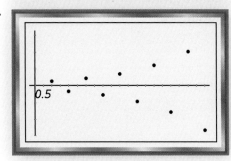

11. Suppose a ball dropped from a height of 11 feet bounces up to 60% of its previous height after each bounce.

   a. Find an explicit formula for the maximum height of the ball after the nth bounce.

   b. Find the height of the ball, to the nearest inch, after the tenth bounce.

12. In the figure at the right, the midpoints of the sides of the largest equilateral triangle have been connected to create the next smaller equilateral triangle, and this process has been continued. If the side length of the largest triangle is s, then the side length of the next smaller triangle is $\frac{1}{2}s$. The sides of the consecutively smaller triangles form a geometric sequence.

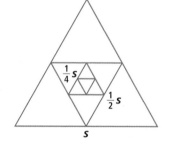

   a. What is the first term of the sequence?

   b. What is the constant multiplier?

APPLYING THE MATHEMATICS

13. Willie Savit invested \$5000 in an account at 2.25% interest compounded annually.

   a. Write an explicit formula for how much money Willie will have in his account.

   b. How much money will Willie have after 20 years?

   c. If Willie had received simple interest instead of compound interest, the value of his investment over the first five years would be as shown at the right. Do the amounts in his account under simple interest form a geometric sequence? Why or why not?

Year	Amount ($)
0	$5000.00
1	$5112.50
2	$5225.00
3	$5337.50
4	$5450.00

14. The fourth term of a geometric sequence is 1. The constant multiplier is $\frac{1}{5}$.

   a. What is the seventh term?   **b.** What is the first term?

In 15 and 16, the first few terms of a geometric sequence are given. Find the next two terms. Then, find an explicit formula for the nth term of the geometric sequence.

15. 14, 28, 56, 112, ...   16. 9, 3, 1, ...

17. **a.** Write a recursive formula for the geometric sequence whose first four terms are 23, –23, 23, –23, ...

   b. Write an explicit formula for the sequence in Part a.

18. y, xy^2, x^2y^3, x^3y^4, and x^4y^5 are the first five terms of a sequence s.

   a. Find each ratio.

     i. $\dfrac{s_2}{s_1}$ **ii.** $\dfrac{s_3}{s_2}$ **iii.** $\dfrac{s_4}{s_3}$ **iv.** $\dfrac{s_5}{s_4}$

   b. Could s be a geometric sequence? Explain why or why not.

19. In the figure at the right, the midpoints of the sides of each square are connected to form the next smaller square. The ratios of the areas of consecutive shaded regions are equal. This is called a Baravelle Spiral. Assume that the side of the largest square has length 1 unit.

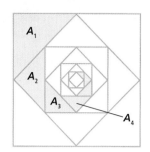

a. Find the area of region A_1.

b. The areas of the consecutively smaller and smaller regions $A_2, A_3, A_4, \ldots$ form a geometric sequence A. Find an explicit formula for A_n.

REVIEW

20. A bank account has an annual interest rate of 3.8% compounded daily. **(Lesson 7-4)**

a. If $10,000 is placed in this account, calculate the value of the account 5 years from now.

b. What is the annual percentage yield of this account?

c. If $10,000 is placed into an account that compounds interest annually at a rate equal to the annual percentage yield in Part b, what will the value of this new account be in 5 years? How does this amount compare to your answer to Part a?

21. Suppose N varies directly as a and inversely as the square of b and the third power of c. Using negative exponents, write a formula for N that does not contain a fraction. **(Lessons 7-3, 2-8)**

22. Rewrite $y = x^2 - 46x + 497$ in vertex form. **(Lesson 6-5)**

23. A circle and a rectangle have the same area. The circumference of the circle is 20 cm and one side of the rectangle equals the diameter of the circle. Determine whether the perimeter of the rectangle is longer or shorter than the circle's circumference. **(Lesson 6-2)**

EXPLORATION

24. a. Refer to the chain letter example from the lesson. The number of e-mails actually sent after the first generation seldom matches the numbers in the table. Why is that?

b. Suppose only 2 of 4 people who receive an e-mail pass it along to 4 others. What, then, is the minimum number of generations it will take for the e-mail to reach 100 million people?

Lesson

7-6

nth Roots

Vocabulary

cube root

nth root

▶ **BIG IDEA** If y is the nth power of x, then x is an nth root of y. Real numbers may have 0, 1, or 2 real nth roots.

Geometric Sequences in Music

A piano tuner adjusts the tension on the strings of a piano so the notes are at the proper pitch. Pitch depends on frequency, measured in hertz or cycles per second. It is common today to tune the A above middle C to 440 hertz. Pythagoras and his followers discovered that a note has exactly half the frequency of the note one octave higher. Thus, the A below middle C is tuned to a frequency of 220 hertz. In most music today, an octave is divided into twelve notes as shown below.

Mental Math

Determine whether the discriminant of the quadratic equation is positive, negative, or zero. Then, tell how many real roots the equation has.

a. $5x^2 + 2x - 1 = 0$

b. $-t^2 - 4t + 1 = 0$

c. $2h^2 + 3h + 2 = 0$

d. $s^2 - 6s + 9 = 0$

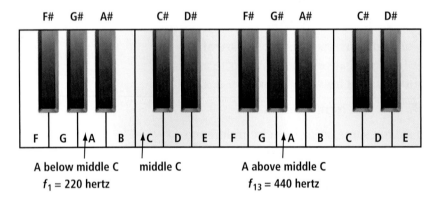

A below middle C
$f_1 = 220$ hertz

middle C

A above middle C
$f_{13} = 440$ hertz

In order for a musical piece to sound much the same in any key, notes in the scale are tuned so that ratios of the frequencies of consecutive notes are equal. To find these frequencies, let $f_1 = $ the frequency of the A below middle C and $f_n = $ the frequency of the nth note in the scale. Then f_{n+1} is the frequency of the next higher note. The frequency of A *above* middle C is f_{13} because there are 12 notes in each octave.

Let $r = $ the ratio of the frequencies of consecutive notes. Then for all integers $n \geq 1$,

$$\frac{f_{n+1}}{f_n} = r.$$

Multiply both sides of this equation by f_n.

$$f_{n+1} = rf_n, \text{ for all integers } n \geq 1$$

Together with the known value $f_1 = 220$, this is a recursive formula for a geometric sequence. It indicates that the frequencies of consecutive notes on a piano (when in tune) are the elements of a geometric sequence. The first term is $f_1 = 220$ and the 13th term of this sequence is $f_{13} = 440$.

From the explicit formula for the nth term of a geometric sequence, substituting 13 for n,

$$f_{13} = 220r^{13-1} = 220r^{12}.$$

To find r, substitute 440 for f_{13}. $\qquad 440 = 220r^{12}$

Divide both sides by 220. $\qquad 2 = r^{12}$

The ratio of the frequencies of consecutive in-tune keys on a piano is called a *12th root* of 2.

A piano being tuned

What Is an *n*th Root?

Recall that x is a square root of t if and only if $x^2 = t$. Similarly, x is a **cube root** of t if and only if $x^3 = t$. For instance, 4 is a cube root of 64 because $4^3 = 64$. Square roots and cube roots are special cases of the following more general idea.

> ### Definition of *n*th Root
>
> Let n be an integer greater than 1. Then b is an **nth root** of x if and only if $b^n = x$.

For example, $-\frac{1}{3}$ is a 5th root of $-\frac{1}{243}$ because $\left(-\frac{1}{3}\right)^5 = -\frac{1}{243}$.

There are no special names for nth roots other than *square roots* (when $n = 2$) and *cube roots* (when $n = 3$). Other nth roots are called *fourth roots, fifth roots*, and so on. In the piano-tuning example above, $r^{12} = 2$. This is why r is a 12th root of 2. And because $(-r)^{12} = r^{12}$ for all real numbers r, there is a negative number that is also a 12th root of 2.

Example 1

Approximate the real 12th roots of 2 to find the ratio r of the frequencies of consecutive in-tune notes, to the nearest hundred-thousandth.

Solution The real 12th roots of 2 are the real solutions to $x^{12} = 2$. So they are the x-coordinates of the points of intersection of $y = x^{12}$ and $y = 2$. Graph these functions using a graphing utility.

(continued on next page)

From the graph at the right, you can see that that there are two real 12th roots of 2. This calculator shows that the real 12th roots of 2 are approximately –1.05946 and 1.05946. Only the positive root has meaning in this context. So, the ratio r of the frequencies of consecutive in-tune keys is about 1.05946.

Check Use a CAS in real-number mode to solve $x^{12} = 2$.

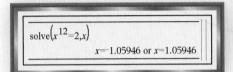

$$\text{solve}\left(x^{12}=2,x\right)$$
$$x=-1.05946 \text{ or } x=1.05946$$

The positive value $x = 1.05946$ checks.

 QY1

▶ QY1

If the A below middle C is tuned to 220 cycles per second, find the frequency of the D above middle C, the 5th note above this A.

How Many Real *n*th Roots Does a Real Number Have?

The number of real *n*th roots of a real number k is the number of points of intersection of the line $y = k$ with the power function $y = x^n$. The number of intersections is determined by whether the value of n is odd or even, and whether the real number k is positive or negative, as illustrated in the graphs below.

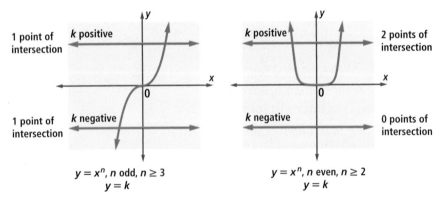

Each intersection determines an *n*th root of k, suggesting the following theorem.

Number of Real Roots Theorem

Every positive real number has: 2 real *n*th roots, when *n* is even.
1 real *n*th root, when *n* is odd.

Every negative real number has: 0 real *n*th roots, when *n* is even.
1 real *n*th root, when *n* is odd.

Zero has: 1 real *n*th root.

For instance, –4 has no real square roots, 4th roots, or 6th roots. It has one real cube root, one real 5th root, and one real 7th root.

Roots and Powers

One reason that powers are important is that the positive nth root of a positive number x is a power of x. The power is directly related to the root: the square root is the $\frac{1}{2}$ power; the cube root is the $\frac{1}{3}$ power; and so on. The general property is stated in the following theorem and can be proved using properties of powers you already know.

$\frac{1}{n}$ Exponent Theorem

When $x \geq 0$ and n is an integer greater than 1, $x^{\frac{1}{n}}$ is an nth root of x.

Proof By the definition of nth root, b is an nth root of x if and only if $b^n = x$.

Suppose $b = x^{\frac{1}{n}}$.

Then $b^n = \left(x^{\frac{1}{n}}\right)^n$ Raise both sides to the nth power.

$ = x^{\left(\frac{1}{n} \cdot n\right)}$ Power of a Power Postulate

$ = x^1$

$ = x.$

Thus, $x^{\frac{1}{n}}$ is an nth root of x.

Mathematicians could decide to let the symbol $x^{\frac{1}{n}}$ be any of the nth roots of x. However, to ensure that $x^{\frac{1}{n}}$ has exactly one value, we restrict the base x to be a nonnegative real number and let $x^{\frac{1}{n}}$ stand for the *unique nonnegative* nth root. For example, $x^{\frac{1}{2}}$ is the positive square root of x, and $2^{\frac{1}{4}}$ is the positive fourth root of 2.

Pay close attention to parentheses when applying the $\frac{1}{n}$ Exponent Theorem. Do not consider negative bases with these exponents because there are properties of powers that do not apply to them. You may use your calculator to find the nth roots of a nonnegative number b by entering b^(1÷n), as in the next Example. You will read more about how your calculator interprets numbers like $(-8)^{\frac{1}{3}}$ in Lesson 7-7.

GUIDED

Example 2
Approximate all real solutions to $x^5 = 77$ to the nearest thousandth.

(continued on next page)

Solution By the definition of nth root, the real solutions of $x^5 = 77$ are the real __?__ roots of __?__. So, one solution is $x = 77$ __?__.

Enter $77 \wedge (1 \div 5)$ into a calculator. The result, to the nearest thousandth is __?__. So, $77^{\frac{1}{5}} \approx$ __?__. By the Number of Real Roots Theorem, $x \approx$ __?__ is the only real solution because 5 is a(n) __?__ number.

 QY2

▶ QY2

Approximate the real solution to $4n^3 = -100$ to the nearest hundredth.

Nonreal nth Roots

Some of the nth roots of a real number are not real.

Example 3

Use the `cSolve` or `Solve` command on a CAS to find all 4th roots of 81.

Solution Set a CAS to complex-number mode. z is a 4th root of 81 if and only if $z^4 = 81$. So solve $z^4 = 81$. One CAS shows four solutions: $z = 3i, -3i, -3,$ or 3.

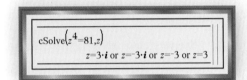

Check Verify that each solution satisfies $z^4 = 81$.

$$3^4 = 3 \cdot 3 \cdot 3 \cdot 3 = 81$$
$$(-3)^4 = (-3)(-3)(-3)(-3) = 81$$
$$(3i)^4 = 3^4 \cdot i^4 = 81 \cdot 1 = 81$$
$$(-3i)^4 = (-3)^4 \cdot i^4 = 81 \cdot 1 = 81$$

So, $3, -3, 3i,$ and $-3i$ are fourth roots of 81.

It can be proved that every nonzero real number has n distinct nth roots. In a later mathematics course you will learn how to find them.

Questions

COVERING THE IDEAS

In 1–3, refer to the discussion of the musical scale at the start of this lesson.

1. What is the exact ratio of the frequencies of consecutive notes in this scale?

2. What is the frequency of E above the middle C, to the nearest hundredth?

3. In 1879, the Steinway & Sons piano manufacturing company tuned the note A above middle C to 457.2 hertz. Given this frequency, what is the frequency of A below middle C?

4. **Fill in the Blank** Let n be an integer greater than 1. Then x is an nth root of d if and only if ___?___.

In 5–7, find the positive real root without a calculator.

5. fourth root of 81   6. cube root of 64   7. fifth root of 32

In 8–10, suppose the graphs of $y = x^n$ and $y = k$ are drawn on the same set of axes.

8. How are the points of intersection related to the nth roots of k?

9. If n is odd and $k > 0$, then at how many points do the graphs intersect?

10. If $k < 0$ and the graphs do not intersect, then is n even or odd?

11. Approximate all real solutions of $m^7 = 5$ to the nearest hundredth.

In 12 and 13, use a CAS to find all solutions of the equation.

12. $x^3 = 1331$ 13. $x^4 = 20{,}736$

14. **a.** Find two solutions to $x^4 = \frac{81}{625}$ in your head.

  b. Your answers to Part a are nth roots of $\frac{81}{625}$. What is n?

15. **True or False** $x^{\frac{1}{n}}$ is defined for $x \geq 0$ and any real number n.

16. Explain why $(-7)^4 = 2401$, but $2401^{\frac{1}{4}} \neq -7$.

17. **Multiple Choice** Which of the following is *not* a 4th root of 6561? Justify your answer.

  A $6561^{\frac{1}{4}}$ **B** -9 **C** $9i$ **D** $6561^{-\frac{1}{4}}$

APPLYING THE MATHEMATICS

18. In the nautilus shell discussed on the first page of the chapter, the eleventh chamber is about twice as long as the smallest chamber. Assuming that the ratio r of the lengths of two adjacent chambers is constant, estimate r to the nearest hundredth.

In 19 and 20, use the Compound Interest Formula and nth roots to determine a rate of growth.

19. In an old animated show, the main character travels to the future and finds out he is a billionaire. Assume that he started with $563 in his account and after 248 years the account held $1,000,000,000. If the interest rate is constant for the entire time period, what is the annual percentage yield?

20. Suppose a house was purchased in 1990 for $100,000.

  a. If its value is increasing by r% each year, then what is its value n years after 1990?

  b. If its value in 2005 was $180,000, then estimate r to the nearest tenth.

21. Give a value of x that makes the inequality $x^{\frac{1}{2}} > x$ true.

22. a. Verify that $i\sqrt{3}$ is a 4th root of 9.
    b. Why is $9^{\frac{1}{4}} \neq i\sqrt{3}$?

Fill in the Blank In 23 and 24, which symbol, $<$, $=$, or $>$, makes it a true statement?

23. $(16.1)^{\frac{1}{4}}$ __?__ 2

24. $0.25^{\frac{1}{2}}$ __?__ $0.25^{\frac{1}{3}}$

REVIEW

25. A Baravelle spiral is shown at the right. The ratios of the areas of consecutive shaded regions are equal. Suppose the area of region $A_1 = 32$ square units and the area of region $A_4 = \frac{1}{2}$ square unit. What is the ratio of consecutive areas? **(Lesson 7-5)**

In 26–28, determine whether the sequence is geometric. If it is, write an explicit formula for its nth term. **(Lesson 7-5)**

26. 4, 9, 16, 25, …   27. 5, 10, 20, 40, …   28. 6, –2, $\frac{2}{3}$, $-\frac{2}{9}$, …

29. **True or False** If two savings accounts have the same published interest rate, then the one with more compoundings per year will always have a higher annual percentage yield. **(Lesson 7-4)**

30. **Multiple Choice** Which of the following graphs show y as a function of x? There may be more than one correct answer. **(Lesson 1-4)**

A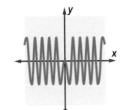

B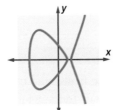

C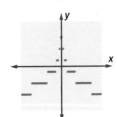

D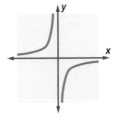

EXPLORATION

31. Use a CAS to find all the complex solutions to $x^8 = 1$.
    a. Choose any two solutions and multiply them. What do you notice?
    b. Add all of the solutions. What do you notice?
    c. Repeat Parts a and b for the solutions for $x^9 = 1$. Make two conjectures about your findings.

QY ANSWERS

1. about 293.660 Hz

2. $n \approx -2.92$

Lesson 7-7 Positive Rational Exponents

▶ **BIG IDEA** The expression $x^{\frac{m}{n}}$ is only defined when x is nonnegative and stands for the positive nth root of the mth power of x, or, equivalently, the mth power of the positive nth root of x.

The Meaning of Positive Rational Powers

In Lesson 7-6, you learned that $x^{\frac{1}{n}}$ stands for the positive nth root of x. For instance, $64^{\frac{1}{6}}$ is the positive 6th root of 64. In this lesson, we ask what $x^{\frac{m}{n}}$ means when m and n are positive integers. For instance, what does $64^{\frac{5}{6}}$ mean? The answer can be found by rewriting the fraction $\frac{5}{6}$ and using the Power of a Power Postulate.

$$64^{\frac{5}{6}} = 64^{\left(\frac{1}{6} \cdot 5\right)} \qquad \frac{5}{6} = \frac{1}{6} \cdot 5$$
$$= \left(64^{\frac{1}{6}}\right)^5 \qquad \text{Power of a Power Postulate}$$

Thus, $64^{\frac{5}{6}}$ is the 5th power of the positive 6th root of 64. With this interpretation, $64^{\frac{5}{6}} = \left(64^{\frac{1}{6}}\right)^5 = 2^5 = 32$.

Notice also that $64^{\frac{5}{6}} = 64^{\left(5 \cdot \frac{1}{6}\right)} = (64^5)^{\frac{1}{6}}$.

So $64^{\frac{5}{6}}$ is also the positive 6th root of the 5th power of 64. With this interpretation, $64^{\frac{5}{6}}$ can be simplified as follows:

$$64^{\frac{5}{6}} = (64^5)^{\frac{1}{6}} = (1{,}073{,}741{,}824)^{\frac{1}{6}} = 32.$$

In general, when an exponent is a simple fraction, the numerator is the power and the denominator is the root. The proof is a generalization of the argument above.

Rational Exponent Theorem

For any nonnegative real number x and positive integers m and n,

$x^{\frac{m}{n}} = \left(x^{\frac{1}{n}}\right)^m$, the mth power of the positive nth root of x, and

$x^{\frac{m}{n}} = (x^m)^{\frac{1}{n}}$, the positive nth root of the mth power of x.

Mental Math

Tyra is painting her bedroom. The room is 8 feet by 12 feet and has 10-foot ceilings. One quart of paint covers about 90 square feet. How many quarts of paint does Tyra need if she wants to paint

a. only the largest wall?

b. all four walls, not including a 24-square-foot window and a 28-square-foot doorway?

Proof

$$x^{\frac{m}{n}} = x^{\frac{1}{n} \cdot m} \qquad \frac{m}{n} = \frac{1}{n} \cdot m$$

$$= \left(x^{\frac{1}{n}}\right)^m \qquad \text{Power of a Power Postulate}$$

Also, $\quad x^{\frac{m}{n}} = x^{m \cdot \frac{1}{n}} \qquad \frac{m}{n} = m \cdot \frac{1}{n}$

$$= \left(x^m\right)^{\frac{1}{n}}. \qquad \text{Power of a Power Postulate}$$

To simplify an expression with a rational exponent you can find powers first or roots first. By hand, it is usually easier to find the root first because you end up working with smaller numbers and fewer digits. With a calculator, you can work more directly.

 QY1

▸ QY1

a. Estimate $7^{\frac{2}{5}}$ by using the key sequence $7^\wedge(2 \div 5)$.

b. Is $7^{\frac{2}{5}}$ greater than, equal to, or less than $7^{0.4}$?

Properties of Positive Rational Exponents

The exponent gives information about the value of each power relative to the base.

Activity 1

Step 1 Complete the following chart for powers of 8.

Rational Power	8^0	$8^{\frac{1}{3}}$	$8^{\frac{1}{2}}$	$8^{\frac{3}{4}}$	8^1	$8^{\frac{5}{4}}$	$8^{\frac{3}{2}}$	$8^{\frac{5}{3}}$	8^2
Decimal Power	8^0	$8^{0.\overline{3}}$	?	$8^{0.75}$	8^1	?	?	?	8^2
Value	1	?	?	?	8	?	?	?	64

Step 2 When the exponent n is between 0 and 1, between what values is 8^n?

Step 3 When the exponent n is between 1 and 2, between what values is 8^n?

Step 4 As the exponent n gets larger, what happens to the value of 8^n?

Step 5 Without calculating, predict between what values $8^{\frac{7}{4}}$ will be. Explain your answer.

In general, if the base is greater than 1, then the greater the exponent, the greater the value. Thus, even without calculating, you can conclude that $8^{\frac{7}{5}}$ is between 8 and 64 because $\frac{7}{5} = 1.4$ is between 1 and 2. In Question 19, you are asked to examine what happens if the base is less than 1 but greater than zero.

 QY2

▸ QY2

Which is smaller, $47^{\frac{3}{4}}$ or $47^{0.78}$?

The properties of powers in Lesson 7-2 hold for all positive exponents.

Example 1

Suppose $y \geq 0$. Simplify $\left(64y^9\right)^{\frac{4}{3}}$.

Solution

$$\left(64y^9\right)^{\frac{4}{3}} = (64)^{\frac{4}{3}}\left(y^9\right)^{\frac{4}{3}} \qquad \text{Power of a Product Postulate}$$

$$= 64^{\frac{4}{3}} \cdot y^{\frac{36}{3}} \qquad \text{Power of a Power Postulate}$$

$$= \left(64^{\frac{1}{3}}\right)^4 \cdot y^{12} \qquad \text{Rational Exponent Theorem}$$

$$= 4^4 \cdot y^{12} \qquad \tfrac{1}{n} \text{ Exponent Theorem}$$

$$= 256y^{12} \qquad \text{Arithmetic}$$

Check This CAS display shows that $\left(64y^9\right)^{\frac{4}{3}} = 256y^{12}$.

Solving Equations with Positive Rational Exponents

Properties of powers can be used to solve equations with positive rational exponents. To solve an equation of the form $x^{\frac{m}{n}} = k$, raise each side of the equation to the $\frac{n}{m}$ power. This can be done because in general, if $a = b$, then $a^n = b^n$.

Example 2

Solve $x^{\frac{3}{4}} = 7.2$ for x.

Solution

$$\left(x^{\frac{3}{4}}\right)^{\frac{4}{3}} = (7.2)^{\frac{4}{3}} \qquad \text{Raise both sides to the } \tfrac{4}{3} \text{ power.}$$

$$x^1 = (7.2)^{\frac{4}{3}} \qquad \text{Power of a Power Postulate}$$

$$x \approx 13.9 \qquad \text{Arithmetic}$$

Check Solve on a CAS as shown at the right.

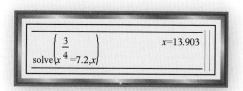

 QY3

▶ **QY3**

Find a solution to the nearest thousandth for $x^{\frac{8}{5}} = 50$.

Applications of Rational Exponents

Rational exponents have many applications, including growth situations, investments, and radioactive decay.

Example 3

The base price of a convertible sports car was $4037 in 1963 and $51,390 in 2006. What was the average annual percent increase in price over these years?

A 1963 Austin-Healey Model 3000

Solution Let p_t be the price of the sports car where $t = 1$ represents 1963 and $t = 44$ represents 2006. Then $p_1 = 4037$ and $p_{44} = 51,390$. Use a geometric sequence P with $p_t = p_1 r^{t-1}$ to model this situation.

$p_{44} = p_1 r^{44-1}$	Geometric sequence model
$51,390 = 4037 r^{43}$	Substitution
$12.730 \approx r^{43}$	Divide both sides by 4037.
$(12.730)^{\frac{1}{43}} \approx r$	Raise each side to the $\frac{1}{43}$ power.
$1.0609 \approx r$	Arithmetic

The constant ratio r represents the previous year's price plus the increase of $0.0609 = 6.09\%$. So, the price of the car increased by about 6.09% per year from 1963 to 2006.

Why Don't We Use Rational Exponents with Negative Bases?

Difficulties arise when working with noninteger rational exponents when the base is negative.

Activity 2

MATERIALS CAS (optional)

Set a CAS or graphing calculator to real-number mode and use the standard window.

Step 1 **a.** Do you think the graph of $y = \left(x^{\frac{1}{6}}\right)^2$ will be the same as the graph of $y = x^{\frac{1}{3}}$? Why or why not?

b. Graph $y = x^{\frac{1}{3}}$.

c. Clear the window and graph $y = \left(x^{\frac{1}{6}}\right)^2$. Was your prediction correct?

Step 2 **a.** Repeat Step 1 with $y = \left(x^{10}\right)^{\frac{1}{2}}$ and $y = x^5$.

b. Predict what the graph of $y = \left(x^{\frac{1}{2}}\right)^{10}$ will look like. Clear the graph screen and check your prediction by graphing $y = \left(x^{\frac{1}{2}}\right)^{10}$.

Step 3 Predict what the graph of $y = x^{0.5} \cdot x^{0.5}$ will look like. Then clear the screen and graph the equation to check your prediction.

Step 4 Based on these examples, what properties of powers do not hold for noninteger rational exponents with negative bases? Compare your results with the others in your class and discuss any differences.

The results of Activity 2 illustrate that to keep the properties of powers valid, the Rational Exponent Theorem requires that the base be a nonnegative real number.

Questions

COVERING THE IDEAS

In 1 and 2, write as a power of x.

1. the 3rd power of the 7th root of x
2. the fifth root of the square of x
3. **a.** Rewrite $10{,}000{,}000^{\frac{4}{7}}$ in two ways as a power of a power of $10{,}000{,}000$.
   b. Which way is easier to calculate mentally?
   c. Calculate $10{,}000{,}000^{\frac{4}{7}}$.

In 4–6, simplify without a calculator.

4. $16^{\frac{3}{4}}$
5. $8^{\frac{5}{3}}$
6. $49^{\frac{3}{2}}$

In 7–9, evaluate with a calculator.

7. $243^{\frac{3}{5}}$
8. $(1.331)^{\frac{4}{3}}$
9. $169^{1.50}$

In 10–12, suppose that the value of each variable is nonnegative. Simplify.

10. $\left(27x^9\right)^{\frac{2}{3}}$
11. $C^{\frac{3}{4}} \cdot C^{\frac{8}{6}}$
12. $\frac{7}{8}y^{\frac{7}{8}} \cdot \frac{8}{7}y^{\frac{8}{7}}$

In 13–15, solve.

13. $R^{\frac{2}{5}} = 100$
14. $j^{\frac{4}{3}} = 3^4$
15. $s^{\frac{4}{7}} - 10 = 0$

16. **True or False** For all real numbers x, $x^{\frac{4}{14}} = x^{\frac{2}{7}}$. Explain your answer.

Fill in the Blank In 17 and 18, complete with >, <, or =.

17. When $y > 1$, $y^{\frac{3}{4}}$? $y^{\frac{4}{3}}$.
18. When $0 < z < 1$, $z^{\frac{5}{6}}$? $z^{\frac{3}{4}}$.

19. Refer to Activity 1.
    a. Complete the table at the right.
    b. As the exponent n gets larger, what happens to the value of $\left(\frac{1}{4}\right)^n$?
    c. Without calculating, predict values that $\left(\frac{1}{4}\right)^{\frac{5}{3}}$ will be between. Explain your answer.
    d. **Fill in the Blank** Complete the following statement: If the base is smaller than 1 (but greater than zero), as the exponent gets larger, the value gets ___?___.

20. Refer to Example 3. Estimate what the base price of the sports car was in 2007.

APPLYING THE MATHEMATICS

21. Suppose n is a positive integer. Write $32^{\frac{n}{5}}$ as an integer power of an integer.

22. **Fill in the Blanks** This question gives one reason why rational exponents are used only with positive bases.
    a. If $(-27)^{\frac{1}{3}}$ were to equal the cube root of -27, then $(-27)^{\frac{1}{3}} = $ ___?___.
    b. If $(-27)^{\frac{2}{6}}$ follows the Rational Exponent Theorem, then
       $(-27)^{\frac{2}{6}} = \left((-27)^2\right)^{\frac{1}{6}} = $ ___?___; and $(-27)^{\frac{2}{6}} = \left(-27^{\frac{1}{6}}\right)^2 = $ ___?___.
    c. In this question, does $(-27)^{\frac{1}{3}} = (-27)^{\frac{2}{6}}$?
    d. Check the answers to this question on a CAS. Is there a difference?
    e. Reread the Rational Exponent Theorem's hypothesis and explain why the base is restricted to nonnegative numbers.

In 23–25, apply the Power of a Quotient Theorem, $\left(\frac{a}{b}\right)^m = \frac{a^m}{b^m}$, to simplify.

23. $\left(\frac{16}{81}\right)^{\frac{3}{4}}$ 24. $\left(\frac{125}{27}\right)^{\frac{2}{3}}$ 25. $(0.00032)^{\frac{3}{5}}$

26. The diameter D of the base of a tree of a given species roughly varies directly with the $\frac{3}{2}$ power of its height h.
    a. Suppose a young sequoia 6 meters tall has a base diameter of 20 centimeters. Find the constant of variation.
    b. The most massive living tree is a California sequoia called General Sherman. Its base diameter is about 11.1 meters. According to the variation in Part a, about how tall is General Sherman?

Rational Power	Decimal Power	Value
$\left(\frac{1}{4}\right)^0$	$\left(\frac{1}{4}\right)^0$	1
$\left(\frac{1}{4}\right)^{\frac{1}{4}}$	?	?
$\left(\frac{1}{4}\right)^{\frac{1}{2}}$	?	?
$\left(\frac{1}{4}\right)^{\frac{2}{3}}$	?	?
$\left(\frac{1}{4}\right)^1$	$\left(\frac{1}{4}\right)^1$	0.25
$\left(\frac{1}{4}\right)^{\frac{4}{3}}$	?	?
$\left(\frac{1}{4}\right)^{\frac{3}{2}}$	?	?
$\left(\frac{1}{4}\right)^{\frac{7}{4}}$	?	?
$\left(\frac{1}{4}\right)^2$	$\left(\frac{1}{4}\right)^2$	0.063

GENERAL SHERMAN

General Sherman

27. Recall from Lesson 2-3 the ancient rodent *Phoberomys*. There you used the direct variation equation $w = kd^3$ to estimate the weight of the rodent given its femur diameter d. This equation actually gives an overestimate of *Phoberomys*'s weight. Scientists have concluded that a better model is $w = kd^{2.5}$. Use this newer model to estimate how the weight of the rodent compares to the weight of a modern guinea pig if the *Phoberomys's* femur diameter is 18 times that of the guinea pig. How does this compare to your estimate in Lesson 2-3?

REVIEW

28. In Byzantine music theory, an octave is divided into 72 notes. What is the ratio of the frequency of a note to the frequency of the note below it in this system? Assume that the ratios between consecutive notes are equal. **(Lesson 7-6)**

29. Verify that $-1 - \sqrt{3}\,i$ is a third root of 8. **(Lessons 7-6, 6-9)**

30. **Fill in the Blanks** Write $<$, $=$, or $>$ in each blank. Suppose the geometric sequence with formula $g_n = ar^{n-1}$ has $g_1 < 0$ and $g_2 > 0$. Then a __?__ 0 and r __?__ 0. **(Lesson 7-5)**

31. If you graph the function $y = x^n$ for $x \leq 0$, what is the range when n is

    a. even?   b. odd? **(Lesson 7-1)**

32. Suppose $f(x) = 3x^2 + 5$. Write an expression for $f(8 - 2x)$ in the standard form of a quadratic. **(Lessons 6-1, 1-3)**

33. Find a transformation T such that $T \circ (S_3 \circ S_{0.5,4}) = I$. **(Lessons 4-7, 4-5, 4-4)**

EXPLORATION

34. Sandwiching irrational powers, such as $x^{\sqrt{2}}$ or x^π, between two close rational powers is one way to give a meaning to irrational powers of nonnegative numbers. Set a CAS to approximate mode so your answers will be decimals.

    a. Evaluate several irrational powers such as $5^{\sqrt{2}}$, $7^{\sqrt{6}}$, and $\sqrt{2}^{\sqrt{3}}$ and some others of your choosing.

    b. Recall that $1.41 < \sqrt{2} < 1.42$. Show that $5^{1.41} < 5^{\sqrt{2}} < 5^{1.42}$.

    c. Given that $\sqrt{6} \approx 2.4494$, is it true that $7^{\sqrt{6}}$ is between $7^{2.449}$ and $7^{2.450}$?

Lesson

7-8

Negative Rational Exponents

▶ **BIG IDEA** The expressions $x^{\frac{m}{n}}$ and $x^{-\frac{m}{n}}$ are reciprocals.

For mammals, a typical relationship between body weight w in kilograms and resting heart rate r in $\frac{beats}{minute}$ is modeled by $r = 241w^{-\frac{1}{4}}$. So far in this chapter you have deduced theorems from the postulates of powers that provide meaning for all positive rational exponents. Now you will use the properties of powers to determine the meaning of negative rational exponents, enabling you to work with equations such as this resting heart rate formula.

Mental Math

Calculate.

a. $16 \cdot 256^{\frac{1}{4}}$

b. $16^{\frac{1}{4}} \cdot 256^{\frac{1}{4}}$

c. $(16 \cdot 256)^{\frac{1}{4}}$

Evaluating Powers with Negative Rational Exponents

Consider the power $x^{-\frac{m}{n}}$. Because $-\frac{m}{n} = -1 \cdot m \cdot \frac{1}{n}$, and the factors can be multiplied in any order, you have the choice of taking the reciprocal, the mth power, or the nth root first.

GUIDED

Example 1

Evaluate $\left(\frac{16}{625}\right)^{-\frac{3}{4}}$ in three different ways.

Solution

Method 1: First find the reciprocal of $\frac{16}{625}$, then take the fourth root, and then calculate the third power.

$$\left(\frac{16}{625}\right)^{-\frac{3}{4}} = \left(\left(\left(\frac{16}{625}\right)^{-1}\right)^{\frac{1}{4}}\right)^3 = \left(\left(\underline{\ ?\ }\right)^{\frac{1}{4}}\right)^3 = \left(\underline{\ ?\ }\right)^3 = \underline{\ ?\ }$$

Method 2: First take the fourth root of $\frac{16}{625}$, then calculate the third power, and then find the reciprocal.

$$\left(\frac{16}{625}\right)^{-\frac{3}{4}} = \left(\left(\left(\frac{16}{625}\right)^{\frac{1}{4}}\right)^3\right)^{-1} = \left(\left(\underline{\ ?\ }\right)^3\right)^{-1} = \left(\underline{\ ?\ }\right)^{-1} = \underline{\ ?\ }$$

Method 3: First calculate the third power of $\frac{16}{625}$, then take the fourth root, and then find the reciprocal.

$$\left(\frac{16}{625}\right)^{-\frac{3}{4}} = \left(\left(\left(\frac{16}{625}\right)^3\right)^{\frac{1}{4}}\right)^{-1} = \left(\left(\underline{\ ?\ }\right)^{\frac{1}{4}}\right)^{-1} = \left(\underline{\ ?\ }\right)^{-1} = \underline{\ ?\ }$$

Example 2

Use the formula $r = 241w^{-\frac{1}{4}}$ to estimate a 180-lb person's resting heart rate in beats per minute.

Solution In the formula, w is in kilograms, so convert the weight in pounds to kilograms.

$$180 \text{ pounds} \cdot \frac{1 \text{ kilogram}}{2.2 \text{ pounds}} \approx 81.8 \text{ kilograms}$$

Substitute $w = 81.8$ into the formula and evaluate using a calculator.

$$r = 241(81.8)^{-\frac{1}{4}}$$

$$r \approx 80$$

So, a normal resting heart rate for a 180-lb person is about 80 beats per minute. Note, however, that resting heart rate is affected by many factors and that there is a wide range of variability in human heart rates. Any rate in the range $50 \leq r \leq 100$ might be considered normal for an individual.

Diet, fitness, and other health-related issues can all affect a person's resting heart rate.

Solving Equations Involving Negative Rational Exponents

The ideas used in Lesson 7-7 to solve equations with positive rational exponents can be used with negative rational exponents as well. For example, if you have a savings goal of G dollars and you already have saved S dollars, then a formula relating S and G to the annual percentage yield r needed over time t in years to reach the savings goal is

$$S = G(1 + r)^{-t}.$$

Example 3

Suppose Jett has \$50,000 of the \$150,000 he hopes to have for his child's college education in $12\frac{1}{2}$ years. Find the interest rate Jett needs in order to meet his savings goal.

Solution Substitute $G = 150{,}000$, $S = 50{,}000$, and $t = 12\frac{1}{2} = \frac{25}{2}$ into the formula above.

$$150{,}000(1 + r)^{-\frac{25}{2}} = 50{,}000$$

$$(1 + r)^{-\frac{25}{2}} = \frac{1}{3}$$

(continued on next page)

The reciprocal of $-\frac{25}{2}$ is $-\frac{2}{25}$, so raise each side to the $-\frac{2}{25}$ power.

$$\left((1+r)^{-\frac{25}{2}}\right)^{-\frac{2}{25}} = \left(\frac{1}{3}\right)^{-\frac{2}{25}}$$

$$(1+r) = \left(\frac{1}{3}\right)^{-\frac{2}{25}} \approx 1.092$$

$$r \approx 0.092$$

Jett needs to find an investment with an APY of about 9.2% to meet his savings goal.

Check Check by substitution.

Is $150{,}000(1+r)^{-\frac{25}{2}} \approx 50{,}000$? $150{,}000(1.092)^{-\frac{25}{2}} \approx$ 49,924. This is close enough to 50,000 given the estimate, so it checks.

Questions

COVERING THE IDEAS

In 2008, the average in-state public 4-year college tuition was over $6,000 per year. For private institutions it was over $23,000.

In 1–3, evaluate without using a calculator.

1. $1000^{-\frac{1}{3}}$

2. $16^{-\frac{3}{2}}$

3. $\left(\frac{625}{81}\right)^{-\frac{3}{4}}$

4. Tell whether or not the expression equals $a^{-\frac{2}{3}}$ for $a > 0$.

   a. $\left(-a^{\frac{1}{3}}\right)^2$ b. $\dfrac{1}{(a^2)^{\frac{1}{3}}}$ c. $\left((a^{-1})^{-2}\right)^{-\frac{1}{3}}$

   d. $-a^{\frac{2}{3}}$ e. $\left(a^{\frac{1}{3}}\right)^{-2}$ f. $\dfrac{1}{a^{\frac{2}{3}}}$

In 5–7, approximate to the nearest thousandth.

5. $75^{-\frac{1}{2}}$

6. $20 \cdot 4.61^{-\frac{3}{5}}$

7. $8^{-0.056}$

In 8 and 9, refer to Example 2.

8. An adult mouse weighs about 20 grams. Estimate its heart rate in beats per minute.

9. If the resting heart rate of an Asian elephant is about 30 beats per minute, estimate its weight.

In 10–12, solve without using a calculator.

10. $r^{-\frac{1}{3}} = 3$

11. $w^{-\frac{2}{5}} = 4$

12. $4z^{-\frac{3}{2}} = \dfrac{1}{16}$

Asian elephants may live as long as 60 years.

13. Solve $\dfrac{x^{-\frac{2}{3}}}{2} - 11 = 0$ to the nearest thousandth.

In 14 and 15, refer to Example 3.

14. a. Rewrite the formula for G in terms of S.

    b. Jett adopts another child and now needs to save \$300,000 in 14.5 years. What APY is needed to meet this goal?

15. A \$2.00 hamburger in 2007 cost as little as 5¢ in 1923. What yearly percentage growth in the price of the hamburger does this represent?

APPLYING THE MATHEMATICS

In 16–18, True or False. Justify your answer.

16. The value of an expression with a negative exponent is always less than zero.

17. If $y = x^{-\frac{1}{5}}$ and $z = -5$, then $y^z = x$.

18. $\left(125^{-\frac{1}{3}}\right)^0 = 0$

19. Find t if $\left(\dfrac{99}{100}\right)^{-\frac{1}{2}} = \left(\dfrac{100}{99}\right)^t$.

20. The amount F of food in grams that a mouse with body mass m must eat daily to maintain its mass is estimated by $F = km^{\frac{2}{3}}$. For a 25-gram mouse, suppose $k = 11.7$. Find the amount of food in grams that this mouse must eat daily to maintain its weight.

In 21–24, rewrite each expression in the form ax^n. Check your answer by substituting a value for x in both the original and the rewritten expression.

21. $\left(x^{-4}\right)^{-\frac{1}{4}}$

22. $\left(36x^{-2}\right)^{-\frac{3}{2}}$

23. $\dfrac{x}{6x^{-\frac{4}{3}}}\left(3x^{\frac{1}{3}}\right)$

24. $\dfrac{-\frac{5}{6}x^{-\frac{5}{6}}}{\frac{1}{6}x^{\frac{1}{6}}}$

Baby mice raised in captivity without their mother may be fed milk or formula.

REVIEW

In 25–27, simplify without a calculator. (Lesson 7-7)

25. $100{,}000{,}000{,}000^{\frac{5}{11}}$

26. $12 \cdot 81^{\frac{3}{4}}$

27. $0.064^{\frac{2}{3}}$

28. German astronomer Johannes Kepler (1571–1630) made observations on planetary orbits. His results became known as Kepler's laws of planetary motion. Kepler's third law states that the ratio of the squares of the periods of any two planets equals the ratio of the cubes of their mean distances from the Sun. (The period of a planet is the length of time it takes the planet to go around the Sun.) If the periods of any two planets are t and T, and their mean distances from the sun are d and D, respectively, then $\frac{T^2}{t^2} = \frac{D^3}{d^3}$. Find the ratio of the periods, $\frac{T}{t}$. (**Lesson 7-7**)

Johannes Kepler

29. Use a graph to explain why negative real numbers have no real nth roots when n is even. (**Lessons 7-6, 7-1**)

30. Cassandra is playing a board game where she rolls two 6-sided dice each turn, and then she adds the two rolls to find how many squares she should move forward. In 50 turns, Cassandra moved forward 2 squares five times. Cassandra thinks the dice might be unfair.

    a. What was her relative frequency of rolling two 1s with these dice?

    b. What seems to be the probability of rolling two 1s with these dice? Is Cassandra's concern valid? (**Lesson 7-6**)

31. Simplify $\left(\frac{a}{b}\right)^3 \cdot \frac{a^2 b}{\left(2ab^2\right)^2}$. (**Lesson 7-2**)

32. Expand $(2x + 3y)^2$. (**Lesson 6-1**)

EXPLORATION

33. Research the average adult weight and heart rate of a mouse and of an Asian elephant. Compare these data to your answers for Questions 8 and 9. How accurate does the model $r = 241w^{-\frac{1}{4}}$ seem to be for predicting the heart rates of these two mammals? Does the weight of the mammal seem to affect the accuracy of the model?

Chapter 7 Projects

1 Financing Post-High School Education

a. Select a college or post-secondary school you have heard about or are interested in attending. Find out its yearly tuition for each year of the past decade.

b. Based on the data in Part a, estimate what tuition will cost during the years you would attend. Explain your answer.

c. Suppose that 15 years ago, a benefactor set up an account for your education. This benefactor made a single deposit of $20,000 that has been earning 6% interest compounded monthly. Since the deposit 15 years ago, no money has been added to or taken from the account. Will this account be sufficient to cover tuition for all four years?

d. Find the smallest annual interest rate for the account to be sufficient to pay for the tuition for the years you will attend.

In 2008 Harvard College, the undergraduate school of Harvard University, adjusted its financial aid program to become more affordable for middle-income families.

2 Powers of 10

Several sites on the Internet show pictures of objects of sizes from 10^{-15} m long to 10^{20} m long, that is, from the smallest particles to the largest multigalactic structures in the universe. Make a table similar to these sites using mass rather than length, from 10^{-30} kg, the mass of an electron, to 10^{30} kg, the mass of our solar system. For various powers of 10, fill in an object that has a mass close to that number in kilograms. For example, the dry mass of an adult water flea is about 10^{-7} kg. The one shown above is carrying eggs.

3 Sums of Powers

In 1637, the French mathematician and lawyer Pierre Fermat asserted that there were no positive integers x, y, and z for which $x^n + y^n = z^n$ for any integer $n > 2$. He wrote that he had a proof but it would not fit in the margin of the book he was writing in. Fermat's "Last Problem," as this assertion was sometimes called, was finally solved by Princeton mathematician Andrew Wiles in 1995. In the over 350 years people worked on this problem, many related problems were posed and solved. Write about efforts to find positive integer solutions to the following equations.

$$w^4 + x^4 + y^4 = z^4$$
$$v^5 + w^5 + x^5 + y^5 = z^5$$

4 The $200,000 Cell Phone

Suppose that, instead of paying $55 per month for cell phone charges, you invest that amount at a 6% interest rate compounded monthly until you retire 50 years from now. A periodic payment (*Pmt*) like this one is called an *annuity* and the amount you will have at retirement is the *future value* (*FV*) of the annuity. A formula for the future value of the annuity is related to the compound interest formula. If i = the interest rate at each compounding and n = the number of compoundings, $FV = Pmt \cdot \left[\dfrac{(1 + i)^n - 1}{i} \right]$.

In this case, you would have

$$FV = 55 \cdot \left[\frac{(1 + 0.005)^{600} - 1}{0.005} \right] \text{ or } \$208{,}295.51$$

at retirement.

a. Use your average cell phone bill or that of a friend. Calculate the future value of those payments 50 years from now if invested at an interest rate from your local bank.

b. Many household expenses recur monthly, such as utility and satellite or cable TV bills. Pick one of these recurring expenses and calculate the future value of that annuity 50 years from now.

c. Find the price of a cup of coffee at a local coffee shop and the price of your favorite car. Assume that car prices increase 3% per year and that you can invest money at a 5% interest rate compounded monthly. Can you buy the car when you retire 50 years from now if you give up a daily cup of coffee?

5 Family of Equations

Graph at least 10 members of the family of equations of the form $ax + by = c$, where a, b, and c are consecutive terms in a geometric sequence. Sketch the graphs. What pattern do you see? You may have to experiment to determine the values to use in the window you have chosen. How can you explain the results?

Chapter 7

Summary and Vocabulary

- When $x > 0$, the expression x^m is defined for any real number m. This chapter covers the meanings and properties of x^m when m is a positive or negative rational number.

- In previous courses, you have learned basic properties of **powers**. For any nonnegative **bases** and nonzero real **exponents**, or any nonzero bases and integer exponents:

Product of Powers Postulate	$x^m \cdot x^n = x^{m+n}$
Power of a Power Postulate	$(x^m)^n = x^{mn}$
Power of a Product Postulate	$(xy)^n = x^n y^n$

- The following theorems can be deduced from these postulates. For any positive bases and real number exponents and any nonzero bases and integer exponents:

Quotient of Powers Theorem	$\dfrac{x^m}{x^n} = x^{m-n}$
Power of a Quotient Theorem	$\left(\dfrac{x}{y}\right)^m = \dfrac{x^m}{y^m}$
Zero Exponent Theorem	$x^0 = 1$
Negative Exponent Theorem	$x^{-m} = \dfrac{1}{x^m}$
Exponent Theorem	$x^{\frac{1}{n}}$ is the positive solution to $b^n = x$.
Rational Exponent Theorem	$x^{\frac{m}{n}} = (x^m)^{\frac{1}{n}} = \left(x^{\frac{1}{n}}\right)^m$

- From these properties, we see that $x^{\frac{1}{n}}$ is the positive root of x and that $x^{\frac{m}{n}}$ is the mth power of the positive **nth root** of x and the positive nth root of the mth power of x. These properties are not always true when $x < 0$, so we do not define x^m when $x < 0$ and m is not an integer. You can use these properties to simplify expressions and to solve equations of the form $x^n = b$. To solve such an equation, raise each side of the equation to the $\frac{1}{n}$ power.

Vocabulary

Lesson 7-1
powering, exponentiation
*base
*exponent
*power
*nth-power function
identity function
squaring function
cubing function

Lesson 7-4
principal
compounding daily
annual percentage yield, APY

Lesson 7-5
*geometric sequence, exponential sequence
constant multiplier
constant ratio

Lesson 7-6
cube root, nth root

○ Equations involving powers are found in many fields, including investments, science, and music. In the **General Compound Interest Formula** $A = P\left(1 + \frac{r}{n}\right)^{nt}$, A is the value of an investment of P dollars earning interest at a rate r compounded n times per year for t years. When P, r, n, and t are given, you can solve for A; when A, r, n, and t are known, you can solve for P. Because of the multitude of ways to calculate interest, federal law requires institutions to advertise the **annual percentage yield (APY)** of their accounts.

○ A **geometric sequence** is a sequence in which the ratios of consecutive terms are constant; each term is a constant multiple r of the preceding term. In symbols, given g_1, then for all $n \geq 2$, $g_n = rg_{n-1}$. The nth term of a geometric sequence can be found explicitly using the formula $g_n = g_1 r^{n-1}$.

Postulates and Theorems

Probability of Repeated Independent
  Events (p. 454)
Product of Powers Postulate (p. 459)
Quotient of Powers Theorem (p. 460)
Power of a Product Postulate (p. 460)
Power of a Quotient Theorem (p. 461)
Zero Exponent Theorem (p. 461)
Power of a Power Postulate (p. 462)
Negative Exponent Theorem (p. 467)
Annual Compound Interest Formula
  (p. 473)

General Compound Interest Formula
  (p. 473)
Recursive Formula for a Geometric
  Sequence (p. 479)
Explicit Formula for a Geometric
  Sequence (p. 481)
Number of Real Roots Theorem
  (p. 488)
$\frac{1}{n}$ Exponent Theorem (p. 489)
Rational Exponent Theorem (p. 493)

Chapter 7 Self-Test

Take this test as you would take a test in class. You will need a calculator. Then use the Selected Answers section in the back of the book to check your work.

1. A student writes $(3 + 4)^{\frac{1}{2}} = 3^{\frac{1}{2}} + 4^{\frac{1}{2}}$. Explain why this sentence is not true.

In 2–4, write as a whole number or simple fraction.

2. 7^{-2} 3. $(214{,}358{,}881)^{\frac{1}{8}}$ 4. $\left(\frac{343}{27}\right)^{-\frac{4}{3}}$

5. Write without an exponent: $\frac{7.3 \cdot 10^3}{10^{-4}}$.

In 6 and 7, simplify. Assume $x > 0$ and $y > 0$.

6. $\left(1728 x^9 y^{27}\right)^{\frac{1}{3}}$ 7. $\frac{84 x^{21} y^5}{6 x^3 y^7}$

8. **True or False** Justify your answer.

   a. $-5^{\frac{1}{3}} < 5^{-\frac{1}{3}}$ b. $3^{-6.4} < 3^{-6.5}$

9. Solve $x^{\frac{3}{2}} = 0.8$ for x. Round your answer to the nearest thousandth.

10. Approximate $4.26^{\frac{1}{6}}$ to the nearest hundredth.

11. A graph of a power function, $y = x^n$, is shown at the right. Is n even or odd? Justify your answer.

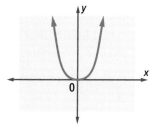

12. Consider this sequence, which gives the maximum height of a bouncing ball after the nth bounce:

$$\begin{cases} h_1 = 5 \\ h_n = 0.85 h_{n-1}, \text{ for integers } n \geq 2. \end{cases}$$

   a. Find the height of the ball after the 1st, 2nd, 3rd, and 4th bounces.

   b. Write an explicit formula for this sequence.

13. **True or False** The range of all power functions is the set of real numbers. Justify your answer.

In 14 and 15, the first few terms of a geometric sequence are given.

   a. Find the next two terms.

   b. Write a recursive formula for the sequence.

   c. Write an explicit formula for the nth term of the sequence.

14. 3, 0.6, 0.12, 0.024, …

15. −4, −12, −36, −108, …

16. A quiz has 10 multiple choice questions. Each question has three choices. What is the probability of guessing all 10 questions correctly?

17. The formula $m = 1.23 \cdot (1.2)^b$ gives the approximate number of minutes m that b bacon slices will take to cook in a typical microwave. How long will it take 2 slices of bacon to cook? Give your answer in minutes and seconds.

18. Identify all real 10th roots of 1024.

19. **True or False** $625^{\frac{1}{4}} = -5$. Justify your answer.

20. A bank account pays 4.25% annual interest. Suppose you deposit $500 in the account and then do not deposit or withdraw any money for 6 years.

   a. How much will you have in the account after 6 years if the interest is compounded annually?

   b. What is the APY if the interest is compounded quarterly?

21. A zero-coupon bond paying 6.8% interest compounded monthly for 7 years has matured, giving the investor $7854. How much did the investor pay for the bond 7 years ago?

Chapter 7 Chapter Review

SKILLS
PROPERTIES
USES
REPRESENTATIONS

SKILLS Procedures used to get answers

OBJECTIVE A Evaluate b^n when $b > 0$ and n is a rational number. (Lessons 7-2, 7-3, 7-6, 7-7, 7-8)

In 1–6, write as a simple fraction or whole number. Do not use a calculator.

1. 17^0

2. 2^{-3}

3. $\left(\frac{5}{3}\right)^{-1}$

4. $64^{-\frac{1}{2}}$

5. $8^{\frac{4}{3}}$

6. $1{,}000{,}000^{\frac{1}{3}}$

In 7–9, approximate to the nearest hundredth.

7. $123^{\frac{2}{3}}$

8. $5 \cdot 8^{\frac{1}{4}}$

9. $7^{1.5}$

OBJECTIVE B Simplify expressions or solve equations using properties of exponents. (Lessons 7-2, 7-3, 7-7, 7-8)

In 10–12, solve.

10. $13^4 \cdot 13^{12} = 13^x$

11. $\frac{2^6}{2^{-3}} = 2^y$

12. $(7^z)^3 = 7^9$

In 13–18, simplify. Assume all variables represent positive numbers.

13. $(-2x^3)^2$

14. $(-2x^2)^3$

15. $\left(\frac{p}{q}\right)^7 \left(\frac{5q}{3p}\right)^3$

16. $\frac{21b}{(7b^{-5})(6b^5)}$

17. $\frac{-16x^8 y^{\frac{5}{2}}}{4x^2 y^{\frac{1}{2}}}$

18. $\frac{(r^3 s^2)^{\frac{1}{2}}}{r^2 s^3}$

OBJECTIVE C Describe geometric sequences explicitly and recursively. (Lesson 7-5)

In 19 and 20, the first few terms of a geometric sequence are given.

a. Find an explicit formula for the nth term.

b. Find a recursive formula for the sequence.

c. Find the 16th term.

19. 4, –12, 36, –108, 324, …

20. 2, 0.5, 0.125, 0.03125, …

21. Find the 25th term of a geometric sequence whose first term is 4 and whose constant multiplier is 1.075. Express your answer

a. exactly.

b. to the nearest thousandth.

22. **Multiple Choice** Which of the following could be the first three terms of a geometric sequence?

A 9, 3, –6, …

B $6\frac{2}{3}, 66\frac{2}{3}, 666\frac{2}{3}, \dots$

C $\frac{4}{7}, \frac{11}{7}, \frac{18}{7}, \dots$

D 0.4, 0.16, 0.64, …

In 23–26, give the first four terms of the geometric sequence described.

23. constant ratio of 3, first term 7

24. first term is $\frac{2}{3}$, fourth term is $\frac{16}{81}$

25. $\begin{cases} t_1 = 6 \\ t_n = \frac{2}{3} t_{n-1}, \text{ for integers } n \geq 2 \end{cases}$

26. $\begin{cases} h_1 = -4 \\ h_{n+1} = -1.5 h_n, \text{ for integers } n \geq 1 \end{cases}$

OBJECTIVE D Find all real solutions to equations of the form $x^n = b$, where $x \geq 0$ and n is a rational number. (Lessons 7-6, 7-7, 7-8)

In 27–35, find all real solutions.

27. $12x^2 = 432$

28. $33 = a^3$

29. $y^4 = 16$

30. $p^{-2} = 16$

31. $6 = y^{\frac{1}{3}}$

32. $7q^{-\frac{3}{5}} = 9$

33. $x^{\frac{4}{9}} = 12$

34. $r^{-\frac{3}{2}} = \frac{1}{8}$

35. $1.55 = m^{\frac{1}{6}}$

PROPERTIES Principles behind the mathematics

OBJECTIVE E Recognize properties of nth powers and nth roots. (Lessons 7-2, 7-6, 7-7, 7-8)

In 36–38, True or False. Justify your answer.

36. $-5 = (390{,}625)^{\frac{1}{8}}$

37. $\pi^{-5.3} < \pi^{-5.4}$

38. $t^{-\frac{3}{4}} = \frac{1}{(t^3)^{\frac{1}{4}}}$ $(t > 0)$

39. a. Identify all the square roots of 121.
    b. Simplify $121^{\frac{1}{2}}$.

40. **True or False** $3i$ is a 4th root of 81.

41. **True or False** For all $x > 1$, $x^{\frac{1}{4}} < x$.

42. Suppose $0 < v < 1$. Arrange from least to greatest: $v, v^{\frac{3}{2}}, v^{-3}, v^{\frac{1}{3}}, v^{-\frac{5}{3}}$.

In 43–46, use properties A–D below. Assume $R > 0$, $m \neq 0$, and $n \neq 0$. Identify the property or properties that justify the equality.

A $R^0 = 1$

B $R^n = \frac{1}{R^n}$

C $R^{\frac{1}{n}}$ is the positive solution to $x^n = R$.

D $R^{\frac{m}{n}} = (R^m)^{\frac{1}{n}} = \left(R^{\frac{1}{n}}\right)^m$

43. $(3.456)^{6-6} = 1$

44. $\left(q^{\frac{1}{4}}\right)^4 = q$

45. $(81)^{-\frac{1}{4}} = \frac{1}{3}$

46. $\left(\frac{1}{b}\right)^{-\frac{7}{8}} = (b^7)^{\frac{1}{8}}$

47. a. For what integer values of n does the equation $x^n = 19$ have exactly one real solution?

    b. How many solutions does it have for other nonzero integer values of n?

48. Explain why rational exponents are not defined for negative bases, using $(-125)^{\frac{1}{3}}$ and $(-125)^{\frac{2}{6}}$ as examples.

49. **Fill in the Blank** The positive nth root of a positive number equals the __?__ power of that number.

USES Real-world applications of mathematics

OBJECTIVE F Solve real-world problems that can be modeled by expressions with nth powers or nth roots. (Lessons 7-1, 7-6, 7-7, 7-8)

50. To qualify for a quiz show, a person must answer all questions correctly in three categories: literature, science, and current events. Suppose a person estimates that the probability of getting one question correct in literature is ℓ, in science is s, and in current events is c. If the person is asked 3 literature, 3 science, and 4 current events questions, what is the probability that the person gets all the questions right?

51. The Merchandise Mart in Chicago has about $4 \cdot 10^6$ ft^2 of floor space. This area is what percent of the $6.6 \cdot 10^6$ ft^2 of floor space in the Pentagon in Washington DC?

52. The intensity I of light varies inversely with the square of the distance d from the light source. Write a formula for I as a function of d using

    a. a positive exponent.

    b. a negative exponent.

In 53 and 54, use this information. Kepler's third law states that the ratio of the squares of the periods of any two planets equals the ratio of the cubes of their mean distances from the Sun. If the periods of the planets are t and T and their mean distances from the Sun are d and D, respectively, then $\frac{T^2}{t^2} = \frac{D^3}{d^3}$.

53. Find the ratio $\frac{D}{d}$ of the distance.

54. Kepler used his third law to determine how far planets were from the Sun. He knew that for Earth, $t \approx 365$ days and $d \approx 150{,}000{,}000$ km. He also knew that for Mars, $T \approx 687$ days. Use this information to find D, the mean distance from Mars to the Sun.

In 55 and 56, use this information about similar figures. If A_1 and A_2 are the surface areas of two similar figures and V_1 and V_2 are their volumes, then $\frac{A_1}{A_2} = \left(\frac{V_1}{V_2}\right)^{\frac{2}{3}}$.

55. Two similar figures have volumes 36 cm^3 and 48 cm^3. What is the ratio of the amounts of paint needed to cover their surfaces?

56. Solve the formula for $\frac{V_1}{V_2}$.

OBJECTIVE G Apply the compound interest formulas. (Lesson 7-4)

57. Deion wants to invest now in a bond that will give him $5000 in 7 years. The bond pays 5.25% interest compounding monthly. How much must Deion invest?

In 58 and 59, Camila put $10,000 in a 6-year 4.825% savings certificate in which interest is compounded daily 365 days per year.

58. What is the APY of Camila's savings certificate?

59. a. How much interest will she earn during the entire 6-year period?

    b. How much interest will she earn during the sixth year?

In 60 and 61, Brooklyn now has $7231 in an account earning interest at a rate of 3.125% compounded quarterly.

60. Assuming she made no deposits or withdrawals in the past four years, how much money was in the account 4 years ago?

61. How much interest did she earn during the past two years?

OBJECTIVE H Solve real-world problems involving geometric sequences. (Lesson 7-5)

62. Suppose a ball bounces up to 77% of its previous height after each bounce, and the ball is dropped from a height of 8 m.

    a. Find an explicit formula for the height after the nth bounce.

    b. Find the height of the ball after the tenth bounce, to the nearest decimeter.

63. Raul set a copy machine to reduce images to 82% of their original size.

    a. If the original image was 20 cm by 25 cm, what are the dimensions of the copy?

    b. If each time Raul made a copy he used the copy as the preimage for the next copy, what are the dimensions of the fifth copy?

64. A weight on a pendulum moves 120 cm on its first swing. On each succeeding swing back or forth it moves 95% of the distance of the previous swing. Write the first four terms of the sequence of swing lengths.

REPRESENTATIONS Pictures, graphs, or objects that illustrate concepts

OBJECTIVE I Graph nth power functions. (Lesson 7-1)

In 65 and 66, an nth power graph is drawn. Write an equation for each function.

65.

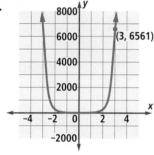

66.

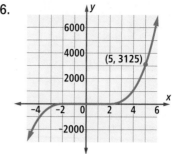

In 67 and 68, a function is given.
a. Graph the function.
b. Identify its domain and range.
c. Describe any symmetries of the graph.

67. $y = x^5$

68. $y = x^4$

69. Use a graph to explain why the equation $x^6 = -5$ has no real solutions.

Chapter

8

Inverses and Radicals

Contents

Year	1940	1950	1960	1970	1980	1990	2000
Median Home Value (unadjusted dollars)	2938	7354	11,900	17,000	47,200	79,100	119,600

U.S. Median Home Values, 1940–2000

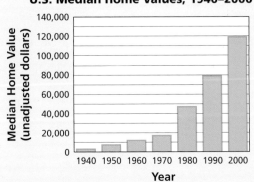

Source: http://www.census.gov/hhes/www/housing/census/historic/values.html

Can you imagine what it would be like to pay these prices: a movie ticket for $0.30, a new car for $800, or a house for $3000? It may be hard to believe, but those were typical prices in the United States in 1940.

Since 1940, the U.S. Census has calculated the median values of single-family homes in the United States. These values are unadjusted for inflation and are shown at the left.

Economists and statisticians analyze price data like these and calculate their rate of change over time using a *geometric mean.* Calculating this mean requires finding *n*th roots or, using *radical notation,* finding $\sqrt[n]{x}$.

You have seen radical notation before. For example, to find the side length of a cube with a volume of 64 cm³, you find the cube root of 64, or $\sqrt[3]{64} = 4$ cm. To check the solution, you cube the answer: $4^3 = 64$. The cubing function $y = x^3$ and the "uncubing" function $y = \sqrt[3]{x}$ are examples of *inverse functions;* each function undoes what the other does.

In this chapter, you will study inverses of functions. We begin with the general idea of following one function with another, using the output of the first as the input of the second. After this, you will learn about the properties and graphs of inverse functions, evaluating expressions with radical notation, and other applications of *n*th roots.

8-1

Composition of Functions

▶ **BIG IDEA** The function that results from following one function by another is called the *composite of the two functions*; the operation is called *composition*.

Activity

The Lee Muns Car Dealership offers two incentives to car buyers. They give a 12% discount off the $33,000 sticker price of a new car, as well as a $2000 rebate for an end-of-the-year clearance. If you were given a choice, which incentive would you take first?

Step 1 Take the discount first and then the rebate. Note that a 12% discount means that you pay 88% of the selling price. How much would you pay for the car if you took the discount first?

Step 2 Take the rebate first then apply the discount. How much would you then pay for the car?

Step 3 Which method gives you a lower price for the car? How much money would you save over the other option?

Mental Math

Write in $a + bi$ form.

a. $\dfrac{6 + 4i}{4}$

b. $5i(2i + i^2)$

c. $(3 + 2i)(3 - 2i)$

d. $(k + ri)(k - ri)$

The Composite of Two Functions

Will calculating the discount before the rebate always result in a lower price? Will the difference always be the same? To answer these questions, you can use algebra. Let x be the sticker price, d the discount function, and r the rebate function. If you take the discount first, then the price after the discount will be $d(x)$. If you then take the rebate, you can find the final price by taking the output of the discount function and using it as input to the rebate function. This is given by the expression $r(d(x))$, read "r of d of x."

When the supply of available new cars exceeds customer demand, auto dealers may lower the price to increase demand.

Example 1

Consider the situation in the Activity. Let $r(x) = x - 2000$ and let $d(x) = 0.88x$.

a. Write an equation for $p(x) = r(d(x))$, that is, the price $p(x)$ if the sticker price is x and discount is taken first, then the rebate.

b. Write an equation for $q(x) = d(r(x))$, that is, the price if the rebate is taken first, then the discount.

c. Is it true for all x that $p(x) < q(x)$?

Solution

a. $p(x) = r(d(x)) = r(0.88x)$ Apply the formula for d.

      $= 0.88x - 2000$ Apply the formula for r.

b. $q(x) = d(r(x)) = d(x - 2000)$ Apply the formula for r.

      $= 0.88(x - 2000)$ Apply the formula for d.

      $= 0.88x - 1760$ Distributive Property

c. Notice that the selling price is $p(x) = r(d(x)) = 0.88x - 2000$ when the discount is given before the rebate, or $q(x) = d(r(x)) = 0.88x - 1760$ when the rebate is given before the discount.

Both $p(x)$ and $q(x)$ are linear equations with the same slope, so the graphs of the functions p and q are parallel lines. The y-intercept of $q(x)$ is $240 more than the y-intercept of $p(x)$. Hence, $p(x) < q(x)$ for all x. That is, with a 12% discount and a $2000 rebate, the final selling price is always $240 less when the discount is taken first, no matter what the sticker price is.

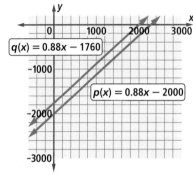

 QY1

> **QY1**
>
> Find $r(d(50,000))$ and $d(r(50,000))$.

Example 1 involves the same ideas you saw in Chapter 4 with transformations. Recall that the result of applying one transformation T after another S is called the composite of the two transformations, written $T \circ S$. Likewise, the function that maps x onto $r(d(x))$ is called the *composite* of the two functions d and r, and is written $r \circ d$.

The composite of two functions is a function. You can describe any function if you know its domain and a rule for obtaining its values. Thus, you define the composite of two functions by indicating its rule and domain.

Definition of Composite

The **composite $g \circ f$** of two functions f and g is the function that maps x onto $g(f(x))$, and whose domain is the set of all values in the domain of f for which $f(x)$ is in the domain of g.

The operation signified by the small circle $\circ$ is called **function composition**, or just **composition**. $g \circ f$ is read "the composite f followed by g" "the composite g following f," or "the composite f then g."

Ways of Writing a Composite

The ways of writing a composite are the same as those used for composites of transformations. Consider the composite $g \circ f$. You can describe the rule for a composite in two ways.

Mapping notation: $g \circ f : x \rightarrow g(f(x))$

$f(x)$ notation: $g \circ f(x) = g(f(x))$

In Example 1, you wrote $r(d(x))$. You could have written this as $r \circ d(x)$.

 QY2

▸ **QY2**

Write "the composite r then d of x" in

a. mapping notation.

b. $f(x)$ notation.

Is Function Composition Commutative?

From the Activity, $r(d(33{,}000)) = 27{,}040$, and $d(r(33{,}000)) = 27{,}280$. This one example is enough to show that *composition of functions is not commutative*. Here is another example.

GUIDED

Example 2

Let $f(x) = x - 2$ and $g(x) = 3x^2$. Calculate

a. $f \circ g(x)$.   b. $g \circ f(x)$.

Solution

a. $f \circ g$ means first square x and multiply by 3, then subtract 2 from the result.

$f \circ g(x) = f(g(x)) = f(\underline{3x^2}) = \underline{3x^2} - 2$

b. To evaluate $g \circ f(x)$, first subtract 2 from x, then square the result and multiply by 3.

$g \circ f(x) = g(f(x)) = g(\underline{\ ?\ }) = 3(\underline{\ ?\ })^2 = 3(\underline{\ \ ?\ \ }) = \underline{\ \ ?\ \ }$

Finding the Domain of a Composite of Functions

The domain for a composite is the largest set for which the composition is defined. That is, the domain can include only those values of x for which the first function is defined and that are paired with values that are in the domain of the second function.

Example 3

Let x be a real number. Let $s(x) = \frac{1}{x^2}$, and let f be given by $f(x) = x^2 - 64$. What is the domain of $s \circ f$?

Solution There are no restrictions on the domain of f, but 0 is not in the domain of s. Thus, $f(x)$ cannot be 0. Because $f(x) \neq 0$, $x^2 - 64 \neq 0$; thus, $x^2 \neq 64$. So, x cannot equal 8 or -8. The domain of $s \circ f$ is the set of real numbers other than 8 or -8.

Questions

COVERING THE IDEAS

In 1 and 2, refer to the rebate and discount functions in the Activity.

1. **a.** If the sticker price is \$25,000 and the rebate is calculated first, what is the selling price of the car?

    b. If the sticker price is \$25,000 and the discount is calculated first, what is the selling price of the car?

2. Suppose the car dealer changes the rebate to \$3000 and the discount to 15%.

    a. Which gives a lower selling price, taking the rebate first or taking the discount first? Justify your answer.

    b. Will the difference in selling price between taking the discount first versus taking the rebate first always be the same? Explain why or why not.

In 3 and 4, if $f(x) = 2x^3$ and $g(x) = 6x - 4$, evaluate the expression.

3. **a.** $f(g(-3))$ **b.** $f \circ g(-3)$

4. **a.** $f \circ g(5)$ **b.** $f(g(5))$

5. Suppose $f(x) = x + 1$ and $g(x) = x - 3$. Is composition of f and g commutative? Justify your answer.

In 6 and 7, let $f(a) = a^2 + a + 1$ and $g(a) = -4a$.

6. Evaluate each expression.

    a. $g(f(-3))$ **b.** $f(g(-3))$ **c.** $g(f(3))$ **d.** $f(g(3))$

7. Find an expression in standard form for each.

    a. $g(f(a))$ **b.** $f(g(a))$

8. Use a CAS to define the functions f and g in Example 2. Evaluate each expression and compare your results with the answers in the example.

    a. $f(g(x))$ **b.** $g(f(x))$

Define $f(x)=x-2$	Done
Define $g(x)=3 \cdot x^2$	Done

9. Refer to Example 3.

 a. What is the domain of $f \circ s$?

 b. Is the domain of $f \circ s$ the same as the domain of $s \circ f$?

APPLYING THE MATHEMATICS

In 10 and 11, let $g(x) = \frac{1}{x}$ and $h(x) = x^4 - 16$.

10. Find the domain of $g \circ h$.   11. Find the domain of $h \circ g$.

12. Suppose a state has a sales tax of 5% and you buy something at a discount of 20%. Let P be the original price of what you have bought.

 a. Write a formula for $f(P)$, the price after the discount.

 b. Write a formula for $g(P)$, the price after tax.

 c. Which gives you a better final price, discount first or tax first? Explain why.

13. Jarrod earns some extra cash by chauffeuring on the weekends. He rents a limousine from Executive Rentals. According to their rental agreement, Jarrod pays Executive a $300 rental fee plus 40% of his remaining proceeds. Luxury Rental makes Jarrod a different offer. They would charge him $300, but they tell him he will pay them 40% of all collected proceeds *before* deducting the $300 fee. Use function composition to show which offer Jarrod should take.

14. Let $g(x) = x^{\frac{2}{3}}$, where x is a real number. The function g can be rewritten as the composite of two functions m and h.

 Define m and h so that $h(m(x)) = m(h(x)) = x^{\frac{2}{3}}$.

15. Composite functions can be used to describe relationships between functions of variation. Suppose w varies inversely as z, and z varies directly as the fourth power of x.

 a. Give an equation for w in terms of z. Use your equation to describe a function $f : z \rightarrow w$.

 b. Give an equation for z in terms of x. Use your equation to describe a function $g : x \rightarrow z$.

 c. Give an equation for w in terms of x. Use your equation to describe a function $h : x \rightarrow w$.

 d. Use words to describe how w varies with x.

 e. How are f, g, and h related?

16. Let $f(x) = x^2$. What is $f(f(f(x)))$?

17. Consider $t(x) = \frac{1}{x^2}$.

    a. Simplify $t(t(x))$. b. When is $t(t(x))$ undefined?

REVIEW

18. Nancy throws a ball upward at a velocity of $18 \frac{m}{sec}$ from the top of a cliff 35 m above sea level. **(Lesson 6-4)**

    a. Write an equation to describe the height h of the ball after t seconds.

    b. What is the maximum height of the ball?

19. Find the inverse of the matrix $\begin{bmatrix} a & -7 \\ 0 & 18 \end{bmatrix}$, $a \neq 0$. **(Lesson 5-5)**

20. Consider the transformation $T : (x, y) \rightarrow (y, x)$. **(Lesson 4-6)**

    a. Let $A = (-4, 0)$, $B = (-1, 5)$, and $C = (2, 5)$. Graph the image of $\triangle ABC$ under T.

    b. What transformation does T represent?

21. **True or False** The line with the equation $\pi = x$ is the graph of a function. **(Lesson 1-4)**

22. The graph at the right shows the height of a flag on a 12-meter pole as a function of time. **(Lesson 1-4)**

    a. Describe what is happening to the flag.

    b. Why are there some horizontal segments on the graph?

    c. What is the domain of the function?

    d. What is its range?

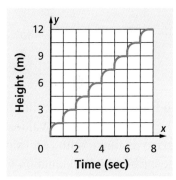

23. a. What number is the additive inverse of $-\frac{13}{4}$?

    b. What number is the multiplicative inverse of $-\frac{13}{4}$?

    (Previous Course)

EXPLORATION

24. Let $f(x) = \frac{1}{1-x}$.

    a. Find $f(f(x))$, $f(f(f(x)))$, and $f(f(f(f(x))))$.

    b. What is the relationship between $f(x)$ and $f(f(f(f(x))))$?

    c. Use your answers to Parts a and b to find $f(f(f(f(f(f(f(f(x))))))))$.

Inverses of Relations

▶ **BIG IDEA** Every relation has an *inverse*. Some of these inverses are functions.

In 1929, astronomers Edwin Hubble and Milton Humason announced their discovery that other galaxies are moving away or *receding* from our galaxy, the Milky Way. Galaxies that are farther away appear to be receding faster. The relationship between distance and recession speed is given in the table below. Distance is given in megaparsecs (Mpc), with 1 Mpc ≈ 3.26 million light years. Speed is given in kilometers per second.

$x =$ Distance (Mpc)	$y =$ Speed $\left(\frac{km}{sec}\right)$
0	0
5	350
10	700
15	1050
20	1400

In this table, the left column has values of the domain variable x and the right column has values of the range variable y. The function that maps distance onto recession speed is called *Hubble's Law* and is described by the equation $y = 70x$.

Astronomers use Hubble's Law to estimate the distance to a galaxy based on its apparent recession speed, which is measured by shifts in the spectrum of light from the galaxy. In other words, the speed becomes the domain variable x and the distance becomes the range variable y. The function mapping the speed onto the distance can be described by the equation $x = 70y$. Solving this equation for y, $y = \frac{x}{70}$.

Galaxies NGC 2207 and IC 2163 are 140 million light years away from us in the direction of the Canis Major constellation. They are expected to meld in 500 million years.

$x =$ Speed $\left(\frac{km}{sec}\right)$	$y =$ Distance (Mpc)
0	0
350	5
700	10
1050	15
1400	20

The relations with equations $y = 70x$ and $y = \frac{1}{70}x$ are related. The ordered pairs of each one are found by switching the values of x and y in the other. Two relations that have this property are called *inverse relations*.

> ### Definition of Inverse of a Relation
>
> The **inverse of a relation** is the relation obtained by switching the coordinates of each ordered pair in the relation.

Example 1

Let $f = \{(-3, 2), (-1, 4), (0, 3), (2, 8), (5, 7), (7, 8)\}$. Find the inverse of f.

Solution Switch the coordinates of each ordered pair. *Call the inverse g.* Then $g = \{(2, -3), (4, -1), (3, 0), (8, 2), (7, 5), (8, 7)\}$.

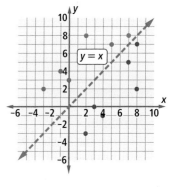

The blue dots at the right are a graph of the function f from Example 1. The red dots are a graph of the inverse g.

Recall that the points (x, y) and (y, x) are reflection images of each other over the line with equation $y = x$. That is, reflection over this line switches the coordinates of the ordered pairs. The graphs of any relation and its inverse are reflection images of each other over the line $y = x$.

The Domain and Range of a Relation and Its Inverse

Recall that the domain of a relation is the set of possible values for the first coordinate, and the range is the set of possible values for the second coordinate. Because the inverse is found by switching the first and second coordinates, the domain and range of the inverse of a relation are the range and domain, respectively, of the original relation. For instance, in Example 1,

$$\text{domain of } f \;=\; \text{range of } g \;=\; \{-3, -1, 0, 2, 5, 7\},$$
$$\text{range of } f \;=\; \text{domain of } g \;=\; \{2, 4, 3, 8, 7\}.$$

The theorem on the next page summarizes the ideas you have read so far in this lesson.

Inverse-Relation Theorem

Suppose f is a relation and g is the inverse of f. Then:

1. If a rule for f exists, a rule for g can be found by switching x and y in the rule for f.

2. The graph of g is the reflection image of the graph of f over the line with equation $y = x$.

3. The domain of g is the range of f, and the range of g is the domain of f.

Caution! The word *inverse*, when used in the term *inverse of a relation*, is different than its use in the phrase *inverse variation*.

Determining Whether the Inverse of a Function Is a Function

The inverse of a relation is a relation. But the inverse of a function is not always a function. In Example 1, g is not a function because it contains the two pairs $(8, 2)$ and $(8, 7)$.

GUIDED

Example 2

Consider the function with domain the set of all real numbers and equation $y = x^2$.

a. What is an equation for the inverse of this function?

b. Graph the function and its inverse on the same coordinate axes.

c. Is the inverse a function? Why or why not?

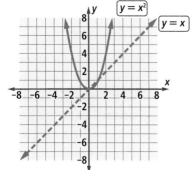

Solution

a. To find an equation for the inverse, switch x and y in the rule for the function. The inverse of the function with equation $y = x^2$ has equation __?__ = __?__.

b. The graph of __?__ is the parabola at the right. The graph of its inverse is the reflection image of the parabola over the line with equation __?__. Copy the graph and add a graph of the inverse to it.

c. Graphs of functions in rectangular coordinates do not include two points with the same first coordinate. The inverse (is/is not) a function because __?__.

In a function, no two points have the same first coordinate. So, if the inverse of a function is a function, no two points of the original function can have the same *second* coordinate. This idea implies that if any horizontal line intersects the graph of a function on a rectangular grid at more than one point, then the function's inverse is not a function. This is called the **horizontal-line test** to check whether the inverse of a function is a function. Notice that the graphs of the functions in Examples 1 and 2 do not pass the horizontal-line test.

Example 3
Two functions are graphed.

i.
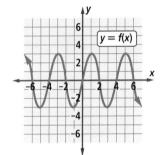

ii.
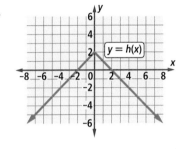

a. Tell whether the inverses of the graphed functions are functions. Explain your reasoning.

b. State the domain and range of each function.

c. State the domain and range of each inverse.

Solution

a. Apply the horizontal-line test. For each function, there is a horizontal line that intersects the graph of the function more than once. Neither the inverse of f nor the inverse of h is a function. Both graphs fail the horizontal-line test.

b. Refer to the graphs above. The domain of f is the set of all real numbers. The range is $\{y \mid -3 \leq y \leq 3\}$. The domain of h is the set of all real numbers. The range is $\{y \mid y \leq 2\}$.

c. Use the Inverse-Relation Theorem. The domain of the inverse of f is the range of f, or $\{x \mid -3 \leq x \leq 3\}$; the range is the domain of f, or the set of all real numbers. The domain of the inverse of h is $\{x \mid x \leq 2\}$; the range is the set of all real numbers.

Questions

1. How can the inverse of a relation be found from the coordinates of the points in the relation?

2. Refer to the functions relating distance and speed at the beginning of the lesson.

    a. At what distance away from us is a galaxy moving at roughly 5000 km/sec?

    b. The photograph at the right shows the disk galaxy NGC 5866 as it appears from Earth, edge-on. NGC 5866 is about 44 million light years, or 13.5 megaparsecs, from the Milky Way. How fast is NGC 5866 receding?

This photograph was taken by the Hubble Space Telescope (named in the astronomer's honor) in June, 2006.

In 3 and 4, let $f = \{(-4, -9), (-3, -7), (0, -1), (3, 5), (4, 7), (6, 11)\}$.

3. a. Find the inverse of f.

    b. Graph f and its inverse on the same set of axes.

    c. How are the two graphs related?

    d. Write an equation for the function f.

    e. Write an equation for the inverse of f.

4. Give the elements of each set.

    a. the domain of f

    b. the range of f

    c. the domain of the inverse of f

    d. the range of the inverse of f

5. Explain how the graphs of any relation f, its inverse, and the line with equation $y = x$ are related.

In 6 and 7, give an equation for the inverse of the relation.

6. $y = 3x$

7. $y = \frac{1}{5}x - 2$

8. Refer to Example 2.

    a. Write an equation for the inverse of $y = x^2$.

    b. Find two points other than those mentioned in the lesson that show the inverse is not a function.

9. The graph of function q is given at the right. Explain if there is a way to tell in advance whether the inverse of q is a function.

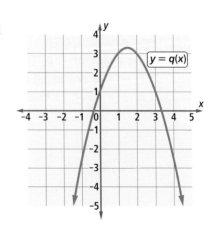

In 10–12, is the inverse of the graphed function a function? Explain.

10.

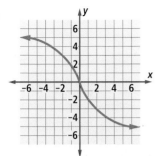

11.

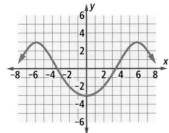

12.

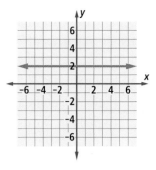

APPLYING THE MATHEMATICS

13. Describe the inverse of the doubling function $D: x \to 2x$.

14. Consider the function f with equation $f(x) = -3$.
   a. Give 5 ordered pairs in f.
   b. Find 5 ordered pairs in the inverse of f.
   c. Draw graphs of f and its inverse on the same set of axes.
   d. What transformation maps the graph of f onto the graph of its inverse?

15. a. Draw the graph of $y = 3x + 7$.
   b. Find an equation for its inverse and solve it for y.
   c. Graph the inverse on the same set of axes.
   d. Is the inverse a function? Explain your answer.
   e. How are the slopes of the graphs of the function and its inverse related?

16. a. Graph the inverse of the absolute-value function $y = |x|$.
   b. Is the inverse a function? Explain why or why not.

17. In 2008, one U.S. dollar ($) was worth about 0.64 euros (€). An item costing $1 cost €0.64. Let U be the cost of an item in U.S. dollars and let E be the cost of an item in euros.
   a. Write an equation for U in terms of E.
   b. Write an equation for E in terms of U.
   c. Are the two functions you found in Parts a and b inverses of each other? Explain your answer.

REVIEW

18. Let c = the cost of a new car. Then $t(c) = 1.07c$ is the cost after a state sales tax of 7%, and $r(c) = c - 3000$ is the cost after a manufacturer's rebate of $3000. **(Lesson 8-1)**

    a. Evaluate $t(r(29{,}350))$.

    b. Find a formula for $t(r(c))$.

    c. Explain in words what $t(r(c))$ represents.

    d. Evaluate $r(t(29{,}350))$.

    e. Find a formula for $r(t(c))$.

    f. Explain in words what $r(t(c))$ represents.

    g. If you were the State Treasurer, which would you prefer, $t(r(c))$ or $r(t(c))$, and why?

19. Let $f(x) = 5x - 7$, $g(x) = \frac{1}{5}(x + 12)$, and $h(x) = \frac{1}{5}(x + 7)$.
    (Lesson 8-1)

    a. Find $f \circ g(3)$.   b. Find $g \circ f(3)$.   c. Find $f \circ h(x)$.

    d. What is another name for the function $f \circ h$?

20. Given $r(x) = 2x^2 - 1$ and $s(x) = x^2 + 3$, find $r \circ s(x)$. **(Lesson 8-1)**

21. **True or False** For all real numbers x, $|-3x| = 3x$. Justify your answer. **(Lesson 6-2)**

22. **True or False** Two matrices A and B are inverses if $AB = \begin{bmatrix} 1 & 0 \\ 0 & 1 \end{bmatrix}$. Explain why or why not. **(Lesson 5-5)**

23. In this figure, $\triangle ADB \sim \triangle AEC$. If $DE = 12$, find BC and CE. **(Previous Course)**

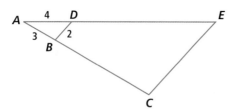

EXPLORATION

24. Some functions are their own inverses. One such function has the equation $y = \frac{1}{x}$.

    a. Find at least one more function that is its own inverse.

    b. Graph $y = \frac{1}{x}$ and your answer(s) to Part a. What property do the graphs share?

8-3 Properties of Inverse Functions

Vocabulary

inverse function, f^{-1}

▶ **BIG IDEA** When a function has an inverse, the composite of the function and its inverse in either order is the identity function.

Using an Inverse Function to Decode a Message

In Chapter 5, you used matrices to code and decode messages. You can use functions to do this too. For instance, suppose you and your friend have agreed on a key that pairs numbers with the letters they stand for and an encoding function $f(x) = 3x + 11$. This means that your friend began with a number code for a letter (the input), multiplied that number by 3, and then added 11 to get the output. For example, if $A = 65$ it is encoded as $3 \cdot 65 + 11 = 206$. Suppose your friend now sends you this coded message:

212 248 215 218 260 107 206 257 218 107 212 248 248 239

To decode the message, you "undo" your friend's coding; subtract 11 and divide the result by 3 to get the original input: $\frac{206 - 11}{3} = 65$. This "undoing" function is the **inverse function** of f. In Question 13, you will use this inverse function to decode the above message.

Mental Math

Give an equation for the graph of the parabola $y = -2x^2$ under the given translation.

a. $T_{1,3}$

b. $T_{-2,-2}$

c. $T_{0.3, 5.4}$

Formulas for Inverses Using f(x) Notation

There are many ways to obtain a formula for the decoding (inverse) function described above. Here is one way.

Start with the original function equation. $\qquad f(x) = 3x + 11$

Replace $f(x)$ with y. $\qquad y = 3x + 11$

Use the Inverse-Relation Theorem and switch x and y in the equation for f. This represents the decoding function in which y is the input from your friend and x is the output. $\qquad x = 3y + 11$

Solve for y. For a given output x from your friend's encoding function, you can use this function to calculate the original input y. $\qquad y = \frac{x - 11}{3}$

Substitute $g(x)$ for y. This describes the inverse, g. $\qquad g(x) = \frac{x - 11}{3}$

Composites of a Function and Its Inverse

Activity 1

MATERIALS CAS

Work with a partner to complete the following steps. One of you should work with $f(x) = 12x - 4$, the other with $f(x) = \frac{1}{3} + \frac{x}{12}$.

Step 1 Using the method described on the previous page, find the inverse of your function. Call your inverse g and write a formula for it.

Step 2 On your CAS, define your functions f and g. Evaluate $f(g(x))$ and $g(f(x))$.

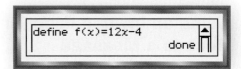

Step 3 Compare your results with those of your partner.

Your results for $f(g(x))$ and $g(f(x))$ in Activity 1 hold for any functions f and g that are inverses. The converse is also true. Both are summarized in the following theorem.

> ### Inverse Functions Theorem
>
> Two functions f and g are inverse functions if and only if:
>
> (1) For all x in the domain of f, $g(f(x)) = x$, and
>
> (2) for all x in the domain of g, $f(g(x)) = x$.

Proof We must prove a statement and its converse. So the proof has two parts.

    (I) *The "only if" part:* Suppose f and g are inverse functions. Let (a, b) be any ordered pair in the function f. Then $f(a) = b$. By the definition of inverse, the ordered pairs in g are the reverse of those in f, so (b, a) is an ordered pair in the function g. Thus $g(b) = a$. Now take the composites.

    For any number a in the domain of f: $g(f(a)) = g(b) = a$.

    For any number b in the domain of g: $f(g(b)) = f(a) = b$.

    (II) *The "if" part:* Suppose (1) and (2) in the statement of the theorem are true. Again let (a, b) be any point on the function f. Then $f(a) = b$, and so $g(f(a)) = g(b)$. But using (1), $g(f(a)) = a$, so, by transitivity, $g(b) = a$. This means that (b, a) is in the function g. So g contains all the points obtained by reversing the coordinates in f.

    By the same reasoning we can show that f contains all points obtained by reversing the coordinates of g. Thus, f and g are inverse functions.

 QY1

> ▶ **QY1**
>
> If $f(x) = 2x + 6$ and $g(x) = \frac{1}{2}x - 6$, are f and g inverses of each other?

Notation for Inverse Functions

Recall that an *identity function* is a function that maps each object in its domain onto itself. Another way of stating the Inverse Functions Theorem is to say that the composite of two functions that are inverses of each other is the identity function I with equation $I(x) = x$.

When an operation on two elements of a set yields an identity element for that operation, then we call the elements *inverses*. For example, 2 and $\frac{1}{2}$ are multiplicative inverses because $2 \cdot \frac{1}{2} = 1$, the identity element for multiplication. This is the reason that we call g the inverse of f, and f the inverse of g: $f(g(x)) = g(f(x)) = I(x)$, the identity function.

The multiplicative inverse of a number x is designated by x^{-1}. Similarly, when a function f has an inverse, we designate the **inverse function** by the symbol f^{-1}, read "f inverse." For instance, for the functions in Activity 1 you showed that for all x, $f(f^{-1}(x)) = f^{-1}(f(x)) = x$, which is read "$f$ of f inverse of x equals f inverse of f of x equals x."

Example

Let $h : x \rightarrow \dfrac{8(x - 2) + 5}{3}$. Find a rule for h^{-1}.

Solution Use a process like that shown for the function f at the beginning of the lesson. From the given information,

$$h(x) = \frac{8(x - 2) + 5}{3}.$$

Simplify. $= \dfrac{? - ?}{3}$

Substitute y for $h(x)$.

$$y = \underline{\quad ? \quad}$$

An equation for the inverse is found by switching x and y.

$$x = \underline{\quad ? \quad}$$

Solve this equation for y.

$$3x = \underline{\ ?\ } - \underline{\ ?\ }$$

$$3x + \underline{\ ?\ } = \underline{\ ?\ }$$

$$y = \underline{\quad ? \quad}$$

So $h^{-1}(x) = \underline{\quad ? \quad}$.

 QY2

> ▶ **QY2**
>
> "Halving" is an inverse for the "doubling" function. What is an inverse for the squaring function?

In earlier lessons, you worked with the functions with equations $y = x^n$, "taking the nth power," and $y = x^{\frac{1}{n}}$, "taking the nth root." For all $x \geq 0$, it is reasonably easy to show that these functions are inverse functions.

Power Function Inverse Theorem

If $f(x) = x^n$ and $g(x) = x^{\frac{1}{n}}$ and the domains of f and g are the set of *nonnegative* real numbers, then f and g are inverse functions.

Proof First, show that $f \circ g(x) = x$ for all x in the domain of g.

Substitute. $\quad f \circ g(x) = f\left(x^{\frac{1}{n}}\right) \quad$ Definition of g

Because x is a nonnegative number, $x^{\frac{1}{n}}$ is always defined and you can apply f.

$$= \left(x^{\frac{1}{n}}\right)^n \quad \text{Definition of } f$$
$$= x^1 = x \quad \text{Power of a Power Postulate}$$

Now you need to show that $g \circ f(x) = x$ for all x in the domain of f. You are asked to do this in Question 12.

An instance of the Power Function Inverse Theorem is illustrated by the graphs of $f(x) = x^6$ and $f^{-1}(x) = x^{\frac{1}{6}}$ at the left below. The graphs are reflection images of each other over the line with equation $y = x$. Notice also that the domain of these functions is the set of nonnegative real numbers.

The function $h(x) = x^6$ with domain the set of *all* real numbers is graphed at the right below. Its inverse is not a function. Notice that the graph of h does not pass the horizontal-line test.

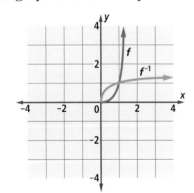

$f(x) = x^6$
Domain of f:
set of nonnegative real numbers

$f^{-1}(x) = x^{\frac{1}{6}}$
Domain of f^{-1}:
set of nonnegative real numbers

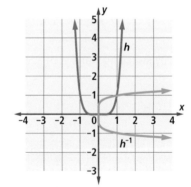

$h(x) = x^6$
Domain of h:
set of real numbers

The inverse of h
is not a function.

A CAS can help you find equations for inverse relations and to limit domains of functions so that their inverses are also functions.

MATERIALS CAS

Use a CAS to find an equation for the inverse of $y = f(x) = x^6$.

Step 1 Clear all variables in a CAS. Define $f(x) = x^6$.

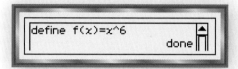

```
define f(x)=x^6
                        done
```

Step 2 Switch x and y and solve $x = f(y)$ for y.

Step 3 Your display should show an equation for the inverse of f. Is this inverse a function? If not, on what domain of f is f^{-1} a function?

Questions

COVERING THE IDEAS

1. Suppose f is a function. What does the symbol f^{-1} represent, and how is it read?

2. Refer to Activity 1. Show that $f(g(2)) = g(f(2))$.

3. Let $f(x) = 5 - 2x$ and $g(x) = \frac{x-5}{-2}$. Are f and g inverses of each other? Justify your answer.

4. The function $h : x \rightarrow 5x + 7$ has an inverse h^{-1} which is a function.
   a. Find a formula for $h^{-1}(x)$.
   b. Check your answer to Part a by finding $h \circ h^{-1}(x)$ and $h^{-1} \circ h(x)$.

5. Consider the "adding 15" function. Call it A.
   a. What is a formula for $A(x)$?
   b. Give a formula for $A^{-1}(x)$.
   c. What is an appropriate name for A^{-1}?
   d. What is $A^{-1} \circ A(63)$?

In 6 and 7, an equation for a function f is given. The domain of f is the set of real numbers. Is the inverse of f a function? If it is, find a rule for f^{-1}. If not, explain why not.

6. $f(x) = 21x$

7. $f(x) = 21$

8. **Fill in the Blank** For any function f that has an inverse, if x is in the range of f, then $f \circ f^{-1}(x) = \underline{\ ?\ }$.

9. If $f(x) = x^7$ with domain the set of nonnegative real numbers, give a formula for $f^{-1}(x)$.

10. Let $m(x) = \frac{x}{5}$. This function could be named "dividing by 5."
    a. Find an equation for m^{-1}.
    b. Give an appropriate name to m^{-1}.

11. a. Use a CAS to find an equation for the inverse of
    $$f(x) = -\frac{(x+12)^3}{2}.$$
    b. Is the inverse of f a function? If so, what is its domain? If not, over what domain of f is the inverse a function?

12. Prove the second part of the Power Function Inverse Theorem in this lesson.

13. Recall the code your friend sent you at the beginning of the lesson. Use the following table to decode the message. A space is represented by 32.

Code	65	66	67	68	69	70	71	72	73	74	75	76	77
Letter	A	B	C	D	E	F	G	H	I	J	K	L	M
Code	78	79	80	81	82	83	84	85	86	87	88	89	90
Letter	N	O	P	Q	R	S	T	U	V	W	X	Y	Z

APPLYING THE MATHEMATICS

14. The function S with $S(n) = (n-2) \cdot 180°$ maps the number of sides of a convex polygon onto the sum of the measures of its interior angles.
    a. Find a formula for $S^{-1}(n)$.
    b. Use the formula to determine the number of sides of a polygon when the sum of the interior angle measures of the polygon is 3780.

15. Let $g(x) = x^{\frac{2}{3}}$ and $h(x) = x^{\frac{3}{2}}$ with $x \geq 0$.
    a. Graph g and h on the same set of axes.
    b. Describe how the graphs are related.
    c. Find $h(g(x))$ and $g(h(x))$.
    d. **Fill in the Blank** The answer to Part c means that since $64^{\frac{3}{2}} = 512$, $\underline{\ ?\ }$.
    e. Are g and h inverses? Why or why not?

In 16–18, find an equation for the inverse of the function with the given equation in the form $y =$ ___?___.

16. $y = \frac{1}{x}$

17. $y = \frac{1}{2}x + 7$

18. $y = x^3 - 2$

19. If f is any function that has an inverse, what is $(f^{-1})^{-1}$? Explain your answer in your own words.

20. a. Explain why the inverse of $g(x) = x^2$ is not a function when the domain of g is the set of all real numbers.

   b. Split the graph of $g(x) = x^2$ along its line of symmetry. For each half,

      i. State the domain and the corresponding range.

      ii. Give a formula for the inverse of that half.

21. **Multiple Choice** Consider the absolute-value function $y = |x| - 1$ graphed below. Which of the following domains gives a function whose inverse is also a function?

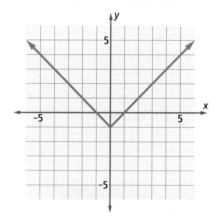

   A $\{x \mid -2 \le x \le 5\}$ B $\{x \mid x \ge 1\}$ C $\{x \mid x \ge -4.5\}$

REVIEW

22. Sandy makes $32,000 per year plus $24 per hour for each hour over 2000 hours that she works per year. An equation for her total income $I(h)$ when she works h hours is $I(h) = 32{,}000 + 24(h - 2000)$, for $h > 2000$. **(Lesson 8-2)**

   a. Find a rule for the inverse of I.

   b. What does the inverse represent?

23. **Fill in the Blanks** Suppose $f : x \to 2x + 5$ and $g : x \to x^{-2}$. **(Lessons 8-2, 8-1)**

   a. $f \circ g(-2) =$ ___?___

   b. $g \circ f(-2) =$ ___?___

   c. $f(g(x)) =$ ___?___

   d. $g(f(x)) =$ ___?___

24. When a beam of light in air strikes
    the surface of water it is *refracted*, or
    bent. Below are the earliest known
    data (translated into modern units) on
    the relation between *i*, the measure of
    the angle of incidence in degrees, and
    r, the measure of the angle of refraction
    in degrees. The measurements are recorded
    in *Optics* by Ptolemy, a Greek scientist
    who lived in the second century BCE.
    (Lessons 6-6, 3-5)

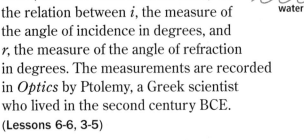

i	10°	20°	30°	40°	50°	60°	70°	80°
r	8°	15.5°	22.5°	29°	35°	40.5°	44.5°	50°

a. Draw a scatterplot of these data.

b. Fit a quadratic model to these data.

c. Fit a linear model to these data.

d. Which model seems more appropriate?
   Explain your decision.

25. Line ℓ is parallel to the line with equation $y = \frac{1}{4}x + 7$. **(Lesson 3-1)**

a. What is the slope of ℓ?

b. What numbers are possible y-intercepts of ℓ?

EXPLORATION

26. The *Shoe and Sock Theorem*: Let *g* be the function with the
    rule "put your sock on the input." Let *f* be the function with the
    rule "put your shoe on the input."

a. Explain in words how to find the output of $f \circ g$ (your foot).

b. What are the rules for g^{-1} and f^{-1}?

c. What is the result of applying $f \circ g$ followed by $(f \circ g)^{-1}$ to
   your foot?

d. What two steps must be taken to apply $(f \circ g)^{-1}$ to your foot?
   In what order must the steps be taken?

e. Translate your answer to Part d into symbols: $(f \circ g)^{-1} = $
   ___?___ .

f. Now let $f(x) = x^5$ and $g(x) = 7x - 19$. Find formulas for $f^{-1}(x)$
   and $g^{-1}(x)$ and use them to find a formula for $(f \circ g)^{-1}(x)$.

QY ANSWERS

1. No; $f(g(x)) = f\left(\frac{1}{2}x - 6\right)$
   $= 2\left(\frac{1}{2}x - 6\right) + 6 = x - 12$
   $+ 6 = x - 6 \neq x$

2. the square root function

Lesson
8-4

Radical Notation for *n*th Roots

Vocabulary

radical sign, $\sqrt{}$

$\sqrt[n]{x}$ when $x \geq 0$

geometric mean

> ▶ **BIG IDEA** For any integer *n*, the largest real *n*th root of *x* can be represented either by $x^{\frac{1}{n}}$ or by $\sqrt[n]{x}$.

As you have learned, the positive *n*th root of a positive number *x* can be written as a power of *x*, namely as $x^{\frac{1}{n}}$. This notation allows all the properties of powers to be used with these *n*th roots. Another notation, using the **radical sign** $\sqrt{}$, can be used to represent all of the positive *n*th roots and some other numbers. However, it is more difficult to see the properties of powers with this notation.

Recall that when *x* is positive, $\sqrt{x}$ stands for the positive square root of *x*. Because $x^{\frac{1}{2}}$ also stands for this square root when *x* is positive, $\sqrt{x} = x^{\frac{1}{2}}$. Similarly the *n*th root of a positive number can be written in two ways.

Definition of $\sqrt[n]{x}$ when $x \geq 0$
When *x* is nonnegative and *n* is an integer ≥ 2, then $\sqrt[n]{x} = x^{\frac{1}{n}}$.

Thus, when *x* is positive, $\sqrt[n]{x}$ is the positive *n*th root of *x*. When $n = 2$, we do not write $\sqrt[2]{x}$, but use the more familiar symbol $\sqrt{x}$.

 QY1

$\sqrt[n]{x}$ is used to calculate a type of average called a *geometric mean*.

The Geometric Mean

You know several ways to describe a set of values with a single statistic such as an average test score or a median housing price. Such values are called *measures of center,* or *measures of central tendency.*

Suppose a data set has *n* values. If you add the values and divide by *n,* you have calculated the *arithmetic mean* of the set. This is commonly called *the average* of the set, but there are other averages. If, instead of adding, you *multiply* the numbers in the set, and instead of dividing you *take the nth root* of the product of the items in the list, you obtain the **geometric mean**.

Mental Math

Vanessa runs an apple orchard. She sells apples for $1.50 per pound, rounded to the nearest half pound.

a. How much does she charge for 3.2 pounds of apples?

b. How much does she charge for 4.6 pounds of apples?

c. Suppose one apple weighs about a third of a pound. About how many apples can you buy for $6?

d. About how many apples can you buy for $10?

> ▶ **QY1**
>
> Solve for *x*:
> $\sqrt[7]{78{,}125} = 78{,}125^{x}$.

The geometric mean may be used when numbers are quite dispersed, to keep one very large number from disproportionately affecting the measure of center. For this reason, the geometric mean is the standard measure of center for data about pollutants and contaminants. The geometric mean is also used to compute an overall rate of percent increase or decrease, as you will see in the Questions.

Example 1

In a water-quality survey, 8 samples give the following levels of *E. coli* bacteria.

Sample	1	2	3	4	5	6	7	8
Count (per 100 mL)	2	54	145	38	597	1152	344	87

Use the geometric mean to compute an average level of bacteria.

Solution Because there are eight numbers, the geometric mean is the 8th root of their product.

$$\sqrt[8]{2 \cdot 54 \cdot 145 \cdot 38 \cdot 597 \cdot 1152 \cdot 344 \cdot 87} \approx 102.6.$$

A typical sample has about 103 bacteria per 100 mL.

 QY2

▶ **QY2**

A ninth sample is taken with an *E. coli* count of 436 per 100 mL. Estimate the average number of bacteria across all nine samples.

Which *n*th Root Does $\sqrt[n]{x}$ Represent?

The symbol $\sqrt[n]{x}$, like $x^{\frac{1}{n}}$, does not represent all *n*th roots of x. When x is positive and n is even, x has two real *n*th roots, but only the *positive* real root is denoted by $\sqrt[n]{x}$. Thus, although 3, –3, 3i, and –3i are all fourth roots of 81, $\sqrt[4]{81} = 81^{\frac{1}{4}} = 3$ only. The negative real fourth root can be written $-81^{\frac{1}{4}}$ or $-\sqrt[4]{81}$, both of which equal –3. Note that the four complex solutions to $x^4 = 81$ are $\pm \sqrt[4]{81}$ and $\pm \sqrt[4]{-81}$, that is, ±3 and ±3i.

Example 2

a. Set a CAS to real-number mode. Evaluate $\sqrt[4]{6561}$ and explain its meaning.

b. Set a CAS to complex-number mode. Find all fourth roots of 6561.

Solution

a. $\sqrt[4]{6561} = \underline{\ ?\ }^{\frac{1}{4}}$ represents the (positive/negative) number whose $\underline{\ ?\ }$ power is 6561. $(6561)^{\frac{1}{4}} = \underline{\ ?\ }$ because $\underline{\ ?\ }^4 = 6561$. So, $\sqrt[4]{6561} = \underline{\ ?\ }$.

b. On a CAS, solve the equation $\underline{\ ?\ }$ in complex-number mode. The solutions are $\underline{\ ?\ }$, $\underline{\ ?\ }$, $\underline{\ ?\ }$, $\underline{\ ?\ }$. These solutions are all the $\underline{\ ?\ }$ roots of 6561.

An early form of the $\sqrt{\ }$ symbol (which looked like a check mark) first appeared in the 1500s, and René Descartes modified it in the early 1600s into the form we use today. The origin of the word *radical* is the Latin word *radix*, which means "root."

Albert Girard suggested the symbol $\sqrt[n]{\ }$, and it first appeared in print in 1690. As you know from Chapter 7, most graphing calculators let you enter the rational power form x^(1/n) to find $\sqrt[n]{x} = x^{\frac{1}{n}}$. Some graphing calculators also let you enter radical forms of roots as shown in Example 2.

Radicals for Roots of Powers

Because radicals are powers, all properties of powers listed in Chapter 7 apply to radicals. In particular, because $\sqrt[n]{x} = x^{\frac{1}{n}}$ for $x > 0$, the mth powers of these numbers are equal. That is, $(\sqrt[n]{x})^m = (x^{\frac{1}{n}})^m$, which equals $x^{\frac{m}{n}}$. If x is replaced by x^m in the definition of $\sqrt[n]{x}$, the result is $\sqrt[n]{x^m} = (x^m)^{\frac{1}{n}}$, which also equals $x^{\frac{m}{n}}$. Thus, there are two radical expressions equal to $x^{\frac{m}{n}}$.

> **Root of a Power Theorem**
>
> For all positive integers $m \geq 2$ and $n \geq 2$,
> $$\sqrt[n]{x^m} = (\sqrt[n]{x})^m = x^{\frac{m}{n}} \text{ when } x \geq 0.$$

 QY3

Notice that $x \geq 0$ in both the Root of a Power Theorem and QY3. This is because the Root of a Power Theorem only applies to positive bases.

 QY4

▶ **QY3**

Suppose $x \geq 0$. Use the Root of a Power Theorem to write $\sqrt[6]{x^{18}}$ in two different ways.

▶ **QY4**

Show that $\sqrt[6]{x^{18}} \neq x^3$ when $x = -2$.

Roots of Roots

Consider the sequence 625, 25, 5, $\sqrt{5}$, ... in which each number is the square root of the preceding number. You can define this sequence recursively.

$$\begin{cases} s_1 = 625 \\ s_n = \sqrt{s_{n-1}}, \text{ for integers } n \geq 2 \end{cases}$$

So, $s_2 = \sqrt{625} = 25$

$s_3 = \sqrt{\sqrt{625}} = \sqrt{25} = 5$

$s_4 = \sqrt{\sqrt{\sqrt{625}}} = \sqrt{\sqrt{25}} = \sqrt{5} \approx 2.24$

 QY5

Rewriting the radicals as rational exponents provides a way to deal with roots of roots.

> ▶ **QY5**
>
> Is this a geometric sequence? Why or why not?

Example 3

Rewrite $\sqrt{\sqrt{\sqrt{6}}}$ using rational exponents. Is this expression an *n*th root of 6? Justify your answer.

Solution $\sqrt{\sqrt{\sqrt{6}}} = \left(\left(6^{\frac{1}{2}}\right)^{\frac{1}{2}}\right)^{\frac{1}{2}} = \left(6^{\frac{1}{2}}\right)^{\frac{1}{4}} = 6^{\frac{1}{8}}$ by the

Power of a Power Postulate. So, by the

definition of nth root, $\sqrt{\sqrt{\sqrt{6}}}$ is the positive 8th root of 6.

Check Enter $\sqrt{\sqrt{\sqrt{6}}}$ on a calculator to see that $\sqrt{\sqrt{\sqrt{6}}} = 6^{\frac{1}{8}}$.

In general, when $x > 0$, it is more common to write $x^{\frac{1}{8}}$ as the single radical $\sqrt[8]{x}$, rather than as $\sqrt{\sqrt{\sqrt{x}}}$. But if you only have a square root key on your calculator, then it is nice to know you can calculate 8th roots.

Questions

1. Who is credited with first using the radical symbol in its current form?

2. Who is credited with first using the symbol $\sqrt[n]{}$?

3. **True or False** $\sqrt[n]{x} = (x)^{\frac{1}{n}}$ for all x.

4. a. How is the geometric mean of n numbers calculated?
   b. Identify a situation in which the geometric mean is the preferred measure of center.

5. The table at the right gives the masses of the eight major planets as a ratio with Earth's mass. Find the geometric mean of these masses.

Planet	Planet's Mass / Earth's Mass
Mercury	0.06
Venus	0.82
Earth	1
Mars	0.11
Jupiter	318
Saturn	95
Uranus	14.5
Neptune	17.2

The Mars rover Spirit took this photo from the eastern edge of the plateau called Home Plate.

6. Evaluate without a calculator.
   a. $\sqrt{64}$
   b. $\sqrt[3]{64}$
   c. $\sqrt[6]{64}$
   d. $\sqrt[10]{64^{10}}$

In 7 and 8, use a calculator to approximate to the nearest hundredth.

7. $\sqrt[17]{7845}$

8. $\sqrt[8]{1 \cdot 2 \cdot 3 \cdot 4 \cdot 5 \cdot 6 \cdot 7 \cdot 8}$

9. **Fill in the Blank** Complete the following statement of the Root of a Power Theorem. For all positive integers m and n with $m \geq 2$ and $n \geq 2$, when $x \geq 0$, $x^{\frac{m}{n}} = \underline{\quad?\quad}$.

10. Refer to Example 2. Write all the complex fourth roots of 625.

11. a. Find all the complex fourth roots of 14,641.
    b. Which root in Part a is $\sqrt[4]{14{,}641}$?

In 12–14, write as a single power using a rational exponent. Assume all variables are positive.

12. $\sqrt[6]{z^{10}}$ 13. $\sqrt[5]{c^{15}}$ 14. $\left(\sqrt[14]{t}\right)^7$

15. Rewrite $\sqrt{\sqrt{\sqrt{x}}}$ with a rational exponent, for $x \geq 0$.

16. Rewrite $\sqrt{\sqrt{\sqrt{81}}}$ using a single radical sign.

APPLYING THE MATHEMATICS

Multiple Choice In 17 and 18, which of the expressions is not equivalent to the other two?

17. **A** $6^{\frac{12}{36}}$ **B** $\sqrt[12]{6^2}$ **C** $\sqrt[12]{36}$

18. **A** $p^{\frac{3}{2}}$ **B** $\left(\sqrt[4]{p^2}\right)^6$ **C** $\left(\sqrt{\sqrt{p}}\right)^6$

19. From the data on page 514, you can compute that the unadjusted median home value in the United States in 2000 was more than 40 times the median value in 1940. The decade-to-decade percentage increase in housing values is summarized in the table below.

Years	1940–1950	1950–1960	1960–1970	1970–1980	1980–1990	1990–2000
% Increase (rounded)	150	62	43	178	68	51
Size-change factor	2.50	?	?	?	?	?

   a. Fill in the table by converting each percent into a size-change factor.

   b. Compute the geometric mean of the size-change factors you found in Part a, and determine the average percentage increase per decade in home values from 1940 to 2000.

20. Recall from Chapter 7 the Baravelle spiral, in which squares are created by connecting midpoints of sides of larger squares. In the Baravelle spiral shown below, the leg length of right triangle A_1 is 1 unit.

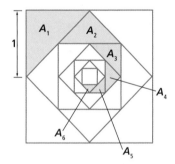

   a. What are the lengths of the hypotenuses of right triangles A_1, A_2, and A_3?

   b. The lengths of the hypotenuses $L_1, L_2, L_3, \ldots$ are a geometric sequence. Write an explicit formula for L_n, the sequence of hypotenuse lengths in this Baravelle spiral.

21. A sphere has volume $V \text{ in}^3$, where $V = \frac{4}{3}\pi r^3$. Express the length of the radius in terms of the volume

    a. using radical notation.

    b. using a rational exponent.

22. Consider the formula $m = 1.23x^3 b$ where m, x, and b are all positive. Solve the formula for x using radical notation.

23. Use the expression $\sqrt[4]{\sqrt[4]{43,046,721z^{48}}}$.

    a. Rewrite the expression with a rational exponent.

    b. Rewrite the expression as one radical.

    c. Evaluate the expression when $z = 2$.

In 24 and 25, write each expression in simplest radical form using no fraction exponents. Assume all variables are positive.

24. $\sqrt{\sqrt{y^{\frac{1}{4}}}}$

25. $\dfrac{\sqrt{k^{\frac{1}{2}}}}{\sqrt{\sqrt{k}}}$

REVIEW

26. Let $h(x) = x^2$ and $k(x) = x^{-\frac{1}{2}}$. **(Lessons 8-3, 8-1, 7-8, 7-2)**

    a. Find $h \circ k(x)$.

    b. Are h and k inverses of each other? How can you tell?

27. The height of a baseball that has been hit is described by the parabola $h(x) = -0.00132x^2 + 0.545x + 4$, where x is the distance in feet from home plate. The effect of a 15-mph wind in the direction the ball is traveling is given by $w(x) = 0.9x$. **(Lessons 8-1, 6-4)**

    a. Which of $w(h(x))$ or $h(w(x))$ represents this situation? (*Hint:* The wind is blowing before the ball is pitched and hit.)

    b. Write a formula for your answer to Part a.

    c. How far from home plate will the ball land if there is no wind?

    d. How far from home plate will the ball land if there is a 15-mph wind in the direction of travel?

In 28 and 29, write without an exponent. Do not use a calculator. (Lesson 7-8)

28. $9^{-\frac{3}{2}}$

29. $-\dfrac{1}{5}^{-3}$

30. Write the reciprocal of $3 + 2i$ in $a + bi$ form.
    (Lesson 6-9)

31. If $g(t) = \dfrac{\frac{t-4}{t+3}}{\frac{t-2}{t+1}}$, what values of t are not in the domain

    of g? (Lesson 1-4)

32. A math text from 1887 gives the following apothecary weights: 3 scruples = 1 dram, 8 drams = 1 ounce. How many scruples are in two and a half ounces? (Previous Course)

In colonial times, apothecaries provided medical treatment, prescribed medicine, and even performed surgery.

EXPLORATION

33. In Parts a–c, use a calculator to estimate to the nearest hundredth.

    a. $\sqrt{1 + \sqrt{1 + \sqrt{1}}}$

    b. $\sqrt{1 + \sqrt{1 + \sqrt{1 + \sqrt{1}}}}$

    c. $\sqrt{1 + \sqrt{1 + \sqrt{1 + \sqrt{1 + \sqrt{1}}}}}$

    d. Use a CAS to solve the equation $x = \sqrt{1 + x}$ such that $x \geq 0$.

    e. Approximate your solution in Part d to the nearest hundredth, and compare it to your answers in Parts a, b, and c. What do you think is happening?

Lesson 8-5

Products with Radicals

> **BIG IDEA** The product of the *n*th roots of nonnegative numbers is the *n*th root of the product of the numbers.

Activity

MATERIALS CAS

Clear variables *a* and *b* and set the CAS to real-number mode.

Step 1 Enter the following expressions, with $a > 0$ and $b > 0$, into the CAS and record the results.

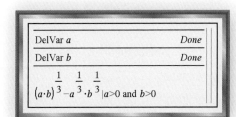

DelVar *a*	Done
DelVar *b*	Done
$(a \cdot b)^{\frac{1}{3}} - a^{\frac{1}{3}} \cdot b^{\frac{1}{3}} \mid a{>}0 \text{ and } b{>}0$	

a. $(a \cdot b)^{\frac{1}{3}} - a^{\frac{1}{3}} \cdot b^{\frac{1}{3}}$

b. $(a \cdot b)^{\frac{1}{12}} - a^{\frac{1}{12}} \cdot b^{\frac{1}{12}}$

c. $a^{\frac{1}{4}} \cdot b^{\frac{1}{4}} - (a \cdot b)^{\frac{1}{4}}$

d. $a^{\frac{1}{7}} \cdot b^{\frac{1}{7}} - (a \cdot b)^{\frac{1}{7}}$

Step 2 Based on the results from Step 1, make a conjecture:
$(a \cdot b)^{\frac{1}{n}} - a^{\frac{1}{n}} \cdot b^{\frac{1}{n}} = \underline{\ \ ?\ \ }$.

Step 3 Based on your conjecture in Step 2, make another conjecture:
$(a \cdot b)^{\frac{1}{n}} = \underline{\ \ ?\ \ }$.

The results of the Activity suggest a property of *n*th roots. This property can also be derived another way. Recall that for all nonnegative numbers *a* and *b*,

$$\sqrt{ab} = \sqrt{a} \cdot \sqrt{b}.$$

Rewriting the above equation with rational powers instead of radicals you can see that this property of radicals is a special case of the Power of a Product Postulate, $(ab)^m = a^m \cdot b^m$.

$$(ab)^{\frac{1}{2}} = a^{\frac{1}{2}} \cdot b^{\frac{1}{2}}$$

If you now let $m = \frac{1}{n}$ in the Power of a Product Postulate, you obtain a theorem about the product of *n*th roots.

Mental Math

Simplify.

a. $\dfrac{w^{17}}{w^{12}}$

b. $\dfrac{x^3}{x^6} \cdot x^5$

c. $4y\left(\dfrac{3y}{y^4}\right)^2$

d. $\left(8wz^2 + \dfrac{z^3}{2w^2}\right)w^3$

> ### Root of a Product Theorem
>
> For any nonnegative real numbers x and y, and any integer $n \geq 2$,
>
> $$(xy)^{\frac{1}{n}} = x^{\frac{1}{n}} \cdot y^{\frac{1}{n}}. \qquad \text{power form}$$
>
> $$\sqrt[n]{xy} = \sqrt[n]{x} \cdot \sqrt[n]{y}. \qquad \text{radical form}$$

Multiplying Radicals

You can use the Root of a Product Theorem to multiply nth roots.

Example 1

Calculate $\sqrt[3]{50} \cdot \sqrt[3]{20}$ without a calculator.

Solution Rewrite the given expression using the Root of a Product Theorem.

$$\sqrt[3]{50} \cdot \sqrt[3]{20} = \sqrt[3]{50 \cdot 20} \qquad \text{Root of Product Theorem}$$

$$= \sqrt[3]{1000} \qquad \text{Arithmetic}$$

$$= 10 \qquad \text{Definition of cube root}$$

Check Use a calculator to find 3-place decimal approximations for $\sqrt[3]{50}$ and $\sqrt[3]{20}$.

$$\sqrt[3]{50} \approx 3.684 \qquad \sqrt[3]{20} \approx 2.714$$

Multiply the decimals.

$3.684 \cdot 2.714 = 9.998376$, close enough given the estimates.

GUIDED

Example 2

Assume that $x \geq 0$. Perform the multiplication: $\sqrt[4]{5x} \cdot \sqrt[4]{125x^3}$.

Solution 1 Rewrite using the Root of a Product Theorem.

$$\sqrt[4]{5x} \cdot \sqrt[4]{125x^3} = \sqrt[4]{\underline{?}}$$

Now use the Root of a Product Theorem to rewrite again.

$$= \sqrt[4]{\underline{?}} \cdot \sqrt[4]{\underline{?}}$$

$$= \underline{?}$$

Solution 2 Convert to rational exponents.

$$\sqrt[4]{5x} \cdot \sqrt[4]{125x^3} = (5x)^{\frac{1}{4}} \cdot (\underline{\quad?\quad})^{\frac{1}{4}}$$

$$= (\underline{\quad?\quad})^{\frac{1}{4}}$$

$$= 625^{\frac{1}{4}} \cdot (\underline{\quad?\quad})^{\frac{1}{4}}$$

$$= 5x$$

Simplifying Radicals

In Example 2, you used the Root of a Product Theorem to rewrite an nth root as a product. For instance, $\sqrt[3]{240}$ can be rewritten several ways:

$$\sqrt[3]{240} = \sqrt[3]{2 \cdot 120} = \sqrt[3]{2} \cdot \sqrt[3]{120};$$

$$\sqrt[3]{240} = \sqrt[3]{8 \cdot 30} = \sqrt[3]{8} \cdot \sqrt[3]{30};$$

$$\sqrt[3]{240} = \sqrt[3]{16 \cdot 15} = \sqrt[3]{16} \cdot \sqrt[3]{15}.$$

Because 8 is a perfect cube, the second form shows that $\sqrt[3]{240} = 2\sqrt[3]{30}$. Some people call $2\sqrt[3]{30}$ the *simplified form* of $\sqrt[3]{240}$. In general, to simplify an nth root, rewrite the expression under the radical sign as a product of perfect nth powers and other factors. Then apply the Root of a Product Theorem.

Example 3

Suppose $r \geq 0$ and $s \geq 0$. Simplify the expression $\sqrt[3]{64r^6s^{15}}$.

Solution 1 Because it is a third root, identify perfect third powers in the expression under the radical.

$$64 = 4^3, r^6 = (r^2)^3, \text{ and } s^{15} = (s^5)^3$$

Rewrite and simplify.

$$\sqrt[3]{64r^6s^{15}} = \sqrt[3]{4^3(r^2)^3(s^5)^3} \qquad \text{Power of a Power Property}$$

$$= \sqrt[3]{4^3} \cdot \sqrt[3]{(r^2)^3} \cdot \sqrt[3]{(s^5)^3} \quad \text{Root of a Product Theorem}$$

$$= 4r^2s^5 \qquad\qquad \text{Root of a Power Theorem}$$

Solution 2 Rewrite using rational exponents.

$$\sqrt[3]{64r^6s^{15}} = (64r^6s^{15})^{\frac{1}{3}} \qquad \text{Definition of } \sqrt[n]{x}$$

$$= 64^{\frac{1}{3}} \cdot r^{\frac{6}{3}} \cdot s^{\frac{15}{3}} \qquad \text{Power of a Product and Power of a Power Postulates}$$

$$= 4r^2s^5 \qquad\qquad \text{Arithmetic}$$

STOP QY1

> ▶ **QY1**
>
> Suppose $a \geq 0$ and $b \geq 0$. Simplify $\sqrt{16a^4b^{10}}$.

Sometimes when you try to simplify a radical, some irreducible portions remain, as in Example 4. Then the new expression may be more complicated than the given expression.

Example 4

Suppose $x \geq 0$. Rewrite $\sqrt[3]{120x^4}$ with a smaller power of x inside the radical.

Solution 1   Keep as a radical.

$$\sqrt[3]{120x^4} = \sqrt[3]{2^3 \cdot 15 \cdot x^3 \cdot x}$$
$$= \sqrt[3]{2^3 \cdot x^3} \cdot \sqrt[3]{15x}$$
$$= 2x\sqrt[3]{15x}$$

Solution 2   Convert to rational exponents.

$$\sqrt[3]{120x^4} = (120x^4)^{\frac{1}{3}}$$
$$= 120^{\frac{1}{3}} \cdot x^{\frac{4}{3}}$$
$$= 8^{\frac{1}{3}} \cdot 15^{\frac{1}{3}} \cdot x \cdot x^{\frac{1}{3}}$$
$$= 2x(15x)^{\frac{1}{3}}$$

 STOP QY2

> ▶ **QY2**
>
> Suppose $w \geq 0$.
> Rewrite $\sqrt[5]{w^{17}}$ with a smaller power of w inside the radical.

Questions

COVERING THE IDEAS

1. State the Root of a Product Theorem.

2. **True or False** $\sqrt{50} \cdot \sqrt{3} = \sqrt{15} \cdot \sqrt{10}$

In 3 and 4, multiply and simplify.

3. $\sqrt[4]{1000} \cdot \sqrt[4]{100,000}$ 4. $\sqrt[3]{9} \cdot \sqrt[3]{81}$

5. Write three different expressions equal to $\sqrt[3]{250}$.

In 6 and 7, find a and b. Assume $a > 0$ and $b > 0$.

6. $\sqrt{360} = \sqrt{a} \cdot \sqrt{10} = b\sqrt{10}$ 7. $\sqrt[3]{297} = \sqrt[3]{a} \cdot \sqrt[3]{11} = b\sqrt[3]{11}$

In 8 and 9, simplify the radicals.

8. $\sqrt[3]{1250}$ 9. $\sqrt{98} \cdot \sqrt{14}$

In 10–12, assume all variables are nonnegative. Simplify or rewrite with a smaller power of the variable inside the radical.

10. a. $\sqrt{144x^4}$

   b. $\sqrt{144x^5}$

11. a. $\sqrt[3]{64y^{18}}$

   b. $\sqrt[3]{\dfrac{y^{25}}{8}}$

12. $\sqrt[4]{1250y^3p^{17}}$

APPLYING THE MATHEMATICS

In 13 and 14, simplify the expression.

13. $\sqrt[3]{12} \cdot \sqrt[3]{18}$

14. $\sqrt[4]{144 \cdot 10^3} \cdot \sqrt[4]{9 \cdot 10^5}$

In 15 and 16, identify which expression is greater.

15. $\sqrt[3]{5} + \sqrt[3]{5}$ or $\sqrt[3]{10}$

16. $\sqrt[3]{600{,}000}$ or 100

17. Ali Baster simplifies $\sqrt[3]{4} \cdot \sqrt[6]{5}$ to $\sqrt[6]{20}$.

   a. Why is Ali's result incorrect?

   b. Solve $\sqrt[3]{4} \cdot \sqrt[6]{5} = \sqrt[6]{n}$ for n.

In 18 and 19, assume all variables are positive. Rewrite with a simpler expression inside the radical.

18. $\sqrt[3]{640x^{12}y^{11}}$

19. $\sqrt{4x^2 + 4y^2}$

REVIEW

20. Since 2000, the price of gasoline has been quite volatile. Here are the changes in average price for the years 2000–2006. Follow the steps to find the average annual price increase.

Year	Percent Increase From Preceding Year	Size-Change Factor
2000	28.5	?
2001	–3.6	?
2002	–6.5	?
2003	16.5	?
2004	18.2	?
2005	21.9	?
2006	12.9	?

   a. Fill in the table above by converting each percent into a size-change factor. (Previous Course)

   b. Compute the geometric mean of the factors you found in Part a to determine the average annual percentage increase. (Lesson 8-4)

21. Suppose $f(x) = \sqrt[4]{x}$. (**Lesson 8-4**)

    a. If x increases from 1 to 2, by how much does $f(x)$ increase?

    b. If x increases from 11 to 12, by how much does $f(x)$ increase?

22. **True or False** The inverse of the power function f with $f(x) = x^a$ is $f^{-1}(x) = x^{-a}$. Explain your reasoning. (**Lesson 8-3**)

23. a. Find an equation for the inverse of the linear function L defined by $L(x) = mx + b$.

    b. How are the slopes of the function and its inverse related?

    c. When is the inverse of L not a function? (**Lessons 8-2, 3-1**)

24. Solve for real values of y and check: $\dfrac{y^{-3}}{y} = \dfrac{1}{81}$. (**Lesson 7-3**)

25. Write $\dfrac{2 + 3i}{7 - 6i}$ in $a + bi$ form. (**Lesson 6-9**)

26. a. Multiply $(2 - \sqrt{5}) \cdot (20 + 10\sqrt{5})$. (**Lesson 6-1**)

    b. Your answer to Part a should be an integer. Tell how you could know that in advance.

EXPLORATION

27. The diagram at the right can be used to compare the arithmetic mean and geometric mean of two positive numbers. Suppose that $AC = x$ and $BC = y$.

    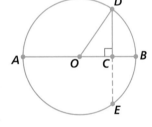

    a. Find the length of the radius of circle O in terms of x and y.

    b. Find CD in terms of x and y. Recall from geometry the Secant Length Theorem: $AC \cdot CB = CD \cdot CE$. In this situation, $CE = CD$.

    c. In $\triangle OCD$, which side corresponds to the arithmetic mean of x and y?

    d. In $\triangle OCD$, which side corresponds to the geometric mean of x and y?

    e. Which of the segments you named in Parts c and d must be longer, and why?

    f. In what situation can the two means be equal?

Lesson 8-6

Quotients with Radicals

Vocabulary

rationalizing the denominator

conjugate

▶ **BIG IDEA**   A fraction with a denominator of the form $a + b\sqrt{c}$ can be rewritten without any square root in the denominator.

In the book *The Phantom Tollbooth* (Juster, 1961), Milo and his traveling companions Humbug and Tock encounter the sign below, which gives the distance to the land of Digitopolis.

DIGITOPOLIS	
5	Miles
1,600	Rods
8,800	Yards
26,400	Feet
316,800	Inches
633,600	Half Inches
AND THEN SOME	

The characters argue about which distance to travel. Humbug wants to use miles because he believes the distance is shorter, while Milo wants to travel by half inches because he thinks it will be quicker.

The joke is that all the distances are the same. For example, 1 rod (a measure used in surveying) is equal to 16.5 ft, so $\frac{26,400 \text{ ft}}{16.5 \frac{\text{ft}}{\text{rod}}} = 1600$ rods. The point is that there are many ways to express one quantity or one number. You saw this when simplifying radicals in the previous lesson. Now we apply this idea to find different ways of writing quotients with radicals.

Rationalizing When the Denominator Is a Radical

Think about how you might approximate the value of $\frac{1}{\sqrt{2}}$ without a calculator. Using long division is no help because dividing 1 by $\sqrt{2}$ requires you to calculate $1.414213\ldots\overline{)1.00}$, which cannot be done by hand. Instead, a process called **rationalizing the denominator** is used to write an equivalent form of the number without a radical in the denominator. *Rationalizing* means rewriting the fraction so that its denominator is a rational number.

Mental Math

At a certain time of day, a tree casts an 18 ft shadow.

a. A nearby 8 ft light pole casts a 12 ft long shadow. How tall is the tree?

b. How long is the shadow cast by the nearby 3 ft mailbox?

c. When Trey gets the mail, his shadow is twice as long as the mailbox's shadow. How tall is he?

Activity

MATERIALS CAS

Step 1 Enter the following expressions into a CAS and record the results. Do not use decimal approximations.

a. $\dfrac{1}{\sqrt{2}}$　　　b. $\dfrac{3}{\sqrt{5}}$　　　c. $\dfrac{13}{\sqrt{7}}$

Step 2 Approximate each original expression in Step 1 and the resulting CAS expression to 3 decimal places. Do the pairs of numbers appear to be equal?

Step 3 Rewrite $\dfrac{1}{\sqrt{a}}$ ($a > 0$) without a radical in the denominator.

The results of the Activity suggest a method for rationalizing denominators of fractions whose denominators are square roots. In general, $\dfrac{a}{\sqrt{x}} = \dfrac{a}{\sqrt{x}} \cdot \dfrac{\sqrt{x}}{\sqrt{x}} = \dfrac{a\sqrt{x}}{x}$ for $x > 0$. This works because $\dfrac{\sqrt{x}}{\sqrt{x}} = 1$ and $\sqrt{x} \cdot \sqrt{x} = x$ for all real numbers x.

 QY

Because of technology, rationalizing denominators to obtain close approximations of quotients is no longer necessary. However, rationalizing denominators is still a useful process because not all technologies put results with radicals in the same form. You can expect to see different but equivalent forms of rationalized expressions depending upon the technology you use.

You can also rationalize denominators involving variable expressions.

> **▶ QY**
>
> Rewrite $\dfrac{26}{\sqrt{13}}$ by rationalizing the denominator.

GUIDED

Example 1

Rationalize the denominator of $\dfrac{3}{\sqrt{12x}}$, where $x > 0$.

Solution 1 Simplify the denominator first, then rationalize.

$$\dfrac{3}{\sqrt{12x}} = \dfrac{3}{\sqrt{3 \cdot 4 \cdot x}} = \dfrac{3}{\sqrt{3} \cdot \sqrt{?} \cdot \sqrt{?}} = \dfrac{3}{? \cdot \sqrt{?}} \cdot \dfrac{?}{?} = \dfrac{?}{?}$$

Solution 2 Rationalize first, then simplify.

$$\dfrac{3}{\sqrt{12x}} \cdot \dfrac{?}{?} = \dfrac{?}{12x} = \dfrac{?}{?}$$

Rationalizing When the Denominator Is a Sum Containing a Radical

▶ **READING MATH**

The word *conjugate* comes from the Latin prefix *co-* meaning "together with," and the Latin verb *jugare*, meaning "to join" or "to connect." In algebra, two complex numbers or radical expressions that are conjugates are joined together as a pair.

Now consider a fraction $\dfrac{n}{a+\sqrt{b}}$, in which a radical in the denominator is added to another term. In this form you cannot easily separate the rational and irrational parts. However, to rationalize the denominator, you can use a technique similar to the one you used in Lesson 6-9 to divide complex numbers.

Recall that to write $\dfrac{1}{5+2i}$ in $a+bi$ form, you multiply both numerator and denominator by the complex conjugate of the denominator, $5-2i$. To rationalize a fraction with a denominator of the form $a+\sqrt{b}$, multiply both numerator and denominator by the **conjugate** $a-\sqrt{b}$. The product has a denominator with no radical terms, because $(x+y)(x-y) = x^2 - y^2$.

Example 2

Write $\dfrac{3}{4+\sqrt{7}}$ in $a+b\sqrt{c}$ form.

Solution 1 The conjugate of $4+\sqrt{7}$ is $4-\sqrt{7}$.

$$\frac{3}{4+\sqrt{7}} \cdot \frac{4-\sqrt{7}}{4-\sqrt{7}} = \frac{3(4-\sqrt{7})}{4^2-\sqrt{7}^2} = \frac{3(4-\sqrt{7})}{16-7} = \frac{3(4-\sqrt{7})}{9} = \frac{4-\sqrt{7}}{3}$$

$$= \frac{4}{3} - \frac{\sqrt{7}}{3}$$

Solution 2 Use a calculator to multiply the numerator and denominator by the conjugate. The result can be rewritten in $a+b\sqrt{c}$ form as $\dfrac{4}{3} - \dfrac{\sqrt{7}}{3}$.

Check Estimate the original and final expressions with decimals.

$$\frac{3}{4+\sqrt{7}} \approx 0.45142$$

$$\frac{4}{3} - \frac{\sqrt{7}}{3} \approx 0.45142$$

It checks.

Questions

COVERING THE IDEAS

1. Verify that the indicated numbers on the DIGITOPOLIS sign are equivalent.
   a. the number of inches and the number of yards
   b. the number of yards and the number of miles

2. Why is it impossible to do long division with $\sqrt{73}$ as a divisor?

3. Estimate the value of $\frac{\sqrt{2}}{2}$ to the nearest tenth.

4. What does the term *rationalize the denominator* mean?

5. If $a > 0$, rewrite $\frac{b}{\sqrt{a}}$ without a radical in the denominator.

6. Are $\frac{2}{\sqrt{17}}$ and $\frac{2\sqrt{17}}{17}$ equal? Justify your answer.

In 7 and 8, rationalize the denominator.

7. $\frac{11}{\sqrt{3}}$

8. $\frac{4}{\sqrt{34}}$

In 9 and 10, a fraction is given.

a. Tell what you would multiply the fraction by to rationalize the denominator.

b. Rationalize the denominator and write the result in $a + b\sqrt{c}$ form.

9. $\frac{7}{3 - \sqrt{5}}$

10. $\frac{16}{\sqrt{12t} + 8}$

APPLYING THE MATHEMATICS

In 11–15, rationalize the denominator of each expression. Assume all variables are positive.

11. $\frac{3}{2\sqrt{3}}$

12. $\frac{47}{\sqrt{47}}$

13. $\frac{3x}{\sqrt{9x^5}}$

14. $\frac{5 - \sqrt{12}}{5 + \sqrt{12}}$

15. $\frac{4}{\sqrt{n} - 6}$

16. As pictured at the right, the largest square piece of wood that can be cut out of a circular log with diameter d has a side of length $\frac{d}{\sqrt{2}}$. If the radius of a log is 17 in., what is the side length of the largest square piece of wood that can be cut from it? Give your answer in both rationalized form and to the nearest hundredth of an inch.

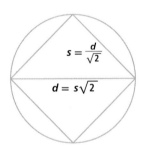

$s = \frac{d}{\sqrt{2}}$

$d = s\sqrt{2}$

17. Recall from geometry the ratio of the sides of a 30-60-90 triangle as shown in the diagram at the right.

a. If the length of the longer leg of the triangle is 8, find the length of the hypotenuse in rationalized form.

b. If the length of the longer leg of the triangle is a, write rationalized expressions for the lengths of the other two sides of the triangle in terms of a.

c. What is the missing x-coordinate of the point on the circle at the far right? Rationalize the denominator of your answer.

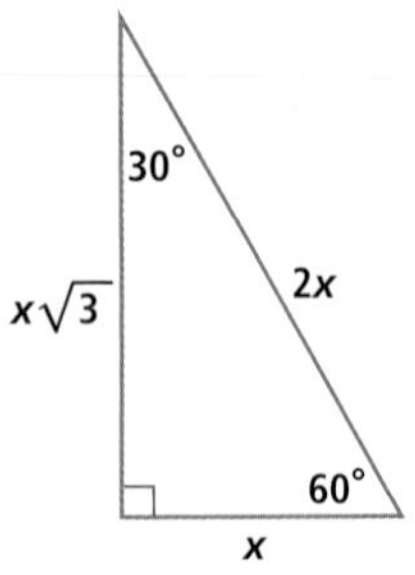

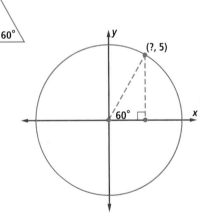

18. In the equilateral triangle at the right, find the ratio of s to h. Write your answer with a rationalized denominator.

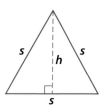

REVIEW

In 19 and 20, simplify the expression. Assume a and b are nonnegative real numbers. (**Lesson 8-5**)

19. $\sqrt[3]{54a^7}$

20. $\sqrt[4]{64a^3b^2}\sqrt{8a^6b}$

True or False In 21 and 22, assume all variables are positive. Justify your answer. (**Lesson 8-4**)

21. $\sqrt{xy^{\frac{1}{4}}} = \sqrt[8]{x^4y}$

22. $\sqrt[3]{m^{\frac{1}{4}}n} = \sqrt[36]{m^3n^{12}}$

23. Suppose $g : x \to x^{\frac{1}{3}}$, and $x > 0$. (**Lessons 8-3, 8-2, 8-1**)
    a. Find an equation for g^{-1}.
    b. If $h : x \to 2x$, find $g \circ h(x)$.

24. Explain why the inverse of $f : x \to 7x^{38}$ is not a function. (**Lessons 8-2, 7-1**)

25. Let f be a function defined by the table below. Is the inverse of f a function? Why or why not? (**Lesson 8-2**)

x	1	3	4	8	11	12
f(x)	3	5	13	5	-2	4

In 26 and 27, assume p and q are positive numbers. Write each expression without negative exponents. (**Lesson 7-8**)

26. $p^{-\frac{1}{5}}q^{\frac{3}{8}}$

27. $\dfrac{p^{-\frac{2}{3}}}{p}$

In 28 and 29, write an expression to describe the situation. (**Lesson 1-1**)

28. Kim collects teacups. She now has 14 teacups and buys one new teacup per month. How many teacups will she have after m months?

29. You buy G granola bars at d dollars per bar. How much did you spend?

EXPLORATION

30. A friend tries to rationalize $\dfrac{1}{\sqrt[3]{x}}$ by performing the following multiplication:
    $$\dfrac{1}{\sqrt[3]{x}} \cdot \dfrac{\sqrt[3]{x}}{\sqrt[3]{x}}.$$
    a. Explain why this method will not work.
    b. Devise a method to rationalize the denominator of $\dfrac{1}{\sqrt[3]{x}}$.
    c. Rewrite $\dfrac{1}{\sqrt[3]{x}}$ as a rational power of x.

QY ANSWER

$\dfrac{26}{\sqrt{13}} = \dfrac{26}{\sqrt{13}} \cdot \dfrac{\sqrt{13}}{\sqrt{13}} = \dfrac{26\sqrt{13}}{13}$

$= 2\sqrt{13}$

Lesson 8-7

Powers and Roots of Negative Numbers

Vocabulary

$\sqrt[n]{x}$ when $x < 0$

▶ **BIG IDEA** Great care must be taken when dealing with nth roots of negative numbers. When the nth roots are not real, then the Root of a Product Theorem may not be true.

You have already calculated some powers and some square roots of negative numbers. First we review the powers.

Integer Powers of Negative Numbers

Activity

Work in pairs. Assign one pair of functions to each partner.

a. $y = (-2)^x$ b. $y = (-6)^x$
    $y = (-7)^x$ $y = (-5)^x$

Step 1 Enter your functions into a calculator and generate a table of values starting at $x = 2$ with an increment of 2.

Step 2 Scroll through the table of values. Compare the results for all four functions with your partner. Describe any patterns you see.

Step 3 Generate a table of values starting at $x = 1$ with an increment of 2. Scroll through the table of values. Compare the results for all four functions with your partner. Describe any patterns you see.

Step 4 Complete the following generalization, choosing the correct words: ___?___ (Even/Odd) integer exponents with negative bases produce ___?___ (positive/negative) powers, while ___?___ (even/odd) integer exponents with negative bases produce ___?___ (positive/negative) powers.

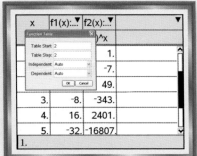

The Activity shows that positive integer powers of negative numbers alternate between positive and negative numbers. Even exponents produce positive numbers, while odd exponents produce negative numbers. The same is true if zero and negative powers are considered, because $(-x)^{-n}$ is the reciprocal of $(-x)^n$.

Mental Math

Miguel is training to run a 3-mile race. In how much time must he run each mile on average if he wants to finish the race in

a. 30 minutes?

b. 28 minutes?

c. 26 minutes?

Integer powers of negative numbers satisfy the order of operations. For example, $-6^4 = -1296$ because the power is calculated before taking the opposite. However, $(-6)^4 = 1296$, so $-6^4 \neq (-6)^4$. All the properties of *integer* powers of positive bases that you studied in Chapter 7 also apply to integer powers of negative bases.

> **GUIDED**
>
> ### Example 1
> Write without exponents: $(-3)^5(-3)^{-2}$.
>
> **Solution** Use the Product of Powers Postulate.
>
> $(-3)^5(-3)^{-2} = (-3)^{\underline{}\,?\,\underline{} + \underline{}\,?\,\underline{}} = (-3)^{\underline{}\,?\,\underline{}} = \underline{\,?\,}$

Are There Noninteger Powers of Negative Numbers?

Many times in this course, you have seen that powers and roots of negative numbers do not have the same properties that powers and roots of positive numbers do. Here are some of the properties that are different.

(1) When x is positive, $\sqrt{x}$ is a real number, but $\sqrt{-x}$ is a pure imaginary number.

(2) If both x and y are negative, $\sqrt{x} \cdot \sqrt{y} \neq \sqrt{xy}$. The left side is the product of two imaginary numbers and is negative; the right side is positive. For example, $\sqrt{-3} \cdot \sqrt{-2} \neq \sqrt{(-3)(-2)}$ because $i\sqrt{3} \cdot i\sqrt{2} = -\sqrt{6}$ and $\sqrt{(-3)(-2)} = \sqrt{6}$.

(3) If x is negative, then $x^n \neq \sqrt{x^{2n}}$ for positive integers n. Again, the left side is negative and the right side is positive. For example, $(-2)^3 \neq \sqrt{(-2)^6}$ because $(-2)^3 = -8$ and $\sqrt{(-2)^6} = \sqrt{64} = 8$.

These examples indicate that powers and roots of negative numbers have to be dealt with very carefully. For this reason, we do not define x^m when x is negative and m is not an integer. *Noninteger powers of negative numbers are not defined in this book.* That is, an expression such as $(-3)^{\frac{1}{2}}$ is not defined. However, we allow square roots of negative numbers to be represented by a radical.

 QY

>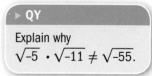
>
> ▶ QY
>
> Explain why $\sqrt{-5} \cdot \sqrt{-11} \neq \sqrt{-55}$.

The Expression $\sqrt[n]{x}$ When x Is Negative and n Is Odd

When x is positive, the radical symbol $\sqrt[n]{x}$ stands for its unique positive nth root. It would be nice to use the same symbol for an nth root of a negative number. This can be done for odd roots of negative numbers. If a number is negative, then it has exactly one real odd root. For instance, –27 has one real cube root, namely –3. Consequently, it is customary to use the symbol $\sqrt[n]{x}$ when x is negative, provided n is odd.

Definition of $\sqrt[n]{x}$ when $x < 0$

When x is negative and n is an odd integer > 2, $\sqrt[n]{x}$ stands for the real nth root of x.

For instance, because $(-5)^3 = -125$, $\sqrt[3]{-125} = -5$.
Because $-100{,}000 = (-10)^5$, you can write $\sqrt[5]{-100{,}000} = -10$.

To evaluate nth roots of negative numbers without a calculator, you can use numerical or graphical methods.

Example 2
Evaluate $\sqrt[3]{-64}$.

Solution 1 $\sqrt[3]{-64}$ represents the real 3rd root of –64, so you can solve $x^3 = -64$. Because $(-4)^3 = -64$, $\sqrt[3]{-64} = -4$.

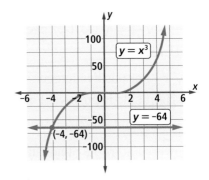

Solution 2 Graph $y = x^3$ and $y = -64$. The x-coordinate of the point of intersection is the real 3rd root of –64, as shown at the right. So, $\sqrt[3]{-64} = -4$.

Notice that the graph of the function f with equation $y = x^3$ or $f(x) = x^3$ verifies that every real number has exactly one real 3rd root, because any horizontal line intersects the graph only once. Therefore, –4 is the only real 3rd root of –64.

The Expression $\sqrt[n]{x}$ When x Is Negative and n Is Even

Square roots of negative numbers are not real numbers, and they do not satisfy all the properties of square roots of positive numbers. However, the radical form $\sqrt{x}$ is used when $x < 0$. When x is negative, $\sqrt{x} = i\sqrt{-x}$. For example, $\sqrt{-7} = i\sqrt{7}$. Other even roots of negative numbers (4th roots, 6th roots, 8th roots, and so on) are also not real, but they are not written using radicals. So, the nth-root expression $\sqrt[n]{x}$ is not defined when x is negative and n is an even integer greater than 2.

Here is a summary of our use of the $\sqrt[n]{}$ symbol when $n > 2$:

(1) When $x \geq 0$, $\sqrt[n]{x}$ is defined for any integer $n > 2$. It equals the positive real nth root of x.

(2) When $x < 0$, $\sqrt[n]{x}$ is defined only for odd integers $n \geq 3$. It equals the negative real nth root of x.

This summary may seem unnecessarily detailed, but it allows you to handle expressions with radical signs in much the same way that square roots are handled, as long as the expressions stand for real numbers.

> **nth Root of a Product Theorem**
>
> When $\sqrt[n]{x}$ and $\sqrt[n]{y}$ are defined and are real numbers, then $\sqrt[n]{xy}$ is also defined and $\sqrt[n]{xy} = \sqrt[n]{x} \cdot \sqrt[n]{y}$.

Example 3

Simplify $\sqrt[5]{-640}$. Leave your answer in radical form.

Solution Look for a perfect fifth power that is a factor of -640.

$-640 = -32 \cdot 20 = (-2)^5 \cdot 20$

So, $\sqrt[5]{-640} = \sqrt[5]{-32} \cdot \sqrt[5]{20}$

$\qquad\qquad = -2 \cdot \sqrt[5]{20}$.

Questions

COVERING THE IDEAS

1. Let $f(x) = (-9)^x$.
   a. Give three values of x that produce a positive value of $f(x)$.
   b. Give three values of x that produce a negative value of $f(x)$.

2. Calculate $(-8)^n$ for all integer values of n from -3 to 3.

3. Tell whether the number is positive or negative.

    a. $(-3)^4$ **b.** -3^4 **c.** $(-4)^{-3}$ **d.** $(-4)^3$

4. **True or False**

    a. $(-x)^{10} = -x^{10}$ **b.** $(-x)^9 = -x^9$

In 5 and 6, write as a single power.

5. $(-2)^6(-2)^{-3}$ 6. $((-4)^5)^6$

7. Calculate $\sqrt{-4} \cdot \sqrt{-9}$.

In 8–10, evaluate.

8. $\sqrt[3]{-125x^9}$ 9. $\sqrt[11]{-1}$ 10. $\sqrt[9]{-512 \cdot 10^{27}}$

In 11–13, simplify.

11. $\sqrt[3]{-\dfrac{64y^{27}}{27}}$ 12. $\sqrt[9]{-10^{63}}$ 13. $\sqrt[7]{1280q^{23}}$

14. What are the domain and the range of the real function with equation $y = \sqrt[6]{x}$?

15. The graphs of $y_1 = x^3$ and $y_2 = -27$ are shown below. What is the significance of the point of intersection of y_1 and y_2?

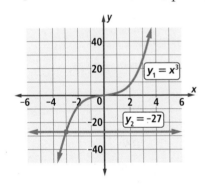

16. Simplify.

    a. $\sqrt[5]{-32} + \sqrt[4]{16}$ **b.** $\sqrt[5]{-32p^{10}} + \sqrt[4]{16p^{16}}$

17. Explain why the graphs of $y = \sqrt[3]{x}$ and $y = \sqrt[18]{x^6}$ are not the same.

18. Let $f : x \rightarrow \sqrt[5]{x}$ and $g : x \rightarrow \sqrt[15]{x}$. Find $f \circ g(x)$.

19. **a.** Show that $2 + 2i$ is a 4th root of -64.

    b. Show that $-2 - 2i$ is a 4th root of -64.

    c. Show that $2 - 2i$ is a 4th root of -64.

    d. Find the one other 4th root of -64 and verify your finding.

    e. How are Parts a–d related to the fact that $\sqrt[4]{-64}$ is not defined?

REVIEW

In 20 and 21, rationalize the denominator and simplify. (Lesson 8-6)

20. $\dfrac{5}{\sqrt{7}}$

21. $\dfrac{2-\sqrt{3}}{2+\sqrt{3}}$

22. a. Find the geometric mean of 2, 4, 8, 16, and 32.
    b. Generalize Part a. (Lesson 8-4)

23. Evaluate $\sqrt[3]{\sqrt{4096}}$. (Lesson 8-4)

24. a. Simplify without using a calculator:
    $(\sqrt{7}-\sqrt{13})(\sqrt{7}+\sqrt{13})$.
    b. Check by approximating $\sqrt{7}$ and $\sqrt{13}$ by decimals with a calculator and multiplying the decimals. (Lessons 8-5, 6-2)

25. Solve $r^{-\frac{2}{3}}=64^{-1}$ for r. (Lesson 7-8)

26. A rectangle has vertices at $(2, 0)$, $(6, 4)$, $(4, 6)$, and $(0, 2)$. What is its area? (Previous Course)

In 27 and 28, use the fact that the population in the United States was about 2.96×10^8 in 2005. (Lesson 7-2)

27. With a land area of about 3.5×10^6 mi², what was the average number of people per square mile?

28. In 2005, people in the U.S. consumed about 2.78×10^{10} pounds of beef. About how much beef was consumed per person in the U.S. in 2005?

EXPLORATION

29. Use your results from Part a to answer the other parts.
    a. Sketch the graphs of $y=\sqrt[3]{x^3}$, $y=\sqrt[4]{x^4}$, $y=\sqrt[5]{x^5}$, and $y=\sqrt[6]{x^6}$.
    b. For what values of n does $\sqrt[n]{x^n}=x$ for every real number x?
    c. For what values of n does $\sqrt[n]{x^n}=|x|$ for every real number x?
    d. Does $y=\sqrt{x^2}$ follow the pattern of other nth root of nth power functions?

Solving Equations with Radicals

▸ **BIG IDEA** To solve an equation of the form $x^{\frac{m}{n}} = k$, where m and n are integers, take the nth power of both sides. But be careful that you do not change the number of solutions to the equation in the process.

Remember that to solve an equation with a single rational power, such as $x^{\frac{4}{5}} = 10$, you can raise both sides to the power of the reciprocal of that exponent.

$$\left(x^{\frac{4}{5}}\right)^{\frac{5}{4}} = 10^{\frac{5}{4}}$$

So $x = 10^{\frac{5}{4}} \approx 17.78$. This checks because $17.78^{\frac{4}{5}} \approx 10$.

Solving an Equation with a Single Radical

Similarly, because the radical $\sqrt[n]{}$ involves an nth root, you can solve an equation containing only this single radical by raising both sides to the nth power.

Mental Math

Find an equation for the inverse of the function and tell whether the inverse is a function.

a. $y = 7x$

b. $3x + y = 4.5$

c. $y = 13$

Example 1

Imagine spinning a ball on a string around in a circle. If the string breaks, the ball will follow a straight-line path in the direction it was traveling at the time of the break as shown below.

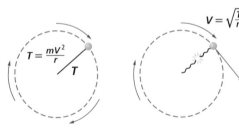

The ball will travel at a velocity V given by the formula $V = \sqrt{\frac{Tr}{m}}$, where T is the tension on the string (in newtons), m is the mass of the ball (in kilograms), and r is the length of the string (in meters). What tension is needed to allow a 5-kilogram ball on a 2-meter string to achieve a velocity of 10 $\frac{\text{meters}}{\text{second}}$?

Solution Here, $V = 10 \frac{m}{s}$, $m = 5$ kg, and $r = 2$ meters. Substitute the given values into the formula and use a CAS to solve for T.

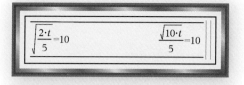

Enter the equation.

Square both sides.

Multiply each side by 5.

Divide each side by 2.

The string tension needs to be 250 newtons.

Check Use the `solve` command. It checks.

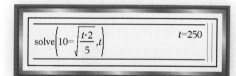

 QY

Extraneous Solutions

There is a major difficulty that may occur when taking an nth power to solve equations with radicals. The new equation may have more solutions than the original equation does. So you must be careful to check each solution in the original equation. If a solution to a later equation does not check in the original equation, it is called an **extraneous solution**, and it is not a solution to the original equation.

> ▶ QY
>
> A 5-kilogram ball traveling 5 meters per second was attached to a string with a tension of 250 newtons when the string broke. How long was the string?

Example 2

Greg and Terrance both solve the equation $5 - \sqrt[6]{x} = 8$, and each one concludes that there are no real solutions. Complete Solution 1 to see Terrance's approach and Solution 2 to see Greg's.

Solution 1 Solve $5 - \sqrt[6]{x} = 8$.

$\underline{\quad ? \quad} = 3$ Add $\underline{\;?\;}$ to both sides.

$(\underline{\;?\;})^6 = (3)^6$ Raise both sides to 6th power.

$x = \underline{\;?\;}$ Definition of $\underline{\;?\;}$ power, arithmetic

Check Substitute $x = \underline{\;?\;}$ into the original equation.

Does $5 - \sqrt[6]{\underline{\;?\;}} = 8?$

$5 - \underline{\;?\;} = 8?$ No.

So, $\underline{\quad ? \quad}$ is not a solution. It is extraneous.
The sentence $5 - \sqrt[6]{x} = 8$ has $\underline{\;?\;}$ real solutions.

Solution 2 Solve $5 - \sqrt[6]{x} = 8$.

$-\sqrt[6]{x} = 3$ Add –5 to both sides.

The left side of the equation is always a negative number because $\underline{\quad ? \quad}$. Therefore, the left side cannot equal positive 3. Thus, there are $\underline{\;?\;}$ real solutions.

In real-number mode, a CAS solution to the equation in Example 2 is shown below. It means that there are no real solutions to this equation.

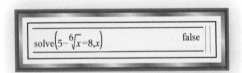

$$\text{solve}\left(5 - \sqrt[6]{x} = 8, x\right) \qquad \text{false}$$

Equations from the Distance Formula

The Pythagorean Distance Formula $d = \sqrt{(x_1 - x_2)^2 + (y_1 - y_2)^2}$ can lead to equations involving square roots. Although the equations may look quite complicated, they can be solved with the same approaches as simpler equations.

Example 3

Find coordinates for the two points on the line with equation $y = 7$ that are 9 units away from the point $(-3, 2)$.

Solution Draw a picture like the one at the right. Let $(x, 7)$ be one of the points you want.

Because the distance from $(x, 7)$ to $(-3, 2)$ is 9,

$$\sqrt{(x - (-3))^2 + (7 - 2)^2} = 9.$$

Simplify. $\sqrt{(x + 3)^2 + 25} = 9$

Now square both sides. $(x + 3)^2 + 25 = 81$

Subtract 25 from both sides. $(x + 3)^2 = 56$

Take the square root of both sides. $|x + 3| = \sqrt{56}$

So, $x + 3 = \sqrt{56}$ or $x + 3 = -\sqrt{56}$

$x = \sqrt{56} - 3$ or $x = -\sqrt{56} - 3$

$x \approx 4.5$ or $x \approx -10.5$.

The two points are exactly $(\sqrt{56} - 3, 7)$ and $(-\sqrt{56} - 3, 7)$, or approximately $(4.5, 7)$ and $(-10.5, 7)$.

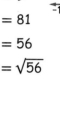

Questions

COVERING THE IDEAS

1. Refer to Example 1. A 2-kilogram ball traveling at $8\frac{m}{s}$ is attached to a string with a tension of 100 newtons. How long is the string?

In 2–5, find all real solutions.

2. $\sqrt[3]{d} = 8$

3. $12 = 2\sqrt[4]{m}$

4. $25 + \sqrt[5]{g} = 10$

5. $3 - \frac{1}{2}\sqrt[6]{w} = -5$

6. What is an extraneous solution to an equation?

In 7 and 8, find all real solutions.

7. $\sqrt{8x - 3} = 4$

8. $12 - \sqrt[8]{3x - 1} = 15$

9. Find the two points on the line with equation $y = -2$ that are 4 units away from the point $(1, -1)$.

10. Find the two points on the line with equation $y = x$ that are 3 units away from the point $(5, 7)$.

APPLYING THE MATHEMATICS

11. Patty Packer is designing rectangular boxes like the one at the right, with square bases and a surface area of 18.2 square feet. The formula $s = -h + \sqrt{h^2 + 9.1}$ gives the base-side length s in terms of height h.

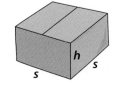

   a. Find s when $h = 2$ feet.
   b. Find h when $s = 2$ feet.
   c. Derive the formula.

12. A sphere has a radius of r centimeters. A new sphere is created with a diameter 3 centimeters less than the original. If the new sphere has a volume of 1500 cm³, what is the radius of the original sphere?

13. Recall the U.S. median home values from page 514. Below is a similar table with home-value data for the state of Texas from 1950 to 2000.

Year	1950	1960	1970	1980	1990	2000
Median Home Value (unadjusted dollars)	5805	8800	12,000	39,100	59,600	82,500

   a. Calculate decade-to-decade percentage growth and the size-change factor for each of the five decades and record your results in a table as shown below. The first decade is done for you.

Years	1950–1960	1960–1970	1970–1980	1980–1990	1990–2000
% Increase	52	?	?	?	?
Size-change Factor	1.52	?	?	?	?

   b. Compute the geometric mean of the size-change factors in Part a, and determine the average increase per decade, expressed as a whole percent, in Texas home values from 1950 to 2000.
   c. The decade-to-decade percentage growth in Texas home values from 1940 to 2000 is ≈ 91%. If this continues, estimate the median Texas home value in 2010.

Ken Warby holds the world water speed record. His boat, the *Spirit of Australia*, traveled at 317.6 mph on October 8, 1978.

14. When traveling at a fast rate, a ship's speed s (in knots) varies directly as the seventh root of the power p (in horsepower) generated by the engine. Suppose the equation $s = 6.5\sqrt[7]{p}$ describes the situation for a particular ship. If the ship is traveling at a speed of 15 knots, about how much horsepower is the engine generating?

In **15** and **16**, find all real solutions.

15. $5\sqrt[3]{x} - 8 = 13\sqrt[3]{x}$

16. $\sqrt{y^2 - 9} = 2\sqrt{y - 3}$

REVIEW

In **17** and **18**, simplify each expression. Assume variables are nonnegative. **(Lessons 8-7, 8-5)**

17. $\sqrt[5]{-32y^{15}}$

18. $\sqrt{45a^5}\sqrt{5b^8}$

19. Give a counterexample to the statement $(x^4)^{\frac{1}{4}} = x$.

True or False In **20** and **21**, assume $t > 0$. Justify your answer. **(Lesson 8-6)**

20. $\dfrac{1}{\sqrt{t}} = \dfrac{\sqrt{t}}{t}$

21. $\dfrac{\sqrt{3t}}{\sqrt{t}} = \dfrac{3\sqrt{t}}{\sqrt{3t}}$

22. **Multiple Choice** If a and k are positive, which of the following values for x is a solution to $a(x + h)^n = k$? **(Lesson 8-6)**

A $\left(\dfrac{a+h}{k}\right)^{-\frac{1}{n}}$

B $\sqrt[n]{\dfrac{k}{a}} - h$

C $\sqrt[n]{\dfrac{k-a}{h}}$

D $\sqrt[n]{\dfrac{k}{a}} + h$

23. Let $u(x) = \sqrt[3]{x}$ and $v(x) = x^6$. **(Lessons 8-4, 8-1)**

  a. Find an equation for $u \circ v$. What is the domain of $u \circ v$?

  b. Find an equation for $v \circ u$. What is the domain of $v \circ u$?

24. Explain why the inverse of the function $f : x \to x(x + 2)$ is not a function. **(Lessons 8-3, 8-2)**

25. Write a system of inequalities whose solution is the shaded region graphed at the right. **(Lesson 5-8)**

26. In 2000, one estimate of known oil reserves worldwide was 1.017×10^{12} barrels (1020 gigabarrels) while annual consumption was estimated to be 2.80×10^{10} barrels. If oil consumption remains constant, how many years after 2000 will known oil reserves last? **(Lesson 7-2)**

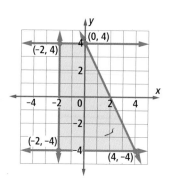

EXPLORATION

27. If you flip a fair coin n times, the expected difference between the number of heads and tails is the integer closest to $\sqrt{\dfrac{2n}{\pi}}$.

  a. How many times would you have to flip a coin to get an expected difference of 8?

  b. Test this formula by using a spreadsheet to simulate flipping a coin the number of times you calculated in Part a. Do multiple simulations and create a bar chart. Does the formula appear to work for your value of n?

QY ANSWER

$\frac{1}{2}$ m

Chapter 8 Projects

1 Roots and Surds

Some graphing utilities produce a graph like the top one below for $y = x^{\frac{1}{3}}$, while others produce a graph like the bottom one below.

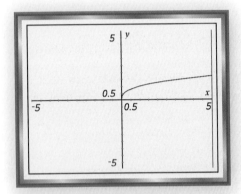

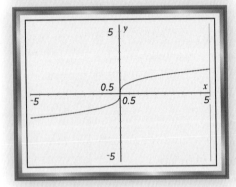

2 Caesar Cipher

Early methods of encoding messages relied on simple or shift ciphers. Research the Caesar cipher. How is the Caesar cipher different from the type of shift cipher you used in this chapter? How does the security of messages encoded with a Caesar cipher compare to the security of the messages used in this chapter?

a. Check which graph your CAS or graphing utility creates for this function. Can both graphs be correct? Why or why not?

b. Investigate the term *surd* and how it relates to these graphs. Summarize your findings, explaining the contexts in which both graphs can be correct.

3 Properties of Irrational Numbers

On page S1, at the back of the book, the field properties of real numbers are listed. Investigate which of these properties apply to the set of irrational numbers. Include examples and counterexamples to illustrate your conclusions.

4 Square Roots of Pure Imaginary Numbers

a. Find a square root of i, and hence a fourth root of –1, by solving the equation $i = (a + bi)^2$ for a and b. It will help to remember that $i = 0 + i$, and that two complex numbers are equal if and only if their real parts are equal and their imaginary parts are equal. This allows you to set up a system to solve for a and b. Write the square root of i in radical form.

b. We know that 4 has another square root besides the one denoted $\sqrt{4}$. Similarly, i has another square root. Use what you know about the other square root of 4 to hypothesize about the other square root of i. Check your hypothesis by squaring your result.

c. Find the square roots of some other nonreal numbers.

5 Square Roots of Matrices

For 2×2 matrices A and B, we say that A is a square root of B if $A^2 = B$.

a. Show that $\begin{bmatrix} 1 & 2 \\ 0 & -1 \end{bmatrix}$ and $\begin{bmatrix} -1 & 7 \\ 0 & 1 \end{bmatrix}$ are square roots of $\begin{bmatrix} 1 & 0 \\ 0 & 1 \end{bmatrix}$.

b. If X is an invertible 2×2 matrix, use the associative property of matrix multiplication to show that for any 2×2 matrix A, $(XAX^{-1})(XAX^{-1}) = XA^2X^{-1}$.

c. Use Part b to find several other square roots of $\begin{bmatrix} 1 & 0 \\ 0 & 1 \end{bmatrix}$. How many do you think there are?

6 America's Cup Yachts

One of the most prestigious international sailing competitions is the America's Cup. In order to compete, yachts must adhere to the International America's Cup Class (IACC) yacht design rules. In 2003, one of the rules was that the yacht length L (in meters), sail area S (in square meters), and displacement D (in cubic meters) had to satisfy the inequality

$$\frac{L + 1.25\sqrt{S} - 9.8\sqrt[3]{D}}{0.686} \leq 24.000 \text{ meters}$$

The displacement D is the volume of water displaced by the yacht. It is calculated by taking the mass (in kilograms) of the yacht and dividing it by 1025. The mass of the yacht must be no greater than 24,000 kg, rounded to the nearest 20 kg. The boats have evolved over the years and now tend toward the maximum allowable displacement.

a. Assuming the maximum displacement, solve the inequality for L in terms of S and graph the function mapping S onto the maximum value of L that would be allowed.

b. Using your answer to Part a, find some pairs of dimensions of the length and sail area that would allow a boat to abide by the formula given the maximum displacement.

c. Research the dimensions of various boats (perhaps winners) in the competition to verify that they complied with the formula.

Chapter 8 Summary and Vocabulary

● Every relation has an **inverse** that can be found by switching the coordinates of its ordered pairs. The graphs of any relation and its inverse are reflection images of each other over the line $y = x$.

● Inverses of some functions are themselves functions. A real function graphed on the coordinate plane has an inverse if and only if no horizontal line intersects the function's graph in more than one point. In general, two functions f and g are inverses of each other if and only if $f \circ g$ and $g \circ f$ are defined, $f(g(x)) = x$ for all values of x in the domain of g, and $g(f(x)) = x$ for all values of x in the domain of f.

● Consider the function f with domain the set of all real numbers and equation of the form $y = f(x) = x^n$. If n is an odd integer ≥ 3, its inverse is the nth root function with equation $x = y^n$ or $y = \sqrt[n]{x}$. These two functions are graphed below at the left. When n is an even integer ≥ 2, the inverse of $y = f(x) = x^n$ is not a function. However, if the domain of f is restricted to the set of nonnegative real numbers, the inverse of f is the nth root function with equation $y = \sqrt[n]{x} = x^{\frac{1}{n}}$. The restricted function and its inverse are graphed below at the right.

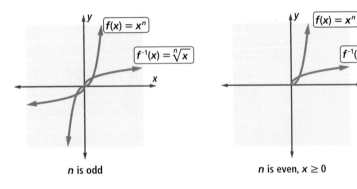

n is odd n is even, $x \geq 0$

That is,

1. When $x \geq 0$, $\sqrt[n]{x}$ is defined for any integer $n > 2$. It equals the positive nth root of x.

2. When $x < 0$, $\sqrt[n]{x}$ is defined only for odd integers $n \geq 3$. It equals the real nth root of x, a negative number.

▷ All properties of powers listed in Chapter 7 apply to radicals when they stand for real numbers. They lead to the following theorems for any real numbers x and y and integers m and n for which the symbols are defined and stand for real numbers.

Root of a Power Theorem: $x^{\frac{m}{n}} = \sqrt[n]{x^m} = (\sqrt[n]{x})^m$

Root of a Product Theorem: $\sqrt[n]{xy} = \sqrt[n]{x} \cdot \sqrt[n]{y}$

▷ These properties are helpful in simplifying radical expressions and in solving equations with radicals. To solve an equation involving an nth root, you need to raise each side to the nth power. When you do this, you may gain **extraneous solutions**. Always check every possible answer in the original equation to make sure that extraneous solutions have not been included.

▷ Radicals appear in many formulas. For example, the nth root of the product of n numbers is the geometric mean of the numbers. When a radical appears in the denominator of a fraction, multiplying both the numerator and denominator by a well chosen number can make the new denominator rational. To **rationalize** a fraction with a denominator of the form $a + \sqrt{b}$, multiply both numerator and denominator by the **conjugate** $a - \sqrt{b}$.

Theorems

Inverse-Relation Theorem (p. 524)
Inverse Functions Theorem (p. 530)
Power Function Inverse Theorem
  (p. 532)

Root of a Power Theorem (p. 539)
Root of a Product Theorem (p. 546)
nth Root of a Product Theorem
  (p. 559)

Chapter

8   Self-Test

Take this test as you would take a test in class. You will need a calculator and graph paper. Then use the Selected Answers section in the back of the book to check your work.

In 1–4, let $f(x) = 4 - x^2$ and $g(x) = 3x + 7$.

1. **Fill in the Blank** $f \circ g : -1 \to$ ___?___ .

2. Write a formula for $f(g(x))$.

3. Write an equation for $g \circ f$.

4. Are f and g inverses? Justify your answer.

5. **a.** Find an equation for the inverse of the function r defined by $r(x) = \frac{1}{2}x + 3$.

   b. Is the inverse a function? Justify your answer.

In 6 and 7, refer to the function graphed below.

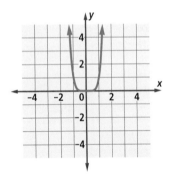

6. Graph the inverse of the function.

7. How can you restrict the domain of the original function so that the inverse is also a function?

8. **True or False** $\sqrt[4]{16} = -2$. Justify your answer.

9. The formula $r = \sqrt{\dfrac{S}{4\pi}}$ gives the radius r of a sphere with surface area S. What is the surface area, to the nearest square centimeter, of a spherical balloon with radius 12 cm?

In 10–12, simplify and rationalize all denominators. Assume $x > 0$ and $y > 0$.

10. $\sqrt[3]{-216x^{12}y^3}$

11. $\sqrt[4]{\dfrac{32x^{12}}{x^4}}$

12. $\dfrac{3x^2}{\sqrt{81x}}$

13. The top of a 15-foot ladder rests against a window ledge on the side of a building. The height h of the window ledge above the ground is equal to the distance from the bottom of the ladder to the base of the building. On a test, students were asked to find h. Three students' answers are below:

    Deon: $h = \sqrt{\dfrac{225}{2}}$

    Tia: $h = \dfrac{15}{2}$

    Carmen: $h = \dfrac{15\sqrt{2}}{2}$

    Which answer(s) is (are) correct? Justify your answer.

14. Rewrite $\sqrt[5]{\sqrt[3]{13}}$ as a power with a simple fraction exponent in lowest terms.

15. Rationalize the denominator and simplify $\dfrac{24}{8 - 2\sqrt{3}}$.

16. Use the formula $r = \sqrt[3]{\frac{3V}{4\pi}}$ to approximate the radius r of a sphere with volume $V = 268$ cubic inches to the nearest hundredth.

In 17 and 18, solve for x.

17. $25 = 4\sqrt{7x}$

18. $17 + \sqrt[3]{2x + 7} = 20$

19. For what values of n is the inverse of the function f with equation $y = f(x) = x^n$ also a function?

20. Give the domain and range of the function $y = \sqrt[6]{x}$, if x and y are real numbers.

21. Six samples taken in a water quality survey have the following levels of *E. coli* bacteria.

Sample	1	2	3	4	5	6
Count (per 100 mL)	27	32	144	46	597	1092

Use the geometric mean to compute an average level of bacteria across the samples.

22. Eddie is buying a new pair of eyeglasses. He has a store coupon for $45 off a new pair of glasses. The glasses he chooses have a starting price of P dollars, but they are on sale for 30% off. For what values of P will Eddie get a better deal if he can use the coupon before the discount is taken?

Chapter 8 — Chapter Review

SKILLS Procedures used to get answers

OBJECTIVE A Find values and rules for composites of functions. (Lesson 8-1)

1. When applying $f \circ g$, which function is applied last?

In 2–4, let $p(x) = x^2 + x + 1$ and $q(x) = x - 6$.

2. a. Find $p(q(5))$.
   b. Find $p(q(x))$.

3. a. Find $q(p(5))$.
   b. Find $q(p(x))$.

4. The function $p \circ q$ maps –10 onto what number?

In 5 and 6, rules for functions f and g are given. Does $f \circ g = g \circ f$? Justify your response.

5. $f: x \to -\frac{3}{8}x$; $g: x \to -\frac{8}{3}x$

6. $f(x) = 3\sqrt{x}$; $g(x) = \frac{9}{x}, x > 0$

7. If $h(x) = x^{\frac{2}{3}}$, find an expression for $h(h(x))$.

8. If $r(x) = \frac{2x + 1}{x}$, find an expression for $r(r(x))$.

OBJECTIVE B Find the inverse of a function. (Lessons 8-2, 8-3)

9. A function has equation $y = 6x - 3$. Write an equation for its inverse in slope-intercept form.

10. A function has equation $y = \sqrt{x^2}$. What is an equation for its inverse?

11. **Fill in the Blank** If $f: x \to 7x + 13$, then $f^{-1}: x \to$ ___?___ .

12. Show that $f: x \to 3x + 2$ and $g: x \to \frac{1}{3}x - 2$ are not inverse functions.

13. **Fill in the Blank** If $g(t) = -t^2$ for $t \leq 0$, then $g^{-1}(t) =$ ___?___ .

14. **Multiple Choice** Suppose $f(x) = x^5$. Then $f^{-1}(x) = x^k$, where $k =$

   A –5.   B $\frac{1}{5}$.
   C $-\frac{1}{5}$.   D 5.

OBJECTIVE C Evaluate radicals. (Lessons 8-4, 8-7)

In 15–18, write as a whole number or simple fraction.

15. $\sqrt[5]{243}$

16. $\sqrt[3]{-27}$

17. $\sqrt[3]{\left(\frac{64}{343}\right)^2}$

18. $\left(\sqrt{15} + \sqrt{13}\right)\left(\sqrt{15} - \sqrt{13}\right)$

In 19–22, approximate to the nearest hundredth.

19. $\sqrt[3]{3}$

20. $\sqrt[4]{81 + 16}$

21. $4\sqrt[3]{-75}$

22. $\sqrt[9]{\sqrt{365}}$

OBJECTIVE D Rewrite or simplify expressions with radicals. (Lessons 8-5, 8-6, 8-7)

In 23–30, simplify. Assume that all variables are positive.

23. $\sqrt{b^8}$

24. $\sqrt[3]{57r^3}$

25. $\sqrt[6]{8} \cdot \sqrt[6]{8}$

26. $\sqrt[3]{-60n^{12}}$

27. $\sqrt[7]{-u^{14}v^{28}}$

28. $\sqrt{3x^3} \cdot \sqrt{6x}$

29. $\sqrt{\sqrt{\sqrt{k}}}$

30. $\sqrt{N} \cdot \sqrt[3]{N} \cdot \sqrt[6]{N}$

In 31–34, rationalize the denominator and simplify, if possible.

31. $\dfrac{13}{\sqrt{13}}$

32. $\dfrac{8}{\sqrt{2}}$

33. $\dfrac{7}{\sqrt{3} - 1}$

34. $\dfrac{p}{p + \sqrt{q}}$ $(p > 0, q > 0)$

OBJECTIVE E Solve equations with radicals. (Lesson 8-8)

In 35–40, find all real solutions. Round answers to the nearest hundredth where necessary.

35. $\sqrt[3]{y} = 2.5$

36. $13 = 11\sqrt[4]{f}$

37. $12 = \frac{1}{4}\sqrt{16 - y}$

38. $\sqrt[3]{x - 1} - 9 = 27$

39. $18 + \sqrt[6]{64n} = 12$

40. $\sqrt{6x} + 3\sqrt{6x} = 12$

PROPERTIES Principles behind the mathematics

OBJECTIVE F Apply properties of the inverse of a function. (Lessons 8-2, 8-3)

In 41 and 42, state whether the statement is true or false.

41. If functions f and g are inverses of each other, then $f \circ g(x) = g \circ f(x)$ for all x for which these functions are defined.

42. When the domain of f is the set of positive real numbers, then the inverse of $y = x^6$ has equation $y = \sqrt[6]{x}$.

43. Suppose the domain of a linear function L is $\{x \mid x \le 0\}$ and the range is $\{y \mid y \ge -6\}$. What are the domain and range of L^{-1}?

In 44 and 45, suppose f and g are inverses of each other.

44. If (ℓ, m) is a point on the graph of f, what point must be on the graph of g?

45. If the domain of g is the set of all positive integers, what can you conclude about the domain or range of f?

OBJECTIVE G Apply properties of radicals and nth root functions. (Lessons 8-4, 8-5, 8-7)

46. If x is negative, for what values of n is $\sqrt[n]{x}$ a real number?

47. **Multiple Choice** Which expression is not defined?

A $\sqrt[3]{625}$ B $\sqrt[3]{-625}$

C $\sqrt[4]{625}$ D $\sqrt[4]{-625}$

48. Explain why the statement $\sqrt[7]{a} = a^{\frac{1}{7}}$ is not true for all real numbers a.

49. For what values of x is $\sqrt[5]{x^5} = x$?

50. Give a counterexample to this statement: For all real numbers x, $\sqrt[8]{x^8} = x$.

In 51 and 52, tell whether the statement $\sqrt[n]{a} \cdot \sqrt[n]{b} = \sqrt[n]{ab}$ is true for given conditions. Justify your answer.

51. a and b are negative, $n = 2$

52. a and b are negative, $n = 3$

USES Applications of mathematics in real-world situations

OBJECTIVE H Solve real-world problems that can be modeled by composite functions. (Lesson 8-1)

53. An electronics store is having a 25%-off sale on flat-screen televisions. The television Amber wants to buy has a sticker price of $1200, and the sales tax in the state is 9%.

a. How much will Amber pay for the television if the discount is taken before the tax is calculated?

b. How much will she pay if the tax is calculated first?

c. Would Amber get a better deal if the tax was calculated first or if the discount was taken first? Explain your answer.

54. A group goes to a restaurant with a $15-off coupon. The restaurant bill comes to b dollars before the tip and before using the coupon. The group wants to tip the server 20%.

a. Find an expression for $f(b)$, the total cost if the coupon is used before the tip is calculated.

b. Find an expression for $g(b)$, the total cost if the tip is calculated before the coupon is used.

c. Restaurants typically urge patrons to tip on the full bill before any discount is applied. Why do you think restaurants do this?

OBJECTIVE I Solve real-world problems that can be modeled by equations with radicals. (Lessons 8-4, 8-8)

55. The maximum distance d you can see from the top of a building with height h is approximated by the formula $d = k\sqrt{h}$. Apartment buildings A and B are 9 and 16 stories high, respectively. If these two apartment buildings have the same height per floor, about how many times farther can you see from the top of apartment B than the top of apartment A?

In 56 and 57, use the following information: To find the speed s (in mph) that a certain car was traveling on a typical dry road, suppose that police use the formula $s = 2\sqrt{5L}$, where L is the length of the skid marks in feet.

56. The car skidded 35 feet before stopping. According to the formula, how fast was the car going?

57. About how far would this car be expected to travel if it skids from 55 mph to a stop?

58. The U.S. Consumer Price Index (CPI) estimates the price of goods and services over time. In the table below, y is the percent change in the CPI from the previous year to the indicated year.

Year	2001	2002	2003	2004	2005	2006	2007
y (percent)	2.8	1.6	2.3	2.7	3.4	3.2	2.8

a. Add a third row to the table in which you convert each percent to a size-change factor. For example, the factor for 2001 is 1.028.

b. Compute the geometric mean of the size-change factors you found in Part a, and determine the average CPI percentage increase over this seven-year time frame.

59. The diameter of a spherical balloon varies directly as the cube root of its volume. If one balloon holds 7 times as much air as a second balloon, how do their diameters compare?

REPRESENTATIONS Pictures, graphs, or objects that illustrate concepts

OBJECTIVE J Make and interpret graphs of inverses of relations and functions. (Lessons 8-2, 8-3)

60. Use the graphs below.

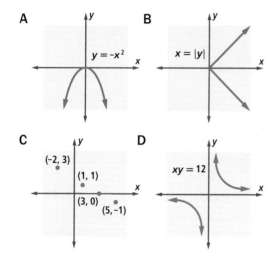

A $y = -x^2$

B $x = |y|$

C $(-2, 3)$, $(1, 1)$, $(3, 0)$, $(5, -1)$

D $xy = 12$

a. **Multiple Choice** Which is the graph of a function whose inverse is not a function?

b. How can you restrict the domain of the function in your answer to Part a so that its inverse is a function?

61. Let $f = \{(-3, 6), (-2, 5), (-1, 2), (0, 3)\}$
   a. Graph f^{-1}.
   b. What transformation maps f onto f^{-1}?

62. Graph the inverse of the function with equation $y = \sqrt{x^2}$.

63. a. Graph the inverse of the function graphed below.

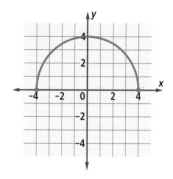

b. Is the inverse a function? Why or why not?

64. Draw a graph of a function with domain $\{x \mid -1 < x < 1\}$ that has an inverse which is not a function.

65. Let $g(x) = x^3$.
   a. Graph $y = g(x)$ and $y = g^{-1}(x)$.
   b. What is the domain of g^{-1}?

In 66 and 67, an equation for a function is given.
a. Graph the function.
b. State the domain and range of the function.
66. $h(x) = \sqrt[6]{x}$ 67. $f(x) = \sqrt[7]{x}$

Chapter 9

Exponential and Logarithmic Functions

Contents

Commercial portable cell phones were first produced in 1984. The table at the left below gives the estimated number of cell phone subscribers in the United States from 1985 to 2007. These data are graphed below.

Year	Subscribers (millions)	Year	Subscribers (millions)
1985	0.3	1997	55.3
1986	0.7	1998	69.2
1987	1.2	1999	86.0
1988	2.1	2000	109.5
1989	3.5	2001	128.4
1990	5.3	2002	140.8
1991	7.6	2003	158.7
1992	11.0	2004	182.1
1993	16.0	2005	207.9
1994	24.1	2006	233.0
1995	33.8	2007	255.4
1996	44.0		

Source: CTIA—The Wireless Association

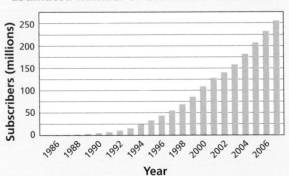

Estimated Number of Cell Phone Subscribers

The number of subscribers S (in millions) is closely modeled by the piecewise defined function

$$S = \begin{cases} 0.3(1.547)^{t-1984} & \text{for 1985–1996} \\ 45.1(1.239)^{t-1996} & \text{for 1997–2001}, \\ 125.5(1.129)^{t-2001} & \text{for 2002–2007} \end{cases}$$

where t is the year. Each equation in the model defines an *exponential function*, so called because the independent variable is in the exponent. Three different equations are needed because the average yearly growth rate of cell phone subscriptions in the United States changed from about 54.7% in the decade following their introduction to about 12.9% after 2002. In Lesson 9-4, you will learn how to find these equations.

Exponential functions have a variety of important applications. They model situations of growth, such as compound interest and population growth. They also model situations of decay, as found in carbon dating and depreciation.

The inverses of exponential functions, the *logarithmic functions*, describe the decibel scale of sound intensity, as well as the pH scale of acidity. They are also useful in solving problems involving exponential functions.

Lesson
9-1
Exponential Growth

Vocabulary

exponential function

exponential curve

growth factor

▶ **BIG IDEA** Exponential functions model situations of constant growth.

A Situation of Exponential Growth

When a species is introduced to a new environment, it often has no natural predators and multiplies quickly. This situation occurred in Australia in 1859, when a landowner named Thomas Austin released 24 rabbits for hunting. The rabbits reproduced so quickly that within 20 years they were referred to as a "grey carpet" on the continent, and drove many native plant and animal species to extinction. Because a pair of rabbits can produce an average of 7 surviving baby rabbits a year when in a dense environment, you can estimate that the rabbit population multiplied by a factor of $\frac{7}{2}$, or 3.5, each year.

To model this situation, let $r_0 = 24$ be the initial population of rabbits in 1859. This is similar to using h_0 for initial height and v_0 for initial velocity in previous formulas. Then let $r_n =$ the number of rabbits n years after 1859. A recursive formula for a sequence modeling this situation is

$$\begin{cases} r_0 = 24 \\ r_n = 3.5r_{n-1}, \text{ for } n \geq 1 \end{cases}.$$

This is a geometric sequence with starting term 24 and constant ratio 3.5. The table below shows the population r_n predicted by the model for $n = 1, 2, 3, \ldots, 10$ years after 1859, rounded to the nearest whole number of rabbits. The first four ordered pairs are graphed below.

Mental Math

Sahar is preparing for a math test. She plans to study on the four days before the test, and each day she will study $1\frac{1}{2}$ times as long as the day before. On the first day, she studies 24 minutes. How many hours and minutes is she planning to study on the fourth day?

n	r_n	n	r_n
0	24	6	44,118
1	84	7	154,414
2	294	8	540,450
3	1,029	9	1,891,575
4	3,602	10	6,620,514
5	12,605		

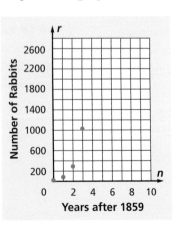

A female rabbit can give birth to several litters in one year, with up to 12 baby rabbits per litter.

An explicit formula for this sequence is $r_n = 24(3.5)^n$ for $n \geq 0$. By representing the population as the function f with equation $r = f(t) = 24(3.5)^t$, you can estimate the population r at any real number of years $t \geq 0$.

 QY1

In the graph at the left below, $r = 24(3.5)^t$ is plotted for values of t from 0 to 3.75, increasing by 0.25. The middle graph shows values for t from 0 to 3.8, increasing by 0.02.

> ▶ **QY1**
>
> Using the equation $r = 24(3.5)^t$, estimate the population of rabbits $\frac{1}{2}$ year after their introduction.

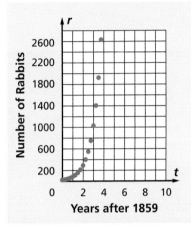

Years after 1859

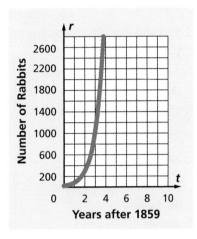

Years after 1859

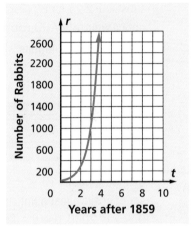

Years after 1859

Because time is continuous when measuring population growth, you can think of the function with equation $r = f(t) = 24(3.5)^t$ as being defined for all real nonnegative values of t, as graphed above at the right. However, the equation has meaning for any real number t. Using the set of real numbers as the domain results in the function graphed at the right below.

This graph shows an *exponential curve*. The shape of an exponential curve is different from the shape of a parabola, a hyperbola, or an arc of a circle. The range of the function f is the set of positive real numbers. Its graph never intersects the t-axis, but gets closer and closer to it as t gets smaller and smaller. Thus, the t-axis is a *horizontal asymptote* to the graph. Substituting $t = 0$ into the equation gives an r-intercept of 24. This represents the number of rabbits present when they were first introduced.

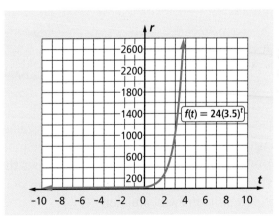

$f(t) = 24(3.5)^t$

Example 1

Use a CAS and the equation $f(t) = 24(3.5)^t$.

a. Estimate the number of rabbits after 2 years and 3 months.

b. Estimate the time to the nearest 0.01 year when there were 5000 rabbits.

c. Use a graph of f to estimate $f\left(-\frac{1}{2}\right)$.

Solution

a. Two years and 3 months equals 2.25 years. In the formula for $f(t)$, substitute 2.25 for t: $f(2.25) = 24(3.5)^{2.25} \approx 402.13$. There were approximately 400 rabbits 2 years 3 months after the release.

b. There were 5000 rabbits when $f(t) = 5000$. So, solve the equation $5000 = 24(3.5)^t$ for t on a CAS. The population reaches 5000 rabbits after about 4.26 years.

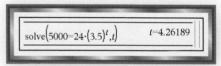

solve$\left(5000=24\cdot(3.5)^t, t\right)$ $t=4.26189$

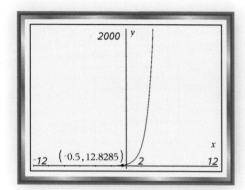

c. Use the window $-12 \le x \le 12$. Since the range of f is the set of positive real numbers, use the window dimension $0 \le y \le 2000$. Use the TRACE feature. $f\left(-\frac{1}{2}\right) \approx 13$

 QY2

What Is an Exponential Function?

The equation $f(t) = 24(3.5)^t$ defines a function in which the independent variable t is the exponent, so it is called an *exponential function*.

> ▶ **QY2**
>
> According to this model, in what year did the rabbit population reach 21 million, the approximate human population of Australia in 2008?

Definition of Exponential Function

The function f defined by the equation $f(x) = ab^x$ ($a \ne 0$, $b > 0$, $b \ne 1$) is an **exponential function**.

The graph of an exponential function is called an **exponential curve**. This particular exponential curve models *exponential growth*; that is, as time increases, so does the population of rabbits. The accelerating increase in the number of rabbits is typical of exponential growth situations.

In the equation $y = ab^x$, with $a > 0$, b is the **growth factor**; it corresponds to the constant ratio r in a geometric sequence. The rabbit situation involves the growth factor $b = 3.5$. In general, when $b > 1$, exponential growth occurs.

The compound interest formula $A = P(1 + r)^t$, when P and r are fixed, also defines an exponential function of t. In this case, A is the dependent variable and $1 + r$ is the growth factor. Since $r > 0$, $1 + r$ is greater than one, and compound interest yields exponential growth. The table below shows how geometric sequences and compound interest are modeled by exponential functions.

	Formula	Independent Variable	Dependent Variable	Starting Value	Growth Factor
Geometric Sequence	$g_n = g_1 r^{n-1}$	n	g_n	g_0 or g_1 = first term	r
Compound Interest	$A = P(1 + r)^t$	t	A	P = principal	$1 + r$
Exponential Function	$y = ab^x$	x	y	$a = y$-intercept	b

GUIDED

Example 2

The speed of a supercomputer is measured in *teraflops*, or trillions of "floating point operations" per second. In 2005, the Blue Gene/L supercomputer recorded a speed of 280.6 teraflops. Over the last 30 years, the speed of the fastest supercomputers has been growing at about 78% per year. Suppose that this growth rate continues, and let $C(x)$ = the speed in teraflops of the fastest supercomputer x years after 2005.

a. Write a formula for $C(x)$.

b. Use your formula to predict how long will it take for the fastest supercomputer speed to double the Blue Gene/L record to 561.2 teraflops.

c. Predict how many more years it will take for the fastest supercomputer speed to double again to 1122.4 teraflops.

Blue Gene/L at Livermore National Laboratory

Solution

a. Model this constant growth situation with an exponential function $C(x) = ab^x$. The initial speed $a =$ ___?___. An annual growth rate of 78% means that each year the computer speed is 178% of the previous year's speed, so $b =$ ___?___. A model is $C(x) =$ ___?___.

b. Solve $561.2 =$ ___?___ (___?___)^x on a CAS to get $x \approx$ ___?___. It will take about ___?___ years for the speed to double.

c. Solve $1122.4 =$ ___?___. So, $x \approx$ ___?___. Because ___?___ $- 1.2 =$ ___?___, it will take about ___?___ more years for the speed to double a second time.

solve($561.2 = 280.6 \cdot 1.78^x, x$)

Parts b and c of Example 2 demonstrate that, with an exponential growth model, the computing speed doubles in the same amount of time regardless of when you start. This constant doubling time is a general feature of exponential growth.

Questions

COVERING THE IDEAS

In 1–3, use the rabbit population model from Example 1.

1. About how many rabbits were there 6.2 years after they were introduced to Australia?

2. After about how many years were 100 million rabbits present?

3. Suppose that Thomas Austin had released only 10 rabbits for hunting. At the same annual growth factor of 3.5, about how many rabbits would there then have been after 5 years?

4. Define *exponential function*.

5. **Multiple Choice** Which is an equation for an exponential function?

    A $y = x^{3.04}$ **B** $y = 3.04x$ **C** $y = 3.04^x$

6. **Multiple Choice** Which graph below best shows exponential growth? Explain your answer.

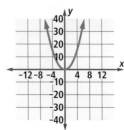 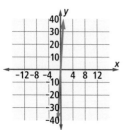 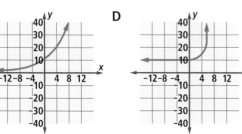

7. Let f be a function with $f(x) = 4 \cdot 2^x$.
    a. Graph $y = f(x)$.
    b. Approximate $f(-1.4)$.
    c. **True or False** f is an exponential function. Explain.

8. Consider the exponential curve with equation $y = ab^x$, where $b > 1$.
    a. **Fill in the Blank** The y-intercept is __?__.
    b. **Fill in the Blank** The constant growth factor is __?__.
    c. Which line is an asymptote to the graph?

In 9 and 10, refer to the situation and function $C(x) = 280.6(1.78)^x$ from Example 2.

9. a. Find $C(-10)$.
    b. In terms of the situation, what does $C(-10)$ represent?

10. a. Use the model to estimate the computing speed of the fastest supercomputer in the year 2010.
    b. Some researchers believe that a supercomputer capable of 10^4 teraflops could simulate the human brain. If current trends continue, when would such computers be possible?

11. Consider the exponential function with equation $A = P(1 + r)^t$.

    a. Name the independent and dependent variables.

    b. What is the growth factor?

APPLYING THE MATHEMATICS

12. Refer to the function $C(x) = 280.6(1.78)^x$.

    a. Find the average rate of change between $x = -4$ and $x = -2$.

    b. Find the average rate of change between $x = 2$ and $x = 4$.

    c. What conclusions can you draw from your answers in Parts a and b?

13. In 2000, the population of Canada was about 31.1 million people, and the population of Morocco was about 30.2 million people. Over the period 1950–2000, the population of Canada grew at an average rate of about 1.16% annually, while the population of Morocco grew at an average rate of about 2.38% annually. Let the function C with equation $C(x) = a \cdot b^x$ represent the population of Canada x years after 2000, in millions, and let the function M with $M(x) = c \cdot d^x$ represent the population of Morocco x years after 2000, in millions.

Morocco is on the northwest side of Africa and its area is slightly larger than California.

    a. Determine the values of a, b, c, and d and write formulas for $C(x)$ and $M(x)$.

    b. Use your formulas from Part a to estimate the populations of Canada and Morocco in 1995.

    c. Graph $y = C(x)$ and $y = M(x)$ on the same set of axes for $0 \le x \le 50$.

    d. Make a prediction comparing the future populations of Canada and Morocco if current trends continue.

14. In September of 2004, the online user-edited encyclopedia *Wikipedia* contained about 1,000,000 articles. In January of 2001, the month it was launched, it contained 617 articles. This means it has grown on average about 18% per month since its launch date. Let $f(x) = ab^x$ represent the number of articles on *Wikipedia* x months after January of 2001 ($x = 0$ represents January of 2001).

    a. From the given information, find the values of a and b.

    b. According to this model, how many articles were there in September of 2004? Explain why the answer is not exactly 1,000,000.

    c. According to this model, how many months did it take for the number of *Wikipedia* articles to double from 1,000,000 to 2,000,000?

15. In 1993, a sample of fish caught in a Mississippi River pool on the Missouri-Illinois border included 1 bighead carp. In 2000, a same-size sample from the same pool included 102 such fish. Other samples taken during this period support an exponential growth model. Let $f(x) = ab^x$ represent the number of carp caught x years after 1993.

    a. Use the information given to determine the value of a.

    b. Find an approximate value for b.

    c. According to your model, how many bighead carp would be in a similar sample caught in 2014?

Hypophthalmichthys nobilis, the bighead carp, is an invasive species of fish that was first introduced in 1986.

REVIEW

16. Suppose $f(x) = 5x - 6$. **(Lessons 8-3, 8-2)**

    a. Find an equation for f^{-1}.

    b. Graph $y = f(x)$ and $y = f^{-1}(x)$ on the same axes.

    c. **True or False** The graphs in Part b are reflection images of each other.

17. The matrix $\begin{bmatrix} 3 & 3 & -3 \\ -1 & -2 & 3 \end{bmatrix}$ represents triangle *TRI*. **(Lessons 4-10, 4-1)**

    a. Give the matrix for the image of $\triangle TRI$ under $T_{1,-2}$.

    b. Graph the preimage and image on the same set of axes.

18. Liberty Lumber sells 6-foot long 2-by-4 boards for $1.70 each, and 8-foot long 2-by-6 boards for $2.50 each. Last week they sold $500 worth of these boards. Let x be the number of 2-by-4s sold and y be the number of 2-by-6s sold. **(Lesson 3-2)**

    a. Write an equation relating x and y.

    b. If 200 2-by-4s were sold, how many 2-by-6s were sold?

19. Suppose a new car costs $28,000 in 2009. Find its value one year later, in 2010, if **(Previous Course)**

    a. the car is worth 82% of its purchase price.

    b. the car depreciated 20% in value.

    c. the value of the car depreciated r%.

EXPLORATION

20. The Australian rabbit plague was initiated by introducing 24 rabbits into the country.

    a. Suppose 12 rabbits, not 24, had been introduced in 1859. How, then, would the number of rabbits in later years have been affected?

    b. Answer Part a if 8 rabbits had been introduced.

    c. Generalize Parts a and b.

QY ANSWERS

1. 45

2. 1869

Lesson
9-2
Exponential Decay

Vocabulary

exponential decay

depreciation

half-life

▶ **BIG IDEA** When the constant growth factor in a situation is between 0 and 1, exponential decay occurs.

In each exponential growth situation in Lesson 9-1, the growth factor b in $f(x) = ab^x$ was greater than 1, so the value of $f(x)$ increased as x increased. When the growth factor b is between 0 and 1, the value of $f(x)$ *decreases* as x increases. The situation then is an instance of **exponential decay**.

Depreciation as an Example of Exponential Decay

Automobiles and other manufactured goods that are used over a number of years often decrease in value. This decrease is called **depreciation**. The function that maps the year onto the value of the car is an example of an exponential decay function.

Mental Math

Imagine the graph of $y = 0.5x^2$. How many lines

a. are asymptotes to this graph?

b. are lines of symmetry of this graph?

c. intersect the graph in two points?

d. are neither horizontal nor vertical and intersect the graph in exactly one point?

e. are horizontal and intersect the graph in exactly one point?

GUIDED

Example 1

Suppose a new SUV costs $36,025 and depreciates 12% each year.

a. Write an equation that gives the SUV's value when it is t years old.
b. Predict the SUV's value when it is 4 years old.

Solution

a. If the vehicle loses 12% of its value annually, it keeps $100\% - 12\% = $ ___?___% of its value. Because each year's value is a constant multiple of the previous year's value, the value of the car after t years can be modeled by an exponential function V with equation $V(t) = ab^t$. $V(t)$ is the value of the car when it is t years old. a is the original value, so $a = $ ___?___.
b is the constant multiplier or ratio, so $b = $ ___?___.
So, $V(t) = $ ___?___.

b. When the car is 4 years old, $t = $ ___?___.
Substitute. $V(\underline{\quad?\quad}) = \underline{\quad?\quad} \cdot 0.88^{\underline{\quad?\quad}} \approx \underline{\quad?\quad}$.
The car's value will be about ___?___ after 4 years.

Chapter 9

Half-Life and Radioactive Decay

In Lesson 9-1, you saw that when a quantity grows exponentially, its doubling time is constant. If a substance decays exponentially, the amount of time it takes for half of the atoms in this substance to decay into another matter is called its **half-life**. Half-life is an important feature of some chemical processes and of radioactive decay. The half-life of a radioactive substance can be as short as a small fraction of a second, as in the 0.002-second half-life of hassium-265, or as long as billions of years, as in the 4.47 billion-year half-life of uranium-238, a naturally occurring radioactive element.

Example 2

Carbon-14 (sometimes written as ^{14}C) has a half-life of 5730 years. This means that in any 5730-year period, half of the carbon-14 decays and becomes nitrogen-14, and half remains. Suppose that an object contains 100 g of carbon-14. Let a_n be the number of grams of carbon-14 remaining after n half-life periods.

a. How many grams of carbon-14 remain after 4 half-life periods?
b. How many years is 4 half-life periods?
c. Write a recursive formula for a_n.
d. Write an explicit formula for a_n.
e. Write a function for the amount of carbon-14 remaining after any real number of half-life periods x.

Solution

a. Make a table of values of a_n for $n = 0, 1, 2, 3,$ and 4, as at the right. The initial amount of carbon-14 is $a_0 = 100$. Each successive value of a_n is one-half the previous value.

6.25 grams of carbon-14 remain after 4 half-life periods.

n	a_n
0	100
1	50
2	25
3	12.5
4	6.25

b. One half-life period = 5730 years, so,
4 half-life periods = 4 • 5730 = 22,920 years.

c. The amount a_n of carbon-14 remaining after n half-life periods forms a geometric sequence with a constant ratio of $\frac{1}{2}$.
A recursive formula for this sequence is
$$\begin{cases} a_0 = 100 \\ a_n = \frac{1}{2}a_{n-1}, \text{ for integers } n \geq 1. \end{cases}$$

d. Use the Explicit Formula for a Geometric Sequence from Chapter 7.
An explicit formula for this sequence is
$$a_n = 100 \cdot \left(\frac{1}{2}\right)^n \text{ for integers } n \geq 0.$$

e. Use the equation $f(x) = ab^x$ with $a = 100$ and $b = \frac{1}{2}$.

$$f(x) = 100 \cdot \left(\frac{1}{2}\right)^x$$

During a plant or animal's life, the carbon-14 in its body is naturally replenished from the environment. Once it dies, the amount of carbon-14 decays exponentially as described in Example 2. Archeologists and historians use this fact to estimate the date at which an ancient artifact made of organic material was created.

GUIDED

Example 3

Suppose a sample of a piece of parchment began with 100% pure carbon-14. Use the information about carbon-14 from Example 2.

a. Write an equation for the percent of carbon-14 remaining in the sample after x half-life periods.

b. Graph your equation from Part a and use it to find the age of the sample to the nearest century if it contains 80% of its original carbon-14.

Parchment is an organic material, usually made of animal skin.

Solution

a. Use the exponential equation $f(x) = ab^x$. Rewrite the initial amount as a whole number: $a = \underline{\quad?\quad}\% = \underline{\quad?\quad}$.
   Let $f(x)$ be the amount of carbon-14 remaining in the sample after x half-life periods. Then $f(x) = \underline{\quad?\quad} \cdot \underline{\quad?\quad}{}^x$.

b. Graph f on a graphing utility. Trace on the graph where $y \approx \underline{\quad?\quad}$ and record the value of x. $x \approx \underline{\quad?\quad}$. So, the piece of parchment is $\approx \underline{\quad?\quad} \cdot 5730 \approx \underline{\quad?\quad}$ years old to the nearest century if it contains 80% of its original carbon-14.

Growth versus Decay

The examples of exponential growth from Lesson 9-1 and the examples of exponential decay from this lesson fit a general model called the *Exponential Change Model*.

Exponential Change Model

If a positive quantity a is multiplied by b (with $b > 0$, $b \neq 1$) in each unit period, then after a period of length x, the amount of the quantity is ab^x.

Activity

MATERIALS dynamic graphing application or graphing utility

Explore how changing a and b affect the graphs of the family of functions with equations $f(x) = ab^x$. You can use a dynamic graphing application provided by your teacher or graph several instances of the family on a graphing utility.

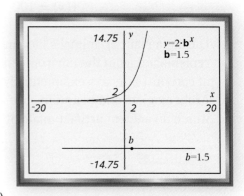

Step 1 Graph $f(x) = ab^x$ for a constant value of $a > 0$ and different values of b between 1 and 4.

    a. What features of the graph change as b changes?

    b. What features of the graph stay the same?
        (Hint: Why are these called exponential *growth* functions?)

Step 2 Now graph $f(x) = ab^x$ for the same a as in Step 1 and different values of b between 0 and 1.

    a. What features of the graph change as b changes?

    b. What features of the graph stay the same? (*Hint:* Why are these called exponential *decay* functions?)

Step 3 How are the graphs, where $0 < b < 1$, different from the graphs where $b > 1$? How are the two kinds of graphs similar?

Step 4 Now graph $f(x) = ab^x$ for a constant value of $b > 0$ and some values of a between 1 and 4. How does a affect the graph of $y = ab^x$?

Step 5 Copy and complete the chart below to summarize the features of exponential growth and exponential decay functions.

Property	Exponential Decay $a > 0, 0 < b < 1$	Exponential Growth $a > 0, b > 1$
Domain	set of real numbers	set of real numbers
Range	?	?
y-intercept	?	?
x-intercept	?	?
Horizontal asymptote	?	?
As x increases, y (increases/decreases).	?	?

Step 6 Sketch a graph showing exponential growth and a graph showing exponential decay, labeling $(0, a)$ on both graphs.

Notice that exponential growth and decay graphs have many features in common. This is not surprising because every exponential growth curve is the reflection image of an exponential decay curve over the *y*-axis. You are asked about this property in Questions 7 and 14.

Questions

COVERING THE IDEAS

1. Refer to Example 1. If the depreciation model continues to be valid, what would the SUV be worth when it is 10 years old?

2. Suppose a car purchased for $16,000 decreases in value by 8% each year.
   a. What is the yearly growth factor?
   b. How much is the car worth after 2 years?
   c. How much is the car worth after *x* years?

3. Define *half-life*.

In 4 and 5, refer to Example 3.

4. What percent of an artifact's carbon-14 would remain after 8000 years?

5. In 1996, a human skeleton, nicknamed Kennewick Man, was found in the Columbia River in the state of Washington. Examination showed that the skeleton had about 32% of its original carbon-14 in its bones. In about what year did Kennewick Man die?

6. **Fill in the Blank** If $y = ab^x$, $a > 0$, and $0 < b < 1$, then y ___?___ as *x* increases.

7. a. Sketch a graph $y = 4^x$, when $-3 \leq x \leq 3$.
   b. Sketch a graph $y = \left(\frac{1}{4}\right)^x$, when $-3 \leq x \leq 3$.
   c. Explain how the graphs in Parts a and b are related to each other.

8. **True or False** The graph of every exponential function with equation $f(x) = ab^x$ has *x*-intercept *a*. Explain your answer.

9. **True or False** The *x*-axis is an asymptote for the graph of an exponential function. Explain your answer.

APPLYING THE MATHEMATICS

10. According to recent estimates, in the future the population of Japan is predicted to *decrease* by 0.9% per year. In 2007, the population of Japan was about 127,433,000 people. Let $J(x)$ be the population (in millions) of Japan x years after 2007. Assume that J is an exponential function with domain $0 \leq x \leq 45$.

Tokyo, Japan is one of the most densely populated cities in the world.

   a. Find a formula for $J(x)$.

   b. Make a table of values of $J(x)$ every 5 years from 2007 until 2052.

   c. According to this model, in what year would the population of Japan first fall below 110 million people?

   d. Assume the population of Japan in 2008 is about 127,288,000. How well does your model predict this number? Check your model's validity with data for more recent years, if available on the Internet.

11. Suppose that a new car costs $20,000 and one year later is worth $18,000. Let N be the value of the car after t years.

   a. Assume the depreciation is exponential. Make a table of values for $t = 1, 2, 3, 4$.

   b. Repeat Part a if the depreciation is linear.

   c. Suppose you lease this car for 6 months and then decide to buy it (called *buying out the lease*). The balance to pay for the car is partly based on the car's existing value; that is, a higher value means you pay more for the car. On this basis, do you prefer that the dealer use a linear or exponential depreciation model? Explain your answer.

12. Sam Dunk has a 78% probability of making a free throw on one attempt. Assume that success or failure on one attempt does not affect the probability of success or failure on the next attempt.

   a. Find the probability that Sam makes 2 free throws in a row.

   b. Find the probability that Sam makes 3 free throws in a row.

   c. Let $f(n)$ be the probability that Sam makes n free throws in a row from the beginning of a game. Write a formula for $f(n)$.

   d. Find the smallest positive value of n such that $f(n) < 0.5$.

   e. What does your answer to Part d mean in the context of Sam Dunk's free throws?

13. In 1965, the computer scientist Gordon Moore first noticed that the size, speed, and cost of computing elements change exponentially over time. In 1981, a 5-megabyte hard drive cost about $1700, or $340 per megabyte. The cost per megabyte has decreased exponentially with a half-life of about 1.2 years since then. Let $P(x)$ be the cost of one megabyte of hard-drive storage x half-life periods after 1981.

 a. Write an equation to for the exponential function P.

 b. Copy and fill in the table below.

x	-2	-1	0	1	2	3
P(x)	?	?	?	?	?	?

 c. The year 2020 corresponds to how many half-life periods since 1981? Use your model to predict the cost of hard-drive storage in 2020.

 d. According to your model, when will the cost of hard-drive storage first be less than $0.01 per *gigabyte*? (1 gigabyte = 1000 megabytes)

 e. Is it realistic that one gigabyte of memory could cost less than a penny? Give a reason for your answer.

14. Suppose $g(x) = ab^x$ is an exponential growth function. Let $h(x) = g(-x)$ for all values of x. Use properties of powers to show that h is an exponential decay function. How are the graphs of h and g related?

REVIEW

15. Consider the function graphed at the right. **(Lessons 9-1, 1-2)**

 a. **Multiple Choice** Which could be an equation for the graph?

 A $f(x) = 10^{-x}$ B $g(x) = 10x$

 C $h(x) = \left(\frac{1}{x}\right)^{10}$ D $j(x) = 10^x$

 b. What is the domain of the function that answers Part a?

 c. What is the range of the function that answers Part a?

16. In 2005 the population of Dhaka, Bangladesh, was 12,576,000, and the growth rate was 3.6% per year. If that growth rate continues, what will the population of Dhaka be in 2015? **(Lesson 9-1)**

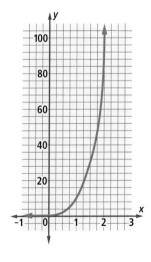

17. Shilah is raising guppies, a popular species of freshwater aquarium fish. To raise guppies, it is important to have plenty of aquarium space, ample food, and places for babies to hide (or the guppy mothers may eat them). Shilah began with 12 guppies and the population increased at a biweekly (every two weeks) rate of 33%. (**Lessons 9-1, 7-2**)

   a. How many guppies did Shilah have after 10 weeks?

   b. What is an equation for the number g of guppies after w weeks?

18. The inflation rate, a rate at which the average price of goods increases, is reported monthly by the U.S. government. Suppose a monthly rate of 0.3% was reported for January. Assume this rate continues for a year. (**Lessons 9-1, 5-1**)

   a. What is the inflation rate for the year?

   b. The value 0.3% has been rounded. The actual value could range from 0.25% up to, but not including, 0.35%. Write an inequality for r, the annual inflation rate, based on those two extreme values.

19. Give the decimal approximation to the nearest tenth. (**Lessons 7-7, 7-2**)

   a. 5^3 b. $5^{\frac{5}{2}}$ c. $5^{\sqrt{7}}$

20. Write $\dfrac{10^{12.7}}{10^{9.3}}$ as a power of ten and as a decimal to the nearest thousandth. (**Lessons 7-2**)

21. Simplify $\dfrac{tr^{n+1}}{r}$. (**Lesson 7-2**)

EXPLORATION

22. Often we think of populations as *growing* exponentially, but as Question 10 shows, a country's population may *decrease* exponentially.

   a. Find some countries in the world whose population has been decreasing.

   b. List some factors that would explain their decreasing population.

23. In this lesson you have seen several cases in which scientists found an object with a certain percent of its carbon-14 left. Research how scientists find out how much carbon-14 was originally present.

Lesson

9-3

Continuous Compounding

Vocabulary

e

compounded continuously

▶ **BIG IDEA** The more times that a given interest rate is compounded in a year, the larger the amount an account will earn. But there is a limit to the amount earned, and the limit is said to be the result of *continuous compounding*.

Recall the General Compound Interest Formula,

$$A = P\left(1 + \frac{r}{n}\right)^{nt},$$

which gives the amount A that an investment is worth when principal P is invested in an account paying an annual interest rate r and the interest is compounded n times per year for t years.

Suppose you put $1 into a bank account. If the bank were to pay you 100% interest compounded annually, your money would double in one year because the interest would equal the amount in the account. Using the formula with $P = 1$ dollar, $r = 100\% = 1$, $n = 1$, and $t = 1$ year, we have

$$A = 1\left(1 + \frac{1}{1}\right)^1 = 2 \text{ dollars.}$$

Now suppose you put $1 into a bank account and the bank paid 100% compounded *semiannually*. This means that the bank pays 50% interest twice a year. Now you receive $0.50 in interest after six months, giving a total of $1.50. Then, after another six months you receive $0.75 in interest on the $1.50, giving a total after one year of $2.25. This agrees with what you would compute using the General Compound Interest Formula with $n = 2$. Then,

$$A = \left(1 + \frac{1}{2}\right)^2 = \left(\frac{3}{2}\right)^2 = \frac{9}{4} = 2.25 \text{ dollars.}$$

You are asked to explore what happens as n gets larger in the following Activity.

Mental Math

Consider the largest sphere that will fit in a cube with side length 6.

a. Find the exact volume of the sphere.

b. How many spheres of diameter 3 will fit in the cube without overlapping?

Activity

Complete a table like the one on the next page to show the value of $1 at the end of one year ($t = 1$) after an increasing number n of compounding periods per year at a 100% annual rate.

(continued on next page)

Compounding Frequency	$n =$ Number of Compoundings per Year	$P\left(1+\frac{r}{n}\right)^{nt}$	A
Annually	1	$1\left(1+\frac{1}{1}\right)^{1}$	$2.00
Semiannually	2	$1\left(1+\frac{1}{2}\right)^{2}$	$2.25
Quarterly	4	?	$2.44141
Monthly	?	?	?
Daily	?	?	?
Hourly	?	?	?
By the Minute	?	?	?

The Activity shows that the more frequently a bank compounds, the more your earnings will be. The sequence of values for the total amount gets closer and closer to the number **e**, which is approximately equal to 2.71828. We say that e is the value of $1 after one year invested at 100% interest **compounded continuously**.

The number e is named after Euler, who proved that the sequence of numbers of the form $\left(1+\frac{1}{n}\right)^{n}$ approaches this particular number as n increases. Like π, e is an irrational number that can be expressed as an infinite, nonrepeating decimal. The following are the first 50 digits of the decimal expansion for e, and a graph of $y = e^x$ is shown at the right.

$e \approx 2.71828\,18284\,59045\,23536\,02874\,71352\,66249\,77572\,47093\,69995\ldots$

Like π, an approximation to e is stored on virtually every calculator.

 QY1

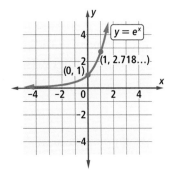

▶ QY1

What value does your calculator display for e? (Hint: Look for a key or menu selection labeled e^x, and evaluate e^1.)

Interest Compounded Continuously

Of course, savings institutions do not pay 100% interest. But the number e appears regardless of the interest rate.

Consider an account in which 4% interest is paid on $1 for one year. The table at the right shows some values of A for different compounding periods. As n increases, the total amount gets closer and closer to $1.040810…, the decimal value of $e^{0.04}$. Furthermore, in t years of continuous compounding at this rate, the dollar would grow to $(e^{0.04})^t$, or $e^{0.04t}$. So, if an amount P were invested, the amount would grow to $Pe^{0.04t}$.

Compounding Frequency	$P\left(1+\frac{r}{n}\right)^{nt}$	A
Annually	$1\left(1+\frac{0.04}{1}\right)^{1}$	$1.04
Quarterly	$1\left(1+\frac{0.04}{4}\right)^{4}$	$1.040604
Daily	$1\left(1+\frac{0.04}{365}\right)^{365}$	$1.040808

In general, for situations where interest is compounded continuously, the general Compound Interest Formula can be greatly simplified.

> ## Continuously Compounded Interest Formula
>
> If an amount P is invested in an account paying an annual interest rate r compounded continuously, the amount A in the account after t years is
>
> $$A = Pe^{rt}.$$

 QY2

Example 1

Talia invested $3,927.54 in a zero-coupon bond paying 5.5% compounded semiannually. After 30 years the value of the bond will be $20,000. How much would Talia's investment have been worth after 30 years if interest were compounded continuously instead of twice per year?

Solution Use $A = Pe^{rt}$, with $P = 3927.54$, $r = 0.055$, and $t = 30$.

$A = 3927.54e^{0.055(30)}$

$A = 3927.54e^{1.65} \approx 20{,}450.62$

After 30 years, the bond would be worth $20,450.62. This is $450.62 more than a bond with semiannual compounding would earn.

> ▶ **QY2**
>
> What will an account with an initial deposit of $1000 be worth after 3 years if interest is compounded continuously at 5% annual interest rate?

Other Uses of the Number e

Many formulas for continuous growth and decay are written using e as the base because the exponential function $y = e^x$ has special properties that make it particularly suitable for applications. We do not study those properties here, but you will learn them if you study calculus. You do not need to know these properties to find values of the function.

GUIDED

Example 2

A *capacitor* is an electrical device capable of storing an electric charge and releasing that charge very quickly. For example, a camera uses a capacitor to provide the energy needed to operate an electronic flash. The percent Q of charge in the capacitor t seconds after a flash begins is given by the formula $Q = Pe^{-10.0055t}$, where P is the initial percent of charge in the capacitor. If 44.8% of the charge is left 0.08 second after a flash begins, to what percent was the capacitor originally charged?

(continued on next page)

Solution We need to find P. When $t = 0.08$, $Q = \underline{\quad?\quad}$.
Substitute these values and solve for P.

$$Q = Pe^{\underline{\quad?\quad}t}$$
$$\underline{\quad?\quad} = Pe^{(\underline{\quad?\quad})(\underline{\quad?\quad})}$$
$$\underline{\quad?\quad} \approx P(\underline{\quad?\quad})$$
$$\underline{\quad?\quad} \approx P$$

The capacitor was originally charged to $\underline{\ ?\ }$% of capacity.

Formulas such as $A = 3000e^{0.05t}$ and $Q = Pe^{-10.0055t}$ model exponential growth and decay, respectively. Models like these are often described using function notation. Let C be the initial amount, and let r be the growth factor by which this amount continuously grows or decays per unit time t. Then $N(t)$, the amount at time t, is given by the equation

$$N(t) = Ce^{rt}.$$

This equation can be rewritten as $N(t) = C(e^r)^t$. So it is an exponential equation of the form

$$y = ab^x,$$

where $a = C$, $x = t$, and the growth factor $b = e^r$. If r is positive, then $e^r > 1$ and there is exponential growth. If r is negative, then $0 < e^r < 1$ and there is exponential decay.

Questions

COVERING THE IDEAS

1. **Fill in the Blank** Like π, the number e is a(n) $\underline{\quad?\quad}$ number.

2. Approximate e to the nearest ten-billionth.

3. Use the General Compound Interest Formula.
   a. What is the value of $1 invested for one year at 100% interest compounded daily, to the nearest hundredth of a cent?
   b. As n increases, the value in Part a becomes closer and closer to what number?

4. Approximate $e^{0.055}$ to the nearest hundred-thousandth.

5. Suppose \$1000 is invested at 6% interest compounded continuously.
   a. What is its value at the end of one year?
   b. What is its value at the end of 3.5 years?

6. Suppose \$8000 is invested at an annual interest rate of 4.5% for 15 years.
   a. How much will be in the account if the interest compounds continuously?
   b. How much will be in the account if the interest compounds annually?

7. Use the formula $Q = Pe^{-10.005t}$ given in Example 2. What is the initial charge of a capacitor if 30% of its charge remains after 0.1 second?

8. Consider the function N with $N(t) = Ce^{rt}$.
   a. What does C represent?
   b. What does e^r represent?
   c. What are the domain and range of N?
   d. How can you determine whether N models exponential growth or exponential decay?

APPLYING THE MATHEMATICS

9. Equations of three functions are shown below.
   (i) $y_1 = e^x$ (ii) $y_2 = \left(\frac{1}{e}\right)^{-x}$ (iii) $y_3 = e^{-x}$
   a. Determine whether each function is increasing, decreasing, or neither on the set of real numbers.
   b. Check your answer in Part a by graphing each function for $-2 \le x \le 2$.
   c. Explain why two of the graphs in Part b coincide.
   d. Which of the functions describe(s) exponential growth?
   e. Which of the functions describe(s) exponential decay?

10. **Fill in the Blank** Write $>$, $<$, or $=$: π^e ___?___ e^{π}.

11. A machine used in an industry depreciates so that its value $N(t)$ after t years is given by $N(t) = Ce^{-0.35t}$.
    a. What is the annual rate r of depreciation of the machine?
    b. If the machine is worth \$90,000 after 4 years, what was its original value?

12. The amount L of americium, a radioactive substance, remaining after t years decreases according to the formula $L = Be^{-0.0001t}$. If 5000 micrograms are left after 2500 years, how many micrograms of americium were present initially?

13. In the 1980s, researchers developed a special "inhibited growth" model to predict the growth of the U.S. population, using data from the 1960s and 1970s. The model predicts the population $p(y)$ (in millions) of the U.S. y years after 1960. An equation for the model is

$$p(y) = \frac{64{,}655.6}{179.3 + 181.3e^{-0.02676y}}.$$

   a. According to this model, what is the predicted U.S. population in 2100?

   b. Let y = the number of years since 2000. Then a newer model for the U.S. population is $q(y) = 281.4e^{0.0112y}$. According to this model, what population is predicted for 2100?

   c. Which answer is greater, the answer for Part a or Part b? Can you think of any reason to account for such differences?

   d. Research last year's U.S. population and compare it to the values predicted by the two models.

REVIEW

14. Nobelium was discovered in 1958 and named after Alfred Nobel. Nobelium-255 (^{255}No) has a half-life of 3 minutes. Suppose 100% of nobelium-255 is present initially. (**Lesson 9-2**)

   a. Make a table of values showing how much nobelium-255 will be left after 1, 2, 3, 4, and 5 half-life periods.

   b. Write a formula for the percent A of nobelium-255 left after x half-life periods.

   c. Write a formula for the percent A of nobelium-255 left after t minutes.

15. Let $f(x) = 16^x$. (**Lessons 9-1, 8-2, 7-7, 7-3**)

   a. Evaluate $f(-3)$, $f(0)$, and $f\left(\frac{3}{2}\right)$.

   b. Identify the domain and range of f.

   c. Give an equation for the reflection image of the graph of $y = f(x)$ over the line with equation $y = x$.

16. Rationalize the denominator of $\dfrac{2\sqrt{5}+1}{2\sqrt{5}-1}$. (**Lesson 8-6**)

17. Solve $5k^3 = 27$. (**Lessons 8-4, 7-6**)

18. Use the diagram at the right. Midpoints of a 12-by-16 rectangle have been connected to form a rhombus. Then midpoints of the rhombus are connected to form a rectangle, and so on. (**Lesson 7-5, Previous Course**)

   a. List the perimeters of the first 6 rectangles.

   b. What kind of sequence is formed by these perimeters?

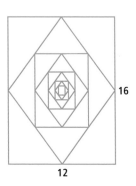

16

12

19. For what values of a does the equation $ax^2 + 8x + 7 = 0$ have no real solutions? (Lesson 6-10)

20. Choose all that apply. What kind of number is $\sqrt{-25}$?
    (Lessons 6-8, 6-2)

    A rational   B irrational   C imaginary

21. The table below gives the average height in centimeters for girls of various ages in the U.S. (Lesson 3-5)

Age	2	4	6	8	10
Height (cm)	85	101	115	127	138

    a. Let x be age and y be height. Find an equation for the line of best fit for this data set.
    b. Use the equation in Part a to predict the height of a girl at age 12.

EXPLORATION

22. Another way to get an approximate value of e is to evaluate the infinite sum
$$1 + \frac{1}{1!} + \frac{1}{2!} + \frac{1}{3!} + \dots .$$
    (Recall that $n!$ is the product of all integers from 1 to n inclusive.)

    a. Use your calculator to calculate each of the following to the nearest thousandth.
    $$1 + \frac{1}{1!} + \frac{1}{2!} + \frac{1}{3!}$$
    $$1 + \frac{1}{1!} + \frac{1}{2!} + \frac{1}{3!} + \frac{1}{4!}$$
    $$1 + \frac{1}{1!} + \frac{1}{2!} + \frac{1}{3!} + \frac{1}{4!} + \frac{1}{5!}$$
    b. How many terms must you add to approximate $e = 2.71828\dots$ to the nearest thousandth? What is the last term you need to add to do this?

Lesson
9-4

Fitting Exponential Models to Data

▶ **BIG IDEA**   When a set of data points in a situation seems to be showing exponential growth or decay, exponential regression can fit an exponential function to the data points.

After a person takes medicine, the amount of drug left in the person's body decreases over time. When testing a new drug, a pharmaceutical company develops a mathematical model to quantify this relationship. To find such a model, suppose a dose of 1000 mg of a certain drug is absorbed by a person's bloodstream. Blood samples are taken every five hours, and the amount of drug remaining in the body is calculated.

Possible data from an experiment are shown in the table and scatterplot below. The scatterplot suggests that an exponential model might be appropriate. Exponential models can be fit to data using methods similar to those that you used to find linear and quadratic models in earlier chapters.

Drug Absorption Data	
Hours Since Drug was Administered	Amount of Drug in Body (mg)
0	1000
5	550
10	316
15	180
20	85
25	56
30	31

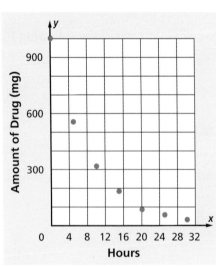

Finding Exponential Models for Data Believed to Be Exponential

As you know, exponential functions have the form $y = ab^x$, where a is the value of y when $x = 0$ and b is the growth factor during each unit period of time.

Example 1

Find an exponential model to fit the drug absorption data using the initial condition and one other point in the table.

Solution You need to find a and b in the equation $y = ab^x$. The initial condition occurs when $x = 0$, so $1000 = ab^0$.
Since $b^0 = 1$, $a = 1000$. So the equation is of the form $y = 1000b^x$.
Choose another point from the table and substitute. We chose $(20, 85)$.

$85 = 1000b^{20}$ Substitute.

$0.085 = b^{20}$ Divide each side by 1000.

$0.884 \approx b$ Take the 20th root of each side
(or raise each side to the $\frac{1}{20}$ power).

So, one model for the data is $y = 1000 \cdot (0.884)^x$.

Check Graph this model on a graphing utility. It looks like an exponential decay function with y-intercept 1000, as you would expect from the data.

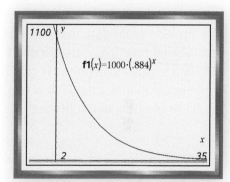

You can also use your graphing utility to find an exponential regression model.

Activity

Step 1   Enter the drug absorption data into your calculator. Apply the exponential regression option to find a and b and then write an equation to model these data.

Step 2   Your calculator probably gives a and b to many digits. The accuracy of the experimental data suggests that a more sensible model rounds a and b to three digits each. Rewrite your equation after rounding.

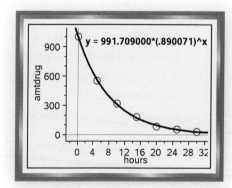

Step 3   Use your regression model from Step 2 and the two-point model $y = 1000 \cdot (0.884)^x$ from Example 1 to estimate the amount of drug left in a body after 5 hours. Which model better fits the actual 550 mg value from the experiment?

Step 4   Compare the accuracy of the two models by making a table of values for each equation and comparing these values to the actual data. Which model appears to be more accurate overall?

Recall that, by design, regression is intended to fit a model as close as possible to *all* the data in a set. The Activity shows that a regression model is indeed a better fit to the drug absorption data than a model based on only two points.

Deciding Whether an Exponential Model Is Appropriate

For some data you may not be sure that an exponential model is appropriate. In that case, consider two things. First, look at a scatterplot of your data to see if it has the general shape of an exponential function. This is a quick way to check if the growth factor between various data points is relatively constant. Second, find an exponential model, then look at a table of values or the graph of your model to see how well it fits the data.

Example 2

Below are the U.S. box office gross for the first eleven weekends of the release of the movie *Mission Impossible III* in 2006.

Weekend	1	2	3	4	5	6	7	8	9	10	11
U.S. Box Office Gross (millions of dollars)	47.70	25.00	11.35	7.00	4.68	3.02	1.34	0.72	0.49	0.31	0.20

a. Fit an exponential model for box office gross based on the number of weekends the movie has been out and determine if the exponential model is appropriate.

b. Use the model to estimate the movie's gross in its 12th weekend.

Solution

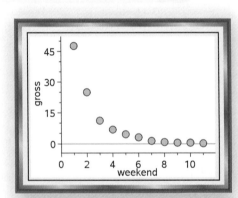

a. The scatterplot appears to have an exponential shape so an exponential model seems appropriate. Perform exponential regression to obtain

$$y = 69.617842 \cdot (0.57873986)^x.$$

Round a and b based on the actual data's accuracy.

$$y = 69.6 \cdot 0.579^x$$

When it is added to the scatterplot, the graph of the model closely follows the pattern of the data points. It appears to be a good model.

Exponential decay models the data closely, but could a parabola fit the data better?

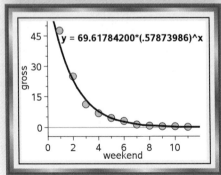

Graph the data together with the graph of a quadratic regression equation.

Exponential decay is a better fit. This makes sense because you know that movie gross earnings typically continues to decrease each successive week after release. Gross earnings do not reach a minimum and then continue to climb indefinitely, so a quadratic model is illogical.

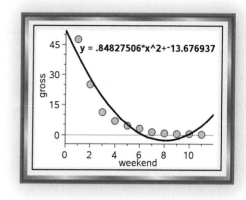

b. Substitute $x = 12$ in for x in the model.

$$y = 69.6 \cdot 0.579^{12}$$

$$y \approx 0.099$$

According to this model, the movie grossed about $99,000 in its 12th week.

The weekend box office gross earnings of many movies decline according to an exponential pattern. This helps theater managers estimate how long they should continue to show a movie.

Questions

COVERING THE IDEAS

In 1 and 2, refer to the drug absorption data given at the beginning of the lesson.

1. a. Use the point $(10, 316)$ and the method of Example 1 to find an equation to model the data.

   b. Use your equation from Part a to predict when there will be less than 10 mg of the drug left in the patient's body.

2. Use the model developed in the Activity to predict when the amount of the drug in the patient's body will fall below 10 mg.

3. Can you use the method of Example 1 to find an equation to model the data in Example 2? Why or why not?

4. **Multiple Choice** Which scatterplot(s) below show(s) data that could be reasonably represented by an exponential model?

A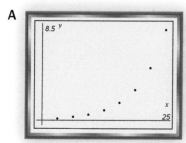

B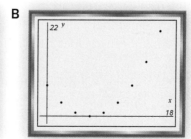

C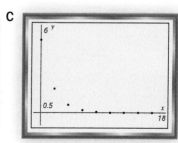

5. Recall the chart of cell phone subscriptions from the opening of this chapter. The table at the right presents the data for seven years.

    a. Find an exponential model for the growth of cell phone subscriptions.

    b. Compare cell phone growth and population growth. The population of the United States was 301.1 million in 2007 and growing exponentially at 0.89% per year. In which year does your model first predict that there will be at least one cell phone subscription per person?

Year	Subscribers (millions)
2001	128.4
2002	140.8
2003	158.7
2004	182.1
2005	207.9
2006	233.0
2007	255.4

APPLYING THE MATHEMATICS

6. The gross earnings for the first three weekends of a popular movie are given in the table at the right.

    a. Find an exponential model for the decline in weekend gross.

    b. Studio executives want to pull the movie from theaters before its weekend gross drops below $1 million. How many weekends should the studio expect to keep the movie in theaters?

Weekend	Gross earnings (millions of dollars)
1	135
2	62
3	35

7. Deven dropped 3728 pennies on the floor. He picked up all the coins showing heads and set them aside. Then he counted the tails, put them in a container, mixed them up, and dropped them. He repeated this several times and made a table of the results as shown at the right. Unfortunately, he forgot to record a couple of entries.

    a. Create an exponential model of the data using the 5 completed row entries.

    b. Use the model to estimate in the missing data.

    c. Deven's brother Tyler said the model looked like a half-life model for radioactive elements. Explain whether Tyler is correct.

Drop Number	Number of Tails
1	1891
2	?
3	470
4	229
5	118
6	?
7	27

8. The table at the right contains the cumulative monthly English-language article total for the website *Wikipedia*. The timetable ranges from March, 2005 to February, 2006.

    a. Develop an *exponential* model for the number y of English-language articles in *Wikipedia* in month x. Does this model seem reasonable based on the graph?

    b. Develop a *quadratic* model for the number y of English-language articles in *Wikipedia* in month x. Does this model seem reasonable based on the graph?

    c. Develop a *linear* model for the number y of English-language articles in *Wikipedia* in month x. Does this model seem reasonable based on the graph?

    d. Use each model to predict how many articles will be on *Wikipedia* in the current year. Check your results on the Internet to determine the most accurate model.

Month	Articles (thousands)
1	507
2	539
3	573
4	609
5	655
6	703
7	744
8	791
9	836
10	886
11	942
12	992

REVIEW

9. **a.** Graph $y = e^{5x}$ for $-2 < x < 2$. Label the coordinates of three points.

   b. State the domain and range of this function. **(Lesson 9-3)**

10. Rumors and fads spread through a population in a process known as *social diffusion*. Social diffusion can be modeled by $N = Ce^{kt}$, where N is the number of people who have heard the rumor after t days. Suppose four friends start a rumor and two weeks later 136,150 people have heard the rumor. **(Lesson 9-3)**

   a. In this situation, what is the value of C?

   b. What is the value of k?

   c. Graph the growth of the rumor during the first two weeks.

   d. How many people heard the rumor after 10 days?

   e. How long will it take for one million people to have heard the rumor?

11. Anne put $4800 in an account with a 3.125% annual interest rate. What will be her balance if she leaves the money untouched for four years compounded

   a. annually?　　　　　　　　**b.** daily?

   c. continuously? **(Lessons 9-3, 7-4)**

12. Consider the sequence defined by
$$\begin{cases} h_1 = 8.375 \\ h_n = 0.8h_{n-1}, \text{ for integers } n \geq 2 \end{cases} \cdot \textbf{(Lessons 9-2, 9-1, 7-5, 3-1)}$$

   a. List the first five terms of this sequence.

   b. Which phrase best describes this sequence: *exponential growth, exponential decay, constant increase,* or *constant decrease*?

In 13 and 14, simplify without using a calculator. (Lessons 7-8, 7-7, 7-3)

13. $\left(\frac{1}{2}\right)^{-4}$　　　　　　　　14. $32^{\frac{6}{5}}$

15. Simplify $(x + y)^2 - (x - y)^2$. **(Lesson 6-1)**

16. Suppose t varies inversely with r, and $t = 24$ when $r = 24$. Find t when $r = 6$. **(Lesson 2-2)**

EXPLORATION

17. Find weekend box office gross data for a movie that has only been out three or four weeks. Develop a model for its decline in gross and predict the gross for future weekends. Track your predictions to see how accurate your model is. (You may want to update your model as new data comes in.) What factors may cause a movie's weekend gross to not follow an exponential model?

Lesson
9-5 Common Logarithms

Vocabulary

logarithm of x to the base 10,
   log of x to the base 10,
   log base 10 of x

common logarithm,
   common log

logarithm function to the
   base 10, common
   logarithm function

▶ **BIG IDEA** When a number is written as a power of 10, the exponent in the expression is the logarithm of the number to the base 10.

Whenever there is a situation described by an equation of the form $y = b^x$, you may know the values of y and b, and want to find x. For instance, people may want to know when a population will reach a certain level, or people may want to know the age of an ancient Egyptian mummy. Recall that an equation for the inverse of a function can be found by switching x and y. So, $x = b^y$ is an equation for the inverse of the exponential function. You can use this equation and what you know about exponential functions to solve exponential equations.

Mental Math

Find the exact geometric mean of the arithmetic mean, median, and mode of the data set {1, 1, 2, 3, 5}.

Logarithms to the Base 10

Consider the equation $x = b^y$ when $b = 10$. Then $x = 10^y$, and we say that y is the *logarithm of x to the base 10*.

Definition of Logarithm of x to the Base 10

y is the **logarithm of x to the base 10**, or the **log of x to the base 10**, or the **log base 10 of x**, written $y = \log_{10} x$, if and only if $10^y = x$.

For example, $2 = \log_{10} 100$, because $10^2 = 100$. In the table below are some other powers of 10 and the related logs to the base 10.

Exponential Form	Logarithmic Form
$10^7 = 10{,}000{,}000$	$\log_{10} 10{,}000{,}000 = 7$
$10^{-3} = 0.001$	$\log_{10} 0.001 = -3$
$10^{\frac{1}{2}} = \sqrt{10}$	$\log_{10} \sqrt{10} = \frac{1}{2}$
$10^{-\frac{1}{4}} = \frac{1}{\sqrt[4]{10}}$	$\log_{10} \frac{1}{\sqrt[4]{10}} = -\frac{1}{4}$
$10^a = b$	$\log_{10} b = a$

▶ **READING MATH**

The word *logarithm* is derived from the Greek words *logos*, meaning "reckoning," and *arithmos*, meaning "number," which is also the root of the word *arithmetic*. A *logarithm* is literally a "reckoning number." Logarithms were used for hundreds of years for calculating (or "reckoning") before electronic calculators and computers were invented.

Logarithms are exponents. The logarithm of x is the exponent to which the base is raised to get x. Logarithms to the base 10 are called **common logarithms**, or **common logs**. We often write common logs without indicating the base. That is, $\log x$ means $\log_{10} x$.

Evaluating Common Logarithms

Common logarithms of powers of 10 can be found quickly without a calculator.

GUIDED

Example 1

Evaluate.

   a. log 100   b. log 0.00001   c. log 1

Solution First write each number as a power of ten. Then apply the definition of a common logarithm. Remember that the logarithm is the exponent.

   a. Because $100 = 10^{\underline{?}}$, $\log_{10} 100 = \underline{\ ?\ }$.

   b. You need to find n such that $10^n = 0.00001$.
        Because $0.00001 = 10^{\underline{?}}$, $\log 0.00001 = \underline{\ ?\ }$.

   c. $10^0 = 1$, so $\log \underline{\ ?\ } = \underline{\ ?\ }$.

 QY1

You can use the logarithm function in your calculator to evaluate $\log_{10} x$ for any positive number x.

▶ **QY1**

   a. What is $\log \sqrt{10}$?

   b. What is the common logarithm of 0.001?

Example 2

Estimate to the nearest hundred-thousandth.

   a. $\log \sqrt{2}$ b. $\log 5$

Solution Use your calculator. You will get a display like that at the right. Round to five decimal places.

   a. $\log \sqrt{2} \approx 0.15051$

   b. $\log 5 \approx 0.69897$

Check Use the definition of logarithm to the base 10: $\log x = y$ if and only if $10^y = x$.

   a. Does $10^{0.15051} \approx \sqrt{2}$? Use the power key on your calculator. It checks.

   b. Does $10^{0.69897} \approx 5$? Yes, it checks.

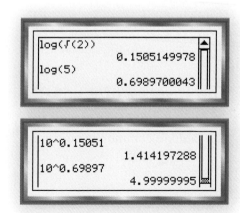

QY2

In general, the common logarithm of 10^x is x. That is, $\log_{10}(10^x) = x$. This is why we say a logarithm is an exponent.

The larger a number, the larger its common logarithm. Because of this, you can estimate the common logarithm of a number without a calculator. Either compare the number to integer powers of 10 or write the number in scientific notation to determine between which two consecutive integers the logarithm falls.

Example 3

Between which two consecutive integers is log 5673?

Solution 1 5673 is between $1000 = 10^3$ and $10{,}000 = 10^4$, so log 5673 is between 3 and 4.

Solution 2 $5673 = 5.673 \cdot 10^3$. This indicates that log 5673 is between 3 and 4.

Check A calculator gives log $5673 \approx 3.753\ldots$. It checks.

Solving Logarithmic Equations

You can solve logarithmic equations using the definition of common logarithms.

Example 4

Solve for x.

a. $\log x = \frac{1}{2}$ b. $\log x = 0.71$

Solution

a. $\log x = \frac{1}{2}$ if and only if $10^{\frac{1}{2}} = x$.
    So, $x = 10^{\frac{1}{2}} = \sqrt{10} \approx 3.162$.

b. $\log x = 0.71$ if and only if $10^{0.71} = x$. So, $x = 10^{0.71} \approx 5.129$.

The Inverse of $y = 10^x$

The inverse of the exponential function f defined by $f(x) = 10^x$ is related to common logarithms. You will find the inverse in the Activity on the next page.

Activity

MATERIALS graph paper

Step 1 Fill in the y-values in the table when $y = 10^x$.

x	-2	-1	-0.75	-0.5	-0.25	0	0.25	0.5	0.75	1	2
y	?	?	?	?	?	?	?	?	?	?	?

Step 2 Plot the points $(x, 10^x)$ from the table on graph paper. Connect the points with a smooth curve.

Step 3 Plot the points of the inverse of the relation in the table on the same graph as in Step 2. Connect the points with a smooth curve.

Step 4 Graph both $y = 10^x$ and $y = \log x$ on the same set of axes in the window $-2 \leq x \leq 2$ and $-2 \leq y \leq 2$ on a graphing utility. Compare them to the graphs you made in Steps 2 and 3.

Your graphs from the Activity should look similar to the one at the right. The graph of $y = 10^x$ passes the horizontal-line test, and so its inverse is a function. This means that $y = \log x$ is an equation for a function, the inverse of the exponential function with equation $y = 10^x$.

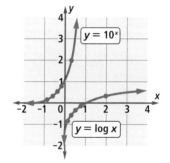

Properties of $y = 10^x$ and $y = \log x$

The inverse of the function f with equation $f(x) = 10^x$ can be described in several ways as shown in the table below.

Ways of Thinking of the Inverse of f with $f(x) = y = 10^x$	Written	Spoken
Switching **x** and **y**	$x = 10^y$	"x equals 10 to the yth power."
Using the language of logs	$y = \log_{10} x$ $y = \log x$	"y equals the log of x to the base 10."
Using function notation	$f^{-1}(x) = \log_{10} x$ $f^{-1}(x) = \log x$	"f inverse of x equals the log of x to the base 10."

The curve defined by these equations is called a *logarithmic curve*. As the graph from the Activity above shows, a logarithmic curve is the reflection image of an exponential curve over the line with equation $y = x$.

The function that maps x onto $\log_{10} x$ for all positive numbers x is called the **logarithm function to the base 10**, or the **common logarithm function**.

Because they are inverses, each property of the exponential function defined by $y = 10^x$ corresponds to a property of its inverse, the common logarithm function defined by $y = \log x$.

Function	Domain	Range	Asymptote	Intercepts
$y = 10^x$	set of real numbers	set of positive real numbers	x-axis ($y = 0$)	y-intercept $= 1$
$y = \log x$	set of positive real numbers	set of real numbers	y-axis ($x = 0$)	x-intercept $= 1$

Questions

COVERING THE IDEAS

1. If $m = \log_{10} n$, what other relationship exists between m, 10, and n?

2. **a.** Write in words how to read the expression $\log_{10} 8$.
   b. Evaluate $\log_{10} 8$ to the nearest ten-thousandth.

In 3–8, evaluate using the definition of common logarithms.

3. $\log 10,000,000$
4. $\log 10^5$
5. $\log 0.0000001$
6. $\log \sqrt[3]{10}$
7. $\log \frac{1}{10}$
8. $\log 10$

9. Between which two consecutive integers is the value of the common logarithm of 100,000,421?

In 10–12, approximate to the nearest thousandth.

10. $\log 3$
11. $\log 0.00309$
12. $\log 309,000$

In 13 and 14, solve for x.

13. $\log x = 5$
14. $\log x = 1.25$

15. Consider the graph of $y = \log_{10} x$.
    a. Name its x- and y-intercepts, if they exist.
    b. Name three points on the graph.
    c. Name the three corresponding points on the graph of $y = 10^x$.

16. What are the domain and range of the common logarithm function?

17. **Fill in the Blank** The functions f and g, with equations $f(x) = 10^x$ and $g(x) = \underline{\ ?\ }$, are inverses of each other.

APPLYING THE MATHEMATICS

18. If $5 \log v = 2$, what is the value of v?

19. If a number is between 10 and 100, its common logarithm is between which two consecutive integers?

20. The common logarithm of a number is -5. What is the number?

21. Evaluate $10^{\log 3.765}$.

22. Explain why for all positive numbers a, $10^{\log a} = a$.

23. If $f(x) = 10^x$ and $g(x) = \log_{10} x$, what is $f \circ g(x)$? Explain your answer.

In 24 and 25, use this information: Most of today's languages are thought to be descended from a few common ancestral languages. The longer the time lapse since a language split from the ancestral language, the fewer common words exist in the descendant language. Let c = the number of centuries since two languages split from an ancestral language. Let w = the fraction of words from the ancestral language that are common to the two descendent languages. In linguistics, the equation $\frac{10}{c} = \frac{2 \log r}{\log w}$ (in which $r = 0.86$ is the index of retention) has been used to relate c and w.

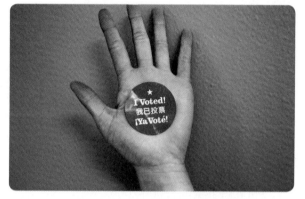

English and Spanish descended from the Indo-European language family. Chinese descended from the Sino-Tibetan family.

24. If about 15% of the words in an ancestral language are common to two different languages, about how many centuries ago did they split from the ancestral language?

25. If it is known that two languages split from an ancestral language about 1500 years ago, about what percentage of the words in the ancestral language are common to the two languages?

REVIEW

26. Find an equation of the form $y = ab^x$ that passes through the points $(0, 3)$ and $(4, 48)$. (Lesson 9-4)

27. A tool and die machine used in a metal working factory depreciates so that its value after t years is given by $N(t) = Ce^{-0.143t}$, where C is its initial value. If after 4 years the machine is worth $921,650$, what was its original value? (Lesson 9-3)

28. Refer to the graph at the right of the cubing function f defined by $f(x) = x^3$. (Lessons 8-3, 7-6, 7-1)

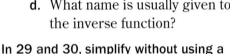

   a. What are the domain and range of this function?
   b. Graph the inverse.
   c. Is the inverse a function? Why or why not?
   d. What name is usually given to the inverse function?

In 29 and 30, simplify without using a calculator. (Lessons 7-6, 7-3, 7-2)

29. $(13^3 \cdot 13^{-6})^2$

30. $\sqrt[5]{8^{15}}$

EXPLORATION

31. In 2005, the three leading international public repositories for DNA and RNA sequence information reached over 100 gigabases (100,000,000,000 bases) of sequence. For perspective, the human genome is about 3 gigabases. The table below presents the approximate size of the international databases in each year from 2000-2005.

Year	2000	2001	2002	2003	2004	2005
Gigabases	11	18	37	53	81	104

   a. Write an equation to predict the number n of bases stored t years after 2000.
   b. What is the growth rate of sequence mapping?
   c. Use your equation in Part a to predict the year that the international databases will hold the amount of information equal to the information contained in the genomes of 1000 people.

QY ANSWERS

1. a. $\frac{1}{2}$ b. -3

2.

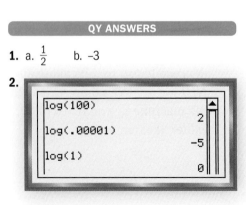

Lesson 9-6 Logarithmic Scales

Vocabulary

logarithmic scale

decibel, dB

▶ **BIG IDEA** A **logarithmic scale** is one in which the numbers are written as the powers of a fixed number, and the exponents of the powers are used as the scale values.

Logarithmic scales are used when all the measures of an attribute are positive and cover a wide range of values from small to very large. Two logarithmic scales you may have heard of are the *decibel scale* that measures relative sound intensity and the *Richter magnitude scale* that measures the intensity of earthquakes.

The Decibel Scale

The decibel scale is based on *watts*. The watt is a measure of power. The quietest sound that a human ear can pick up has an intensity of about 10^{-12} watts per square meter $\left(\frac{W}{m^2}\right)$. The human ear can also hear sounds with an intensity as large as $10^2 \frac{W}{m^2}$. Because the range from 10^{-12} to 10^2 is so large, it is convenient to use a measure that is based on the exponents of 10 in the intensity. This measure of relative sound intensity is the **decibel**, abbreviated **dB**. Because it is based on a ratio of units, the decibel is a dimensionless unit, like angle measure. A formula that relates the sound intensity N in $\frac{W}{m^2}$ to its relative intensity D in decibels is

$$D = 10 \log\left(\frac{N}{10^{-12}}\right).$$

The chart at the right gives the sound intensity in $\frac{W}{m^2}$ and the corresponding decibel values for some common sounds.

Notice that as the decibel values in the right column increase by 10, corresponding intensities in the left column are multiplied by 10. Thus, if the number of decibels increases by 20, the sound intensity multiplies by $10 \cdot 10$ or 10^2. If you increase the number of decibels by 40, you multiply the watts per square meter by $10 \cdot 10 \cdot 10 \cdot 10$, or 10^4.

Mental Math

Simplify.

a. $\dfrac{10^6}{10^2}$

b. $\dfrac{4.3 \cdot 10^4}{4.3 \cdot 10^{-7}}$

c. $\dfrac{2.42 \cdot 10^{12}}{1.21 \cdot 10^3}$

d. $\dfrac{a \cdot 10^m}{b \cdot 10^n}$

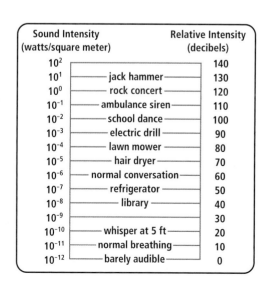

Sound Intensity (watts/square meter)		Relative Intensity (decibels)
10^2		140
10^1	jack hammer	130
10^0	rock concert	120
10^{-1}	ambulance siren	110
10^{-2}	school dance	100
10^{-3}	electric drill	90
10^{-4}	lawn mower	80
10^{-5}	hair dryer	70
10^{-6}	normal conversation	60
10^{-7}	refrigerator	50
10^{-8}	library	40
10^{-9}		30
10^{-10}	whisper at 5 ft	20
10^{-11}	normal breathing	10
10^{-12}	barely audible	0

In general, an increase of n dB multiplies the sound intensity by $10^{\frac{n}{10}}$.

 QY

▶ **QY**

How many times as intense is a sound of 90 decibels as a sound of 60 decibels?

Example 1

Due to a law in France, manufacturers had to limit the sound intensity of their MP3 players to 100 decibels. How many times as intense is a particular MP3 player's maximum 115-decibel intensity than the limited intensity imposed in France?

Solution Use the generalization above. An increase of n dB multiplies the sound intensity by $10^{\frac{n}{10}}$. The difference in the decibel level is $115 - 100 = 15$ dB. So, the maximum sound intensity is $10^{\frac{15}{10}} = 10^{1.5} \approx 32$ times as intense as the limit imposed in France.

Experiments have shown that people perceive a sound that is 10 decibels louder than another sound as only twice as loud. That is, the ear perceives a sound of 80 decibels as only sixteen times as loud as one of 40 decibels, even though in fact the sound is 10^4 times as intense. Even if a person does not feel pain, long or repeated exposure to sounds at or above 85 decibels can cause permanent hearing loss.

Using the formula on the previous page, if you know the intensity of a sound, you can find its relative intensity in decibels.

A decibel is $\frac{1}{10}$ of a bel, a unit named after Alexander Graham Bell (1847–1922), the inventor of the telephone.

Example 2

Grunting while hitting the ball has become a controversial issue in professional tennis. Some people are concerned that such loud sounds are unfair distractions to the opposing player. Serena Williams's grunts have been measured at a sound intensity of $6.31 \cdot 10^{-4} \frac{W}{m^2}$. Find the relative intensity of the sound in decibels.

Solution Substitute $6.31 \cdot 10^{-4}$ for N in the formula $D = 10 \log \frac{N}{10^{-12}}$.

$D = 10 \log\left(\frac{6.31 \cdot 10^{-4}}{10^{-12}}\right) = 10 \log 631{,}000{,}000 \approx 88$

The relative intensity of the grunting is about 88 dB.

Check Refer to the chart on the previous page. Notice that $10^{-4} < 6.31 \cdot 10^{-4} < 10^{-3}$, and the relative intensity of 88 dB falls between 80 and 90 dB. It checks.

To convert from the decibel scale to the $\frac{W}{m^2}$ scale, you can solve a logarithmic equation.

Example 3

The maximum sound intensity of Melody's MP3 player is 115 dB.
What is the maximum sound intensity in watts per square meter?

Solution 1

$115 = 10 \log\left(\dfrac{N}{10^{-12}}\right)$	Substitute 115 for D in the formula.
$11.5 = \log\left(\dfrac{N}{10^{-12}}\right)$	Divide each side by 10.
$10^{11.5} = \dfrac{N}{10^{-12}}$	Definition of common logarithm
$10^{11.5} \cdot 10^{-12} = N$	Multiply each side by 10^{-12}.
$10^{-0.5} = N$	Product of Powers Postulate

$10^{-0.5} \approx 0.316$. So, the maximum sound intensity of the MP3 player is about $0.316 \; \frac{W}{m^2}$.

Check Solve on a CAS. It checks as shown on the calculator display at the right.

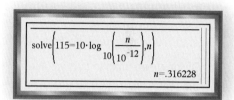

$$\text{solve}\left(115 = 10 \cdot \log_{10}\left(\frac{n}{10^{-12}}\right), n\right)$$

$$n = .316228$$

Earthquake Magnitude Scales

The most popular scale used by the media for describing the magnitudes of earthquakes is the Richter scale, designed by Charles Richter in 1935. Like the decibel scale, the Richter scale is a logarithmic scale. A value x on the Richter scale corresponds to a measured force, or amplitude, of $k \cdot 10^x$, where the constant k depends on the units being used to measure the quake. Consequently, each increase of 1 on the Richter scale corresponds to a factor of 10 change in the amplitude. The table below gives a brief description of the effects of earthquakes of different magnitudes.

Richter Magnitute	Possible Effects
1–2	Usually not felt except by instruments
3	May be felt but rarely causes damage
4	Like vibrations from heavy traffic
5	Strong enough to wake sleepers
6	Walls crack, chimneys fall
7	Ground cracks, houses collapse
8	Few buildings survive, landslides

When $k = 1$, the Richter magnitude is the common logarithm of the force, because x is the common logarithm of 10^x.

GUIDED

Example 4

Just before a 1989 World Series game in San Francisco, California there was an earthquake in the nearby Santa Cruz Mountains that registered 6.9 on the Richter scale. The tsunami that hit twelve Indian Ocean nations in December 2004 was triggered by an earthquake that measured 9.3 on the Richter scale. The force of the 2004 earthquake was how many times the force of the 1989 quake?

The 1989 earthquake in San Francisco caused the third floor of this apartment building to collapse onto this car.

Solution The amplitude of the 1989 World Series earthquake was $k \cdot 10^{\underline{?}}$. The amplitude of the 2004 quake was $k \cdot 10^{\underline{?}}$.

Divide these quantities to compare the amplitudes. Since we are assuming the forces were measured in the same units, the two k-values are the same.

$$\frac{k \cdot 10^{\underline{?}}}{k \cdot 10^{\underline{?}}} = 10^{\underline{?}}.$$

As a decimal, $10^{\underline{?}} \approx \underline{\;?\;}$, indicating that the 2004 earthquake had about $\underline{\;?\;}$ times as much force as the $\underline{\;?\;}$ earthquake.

Some other examples of logarithmic scales include the pH scale for measuring the acidity or alkalinity of a substance, the scales used on radio dials, and the scale for measuring the magnitude (brightness) of stars.

Questions

COVERING THE IDEAS

In 1 and 2, use the formula $D = 10 \log\left(\dfrac{N}{10^{-12}}\right)$ that relates relative intensity D in dB of sound to sound intensity N in $\dfrac{\text{w}}{\text{m}^2}$.

1. Find the relative intensity in decibels of an explosion that has a sound intensity of $2.45 \cdot 10^{-5}$.

2. Find the intensity in $\dfrac{\text{w}}{\text{m}^2}$ of a sound that has a relative intensity of 78 decibels.

In 3–5, refer to the chart of sound intensity levels from the beginning of this lesson.

3. In Example 2, we found that Serena Williams's grunts reach an intensity of 88 decibels. Between which two powers of 10 is the equivalent sound intensity?

Serena Williams

4. If you are near a refrigerator, how many times as intense would the sound have to be in order for it to reach the level emitted by a jackhammer?

5. How many times as intense is a whisper than a noise that is barely audible?

6. Lester Noyes is deciding between two dishwashers for purchase. One dishwasher's rating is 56 decibels (dB) and the other is 59 dB. The salesman says the sound intensity of the 59 dB dishwasher is only 5% more than the other one, and this difference is insignificant. Do you agree? Explain.

7. The grunts of the tennis player Maria Sharapova have been measured at 100 dB. How many times as loud as 88 dB is 100 dB?

8. An earthquake measuring 6.5 on the Richter scale carries how many times as much force as one measuring 5.5?

9. An earthquake measuring 8.2 on the Richter scale carries how many times as much force as one measuring 6.7?

10. If one earthquake's Richter value is 0.4 higher than another, how many times as much force is in the more powerful earthquake?

APPLYING THE MATHEMATICS

In 11 and 12, use this information. In 1979, seismologists introduced the *moment magnitude scale* based on the formula
$M_w = \dfrac{\log M_0}{1.5} - 10.7$ for moment magnitude M_w. Seismic moment M_0 is a measure of the size of the earthquake. Its units are dyne-cm. (One dyne is the force required to accelerate a mass of 1 gram at a rate of 1 centimeter per second squared.)

11. An earthquake had a seismic moment of 3.86×10^{27} dyne-cm. Find its moment magnitude M_w.

12. If the moment magnitude of one earthquake is 1 higher than that of another, how do their seismic moments differ?

In 13–16, use the pH scale for measuring acidity or alkalinity of a substance shown at the right. Its formula is $pH = -\log(H^+)$, where H^+ is the concentration of hydrogen ions in $\frac{\text{moles}}{\text{liter}}$ of the substance.

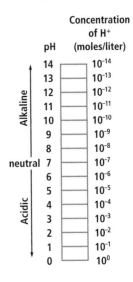

Concentration of H$^+$

pH	(moles/liter)
14	10^{-14}
13	10^{-13}
12	10^{-12}
11	10^{-11}
10	10^{-10}
9	10^{-9}
8	10^{-8}
7	10^{-7}
6	10^{-6}
5	10^{-5}
4	10^{-4}
3	10^{-3}
2	10^{-2}
1	10^{-1}
0	10^{0}

Alkaline, neutral 7, Acidic

13. The pH of normal rain is 5.6 and the pH of acidic rain is 4.3. The concentration of hydrogen ions in acid rain is how many times the concentration of hydrogen ions in normal rain?

14. For a healthy individual, human saliva has an average pH of 6.4.

   a. Is saliva acidic or alkaline?

   b. What is the concentration of hydrogen ions in human saliva?

15. The concentration of hydrogen ions in a typical piece of white bread is $3.16 \times 10^{-6} \frac{\text{mol}}{\text{liter}}$. What is the pH of white bread?

16. The Felix Clean Company manufactures soap. In one advertisement the company claimed that its soap has a tenth of the pH of the competing brand.

   a. What range of values could the pH of the soap have if this were true?

   b. What would happen if the soap did indeed have a pH as claimed in the advertisement?

   c. What did the advertiser probably mean?

In 17 and 18, use this information: In astronomy, the *magnitude* (brightness) m of a star is measured not by the energy I meeting the eye, but by its logarithm. In this scale, if one star has radiation energy I_1 and absolute magnitude m_1, and another star has energy I_2 and absolute magnitude m_2, then $m_1 - m_2 = -2.5 \log\left(\frac{I_1}{I_2}\right)$.

17. The star Rigel in the constellation Orion radiates about 150,000 times as much energy as the Sun. The Sun has absolute magnitude 4.8. If the ratio of intensities is 150,000, find the absolute magnitude of Rigel.

18. Suppose the difference $m_1 - m_2$ in absolute magnitudes of two stars is 5. Find $\frac{I_1}{I_2}$, the ratio of the energies they radiate.

REVIEW

In 19 and 20, explain how to evaluate without using a calculator. (Lesson 9-5)

19. $\log_{10} 1{,}000{,}000$

20. $\log_{10} 10^{-5}$

In 21 and 22, solve. (Lesson 9-5)

21. $\log x = 7$

22. $\log 7 = x$

23. Give two equations for the inverse of the function with the equation $y = 10^x$. **(Lesson 9-5)**

24. The *Haugh unit* is a measure of egg quality that was introduced in 1937 in the *U.S. Egg and Poultry Magazine*. The number U of Haugh units of an egg is given by the formula

$$U = 100 \log\left[H - \tfrac{1}{100}\sqrt{32.2}\,(30W^{0.37} - 100) + 1.9\right],$$

where W is the weight of the egg in grams, and H is the height of the albumen in millimeters when the egg is broken on a flat surface. Find the number of Haugh units of an egg that weighs 58.8 g and for which $H = 6.3$ mm. **(Lesson 9-5)**

25. Is $y = 2^{-x}$ the equation for a function of exponential growth or exponential decay? Explain how you can tell. **(Lesson 9-2)**

26. Consider the graph of the function g at the right. Give the domain and range of g^{-1}. **(Lesson 8-2)**

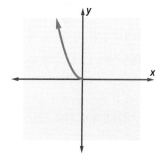

In 27 and 28, assume all variables represent positive real numbers. Simplify. **(Lessons 7-7, 7-2)**

27. $\dfrac{p^5 q^4}{(pm)^3}$

28. $\left(r^2\right)^{\frac{1}{4}} \left(t^{10}\right)^{\frac{3}{5}}$

EXPLORATION

29. The mathematician L.F. Richardson classified conflicts according to their magnitude, the base-10 logarithm of the total number of deaths. For example, a war in which there were 10,000 deaths would have a magnitude of 4 because $10^4 = 10,000$. A gang fight with 10 casualties would have a magnitude of 1 because $10^1 = 1$. Use Richardson's scale to classify the Revolutionary War, the Civil War, World War I, World War II, the Vietnam War, and the Persian Gulf War. Comment on the effectiveness of Richardson's scale in comparing the number of deaths in the wars.

Lesson
9-7

Logarithms to Bases Other Than 10

Vocabulary

logarithm of a to the base b

logarithm function with base b

▶ **BIG IDEA** When a number is written as a power of b, the exponent in the expression is the *logarithm of the number to the base b*.

In Lesson 9-5, you learned about the common logarithm, which allows you to solve for x when $10^x = a$. But how do you solve $b^x = a$ when b is any positive value other than 1? The same question was considered by Leonhard Euler in the 18th century. It led him to the following definition of a logarithm to any positive base other than 1.

Mental Math

Let $P = (3, 5)$. Give the coordinates of the image of P under the given transformation.

a. r_y

b. S_3

c. $S_{-2, -1}$

d. $T_{-2, -1}$

Definition of Logarithm of a to the Base b

Let $b > 0$ and $b \neq 1$. Then x is the **logarithm of a to the base b**, written $x = \log_b a$, if and only if $b^x = a$.

For example, because $3^5 = 243$, you can write $5 = \log_3 243$. This is read "5 is the logarithm of 243 with base 3" or "5 is log 243 to the base 3" or "5 is the log base 3 of 243." At the right are some other powers of 3 and the related logs to the base 3.

Exponential Form	Logarithmic Form
$3^4 = 81$	$\log_3 81 = 4$
$3^3 = 27$	$\log_3 27 = 3$
$3^2 = 9$	$\log_3 9 = 2$
$3^1 = 3$	$\log_3 3 = 1$
$3^{0.5} = \sqrt{3}$	$\log_3 \sqrt{3} = 0.5$
$3^0 = 1$	$\log_3 1 = 0$
$3^{-1} = \frac{1}{3}$	$\log_3\left(\frac{1}{3}\right) = -1$
$3^{-2} = \frac{1}{9}$	$\log_3\left(\frac{1}{9}\right) = -2$
$3^y = x$	$\log_3 x = y$

(STOP) **QY1**

▶ **QY1**

Rewrite $7^4 = 2401$ in logarithmic form.

Example 1

Write the equation $P = 9(1.028)^x$ in logarithmic form.

Solution First rewrite the equation in $b^n = m$ form. Divide both sides by 9.

$$1.028^x = \frac{P}{9}$$

Apply the definition of logarithm. The base is 1.028.

$$\log_{1.028}\left(\frac{P}{9}\right) = x$$

Evaluating Logarithms to Bases Other Than 10

The methods of evaluating logs and solving equations of logs with bases other than 10 are very similar to the methods used with common logarithms. For example, when a is a known power of the base b, $\log_b a$ can be quickly found without a calculator.

GUIDED

Example 2

Evaluate the following.

a. $\log_7 49$ b. $\log_8 2$ c. $\log_4\left(\frac{1}{64}\right)$

Solution

a. Let $\log_7 49 = x$.

$$7^x = 49 \qquad \text{Definition of logarithm}$$
$$7^x = 7^2 \qquad \text{Rewrite 49 as a power of 7.}$$
$$x = \underline{\ ?\ } \qquad \text{Equate the exponents.}$$

So, $\log_7 49 = \underline{\ ?\ }$.   Transitive Property of Equality

b. Let $\log_8 2 = x$.

$$8^x = 2 \qquad\qquad \underline{\ ?\ }$$
$$(2^3)^x = 2^{\underline{\ ?\ }} \qquad \text{Rewrite both sides as powers of 2.}$$
$$2^{\underline{\ ?\ }} = 2^{\underline{\ ?\ }} \qquad \text{Power of a Power Postulate}$$
$$\underline{\ ?\ } = \underline{\ ?\ } \qquad \text{Equate the exponents.}$$
$$x = \underline{\ ?\ } \qquad\qquad \underline{\ ?\ }$$

So, $\log_8 2 = \underline{\ ?\ }$.

c. Let $\log_4\left(\frac{1}{64}\right) = x$.

$$\underline{\ ?\ } = \frac{1}{64}$$
$$\underline{\ ?\ } = 4^{\underline{\ ?\ }}$$
$$\underline{\ ?\ } = \underline{\ ?\ }$$

So, $\log_4\left(\frac{1}{64}\right) = \underline{\ ?\ }$.

In each part of Example 2 you moved from an expression of the form $b^m = b^n$ to one of the form $m = n$. We call this "equating the exponents." When $b \neq 1$ and b is positive, this is a valid process because the exponential function with base b takes on a unique value for each exponent.

When a is not an integer power of b, a CAS can be used to find $\log_b a$. Some CAS allow you to enter any base b when you press the logarithm key. The template used for logs on one CAS is shown at the left below. The CAS response to $\log_2 3.5$ is shown at the right below.

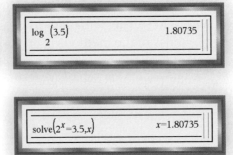

In addition, any CAS may be used to find x when $\log_b a = x$. Simply rewrite the equation in exponential form and solve for x. This method is used to find $\log_2 3.5$ as shown at the right.

 QY2

Graphs of Logarithm Functions

Both the exponential equation $3^y = x$ and the logarithmic equation $y = \log_3 x$ describe the *inverse* of the function with equation $y = 3^x$. These functions are graphed at the right. In general, the **logarithm function with base b**, $g(x) = \log_b x$, is the inverse of the exponential function with base b, $f(x) = b^x$.

Recall that the domain of the exponential function with equation $y = 3^x$ is the set of all real numbers. Consequently, the range of the logarithm function with equation $y = \log_3 x$ is also the set of all real numbers. So logarithms to the base 3 can be negative. However, the range of the exponential function $y = 3^x$ and the domain of the corresponding logarithm function is the set of positive real numbers. This means that in the set of real numbers there is no logarithm of a nonpositive number.

▶ **QY2**

a. Use a CAS to estimate $\log_7 124$ to five decimal places.

b. To what exponential equation is the result of Part a an estimated solution?

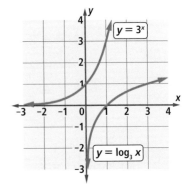

Activity

Work with a partner and use a graphing utility.

Step 1 Each partner should choose a different value of b, with $b > 0$ and $b \neq 1$, and graph $y = \log_b x$. An example is shown at the right.

Step 2 Fill in a chart like the one on the next page to record the domain, range, intercepts, and asymptotes of the graph for your value of b. Then have your partner fill in another row of the chart for his or her value of b.

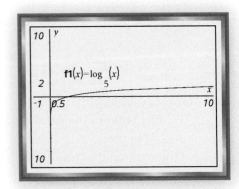

b	Domain	Range	x-intercepts	y-intercepts	Asymptotes
?	?	?	?	?	?

Step 3 Set the window on your graphing utility to square. Each partner should graph $y = b^x$ on the same set of axes as in Step 1. What is the relationship between the graph of $y = \log_b x$ and the graph of $y = b^x$?

Step 4 Both partners should repeat Steps 1–3 for two additional values of b. Choose values that are different from your partner's values, and be sure to choose some noninteger values for b.

Step 5 Summarize your results. What properties do these functions seem to share?

The Activity shows that there are properties shared by all logarithm functions, regardless of their base. Every function with an equation of the form $y = \log_b x$, where $b > 0$ and $b \neq 1$, has the following properties.

1. Its domain is the set of positive real numbers.
2. Its range is the set of all real numbers.
3. Its x-intercept is 1, and there is no y-intercept.
4. The y-axis ($x = 0$) is an asymptote to the graph.
5. Its graph is the reflection image over $y = x$ of the graph of $y = b^x$.

You may wonder why b cannot equal 1 in the definition of the logarithm of a with base b. It is because the inverse of the function with equation $y = 1^x$ is not a function.

 QY3

> ▶ **QY3**
>
> Why can the base b of logarithms never be negative?

Solving Logarithmic Equations

To solve a logarithmic equation, it often helps to use the definition of logarithm to rewrite the equation in exponential form.

Example 3

Solve for h: $\log_4 h = \frac{3}{2}$.

Solution

$4^{\frac{3}{2}} = h$ definition of logarithm

$8 = h$ Rational Exponent Theorem

To solve for the base in a logarithmic equation, apply the techniques you learned in Chapters 7 and 8 for solving equations with nth powers.

Example 4

Find w if $\log_w 1024 = 10$.

Solution

$$w^{10} = 1024 \qquad \text{definition of logarithm}$$
$$(w^{10})^{\frac{1}{10}} = (1024)^{\frac{1}{10}} \quad \text{Raise both sides to the } \tfrac{1}{10} \text{ power.}$$
$$w = 2 \qquad \text{Power of a Power Postulate}$$

Check Does $\log_2 1024 = 10$? Does $2^{10} = 1024$? Yes, it checks.

Questions

COVERING THE IDEAS

1. Logarithms to the base b arose from Euler's attempt to describe the solution(s) to what equation?

2. **Fill in the Blank** Suppose $b > 0$ and $b \neq 1$. When $b^n = m$, $n = \underline{\ ?\ }$.

3. Write the equivalent logarithmic form for $(6\sqrt{3})^8 = 136{,}048{,}896$.

In 4 and 5, write the equivalent exponential form for the sentence.

4. $\log_4 0.0625 = -2$

5. $\log_b a = c$

6. Write the inverse of the function with equation $y = 2^x$ as
   a. an exponential equation. b. a logarithmic equation.

7. State the domain and range of the function defined by $y = \log_3 x$.

8. Write an equation for the asymptote to the graph of $y = \log_b x$, where $b > 0$ and $b \neq 1$.

9. Sketch the graph of $y = \log_4 x$.

In 10–12, write the corresponding exponential form of each logarithmic equation, then solve for the exponent on a CAS.

10. $\log_{121} 1331 = x$

11. $\log_{\frac{1}{4}}\left(\frac{1}{64}\right) = y$

12. $\log_{\sqrt{5}} 625 = z$

In 13–17, write the corresponding exponential form of each logarithmic equation. Then, solve for the given variable.

13. $\log_a 8 = \frac{1}{3}$

14. $\log_7 c = 4$

15. $\log_{47} d = -0.2$

16. $\log_t\left(\frac{1}{6}\right) = -\frac{1}{3}$

17. $\log_w w = 1$

APPLYING THE MATHEMATICS

18. **a.** Make a table of values for $y = 2^x$ with $x = -2, -1, 0, 1, 2$ and a corresponding table of values for $y = \log_2 x$.
    b. Graph $y = 2^x$ and $y = \log_2 x$ on the same set of axes.
    c. **True or False** The domain of $y = 2^x$ is the range of $y = \log_2 x$.

19. *Self-information I* is a measure (in bits) of how the knowledge gained about a certain event adds to your overall knowledge. A formula for self-information is $I = \log_2\left(\frac{1}{x}\right)$, where x is the probability of a certain event occurring.
    a. When flipping a coin, the probability of it landing on tails is 0.5. Find the number of bits this adds to your self-information.
    b. The probability of drawing a card with a diamond on it from a standard deck of cards is 0.25. Find the number of bits this adds to your self-information.
    c. If the self-information added is 2.3 bits, what was the probability of the event?

20. The depreciation of a certain automobile that initially costs $25,000 is given by the formula $N = 25{,}000(0.85)^t$, where N is the current value after t years.
    a. Write this equation in logarithmic form.
    b. How old is a car that has a current value of $9500?

21. **a.** Evaluate $\log_6 216$ and $\log_{216} 6$.
    b. Evaluate $\log_7 49$ and $\log_{49} 7$.
    c. Generalize the results of Parts a and b.

REVIEW

In 22 and 23, refer to the representation of the electromagnetic spectrum below. (Lesson 9-6)

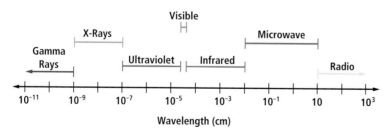

22. What kind of scale is used to depict the wavelengths for the electromagnetic spectrum?

23. The shortest radio wave is how many times the length of the longest gamma ray wave?

In 24 and 25, consider the Palermo Technical Impact Hazard Scale, a logarithmic scale used by astronomers to rate the potential hazard of impact of near-Earth objects (NEO) such as asteroids, comets, and large meteoroids. A Palermo value P is given by the formula $P = \log\left(\dfrac{p_i}{f_B T}\right)$, where $p_i =$ the event's probability, $T =$ the time in years until the event, and $f_B = 0.03E^{-0.8}$ is the annual probability of an impact event with energy E (in megatons of TNT) at least as large as the event in question. (Lesson 9-6)

24. NASA keeps a database of all near-Earth objects. Discovered in 2008, the asteroid 2008 HJ is expected to approach Earth in 2077. For that year, it has an impact probability of $1.3 \cdot 10^{-6}$ and an expected impact energy of 0.408 megaton. Calculate the 2020 Palermo value for 2008 HJ.

25. 2002 RB182 is an asteroid with a 2005 Palermo value of –7.17. If it were to collide with Earth in 2044, it would have an expected energy of 35.6 megatons. What is the probability that it will impact Earth in 2044?

While most meteor showers are caused by comet debris, the Geminid meteor showers are caused by asteroid 3200 Phaethon.

In 26 and 27, evaluate using powers of 10. (Lesson 9-5)

26. $\log 10^7$

27. $\log 0.0001$

28. Thorium-232 (^{232}Th) has a half-life of 1.41×10^{10} years. (Lesson 9-2)

   a. Write an equation giving the percent p of thorium-232 in a substance after n years.

   b. Sketch a graph of your equation from Part a.

   c. Calculate the number of years needed for 99% of the original thorium to be present.

29. Simplify each expression. (Lessons 8-4, 7-2)

   a. $x^3 \cdot x^9$ b. $\dfrac{x^3}{x^9}$ c. $\left(x^3\right)^9$ d. $\sqrt[3]{x^9}$

EXPLORATION

30. Logarithms were invented by English mathematician John Napier (1550–1617), in the early 1600s. Henry Briggs, also from England, first used common logarithms around 1620. In England even today logs to the base 10 are sometimes called Briggsian logarithms. Find more information about one of these mathematicians.

QY ANSWERS

1. $\log_7 2401 = 4$

2. a. 2.47713
   b. $7^x = 124$

3. Exponential functions are not defined for negative bases.

Lesson
9-8
Natural Logarithms

Vocabulary

natural logarithm of m, ln m

▶ **BIG IDEA** When a number is written as a power of the irrational number e, the exponent in the expression is the natural logarithm of the number.

What Are Natural Logarithms?

Any positive number except 1 can be the base of a logarithm. The number e that you studied in Lesson 9-3 is frequently used as a logarithm base in real-world applications.

Logarithms to the base e are called *natural logarithms*. Just as log x (without any base named) is shorthand for $\log_{10} x$, ln x is shorthand for $\log_e x$.

Definition of Natural Logarithm of m

n is the **natural logarithm of m**, written $n = \ln m$, if and only if $m = e^n$.

The symbol ln x is usually read "natural log of x".

Natural logarithms of powers of e can be determined in your head from the definition.

$\ln 1 = \log_e 1 = 0$ because $1 = e^0$.

$\ln e = \log_e e = 1$ because $e = e^1$. That is, $\ln 2.718 \approx 1$ because $e^1 \approx 2.718$.

$\ln e^3 = \log_e e^3 = 3$ because $e^3 = e^3$. That is, $\ln 20.086 \approx 3$ because $e^3 \approx 20.086$.

In general, $\ln(e^x) = x$.

 QY1

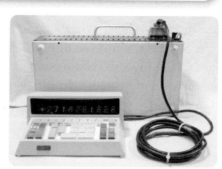

Image courtesy The Old Calculator Museum.

The Wang Model 360E calculator, first developed in 1964, was one of the first calculators capable of computing logarithms.

Mental Math

Find an expression for $f \circ g(x)$ if

a. $f(x) = x^2$ and $g(x) = x^4$

b. $f(x) = x^4$ and $g(x) = x^2$

c. $f(x) = \sqrt{x}$ and $g(x) = 6x - 3$

d. $f(x) = 6x - 3$ and $g(x) = \sqrt{x}$

▶ **QY1**

What is the value of $\ln\left(\frac{1}{e}\right)$?

Evaluating Natural Logarithms

Example 1

Estimate the following to the nearest thousandth using a calculator.

a. ln 100

b. ln 5

Solution Enter as shown at the right.

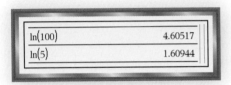

ln(100)	4.60517
ln(5)	1.60944

a. ln 100 ≈ 4.605

b. ln 5 ≈ 1.609

 QY2

▶ **QY2**

Check your answers to Example 1 using powers of e.

Caution! In many computer languages, the natural logarithm function is denoted as $\log(x)$. This can be confusing because in most other places, including on calculators, $\log(x)$ means the common log, with base 10.

The Graph of $y = \ln x$

The function with equation $y = \ln x$ is the inverse of the function with equation $y = e^x$, just as $y = \log x$ is the inverse of $y = 10^x$ and $y = \log_2 x$ is the inverse of $y = 2^x$. The inverse relationship of $y = e^x$ and $y = \ln x$ is displayed in the tables and graphs below. The graph of each function is the reflection image of the other over the line with equation $y = x$.

x	$y = e^x$
−1	0.37
0	1.00
1	$e \approx 2.72$
1.6	4.95

x	$y = \ln x$
0.37	−1
1.00	0
$e \approx 2.72$	1
4.95	1.6

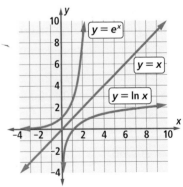

The function f with equation $f(x) = \ln x$ has all the properties held by all other logarithmic functions. In particular, the domain of the natural logarithm function is the set of positive real numbers, and its range is the set of all real numbers.

Applications of Natural Logarithms

Natural logarithms are frequently used in formulas.

Example 2

According to the Beer-Lambert law, if you shine a 10-lumen light into a lake, the light intensity I (in lumens) at a depth of d feet under water is given by $d = -k \cdot \ln\left(\frac{I}{10}\right)$, where k is a measure of the light absorbance of the water. For a lake where $k = 34$, at what depth is the light intensity 6 lumens?

Solution Substitute the known values of k and I and solve for d.

$$k = 34 \text{ and } I = \underline{\ ?\ }$$

So, $d = \underline{\ ?\ } \cdot \ln\left(\frac{?}{10}\right)$

$d \approx \underline{\ ?\ }$

The light intensity is 6 lumens at about $\underline{\ ?\ }$ feet.

Example 3

Let O_1 and O_2 be the temperatures of a cooling object before and after taking t minutes to cool, and let S_1 and S_2 be the temperatures of the surrounding environment before and after those t minutes. Newton's Law of Cooling says that t varies directly as the natural log of the ratio of the differences in temperature between the object and the surrounding area, or $t = k \ln\left(\frac{O_2 - S_2}{O_1 - S_1}\right)$.

The constant of variation k depends on the temperature scale (Celsius or Fahrenheit), the type of container the object is in, the altitude, and other environmental conditions.

a. After bringing it to a boil (212°F), Laura let her soup cool for 15 minutes to 140°F. Use the data to find a value of k if Laura is making her soup in an 80°F kitchen.

b. Suppose Laura heats a pot of vegetable soup to boiling, and lets it cool 15 minutes before serving. What is the serving temperature of the soup if the kitchen is 70°F?

(continued on next page)

Solution

a. Substitute values for O_1, O_2, S_1, S_2, and t into the equation to solve for k. Note that because the kitchen temperatures did not change, $S_1 = S_2 = 80$. Enter the equation into a CAS as shown below. This CAS simplifies the fraction automatically.

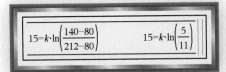

Divide both sides by $\ln\left(\frac{5}{11}\right)$, as shown at the right.

So, $k \approx -19$.

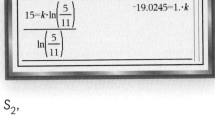

b. From Part a, $T = -19 \ln\left(\frac{O_2 - S_2}{O_1 - S_1}\right)$. Substitute values for O_1, S_1, S_2, and t into the equation from Part a and divide both sides by -19.

$$15 = -19 \ln\left(\frac{O_2 - 70}{212 - 70}\right)$$

$$-\frac{15}{19} = \ln\left(\frac{O_2 - 70}{142}\right)$$

Use the definition of natural logarithm to rewrite the equation.

$$e^{-\frac{15}{19}} = \frac{O_2 - 70}{142}$$

Calculate $e^{-\frac{15}{19}}$ and multiply both sides by 142.

$$0.454(142) \approx O_2 - 70$$

$$134 \approx O_2$$

The serving temperature of the soup is about 134°F.

Check Solve on a CAS as shown at the right.

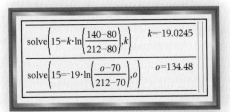

Questions

COVERING THE IDEAS

1. What are logarithms to the base e called?

2. **Multiple Choice** Which of the following is not equivalent to $y = \ln x$? There may be more than one correct answer.

   A $x = \log_e y$ B $e^y = x$ C $y = \log_e x$ D $e^x = y$

In 3 and 4, write an equivalent exponential equation.

3. $\ln 1 = 0$

4. $\ln 1000 \approx 6.908$

In 5 and 6, write in logarithmic form.

5. $e^5 \approx 148.41$

6. $e^{0.5} \approx 1.65$

7. Without using a calculator, tell which is greater, ln 1000 or log 1000. How do you know?

8. Approximate ln 210 to the nearest thousandth.

9. Refer to Example 2. If the light intensity is 4.5 lumens, at what depth is the light in the lake?

10. Does the function with equation $f(x) = \ln x$ have an inverse? If so, what is it?

In 11–13, evaluate without using a calculator.

11. $\ln e$

12. $\ln e^4$

13. $\ln e^{-2}$

14. Refer to Part b of Example 3. What will the temperature of Laura's soup be if it cools for 10 minutes instead of 15?

APPLYING THE MATHEMATICS

15. Refer to Example 3. When cooling hot food in a refrigerator or freezer for service at a later time, caterers want to move through the temperature range from 145°F to 45°F as quickly as possible to avoid pathogen growth. How much longer will it take soup to go from 145° to 45° in a 34°F refrigerator than a 5°F freezer? (Assume $k = -19$.)

16. a. What happens when you try to find $\ln(-5)$ on a calculator?

    b. Use the graph of $y = \ln(x)$ to explain why the calculator displayed what you saw in Part a.

17. In Lesson 9-6 you saw that the decibel scale is a logarithmic scale used to measure relative power intensity, often for sound. Named after John Napier, the *neper* is another measure of relative power intensity based on the natural log scale. A conversion formula between nepers and decibels is 1 neper $= \frac{20}{\ln 10}$ decibels. Find the number of decibels that are equivalent to 3 nepers.

18. How can you use a graph of $y = e^x$ to find the value of $\ln\left(\frac{2}{3}\right)$?

19. a. What is the y-intercept of the graph of $y = \ln x$?

    b. What is the y-intercept of the graph of $y = \log x$?

    c. Prove a generalization about the y-intercepts of the graphs of all equations of the form $y = \log_b x$.

In 20 and 21, suppose ln $x = 9$, ln $y = 3$, and ln $z = 27$. Evaluate.

20. $\ln(xyz)$

21. $\ln \sqrt[y]{\frac{z}{x}}$

REVIEW

22. Let $y = 7^x$. Write an equivalent logarithmic equation. **(Lesson 9-7)**

23. Solve for x: $\log_x 37 = 2$. **(Lesson 9-7)**

24. Use the earthquake moment magnitude scale formula
$M_w = \log \dfrac{M_0}{1.5} - 10.7$. Determine the seismic moment M_0 for the earthquake in Chile, May 22, 1960, where $M_w = 9.5$, the largest (as of January 2007) ever recorded with a seismometer.
(Lesson 9-6)

25. **Multiple Choice** Pick the equation for an exponential decay function. Explain your choice. **(Lesson 9-2)**

    A $f(x) = \dfrac{1^x}{7}$ **B** $f(x) = 7^{-x}$ **C** $f(x) = 7^x$ **D** $f(x) = \left(\dfrac{1}{7}\right)^{-x}$

26. A 1:1000 scale model of the Empire State Building is a little less than $1\frac{1}{2}$ feet tall with a volume of 0.037 cubic foot. From this information, about how many cubic feet does the Empire State Building contain? **(Lesson 2-1)**

EXPLORATION

27. John Napier, the Scottish mathematician who invented logarithms, also invented a calculating device known as *Napier's bones*. Investigate how this device works. Use your findings to simulate and describe Napier's process for any 3-digit-by-2-digit multiplication.

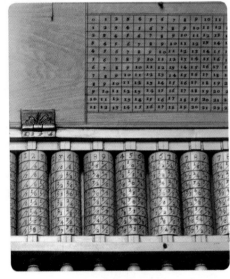

Napier's bones was an early counting device.

Lesson

9-9 Properties of Logarithms

▶ **BIG IDEA** Each basic property of powers corresponds to a basic property of logarithms.

The following properties of powers should be familiar both from Chapter 7 and from your work in previous courses. For all positive numbers x and for all real numbers m and n:

Zero Exponent Theorem	$x^0 = 1$
Product of Powers Postulate	$x^m \cdot x^n = x^{m+n}$
Quotient of Powers Theorem	$\dfrac{x^m}{x^n} = x^{m-n}$
Power of a Power Postulate	$(x^m)^n = x^{mn}$

Mental Math

Put in order from least to greatest.

a. $2 \cdot 10, 2^{10}, 10^2$

b. $3^{16}, \pi^{16}, e^{16}$

c. $0.5, \sqrt{0.5}, \sqrt[3]{0.5}$

Because every logarithm in base b is the exponent n of b^n, properties of logarithms can be derived from the properties of powers. In this lesson you will see five theorems, each related to one of the four properties of powers mentioned above.

The Logarithm of a Power of the Base

Recall that the base b of a logarithm can be any positive number other than 1.

Activity 1

Step 1 Evaluate each of the following:

a. $\log_2 1$	**b.** $\log_2 2$	**c.** $\log_2 2^{-2}$
$\log_3 1$	$\log_3 3$	$\log_2 2^{-1}$
$\log_4 1$	$\log_4 4$	$\log_2 2^0$
$\ln 1$	$\ln e$	$\log_2 2^1$

Step 2 Describe the pattern or write a generalization for each of the three sets of expressions in Step 1.

Steps 1b and 1c of Activity 1 illustrate the following theorem. You already used this theorem in Lessons 9-7 and 9-8 to evaluate logarithms of numbers with various bases.

$\log_b$ of b^n Theorem

For every positive base $b \neq 1$, and any real number n, $\log_b b^n = n$.

Proof

1. Suppose $\log_b b^n = x$.
2. $b^x = b^n$ Definition of logarithm
3. $x = n$ Equate the exponents.
4. $\log_b b^n = n$ Substitution (Steps 1 and 3)

So, if a number can be written as the power of the base, the exponent of the power is its logarithm.

Let $n = 0$ in the $\log_b$ of b^n Theorem. Then $\log_b b^0 = 0$, and because $b^0 = 1$ for $b \neq 0$, $\log_b 1 = 0$. This should agree with what you found in Step 1a of Activity 1.

Logarithm of 1 Theorem

For every positive base $b \neq 1$, $\log_b 1 = 0$.

The Logarithm of 1 Theorem is a special case of the $\log_b$ of b^n Theorem.

The Logarithm of a Product

The Product of Powers Postulate says that in order to multiply two powers with the same base, add their exponents. In particular, for any base b (with $b > 0$, $b \neq 1$) and any real numbers m and n, $b^m \cdot b^n = b^{m+n}$.

The corresponding property of logarithms is about the logarithm of a product of two numbers.

Activity 2

MATERIALS CAS

Step 1 Make a table like the one on the next page.
Use a CAS to evaluate each expression, and then record the result.

A	$\log(3 \cdot 2)$	?
	$\log 3 + \log 2$	?
B	$\log\left(\frac{1}{6} \cdot 216\right)$	?
	$\log \frac{1}{6} + \log 216$	?
C	$\log_5\left(7 \cdot \frac{25}{2}\right)$	?
	$\log_5 7 + \log_5 \frac{25}{2}$	?
D	$\ln(4 \cdot 32)$	?
	$\ln 4 + \ln 32$	?

Step 2 Make a conjecture based on the results of Step 1:
$\log_b (x \cdot y) = \underline{\ ?\ }$.

Activity 2 shows that the logarithm of a product equals the sum of the logarithms of the factors. We state this result as a theorem and prove it using the Product of Powers Postulate.

Logarithm of a Product Theorem

For any positive base $b \neq 1$ and positive real numbers x and y,
$\log_b(xy) = \log_b x + \log_b y$.

Proof Let $x = b^m$ and $y = b^n$, for any $b > 0$, $b \neq 1$. Let $z = xy$.

Then $\log_b x = m$ and $\log_b y = n$.	Definition of logarithm
So $\log_b(xy) = \log_b(b^m \cdot b^n)$	Substitution (Given)
$= \log_b(b^{m+n})$	Product of Powers Postulate
$= m + n$	Log_b of b^n Theorem
$= \log_b x + \log_b y$	Substitution (Step 1)

Example 1

Find $\log_8 128 + \log_8 4$ and check your result.

Solution By the Logarithm of a Product Theorem,

$\log_8 128 + \log_8 4 = \log_8(128 \cdot 4) = \log_8 512$

Because $512 = 8^3$, $\log_8 512 = 3$.

So, $\log_8 128 + \log_8 4 = 3$.

Check Notice that 8, 128, and 4 are all integer powers of 2.

Let $x = \log_8 128$. Then $8^x = 128$, or $2^{3x} = 2^7$. So, $x = \frac{7}{3}$.

Let $y = \log_8 4$. Then $8^y = 4$, or $2^{3y} = 2^2$. So, $y = \frac{2}{3}$.

Consequently, $x + y = \frac{7}{3} + \frac{2}{3} = \frac{9}{3} = 3$. It checks.

Example 2

Write $\log_7 96$ as the sum of two logarithms.

Solution Find two positive integers whose product is 96.

$$\log_7 96 = \log_7 (3 \cdot 32)$$
$$= \log_7 3 + \log_7 32 \quad \text{Logarithm of a Product Theorem}$$

 QY

▶ QY

Write $\log_7 96$ as the sum of a different pair of logarithms than those used in Example 2.

The Logarithm of a Quotient

A Logarithm of a Quotient Theorem follows from the related Quotient of Powers Theorem, $b^m \div b^n = b^{m-n}$.

Logarithm of a Quotient Theorem

For any positive base $b \neq 1$ and for any positive real numbers x and y,

$$\log_b\left(\frac{x}{y}\right) = \log_b x - \log_b y.$$

The proof of the Logarithm of a Quotient Theorem is similar to that of the Logarithm of a Product Theorem. You are asked to complete the proof in Question 22.

GUIDED

Example 3

Recall from Lesson 9-6 that the formula $D = 10 \log\left(\frac{N}{10^{-12}}\right)$ is used to compute the number D of decibels from the sound intensity N measured in watts per square meter. Use the Logarithm of a Quotient Theorem to rewrite this formula without a quotient.

Solution By the Logarithm of a Quotient Theorem,

$$\log\left(\frac{N}{10^{-12}}\right) = \log \underline{\ ?\ } - \log \underline{\ ?\ }$$
$$= \log \underline{\ ?\ } + \underline{\ ?\ }, \text{ because } \log_{10}(10^x) = x.$$

Thus, $D = 10 \log\left(\frac{N}{10^{-12}}\right)$ is equivalent to
$$D = 10(\log \underline{\ ?\ } + \underline{\ ?\ }).$$

Members of marching bands are encouraged to wear earplugs to prevent hearing loss.

The Logarithm of a Power

Recall that $\log_b b^n = n$. In Activity 3 you will explore $\log_b x^n$, where $x \neq b$.

Activity 3

Step 1 Estimate each logarithm to the nearest thousandth.

    a. log 3   **b.** $\log 3^2$

    c. $\log 3^3$ **d.** $\log 3^4$

    e. ln 10   **f.** ln 100

    g. ln 1000   **h.** ln 10,000

Step 2 Make a conjecture based on the results of Step 1: $\log_b x^n = $ ___?___.

Step 3 In Step 1 you should have found that $\log 3^4 = \log 3 + \log 3 + \log 3 + \log 3 = 4 \log 3$. What property of logarithms supports this conclusion?

Activity 3 illustrates the following theorem.

Logarithm of a Power Theorem

For any positive base $b \neq 1$ and for any positive real number x and any real number n, $\log_b(x^n) = n \log_b x$.

You are asked to complete the proof of this theorem in Question 23.

You can use the properties of logarithms to solve equations.

Example 4

Solve for k: $\log k = 2 \log 5 + \log 6 - \log 3$.

Solution Use the properties of logarithms.

$\log k = \log 5^2 + \log 6 - \log 3$ Logarithm of a Power Theorem

$\log k = \log(5^2 \cdot 6) - \log 3$ Logarithm of a Product Theorem

$\log k = \log\left(\frac{5^2 \cdot 6}{3}\right) = \log 50$ Logarithm of a Quotient Theorem, arithmetic

    $k = 50$

Check Evaluate $2 \log 5 + \log 6 - \log 3$ on a CAS. The result, as shown at the right, is log 50. It checks.

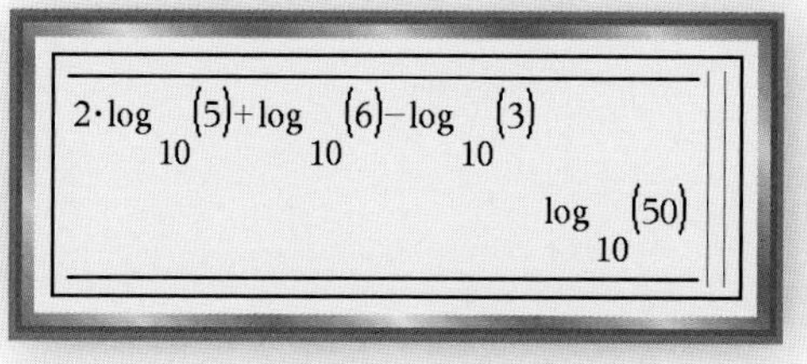

$$2 \cdot \log_{10}(5) + \log_{10}(6) - \log_{10}(3)$$
$$\log_{10}(50)$$

Questions

COVERING THE IDEAS

In 1–3, simplify.

1. $\log_{5.1} 5.1^{15.4}$ 2. $\log_r r^n$ 3. $\log_p 1$

4. Write $\log_7 96$ as the sum of a pair of logarithms different from the pair used in Example 2 and the QY.

In 5–7, an expression is given.
   a. Write it as the logarithm of a single number or as a number without a logarithm.
   b. Name the property or properties of logarithms that you used.

5. $\log_2 49 - \log_2 7$

6. $\log_{64} 3 + \log_{64} 8 - \log_{64} 4$

7. $3 \log_3 \sqrt{9}$

8. Here is the start of a proof to show that $\log x^4 = 4 \log(x)$ without using the Logarithm of a Power Theorem. Begin by realizing $\log x^4 = \log(x \cdot x \cdot x \cdot x)$. Next, use the Logarithm of a Product Theorem. With this hint, complete the rest of the proof.

9. A student enters $\log(64)$ on a CAS and gets $6 \log 2$ as an output. Show that the two expressions are equivalent using the properties of logarithms.

In 10 and 11, rewrite the expression as a single logarithm.

10. $\log 14 - 3 \log 4$ 11. $\log_b x - \log_b y + \frac{1}{2} \log_b z$

True or False In 12–17, if the statement is false, give a counterexample, and then correct the statement to make it true.

12. $\log_b (3x) = 3 \log_b x$ 13. $\ln M - \ln N = \ln\left(\frac{M}{N}\right)$

14. $\frac{\log p}{\log q} = \log\left(\frac{p}{q}\right)$ 15. $\log(M + N) = \log M + \log N$

16. $\log_7\left(\frac{M}{N}\right) = \frac{\log_7 M}{\log_7 N}$ 17. $6 \log_2 T = \log_2(T^6)$

APPLYING THE MATHEMATICS

In 18 and 19, solve for y.

18. $\ln h = \frac{1}{4} \ln x + \ln y$ 19. $\log w = \log\left(\frac{x}{y^4}\right)$

20. In the formula for the decibel scale $D = 10 \log\left(\frac{N}{10^{-12}}\right)$, where N is the sound intensity and D is relative intensity, show that $D = 120 + \log N^{10}$ by using properties of logarithms.

21. Janella used a CAS to expand $\log(a*b)$, where a and b are both positive. The display at the right shows her CAS output. Show that this CAS output is equivalent to the original expression.

$$\text{expand}\left(\log_{10}(a \cdot b) | a > 0 \text{ and } b > 0\right)$$
$$\frac{\log_{10}(a)}{\log_{10}(5) + \log_{10}(2)} + \frac{\log_{10}(b)}{\log_{10}(5) + \log_{10}(2)}$$

22. **Fill in the Blanks** Justify each step in the following proof of the Logarithm of a Quotient Theorem.

Let $x = b^m$, $y = b^n$, and $z = \frac{x}{y}$.

a. $\log_b x = m$ and $\log_b y = n$ a. ___?___

b. $\log_b\left(\frac{x}{y}\right) = \log_b\left(\frac{b^m}{b^n}\right)$ b. ___?___

c. $= \log_b(\underline{\ ?\ })$ c. ___?___

d. $= m - n$ d. ___?___

e. $\log_b\left(\frac{x}{y}\right) = \log_b x - \log_b y$ e. ___?___

23. **Fill in the Blanks** Complete the following proof of the Logarithm of a Power Theorem.

Let $\log_b x = m$.

a. Then, $x = \underline{\ ?\ }$.   definition of logarithm

b. $x^n = \underline{\ ?\ }$ Raise both sides to the nth power.

c. $x^n = \underline{\ ?\ }$ Power of a Power Postulate

d. $x^n = \underline{\ ?\ }$ Commutative Property of Multiplication

e. $\log_b x^n = \underline{\ ?\ }$ definition of logarithm

     $\log_b x^n = n \log_b x$ substitution

REVIEW

24. Simplify without a calculator. **(Lesson 9-7)**

a. $\log_{81} 81$ b. $\log_{81} 9$ c. $\log_{81} 3$

d. $\log_{81} 1$ e. $\log_{81}\left(\frac{1}{81}\right)$ f. $\log_{81}\left(\frac{1}{9}\right)$

In 25 and 26, solve. (Lessons 9-7, 9-5)

25. $\log_x 144 = 2$

26. a. $\log y = -4$ b. $\log(-4) = z$

27. After World War I, harsh reparation payments imposed on Germany caused the value of the deutschmark to decline against foreign currencies and German wholesale prices to increase rapidly. The table at the right shows the wholesale price index in Germany from 1914 to 1922. (**Lesson 9-4**)

    a. Sketch a scatterplot of these data.

    b. Why do these data suggest an exponential model?

    c. Find an exponential equation that models these data for semiannual periods after July 1914.

    d. Use Part c to predict the wholesale price index for January 1923.

    e. The actual wholesale price index for January 1923 was 2785. Does this model over or underpredict the actual wholesale price index?

Date	Wholesale Price Index
July 1914	1.0
Jan. 1919	2.6
July 1919	3.4
Jan. 1920	12.6
Jan. 1921	14.4
July 1921	14.3
Jan. 1922	36.7
July 1922	100.6

28. A student borrows $2000 to go to college. The student loan accrues interest continously at a rate of 3.75% for four years. What is the total loan balance after four years? (**Lesson 9-3**)

29. a. Find the average rate of change between $x = -3$ and $x = -2$ for the function with equation $y = \frac{1}{3}x^3$. (**Lesson 7-1**)

    b. Sketch a graph and use it to explain your answer to Part a.

30. **Multiple Choice** A distributor sells old-time movies from the 1940s and 1950s for $5.99 each plus a shipping and handling charge of $3.89 per order. As a bonus, every fifth movie ordered is free. Which equation gives the correct charge $f(m)$ for m movies? (**Lesson 3-9**)

    A $f(m) = 3.89 + 5.99m - 5.99\left\lfloor \frac{m}{5} \right\rfloor$

    B $f(m) = 3.89 + 5.99m - 5.99\left\lfloor 1 - \frac{m}{5} \right\rfloor$

    C $f(m) = \frac{5.99m}{5} + 3.89$

    D $f(m) = 3.89 + 5.99m - \frac{5.99m}{5}$

EXPLORATION

31. What common logarithms of whole numbers from 1 through 20 can you estimate without a calculator using only log 1, $\log 2 \approx 0.301$, $\log 3 \approx 0.477$, and log 10?

Lesson 9-10

Using Logarithms to Solve Exponential Equations

▶ **BIG IDEA** The solution to the exponential equation $b^x = a$ can be written as $\log_b a$, or it can be written as $\dfrac{\log a}{\log b}$.

Methods for Solving $b^x = a$

You already know a few ways to solve equations of the form $b^x = a$. When a is an integer power of b, you may solve the equation in your head. For instance, to solve $4^x = 64$, you may know that 3 is a solution, because $4^3 = 64$.

Sometimes you may notice that a and b are powers of the same number. For instance,

$$100^x = \sqrt{10}$$

can be solved by noting that both sides of the equation can be written as powers of 10.

$$100^x = (10^2)^x = 10^{2x} \quad \text{and} \quad \sqrt{10} = 10^{0.5}$$

Substitute into the given equation.

$$10^{2x} = 10^{0.5}$$

Equate the exponents and solve the resulting equation for x.

$$2x = 0.5$$

$$x = 0.25$$

Still another way to solve an equation of the form $b^x = a$ is to graph $y = b^x$ and see where the graph intersects the horizontal line $y = a$.

And, of course, you could use a CAS to solve $b^x = a$ directly.

But what happens if a and b are not easily found powers of the same number and a graphing utility or a CAS is not available? Then you can use logarithms to solve these equations.

Mental Math

Solve.

a. $3^x = 81$

b. $5^y + 75 = 200$

c. $4 \cdot 7^z = 196$

d. $2^{w+2} - 16 = 0$

Using Logarithms to Solve $b^x = a$

The process is similar to the process you have used to solve many equations where the unknown is on one side of the equation. Instead of adding the same number to both sides or multiplying both sides by the same number, you take the logarithm of both sides. Then apply the Logarithm of a Power Theorem to solve the resulting equation. When the base of the exponential equation is e, use natural logarithms.

Example 1

At what rate of interest, compounded continuously, do you have to invest your money so that it will double in 12 years?

Solution Use the Continuously Compounded Interest Formula $A = Pe^{rt}$. You want to know the rate at which after 12 years A will equal twice P, so substitute $2P$ for A and 12 for t.

$$A = 2P = Pe^{12r}$$

Since P is not zero, you can divide both sides by P.

$$2 = e^{12r}$$

Because the base in the equation is e, it is easiest to take the natural logarithm of each side.

$$\ln 2 = \ln(e^{12r})$$

$$\ln 2 = 12r \qquad \text{Log}_b \text{ of } b^n \text{ Theorem}$$

$$r = \frac{\ln 2}{12} \approx 0.0578 = 5.78\%$$

Check Choose a value for P and substitute the values back into the formula. We use $P = \$500$.

$A = 500 \cdot e^{(0.0578)(12)} \approx 1000$, or twice the investment. It checks.

 QY

 QY

Why would it be less efficient to use common logarithms in Example 1?

Decay or depreciation problems modeled by exponential equations with negative exponents can be solved in a similar way.

GUIDED

Example 2

Archie Oligist finds the remains of an ancient wood cooking fire and determines that it has lost about 23.2% of the carbon-14 expected for that kind of wood. Recall that the half-life of carbon-14 is about 5730 years.

A radioactive decay model for this situation is $A = Ce^{-\frac{\ln 2}{5730}t}$, where C is the original amount of carbon-14 and A is the amount present after t years. Use the model to estimate the age of the wood.

Solution Substitute values for the variables into the model. Because 23.2% of the carbon-14 has decayed, 76.8% remains and $A = 0.768C$.

$$A = Ce^{-\frac{\ln 2}{5730}t}$$

$0.768C =$ __?__		Substitute for A.
$0.768 =$ __?__		Divide both sides by C.
$\ln 0.768 =$ __?__		Take the natural log of both sides.
$\ln 0.768 =$ __?__		Log_b of b^n Theorem
__?__ $\cdot \ln 0.768 = t$		Solve for t.
$t \approx$ __?__		

The wood is about __?__ years old.

Check Solve on a CAS as shown below. It checks.

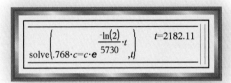

$$\text{solve}\left(.768 \cdot c = c \cdot e^{\frac{-\ln(2)}{5730} \cdot t}, t\right) \qquad t = 2182.11$$

In the previous Examples, exponential equations were solved by taking natural logarithms of both sides because the base in the exponential equation was e. You can use any base of a logarithm to solve an exponential equation.

Example 3

Solve $7^x = 28$ by taking

a. the common logarithm of each side.
b. the natural logarithm of each side.

Solution Solutions for Parts a and b are written side-by-side. First read the solution to Part a (the columns at the left and center). Then read the solution to Part b (the center and right columns). Then reread both solutions by reading across each line.

a. $7^x = 2$		b. $7^x = 28$
$\log 7^x = \log 28$	Take the log of each side.	$\ln 7^x = \ln 28$
$x \log 7 = \log 28$	Logarithm of a Power Theorem	$x \ln 7 = \ln 28$
$x = \dfrac{\log 28}{\log 7}$	Divide both sides by the coefficient of x.	$x = \dfrac{\ln 28}{\ln 7}$
$x \approx 1.7124$	Evaluate with a calculator.	$x \approx 1.7124$

Logarithms to bases 10 and *e* are used in the solutions to Example 3. Any other base for the logarithms could have been used to solve the equation. Because the same results are *always* obtained regardless of the base, you may choose common logarithms, natural logarithms, or logarithms to any other base for a given situation.

Changing the Base of a Logarithm

While most graphing utilities have keys for the natural logarithm LN and common logarithm LOG, some of them do not have a single operation that allows you to find logarithms with bases other than 10 or *e*. However, with the following theorem you can convert logarithms with any base *b*, such as $\log_5 18$, to a ratio of either common logarithms or natural logarithms. The proof is a generalization of the process used in Examples 1–3.

Change of Base Theorem

For all positive real numbers *a*, *b*, and *t*, $b \neq 1$ and $t \neq 1$,
$$\log_b a = \frac{\log_t a}{\log_t b}.$$

Proof Suppose $\log_b a = x$.

Definition of logarithm	$b^x = a$
Take the log base *t* of each side.	$\log_t b^x = \log_t a$
Apply the Logarithm of a Power Theorem.	$x \log_t b = \log_t a$
Divide.	$x = \dfrac{\log_t a}{\log_t b}$
Transitive Property of Equality	$\log_b a = \dfrac{\log_t a}{\log_t b}$

The Change of Base Theorem says that the logarithm of a number to any base is the log of the number divided by the log of the base. Notice that the logarithms on the right side of the formula must have the same base.

Example 4

Approximate $\log_5 18$ to the nearest thousandth.

Solution 1 Use the Change of Base Theorem with common logarithms.
$$\log_5 18 = \frac{\log 18}{\log 5} \approx 1.796$$

Solution 2 Use the Change of Base Theorem with natural logarithms.

$$\log_5 18 = \frac{\ln 18}{\ln 5} \approx 1.796$$

Check By definition, $\log_5 18 \approx 1.796$ is equivalent to $5^{1.796} \approx 18$. The calculator display at the right shows that $5^{1.796} \approx 18$. It checks.

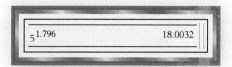

| $5^{1.796}$ | 18.0032 |

Questions

COVERING THE IDEAS

1. Solve $3^x = 243$
   a. in your head.
   b. by taking common logarithms.
   c. by taking natural logarithms.

2. Solve $20^r = 30$ to the nearest thousandth by using logarithms.

In 3 and 4, solve and check.

3. $\log_4 140 = y$

4. $53.75^z = 44$

5. Solve Example 1 using common logarithms.

6. Refer to Example 1. What interest rate, compounded continuously, would it take to triple your money in 12 years?

7. Refer to Example 2. Another artifact found several miles away contains only 48% of its expected carbon-14. About how old is this artifact?

In 8 and 9, approximate the logarithm to the nearest thousandth and check your answer.

8. $\log_2 30$

9. $\log_{0.5} 80$

10. Express as a single logarithm: $\dfrac{\log_8 P}{\log_8 Q}$.

APPLYING THE MATHEMATICS

11. Suppose you invest $250 in a savings account paying 3.25% interest compounded continuously. How long would it take for your account to grow to $300, assuming that no other deposits or withdrawals are made?

12. Suppose a colony of bacteria grows according to $N = Be^{2t}$, where N is the number of bacteria after t hours and B is the initial number in the colony. How long does it take the colony to grow to 10 times its original size?

In 13 and 14, solve.

13. $8^{4y} = 1492$

14. $\log_{12} 313 = -3x$

15. In 1999 the world population was estimated to be 5,995,000,000, and in 2000 the estimate was 6,071,000,000.

  a. Calculate the percentage increase between 1999 and 2000.

  b. Assume the percent increase remains the same. In what year will the world population reach 7 billion? Explain your answer.

  c. In what year will the population be double the 2000 population?

  d. Use the Internet to find the current and previous year's world population, and then calculate the percentage increase. With this growth rate, estimate when the world population will be twice the 2000 population.

16. In the formula $A = C\left(\frac{1}{2}\right)^{\frac{t}{H}}$, A is the amount of carbon left after time t, where C is the initial amount and H is the half-life. In this lesson the formula $A = Ce^{-\frac{\ln 2}{H}t}$ is used. Use the properties of exponents and logarithms to prove that $C\left(\frac{1}{2}\right)^{\frac{t}{H}} = Ce^{-\frac{\ln 2}{H}t}$ for all t.

In 17–19, between which two consecutive whole numbers does each logarithm fall? Answer without using a calculator.

17. $\log_3 21$ 18. $\log_{12} 200$ 19. $\log_7\left(\frac{1}{50}\right)$

REVIEW

In 20 and 21, solve and check. (Lesson 9-9)

20. $\log w = \frac{3}{2}\log 16 - \log 8$ 21. $\log 125 = x \log 5$

22. Find the error in the following proof that $3 < 2$. (*Hint:* What is the sign of $\log\left(\frac{1}{3}\right)$?) (Lessons 9-9, 9-5, 5-1)

$$\frac{1}{27} < \frac{1}{9}$$
$$\log\left(\frac{1}{27}\right) < \log\left(\frac{1}{9}\right)$$
$$\log\left[\left(\frac{1}{3}\right)^3\right] < \log\left[\left(\frac{1}{3}\right)^2\right]$$
$$3 \log\left(\frac{1}{3}\right) < 2 \log\left(\frac{1}{3}\right)$$
$$3 < 2$$

23. For a 3-stage model rocket, the formula $V = c_1 \cdot \ln R_1 + c_2 \cdot \ln R_2 + c_3 \cdot \ln R_3$ is used to find the velocity of the rocket at final burnout. If $R_1 = 1.37$, $R_2 = 1.59$, $R_3 = 1.81$, $c_1 = 2185 \frac{m}{sec}$, $c_2 = 2530 \frac{m}{sec}$, and $c_3 = 2610 \frac{m}{sec}$, find V. (Lesson 9-8)

24. Recall that the relative intensity of sound is given by $D = 10 \log\left(\frac{N}{10^{-12}}\right)$. Find the decibel level of a sound that has an absolute intensity of $3.7 \cdot 10^{-5}$. (**Lesson 9-6**)

In 25 and 26, evaluate, given $g(x) = \log x$. (**Lessons 9-5, 8-3**)

25. $g(0.01)$

26. $g^{-1}(4)$

27. **a.** Graph the triangle represented by $\begin{bmatrix} -5 & 5 & 0 \\ 0 & 0 & 5 \end{bmatrix}$. (**Lessons 4-5, 4-1**)

    b. What kind of triangle is this?

    c. What matrix describes the image of the triangle in Part a under the transformation given by $\begin{bmatrix} 4 & 0 \\ 0 & 1 \end{bmatrix}$?

    d. Graph the image on the same set of axes as in Part a.

    e. Are the triangles in Parts a and c similar?

EXPLORATION

28. The table shows part of the Krumbein *phi* (φ) scale. Geologists use this scale to measure the grain size of the individual particles that make up rocks, soils, and other solids. Notice that the sizes in the first column are all powers of 2. Determine a logarithmic relationship between the size range and the φ-scale numbers in the second column.

Size Range (mm)	φ Scale	Aggregate Name	Other Names
> 256	< -8	Boulder	
64–256	-6 to -8	Cobble	
32–64	-5 to -6	Very coarse gravel	Pebble
16–32	-4 to -5	Coarse gravel	Pebble
8–16	-3 to -4	Medium gravel	Pebble
4–8	-2 to -3	Fine gravel	Pebble
2–4	-1 to -2	Very fine gravel	Granule
1–2	0 to -1	Very coarse sand	
$\frac{1}{2}$–1	1 to 0	Coarse sand	
$\frac{1}{4}$–$\frac{1}{2}$	2 to 1	Medium sand	

Chapter 9 Projects

1 Slide Rules

Battery operated electronic calculators first became available in the early 1970s. Before that time many engineers and scientists used a slide rule to do calculations involving multiplication, division, exponentiation, finding roots, and so on (but not adding and subtracting). Slide rules are based on the principles of common logarithms. Find out how they work. Summarize your findings in a report and demonstrate a few calculations for the class.

2 Non-Native Species

The Australian rabbit plague shows how introducing a species to a new environment can be catastrophic. Other non-native species whose introductions have had or are having disastrous consequences for environments in the United States include the Asian long-horned Beetle, the emerald ash borer, the zebra mussel, and the fungi causing Dutch elm disease (*Ophiostoma ulmi* and related fungi) and sudden oak death syndrome (*Phytophthora ramorum*). Research one of these foreign species invasions to find out when and how the species was introduced, how quickly it has spread, what the consequences are, and how the species' spread is being controlled (if at all).

Damage from an emerald ash borer

3 Car Values

a. Consult a car dealer, insurance agent, books, magazines, or the Internet to learn how automobile depreciation is typically calculated. Is depreciation typically described by a linear, exponential, or some other type of model?

b. Gather data on the book value of two or three cars that interest you. Include only cars whose values you can determine for at least the past five years. Which book values over time, if any, seem to illustrate exponential decay? Find equations that model the value of these cars over time. Use the equations to predict the value of each car five years from now. In your opinion, how reasonable are these estimates? What are some limitations of your mathematical models?

4 How Many Digits Are in That Number?

Logarithms can be used to determine how many digits a number has. While initially this may not seem useful, consider finding the number of digits in the number 2^{5000} when written in base 10. Most calculators will not calculate the value, let alone help you figure out how many digits it has. However, by using logarithms, you can find that this number has 1506 digits. Research how to perform this process and find out why it works.

5 Predicting Cooling Times

You can predict how long it will take a 12-fluid-ounce soft drink to cool to a desired temperature on a particular shelf in your refrigerator using either exponential or logarithmic functions. You will need a thermometer that measures a wide range of temperatures.

a. Measure the temperature T_R in your refrigerator. (Let's say it is $42°F$.) Assume T_R is constant. Fill a ceramic cup with very hot tap water. Measure the temperature T_M of the water. (Say it is $135°F$.) Place the cup in your refrigerator. Make a table like the one below and record your first measurement.

t (min)	T_M (°F)	$T_M - T_R$ (°F)
Time since cooling began	Measured temperature of water	Difference between measured temperature and refrigerator temperature
0	135	$135 - 42 = 93$

b. Measure the temperature of the water periodically. At first you will want to take a measurement every 5 minutes. Later you might wait longer between measurements. Record the temperatures in the table. Continue taking measurements for at least 4 hours. Be sure to take the measurements the same way each time.

c. Plot the ordered pairs $(t, T_M - T_R)$. Describe the shape of the graph. Use the modeling features on a graphing utility or statistics application to fit a reasonable curve to your data. Explain why you chose the model you did.

d. Use technology to fit a logarithmic model to your data. This time, let $T_M - T_R$ be your independent variable and t be your dependent variable. The statistics application might give you a model such as $t = a + b \cdot \ln(T_M - T_R)$, where a and b are parameters for the model. Is this model equivalent to the one you found in Part c? Why or why not?

e. Use your model(s) from Parts c and d to predict how long it takes for a 12-fluid-ounce can of soft drink in your refrigerator to cool from $70°F$ (about room temperature) to $48°F$ (an acceptable drinking temperature). Comment on the answer(s) you get.

6 Finding when $a^x = \log_a x$

a. Use a graphing utility to graph $f(x) = a^x$ and $g(x) = \log_a x$ for several values of a such that $0 < a < 1$.

b. Use a graphing utility to graph $f(x) = a^x$ and $g(x) = \log_a x$ for several values of $a > 1$.

c. Based on the results of Parts a and b, what approximate values of a cause these functions to intersect once? What approximate values of a cause them to intersect twice?

d. Approximate the a-value at which these two functions intersect for the last time.

Chapter 9 Summary and Vocabulary

○ A function with an equation of the form $y = ab^x$, where $b > 0$ and $b \neq 1$, is an **exponential function**. All geometric sequences are also exponential functions. In the formula $A = P\left(1 + \frac{r}{n}\right)^{nt}$, when P, r, and n are given, A is an exponential function of t.

○ The exponential function with base b has an equation of the form $f(x) = ab^x$. Some exponential functions represent **exponential growth** or **decay** situations. In an exponential growth situation, the growth factor b is greater than one. In an exponential decay situation, b is between 0 and 1. Over short periods of time, many populations grow exponentially. The value of many items depreciates exponentially. Quantities that grow or decay exponentially have a constant doubling time or **half-life**, respectively. Real data from these and other contexts can be modeled using exponential functions.

○ When an initial amount of $1.00 is continuously compounded at 100% interest, the value of the investment after one year is $e \approx 2.71828$. Like π, the number e is an irrational number. In general, the formula $A = Pe^{rt}$ can be used to calculate the value A of an investment of P dollars at r% interest **compounded continuously** for t years.

○ The inverse of the exponential function $f: x \to b^x$ is $f^{-1}: b^x \to x$, the **logarithm function with base b**. Thus, $b^x = a$ if and only if $x = \log_b a$. Because exponential and logarithm functions are inverses, their graphs are reflection images of each other over the line $y = x$. Properties of logarithm functions can be derived from the corresponding properties of exponential functions.

Exponential Growth Function
$y = b^x, b > 1$

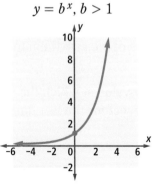

Logarithmic Function
$y = \log_b x, b > 1$

Vocabulary

Lesson 9-1
*exponential function
exponential curve
growth factor

Lesson 9-2
exponential decay
depreciation
half-life

Lesson 9-3
e
compounded continuously

Lesson 9-5
logarithm of x to the
    base 10, log of x to the
    base 10, log base 10 of x
common logarithm,
    common log
logarithmic function to
    the base 10, common
    logarithm function

Lesson 9-6
logarithmic scale
decibel, dB

Lesson 9-7
*logarithm of a to the
    base b
logarithm function with
    base b

Lesson 9-8
*natural logarithm of m,
    ln m

Exponential Growth Function		Logarithmic Function
all real numbers	**Domain**	all positive reals
all positive reals	**Range**	all real numbers
y-intercept is 1, no x-intercept	**Intercepts**	x-intercept is 1, no y-intercept
the x-axis ($y = 0$)	**Asymptotes**	the y-axis ($x = 0$)

◐ Logarithmic functions are used to scale data having a wide range (for example, **decibel** levels or earthquake magnitudes) and to solve equations of the form $b^x = a$, where b and a are positive and $b \neq 1$. One way to solve equations of this type is to take the logarithm of both sides; the solution is $x = \dfrac{\log a}{\log b}$.

◐ The base of a logarithmic function can be any positive real number not equal to 1, but the most commonly used bases are 10 and e. When the base is 10, the values of the log function are called **common logarithms**. When the base is e, the values of the log function are called **natural logarithms**.

◐ The basic properties of logarithms correspond to properties of powers. Let $x = b^m$ and $y = b^n$, and take the logarithms of both sides of each power property. The result is a logarithm property.

Power Property	**Logarithm Property**
$b^0 = 1$	$\log_b 1 = 0$
$b^m \cdot b^n = b^{m+n}$	$\log_b(xy) = \log_b x + \log_b y$
$\dfrac{b^m}{b^n} = b^{m-n}$	$\log_b\left(\dfrac{x}{y}\right) = \log_b x - \log_b y$
$(b^m)^a = b^{am}$	$\log_b(x^a) = a \log_b x$
$b^x = a$	$\log_b a = \dfrac{\log_t a}{\log_t b}$

Theorems

Continuously Compounded Interest
   Formula (p. 597)
Log_b of b^n Theorem (p. 636)
Logarithm of 1 Theorem (p. 636)
Logarithm of a Product Theorem

(p. 637)
Logarithm of a Quotient Theorem
   (p. 638)
Logarithm of a Power Theorem (p. 639)
Change of Base Theorem (p. 646)

Chapter 9 — Self-Test

Take this test as you would take a test in class. You will need a calculator. Then use the Selected Answers section in the back of the book to check your work.

1. What are the domain and range of the function f defined by $f(x) = \log_{13} x$?

2. The half-life of uranium-237 is about 7 days. What percent of the original amount of uranium will remain in an artifact after three weeks?

3. Let $f(x) = 8^x$.

   a. Graph f on the interval $-2 \leq x \leq 2$.

   b. Approximate $f(\pi)$ to the nearest tenth.

   c. Does the graph have any asymptotes? If it does, state the equations for all asymptotes. If it does not, explain why not.

4. a. Give a value of b for which the equation $y = ab^x$ models exponential decay.

   b. What are the domain and range of the function in Part a for your value of b?

In 5–7, explain how you would evaluate each expression exactly without using a calculator.

5. $\log 100{,}000{,}000$

6. $\log_2\left(\frac{1}{16}\right)$

7. $\ln e^{-4}$

8. Rewrite $\log a + 2 \log t - \log s$ as a single logarithm.

9. Without natural predators, the number of a certain species of bird will grow each year by 12%. A colony of 50 birds is started in a predator-free area. What is the expected number of birds of this species after 3 years?

In 10–12, solve. If necessary, round solutions to the nearest hundredth.

10. $\log_x 27 = \frac{3}{4}$

11. $5^y = 40$

12. $\ln(7z) = \ln 3 + \ln 21$

13. **True or False** $\log_5 a + \log_3 b = \log_{15}(ab)$. Justify your answer.

14. Write an equation for the inverse of the function with equation $y = \log_4 x$.

15. The Henderson-Hasselbalch formula $pH = 6.1 + \log\left(\frac{B}{C}\right)$ can be used to find the pH of a patient's blood as a function of the bicarbonate concentration B and the carbonic-acid concentration C. A patient's blood has a bicarbonate concentration of 23 and a pH reading of 7.3. Find the concentration of carbonic acid.

16. To the nearest thousandth, what is $\log_{14} 24.72$?

17. State the general property used in simplifying the expression $\log_{19} 19^{23}$.

18. Consider the function defined by $y = \log_5 x$.

   a. State the coordinates of three points on the graph.

   b. State the domain and range of the function.

   c. Graph the function.

   d. State an equation for its inverse.

   e. Graph the inverse on the same axes you used in Part c.

19. Multiple Choice Assume that the value of an investment grows according to the model $y = I(1.075)^x$, where I is the original investment and y is the amount present after x years. Which graph below could represent this situation?

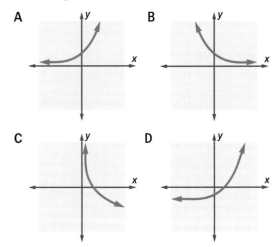

A   B   C   D

20. The Population Reference Bureau reports the following data about the population P of the United States.

Population mid-2007	302,200,000
Births per 1000 population	14
Deaths per 1000 population	8
Rate of natural increase/year	0.6% = 0.006
Projected population, mid-2025	349,400,000

Assume the continuous growth model $P = 302,200,000e^{rt}$, where r is the annual rate of increase and t is the number of years after 2007.

a. Use the model to show that the reported rate of natural increase of 0.6% is *not* the rate leading to the projected population in 2025.

b. Calculate the growth rate to the nearest 0.1% that will give the reported projected population for 2025 of 349,400,000. Show how you arrived at your answer.

21. For the equation $27^x = 14$,

a. find the exact solution.

b. find a decimal solution rounded to the nearest thousandth.

22. The table below shows the total number of German-language articles in *Wikipedia* each month from March 2007 to February 2008.

Month	Articles (thousands)
March 2007	570
April 2007	584
May 2007	599
June 2007	612
July 2007	626
August 2007	640
September 2007	655
October 2007	668
November 2007	681
December 2007	695
January 2008	711
February 2008	726

a. Fit an exponential model to these data.

b. The number of English articles in *Wikipedia* surpassed 1 million in March, 2006. Use your model from Part a to predict when the number of German articles will reach this benchmark.

Chapter 9 Chapter Review

SKILLS
PROPERTIES
USES
REPRESENTATIONS

SKILLS Procedures used to get answers

OBJECTIVE A Determine values of logarithms. (Lessons 9-5, 9-7, 9-8, 9-10)

In 1–8, find the exact value of each logarithm without using a calculator.

1. $\log 10{,}000$
2. $\log 0.00001$
3. $\ln e^8$
4. $\log_4 1024$
5. $\log_{13}(13^{15})$
6. $\ln 1$
7. $\log_{\frac{1}{3}} 27$
8. $\log_9 \sqrt[3]{9}$

In 9–14, approximate each logarithm to the nearest hundredth.

9. $\log 98{,}765$
10. $\ln 10.95$
11. $\ln(-3.7)$
12. $\log 0.003$
13. $\log_7 25$
14. $\log_3 12.3$

OBJECTIVE B Use logarithms to solve exponential equations. (Lesson 9-10)

In 15–22, solve. If necessary, round to the nearest hundredth.

15. $\log_6 5 = t$
16. $\log_{12} 9900 = s$
17. $2000(1.06)^n = 6000$
18. $13 \cdot 2^x = 1$
19. $e^z = 44$
20. $(0.8)^w = e$
21. $11^{a+1} = 1011$
22. $3^{-2a} = 53$

OBJECTIVE C Solve logarithmic equations. (Lessons 9-5, 9-6, 9-7, 9-9)

In 23–30, solve. If necessary, round to the nearest hundredth.

23. $\log_x 33 = \log_{11} 33$
24. $\ln(4y) = \ln 9 + \ln 12$
25. $\log z = 18$
26. $\log x = 3.71$
27. $3 \ln 5 = \ln x$
28. $\log_8 x = \frac{3}{7}$
29. $\log_x 347 = 3$
30. $\log_x 5 = 10$

PROPERTIES Principles behind the mathematics

OBJECTIVE D Recognize properties of exponential functions. (Lessons 9-1, 9-2, 9-3)

31. What are the domain and range of the function f defined by $f(x) = e^x$?

32. What are the domain and range of the function g defined by $g(x) = 2^x$?

33. When does the function $f : x \rightarrow a^x$ describe exponential growth?

34. What must be true about the value of b in the equation $y = ab^x$, if the equation models exponential decay?

35. Write the equation(s) of the asymptote(s) to the graph of $y = 27(1.017)^x$.

36. **Multiple Choice** Which situation does the equation $y = e^{-x}$ describe?

    A constant increase

    B constant decrease

    C exponential growth

    D exponential decay

OBJECTIVE E Recognize properties of logarithmic functions. (Lessons 9-5, 9-7, 9-8)

37. What is the inverse of f, when $f(x) = e^{-x}$?

38. Give an equation of the form $y = \underline{\ ?\ }$ for the inverse of the function with equation $y = \log_3 x$.

39. **True or False** The domain of the log function with base 12 is the range of the exponential function with base 12.

40. **True or False** Negative numbers are not included in the domain of f when $f(x) = \log_b x$.

OBJECTIVE F Apply properties of logarithms. (Lessons 9-9, 9-10)

In 41–44, write in exponential form.

41. $\log_3\left(\frac{1}{243}\right) = -5$ 42. $\ln 23.14 \approx \pi$

43. $\log m = n$ 44. $\log_b p = q$

In 45–48, write in logarithmic form.

45. $10^{-1.8} \approx 0.01585$ 46. $e^5 \approx 148.413$

47. $x^y = z, x > 0, x \neq 1$ 48. $4^a = 18$

In 49–56, rewrite the expression as a whole number or a single logarithm and state the theorem or theorems you used.

49. $\ln 17 + \ln 12$ 50. $\log 50 - \log 5$

51. $-2 \log_{12} 11$ 52. $\ln e$

53. $\log_{107} 107^{79}$ 54. $\log_{6.3} 1$

55. $\log a - 3 \log b$

56. $\log u + \log v + 0.7 \log w$

USES Applications of mathematics in real-world situations

OBJECTIVE G Create and apply exponential growth and decay models. (Lessons 9-1, 9-2, 9-3, 9-10)

57. In 2005 the population of the Tokyo-Yokohama region in Japan was about 35.327 million, the largest metropolitan area in the world. The average annual growth rate was 0.43%. Assuming this growth rate continues, find the population of the Tokyo-Yokohama area in 2020.

58. In 2005 the sixth largest metropolitan area in the world was that of Mumbai, India, with 18.202 million people. Mumbai was growing at an average rate of 1.96% annually. Suppose this rate continues indefinitely.

   a. Find the population of this area in 2020.

   b. In what year will the population of Mumbai reach 30 million?

59. Refer to Questions 57 and 58. Estimate the year in which Mumbai's population will first exceed Tokyo-Yokohama's population.

60. The population of a certain strain of bacteria grows according to $N = C \cdot 3^{0.593t}$, where t is the time in hours. How long will it take for 30 bacteria to increase to 500 bacteria?

61. The amount A of radioactivity from a nuclear explosion is given by $A = Ce^{-0.2t}$, where t is measured in days after the explosion. What percent of the original radioactivity is present 9 days after the explosion?

62. Strontium-90 (^{90}Sr) has a half-life of 29 years. If there was originally 25 grams of ^{90}Sr,

   a. how much strontium will be left after 87 years?

   b. how much strontium will be left after t years?

63. A new car costing $28,000 is predicted to depreciate at a rate of 14% per year. About how much will the car be worth in six years?

OBJECTIVE H Fit an exponential model to data. (Lesson 9-4)

64. Find an equation for the exponential function $f: x \rightarrow ab^x$ passing through $(0, 1.2)$ and $(3, 25)$.

65. A bacteria population was counted every hour for a day with the following results.

Hour h	1	2	3	4	5	6	7
Population p (hundreds)	5	13	25	49	103	211	423

   a. Construct a scatterplot of these data.

   b. Fit an exponential model to these data.

   c. Use your model to estimate the population at the 11th hour.

66. A hypothetical new substance, mathium, was manufactured and experiments showed that it decayed at the following rate.

Days	Amount Present (g)
1	1156
2	907
3	715
4	660
5	432
6	340
7	273
8	210
9	168
10	129

a. Construct a scatterplot of these data.

b. From the data in the scatterplot, what is the approximate half-life of this new substance? Explain your answer.

c. Fit an exponential model to these data.

d. On the 20th day, how much of the substance will be present?

OBJECTIVE I Apply logarithmic scales, models, and formulas. (Lessons 9-6, 9-8)

In 67–69, use the formula $D = 10 \log\left(\dfrac{I}{10^{-12}}\right)$ to convert sound intensity I in $\dfrac{W}{m^2}$ into relative intensity D in decibels.

67. Find D when $I = 3.88 \cdot 10^9$.

68. What sound intensity corresponds to a relative intensity of 80 decibels?

69. How many times as intense is a 60 dB sound as a 20 dB sound?

70. Baking soda has a pH value of 8, while pure water has a pH value of 7. How many times as acidic is water than baking soda?

71. The boiling point T of water in degrees Fahrenheit at barometric pressure P in inches Hg (inches of mercury) is given by the model

$$T = 49.161 \cdot \ln P + 44.932.$$

At what temperature does water boil in Colorado if the average barometric pressure is 27 inches Hg?

REPRESENTATIONS   Pictures, graphs, or objects that illustrate concepts

OBJECTIVE J Graph exponential functions. (Lessons 9-1, 9-2)

72. Graph $y = 3^x$ using at least five points.

73. Graph $y = \left(\dfrac{1}{3}\right)^x$ using at least five points.

74. Graph $g(x) = \left(\dfrac{1}{5}\right)^x$ and $h(x) = \left(\dfrac{1}{5}\right)^{2x}$ on the same set of axes.

a. Which function has greater values when $x > 0$?

b. Which function has greater values when $x < 0$?

75. Below are the graphs of the equations $y = 2^x$ and $y = 3^x$.

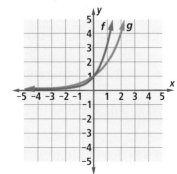

a. Which equation corresponds to the graph of f?

b. Which equation corresponds to the graph of g?

c. Describe how the graph of $y = e^x$ is related to the graphs of f and g.

76. **Multiple Choice** Which graph below represents exponential decay?

A

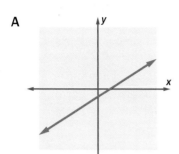

B

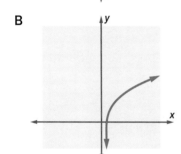

C

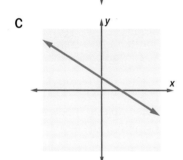

D

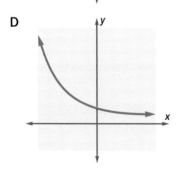

OBJECTIVE K Graph logarithmic curves.
(Lessons 9-5, 9-7)

77. **a.** Graph $y = 4^x$ using at least five points.

   b. Use the results of Part a to plot at least five points on the graph of $y = \log_4 x$.

78. **a.** Plot $y = 10^x$ and $y = \log_{10} x$ on the same set of axes.

   b. Identify all intercepts of these curves.

79. **a.** Graph $y = \ln x$ using at least five points.

   b. Give an equation for the inverse of this function.

80. The graph below has the equation $y = \log_Q x$. Find Q.

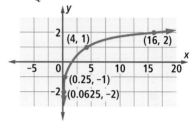

81. What is the x-intercept of the graph of $y = \log_b x$, where $b > 1$?

Basic Ideas of Trigonometry

Contents

The word *trigonometry* is derived from Greek words meaning "triangle measure," and its study usually begins by examining relationships between sides and angles in right triangles. These ideas originated thousands of years ago. As early as 1500 BCE, the Egyptians had sun clocks. Using their ideas, the ancient Greeks created sundials by erecting a *gnomon*, or staff, in the ground. The shadows and the gnomon created triangles that could be used to measure the angle of the Sun at any time. With these measurements, ancient mathematicians could measure the length of a year by seeing when the angles of the Sun repeated themselves.

Using geometry and trigonometry, the ancient Greeks were also able to measure the circumference of Earth to within 1 percent of its actual value. Additionally, the astronomers Hipparchus and Ptolemy used trigonometry to estimate the distance from Earth to the Moon to be about 250,000 miles, impressively close to the current estimated value of 238,000 miles. To arrive at this estimate, Hipparchus and Ptolemy used observations of *parallax*, the apparent motion of the Moon against the background of the stars when observed from different positions. The same method is still used today. The satellite *Hipparcos*, (the Greek version of the Latin name Hipparchus), collected parallax data from 1989 until 1993 that has established the distances from Earth to over 120,000 stars.

Trigonometry is also used to describe wave-like patterns. For instance, when a spacecraft is launched from Cape Canaveral, tracking its position with respect to the equator produces a graph like the curve below. This curve can be described with trigonometric functions.

This chapter introduces the three fundamental trigonometric functions, *sine, cosine,* and *tangent*, as ratios of sides in right triangles. It explores ancient astronomy by recreating the computations of Earth's circumference and the distances to the Moon and to a nearby star. Many real-world uses of trigonometry, including Global Positioning Systems (GPS), are based on solving triangles. Beyond right-triangle relationships, you will study extensions of the sine and cosine functions to the domain of real numbers.

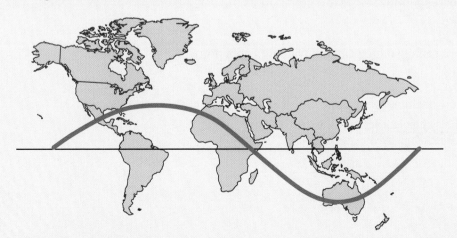

Lesson

10-1

Three Trigonometric Functions

Vocabulary

right-triangle definitions of
  sine (sin), cosine (cos),
  and tangent (tan)

sine function

cosine function

tangent function

angle of elevation

▶ **BIG IDEA**   The sine, cosine, and tangent of an acute angle are each a ratio of particular sides of a right triangle with that acute angle.

Suppose a flagpole casts a 22-foot shadow when the Sun is at an angle of 39° with the ground. What is the height of the pole?

The height of the pole is determined by the given information because you know the measures of two angles of the right triangle (the 39° angle and the right angle) and the side they include.

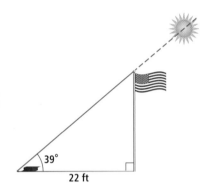

The ASA Congruence Theorem indicates that all triangles with these measurements are congruent.

Problems like the one above led to the development of *trigonometry*. Consider the two right triangles ABC and $A'B'C'$, with $\angle A \cong \angle A'$.

Mental Math

Picture two similar triangles, $\triangle BOW$ and $\triangle TIE$. Are the following ratios equal?

a. $\frac{BO}{TI}$ and $\frac{OW}{IE}$

b. $\frac{WB}{ET}$ and $\frac{TI}{BO}$

c. $\frac{OB - OW}{WB}$ and $\frac{TI - IE}{TE}$

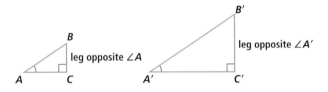

By the AA Similarity Theorem, these triangles are similar, so ratios of the lengths of corresponding sides are equal. In particular,

$$\frac{B'C'}{BC} = \frac{A'B'}{AB}.$$

Exchanging the means produces an equivalent proportion.

$$\frac{B'C'}{A'B'} = \frac{BC}{AB}$$

Look more closely at these two ratios:

$$\frac{B'C'}{A'B'} = \frac{\text{length of the leg opposite } \angle A'}{\text{length of the hypotenuse of } \triangle A'B'C'}$$

and $$\frac{BC}{AB} = \frac{\text{length of the leg opposite } \angle A}{\text{length of the hypotenuse of } \triangle ABC}.$$

Thus, in every right triangle with an angle congruent to $\angle A$, the ratio of the length of the leg opposite that angle to the length of the hypotenuse of the triangle is the same.

Three Trigonometric Ratios

In similar right triangles, any other ratio of corresponding sides is also constant. These ratios are called *trigonometric ratios*. There are six possible trigonometric ratios. All six have special names, but three of them are more important and are defined here. The Greek letter θ (theta) is customarily used to refer to an angle or to its measure.

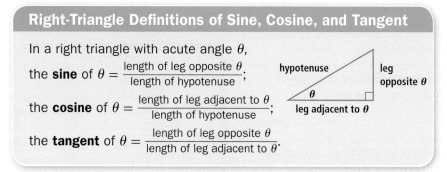

Right-Triangle Definitions of Sine, Cosine, and Tangent

In a right triangle with acute angle θ,

the **sine** of $\theta = \dfrac{\text{length of leg opposite } \theta}{\text{length of hypotenuse}}$;

the **cosine** of $\theta = \dfrac{\text{length of leg adjacent to } \theta}{\text{length of hypotenuse}}$;

the **tangent** of $\theta = \dfrac{\text{length of leg opposite } \theta}{\text{length of leg adjacent to } \theta}$.

> ▶ **READING MATH**
>
> Trigonometry, like algebra and geometry, is a branch of mathematics. The word *trigonometry* comes from the Greek words *trigon*, meaning "triangle" and *metron*, meaning "measure," so trigonometry means "triangle measure," that is, measuring the sides and angles of triangles.

Following a practice begun by Euler, we use the abbreviations sin θ, cos θ, and tan θ to stand for these ratios, and abbreviate them as shown below.

$$\sin \theta = \frac{\text{opposite}}{\text{hypotenuse}} = \frac{\text{opp.}}{\text{hyp.}}$$

$$\cos \theta = \frac{\text{adjacent}}{\text{hypotenuse}} = \frac{\text{adj.}}{\text{hyp.}}$$

$$\tan \theta = \frac{\text{opposite}}{\text{adjacent}} = \frac{\text{opp.}}{\text{adj.}}$$

 QY

Sine, Cosine, and Tangent Functions

The three correspondences that map an angle measure θ in a right triangle onto each right-triangle ratio above define functions called the **sine**, **cosine**, and **tangent functions**.

$$sin: \theta \rightarrow \sin \theta \qquad cos: \theta \rightarrow \cos \theta \qquad tan: \theta \rightarrow \tan \theta$$

From their right-triangle definitions, the domain of each of these functions is the set of all possible acute angle measures, that is, $\{\theta \mid 0° < \theta < 90°\}$. However, in a later lesson we will extend the domain so that θ can have any degree measure, positive, negative, or zero.

> ▶ **QY**
>
> Write the ratios sin θ, cos θ, and tan θ when $\theta = \angle A$ in $\triangle ABC$ on the previous page.

In the early history of trigonometry, mathematicians calculated values of the sine, cosine, and tangent functions and published the values in tables.

Today, people use calculators to find these values, with built-in programs using formulas derived from calculus. Most calculators allow you to enter angle measures in *degrees* or *radians*. You will learn about radians in Lesson 10-9. For now, make sure that any calculator you use is set to degree mode.

Activity

MATERIALS ruler and protractor or DGS

See how close you can measure sides to obtain values of the trigonometric functions.

Step 1 Draw a 39° angle and label it *PQR*.

Step 2 Draw a perpendicular from *P* to $\overrightarrow{QR}$ to form a right triangle *PQR*, where ∠*R* is the right angle.

Step 3 Measure the sides of △*PQR* as accurately as you can.

Step 4 Use the measures in Step 3 to calculate ratios to the nearest thousandth to estimate the sine, cosine, and tangent of 39°.

Step 5 Set your calculator to degree mode. Find sin 39°, cos 39°, and tan 39° to the nearest thousandth.

Step 6 Subtract to calculate the error in each of the estimates you found by measuring. Divide each error by the calculator value to determine the *relative error* of the estimate. Consider yourself to have measured well if your relative error is less than 3%.

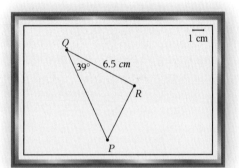

In this book, we usually give values of the trigonometric functions to the nearest thousandth. But when trigonometric values appear in long calculations, we do not round the calculator values until the end.

Using Trigonometry to Find Sides of Right Triangles

Example 1

Find the height of the flagpole mentioned in the first paragraph of this lesson.

Solution With respect to the 39° angle, the adjacent leg is known and the opposite leg is needed. Use the tangent ratio to set up an equation. Let *x* be the height of the flagpole.

$\tan 39° = \dfrac{\text{opposite}}{\text{adjacent}}$

$\tan 39° = \dfrac{x}{22}$

Solve for x.

$x = 22 \cdot \tan 39°$

From the Activity, we know that $\tan 39° \approx 0.810$. Substitute.

$x \approx 22(0.810) = 17.82$.

The flagpole is about 18 feet high.

Check Recall from geometry that within a triangle, longer sides are opposite larger angles. We have found that the side opposite the 39° angle is about 18 feet long. The angle opposite the 22-foot side has measure 51°, which is larger than 39°. So the answer makes sense.

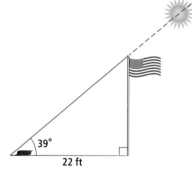

The **angle of elevation** of an object, such as the Sun above the horizon or the peak of a mountain from its base, is the angle between the horizontal base of the object and the observer's line of sight to the object. If you know the angle of elevation of an object, you can use trigonometry to find distances that would otherwise be difficult to find.

angle of elevation

GUIDED

Example 2

The Zephyr Express chair lift in Winter Park, Colorado, has a vertical rise of about 490 meters. Suppose the lift travels at an average 16.7° angle of elevation. How many meters long is the ride?

Solution A diagram of this situation is given below.

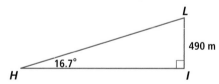

A chairlift travels in a continuous loop, so riders must get on and off while it is moving. Many first-time riders fall on their way off.

You know the length of the side opposite the 16.7° angle and want to find the length of the hypotenuse LH. So use the sine ratio.

$\sin \underline{\ ?\ } = \dfrac{\text{opp.}}{\text{hyp.}}$

$\qquad = \dfrac{?}{LH}$ Substitution

$LH = \dfrac{?}{?}$ Solve for LH.

$LH \approx \underline{\ ?\ }$ Calculate.

So, the ride up the chair lift is about $\underline{\ ?\ }$ meters long.

Check Use the Pythagorean Theorem to calculate HI.

$HI = \sqrt{\underline{\ ?\ }^2 - \underline{\ ?\ }^2} \approx \underline{\ ?\ }$ meters. Using the known angle, the cosine ratio relates HI and LH. $\cos \underline{\ ?\ } \approx \underline{\ ?\ }$ and $\dfrac{HI}{LH} \approx \underline{\ ?\ }$. It checks.

Drawing Auxiliary Lines to Create Right Triangles

When you wish to find the length of a segment and no right triangle is given, you can sometimes draw an auxiliary line to create a right triangle.

Example 3

Each side in the regular pentagon *VIOLA* is 7.8 cm long. Find the length of diagonal $\overline{VO}$.

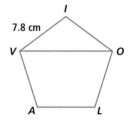

Solution Create a right triangle by drawing the segment perpendicular to $\overline{VO}$ from *I*. Call the intersection point *P*, as in the drawing at the right. Recall that each angle in a regular pentagon has measure

$$\frac{180(5-2)}{5} = 108°. \text{ So } m\angle VIO = 108°.$$

$\overline{IP}$ bisects $\angle VIO$. So $m\angle VIP = 54°$. Since the hypotenuse of $\triangle VIP$ is known and the opposite leg, *VP*, is needed, use the sine ratio.

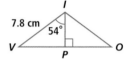

$$\sin 54° = \frac{\text{opposite}}{\text{hypotenuse}} = \frac{VP}{7.8}$$

$$VP = 7.8 \cdot \sin 54° \approx 6.3$$

The perpendicular $\overline{IP}$ bisects $\overline{VO}$, so the diagonal is about 2(6.3), or 12.6 cm long.

Using Special Right Triangles to Find Sines and Cosines

When θ is 30°, 45°, or 60°, you can use properties of 45-45-90 triangles or 30-60-90 triangles to find exact values of sin θ and cos θ.

Example 4

Use a 30-60-90 triangle to find cos 30°.

Solution Draw a 30-60-90 triangle. Label the 30° and 60° angles and let the length of the hypotenuse be *s*. Recall from geometry that since a 30-60-90 triangle is half of an equilateral triangle, if the length of the hypotenuse is *s*, the lengths of the legs are $\frac{s}{2}$ and $\frac{s}{2}\sqrt{3}$.

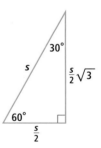

So, the length of the leg opposite the 30° angle is $\frac{s}{2}$, and the length of the leg adjacent to the 30° angle is $\frac{s}{2}\sqrt{3}$.

Then, by the right-triangle definition of cosine, $\cos 30° = \frac{\frac{s}{2}\sqrt{3}}{s} = \frac{\sqrt{3}}{2}$.

Using similar methods, you can find all the sine and cosine values below. You will use these special values of sine and cosine often in later mathematics courses.

$$\sin 30° = \frac{1}{2} \qquad \sin 60° = \frac{\sqrt{3}}{2} \qquad \sin 45° = \frac{\sqrt{2}}{2}$$

$$\cos 30° = \frac{\sqrt{3}}{2} \qquad \cos 60° = \frac{1}{2} \qquad \cos 45° = \frac{\sqrt{2}}{2}$$

Questions

COVERING THE IDEAS

1. **Multiple Choice** The fact that different angles with the same measure have the same sine value is due to a property of which kinds of triangles?

   A   congruent triangles   B   isosceles triangles

   C   right triangles   D   similar triangles

2. **Fill in the Blanks** Refer to $\triangle JMS$ at the right.

   a. $\overline{JS}$ is the ___?___ of the triangle.

   b. ___?___ is the leg opposite $\angle J$.

   c. ___?___ is the leg adjacent to $\angle J$.

   d. $\dfrac{MJ}{JS} = \underline{}\ J$

   e. $\dfrac{MS}{MJ} = \underline{}\ J$

   f. $\dfrac{MS}{JS} = \underline{}\ J$

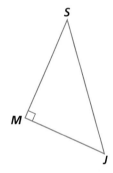

Fill in the Blanks In 3 and 4, use $\triangle ABC$ at the right. Answer with expressions involving a, b, and/or c.

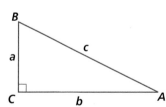

3. a. $\sin A = \underline{}$ b. $\cos A = \underline{}$ c. $\tan A = \underline{}$

4. a. $\sin B = \underline{}$ b. $\cos B = \underline{}$ c. $\tan B = \underline{}$

5. Suppose a construction crane casts a shadow 21 meters long when the Sun is 78° above the horizon. How high is the crane?

6. Suppose that a tree near your school casts a 32-foot shadow from its base when the Sun is 72° above the horizon. How tall is the tree?

7. Refer to Example 3. Use a trigonometric ratio to find IP to the nearest hundredth.

8. a. Use a 30-60-90 triangle to show that $\cos 60° = \frac{1}{2}$.

   b. Use a 45-45-90 triangle to show that $\sin 45° = \frac{\sqrt{2}}{2}$.

APPLYING THE MATHEMATICS

9. The Americans with Disabilities Act specifies that the angle of elevation of ramps can be no greater than about 4.76°. The door pictured at the right is 2 feet off the ground.

   a. What is the length of the shortest ramp that will meet the code?

   b. How far from the building will the ramp in Part a extend?

10. Explain why tan 45° = 1. (*Hint:* Draw a right triangle.)

11. a. Fill in the table.

    b. What is the relationship between the angles in each pair?

sin 30° = ?	sin 40° = ?
cos 60° = ?	cos 50° = ?
sin 15° = ?	sin 9.3° = ?
cos 75° = ?	cos 80.7° = ?

    c. What is the relationship between the sine and cosine values you calculated in Part a?

    d. Generalize your answers to Parts b and c as a conjecture.

    e. Test your answer to Part d by finding sine and cosine of other angle pairs.

12. The Great Pyramid of Giza had a square base with side length 230 meters and a height of 146.6 meters when first built around 2570 BCE. Due to erosion, the pyramid's height has decreased, but its side lengths are still nearly the same. The pyramid casts a 30.5-meter shadow when the Sun is at a 43.3° angle of elevation, as shown below. What is the current height of the pyramid to the nearest meter?

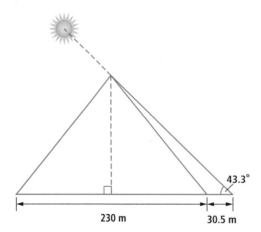

43.3°

230 m   30.5 m

13. A regular octagon has diagonals of three different lengths. What is the length of the shortest diagonal if a side of the regular octagon has length 1 unit?

14. In $\triangle TRI$ at the right, $m\angle I = 90°$. Let $m\angle R = \theta$.

   a. Find $\sin \theta$, $\cos \theta$, and $\tan \theta$.

   b. Use your answer from Part a to prove that $\dfrac{\sin \theta}{\cos \theta} = \tan \theta$.

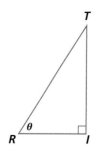

15. Use the figure at the right.

   a. Prove that the area of $\triangle ABC$ is $\frac{1}{2}ab \sin C$. (*Hint:* Find h.)

   b. Find the area of $\triangle ABC$ if $m\angle C = 20.3°$, $a = 72$, and $b = 75$.

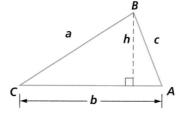

16. If it takes 34.66 years for an investment to double under continuous compounding, how long would it take to be multiplied by a factor of 2.5? (**Lessons 9-10, 9-3**)

17. Solve for z: $9\sqrt[3]{z - 5} = 27$. (**Lesson 8-8**)

18. Write two different expressions equal to $\dfrac{x^2}{x\sqrt{8}}$, $x \neq 0$. At least one expression should have a rational denominator. (**Lesson 8-6**)

19. Suppose h is a function with an inverse. Simplify. (**Lesson 8-3**)

   a. $h(h^{-1}(\pi))$

   b. $h^{-1}(h(\pi))$

20. One group of students ordered 5 hamburgers and 5 veggie burgers and paid \$66.00. Another group ordered 4 hamburgers and 7 veggie burgers and paid \$71.55. At these prices, what was the cost of one hamburger? (**Lesson 5-4**)

True or False. In 21 and 22, refer to the figure at the right where $j \parallel k$. Explain your answer. (**Previous Course**)

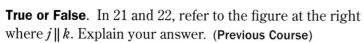

21. $\angle 1 \cong \angle 7$

22. $\angle 2 \cong \angle 6$

23. In the figure at the right, $m\angle HTC = x°$ and $\overline{TH} \parallel \overline{BC}$. (**Previous Course**)

   a. Find $m\angle BTC$.

   b. Find $m\angle C$.

EXPLORATION

24. a. As θ increases from $0°$ to $90°$, what happens to the value of $\sin \theta$? Does it increase? Does it decrease? Does it shift back and forth? Use values obtained from a calculator to find the pattern. Then use a geometrical argument and the definition of $\sin \theta$ to explain why $\sin \theta$ acts as it does.

   b. What happens to the value of $\cos \theta$ over the same interval?

   c. What happens to the value of $\tan \theta$?

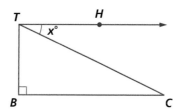

QY ANSWER

$\sin \theta = \dfrac{BC}{AB}$; $\cos \theta = \dfrac{AC}{AB}$;

$\tan \theta = \dfrac{BC}{AC}$.

Lesson

10-2 More Right-Triangle Trigonometry

Vocabulary

inverse sine function, $\sin^{-1}$

inverse cosine function, $\cos^{-1}$

inverse tangent function, $\tan^{-1}$

angle of depression

▶ **BIG IDEA** If you know two sides of a right triangle, you can use inverse trigonometric functions to find the measures of the acute angles.

In the last lesson, you used sines, cosines, and tangents of angles to find lengths of sides in right triangles. In this lesson, you will see how to use these ratios to find the measures of angles in right triangles.

Finding an Angle from a Trigonometric Ratio

Given an angle measure x, you can use a calculator to find its sine, cosine, and tangent. That is, you can find y when $y = \sin x$, $y = \cos x$, or $y = \tan x$. Now, instead of knowing the angle, suppose you know its sine, cosine, or tangent. That is, suppose you know y in one of these situations. Can you find x?

The function that maps $\sin x$ onto x is the *inverse* of the function that maps x onto $\sin x$. Appropriately, this function is called the **inverse sine function**. On a calculator, this function is denoted by the symbol **$\sin^{-1}$**. Like the inverse of any function, when composed with the original function, the result is the identity. That is, for any angle θ between $0°$ and $90°$, $\sin^{-1}(\sin \theta) = \theta$.

Mental Math

Between which two consecutive whole numbers does each value fall?

a. $\log 15{,}823{,}556$

b. $\ln e^{8.76}$

c. $\log_{18} 7$

d. $\log_5 620 + \log_9 77$

Example 1

Consider a right triangle *TOP* in which the hypotenuse $\overline{TP}$ has length 25 and $TO = 17$. What is $m\angle TPO$?

Solution Let θ be the unknown angle measure. Draw a figure, as shown at the right. From the figure, notice that $\sin \theta = \dfrac{\text{opp.}}{\text{hyp.}} = \dfrac{17}{25} = 0.68$. To solve for θ, apply the inverse sine function to each side of the equation. Make sure that your calculator is in degree mode.

$$\sin \theta = 0.68$$
$$\sin^{-1}(\sin \theta) = \sin^{-1}(0.68)$$
$$\theta \approx 42.84°$$
$$m\angle TPO \approx 43°$$

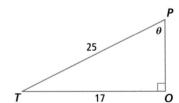

The **inverse cosine function** (denoted **cos⁻¹**) and the **inverse tangent function (tan⁻¹)** can be used in a similar way.

 QY1

 GUIDED

Example 2

Find the measures of the acute angles of the 5-12-13 right triangle below at the right.

Solution The two acute angles of this triangle are ∠B and __?__.

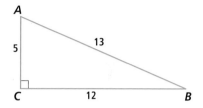

Find the measure of ∠B by using any of the trigonometric ratios.

$$\underline{}\,B = \frac{?}{?}$$

Apply the __?__⁻¹ function to both sides of the equation.

$$\underline{}^{-1}(\underline{}B) = \underline{}^{-1}\!\left(\frac{?}{?}\right)$$

Simplify. $m\angle B \approx \underline{}°$

To find m∠A you could use another ratio, but applying the Triangle-Sum Theorem may be quicker.

$$m\angle A + m\angle B + m\angle C = 180°$$
$$m\angle A + \underline{} + 90° \approx 180°$$
$$m\angle A \approx \underline{}$$

Check Use another ratio to find m∠B and see if you get the same answers.

 QY2

Finding Angles of Elevation and Depression

One of the most important applications of trigonometry is to find distances and angle measures that are difficult or impossible to measure directly. In the last lesson, you used the trigonometric functions to find distances. The inverse trigonometric functions can be used to find angles.

Example 3

Suppose a cell tower is anchored to the ground by a supporting wire. The wire is attached at a point on the tower 120 feet above the ground and is attached to the level ground 150 feet from the base of the tower. What is the angle of elevation of the wire?

Solution In the drawing, $BC = 120$ feet, $AC = 150$ feet, and $\frac{BC}{AC} = \tan \theta$, where θ is the angle of elevation.

Then, $\tan \theta = \frac{120}{150} = 0.8$ and $\theta = \tan^{-1}(0.8) \approx 38.660$.

So, the angle of elevation of the wire is about 39°.

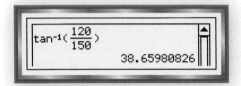

$$\tan^{-1}(\frac{120}{150})$$
$$38.65980826$$

Check Does $\tan 39° = 0.8$? $\tan(39°) \approx 0.8098$.
Yes, it checks.

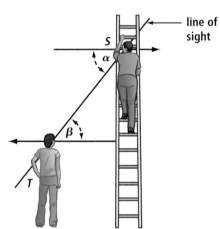

In the figure at the right, β (the Greek letter beta) represents the angle of elevation as person T looks up at person S. If S looks down at T, the angle α (the Greek letter alpha) between S's line of sight and the horizontal is called the **angle of depression**. The line of sight between T and S is a transversal for the parallel horizontal lines. Thus, β and α are alternate interior angles and are congruent. *So, the angle of elevation is equal to the angle of depression.*

GUIDED

Example 4

A surveyor standing on a bridge points her scope towards an assistant standing on level ground 110.56 feet from the base of the bridge. Her surveying laser measures a direct (slant) distance to the assistant of 125.75 feet. To the nearest hundredth of a degree, find the angle of depression of the scope.

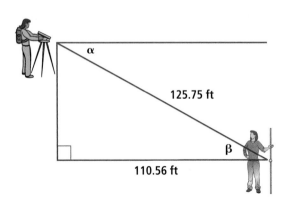

Solution The angle of depression α is not inside the triangle, so you cannot use it directly to set up a trigonometric ratio. However, the angle of depression α is equal to the angle of ___?___. So the angle of depression can be found using the ___?___ ratio.

$$\cos \underline{\ ?\ } = \frac{?}{?}$$ definition of cosine

$$\cos^{-1}(\cos \underline{\ ?\ }) = \cos^{-1}\underline{\ ?\ }$$ Apply the inverse cosine function.

$$m\angle\beta \approx \underline{\ ?\ }$$

Since $m\angle\alpha = m\angle\beta$, the angle of depression is about $\underline{\ ?\ }$ degrees.

Questions

COVERING THE IDEAS

1. Write a key sequence for your calculator to find θ if $\sin\theta = 0.475$.

In 2 and 3, approximate to three decimal places.

2. $\cos^{-1}(0.443)$

3. $\sin^{-1}\left(\dfrac{\sqrt{2}}{2}\right)$

4. Refer to $\triangle ABC$ at the right. Find θ to the nearest degree.

5. Refer to Example 2.
   a. Find $m\angle B$ using the tangent ratio.
   b. Find $m\angle A$ using the sine ratio.

6. Find the measures of the acute angles of an 8-15-17 right triangle to the nearest thousandth.

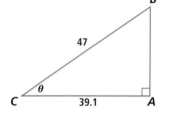

7. **Fill in the Blank** The angle of depression is the angle made between the line of sight to an object and the $\underline{\ ?\ }$.

8. Suppose a statue 15.5 meters high casts a shadow 21 meters long. What is the angle of elevation of the Sun?

9. Explain why the angle of depression from a point R to a point S equals the angle of elevation from S to R.

10. Refer to Example 4. Suppose the assistant stands 75 feet from the base of the bridge and the direct distance between the surveyor and the assistant is 120 feet. What is the angle of depression?

APPLYING THE MATHEMATICS

11. According to one railway company's guidelines, industrial railroad tracks must be built with a 1.5% grade or less. A 1.5% grade means that the track rises 1.5 feet vertically for every 100 feet horizontally.
    a. What is the slope of a line with a 1.5% grade?
    b. What is the tangent of the angle of elevation of a 1.5% grade?
    c. What angle does a track with a 1.5% grade make with the horizontal?

This train, on White Pass in Alaska, will rise approximately 3000 feet over just 20 miles.

More Right-Triangle Trigonometry   **673**

12. Consider the line with equation $y = \frac{2}{3}x + b$.
   a. What is the slope of the line?
   b. What is the tangent of the acute angle that the line makes with x-axis?
   c. What is the measure of the acute angle that the line makes with the x-axis, to the nearest degree?
   d. What is the tangent of the acute angle that the line with equation $y = mx + b$ makes with the x-axis?

13. Haylee has the plans to build a skateboard launch ramp with dimensions as shown in the drawing at the right. What is the angle of elevation of the ramp?

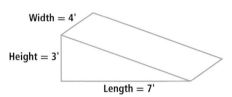

Width = 4'
Height = 3'
Length = 7'

14. Explain why the domain of the inverse sine function cannot include a number greater than 1.

15. Dawn Hillracer, an avid skier, rides 1600 meters on a chair lift to the top of a slope. If the lift has a 450-meter vertical rise, at what average angle of elevation does it ascend the slope?

REVIEW

16. Chinese legend tells of a General Han Hsin, who used a kite during battle to work out the distance between his army and a castle so he could dig tunnels under the walls. Suppose the kite's string was 200 meters long and at an angle of 62° with the ground when the kite was above the courtyard of a castle. If the tunnels are dug straight, how long should the tunnels be in this case? Round to the nearest meter. (**Lesson 10-1**)

17. Rationalize the denominator of $\frac{11}{\sqrt{8}}$. (**Lesson 8-6**)

18. If an isosceles right triangle has a leg of length z, how long is its hypotenuse? (**Lesson 8-5, Previous Course**)

19. Solve for x: $rx^2 + sx + t = 0$. (**Lesson 6-7**)

In 20 and 21, use triangle _DON_ at the right. (Lessons 4-7, 4-6, 4-4)

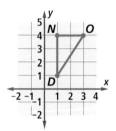

20. a. Find the coordinates of the vertices of $r_y(\triangle DON)$.
   b. Find the coordinates of the vertices of $r_x \circ r_y(\triangle DON)$.

21. Find DO.

EXPLORATION

22. a. Graph the $\sin^{-1}$ function over the domain $\{x \mid 0 < x < 1\}$. What is the range of the $\sin^{-1}$ function over this domain?
   b. Graph $\sin \circ \sin^{-1}$ over the domain $\{x \mid 0 < x < 1\}$. What is the range of this function over this domain?

QY ANSWERS

1. a. $\theta \approx 19°$
   b. $\theta \approx 54°$
   c. $\theta \approx 55°$

2. $\frac{5}{13}$

Lesson
10-3
Trigonometry, Earth, Moon, and Stars

▶ **BIG IDEA** Trigonometry has been used to estimate with great accuracy very large distances, such as are found in the solar system and in our galaxy.

Over 2000 years ago the ancient Greeks observed that when a ship arrived in port, the mast was the first part of the ship that could be seen. They concluded that Earth's surface must be curved. Using trigonometry and without modern equipment, the ancient Greeks were able to estimate the circumference of Earth and the distance from Earth to our Moon with remarkable accuracy. Their methods were so powerful that the same ideas are still used today to measure interstellar distances. One of these ideas is that of *parallax*.

Activity 1

Work with a partner.

Step 1 Have your partner stand about eight feet in front of you. Extend your arm in front of your face and hold up one thumb. Close your left eye. Then open your left eye and close your right eye.

Your thumb and your partner both appear to jump to new locations. Which jumps more? Does the background behind your partner jump as well?

Take four steps back and repeat the experiment. What moves now?

Step 2 When you switched eyes, you changed the position from which you viewed your thumb and partner by about three inches. As a result, both your thumb and your partner appeared to move across the background behind them. This illusion of motion created by changing viewing positions is called *parallax*.

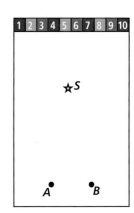

Copy the parallax diagram at the right. Draw lines from the observation sites A and B through the star S to the row of numbered squares in the background. The line from A indicates which square the star appears to obscure when viewed from the left site. That square marks the *apparent position* of the star from site A. Record that square, and the square obscured from site B.

(continued on next page)

Step 3 As you noticed with your partner and your thumb, objects that are closer to the observer appear to move sideways more than objects that are far away.

    a. On the diagram you copied in Step 2, draw three more stars between the two observation sites and the shaded background.

    b. Draw lines to identify each star's apparent position from each site and record the results.

    c. For each pair of observations, determine how many squares the star "jumped." How does that number relate to the angle formed by site *A*, the star, and site *B*?

    d. Write a sentence explaining why closer objects seem to jump sideways more than more distant objects.

In Activity 1 you measured the parallax effect by the magnitude of background movement. Another way is to measure $\angle ASB$. The larger the measure of this angle, the closer the star is to *A* and *B*. In practice, astronomers use half this angle and call it the **parallax angle**.

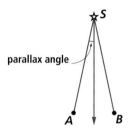

Astronomers measure parallax angles with special equipment, including satellites and telescopes with scales for measuring angles. You and a partner can simulate what they do by making a larger version of the parallax diagram used in Activity 1. One partner takes the place of the star, and the other makes observations from two sites.

Activity 2

MATERIALS whiteboard or butcher paper, ruler, string

Step 1 **a.** Mark points *A* and *B* on the floor 4 feet apart so that $\overline{AB}$ is parallel to, and several feet from, a whiteboard or wall covered in butcher paper.

    b. Draw equally spaced vertical lines on the board. Use a ruler to accurately space the lines 1 inch apart for every foot between $\overline{AB}$ and the board. (For example, if $\overline{AB}$ is 16 feet from the board, the lines should be 16 inches apart.)

Step 2 The star needs to be equidistant from the two observation sites. (In astronomical parallax, the variations in Earth's position are so small compared to the distances being measured that this condition is easy to meet.) Use string to construct the perpendicular bisector $\overleftrightarrow{MS}$ of $\overline{AB}$. Have your partner, the star, stand on $\overleftrightarrow{MS}$ between $\overline{AB}$ and the background. Observe your partner's head from *A* and from *B* by closing one eye and noting with the other eye the line on the board that appears to be closest to your partner's head. Close the same eye each time. Record the number of spaces between the observed board lines.

Step 3 Convert your range of spaces to an angle. If *A* and *B* are 16 feet from the board, each space corresponds to an angle of about 5°. So if you observed a 3.5-space difference from the two sites, this corresponds to a 5 • 3.5 = 17.5° angle. Half this angle (8.25°) is the parallax angle shown in the diagram at the right. Copy the diagram and write your angle measurement in the "Par ∠" space. If you are not able to do the physical experiment, use the lines in the diagram shown here to estimate a parallax angle.

Step 4 △*ASB* is isosceles. Use this fact to fill in the measures of all other labeled angles.

Step 5 Use trigonometry to find *MS*.

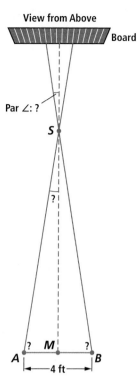

In Activity 2, *MS* is the distance from the observation sites line to the star. Using trigonometry, you were able to determine the distance to this remote object without having to go there and without sending anything to it.

The Distance to the Moon

Since ancient times, astronomers have used parallax to measure distances from Earth to celestial objects: the Moon, the planets, and in the last 200 years, nearby stars. In the second century BCE, Hipparchus used data recorded during a solar eclipse to estimate the distance from Earth to the Moon. Although the books describing his exact procedure have been lost, we know he used estimates of the Moon's parallax angle in his calculations. Now that you have seen how parallax works, you can use trigonometry to make your own estimates of the distance from Earth to the Moon.

Activity 3

Step 1 Data from observers of an eclipse in two Greek cities, Alexandria and Hellespont, led Hipparchus to approximate the Moon's parallax angle as 0.11°. This is the measure of ∠*ANM* in the diagram at the right.

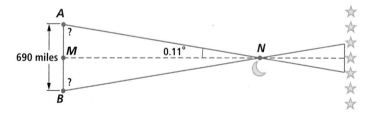

Because the Moon, represented by point *N*, is far away from Earth, *AN* ≈ *BN*, so △*ANB* is isosceles with base $\overline{AB}$. Let *M* be the midpoint of $\overline{AB}$. Approximate the measures of the missing angles.

Step 2 Given that *AB* = 690 miles, use trigonometry to compute the distance *MN* from Earth to the Moon.

The actual distance from Earth to the Moon is about 238,000 miles. The data you used in Activity 3 leads to an underestimate by about 25%, which means that 0.11° is probably not the measure of the parallax angle when the Moon is exactly equidistant from the two cities. When Hipparchus did his calculations without modern trigonometry, he overestimated the distance by about 5%. The inaccuracy of Hipparchus' calculations may be due to the difficulty in verifying the data he worked with, and to the fact that none of the observers of his time had telescopes.

The Distance to a Star

To find distances to stars, you need to use observation sites that are much farther apart than 690 miles, and even then the parallax angles are extremely small. So while the ancient Greeks knew that the stars were extremely far away, they did not have an accurate way to measure those distances.

The first successful attempt to measure the distance to a star was by Friedrich Wilhelm Bessel in 1838, when he used the opposite sides of Earth's orbit around the Sun to measure a parallax angle to the star 61 Cygni. (This technique is called *annual parallax*.)

Activity 4

The angle measured by Bessel was 0.314 arc-second; one arc-second is $\frac{1}{3600}$ degree. Earth's orbit is roughly circular with a diameter of 186,000,000 miles.

Step 1 Draw a diagram and compute the distance from Earth to 61 Cygni.

Step 2 One light-year is the distance light can travel in one Earth year, approximately $5.88 \cdot 10^{12}$ miles. How many light-years away is 61 Cygni?

Friedrich Wilhelm Bessel has an asteroid named for him, 1552 Bessel.

The Circumference of Earth

Hipparchus's original parallax computations gave the distance from Earth to the Moon in terms of Earth's radius. The radius of Earth was already known in Hipparchus's time, having been computed by Eratosthenes, a mathematician who lived in the third century BCE. Eratosthenes lived in Alexandria, Egypt, an important location of learning at the time. Eratosthenes found Earth's radius from its circumference. Activity 5 recreates his measurement of Earth's circumference.

Activity 5

Step 1 Eratosthenes knew that at noon on a particular day each year, in Cyene (an ancient city on the Nile River near what is now Aswan, Egypt), the Sun would shine directly overhead, even reflecting from the water in a deep well. In Alexandria on the same day each year, the Sun at noon was *not* directly overhead. In that city at noon, a 10-meter pole would cast a shadow approximately 1.27 meters long. Copy the diagram at the right, and use trigonometry to compute θ.

10 m

1.27 m

Step 2 Because the Sun is very far away from Earth, the rays reaching Earth are essentially parallel. If the Sun's rays strike Earth at different angles, it can only be because Earth's surface is curved. Copy the diagram at the right, substituting the angle measure θ you found in Step 1 and including the measures of all other angles in the figure. What is the degree measure of $\overset{\frown}{AC}$?

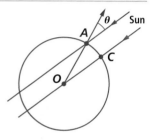

Step 3 Your work in Step 2 shows that the measure of $\overset{\frown}{AC}$ is $\frac{\theta}{360}$ of Earth's circumference. From caravans of merchants traveling between the two cities, Eratosthenes knew that Alexandria and Cyene were approximately 5,000 *stadia* apart. Although the exact correspondence is not known, scholars believe that 1 kilometer is approximately 6.27 stadia. Convert 5000 stadia into kilometers.

Step 4 If x is Earth's circumference and d is the distance between Alexandria and Cyene, then $\frac{\theta}{360}x = d$. Solve this equation for x using the values of θ and d that you computed in Steps 1 and 3.

The lack of modern technology at the time of Eratosthenes and Hipparchus makes their achievements even more amazing. Without using telescopes, much less calculators and computers, they were able to predict huge distances with incredible precision. The success of their techniques is a testament not just to their persistence but to the brilliance of their work and the power of mathematics in helping us understand the world around us.

Questions

COVERING THE IDEAS

1. What is meant by *parallax*?

2. If an object has a parallax angle of 6.5° from two sites 4 feet apart, about how many feet away is the object?

3. In recreating Hipparchus's computation of the Moon's distance from Earth, what did we assume about the two observation sites? What justifies this assumption?

4. Refer to Activity 4. The ancient Egyptians marked the seasons by the location of the brightest star in the night sky, Sirius. Modern observations show that Sirius has a parallax angle of 0.38 arc-second when measured from opposite ends of Earth's orbit.

   a. Compute the distance from Sirius to Earth in miles.

   b. Use your result from Part a to compute the distance between Earth and Sirius in light-years. (Sirius is one of the closest stars to Earth.)

Hipparchus of Rhodes is considered the Father of Astronomy and was the first person to systematically survey the sky.

APPLYING THE MATHEMATICS

5. Depending on conditions, an unaided human eye can distinguish objects as little as $\frac{1}{60}$ degree apart. For apparent motion, however, the human eye can detect only differences of about 1° or greater. Suppose that when you close one eye and open the other, an object appears to jump 1°, for a parallax angle of 0.5°. If your eyes are 3 inches apart, how many feet away is the object?

6. The nearest star to Earth is Proxima Centauri, 4.22 light-years away. Using the fact that 1 light-year $= 5.88 \times 10^{12}$ miles, compute the parallax angle you would observe for Proxima Centauri, using the endpoints of Earth's orbit as the two observation sites.

REVIEW

In 7 and 8, use the following information. The top of the Rock of Gibraltar is 426 meters above sea level. The Strait of Gibraltar, the body of water separating Gibraltar from the African continent, is 14 kilometers wide at its narrowest point. (Lessons 10-2, 10-1)

7. If the captain of the ship HMS Pinafore took a sighting on the Rock of Gibraltar and recorded a 14° angle of elevation, how far was the Pinafore from Gibraltar?

8. To avoid running into the African shore, what is the minimum angle of elevation possible?

9. In the triangle at the right, find each value. (**Lesson 10-1**)

a. $\sin \theta$ b. $\cos \theta$ c. $\tan \theta$

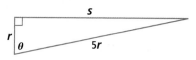

10. a. Solve the system $\begin{cases} y = x^2 - x - 2 \\ y = 3 \end{cases}$ by graphing.

b. Check your work using some other method. (**Lessons 5-3, 5-2**)

11. Find the coordinates of each point. (**Lesson 4-8**)

a. $R_{90}(1, 0)$ b. $R_{180}(1, 0)$

c. $R_{270}(1, 0)$ d. $R_{-90}(1, 0)$

12. After an initial 5000-mile break-in period, Mr. Euler changed the oil in his car and, thereafter changed the oil every 3000 miles. (**Lessons 3-8, 3-6, 3-1**)

a. What did the car's odometer read at the time of the second oil change?

b. What will the car's odometer read at the nth oil change after the break-in period?

c. How many times will Mr. Euler have changed his oil when the odometer reads 67,000?

13. State the Quadrants (I, II, III, or IV) in which (x, y) may be found if

a. x is negative and y is negative.

b. x is negative and y is positive.

c. $x = y$ and $xy \neq 0$. (**Previous Course**)

EXPLORATION

14. Let x be is the parallax angle for a star as seen from the endpoints of Earth's orbit, and let d be the star's distance from Earth in light-years.

a. Write a formula for d in terms of x.

b. Graph your equation from Part a using values of x between $0°$ and $\left(\frac{1}{3600}\right)°$. Describe the graph clearly.

c. Using a table or a graph, compare the values of $f(x) = \tan x$ and $g(x) = \frac{\pi x}{180}$ for x between $0°$ and $\left(\frac{1}{3600}\right)°$.

d. Use your observation in Part c to rewrite your formula in Part a. Does your rewritten formula hold for larger values of x, such as $20°$?

10-4

The Unit-Circle Definition of Cosine and Sine

Vocabulary

unit circle

unit-circle definition of cosine and sine

▶ **BIG IDEA** Every point P on the unit circle has coordinates of the form $(\cos \theta, \sin \theta)$, where θ is the magnitude of a rotation that maps $(1, 0)$ onto P.

In a right triangle, the two angles other than the right angle each have a measure between $0°$ and $90°$. So the definitions of sine, cosine, and tangent given in Lesson 10-1 only apply to measures between $0°$ and $90°$. However, the sine, cosine, and tangent functions can be defined for all real numbers. To define cosines and sines for all real numbers, we use rotations with center $(0, 0)$.

Mental Math

Let g be a geometric sequence with the formula $g_n = 120(0.75)^{n-1}$.

a. What is the second term of the sequence?

b. If the sequence models the height in inches of a dropped ball after the nth bounce, from what height in feet was the ball dropped?

c. Could this sequence model the number of people who have heard a rumor?

Activity

MATERIALS compass, protractor, graph paper

Work with a partner.

Step 1 Draw a set of coordinate axes on a piece of graph paper. Let each side of a square on your coordinate grid have length 0.1 unit. With the origin as the center, use a compass to draw a circle with radius 1. Label the positive x-intercept of the circle as A_0. Your circle should look like the one below at the right.

Step 2 a. With a protractor, locate the image of $A_0 = (1, 0)$ under R_{20}. Label this point A_{20}.

　　　b. Use the grid to estimate the x- and y-coordinates of A_{20}.

　　　c. Use a calculator to find $\cos 20°$ and $\sin 20°$.

Step 3 a. With a protractor, locate $R_{40}(1, 0)$. Label it A_{40}.

　　　b. Use the grid to estimate the x- and y-coordinates of A_{40}.

　　　c. Use a calculator to find $\cos 40°$ and $\sin 40°$.

Step 4 a. Locate $R_{75}(1, 0)$. Label it A_{75}.

　　　b. Estimate the x- and y-coordinates of this point.

　　　c. Use a calculator to evaluate $\cos 75°$ and $\sin 75°$.

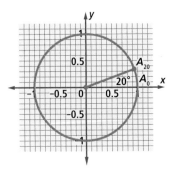

Step 5 a. Look back at your work for Steps 2–4. What relationship do you see between the x- and y-coordinates of $R_\theta(1, 0)$, cos θ, and sin θ?

    b. Use your answer to Step 5 Part a to estimate the values of cos 61° and sin 61° from your figure without a calculator.

    c. What is the relative error between your predictions in Step 5 Part b and the actual values of cos 61° and sin 61°? Were you within 3% of the actual values?

The Unit Circle, Sines, and Cosines

The circle you drew in the Activity is a *unit circle*. The **unit circle** is the circle with center at the origin and radius 1 unit. If the point (1, 0) on the circle is rotated around the origin with magnitude θ, then the image point (x, y) is also on the circle. The coordinates of the image point can be found using sines and cosines, as you should have discovered in the Activity.

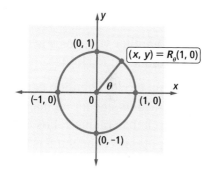

Example 1

What are the coordinates of the image of (1, 0) under R_{70}?

Solution Let $A = (x, y) = R_{70}(1, 0)$. In the figure at the right, $OA = 1$ because the radius of the unit circle is 1. Draw the segment from A to $B = (x, 0)$. $\triangle ABO$ is a right triangle with legs of length x and y, and hypotenuse of length 1. Now use the definitions of sine and cosine.

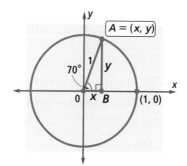

$$\cos 70° = \frac{adj}{hyp} = \frac{x}{1} = x$$

$$\sin 70° = \frac{opp}{hyp} = \frac{y}{1} = y$$

The first coordinate is cos 70°, and the second coordinate is sin 70°. Thus, $(x, y) = (\cos 70°, \sin 70°) \approx (0.342, 0.940)$. That is, the image of (1, 0) under R_{70} is (cos 70°, sin 70°), or about (0.342, 0.940).

Check Use the Pythagorean Theorem with cos 70° and sin 70° as the lengths of the legs.

Is $(0.342)^2 + (0.940)^2 \approx 1^2$? $0.117 + 0.884 = 1.001 \approx 1$, so it checks.

The idea of Example 1 can be generalized to define the sine and cosine of any magnitude θ. Since any real number can be the magnitude of a rotation, this definition enlarges the domain of these trigonometric functions to be the set of all real numbers.

Unit-Circle Definition of Cosine and Sine

Let R_θ be the rotation with
center (0, 0) and magnitude θ.
Then, for any θ, the point
(cos θ, sin θ) is the image
of (1, 0) under R_θ.

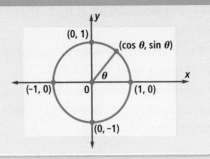

Stated another way, cos θ is the x-coordinate of $R_\theta(1, 0)$; sin θ is the
y-coordinate of $R_\theta(1, 0)$.

This unit-circle definition agrees with the right-triangle definition
of cosine and sine for all magnitudes between $0°$ and $90°$. Since the
unit circle has radius 1, cos θ is the ratio of the side adjacent to θ to
the hypotenuse, and sin θ is the ratio of the side opposite θ to the
hypotenuse.

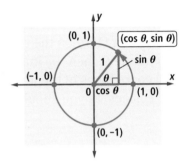

Cosines and Sines of Multiples of 90°

Sines and cosines of multiples of $90°$ can be found from the unit-
circle definition without using a calculator.

Example 2

Explain how to use the unit circle to find

a. cos 90°.   b. sin (−180°).

Solution

a. Think: cos 90° is the x-coordinate of $R_{90}(1, 0)$.
R_{90} (1, 0) = (0, 1). So cos 90° = 0.

b. $R_{-180}(1, 0) = (−1, 0)$. Since sin(−180°) is
the y-coordinate of this point, sin(−180°) = 0.

Check Check these values on your calculator.

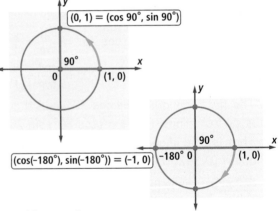

As you saw in your study of geometry and in Chapter 4, if a rotation
of magnitude x is followed by a rotation of magnitude y with the
same center, then the composite transformation is a rotation with
magnitude $x + y$. That is, $R_x \circ R_y = R_{x+y}$. Furthermore, rotations of
multiples of $360°$ are the identity transformation. Consequently, if
you add or subtract integer multiples of $360°$ from the magnitude of
a rotation, the rotation is the same. This means that you can add or
subtract integer multiples of $360°$ from the arguments of the sine or
cosine functions, and the value remains the same.

Example 3

a. Find sin 630°.

b. Find cos –900°.

Solution Add or subtract multiples of 360° from the argument until you obtain a value from 0° to 360°.

a. $630° - 360° = 270°$. So R_{630} equals one complete revolution followed by a 270° rotation, and R_{630} and R_{270} have the same images. $R_{630}(1, 0) = R_{270}(1, 0) = (0, -1)$. So sin 630° = –1.

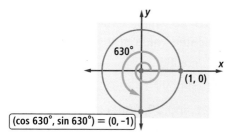

(cos 630°, sin 630°) = (0, –1)

b. Add $3 \cdot 360°$ to –900° to obtain a magnitude from 0° to 360°. $-900° + 3 \cdot 360° = 180°$. $R_{-900} = R_{180}$. So cos –900° = cos 180° = –1.

(cos(–900°), sin(–900°)) = (–1, 0)

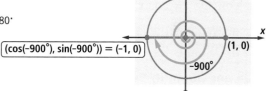

Questions

COVERING THE IDEAS

1. **Fill in the Blanks** If $(1, 0)$ is rotated θ degrees around the origin,

   a. $\cos \theta$ is the __?__-coordinate of its image.

   b. $\sin \theta$ is the __?__-coordinate of its image.

2. **True or False** The image of $(1, 0)$ under R_{23} is $(\sin 23°, \cos 23°)$.

3. **Fill in the Blanks** $R_0(1, 0) = $__?__, so $\cos 0° = $__?__ and $\sin 0° = $__?__.

4. Explain how to use the unit circle to find sin 180°.

In 5–7, use the unit circle to find the value.

5. cos 90°   6. sin (–90°)   7. cos 270°

8. If $(1, 0)$ is rotated –42° about the origin, what are the coordinates of its image, to the nearest thousandth?

9. **Fill in the Blanks**

   a. A rotation of 540° equals a rotation of 360° followed by __?__.

   b. The image of $(1, 0)$ under R_{540} is __?__.

   c. Evaluate sin 540°.

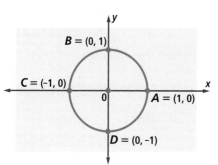

In 10–12, suppose $A = (1, 0)$, $B = (0, 1)$, $C = (-1, 0)$, and $D = (0, -1)$. Which of these points is the image of $(1, 0)$ under the stated rotation?

10. R_{450} 11. R_{540} 12. R_{-720}

In 13 and 14, evaluate without using a calculator.

13. cos 450° and sin 450°

14. cos(-720°) and sin(-720°)

APPLYING THE MATHEMATICS

In 15–20, which letter on the figure at the right could stand for the indicated value of the trigonometric function?

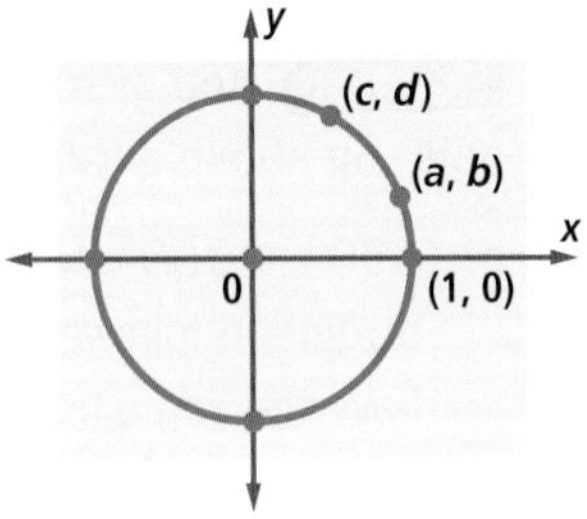

15. cos 80°

16. sin 80°

17. cos(-280°)

18. sin 800°

19. cos 380°

20. sin(-340°)

In 21 and 22, find a solution to the equation between 0° and 360°. Then check your answer by using a calculator to approximate both sides of the equation to the nearest thousandth.

21. cos 392° = cos x

22. sin(-440°) = sin y

23. a. What is the largest possible value of cos θ?

  b. What is the smallest possible value of sin θ?

In 24 and 25, verify by substitution that the statement holds for the given value of θ.

24. $(\cos \theta)^2 + (\sin \theta)^2 = 1$; $\theta = 7290°$

25. $\sin \theta = \sin(180° - \theta)$; $\theta = -270°$

REVIEW

26. If an object has a parallax angle of 3° from two sites 100 meters apart, about how far away is the object? (Lesson 10-3)

27. Why must the observation sites be very far apart to determine the distance to a star by parallax? (Lesson 10-3)

28. A private plane flying at an altitude of 5000 feet begins its descent along a straight line to an airport 5 miles away. At what constant angle of depression does it need to descend? (Lesson 10-2)

29. A submarine commander took a sighting at sea level of the aircraft carrier USS Enterprise, the tallest ship in the U.S. Navy at 250 feet. He knew that the top of the ship was about 210 feet above sea level, and he noted that the angle of elevation to the top of the mast was 4°. How far from the Enterprise was the submarine? (Lesson 10-1)

30. Use the distance formula $d = \sqrt{(x_1 - x_2)^2 + (y_1 - y_2)^2}$ to find the distance between (-2, 3) and (2, -7). (Lesson 4-4)

EXPLORATION

31. In Chapter 4, you used the Matrix Basis Theorem to develop rotation matrices for multiples of 90°. Use that theorem and the unit circle to produce a rotation matrix for any magnitude θ.

Lesson 10-5

Relationships among Sines and Cosines

Vocabulary

identity

tangent of θ
(for all values of θ)

> ▶ **BIG IDEA** Many properties of sines and cosines follow logically from the definition $(\cos\theta, \sin\theta) = R_\theta(1, 0)$ and properties of the unit circle.

With a calculator, it is easy to determine the values of $\cos\theta$ and $\sin\theta$ for any value of θ. But how can you check that you are correct? If θ is a multiple of $90°$, you can find $\cos\theta$ and $\sin\theta$ by visualizing the exact location of $R_\theta(1, 0)$. If $0 < \theta < 90°$, then $R_\theta(1, 0)$ is in the first quadrant and you can estimate the values by drawing a right triangle. For other values of θ, you can use the symmetry of the unit circle.

Determining the Signs of cos θ and sin θ

When θ is not a multiple of $90°$, $(\cos\theta, \sin\theta)$ is in one of the four quadrants. As the figure below at the left shows, each quadrant is associated with one-fourth of the interval $0° < \theta < 360°$.

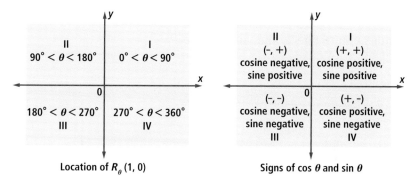

Location of R_θ (1, 0)   Signs of cos θ and sin θ

The quadrants enable you to determine quickly whether $\cos\theta$ and $\sin\theta$ are positive or negative. Refer to the figure above at the right. Because $\cos\theta$ is the first or x-coordinate of the image, it is positive when $R_\theta(1, 0)$ is in Quadrant I or IV and negative when $R_\theta(1, 0)$ is in Quadrant II or III. Because $\sin\theta$ is the second or y-coordinate of $R_\theta(1, 0)$, $\sin\theta$ is positive when $R_\theta(1, 0)$ is in Quadrant I or II and negative when $R_\theta(1, 0)$ is in Quadrant III or IV.

You do *not* need to memorize this information. You can always rely on the definition of $\cos\theta$ and $\sin\theta$ or visualize $R_\theta(1, 0)$ on the unit circle.

STOP QY

Mental Math

Jamal is packing bulk auto parts into boxes to ship to automotive stores. He can only send full boxes. How many parts will Jamal have left over if he has

a. 672 windshield-wiper blades and 160 fit in a box?

b. 223 fan belts and 24 fit in a box?

c. 17 fuel pumps and he can use boxes that fit 4 or 5 pumps?

> ▶ **QY**
>
> A value of $\cos\theta$ or $\sin\theta$ is given. Identify the quadrant of $R_\theta(1, 0)$ and tell whether the value is positive or negative.
>
> **a.** $\cos 212°$
> **b.** $\sin 212°$
> **c.** $\cos -17°$
> **d.** $\sin -17°$

Determining sin θ from cos θ, and Vice Versa

Relationships that are true for all values of variables in a domain are called **identities**. The basic identity relating sin θ and cos θ comes from the Pythagorean Distance Formula. So it is called the *Pythagorean Identity*.

> ### Pythagorean Identity Theorem
>
> For all θ, $(\cos \theta)^2 + (\sin \theta)^2 = 1$.

Proof For any θ, $(\cos \theta, \sin \theta)$ is a point on the unit circle. Let $A = (\cos \theta, \sin \theta)$ and $O = (0, 0)$. By the Pythagorean Distance Formula,

$$OA = \sqrt{(\cos \theta - 0)^2 + (\sin \theta - 0)^2}.$$

But $OA = 1$ for every point A on the circle. Substituting 1 for OA and squaring both sides,

$$(\cos \theta)^2 + (\sin \theta)^2 = 1.$$

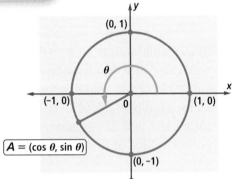

To avoid having to write parentheses, $(\sin \theta)^n$ is written $\sin^n \theta$, and powers of the other trigonometric functions are written similarly. With this notation, the Pythagorean Identity is:

$$\text{For all } \theta, \cos^2 \theta + \sin^2 \theta = 1.$$

The Pythagorean Identity enables you to determine sin θ if cos θ is known, and vice versa. It also enables you to check sine and cosine values that you have obtained.

Example 1

Refer to the unit circle at the right.

a. Approximate the coordinates of point C to the nearest thousandth.

b. Verify that your coordinates in Part a satisfy the Pythagorean Identity.

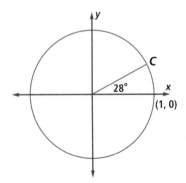

Solution

a. Since $C = R_{28}(1, 0)$, $C = (\cos 28°, \sin 28°) \approx (0.883, 0.469)$.

b. Here $\theta = 28°$.
$$\cos^2 28° + \sin^2 28° \approx 0.883^2 + 0.469^2 = 0.99965 \approx 1$$

Using Symmetry to Find Sines and Cosines

A circle is reflection-symmetric to any line through its center. For this reason, once you know the coordinates of one point on the circle, you can find the coordinates of many other points.

Second-Quadrant Values

When θ is between $90°$ and $180°$, $R_\theta(1, 0)$ is in Quadrant II. Every point on the unit circle in Quadrant II is the image of a point on the circle in Quadrant I under a reflection over the y-axis. For instance, the point $(\cos 150°, \sin 150°)$ is the reflection image over the y-axis of the point $(\cos 30°, \sin 30°)$, which is in the first quadrant. Notice that the acute angles determined by these two points and the x-axis are congruent.

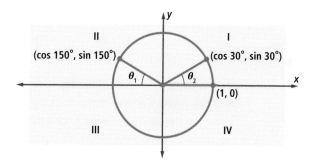

Recall that under r_y, the image of (x, y) is $(-x, y)$. The first coordinates of these points are opposites. Thus, $\cos 150° = -\cos 30° = -\dfrac{\sqrt{3}}{2}$.

The second coordinates are equal, so $\sin 150° = \sin 30° = \dfrac{1}{2}$.

Third-Quadrant Values

When a point is in Quadrant I, rotating it $180°$ gives a point in Quadrant III. Thus, to find $\sin \theta$ or $\cos \theta$ when $180° < \theta < 270°$, think of the angle with measure $\theta - 180°$.

Example 2

Show why $\sin 222° = -\sin 42°$.

Solution Make a sketch. $P' = (\cos 222°, \sin 222°)$ is the image of $P = (\cos 42°, \sin 42°)$ under R_{180}. Because the image of (x, y) under a rotation of $180°$ is $(-x, -y)$, P' has coordinates $(-\cos 42°, -\sin 42°)$. Thus, $\sin 222° = -\sin 42°$.

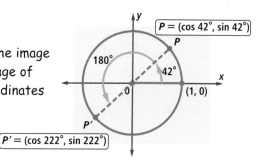

Check Using a calculator, $\sin 222° \approx -0.669$ and $-\sin 42° \approx -0.669$, so it checks.

Fourth-Quadrant Values

Points in Quadrant IV are reflection images over the x-axis of points in Quadrant I.

Example 3

Find an exact value for $\cos 315°$.

Solution $\cos 315°$ is the first coordinate of a point in Quadrant IV, so the cosine is positive. Reflect $(\cos 315°, \sin 315°)$ over the x-axis. Since $360° - 315° = 45°$, the image point is $(\cos 45°, \sin 45°)$. Since the first coordinates of these points are equal, $\cos 315° = \cos 45° = \dfrac{\sqrt{2}}{2}$.

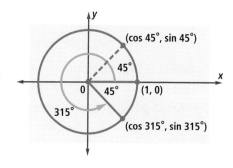

(continued on next page)

> **Check** A calculator shows $\cos 315° \approx 0.707$. Since 0.707 is an approximation for $\frac{\sqrt{2}}{2}$, it checks.

Activity

Step 1 Draw a good-sized copy of the figure of Example 1. Reflect C over the y-axis. Call the image C_2 (since it is in Quadrant II). Using the answer to Example 1, give the coordinates of C_2 to the nearest thousandth.

Step 2 Explain why R_{152} maps $(1, 0)$ onto C_2. Estimate $\cos 152°$ and $\sin 152°$ to the nearest thousandth.

Step 3 Reflect the original point C over the x-axis. Call the image C_4 (since it is in Quadrant IV). Using the answer to Example 1, give the coordinates of C_4 to the nearest thousandth.

Step 4 What is the magnitude θ of a rotation that maps $(1, 0)$ onto C_4? Use your answer to obtain $\sin \theta$ and $\cos \theta$ for this particular θ.

Step 5 Let $C_3 = R_{180}(C)$. Add C_3 to your figure and use it to obtain $\sin \theta$ and $\cos \theta$ for another value of θ.

In Lesson 10-4, you learned that if you know $\sin x$, you also know $\sin (x + n \cdot 360)$ because you can add or subtract multiples of $360°$ from the argument without changing the value of the sine. With the techniques in this lesson, if you know $\sin x$, you can use the Pythagorean Identity, reflections, and rotations to obtain the sines and cosines of many other arguments between $0°$ and $360°$.

Questions

COVERING THE IDEAS

Fill in the Blank In 1–3, choose one of the following: *is always positive, is always negative,* or *may be positive or negative.*

1. If $R_\theta(1, 0)$ is in Quadrant II, then $\cos \theta$ __?__.
2. When $180° < \theta < 360°$, $\sin \theta$ __?__.
3. When $0° > \theta > -90°$, $\cos \theta$ __?__.
4. When $\cos \theta$ is negative, in which quadrant(s) can $R_\theta(1, 0)$ be?

In 5 and 6, a trigonometric value is given. Draw the corresponding point on the unit circle. Then, without using a calculator, state whether the value is positive or negative.

5. $\sin 271°$
6. $\cos 200°$

7. Suppose $\sin x = \frac{5}{13}$. Use the unit circle to
   a. find the two possible values of $\cos x$.
   b. explain why $\sin(-x) = -\frac{5}{13}$.

In 8 and 9, the given statement is true. Use the unit circle and transformations to explain why the statement is true. Then, verify the statement with a calculator.

8. $\cos 130° = \cos 230°$ 9. $\sin 295° = -\sin 65°$

10. **Fill in the Blanks** Copy and complete with *positive* or *negative*.
    If $\angle B$ is obtuse, then $\cos B$ is ___?___ and $\sin B$ is ___?___.

APPLYING THE MATHEMATICS

In 11–14, find the exact value without a calculator.

11. $\cos 315°$ 12. $\sin 135°$ 13. $\sin(-120°)$ 14. $\cos 930°$

15. Refer to the unit circle at the right. Find θ to the nearest degree.

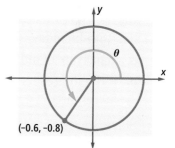

16. Given that $0° < x < 180°$ and $\cos x = -0.433$, find x to the nearest hundredth of a degree.

In 17–20, use this information. The tangent of θ, tan θ, is defined for all values of θ as $\frac{\sin \theta}{\cos \theta}$. This agrees with its right-triangle definition when $0° < \theta < 90°$.

17. a. Use your calculator to evaluate $\sin 300°$, $\cos 300°$, and $\tan 300°$.
    b. Verify that $\frac{\sin 300°}{\cos 300°} = \tan 300°$.

18. What is the sign of tan θ when $90° < \theta < 180°$? Justify your answer using the definition of tan θ.

19. Without a calculator, evaluate $\tan 900°$.

20. Without a calculator, give the exact value of $\tan 135°$.

21. Use the unit circle to explain why, for all θ, $\sin \theta = \sin(180° - \theta)$.
    (Hint: What is the image of a point when reflected over the y-axis?)

REVIEW

22. Without a calculator, evaluate $\sin 90°$ and $\cos 90°$. **(Lesson 10-4)**

23. **True or False** When measuring an object's distance using the parallax effect, a parallax angle of $110°$ is possible. **(Lesson 10-3)**

24. In the picture at the right, a person is standing on a cliff looking down at a boat. Which angle, θ or α, is the angle of depression? **(Lesson 10-2)**

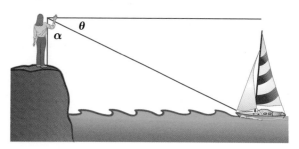

25. Before 1992, one national building code specified that stairs in homes should be built with an $8\frac{1}{4}$-inch maximum riser height and 9-inch minimum tread. Falls are a leading cause of nonfatal injuries in the United States. In an effort to reduce the number of falls, the building code was changed in 1992 to require a 7-inch maximum riser and an 11-inch minimum tread.
(**Lesson 10-2**)

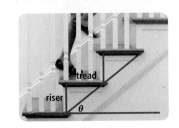

a. By how many degrees did this change decrease the angle θ the stairs make with the horizontal?

b. Why do you think the writers of the code thought the new stairs would be safer?

26. Cierra's grandmother put $10,000 into a college account at Cierra's birth. The money is invested so she will have $30,000 on her 18th birthday. (**Lessons 9-10, 9-3, 7-4**)

a. What rate of interest compounded annually will allow this?

b. If the interest is compounded continuously, what rate will be required?

27. Solve $3(x - 5)^6 = 12{,}288$ for x. (**Lesson 7-6**)

28. Determine whether the triangles in each pair are congruent. (**Previous Course**)

a. 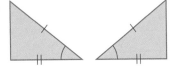   b.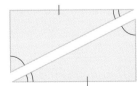

c. 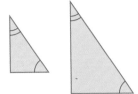   d.

QY ANSWERS

a. III; negative

b. III; negative

c. IV; positive

d. IV; negative

EXPLORATION

29. a. Copy and complete the table at the right using a calculator or spreadsheet. Round answers to 6 decimal places.

b. You should find that $(\sin \theta - \tan \theta)$ gets closer and closer to 0. Explain why this happens.

θ	$\sin \theta$	$\tan \theta$	$\sin \theta - \tan \theta$
10°	?	?	?
5°	?	?	?
2°	?	?	?
1°	?	?	?
0.5°	?	?	?
0.1°	?	?	?

10-6

The Cosine and Sine Functions

Vocabulary

periodic function, period

sine wave

sinusoidal

▶ **BIG IDEA** The graphs of the cosine and sine functions are sine waves with period 2π.

Remember that when (1, 0) is rotated θ degrees around the origin, its image is the point (cos θ, sin θ). The correspondence $\theta \rightarrow \cos \theta$ is the cosine function, with domain the set of real numbers. The values of this function are the first coordinates of the images of (1, 0) under rotations about the origin.

Similarly, the correspondence $\theta \rightarrow \sin \theta$ is the sine function. The values of this function are the second coordinates of the images of (1, 0) under R_θ.

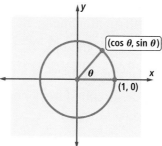

Mental Math

Which is longer?

a. the side of a regular octagon or its shortest diagonal

b. the leg opposite a 40° angle in a right triangle or the other leg

c. the diagonal of a square or the diameter of a circle inscribed in it

d. the diagonal of a square or the diameter of a circle circumscribed around it

A Graph of the Cosine Function

To imagine the graph of $y = \cos \theta$ as θ increases from 0, think of a point moving around the unit circle counterclockwise from (1, 0). As the point moves halfway around the circle, its first coordinate decreases from 1 to –1. As the point continues to move around the circle, its first coordinate increases from –1 to 1. The Activity provides more detail.

Activity

MATERIALS calculator

Set a calculator to degree mode.

Step 1 Make a table of values of cos θ for values of θ in the interval $0° \leq \theta \leq 360°$, in increments of 15°. You will need 25 rows, not 10 as shown at the right. Round the cosines to the nearest hundredth. The first few pairs in the table are shown.

Step 2 Graph the points you found in Step 1. Plot θ on the horizontal axis and cos θ on the vertical axis. Connect the points with a smooth curve.

Step 3 Describe two patterns you notice in your graph.

Step 4 What is the largest value that cos θ can have? What is the smallest value that cos θ can have?

θ	cos θ
0°	1.00
15°	0.97
30°	0.87
45°	0.71
60°	?
75°	?
90°	?
⋮	⋮
345°	?
360°	?

Recall that as θ takes on values greater than 360°, cos θ repeats its values. So the graph of $y = \cos \theta$ repeats every 360°. Below is a graph of this function when $-360° \leq \theta \leq 720°$.

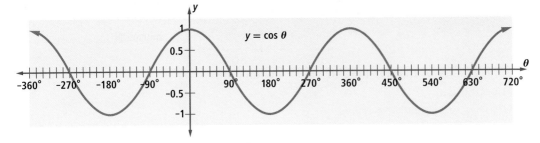

The Graph of the Sine Function

The graph of the sine function is constructed by a similar process, using the second coordinate of the rotation image of $(1, 0)$ as the dependent variable. For instance, $R_{60}(1, 0) = \left(\frac{1}{2}, \frac{\sqrt{3}}{2}\right)$, so $\sin 60° = \frac{\sqrt{3}}{2}$ and the point $\left(60°, \frac{\sqrt{3}}{2}\right)$ is on the graph of the sine function. Below is a graph of $y = \sin \theta$. Notice that the graph of the sine function looks congruent to the graph of the cosine function.

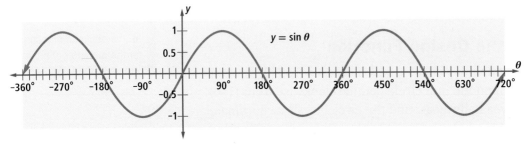

Properties of the Sine and Cosine Functions

A function is a **periodic function** if its graph can be mapped onto itself by a horizontal translation. Algebraically, this means that a function f is periodic if there is a positive number p such that $f(x + p) = f(x)$ for all values of x. The smallest positive number p with this property is called the **period** of the function. Both the sine and cosine functions are periodic because their values repeat every 360°. That is, for all θ, $\sin(\theta + 360°) = \sin \theta$ and $\cos(\theta + 360°) = \cos \theta$. This means that under a horizontal translation of magnitude 360°, the graph of $y = \sin \theta$ coincides with itself. Similarly, under this translation, the graph of $y = \cos \theta$ coincides with itself.

Notice that each of these functions has range $\{y|-1 \leq y \leq 1\}$. Also, each function has infinitely many x-intercepts, but still only one y-intercept. These and other properties of sine and cosine functions are summarized in the table on the next page.

	Cosine Function	Sine Function
Domain	set of all real numbers	set of all real numbers
Range	$\{y \mid -1 \leq y \leq 1\}$	$\{y \mid -1 \leq y \leq 1\}$
-intercepts	odd multiples of 90° $\{\ldots, -90°, 90°, 270°, 450°, \ldots\}$	even multiples of 90° $\{\ldots, -180°, 0°, 180°, 360°, \ldots\}$
Period	360°	360°
-intercept	1	0

The graph of the cosine function can be mapped onto the graph of the sine function by a horizontal translation of 90°. So the graphs of $y = \cos \theta$ and $y = \sin \theta$ are congruent. Both graphs are called *sine waves*.

Definition of Sine Wave

A **sine wave** is a graph that can be mapped onto the graph of the sine function $s: \theta \rightarrow \sin \theta$ by any composite of translations, scale changes, or reflections.

Because the graph of the cosine function $c: \theta \rightarrow \cos \theta$ is a translation image of the graph of $s: \theta \rightarrow \sin \theta$, its graph is a sine wave. Situations that lead to sine waves are said to be **sinusoidal**.

Example

The Great Ferris Wheel built in 1893 for the Columbian Exposition in Chicago had a 125-foot radius and a center that stood 140 feet off the ground. A ride on the wheel took about 20 minutes and allowed the rider to reach the top of the wheel twice. Assume that a ride began at the bottom of the wheel and did not stop. Which sine wave below models the rider's height h off the ground t minutes after the ride began? Explain your choice.

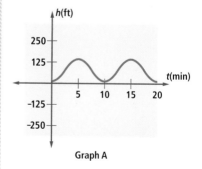

Graph A

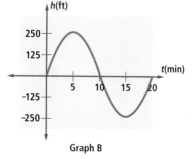

Graph B

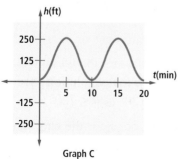

Graph C

(continued on next page)

Solution Find the minimum and maximum height of a ride on the Ferris wheel.

The minimum height is the difference between the height of the center of the wheel and its radius, or 140 ft − 125 ft = 15 ft. The maximum height is then 250 ft + 15 ft = 265 ft, the diameter of the wheel plus the wheel's height off the ground.

Find when the minimum and maximum height of a ride occurred.

The maximum height occurred twice in 20 minutes. A complete revolution took 10 minutes, so, the rider is 15 feet high at t = 0, 10, and 20 minutes. Without stops, the rider reached the top at t = 5 and 15 minutes. So, graph C is the correct graph.

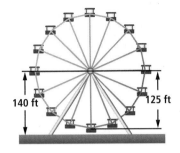

140 ft 125 ft

Sine waves occur frequently in nature: in ocean waves, sound waves, and light waves. Also, the graph of average daily temperatures for a specific location over the year often approximates a sine wave. The voltages associated with alternating current (AC), the type used in electrical transmission lines, have sinusoidal graphs. Sine waves can be converted to electrical signals and then viewed on an oscilloscope.

An oscilloscope can be used to test electronic equipment.

Questions

COVERING THE IDEAS

1. What function maps θ onto the first coordinate of the image of (1, 0) under R_θ?

2. What function maps θ onto the second coordinate of the image of (1, 0) under R_θ?

In 3–5, consider the cosine function.

3. a. **Fill in the Blanks** As θ increases from 0° to 90°, cos θ decreases from __?__ to __?__.

   b. **Fill in the Blanks** As θ increases from 90° to 180°, cos θ decreases from __?__ to __?__.

   c. As θ increases from 180° to 270°, does the value of cos θ increase or decrease?

4. Name two points on the graph of the function when $\theta > 360°$.

5. How many solutions are there to the equation cos $\theta = 0.5$ if $-720° \le \theta \le 720°$?

In 6–8, consider the sine function.

6. Explain why its period is 360°.

7. Name all θ-intercepts between –360° and 360°.

8. How many solutions are there to the equation sin $\theta = 2$ if $-720° \le \theta \le 720°$?

In 9–11, true or false.

9. The graph of the cosine function is called a cosine wave.

10. The graph of the sine function is the image of the graph of the cosine function under a horizontal translation of 180°.

11. The ranges of the sine and cosine function are identical.

12. Refer to the Example.

   a. Describe the Ferris Wheel ride shown by graph A.

   b. Why does graph B not describe a Ferris Wheel ride?

APPLYING THE MATHEMATICS

13. Use the graphs of the sine and cosine functions.

   a. Find two values of θ, one positive and one negative, such that $\cos \theta > \sin \theta$.

   b. Name two values of θ for which $\cos \theta = \sin \theta$.

14. Consider these situations leading to periodic functions. What is the period?

   a. days of the week

   b. the ones digit in the successive integers in base 10

In 15–18, part of a function is graphed. Does the function appear to be periodic? If so, what is the period? If not, why not?

15.

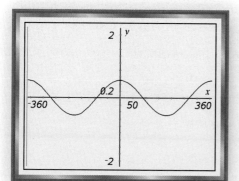

16.

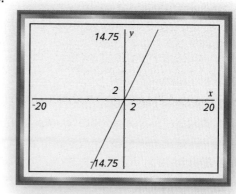

17.

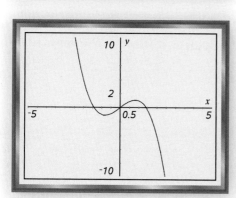

18.

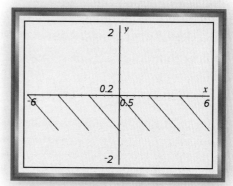

19. Below is a table of average monthly high temperatures T (all in degrees Fahrenheit) for Phoenix, Arizona.

Jan.	Feb.	Mar.	Apr.	May	June	July	Aug.	Sept.	Oct.	Nov.	Dec.
66	70	75	84	93	103	105	103	99	88	75	66

   a. Explain why these data could be modeled by a sine wave.
   b. Estimate the domain and range of a sinusoidal function that models these data.

20. a. Graph $y = \sin x$ and $y = \sin(180° - x)$.
    b. What identity is suggested by the graphs?

21. a. Graph the function with equation $y = \sin x + \cos x$. Does this appear to be a periodic function? If so, what is its period?
    b. What are the domain and range of this function?

REVIEW

22. **Multiple Choice** Which of the following is equal to $\sin(-45°)$?
    (Lesson 10-5)

    A $\sin 45°$ B $\sin 135°$
    C $\sin 405°$ D $\sin 675°$

In 23 and 24, give the exact value without using a calculator.
(Lesson 10-4)

23. $\cos 270°$ 24. $\tan 180°$

25. On Mars, the height h in meters of a thrown object at time t seconds is given by $h = -1.86t^2 + v_0 t + h_0$. A space traveler standing on a 47-meter high Martian cliff tosses a rock straight up with an initial velocity of $15 \frac{m}{sec}$. (Lessons 6-7, 6-4)

    a. Write an equation to describe the height of the rock at time t.
    b. Graph your equation in Part a.
    c. What is the maximum height of the rock to the nearest meter?
    d. To the nearest tenth of a second, when does the rock hit the Martian ground?

Cape Verde juts out from the walls of Victoria Crater on Mars.

In 26 and 27, $A = \begin{bmatrix} -72 & -27 \\ 8 & 3 \end{bmatrix}$.

26. **a.** Find det A.

    b. Does A^{-1} exist? If so, find it. If not, explain why it does not exist. **(Lesson 5-5)**

27. **a.** Find an equation for the line through the two points represented by matrix A.

    b. What kind of variation is described by the answer to Part a? **(Lessons 4-1, 3-4, 2-1)**

28. Approximate QW to the nearest hundredth. **(Lesson 4-4)**

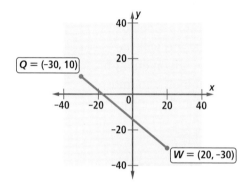

EXPLORATION

29. Oscilloscopes can be used to display sound waves. Search the Internet to find some sites that simulate oscilloscope output for different sounds. Do additional research about sound waves and report on the following.

    a. the oscilloscope patterns for at least two sound waves (for example, whistling a tune, middle C on the piano)

    b. the meaning of frequency and amplitude of a sound wave

    c. the effect on sound tone as a result of changes in amplitude and frequency

Lesson
10-7
The Law of Sines

Vocabulary

solving a triangle

▶ **BIG IDEA** Given AAS or ASA in a triangle, the Law of Sines enables you to find the lengths of the remaining sides.

One of the most important applications of trigonometry is to find unknown or inaccessible lengths or distances. In previous lessons, you learned to use trigonometric ratios to find unknown sides or angles of *right* triangles. In this lesson and the next, you will see how to find unknown sides or angles in *any* triangle, given enough information to determine the triangle. Using trigonometry to find all the missing measures of sides and angles of a triangle is called **solving the triangle**.

Mental Math

How many lines of symmetry does the graph of each equation have?

a. $y = -x^2$

b. $y = -x^3$

c. $y = -x$

Solutions to cos θ = k When 0° < θ < 180°

Although the sine and cosine functions are periodic over the domain of real numbers, it is important to remember that if θ is an angle in a triangle, then $0° < \theta < 180°$.

When $0° < \theta < 180°$, the equation $\cos \theta = k$ has a unique solution. To see why, consider the graph of $y = \cos \theta$ on this interval. For any value of k between –1 and 1, the graph of $y = k$ intersects $y = \cos \theta$ at a single point. The θ-coordinate of this point is the solution to $\cos \theta = k$.

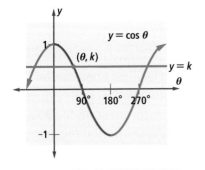

STOP QY1

▶ **QY1**

Solve $\cos \theta = -\frac{1}{2}$ when $0° < \theta < 180°$ using the inverse cosine function.

Solutions to sin θ = k When 0° < θ < 180°

The situation is different for the equation $\sin \theta = k$. On the interval $0° < \theta < 180°$, for any value of k between 0 and 1, the graph of $y = k$ intersects $y = \sin \theta$ in two points. In the graph at the right, we call these points (θ_1, k) and (θ_2, k). The numbers θ_1 and θ_2 are the solutions to $\sin \theta = k$.

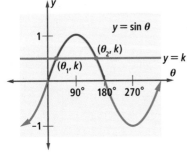

The points (θ_1, k) and (θ_2, k) are reflection images of each other over the vertical line with equation $\theta = 90°$. This is because when $0 < k < 1$, the two solutions to $\sin \theta = k$ between $0°$ and $180°$ are supplementary angles.

> **Supplements Theorem**
>
> For all θ in degrees, $\sin \theta = \sin(180° - \theta)$.

The Supplements Theorem allows you to solve equations of the form $\sin \theta = k$ without graphing.

Example 1

Find all solutions to $\sin \theta = 0.842$ in the interval $0° < \theta < 180°$.

Solution Use the inverse sine function to find a solution between $0°$ and $90°$. $\theta_1 = \sin^{-1} 0.842 \approx 57.4°$

The second solution is the supplement of the first.
$\theta_2 = 180° - \theta_1 \approx 180° - 57.4° = 122.6°$
So, when $\sin \theta = 0.842$, $\theta \approx 57.4°$ or $122.6°$.

Check $\sin 57.4° \approx 0.842$ and $\sin 122.6° \approx 0.842$. They check.

STOP QY2

The Law of Sines

Activity

MATERIALS ruler, protractor
Work with a partner.

Step 1 Each partner should draw a different triangle *ABC* on a sheet of notebook paper. Measure the side lengths to the nearest tenth of a centimeter and the angles as accurately as you can.

Step 2 Let *a*, *b*, and *c* be the lengths of the sides opposite angles *A*, *B*, and *C*, respectively. Use a calculator to compute $\frac{\sin A}{a}$, $\frac{\sin B}{b}$, and $\frac{\sin C}{c}$ to the nearest hundredth.

Step 3 Compare your results to those of your partner. What do you notice?

The results of the Activity are instances of a theorem that enables a triangle to be solved when the measures of two angles and one side are known.

> ▶ **QY2**
>
> **a.** Solve $\sin \theta = -\frac{1}{2}$ when $0° < \theta < 180°$ using the inverse sine function or the graph of $y = \sin x$.
>
> **b.** Solve $\sin \theta = \frac{1}{2}$ when $0° < \theta < 180°$ using the inverse sine function or the graph of $y = \sin x$.

Law of Sines Theorem

In any triangle ABC, $\frac{\sin A}{a} = \frac{\sin B}{b} = \frac{\sin C}{c}$.

The Law of Sines states that in any triangle, the ratios of the sines of its angles to the lengths of the sides opposite them are equal. A proof of the theorem is given below. You are asked in Question 13 to fill in the missing information.

Proof Recall from Question 15 of Lesson 10-1 that the area of a triangle is $\frac{1}{2}$ the product of two sides and the sine of the included angle. Consequently, area($\triangle ABC$) $= \frac{1}{2}ab \sin C$, area($\triangle ABC$) $= \frac{1}{2}ac \sin B$, and area($\triangle ABC$) $= \frac{1}{2}bc \sin A$. So, by the ___?___, $\frac{1}{2}ab \sin C = \frac{1}{2}ac \sin B = \frac{1}{2}bc \sin A$. Multiply all three parts of this equation by 2. The result is ___?___. Divide all three parts by abc. You get ___?___. Simplify the fractions to get $\frac{\sin C}{c} = \frac{\sin B}{b} = \frac{\sin A}{a}$.

Example 2

Two forest rangers are in their stations, S and T, 30 miles apart. On a certain day, the ranger at S sees a fire at F, at an angle of $38°$ with segment $\overline{ST}$. The ranger at T sees the same fire at an angle of $64°$ with $\overline{ST}$. Find the distance from station T to the fire.

Solution Let s be the desired distance. The angle opposite s is $\angle S$, with measure $38°$. To use the Law of Sines, you need the measures of another angle and its opposite side. Because the sum of the measures of the angles in a triangle is $180°$, $\angle F$ has measure $78°$. Now there is enough information to use the Law of Sines.

$$\frac{\sin S}{s} = \frac{\sin F}{f}$$

$$\frac{\sin 38°}{s} = \frac{\sin 78°}{30} \qquad \text{Substitution}$$

$$s = \frac{30 \sin 38°}{\sin 78°} \qquad \text{Solve for } s.$$

$$s \approx \frac{30(0.616)}{0.978} \approx 18.9$$

The fire is about 19 miles from station T.

The Law of Sines also can be used to find lengths in triangles when you know two angles and an adjacent side, the AAS (Angle-Angle-Side) condition from geometry.

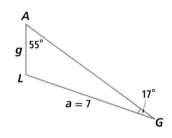

GUIDED

Example 3

In △*ALG* at the right, m∠*A* = 55°, m∠*G* = 17°, and *a* = 7. Find *g*.

Solution

$$\frac{\sin A}{a} = \frac{\sin G}{g}$$ Law of Sines

$$\underline{\quad ? \quad} = \underline{\quad ? \quad}$$ Substitution

$$g \approx \underline{\quad ? \quad}$$ Solve for *g*.

Ptolemy knew of the Law of Sines in the 2nd century CE. This led eventually to *triangulation*, the process of dividing a region into triangular pieces, making a few accurate measurements, and using trigonometry to determine most of the unknown distances. This made it possible for reasonably accurate maps of parts of Earth to be drawn well before the days of artificial satellites. Now, Global Positioning Systems (GPS) use triangulation along with accurate data about the contours of Earth to produce highly accurate maps and to calculate the precise latitude and longitude of GPS devices.

Questions

COVERING THE IDEAS

1. Explain why the equation sin θ = *k* for 0° < θ < 180° and 0 < *k* < 1 has two solutions.

2. Solve sin θ = 0.954 for 0° < θ < 180° by graphing.

3. Since sin 45° = $\frac{\sqrt{2}}{2}$, for what other value of θ with 0° < θ < 180° does sin θ = $\frac{\sqrt{2}}{2}$?

4. Explain how the Activity illustrates the Law of Sines.

5. Write a description of the Law of Sines as if you were explaining it to a friend.

6. Refer to Example 2. Find the distance from the fire to station *S*.

In 7 and 8, find *r*.

7.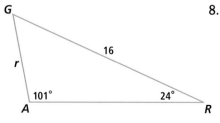

8.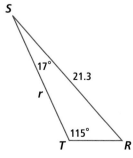

APPLYING THE MATHEMATICS

9. Refer to $\triangle BSG$ at the right.

   a. Why is the statement $\sin 120° = \dfrac{21}{BG}$ incorrect?

   b. Can you use the Law of Sines to find BG? Explain your answer.

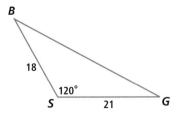

10. Refer to the picture below. A rock at C and a house at D are 100 feet apart. A tree is across the river at B. $m\angle BCD = 47°$ and $m\angle BDC = 80°$. Find the distance across the river from the house to the tree.

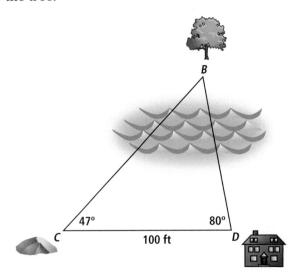

11. When a beam of light traveling through air strikes the surface of a diamond, it is *refracted*, or bent, as shown at the right. The relationship between α and θ is known as Snell's Law,

$$\frac{\sin \alpha}{\text{speed of light in air}} = \frac{\sin \theta}{\text{speed of light in diamond}}.$$

   The speed of light in air is about $3 \cdot 10^8$ meters per second.

   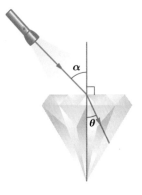

   a. If $\alpha = 45°$ and $\theta = 17°$, estimate the speed of light in diamond.

   b. Diamonds are identified as authentic by their *refractive index*. The refractive index of an object is the ratio of the speed of light in air to the speed of light in that object. Find the refractive index for a diamond.

   c. Cubic zirconia, a lab-created alternative to diamond, has a refractive index of 2.2. Estimate the speed of light through cubic zirconia.

12. Because surveyors cannot get inside a mountain, every mountain's height must be measured indirectly. Refer to the diagram at the right. Assume all labeled points lie in a single plane.

   a. Find the measures of $\angle ABD$ and $\angle ADB$ without trigonometry.

   b. Find BD.

   c. Find DC, the height of the mountain.

13. Fill in the blanks in the proof of the Law of Sines on page 702.

Fill in the blanks in the proof of the Law of Sines on page 702.

REVIEW

14. **Fill in the Blank** A function is periodic if its graph can be mapped onto itself under a ___?___. (**Lesson 10-6**)

15. Name three x-intercepts of the graph of $y = \cos x$. (**Lesson 10-6**)

Fill in the Blank In 16 and 17, complete each statement with a trigonometric expression to make the equation true for all θ. (**Lesson 10-5**)

16. $\cos^2 \theta +$ ___?___ $= 1$ 17. $\cos \theta =$ ___?___

18. Estimate to the nearest thousandth. (**Lessons 9-8, 9-7, 9-5**)

   a. $\log 49$ b. $\ln 49$ c. $\log_2 49$

EXPLORATION

19. Use a DGS to draw a circle with diameter $\overline{AB}$ and mark points $C, D, E,$ and F not equally spaced on the circle.

   a. Draw $\triangle ABC$ and measure to find $\frac{\sin A}{a}$, $\frac{\sin B}{b}$, and $\frac{\sin C}{c}$.

   b. Draw $\triangle DEF$ and measure to find $\frac{\sin D}{d}$, $\frac{\sin E}{e}$, and $\frac{\sin F}{f}$.

   c. What do you notice about your results in Parts a and b? What is the connection between the ratios and the length of the diameter?

20. In QY 2, you were asked to solve $\sin \theta = 0.5$ using its inverse function or the graph of $y = \sin x$.

   a. Solve $\sin \theta = 0.5$ on a CAS.

   b. One CAS gives the results at the right.
      $\theta = 360 \cdot (n1 + .416667)$ or
      $\theta = 360 \cdot (n1 + .083333)$
      Interpret what the CAS display means.

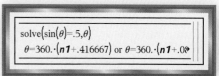

1. $\theta = 120°$

2. a. There are no solutions.
   b. $\theta = 30°$ or $150°$

Lesson 10-8
The Law of Cosines

▶ **BIG IDEA** Given SAS or SSS in a triangle, the Law of Cosines enables you to find the lengths of the remaining sides or measures of angles of the triangle.

The Law of Sines enables you to solve a triangle if you know two angles and a side of the triangle (the ASA or AAS conditions). However, if you know the measures of two sides and the included angle (the SAS condition) or the measures of three sides of a triangle (the SSS condition), the Law of Sines cannot be used to solve the triangle. Fortunately, you can find other measures in the triangle using the *Law of Cosines*.

Law of Cosines Theorem

In any triangle ABC,
$c^2 = a^2 + b^2 - 2ab \cdot \cos C$.

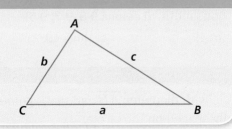

Mental Math

True or False

a. The length of your hair varies directly with the amount of time since your last haircut.

b. The length of your hair varies directly with the number of haircuts you've had this year.

c. The amount of money you spend on haircuts each year varies jointly with the price of each haircut and the number of haircuts you have.

Proof Set up $\triangle ABC$ on a coordinate plane so that $C = (0, 0)$ and $A = (b, 0)$. To find the coordinates of point B, notice that B is the image of $(\cos C, \sin C)$ under a size change of magnitude a. Thus, $B = (a \cos C, a \sin C)$. Recall the Pythagorean Distance Formula for the distance d between

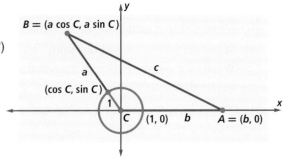

(x_1, y_1) and (x_2, y_2), $d = \sqrt{|x_2 - x_1|^2 + |y_2 - y_1|^2}$.
Use it to find c and square the result.

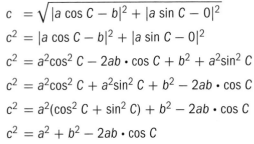

$c = \sqrt{|a \cos C - b|^2 + |a \sin C - 0|^2}$ Pythagorean Distance Formula

$c^2 = |a \cos C - b|^2 + |a \sin C - 0|^2$ Square both sides.

$c^2 = a^2\cos^2 C - 2ab \cdot \cos C + b^2 + a^2\sin^2 C$ Expand the binomials.

$c^2 = a^2\cos^2 C + a^2\sin^2 C + b^2 - 2ab \cdot \cos C$ Commutative Property of Addition

$c^2 = a^2(\cos^2 C + \sin^2 C) + b^2 - 2ab \cdot \cos C$ Distributive Property

$c^2 = a^2 + b^2 - 2ab \cdot \cos C$ Pythagorean Identity

The Law of Cosines applies to any two sides of a triangle and their included angle. So it is also true that in $\triangle ABC$,

$$a^2 = b^2 + c^2 - 2bc \cos A \text{ and } b^2 = a^2 + c^2 - 2ac \cos B.$$

In words the Law of Cosines says that in any triangle, the sum of the squares of two sides minus twice the product of these sides and the cosine of the included angle equals the square of the third side.

Using the Law of Cosines to Find a Length

With the Law of Cosines, finding the length of the third side of a triangle when two sides and the included angle are known requires only substitution.

Example 1

Two straight roads meet in Canton at a 27° angle. Anton is 7 miles down one road, and Banton is 8 miles down the other. How far apart are Anton and Banton?

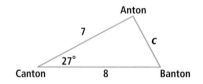

Solution 1 Because this is an SAS situation, use the Law of Cosines. Let c be the distance between Anton and Banton.

$$c^2 = a^2 + b^2 - 2ab \cos C$$

Substitute. $\quad c^2 = 8^2 + 7^2 - 2 \cdot 8 \cdot 7 \cdot \cos 27°$

$$c^2 \approx 13.207$$

$$c \approx \pm 3.63$$

Because lengths of sides of triangles must be positive numbers, only the positive solution makes sense in this situation. So $c \approx 3.63$. **Anton and Banton are about 3.6 miles apart.**

Solution 2 Enter the equation into a CAS and then take the square root of the result.

This display shows both exact and approximate solutions. They agree with Solution 1.

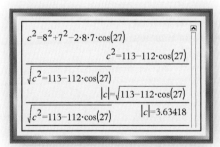

You can also use the Law of Cosines to find a length in triangles that meet the SsA condition, where the lengths of two sides are known, along with the measure of the angle opposite the longer side.

Example 2

In $\triangle ABC$, $c = 6$ cm, $a = 8$ cm, and $m\angle A = 30°$. Find b.

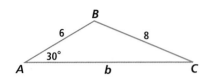

Solution This SsA situation indicates that you can use the Law of Cosines.

$$a^2 = b^2 + c^2 - 2bc \cos A$$
$$8^2 = b^2 + 6^2 - 2 \cdot 6 \cdot b \cdot \cos 30°$$

Solve on a CAS.

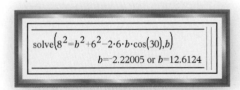

$$\text{solve}\left(8^2 = b^2 + 6^2 - 2 \cdot 6 \cdot b \cdot \cos(30), b\right)$$
$$b = -2.22005 \text{ or } b = 12.6124$$

Because side lengths must be positive numbers, $b \approx 12.6$ cm.

Using the Law of Cosines to Find an Angle

You can also use the Law of Cosines to find the measure of any angle in a triangle when you know the lengths of all three sides (the SSS condition).

GUIDED

Example 3

A city wants to build a grass-covered playground on a small triangle-shaped lot with boundaries of length 12 m, 14 m, and 20 m. Through what angle measure should an automatic sprinkler be set to water the grass if it is placed at the corner C?

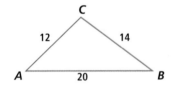

Solution In this SSS situation, you want to find $m\angle C$. Use the Law of Cosines.

Let $a = \underline{\ ?\ }$, $b = \underline{\ ?\ }$, and $c = \underline{\ ?\ }$.

$$c^2 = a^2 + b^2 - 2ab \cos C$$

Substitute. $$\underline{\ ?\ }^2 = \underline{\ ?\ }^2 + \underline{\ ?\ }^2 - 2 \cdot \underline{\ ?\ } \cdot \underline{\ ?\ } \cos C$$

$$\underline{\ ?\ } = \underline{\ ?\ } - \underline{\ ?\ } \cos C$$

Solve for $\cos C$. $\underline{\ ?\ } = \underline{\ ?\ } \cos C$

$$\cos C \approx \underline{\ ?\ }$$

Apply $\cos^{-1}$ to both sides. $m\angle C \approx \underline{\ ?\ }$

The sprinkler should be set to cover an angle of about $\underline{\ ?\ }$.

▶ **QY**

Check your solution to Example 3 by substituting a, b, c, and your solution for $m\angle C$ into $a^2 + b^2 - 2ab \cos C = c^2$.

STOP QY

The Law of Cosines ($a^2 + b^2 - 2ab\cos C = c^2$) is the Pythagorean Theorem ($a^2 + b^2 = c^2$) with an extra term, $-2ab\cos C$. Consider three different triangles:

- If $\angle C$ is acute, as in Example 1, then $\cos C$ is positive and the extra term, $-2ab\cos C$, is negative. So $c^2 < a^2 + b^2$.

- If $\angle C$ is obtuse, as in Example 3, then $\cos C$ is negative and the extra term, $-2ab\cos C$, is positive. So $c^2 > a^2 + b^2$.

- If $\angle C$ is a right angle, then the extra term, $-2ab\cos 90°$, is equal to 0, and $c^2 = a^2 + b^2$.

This shows that the Law of Cosines is a generalization of the Pythagorean Theorem.

The last two lessons can be summarized as follows. If you need to find a side or an angle of a triangle, use the simplest methods possible before using trigonometry. If you still have sides or angles to find:

- Use right-angle trigonometric ratios when the missing side or angle is part of a right triangle.

- Use the Law of Sines for triangles meeting the ASA or AAS conditions.

- Use the Law of Cosines for triangles meeting the SAS, SSS, or SsA conditions.

Questions

COVERING THE IDEAS

In 1–3, according to the Law of Cosines, what expression is equal to the following in $\triangle PQR$?

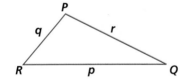

1. $p^2 + r^2 - 2pr\cos Q$ 2. $r^2 + q^2 - 2rq\cos P$ 3. r^2

4. **Multiple Choice** Which of the following describes the Law of Cosines?

   A The third side of a triangle equals the sum of the squares of the other two sides minus the product of the two sides and the included angle.

   B The square of the third side of a triangle equals the sum of the squares of the other two sides minus the product of the two sides and the cosine of the included angle.

   C The square of the third side of a triangle equals the sum of the squares of the other two sides minus twice the product of the two sides and the cosine of the included angle.

   D none of the above

5. In $\triangle ABC$, m$\angle A = 27.5°$, $AB = 10$, and $AC = x$. Write an expression for the length of BC.

6. Refer to Example 1. Danton is a town on the road between Canton to Banton, 5 miles from Canton, with a straight road connecting Danton and Anton. What is the direct distance from Danton to Anton?

7. Check the answer to Example 2 by following these steps.

   Step 1 Solve $\frac{\sin A}{a} = \frac{\sin C}{c}$ for m$\angle C$.

   Step 2 Find m$\angle B$.

   Step 3 Solve $\frac{\sin A}{a} = \frac{\sin B}{b}$ for b.

8. Refer to Guided Example 3. Find m$\angle B$.

9. For $\triangle ABC$, use the Law of Cosines to prove that, if $\angle C$ is acute, then $a^2 + b^2 > c^2$.

APPLYING THE MATHEMATICS

10. In $\triangle ABC$, m$\angle A = 53°$, m$\angle B = 102°$, and m$\angle C = 25°$. Explain why this triangle cannot be solved.

11. In $\triangle ABC$, use the Law of Cosines to get a formula for $\cos A$ in terms of a, b, and c.

12. An airplane flies the 261 miles from Albany, New York, to Buffalo, New York. It then changes its direction by turning 26° left and flies a distance of 454 miles to Chicago, Illinois. What is the direct distance from Albany to Chicago?

13. Refer to the triangle at the right.
    a. Find a.
    b. Find θ.

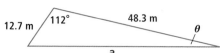

14. The diagonals of a rectangle are 40 centimeters long and intersect at an angle of 28.5°. How long are the sides of the rectangle?

15. The sides of a rectangle are 4 inches and 7 inches. Find the measure of the acute angle formed by the diagonals.

16. At a criminal trial, a witness gave the following testimony: "The defendant was 25 feet from the victim. I was 65 feet from the defendant and about 100 feet from the victim when the robbery occurred. I saw the whole thing."
    a. Draw a triangle to represent the situation.
    b. Use the Law of Cosines to show that the testimony has errors.
    c. How else could you know that the testimony has errors?

17. At a track meet, an electronic device measures the distance a discus travels to the nearest centimeter. The device is placed as shown in the diagram at the right. It first measures the distance p to the discus circle. After the athlete throws the discus, the device measures angle α and distance d and calculates the length w of the throw to the nearest centimeter. Suppose $p = 3.2$ m, $\alpha = 147.207°$ and $d = 47.40$ meters. How long was the throw?

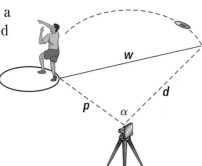

REVIEW

18. In $\triangle DOG$, $OG = 42$, $m\angle D = 118°$, and $m\angle G = 27°$. Find DO and DG. (Lesson 10-7)

19. Does the graph of the function $f(x) = \sin x$ have any lines of symmetry? If so, give an equation for one such line. (Lesson 10-6)

20. a. Is the relationship graphed at the right a function? Why or why not? (Lessons 10-6, 8-2, 1-2)
    b. Is the relation periodic? If so, what is the period?
    c. Is the inverse of the relation a function? Why or why not?

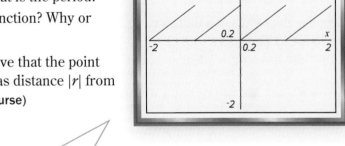

21. Use the Pythagorean Identity to prove that the point with coordinates $(r \cos \theta, r \sin \theta)$ has distance $|r|$ from the origin. (Lesson 10-5, Previous Course)

22. Find the area of the triangle at the right. (Lesson 10-1)

In 23 and 24, refer to circle O at the right. (Previous Course)

23. If $\theta = 85°$, what fraction of the circle's area is the area of the shaded sector?

24. If the length of $\overset{\frown}{AB}$ is $\frac{9}{24}$ of the circumference of the circle, find θ.

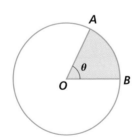

EXPLORATION

25. "You, who wish to study great and wonderful things, who wonder about the movement of the stars, must read these theorems about triangles. Knowing these ideas will open the door to all of astronomy and to certain geometric problems." This quotation is from *De Triangulis Omnimodis* by Regiomontanus. Find out more about this 15th-century mathematician and his work in trigonometry.

QY ANSWERS

$14^2 + 12^2 - 2 \cdot 14 \cdot 12$
$\cos 100° \approx 398 \approx 400 =$
20^2. It checks.

Lesson

10-9

Radian Measure

Vocabulary

radian

▶ **BIG IDEA** The *radian* is an alternate unit of angle measure defined so the length of an arc in a unit circle is equal to the radian measure of that arc.

So far in this chapter you have learned to evaluate $\sin x$, $\cos x$, and $\tan x$ when x is given in degrees. Angles and magnitudes of rotations may also be measured in *radians*. In calculus and some other areas of mathematics, radians are used more often than degrees.

What Is a Radian?

Because the radius of a unit circle is 1, the circumference of a unit circle is 2π. Thus, on a unit circle, a $360°$ arc has length 2π. Similarly, a $180°$ arc has length $\frac{1}{2}(2\pi) = \pi$, and a $90°$ arc has length $\frac{1}{4}(2\pi) = \frac{\pi}{2}$.

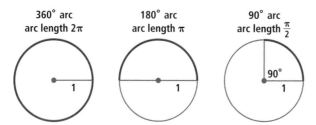

The radian is a measure created so that the arc *measure* and the arc *length* are the same number.

Definition of Radian

The **radian** is a measure of an angle, arc, or rotation such that π radians = 180 degrees.

Notice that a $180°$ arc on the unit circle has measure π radians, and this arc has length π. A $90°$ angle has measure $\frac{\pi}{2}$ radians, and its arc has length $\frac{\pi}{2}$. In general, *the measure of an angle, arc, or rotation in radians equals the length of its arc on the unit circle.*

 QY

Mental Math

Ronin just returned from a trip and has several kinds of currency in his wallet. Assume there are 1.5 U.S. dollars in 1 euro, 100 Japanese yen in 1 U.S. dollar, and 0.75 South African rand in 10 Japanese yen.

a. How many euros can Ronin get for his $36?

b. About how many euros can he get for his 1700 yen?

c. How many U.S. dollars can he get for his 120 rand?

▶ **QY**

What is the length of a $60°$ arc on the unit circle?

Conversion Factors for Degrees and Radians

The definition of radian can be used to create conversion factors for changing degrees into radians, and vice versa. Begin with the equation

$$\pi \text{ radians} = 180°.$$

Dividing each side by π radians gives $\dfrac{\pi \text{ radians}}{\pi \text{ radians}} = \dfrac{180°}{\pi \text{ radians}}$.

So, $$1 = \dfrac{180°}{\pi \text{ radians}}.$$

Similarly, dividing each side by 180° gives $\dfrac{\pi \text{ radians}}{180°} = \dfrac{180°}{180°}$,

so, $$\dfrac{\pi \text{ radians}}{180°} = 1.$$

> ### Conversion Factors for Degrees and Radians
>
> To convert radians to degrees, multiply by $\dfrac{180°}{\pi \text{ radians}}$.
>
> To convert degrees to radians, multiply by $\dfrac{\pi \text{ radians}}{180°}$.

You may be wondering how big a radian is. Example 1 gives an answer.

Example 1

Convert 1 radian to degrees.

Solution Because radians are given, multiply by the conversion factor with radians in the denominator.

$$\begin{aligned}
1 \text{ radian} &= 1 \text{ radian} \cdot \dfrac{180°}{\pi \text{ radians}} \\
&= \dfrac{180°}{\pi} \\
&\approx 57.3°
\end{aligned}$$

Notice that one radian is much larger than one degree.

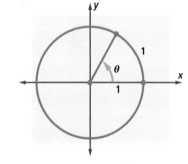

Because the measure of an angle in radians equals the length of its arc on the unit circle, the angle of 1 radian in Example 1 determines an arc of length 1. A unit circle has circumference $2\pi \approx 6.28$, so an arc length of 1 is a little less than $\frac{1}{6}$ of the circle's circumference. So the angle $\theta = 1$ radian measures a little less than $\frac{1}{6}$ of 360°, or a little less than 60°.

GUIDED

Example 2

a. Convert 45° to its exact radian equivalent.

b. Convert $\frac{2}{3}\pi$ radians to its exact degree equivalent.

Solution

a. Multiply 45° by one of the conversion factors. Because you want radians, choose the ratio with radians in the numerator.

$$45° \cdot \underline{\quad ? \quad} = \underline{\ ? \ } \text{ radians}$$

$$= \underline{\ ? \ } \text{ radian}$$

$$45° = \underline{\ ? \ } \text{ radian, exactly.}$$

b. Multiply $\frac{2}{3}\pi$ radians by one of the conversion factors. Because you want degrees, choose the ratio with degrees in the numerator.

$$\frac{2}{3}\pi \cdot \underline{\quad ? \quad} = \underline{\quad ? \quad}$$

$$= \underline{\ ? \ }$$

$$\frac{2}{3}\pi = \underline{\ ? \ }, \text{ exactly.}$$

Radian expressions are often left as multiples of π because this form gives an exact value. Usually in mathematics, the word *radian* or the abbreviation *rad* is omitted. In trigonometry, when no degree symbol or other unit is specified, we assume that the measure of the angle, arc, or rotation is radians.

$\theta = 2°$ means
  "the angle (or the arc or rotation) θ has measure 2 degrees."

$\theta = 2$ means
  "the angle (or the arc or rotation) θ has measure 2 radians."

Refer to Guided Example 2. Because $\frac{\pi}{4} = 45°$, you can conclude that $\frac{3\pi}{4} = 3 \cdot \frac{\pi}{4} = 3 \cdot 45° = 135°$.

Similarly, $\frac{5\pi}{4} = 5 \cdot 45° = 225°$.

The diagram at right shows some common equivalences of degrees and radians.

In general, the multiples of π and the simplest fractional parts of π ($\frac{\pi}{2}, \frac{\pi}{3}, \frac{\pi}{4}, \frac{\pi}{6}$, and their multiples) correspond to those angle measures that give exact values of sines, cosines, and tangents.

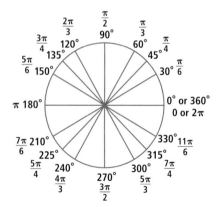

Trigonometric Values in Radians

Every scientific calculator can evaluate sin θ, cos θ, and tan θ, where θ is in radians.

Example 3

Set your calculator to radian mode. Evaluate

a. $\cos 2.$ b. $\tan \frac{3\pi}{4}.$

Solution

a. Enter cos __?__. You should find $\cos$ __?__ $\approx$ __?__.

b. Enter tan __?__. You should find $\tan$ __?__ $=$ __?__.

Check

a. 1 radian $\approx$ __?__ °, so 2 radians $\approx$ __?__ °.

$\cos$ __?__ ° $\approx$ __?__. It checks.

b. $\frac{3\pi}{4}$ radians $=$ __?__ °, and $\tan$ __?__ ° $=$ __?__. It checks.

Graphs of the Sine and Cosine Functions Using Radians

The cosine and sine functions are graphed below with x measured in radians rather than degrees. Notice that each function is (still) periodic, but that each period is 2π rather than 360°.

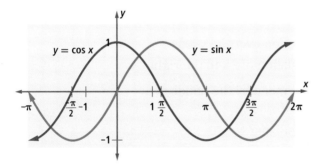

Questions

1. A circle has a radius of 1 unit. Give the length of an arc with the given measure.

   a. 360° b. 180° c. 90° d. 110°

2. **Fill in the Blanks**

   a. π radians $=$ __?__ °

   b. 1 radian $=$ __?__ °

   c. __?__ radian(s) $= 1°$

In 3–6, convert the radian measure to degrees.

3. 8π 4. $\frac{11\pi}{2}$ 5. $\frac{14\pi}{45}$ 6. $-\frac{5\pi}{4}$

In 7–10, convert to radians. Give your answer as a rational number times π.

7. 90°   **8.** 15°   **9.** 225°   **10.** 330°

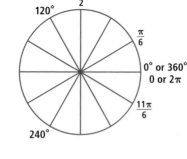

11. a. Explain the different meanings of sin 6 and sin 6°.

  b. Evaluate sin 6 and sin 6°.

12. Six equally-spaced diameters are drawn on a unit circle, as shown at the right. Copy and complete the diagram, giving equivalent measures in degrees and radians at the end of each radius.

In 13 and 14,
   a. evaluate on a calculator in radian mode.
   b. check your answer using degrees.

13. $\cos \frac{3\pi}{2}$ **14.** $\tan \frac{\pi}{6}$

In 15 and 16, suppose x is in radians.

15. What is the period of the function $y = \sin x$?

16. Name three x-intercepts of the function $y = \cos x$ on the interval $-2\pi < x < 2\pi$.

APPLYING THE MATHEMATICS

17. Multiple Choice Suppose x is in radians. Which transformations map the graph of $y = \cos x$ onto itself?

  A reflection over the x-axis   **B** reflection over the y-axis
  C translation of π to the right   **D** translation of 2π to the right

18. What is the measure of the obtuse angle made by the hands of a clock at 1:30

  a. in degrees?   **b.** in radians?

In 19 and 20, find the exact values.

19. $\sin\left(\frac{9\pi}{2}\right)$ **20.** $\cos\left(\frac{11\pi}{4}\right)$

In 21–23, use the following relationship between radian measure and arc length: In a circle of radius r, a central angle of x radians has an arc of length rx.

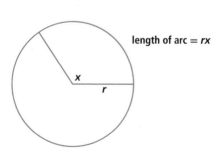

length of arc = rx

21. a. How long is the arc of a $\frac{\pi}{4}$-radian angle in a circle of radius 10?

  b. How long is a 45° arc in a circle of radius 20?

22. On a circle of radius 2 meters, find the length of a 75° arc.

23. How long is the arc of a $\frac{2\pi}{3}$-radian angle in a circle of radius 8 feet?

REVIEW

24. In $\triangle ABC$, m$\angle B = 16°$, $b = 10$, and $c = 24$. Explain why there are two possible measures of $\angle C$, and find both of them. (**Lesson 10-8**)

25. Suppose $0° < \theta < 180°$. Solve $\sin \theta = 0.76$. (**Lesson 10-7**)

26. In $\triangle MAP$, $m = 22$, m$\angle M = 149°$, and m$\angle P = 23°$. Find the lengths of p and a. (**Lesson 10-7**)

In 27 and 28, evaluate the expression without using a calculator. (**Lesson 10-5**)

27. $(\sin 450°)^2 + (\cos 450°)^2$ **28.** $\tan 135°$

29. The newspaper article at the right is from the *Detroit Free Press*, January 29, 1985. Using the drawing below, explain why the construction technique leads to an angle of 26.5°. (**Lesson 10-2**)

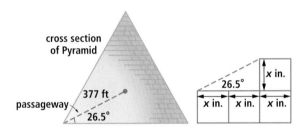

30. One of Murphy's Laws says that the relative frequency r of rain is inversely proportional to the diameter d of the umbrella you are carrying. (**Lesson 2-2**)

   a. Write an equation relating r and d.

   b. If the probability of rain is $\frac{1}{4}$ when you are carrying a 46″-diameter umbrella, what is the probability when you are carrying a 36″-diameter umbrella?

EXPLORATION

31. When x is in radians, $\sin x$ can be estimated by the formula

$$\sin x = x - \frac{x^3}{6} + \frac{x^5}{120} - \frac{x^7}{5040}.$$

   a. How close is the value of this expression to $\sin x$ when $x = \frac{\pi}{4}$?

   b. To get greater accuracy, you can add $\dfrac{x^9}{362,880}$ to the value you got in Part a. Where does the 362,880 come from?

   c. Add $\dfrac{\left(\frac{x}{4}\right)^9}{362,000}$ to your answer to Part a. How close is this value to $\sin \frac{\pi}{4}$?

Astronomer Solves Riddle of Pyramid

United Press International

   A Navy astronomer has devised a surprisingly simple explanation for the angle of a descending passageway in the Great Pyramid of Cheops in Egypt.

   In the early 19th century, English astronomer John Herschel suggested the 377–foot–long passageway was built at its angle of 26.5230 degrees to point at the North Star, making the pyramid an astronomical observatory as well as a tomb for Cheops.

   Richard Walker, a U.S. Naval Observatory astronomer based at Flagstaff, Ariz., checked Herschel's idea and found that, because of the wobble of the Earth's axis in its orbit around the sun, no prominent star could have been seen from the base of the passageway built in 2800 BC when the pyramid was built.

   Then why was the passageway inclined at an angle of 26.5 degrees?

   According to Walker's report, the angle merely was the result of the construction technique.

   By placing three stones of equal length horizontally and then placing a fourth stone of equal size on the top of the third horizontal stone, Walker determined that the angle from the top stone to the bottom stone at the other end is 26.5 degrees.

QY ANSWER

$\dfrac{\pi}{3}$

Chapter 10 Projects

The roof of the Kresge Auditorium on the Massachusetts Institute of Technology campus is a spherical triangle.

1 Area Under a Sine Curve

a. Draw a graph of the equation $y = \sin x$ from $x = 0$ to $x = \pi$ radians on graph paper. Let one gridline equal 0.1 unit on each axis.

b. Using the scale of your graph, what is the area of each square?

c. How many whole squares are between the sine curve and the x-axis? Estimate the number of whole squares you can make from the remaining partial squares. Add these to estimate the total number of squares between the sine curve and the x-axis.

d. Calculate $\frac{\text{area}}{\text{square}} \cdot$ (number of squares) to estimate the total area under the graph of $y = \sin x$ from $x = 0$ to $x = \pi$. The final answer should be surprisingly simple.

e. Predict the area under the curve $y = \cos x$ from $x = 0$ to $x = \frac{\pi}{2}$. Devise a method to test your prediction and carry it out.

f. Summarize what you found.

2 Spherical Trigonometry

When three great circles on a sphere intersect at 3 different points, they determine regions called *spherical triangles*. Research the Law of Sines and Law of Cosines for spherical triangles.

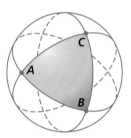

3 Amplitudes, Periods, and the Tangent Function

Set a graphing utility to radian mode.

a. Graph $y = A\sin(x)$ and $y = A\cos(x)$ for several values of A. How does the value of A appear to affect the graphs of these functions?

b. Choose a value of A and graph $y = A\sin(Bx)$ and $y = A\cos(Bx)$ for $B = \{-2\pi, -\pi, 0, \pi, 2\pi\}$. How does the value of B appear to affect the graphs of these functions?

c. Graph $y = \tan(x)$. Determine the period. Describe all asymptotes.

d. Do your conjectures in Parts a and b apply to the graph of $y = A\tan(Bx)$? Explain your answer.

4 The Pendulum Swings

a. Using a motion detecting device, record the distance between a pendulum and the device at different times.

b. From this data, determine a curve to model the situation. What is the period of the pendulum swing?

c. Repeat the process for pendulums of different lengths. How does the period depend on the pendulum length?

5 Benjamin Banneker

Benjamin Banneker was one of the first African-American mathematicians. Among other things, he was responsible for surveying Washington, D.C., when it was first being developed. Find out more about Banneker and how he used mathematics, particularly trigonometry, in his work.

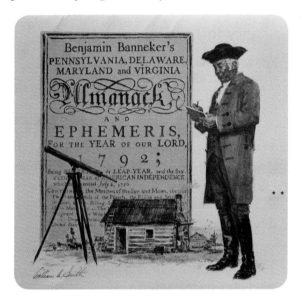

6 The Great Trigonometric Survey

In the 19th century a trigonometric survey was commissioned in India. Initially, William Lambton and George Everest participated in the work. Research this significant mapping project and find out how the Himalayan Mountains were measured. Summarize your findings in a report.

Chapter 10 Summary and Vocabulary

▶ Trigonometry is the study of relationships between sides and angles in triangles. In a right triangle, three important trigonometric ratios are the **sine**, **cosine**, and **tangent** of an acute angle θ, defined as follows for $0° < \theta < 90°$:

$$\sin \theta = \frac{\text{leg opposite } \theta}{\text{hypotenuse}};$$

$$\cos \theta = \frac{\text{leg adjacent to } \theta}{\text{hypotenuse}};$$

$$\tan \theta = \frac{\text{leg opposite } \theta}{\text{leg adjacent to } \theta}.$$

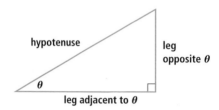

▶ The sine, cosine, and tangent ratios are frequently used to find lengths in situations involving right triangles. Angle measures are found using inverses of the trigonometric functions: $\sin^{-1}$, $\cos^{-1}$, and $\tan^{-1}$. Applications include finding **angles of elevation**, **depression**, and **parallax**.

▶ The trigonometric ratios can be generalized to find sines, cosines, and tangents for any real number θ. Every point on a **unit circle** is a rotation image of the point $(1, 0)$ about the origin with magnitude θ. $\cos \theta$ is the x-coordinate of $R_\theta (1, 0)$, and $\sin \theta$ is the y-coordinate of $R_\theta (1, 0)$.

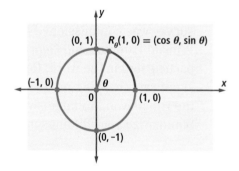

Vocabulary

10-1
right-triangle definitions of sine (sin), cosine (cos) and tangent (tan)
*sine function
*cosine function
*tangent function
*angle of elevation

10-2
*inverse sine function, $\sin^{-1}$
*inverse cosine function, $\cos^{-1}$
*inverse tangent function, $\tan^{-1}$
*angle of depression

10-3
*parallax angle

10-4
*unit circle
unit-circle definition of cosine and sine

10-5
identity
tangent of θ (for all values of θ)

10-6
*periodic function, period
*sine wave
sinusoidal

10-7
*solving a triangle

10-9
*radian

● The mappings $\theta \to \cos \theta$ and $\theta \to \sin \theta$ are functions whose domains are the set of real numbers and whose ranges are $\{y\mid -1 \leq y \leq 1\}$. When θ is in degrees, the graphs of these functions are **sine waves** with **period** $360°$. When θ is in **radians**, the period is 2π, because radians are defined such that π radians $= 180°$.

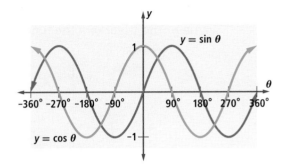

● The Law of Cosines and the Law of Sines relate sides and measures of angles in triangles. In any triangle ABC,

$$\frac{\sin A}{a} = \frac{\sin B}{b} = \frac{\sin C}{c} \text{ (Law of Sines)}$$

$$c^2 = a^2 + b^2 - 2ab \cos C \text{ (Law of Cosines)}.$$

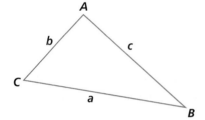

These theorems can be used to **solve a triangle**, that is, to find unknown sides and angle measures in triangles. The Law of Cosines is most useful when an SAS, SSS, or SsA condition is given; the Law of Sines can be used in all other situations that determine triangles.

Theorems and Properties

Pythagorean Identity Theorem (p. 688)
Supplements Theorem (p. 701)
Law of Sines Theorem (p. 702)
Law of Cosines Theorem (p. 706)

Chapter

10 Self-Test

Take this test as you would take a test in class. You will need a calculator. Then use the Selected Answers section in the back of the book to check your work.

1. Let $\theta = 17.4°$. Approximate to the nearest thousandth.

   a. $\cos \theta$ **b.** $\sin \theta$

2. Suppose $0 < \theta < 90°$ and $\tan \theta = 0.64$. Approximate θ to the nearest thousandth of a degree.

3. Use the triangle below. Evaluate.

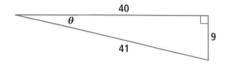

   a. $\cos \theta$ **b.** $\tan \theta$

4. If $\sin \theta = 0.280$ and $90° < \theta < 180°$, what is $\cos \theta$?

5. **a.** What are the exact coordinates of $R_{-423}(1, 0)$?

   b. Justify your answer to Part a.

6. For what value of x such that $0° < x \le 180°$ and $x \ne 17°$ does $\sin 17° = \sin x$?

In **7** and **8**, refer to the unit circle below. Name the letter that could be equal to the value of the trigonometric function.

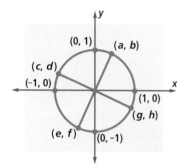

7. $\cos 70°$ 8. $\sin 250°$

9. Ten minutes ago, Robin Hood spied the Sheriff of Nottingham 250 meters down the road. In order to sneak around, Robin left the road at a 35° angle and traveled 175 meters into the forest as shown below. If the Sheriff has not moved, how far is Robin Hood from the Sheriff now, to the nearest meter?

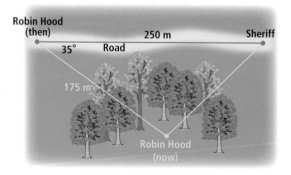

10. In $\triangle SLR$, $m\angle L = 12°$, $s = 425$, and $\ell = 321$. Approximate $m\angle S$ to the nearest 0.1 degree.

In 11 and 12, consider the function graphed below.

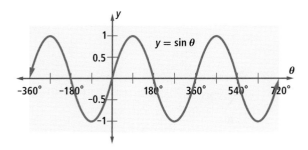

11. What is the period of this function?

12. Name 2 intervals on which the function increases as θ increases.

13. A parallelogram has sides of length 20 and 30. If the shorter diagonal has length 15, find the measures of the angles of the parallelogram.

14. The star Betelgeuse has a parallax angle of about 0.008 arc-second when viewed from the endpoints of Earth's orbit. How many light-years away is it? Use the facts that 1 arc-second is $\frac{1}{3600}$ degree, 1 light-year is $5.88 \cdot 10^{12}$ miles, and Earth's orbit has a radius of about 93,000,000 miles.

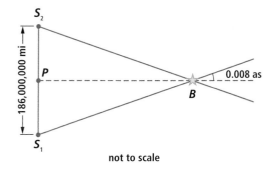

not to scale

15. Convert to radians. Give your answer as a rational number times π.

   a. 60°   **b.** 25°

16. Convert to degrees. Round your answer to the nearest tenth.

   a. 4π radians   **b.** $\frac{\pi}{7}$ radian

17. A photographer is in a hot air balloon 1500 feet above the ground. She tilts her camera down 36° from horizontal to aim it at an African elephant.

   a. How far away from the elephant is the photographer?

   b. How far away from the elephant is her assistant, who is standing directly beneath the balloon?

Chapter 10 · Chapter Review

SKILLS
PROPERTIES
USES
REPRESENTATIONS

SKILLS Procedures used to get answers

OBJECTIVE A Approximate values of trigonometric functions using a calculator. (Lessons 10-1, 10-9)

In 1–6, evaluate to the nearest thousandth.

1. $\sin 34°$
2. $\cos^2 125°$
3. $\sin\left(-\dfrac{\pi}{8}\right)$
4. $\cos 3$
5. $\sin\left(\dfrac{8\pi}{5}\right)$
6. $\tan 167°$

OBJECTIVE B Determine the measure of an angle given its sine, cosine, or tangent. (Lessons 10-2, 10-7)

In 7–10, find all θ between 0° and 180° satisfying the given equation.

7. $\sin \theta = 0.5$
8. $\cos \theta = \dfrac{\sqrt{2}}{2}$
9. $\cos \theta = 0$
10. $\tan \theta = 1$

OBJECTIVE C Find missing side lengths and angle measures of a triangle using the Law of Sines or the Law of Cosines. (Lessons 10-7, 10-8)

In 11–15, use the Law of Sines or the Law of Cosines to solve for the variable. Round your answers to the nearest tenth.

11.

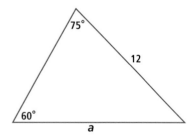

12.

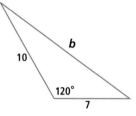

13.

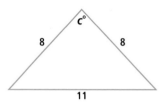

14.

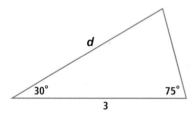

15.

OBJECTIVE D Convert angle measures from radians to degrees or from degrees to radians. (Lesson 10-9)

In 16–19, convert to radians. Express your answers in terms of π.

16. $30°$
17. $-105°$
18. $360°$
19. $405°$

In 20–23, convert the radian measure to degrees. Round your answers to the nearest tenth.

20. π
21. $\dfrac{3\pi}{2}$
22. $-\dfrac{\pi}{8}$
23. $\dfrac{7\pi}{6}$

PROPERTIES Principles behind the mathematics

OBJECTIVE E Identify and use the definitions of sine, cosine, and tangent. (Lessons 10-1, 10-4, 10-5)

Multiple Choice In 24–27, use the diagram below. Identify the given trigonometric function as one of the following ratios.

A $\dfrac{AB}{BC}$ B $\dfrac{AC}{BC}$

C $\dfrac{AB}{AC}$ D $\dfrac{AC}{AB}$

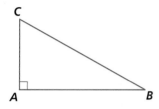

24. $\sin B$ 25. $\tan B$

26. $\cos C$ 27. $\tan C$

In 28–30, use the triangle below. Evaluate each trigonometric function.

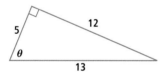

28. $\sin \theta$ 29. $\cos \theta$ 30. $\tan \theta$

31. **True or False** $\sin \dfrac{4\pi}{7} = \tan \dfrac{4\pi}{7} \cdot \cos \dfrac{4\pi}{7}$. Justify your answer without a calculator.

32. **Multiple Choice** What is the image of $(1, 0)$ under R_{-600}?

A $(-\sin 600°, -\cos 600°)$

B $(-\cos 600°, -\sin 600°)$

C $(\sin -600°, \cos -600°)$

D $(\cos -600°, \sin -600°)$

33. **Fill in the Blanks** Because $R_{253}(1, 0) \approx$ $(-0.292, -0.956)$, $\cos 253° \approx$ __?__ and $\sin 253° \approx$ __?__.

OBJECTIVE F Identify and use theorems relating sines and cosines. (Lessons 10-5, 10-7).

34. **True or False** For all real numbers θ, $\sin \theta - \sin(180° - \theta) = 0$. Justify your answer.

35. **True or False** For all real numbers θ, $\cos \theta - \cos(180° - \theta) = 0$. Justify your answer.

36. Suppose $0° \leq \theta < 90°$. If $\sin \theta = \dfrac{3}{5}$, use the Pythagorean Identity to determine $\cos \theta$.

37. Suppose $\sin x = 0.25$. What are all possible values of $\cos x$?

USES Applications of mathematics in real-world situations

OBJECTIVE G Solve real-world problems using the trigonometry of right triangles. (Lessons 10-1, 10-2, 10-3)

38. A tower is anchored by a 250-foot guy wire attached to the tower at a point 150 feet above the ground. What is the angle of elevation of the wire, to the nearest degree?

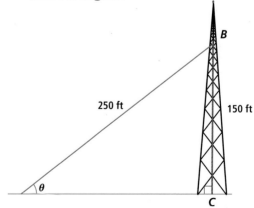

39. A wheelchair ramp is built with slope $\frac{1}{12}$. To the nearest tenth of a degree, what angle does the ramp make with the horizontal?

40. A ship sails 560 kilometers at an angle 35° clockwise from north. How far east of its original position is the ship?

41. Neil and Buzz are exploring the Moon. At one moment, the Sun is shining directly over Neil. Meanwhile, Buzz has just set up a 10-meter tall flagpole 150 kilometers away from Neil. The flagpole casts a shadow that is 86 centimeters long. Using this information, compute the circumference of the Moon to the nearest 100 kilometers.

OBJECTIVE H Solve real-world problems using the Law of Sines or Law of Cosines. (Lessons 10-7, 10-8)

42. Observers in two ranger stations 8 miles apart spot a fire. The observer in station A spots the fire at an angle of 44° with the line between the two stations, while the observer in station B spots the fire at a 105° angle with the same line. Which station is closer to the fire, and how far away is the fire from this station?

43. Pictured below is the top view of a chandelier with 10 spokes equally spaced around a central point in the same plane. If each spoke is 50 cm long, what is the perimeter of the chandelier?

50 cm

44. Two observers are in lighthouses 25 miles apart, as shown below. The observer in lighthouse A spots a ship in distress at an angle of 20° with the line between the lighthouses. The observer in lighthouse B spots the ship at an angle of 30° with that line. How far is the ship from each lighthouse?

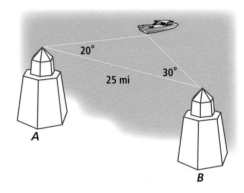

20°
30°
25 mi
A
B

45. The White House, the Washington Monument, and the Lincoln Memorial form the triangle shown below. To the nearest degree, find the angle θ between the Lincoln Memorial and the White House at the Washington Monument.

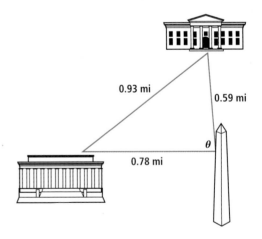

0.93 mi
0.59 mi
θ
0.78 mi

REPRESENTATIONS Pictures, graphs, or objects that illustrate concepts

OBJECTIVE I Use the properties of a unit circle to find values of trigonometric functions. (Lessons 10-4, 10-5)

In 46–51, use the fact that
$$\sin 15° = \cos 75° = \frac{\sqrt{6} - \sqrt{2}}{4},$$
and $\sin 75° = \cos 15° = \frac{\sqrt{6} + \sqrt{2}}{4}$.

Evaluate without a calculator. Draw a picture to explain your result.

46. $\sin 345°$ **47.** $\cos 105°$

48. $\sin 285°$ **49.** $\cos 165°$

50. $\sin(-15°)$ **51.** $\cos(-435°)$

In 52 and 53, use the diagram below.

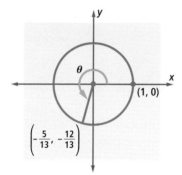

52. What is the value of $\sin \theta$?

53. Find θ to the nearest radian.

In 54–56, use the unit circle below. Which letter could stand for the given number?

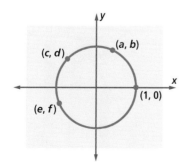

54. $\sin 67°$ **55.** $\cos 560°$

56. $\sin(-222°)$

OBJECTIVE J Identify properties of the sine and cosine functions using their graphs. (Lesson 10-6)

57. a. Graph the sine function on the interval $0° \leq \theta \leq 360°$.

    b. State the domain and range of the sine function.

58. a. Graph the cosine function on the interval $-2\pi \leq \theta \leq 2\pi$.

    b. At what points does the graph of the cosine function intersect the x-axis?

    c. What is the period of the cosine function?

59. The graph of $y = \cos x$ is the image of the graph of $y = \sin x$ under what translation?

In 60 and 61, use the graph below.

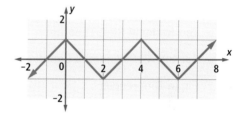

60. What is the period of this graph?

61. Is this graph a sine wave? Why or why not?

Chapter

11 Polynomials

Contents

The shape of the roller coaster track in the picture below cannot be modeled by any of the types of functions you have studied so far. However, using techniques similar to those you used to fit quadratic models to data, an equation can be found for a curve that approximates the track.

Let the origin be the left most point of the roller coaster in the photo, and let $H(x)$ be the height of the track in the picture at a horizontal distance x from the origin. Then the track as seen in this picture can be modeled by

$$H(x) = 0.001x^4 + 0.010x^3 + 0.060x^2 - 0.564x - 0.011.$$

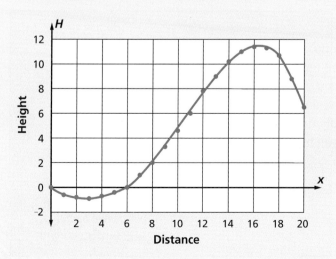

The expression on the right side of this equation is a *polynomial* and the equation is a *polynomial equation*. The function H described by this equation is a *polynomial function*. For any finite set of points, no two of which are on the same vertical line, there is a polynomial function whose graph contains those points. In this chapter, you will study situations that lead to polynomial functions. You will learn how to graph and analyze them, and how to describe data with a polynomial model.

Lesson

11-1

Introduction to Polynomials

Vocabulary

degree of a polynomial

term of a polynomial

polynomial in x of degree n

standard form of a polynomial

coefficients of a polynomial

leading coefficient

constant term, constant

polynomial function

▶ **BIG IDEA** *Polynomials* are a common type of algebraic expression that arise from many kinds of situations, including those of multiple investments compounded over different lengths of time.

You are likely to have studied polynomials in a previous course. This lesson reviews some of the terminology that is used to describe them.

Vocabulary Used with Polynomials

The expression

$$-0.001x^4 + 0.010x^3 + 0.060x^2 - 0.564x - 0.011$$

from the previous page is a *polynomial in the variable x*. When the polynomial is in only one variable, the largest exponent of the variable is the **degree of the polynomial**. The polynomial above has degree 4. The expressions $-0.001x^4$, $0.010x^3$, $0.060x^2$, $-0.564x$, and -0.011 are the **terms of the polynomial**. A polynomial is the sum of its terms.

Definition of Polynomial in x of Degree n

A **polynomial in x of degree n** is an expression of the form $a_nx^n + a_{n-1}x^{n-1} + a_{n-2}x^{n-2} + \dots + a_1x^1 + a_0$, where n is a nonnegative integer and $a_n \neq 0$.

The **standard form** of the general nth-degree polynomial is the one displayed in the definition. Notice that the terms are written in descending order of exponents. The numbers a_n, a_{n-1}, a_{n-2}, ..., a_0 are the **coefficients of the polynomial**, with **leading coefficient** a_n. The number a_0 is the **constant term**, or simply the **constant**. For instance, the standard form of a 4th-degree polynomial is

$$a_4x^4 + a_3x^3 + a_2x^2 + a_1x^1 + a_0.$$

It has coefficients a_4, a_3, a_2, a_1, and a_0, with leading coefficient a_4.

 QY1

Mental Math

Give an example of the following or say that it does not exist.

a. a number without a multiplicative inverse

b. a matrix without an inverse

c. a relation without an inverse

d. a function whose inverse is not a function

▶ **QY1**

What is the constant term of the 4th-degree polynomial $a_4x^4 + a_3x^3 + a_2x^2 + a_1x^1 + a_0$?

GUIDED

Example 1

a. List all the coefficients of $x^2 - 16x^4 + 3x^2 + 96$.

b. What are the degree and leading coefficient of $x^2 - 16x^4 + 3x^2 + 96$?

c. Rewrite the polynomial in standard form.

Solution

a. First, combine like terms.

$$x^2 - 16x^4 + 3x^2 + 96 = (\underline{\ ?\ } + \underline{\ ?\ })x^2 - \underline{\ ?\ }x^4 + \underline{\ ?\ }$$
$$= \underline{\ ?\ }x^2 - \underline{\ ?\ }x^4 + \underline{\ ?\ }$$

So, by the definition on the previous page, $a_4 = \underline{\ ?\ }$, $a_3 = \underline{\ ?\ }$, $a_2 = \underline{\ ?\ }$, $a_1 = \underline{\ ?\ }$, and $a_0 = \underline{\ ?\ }$.

b. The largest exponent is $\underline{\ ?\ }$. So, the degree of the polynomial is $\underline{\ ?\ }$. The coefficient of the x^4 term is $\underline{\ ?\ }$. Thus, the leading coefficient is $\underline{\ ?\ }$.

c. In standard form, the terms are in descending order. So, the standard form of this polynomial is

$$\underline{\ ?\ }x^4 + \underline{\ ?\ }x^2 + \underline{\ ?\ }.$$

Check Enter $x^2 - 16x^4 + 3x^2 + 96$ into a CAS. It automatically puts the polynomial in standard form.

Polynomials can be classified by their degree. Those of degree 1 through 5 have special names.

Degree	Polynomial Name	Example
1	Linear	$mx + b$
2	Quadratic	$ax^2 + bx + c$
3	Cubic	$ax^3 + bx^2 + cx + d$
4	Quartic	$ax^4 + bx^3 + cx^2 + dx + e$
5	Quintic	$ax^5 + bx^4 + cx^3 + dx^2 + ex + f$

You can think of nonzero constants such as 5, $\pi + 2$, or a_0 as polynomials of degree 0. This is because a constant k can be written $k \cdot x^0$, which is a polynomial of degree 0. However, the constant 0 is not assigned a degree because its leading coefficient is zero.

Polynomial Functions and Graphs

A **polynomial function** is a function of the form $P: x \rightarrow P(x)$, where $P(x)$ is a polynomial. Polynomial functions of degree 1 are the linear functions and have graphs that are lines.

Polynomial functions of degree 2 are the quadratic functions and have graphs that are parabolas. For linear and quadratic functions, the coefficients of the polynomial help identify key points or properties of the graph such as slope, vertices, or intercepts. For polynomials of higher degree, the connection between the polynomial's coefficients and its graph is not as simple.

Example 2

Consider the polynomial function P with equation
$P(x) = x^4 + 8x^3 + 20x^2 + 16x$.

a. What is $P(-1)$?

b. Graph this function in the window $-5 \leq x \leq 2, -5 \leq y \leq 4$.

Solution

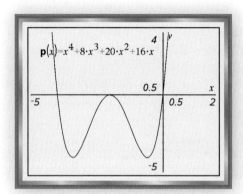

a. Substitute -1 for x.
$$P(-1) = (-1)^4 + 8(-1)^3 + 20(-1)^2 + 16(-1)$$
$$= 1 - 8 + 20 - 16$$
$$= -3$$

b. A graph of P is shown at the right. The curve is related to the graph of $y = x^4$. The extra terms are responsible for the waves and translation off of the origin.

You will further explore relationships between coefficients and graphs in Lesson 11-5.

Operations on Polynomials and Polynomial Functions

Sums, differences, products, and powers of polynomials are themselves polynomials. The degree of the result of operations with polynomials depends on the degrees of the polynomials.

Activity 1

MATERIALS CAS

Let $f(x) = 2x^2 + 3x + 4$ and $g(x) = 5x^3 + 1$.

Step 1 Evaluate each expression. Write your answer in standard form and give the degree.

    a. $f(x) + g(x)$ **b.** $f(x) - g(x)$

Step 2 Without multiplying the polynomials, predict

    a. the degree of $f(x) \cdot g(x)$.

    b. the leading coefficient of $f(x) \cdot g(x)$.

Step 3 a. Define the functions f and g on a CAS. Expand $f(x) \cdot g(x)$ and write the result in standard form.

b. Check your answers to Parts a and b of Step 2.

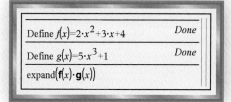

Define $f(x)=2 \cdot x^2+3 \cdot x+4$ *Done*

Define $g(x)=5 \cdot x^3+1$ *Done*

expand$(f(x) \cdot g(x))$

Step 4 Generalize the results of Steps 2 and 3:

The degree of the product of two polynomials is the ___?___ of the degrees of the polynomials. The leading coefficient of the product of two polynomials is the ___?___ of the leading coefficients of the polynomials.

The composite of two polynomial functions is a polynomial function.

Activity 2

MATERIALS CAS

Use the polynomials $f(x)$ and $g(x)$ from Activity 1.

Step 1 Write an unsimplified expression for $f(g(x))$ and use it to predict

a. the degree of $f(g(x))$.

b. the leading coefficient of $f(g(x))$.

Step 2 a. Expand $f(g(x))$ on a CAS and write the result in standard form.

b. Check your answers to Parts a and b of Step 1.

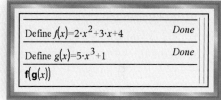

Define $f(x)=2 \cdot x^2+3 \cdot x+4$ *Done*

Define $g(x)=5 \cdot x^3+1$ *Done*

$f(g(x))$

Step 3 Generalize the results of Steps 1 and 2.

GUIDED

Example 3

Let $f(x) = 3x^2 + 2x + 1$ and $g(x) = 4x - 5$. Predict

a. the degree of $f(x) \cdot g(x)$.

b. the leading coefficient of $f(x) \cdot g(x)$.

c. the degree of $f(g(x))$.

d. the leading coefficient of $f(g(x))$.

Solution

a. Because the degree of f is 2 and the degree of g is 1, **the degree of** $f(x) \cdot g(x)$ is ___?___ + ___?___ = ___?___.

b. Because the leading coefficient of f is 3 and the leading coefficient of g is 4, **the leading coefficient of** $f(x) \cdot g(x)$ is ___?___ · ___?___ = ___?___.

c. Because the degree of f is 2 and the degree of g is 1, **the degree of** $f(g(x))$ is ___?___ · ___?___ = ___?___.

d. Because the leading coefficient of f is 3, the leading coefficient of g is 4, and the degree of f is 2, **the leading coefficient of** $f(g(x))$ is ___?___ · ___?___^{2} = ___?___.

Savings and Polynomials

Recall that the total value over time of a single investment compounded at a constant rate can be described by an exponential function. If, instead, you have multiple investments compounded over different lengths of time, then the total value can be described by a polynomial function. Example 4 illustrates such a situation.

Example 4

Anita Loan and her family start saving for college when she graduates from eighth grade. At the end of each summer, they deposit money into a savings plan with an annual percentage yield (APY) of 5.8%. Anita is planning to go to college in the fall following her graduation from high school. How much money will be in the account when she leaves for college, if no other money is added or withdrawn?

End of Summer after:	Amount Deposited ($)
8th grade	1200
9th grade	850
10th grade	975
11th grade	1175
12th grade	1300

Solution The money deposited at the end of the summer after 8th grade earns interest for 4 years, so it is worth $1200(1.058)^4$ when Anita goes to college. Similarly, the amount deposited at the end of the summer after 9th grade is worth $850(1.058)^3$ because it only earns interest for 3 years. Adding the values at the end of each summer gives the total amount that will be in Anita's account. Notice that, because the last deposit does not earn any interest, the last term is not multiplied by a power of 1.058.

$$1200(1.058)^4 \quad + \quad 850(1.058)^3 \quad + \quad 975(1.058)^2 \quad + \quad 1175(1.058)^1 \quad + \quad 1300$$

| End of summer after 8th grade | End of summer after 9th grade | End of summer after 10th grade | End of summer after 11th grade | End of summer after 12th grade |

Evaluating this expression shows that Anita will have about $6144.74 in her account when she leaves for college.

In Example 4, you could replace 1.058 with x. Then when Anita goes to college she will have (in dollars)

$$1200x^4 + 850x^3 + 975x^2 + 1175x + 1300.$$

If you let $x = 1 + r$, where r is the APY, evaluating this expression gives the amount in the account for any APY. You could use the expression to compare the total savings for different rates, or to compute the rate required to obtain a certain total. Since the first deposit earns interest for 4 years, the polynomial has degree 4.

 QY2

> ▸ **QY2**
>
> How much would Anita have if the APY in the account is 4.95%?

Questions

COVERING THE IDEAS

In 1–3, tell whether the expression is a polynomial. If it is, state its degree and its leading coefficient. If it is not a polynomial, explain why not.

1. $17 + 8y$ 2. $14x^3 + 5x^{-1}$ 3. $14x^3 + 12x^2 + 6x^5 + 3$

4. Refer to the definition of a polynomial of degree n. State each value for the polynomial $-x^6 + 16x^4 + 3x^3 + \frac{4x}{7} - 17$.
   a. n b. a_n c. a_{n-1} d. a_0
   e. a_1 f. a_2 g. a_4

5. Write the standard form of a quartic polynomial in the variable t.

6. **True or False** The number π is a polynomial.

7. Consider the polynomial function P with equation $P(x) = x^3 - 6x^2 + 3x + 10$.
   a. Evaluate $P(4)$.
   b. Graph P in the window $-2 \le x \le 6$, $-30 \le y \le 30$.

In 8 and 9, let $f(x) = 2x^4 - 1$ and $g(x) = \frac{3}{2}x^2 + 4x$. An expression is given.
   a. Predict its degree.
   b. Predict its leading coefficient.
   c. Expand the expression and write the result in standard form.

8. $g(x) \cdot f(x)$ 9. $g(f(x))$

10. Refer to Example 4. Suppose that in successive summers beginning after high school graduation, Javier put $1250, $750, $2250, $3500, and $3300 into a bank account.

   a. Assume Javier goes to graduate school in the fall immediately after finishing 4 years of college and that the annual percentage yield is r. If no other money is added or withdrawn, how much is in his account when he goes to graduate school? Express your answer in terms of x, where $x = 1 + r$.
   b. Evaluate your answer to Part a when $r = 4.5\%$.
   c. What is the degree of the polynomial you found in Part a?

11. Suppose $f(x)$ is a polynomial of degree 3, and $g(x)$ is a polynomial of degree 5.
   a. What is the degree of $f(x) + g(x)$?
   b. What is the degree of $f(x) \cdot g(x)$?
   c. What is the degree of $f(g(x))$?

APPLYING THE MATHEMATICS

12. Let $q(x) = 4x^2 - 3$.

 a. If $p(x) + q(x)$ has degree 3, what do you know about the degree of $p(x)$?

 b. If $r(x) \cdot q(x)$ has degree 8, what do you know about the degree of $r(x)$?

 c. If $s(x) \cdot q(x)$ has a leading coefficient of -28, what do you know about the leading coefficient of $s(x)$?

13. Recall the formula for the height h in feet of an object thrown upward: $h = -\frac{1}{2}gt^2 + v_0t + h_0$, where t is the number of seconds after being thrown, h_0 is the initial height in feet, v_0 is the initial velocity in $\frac{ft}{sec}$, and g is the acceleration due to gravity ($32\frac{ft}{sec^2}$ on Earth). This formula describes a polynomial function in t.

 a. What is the degree of this polynomial?

 b. What is the leading coefficient?

 c. Suppose a ball is thrown upward from the ground with initial velocity $55\frac{ft}{sec}$. Find its height after 1.6 seconds.

14. Consider $f(x) = 4^x$ and $g(x) = x^4$.

 a. Which of f or g is a polynomial function?

 b. Which of f or g is an exponential function?

 c. Explain how to tell the difference between an exponential function and a polynomial function.

15. The whole number 45,702 can be written as the polynomial function $P(x) = 4x^4 + 5x^3 + 7x^2 + 0x^1 + 2$, with $x = 10$.

 a. Verify that $P(10) = 45{,}702$.

 b. What is the base-10 value of the base-8 number 45,702?

16. In Hebrew, the number 18 stands for life and, for this reason, one custom is to give a child $18 each year to save until his or her 18th birthday.

 a. Suppose a child is given $18 on each birthday (including the day he or she is born) until his or her 18th birthday, and that these gifts are put into an account with an annual yield of r. (No money is given on the 18th birthday itself.) Write a polynomial expression to give the total amount in the account on the child's 18th birthday. Let $x = 1 + r$.

 b. Evaluate your answer to Part a for an APY of 4.2%.

REVIEW

17. Of the sequences A and B below, one is arithmetic, the other geometric. (**Lessons 7-5, 3-8, 3-6**)

   A: 49, 7, 1, $\frac{1}{7}$, … B: 47, 36, 25, 14, …

   a. Write the next two terms of each sequence.

   b. Write an explicit formula for the geometric sequence.

   c. Write an explicit formula for the arithmetic sequence.

   d. Which sequence might model the successive maximum heights of a bouncing ball?

18. Solve this system of equations: $\begin{cases} r = 5t \\ s = r - 7 \\ t = r + 2s \end{cases}$. (**Lesson 5-3**)

19. A cat stalking a mouse creeps forward for 2 seconds at 0.5 $\frac{\text{ft}}{\text{sec}}$ then stops for 2 seconds. The cat springs forward at a rate of 10 $\frac{\text{ft}}{\text{sec}}$, but is stopped after 1 second by the refrigerator that the mouse ran under. Graph the situation, plotting time on the horizontal axis and distance on the vertical axis. (**Lesson 3-4**)

In 20 and 21, refer to the four equations below. (**Lessons 2-6, 2-5, 2-4**)

   A $y = kx$ **B** $y = kx^2$ **C** $y = \frac{k}{x}$ **D** $y = \frac{k}{x^2}$

20. Which equations have graphs that are symmetric to the y-axis?

21. The graph of which equation is a parabola?

22. The weight of a body varies inversely with the square of the distance from the center of Earth. If Deja weighs 110 pounds on the surface of Earth, how much will she weigh in space, 6000 miles from the surface? (The radius of Earth is approximately 4000 miles.) (**Lesson 2-2**)

23. Refer to kite $FLYR$ at the right. (**Previous Course**)

   a. Find LY.   b. Find m$\angle R$.

EXPLORATION

24. Suppose $f(x)$ and $g(x)$ are polynomials of degree 4. Justify your answers with examples or proofs.

   a. What are the possible degrees of $f(x) + g(x)$?

   b. What are the possible degrees of $f(x) \cdot g(x)$?

   c. What are the possible degrees of $f(g(x))$?

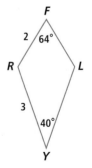

QY ANSWERS

1. a_0

2. $6045.48

Lesson

11-2

Multiplying Polynomials

Vocabulary

monomial, binomial, trinomial

degree of a polynomial in
several variables

▶ **BIG IDEA**   The product of two or more polynomials is a polynomial whose degree is the sum of the degrees of the factors.

Classifying Polynomials by the Number of Terms

In Lesson 11-1, you saw that polynomials can be classified by their degree. They can also be classified according to the number of terms they have after combining like terms. A **monomial** is a polynomial with one term, a **binomial** is a polynomial with two terms, and a **trinomial** is a polynomial with three terms. Below are some examples.

monomials: $-7, x^2, 3y^4$
binomials: $x^2 - 11, 3y^4 + y, 12a^5 + 4a^3$
trinomials: $x^2 - 5x + 6, 10y^6 - 9y^5 + 17y^2$

Notice that monomials, binomials, and trinomials can be of any degree. No special name is given to polynomials with more than three terms.

When a polynomial in one variable is added to or multiplied by a polynomial in another variable, the result is a polynomial in several variables. The **degree of a polynomial in several variables** is the largest sum of the exponents of the variables in any term. For instance, $x^3 + 8x^2y^3 + xy^2$ is a trinomial in x and y of degree 5. Notice that the sum of the exponents in the middle term is 5, while in both the first and last terms the sum of the exponents is 3.

The Extended Distributive Property

The product of a monomial and a binomial can be found using the Distributive Property, which says that for all numbers a, b, and c, $a(b + c) = ab + ac$. So, to multiply a monomial by a binomial, multiply the monomial by each term of the binomial and then add the products.

Repeated application of the Distributive Property allows you to find the product of any two polynomial factors. In general, if one polynomial has m terms and the second n terms, there will be mn terms in their product before combining like terms.

Mental Math

A square park is 1 block on a side. A person walks from the midpoint of one side to the midpoint of the next, and so on, until he returns to his starting point. How far has he walked?

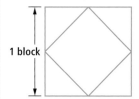

1 block

▶ **READING MATH**

The prefixes *mono-, bi-, tri-,* and *poly-* mean "one," "two," "three," and "many," respectively. These prefixes are used in many common English words, such as monopoly, bicycle, and tricycle, and in geometric terms such as triangle and polyhedron.

Example 1

Expand $(2x^3 + 3x^2 - 2)(5x^2 + 4)$ and write your answer in standard form.

Solution 1 Expand on a CAS.

Solution 2 Use the Distributive Property by treating $(2x^3 + 3x^2 - 2)$ as a single unit.

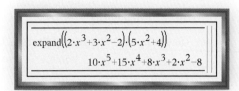

$$(2x^3 + 3x^2 - 2)(5x^2 + 4)$$
$$= (2x^3 + 3x^2 - 2)(5x^2) + (2x^3 + 3x^2 - 2)(4)$$

Now use the Distributive Property to expand each product on the right side.

$$= 2x^3 \cdot 5x^2 + 3x^2 \cdot 5x^2 + -2 \cdot 5x^2 + 2x^3 \cdot 4 + 3x^2 \cdot 4 + -2 \cdot 4$$
$$= 10x^5 + 15x^4 - 10x^2 + 8x^3 + 12x^2 - 8$$

There are six terms. Combine like terms and write in standard form.

$$= 10x^5 + 15x^4 + 8x^3 + 2x^2 - 8$$

 QY

Notice that in Example 1 each of the terms of the trinomial $2x^3 + 3x^2 - 2$ is multiplied by each of the terms of the binomial $5x^2 + 4$. We call this generalization of the Distributive Property the *Extended Distributive Property*.

> ▸ **QY**
>
> Check Example 1 by letting $x = 2$ in both the given expression and the answer.

Extended Distributive Property

To multiply two polynomials, multiply each term in the first polynomial by each term in the second and add the products.

The Extended Distributive Property is applied several times when multiplying more than two polynomials. Because multiplication is associative and commutative, one way to multiply three polynomials is to start by multiplying any two of the polynomials and then multiplying their product by the remaining polynomial.

Used together, the Extended Distributive Property and the Associative Property of Multiplication let you multiply any number of polynomials in any order.

Example 2

a. Find the volume of the large box by multiplying its dimensions.

b. Find the volume of the large box by adding the volumes of each of the small boxes.

c. Show that the answers to Parts a and b are equal.

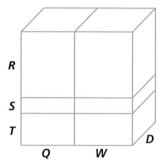

Solution

a. The box has width $Q + W$, height $R + S + T$, and depth D. Its volume is the product of its dimensions.

$$\text{Volume} = (Q + W)(R + S + T)D$$

b. There are six small boxes, each with depth D. The volume of the big box is the sum of the volumes of the 6 smaller boxes (2 smaller boxes in each of 3 layers), from the top, left to right:

$$\text{Volume} = QRD + WRD + QSD + WSD + QTD + WTD$$

c. The two expressions in Parts a and b must be equivalent because they represent the same volume. To show this, you can expand the product from Part a. Because of the Associative Property of Multiplication, either $Q + W$ can be multiplied by $R + S + T$ first, or $R + S + T$ can be multiplied by D first. We begin by multiplying by D first.

$$\text{Volume} = (Q + W)(RD + SD + TD)$$

Now apply the Distributive Property, distributing $(Q + W)$ over the trinomial $RD + SD + TD$.

$$\text{Volume} = (Q + W)RD + (Q + W)SD + (Q + W)TD$$

Apply the Distributive Property again.

$$\text{Volume} = QRD + WRD + QSD + WSD + QTD + WTD$$

Notice how each term of the expanded form is the product of a term from $Q + W$, a term from $R + S + T$, and the term D.

Finding Terms of Products without Finding the Entire Product

In Example 1, you found that the product $(2x^3 + 3x^2 - 2)(5x^2 + 4)$ of two polynomials is equal to the polynomial $10x^5 + 15x^4 + 8x^3 + 2x^2 - 8$. All three polynomials were written in standard form. Notice the leading term $10x^5$ of the product is the product of the leading terms of the polynomial factors. Also, the last term –8 of the product is the product of the last terms of the factors.

In general, the leading term of the product of n polynomials written in standard form is the product of the leading terms of the polynomial factors, and the last term is the product of the last terms of the factors.

GUIDED

Example 3

Without expanding, find the leading term, the last term, and the coefficient of the term with x^3 of the product $(5x^2 + 2)(4x^2 + 8)(11x - 3)$ when written in standard form.

Solution The leading term is the product of the leading terms of the factors.

The leading term of the product is
$\underline{\quad?\quad} \cdot \underline{\quad?\quad} \cdot \underline{\quad?\quad} = \underline{\quad?\quad}$.

The last term is the product of the last terms of the factors.

The last term of the product is
$\underline{\quad?\quad} \cdot \underline{\quad?\quad} \cdot \underline{\quad?\quad} = \underline{\quad?\quad}$.

A term with x^3 will arise from multiplying $5x^2$ from the first factor, 8 from the second factor, and $11x$ from the third factor. The only other term with x^3 will arise from multiplying 2 from the first factor, $\underline{\quad?\quad}$ from the second factor, and $\underline{\quad?\quad}$ from the third factor. The first product is $\underline{\quad?\quad} x^3$; the second is $\underline{\quad?\quad} x^3$. So, after combining like terms, the term with x^3 in the product is $528x^3$.

Check Expand on a CAS and check the leading and last terms and the coefficient of x^3 in the product.

$$\text{expand}\left(\left(5{\cdot}x^2{+}2\right){\cdot}\left(4{\cdot}x^2{+}8\right){\cdot}\left(11{\cdot}x{-}3\right)\right)$$

Applications of Polynomials

A classic problem in mathematics is to find the maximum volume of an open box like the one in Example 4.

Example 4

A rectangular piece of cardboard measuring 16 inches by 20 inches is to be folded into an open box after cutting squares of side length x from each corner. Let $V(x)$ be the volume of the box.

a. Write a polynomial formula for $V(x)$ in standard form.

b. Use the graph of the function V to find the maximum possible volume.

(continued on next page)

Solution

a. Draw a diagram.

When the cardboard is folded up, the dimensions of the box are $(20 - 2x)$ inches long by $(16 - 2x)$ inches wide by x inches high. The volume is the product of these dimensions.

$$V(x) = (20 - 2x)(16 - 2x)(x)$$

Use a CAS to expand this product. The CAS will automatically write the product in standard form.

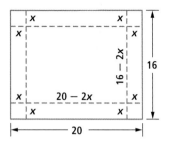

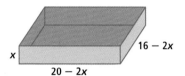

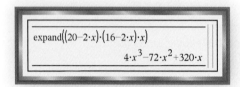

So $V(x) = 4x^3 - 72x^2 + 320x$.

b. Because dimensions of a box are positive, $x > 0$, $20 - 2x > 0$ and $16 - 2x > 0$. Solving these inequalities for x, we have $x > 0$, $x < 10$, and $x < 8$, which means that $0 < x < 8$ is the largest domain for V in this situation. Graphing V over this domain shows that the largest possible volume is approximately 420.1 in^3, which occurs when $x \approx 2.94$ in. You can substitute this value for x to find that the box dimensions for this volume are 14.12 inches long by 10.12 inches wide by 2.94 inches high.

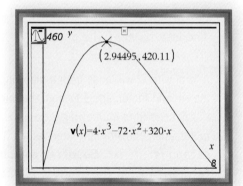

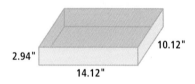

Questions

COVERING THE IDEAS

In 1–6, a polynomial is given.
   a. State whether the polynomial is a monomial, a binomial, a trinomial, or none of these.
   b. Give its degree.

1. $w^7 - w^5 z^3 + \left(\frac{3}{2}\right) z^5$ 2. $6a \cdot 2a \cdot \left(\frac{1}{3}\right) a$ 3. $x^3 - x$

4. $4x^3 - 6x^2 - 5$ 5. 3^2 6. $182x^4 w t^2$

7. Give an example of a 5th-degree binomial in two variables.

In 8 and 9, multiply by hand and write in standard form.

8. $\left(5x^2 + \frac{1}{2}x - 2\right)\left(x + \frac{1}{5}\right)$

9. $(b^2 + 1)(2b - 3)(5b)$

10. Polynomial P is the product in standard form of
$(12x^3 + 7)\left(3x^2 + \frac{3}{8}x - 1\right)(7x^2 - x)$.
Without expanding, write the first and last terms of P.

In 11 and 12, refer to Example 4.

11. **a.** Use $V(x) = 4x^3 - 72x^2 + 320x$ to complete the table below
for $x = 0$ to 10.

x	1	2	3	4	5	6	7	8	9	10
$V(x)$	252	?	420	?	?	192	?	?	-36	?

b. For what integer value of x from 0 to 10 is $V(x)$ largest?

12. Give a reasonable domain for V if a new box has volume
$V(x) = (15 - 2x)(18 - 2x)x$.

APPLYING THE MATHEMATICS

13. Expand and write in standard form.
    a. $(x^2 - 3x + 3)(x^2 + 3x + 3)$ **b.** $(x^2 - 4x + 4)(x^2 + 4x + 4)$
    c. $(x^2 - 5x + 5)(x^2 + 5x + 5)$ **d.** $(x^2 - 6x + 6)(x^2 + 6x + 6)$
    e. Based on your answers to Parts a–d, what do you predict the
    expanded form of $(x^2 - nx + n)(x^2 + nx + n)$ will be for any
    positive integer n?

14. A box measures 12 inches by 18 inches by 10 inches. A y-foot
long roll of wrapping paper is x feet wide. Assuming no overlap,
how much wrapping paper will be left after wrapping the box?

15. A piece of material is in the shape of an equilateral triangle. Each
side measures 15 inches. Kites with sides x and $x\sqrt{3}$ are cut from
each corner. Then the flaps are folded up to form an open box.
    a. Write a formula for the volume $V(x)$ of the box as a product
    of polynomials.
    b. Write $V(x)$ in standard form.
    c. Find $V(4)$.
    d. Find the maximum possible value of $V(x)$.

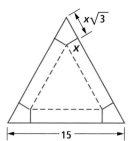

16. Melissa knows that $347 = 3 \cdot 10^2 + 4 \cdot 10 + 7$. Explain how
Melissa can use what she knows together with the Extended
Distributive Property to multiply any 3-digit number
$a \cdot 10^2 + b \cdot 10 + c$ in base 10, where a, b, and c are single
digits, by any 4-digit number in base 10.

17. A town's zoning ordinance shows an aerial view of a lot like the one at the right. Distances a, b, and c are the minimum setbacks allowed to the street, the side lot lines, and the rear lot line, respectively. If a rectangular lot has 75 feet of frontage and is 150 feet deep, what is the maximum ground area possible for a one-story house in terms of a, b, and c?

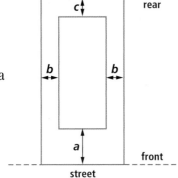

18. Find a monomial and a binomial whose product is $2p^2q + 4p$.

19. Find two binomials whose product is $2y^2 + 15y + 7$.

20. Find a binomial and a trinomial whose product is $3a^2 + 7ab + 2b^2 + 3a + b$.

REVIEW

21. Nancy invested different amounts at an APY of r. On the fifth anniversary of her initial investment her savings were $872x^5 + 690x^4 + 737x^3 + 398x^2 + 1152x + 650$ dollars, where $x = 1 + r$. (Lesson 11-1)

    a. What is the degree of the polynomial?

    b. If Nancy invested at an APY of 5.125%, how much did Nancy have in the account on the fifth anniversary?

22. During the early part of the twentieth century, the deer population of the Kaibab Plateau in Arizona grew rapidly. Later, the increase in population depleted the food supply and the deer population declined quickly. The number $N(t)$ of deer from 1905 to 1930 is approximated by $N(t) = -0.125t^5 + 3.125t^4 + 4000$, where t is the time in years after 1905. This function is graphed at the right. (Lessons 11-1, 1-4)

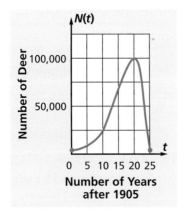

    a. What is the degree of this polynomial function?

    b. Estimate the deer population in 1905.

    c. Estimate the deer population in 1930.

    d. Over what time period was the deer population increasing?

23. a. Graph $c(x) = \cos x$ when $0° \leq x \leq 360°$. (Lesson 10-6)

    b. For what values of x in this domain does $c(x) = 0$?

24. Solve $3x^2 - 16x - 64 = 0$. (Lesson 6-7)

25. Expand $(x + 7)^2$. (Lesson 6-1)

EXPLORATION

26. a. Find the products below.

    $(x - 1)(x + 1)$ $(x - 1)(x^2 + x + 1)$ $(x - 1)(x^3 + x^2 + x + 1)$

    b. State a general rule about polynomials whose product is $x^n - 1$, where $n = 2, 3, 4, 5\ldots$.

Lesson

11-3

Quick-and-Easy Factoring

Vocabulary

factoring a polynomial

factored form of a polynomial

greatest common
  monomial factor

prime polynomial,
  irreducible polynomial

▶ **BIG IDEA** Some polynomials can be factored into polynomials of lower degree; several processes are available to find factors.

A polynomial is, by definition, a sum.

$$P(x) = a_n x^n + a_{n-1} x^{n-1} + a_{n-2} x^{n-2} + \ldots + a_2 x^2 + a_1 x^1 + a_0, a_n \neq 0$$

However, polynomials are sometimes written as a product. For instance, recall the polynomial $4x^3 - 72x^2 + 320x$ from Lesson 11-2 that represents the volume of a box with sides of length x, $20 - 2x$, and $16 - 2x$. A *factored form* of this polynomial is $x(20 - 2x)(16 - 2x)$. By factoring out a 2 from each binomial in the factored form you can obtain another factored form. The expressions below are all equivalent.

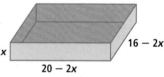

x $16 - 2x$ $20 - 2x$

$$4x^3 - 72x^2 + 320x \qquad (16 - 2x)(20 - 2x)x \qquad 4x(8 - x)(10 - x)$$

The process of rewriting a polynomial as a product of two or more factors is called **factoring the polynomial**, or writing the polynomial in **factored form**. Factoring undoes multiplication. Most factoring is based on three properties:

- Distributive Property (common monomial factoring)

- Special Factoring Patterns (difference of squares, etc.)

- Factor Theorem (for polynomials)

You may have used some of these properties to factor polynomials in earlier courses. The Factor Theorem is introduced in the next lesson. This lesson discusses how to apply the other two properties when factoring polynomials.

Mental Math

Let $\sin x = 0.15$. Find each of the following.

a. $\sin(-x)$

b. $\sin(180° - x)$

c. $\sin(180° + x)$

d. $\sin(360° + x)$

Common Monomial Factoring

Similar to the greatest common factor of a set of numbers, the **greatest common monomial factor** of the terms of a polynomial is the monomial with the greatest coefficient and highest degree that evenly divides all the terms of the polynomial. Common monomial factoring applies the Distributive Property.

Example 1

Factor $5x^3 - 15x^2$.

Solution Look for the greatest common monomial factor of the terms. The greatest common factor of 5 and 15 is 5. x^2 is the highest power of x that divides each term. So, $5x^2$ **is the greatest common monomial factor of $5x^3$ and $-15x^2$.** Now apply the Distributive Property.

$$5x^3 - 15x^2 = 5x^2(x - 3)$$

$5x^2(x - 3)$ is completely factored; it cannot be factored further.

Check Expand $5x^2(x - 3)$. $5x^2(x - 3) = 5x^2 \cdot x - 5x^2 \cdot 3 = 5x^3 - 15x^2$. It checks.

Special Factoring Patterns

CAS machines have built-in operations for factoring polynomials. In this Activity, you will use a CAS to discover some special factoring patterns.

Activity

MATERIALS CAS

Work with a partner.

Step 1 Factor the following expressions on a CAS.

a. $x^2 - 24x + 144$ b. $x^2 - 49$

c. $x^2 - 64$ d. $x^2 + 8x + 16$

e. $4x^2 - 81$ f. $x^2 - 6x + 9$

g. $x^2 + 10x + 25$ h. $36x^2 - 1$

```
factor(x²-24x+144)
```

Step 2 Based on the results of Step 1, sort the polynomials into groups. Describe how you factor the polynomials in each group.

Step 3 Use your descriptions in Step 2 to factor the following.

a. $x^2 - y^2$ b. $x^2 - 2xy + y^2$ c. $x^2 + 2xy + y^2$

The three factoring relationships in the Activity are common enough to be worth knowing, and are summarized on the next page.

Difference of Squares Factoring Theorem

For all a and b,

$$a^2 - b^2 = (a + b)(a - b).$$

Binomial Square Factoring Theorem

For all a and b,

$$a^2 + 2ab + b^2 = (a + b)^2;$$
$$a^2 - 2ab + b^2 = (a - b)^2.$$

Be aware that many polynomials do not factor into binomials involving integers, or do not factor at all. For example, the sum of two squares, $a^2 + b^2$, cannot be factored over the set of polynomials with real coefficients.

GUIDED

Example 2

Factor each polynomial.

a. $9x^6 - 100y^2$

b. $p^2 - 18p + 81$

Solution

a. $9x^6 = (3x^3)^2$ and $100y^2 = (10y)^2$, so this polynomial is a difference of squares. Use difference of squares factoring.

$$9x^6 - 100y^2 = (\underline{})^2 - (\underline{})^2$$
$$= (\underline{} + \underline{})(\underline{} - \underline{})$$

b. This polynomial is in binomial square form with $a = p$ and $b = \pm9$. In order for the signs to agree with the second Binomial Square Factoring pattern, you need $b = 9$. So,

$$p^2 - 18p + 81 = p^2 - 2 \cdot \underline{} \cdot \underline{} + \underline{}^2$$
$$= (\underline{})^2$$

Trial and Error

When there is no common monomial factor and the polynomial does not fit any of the special factoring patterns discussed above, a CAS can be used to factor. Additionally, some polynomials are simple enough that you can guess their factors and check by multiplying. Before CAS, some algebra students spent many hours learning strategies to guess more efficiently.

Example 3

Factor $x^2 - 2x - 15$.

Solution You want solutions to $x^2 - 2x - 15 = (x + \underline{\;?\;})(x + \underline{\;?\;})$.

Because -15 is the product of the two missing numbers on the right side, begin by considering the factors of -15. The factors are either -15 and 1, 15 and -1, 5 and -3, or -5 and 3. Try each combination by expanding until you find one that checks:

$$(x - 15)(x + 1) = x^2 - 14x - 15 \qquad \text{Not correct}$$
$$(x + 15)(x - 1) = x^2 + 14x - 15 \qquad \text{Not correct}$$
$$(x + 5)(x - 3) = x^2 + 2x - 15 \qquad \text{Not correct}$$
$$(x - 5)(x + 3) = x^2 - 2x - 15 \qquad \text{Correct!}$$

So, $x^2 - 2x - 15 = (x - 5)(x + 3)$.

Prime Polynomials

How do you know when a polynomial is factored completely? When you factor a number such as $12 = 2 \cdot 2 \cdot 3$, the number is factored completely when all the factors are prime. Similarly, a polynomial is factored completely when all the factors are *prime polynomials*. A polynomial is **prime**, or **irreducible**, over a set of numbers if it cannot be factored into polynomials of lower degree whose coefficients are in the set. Writing a polynomial as a product of factors with coefficients in a set is called *factoring over* that set.

Example 4

a. Is $x^2 - 14$ prime over the integers?

b. Is $x^2 - 14$ prime over the real numbers?

Solution

a. If $x^2 - 14$ factors over the integers, then you can find integers a and b so that $x^2 - 14 = (x + a)(x + b)$. Use trial and error to test values of a and b when $ab = -14$.

The integer factors of -14 are 1 and -14, -1 and 14, 2 and -7, or -2 and 7. Expand each combination.

$$(x + 1)(x - 14) = x^2 - 13x - 14$$
$$(x - 1)(x + 14) = x^2 + 13x - 14$$
$$(x + 2)(x - 7) = x^2 - 5x - 14$$
$$(x - 2)(x + 7) = x^2 + 5x - 14$$

None of these products expand to equal $x^2 - 14$.

So, $x^2 - 14$ is prime over the set of integers.

b. If $x^2 - 14$ factors over the real numbers, then you can find real numbers a and b so that $x^2 - 14 = (x + a)(x + b)$. Because a and b do not need to be integers, you can use Difference of Squares factoring.

$$x^2 - 14 = x^2 - \left(\sqrt{14}\right)^2 = \left(x + \sqrt{14}\right)\left(x - \sqrt{14}\right)$$

This factorization shows that $x^2 - 14$ is not prime over the set of real numbers.

By default, most CAS machines factor over the rationals. However, a CAS will also factor over the real numbers, as shown below. Different machines may have different commands for factoring over the reals. Some machines factor over the reals when ",x" is entered after a polynomial in x. The machine output pictured at the left below uses an `rfactor` command to factor over the reals.

Note that when a polynomial is prime over a set, a CAS command to factor over that set will produce an output identical to the original polynomial. For example, the screen at the right below shows that $x^2 - 14$ is prime over the rationals.

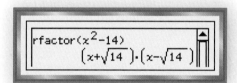
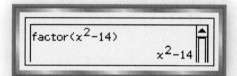

Questions

COVERING THE IDEAS

1. **Fill in the Blanks** Copy and complete:
   $21m^3n + 35m^2n - 14mn^2 = 7mn(\underline{\ ?\ } + \underline{\ ?\ } + \underline{\ ?\ })$

In 2 and 3, factor out the greatest common monomial. Check with a CAS.

2. $25y^4 - 50y$

3. $90x^3y^2 + 270xy^2 + 180x^2y^2$

In 4–9, a polynomial is given.
   a. Tell whether the polynomial is a binomial square, a difference of squares, or neither.
   b. Factor over the real numbers, if possible.

4. $a^2 - b^2$

5. $x^2 - 2xy + y^2$

6. $r^2 - 121$

7. $49x^2 - 144b^2$

8. $25b^2 - 70bc + 49c^2$

9. $y^2 + 25$

10. Check your solutions to Example 2 by expanding your answers.

11. a. Factor $12t^3 - 12t$ into linear factors.

    b. Check by multiplying.

In **12** and **13**, factor completely over the rational numbers and check your answer.

12. $64s^2 - 49$

13. $x^4 - 10x^2 + 9$

In **14–18**, factor the polynomial completely over the rational numbers and state which factoring method(s) you used. If it is not factorable, write "prime over the rational numbers."

14. $3x^2 + 24x + 36$

15. $-1 - 10y - 25y^2$

16. $b^2 - 3b + 8$

17. $12x^3 - 75x$

18. a. $y^2 + 5y - 6$

    b. $y^2 - 5y - 6$

    c. $y^2 + 5y + 6$

    d. $y^2 - 5y + 6$

19. a. Is $x^2 - 30$ prime over the integers? If not, factor it.

    b. Is $x^2 - 30$ prime over the reals? If not, factor it.

APPLYING THE MATHEMATICS

20. $48^2 = 2304$. Use this information to factor $2303 = 48^2 - 1$.

21. **Multiple Choice** Which of the following is a perfect square trinomial?

    A $16x^2 + 24x + 9$

    B $q^4 - 16q^3 + 64$

    C $t^2 - 64$

    D $4x^2 + 16x + 25$

22. a. Write $x^4 - 16$ as the product of two binomials.

    b. Write $x^4 - 64$ as the product of three binomials.

23. One factor of $12x^2 + x - 35$ is $(3x - 5)$. Find the other factor.

In **24** and **25**, factor the polynomial by trial and error and check with a CAS.

24. $8z^2 + 6z + 1$

25. $6q^2 + 2q - 4$

In **26** and **27**, a polynomial is given.

   a. First factor out the greatest common monomial factor. Then complete the factorization.

   b. Check by multiplying.

26. $242a^4b^2 - a^2b^4$

27. $5x^3 - 10x^2 - 175x$

28. **Multiple Choice** Which is a factorization of $a^2 + 100$ over the complex numbers?

    A $(a + 10)(a + 10)$

    B $(a + 10i)(a + 10i)$

    C $(a + 10i)(a - 10i)$

    D $(a + 10i)^2$

29. Consider the expression $\dfrac{4x^2 - 9}{2x^2 + x - 3}$.

    a. Factor the numerator and denominator.

    b. Because binomials are numbers, they may be multiplied and divided in the same way numbers are. Simplify your answer to Part a.

    c. Using your factored expression in Part a, determine the domain of the function $f(x) = \dfrac{4x^2 - 9}{2x^2 + x - 3}$. Explain your reasoning.

REVIEW

In 30 and 31, consider a closed rectangular box with dimensions $2h$, $h + 2$, and $2h + 3$. Write a polynomial in standard form for each measure.

30. $S(h)$, the surface area of the box (**Lesson 11-2**)

31. $V(h)$, the volume of the box (**Lesson 11-2**)

32. Give an example of a cubic binomial. (**Lesson 11-1**)

33. The lateral height of a cone is 9 centimeters and its height is h. (**Lessons 11-1, 1-7**)

    a. Write a formula for the volume of the cone in terms of r and h.

    b. Write a formula for the radius r of the cone in terms of h.

    c. Substitute your expression for r in Part b into your formula in Part a.

    d. **True or False** The volume of this cone is a polynomial function of h.

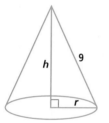

34. Write $\log_7 99 - \log_7 33$ as a logarithm of a single number. (**Lesson 9-9**)

35. Evaluate $\log_{13}\left(\dfrac{1}{13}\right)$. (**Lesson 9-7**)

EXPLORATION

36. Consider the polynomial $P(x) = \left(x - \dfrac{2}{3}\right)\left(x + \dfrac{3}{4}\right)$.

    a. Write $P(x)$ in expanded form.

    b. Use a CAS to factor the expanded form.

    c. The answer to Part b does not look like the form you started with. What did the CAS do to the expanded form before it factored it?

    d. Use the ideas in Parts a–c to factor $x^2 - \dfrac{5}{6}x + \dfrac{1}{6}$.

Lesson
11-4

The Factor Theorem

Vocabulary

zero of a polynomial, root of a polynomial

▶ **BIG IDEA**   If $P(x)$ is a polynomial, then a is a solution to the equation $P(x) = 0$ if and only if $x - a$ is a factor of $P(x)$.

Zeros of a Polynomial

Consider an equation in which there is a polynomial on each side, such as

$$14x^3 + 3x - 10 = 5x - 9x^2 + 5.$$

Add the opposite of $5x - 9x^2 + 5$ to each side. Then

$$14x^3 + 9x^2 - 2x - 15 = 0.$$

In this way, every equality of two polynomials in x can be converted into an equivalent equation of the form $P(x) = 0$. More generally, any equation involving two expressions in x can be converted into an equation of the form $f(x) = 0$.

As you know, when f is a function, then a solution to the equation $f(x) = 0$ is called a *zero* of f. When P is a polynomial function, then a zero of P is also called a **zero** or **root of the polynomial** $P(x)$. For example, when $P(x) = 3x - 18$, then 6 is a zero, or root, of $P(x)$ because $3 \cdot 6 - 18 = 0$. A zero of $P(x)$ is an x-intercept of the graph of P, as shown at the right.

The factors of a polynomial $P(x)$ are connected to its zeros, or x-intercepts. Understanding this connection can help you to solve equations with polynomials and to understand their algebraic structure.

Recall that a product of numbers equals 0 if and only if one of the factors equals 0. We call this result the *Zero-Product Theorem*.

Zero-Product Theorem

For all a and b, $ab = 0$ if and only if $a = 0$ or $b = 0$.

For instance, if $f(x) = 0$ and $f(x) = g(x) \cdot h(x)$, then $g(x) = 0$ or $h(x) = 0$. This is one reason for manipulating an equation so that 0 is on one side.

Mental Math

Let h be a sequence with explicit formula $h_n = h_0 \cdot r^n$ that models the maximum height of a ball after n bounces. Are the following *always*, *sometimes but not always*, or *never* true?

a. r is greater than 1.

b. r is less than zero.

c. h_0 is greater than h_3.

d. h_4 is greater than h_3.

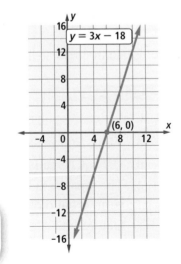

Example 1

In Example 4 of Lesson 11-2, the volume $V(x)$ of the box shown at the right is given by $V(x) = x(20 - 2x)(16 - 2x) = 4x^3 - 72x^2 + 320x$. Find the zeros of V.

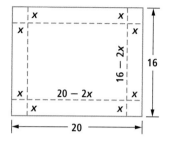

Solution To find the zeros, solve $x(20 - 2x)(16 - 2x) = 0$. By the Zero-Product Theorem, at least one of these factors must equal zero.

Either $x = 0$ or $20 - 2x = 0$ or $16 - 2x = 0$.
So, $x = 0$ or $x = 10$ or $x = 8$.

The zeros are 0, 8, and 10.

Check Notice that the zeros are the three values of x for which one of the dimensions of the box is 0, so there would be no volume.

Activity 1

MATERIALS CAS

Work with a partner and use a graphing utility. One partner should work with $p(x) = x^3 - 3x^2 - 13x + 15$, the other with $q(x) = x^4 - x^3 - 24x^2 + 4x + 80$.

Step 1 Graph your polynomial equation in the window $-10 \leq x \leq 10$; $-100 \leq y \leq 100$.

Step 2 Find the x-intercepts of your graph.

Step 3 Factor your polynomial on a CAS.

Step 4 Compare your results with your partner's. How are the x-intercepts of each graph related to the factors of each polynomial?

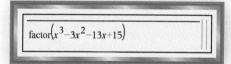

Step 5 Use your answer to Step 4 to write a polynomial whose graph has x-intercepts 4, 5, and –1.

The results of Activity 1 show that the zeros of a polynomial function correspond to the polynomial's linear factors. For example, the zeros of p are –3, 1, and 5, and the factors of $p(x)$ are $x + 3$, $x - 1$, and $x - 5$. The following theorem generalizes this relationship.

Factor Theorem

$x - r$ is a factor of $P(x)$ if and only if $P(r) = 0$, that is, r is a zero of P.

Proof If $x - r$ is a factor of $P(x)$, then for all x, $P(x) = (x - r) \cdot Q(x)$, where $Q(x)$ is some polynomial. Substitute r for x in this formula.

$$P(r) = (r - r) \cdot Q(r) = 0 \cdot Q(r) = 0.$$

Proving the other direction of the theorem, that $P(r) = 0$ implies that $x - r$ is a factor of $P(x)$, requires more work. Consider the special case where $r = 0$. Write $P(x)$ in general form:

$$P(x) = a_n x^n + a_{n-1} x^{n-1} + \dots + a_1 x + a_0.$$

Then $P(r) = P(0) = a_n 0^n + a_{n-1} 0^{n-1} + \dots + a_1 0 + a_0$.

So $P(0) = a_0$, the constant term.

Thus, when $r = 0$ and $P(0) = 0$, $a_0 = 0$.

So, x is a factor of every term of $P(x)$, and we can write

$$P(x) = x(a_n x^{n-1} + a_{n-1} x^{n-2} + \dots + a_1).$$

So $x = x - 0$ is a factor of $P(x)$.

If $r \neq 0$ and $P(r) = 0$, then the graph of $y = P(x)$ contains the point $(r, 0)$. Think of the graph of $y = P(x)$ as a translation image r units to the right of the graph of a polynomial function G that contains $(0, 0)$. Because $G(0) = 0$, the case above applies to $G(x)$, so $G(x) = x \cdot H(x)$ for some polynomial $H(x)$. By the Graph-Translation Theorem, $P(x)$ can be formed by replacing x in $G(x)$ by $x - r$. Therefore, $P(x) = (x - r) \cdot H(x - r)$, and so $x - r$ is a factor of $P(x)$.

 QY1

▶ **QY1**

5 is a zero of q in Activity 1. What factor of $q(x)$ is associated with this zero?

Finding Zeros by Factoring

Both the Zero-Product Theorem and the Factor Theorem provide methods for finding zeros of polynomials by factoring.

Example 2

Find the roots of $P(x) = x^4 - 14x^2 + 45$ by factoring.

Solution 1 Use a CAS command to factor $P(x)$ over the rationals, as shown at the right.

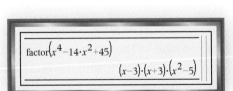

The roots are the solutions to $(x - 3)(x + 3)(x^2 - 5) = 0$.

Use the Zero-Product Theorem.

$$x - 3 = 0 \quad \text{or} \quad x + 3 = 0 \quad \text{or} \quad x^2 - 5 = 0$$
$$x = 3 \quad \text{or} \quad x = -3 \quad \text{or} \quad x = \sqrt{5} \text{ or } -\sqrt{5}$$

So, the roots of $P(x)$ are $3, -3, \sqrt{5}$ and $-\sqrt{5}$.

Solution 2 Use a CAS command to factor $P(x)$ over the reals. Notice the difference in the command in the screenshot at the right. This shows

$$P(x) = (x - 3)(x + 3)(x + \sqrt{5})(x - \sqrt{5}).$$

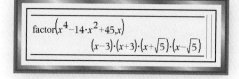

$$\text{factor}\left(x^4 - 14 \cdot x^2 + 45, x\right)$$
$$(x-3) \cdot (x+3) \cdot \left(x + \sqrt{5}\right) \cdot \left(x - \sqrt{5}\right)$$

Now use the Factor Theorem.

The factor $x - 3$ means that 3 is a root of P.
The factor $x + 3 = x - (-3)$ means that –3 is a root of P.
The factor $x - \sqrt{5}$ means that $\sqrt{5}$ is a root of P.
The factor $x + \sqrt{5} = x - \left(-\sqrt{5}\right)$ means that $-\sqrt{5}$ is a root of P.

So, the roots of $P(x)$ are 3, –3, $\sqrt{5}$ and $-\sqrt{5}$.

 QY2

> ▶ **QY2**
>
> Use the factorization of $x^2 - 225$ to find the roots of $f(x) = x^2 - 225$.

Finding Equations from Zeros

The Factor Theorem also says that if you know the zeros of a polynomial function, then you can determine the polynomial's factors.

Example 3

A polynomial function p with degree 4 and a leading coefficient of 1 is graphed at the right. Find the factors of $p(x)$ and use them to write a formula for $p(x)$.

Solution From the graph, the zeros appear to be –6, –2, –1, and 2.

By the Factor Theorem, the factors are $x - (-6)$, __?__, __?__, and $x - 2$. Therefore, $p(x) = (x + 6)(\underline{\;\;?\;\;})(\underline{\;\;?\;\;})(x - 2)$.

In standard form, $p(x) = x^4 + 7x^3 + 2x^2 - 28x - 24$.

Check Graph $y = p(x) = x^4 + 7x^3 + 2x^2 - 28x - 24$ in the window $-8 \leq x \leq 4$; $-100 \leq y \leq 50$. Does it match the given graph?

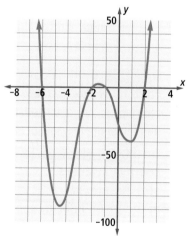

As you saw in Example 3, given a set of zeros, you can create a polynomial function with those zeros by multiplying the associated factors. However, if you do not restrict the degree or leading coefficient, different polynomial functions can have the same zeros.

Activity 2

Step 1 Find the zeros of each polynomial function below without graphing.

    a. $f(x) = x(x + 4)(x - 3)$ **b.** $g(x) = 4x(x + 4)(x - 3)$

    c. $h(x) = x^2(x + 4)(x - 3)$ **d.** $k(x) = -\frac{1}{4}x^2(x + 4)(x - 3)^3$

Step 2 Graph the four functions from Step 1 in the window $-5 \le x \le 5$; $-100 \le y \le 100$. How are their graphs similar? Describe any differences.

Step 3 Find an equation for a polynomial function whose zeros are $0, -4$, and 3, but whose graph is different from the graphs in Step 2.

Activity 2 illustrates that a polynomial can be transformed in at least two ways to produce another polynomial with the same zeros:

- The original polynomial can be multiplied by a constant factor k.

- One or more factors of the original polynomial can be raised to a different positive integer power.

For example, every polynomial of the form $kx^a(x + 4)^b(x - 3)^c$ has the same zeros as the polynomial $x(x + 4)(x - 3)$.

Example 4

Find the general form of a polynomial function P whose only zeros are $-\frac{7}{2}, \frac{1}{4}$, and 2.

Solution By the Factor Theorem, $x - \left(-\frac{7}{2}\right)$, $x - \frac{1}{4}$, and $x - 2$ are factors of $P(x)$, and because P has no other zeros, these factors are the only factors. Any of these factors can be raised to any positive integer power, and the entire polynomial could be multiplied by any nonzero constant.

So, $P(x) = k\left(x + \frac{7}{2}\right)^a\left(x - \frac{1}{4}\right)^b(x - 2)^c$ where $k \ne 0$ and a, b, c are positive integers.

From the given information in Example 4, you know the degree of $P(x)$ is at least three. However, you have no information about the exponents a, b, and c except that they are positive integers. Therefore, you cannot be sure of the degree of $P(x)$.

Questions

COVERING THE IDEAS

1. State the Zero-Product Theorem.

In 2 and 3, find the roots of the equation.

2. $(x - 5)(6x + 33) = 0$ 3. $\left(-\frac{3}{5}k + 12\right)(k^2 - 2)(k + 1) = 0$

4. If $f(x) = 4x(x - 2)(x + 16)$, find the zeros of f.

5. **Multiple Choice** If the graph of a polynomial function intersects the x-axis at $(5, 0)$ and $(-2, 0)$, then which of the following must be two of the polynomial's linear factors?

   A $x - 5$ and $x - 2$ B $x + 5$ and $x - 2$

   C $x - 5$ and $x + 2$ D $x + 5$ and $x + 2$

6. Suppose that P is a polynomial function and $P(1.7) = 0$. According to the Factor Theorem, what can you conclude?

In 7 and 8, an equation for a polynomial function is given.
   a. Factor the polynomial.
   b. Find the zeros of the function.

7. $r(t) = 2t^3 - t^2 - 21t$ 8. $w(b) = -b^4 + 4b^3 + 11b^2 - 30b$

9. **a.** Write the general form of an equation for a polynomial function whose only zeros are 5.6 and –2.9.

   b. Graph two different third-degree polynomial functions satisfying the condition in Part a.

10. **Multiple Choice** A polynomial has real zeros 4, –2, and 6. What must the degree of the polynomial be?

    A 2   B 3   C 8

    D 12   E an integer ≥ 3

APPLYING THE MATHEMATICS

11. The graph of a polynomial function is shown at the right. There are no other x-intercepts and the degree of the function is 4. Give an equation for the function.

12. **a.** Is it possible to have a polynomial with integer coefficients that has 3 and $\frac{3}{5}$ as zeros, and no other zeros? Justify your answer.

    b. Is it possible to have a polynomial with integer coefficients that has 3 and $\sqrt{5}$ as zeros, and no other real zeros? Justify your answer.

13. Consider the polynomial function $P(x) = x^3 + 5x^2 + 3x$.

    a. Factor $P(x)$ over the rationals.

    b. Use the factors you found in Part a to find the zeros of the function.

    c. Your answer to Part b should suggest that there are other factors. Use a CAS to factor over the reals and compare your result to your factorization in Part a.

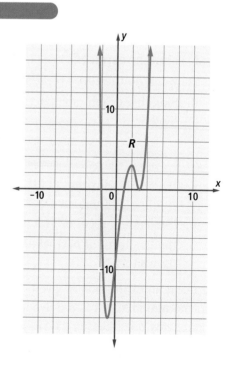

14. Find all possible values of a such that $x - 2$ is a factor of $x^3 + x^2 + ax + 4$.

15. The graph of the polynomial function with equation $y = \frac{1}{2}(x^2 + 1) \cdot (x + 2)^2(x - 3)$ has only two x-intercepts: -2 and 3. Why does the factor $x^2 + 1$ not change the number or location of the x-intercepts?

16. Let $q(x) = 3x^2 - 7x - 6$.
    a. Graph q. Where do the roots of $q(x)$ appear to be located?
    b. Use the quadratic formula to solve $3x^2 - 7x - 6 = 0$.
    c. Use your result from Part b to factor $q(x)$.

17. Let $p(x) = k(x - 2)^2(x + 3)$, where $k \neq 0$.
    a. What are the roots of $p(x)$?
    b. If $p(-2) = 4$, find the value of k.

18. Let $q(m)$ be a polynomial, and let $q(m) = 0$ when $m = -3$, $m = 1$, and $m = 4$.
    a. Write a possible 3rd-degree equation for q.
    b. Write a possible equation for q that has degree 4.
    c. Graph your equations from Parts a and b. Describe any similarities and differences you see.

19. A manufacturer determines that n employees on a production line will produce $f(n)$ units per month where $f(n) = 80n^2 - 0.1n^4$.

    a. Factor the polynomial over the reals.
    b. Find the zeros of f.
    c. What do the zeros represent in this situation?
    d. Sketch a graph of f. Give a reasonable domain for this model.

REVIEW

20. Consider the polynomial $p(x) = (2x - 3)(x^2 + 4)(10 - x)$. Without expanding the polynomial, find
    a. the leading term of $p(x)$ when written in standard form.
    b. the last term of $p(x)$ when written in standard form.
    (Lesson 11-2)

In 21–23, an expression is given. Tell whether the expression is a polynomial in x. If it is a polynomial in x, give its degree. If not, indicate why not. (Lesson 11-1)

21. $x^2y + xy^2 - \frac{1}{x}y^3$ 22. $\log(x^2y + xy^2)$ 23. $\sqrt{3}x^2y + \frac{\pi}{2}z$

In 24 and 25, use the graph below of the sine function.

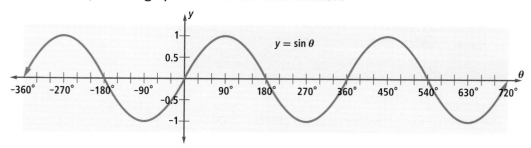

24. What is the period of this function? (**Lesson 10-6**)

25. **Fill in the Blanks** As θ increases from $90°$ to $180°$, the value of $\sin \theta$ decreases from __?__ to __?__. (**Lesson 10-6**)

26. In the general compound interest formula $A = P\left(1 + \frac{r}{n}\right)^{nt}$, as the value of n increases while r and t are kept constant, the value of A gets closer and closer to what number? (**Lesson 9-3**)

27. a. **True or False** $\sqrt{12} \cdot \sqrt{3} = \sqrt{2} \cdot \sqrt{18}$. Justify your answer.
    b. **True or False** $\sqrt{-12} \cdot \sqrt{-3} = \sqrt{2} \cdot \sqrt{18}$. Justify your answer. (**Lessons 8-7, 8-5**)

28. A student, removing the bolts from the back of a large cabinet in a science lab, knew that it was easier to turn a bolt with a long wrench than with a short one. The student decided to investigate the force required with wrenches of various lengths, and obtained the data at the right. (**Lessons 2-7, 2-6, 2-2**)
    a. Graph these data points.
    b. Which variation equation is a better model for this situation, $F = \frac{k}{L}$ or $F = \frac{k}{L^2}$? Justify your answer.
    c. How much force would be required to turn one of these bolts with a 12-inch wrench?

Length of Wrench (in.) L	Force (lb) F
3	120
5	72
6	60
8	45
9	40

EXPLORATION

29. If $P(x) = (x - 5)(x - 3)^2(x + 1)^3$, 3 is called a *zero of multiplicity 2* and –1 is called a *zero of multiplicity 3*. Give equations for several different polynomial functions whose only real zeros are 5, 3, and –1, and graph them. How does the multiplicity of a zero affect the way the graph looks at the zero?

QY ANSWERS

1. $x - 5$

2. The factorization is $(x + 15)(x - 15)$. The roots are 15 and –15.

Lesson

11-5

The Rational-Root Theorem

▶ **BIG IDEA** The Rational-Root Theorem gives a criterion that any rational root of a polynomial equation must satisfy, and typically limits the number of rational numbers that need to be tested to a small number.

Remember that a *rational number* is a number that can be written as a quotient of two integers, that is, as a simple fraction. Any integer n is a rational number, because n can be written as $\frac{n}{1}$. Zero is a rational number because, for example, $0 = \frac{0}{26}$. The number 10.32 is rational because $10.32 = 10 + \frac{3}{10} + \frac{2}{100} = \frac{1032}{100}$. An *irrational number* is a real number that is not rational. The irrational numbers are infinite nonrepeating decimals, including $\sqrt{60}$, $\pi + 5$, and e. Nonreal complex numbers such as $3i$ and $i - 4$ are neither rational nor irrational.

Every *real* zero of a function f corresponds to an x-intercept of the graph of $y = f(x)$. But it can be difficult to tell from a graph which zeros are rational. However, there is a theorem that details the possible rational zeros that a polynomial function can have. The following Activity is about this theorem.

Mental Math

a. Which is the better value, an 8-oz box of pasta for $2.50 or a 12-oz box of pasta for $3.50?

b. Should you buy a $30 sweater using a 30%-off coupon or using a $10-off coupon?

c. Do you save more when a $25 instant rebate on contact lenses is taken before tax is calculated or after tax is calculated?

d. Which will pay more interest, an account with 3.5% interest compounded annually, or an account with continuously compounded interest with an APY of 3.5%?

Activity

Let $Q(x) = (2x - 3)(9x + 4)$ and $P(x) = (2x - 3)(9x + 4)(5x + 7)$.

Step 1 Solve $Q(x) = 0$ and $P(x) = 0$ for x.

Step 2 Without expanding the polynomials, find the first and last terms of $Q(x)$ and $P(x)$ when they are written in standard form.

Step 3 Describe the connection between the denominators of the roots of the polynomial equation $Q(x) = 0$ and the coefficient of the first term of the expanded polynomial $Q(x)$. Repeat for $P(x)$.

Step 4 Describe the connection between the numerators of the roots of the polynomial equation $Q(x) = 0$ and the constant term of the expanded polynomial $Q(x)$. Repeat for $P(x)$.

A generalization of the results of the Activity is called the *Rational-Root Theorem*, or *Rational-Zero Theorem*.

> ### Rational-Root (or Rational-Zero) Theorem
>
> Suppose that all the coefficients of the polynomial function described by
>
> $$f(x) = a_n x^n + a_{n-1} x^{n-1} + \ldots + a_2 x^2 + a_1 x + a_0$$
>
> are integers with $a_n \neq 0$ and $a_0 \neq 0$. If $\frac{p}{q}$ is a root of $f(x)$ in lowest terms, then p is a factor of a_0 and q is a factor of a_n.

Stated another way, the Rational-Root Theorem says that if a simple fraction in lowest terms (a rational number) is a root of a polynomial function with integer coefficients, then the numerator of the rational root is a factor of the constant term of the polynomial, and the denominator of the rational root is a factor of the leading coefficient of the polynomial. A proof of this theorem is left to a later course.

Identifying Possible and Actual Rational Roots

Notice that the Rational-Root Theorem gives a way to decide the *possible* rational roots of a polynomial. It does not determine which of these possible roots are actual roots of the polynomial.

> ### Example 1
>
> Apply the Rational-Root Theorem to *identify* possible rational roots of $f(x)$ when $f(x) = 4x^4 + 3x^3 + 4x^2 + 11x + 6$.
>
> **Solution** Let $\frac{p}{q}$ in lowest terms be a rational root of $f(x)$. Then p is a factor of 6 and q is a factor of 4.
>
> So, p equals $\pm1, \pm2, \pm3$, or ±6, and q equals $\pm1, \pm2$, or ±4. Now take all possible quotients $\frac{p}{q}$. It looks like there are as many as $8 \cdot 6 = 48$ possible quotients, but actually there are fewer because many of them, such as $\frac{6}{2}$ and $\frac{-3}{-1}$, are equal. So, the possible rational roots are $\pm1, \pm2, \pm3, \pm6, \pm\frac{1}{2}, \pm\frac{3}{2}, \pm\frac{1}{4}$, and $\pm\frac{3}{4}$.

🛑 **QY1**

Example 1 shows that there can be several possible rational roots for a polynomial. You can test each possible root by hand, but with a graphing utility, a CAS, the Rational-Root Theorem, and the Factor Theorem, you can greatly reduce the time it takes to identify roots.

> ▶ **QY1**
>
> $\frac{2}{4}$ and $\frac{6}{2}$ are both possible rational roots of $f(x)$ in Example 1. Why do they not appear in this form on the list in the solution to Example 1?

Example 2

Use the Rational-Root Theorem to *find* all rational roots of
$f(x) = 4x^4 + 3x^3 + 4x^2 + 11x + 6$ from Example 1.

Solution The possible rational roots of
$f(x)$ are $\pm 1, \pm 2, \pm 3, \pm 6, \pm\frac{1}{2}, \pm\frac{3}{2}, \pm\frac{1}{4}$, and $\pm\frac{3}{4}$.

The smallest possible rational root is –6, and the largest is 6,
so graph f on the interval $-6 \le x \le 6$ to see where rational
roots might be.

The graph at the right shows that there are two real roots on
the interval $-1 \le x \le 0$.

The possible rational roots for $f(x)$ in that interval are
$-\frac{1}{4}, -\frac{1}{2}, -\frac{3}{4}$, and –1. Test these values in the equation for f.

$$f\left(-\frac{1}{4}\right) \approx 3.47 \qquad f\left(-\frac{1}{2}\right) \approx 1.38$$

$$f\left(-\frac{3}{4}\right) = 0 \qquad f(-1) = 0$$

So, –1 and $-\frac{3}{4}$ are the only rational roots of $f(x)$.

Check Use a graphing utility to find the x-intercepts of the
graph of f. The answers check.

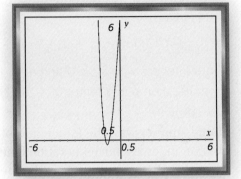

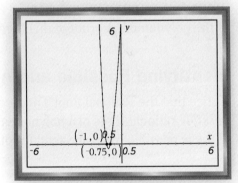

Irrational Roots

The Rational-Root Theorem identifies all possible rational roots of a
polynomial with integer coefficients. So, any real roots that are not
identified must be irrational.

Example 3

In a Hans Magnus Enzensberger book, a boy named Robert argues with a
sprite about $\sqrt{2}$. The sprite tries to convince Robert that $\sqrt{2}$ is irrational
by showing him that its decimal expansion is infinite. How could the sprite
show that $\sqrt{2}$ is irrational using the Rational-Root Theorem?

Solution $\sqrt{2}$ is a solution to the equation $x^2 = 2$ and a root
of $x^2 - 2 = 0$. By the Rational-Root Theorem, if $\frac{a}{b}$ is a rational root
of $x^2 - 2 = 0$, then a is a factor of 2 and b is a factor of 1.

Thus, the only possible rational roots of $x^2 - 2 = 0$ are $\pm\frac{2}{1}$ and $\pm\frac{1}{1}$, that is, 2, –2, 1, and –1.

Substitute to see if any of these numbers is a root of the equation $x^2 - 2 = 0$.

$$1^2 - 2 = -1 \neq 0 \quad (-1)^2 - 2 = -1 \neq 0$$
$$2^2 - 2 = 2 \neq 0 \quad (-2)^2 - 2 = 2 \neq 0$$

None of the possible rational roots is a root of the equation. Thus, $x^2 - 2 = 0$ has no rational roots, so $\sqrt{2}$ must be irrational.

 QY2

> ▶ QY2
>
> What equation can you consider in order to prove that $\sqrt[3]{9}$ is irrational?

Questions

COVERING THE IDEAS

1. Suppose that $\frac{7}{5}$ is a root of a polynomial equation with integer coefficients. What can you say about the leading coefficient and the constant term of the polynomial?

2. Does the Rational-Root Theorem apply to finding roots of $Q(x) = 10x^4 - 4\sqrt{2}x + 9$? Explain your response.

In 3 and 4, a polynomial equation is given.
   a. Use the Rational-Root Theorem to list the possible rational roots.
   b. Find all of the rational roots.

3. $8x^3 - 4x^2 + 44x + 24 = 0$

4. $f(n) = -n^2 + 3n^3 + 5n^5 - 8n + 12 - 11n^4$

5. The graph of a polynomial function with equation $R(x) = 7x^4 - 2x^3 - 42x^2 - 61x - 14$ is shown at the right.
   a. Using the Rational-Zero Theorem, list all possible rational zeros of R.
   b. Which possible rational zeros of R are actual zeros? How did you decide which roots to test?

6. Prove that $\sqrt[3]{9}$ is irrational.

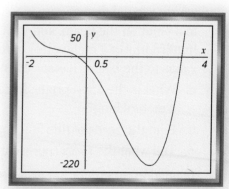

APPLYING THE MATHEMATICS

7. Consider the function f, where $f(n) = 2n^4 + 5n^3 + 12$.
   a. List all possible rational zeros of this function.
   b. Graph f and explain why it has no rational zeros.

8. Explain how the Rational-Root Theorem and the Factor Theorem can be used to factor a polynomial equation with only rational roots.

9. When a polynomial equation with integer coefficients has a root of the form $a + \sqrt{p}$, where p is not a perfect square, then $a - \sqrt{p}$ must also be a root of the equation.

   a. Use this fact to find a polynomial equation with integer coefficients in which $a + \sqrt{p}$ is a root.

   b. If p is a prime number and a is an integer, show that $a + \sqrt{p}$ is irrational.

10. Let $f(x) = a_3x^3 + a_2x^2 + a_1x + a_0$ and
    $g(x) = 7a_3x^3 + 7a_2x^2 + 7a_1x + 7a_0$,
    where a_3, a_2, a_1, and a_0 are integers with $a_3 \neq 0$ and $a_0 \neq 0$.

    a. How are the possible rational zeros of these functions related? Explain your reasoning.

    b. Let $f(x)$ be defined as in Part a and $h(x) = k \cdot f(x)$, where k is a nonzero constant. How are the possible rational zeros of f and h related?

REVIEW

11. A horizontal beam has its left end built into a wall, and its right end resting on a support, as shown at the right. The beam is loaded with weight uniformly distributed along its length. As a result, the beam sags downward according to the equation $y = -x^4 + 24x^3 - 135x^2$, where x is the distance (in meters) from the wall to a point on the beam, and y is the distance (in hundredths of a millimeter) of the sag from the x-axis to the beam. **(Lesson 11-3)**

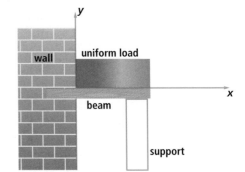

    a. What is the appropriate domain for x if the beam is 9 meters long?

    b. Find the zeros of this function.

    c. Tell what the roots represent in this situation.

12. Insulation tubing for hot water pipes is shaped like a cylindrical solid of outer radius R from which another cylindrical solid of inner radius r has been removed. The figure at the right shows a piece of insulation tubing. **(Lessons 11-1, 1-7, Previous Course)**

    a. Suppose the piece of tubing has length L. Find a formula for the volume V of the tube in terms of r, R, and L.

    b. Suppose $R = 1$ inch and $L = 6$ feet. Write a formula for V as a function of r.

    c. What is the degree of the polynomial in Part b?

13. The building code in one state specifies that accessibility ramps into public swimming pools must not drop more than one inch for every horizontal foot. What is the maximum angle of depression a ramp can make with the surface of the water? (**Lesson 10-2**)

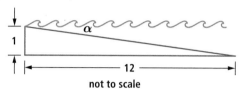

not to scale

14. Find the roots of $z^2 + 2z + 6 = 0$. (**Lesson 6-10**)

15. Triangle MAP is shown below. (**Lessons 4-7, 4-6**)

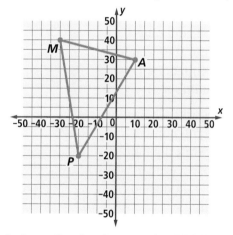

  a. Graph the reflection image of $\triangle MAP$ over the x-axis. Label the vertices M', A', and P', respectively.

  b. Graph the reflection image of $\triangle M'A'P'$ over the y-axis. Label the vertices M'', A'', and P'', respectively.

  c. What single transformation maps $\triangle MAP$ onto $\triangle M''A''P''$?

EXPLORATION

16. Find information about *Descartes' Rule of Signs* for polynomials, and then explain what it tells you about $P(x)$ from the Activity in this lesson.

Lesson
11-6

Solving All Polynomial Equations

Vocabulary

double root, root of multiplicity 2

multiplicity

▶ **BIG IDEA** Every polynomial equation of degree $n \geq 1$ has exactly n solutions, counting multiplicities.

What Types of Numbers Are Needed to Solve Polynomial Equations?

An exact solution to any linear equation with real coefficients is always a real number, but exact solutions to quadratic equations with real coefficients sometimes are not real. It is natural to wonder whether exact solutions to higher-order polynomial equations require new types of numbers beyond the complex numbers.

As early as 2000 BCE, the Babylonians developed algorithms to solve problems using quadratic polynomials. The study of polynomials progressed to cubics and quartics, with much early work done in Persia, China, and Italy. In 1535, Niccolo Tartaglia discovered a method for solving all cubic equations. Soon after that, Ludovico Ferrari discovered how to solve any quartic equation.

Surprisingly, no numbers beyond complex numbers are needed to solve linear, quadratic, cubic, and quartic polynomial equations. However, when over 250 years passed without finding a general solution to every *quintic* (5th-degree) polynomial equation, mathematicians began wondering whether new numbers might be needed. At last, in 1824, the Norwegian mathematician Niels Henrik Abel proved that there is no general formula for solving quintic polynomial equations. Shortly after that, Evariste Galois deduced that there is no general formula involving complex numbers for finding roots of all polynomials of degree five or higher.

But how do we know that new numbers beyond the complex numbers were not necessary? In 1797, at the age of 18, Carl Gauss offered a proof of the following theorem, the name of which indicates its significance.

Fundamental Theorem of Algebra

Every polynomial equation $P(x) = 0$ of any degree ≥ 1 with complex number coefficients has at least one complex number solution.

Mental Math

How many lines of symmetry does each of the following have at a minimum?

a. a rhombus

b. a rectangle

c. an isosceles triangle

d. a circle

Carl Friedrich Gauss is often called "the prince of mathematicians."

From the Fundamental Theorem of Algebra and the Factor Theorem it is possible to prove that *every solution to a polynomial equation with complex coefficients is a complex number.* (Remember that complex numbers include the real numbers.) Thus, no new type of number is needed to solve higher-degree polynomial equations. So, for instance, the solutions to $x^5 + 2x^4 - 3ix^2 + 3 + 7i = 0$ are complex numbers.

How Many Complex Solutions Does a Given Polynomial Equation Have?

It is easy to show that every solution to a polynomial equation is complex when the degree of the polynomial is small. For example, if $a \neq 0$, the linear equation $ax + b = 0$ has one root: $x = -\frac{b}{a}$. Therefore, every linear polynomial has exactly one complex root.

The quadratic equation $ax^2 + bx + c = 0$, $a \neq 0$, has two solutions, given by the quadratic formula: $x = \frac{-b \pm \sqrt{b^2 - 4ac}}{2a}$. When the discriminant $b^2 - 4ac > 0$, the roots are real numbers. When $b^2 - 4ac < 0$, the roots are nonreal complex numbers. When the discriminant $b^2 - 4ac = 0$, the two roots are equal, and the root is called a **double root**, or a **root of multiplicity 2**. For instance, consider $x^2 - 10x + 25 = 0$. Here $b^2 - 4ac = (-10)^2 - 4 \cdot 1 \cdot 25 = 0$, so there is a double root. Because $x^2 - 10x + 25 = (x - 5)^2$, that double root is 5. So every quadratic has two complex roots, although the two roots might not be distinct.

In general, the **multiplicity** of a root r in an equation $P(x) = 0$ is the highest power of the factor $x - r$. For example, in $17(x - 4)(x - 11)^3 = 0$, 11 is a root of multiplicity 3, and 4 is a root of multiplicity 1. So $P(x) = 0$ has four roots altogether.

 QY1

▶ **QY1**

State the multiplicity of each root of $4(x - 3)^2 \cdot (x + 5)^4(x - 17) = 0$. How many roots does the equation have altogether?

Activity

Consider the polynomial function P defined by $P(x) = x^3 + 2x^2 - 14x - 3$.

Step 1 Verify that 3 is a zero of P. What theorem justifies the conclusion that $x - 3$ is a factor of $P(x)$?

Step 2 If $P(x) = (x - 3)Q(x)$, what is the degree of $Q(x)$?

Step 3 Without finding $Q(x)$ explicitly, how many roots does the equation $Q(x) = 0$ have? Why are these roots also roots of $P(x) = 0$?

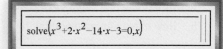

Step 4 Check your answer to Step 3 by solving $P(x) = 0$ on a CAS.

The Fundamental Theorem of Algebra only guarantees the existence of a single complex (possibly real) root r_1 for any polynomial $P(x)$. However, using the Factor Theorem, you can rewrite $P(x) = (x - r_1) \cdot Q(x)$, where the degree of $Q(x)$ is one less than the degree of $P(x)$. By applying the Fundamental Theorem of Algebra to $Q(x)$, you get another root r_2, and therefore another factor. So you could write $P(x) = (x - r_1)(x - r_2)Q_2(x)$, where the degree of $Q_2(x)$ is *two* less than the degree of $P(x)$. This process can continue until you "run out of degrees," that is, until $Q(x)$ is linear.

For example, if you start with a 4th degree polynomial $P(x)$, you can factor it as

$$\begin{aligned} P(x) &= (x - r_1)Q_1(x) \\ &= (x - r_1)(x - r_2)Q_2(x) \\ &= (x - r_1)(x - r_2)(x - r_3)Q_3(x), \end{aligned}$$

and now $Q_3(x)$ is linear. So, $P(x) = 0$ has four roots: $r_1, r_2, r_3,$ and the single root of $Q_3(x) = 0$. These four roots are *not* necessarily four different real numbers. In fact, some or all of them may be nonreal complex numbers. This conclusion is summarized in the following theorem.

> **Number of Roots of a Polynomial Equation Theorem**
>
> Every polynomial equation of degree n has exactly n roots, provided that multiple roots are counted according to their multiplicities.

 QY2

Finding Real Solutions to Polynomial Equations

The Number of Roots of a Polynomial Equation Theorem tells you how many roots a polynomial equation $P(x) = k$ has. It does not tell you how to find the roots, nor does it tell you how many of the roots are real. To answer these questions, you can apply the methods studied in this chapter for finding and analyzing zeros of polynomial functions.

> ▸ **QY2**
>
> How many roots does each equation have?
>
> **a.** $5x^{12} - 64x^3 + 4x^2 + 1 = 0$
>
> **b.** $4ex^5 + 3ix^2 - (2 + i)x + 4 = 0$

Example

Consider the polynomial function P defined by
$P(x) = x^5 - x^4 - 21x^3 - 37x^2 - 98x - 24.$

a. How many real zeros does P have?

b. How many nonreal complex zeros does P have?

Solution 1

a. Solve $P(x) = 0$ on a CAS in real-number mode. The CAS shows that P has three real zeros. One zero, 6, is rational, but the other two real zeros, $-2 \pm \sqrt{3}$, are irrational.

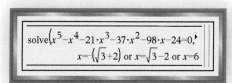

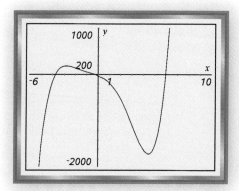

Confirm with a graph of $y = P(x)$ that the three x-intercepts correspond to the three real zeros.

b. Because the degree of $P(x)$ is five, there are five zeros altogether by the Number of Roots of a Polynomial Equation Theorem. Since there are three real zeros, there are two nonreal complex zeros. Solve in complex-number mode on a CAS to find all five real and nonreal zeros. There are two nonreal zeros, $-\frac{1}{2} \pm \frac{\sqrt{15}}{2}i$. So, of the five zeros of $P(x)$, one is rational, two are irrational, and two are nonreal complex.

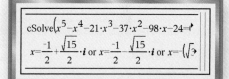

Check Factor $P(x)$ on a CAS in complex mode. Part of the solution line is shown at the right. The entire output is

$$\frac{(x - 6)(x + \sqrt{3} + 2)(x - \sqrt{3} + 2)(2x - (-1 + \sqrt{15}\,i))(2x + 1 + \sqrt{15}\,i)}{4}.$$

When the constant $\frac{1}{4}$ is factored out, there are five factors that can be set equal to zero and solved for x. Solving shows that **one root of P(x) is rational, two are irrational, and two are nonreal. It checks.**

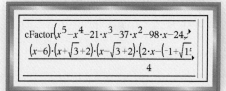

Questions

COVERING THE IDEAS

1. State the Fundamental Theorem of Algebra.

2. Zelda is confused. The polynomial $x^3 - 25x$ is supposed to have three complex roots, but the roots 0, 5, and –5 are all real numbers. Resolve Zelda's confusion.

3. **Multiple Choice** The equation $\pi w^3 - 3iw + \frac{1}{17} = 0$ has how many complex solutions?

   A none   **B** two   **C** three   **D** four

4. Who first proved the Fundamental Theorem of Algebra?

In 5 and 6, $a \neq 0$. Solve for x.

5. $ax + b = 0$

6. $ax^2 + bx + c = 0$

In 7–9, an equation is given. Find and classify all solutions as rational, irrational, or nonreal. Then, identify any multiple roots.

7. $x^2 + 16x + 64 = 0$

8. $y^3 - 8y = 0$

9. $(z - 3)^4(7z - 22)(z^2 - 6z + 25) = 0$

10. a. The equation $x^3 + x^2 - 8x - 12 = 0$ has only two roots, as the CAS screen shows at the right. However, the Number of Roots of a Polynomial Equation Theorem says that a cubic equation has three roots. How is this possible?

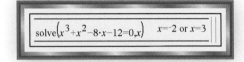

b. Without using a CAS, give a factorization of $x^3 + x^2 - 8x - 12$ that explains the result in Part a.

11. How many complex roots does the equation $x^5 + 14ix = 0$ have?

APPLYING THE MATHEMATICS

12. Consider the polynomial function P defined by $P(x) = 4x^5 + x^3 - 5x^2 - 18x + 10$ graphed at the right.

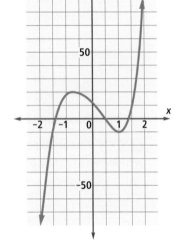

a. How many real zeros does P have?

b. How many nonreal zeros does it have?

c. The graph shows that the real zeros are approximately -1.4, $\frac{1}{2}$, and 1.4. Which of these is an exact zero?

d. How many irrational zeros does this function have?

In 13 and 14, solve the equation.

13. $-4ix + 8 + 16i = 0$

14. $ix^2 + 10x + 11i = 0$

15. Consider $f(x) = x^5 - 6x^4 - 5x^3 + 42x^2 + 40x + k$, where k is a real number. The case where $k = 0$ is graphed at the right.

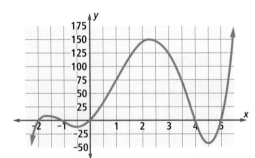

a. What effect does changing the value of k have on the graph of $y = f(x)$?

b. Find a value of k for which f has exactly two nonreal zeros.

c. For what values of k (approximately) does f have exactly four nonreal zeros?

16. The 4th roots of -16 are nonreal complex numbers. If z is a 4th root of -16, then $z^4 = -16$ and $z^4 + 16 = 0$.

a. How many complex solutions does $z^4 + 16 = 0$ have?

b. Use a CAS to find all of the complex fourth roots of -16.

c. By similar logic as in Parts a and b, -8 has __?__ complex cube roots. Find them.

REVIEW

17. A 3rd-degree polynomial with leading coefficient 3 has roots $-4, \frac{2}{3}$, and 5. Find an equation for the polynomial. (**Lesson 11-4**)

18. The sum of the cube and the square of a number is 3. To the nearest thousandth, what is the number? (**Lessons 11-3, 11-1**)

19. Myron makes a bread pan by cutting squares of side length w from each corner of a 5-inch by 10-inch sheet of aluminum and folding up the sides. Let $V(w)$ represent the volume of the pan. (**Lesson 11-2**)
    a. Find a polynomial formula for $V(w)$.
    b. Can Myron's process yield a pan with volume of 25 in³?
    c. What value of w produces the pan with the greatest volume?

20. a. Which is more help in finding the distance c in the diagram at the right, the Law of Sines or the Law of Cosines?
    b. Find c. (**Lesson 10-8**)

In 21 and 22, solve. (**Lessons 10-6, 9-5**)

21. $2 \log x = 4$ 22. $2 \sin x = 4$

23. a. State an equation for the image of the graph of $y = 3x^2$ under $T_{-3, 4}$.
    b. Graph the image in Part a. (**Lesson 6-3**)

24. The wind force on a vertical surface varies jointly as the area A of the surface and the square of the wind speed S. The force is 340 newtons on a vertical surface of 1 m² when the wind blows at $18 \frac{m}{\sec}$. Find the force exerted by a $35 \frac{m}{\sec}$ wind on a vertical surface of area 2 m². (*Note:* One newton equals one $\frac{\text{kilogram-meter}}{\sec^2}$.) (**Lesson 2-9**)

25. Make a spreadsheet to create the table of values for $y = 4x^2 - 13x$ as shown at the right. Use the table to find y when $x = 3.7$. (**Lesson 1-6**)

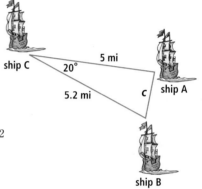

EXPLORATION

26. **Fill in the Blanks** Because $1^n = 1$ for all n, $x^n = 1$ has the solution 1 for all n. Thus, for all n, by the Factor Theorem, $x^n - 1$ has the factor $x - 1$.
    a. $x^2 - 1$ is the product of $x - 1$ and ___?___.
    b. $x^3 - 1$ is the product of $x - 1$ and ___?___.
    c. $x^4 - 1$ is the product of $x - 1$ and ___?___.
    d. Generalize Parts a–c.

Lesson

11-7 Finite Differences

Vocabulary

method of finite differences

▶ **BIG IDEA** Given n points in the plane, no two of which have the same first coordinate, it is possible to determine whether there is a polynomial function of degree less than $n − 1$ that contains all the points.

When you find an equation of a line through two data points, you have an *exact model* for the data. When you find an equation for a regression line through a set of data points that are roughly linear, you have an *approximate model* for the data. Similarly, you can find exact and approximate quadratic and exponential models. This lesson is about finding exact polynomial models through data points.

Mental Math

Factor.

a. $c^{14} − 100$

b. $r^2 − 10r + 25$

c. $\pi x^4 − 13\pi x^3$

d. $4p^2 + 12pq + 9q^2$

Consider the data points and their scatterplot at the right. The graph looks like part of a parabola, perhaps with its vertex at the origin, so it is reasonable to think that a quadratic polynomial function models these data. But the graph also looks somewhat like an exponential function, translated down one unit.

s	0	10	20	30	40
d	0	8	33	75	134

Differences between Values of Polynomial Functions

It is possible to determine that a quadratic function is an exact model for these data. The determination relies on finding differences between certain values of the function.

Consider the spreadsheet at the right. Columns A and B show values of x and y, respectively, for the linear function with equation $y = 2x + 7$ when $x = 1, 2, 3, 4, 5,$ and 6.

The values in the cells of column C are the differences between consecutive values in the cells of column B.

C2 = B2 − B1 = 11 − 9 = 2;
C3 = B3 − B2 = 13 − 11 = 2;
C4 = B4 − B3 = 15 − 13 = 2; and so on.

Notice that all these differences are 2, which is the slope of the line with equation $y = 2x + 7$. So, the spreadsheet shows the constant increase of the linear polynomial $2x + 7$.

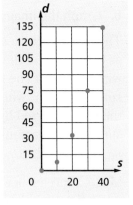

Activity 1

Step 1 **a.** Make a spreadsheet to show x- and y-values for the quadratic polynomial function with equation $y = 4x^2 - 5x - 3$, for $x = 1$ to 7. The first six rows of our spreadsheet are shown at the right.

b. Define a third column showing the difference between consecutive y-values.

c. Define a fourth column showing the difference between consecutive cells of the third column.

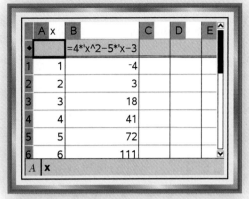

Step 2 **a.** Make another spreadsheet and repeat Step 1 for the cubic polynomial function with equation $y = x^3 - 3x^2 + 4x - 5$.

b. Define a fifth column that finds the difference between consecutive terms of the fourth column.

Step 3 Make a conjecture about what will happen if you calculate four consecutive sets of differences of y-values with $y = -2x^4 + 8x^3 + 11x^2 - 3x$ for $x = 1$ to 7.

The Polynomial-Difference Theorem

Activity 1 shows that if you evaluate a polynomial of degree n for consecutive integer values of x and take differences between consecutive y-values, then after n sets of differences you get a constant difference. You can see the results of the calculations for the polynomial of Step 3 at the right.

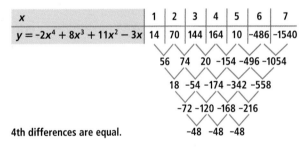

4th differences are equal.

Consider again the linear function $y = 2x + 7$. Instead of using consecutive integers for x-values, use the arithmetic sequence –5, –1, 3, 7, 11, ...

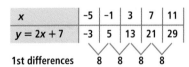

1st differences

Again the 1st differences are equal. Each of these examples is an instance of the following theorem.

Polynomial-Difference Theorem

$y = f(x)$ is a polynomial function of degree n if and only if, for any set of x-values that form an arithmetic sequence, the nth differences of corresponding y-values are equal and the $(n-1)$st differences are not equal.

The Polynomial-Difference Theorem provides a technique to determine whether a polynomial function of a particular degree can be an exact model for a set of points. The technique is called the **method of finite differences**. From a table of y-values corresponding to an arithmetic sequence of x-values, take differences of consecutive y-values and continue to take differences of the resulting y-value differences as needed. Only if those differences are eventually constant is the function a polynomial function, and the number of the differences indicates the polynomial function's degree.

Example 1

Consider the data from the beginning of the lesson. Use the method of finite differences to determine the degree of the polynomial function mapping s onto d.

Solution Notice that the values of the independent variable s form an arithmetic sequence, so the Polynomial-Difference Theorem applies. Calculate the differences.

s	0	10	20	30	40
d	0	8	33	75	134

1st differences are not equal.   8 25 42 59

2nd differences are equal.   17 17 17

d is a 2nd-degree polynomial function of s because the 2nd differences are equal.

The Polynomial-Difference Theorem does *not* generalize to nonpolynomial functions.

Activity 2

Step 1   Make a spreadsheet to show x- and y-values of $y = 3^x$ for $x = 1$ to 7.

Step 2   Analyze the data using the method of finite differences through at least 3rd differences. Describe what you observe.

Activity 2 shows that when the method of finite differences is used with functions other than polynomial functions, the differences do not become constant.

Example 2

Consider the sequence a defined by the recursive formula
$$\begin{cases} a_1 = 3 \\ a_n = 2a_{n-1} + 1, \text{ for integers } n \geq 2. \end{cases}$$

a. Identify the first seven terms of this sequence.

b. Use the method of finite difference to decide if there is an explicit polynomial formula for this sequence.

Solution

a. From the recursive definition, the sequence is $3, 7, \underline{\ ?\ }, \underline{\ ?\ }, \underline{\ ?\ }, \underline{\ ?\ }, \dots$.

b. Take differences between consecutive terms.

The pattern of differences appears to repeat and (will/will not) eventually give constant differences. So, there (is/is not) an explicit $\underline{\ ?\ }$ formula for this sequence.

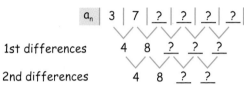

Questions

In 1 and 2, refer to Activity 1.

1. How many sets of differences did you need to take for the cubic function before the values were equal?

2. How many sets of differences would you need to take for the function with equation $y = 2x^2 + 3x^5 + x$?

3. If a set of x-values is an arithmetic sequence, the 11th differences of corresponding y-values are not all equal, but the 12th differences are equal, is y a polynomial function of x? If so, what is its degree? If not, why not?

4. a. According to the Polynomial-Difference Theorem, how many sets of differences will it take to get equal differences when $y = 5x^7 - 4x^5 + 6x^2$?

   b. Check your answer to Part a by making a spreadsheet to calculate y and the finite differences for $x = -2, -1, 0, 1, 2, 3, 4,$ and 5.

5. Tanner calculated three sets of finite differences for $y = x^2$ beginning with the table of values at the right. While Tanner knows this is a polynomial function, he did not get a constant set of differences. Explain why.

x	-3	0	1	4	5	7
y	9	0	1	16	25	49

6. Consider the sequence $\begin{cases} a_1 = 0.3 \\ a_n = a_{n-1} - 1 \text{ for integers } n \geq 2 \end{cases}$.

   a. Generate the first seven terms of the given sequence.

   b. Tell whether the sequence can be described explicitly by a polynomial function of degree less than 4.

In 7 and 8, use the data points listed in the table.

   a. Determine if y is a polynomial function of x of degree ≤ 5.

   b. If the function is a polynomial, state its degree.

7.

x	-3	-2	-1	0	1	2	3	4	5
y	67.5	16	1.5	0	-0.5	0	13.5	64	187.5

8.

x	0	1	2	3	4	5	6	7	8	9
y	1	1	0	-1	0	7	28	79	192	431

APPLYING THE MATHEMATICS

9. Consider the following sequence of sums of products of three consecutive integers.

   $f(1) = 1 \cdot 2 \cdot 3 = 6$

   $f(2) = 1 \cdot 2 \cdot 3 + 2 \cdot 3 \cdot 4 = 30$

   $f(3) = 1 \cdot 2 \cdot 3 + 2 \cdot 3 \cdot 4 + 3 \cdot 4 \cdot 5 = 90$

   $f(4) = 1 \cdot 2 \cdot 3 + 2 \cdot 3 \cdot 4 + 3 \cdot 4 \cdot 5 + 4 \cdot 5 \cdot 6 = 210$

   $f(5) = 1 \cdot 2 \cdot 3 + 2 \cdot 3 \cdot 4 + 3 \cdot 4 \cdot 5 + 4 \cdot 5 \cdot 6 + 5 \cdot 6 \cdot 7 = 420$

   $f(6) = 1 \cdot 2 \cdot 3 + 2 \cdot 3 \cdot 4 + 3 \cdot 4 \cdot 5 + 4 \cdot 5 \cdot 6 + 5 \cdot 6 \cdot 7 + 6 \cdot 7 \cdot 8 = 756$

   $f(7) = 1 \cdot 2 \cdot 3 + 2 \cdot 3 \cdot 4 + 3 \cdot 4 \cdot 5 + 4 \cdot 5 \cdot 6 + 5 \cdot 6 \cdot 7 + 6 \cdot 7 \cdot 8 + 7 \cdot 8 \cdot 9 = 1260$

   Determine whether or not there is a polynomial function f of degree less than 5 that models these data exactly.

10. Let $f(n)$ = the sum of the 4th powers of the integers from 1 to n.

    $f(1) = 1^4 = 1$

    $f(2) = 1^4 + 2^4 = 17$

    $f(3) = 1^4 + 2^4 + 3^4 = 98$

    $f(4) = 1^4 + 2^4 + 3^4 + 4^4 = 354$, and so on.

    a. Find $f(5), f(6), f(7)$, and $f(8)$.

    b. According to the Polynomial-Difference Theorem, what is the degree of the polynomial $f(n)$?

11. a. Fill in the table of values for the linear function with equation $y = mx + b$.

x	0	1	2	3	4
y	?	?	?	?	?

    b. Find the first differences for the table in Part a and explain your results.

12. If the second differences for $y = kx^2$ all equal $\frac{2}{9}$, what is the value of k?

REVIEW

13. Consider the function f with equation $f(x) = 3x^4 - 10x^2 - 8x + 10$. (Lessons 11-6, 11-5)

    a. How many roots does $f(x)$ have? Justify your answer.

    b. According to the Rational-Root Theorem, what are the possible rational roots?

    c. Find all roots of $f(x)$. Write all rational roots as fractions. Approximate all irrational roots to the nearest hundredth.

14. Find all the roots of $z^4 - 1 = 0$ by factoring and using the Zero-Product Theorem. (Lessons 11-6, 11-3)

15. Let $P(x) = (x - 1)(-x^2 + 3x + 3)$.

    a. Rewrite $P(x)$ in standard form.

    b. How many x-intercepts does the graph of $y = P(x)$ have? Find the exact value of the largest of these. (Lessons 11-4, 11-1, 6-7)

16. Refer to the figure at the right. A boat sails 20 miles from A to B, then turns $175°$ as indicated and sails 12 miles to point C. How far is A from C? (Lesson 10-8, Previous Course)

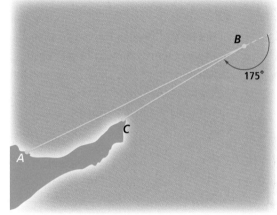

17. Solve the system $\begin{cases} 4x - 2y + 3z = 1 \\ 8x - 3y + 5z = 4 \\ 7x - 2y + 4z = 5 \end{cases}$ using matrices.

    (Lesson 5-6)

18. **Multiple Choice** Determine which equation describes the relationships graphed below, where k is a constant. (Lesson 2-8)

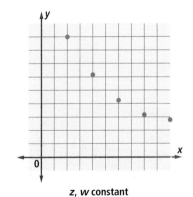

z, w constant

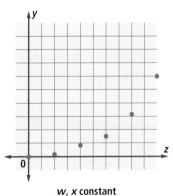

w, x constant

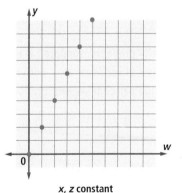

x, z constant

A $y = \dfrac{kwz}{x^2}$ B $y = kwxz$ C $y = \dfrac{kwz^2}{x}$ D $y = \dfrac{kwx^2}{z}$

EXPLORATION

19. Find a sequence for which the 3rd differences are all 30.

11-8

Modeling Data with Polynomials

> ▶ **BIG IDEA** Given *n* points in the plane, no two of which have the same first coordinate, it is generally possible to find an equation for a polynomial function that contains all the points.

In Lesson 11-7, you saw how to determine whether a polynomial formula exists for a given function. Now, if the formula exists, you will see how to find the coefficients of the polynomial.

Mental Math

Consider the equation $x^2 + bx + 4 = 0$. Give a value of *b* so that

a. the equation has no real roots.

b. the equation has two irrational roots.

c. the equation has two different rational roots.

d. the equation has a double root.

Example 1

In the land of Connectica, it was deemed that all cities built must be joined to every other city directly by a straight road. When the land was first developed, there were three cities and only three roads were needed. After a few years there were more cities and the government found itself having to build more and more roads.

Number of Cities	3	4	5	6	7
Number of Roads	3	6	10	15	21

The city planners wondered if there is a polynomial formula relating *r*, the number of roads needed, with *x*, the number of cities built. Does such a formula exist? If so, find the formula.

Solution Use the method of finite differences to determine the degree of a polynomial that will fit the data, if it exists.

Because the 2nd differences are equal, there is a quadratic polynomial that is an exact model for these five data points. That is,

$$r = ax^2 + bx + c.$$

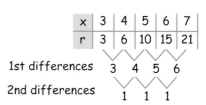

x	3	4	5	6	7
r	3	6	10	15	21

1st differences   3   4   5   6

2nd differences   1   1   1

Now you need to find values of the coefficients a, b, and c. As in Lesson 6-6, we find a, b, and c by solving a system of equations.

First, substitute three known ordered pairs (x, r) into the above equation. We choose $(3, 3)$, $(4, 6)$, and $(5, 10)$.

Substitute the ordered pairs into the equation to get the following system.

$$\begin{cases} 3 = a \cdot 3^2 + b \cdot 3 + c \\ 6 = a \cdot 4^2 + b \cdot 4 + c \\ 10 = a \cdot 5^2 + b \cdot 5 + c \end{cases}$$

You may solve this system by using linear combinations or by using matrices. Our matrix solution is shown below. You see that $a = \frac{1}{2}$, $b = -\frac{1}{2}$, and $c = 0$. So, $r = \frac{1}{2}x^2 - \frac{1}{2}x$ models the data from the city planners.

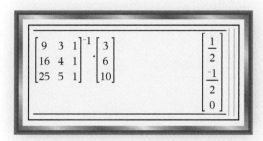

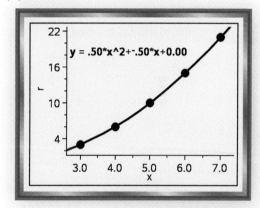

Check Enter the data into a CAS and run a quadratic regression. We obtain the results at the right. **It checks.**

STOP **QY1**

> **QY1**
>
> How many roads would be needed for 10 cities in Connectica?

Finding Higher-Degree Polynomials

The idea of Example 1 can be used to find a polynomial function of degree greater than 2 that exactly fits some data. Consider the sequence of squares shown on the next page.

Let S be the total number of squares of any size that can be found on an n-by-n checkerboard.

For example, on the 3-by-3 checkerboard there are:

9 squares this size: ☐, 4 squares this size: ⊞, and 1 square this size: ⊞ .

So, when $n = 3$, $S = 9 + 4 + 1 = 14$. There are 14 squares on a 3×3 board. The numbers of squares for $n = 1$ through 6 are given on the next page.

n	1	2	3	4	5	6
S	1	5	14	30	55	91

Example 2

How many squares are on an $n \times n$ checkerboard?

Solution First, use the method of finite differences to determine whether a polynomial model fits the data.

n	1	2	3	4	5	6
S	1	5	14	30	55	91
1st differences		4	?	?	?	?
2nd differences			?	?	?	?
3rd differences				?	?	?

The 3rd differences are constant. Therefore, **the data can be modeled by a polynomial function of degree __?__.**

Now use a system of equations to find a polynomial model. You know that the polynomial is of the form $S = an^3 + bn^2 + cn + d$. Substitute $n = 4$, 3, 2, then 1 and the corresponding values of S into the equation and solve the resulting system.

$$\begin{cases} 30 = a(4)^3 + b(4)^2 + c(4) + d \\ 14 = a(3)^3 + b(3)^2 + c(3) + d \\ 5 = a(2)^3 + b(2)^2 + c(2) + d \\ 1 = a(1)^3 + b(1)^2 + c(1) + d \end{cases}$$

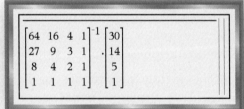

$$\begin{bmatrix} 64 & 16 & 4 & 1 \\ 27 & 9 & 3 & 1 \\ 8 & 4 & 2 & 1 \\ 1 & 1 & 1 & 1 \end{bmatrix}^{-1} \cdot \begin{bmatrix} 30 \\ 14 \\ 5 \\ 1 \end{bmatrix}$$

Thus, $a =$ __?__, $b =$ __?__, $c =$ __?__, and $d =$ __?__.

So, a formula for S in terms of n is $S =$ ____?____.

Check Use *cubic regression* on a CAS to find the model.

Modeling a Finite Set of Points

In Examples 1 and 2, you were asked to find a polynomial function model that related two variables, and in each case the independent variable could take on an infinite number of values. Consequently, you need a way to prove that the polynomial function works for all values. Sometimes you can appeal to the definition of the function (see Question 10). At other times you may need to use methods beyond the scope of this course.

However, when you need a model only to fit a finite number of values in a function, you can always find a polynomial model. Such a model exists even when the data do not follow a simple pattern.

Year	Population
1900	1,850,000
1920	2,284,000
1940	1,890,000
1960	1,698,000
1980	1,428,000
2000	1,537,000

The population of Manhattan Island (part of New York City) has gone up and down over the past 100 years. At the right are a table and a graph showing the population every 20 years from 1900 to 2000.

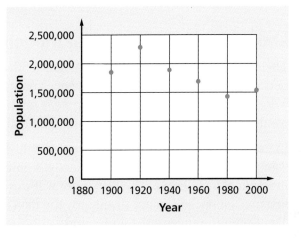

Let x = the number of 20-year periods since 1900. Let $P(x)$ be the population (in millions) of Manhattan at these times. Here is how to find a polynomial expression for $P(x)$ that contains the six points in the table.

Use the methods of Examples 1 and 2 and find 1st, 2nd, 3rd, 4th, and 5th differences. When you reach the 5th differences, there is only one value. So all the 5th differences are equal! Consequently, by the Polynomial-Difference Theorem, *for these six points* there is a polynomial of degree 5 that fits:

$$P(x) = ax^5 + bx^4 + cx^3 + dx^2 + ex + f.$$

To find a, b, c, d, e, and f, you need to solve a system of six linear equations in six variables. These equations arise from the six given data points. For instance, for the year 1940, $x = 2$, the number of 20-year periods since 1900, and $P(x) = 1.890$, the population of Manhattan in millions. Substituting 2 for x in the general formula above,

$$P(2) = 1.89 = a \cdot 2^5 + b \cdot 2^4 + c \cdot 2^3 + d \cdot 2^2 + e \cdot 2 + f$$
$$= 32a + 16b + 8c + 4d + 2e + f.$$

One World Trade Center, Manhattan, New York City

Similarly, you can find the other five equations. When the system is solved, it turns out that, to four decimal places, $a = 0.0171$, $b = -0.2252$, $c = 1.0962$, $d = -2.3823$, $e = 1.9282$, and $f = 1.85$. So,

$$P(x) = 0.0171x^5 - 0.2252x^4 + 1.0962x^3 - 2.3823x^2 + 1.9282x + 1.85.$$

Unfortunately, polynomial regression on most calculators stops at a 4th-degree polynomial. But some computer spreadsheets and data-analysis applications allow you to find regression equations to 5th and higher degrees. We found the equation for $P(x)$ using a spreadsheet. Although this polynomial model fits the data, because the decimal coefficients are rounded, evaluating the polynomial $P(x)$ will not produce exact populations.

 QY2

In general, if you have n points of a function, there exists a polynomial formula of some degree less than n that models those n points exactly.

Limitations of Polynomial Modeling

The model for Manhattan's population is a 5th-degree polynomial because you could take finite differences only five times. If there were more data, then the 5th differences might not be equal and you would need a larger degree polynomial to fit all the points.

> **QY2**

Substitute 3 for x in the formula for the population of Manhattan.
What year's population of Manhattan should you obtain?
How close is your result to that year's population?

In general, if you have n data points through which you want to fit a polynomial function exactly, a polynomial function of any degree $n - 1$ *or more* will work. For instance, suppose you are given only the data at the right. The first differences are 1 and 2, and there is only one 2nd difference. So, there is a quadratic function that fits the data. That function has the formula $y = \frac{x^2 - x + 2}{2}$. If (4, 7) and (5, 11) are two more given data points, then the second differences are still equal and the same formula models all the data.

x	1	2	3
y	1	2	4
1st differences		1	2
2nd difference			1

x	1	2	3	4	5
y	1	2	4	7	11
1st differences		1	2	3	4
2nd differences			1	1	1

However, if (4, 8) and (5, 15) are the next data points, then the second differences are no longer equal. The 3rd differences are equal, so there is a 3rd-degree polynomial equation modeling the data: $y = \frac{x^3 - 3x^2 + 8x}{6}$.

In this way, you can see that
$y = \frac{x^2 - x + 2}{2}$ and $y = \frac{x^3 - 3x^2 + 8x}{6}$ are only two of many polynomial formulas fitting the original three data points (1, 1), (2, 2), and (3, 4).

x	1	2	3	4	5
y	1	2	4	8	15
1st differences		1	2	4	7
2nd differences			1	2	3
3rd differences				1	1

Since many different polynomial functions can fit a set of n data points, polynomial models based on only a few points are usually not good models for making predictions. However, polynomial models do provide an efficient way to store data because the model fits all the given points exactly.

Questions

COVERING THE IDEAS

In 1–3, refer to Example 1.

1. How did the planners know what degree polynomial model would fit the data?

2. Show that the model is correct for $x = 6$.

3. Use the model to calculate how many roads would be needed for 20 cities.

In 4 and 5, refer to Example 2.

4. How many squares of any size are on a standard 8-by-8 checkerboard?

5. How many more squares of any size are created when you turn an 8-by-8 checkerboard into a 9-by-9 checkerboard?

6. Consider the data at the right.
   a. Determine the lowest possible degree of a polynomial function that fits these data.
   b. Find a formula for the function.

x	0	1	2	3	4	5
y	1	2	7	46	173	466

7. Suppose the data in the table at the right are modeled by a formula of the form $y = ax^2 + bx + c$. What three equations are satisfied by a, b, and c?

x	2	5	8
y	12	60	162

8. Refer to the formula modeling Manhattan's population. Verify that the model accurately stores the population for the given year.
   a. 1920
   b. 2000

APPLYING THE MATHEMATICS

9. Ronin did not use matrices to solve the system of Example 1. Instead he reordered the equations so that the largest coefficients are on the top line as shown below, then repeatedly subtracted each equation from the one above it. Use this method to solve the system and check the solution to Example 1.

$$\begin{cases} 25a + 5b + c = 10 \\ 16a + 4b + c = 6 \\ 9a + 3b + c = 3 \end{cases}$$

10. Example 1 shows that $r = \frac{1}{2}x^2 - \frac{1}{2}x$ fits the values of the function for integers x with $1 \le x \le 6$. Check that this formula works for any integer x by going through the following steps.

    a. Let $r = f(x)$ and calculate $f(n)$ and $f(n-1)$.

    b. The expression $f(n) - f(n-1)$ is the difference between connecting n cities and connecting $n-1$ cities. Find an expression for $f(n) - f(n-1)$ in terms of n.

    c. Interpret your answer to Part b by referring back to the meaning of $f(n)$.

11. The employees at Primo's Pizzeria like to cut pizza into oddly-shaped pieces. In so doing, they noticed that there is a maximum number of pieces that can be formed from a given number of cuts. Write a polynomial model for finding the maximum number of pieces from the number of cuts.

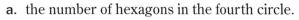

Number of Cuts	0	1	2	3	4
Maximum Number of Pieces	1	2	4	7	11

12. Recall that a *tessellation* is a pattern of shapes that covers a surface completely without overlaps or gaps. One cross section of a honeycomb is a tessellation of regular hexagons, with three hexagons meeting at each vertex. One way to construct the tessellation is to start with 1 hexagon then surround it with 6 more hexagons and then surround these with another "circle" of 12 hexagons, and so on. If this pattern were to continue, find

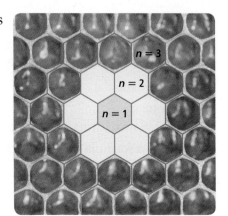

    a. the number of hexagons in the fourth circle.

    b. the total number of hexagons in the first four circles.

    c. a polynomial equation which expresses the *total* number of hexagons h as a function of the number of circles n.

    d. the total number of hexagons in a honeycomb with 10 circles.

13. Refer to the roller coaster on the first page of the chapter. The table below gives the roller coaster's height $H(x)$ in the picture for different values of horizontal distance x along the picture.

x	0	4	8	12	16
$H(x)$	0	–0.7	2	7.8	11.4

    a. Use the method of Examples 1 and 2 to fit a polynomial function to these data points.

    b. Graph the polynomial you find to see how close its graph is to the picture.

14. The drawings given in Example 1 look similar to the drawings you saw in geometry when you learned that the number d of diagonals in a polygon with n sides is given by $d = \frac{n(n-3)}{2}$.

   a. Show that the formula for the number of diagonals and the formula for the number of roads are not equivalent.

   b. Subtract the polynomial for the number of diagonals from the polynomial for the number of connections. What is the difference? What does the difference mean?

REVIEW

15. Consider the data at the right. (**Lesson 11-7**)

   a. Can the data be modeled by a polynomial function?

   b. If so, what degree is the polynomial?

r	1	2	3	4	5	6	7	8	9
t	3	11	31	69	131	223	351	521	739

16. How much larger is the volume of a cube with side $x + 1$ than the volume of a cube with side $x - 1$? (**Lesson 11-2**)

17. May Vary invested different amounts each year for the past 6 years as summarized in the table at the right. Each amount is invested at 4.6% APY. What is the total current value of her investments? (**Lesson 11-1**)

Years Ago	Amount ($)
5	7100
4	6600
3	5800
2	5100
1	6000
0	5500

18. Describe three properties of the graph of $f(x) = \sin x$. (**Lesson 10-6**)

19. Express as a logarithm of a single number. (**Lesson 9-9**)

   a. $\log_7 36 - \log_7 6$ b. $4 \log 3$ c. $\frac{1}{6} \log 64$

20. Ivan Speeding was driving 25% faster than the speed limit. By what percent must he reduce his speed to be driving at the speed limit? (**Previous Course**)

EXPLORATION

21. In the last part of this lesson, the functions with equations $y = \frac{x^2 - x + 2}{2}$ and $y = \frac{x^3 - 3x^2 + 8x}{6}$ are shown to fit the points $(1, 1)$, $(2, 2)$, and $(3, 4)$.

   a. Explain why $y = \frac{x^3 - 3x^2 + 8x}{6} + (x - 1)(x - 2)(x - 3)$ describes another 3rd degree polynomial function that contains these points.

   b. Find an equation for a polynomial function of 4th degree that contains these points.

   c. Explain how you could find as many different polynomials as desired that contain these three points.

Chapter 11 Projects

1 Proving That Certain *n*th Roots Are Irrational

In Lesson 11-5, you proved that $\sqrt{2}$ and $\sqrt[3]{9}$ are irrational.

a. Use the same idea to prove that the following numbers are irrational.

   i. $\sqrt{37}$ **ii.** $\sqrt[3]{7}$ **iii.** $\sqrt[5]{2}$

b. Prove that some other number of your choosing is irrational.

c. Explain why this process does not work to prove that $\sqrt{49}$ is irrational.

2 Polynomial Inequalities

a. Consider $P(x) = 3x^2 - 20$. Solve $P(x) = 0$. Using a graph, solve $P(x) < 0$ and $P(x) > 0$.

b. Describe how to solve the inequalities in Part a algebraically. Check your method for other polynomials of the form $ax^2 + b$.

c. Suppose $Q(x) = 2x^2 + x - 6$. Solve $Q(x) = 0$. Using a graph, solve $Q(x) < 0$ and $Q(x) > 0$.

d. Describe how to solve the inequalities in Part c algebraically. Check your method for other polynomials of the form $ax^2 + bx + c$.

e. Extend your strategy in Part d to describe how to solve $0 < ax^3 + bx^2 + cx + d$.

3 Sums of Products

a. Use the method of finite differences to find an explicit formula for the sequence a of sums of squares shown below.
$$a_1 = 1^2 = 1$$
$$a_2 = 1^2 + 2^2 = 5$$
$$a_3 = 1^2 + 2^2 + 3^2 = 14$$
$$a_4 = 1^2 + 2^2 + 3^2 + 4^2 = 30, \text{ and so on.}$$

b. Use the method of finite differences to find an explicit formula for the sequence b of sums of products of consecutive integers shown below.
$$b_1 = 3 \cdot 4 = 12$$
$$b_2 = 3 \cdot 4 + 4 \cdot 5 = 32$$
$$b_3 = 3 \cdot 4 + 4 \cdot 5 + 5 \cdot 6 = 62$$
$$b_4 = 3 \cdot 4 + 4 \cdot 5 + 5 \cdot 6 + 6 \cdot 7 = 104,$$
and so on.

c. Use the method of finite differences to find an explicit formula for the sequence c of sums of products of integers that differ by 3 shown below.
$$c_1 = 2 \cdot 5 = 10$$
$$c_2 = 2 \cdot 5 + 3 \cdot 6 = 28$$
$$c_3 = 2 \cdot 5 + 3 \cdot 6 + 4 \cdot 7 = 56$$
$$c_4 = 2 \cdot 5 + 3 \cdot 6 + 4 \cdot 7 + 5 \cdot 8 = 96,$$
and so on.

d. What do all the polynomial formulas in Parts a–c have in common?

e. Write another sequence of sums of products that you think has an explicit formula that shares the characteristic you gave in Part d. Use the method of finite differences to check your prediction.

4 Synthetic Division

A polynomial can be divided by a linear binomial $x - r$ using a process called *synthetic division*. To perform synthetic division, start by writing the coefficients of the polynomial to be divided in a line across the page. Multiply the first coefficient by r and add the second coefficient. Multiply the resulting sum by r and add the next coefficient. Repeat this process until the final coefficient has been added. The sums you find in each step are the coefficients of the quotient, and the final sum is the remainder. For example, to divide $P(x) = 2x^4 - 9x^3 + 4x - 7$ by $x - 5$, you would write the coefficients 2, -9, 0, 4, and -7. Then follow the arrows to perform the calculation. The quotient $Q(x)$ is $2x^3 + x^2 + 5x + 29$ with a remainder R of 138.

$$
\begin{array}{ccccc}
2 & -9 & 0 & 4 & -7 \\
\text{multiply by 5} & \text{multiply by 5} & \text{multiply by 5} & \text{multiply by 5} & \\
+\downarrow \quad 10 & 5 & 25 & 145 & \\
\hline
2 & 1 & 5 & 29 & 138
\end{array}
$$

a. Verify that the division was done correctly by using a CAS to show that $P(x) = Q(x) \cdot (x - 5) + R$.

b. Find the value of $P(5)$ by substituting $x = 5$ into the expression for $P(x)$ in Part a.

c. When a polynomial is divided by $x - r$, what does the remainder tell you about the value of the polynomial at r?

d. Use synthetic division to divide $G(x) = 3x^4 + 2x^3 - 20x^2 - 3x + 12$ by $x - 7$ and evaluate $G(7)$.

5 Effect of Coefficients on $y = ax^4 + bx^3 + cx^2 + dx + e$

Use a DGS or CAS to create a dynamic graph of $y = ax^4 + bx^3 + cx^2 + dx + e$, where a, b, c, d, and e are real numbers and can be varied. Explore several possible values of each coefficient to decide how each affects the graph of the function. Summarize your findings in a report.

6 Factoring and Solving Cubic Polynomial Functions

Factoring the general cubic polynomial function has a rich history. While a general formula for factoring any cubic polynomial in one variable exists, there are a few special forms that do not require the general formula.

a. Find out how to factor $x^3 - a^3$ and $x^3 + a^3$, where a is any real number.

b. Find out how to solve $x^3 + mx = n$, where m and n are any real numbers.

c. Research the history of solving the general cubic polynomial equation
$$a_3x^3 + a_2x^2 + a_1x + a_0 = 0.$$

Summarize your research into a brief report that includes information about the person who first created the general formula.

Mathematicians have developed a method for solving cubic equations with origami.

Chapter 11 Summary and Vocabulary

A **polynomial in x of degree n** is an expression that can be written in the standard form

$$a_n x^n + a_{n-1} x^{n-1} + \ldots + a_2 x^2 + a_1 x + a_0,$$

where $a_n \neq 0$. a_n is called the **leading coefficient**, and a_0 is the **constant term**. Polynomials written in factored form can be rewritten in standard form using the Extended Distributive Property. **Polynomial functions** include linear, quadratic, and functions of higher degree such as the ones graphed below.

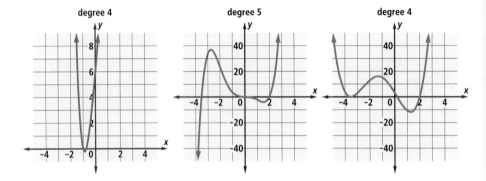

When a polynomial $P(x)$ of degree n is set equal to zero, the resulting equation has n **roots**, or **zeros**. The Fundamental Theorem of Algebra guarantees that $P(x) = 0$ has at least one complex root. The Factor Theorem states that if r is a root of $P(x) = 0$, then $P(x)$ can be factored as $P(x) = (x - r) \cdot Q(x)$. From this we can deduce that a polynomial function of degree n has exactly n complex roots, although some may be **multiple roots**.

If the degree of a polynomial function is less then 5, formulas such as the Quadratic Formula can be used to find exact zeros. For polynomials of degree 5 or higher, such formulas do not exist. The Rational-Root Theorem provides a way to identify all the possible rational roots of a polynomial with integer coefficients. A CAS can be used to find exact or approximate roots, or to factor or expand polynomials.

Vocabulary

Lesson 11-1
degree of a polynomial
term of a polynomial
*polynomial in x
  of degree n
standard form of
  a polynomial
coefficients of a polynomial
leading coefficient
constant term, constant
polynomial function

Lesson 11-2
monomial, binomial,
  trinomial
degree of a polynomial
  in several variables

Lesson 11-3
factoring a polynomial
factored form of
  a polynomial
greatest common
  monomial factor
*prime polynomial,
  irreducible polynomial

Lesson 11-4
zero of a polynomial, root
  of a polynomial

Lesson 11-6
double root, root
  of multiplicity 2
multiplicity

Lesson 11-7
method of finite differences

◖ Polynomials arise directly from compound-interest situations, orbits, and questions about volume. They describe many numerical patterns. They can model many other real-world situations based on finite sets of data points. The degree of the model's polynomial can be found by the **method of finite differences**, and the coefficients of the polynomial can be found by solving a system of linear equations or polynomial regression.

Theorems and Properties

Extended Distributive Property (p. 739)
Difference of Squares Factoring Theorem (p. 747)
Binomial Square Factoring Theorem (p. 747)
Zero-Product Theorem (p. 752)
Factor Theorem (p. 753)
Rational-Root (or Rational-Zero) Theorem (p. 761)
Fundamental Theorem of Algebra (p. 766)
Number of Roots of a Polynomial Equation Theorem (p. 768)
Polynomial-Difference Theorem (p. 773)

Chapter

11 Self-Test

Take this test as you would take a test in class. Use the Selected Answers section in the back of the book to check your work.

1. Khalil has a 30-inch by 40-inch rectangular piece of cardboard. He forms a box by cutting out squares with sides of length x from each corner and folding up the sides. Write a polynomial formula for the volume $V(x)$ of the box.

2. **Multiple Choice** Which term appears in the expansion of $(a + b)(c + d + e)(f + g)$?

   A acf B bde

   C af^2 D ad^2g

3. Write $(z + 5)(z^2 - 3z + 1)$ in the standard form of a polynomial.

4. Give an example of a binomial with degree 4.

In 5 and 6, use these facts: When Francesca turned 16, she began saving money from her summer jobs. After the first summer, she saved $680. After the second summer, she saved $850. After the third, she saved $1020, and after the following two summers she saved $1105 and $935, respectively. Francesca invested all this money at an annual percentage yield r and did not deposit or withdraw any other money.

5. If $x = (1 + r)$, write a polynomial in terms of x that gives the final amount of money in Francesca's account the summer after her 20th birthday.

6. How much money would she have the summer after her 20th birthday if she were able to invest all the money at an APY of 5.1%?

In 7 and 8, consider the polynomial function P where $P(x) = x^5 - 17x^2 + 4x^6 + 11$.

7. a. What is the degree of the polynomial?

   b. How many complex roots does the polynomial have?

8. State all rational roots of $P(x) = 0$ that are possible according to the Rational-Root Theorem.

In 9 and 10, consider the polynomial function with equation $y = x^5 - 11x^3 + 12x^2 - 35x - 12$. A graph of the function is given below.

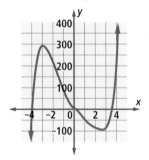

9. a. How many real zeros does the function have? Assume no multiplicities are greater than 1.

   b. How many nonreal zeros does the function have?

10. a. List all possible rational zeros of the function according to the Rational-Root Theorem.

   b. Using the results of Part a and the graph, between what pairs of consecutive integers must irrational zeros be located?

In 11 and 12, consider the polynomial function g where $g(x) = (x - 2)^3(11x + 37)(x^2 - 7)$.

11. Find the zeros of g.

12. What is the multiplicity of each zero you found in Question 11?

13. **Multiple Choice** Suppose $x - r$ and $x - s$ are factors of a quadratic polynomial $P(x)$. Which of the following is *not* true for all x?

  A $P(r) = 0$ B $k(x - r)(x - s) = P(x)$

  C $P(s) = 0$ D $(x - r)(x - s) = 0$

14. Use the Rational-Root Theorem to prove that $\sqrt{21}$ is an irrational number.

In 15 and 16, consider the polynomial function r with $r(y) = y^5 - 5y^3 - 27y^2 + 135$.

15. a. Factor $r(y)$ over the rationals.

  b. Factor $r(y)$ over the complex numbers.

16. Find all complex zeros of r.

17. Factor completely: $10uv^2w + 24vw$.

18. A polynomial function p has a root at $x = -2$ and a double root at $x = 6$. Write a possible equation for p and graph it.

19. A function f produces the table of values below.

n	1	2	3	4	5	6
f(n)	0	6	24	60	120	210

  a. Can f be modeled by a polynomial?

  b. If so, what is the smallest possible degree of the polynomial? If not, why not?

20. Find an equation for a polynomial function that describes the data points below.

t	-2	-1	0	4	2	3	4
r	12	4	0	0	4	12	24

Chapter 11 Chapter Review

SKILLS Procedures used to get answers

OBJECTIVE A Use the Extended Distributive Property to multiply polynomials.
(Lesson 11-2)

In 1 and 2, write the polynomial in standard form.

1. $(x + 1)(x^2 - 2x + 3)$

2. $(a + 2)(a + 3)(a + 4)$

In 3 and 4, expand and combine like terms.

3. $(x - y)^2(x + y)$ 4. $(r + s + t)(r - s + t)$

OBJECTIVE B Factor polynomials.
(Lesson 11-3)

Fill in the Blanks In 5 and 6, complete the factoring.

5. $15a^7b^{11} - 40a^{13}b^6 = \underline{\ ?\ }(3b^5 - \underline{\ ?\ })$

6. $z^2 + \underline{\ ?\ } + 81 = (z + \underline{\ ?\ })^2$

In 7–10, factor over the set of integers by hand and check with a CAS.

7. $x^2 - 6x + 9$ 8. $9m^2 - 48m + 64$

9. $p^4q^4 - 16$ 10. $x^2 + 5x + 6$

In 11–14, factor over the reals, if possible.

11. $a^2 + b^2$ 12. $y^2 + 7y + 10$

13. $37 + 3t - t^2$ 14. $z^2 - 17$

OBJECTIVE C Find zeros of polynomial functions by factoring. (Lessons 11-4, 11-6)

In 15–17, solve the equation and identify any multiple roots.

15. $0 = 6n(n + 3)(8n - 7)$

16. $0 = (t + 13)^3(2t - 3)^2$

17. $x^3 + 12x^2 + 36x = 0$

In 18 and 19, find the exact zeros of the polynomial function.

18. $f(x) = x(x - \pi)(2x + 1)$

19. $P(a) = a^4 - 25a^2$

OBJECTIVE D Determine an equation for a polynomial function from data points.
(Lessons 11-7, 11-8)

20. Consider the polynomial function of smallest degree that models the data points below.

x	1	2	3	4	5	6
y	1	5	14	30	55	91

a. What is the degree?

b. **Multiple Choice** Which system of equations could be solved to find the coefficients of the polynomial?

i. $\begin{cases} 64a + 16b + 4c + d = 30 \\ 27a + 9b + 3c + d = 14 \\ 8a + 4b + 2c + d = 5 \\ a + b + c + d = 1 \end{cases}$

ii. $\begin{cases} 4x^3 + 4x^2 + 4x + 4 = 0 \\ 3x^3 + 3x^2 + 3x + 3 = 0 \\ 2x^3 + 2x^2 + 2x + 2 = 0 \\ x^3 + x^2 + x + 1 = 0 \end{cases}$

iii. $\begin{cases} a + b + c + d = 5 \\ a + 2b + c + d = 19 \\ a + 2b + 3c + d = 43 \\ a + 2b + 3c + 4d = 77 \end{cases}$

iv. none of these

c. Determine an equation for the polynomial function.

In 21–23, can the given relation be described by a polynomial function of degree ≤ 3? If so, find an equation for the function. If not, explain why not. Assume the first variable is the independent variable.

21.

x	1	2	3	4	5	6
y	1	1	3	7	15	31

22.

x	1	2	3	4	5	6
y	56	58	62	74	100	146

23. the sequence defined by
$$\begin{cases} a_1 = 4 \\ a_n = 2a_{n-1} - 1, \text{ for integers } n \geq 2 \end{cases}$$

24. The graph of a function G below contains $(-2, 0)$, $(0, 6)$, $(1, 0)$, and $(3, 0)$. Suppose $G(x) = a_3x^3 + a_2x^2 + a_1x + a_0$.

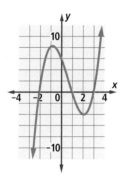

a. What is the value of a_0?

b. Find the values of a_1, a_2, and a_3.

c. Give the value of $G(5)$.

PROPERTIES Principles behind the mathematics

OBJECTIVE E Describe attributes of polynomials. (Lessons 11-1, 11-2, 11-3)

In 25 and 26, state: a. the degree and b. the leading coefficient of the polynomial.

25. $18c^6 + 9c^4 + 2c + 15$

26. $3 + 2d - 2d^5 - 35d^9$

Multiple Choice In 27–29, choose the term that applies to the polynomial.

A monomial 　 B binomial

C trinomial 　 D none of these

27. $p^7 - 8$ 　 28. $17r^4t^5$ 　 29. $\dfrac{\pi}{h^2}$

30. Give an example of a trinomial with degree 7.

31. a. Is $x^2 - 42$ prime over the rationals? If not, factor it.

   b. Is $x^2 - 42$ prime over the reals? If not, factor it.

OBJECTIVE F Apply the Zero-Product Theorem, Factor Theorem, and Fundamental Theorem of Algebra. (Lessons 11-4, 11-6)

In 32 and 33, explain why the Zero-Product Theorem cannot be used directly on the given equation.

32. $(x + 11)(x + 6) = 4$

33. $3(m + 2) - (2m - 6) = 0$

34. Suppose $g(t) = (t + 4)^3(t - 5)^a$. If g has degree 8, what is the value of a?

35. **True or False** If $(r - \pi)$ is a factor of a polynomial function V, then $V(\pi) = 0$.

36. **Multiple Choice** Suppose $p(x)$ is a polynomial, $p(r) = 0$, $p(s) = 0$, and $p(t) = 5$. Which of the following is *not* true?

A $p(s) \cdot p(r) = 0$

B $k(x - r)(x - s)(x - t) = p(x)$

C r and s are intercepts of the graph of $p(x)$.

D r and s are roots of the equation $p(x) = 0$.

37. Suppose a 7th-degree polynomial equation has 5 real roots and 2 irrational roots.

a. How many rational roots does the equation have?

b. How many nonreal complex roots does it have?

OBJECTIVE G Apply the Rational-Root Theorem. (Lesson 11-5)

38. **True or False** By the Rational-Root Theorem, $P(x) = 11x^2 - 5x + 3$ could have a rational root at $x = -\frac{11}{3}$.

39. a. List all possible rational roots of $R(x) = 2x^4 - 7x^3 + 5x^2 - 7x + 3$, according to the Rational-Root Theorem.

    b. Find the rational zeros of R.

40. Use the Rational-Root Theorem to factor $z^4 - z^3 - 19z^2 - 11z + 30$ over the integers.

41. Consider $f(n) = 10n^5 - 3n^2 + n - 6$.

    a. List all possible rational zeros of f, according to the Rational-Root Theorem.

    b. Use a graph to explain why f has exactly one irrational zero.

42. Prove that $25^{\frac{1}{3}}$ is an irrational number.

USES Applications of mathematics in real-world situations

OBJECTIVE H Use polynomials to model real-world situations. (Lessons 11-1, 11-8)

43. The number G of games needed for n tic-tac-toe players to play every other player twice is given in the following table.

n	2	3	4	5	6
G	2	6	12	20	30

    a. Find a polynomial function relating n and G.

    b. Assume there are 24 students in your class. How many games will your class play if each student plays every other student twice?

44. Consider $75x^3 + 100x^2 + 150x + 250$.

    a. Write a question involving money that could be answered by evaluating this expression.

    b. Answer your question in Part a.

45. Gary's grandmother put \$250 in a savings account each year starting on Gary's tenth birthday. The account compounds at an APY of r.

    a. Write a polynomial in x, where $x = 1 + r$, that represents the value of the account on Gary's fifteenth birthday.

    b. If the account pays 6% interest annually, calculate how much money Gary would have on his fifteenth birthday.

    c. If Gary had \$1700 on his fifteenth birthday, what rate of interest did he earn on his account?

46. Recall that when a beam of light in the air strikes the surface of water, it is refracted. Below are the earliest known data on the relation between i, the angle of incidence in degrees, and r, the angle of refraction in degrees, recorded by Ptolemy in the 2nd century CE.

i	10	20	30	40	50	60	70	80
r	8	15.5	22.5	29	35	40.5	45.5	50

    a. Can these data be modeled by a polynomial function?

    b. If so, what is the degree of the function? If not, explain why not.

OBJECTIVE I Use polynomials to describe geometric situations. (Lesson 11-2)

47. A student forms a small open box out of a 3-inch by 5-inch index card by cutting squares of side length s out of the corners and folding up the sides. Write a polynomial for the volume $V(s)$ of the box.

In 48 and 49, an open box is formed out of a 50-cm by 80-cm sheet of cardboard by removing squares of side length x cm from each corner and folding up the four sides.

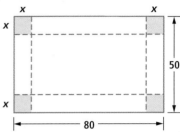

48. Write an expression for the area of the bottom of the box.

49. Write a formula for the volume $V(x)$ of the box.

50. A right circular cone has a slant height $s = 23$.

a. Express its radius r in terms of its altitude h.

b. Use the results of Part a to express the volume V of this cone as a polynomial function in h.

REPRESENTATIONS Pictures, graphs, or objects that illustrate concepts

| **OBJECTIVE J Graph polynomial functions.** (Lessons 11-1, 11-4, 11-5)

In 51 and 52, an equation for a function is given. Graph the function. Then, use the graph to factor the polynomial.

51. $f(x) = 3x^3 - x^2 - 20x - 12$

52. $g(x) = x^3 - x^2 - 20x$

53. A polynomial function h with degree 4 has zeros at –1, 0, 1, and 3.

a. Write a possible equation for h in factored form.

b. Suppose that the leading coefficient of $h(x)$ is 3. Find an equation for h.

c. Graph your equation in Part b.

54. A polynomial function f of degree 3 has zeros at –4, –1, and 5.

a. Write an equation for one function satisfying these conditions.

b. Graph the function in Part a.

c. Write the general form of an equation for f.

d. What do the graphs of all functions with equations of the form in Part c have in common with the graph in Part b?

| **OBJECTIVE K Estimate zeros of polynomial functions using graphs.** (Lessons 11-4, 11-5)

55. A 4th-degree polynomial function with equation $y = g(x)$ and integer zeros is graphed at the right. List the four zeros.

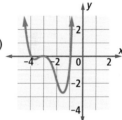

56. Refer to the graph of a function at the right. Name two pairs of consecutive integers between which a zero of f must occur.

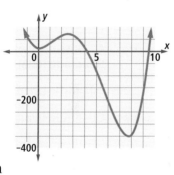

57. Let $p(x) = 2x^4 + 3x^2 + 2x - 6$.

a. List all the possible rational roots of $p(x)$, according to the Rational-Root Theorem.

b. A graph of p is shown at the right. Based on the graph, which possible rational roots from Part a should you test? Explain.

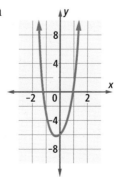

c. Find all rational roots of $p(x)$.

Chapter

12 Quadratic Relations

Contents

The general quadratic equation in two variables x and y is the equation

$$Ax^2 + Bxy + Cy^2 + Dx + Ey + F = 0,$$

where A, B, C, D, E, and F are real numbers, and at least one of A, B, or C is not zero.

The set of ordered pairs that satisfy a sentence equivalent to one in the above form is called a *quadratic relation* in two variables.

Quadratic relations have connections with a wide variety of ideas you have already studied. Those ideas include the parabolas you studied in Chapters 2 and 6 and the hyperbolas you saw in Chapter 2. Quadratic relations also describe the orbits of comets, satellites, and planets, and the shapes of communication receivers and mirrors used in car headlights.

Quadratic relations may also be defined geometrically as the intersection of a plane and a *double cone*. Such cross sections of a double cone are called *conic sections*, or simply *conics*. They include hyperbolas, parabolas, and ellipses. In this description, circles are special cases of ellipses.

In this chapter, you will study quadratic relations both algebraically, as equations and inequalities, and geometrically, as figures with certain properties. You will also see some of the many situations these relations model.

Hyperbola:
A plane intersects
both cones.

Parabola:
A plane intersects one cone
and is parallel to the cone's edge.

Ellipse:
A plane intersects one cone
but is not parallel to the cone's edge.

Lesson
12-1

Parabolas

Vocabulary

parabola

focus, directrix

axis of symmetry

vertex of a parabola

paraboloid

▶ **BIG IDEA**   From the geometric definition of a parabola, it can be proved that the graph of the equation $y = ax^2$ is a *parabola*.

What Is a Parabola?

In Chapter 6 you were told that the path of a shot or tossed object, such as a fly ball in baseball, is part of a *parabola*.

But how do we know this? In order to determine whether a curve is a parabola, a definition of *parabola* is necessary. Here is a geometric definition.

Mental Math

Which expression or equation is not equivalent to the others?

a. $\log_{10} 1742$, $\log_{16} 1742$, $\log 1742$

b. $\log_3 100$, $2 \log_3 10$, $\log_3 20$

c. $r = e^{3t}$, $r = 3e^t$, $\ln r = 3t$

d. $\dfrac{\ln 2}{\ln p}$, $\dfrac{\log 2}{\log p}$, $\dfrac{\log_6 2}{\log_6 p}$

Definition of Parabola

Let ℓ be a line and F be a point not on ℓ. A **parabola** is the set of all points in the plane of ℓ and F equidistant from F and ℓ.

To understand the definition of parabola, recall that the distance from a point P to a line ℓ is the length of the perpendicular from P to ℓ. In the diagram at the right below, four points on a parabola, V, P_1, P_2, and P_3, are identified. Note that each is equidistant from F and the line ℓ. For example, $\overline{P_1Q_1} \perp \ell$ and $P_1Q_1 = P_1F$. Also, $\overline{P_2Q_2} \perp \ell$ and $P_2Q_2 = P_2F$, and so on.

F is the **focus** and ℓ is the **directrix** of the parabola. Thus, a parabola is the set of points in a plane equidistant from its focus and its directrix. Neither the focus nor directrix is on the parabola. The line through the focus perpendicular to the directrix is the **axis of symmetry** of the parabola. The point V on the parabola and on the axis of symmetry is the **vertex of the parabola**.

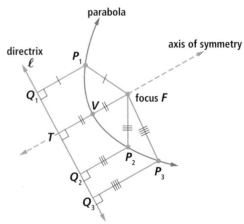

QY1

Drawing a Parabola

You can draw as many points on a parabola as you wish using only a compass and a straightedge.

▶ **QY1**

FV is equal to what other distance shown in the diagram on the previous page?

Activity

MATERIALS compass, straightedge

Step 1 Begin with a blank sheet of paper. Draw a directrix ℓ and a focus *F* not on ℓ. With *H* on ℓ, construct $\overleftrightarrow{FH}$ perpendicular to ℓ. Then construct the midpoint of $\overline{FH}$ and label it *V* for vertex. Your drawing should resemble the one at the right.

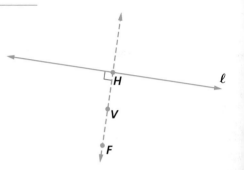

Step 2 Let $FH = d$. Construct a line segment parallel to ℓ of length 2*d* through *F*, where *F* is the midpoint. Label the endpoints of this segment P_1 and P_2.

Step 3 **a.** Construct perpendicular line segments from P_1 to ℓ and P_2 to ℓ. Note that P_1 and P_2 are vertices of squares with $\overline{FH}$ as the common side.
   b. How does FP_1 compare to the perpendicular distance from P_1 to ℓ? What does this tell you about P_1?
   c. How does FP_2 compare to the perpendicular distance from P_2 to ℓ? What does this tell you about P_2?

Step 4 Construct a circle with center *F* and any radius $r > \frac{d}{2}$.

Step 5 **a.** Construct a line *n* parallel to ℓ that is distance *r* from ℓ. (*Hint:* Use your compass to measure *r*.) Label the two intersections of this line and the circle P_3 and P_4.
   b. Are P_3 and P_4 on the parabola? Justify your answer.

Step 6 Find two more points by repeating Steps 4 and 5 for a different $r > \frac{d}{2}$.

Step 7 You have found seven points on the parabola. Connect them with a smooth curve to sketch part of a parabola.

Equations for Parabolas

Suppose that you know the coordinates of the focus and an equation for the directrix of a parabola. You can find an equation for the parabola by using the definition of parabola and the Pythagorean Distance Formula.

Example

Find an equation for the parabola with focus $F = (0, 2)$ and directrix $y = -2$.

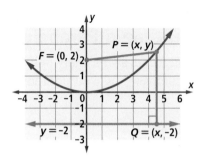

Solution Sketch the given information. Let $P = (x, y)$ be any point on the parabola. Because the directrix is a horizontal line, the distance from a point on the parabola to the directrix is measured along a vertical line. Let Q be the point on the directrix and on the vertical line through P. If $P = (x, y)$, then $Q = (x, -2)$.

$PF = PQ$	definition of parabola
$\sqrt{(x - 0)^2 + (y - 2)^2} = \sqrt{(x - x)^2 + (y - (-2))^2}$	Pythagorean Distance Formula
$x^2 + (y - 2)^2 = (y + 2)^2$	Square both sides.
$x^2 + y^2 - 4y + 4 = y^2 + 4y + 4$	Expand.
$x^2 - 4y = 4y$	Add $-y^2 - 4$ to both sides.
$x^2 = 8y$	Add $4y$ to both sides.
$y = \frac{1}{8}x^2$	Solve for y.

An equation for the parabola is $y = \frac{1}{8}x^2$.

Check Pick any point on $y = \frac{1}{8}x^2$. We use $A = (12, 18)$. Now show that A is equidistant from $F = (0, 2)$ and $y = -2$.

$$AF = \sqrt{(12 - 0)^2 + (18 - 2)^2} = \sqrt{12^2 + 16^2} = \sqrt{400} = 20$$

The distance from A to $y = -2$ is the distance from $(12, 18)$ to $(12, -2)$, which is 20, also. So, A is on the parabola with focus $(0, 2)$ and directrix $y = -2$.

In the Example, if you replace the focus by $\left(0, \frac{1}{4}\right)$ and replace the directrix by $y = -\frac{1}{4}$, the equation for the parabola is $y = x^2$.

If a is nonzero, $(0, 2)$ is replaced by $\left(0, \frac{1}{4a}\right)$, and the directrix is replaced by $y = -\frac{1}{4a}$, then the parabola has equation $y = ax^2$. The derivation of both of these equations uses the same steps as the Example, and demonstrates the following theorem.

Focus and Directrix of a Parabola Theorem

For any nonzero real number a, the graph of $y = ax^2$ is the parabola with focus at $\left(0, \frac{1}{4a}\right)$ and directrix at $y = -\frac{1}{4a}$.

 QY2

Recall that because the image of the graph of $y = ax^2$ under the translation $T_{h,k}: (x, y) \rightarrow (x + h, y + k)$ is the graph with equation $y - k = a(x - h)^2$, the graph of any quadratic equation of the form $y = a(x - h)^2 + k$ or $y = ax^2 + bx + c$ is also a parabola. You can find the focus and directrix of a parabola with equation $y = a(x - h)^2 + k$ by applying the appropriate translation to the focus and directrix of $y = ax^2$.

When $a < 0$, you have learned that the parabola opens down. In this case, when the vertex is $(0, 0)$, the directrix is above the x-axis, and the focus is below.

If a parabola is rotated in space around its axis of symmetry it creates a 3-dimensional **paraboloid**. The focus of a paraboloid is the focus of the rotated parabola. Paraboloids are common in modern technology. The shape of a satellite receiving dish is based on a paraboloid. Residents of a wheat-growing commune in southern China use a tiled solar reflector in the shape of a paraboloid. A teapot is placed at the focus of the paraboloid. Sunlight is reflected toward the teapot, boiling the water in 20 minutes without burning any wood, which is a precious resource. Cooking with a Dutch oven follows the same principle.

▶ **QY2**

Find the focus and directrix of the parabola with equation $y = -\frac{1}{6}x^2$.

Questions

COVERING THE IDEAS

1. a. Can the focus of a parabola be a point on the directrix? Why or why not?
   b. Can the vertex be on the directrix? Why or why not?

True or False In 2–4, refer to the parabola at the right with focus F and directrix ℓ. P_1, P_2, P_3, and P_4 are points on the parabola.

2. $P_3F = FG_3$
3. $FG_1 = FG_2$
4. The focus of this parabola is its vertex.

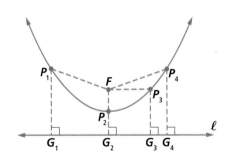

5. Refer to the Example.

 a. Graph the parabola with equation $y = \frac{1}{8}x^2$.

 b. Name its focus, vertex, and directrix.

 c. Verify that the point $(5, 3.125)$ is equidistant from the focus and directrix, and therefore is a point on the parabola $y = \frac{1}{8}x^2$.

 d. Find another point on the graph of $y = \frac{1}{8}x^2$. Show that it is equidistant from the focus and directrix.

6. Verify that the graph of $y = -x^2$ is a parabola with focus $\left(0, -\frac{1}{4}\right)$ and directrix $y = \frac{1}{4}$ by choosing a point on the graph and showing that two appropriate distances are equal.

7. Let $F = (0, -3)$ and ℓ be the line with equation $y = 3$. Write an equation for the set of points equidistant from F and ℓ.

8. a. Using graph paper, follow the steps of the Activity to draw five points that are on the parabola with focus $F = (0, 1)$ and directrix defined by $y = -1$. You do not need to construct with a ruler and compass.

 b. Refer to the Example. Find an equation for the parabola you drew in Part a.

 c. Verify that the points you drew in Part a are on the graph of the function defined in Part b.

9. Give the focus and directrix of the parabola with equation $y = -2xz^2$.

10. What is a paraboloid?

APPLYING THE MATHEMATICS

11. What are the focus and directrix of $y - 6 = (x + 5)^2$?

12. Prove the Focus and Directrix of a Parabola Theorem.

In 13 and 14, an equation for a parabola is given.
 a. Tell whether the parabola opens up or down.
 b. Give the focus of the parabola.

13. $y = -5x^2$

14. $y = \frac{1}{4}x^2$

15. a. Find an equation for the parabola with focus $(5, 0)$ and directrix $x = -5$.

 b. Give the coordinates of three points on this parabola, including the vertex.

The Municipal Asphalt Plant in New York City has a parabolic shape. This landmark facility is used for community sports, fitness, and recreation.

REVIEW

16. An isotope has a half-life of 135 seconds. How long will it take 75 mg of this isotope to decay to 20 mg? (**Lesson 9-2**)

17. Solve for y. (**Lessons 8-8, 6-2**)
    a. $y^2 = 13$
    b. $(y + 3)^2 = 13$
    c. $\sqrt{y} = 13$
    d. $\sqrt{y + 3} = 13$

18. Determine whether the set of points (x, y) satisfying the given equation describes y as a function of x. Explain your answer. (**Lessons 8-2, 2-5**)
    a. $y = (x + 1)^2$
    b. $x = (y + 1)^2$

19. Simplify. (**Lessons 7-8, 7-7**)
    a. $\left(\frac{16}{49}\right)^{\frac{1}{2}}$
    b. $(0.0001)^{-\frac{3}{4}}$

20. Suppose the transformation $T_{-3,2}$ is applied to the parabola with equation $y = \frac{5}{9}x^2$. Find an equation for its image. (**Lesson 6-3**)

21. **Fill in the Blank** If x is a real number, then $\sqrt{x^2} = \underline{\quad?\quad}$.
    (**Lesson 6-2**)

    A x B $-x$ C $|x|$ D none of these

EXPLORATION

22. Parabolas can be formed without equations or graphs. Follow these steps to see how to make a parabola by folding paper.

    a. Start with a sheet of unlined paper. Fold it in half as shown at the right. Cut or tear along the fold to make two congruent pieces. On one piece mark a point P about one inch above the center of the lower edge. Fold the paper so that the lower edge touches P, and crease well as shown at the right. Then unfold the paper. Repeat 10 to 15 times, each time folding so that a different point on the bottom edge of the paper aligns with P. The creases represent the *tangents* to a parabola. (A *tangent* to a parabola is a line not parallel to its line of symmetry that intersects the parabola in exactly one point.) Where are the focus and directrix of this parabola?

    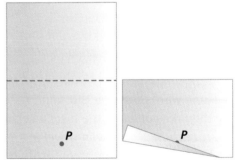

    b. On the other piece of paper mark a point Q approximately in the center. Repeat the procedure used in Part a. Where are the focus and directrix for this parabola?

    c. **Fill in the Blank** The two parabolas formed in Parts a and b illustrate the property that as the distance between the focus and directrix increases, the parabola __?__.

Lesson
12-2
Circles

▶ **BIG IDEA** From the geometric definition of a circle, an equation for any circle in the plane can be found.

As you know, a **circle** is the set of all points in a plane at a given distance (its **radius**) from a fixed point (its **center**). When a person throws a pebble into a calm body of water, *concentric circles* soon form around the point where the pebble hit the water. **Concentric circles** have the same center, but different radii. When an earthquake or mass movement of land (such as a volcanic eruption) occurs, various kinds of *seismic waves* radiate in roughly concentric circles from the *focus*, the point below Earth's surface where the earthquake began.

After reaching the surface, seismic waves travel along the ground in concentric circles around the *epicenter*, or the point on Earth's surface above the focus. Warning systems for tsunamis and other natural disasters are triggered by measuring seismic waves with an instrument called a *seismograph*.

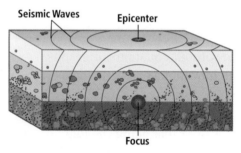

Seismic waves travel at the same speed in all directions. For earthquakes on land, the fastest seismic waves, called *P-waves* or *compression waves*, can travel at speeds of up to $8\frac{\text{km}}{\text{sec}}$. So the points on Earth's surface that a compression wave reaches at a given time make an approximate circle whose center is the epicenter of a quake.

Equations for Circles

You can find the equation for any circle using the definition of circle and the Pythagorean Distance Formula.

Mental Math

Suppose Aaron throws a rock off a cliff and its height h in meters t seconds after he throws it is given by $h = -4.9t^2 + 4.5t + 25$.

a. Did Aaron throw the rock upward or downward?

b. What was the rock's initial velocity?

c. How high is the cliff?

d. Will the rock have hit the ground after 10 seconds?

Example 1

Suppose the compression waves of an earthquake travel at a speed of $8\frac{km}{sec}$. Find an equation for the set of points that are reached by the compression waves in 7 seconds.

Solution Let the unit of a graph equal one kilometer, and the origin (0, 0) represent the epicenter of the quake. In 7 seconds, a compression wave travels about $8\frac{km}{sec}$ å• 7 sec = 56 km. So, the circle has radius 56.

Let (x, y) be any point on the circle. By the definition of circle, the distance between (x, y) and $(0, 0)$ is 56.

Use the Pythagorean Distance Formula.

$$\sqrt{(x-0)^2 + (y-0)^2} = 56$$
$$(x-0)^2 + (y-0)^2 = 56^2 \quad \text{Square both sides.}$$
$$x^2 + y^2 = 3136 \quad \text{Simplify.}$$

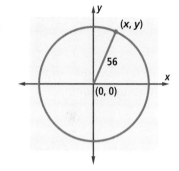

The equation you found in Example 1 is a quadratic equation in x and y. This is also true of equations for circles not centered at the origin.

GUIDED

Example 2

Find an equation for the circle with radius $\frac{27}{2}$ and center $\left(12, \frac{17}{2}\right)$.

Solution Let (x, y) be any point on the circle.

Use the Distance Formula. $\quad \sqrt{(x - \underline{\ ?\ })^2 + (y - \underline{\ ?\ })^2} = \underline{\ ?\ }$

Square both sides. $\quad (x - \underline{\ ?\ })^2 + (y - \underline{\ ?\ })^2 = \underline{\ ?\ }$

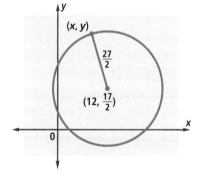

 QY

Example 2 can be generalized to determine an equation for *any* circle. Let (h, k) be the center of a circle with radius r, and let (x, y) be any point on the circle. Then, by the definition of a circle, the distance between (x, y) and (h, k) equals r.

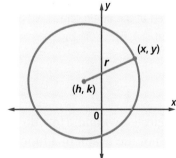

▶ **QY**

The point (25.5, 8.5) satisfies the equation in Example 2. Check your answer to Example 2 by showing that this point is $\frac{27}{2}$ units from the center of the circle.

By the Distance Formula, $\sqrt{(x-h)^2 + (y-k)^2} = r$.

Squaring gives an equation without radicals: $(x-h)^2 + (y-k)^2 = r^2$.

This proves the following theorem.

Circle Equation Theorem

The circle with center (h, k) and radius r is the set of points (x, y) that satisfy $(x-h)^2 + (y-k)^2 = r^2$.

When the center of a circle is the origin, $(h, k) = (0, 0)$ and the equation becomes

$$x^2 + y^2 = r^2.$$

Graphing a Circle

By the Graph-Translation Theorem, the translation $T_{h, k}$ maps the circle with equation $x^2 + y^2 = r^2$ onto a circle with equation $(x-h)^2 + (y-k)^2 = r^2$.

Example 3

a. Find the center and radius of the circle with equation $(x + 4)^2 + (y - 5)^2 = 49$.

b. Graph this circle.

c. What translation maps the circle with equation $x^2 + y^2 = 49$ onto the circle you drew in Part b?

Solution

a. By the Circle Equation Theorem, the center $(h, k) = (-4, 5)$ and the radius $r = \sqrt{49} = 7$.

b. You can make a quick sketch of this circle by locating the center and then four points on the circle whose distance from the center is 7, as illustrated at the right.

c. The preimage circle is centered at the origin and the image circle is centered at $(-4, 5)$, so $T_{-4, 5}$ translates $x^2 + y^2 = 49$ onto $(x + 4)^2 + (y - 5)^2 = 49$.

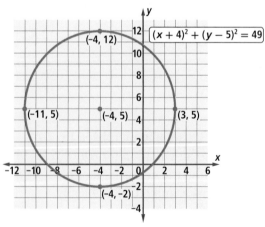

In Example 3, you graphed the circle by hand. How do you graph it on a graphing utility? A circle is not the graph of a function, and most graphing utilities will only graph functions. However, you can divide the circle in half with a horizontal line. Each half-circle, or **semicircle**, is the graph of a function, and both can be graphed in the same window to form a circle.

Example 4

a. Solve $(x + 4)^2 + (y - 5)^2 = 49$ for y.

b. Graph the circle with the equation in Part a on a graphing utility.

Solution

a. Solve for y on a CAS. One CAS solution is shown below.

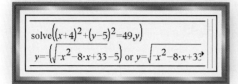

b. Copy the two equations into a graphing utility. Graph
    both on the same screen, using a square window. The
    graph of each equation is half of a circle, but due to
    the graphing utility's limitations, the circle's graph is not
    completely accurate.

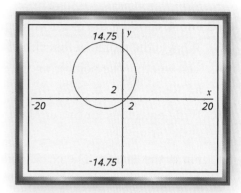

If you know an equation for a circle and one coordinate of a point on
the circle, you can determine the other coordinate of that point.

GUIDED

Example 5

Refer to the circle in Examples 3 and 4. Find the y-coordinate of each point
with x-coordinate 1.

Solution Substitute __?__ for x in the equation for the circle
$(x + 4)^2 + (y - 5)^2 = 49$ and solve for y.

$$(\underline{} + 4)^2 + (y - 5)^2 = 49 \qquad \text{Substitute.}$$

$$(y - 5)^2 = \underline{} \qquad \text{Simplify. Then subtract } \underline{}$$
from both sides.

$$|\underline{}| = \sqrt{\underline{}} \qquad \text{Take the square root of}$$
both sides and apply the
Absolute Value-Square Root
Theorem.

$$y - \underline{} = \underline{} \text{ or } y - \underline{} = \underline{} \qquad \text{Definition of absolute value}$$

$$y = \underline{} \qquad \text{or} \quad y = \underline{} \qquad \text{Add } \underline{} \text{ to both sides of}$$
each equation.

Check Solve the original equation for the circle on a CAS for
y such that $x = 1$. One solution is shown at the right. If yours
looks different, check to see that it is equivalent to this one.

solve$\left((x+4)^2+(y-5)^2=49,y\right)|x=1$
$y=\left(2\cdot\sqrt{6}-5\right)$ or $y=2\cdot\sqrt{6}+5$

Questions

COVERING THE IDEAS

1. Suppose the epicenter for an earthquake is at the origin. If 1 unit on the graph represents 1 km, find an equation describing the set of points (x, y) reached by compression waves in 1 minute if the waves travel at $8\frac{km}{sec}$.

2. Repeat Question 1 if the epicenter is at $(40, 22)$.

In 3 and 4, consider the circle with equation $x^2 + y^2 = 34^2$.

3. What is the radius of this circle?

4. Tell whether the point is on the circle.

   a. $(0, 0)$ b. $(-34, 0)$ c. $(0, 34)$ d. $(\sqrt{34}, 0)$

5. Write equations for two circles that are concentric and centered at the origin.

6. **Fill in the Blanks** The circle with equation $(x - h)^2 + (y - k)^2 = r^2$ has center __?__ and radius __?__.

In 7 and 8, an equation for a circle is given. State the center and radius of the circle, sketch the circle, and solve the equation for y.

7. $x^2 + y^2 = 121$ 8. $(x + 7)^2 + (y + 3)^2 = 25$

9. Refer to the circle from Example 2. Find the y-coordinates of all points on the circle

   a. where $x = -6$.   b. where $x = 800$.

10. Find an equation for the circle with center $\left(-\frac{1}{2}, \frac{5}{2}\right)$ and radius 8.

APPLYING THE MATHEMATICS

11. Planets in our solar system travel in elliptical orbits about the Sun, but these orbits are nearly circular. In the late 1990s, as astronomers searched for solar systems similar to ours, they discovered other planets with almost circular orbits. One example, discovered in 2002, is a planet orbiting the star Tau Gruis. This planet is 2.5 astronomical units (AU) away from Tau Gruis. (1 AU is the average distance between Earth and the Sun.)

    a. Write an equation for the orbit of this planet about Tau Gruis.

    b. If Earth's orbit is assumed to be circular, it revolves about the Sun at a speed of approximately 67,062 miles per hour. If it takes 3.5 years for the planet orbiting Tau Gruis to make one revolution, about how fast is the planet moving?

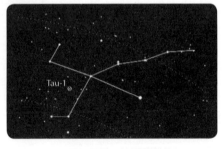

Tau Gruis, visible from the Southern Hemisphere, is located about 100 light-years from Earth.

12. The equation of the circle in Example 4 is equivalent to an equation of the form $Ax^2 + Bxy + Cy^2 + Dx + Ey + F = 0$. What are the values of $A, B, C, D, E,$ and F?

13. A circle centered at the origin has radius 5.
    a. Find the equation of its image under the translation $T_{4, -1}$.
    b. Identify one point on the circle. Verify that its image satisfies the image circle of Part a.

14. To locate the epicenter of an earthquake, data about waves originating from the epicenter are collected by earthquake research stations. Three circles, each centered at a research station with a radius equal to the distance from the epicenter to the particular station, allow one to determine the epicenter location. The location is given by a unique intersection shared by all three circles. Suppose that:

Earthquake data allow a research station to calculate its distance from an epicenter, but not its direction.

   Station 1 is 7 miles from the epicenter and located at $(6, 1)$.
   Station 2 is 5 miles from the epicenter and located at $(11, 8)$.
   Station 3 is 4 miles from the epicenter and located at $(2, 8)$.
    a. Graph the three circles defined above.
    b. Find the epicenter of the earthquake.

REVIEW

15. A parabola has focus $(0, -1)$ and directrix $y = 1$. **(Lesson 12-1)**
    a. What is its vertex? Is it a minimum point or a maximum point?
    b. Give an equation for the parabola.
    c. Give an equation for its axis of symmetry.

16. Give the focus and directrix of the parabola $y = \frac{2}{3}x^2$. **(Lesson 12-1)**

17. What are the zeros of the 4th-degree polynomial function graphed at the right? **(Lesson 11-4)**

18. In 2007, Karla deposited $5000 in a retirement account paying interest compounded continuously. If no additional deposits or withdrawals are made, when Karla retires in 2040 the account will be worth $8172.40. What is the annual percentage yield of this account? **(Lessons 9-10, 9-1, 7-4)**

19. Find an equation for the line containing the origin and $(-3, 7)$. **(Lesson 3-4)**

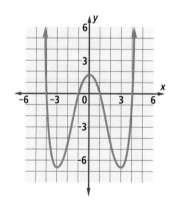

EXPLORATION

20. A *lattice point* is a point with integer coordinates. If possible, find an equation for a circle that passes through
    a. no lattice points.
    b. exactly one lattice point.
    c. exactly two lattice points.
    d. exactly three lattice points.
    e. more than ten lattice points.

Lesson

12-3

Semicircles, Interiors, and Exteriors of Circles

Vocabulary

interior, exterior of a circle

> ▶ **BIG IDEA** An inequality describing the points (x, y) in the *interior* or *exterior* of a circle can be found by replacing the equal sign in a circle's equation by $<$ or $>$.

Semicircles

In Lesson 12-2 you used two semicircles to create a graph of a circle on a graphing utility. In everyday life, semicircles often occur in design and architecture, as shown in the photo at the right of the Arc de Triomphe du Carrousel, located in Paris, France.

Mental Math

Suppose $P(x) = (x + 2)(x - 7)^2(2x + 3)$.

a. Give the degree of $P(x)$.

b. Determine all the roots of $P(x)$.

c. Determine all the roots with multiplicity greater than 1.

Example 1

An architect is designing a 6-foot wide and 8-foot tall passageway with a semicircular arch at its top.

a. Suppose the x-axis represents the floor with the origin at the center of the passageway. What is an equation for the semicircle representing the arch?

b. How high is the arch above a point on the ground 2 feet from the center of the passageway?

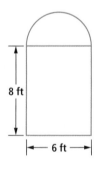

8 ft

6 ft

Solution

a. Let 1 unit in the coordinate system equal 1 foot. Then the circle whose top half is the arch has center $(0, 8)$ and diameter 6, so its radius is 3. Therefore, its equation is $x^2 + (y - 8)^2 = 3^2$. Solve this equation for y.

$$x^2 + (y - 8)^2 = 9$$
$$(y - 8)^2 = 9 - x^2$$

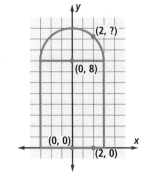

$$y - 8 = \pm\sqrt{9 - x^2}$$
$$y = \pm\sqrt{9 - x^2} + 8$$

Since we want the upper semicircle, $y = \sqrt{9 - x^2} + 8$ is the equation for the semicircle.

b. Since we have an equation for the semicircle, we can easily use it to find the height of the arch at any point. At a point 2 feet along the ground from the center of the passageway, $x = 2$ or $x = -2$. Substitute either x-value into the equation for the semicircle to find the corresponding y-value, which is the height of the arch above this point.

$$y = \sqrt{9 - 2^2} + 8$$
$$= \sqrt{5} + 8 \approx 10.24$$

The arch is about 10.24 feet high 2 feet from the center of the passageway.

Check Graph $y = \sqrt{9 - x^2} + 8$ using technology. Use the trace feature to estimate the y-value when $x = 2$. It checks.

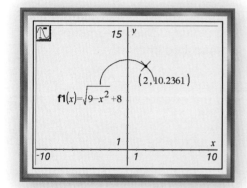

In general, when the equation $(x - h)^2 + (y - k)^2 = r^2$ is solved for y, the result is a pair of equations in the form $y = \pm\sqrt{r^2 - (x - h)^2} + k$. The equation with the positive square root describes the upper semicircle, and the equation with the negative square root describes the lower semicircle.

 QY

▶ **QY**

Write an equation for the lower semicircle of the circle in Example 1.

Interiors and Exteriors of Circles

Every circle separates the plane into two regions. The region inside the circle is called the **interior of the circle**. The region outside the circle is called the **exterior of the circle**. The circle itself is the boundary between these two regions and is not part of either region.

Regions bounded by concentric circles are often used in target practice. Consider the target shown at the right.

To describe the colored regions mathematically, you can place the target on a coordinate system with the origin at the target's center. Notice that the region worth 100 points is the interior of a circle with radius 2. All points in this region are less than 2 units from the origin. Thus if (x, y) is a point in the 100-point region, then $\sqrt{x^2 + y^2} < 2$.

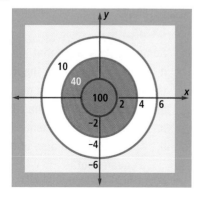

The expressions on both sides of this inequality are positive. Recall that whenever a and b are positive and $a < b$, then $a^2 < b^2$. Thus, when both sides of the inequality are squared, the sentence becomes $x^2 + y^2 < 4$, which also describes the 100-point region.

Similarly, the points in the region worth less than 100 points constitute the exterior of the circle with radius 2. All (x, y) in this region satisfy the sentence $\sqrt{x^2 + y^2} > 2$, or $x^2 + y^2 > 4$.

The two instances above are generalized in the following theorem.

Interior and Exterior of a Circle Theorem

Let C be the circle with center (h, k) and radius r. Then the interior of C is described by

$$(x - h)^2 + (y - k)^2 < r^2$$

and the exterior of C is described by

$$(x - h)^2 + (y - k)^2 > r^2.$$

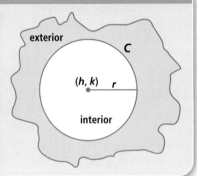

If $\geq$ or $\leq$ is used instead of $>$ or $<$ in the theorem above, the boundary (the circle itself) is included.

Example 2

Graph the points satisfying $(x + 2)^2 + (y - 1)^2 \geq 7$.

Solution The sentence represents the union of a circle, with center at $(-2, 1)$ and radius $\sqrt{7}$, and its exterior. The shaded region and the circle at the right make up the graph.

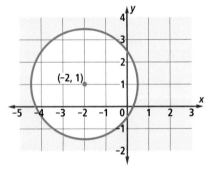

Check Test a specific point. The point $(0, 0)$ is not in the exterior, so it should not satisfy the inequality. This is the case, because $(0 + 2)^2 + (0 - 1)^2 = 5$ and $5 < 7$.

GUIDED

Example 3

Some scholars argue that King Arthur's round table had a hole in it like the one shown at the right. An estimate of the diameter of the table is 18 feet. (The painting is not to scale.) Suppose the inner circle of the table has diameter 5 feet.

a. Write a system of inequalities describing the tabletop.

b. Approximate the area of the tabletop to the nearest tenth of a square foot.

Solution

a. Sketch a top view of the table, and superimpose a coordinate system with the center of the table at the origin.

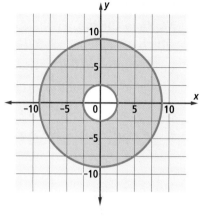

The inner circle has center __?__, radius __?__ ft, and equation __?__. The outer circle has center __?__, radius __?__ ft, and equation __?__.

The tabletop is the intersection of the interior of the outer circle and the exterior of the inner circle, also including the circles themselves.

So, a system of inequalities describing the tabletop

is $\begin{cases} \underline{\quad ? \quad} \geq \underline{\quad ? \quad} \\ \underline{\quad ? \quad} \leq \underline{\quad ? \quad} \end{cases}$.

b. The area of the tabletop is the difference of areas of the two circles. The area of the inner circle is __?__ ft². The area of the outer circle is __?__ ft². So, the area of the region bounded by these circles is __?__ − __?__ = __?__ ft² or about __?__ ft².

Questions

COVERING THE IDEAS

1. Sketch the graph of each equation by hand on separate axes.
   a. $x^2 + y^2 = 25$ b. $y = \sqrt{25 - x^2}$ c. $y = -\sqrt{25 - x^2}$

2. Graph $x^2 + (y - 4)^2 \leq 25$.

3. Use the equation for the arch found in Example 1. To the nearest inch, how high is the arch above a point on the ground 2 feet 6 inches from the center of the passage?

4. A bridge over water has a semicircular arch with radius 4 m set on pillars that extend 6 m above the water. How many meters above the water is the arch at a point 1.5 m from one of the pillars?

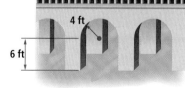

5. At the right, C is a circle. What is the shaded region called?

6. Refer to the target in this lesson. Write a system of inequalities to describe the set of points (x, y)
   a. in the 10-point region of the target.
   b. in the 40-point region of the target.

7. Write an inequality that describes all the points in the shaded region at the right.

8. **Multiple Choice** Given a circle with center (h, k) and radius r, which of sentences A to D below describes

   a. the union of the circle with its interior?

   b. the exterior of the circle?

   A $(x - h)^2 + (y - k)^2 > r^2$ B $(x - h)^2 + (y - k)^2 < r^2$

   C $(x - h)^2 + (y - k)^2 \geq r^2$ D $(x - h)^2 + (y - k)^2 \leq r^2$

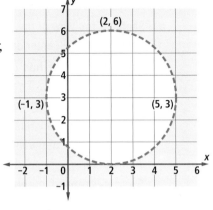

APPLYING THE MATHEMATICS

9. An architect designed a semicircular tunnel so that two trucks can pass each other with a 3-foot clearance between them and a 2-inch vertical clearance with the ceiling.

   a. If the trucks are 8 feet wide and 12 feet tall, what is the smallest possible radius of the semicircle?

   b. What is the tallest 8-foot wide vehicle that can drive down the center of the tunnel?

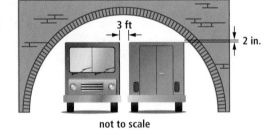

not to scale

10. Graph the following system of inequalities and describe the graph.

$$\begin{cases} x^2 + y^2 \geq 4 \\ (x - 5)^2 + (y - 11)^2 \geq 16 \end{cases}$$

11. In sumo wrestling, participants wrestle in a *dohyo*, a circular arena with a 4.55-meter diameter centered inside a square with side length 6.7 meters. Pushing your opponent out of the circle is one way to win the match. Sometimes the area between the circle and square is layered with fine sand to assist in determining when a wrestler falls outside of the circle. Put the origin at the center of the arena and describe the sandy region with a system of 5 inequalities.

REVIEW

In 12 and 13, define the term. (Lessons 12-2, 12-1)

12. circle   13. parabola

14. Find an equation for the circle with center at $(3, 6)$ and radius 2. (Lesson 12-2)

15. A circle with center at the origin passes through the point $(5, 12)$. **(Lesson 12-2)**

    a. Find the radius of the circle.

    b. Find an equation for the circle.

16. Identify the focus and directrix of the parabola described by $y = \frac{1}{800} x^2$. **(Lesson 12-1)**

17. Expand and simplify. **(Lessons 11-2, 6-1)**

    a. $(x + 8)^2 + y^2$ b. $(x + y)^2 + 8^2$ c. $(x + y + 8)^2$

18. A 3×5 drawing is enlarged to 6×10 by using a size change. **(Lesson 4-4)**

    a. What is a matrix for the size change?

    b. Suppose $A = (1, 2.3)$, $B = (0.3, 2)$, and $C = (2.7, 4.3)$ are three points on the smaller drawing. Write a matrix representing the locations of A', B', and C' on the image.

    c. Find the distance between A and B.

    d. Find the distance between A' and B'.

In 19 and 20, the graph of a function is given. State a. the function's domain, and b. its range. (Lesson 1-4)

19.
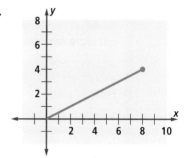

20.
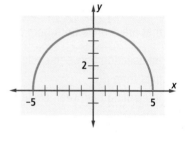

21. Caleb has three scraps of wood with lengths 1.5 feet, 2.3 feet, and 4 feet. Without cutting them, can he use these scraps as the three sides of a triangular flower planter? Why or why not? **(Previous Course)**

EXPLORATION

22. a. A target with five circles all with center $(0, 0)$ is shown at right. If the largest circle has radius 1 and the areas of the five nonoverlapping regions are all equal, find the radii and equations for the five circles.

    b. Generalize Part a.

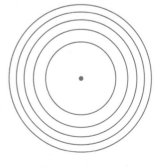

Lesson

12-4 Ellipses

Vocabulary

ellipse

foci, focal constant of an ellipse

standard position for an ellipse

standard form of an equation for an ellipse

major axis, minor axis

center of an ellipse

semimajor axes, semiminor axes

> ▶ **BIG IDEA** From the geometric definition of an ellipse, an equation for any ellipse symmetric to the x- and y-axes can be found.

On the first page of the chapter, we noted that when a plane intersects a cone but is not parallel to the cone's edge, the intersection is a figure called an *ellipse*. Drawing an ellipse freehand is challenging. Identifying several points on the ellipse and connecting them with a smooth curve simplifies the task. The following Activity provides one method for sketching an ellipse.

Activity

MATERIALS conic graph paper with 8 units between the centers of the circles

Step 1 Label the centers of the circles as F_1 and F_2. Plot the two points of intersection of the circles that are 5 units from both F_1 and F_2.

Step 2 **a.** Mark the two points of intersection between circles that are 4 units from F_1 and 6 units from F_2.

 b. Mark the two points of intersection between circles that are 6 units from F_1 and 4 units from F_2.

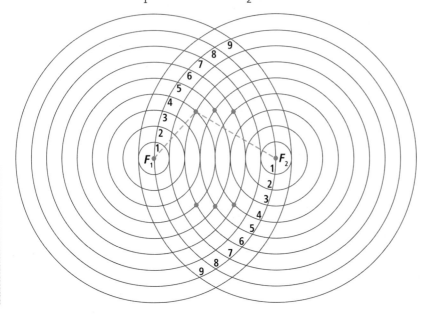

Mental Math

At the state championship basketball game, the Beatums are beating the Underdogs. The score is 84-52. If the Beatums do not score any more points,

a. how many 2-point shots must the Underdogs make to win?

b. how many 3-point shots must they make to win?

c. how many 2-point shots must they make to win if they also make five 3-point shots?

Step 3 Repeat the process of Step 2 plotting the intersections of circles that are x units from F_1 and $(10 - x)$ units from F_2 until you have 16 points plotted.

Step 4 Starting on the left and moving clockwise, label the points $P_1, P_2, ..., P_{16}$. Connect the dots with a smooth curve to form an ellipse.

What Is an Ellipse?

The ellipse in the Activity is determined by the two points F_1 and F_2, called its *foci* (pronounced "foe sigh," plural of focus), and a number called the *focal constant*. The focal constant is the constant sum of the distances from any point P on the ellipse to the foci. The focal constant of the ellipse in the Activity is 10. That is, $P_nF_1 + P_nF_2 = 10$ for all points P_n on the curve. Every ellipse can be determined in this way.

> ### Definition of Ellipse
>
> Let F_1 and F_2 be any two points in a plane and let d be a constant with $d > F_1F_2$. Then the **ellipse** with **foci** F_1 and F_2 and **focal constant** d is the set of points P in the plane for which $PF_1 + PF_2 = d$.

The equation $PF_1 + PF_2 = d$ is said to define the ellipse. For any point P on the ellipse, the focal constant $PF_1 + PF_2$ has to be greater than F_1F_2 because of the Triangle Inequality. That is why $d > F_1F_2$.

Equations for Some Ellipses

To find an equation for the ellipse in the Activity, consider a coordinate system with $\overleftrightarrow{F_1F_2}$ as the x-axis and with the origin midway between the foci on the axis. Then the foci are $F_1 = (-4, 0)$ and $F_2 = (4, 0)$. This is the *standard position* for the ellipse.

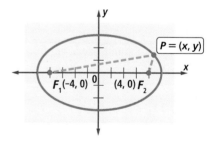

If $P = (x, y)$ is on the ellipse, then because the focal constant is 10,

$$PF_1 + PF_2 = 10.$$

So, by the Distance Formula,

$$\sqrt{(x + 4)^2 + (y - 0)^2} + \sqrt{(x - 4)^2 + (y - 0)^2} = 10,$$

or

$$\sqrt{(x + 4)^2 + y^2} + \sqrt{(x - 4)^2 + y^2} = 10.$$

This equation for the ellipse is quite involved. Surprisingly, to find an equation for this ellipse, and all others with their foci on an axis, it is simpler to begin with a more general case. The resulting equation is well worth the effort it takes to derive it, and so on the next page we state it as a theorem. You are asked to justify the numbered steps in the proof of this theorem in Question 14.

Equation for an Ellipse Theorem

The ellipse with foci $(c, 0)$ and $(-c, 0)$ and focal constant $2a$ has equation $\dfrac{x^2}{a^2} + \dfrac{y^2}{b^2} = 1$, where $b^2 = a^2 - c^2$.

Proof Let $F_1 = (-c, 0)$, $F_2 = (c, 0)$, and $P = (x, y)$. By the definition of an ellipse,
$$PF_1 + PF_2 = 2a.$$

1. $\sqrt{(x + c)^2 + y^2} + \sqrt{(x - c)^2 + y^2} = 2a$

2. $\sqrt{(x - c)^2 + y^2} = 2a - \sqrt{(x + c)^2 + y^2}$

3. $(x - c)^2 + y^2 = 4a^2 - 4a\sqrt{(x + c)^2 + y^2} + (x + c)^2 + y^2$

4. $-2cx = 4a^2 - 4a\sqrt{(x + c)^2 + y^2} + 2cx$

5. $4a\sqrt{(x + c)^2 + y^2} = 4a^2 + 4cx$

6. $a\sqrt{(x + c)^2 + y^2} = a^2 + cx$

7. $a^2((x + c)^2 + y^2) = a^4 + 2a^2cx + c^2x^2$

8. $a^2x^2 + a^2c^2 + a^2y^2 = a^4 + c^2x^2$

9. $(a^2 - c^2)x^2 + a^2y^2 = a^2(a^2 - c^2)$

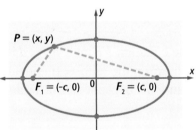

10. Because $c > 0$, $F_1F_2 = 2c$, and $2a > F_1F_2$, you can conclude that $2a > 2c > 0$, so $a > c > 0$. Thus, $a^2 > c^2$, and $a^2 - c^2$ is not negative. So $a^2 - c^2$ can be considered as the square of some real number, say b. Let $b^2 = a^2 - c^2$ and substitute to get
$$b^2x^2 + a^2y^2 = a^2b^2.$$

11. $\dfrac{x^2}{a^2} + \dfrac{y^2}{b^2} = 1$

The Standard Form of an Equation for an Ellipse in Standard Position

An ellipse centered at the origin with its foci on an axis is in **standard position**, and $\dfrac{x^2}{a^2} + \dfrac{y^2}{b^2} = 1$ is in the **standard form** of an equation for this ellipse. Even without knowing the geometric interpretation of this equation, analyzing the formula can tell you a lot about the shape of its graph.

GUIDED

Example 1

Given the equation $\frac{x^2}{9} + \frac{y^2}{4} = 1$, describe

a. the x- and y-intercepts of its graph.

b. the possible x- and y-values.

c. the symmetries of the graph, if any.

Solution

a. The x-intercepts are the x-values for which $y = 0$. **Substitute 0 for y** to get $\frac{x^2}{9} + \frac{0^2}{4} = 1$. Then solve for x to get $x = \underline{}$. Similarly, the y-intercepts are the y-values for which $x = \underline{}$. Substitute and solve for y to get $y = \underline{}$.

b. Now $y^2 \geq 0$ and $\frac{x^2}{9} \leq 1$, so $x^2 \leq 9$. Therefore, $\sqrt{x^2} \leq 3$ and $|x| \leq 3$, so $-3 \leq x \leq 3$. Similarly, $x^2 \geq 0$ and $\frac{y^2}{4} \leq \underline{}$, so $\underline{} \leq y \leq \underline{}$.

c. Recall that $r_y : (x, y) \rightarrow (-x, y)$. If you replace x with $-x$ in the equation, it becomes $\frac{(-x)^2}{9} + \frac{y^2}{4} = 1$. Because $x^2 = (-x)^2$, this equation is equivalent to the original. **Therefore, the graph is symmetric to the** $\underline{}$.

To check for symmetry over the x-axis, recall that $r_x : (x, y) \rightarrow \underline{}$. So, if you replace $\underline{}$ with $\underline{}$ in the original equation, it becomes $\underline{}$, which is equivalent to the original. **Therefore, the graph is symmetric to the** $\underline{}$ **as well.**

Generalizing from Example 1, the intercepts of the ellipse with equation $\frac{x^2}{a^2} + \frac{y^2}{b^2} = 1$ are $(a, 0)$, $(-a, 0)$, $(0, b)$ and $(0, -b)$. The possible x- and y-values indicate that the entire ellipse is contained in the rectangle $\{(x, y): -a \leq x \leq a \text{ and } -b \leq y \leq b\}$. These facts can help you sketch a graph of an ellipse by hand.

Consider the ellipse at the right. The segments $\overline{A_1 A_2}$ and $\overline{B_1 B_2}$ are, respectively, the **major** and **minor axes** of the ellipse. The major axis contains the foci and is always longer than the minor axis. The axes lie on the symmetry lines and intersect at the **center** O of the ellipse. Each segment $\overline{OA_1}$ and $\overline{OA_2}$ is a **semimajor axis** of the ellipse. The segments $\overline{OB_1}$ and $\overline{OB_2}$ are the **semiminor axes** of the ellipse. The diagram illustrates the following theorem. It applies to all ellipses centered at the origin with foci on one of the coordinate axes.

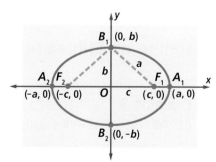

Length of Axes of an Ellipse Theorem

In the ellipse with equation $\frac{x^2}{a^2} + \frac{y^2}{b^2} = 1$, $2a$ is the length of the horizontal axis, and $2b$ is the length of the vertical axis.

The length of the major axis is the focal constant. If $a > b$, then the major axis is horizontal and $(c, 0)$ and $(-c, 0)$ are the foci. The focal constant is $2a$, the length of the semimajor axis is a, and by the Pythagorean Theorem $b^2 = a^2 - c^2$ (as in the ellipse on the previous page).

If $b > a$, then the major axis is vertical. So the foci are $(0, c)$ and $(0, -c)$, the focal constant is $2b$, and $a^2 = b^2 - c^2$.

 QY

> ▶ QY
>
> In the ellipse of Example 1, what are the lengths of the major and minor axes?

GUIDED

Example 2

Find the endpoints of the major and minor axes, the foci, and an equation for the ellipse in the Activity in standard position.

Solution Position a coordinate system on the ellipse, with origin at the center of the ellipse and the foci on the x-axis. Let the radius of the smallest circle on the conic graph paper be 1 unit on the axes scales.

The major axis has length __?__, so $a =$ __?__.

The minor axis has length __?__, so $b =$ __?__.

Therefore, in standard position, the endpoints of the major axis of the ellipse are (__?__, 0) and (__?__, 0), and the endpoints of the minor axis are (0, __?__) and (0, __?__).

Since $a > b$, $b^2 = a^2 - c^2$. So $9 = 25 - c^2$ and $c = 4$.

The foci for the ellipse are (__?__, 0) and (__?__, 0).

An equation for this ellipse is $\frac{x^2}{?} + \frac{y^2}{?} = 1$.

Graphing an Ellipse in Standard Form

Example 3

Consider the ellipse with equation $\frac{x^2}{9} + \frac{y^2}{10} = 1$.

a. Identify the endpoints and the lengths of the major and minor axes.

b. Graph the ellipse.

Solution

a. $a^2 = 9$ and $b^2 = 10$. So $a = 3$ and $b = \sqrt{10}$. Because $b > a$, the foci of the ellipse are on the y-axis. The endpoints of the major axis are $(0, \sqrt{10})$, and $(0, -\sqrt{10})$. The endpoints of the minor axis are $(3, 0)$, and $(-3, 0)$. The length of the major axis is $2\sqrt{10}$ and the length of the minor axis is 6.

b. Plot the four axis endpoints, then sketch the rest of the ellipse. A graph of the ellipse is at the right.

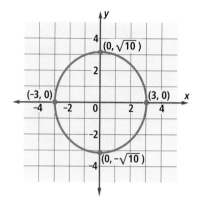

Questions

COVERING THE IDEAS

In 1 and 2, refer to the ellipse in the Activity.

1. What is the focal constant?

2. What is the length of each
   a. semimajor axis?   b. semiminor axis?

3. On the ellipse at the right, $OA = OB$, $OD = OC$ and $\overline{AB} \perp \overline{CD}$. Identify its
   a. foci.   b. major axis.   c. minor axis.

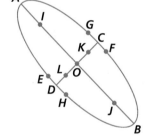

In 4–8, consider the ellipse with equation $\frac{x^2}{a^2} + \frac{y^2}{b^2} = 1$. Assume $b >$ a. Identify the following.

4. the center of the ellipse

5. the endpoints of the major and minor axes

6. the lengths of the semimajor and semiminor axes

7. the possible values of x and the possible values of y

8. the x- and y-intercepts

9. Sketch a graph of the ellipse with equation $\frac{x}{16} + y^2 = 1$.

10. Write an equation in standard form for the ellipse with focal constant 17 and foci $(8, 0)$ and $(-8, 0)$.

APPLYING THE MATHEMATICS

11. Refer to the ellipse graphed at the right.
    a. Find an equation for the ellipse.
    b. Write an inequality describing the points in the interior of the ellipse.

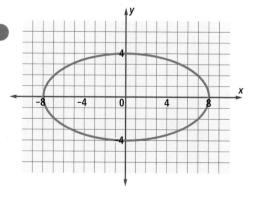

12. **a.** Write an equation describing the set of points (x, y) whose distances from $(-1, -1)$ and $(1, 1)$ add up to 4.

   b. Use a CAS to solve your equation in Part a for y.

   c. Graph the equation on a graphing utility and sketch the result.

13. **a.** Find the foci and focal constant of the ellipse with equation $\frac{x^2}{100} + \frac{y^2}{100} = 1$.

   b. What is special about this ellipse?

14. Write justifications for the statements in the proof of the Equation for an Ellipse Theorem.

15. The orbits of the planets are nearly elliptical with the Sun at one focus. The shape of Mercury's orbit can be approximated by the equation $\frac{x^2}{1295} + \frac{y^2}{1240} = 1$, where x and y are in millions of miles.

   a. What is the farthest Mercury gets from the Sun?

   b. What is the closest Mercury gets to the Sun?

   c. How far apart are the two foci?

16. **a.** Using two thumbtacks and a piece of string, draw a curve as shown at the right.

   b. Explain why the curve is an ellipse.

   c. What part of your equipment represents the focal constant of the ellipse?

17. A *superellipse* has equation $\left|\frac{x}{a}\right|^n + \left|\frac{y}{b}\right|^n = 1$, for nonzero a and b and $n > 0$.

   a. Sketch a graph of the superellipse for $a = 2$, $b = 3$, and $n = 1$.

   b. Identify the major and minor axis on your sketch in Part a.

   c. List the x- and y- intercepts of the graph.

   d. Repeat Part a when $a = 2$, $b = 3$, and $n = 3$.

   e. How are your graphs in Parts a and d related? How are they different?

REVIEW

18. The figure at the right shows a cross section of a tunnel with diameter 40 feet. A rectangular sign reading "Do Not Pass" must be placed at least 16 feet above the roadway. **(Lesson 12-3)**

   a. Find the length BE of the beam that supports the sign.

   b. If the sign is 16 feet long, what is its maximum height?

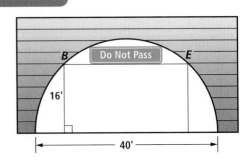

19. Refer to the diagram at the right. Circle A is centered at the origin with radius 3. Circle B is tangent to both axes and has its center on circle A. (Lessons 12-3, 12-2, Previous Course)

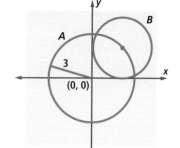

a. Write an equation for circle A.

b. Write an equation for circle B.

c. Find an inequality describing the interior of circle A.

d. Find the circumference of each circle.

e. Find the area of each circle.

20. Use $\triangle PQR$ at the right. Find the length of $\overline{QR}$ to the nearest tenth. (Lesson 10-8)

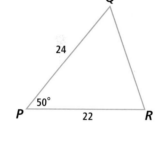

In 21 and 22, find b. (Lesson 10-7)

21.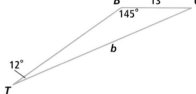

22.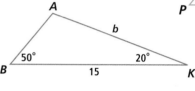

23. What is the image of (x, y) under the scale change $S_{7, 8}$? (Lesson 4-5)

EXPLORATION

24. Over the years, many devices were invented to help drafters draw ellipses by hand. One such tool is the *carpenter's trammel*. Research the carpenter's trammel and explain how it is used to draw an ellipse.

QY ANSWER

major axis, 6; minor axis, 4

Lesson

12-5

Relationships between Ellipses and Circles

▶ **BIG IDEA** Ellipses are stretched circles; circles are ellipses whose major and minor axes have the same length.

In some ellipses, the major axis is much longer than the minor axis. In others, the two axes are almost equal in length. The diagrams below illustrate three of the possible cases. All three ellipses have the same focal constant. You can see that when the focal constant stays the same, the positions of the foci in an ellipse affect its shape. The closer the foci are to the origin, the rounder the ellipse appears.

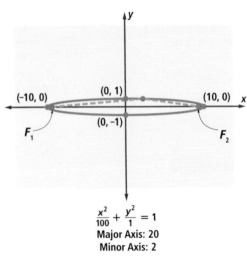

$$\frac{x^2}{100} + \frac{y^2}{1} = 1$$
Major Axis: 20
Minor Axis: 2

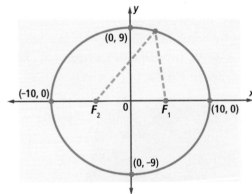

$$\frac{x^2}{100} + \frac{y^2}{81} = 1$$
Major Axis: 20
Minor Axis: 18

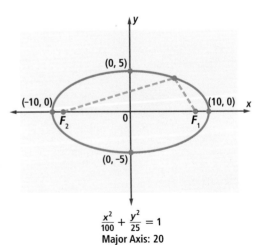

$$\frac{x^2}{100} + \frac{y^2}{25} = 1$$
Major Axis: 20
Minor Axis: 10

Circles as Special Ellipses

A special kind of ellipse results if the major and minor axes are equal. At the right is an ellipse with major axis 20 and minor axis 20. It has equation

$$\frac{x^2}{100} + \frac{y^2}{100} = 1,$$

which can be rewritten as

$$x^2 + y^2 = 100.$$

So this ellipse is a circle.

This can be generalized. Consider the standard form of an equation for an ellipse,

$$\frac{x^2}{a^2} + \frac{y^2}{b^2} = 1.$$

If the major and minor axes each have length $2r$, then $2a = 2r$ and $2b = 2r$, so you may substitute r for both a and b to get $\frac{x^2}{r^2} + \frac{y^2}{r^2} = 1$.

Multiply both sides by r^2 to get $x^2 + y^2 = r^2$.

This is an equation for the circle with center at the origin and radius r. So, a circle is a special kind of ellipse whose major and minor axes are equal in length.

 QY1

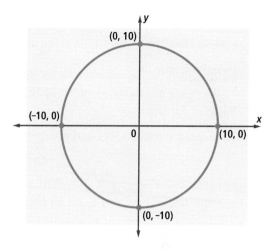

▶ **QY1**

Find an equation of the ellipse with center at the origin and major and minor axes of length 14.

Ellipses as Stretched Circles

An ellipse can be thought of as a stretched circle. The transformation that stretches and shrinks figures is the *scale change*, which you studied in Lesson 4-5.

Consider the unit circle with equation $x^2 + y^2 = 1$ and a scale change with horizontal magnitude 4 and vertical magnitude 2. The x- and y-intercepts of the unit circle and their images under this scale change are graphed at the right.

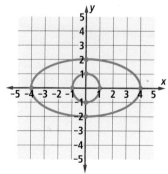

$$
\begin{aligned}
S_{4,2}: & \quad (1, 0) \rightarrow (4, 0) \\
S_{4,2}: & \quad (0, 1) \rightarrow (0, 2) \\
S_{4,2}: & \quad (-1, 0) \rightarrow (-4, 0) \\
S_{4,2}: & \quad (0, -1) \rightarrow (0, -2)
\end{aligned}
$$

From these four points you can see that the image of the unit circle under this scale change is not a circle. It appears to be an ellipse with foci on the x-axis.

Example

Find an equation for the image of the unit circle $x^2 + y^2 = 1$ under $S_{4, 2}$.

Solution To find an equation for the image of the circle, let (x', y') be the image of point (x, y) on the circle. Then $(x', y') = (4x, 2y)$. So, $x' = 4x$ and $y' = \underline{\ ?\ }$.

Solve these equations for x and y.

$$x = \underline{\ ?\ } \text{ and } y = \underline{\ ?\ }$$

You know that $x^2 + y^2 = 1$. Substitute the expressions for x and y involving x' and y' into $x^2 + y^2 = 1$ to get an equation for the image.

$$(\underline{\ ?\ })^2 + (\underline{\ ?\ })^2 = 1.$$

Now let (x, y) be a point on the image. Rewrite the equation for the image using x and y in place of x' and y'.

$$(\underline{\ ?\ })^2 + (\underline{\ ?\ })^2 = 1$$

The equation you have written is for an ellipse with a minor axis of length $\underline{\ ?\ }$ and a major axis of length $\underline{\ ?\ }$.

Check Substitute some points known to be on the image into the equation. Do their coordinates satisfy your equation for the ellipse?

Try $(4, 0)$: $\frac{4^2}{?} + \frac{0^2}{?} = \underline{\ ?\ } + \underline{\ ?\ } = \underline{\ ?\ }$.

Try $(0, -2)$: $\frac{0^2}{?} + \frac{(-2)^2}{?} = \underline{\ ?\ } + \underline{\ ?\ } = \underline{\ ?\ }$. It checks.

The procedure you followed in the Example can be repeated using a in place of 4 and b in place of 2. This shows that any ellipse in standard form can be thought of as a scale-change image of the unit circle.

Circle Scale-Change Theorem

The image of the unit circle with equation $x^2 + y^2 = 1$ under $S_{a, b}$ is the ellipse with equation $\left(\frac{x}{a}\right)^2 + \left(\frac{y}{b}\right)^2 = 1$.

The previous theorem is a special case of a more general Graph Scale-Change Theorem, which is analogous to the Graph-Translation Theorem you studied in Chapter 6.

Graph Scale-Change Theorem

In a relation described by a sentence in x and y, the following two processes yield the same graph:

1. replacing x by $\frac{x}{a}$ and y by $\frac{y}{b}$;
2. applying the scale change $S_{a,b}$ to the graph of the original relation.

 QY2

▶ QY2

Find an equation for the image of $x^2 + y^2 = 1$ under $S_{6,14}$.

A Formula for the Area of an Ellipse

Consider the figure below at the left. If each grid square has area 1, the area of the figure is equal to the number of grid squares inside the figure. Now suppose the scale change $S_{a,b}$ is applied to the figure and the grid. The result is the figure at the right below. Each grid square is transformed into a rectangle with length a, width b, and area ab. Since the area of each rectangle is ab times the area of one grid square, the area of the transformed figure is ab times the area of its preimage.

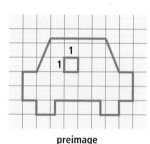

preimage

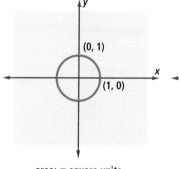

image under $S_{a,b}$

This illustrates that in general, the area of the image of a figure under the scale change $S_{a,b}$ is ab times the area of the preimage. This fact can be used to derive a formula for the area of an ellipse. The area of the unit circle is $\pi(1)^2 = \pi$, and the ellipse with equation $\frac{x^2}{a^2} + \frac{y^2}{b^2} = 1$ is the image of the unit circle under $S_{a,b}$. So, the area of this ellipse is $\pi(ab) = \pi ab$.

area: π square units

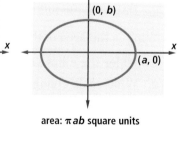

area: πab square units

Area of an Ellipse Theorem

An ellipse with semimajor and semiminor axes of lengths a and b has area $A = \pi ab$.

 QY3

▶ QY3

The Statuary Hall gallery in the United States Capitol is an elliptical chamber. The gallery is about 83 feet wide and about 96 feet long. Find its floor area.

Questions

COVERING THE IDEAS

1. **Fill in the Blank** A circle is an ellipse in which the major and minor axes

   A   are parallel.
   B   are perpendicular
   C   are of equal length.
   D   coincide.

True or False In 2 and 3, indicate whether the statement is true or false. If false, draw a counterexample.

2. All ellipses are circles.
3. All circles are ellipses.

4. **Fill in the Blank** All three ellipses below are the same width. In which ellipse are the foci farthest apart? Explain your answer.

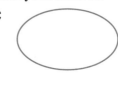

A   B   C

In 5 and 6, consider the circle $x^2 + y^2 = 1$ and the scale change $S_{5,\frac{1}{2}}$.

5. a. Find the image of $(1, 0)$ under $S_{5,\frac{1}{2}}$.

   b. Find the image of $(0, 1)$ under $S_{5,\frac{1}{2}}$.

   c. Write an equation of the image of the circle under $S_{5,\frac{1}{2}}$.

6. What is the area of the image?

7. Consider the ellipse drawn at the right.

   a. What scale change maps the unit circle onto this ellipse?

   b. Write an equation for the ellipse.

   c. Find its area.

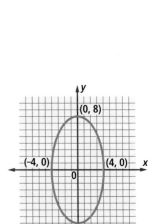

APPLYING THE MATHEMATICS

8. **a.** Write an equation for an ellipse in which the semimajor and semiminor axes each have length 5 and the distance between the foci is 0.

   b. What relationship between ellipses and circles does this illustrate?

9. **a.** Give an example of a scale change $S_{h,k}$ such that the unit circle and its image under the scale change are similar.

   b. Give an example of a scale change $S_{h,k}$ such that the unit circle and its image under the scale change are not similar.

10. Use the method of the Example to prove that the image of the unit circle under the scale change $S_{a,b}$ is the ellipse $\frac{x^2}{a^2} + \frac{y^2}{b^2} = 1$.

11. The orbit of the planet Mars is in the shape of an ellipse whose minor axis is 282.0 million miles long and whose major axis is 283.3 million miles long. What is the area of this ellipse?

In 12–14, use this definition: The *eccentricity* of an ellipse is the ratio of the distance 2c between its foci to the length 2a of its major axis.

12. **a.** Write a formula for the eccentricity e of an ellipse.

   b. What is the eccentricity of the ellipse in the Example?

13. An ellipse's major axis is 12 units long. If its eccentricity is $\frac{1}{3}$, how long is its minor axis?

14. **a.** What is the eccentricity of a circle?

   b. Is there a maximum possible value for the eccentricity of an ellipse? Why or why not?

15. Write the equations of four different ellipses with area 24π.

REVIEW

16. In 1937, a Whispering Gallery was constructed in the Museum of Science and Industry in Chicago. The Gallery was constructed in the form of an *ellipsoid* (an ellipse rotated around its major axis). When a visitor located at one focus whispers, the sound reflects directly to the focus at the other end of the gallery. The width of an ellipse in the plane of the foci is 13 feet 6 inches, and the length of the ellipse is 47 feet 4 inches.

   a. Find an equation that could describe this ellipse.

   b. How far are the foci from the endpoints of the major axis of this ellipse? (**Lesson 12-4**)

Matching In 17–21, match each equation with the best description. A letter may be used more than once. Do not graph. **(Lessons 12-4, 12-3, 12-2)**

17. $(x - 2)^2 + y^2 < 36$

18. $\dfrac{x^2}{48} + \dfrac{y^2}{64} > 1$

19. $x^2 + 9y^2 = 121$

20. $\dfrac{x^2}{16} + \dfrac{y^2}{81} = 4$

21. $x^2 + y^2 = 25$

  i   circle

  ii ellipse

  iii interior of a circle

  iv interior of an ellipse

  v   exterior of a circle

  vi exterior of an ellipse

22. At the right is a top view of a circular fountain surrounded by a circular flower garden. The distance from the center of the fountain to the outside edge of the garden is 65 feet. It is 45 feet from the outside edge of the garden to the fountain. If the center of the fountain is the origin, write a system of inequalities to describe the set of points in the flower garden. **(Lesson 12-3)**

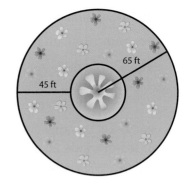

In 23 and 24, solve. **(Lesson 6-2)**

23. $|2n + 1| = 0.5$

24. $\sqrt{x^2} = \sqrt{40}$

In 25 and 26, consider the line ℓ with equation $y = -\dfrac{3}{7}x + 6$ and the point $P = (5, 2)$. Find an equation for the line through P

25. parallel to ℓ. **(Lesson 4-10)**

26. perpendicular to ℓ. **(Lesson 4-9)**

EXPLORATION

27. **a.** Cut out a paper circle and make a dot anywhere in its interior. Fold a point on the circle onto the dot then unfold. Repeat with other points on the circle. What shape is outlined by the crease lines (shown as segments in the diagram)?

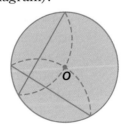

    b. Use a new circle. Place the dot at the center of the circle and repeat the activity in Part a.

    c. Explain your results in Parts a and b.

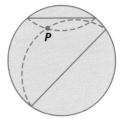

QY ANSWERS

1. $\dfrac{x^2}{49} + \dfrac{y^2}{49} = 1$ or
$x^2 + y^2 = 49$

2. $\dfrac{x^2}{36} + \dfrac{y^2}{196} = 1$

3. $1992\pi \approx 6258$ ft^2

Lesson 12-6

Equations for Some Hyperbolas

Vocabulary

hyperbola

foci, focal constant of
 a hyperbola

vertices of a hyperbola

standard position of
 a hyperbola

standard form of an equation
 for a hyperbola

▶ **BIG IDEA**   From the geometric definition of a hyperbola, an equation for any *hyperbola* symmetric to the *x*- and *y*-axes can be found.

The edges of the silhouettes of each of the towers pictured at the right are parts of *hyperbolas*. Structures with this shape are able to withstand higher winds and require less material to build than any other form.

Mental Math

Suppose a function *f* contains the points **(4, 17), (9, 12),** and **(13, 13).**

a. Find the rate of change from (4, 17) to (9, 12).

b. Find the rate of change from (9, 12) to (13, 13).

c. Could the graph of *f* be a line?

What Is a Hyperbola?

Like an ellipse, a hyperbola is determined by two foci and a focal constant. However, instead of a constant sum of distances from the foci, a point on a hyperbola must be at a *constant difference* of distances from the foci. The following Activity shows one way to find points on a hyperbola.

Activity

MATERIALS conic graph paper with 6 units between the centers of the circles

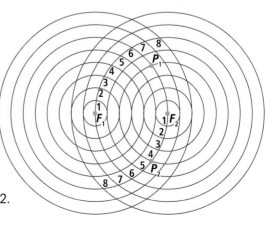

Step 1 Copy the foci and points P_1 and P_2 at the right. Find P_1F_1, P_2F_1, P_1F_2, and P_2F_2, then calculate $P_1F_1 - P_1F_2$ and $P_2F_1 - P_2F_2$. Do both differences equal the same constant?

Step 2 Plot two more points P_n such that $P_nF_1 = 8$ and $P_nF_2 = 6$, and then two more such that $P_nF_1 = 7$ and $P_nF_2 = 5$. Continue this process to find four more points such that $P_nF_1 - P_nF_2$ is always 2.

(continued on next page)

Step 3 Repeat Step 2, plotting ten points P_n such that $P_nF_2 - P_nF_1 = 2$.

Step 4 Draw a smooth curve through the points you plotted in Step 2, and another through the points you plotted in Step 3. These are two branches of a hyperbola. The branches do not intersect.

In general, if d is a positive number less than F_1F_2, the set of all points P such that $|PF_1 - PF_2| = d$ is a hyperbola. The absolute value means that the hyperbola has two branches, one from $PF_1 - PF_2 = d$, and the other from $PF_1 - PF_2 = -d$. The absolute value function allows both branches to be described with one equation.

Definition of Hyperbola

Let F_1 and F_2 be any two points and d be a constant with $0 < d < F_1F_2$. Then the **hyperbola** with **foci** F_1 and F_2 and **focal constant** d is the set of points P in a plane that satisfy $|PF_1 - PF_2| = d$.

The **vertices** V_1 and V_2 of the hyperbola are the intersection points of $\overleftrightarrow{F_1F_2}$ and the hyperbola.

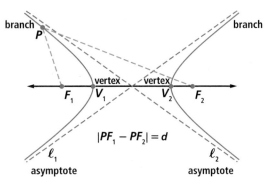

While it may look like each branch of the hyperbola is a parabola, this is not the case. Each branch of a hyperbola has *asymptotes*. In the figure at the right, ℓ_1 and ℓ_2 are asymptotes. The farther points on the hyperbola are from a vertex of the hyperbola, the closer they are to an asymptote, without ever touching. In contrast, parabolas do not have asymptotes.

The Standard Form of an Equation for a Hyperbola

A hyperbola is in **standard position** if it is centered at the origin with its foci on an axis. An equation for a hyperbola in standard position resembles the standard form of an equation for an ellipse.

Equation for a Hyperbola Theorem

The hyperbola with foci $(c, 0)$ and $(-c, 0)$ and focal constant $2a$ has equation $\frac{x^2}{a^2} - \frac{y^2}{b^2} = 1$, where $b^2 = c^2 - a^2$.

Proof The proof is almost identical to the proof of the
Equation for an Ellipse Theorem in Lesson 12-4.
Let $P = (x, y)$ be any point on the hyperbola with
foci $F_1 = (-c, 0)$ and $F_2 = (c, 0)$ and focal
constant $2a$. Then, by the definition of a hyperbola,

$$|PF_1 - PF_2| = 2a.$$

By the definition of absolute value, you know that this
equation is equivalent to

$$PF_1 - PF_2 = \pm 2a.$$

Now substitute $P = (x, y)$, $F_1 = (-c, 0)$, and $F_2 = (c, 0)$ into the
Pythagorean Distance Formula to get

$$\sqrt{(x + c)^2 + (y - 0)^2} - \sqrt{(x - c)^2 + (y - 0)^2} = \pm 2a.$$

Do algebraic manipulations similar to those in Steps 1–9 of the proof in
Lesson 12-4, and the same equation in Step 9 results.

$$(a^2 - c^2)x^2 + a^2y^2 = a^2(a^2 - c^2)$$

Then in Step 10, for hyperbolas, $c > a > 0$, so $c^2 > a^2$. Thus, $c^2 - a^2$ is
positive and you can let $b^2 = c^2 - a^2$. So $-b^2 = a^2 - c^2$. This accounts
for the minus sign in the equation.

$$\frac{x^2}{a^2} - \frac{y^2}{b^2} = 1$$

The equation $\frac{x^2}{a^2} - \frac{y^2}{b^2} = 1$ is the **standard form of an equation for
a hyperbola**.

Example 1

Find an equation for the hyperbola with foci F_1 and F_2,
where $F_1F_2 = 10$ and $|PF_1 - PF_2| = 8$, on a rectangular
coordinate system in standard position.

Solution Use the Equation for a Hyperbola Theorem.
You are given $F_1F_2 = 10$, so $2c = 10$, and $c = 5$.
The focal constant is 8, so $2a = 8$, and $a = 4$.
Now, $b^2 = 5^2 - 4^2 = 9$. Thus, an equation for
this hyperbola is $\frac{x^2}{16} - \frac{y^2}{9} = 1$.

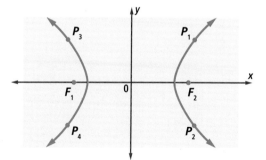

Asymptotes of a Hyperbola in Standard Position

To find equations for the asymptotes of the hyperbola with
equation $\frac{x^2}{a^2} - \frac{y^2}{b^2} = 1$, it helps to examine the special case when a
and b both equal 1. (This is like examining the unit circle to learn
about ellipses.)

Then $x^2 - y^2 = 1$. The hyperbola with this equation is symmetric to both axes. Consequently, each point on the hyperbola in the first quadrant has reflection images on the hyperbola in other quadrants. The graph at the right shows the reflection images of A, B, C, and D over the x-axis and the y-axis.

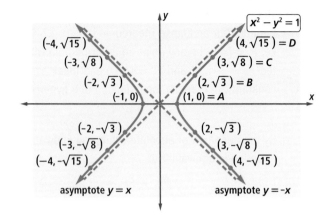

$$A = (1, 0)$$
$$B = (2, \sqrt{3}) \approx (2, 1.73)$$
$$C = (3, \sqrt{8}) \approx (3, 2.83)$$
$$D = (4, \sqrt{15}) \approx (4, 3.87)$$

The lines $y = -x$ and $y = x$ appear to be the asymptotes of $x^2 - y^2 = 1$. We can verify the equations for the asymptotes algebraically.

When $x^2 - y^2 = 1$,
$$y^2 = x^2 - 1.$$
So
$$y = \pm \sqrt{x^2 - 1}.$$

As values of x get larger, $\sqrt{x^2 - 1}$ becomes closer to $\sqrt{x^2}$, which is $|x|$. However, because $\sqrt{x^2 - 1} \neq \sqrt{x^2}$, the curve $x^2 - y^2 = 1$ never intersects the lines with equations $y = x$ or $y = -x$. So, y gets closer to x or $-x$ but never reaches it.

According to the Graph Scale-Change Theorem, the scale change $S_{a,b}$ maps $x^2 - y^2 = 1$ onto $\frac{x^2}{a^2} - \frac{y^2}{b^2} = 1$. Under the same scale change, the asymptotes $y = \pm x$ of $x^2 - y^2 = 1$ are mapped onto the lines with equations $\frac{y}{b} = \pm \frac{x}{a}$. These lines are the asymptotes of $\frac{x^2}{a^2} - \frac{y^2}{b^2} = 1$.

Asymptotes of a Hyperbola Theorem

The asymptotes of the hyperbola with equation $\frac{x^2}{a^2} - \frac{y^2}{b^2} = 1$ are $\frac{y}{b} = \pm \frac{x}{a}$, or $y = \pm \frac{b}{a}x$.

 QY

▸ **QY**

What are the asymptotes of the hyperbola in Example 1?

Graphing a Hyperbola with Equation in Standard Form

To graph $\frac{x^2}{a^2} - \frac{y^2}{b^2} = 1$ by hand, notice that $(a, 0)$ and $(-a, 0)$ satisfy the equation. These are the vertices of the hyperbola. When $x = 0$, y is not a real number, so the hyperbola does not intersect the y-axis. Use the asymptotes to make an accurate sketch of the graph. Remember that the asymptotes are not part of the hyperbola.

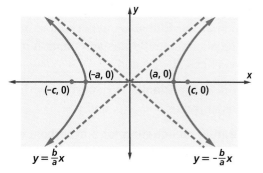

Example 2

Graph the hyperbola with equation $\frac{x^2}{16} - \frac{y^2}{36} = 1$.

Solution The equation is in standard form. So, $a^2 = 16$ and $a = 4$. The vertices are $(4, 0)$ and $(-4, 0)$. The asymptotes are $\frac{y}{6} = \pm \frac{x}{4}$, or $y = \pm \frac{3}{2}x$. Carefully graph the vertices and asymptotes. Then sketch the hyperbola.

Check Solve $\frac{x^2}{16} - \frac{y^2}{36} = 1$ for y on a CAS.
One CAS solution is shown below.

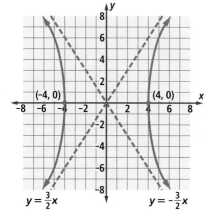

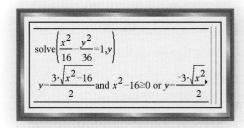

The complete solution is

$$y = \frac{3 \cdot \sqrt{x^2 - 16}}{2} \text{ and } x^2 - 16 \geq 0 \text{ or } y = \frac{-3 \cdot \sqrt{x^2 - 16}}{2} \text{ and } x^2 - 16 \geq 0.$$

So $y = \frac{3\sqrt{x^2 - 16}}{2}$ or $y = -\frac{3\sqrt{x^2 - 16}}{2}$.

Graph both equations on the same axes on a graphing utility.

Although the graphing utility may have trouble graphing values close to the vertices of the hyperbola, the output closely resembles the hand-drawn solution.

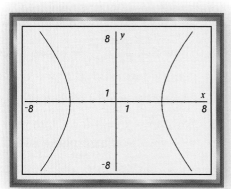

Questions

COVERING THE IDEAS

1. **Fill in the Blanks** A hyperbola with foci $(c, 0)$ and $(-c, 0)$ and focal constant $2a$ has an equation of the form ___?___ and vertices at ___?___ and ___?___.

2. **Fill in the Blanks** A hyperbola with equation $\frac{x^2}{a^2} - \frac{y^2}{b^2} = 1$ has asymptotes $y =$ ___?___ and $y =$ ___?___.

In 3 and 4, an equation for a hyperbola is given. Identify its vertices, its foci, and its asymptotes.

3. $1 = x^2 - y^2$

4. $\frac{x^2}{7^2} - \frac{y^2}{3^2} = 1$

5. **True or False** The focal constant of a hyperbola equals the distance between the foci.

6. **True or False** If F_1 and F_2 are the foci of a hyperbola, then $\overleftrightarrow{F_1F_2}$ is a line of symmetry for the curve.

7. What does the phrase "$\sqrt{x^2 - 1}$ is close to $\sqrt{x^2}$ for large values of x" mean?

8. Consider the hyperbola with equation $\frac{x^2}{25} - \frac{y^2}{64} = 1$.
   a. Name its vertices and state equations for its asymptotes.
   b. Graph the hyperbola.

APPLYING THE MATHEMATICS

9. Explain why $y = |x|$ is not an equation describing the asymptotes of $x^2 - y^2 = 1$.

10. Write an equation for the hyperbola with vertices at $(4, 0)$ and $(-4, 0)$ and one focus at $(7, 0)$.

11. The point $(-6, 3)$ is on a hyperbola with foci $(4, 0)$ and $(-4, 0)$.
   a. Find the focal constant of the hyperbola.
   b. Give an equation for this hyperbola in standard form. (*Hint:* Find b using $b^2 = c^2 - a^2$.)
   c. Graph this hyperbola.

12. Show that $\frac{x^2}{91} - \frac{y^2}{49} = 1$ is equivalent to an equation of the general form $Ax^2 + Bxy + Cy^2 + Dx + Ey + F = 0$ by finding the values of $A, B, C, D, E,$ and F.

13. Solve $x^2 - y^2 = 1$ for y. Use your solution to graph $x^2 - y^2 = 1$ on a graphing utility.

REVIEW

14. In Australia, a type of football is played on elliptical fields. One such field has a major axis of length 185 meters and minor axis of length 155 meters. Surrounding it is an elliptical fence with major axis of length 187 meters and minor axis of length 157 meters. The 1-meter wide track between the fence and the field is to be covered with turf. Find the area of the track. **(Lesson 12-5)**

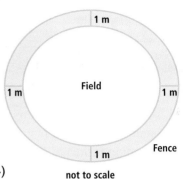

1 m

1 m   Field   1 m

1 m   Fence

not to scale

In 15 and 16, graph the ellipse with the given equation. **(Lesson 12-4)**

15. $\frac{x^2}{4} + \frac{y^2}{25} = 1$

16. $\frac{x^2}{9} + y^2 = 1$

17. Standard Quonset huts are semicircular with a diameter of 20 feet and a length of 48 feet. **(Lesson 12-2)**

   a. Inside the hut, how close to either side of the hut could a 6-foot soldier stand upright?

   b. What is the volume of a hut?

During World War II, easy-to-build Quonset huts were used as barracks for troops.

18. An auto dealer is having a Fourth of July extravaganza. The dealership plans to be open for 72 hours straight. Suppose the dealer has 100 new cars on the lot and is able to sell an average of 4 cars every 3 hours. **(Lesson 3-1)**

   a. Let h be the number of hours the car dealership has been open and let C be the number of cars remaining on the lot. Find three other pairs of values that satisfy this relation and complete the table.

h	0	3	?	?	?
C	100	96	?	?	?

   b. Write a formula for the number of cars C on the lot as a function of the number of hours h the sale has been on.

   c. After how many hours will there be only 60 cars left?

   d. If the dealership is able to maintain the pace of 4 cars sold every 3 hours, will the dealer sell all the cars on the lot during the sale? How can you tell?

EXPLORATION

19. The words *ellipsis* and *hyperbole* have literary meanings. What are these meanings?

20. In *Round the Moon*, a novel written by Jules Verne in 1870, a group of men launch a rocket to the Moon. During the journey they argue whether the rocket trajectory is hyperbolic or parabolic. Because each curve is infinite, the men believe they are doomed to travel infinitely through space. Find out on which trajectory modern day rockets travel and whether or not the men had reason to worry.

QY ANSWER

$y = \pm \frac{3}{4}x$

Equations for Some Hyperbolas   **837**

Lesson

12-7

A General Equation for Quadratic Relations

Vocabulary

conic section

standard form of an equation for a quadratic relation

▶ **BIG IDEA** Equations for specific parabolas, ellipses, circles, and hyperbolas are all special cases of one general equation form.

The Conic Sections

On the opening page of this chapter, you read that parabolas, hyperbolas, and ellipses can all be formed when a plane and a double cone intersect. A double cone is formed by rotating a line (in space) about a line it intersects (its axis). Any of the possible images of the line is an *edge* of the cone. The intersection of a plane and a double cone is called a **conic section**. Let k be the measure of the acute angle between the axis of the double cone and its edge. Let θ be the measure of the smallest angle between the axis and the intersecting plane. The following three possible relationships between θ and k determine the three types of conic sections.

$k < \theta < 90°$
ellipse

$\theta = k$
parabola

$\theta < k$
hyperbola

The Standard Form of an Equation for a Quadratic Relation

You have now seen equations for all three different types of conics. Here are some of these equations in standard form, including the hyperbola form $xy = k$ that you will examine in this lesson.

$y = ax^2 + bx + c$	parabola
$(x - h)^2 + (y - k)^2 = r^2$	circle (special ellipse)
$\dfrac{x^2}{a^2} + \dfrac{y^2}{b^2} = 1$	ellipse
$\dfrac{x^2}{a^2} - \dfrac{y^2}{b^2} = 1$ or $xy = k$	hyperbola

Mental Math

Serena is cutting a pie into 8 wedge-shaped pieces.

a. If she wants all the pieces to be the same size, what is the measure in degrees of the angle of each wedge?

b. If she wants each of four of the pieces to be twice as big as each of the other four, what is the measure in degrees of the angle of the larger wedges?

c. If she wants each of two of the pieces to be twice as big as each of the other six, what is the measure in degrees of the angle of the larger wedges?

Although the equations for a hyperbola and an ellipse look similar, the others look different. However, all these equations contain only constant terms or terms with x^2, xy, y^2, x, or y. Thus, all the conic sections are special types of relations with polynomial equations of 2nd degree. These are called *quadratic relations*. The **standard form of an equation for a quadratic relation** is

$$Ax^2 + Bxy + Cy^2 + Dx + Ey + F = 0,$$

where A, B, C, D, E, and F are real numbers, and at least one of A, B, or C is nonzero.

Example 1

Show that the circle with equation $(x - 5)^2 + (y - 1)^2 = 12$ is a quadratic relation.

Solution Rewrite the equation in the standard form of an equation for a quadratic relation. First expand the squares of the binomials and combine like terms.

$$x^2 - 10x + y^2 - 2y + 26 = 12$$

Then add –12 to both sides and use the Commutative Property of Addition to reorder the terms so that they are in the order x^2, xy, y^2, x, y, and constants.

$$x^2 + 0xy + y^2 - 10x - 2y + 14 = 0$$

This is in standard form with $A = 1$, $B = 0$, $C = 1$, $D = -10$, $E = -2$, and $F = 14$. Because at least one of A, B, or C is nonzero, this is a quadratic relation.

 QY1

▶ QY1

If you put the equation $xy = 25$ for a hyperbola into standard form, what are the values of A, B, C, D, E, and F?

Generating More Hyperbolas from Inverse Variation

Recall Anna and Jenna Lyzer's experiment to verify the Law of the Lever in Lesson 2-7. With Anna seated at a fixed point on one side of a seesaw's pivot, her friends with weights w took turns balancing her while Jenna recorded their distances d feet from the pivot. Anna and Jenna found that the inverse-variation function with equation $d = \frac{119}{w}$ is a good model.

The Lyzers' equation is one instance of the general inverse-variation function, $y = \frac{k}{x}$. In Chapter 2 we claimed that the graph of $y = \frac{k}{x}$ is a hyperbola. Now you can prove it by showing that it satisfies the geometric definition of hyperbola given in the last lesson.

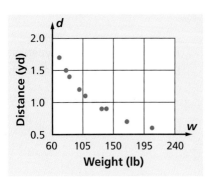

If $k > 0$, each branch of the graph is reflection-symmetric over the line $y = x$, and so the foci must be on the line $y = x$. Because the graph of $y = \frac{k}{x}$ is rotation-symmetric about the origin, the foci are also rotation-symmetric about the origin. Example 2 shows how an equation of the form $y = \frac{k}{x}$ arises when the foci meet these criteria.

Example 2

Find an equation of the form $y = \frac{k}{x}$ for the hyperbola with foci $F_1 = (c, c)$ and $F_2 = (-c, -c)$ and focal constant $2c$.

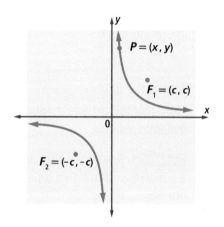

Solution Let $P = (x, y)$ be a point on the hyperbola. Then, by the definition of hyperbola, one branch of the curve is the set of points P such that

$$PF_1 - PF_2 = 2c.$$

Use the Distance Formula with $F_1 = (c, c)$, and $F_2 = (-c, -c)$.

$$\sqrt{(x - c)^2 + (y - c)^2} - \sqrt{(x + c)^2 + (y + c)^2} = 2c$$

Now proceed as in Lesson 12-4 with the derivation of an equation for an ellipse. You may find your CAS helpful as you follow the steps. First add $\sqrt{(x + c)^2 + (y + c)^2}$ to both sides.

$$\sqrt{(x - c)^2 + (y - c)^2} = 2c + \sqrt{(x + c)^2 + (y + c)^2}$$

Square both sides. Notice that the right side is a binomial.

$$(x - c)^2 + (y - c)^2 = 4c^2 + 4c\sqrt{(x + c)^2 + (y + c)^2} + (x + c)^2 + (y + c)^2$$

Expand the squares of the binomials, combine like terms, and simplify.

$$-4cx - 4cy - 4c^2 = 4c\sqrt{(x + c)^2 + (y + c)^2}$$

Divide by $-4c$; then square both sides.

$$(x + y + c)^2 = (x + c)^2 + (y + c)^2$$

The other branch of the hyperbola also satisfies this equation. Expand both sides again and simplify.

$$x^2 + 2xy + 2cx + y^2 + 2cy + c^2 = x^2 + 2cx + c^2 + y^2 + 2cy + c^2$$

$$2xy = c^2$$

$$xy = \frac{c^2}{2}$$

$$y = \frac{\frac{c^2}{2}}{x}$$

This is an equation of the form $y = \frac{k}{x}$, where $k = \frac{c^2}{2}$.

The following theorem summarizes attributes of hyperbolas of the form $xy = k$.

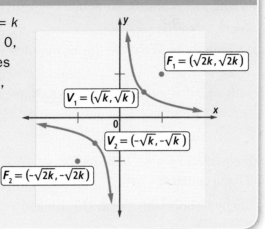

Attributes of $y = \frac{k}{x}$ Theorem

The graph of $y = \frac{k}{x}$ or $xy = k$ is a hyperbola. When $k > 0$, this hyperbola has vertices $(\sqrt{k}, \sqrt{k})$ and $(-\sqrt{k}, -\sqrt{k})$, foci $(\sqrt{2k}, \sqrt{2k})$ and $(-\sqrt{2k}, -\sqrt{2k})$, and focal constant $2\sqrt{2k}$. The asymptotes of the graph are $x = 0$ and $y = 0$.

$F_1 = (\sqrt{2k}, \sqrt{2k})$

$V_1 = (\sqrt{k}, \sqrt{k})$

$V_2 = (-\sqrt{k}, -\sqrt{k})$

$F_2 = (-\sqrt{2k}, -\sqrt{2k})$

 QY2

▶ QY2

According to the theorem above, what are the foci and focal constant of the hyperbola that is the graph of $d = \frac{160}{w}$?

GUIDED

Example 3

Find an equation of the form $y = \frac{k}{x}$ for the hyperbola with foci $F_1 = (8, 8)$ and $F_2 = (-8, -8)$, and focal constant 16.

Solution From the Attributes of $y = \frac{k}{x}$ Theorem, you know that the hyperbola has foci $(\sqrt{2k}, \sqrt{2k})$ and $(\underline{\ ?\ }, \underline{\ ?\ })$.
So $(\underline{\ ?\ }, \underline{\ ?\ }) = (\sqrt{2k}, \sqrt{2k})$. Solve for k.

$$\underline{\ ?\ } = \sqrt{2k}$$
$$\underline{\ ?\ } = 2k$$
$$\underline{\ ?\ } = k$$

Substitute $\underline{\ ?\ }$ for k in $y = \frac{k}{x}$.

An equation for the hyperbola is $y = \underline{\ ?\ }$.

Questions

COVERING THE IDEAS

In 1–4, determine whether the equation is for a quadratic relation. If so, put the equation in standard form of an equation for a quadratic relation. If not, tell why not.

1. $(x - 3)^2 + (y + 7)^2 = 118$ 2. $2x^2 + 4x - 7xy + 8y^2 + 3y = -3$
3. $\pi y - 6xy + y^2 + x^2 = \sqrt{11}$ 4. $-2x^2y + 2xy + 2x^2 + 6x + 5y = 0$

5. Recall Anna and Jenna's equation, $d = \frac{119}{w}$.

   a. What are the foci of the hyperbola with this equation?

   b. What is the focal constant?

6. At the right is a hyperbola with foci A and B. What must be true about $|Q_1A - Q_1B|$ and $|Q_2A - Q_2B|$?

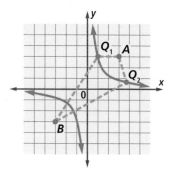

7. The graph of $dw = 240$ is a hyperbola. What are the asymptotes of this hyperbola?

8. Consider the hyperbola with equation $xy = k$ where $k > 0$. Name

   a. its foci.
   b. its asymptotes.

   c. its focal constant.
   d. its vertices.

9. Graph the hyperbola with equation $xy = 8$. Name its foci, vertices, asymptotes, and focal constant.

10. a. Find an equation for the hyperbola with foci at (15, 15) and (–15, –15) and focal constant 30.

    b. Verify that the point (5, 22.5) is on the hyperbola in Part a.

APPLYING THE MATHEMATICS

In 11 and 12, rewrite the equation in the standard form for a quadratic relation, and give the values of A, B, C, D, E, and F.

11. $\frac{x^2}{4} - \frac{y^2}{9} = 1$

12. $y = 5(x + 2)^2 - 11$

13. A hyperbola with perpendicular asymptotes is called a *rectangular hyperbola*. Explain whether or not each of the following is an equation for a rectangular hyperbola.

    a. $xy = 32$
    b. $x^2 - y^2 = 1$
    c. $\frac{x^2}{4} - \frac{y^2}{9} = 1$

14. A car travels the 2.5 miles around the Indianapolis Motor Speedway in t seconds at an average rate of r mph. Racing fans with stopwatches can calculate how fast the car is traveling if they know the value of the constant rt.

    a. What is that value? (*Hint:* Convert miles per second to miles per hour.)

    b. Graph $rt = k$, where k is your answer to Part a.

    c. If a driver completes one lap in 45 seconds, how fast is the driver traveling in mph?

15. Sketch a graph of the inequality.

    a. $xy > 4$
    b. $xy \leq 4$

The Indianapolis Motor Speedway is the largest stadium in the United States. It seats 250,000 people.

REVIEW

16. Consider the hyperbola with the equation $\frac{x^2}{64} - \frac{y^2}{121} = 1$.

    a. What are its foci?   b. Name its vertices.

    c. State equations for its asymptotes. (**Lesson 12-6**)

Multiple Choice In 17–19, choose the set of points that best meets the given condition. (**Lessons 12-6, 12-4, 12-2, 12-1**)

    A circle   **B** ellipse   **C** parabola   **D** hyperbola

17. equidistant from a given focus and directrix

18. satisfy the equation
$$\left| \sqrt{(x+23)^2 + (y-15)^2} - \sqrt{(x-23)^2 + (y-15)^2} \right| = 7$$

19. satisfy the equation $9x^2 + 2y^2 = 71$

20. **Multiple Choice** Which of the following describes the set of points P whose distances from $(7, 2)$ and $(3, 4)$ add up to 12? (**Lesson 12-4**)

    A $(x-7)^2 + (y-2)^2 + (x-3)^2 + (y-4)^2 = 12$

    B $(x+7)^2 + (y+2)^2 + (x+3)^2 + (y+4)^2 = 12$

    C $\sqrt{(x-7)^2 + (y-2)^2} + \sqrt{(x-3)^2 + (y-4)^2} = 12$

    D $\sqrt{(x-7)^2 + (y+2)^2} + \sqrt{5(x+3)^2 + (y+4)^2} = 12$

21. To estimate the distance across a river, Sir Vayer marks point A near one bank, sights a tree at point T growing on the opposite bank, and measures off a distance AB of 100 feet along the bank. At B he sights T again. If $m\angle A = 90°$ and $m\angle B = 76°$, how wide is the river? (**Lesson 10-1**)

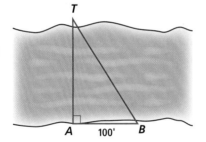

22. a. Evaluate $\log_5 125$ and $\log_{125} 5$ without a calculator.

    b. Evaluate $\log_4 16$ and $\log_{16} 4$ without a calculator.

    c. Generalize Parts a and b. (**Lesson 9-7**)

23. At the zoo, Nigel bought 3 slices of vegetable pizza and 1 small lemonade for \$5.40. Rosa paid \$4.80 for 2 slices of vegetable pizza and 2 small lemonades. What is the cost of a small lemonade? (**Lesson 5-4**)

EXPLORATION

24. In the standard form of a quadratic relation equation, there are six coefficients: A, B, C, D, E, and F. Changing these coefficients affects the type and appearance of the conic section. Search the Internet to find interactive websites that allow you to change the appearance of the conics by varying the coefficients. Make at least three conjectures about how certain types of coefficient changes affect the appearance of the graphs.

QY ANSWERS

1. $A = 0, B = 1, C = 0,$
   $D = 0, E = 0, F = -25$

2. The foci are
   $(\sqrt{320}, \sqrt{320}) = (8\sqrt{5}, 8\sqrt{5})$ and
   $(-\sqrt{320}, -\sqrt{320}) = (-8\sqrt{5}, -8\sqrt{5})$.
   The focal constant
   is $2\sqrt{320} = 16\sqrt{5}$.

Lesson

12-8

Quadratic-Linear Systems

Vocabulary

quadratic system

quadratic-linear system

▶ **BIG IDEA** Solutions to systems with one quadratic and one linear equation in x and y can be found by graphing, substitution, using linear combinations, or using a CAS.

A **quadratic system** is a system that involves polynomial sentences of degrees 1 and 2, at least one of which is a quadratic sentence. One way to solve a quadratic system is to examine the points of intersection of the graphs of the equations.

Examining Quadratic-Linear Systems Geometrically

A quadratic system with at least one linear sentence is called a **quadratic-linear system**. No new properties are needed to solve quadratic-linear systems. Geometrically, the task is to find the intersection of a conic section and a line. For example, in a system of a parabola and a line, there are three possibilities.

Mental Math

Give all points of intersection of the graph of $y = x^2$ and the graph of the given equation.

a. $y = 0$

b. $y = 4$

c. $y = x$

d. $y = -1.5$

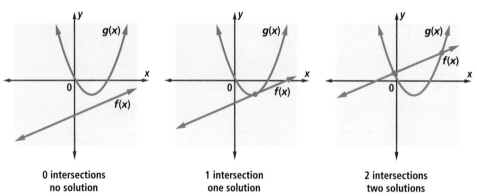

0 intersections
no solution

1 intersection
one solution

2 intersections
two solutions

The following Activity will help you determine the possible numbers of intersection points of a line and the other quadratic relations that you have studied.

Activity

Step 1 Sketch a circle, an ellipse, and a hyperbola on three separate sets of axes.

Step 2 On the graph of the circle, sketch several lines that intersect the circle in various ways. In how many points can a circle and a line intersect?

Step 3 Repeat Step 2 for the ellipse and the hyperbola.

Step 4 Fill in a table like the one at the right with the number of intersections between a line and the indicated quadratic relation. The Line-Parabola cell has been filled in for you.

Quadratic Relation	Number of Intersections with Line
Parabola	0, 1, or 2
Circle	?
Ellipse	?
Hyperbola	?

As you saw in the Activity, a quadratic-linear system may have 0, 1, or 2 solutions. This is because a quadratic equation may have 0, 1, or 2 solutions.

Solving Quadratic-Linear Systems Algebraically

One way to solve a quadratic-linear system is to solve the linear equation for one variable and substitute the resulting expression into the quadratic equation.

Example 1

Find exact solutions to the system $\begin{cases} y + 4x = 10 \\ xy = 4 \end{cases}$.

Solution 1 Geometrically, the solution to this system is the intersection of a line and a hyperbola.

Solve the first sentence for y.

$$y = 10 - 4x$$

Substitute the expression $10 - 4x$ for y in the second sentence.

$$x(10 - 4x) = 4$$

This is a quadratic equation that you can solve by the Quadratic Formula or by factoring. To use the Quadratic Formula, expand the left side and rearrange the equation so one side is zero.

$$10x - 4x^2 = 4$$
$$4x^2 - 10x + 4 = 0$$
$$x = \frac{(-10) \pm \sqrt{100 - 4 \cdot 4 \cdot 4}}{8} = \frac{10 \pm \sqrt{36}}{8}$$
$$x = \frac{10 - 6}{8} \quad \text{or} \quad x = \frac{10 + 6}{8}$$
$$x = \frac{1}{2} \quad \text{or} \quad x = 2$$

(continued on next page)

Now substitute each value into $y = 10 - 4x$ to find y.

When $x = \frac{1}{2}$, $y = 10 - 4\left(\frac{1}{2}\right) = 8$. So, one solution is $\left(\frac{1}{2}, 8\right)$.

When $x = 2$, $y = 10 - 4(2) = 2$. The other solution is $(2, 2)$.

So, the solutions are $\left(\frac{1}{2}, 8\right)$ and $(2, 2)$.

Check Solve both equations for y.

$$y = 10 - 4x \qquad \text{and} \qquad y = \frac{4}{x}$$

Now graph the system with a graphing utility. Zoom in to estimate the coordinates of the intersection points.

The curves intersect at $(0.5, 8)$ and $(2, 2)$. It checks.

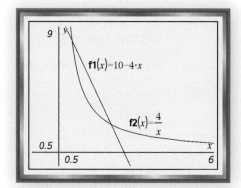

You could also solve Example 1 by using a graphing utility to find the exact coordinates of the intersection points. (See Question 2.)

Example 2 illustrates two solution methods that are quite different from each other. The first solution is algebraic and uses a CAS. The second solution draws on what you have already learned about circles in your study of geometry.

Example 2

At the right are graphs of the equations $x^2 + y^2 = 100$ and $y = -\frac{3}{4}x + \frac{25}{2}$. It appears that they intersect in only one point, $(6, 8)$. (That is, the line is tangent to the circle.) Is this so? Justify your answer.

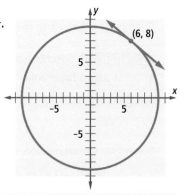

Solution 1 First, check that the graphs intersect at $(6, 8)$.

Does $6^2 + 8^2 = 100$? Yes.

Does $8 = -\frac{3}{4}(6) + \frac{25}{2}$? Yes, because $-\frac{3}{4}(6) + \frac{25}{2} = -\frac{9}{2} + \frac{25}{2} = \frac{16}{2} = 8$.

To find all solutions, solve the system $\begin{cases} x^2 + y^2 = 100 \\ y = -\frac{3}{4}x + \frac{25}{2} \end{cases}$.

Use the solve command on a CAS.

The only solution is $x = 6$ and $y = 8$. So, the only point of intersection is $(6, 8)$.

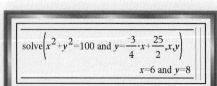

Solution 2 You may remember from your study of geometry that a tangent to a circle is the line perpendicular to the radius at the point of intersection. In this case, a radius from $(0, 0)$ to $(6, 8)$ has slope $\frac{4}{3}$. The line $y = -\frac{3}{4}x + \frac{25}{2}$ has slope $-\frac{3}{4}$. Since the product of the slopes is -1, this line is perpendicular to the radius at $(6, 8)$ and is a tangent.

Inconsistent Quadratic Systems

Like linear systems, quadratic systems can be inconsistent. That is, there can be no solution. One signal for inconsistency is that the solutions to the quadratic system are not real. This means that the graphs of the equations in the system do not intersect.

GUIDED

Example 3

Find the points of intersection of the line $y = x$ and the parabola $y = x^2 + 2$.

Solution 1 Graph the line and parabola, as shown at the right. There are ___?___ points of intersection.

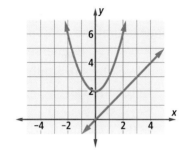

Solution 2 Solve the system $\begin{cases} y = x \\ y = x^2 + 2 \end{cases}$.

Substitute x for y in the second sentence.

$$\underline{} = x^2 + 2$$

Put this equation in standard form so you can use the Quadratic Formula.

$$\underline{} - \underline{} + \underline{} = 0$$

From that formula, $x = \dfrac{? \pm \sqrt{?}}{?}$.

Both solutions to this equation are nonreal complex numbers. So there are ___?___ points of intersection.

Questions

COVERING THE IDEAS

1. How many solutions can a system of one linear and one quadratic equation have?

2. Refer to Example 1. Check the solutions by graphing the system on a graphing utility and finding the exact coordinates of the points of intersection.

3. A graph of the system $\begin{cases} y = 2x \\ 4x^2 + 2y^2 = 48 \end{cases}$ is shown at the right.

   a. How many solutions are there?
   b. Use the graph to approximate the solutions.
   c. Check your answers in Part b.

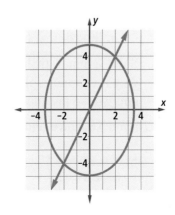

In 4 and 5, a system is given. Estimate the solutions by graphing. Then find exact solutions by substitution.

4. $\begin{cases} mn = 12 \\ n = -\frac{2}{3}m + 6 \end{cases}$

5. $\begin{cases} y = 2x^2 + 5x - 1 \\ y = x + 5 \end{cases}$

In 6 and 7, find all intersection points of the line and the parabola.

6. $y = x + 4$ and $y = x^2$

7. $y = 2x^2 - 4$ and $y = x - 3$

8. a. What term is used to describe a system that has no solutions?

   b. Give equations for such a system involving an ellipse and a line.

In 9 and 10, consider the figure at the right, which suggests that the parabola $y = x^2 + 7x - 10$ and the line $y = -5x - 46$ intersect at or near the point (-6, -16).

9. Check by substitution that this point is on both curves.

10. Solve the system algebraically to verify that this is the only solution.

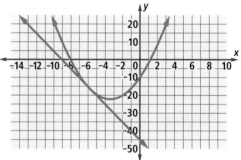

APPLYING THE MATHEMATICS

11. Explain why a quadratic-linear system cannot have an infinite number of solutions.

12. Phillip has 200 meters of fencing material. Next to his barn, he wants to fence in a rectangular region with area 1700 square meters, using a wall of the barn as one of the sides of the enclosure.

    a. Let x be the width of the region and y be its length. Draw a picture of the situation.

    b. Assuming Phillip uses all 200 meters of fencing, write a system of equations that can be solved to find x and y.

    c. Graph your system to estimate the dimensions of this region.

    d. Solve your system from Part b and explain your answer in the context of the problem.

13. Solve a quadratic-linear system and use the solution to explain why it is impossible to have two real numbers whose sum is 8 and whose product is 24.

In 14 and 15, solve and check.

14. $\begin{cases} 4r^2 + s^2 = 9 \\ 3r + s = 1 \end{cases}$

15. $\begin{cases} \dfrac{x^2}{16} + \dfrac{y^2}{36} = 1 \\ y = \frac{1}{2}x \end{cases}$

REVIEW

16. Consider the hyperbola with equation $xy = k$, where $k > 0$. (Lesson 12-7)

   a. Give the coordinates of its foci.

   b. Identify its asymptotes.

   c. Determine the focal constant.

17. Find an equation for the hyperbola with foci at (10, 10) and (–10, –10) and focal constant 20. (Lesson 12-7)

18. Give an equation for a hyperbola that

   a. is the graph of a function.

   b. is not the graph of a function. (Lessons 12-7, 12-6, 1-4)

In 19–21, a polynomial is given.

   a. Tell whether the polynomial is a binomial square, a difference of squares, or a sum of squares.

   b. Factor, if possible, over the integers. (Lesson 11-3)

19. $r^{20} - t^2$

20. $49a^2 - 42ab + 9b^2$

21. $x^2 + 2500$

In 22 and 23, simplify without a calculator. (Lessons 8-7, 7-8, 6-9)

22. $\left(2 - \sqrt{-9}\,\right)\!\left(2 + \sqrt{-9}\,\right)$

23. $\sqrt{-64} + \sqrt[3]{-64} + 64^{-\frac{1}{6}}$

24. Suppose a car rental company charges a flat rate of $25 plus 45 cents for each mile or part of a mile driven. (Lesson 3-9)

   a. Write an equation relating the total charge c to the number m of miles driven.

   b. How much would you pay if you drove 36.3 miles in a rental car?

   c. The company also allows you to pay a flat rate of $65 a day for an unlimited number of miles. How many miles would you have to drive in a day to make this the better deal?

The first U.S. airport location of a car rental company opened at Chicago's Midway Airport in 1932.

EXPLORATION

25. Find equations for a noncircular ellipse and an oblique (slanted) line that have exactly one point of intersection.

Lesson

12-9

Quadratic-Quadratic Systems

▶ **BIG IDEA**   Solutions to systems with two quadratic equations in x and y can be found by graphing, using linear combinations, by substitution, or using technology.

Recall from Lesson 12-2 that an earthquake sends out seismic waves from its epicenter. A single monitoring station can determine that the epicenter lies on a particular circle with the station as its center, but it takes multiple stations combining their information to locate the epicenter.

Mental Math

Give an inequality that represents the situation.

a. The price n of a gallon of gasoline now is more than 4 times the price t of a gallon 10 years ago.

b. The life ℓ of a battery is guaranteed to be at least 30 hours.

c. The average test score s was 86.4 with a 0.5-point margin of error.

d. You can order from the children's menu if your age a is under 7 years.

Example 1

An earthquake monitoring station A determines that the center of a quake is 100 km away. A second station B 50 km west and 30 km south of the first finds that it is 70 km from the quake's center. Find all possible locations of the epicenter in relation to station A.

Solution   Let Station A be located at $(0, 0)$. Then Station B is at $(-50, -30)$. Draw the circles with radii 100 and 70. The graph shows two intersections of the circles. These are the possible locations of the epicenter.

Because the epicenter is 100 km from A, the larger circle has equation $x^2 + y^2 = 10{,}000$. Because the epicenter is 70 km from B, the smaller circle has equation $(x + 50)^2 + (y + 30)^2 = 4900$. These two equations form the following *quadratic-quadratic system*.

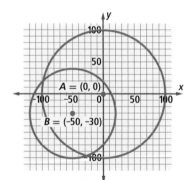

$$\begin{cases} x^2 + y^2 = 10{,}000 \\ (x + 50)^2 + (y + 30)^2 = 4900 \end{cases}$$

In order to find the exact solutions to this system, you can use by-hand methods or a CAS. One CAS solution is partially shown at the right. The full solution, found by scrolling the display, is:

$x = -97.7251$ and $y = 21.2085$ or
$x = -27.2749$ and $y = -96.2085$

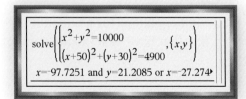

The epicenter is located either about 98 kilometers west and 21 kilometers north of Station A, or about 27 kilometers west and 96 kilometers south of Station A. Information from a third station would help you determine which was the actual location.

 QY

▶ **QY**

Check both solutions to Example 1 in both equations of the system.

In general, a **quadratic-quadratic system** involves two or more quadratic sentences. Geometrically, a quadratic-quadratic system involves curves represented by quadratic relations: circles, ellipses, hyperbolas, and parabolas.

In Lesson 12-8, you looked at ways a line could intersect the different conic sections. Below are examples of the six different ways that a quadratic-quadratic system of a circle and an ellipse can intersect.

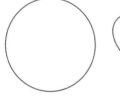

no intersections

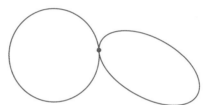

1 intersection

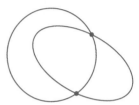

2 intersections

3 intersections

4 intersections

infinite number of intersections

Activity

MATERIALS DGS (optional)

Using the examples above as a guide, fill in a table like the one at the right for the possible numbers of intersections for each quadratic-quadratic system. You may find a DGS helpful.

	Parabola	Circle	Ellipse	Hyperbola
Parabola	?	?	?	?
Circle	?	?	?	?
Ellipse	?	0, 1, 2, 3, 4, or infinitely many	?	?
Hyperbola	?	?	?	?

The results of the Activity can be generalized. Quadratic-quadratic systems may have 0, 1, 2, 3, 4, or infinitely many solutions (if the quadratics are equivalent equations).

Some quadratic-quadratic systems are simple enough to be solved by hand.

Example 2

Find all solutions (x, y) to $\begin{cases} 4x^2 + 3y^2 = 37 \\ 3x^2 - y^2 = 5 \end{cases}$.

Solution This system represents the intersection of an ___?___ and a ___?___. There are 0, 1, 2, 3, or 4 possible solutions. Use the linear combination method to find them.

$$4x^2 + 3y^2 = 37$$
$$\underline{\ ?\ }x^2 - \underline{\ ?\ }y^2 = \underline{\ ?\ } \qquad \text{Multiply the second equation by 3.}$$
$$\underline{\ ?\ }x^2 = \underline{\ ?\ } \qquad \text{Add the equations to eliminate } y.$$
$$x^2 = \underline{\ ?\ }$$
$$x = \pm\ \underline{\ ?\ }$$

Substitute each value of x into one of the given equations.

If $x = \underline{\ ?\ }$, then $4(\underline{\ ?\ })^2 + 3y^2 = 37$, so $y = \pm\ \underline{\ ?\ }$.

If $x = \underline{\ ?\ }$, then $4(\underline{\ ?\ })^2 + 3y^2 = 37$, so $y = \pm\ \underline{\ ?\ }$.

So, this system has 4 solutions: $(\underline{\ ?\ }, \underline{\ ?\ })$, $(\underline{\ ?\ }, \underline{\ ?\ })$, $(\underline{\ ?\ }, \underline{\ ?\ })$, and $(\underline{\ ?\ }, \underline{\ ?\ })$.

Check Solve on a CAS.

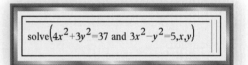

$$\text{solve}\!\left(4x^2 + 3y^2 = 37 \text{ and } 3x^2 - y^2 = 5, x, y\right)$$

The following quadratic-quadratic system involves the intersection of two rectangular hyperbolas.

Example 3

In one month, Willie's Western Wear took in $10,500 from boot sales. Although Willie sold 30 fewer pairs of boots the next month, he still sold $10,800 worth of boots by raising the price $10 per pair. Find the price of a pair of boots in each month.

Solution Let $n =$ the number of pairs of boots sold in the first month. Let $p =$ the price of a pair of boots in the first month.

The equations for total sales in the first and second months, respectively, are:

(1) $\qquad\qquad np = 10{,}500$

(2) $\qquad (n - 30)(p + 10) = 10{,}800$

Civil War era military boots were not the ideal boot for cowboys. So, in the late 1800s, boots were given pointy toes to help feet get into stirrups and higher heels to help feet stay there.

Our goal is to obtain an equation in only one of these variables.

In (1), solve for p. $$p = \frac{10{,}500}{n}$$

In (2), expand. $$np + 10n - 30p - 11{,}100 = 0$$

The two forms of equation (1) allow you to make two substitutions into equation (2). Substitute 10,500 for np and $\frac{10{,}500}{n}$ for p to get an equation in one variable.

$$10{,}500 + 10n - 30\left(\frac{10{,}500}{n}\right) - 11{,}100 = 0.$$

Divide both sides by 10.

$$1050 + n - 3\left(\frac{10{,}500}{n}\right) - 1110 = 0$$

Simplify. $$n - 60 - \frac{31{,}500}{n} = 0$$

Multiply both sides by n. $$n^2 - 60n - 31{,}500 = 0$$

This is a quadratic equation you can solve in many ways. One CAS solution is shown at the right.

The number n of pairs of boots sold can only be positive.

So, $$n = 210.$$

Since the price $p = \frac{10{,}500}{n}$, $p = \frac{10{,}500}{210} = 50.$

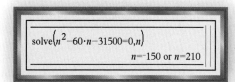

solve$\left(n^2 - 60 \cdot n - 31500 = 0, n\right)$
$n = -150$ or $n = 210$

The boots were priced at $50 a pair the first month, and $p + 10 = \$60$ a pair the second month.

Check With these numbers, 210 pairs of boots were sold the first month at $50/pair, so Willie took in $10,500. That checks. The second month, Willie sold 180 pairs of boots at $60/pair. So he took in $10,800. The solution checks.

Questions

COVERING THE IDEAS

1. A system with two quadratic equations in x and y that are not equivalent can have at most how many solutions?

2. Without doing any algebra or graphing, tell how many solutions the system $x^2 + y^2 = 9$ and $3x^2 + 3y^2 = 48$ has. Explain your response.

3. Find all solutions (x, y) to $\begin{cases} 9x^2 + 4y^2 = 36 \\ -16x^2 + 4y^2 = 36 \end{cases}.$

4. **True or False** The solutions to the system in Example 2 lie on the symmetry lines of both the hyperbola and the ellipse.

5. Consider the circle $x^2 + y^2 = 9$ and the parabola $y = x^2$.
   a. How many intersection points do you expect?
   b. Check your answer to Part a by finding the points of the intersection.

In 6 and 7, refer to Example 3.

6. a. What two substitutions could you make to transform the second equation into an equation in terms of p only?
   b. Check the solution by solving the original system using the substitutions in Part a.

7. In the second month, if Willie had instead raised prices $18 per pair and earned $12,000 from selling 50 fewer pairs of boots than the previous month, what would have been the price of a pair of boots in each month?

APPLYING THE MATHEMATICS

8. The product of two real numbers is 6984. If one number is increased by 4 and the other is decreased by 10, the new product is 6262.
   a. Write a quadratic system that can be solved to find these numbers.
   b. Find the numbers.

In 9–11, solve the system.

9. $\begin{cases} x^2 + y^2 = 25 \\ \quad y = x^2 - 13 \end{cases}$

10. $\begin{cases} \dfrac{x^2}{16} + \dfrac{y^2}{9} = 1 \\ x^2 - y^2 = 7 \end{cases}$

11. $\begin{cases} y = x^2 + 3x - 4 \\ y = 2x^2 + 5x - 3 \end{cases}$

12. Draw a graph of the solution set of the system $\begin{cases} y \geq x^2 \\ x^2 + y^2 \leq 4 \end{cases}$.

13. A fleet of fishing boats locates its nets by sonar. One boat determines that a net is 1000 meters away. A second boat, 200 meters east and 800 meters north of the first, finds that it is 400 meters from the net. A third boat, 1000 meters east and 1100 meters north of the first, finds that it is 500 meters away from the net.

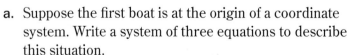

   a. Suppose the first boat is at the origin of a coordinate system. Write a system of three equations to describe this situation.
   b. The graphs of the three equations have one intersection point in common. Solve the system and find the location of the net with respect to the first boat.

14. The altitude h of the 3-4-5 triangle at the right splits the hypotenuse into segments of length x and $5 - x$.

   a. Use the Pythagorean Theorem to create two quadratic relations in terms of x and h.

   b. Solve your system of equations in Part a for x and h.

   c. Which solution(s) make(s) sense in the context of this problem?

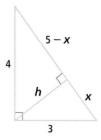

REVIEW

15. The sum of Mr. Hwan's age in years and his baby's age in months is 37. The product of the ages is 300. Solve a system of equations to find their ages. **(Lesson 12-8)**

In 16 and 17, an equation is given. Does the equation represent a quadratic relation? If so, put the equation in standard form for a quadratic relation. If not, explain why not. **(Lesson 12-7)**

16. $x^2 + 5xy^2 = 10$

17. $\frac{1}{2}x - 13y^2 = \sqrt{7}y$

In 18 and 19, graph the equation. **(Lesson 12-6, 12-4)**

18. $\frac{x^2}{9} + \frac{y^2}{36} = 1$

19. $\frac{x^2}{9} - \frac{y^2}{36} = 1$

20. The third and fourth terms of a geometric sequence are 20 and -40. **(Lesson 7-5)**

   a. What is the constant ratio?

   b. What are the first and second terms?

   c. Write an explicit formula for the nth term.

   d. What is the 15th term?

21. Explain why every transformation is a function. **(Lesson 4-7)**

In 22 and 23, multiply the matrices. **(Lesson 4-3)**

22. $\begin{bmatrix} 3 & 5 & 7 \end{bmatrix} \begin{bmatrix} 1 \\ 0 \\ -2 \end{bmatrix}$

23. $\begin{bmatrix} 3 & 0 & 5 \\ -1 & 4 & 2 \end{bmatrix} \begin{bmatrix} 2 & -2 \\ 0 & 1 \\ -3 & 4 \end{bmatrix}$

EXPLORATION

24. In 2002, a new method of code breaking called the XSL (eXtended Sparse Linearization) attack was published. The attack is a type of *algebraic cryptanalysis*. Search the Internet to answer these questions.

   a. What is the purpose of the XSL attack?

   b. How does this system use quadratic systems?

QY ANSWER

$(-97.7)^2 + (21.2)^2 = 9994.73 \approx 10{,}000$;
$(-97.7 + 50)^2 + (21.2 + 30)^2 = 4896.73 \approx 4900$;
$(-27.3)^2 + (-96.2)^2 = 9999.73 \approx 10{,}000$;
$(-27.3 + 50)^2 + (-96.2 + 30)^2 = 4897.73 \approx 4900$

Chapter

12 Projects

1 Reflection Properties of Conics

Parabolas, ellipses, and hyperbolas are used in telescopes, whispering galleries, navigation, satellite dishes, and headlights because they have reflection properties.

a. Explain how the reflection properties of a conic section are used to create whispering galleries. Illustrate your explanation with an accurate drawing.

b. Describe the reflection properties of the remaining conic sections and illustrate them with drawings and examples of their use in the real world.

2 Arches

Semicircular arches were popular with the early Romans. Other kinds of arches have been popular at other times and places. Prepare a brief report on the various kinds of arches that have been used in architecture throughout history. You might include arches from the Roman and Renaissance eras and modern arches such as the Gateway Arch in St. Louis. If possible, give equations for curves that outline these arches.

3 The General Ellipse and Hyperbola

Research the standard, or general, form of a quadratic equation

$$Ax^2 + Bxy + Cy^2 + Dx + Ey + F = 0$$

compared to the standard forms of the equations for the conic sections you studied in this chapter. There are many Internet sites that can help your research.

a. Find out how to interpret $A, B, C, D, E,$ and F to identify which conic section the standard quadratic equation represents.

b. Find out how to transform the standard quadratic equation into the standard forms for ellipses and hyperbolas.

c. Summarize your findings in a report.

4 Orbits of the Planets

The orbits of planets around the Sun are ellipses with the Sun at one focus.

a. Describe these ellipses, giving the major and minor axes for the orbit of each planet and indicating the nearest and farthest distances of each planet to the Sun. (These are called the planet's *perihelion* and *aphelion*, respectively.)

b. Draw two accurate pictures of these orbits, one with the inner four major planets, the other with the outer four major planets.

c. Research the orbit of the dwarf planet Pluto. Add its orbit to your drawing of the outer planets' orbits. How does it differ from the orbits of the other planets?

d. The closeness of an ellipse to a circle is measured by the *eccentricity* of the ellipse. Give the eccentricity of each orbit. Which planets' orbits are closest to being circular?

5 Constructing Conic Sections

The Exploration in Lesson 12-1 showed you how to construct a parabola by using paper folding. This project uses a DGS to mimic paper-folding constructions of parabolas and other conic sections.

a. Construct line m and point P not on the line. Construct a line segment from P to point A on the line. Construct the perpendicular bisector of $\overline{PA}$. Set your DGS to trace the path of the perpendicular bisector of $\overline{PA}$ as you vary the location of A on line m. Explain why this construction results in a parabola. Identify the focus and directrix in the construction.

b. Construct circle C and point P in the exterior of the circle. Construct a line segment from P to point A on the circle. Construct the perpendicular bisector $\overline{PA}$. Set your DGS to trace the path of the perpendicular bisector of $\overline{PA}$ as you vary the location of A on circle C. Which conic section does this construction make? Explain why this construction makes the conic section it does.

c. Modify your construction in Part b by moving P to the interior of the circle. Set your DGS to trace the path of the perpendicular bisector of $\overline{PA}$ as you vary the location of A on circle C. Which conic section does this construction make? Explain why this construction makes the conic section it does.

Chapter 12

Summary and Vocabulary

- In this chapter, you studied **quadratic relations** in two variables, their graphs, and their geometric properties. Equations for all quadratic relations in two variables can be written in the standard form $Ax^2 + Bxy + Cy^2 + Dx + Ey + F = 0$, where not all of A, B, and C are zero. The properties of the various quadratic relations are summarized in the table on the next page.

- **Conic sections** appear naturally as orbits of planets and comets, in paths of thrown objects, as energy waves radiating from the epicenter of an earthquake, and in many manufactured objects such as tunnels, windows, and satellite receiver dishes.

- Systems of equations with quadratic sentences can be solved in much the same way as linear systems, that is, by graphing, substitution, using linear combinations, or using a CAS. A system of one linear and one quadratic equation may have 0, 1, or 2 solutions; a system of two quadratic equations may have 0, 1, 2, 3, 4, or infinitely many solutions.

Theorems and Properties

Focus and Directrix of a Parabola Theorem (p. 801)
Circle Equation Theorem (p. 806)
Interior and Exterior of a Circle Theorem (p. 812)
Equation for an Ellipse Theorem (p. 818)
Length of Axes of an Ellipse Theorem (p. 820)
Circle Scale-Change Theorem (p. 826)
Graph Scale-Change Theorem (p. 827)
Area of an Ellipse Theorem (p. 828)
Equation for a Hyperbola Theorem (p. 832)
Asymptotes of a Hyperbola Theorem (p. 834)
Attributes of $y = \frac{k}{x}$ Theorem (p. 841)

Vocabulary

Lesson 12-1
*parabola
focus, directrix
axis of symmetry
vertex of a parabola
paraboloid

Lesson 12-2
*circle, radius, center
*concentric circles
semicircle

Lesson 12-3
*interior, exterior of a circle

Lesson 12-4
*ellipse
foci, focal constant of
  an ellipse
standard position for
  an ellipse
standard form of an
  equation for an ellipse
*major axis, minor axis
center of an ellipse
semimajor axes,
  semiminor axes

Lesson 12-6
*hyperbola
foci, focal constant of
  a hyperbola
vertices of a hyperbola
standard position of
  a hyperbola
*standard form of an
  equation for a hyperbola

Lesson 12-7
conic section
*standard form of
  an equation for a
  quadratic relation

Lesson 12-8
quadratic system
quadratic-linear system

Lesson 12-9
quadratic-quadratic system

	Quadratic Relation			
	Circle	**Ellipse**	**Parabola**	**Hyperbola**
Geometric Definition	given center C and radius r, set of points P such that $PC = r$	given foci F_1 and F_2 and focal constant $2a$, set of points P such that $PF_1 + PF_2 = 2a$	given focus F and directrix ℓ, set of points P equidistant from F and ℓ	given foci F_1 and F_2 and focal constant $2a$, set of points P such that $\lvert PF_1 - PF_2 \rvert = 2a$
Equation(s) in Standard Form	$(x - h)^2 + (y - k)^2 = r^2$	$\dfrac{x^2}{a^2} + \dfrac{y^2}{b^2} = 1$	$y = ax^2 + bx + c$ or $y - k = a(x - h)^2$	$\dfrac{x^2}{a^2} - \dfrac{y^2}{b^2} = 1$ or $xy = k$
Graph	center: (h, k) radius: r	If $a > b$ foci: $(-c, 0), (c, 0)$ length of major axis (focal constant): $2a$ length of minor axis: $2b$ $b^2 = a^2 - c^2$ If $b > a$ foci: $(0, -c), (0, c)$ length of major axis (focal constant): $2b$ length of minor axis: $2a$ $a^2 = b^2 - c^2$	$y = ax^2$ axis of symmetry: $x = 0$ vertex: $(0, 0)$ focus: $\left(0, \dfrac{1}{4a}\right)$ directrix: $y = -\dfrac{1}{4a}$ $y - k = a(x - h)^2$ axis of symmetry: $x = h$ vertex: (h, k)	$\dfrac{x^2}{a^2} - \dfrac{y^2}{b^2} = 1$ foci: $(-c, 0), (c, 0)$; $b^2 = c^2 - a^2$ asymptotes: $\dfrac{y}{b} = \pm\dfrac{x}{a}$ $xy = k$ foci: $(\sqrt{2k}, \sqrt{2k}), (-\sqrt{2k}, -\sqrt{2k})$ focal constant: $2\sqrt{2k}$ asymptotes: $x = 0, y = 0$
Conic Section				

Chapter

12 Self-Test

Take this test as you would take a test in class. You will need graph paper and a calculator. Use the Selected Answers section in the back of the book to check your work.

1. a. Rewrite $\frac{x^2}{25} - \frac{y^2}{144} = 1$ in the form $Ax^2 + Bxy + Cy^2 + Dx + Ey + F = 0.$

   b. Give the values of A, B, C, D, E, and F.

   c. Identify the conic section represented by the equation in Part a.

In 2 and 3, write an equation or inequality for

2. the interior of the circle with center $(3, -5)$ and radius 7.

3. the image of the graph of $x^2 + y^2 = 1$ under $T_{-2, 5}$.

4. Consider the parabola with equation $y = \frac{1}{11}x^2$.

   a. Give the coordinates of its focus.

   b. Give the coordinates of its vertex.

   c. Write an equation for its directrix.

5. Explain what has to be true about the foci and the major and minor axes of an ellipse for it to be a circle.

6. Ernesto's Eye Extravaganza sold $5600 worth of sunglasses last year. This year, Ernesto's lowered the prices by two dollars, sold seventy more pairs of sunglasses, and took in $5880. Assuming sunglasses are all the same price,

   a. how much is he selling his sunglasses for now?

   b. how many pairs did he sell this year?

In 7–9, refer to the ellipse below.

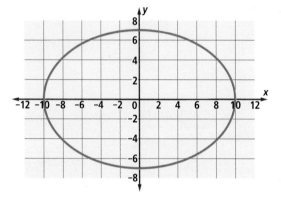

7. Write an equation for the ellipse.

8. Find the area of the ellipse.

9. The ellipse is the image of the circle $x^2 + y^2 = 1$ under what scale change?

10. Graph the conic section with equation $\frac{x^2}{25} - \frac{y^2}{16} = 1$. Identify all its major features.

11. Graph the system $\begin{cases} xy = 2 \\ 2x + 5 = y \end{cases}$ and identify the points of intersection.

12. Solve the system $\begin{cases} y = x^2 - 4x + 3 \\ x^2 + y^2 = 9 \end{cases}$.

13. Consider the equation $xy = -15$.

   a. What type of curve is the graph of this equation?

   b. State the foci and focal constant of the curve.

   c. Does the curve have any lines of symmetry? If so, write an equation for one of them.

14. The elliptically shaped pool shown below is to be surrounded by a tile walkway so that the outer edge of the walkway is also an ellipse. The major axis of the pool has length 15 m, and the minor axis of the pool has length 8 m. The length AB of the major axis of the outer edge of the walkway is 18 m, and the length CD of the minor axis is 11 m. What is the area of the walkway?

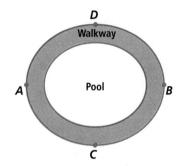

15. Colorado has an area of approximately 104,094 square miles and a perimeter of about 1320 miles. Assuming that Colorado is rectangular, find its dimensions.

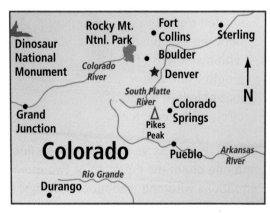

16. Halley's comet has an elliptical orbit with the Sun at one focus. Its closest distance to the Sun is about $9 \cdot 10^7$ km, while its farthest distance is about $5.3 \cdot 10^9$ km.

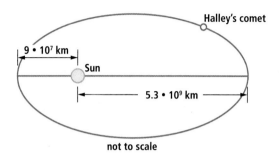

Find the length of the major axis of Halley's comet's orbit.

17. Find equations for a quadratic-quadratic system that has exactly three solutions.

18. Graph the set of points that are 8 units from $(3, 2.5)$ and describe the graph with an equation.

19. The entrance to a cave is a semicircular arch that is 14 feet wide at its base. How far from the center of the opening can a 5'8"-tall spelunker stand upright?

20. Graph the set of points that satisfy the inequality $\frac{x^2}{9} + \frac{y^2}{100} \geq 1$.

Chapter 12 Chapter Review

SKILLS Procedures used to get answers

OBJECTIVE A Rewrite an equation for a conic section in the standard form of a quadratic equation in two variables. (Lesson 12-7)

In 1–6, rewrite the equation in the form $Ax^2 + Bxy + Cy^2 + Dx + Ey + F = 0$. Give the values of A, B, C, D, E, and F.

1. $(x - 7)^2 + (y + 2)^2 = 81$

2. $y = 6(t - 3)^2 + 4$

3. $\dfrac{x^2}{36} - \dfrac{y^2}{16} = 1$

4. $\dfrac{a^2}{8} + \dfrac{b^2}{5} = 1$

5. $y = \sqrt{169 - x^2}$

6. $y = \dfrac{30}{x}$

OBJECTIVE B Write equations for quadratic relations and inequalities for their interiors and exteriors. (Lessons 12-1, 12-2, 12-3, 12-4, 12-6, 12-7)

7. Find an equation for the circle with center at the origin and radius 12.

8. Find an equation for the circle with center at (-3, 6) and diameter 7.

9. Give an equation for the upper semicircle of the circle with equation $a^2 + b^2 = 35$.

10. What inequality describes the interior of the circle with equation $x^2 + y^2 = 64$?

11. What sentence describes the exterior of the circle with equation $x^2 + y^2 = 64$?

In 12 and 13, write an equation for the ellipse satisfying the given conditions.

12. foci: (0, 1) and (0, -1); focal constant: 5

13. The endpoints of the major and minor axes are (4, 0), (-4, 0), (0, 7), and (0, -7).

14. Write an equation for the parabola with focus (0, -3) and directrix $y = 3$.

In 15 and 16, find an equation for a hyperbola satisfying the given conditions.

15. vertices: (2, 2) and (-2, -2)

16. foci: (9, 0) and (-9, 0); focal constant: 11

OBJECTIVE C Find the area of an ellipse. (Lesson 12-5)

In 17 and 18, find the area of the ellipse satisfying the given conditions.

17. It has equation $\dfrac{x^2}{100} + \dfrac{y^2}{49} = 1$.

18. The endpoints of its axes are (3, 0), (-3, 0), (0, 6), and (0, -6).

19. Which has a larger area and by how much: a circle of radius 15 or an ellipse with major and minor axes of lengths 46 and 20?

20. Find the area of the shaded region between an ellipse with major axis of length 7 and minor axis of length 5, and a circle with diameter 5.

OBJECTIVE D Solve systems of one linear and one quadratic equation or two quadratic equations with and without technology. (Lessons 12-8, 12-9)

In 21–26, solve.

21. $\begin{cases} y = x^2 + 8 \\ y = -x^2 + 7x + 8 \end{cases}$

22. $\begin{cases} 6x + y = 24 \\ y = x^2 + 4x - 5 \end{cases}$

23. $\begin{cases} t = r^2 + 2r - 8 \\ t = 2r^2 + 2r - 6 \end{cases}$

24. $\begin{cases} (x - 1)^2 + y^2 = 3 \\ x^2 + (y + 1)^2 = 3 \end{cases}$

25. $\begin{cases} ab = 4 \\ b = 3a + 1 \end{cases}$

26. $\begin{cases} p^2 - q^2 = 4 \\ \dfrac{p^2}{16} + \dfrac{q^2}{9} = 1 \end{cases}$

27. The product of two numbers is 1073. If one number is increased by 3 and the other is decreased by 7, the new product is 960.

a. Write a system of equations representing this situation.

b. Find the numbers.

PROPERTIES Principles behind the mathematics

OBJECTIVE E Identify characteristics of parabolas, circles, ellipses, and hyperbolas. (Lessons 12-1, 12-2, 12-4, 12-6, 12-7)

In 28 and 29, identify the center and radius of the circle with the given equation.

28. $(r + 6)^2 + s^2 = 196$ 29. $x^2 + y^2 = 361$

30. Consider the ellipse with equation $\frac{x^2}{169} + \frac{y^2}{64} = 1$.

a. Name the endpoints of its axes.

b. State the length of its minor axis.

31. Consider the parabola with equation $y = \frac{1}{7}x^2$.

a. Give the coordinates of its focus.

b. Give the coordinates of its vertex.

c. State the equation of its directrix.

32. Consider the hyperbola with equation $\frac{p^2}{25} - \frac{q^2}{9} = 1$.

a. Name its vertices.

b. State equations for its asymptotes.

33. Consider the ellipse with equation $\frac{x^2}{16} + \frac{y^2}{100} = 1$.

a. Give the coordinates of its foci F_1 and F_2.

b. Suppose P is on the ellipse. Find the value of $PF_1 + PF_2$.

34. Identify the asymptotes of the hyperbola with equation $xy = \frac{23}{5}$.

OBJECTIVE F Classify curves as circles, ellipses, parabolas, or hyperbolas using algebraic or geometric properties. (Lessons 12-1, 12-2, 12-4, 12-5, 12-6)

In 35 and 36, consider two fixed points F_1 and F_2 and a constant d. Name the curve formed by the set of points P satisfying the given conditions.

35. $F_1P + F_2P = d$, where $d > F_1F_2$

36. $|F_1P - F_2P| = d$, where $d < F_1F_2$

37. Each figure below shows a double cone intersected by a plane. In Figure b, the plane is parallel to the edge of the cone; in Figure d, the plane is perpendicular to the axis of the cone. Identify the curve produced by each intersection.

a. 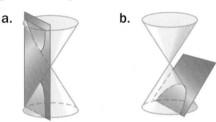   b.

c. 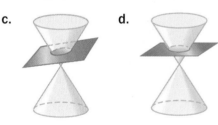   d.

In 38–40, answer true or false.

38. A hyperbola can be considered as the union of two parabolas.

39. All quadratic relations in two variables can be formed by the intersection of a plane and a double cone.

40. The image of an ellipse under a scale change can be a circle.

41. a. What equation describes the image of the circle with equation $x^2 + y^2 = 1$ under the transformation $S : (x, y) \to (5x, 8y)$?

b. What kind of curve is the image in Part a?

USES Applications of mathematics in real-world situations

OBJECTIVE G Use circles and ellipses to solve real-world problems. (Lesson 12-2, 12-3, 12-4, 12-5)

42. An elliptical garden surrounds a circular fountain with diameter 10 feet. The major axis of the garden is 20 feet long, and the minor axis is 10 feet long. What is the area of the garden?

43. A castle is surrounded by a circular moat 15 feet wide. The distance from the center of the castle to the outside of the moat is 500 feet. If the center of the castle is considered the origin, write a system of inequalities to describe the set of points on the surface of the moat.

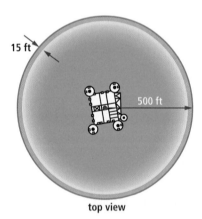

15 ft
500 ft
top view

44. A tent is in the form of half a cylindrical surface. Each cross section is a semiellipse (half an ellipse) with base length 20 feet and height 8 feet. How close to either end can a person 5 feet tall stand straight up?

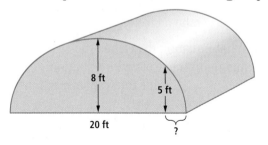

8 ft
5 ft
20 ft
?

45. A moving van 6 ft wide and 12 ft tall is approaching a semicircular tunnel with radius 13 ft.

   a. Explain why the truck cannot pass through the tunnel if it stays in its lane.

   b. Can the moving van fit through the tunnel if it is allowed to drive anywhere on the roadway? Justify your answer.

OBJECTIVE H Use systems of quadratic equations to solve real-world problems. (Lessons 12-8, 12-9)

46. A piece of paper has an area of 93.5 square inches and a perimeter of 39 inches. Find the dimensions of the piece of paper.

47. Suppose the epicenter of an earthquake is 75 miles away from monitoring stations 1 and 2. Station 2 is 30 miles west and 50 miles north of Station 1. Let Station 1 be at the origin of a coordinate system.

   a. Write a system of equations to describe the location (x, y) of the epicenter.

   b. Find the coordinates of all possible locations of the epicenter relative to Station 1.

48. Perla has 150 meters of fencing material and wants to form a rectangular pen with an area of 1300 square meters.

   a. Let w = the width of the pen and ℓ = its length. Write a system of equations to determine w and ℓ.

   b. Solve the system and interpret your answer in the context of the problem.

49. One month Wanda's Wonderful Wagons took in $12,000 from sales of wagons. The next month Wanda sold 40 fewer wagons because she had raised the price by $20. In spite of this, total sales rose to $12,800. Find the price of wagons in each month.

REPRESENTATIONS Pictures, graphs, or objects that illustrate concepts

| **OBJECTIVE I** Graph quadratic relations when given equations for them in standard form, and vice versa. (Lessons 12-2, 12-4, 12-6, 12-7)

In 50–53, sketch a graph of the equation.

50. $\frac{x^2}{25} + \frac{y^2}{64} = 1$ 51. $\frac{x^2}{25} - \frac{y^2}{64} = 1$

52. $xy - 18 = 0$ 53. $x^2 + y^2 = 16$

In 54 and 55, state an equation for the curve.

54.
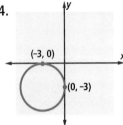

55.
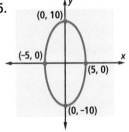

Multiple Choice In 56 and 57, select the equation that best describes each graph.

A $\frac{x^2}{a^2} + \frac{y^2}{b^2} = 1$ **B** $\frac{x^2}{a^2} - \frac{y^2}{b^2} = 1$

C $y = ax^2$ **D** $xy = a$

56.
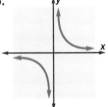

57.
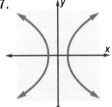

| **OBJECTIVE J** Graph interiors and exteriors of ellipses when given inequalities for them, and vice versa. (Lesson 12-3)

58. Write a system of inequalities to represent the points in the shaded region at the right.

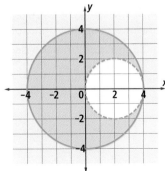

In 59 and 60, sketch a graph of the inequality.

59. $x^2 + (y - 4)^2 > 25$ 60. $\frac{x^2}{16} + \frac{y^2}{4} \le 1$

| **OBJECTIVE K** Interpret representations of quadratic-linear and quadratic-quadratic systems. (Lessons 12-8, 12-9)

In 61 and 62, give equations for

61. a circle and a hyperbola that intersect in exactly three points.

62. two hyperbolas that intersect exactly twice.

63. Someone claims that the sum of two real numbers is 17 and their product is 73. Use equations and a graph to explain why this is impossible.

64. a. Graph $x^2 + y^2 = 5$ and $y = x^2 - 2$ in the same window.

    b. Find the points of intersection to the nearest tenth.

| **OBJECTIVE L** Draw a graph or interpret drawings or graphs of conic sections based on their definitions. (Lessons 12-1, 12-2, 12-4, 12-6)

65. Graph the set of points equidistant from the point (5, 3) and the line $y = -1$.

66. Draw a line ℓ and point F not on ℓ. Sketch 5 points equidistant from F and ℓ.

67. Graph the set of points that are 2.5 units from the origin.

68. Draw two points F_1 and F_2. Sketch 6 more points whose distances from F_1 and F_2 add up to $2F_1F_2$.

69. Graph the set of points whose distances from (0, –3) and (0, 3) add up to 8.

70. Draw two points F_1 and F_2 4 units apart. Sketch 3 points that are 2 units farther from F_1 than from F_2 and 3 points that are 2 units farther from F_2 than from F_1.

71. Graph the set of points whose difference of distances from (–6, 0) and (6, 0) is 5.

Series and Combinations

Contents

This chapter on series and combinations extends many of the theorems, techniques, and ideas from earlier in the book.

When an employee sets aside a regular portion of wages as savings earning interest, or a ball bounces until it comes to rest, the total amount saved or the total distance traveled is the sum of terms of a sequence. The indicated sum of terms of a sequence is called a *series*. Using the mathematics of polynomials and exponential functions that you have already learned, you can derive formulas related to these series and answer questions about them.

An important pattern with surprising connections is *Pascal's Triangle*. Rows 0–7 of this infinite triangular array of numbers are shown below.

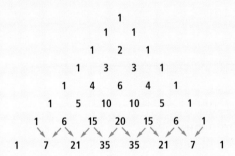

The arrows between rows 6 and 7 show how the *sum* of two elements in one row is an element in the next row. Yet an explicit formula for each element involves *products*.

Pascal's Triangle may have been discovered independently by ancient Chinese, Indian, and Persian mathematicians, who used it to expand the power $(x + y)^n$ of any binomial. The French mathematician Blaise Pascal (1623–1662) later rediscovered the pattern and publicized it as a way to answer questions about probability. The entries in Pascal's Triangle are used to answer counting questions, to compute probabilities, and to help describe statistical distributions.

Lesson

13-1 Arithmetic Series

Vocabulary

series

arithmetic series

Σ, sigma

Σ-notation, sigma notation, summation notation

index variable, index

▶ **BIG IDEA** There are several ways to find sums of the successive terms of an arithmetic sequence.

Sums of Consecutive Integers

There is a story the famous mathematician Carl Gauss often told about himself. When he was in third grade, his class misbehaved and the teacher gave the following problem as punishment:

"Add the whole numbers from 1 to 100."

Gauss solved the problem in almost no time at all. His idea was the following. Let S be the desired sum.

$$S = 1 + 2 + 3 + \ldots + 98 + 99 + 100$$

Using the Commutative Property of Addition, the sum can be rewritten in reverse order.

$$S = 100 + 99 + 98 + \ldots + 3 + 2 + 1$$

Now add corresponding terms in the equations above. The sums $1 + 100$, $2 + 99$, $3 + 98$, ... all have the same value!

So $\quad 2S = \underbrace{101 + 101 + 101 + \ldots + 101 + 101 + 101}_{100 \text{ terms}}$.

Thus, $\quad 2S = 100 \cdot 101$

and $\quad\quad S = 5050$.

Gauss wrote only the number 5050 on his slate, having done all the figuring in his head. The teacher (who had hoped the problem would keep the students working for a long time) was quite irritated. However, partly as a result of this incident, the teacher did recognize that Gauss was extraordinary and gave him some advanced books to read. (You read about Gauss's work in Lesson 11-6 and may recall that he proved the Fundamental Theorem of Algebra at age 18.)

 QY1

Mental Math

Consider the arithmetic sequence defined by $a_n = 3n - 12$.

a. Find a_1, a_2, and a_3.

b. Find $a_1 + a_2 + a_3$.

c. Find a_{101}, a_{102}, and a_{103}.

d. Find $a_{101} + a_{102} + a_{103}$.

▶ **QY1**

Use Gauss's method to add the integers from 1 to 40.

What Is an Arithmetic Series?

Recall that an *arithmetic* or *linear sequence* is a sequence in which the difference between consecutive terms is constant. An arithmetic sequence has the form

$$a_1, a_1 + d, a_1 + 2d, ..., a_1 + (n - 1)d, ... ,$$

where a_1 is the first term and d is the constant difference. For example, the odd integers from 1 to 999 form a finite arithmetic sequence with $a_1 = 1$, $n = 500$, and $d = 2$.

A **series** is an indicated sum of terms of a sequence. For example, for the sequence 1, 2, 3, a series is the indicated sum $1 + 2 + 3$. The addends 1, 2, and 3 are the *terms* of the series. The value, or sum, of the series is 6. In general, the sum of the first n terms of a series a is

$$S_n = a_1 + a_2 + a_3 + ... + a_{n-2} + a_{n-1} + a_n.$$

If the terms of a series form an arithmetic sequence, the indicated sum of the terms is called an **arithmetic series**.

If a is an arithmetic series with first term a_1 and constant difference d, you can find a formula for the value S_n of the series by writing the series in two ways:

Start with the first term a_1 and successively add the common difference d.

$$S_n = a_1 + (a_1 + d) + (a_1 + 2d) + ... + (a_1 + (n - 1)d)$$

Start with the last term a_n and successively subtract the common difference d.

$$S_n = a_n + (a_n - d) + (a_n - 2d) + ... + (a_n - (n - 1)d)$$

Now add corresponding pairs of terms of these two formulas, as Gauss did. Then each of the n pairs has the same sum, $a_1 + a_n$.

$$S_n + S_n = \underbrace{(a_1 + a_n) + (a_1 + a_n) + (a_1 + a_n) + ... + (a_1 + a_n)}_{n \text{ terms}}$$

So $2S_n = n(a_1 + a_n).$

Thus, $S_n = \frac{n}{2}(a_1 + a_n).$

This proves that if $a_1 + a_2 + ... + a_n$ is an arithmetic series, then a formula for the value S_n of the series is $S_n = \frac{n}{2}(a_1 + a_n).$

 QY2

Arithmetic series that involve the sum of consecutive integers from 1 to n lead to a special case of the above formula. In these situations, $a_1 = 1$ and $a_n = n$, so the sum of the integers from 1 to n is $\frac{n}{2}(1 + n)$, or $\frac{n^2 + n}{2}$.

▶ **QY2**

Use the formula for S_n to find the sum of the odd integers from 1 to 999.

GUIDED

Example 1

Part of the lyrics of a popular Christmas carol say, "On the 12th day of Christmas my true love gave to me... ." The 12th-day gifts are listed at the right. How many gifts did the singer receive on the 12th day of Christmas?

12 drummers drumming,
11 pipers piping,
10 lords-a-leaping,
9 ladies dancing,
8 maids-a-milking,
7 swans-a-swimming,
6 geese-a-laying,
5 golden rings,
4 calling birds,
3 French hens,
2 turtle doves, and
a partridge in a pear tree

Solution Find the sum of consecutive integers from 1 to 12.
Substitute 12 for n in the formula for S_n.

$$S_{12} = \frac{?}{2}(1 + \underline{\ ?\ })$$

$$S_{12} = \underline{\ ?\ }$$

The singer received __?__ gifts on the 12th day of Christmas.

The formula $S_n = \frac{n}{2}(a_1 + a_n)$ is convenient if the first and nth terms of the series are known. If the nth term is not known, you can use another formula. Start with the formula for the nth term of an arithmetic sequence.

$$a_n = a_1 + (n - 1)d$$

Substitute this expression for a_n in the right side of the formula for S_n and simplify.

$$S_n = \frac{n}{2}\left[a_1 + (a_1 + (n - 1)d\right]$$
$$S_n = \frac{n}{2}\left[2a_1 + (n - 1)d\right]$$

This argument proves that if $a_1 + a_2 + \ldots + a_n$ is an arithmetic series with constant difference d, then the value S_n of the series can be found using the formula $S_n = \frac{n}{2}\left[2a_1 + (n - 1)d\right]$.

Example 2

An auditorium has 20 rows, with 14 seats in the front row and 2 more seats in each row thereafter. How many seats are there in all?

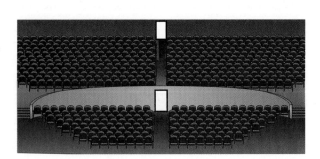

Solution This is an arithmetic-series situation. Because the first term, difference, and number of terms are given, use the formula

$$S_n = \frac{n}{2}(2a_1 + (n - 1)d).$$

In this case, $n = 20$, $a_1 = 14$, and $d = 2$.

So
$$S_{20} = \frac{20}{2}(2 \cdot 14 + (20 - 1)2)$$
$$= \frac{20}{2}(28 + 38) = 660.$$

There are 660 seats in the auditorium.

Check Use the formula $S_n = \frac{n}{2}(a_1 + a_n)$. You need to know how many seats are in the first and last row. In this case,

$a_1 = 14$ and $a_n = a_{20} = 14 + 19 \cdot 2 = 52$.

So $S_{20} = \frac{20}{2}(14 + 52) = \frac{20}{2} \cdot 66 = 660$.

There are 660 seats in the auditorium. It checks.

Summation Notation

The sum of the first six terms of a sequence a_n is

$$a_1 + a_2 + a_3 + a_4 + a_5 + a_6.$$

However, when there are many numbers in the series, this notation is too cumbersome. You can shorten this by writing

$$a_1 + a_2 + \cdots + a_6.$$

It is understood that the terms a_3, a_4, and a_5 are included.

This notation can be shortened even further. In a spreadsheet, suppose you have the sum A1 + A2+ A3 + A4 + A5 + A6. That sum can be written as $\mathsf{SUM(A1:A6)}$. In algebra, the upper-case Greek letter **Σ (sigma)** indicates a sum. In **Σ-notation**, called **sigma notation** or **summation notation,** the above sum is written

$$\sum_{i=1}^{6} a_i.$$

The expression can be read as "the sum of the values of a sub i, for i equals 1 to 6." The variable i under the Σ sign is called the **index variable,** or **index**. It is common to use the letters $i, j, k,$ or n as index variables. (In summation notation, i is *not* the complex number $\sqrt{-1}$.) In this book, index variables have only integer values.

> ▶ **READING MATH**
>
> While you use your index finger to point an object, the *index variable* is used to point to a value. Thus, a_3 points to the third term of the series named a.

Writing Formulas Using Σ-Notation

The two arithmetic series formulas $S_n = \frac{n}{2}(a_1 + a_n)$ and $S_n = \frac{n}{2}(2a_1 + (n-1)d)$ can be restated using Σ-notation. Notice that i is used as the index variable to avoid confusion with the variable n.

Arithmetic Series Formula

In an arithmetic sequence $a_1, a_2, a_3, ..., a_n$ with constant difference d,

$$\sum_{i=1}^{n} a_i = \frac{n}{2}(a_1 + a_n) = \frac{n}{2}(2a_1 + (n-1)d).$$

When $a_i = i$, the sequence is the set of all positive integers in increasing order 1, 2, 3, 4, ... Then

$$\sum_{i=1}^{n} i = \frac{n}{2}(1 + n) = \frac{n(n+1)}{2}.$$

This is a Σ-notation version of Gauss's sum.

 QY3

One advantage of Σ-notation is that you can substitute an expression for a_i. For instance, suppose $a_n = 2n$, the sequence of even positive integers. Then,

$$\sum_{i=1}^{6} a_i = \sum_{i=1}^{6} (2i) = 2 \cdot 1 + 2 \cdot 2 + 2 \cdot 3 + 2 \cdot 4 + 2 \cdot 5 + 2 \cdot 6$$

$$= 2 + 4 + 6 + 8 + 10 + 12$$

$$= 42.$$

The sum of the first six positive even integers is 42.

> ▶ **QY3**
>
> Find $\sum_{i=1}^{40} i$.

GUIDED

Example 3

Consider $\sum_{i=1}^{500} a_i$, where $a_n = 4n + 6$.

a. Write the series without Σ-notation.

b. Evaluate the sum.

Solution

a. Substitute the expression for a_i from the explicit formula and use it to write out the terms of the series.

$$\sum_{i=1}^{500} (4i + 6) = (4 \cdot \underline{\;?\;} + 6) + (4 \cdot \underline{\;?\;} + \underline{\;?\;}) + \underline{\;?\;} + \\ \cdots + \underline{\;?\;}$$

$$= \underline{\;?\;} + \underline{\;?\;} + \underline{\;?\;} + \cdots + \underline{\;?\;}$$

b. This is an arithmetic series. The first term is $\underline{\;?\;}$. The constant difference is $\underline{\;?\;}$. There are $\underline{\;?\;}$ terms in the series.

Use the formula $\sum_{i=1}^{n} a_i = \frac{n}{2}(2a_1 + (n-1)d)$ to evaluate the series.

$$\sum_{i=1}^{500} (4i + 6) = \frac{?}{2}(2 \cdot \underline{\;?\;} + (\underline{\;?\;} - 1)\underline{\;?\;})$$

$$= \underline{\;?\;}$$

Most scientific calculators and CAS have commands to evaluate a series, but the commands and entry styles vary considerably. The entry for one CAS is shown at the right.

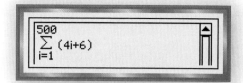

 QY4

Questions

▸ QY4
Check the solution to Example 3 on your calculator or CAS.

COVERING THE IDEAS

In 1 and 2, tell whether what is given is an arithmetic sequence, an arithmetic series, or neither.

1. $26 + 29 + 32 + 35$

2. $35, 32, 29, 26$

3. What problem was Gauss given in third grade, and what is its answer?

4. Find the sum of the integers from 1 to 500.

5. **Fill in the Blank** The symbol Σ is the upper-case Greek letter ___?___ .

6. **Fill in the Blank** In Σ-notation, the variable under the Σ sign is called the ___?___ .

7. Consider the arithmetic sequence with first term a_1 and constant difference d.
   a. Write a formula for the nth term.
   b. Write a formula for the sum S_n of the first n terms using Σ-notation.
   c. Write an equivalent formula to the one you wrote in Part b.

8. Consider the arithmetic series $8 + 13 + 18 + \ldots + 38$.
   a. Write out all the terms of the series. How many terms are there?
   b. What is the sum of all the terms?
   c. Write this series using Σ-notation.

9. Refer to Example 3. Check your answer using the formula $\displaystyle\sum_{i=1}^{n} a_i = \frac{n}{2}(a_1 + a_n)$.

10. **Multiple Choice** $\displaystyle\sum_{i=1}^{5} i^2 = $ ___?___ .

    A $1 + 4 + 9 + 16 + 25$ B 5^2

    C $1 + 2 + \ldots + 5$ D none of these

11. **Multiple Choice** $7 + 14 + 21 + 28 + 35 + 42 + 49 + 56 + 63 =$ ___?___.

A $\displaystyle\sum_{i=7}^{63} i$ B $\displaystyle\sum_{i=7}^{63} (7i)$ C $\displaystyle\sum_{i=1}^{9} (7i)$ D none of these

12. In $\displaystyle\sum_{i=100}^{300} (5i)$, how many terms are added? (*Be careful!*)

In **13** and **14**, evaluate the sum.

13. $\displaystyle\sum_{i=1}^{25} (6i - 4)$

14. $\displaystyle\sum_{n=-1}^{3} 9 \cdot 3^n$

APPLYING THE MATHEMATICS

15. Finish this sentence: The sum of the first n terms of an arithmetic sequence equals the average of the first and last terms multiplied by ___?___.

16. The Jewish holiday Chanukah is celebrated by lighting candles in a *menorah* for eight nights. On the first night, two candles are lit, one in the center and one on the right. The two candles are allowed to burn down completely. On the second night, three candles are lit (one in the center, and two others) and are allowed to burn down completely. On each successive night, one more candle is lit than the night before, and all are allowed to burn down completely. How many candles are needed for all eight nights?

17. Penny Banks decides to start saving money in a Holiday Club account. At the beginning of January she will deposit $100, and each month thereafter she will increase the deposit amount by $25. How much will Penny deposit during the year?

18. a. How many even integers are there from 50 to 100?
    b. Find the sum of the even integers from 50 to 100.

19. a. Write the sum of the squares of the integers from 1 to 100 in Σ-notation.
    b. Evaluate the sum in Part a.

20. a. Translate this statement into an algebraic formula using Σ-notation: The sum of the cubes of the integers from 1 to n is the square of the sum of the integers from 1 to n.
    b. Verify the statement in Part a when $n = 8$

21. Write the arithmetic mean of the n numbers $a_1, a_2, a_3, \ldots, a_n$ using Σ-notation.

REVIEW

22. The function with equation $y = k \cdot 2^x$ contains the point (6, 10).
    a. What is the value of k?
    b. Describe the graph of this function. **(Lesson 9-1)**

23. Consider the geometric sequence 16, –40, 100, –250, … .
    a. Determine the common ratio.
    b. Write the 5th term.
    c. Write an explicit formula for the nth term. **(Lesson 7-5)**

24. Suppose an account pays 5.75% annual interest compounded monthly. **(Lesson 7-4)**
    a. Find the annual percentage yield on the account.
    b. Find the value of a $1700 deposit after 5 years if no other money is added or withdrawn from the account.

25. Find an equation for the parabola with a vertical line of symmetry that contains the points (5, 0), (1, 5), and (3, 8). **(Lesson 6-6)**

26. a. Find the inverse of $\begin{bmatrix} 5 & 2 \\ n & 1 \end{bmatrix}$.
    b. For what value(s) of n does the inverse not exist? **(Lesson 5-5)**

EXPLORATION

27. The number 9 can be written as an arithmetic series: $9 = 1 + 3 + 5$. What other numbers from 1 to 100 can be written as an arithmetic series with three or more positive integer terms?

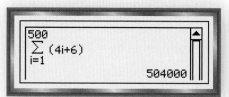

Lesson
13-2

Geometric Series

Vocabulary

geometric series

▶ **BIG IDEA** There are several ways to find the sum of the successive terms of a finite geometric sequence.

Activity

Step 1 Draw a large square on a sheet of paper.

Step 2 Divide the square into two equal parts and shade one of the regions. How much of the square has been shaded?

Step 3 Divide the unshaded half into two equal parts and shade one of the regions. How much of the original square does this region represent? How much of the total square have you shaded? The figure at the right shows one possible result of Steps 1–3.

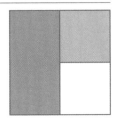

Step 4 Repeat Step 3 three more times. Fill in a table like the one below where n represents the number of shaded regions and S_n is the total fraction of the original large square that is shaded.

n	1	2	3	4	5
g_n (series)	$\frac{1}{2}$	$\frac{1}{2} + \frac{1}{4}$	$\frac{1}{2} + \frac{1}{4} + \frac{1}{8}$	?	?
S_n (value)	$\frac{1}{2}$	$\frac{3}{4}$	?	?	?

Step 5 What is the value of S_6? Describe the nth term of the series and the value of S_n.

Step 6 Will the original square ever be entirely shaded? Explain why or why not.

In the Activity, the terms in the series g form a geometric sequence with first term $\frac{1}{2}$ and constant ratio $\frac{1}{2}$. So the kth term in the sequence is $\left(\frac{1}{2}\right)^k$. For instance, the 5th term in the sequence is $\left(\frac{1}{2}\right)^5 = \frac{1}{32}$, and

$$S_5 = \frac{1}{2} + \frac{1}{4} + \frac{1}{8} + \frac{1}{16} + \frac{1}{32} = \sum_{k=1}^{5} \frac{1}{2^k} = \frac{31}{32}.$$

Mental Math

Is the point (2, 1) included in the set?

a. the circle with equation $x^2 + y^2 = 5$

b. the interior of the circle with equation $x^2 + y^2 = 5$

c. the exterior of the circle with equation $x^2 + y^2 = 3$

d. the interior of the circle with equation $(x - 1)^2 + (y + 1)^2 = 3$

GUIDED

Example 1

a. Write the value of S_{10} from the Activity as a sum of terms of a geometric sequence.

b. Write S_{10} using Σ-notation.

c. Compute exact and approximate values of S_{10} using a calculator or CAS.

Solution

a. The terms in the geometric sequence are the first ten positive integer powers of $\frac{1}{2}$. So, $S_{10} = \frac{1}{2} + \frac{1}{4} + \frac{1}{8} + \underline{\ ?\ } + \underline{\ ?\ } + \underline{\ ?\ } + \underline{\ ?\ } + \underline{\ ?\ } + \underline{\ ?\ } + \underline{\ ?\ }$.

b. There are 10, terms and an expression for the kth term is $\left(\frac{1}{2}\right)^k$.

So $S_{10} = \sum_{\underline{?}}^{\underline{?}} \underline{\ ?\ }$.

c. The exact value is $\underline{\ ?\ }$. An approximate value is $\underline{\ ?\ }$.

An indicated sum of successive terms of a geometric sequence, like the one for S_{10} in Part a of Example 1, is called a **geometric series**. As with arithmetic series, there are formulas for the values of geometric series.

A Formula for the Value of Any Finite Geometric Series

In Example 1, notice that if each term of the sequence is halved, many values are identical to those in the original series:

$$S_{10} = \frac{1}{2} + \frac{1}{4} + \frac{1}{8} + \frac{1}{16} + \ldots + \left(\frac{1}{2}\right)^9 + \left(\frac{1}{2}\right)^{10}$$

$$\frac{1}{2}S_{10} = \quad \frac{1}{4} + \frac{1}{8} + \frac{1}{16} + \ldots + \left(\frac{1}{2}\right)^9 + \left(\frac{1}{2}\right)^{10} + \left(\frac{1}{2}\right)^{11}.$$

Subtracting the second equation from the first yields

$$\frac{1}{2}S_{10} = \frac{1}{2} + \left(\frac{1}{4} - \frac{1}{4}\right) + \left(\frac{1}{8} - \frac{1}{8}\right) + \ldots + \left(\left(\frac{1}{2}\right)^{10} - \left(\frac{1}{2}\right)^{10}\right) - \left(\frac{1}{2}\right)^{11}.$$

That is, $\frac{1}{2}S_{10} = \frac{1}{2} - \left(\frac{1}{2}\right)^{11}$,

and so $S_{10} = 2\left(\frac{1}{2} - \left(\frac{1}{2}\right)^{11}\right) = 1 - \left(\frac{1}{2}\right)^{10} = 1 - \frac{1}{1024} = \frac{1023}{1024}$.

This procedure can be generalized to find the value S_n of any finite geometric series. Let S_n be the geometric series with first term g_1, constant ratio $r \neq 1$, and n terms.

$$S_n = g_1 + g_1 r + g_1 r^2 + \dots + g_1 r^{n-1}$$

$$r S_n = g_1 r + g_1 r^2 + \dots + g_1 r^{n-1} + g_1 r^n \quad \text{Multiply by } r.$$

$$S_n - r S_n = g_1 - g_1 r^n \qquad\qquad \text{Subtract the second equation from the first.}$$

$$(1-r)S_n = g_1(1-r^n) \qquad\quad \text{Use the Distributive Property.}$$

$$S_n = \frac{g_1(1-r^n)}{1-r} \qquad\qquad\quad \text{Divide each side by } 1-r.$$

This proves the following theorem.

Finite Geometric Series Formula

Let S_n be the sum of the first n terms of the geometric sequence with first term g_1 and constant ratio $r \neq 1$. Then $S_n = \dfrac{g_1(1-r^n)}{1-r}$.

The constant ratio r cannot be 1 in this formula. (Do you see why?) But that is not a problem. If $r = 1$, the series is $g_1 + g_1 + g_1 + \dots + g_1$, with n terms, and its sum is $n g_1$.

STOP QY

The formula for a geometric series works even when the constant ratio is negative.

▶ **QY**

Find the sum of the first 10 terms of a geometric series sequence with common ratio $\frac{3}{4}$ and first term 16.

GUIDED

Example 2

a. Write the indicated sum given by $\displaystyle\sum_{k=1}^{5} 27\left(-\frac{1}{3}\right)^{k-1}$.

b. Compute the value of the series in Part a.

Solution

a. The indicated sum is $27(\underline{\ ?\ })^0 + 27(\underline{\ ?\ })^1 + 27(\underline{\ ?\ })^{\underline{?}} +$
$27(\underline{\ ?\ })^{\underline{?}} + \underline{\ ?\ }(\underline{\ ?\ })^{\underline{?}} = \underline{\ ?\ } + \underline{\ ?\ } + \underline{\ ?\ } +$
$\underline{\ ?\ } + \underline{\ ?\ }.$

b. Use the Finite Geometric Series Formula.
$$S_5 = \frac{?(1 - ?^?)}{1 - ?} = \underline{\ ?\ }$$

Check Compute the sum by hand to check that the formula works for negative values of r.

$\underline{\ ?\ } + \underline{\ ?\ } + \underline{\ ?\ } + \underline{\ ?\ } + \underline{\ ?\ } = \underline{\ ?\ }$. It checks.

Geometric Series and Compound Interest

Geometric series arise in compound-interest situations when the equal amounts of money are deposited or invested at regular intervals. The total value of such investments can be found using the Finite Geometric Series Formula.

Example 3

On the day her granddaughter Savanna was born, Mrs. Kash began saving for Savanna's college education by depositing $1000 in an account earning an annual percentage yield of 5.2%. She continued to deposit $1000 each year on Savanna's birthday into the same account at the same interest rate. How much money will be in Savanna's account on her 18th birthday, not including that birthday's payment?

Solution Make a table showing each deposit and its value on Savanna's 18th birthday.

Birthday	Deposit	Value on 18th Birthday
0	$1000	$1000(1.052)^{18}$
1	$1000	$1000(1.052)^{17}$
2	$1000	$1000(1.052)^{16}$
⋮	⋮	⋮
17	$1000	$1000(1.052)^{1}$

The amount in the account is the value of the geometric series $1000(1.052) + 1000(1.052)^2 + ... + 1000(1.052)^{18}$.

The first term is $a = 1000(1.052)$, the ratio is $r = 1.052$, and there are 18 terms. Therefore, the sum is

$$\frac{1000(1.052)(1 - 1.052^{18})}{1 - 1.052} \approx 30{,}153.58.$$

Savanna will have $30,153.58 in the account on her 18th birthday.

Questions

COVERING THE IDEAS

1. Refer to the Activity. What fraction of the square is shaded when $n = 8$?

2. **a.** State a formula for the sum of the first n terms of a geometric sequence with first term g_1 and constant ratio r.
   b. In the formula in Part a, what value can r not have?
   c. Why can r not have the value in Part b? In this situation, what is the value of the series?

In 3–6, a geometric series is given.

    a. **How many terms does the series have?**

    b. **Write the series in Σ-notation.**

    c. **Use the Finite Geometric Series Formula to evaluate the series.**

3. $5 + 10 + 20 + 40 + \ldots + 5 \cdot 2^7$

4. $170 + 17 + 1.7 + 0.17 + 0.017 + 0.0017$

5. $170 - 17 + 1.7 - 0.17 + 0.017 - 0.0017$

6. $a + \frac{1}{2}a + \frac{1}{4}a + \frac{1}{8}a + \ldots + \left(\frac{1}{2}\right)^{16}a$

7. Consider the geometric series in Example 2.

    a. Calculate the following sums.

       i. S_2 ii. S_3 iii. S_4 iv. S_5

    b. Plot the sums S_1 (which is 27), S_2, S_3, S_4, and S_5 on a number line. How is S_n related to S_{n-1} and S_{n-2}?

8. Find the sum of the first 17 terms of the geometric sequence with first term 20 and constant ratio 1.

9. Suppose $500 is deposited into a bank account on July 17 for seven consecutive years and earns an annual percentage yield of 4%.

    a. Write a geometric series that represents the value of this investment on July 17 of the eighth year (before that year's deposit).

    b. Rewrite your answer to Part a in Σ-notation.

    c. How much is in the account on July 17 of the eighth year?

APPLYING THE MATHEMATICS

10. a. A worker deposits $2000 at the end of each year into a retirement account earning an annual percentage yield of 5.1%. Assume no other deposits or withdrawals from the account. To the nearest dollar, how much will the worker have after 40 years, assuming no deposit is made at the end of the 40th year?

    b. A second worker waits ten years before starting to save money for retirement. Assume that this worker saves $A per year for thirty years, also earning an APY of 5.1%. Write an expression for the total amount of money this worker will have saved, including interest, over thirty years, assuming no deposit is made at the end of the 30th year.

    c. How much does the second worker have to save each year in order to have the same amount after 30 years as the first worker has after 40 years?

11. **a.** Complete the table below by expanding or factoring each polynomial on a CAS.

Standard Form	Factored Form
?	$(r-1)(r+1)$
$r^3 - 1$	?
$r^4 - 1$	?
$r^5 - 1$	?
$r^6 - 1$	?

factor(r^3-1)

b. Fill in the Blanks According to the Factor Theorem, $r - 1$ is a factor of a polynomial $P(r)$ if and only if $P(\underline{\,?\,}) = \underline{\,?\,}$. Use this fact to prove that $r - 1$ is a factor of $r^n - 1$ for all n.

c. Use the results of Parts a and b to simplify $\frac{r^n - 1}{r - 1}$ for $n = 2, 3, 4, 5,$ and 6.

12. A ball is dropped from a height of 10 feet. Each bounce returns it to $\frac{4}{5}$ of the height of the previous bounce.

a. Draw a diagram showing the ball's path until it touches the ground for the fourth time.

b. Find the total vertical distance the ball has traveled when it touches the ground for the fourth time.

13. Consider the geometric series $x^7 + x^{12} + x^{17} + \ldots + x^{77}$.

a. What is the common ratio?

b. Use the Finite Geometric Series Formula to find a formula for the value of the series.

c. Verify your result in Part b with a CAS.

14. The number $\frac{1}{3}$ can be approximated by the finite geometric series $0.3 + 0.03 + 0.003 + 0.0003 + \ldots + g_1 r^{n-1}$.

a. Identify g_1 and r for the series.

b. What approximation to $\frac{1}{3}$ occurs when $n = 6$?

c. How far from $\frac{1}{3}$ is the approximation in Part b?

REVIEW

15. Find the sum of the integers from 101 to 200. **(Lesson 13-1)**

16. **a.** How many odd integers are there from 25 to 75?

b. Find the sum of the odd integers from 25 to 75. **(Lesson 13-1)**

17. A math club mails a monthly newsletter. In January, the club mailed newsletters to each of its 325 members. If membership increases by 5 members each month, how many newsletters will it mail for the entire year? **(Lesson 13-1)**

18. Suppose $t_n = -2n + 9$. Find $t_1 + t_2 + t_3 + \ldots + t_{19}$.
    (Lessons 13-1, 3-8)

19. Let $f(x) = 27^x$. (Lessons 9-7, 7-7, 7-3, 7-2, 7-1)
    a. Evaluate $f(-3)$, $f(0)$, and $f\left(\frac{2}{3}\right)$.
    b. Identify the domain and range of f.
    c. Give an equation for the reflection image of the graph of
       $y = f(x)$ over the line $y = x$.

20. a. Identify the type of quadrilateral graphed at the right.
    b. Prove or disprove that the diagonals of this quadrilateral have
       the same length. (Lesson 4-4, Previous Course)

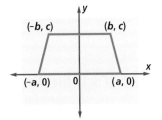

21. Write an equation for the line parallel to $7x + 2y = 13$ and
    containing the point $(6, 5)$. (Lessons 3-4, 3-2)

EXPLORATION

22. Suppose that, because of inflation, a payment of P dollars n years
    from now is estimated to be worth $P(0.96)^n$ today. This worth is
    called the *present value* of a future payment.
    a. A lottery advertising $10,000,000 in winnings actually plans
       to pay the winner $500,000 each year for 20 years, starting
       the day the winner wins the lottery. Write the sum of the
       present values of these payments as a geometric series.
    b. Evaluate the geometric series in Part a. How does the sum
       compare to the advertised jackpot of $10,000,000?

13-3 Using Series in Statistics Formulas

Vocabulary

mean

measure of center, measure of central tendency

absolute deviation

mean absolute deviation, m.a.d.

standard deviation, s.d.

▶ **BIG IDEA** Formulas for certain statistics, such as the *mean*, *mean absolute deviation*, and *standard deviation*, involve sums.

When you used Σ-notation in the previous two lessons, there was a formula for the numbers being added. In many situations, and particularly in statistics, data may be represented with an index variable and no formula. You can think of the data as terms of a sequence. We also say that each datum is an element of a data set. In data sets, different elements can have the same value. For instance, suppose that you have to read 10 short stories this semester in your English class and they have the following numbers of pages:

$$6, 14, 3, 8, 8, 9, 4, 11, 10, 23.$$

Let L_i be the length of the ith story, so $L_1 = 6$, $L_2 = 14$, $L_3 = 3$,

and so on. The sum of these 10 numbers can be represented as $\sum_{i=1}^{10} L_i$.

In this case, $\sum_{i=1}^{10} L_i = 96$, and so the mean of the lengths L_i is

$\dfrac{\sum_{i=1}^{10} L_i}{10} = \dfrac{96}{10} = 9.6$. That is, the mean length of a story is 9.6 pages.

In general, when S is a data set of n numbers $x_1, x_2, x_3, \ldots, x_n$, then

the **mean** of $S = \dfrac{\sum_{i=1}^{n} x_i}{n} = \dfrac{1}{n}\sum_{i=1}^{n} x_i$.

The Greek letter μ (mu, pronounced "mew") is customarily used to represent the mean of a data set.

Deviations and Absolute Deviations

Recall from Lesson 3-5 that the difference between an element of a data set and the mean of the set is called the element's deviation from the mean. If the mean is μ and the element is x_i, then the deviation is

$$x_i - \mu.$$

Mental Math

Roderick has averaged 87 on two advanced algebra tests. What is the minimum score he can receive on the final test if he wants to finish the course with an average test score of at least

a. 90?

b. 70?

c. 95?

For instance, in the data set of story lengths, the deviation of the 6-page story from the mean of 9.6 pages is $6 - 9.6$, or -3.6. That is, the story is 3.6 pages shorter than the mean length. In general, deviation is positive if an element is larger than the mean and negative if an element is smaller than the mean.

The sum of the deviations of the elements of a set from the mean of the set is 0. That is,

$$\sum_{i=1}^{n} (x_i - \mu) = 0.$$

This suggests that the mean is a number on which the data set "balances." For this reason, the mean is a called a **measure of center**, or a **measure of central tendency**, of a data set.

 QY1

Suppose you are estimating the number of beans in a jar and you want to know how close your estimate E is to the actual number A. Then you do not care whether E is larger or smaller than A. You want to know the *absolute deviation* of your estimate from A, or $|E - A|$. The **absolute deviation** of an element of a data set from the mean of the set is

$$|x_i - \mu|.$$

For instance, the absolute deviation of the element 6 from the mean 9.6 is $|6 - 9.6| = |-3.6| = 3.6$.

 QY2

The mean of the absolute deviations is a statistic called the **mean absolute deviation**, or **m.a.d.** The *m.a.d.* is a measure of the *spread* or *dispersion* of a data set. In Σ-notation,

$$m.a.d. = \frac{1}{n} \sum_{i=1}^{n} |x_i - \mu|.$$

> ▶ **QY1**
>
> a. Find the deviations of all 10 elements of the data set of story lengths from the mean of the lengths.
>
> b. Find the mean of the deviations.

> ▶ **QY2**
>
> a. Find the absolute deviations of all 10 elements of the data set of story lengths from the mean of the lengths.
>
> b. Find the mean of these absolute deviations.

Example

You are offered the chance to play one of two games, Game A or Game B. In each game you reach into a jar and pull out a slip of paper. You win the dollar amount written on the slip. In both games there are 10 slips of paper in the jar with a mean value of $50.

Calculate the *m.a.d.* of the values on the slips for each game. Which game gives more spread out results?

Game A Slips				
$49	$49	$49	$49	$50
$50	$51	$51	$51	$51

Game B Slips				
$0	$0	$0	$0	$0
$0	$0	$0	$0	$500

Solution For each game, the mean dollar amount μ is $50. For Game A, there are 8 slips whose amounts deviate by $1 from the mean, and the other two slips have no deviation from the mean. So,

$$\text{m.a.d. for Game A} = \frac{1}{n}\sum_{i=1}^{n}|x_i - \mu| = \frac{1}{10}(8 \cdot 1 + 2 \cdot 0) = 0.8.$$

In Game B, there are 9 amounts that deviate by $50 from the mean, and 1 amount that deviates by $450. So,

$$\text{m.a.d. for Game B} = \frac{1}{n}\sum_{i=1}^{n}|x_i - \mu| = \frac{1}{10}(9 \cdot 50 + 1 \cdot 450) = 90.$$

There is quite a difference in the spread of dollar amounts in the games. The more the values are spread out from the mean, the less likely you are to win an amount of money close to the mean amount. Since 90 is greater than 0.8, the results of Game B are more spread out.

The Standard Deviation

The *m.a.d.* of a data set is relatively easy to calculate, but it does not have as many useful properties as a second measure of spread, the *standard deviation*. The lower-case Greek letter sigma (σ) is often used to denote the standard deviation, as in the following definition.

Definition of Standard Deviation

Let S be a data set of n numbers $\{x_1, x_2, ..., x_n\}$. Let μ be the mean of S. Then the **standard deviation, or s.d.,** of S is given by

$$\text{s.d.} = \sigma = \sqrt{\frac{1}{n}\sum_{i=1}^{n}(x_i - \mu)^2}.$$

This formula looks complicated, so it may help to describe it in words. To use the formula, find the mean of the data set and then square each element's deviation from the mean. Then find the mean of these squared deviations. The square root of this mean is the standard deviation. The mean of the squared deviations is also called the *variance* of the data set. So, the standard deviation of a set is the square root of the variance of the set.

Standard deviation is used in a wide variety of statistical analyses. For example, a readability index is a measure of how difficult written text is to understand. To find one index, a computer program analyzes random paragraphs to find the mean and standard deviation of the number of words per sentence.

Activity

MATERIALS Paper and the first two paragraphs under the Big Idea of Lesson 13-5

Step 1	Make a table like the one at the right and record the number of words in each sentence of each paragraph. When counting words, note that a "word" is any character or group of characters with a space before and after it.
Step 2	Calculate the mean number of words per sentence in each paragraph.
Step 3	Calculate the square of the deviation of each element in the Paragraph 1 data set from its mean. Repeat for the Paragraph 2 data set.
Step 4	Find the mean of the squared deviations for each data set in Step 3. Then calculate the standard deviation for each.
Step 5	Which paragraph from Lesson 13-5 do you think is easier to read and understand? Support your conclusion with your statistics from Steps 2–4.

Number of words		
Sentence Number	Paragraph 1	Paragraph 2
1	?	?
2	?	?
3	?	?
4	?	?
5	?	?

Most spreadsheet programs and calculators calculate standard deviations. One CAS shows the results at the right for the standard deviations of the data sets in the Activity.

stDevPop($\{12,11,10,11,9\}$)	1.0198
stDevPop($\{27,16,19,31,14\}$)	6.52993

Questions

COVERING THE IDEAS

1. Find the mean absolute deviation of the data set $\{2, 3, 3, 4, 5, 6, 6, 6, 6, 7\}$.

2. In 2006, the Miami Heat salaries, in millions of dollars, were approximately 0.4, 0.74, 2.88, 0.15, 0.07, 5.53, 0.41, 0.075, 1.19, 2.5, 20, 0.74, 6.39, 0.41, 0.93, 3.84, 7.61, 8.25, and 1.33. Find the *m.a.d.* of these salaries.

3. Calculate the standard deviation of the data set of story lengths on the first page of this lesson.

4. A person bowls games of 158, 201, 175, and 134. For these scores, calculate
   a. the *m.a.d.*
   b. the *s.d.*

In the 2005–2006 season, Shaquille O'Neal of the Miami Heat was the highest paid NBA player.

In 5–9, suppose that 100 scores are identified as $s_1, s_2, \ldots, s_{100}$.
What does each expression represent?

5. $\displaystyle\sum_{i=1}^{100} s_i$

6. $\displaystyle\frac{1}{100}\sum_{i=1}^{100} s_i$

7. $\displaystyle\frac{1}{100}\sum_{i=1}^{100}(s_i - \mu)$

8. $\displaystyle\frac{1}{100}\sum_{i=1}^{100}|s_i - \mu|$

9. $\displaystyle\sqrt{\frac{1}{100}\sum_{i=1}^{100}(s_i - \mu)^2}$

10. Refer to the two games described in the Example.
    a. Calculate the standard deviation of the amounts in each game
    b. Why would someone want to play Game B?

11. Fundraisers sell 10,000 raffle tickets for $5 each. The raffle officials need to decide whether to have several winners of small amounts or just a few winners of large amounts. One option is to have two prizes worth $10,000 each and two prizes worth $5000 each. A second option is to have two prizes worth $15,000 apiece. In which option are the results more spread out? (Consider any non-winning ticket as a $0 prize.)

APPLYING THE MATHEMATICS

12. **Multiple Choice** A store has two managers and nine employees. Each manager earns $40,000 a year, six employees earn $25,000 a year, and three employees earn $15,000 a year. If each person gets a $1000 raise next year, the standard deviation of the salaries
    A will increase by $1000.   B will increase by $3000.
    C will not change.   D will increase by about 3%.

13. a. Let $x_i = 2i$, for $i = 1, 2, 3, \ldots, 10$. Find the mean and standard deviation of the x_i values.
    b. Let $y_i = 2i + 1$, for $i = 1, 2, 3, \ldots, 10$. Find the mean and standard deviation of the y_i values.

14. Below are the ages of the Democratic and Republican United States Presidents when they were first inaugurated into office, as of 2008.

    Democrats: 43, 46, 47, 48, 49, 51, 52, 54, 55, 55, 56, 60, 61, 65

    Republicans: 42, 46, 49, 50, 51, 51, 51, 52, 54, 54, 54, 55, 55, 56, 61, 62, 64, 69

    Compare the ages at inauguration of Democrats and Republicans by calculating means and standard deviations.

15. Give an example, different from the one in the lesson, of two different data sets that have the same mean but different standard deviations.

While John F. Kennedy was the youngest person elected president, Theodore Roosevelt was the youngest person to become president when William McKinley was assassinated.

16. What would a data set with standard deviation equal to 0 look like?

REVIEW

17. Lotta Moola invests $350 on the first day of every month in an account that earns an annual interest rate of 6% compounded monthly. Assume no other deposits or withdrawals are made.

    a. How much interest will the first $350 deposit earn in 6 months?

    b. How much will be in Lotta's account just after she makes her 7th deposit? (**Lessons 13-2, 11-1, 7-4**)

In 18 and 19, suppose a tennis ball is released from a height of 1 meter above the floor. Each time it hits the floor it bounces to 40% of its previous height. (**Lessons 13-2, 7-5**)

18. Suppose the ball has hit the floor four times. How high will it get on the next bounce?

19. If the ball hits the floor eight times, find the vertical distance it will have traveled.

20. Beginning with 1, how many consecutive positive integers do you have to add in order to total 2701? (**Lesson 13-1**)

In 21–23, evaluate and write your answer in $a + bi$ form. (Lessons 6-9, 6-8)

21. $(1 + i)^2$

22. $\dfrac{-8 + 2i}{i}$

23. $i^4 + i^5 + i^6 + i^7$

24. If 4 thingies and 3 somethings weigh 190 lb, and 6 thingies and 7 somethings weigh 350 lb, what will 2 thingies and 4 somethings weigh? (**Lesson 5-4**)

25. Ivan has test scores of 80, 97, 90, and 88. What must Ivan score on the next test to have

    a. a mean of 90 for the five tests?

    b. a median of 90 for the five tests?

    c. a mode of 90 for the five tests? (**Previous Course**)

EXPLORATION

26. You have used your calculator's statistical regression functions to find equations to model sets of data points. To do this, the regression procedure minimizes the sum of the squares of the deviations of the points from the curve. Search the Internet (for example, search for "sum of squares applet") to find interactive websites that allow you to graphically explore how to minimize a sum of squares.

QY ANSWERS

1. a. –3.6, 4.4, –6.6, –1.6, –1.6, –0.6, –5.6, 1.4, 0.4, 13.4

   b. 0

2. a. 3.6, 4.4, 6.6, 1.6, 1.6, 0.6, 5.6, 1.4, 0.4, 13.4

   b. 3.92

13-4

Subsets and Combinations

Vocabulary

permutation

!, factorial symbol

$n!$

combination

> ▶ **BIG IDEA** Given a set of n objects, there are formulas for the number of ways of choosing r objects where the order or the objects matters, and for the number of ways of choosing r objects without regard to their order.

Permutations

An arrangement of objects where order matters is called a **permutation.** With 3 objects A, B, and C, there are 6 possible permutations: ABC, ACB, BAC, BCA, CAB, and CBA. You can think of these 6 permutations in many ways, such as ways to arrange 3 objects on a shelf or orders in which runners could win medals in an Olympic race.

Example 1

a. Write all the possible orders in which 4 runners A, B, C, and D might finish a race.

b. How many permutations of 4 runners are there?

Solution

a. Make a list as shown below. Assume A finishes first. The left column lists the 6 possible orders of B, C, and D finishing behind A. The next column has B first followed by the 6 possible orders of A, C, and D. The third and fourth columns begin with C and D, respectively.

ABCD	BACD	CABD	DABC
ABDC	BADC	CADB	DACB
ACBD	BCAD	CBAD	DBAC
ACDB	BCDA	CBDA	DBCA
ADBC	BDAC	CDAB	DCAB
ADCB	BDCA	CDBA	DCBA

b. Count the permutations you listed. **There are 24 permutations.** Notice that the number of permutations of 4 objects is 4 times the number of permutations of 3 objects, or 4 · 6.

Mental Math

Simplify.

a. $\{n \mid n < 3\} \cup \{n \mid n \geq -8\}$

b. $\{p \mid p$ is divisible by 2$\} \cap \{p \mid p$ is divisible by 3$\}$

c. the set of all circles $\cup$ the set of all ellipses

d. the set of all rectangles with perimeter 50 $\cap$ the set of all squares with area 225

Special Olympics serves people with intellectual disabilites in over 180 countries.

To list the possible ways in which 5 people could finish a race, you could begin with the list in Example 1. Call the fifth racer E. In each permutation in the list, you can insert E in 5 places: at the beginning, in one of the three middle spots, or at the end. For instance, inserting E into $ABCD$ yields $EABCD$, $AEBCD$, $ABECD$, $ABCED$, or $ABCDE$. This means that the number of permutations of 5 objects is 5 times the number of permutations of 4 objects, or $5 \cdot 24$.

The Factorial Symbol

You may have noticed a pattern. The number of permutations of 2 objects A and B is 2, AB and BA, and $2 = 2 \cdot 1$. The number of permutations of 3 objects A, B, and C is 6, which is $3 \cdot 2$, or $3 \cdot 2 \cdot 1$. The number of permutations of 4 objects is $4 \cdot 6$, or $4 \cdot 3 \cdot 2 \cdot 1$. The number of permutations of 5 objects is $5 \cdot 4 \cdot 3 \cdot 2 \cdot 1$, or 120.

These products of the integers n through 1 are represented by a special symbol, called the *factorial symbol*. The **factorial symbol, !,** is an exclamation point, and $n!$ is read "n factorial."

> ### Definition of Factorial
>
> Let n be any integer ≥ 2. Then $n!$ is the product of the integers from 1 through n.

A generalization of Example 1 can be described using factorials.

> ### Number of Permutations Theorem
>
> There are $n!$ permutations of n distinct objects.

In the order of operations, factorials are calculated before multiplications or divisions. That is, $2 \cdot 5! = 2 \cdot 120 = 240 \neq 10!$.

> ### Activity
>
> **Step 1**   Copy and fill in the table at the right.
>
> **Step 2**   Describe the pattern you see in the table. Use the pattern to write a recursive formula for the sequence $f_n = n!$.

n	3	4	5	6	7	8	9	10
$n!$	?	?	120	?	?	?	362,880	?
$(n-1)!$	2	?	?	?	720	?	?	?
$\frac{n!}{(n-1)!}$	?	?	?	?	?	?	?	?

The pattern in the Activity is a fundamental property of factorials.

> ### Factorial Product Theorem
>
> For all $n \geq 1$, $n! = n \cdot (n-1)!$.

When $n \geq 3$, the theorem follows from the definition of factorial. For the theorem to hold when $n = 2$, we must have $2! = 2 \cdot (2 - 1)!$ $= 2 \cdot 1!$. This means that $1!$ has to equal 1. This makes sense with permutations. If there is only one object, there is only one order. If the theorem is to hold when $n = 1$, then it must be that $1! = 1 \cdot (1 - 1)! = 1 \cdot 0!$. This means that we must have $0! = 1$.

Many calculators and CAS give exact values of $n!$ for small values of n, but for larger values, they give approximations in scientific notation. For instance, when $20!$ is entered, one calculator displays

$$2432902008176640000$$

while another displays 2.4329 E 18, which means 2,432,900,000,000,000,000.

Products of Consecutive Integers

Factorials help you calculate products of consecutive integers, starting at any number.

Example 2

Find $7 \cdot 8 \cdot 9 \cdot 10 \cdot 11 \cdot 12 \cdot 13 \cdot 14 \cdot 15 \cdot 16$ using factorials.

Solution 1 Multiply the given product by a factorial so that the final product is a factorial.

Let $\qquad x = 7 \cdot 8 \cdot 9 \cdot 10 \cdot 11 \cdot 12 \cdot 13 \cdot 14 \cdot 15 \cdot 16$.

Notice that $\quad 6! \cdot x = 1 \cdot 2 \cdot 3 \cdot 4 \cdot 5 \cdot 6 \cdot x = 16!$.

Solving for x, $\quad x = \frac{16!}{6!} = 29{,}059{,}430{,}400$.

Solution 2 Multiply the given product by $\frac{6!}{6!}$. This does not change its value.

$$7 \cdot 8 \cdot 9 \cdot 10 \cdot 11 \cdot 12 \cdot 13 \cdot 14 \cdot 15 \cdot 16$$
$$= \frac{6!}{6!} \cdot 7 \cdot 8 \cdot 9 \cdot 10 \cdot 11 \cdot 12 \cdot 13 \cdot 14 \cdot 15 \cdot 16$$
$$= \frac{16!}{6!} = 29{,}059{,}430{,}400$$

 QY1

> ▸ QY1
>
> Write $22 \cdot 23 \cdot 24$ as a quotient of two factorials.

Subsets and Combinations

You can apply the technique in Example 2 to problems where you are choosing subsets and order does not matter.

Example 3

A committee of 4 people is to be chosen from 10 applicants. In how many different ways can this be done?

Solution Think of the applicants as the set {A, B, C, D, E, F, G, H, I, J}. Each possible committee is a 4-element subset of this set. For instance, two possible committees are {D, C, A, B} and {C, D, F, J}.

Form the committees one person at a time. There are 10 possibilities for the first person. After selecting the first person, there are 9 possibilities for the second person. After selecting the first two people, there are 8 possibilities for the third person. After selecting the first three people, there are 7 possibilities for the fourth person. So it seems that there are $10 \cdot 9 \cdot 8 \cdot 7$ possible committees.

However, this assumes that the order in which the people are chosen makes a difference, but the order of people in a committee does not matter: {B, E, H, I} and {H, B, E, I} are the same committee. In fact, there are $4! = 24$ different orders of the elements B, E, H, and I, all of which form the same committee. So, the answer $10 \cdot 9 \cdot 8 \cdot 7$ is $4!$ times what you need.

The number of committees with 4 people is $\dfrac{10 \cdot 9 \cdot 8 \cdot 7}{4!}$.

Multiply both the numerator and denominator by $6!$.

$$\frac{10 \cdot 9 \cdot 8 \cdot 7}{4!} \cdot \frac{6!}{6!} = \frac{10!}{4! \cdot 6!} = 210$$

So, there are 210 ways to choose a committee of 4 from a set of 10 people.

 QY2

Example 3 can be viewed as a problem in counting subsets. How many subsets of 4 elements are possible from a set of 10 elements? It also can be viewed as a problem in counting *combinations* of objects. How many combinations of 4 objects are possible from 10 different objects?

Any choice of r objects from n objects *when the order of choice does not matter* is called a **combination.** The number of combinations of r objects that can be created from n objects is denoted $_nC_r$. The following theorem connects combinations with counting subsets.

> ▶ **QY2**
>
> How many committees of 3 people can be chosen from 10 applicants?

Combination Counting Formula

The number $_nC_r$ of subsets, or combinations, of r elements that can be formed from a set of n elements is given by the formula

$$_nC_r = \frac{n!}{r!(n-r)!}.$$

Proof There are n choices for the first element in a subset. Once that element has been picked, there are $n - 1$ choices for the second element, and $n - 2$ choices for the third element. This continues until all r elements have been picked. There are $(n - r + 1)$ choices for the rth element.

So, if all possible orders are considered different, there are

$$\underbrace{n(n - 1)(n - 2) \ldots (n - r + 1)}_{r \text{ factors}}$$

ways to choose them. But each subset is repeated $r!$ times with the same elements in various orders. So the number of different subsets is

$$_nC_r = \frac{n(n - 1)(n - 2) \ldots (n - r + 1)}{r!}.$$

Multiplying both numerator and denominator by $(n - r)!$ gives the Combination Counting Formula.

$_nC_r$ is sometimes read "n choose r." Another notation for the number of combinations is $C(n, r)$. Both have the same meaning and are equal to $\frac{n!}{r!(n - r)!}$. Example 3 shows that $_{10}C_4 = C(10, 4) = 210$. Many calculators have keys that enable you to calculate $_nC_r$ directly.

GUIDED

Example 4

How many subsets of 11 elements are possible from a set of 13 elements?

Solution Evaluate $C(13, 11)$. Use the Combination Counting Formula with $n = \underline{\ \ ?\ \ }$ and $r = \underline{\ \ ?\ \ }$.

Then $\frac{n!}{r!(n - r)!} = \frac{?}{? \cdot ?} = \underline{\ \ ?\ \ }$.

Check Use a calculator or CAS. At the right is one way to enter the combination.

nCr(13,11)

Example 5

Given 7 points in a plane, with no 3 of them collinear, how many different triangles can have 3 of these points as vertices?

Solution 1 Because no 3 points are collinear, any choice of 3 points from the 7 points determines a triangle. Use the Combination Counting Formula with $n = 7$ and $r = 3$.

The number of possible triangles is $_7C_3 = \frac{7!}{3!(7 - 3)!} = \frac{7!}{3!4!} = 35$.

Solution 2 Use the idea of the proof of the Combination Counting Formula.

(continued on next page)

The first vertex of the triangle can be chosen in 7 ways. The second vertex can then be chosen in 6 ways. And the third vertex can then be chosen in 5 ways. So, if order mattered, there would be 7 • 6 • 5 = 210 different triangles. But order doesn't matter and each triangle is counted 3! = 6 times. So divide 210 by 6, giving 35 different triangles.

Questions

COVERING THE IDEAS

1. **a.** Write all permutations of the three symbols P, R, M.
   b. Write all combinations of the three symbols P, R, M.

2. How many permutations are there for the 5 vowels A, E, I, O, and U?

3. Give the values of 1!, 2!, 3!, 4!, 5!, 6!, and 7!.

4. Explain why $23! = 23 \cdot 22!$.

5. Explain, in words, the difference between a permutation and a combination.

6. Write $100 \cdot 101 \cdot 102 \cdot 103$ as the quotient of two factorials.

7. How many combinations of r objects can you make from n different objects?

8. What is another way to represent $_nC_r$?

9. A cone of 3 different scoops of ice cream is to be chosen from 5 different flavors. In how many ways can this be done?

10. How many subcommittees of 6 people are possible in a committee of 15?

11. Refer to Example 5. How many different line segments can have 2 of the 7 points as endpoints?

APPLYING THE MATHEMATICS

12. Prove that $(n + 1)! = n!(n + 1)$.

13. Recall that the U.S. Congress consists of 100 senators and 435 representatives.
    a. How many four-person senatorial committees are possible?
    b. How many four-person house committees are possible?
    c. A "conference committee" of 4 senators and 4 representatives is chosen to work out differences in bills passed by the two houses. How many different conference committees are possible?

14. Consider a set of *n* elements.

   a. How many subsets of any number of elements are possible when *n* = 1, 2, 3, 4, 5?

   b. Based on your answers to Part a, make a conjecture about the number of subsets for a set with *k* elements.

15. Dyana sells custom-made tie-dyed T-shirts. A customer chooses 6 dyes from 25 possibilities. Dyana advertises that she offers 150,000 different dye combinations.

   a. Assuming a customer chooses 6 dyes, how many dye combinations are possible? Is the advertisement correct?

   b. How many choices does a customer have if the order of color choice matters?

REVIEW

16. Find the standard deviation of the data set {4, 11, 25, 39, 39, 25, 11, 4}. (**Lesson 13-3**)

17. If two sets of scores have the same mean but the standard deviation of the first set is much larger than that of the second set, what can you conclude? (**Lesson 13-3**)

18. In $\triangle SPX$, $m\angle S = 75°$, $s = 11$, and $x = 9$. Find $m\angle X$. (**Lesson 10-7**)

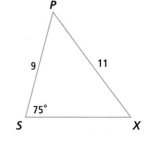

19. a. Find an equation for the inverse of the linear function with equation $y = mx + b$.

   b. How are the slopes of a linear function and its inverse related?

   c. When is the inverse not a function? (**Lessons 8-2, 1-4**)

20. Suppose *y* varies inversely as the cube of *w*. If *y* is 6 when *w* is 5, find *y* when *w* is 11. (**Lesson 2-2**)

21. Give a set of integers whose mean is 12, whose mode is 14, and whose median is 13. (**Previous Course**)

EXPLORATION

22. a. Consider the state name MISSISSIPPI. In how many different ways can you rearrange the letters, if you do not distinguish between letters that are the same?

   b. Repeat Part a using WOOLLOOMOOLOO, the name of a town near Sydney, Australia.

QY ANSWERS

1. $22 \cdot 23 \cdot 24 = \frac{24!}{21!}$

2. $\frac{10 \cdot 9 \cdot 8}{3!} \cdot \frac{7!}{7!} = \frac{10!}{3! \cdot 7!} = 120$

Lesson
13-5

Pascal's Triangle

Vocabulary

Pascal's Triangle

> ▶ **BIG IDEA** The nth row of Pascal's Triangle contains the number of ways of choosing r objects out of n objects without regard to their order, that is, the number of combinations of r objects out of n.

Very often a single idea has applications to many parts of mathematics. On the first page of this chapter, we mentioned *Pascal's Triangle*. Pascal's Triangle is not a triangle in the geometric sense. It is an infinite array of numbers in a triangular shape. The top rows of the triangle are shown below.

				1					row 0
			1		1				row 1
		1		2		1			row 2
	1		3		3		1		row 3
1		4		6		4		1	row 4

1   5   10   10   5   1   row 5
1   6   15   20   15   6   1   row 6
1   7   21   35   35   21   7   1   row 7

Pascal's Triangle

This array seems to have first appeared in the 11th century in the works of Abu Bakr al-Karaji, a Persian mathematician, and Jia Xian, a Chinese mathematician. The works of both of these men are now lost, but 12th-century writers refer to them. Versions of the array were discovered independently by the Europeans Peter Apianus in 1527 and Michael Stifel in 1544. But in the western world the array is known as Pascal's Triangle after Blaise Pascal (1623–1662), the French mathematician and philosopher who discovered many properties relating the numbers in the array. Pascal himself called it the *triangle arithmetique*, which literally translates as the "arithmetical triangle."

Mental Math

Let D be a relation that maps any polynomial onto its degree.

a. Name three ordered pairs in D.

b. Is D a function?

c. What is the range of D?

d. Is the inverse of D a function?

How Is Pascal's Triangle Formed?

Pascal's Triangle is formed in a very simple way. You can think of Pascal's Triangle as a two-dimensional sequence in which each element is determined by a row and its position in that row.

Here is a recursive definition for the sequence: The only element in the top row (row 0) is 1. The first and last elements of all other rows are also 1. If x and y are located next to each other in a row, the element just below and directly between them is $x + y$, as illustrated at the right.

For instance, from row 7 you can get row 8 as follows.

```
1   7   21   35   35   21   7   1      row 7
1   8   28   56   70   56   28   8   1   row 8
```

From this recursive definition, you can obtain any row in the array if you know the preceding row.

Example 1

Write row 9 of Pascal's Triangle.

Solution Begin by listing row 8, as shown above. Apply the recursive definition to generate row 9. Remember that the first and last elements of each row are 1.

```
1   8   28   56   70   56   28   8   1
1   9   36   84   126   126   84   36   9   1
```

 QY1

The elements in the nth row of Pascal's triangle are identified as $\binom{n}{0}$, $\binom{n}{1}$, $\binom{n}{2}$, ..., $\binom{n}{n}$. The top row of the array is called row 0, so in this row $n = 0$. It has one element, its first element, $\binom{0}{0}$. So, $\binom{0}{0} = 1$. The two elements of row 1 are $\binom{1}{0}$ and $\binom{1}{1}$. The three elements of row 2 are $\binom{2}{0}$, $\binom{2}{1}$, and $\binom{2}{2}$. In general, the $(r + 1)$st element in row n of Pascal's triangle is denoted by $\binom{n}{r}$.

 QY2

> ▸ **QY1**
>
> Write row 10 of Pascal's Triangle.

> ▸ **QY2**
>
> Write row 8 of Pascal's Triangle using $\binom{n}{r}$ notation.

On the previous page, we wrote a recursive definition of Pascal's Triangle in words. Now, we can write a recursive definition using the $\binom{n}{r}$ symbol. The recursive rule involves two variables because Pascal's Triangle is a two-dimensional sequence, that is, a sequence in two directions: down and across.

Definition of Pascal's Triangle

Pascal's Triangle is the sequence satisfying

1. $\binom{n}{0} = \binom{n}{n} = 1$, for all integers $n \geq 0$ and

2. $\binom{n+1}{r+1} = \binom{n}{r} + \binom{n}{r+1}$, for $0 \leq r < n$.

Part 1 of the definition gives the "sides" of the triangle. Part 2 is a symbolic way of stating that adding two adjacent elements in one row gives an element in the next row.

Example 2

Find a solution to the equation $\binom{x}{y} = \binom{7}{5} + \binom{7}{6}$.

Solution 1 Apply Part 2 of the definition of Pascal's Triangle. Here $n = 7$ and $r = 5$. Substituting these values into Part 2, we have

$$\binom{7}{5} + \binom{7}{6} = \binom{8}{6}.$$

So, $x = 8$ and $y = 6$.

Solution 2 Find $\binom{7}{5}$ and $\binom{7}{6}$ in Pascal's triangle and add the results.

$\binom{7}{5}$ is the 6th element in row 7. So $\binom{7}{5} = 21$. $\binom{7}{6}$ is the 7th element in row 7, so $\binom{7}{6} = 7$. $21 + 7 = 28$.

Locate where 28 appears in Pascal's Triangle.

28 is the 3rd element in row 8 and the 7th element in row 8. So, $x = 8$ and $y = 2$ is one solution, and $x = 8$ and $y = 6$ is another solution.

Entries in Pascal's Triangle

There is a very close connection between combinations and the elements in the rows of Pascal's Triangle.

Activity

Step 1 Calculate $_4C_0$, $_4C_1$, $_4C_2$, $_4C_3$, and $_4C_4$.

Step 2 How are the results of Step 1 related to Pascal's Triangle?

Step 3 Find $_7C_0$, $_7C_1$, $_7C_2$, $_7C_3$, $_7C_4$, $_7C_5$, $_7C_6$, and $_7C_7$.

Step 4 How are the results of Step 3 related to Pascal's Triangle?

Step 5 Generalize Steps 1–4.

The generalization of the Activity is stated below. It was first proved by the famous English mathematician and physicist Isaac Newton in the 17th century.

Pascal's Triangle Explicit Formula

If n and r are integers with $0 \leq r \leq n$, then $\binom{n}{r} = {}_nC_r = \frac{n!}{r!(n-r)!}$.

Proof To show that $\binom{n}{r} = \frac{n!}{r!(n-r)!}$, it is enough to show that the factorial expression $\frac{n!}{r!(n-r)!}$ satisfies the relationships involving $\binom{n}{r}$ in the recursive definition of Pascal's Triangle.

(1) When $n \geq 0$, does the expression $\frac{n!}{0!(n-0)!}$ equal the expression $\frac{n!}{n!(n-n)!}$ and equal 1? Yes, since $\frac{n!}{0!(n-0)!} = \frac{n!}{0!n!} = \frac{n!}{1 \cdot n!} = 1$, and $\frac{n!}{n!(n-n)!} = \frac{n!}{n!0!} = \frac{n!}{n! \cdot 1} = 1$.

Thus, the formula works for the "sides" of Pascal's triangle.

(2) To prove that the expression $\frac{(n+1)!}{(r+1)!(n-r)!}$ is the sum of the expressions $\frac{n!}{r!(n-r)!}$ and $\frac{n!}{(r+1)!(n-r-1)!}$, use a calculator or CAS. Enter $_nC_r + {}_nC_{r+1}$.

This CAS displays $\frac{(n+1) \cdot n!}{(r+1) \cdot r! \cdot (n-r)!}$.

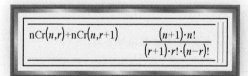

Using the definition of factorial,

$$\frac{(n+1) \cdot n!}{(r+1) \cdot r! \cdot (n-r)!} = \frac{(n+1)!}{(r+1)!(n-r)!}.$$

The right side is an expression for $\binom{n+1}{r+1}$. So, this explicit formula gives the same sequence as the recursive formula that defines Pascal's Triangle.

The result of all this is an exceedingly useful fact: *The elements in row n of Pascal's triangle are the numbers of combinations possible from n things taken 0, 1, 2, ..., n at a time.* So, you do not need to calculate all the rows of Pascal's triangle to get the next row. You can use your knowledge of combinations.

GUIDED

Example 3

Find $\binom{8}{5}$.

Solution 1 Use the Pascal's Triangle Explicit Formula.

$$\binom{8}{5} = \frac{?!}{?!(?-?)!} = \underline{\quad?\quad}$$

Solution 2 $\binom{8}{5}$ is the __?__th element in row __?__ of Pascal's triangle. From the second page of the lesson, it is __?__.

Questions

COVERING THE IDEAS

1. When and where did the array known as Pascal's Triangle first appear?

2. When and where did Pascal live?

3. Explain how entries in a row of Pascal's Triangle can be used to obtain entries in the next row.

4. Write row 11 of Pascal's Triangle.

5. Write row 5 of Pascal's Triangle using $\binom{n}{r}$ notation.

6. **Fill in the Blanks** The element $\binom{16}{8}$ is the __?__ element in row __?__ of Pascal's Triangle.

In 7–12, calculate the element of Pascal's Triangle and give its location in the triangle.

7. $\binom{6}{4}$

8. $\binom{12}{3}$

9. $_{10}C_0$

10. $\binom{8}{8}$

11. $\binom{7}{5}$

12. $_{18}C_{10}$

13. Calculate the 3rd element in the 100th row of Pascal's triangle.

APPLYING THE MATHEMATICS

14. Simplify $\binom{n}{n-1}$.

In 15 and 16, find a solution to the equation.

15. $\binom{9}{4} + \binom{9}{5} = \binom{x}{y}$

16. $\binom{11}{5} + \binom{a}{b} = \binom{12}{5}$

17. **a.** Find the sum of the elements in each of the rows 1 through 6 of Pascal's triangle.

    b. Based on the results of Part a, what do you think is the sum of the elements in row 7?

    c. Write an expression for the sum of the elements in row n of Pascal's Triangle.

18. What sequence is defined by the second element of each row in Pascal's Triangle?

19. Where in Pascal's Triangle can the sequence of triangular numbers 1, 3, 6, 10, 15, ... $\frac{n(n+1)}{2}$, ... be found?

20. Where in Pascal's Triangle can the sequence of series with terms the first n triangular numbers 1, 1 + 3, 1 + 3 + 6, 1 + 3 + 6 + 10, ... be found?

REVIEW

21. How many golf foursomes can be formed from twelve people? **(Lesson 13-4)**

22. Consider drawing cards at random from a standard 52-card deck.

    a. In how many ways can you draw one card?

    b. How many different pairs of cards can you draw if order does not matter?

    c. What are the chances of drawing the four aces on your first 4 draws? **(Lesson 13-4)**

23. Will had a mean of 86 on five tests. After the lowest test score was dropped, his mean was 93. What was the score that was dropped? **(Lesson 13-3)**

24. Give the first 8 terms of the sequence $a_n = \sin(n \cdot 45°)$. **(Lessons 10-1, 1-8)**

25. A square has double the area of a circle. The square has side length 10. What is the radius of the circle? **(Lesson 6-2)**

26. Expand. **(Lesson 6-1)**
   a. $(a + b)^2$
   b. $3(x - 7)^2$
   c. $(3p + 6q)^2 - (3p - 6q)^2$

27. Explain why $\begin{bmatrix} 3 & 1 \\ 6 & 2 \end{bmatrix}$ does not have an inverse. **(Lesson 5-5)**

In 28–30 solve. (Lesson 1-6)

28. $3y + 60 = 5y + 42$

29. $\frac{4}{y} + \frac{8}{y} = 5$

30. $0.05x + 0.1(2x) + 0.25(100 - 3x) = 20$

EXPLORATION

31. There are six elements surrounding each element not on a side of Pascal's Triangle. For instance, around 15 in row 6 are the elements 5, 10, 20, 35, 21, and 6. In 1969, an amazing property about the product of these six elements was discovered by Verner Hoggatt and Walter Hansell of San Jose State University.

1	5	10	10	5	1			row 5
1	6	15	20	15	6	1		row 6
1	7	21	35	35	21	7	1	row 7

   a. Find the product of all the elements surrounding the number 15 in row 6.
   b. Repeat for the number 3 in row 3.
   c. Find the surrounding product for each element not on a side of row 5 of the triangle.
   d. Describe the pattern you see in your products in Parts a through c.
   e. Support your answer to Part d by calculating the product for two elements in row 8 of the triangle.

QY ANSWERS

1. 1 10 45 120 210 252 210 120 45 10 1

2. $\binom{8}{0}$ $\binom{8}{1}$ $\binom{8}{2}$ $\binom{8}{3}$ $\binom{8}{4}$ $\binom{8}{5}$ $\binom{8}{6}$ $\binom{8}{7}$ $\binom{8}{8}$

Lesson

13-6

The Binomial Theorem

Vocabulary

binomial coefficients

▶ **BIG IDEA** The nth row of Pascal's Triangle contains the coefficients of the terms of $(a + b)^n$.

You have seen patterns involving squares of binomials in many places in this book. In this lesson we examine patterns involving the coefficients of higher powers.

Mental Math

Expand.

a. $(r - s)^2$

b. $(7p + 9)^2$

c. $(2ab - 3c)^2$

d. $k^4(5m^2 + n^3)^2$

Activity

MATERIALS CAS

Step 1 Expand the binomials in column 1 on a CAS and record the results in column 2 of a table like the one below.

Power of $(a + b)$	Expansion of $(a + b)^n$	Sum of Exponents of the Variables in Each Term
$(a + b)^0$	1	0
$(a + b)^1$	$a + b$	1
$(a + b)^2$	$a^2 + 2ab + b^2$	2
$(a + b)^3$	?	?
$(a + b)^4$	?	?
$(a + b)^5$	?	?
$(a + b)^6$	?	?

$$\text{expand}\left((a+b)^2\right) \qquad a^2+2\cdot a\cdot b+b^2$$

Step 2 In column 3 of the table above, record the sum of the exponents of the variables in each term of the expansion of $(a + b)^n$.

Step 3 Set up a table like the one at the right. Record the coefficients of the terms in each expansion in column 2.

(continued on next page)

Power of $(a + b)$	Coefficients in the Expansion of $(a + b)^n$	Exponents of a in the Expansion	Exponents of b in the Expansion
$(a + b)^0$	1	0	0
$(a + b)^1$	1, 1	1, 0	0, 1
$(a + b)^2$	1, 2, 1	2, 1, 0	0, 1, 2
$(a + b)^3$	?	?	?
$(a + b)^4$	?	?	?
$(a + b)^5$	?	?	?
$(a + b)^6$	?	?	?

Step 4 What do you notice about the coefficients of the expansions of $(a + b)^n$?

Step 5 Write the coefficients of the expansion of $(a + b)^5$ using $\binom{n}{r}$ notation.

Step 6 Record the exponents of the powers of a and b in each term in the binomial expansions in the rightmost two columns of the table in Step 3.

Step 7 What do you notice about the exponents of a in each expansion of $(a + b)^n$? What do you notice about the exponents of b in each expansion of $(a + b)^n$?

The Activity reveals several properties of the expansion of $(a + b)^n$. Knowledge of these properties makes expanding $(a + b)^n$ easy.

- In each term of the expansion, the sum of the exponents of a and b is n.

- All powers of a occur in decreasing order from n to 0, while all powers of b occur in increasing order from 0 to n.

- If the power of b is r, then the coefficient of the term is $\binom{n}{r} = {_nC_r}$.

As a consequence of these properties, binomial expansions can be written using the $\binom{n}{r}$ symbol.

$$(a + b)^0 = \binom{0}{0}$$

$$(a + b)^1 = \binom{1}{0}a + \binom{1}{1}b$$

$$(a + b)^2 = \binom{2}{0}a^2 + \binom{2}{1}ab + \binom{2}{2}b^2$$

$$(a + b)^3 = \binom{3}{0}a^3 + \binom{3}{1}a^2b + \binom{3}{2}ab^2 + \binom{3}{3}b^3$$

$$\vdots \qquad \qquad \vdots$$

The information above is summarized in a famous theorem that was known to Omar Khayyam, the Persian poet, mathematician, and astronomer, who died around the year 1123. Of course, he did not have the notation we use today. Our notation makes it clear that the nth row of Pascal's triangle contains the coefficients of $(a + b)^n$.

Omar Khayyam

Binomial Theorem

For all complex numbers a and b, and for all integers n and r with $0 \leq r \leq n$,

$$(a + b)^n = \sum_{r=0}^{n} \binom{n}{r} a^{n-r} b^r.$$

A proof of the Binomial Theorem requires mathematical induction, a powerful proof technique beyond the scope of this book. You will see this proof in a later course.

Example 1

Expand $(a + b)^7$.

Solution First, write the powers of a and b in the form of the answer. Leave spaces for the coefficients.

$$(a + b)^7 = \underline{\quad}a^7 + \underline{\quad}a^6b + \underline{\quad}a^5b^2 + \underline{\quad}a^4b^3 +$$
$$\underline{\quad}a^3b^4 + \underline{\quad}a^2b^5 + \underline{\quad}ab^6 + \underline{\quad}b^7$$

Second, fill in the coefficients using $\binom{n}{r}$ notation.

$$(a + b)^7 = \binom{7}{0}a^7 + \binom{7}{1}a^6b + \binom{7}{2}a^5b^2 + \binom{7}{3}a^4b^3 +$$
$$\binom{7}{4}a^3b^4 + \binom{7}{5}a^2b^5 + \binom{7}{6}ab^6 + \binom{7}{7}b^7$$

Finally, evaluate the coefficients, either by referring to row 7 of Pascal's Triangle or by using the formula $\binom{n}{r} = \frac{n!}{r!(n-r)!}$.

$$(a + b)^7 = a^7 + 7a^6b + 21a^5b^2 + 35a^4b^3 + 35a^3b^4 +$$
$$21a^2b^5 + 7ab^6 + b^7$$

The Binomial Theorem can be used to expand a variety of expressions.

Example 2

Expand $(3x - 4y)^3$.

Solution 1 The expansion follows the form of $(a + b)^3$.
$$(a + b)^3 = \binom{3}{0}a^3 + \binom{3}{1}a^2b + \binom{3}{2}ab^2 + \binom{3}{3}b^3.$$

(continued on next page)

Think of $3x$ as a and $-4y$ as b and substitute.

$(3x - 4y)^3 = 1(3x)^3 + 3(3x)^2(-4y) + 3(3x)(-4y)^2 + 1(-4y)^3$
$= 27x^3 - 108x^2y + 144xy^2 - 64y^3$

Solution 2 Expand the binomial on a CAS.

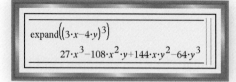

GUIDED

Example 3

Expand $(2x^2 + 1)^4$.

Solution Think of $2x^2$ as a and 1 as b. Then follow the form of $(a + b)^4$.

$(2x^2 + 1)^4 = (\underline{\ ?\ })(2x^2)^{\underline{\ ?\ }} + (\underline{\ ?\ })(2x^2)^{\underline{\ ?\ }} \cdot (1)^{\underline{\ ?\ }} +$
$\qquad (\underline{\ ?\ })(2x^2)^{\underline{\ ?\ }} \cdot (1)^{\underline{\ ?\ }} + (\underline{\ ?\ })(2x^2)^{\underline{\ ?\ }} \cdot (1)^{\underline{\ ?\ }} +$
$\qquad (\underline{\ ?\ }) \cdot (1)^{\underline{\ ?\ }}$
$\qquad = \underline{\ ?\ }x^{\underline{\ ?\ }} + \underline{\ ?\ }x^{\underline{\ ?\ }} + \underline{\ ?\ }x^{\underline{\ ?\ }} + \underline{\ ?\ }x^{\underline{\ ?\ }} + \underline{\ ?\ }$

Check Substitute a value for x in the binomial power and in the expansion. The two results should be equal.

You can also use the Binomial Theorem to quickly find any term in the expansion of a binomial power without writing the full expansion.

Example 4

Find the 8th term in the expansion of $(a + b)^{20}$.

Solution The formula $(a + b)^{20} = \sum_{r=0}^{20} \binom{20}{r} a^{20-r}b^r$ gives the full

expansion of the binomial. Because r starts at 0, the 8th term is when $r = 7$.

$$\binom{20}{7} a^{20-7}b^7 = 77{,}520\,a^{13}b^7$$

The 8th term in the expansion is $77{,}520a^{13}b^7$.

 QY

▶ **QY**

Find the 13th term in the expansion of $(x + 2)^{15}$.

Due to their use in the Binomial Theorem, the numbers in Pascal's Triangle are sometimes called **binomial coefficients.** The Binomial Theorem has a surprising number of applications in estimation, counting problems, probability, and statistics. You will study these applications in the remainder of this chapter.

Questions

COVERING THE IDEAS

1. **a.** Expand $(x + y)^2$.
   b. What are the coefficients of the terms in the expansion of $(x + y)^3$?

In 2–4, expand each binomial power.

2. $(a - 3b)^3$

3. $\left(\frac{1}{2} - m^2\right)^4$

4. $(2x + 5y)^3$

In 5 and 6, find the 5th term in the binomial expansion.

5. $(x + y)^{10}$

6. $(a - 3b)^8$

In 7 and 8, find the second-to-last term in the binomial expansion. (This term is called the *penultimate* term.)

7. $(5 - 2n)^9$

8. $(3j + k)^{12}$

APPLYING THE MATHEMATICS

In 9 and 10, convert to an expression in the form $(a + b)^n$.

9. $\displaystyle\sum_{r=0}^{14} \binom{14}{r} x^{14-r} 2^r$

10. $\displaystyle\sum_{i=0}^{n} \binom{n}{i} y^{n-i} (-3w)^i$

11. Multiply the binomial expansion for $(a + b)^3$ by $a + b$ to check the expansion for $(a + b)^4$.

12. **a.** Multiply and simplify $(a^2 + 2ab + b^2)(a^2 + 2ab + b^2)$.
    b. Your answer to Part a should be a power of $a + b$. Which one? Explain your answer.

In 13 and 14, use this information. The Binomial Theorem can be used to approximate some powers quickly without a calculator. Here is an example.

$$(1.002)^3 = (1 + 0.002)^3$$
$$= 1^3 + 3 \cdot 1^2 \cdot (0.002) + 3 \cdot 1 \cdot (0.002)^2 + (0.002)^3$$
$$= 1 + 0.006 + 0.000012 + 0.000000008$$
$$= 1.006012008$$

Because the last two terms in the expansion are so small, you may ignore them in an approximation. So $(1.002)^3 \approx 1.006$ to the nearest thousandth.

13. Show how to approximate $(1.003)^3$ to the nearest thousandth without a calculator. Check your answer with a calculator.

14. Show how to approximate $(1.001)^4$ to nine decimal places without a calculator.

15. **a.** Evaluate 11^0, 11^1, 11^2, 11^3, and 11^4. How are these numbers related to Pascal's Triangle?

   b. Expand $(10 + 1)^4$ using the Binomial Theorem.

   c. Use the Binomial Theorem to calculate 11^5.

REVIEW

True or False In 16 and 17, explain your reasoning.

16. $\binom{99}{17}$ is an integer. **(Lesson 13-5)**

17. $\dfrac{n!}{(n-2)!}$ is always an integer when $n \geq 2$. **(Lesson 13-4)**

18. Simplify: ${}_9C_0 + {}_9C_1 + {}_9C_2 + {}_9C_3 + {}_9C_4 + {}_9C_5 + {}_9C_6 + {}_9C_7 + {}_9C_8 + {}_9C_9$. **(Lesson 13-4)**

19. Consider the ellipse with equation $\dfrac{x^2}{15} + \dfrac{y^2}{26} = 1$.
    (Lessons 12-5, 12-4)

   a. Give the length of its major axis.

   b. Give the coordinates of the endpoints of its major and minor axes.

   c. Find the coordinates of its foci F_1 and F_2.

   d. If P is a point on this ellipse, find $PF_1 + PF_2$.

   e. Find the area of the ellipse.

20. Paola has been saving to buy a condo for five years. At the beginning of the first year, she placed $2200 in a savings account that pays 3.7% interest annually. At the beginning of the second, third, fourth, and fifth years, she deposited $2350, $2125, $2600, and $2780, respectively, into the same account. At the end of the five years, does Paola have enough money in the account to make a $15,000 down payment? If not, how much more does she need? **(Lesson 11-1)**

In 21–24, solve. **(Lessons 9-9, 9-7, 9-5)**

21. $\log_7 y = 2 \log_7 13$

22. $2 \ln 13 = \ln x$

23. $\log z = 5$

24. $\ln(3x) = \ln 2 + \ln 18$

EXPLORATION

25. The expansion of $(a + b)^3$ has 4 terms.

   a. How many terms are in the expansion of $(a + b + c)^3$?

   b. How many terms are in the expansion of $(a + b + c + d)^3$?

   c. Generalize these results.

QY ANSWER

$\binom{15}{12}x^{15-12} \cdot 2^{12} =$

$1{,}863{,}680x^3$

Lesson 13-7

Probability and Combinations

Vocabulary

trial

binomial experiment

▶ **BIG IDEA** The probability of an event occurring *r* times in *n* trials of a *binomial experiment* can be found by calculating combinations.

Pascal originally conceived of the triangle named in his honor in the context of probability problems. Activity 1 can help you connect Pascal's Triangle and probabilities.

Mental Math

Calculate

a. $_{1000}C_{999}$

b. $_{1000}C_{1000}$

c. $_{1000}C_1$

d. $_{1000}C_2$

Activity 1

MATERIALS penny or other coin

The task is to estimate the probability of getting exactly 2 heads in 4 tosses of the coin by repeating an experiment a large number of times.

Step 1 Make a table with headings as shown at the right. Include ten rows for trials 1 through 10. Flip a coin 4 times and record the results. For example, if the coin came up heads, then tails, then tails, and then heads, write HTTH in the Sequence column and 2 in the Number of Heads column.

Trial	Sequence	Number of Heads
1	?	?
2	?	?
⋮	?	?

Step 2 Repeat Step 1 nine more times until you have filled the entire table. Then tally the number of times you got 0 heads, 1 head, 2 heads, 3 heads, and 4 heads.

Step 3 Combine your results with others in your class. Then compute relative frequencies of each number of heads for the class as a whole. Graph the five ordered pairs, where each ordered pair is of this form: (number of heads in 4 flips, relative frequency of that number of heads). What is your class's relative frequency for getting 2 heads in 4 flips?

Many people are surprised to find out that the relative frequency of getting 2 heads in 4 flips is usually not too close to 50%. But it is not difficult to compute the probability if the coins are fair. Activity 2 explores that computation.

Activity 2

Step 1 Write every possible sequence of four H's and T's. Organize the sequences in a table like the one at the right. Three sequences have been written in the table to get you started. Then count the number of sequences in each cell and write the counts in the bottom row of the table.

n	0	1	2	3	4
Sequences of 4 Tosses with n Heads	?	THTT HTTT ?	HTTH ?	?	?
y = Number of Sequences with n Heads	?	?	?	?	?

Step 2 Graph the five points (n, y), where y is the number of sequences of 4 H's and T's that have n heads.

Step 3 Compare the shape of your graph in Step 2 with the shape of the graph in Step 3 of Activity 1.

You should find that the shapes of the graphs in the Activities are quite similar. In Activity 2 you should have found 6 different sequences of 2 heads and 2 tails:

HHTT, HTHT, HTTH, THHT, THTH, TTHH.

If the coin is fair, for a single toss, the probability $P(H)$ of heads and the probability of $P(T)$ tails each equal $\frac{1}{2}$. So, each of these 6 sequences has the same probability, $\frac{1}{2} \cdot \frac{1}{2} \cdot \frac{1}{2} \cdot \frac{1}{2} = \frac{1}{16}$, regardless of the order of the heads and tails. Therefore, the probability of getting 2 heads in 4 tosses of a fair coin is $6 \cdot \frac{1}{16} = \frac{3}{8} = 0.375$. This should be close to your class's relative frequency from Step 3 of Activity 1.

You can use the same idea to find the probability of obtaining any number of heads in any number of tosses of a fair coin.

Example 1

Suppose a fair coin is flipped 6 times. What is the probability of obtaining exactly 2 heads?

Solution 1 First count the number of sequences of 6 flips with exactly 2 heads. You could list the sequences by hand, but the Combination Counting Formula gives a faster way to count. Number the six flips 1, 2, 3, 4, 5, and 6. Then choosing two flips to be heads is equivalent to choosing a two-element subset of {1, 2, 3, 4, 5, 6}. The number of two-element subsets of a six-element set is

$$_6C_2 = \binom{6}{2} = \frac{6!}{2!(6-2)!} = 15.$$

Now compute the probability of each sequence occurring. (Remember, each sequence has the same probability.)

$$P(H) \cdot P(H) \cdot P(T) \cdot P(T) \cdot P(T) \cdot P(T)$$
$$= P(H)^2 \cdot P(T)^4$$
$$= \left(\frac{1}{2}\right)^2 \left(\frac{1}{2}\right)^4 = \frac{1}{64}.$$

So, the probability of getting exactly 2 heads in 6 tosses of a fair coin is $_6C_2 \cdot P(H)^2 \cdot P(T)^4 = 15 \cdot \frac{1}{64} = \frac{15}{64}$.

Solution 2 The 15 sequences with two heads are shown below.

HHTTTT	HTHTTT	HTTHTT	HTTTHT	HTTTTH
THHTTT	THTHTT	THTTHT	THTTTH	TTHHTT
TTHTHT	TTHTTH	TTTHHT	TTTHTH	TTTTHH

There are two outcomes for each of the flips (*H* or *T*), so the total number of sequences is $2 \cdot 2 \cdot 2 \cdot 2 \cdot 2 \cdot 2 = 2^6 = 64$. Thus, the probability of getting exactly two heads is $\frac{15}{64}$.

Connecting Probabilities with Combinations

Notice that the number of sequences of 2 heads in 6 flips equals the binomial coefficient $\binom{6}{2}$, which is the third number in row 6 of Pascal's Triangle. This is no accident: If an experiment is repeated *n* times , the number of possible sequences with *r* successes (and *n* − *r* failures) is $\binom{n}{r}$, because determining such a sequence is equivalent to picking an *r*-element subset from {1, 2, ..., *n*}.

What happens if the two outcomes are not equally likely? Example 2 addresses this issue.

Example 2

Two generations ago, around 1950, the probability that a birth would be a multiple birth (twins, triplets, etc.) was about $\frac{1}{87}$. Mrs. Pereskier gave birth five times. Three of the births resulted in twins. What is the probability of this happening if multiple births occur at random?

Solution Let *M* be a multiple birth and *S* be a single birth. In this case, $P(M) = \frac{1}{87}$, so $P(S) = \frac{86}{87}$.

(continued on next page)

The Multiple Birth Family Reunion in Mexico is an annual event for families of multiples.

There are $\binom{5}{3} = 10$ ways that 3 of the 5 births could be multiple births.

One of these ways yields the sequence SSMMM, which is what happened in the Pereskier family. The probability of this sequence is

$$P(S) \cdot P(S) \cdot P(M) \cdot P(M) \cdot P(M) = (P(S))^2 \cdot (P(M))^3$$

$$= \left(\frac{86}{87}\right)^2 \cdot \left(\frac{1}{87}\right)^3$$

$$= \frac{7396}{4{,}984{,}209{,}207}$$

$$\approx 0.00000148.$$

There are 10 such sequences possible, so the probability of 3 multiple births in 5 births at that time was about $10 \cdot (0.00000148)$, or 0.0000148, or about 15 in a million.

This is a very low probability, and the actual occurrence of multiple births in some families is much higher than would be expected if they occurred randomly. This is how doctors realized that a tendency towards multiple births runs in some families.

 QY

▶ QY

Refer to Example 2. What was the probability of exactly 1 multiple birth in 5 births?

Binomial Experiments

The situations of Examples 1 and 2 satisfy four criteria.

1. A task, called a **trial,** is repeated n times, where $n \geq 2$.

2. Each trial has outcomes that can be placed in one of only two categories, sometimes called "success" and "failure."

3. The trials are independent, that is, the probability of success on one trial is not affected by the results of earlier trials.

4. Each trial has the same probability of success.

When these four criteria are satisfied, the situation is called a **binomial experiment.** In a binomial experiment, the following properties hold:

•   In n trials, there are $\binom{n}{r}$ possible sequences of r successes and $n - r$ failures.

•   If the probability of success in any one trial is p, then the probability of failure is $q = 1 - p$ because success and failure are the only possible outcomes and they are mutually exclusive. Recall that categories are mutually exclusive if it is impossible for an element to belong to more than one of the categories.

- Because each trial has the same probability of success, the probability of any particular sequence of r successes and $n - r$ failures is $p^r q^{n-r}$.

- Because the trials are independent, the order in which successes and failures occur does not affect their probabilities. Therefore, the probability of *each* sequence of r successes and $n - r$ failures is $p^r q^{n-r}$, and their combined probability is $\binom{n}{r} p^r q^{n-r}$.

This argument proves the following theorem.

Binomial Probability Theorem

Suppose an experiment has an outcome with probability p, so that the probability the outcome does not occur is $q = 1 - p$. Then in n independent repetitions of the experiment, the probability that the outcome occurs r times is $\binom{n}{r} p^r q^{n-r}$.

These probabilities are often called *binomial probabilities* because of their connection with binomial coefficients.

GUIDED

Example 3

Suppose you roll two dice on each of three successive turns in a game. Compute the probability of rolling two sixes 0 times, 1 time, 2 times, and 3 times in those turns.

Solution Organize the computations in a table. Let rolling two sixes be a success and rolling anything else be a failure. If the die is fair, $P(\text{success})$ = probability of rolling a six on one die • probability of rolling a six on the other die = $\frac{1}{6} \cdot \underline{\quad?\quad} = \underline{\quad?\quad}$. To compute the final result, substitute $\underline{\quad?\quad}$ for p and $1 - \underline{\quad?\quad} = \underline{\quad?\quad}$ for q into the Binomial Probability Theorem.

Number of Successes	Number of Failures	Binomial Expression	Probability
0	3	$\binom{3}{0} p^0 q^3$	$\underline{\ ?\ } \approx \underline{\ ?\ }$
1	2	$\binom{3}{1} p^1 q^{\underline{?}}$	$\underline{\ ?\ }$
2	1	$\underline{\ ?\ } p^{\underline{?}} q^{\underline{?}}$	$\underline{\ ?\ }$
3	$\underline{\ ?\ }$	$\underline{\ ?\ }$	$\underline{\ ?\ }$

In Example 3, notice that the expressions in the Binomial Expression column are also the terms in the expansion of $(p + q)^3$. In general, the probability of r successes in n trials is the $p^r q^{n-r}$ term in the expansion of $(p + q)^n$. Because $p + q = 1$, $(p + q)^n = 1^n = 1$. That is, the sum of the probabilities of all possible outcomes computed using the Binomial Probability Theorem is 1.

Questions

COVERING THE IDEAS

In 1 and 2, a fair coin is flipped 5 times. A sequence of 5 H's (heads) and T's (tails) is recorded.

1. a. How many different sequences are possible with exactly 2 H's?
    b. How many different sequences are possible with exactly 2 T's?
    c. What is the probability of flipping exactly 2 heads?

2. a. How many different sequences are possible with exactly 4 T's?
    b. What is the probability of flipping exactly 4 tails?

3. About 1 in 35 births today is a multiple birth. Suppose there are 4 births in a family. What is the probability that exactly 1 of them is a multiple birth?

In 4 and 5, a fair coin is flipped 8 times. Give the probability of each event.

4. getting exactly 8 heads
5. getting exactly 4 heads

6. What is the probability of getting exactly r heads in n tosses of a fair coin?

7. Suppose that a coin is biased so that there is a 55% chance that the coin will show tails when tossed. Find the probability of each event.
    a. The coin shows heads when tossed.
    b. When tossed twice, the coin shows heads the first time and tails the second time.
    c. When tossed twice, the coin shows heads once and tails once.

8. Use the information in Question 3. What is the probability that, of the 3 births in a family today, at least 1 is a multiple birth?

In 9 and 10, suppose a stoplight is red for 45 seconds and green for 30 seconds. Suppose, also, that every day for a full week you get to this stoplight at a random time. What is the probability that the stoplight will be red

9. every day of the full week?
10. 3 of the 5 days of the work week?

The stoplight was invented by Garrett Augustus Morgan, Sr., an African American born in Paris, Kentucky. He received a patent for the stoplight on November 20, 1923.

APPLYING THE MATHEMATICS

11. An O-ring is a circular mechanical seal (usually made of rubber) that generally prevents leakage between two compressed objects. A manufacturer of 14-mm diameter O-rings claims that 97.5% of the O-rings he manufactures are less than 14.5 mm in diameter. In a sample of 10 such O-rings, you find that three O-rings have a diameter greater than 14.5 mm. If the manufacturer's claim is correct, what is the probability of this outcome?

12. Slugger Patty McBattie has a batting average of 0.312. Use this average as her probability of getting a hit in a particular time at bat. In a game where she bats 5 times, what is the probability she gets exactly 2 hits?

In **13** and **14**, suppose you have two minutes left to fill in the last 8 questions on a multiple-choice test. You can eliminate enough answers so your probability of guessing the correct answer to any question is $\frac{1}{4}$.

13. What is the probability you get exactly 5 questions correct?

14. What is the probability you get 2 or more questions correct?

In **15** and **16**, a student is given the quiz at the right.

15. Using R for right and W for wrong, list all possible ways the quiz might be answered. For example, getting all four right is coded RRRR. Assume that an unanswered question is wrong.

16. Assuming that the student guesses on each item and that the probability of guessing the right answer is $\frac{1}{2}$, calculate each probability.

    a. The student gets all 4 correct.

    b. The student gets exactly 2 correct.

    c. The student gets at least 2 correct.

> **QUIZ**
> 1. Which is farther north, Anchorage, Alaska or Helsinki, Finland?
> 2. Is the 1,000,000th decimal place of π 5 or greater?
> 3. Is the area of Central Park in New York greater than 5 times the area of Hyde Park in London?
> 4. Did Euler die before or after Gauss was born?

REVIEW

17. Expand $(a + b)^5$. **(Lesson 13-6)**

18. Use the Binomial Theorem to approximate $(1.001)^{10}$ to fifteen decimal places. **(Lesson 13-6)**

19. The mean of three consecutive terms in a geometric sequence is 35. The first of these terms is 15. What might the other terms be? **(Lessons 13-3, 7-5)**

20. A snail is crawling straight up a wall. The first hour it climbs 16 inches, the second hour it climbs 12 inches, and each succeeding hour it climbs $\frac{3}{4}$ the distance it climbed the previous hour. Assume this pattern holds indefinitely.

    a. How far does the snail climb during the 7th hour?

    b. What is the total distance the snail has climbed in 7 hours? **(Lesson 13-2)**

In 21 and 22, expand and simplify. (Lesson 11-2)

21. $(t^2 - r^2)(t^2 + r^2)$

22. $(4a^2 + 2a + 1)(2a - 1)$

23. In the triangle at the right, find $\frac{\sin \theta}{\cos \theta}$. **(Lesson 10-1)**

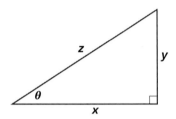

24. Suppose an experiment begins with 120 bacteria and that the population of bacteria doubles every hour.

    a. About how many bacteria will there be after 3 hours?

    b. Write a formula for the number y of bacteria after x hours. **(Lesson 9-1)**

EXPLORATION

25. a. Use a random-number generator in a spreadsheet to simulate 50 trials of tossing 4 coins simultaneously. Record your results.

    b. Calculate the relative frequency of each of the following outcomes: 0 heads, exactly 1 head, exactly 2 heads, exactly 3 heads, 4 heads.

    c. How closely do your results agree with predictions based on the Binomial Probability Theorem? Do you think your random number generator program is a good one? Explain why or why not.

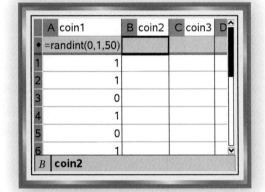

13-8

Lotteries

Vocabulary

lottery

unit fraction

▶ **BIG IDEA** The probability of winning a lottery prize can be calculated using permutations and combinations.

A **lottery** is a game or procedure in which prizes are distributed by pure chance. The simplest lotteries are raffles in which you buy tickets that are put in a bin, and winning tickets are picked from that bin. In recent years, however, more complicated lotteries have been designed. These lotteries pay out large amounts of money to a few individuals in order to attract participants. Today, several countries run lotteries themselves or allow private lotteries. In the United States, about 94% of people live in states that run lotteries, and participants must be 18 years of age or older.

Lotteries are designed to make money, so they always take in more than they pay out. This means that many more people will lose money in a lottery than will win. Still, the possibility, however remote, of winning a huge amount of money is very attractive.

Consider a typical lottery. To participate in the Mega Millions game, which is played across several states, a participant pays $1 and picks five numbers out of the set of consecutive integers {1, 2, 3, ..., 56} and one number (called the "Mega Ball") from the set of consecutive integers {1, 2, 3, ..., 46}. Twice a week, five balls are picked at random from balls numbered 1 through 56, and one Mega Ball is randomly picked from a different set of 46 balls. These balls show the winning numbers. For instance, the winning numbers in one Mega Millions game were 1, 5, 13, 18, 33, and 30. The people who pick all six winning numbers split the grand prize, which is always at least $12 million.

Mental Math

Given that $\sin 52° \approx 0.788$, find

a. $\sin(-52°)$

b. $\sin 232°$

c. $\sin 128°$

d. $\cos 38°$

GUIDED

Example 1

What is the probability of picking the six winning numbers in the Mega Millions lottery?

Solution The number of possible combinations for the first five numbers that could be chosen is $\begin{pmatrix} 56 \\ ? \end{pmatrix} = \frac{56!}{? \cdot ?} = \underline{\quad?\quad}$.

(continued on next page)

There are 46 choices for the Mega Ball number. So there are a total of __?__ · 46 = __?__ possible combinations of numbers to choose. Because each combination has the same chance of being drawn, the probability that a particular combination will appear is $\frac{1}{?}$. So, the probability of picking the six winning numbers is $\frac{1}{?}$ or about __?__.

The answer to Example 1 explains why there is often no winner in a lottery. The chance of choosing all the winning numbers is quite small.

Because the probability of winning the grand prize is so low, most lotteries give smaller prizes to participants who pick almost all of the numbers. For instance, in Mega Millions, there is a *much* smaller prize for picking the five regular winning numbers but not the Mega Ball number. There is an even smaller prize given for picking four of the five regular winning numbers.

Example 2

What is the probability of picking exactly four of the five regular winning numbers in Mega Millions?

Solution Of the five regular numbers, four must be picked from the five winning regular numbers, one must be picked from the 51 incorrect regular numbers that remain, and the Mega Ball number must be chosen incorrectly.

There are $\binom{5}{4}$ different sets of 4 numbers from the 5 winning regular numbers.

There are 51 ways to choose one incorrect number from the remaining regular numbers.

There are 45 non-winning Mega Ball numbers.

Winners are chosen from the 175,711,536 possible combinations found in Example 1. So, the probability of picking four of the five regular winning numbers is

$$\frac{\binom{5}{4} \cdot 51 \cdot 45}{175,711,536} = \frac{11,475}{175,711,536} \approx 0.000065 \approx \frac{1}{15,385}.$$

A **unit fraction** is a simple fraction with 1 in its numerator. In Example 2, the probability 0.000065 is approximated by the unit fraction $\frac{1}{15,385}$ to make it easier to interpret the answer as "about 1 chance in 15,385" of picking four winning balls. This fraction was found by calculating the reciprocal of 0.000065. That reciprocal is about 15,385. The reciprocal of the reciprocal equals the original number.

QY

Because probabilities of winning are so low in many lotteries, people can get discouraged from entering repeatedly. Consequently, states also have lotteries with fewer numbers to match in which more people win, but the payouts are much lower. Some of these lotteries require that three or four digits (from 0 to 9) be matched exactly, in order. Because the order matters, the probability of winning cannot be calculated using combinations.

> ▸ **QY**
>
> A probability of 0.0028 means "about 1 in ___?___."

Example 3

To play Florida's basic "Play 4" lottery, a participant picks four digits, each from 0 to 9, and must match a 4-digit number. What is the probability of matching the four winning digits?

Solution 1 There is a probability of $\frac{1}{10}$ that each digit will be matched. Because these events are independent, the probability of matching all four numbers is $\frac{1}{10} \cdot \frac{1}{10} \cdot \frac{1}{10} \cdot \frac{1}{10}$, or $\frac{1}{10,000}$.

Solution 2 Think of the four digits as forming one number. There are ten thousand numbers from 0000 to 9999. The probability of matching one of these is $\frac{1}{10,000}$.

Some people have computers pick lottery numbers for them. Others study past winning numbers to look for patterns. Still others use their "lucky" numbers or their birthdays. None of these strategies changes the probability of winning. The probability is always tiny, and there is no systematic way to win lotteries like these.

Questions

COVERING THE IDEAS

1. What is a lottery?

In 2 and 3, consider the Mega Millions game.

2. What must a participant do in order to win the grand prize in this lottery?

3. What is the probability of picking only the five regular winning numbers in this lottery?

4. New York's "Take 5" lottery gives participants a chance to win a jackpot by matching five numbers (in any order) from a set of 39. What is the probability of winning this lottery?

5. Refer to Example 3. If Florida's "Play 4" lottery had an option where a participant must match the first three of four numbers (each of which can be any digit from 0 through 9) in exact order, what would be the probability of winning?

APPLYING THE MATHEMATICS

6. Another multi-state lottery, Powerball, is played by picking five numbers from balls numbered 1 to 55 and one additional Powerball number from balls numbered 1 to 42. A smaller prize is awarded to participants who match four of the regular winning numbers and the Powerball number.

   a. In how many ways can a participant win the smaller prize?

   b. What is the probability of winning the smaller prize?

7. The European lottery "Euro Millions" is played by choosing five numbers from a set of 50 and two "star" numbers from a set of nine (the digits 1 to 9). What is the probability of picking all seven winning numbers?

8. Consider New York's "Take 5" lottery in Question 4.

   a. How many different tickets are possible for this lottery? Remember that order does not matter, so all tickets listing the numbers 12, 32, 11, 19, and 6, for instance, are considered the same.

   b. How many tickets from your answer to Part a are losers?

   c. A special promotion advertises that a participant with a losing ticket can enter for a second chance prize. What is the probability of being eligible for this drawing?

9. To play the "Cash 5" lottery in Connecticut, a participant pays $1 and picks five numbers from a set of 35. For 50¢ more, the participant can add a "kicker" number picked from the remaining 30 numbers not already picked. One way to win a kicker prize is to match the kicker number and four out of five of the regular numbers. What is the probability of winning a kicker prize in this way?

10. In the Quinto lottery formerly played in Washington state, a participant picked five cards from a standard 52-card deck. The lottery paid $1000 for a ticket matching four of the five cards, and $20 for a ticket matching three of the five. Does this mean that participants were $\frac{1000}{20} = 50$ times as likely to match three of five winning cards as to match four of five winning cards? Explain your answer.

REVIEW

11. A golden retriever has a litter of 8 puppies. If males and females are equally likely to be in the litter, what is the probability that 4 of the puppies are male and 4 are female? (**Lesson 13-7**)

12. Sam Dunk makes 75% of the free throws he attempts. What is the probability that he will make at least 7 out of 10 free throws? (**Lesson 13-7**)

13. You pick four numbers out of a set of 30. If four numbers are selected at random from this set, what is the probability that the four numbers you picked are selected? (**Lesson 13-7**)

14. Solve $_nC_3 = 84$. (**Lessons 13-4, 11-6**)

In 15 and 16, do not solve the equation.
a. State the number of roots each equation has.
b. State the number of positive roots the equation has. (**Lesson 11-6**)

15. $x^5 + 8x^3 + x = 0$

16. $13t^2 + 6t^7 + it^3 = 14$

17. **Multiple Choice** Choose the equation for an exponential decay function and explain why it is that kind of function. (**Lesson 9-2**)

A $f(x) = \frac{1^x}{5}$ B $f(x) = 5^{-x}$ C $f(x) = 5^x$ D $f(x) = \left(\frac{1}{5}\right)^{-x}$

18. Write the reciprocal of $2 + i$ in $a + bi$ form. (**Lesson 6-9**)

19. Assume that the cost of a spherical ball bearing varies directly as the cube of its diameter. What is the ratio of the cost of a ball bearing 6 mm in diameter to the cost of a ball bearing 3 mm in diameter? (**Lesson 2-3**)

EXPLORATION

20. The term "odds" is often used interchangeably (and often incorrectly) with the term "probability" when discussing probability. Look up information about odds and find out how odds are related to, but are not the same thing as, probability.

Binomial and Normal Distributions

Vocabulary

probability function,
    probability distribution

binomial probability
    distribution, binomial
    distribution

normal distribution

normal curve

standard normal curve

standardized scores

▶ **BIG IDEA**   As the number of trials of a binomial experiment increases, the graphs of the probabilities of each event and the relative frequencies of each event approaches a distribution called a *normal distribution*.

A Binomial Distribution with Six Points

Let $P(n)$ = the probability of n heads in 5 tosses of a fair coin. Then the domain of P is $\{0, 1, 2, 3, 4, 5\}$. By the Binomial Probability Theorem, and because the probability of heads = the probability of tails = $\frac{1}{2}$,

$$P(n) = \left(\frac{1}{2}\right)^5 \binom{5}{n} = \frac{1}{32}\binom{5}{n}.$$

Mental Math

Tell whether each graph is the graph of a function.

a. the image of $y = \sin x$ under R_{90}

b. the image of $y = \cos x$ under $r_{x\text{-axis}}$

c. the image of $x = 12$ under $T_{2, 8}$

d. the image of $x = 12$ under R_3

GUIDED

Example 1

a.   Copy and complete the table shown below.

b.   Graph and label the coordinates of the six points $(n, P(n))$.

Solution

a.   Use the formula for $P(n)$ given above to fill in the second row.

n = Number of Heads	0	1	2	3	4	5
P(n) = Probability of n Heads in 5 Tosses of a Fair Coin	$\frac{1}{32}\binom{5}{0} = \frac{1}{32}$	$\frac{1}{32}\binom{5}{?} = \underline{\ ?\ }$	?	?	?	?

b.   The points are graphed below. Fill in the missing coordinates with values from the table in Part a.

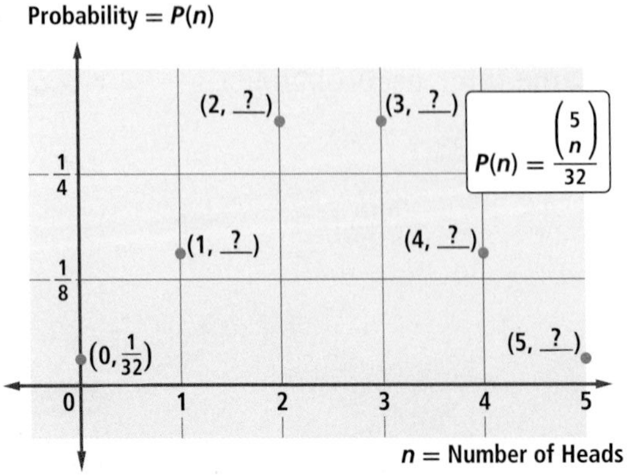

$$P(n) = \frac{\binom{5}{n}}{32}$$

P is a *probability function*. A **probability function**, or **probability distribution**, is a function that maps a set of events onto their probabilities. Because the function P results from calculations of binomial probabilities, it is called a **binomial probability distribution,** or simply a **binomial distribution.**

A Binomial Distribution with Eleven Points

If a fair coin is tossed 10 times, the possible numbers of heads are 0, 1, 2, ..., 10, so there are 11 points in the graph of the corresponding probability function. Again, by the Binomial Probability Theorem with equally likely events, the probability $P(x)$ of tossing x heads is given by

$$P(x) = \left(\frac{1}{2}\right)^{10}\binom{10}{x} = \frac{1}{1024}\binom{10}{x}.$$

The 11 probabilities are easy to calculate because the numerators in the fractions are the numbers in the 10th row of Pascal's triangle. That is, they are binomial coefficients.

n = Number of Heads	0	1	2	3	4	5	6	7	8	9	10
$P(x)$ = Probability of x Heads	$\frac{1}{1024}$ ≈ 0.001	$\frac{10}{1024}$ ≈ 0.01	$\frac{45}{1024}$ ≈ 0.04	$\frac{120}{1024}$ ≈ 0.12	$\frac{210}{1024}$ ≈ 0.21	$\frac{252}{1024}$ ≈ 0.25	$\frac{210}{1024}$ ≈ 0.21	$\frac{120}{1024}$ ≈ 0.12	$\frac{45}{1024}$ ≈ 0.04	$\frac{10}{1024}$ ≈ 0.01	$\frac{1}{1024}$ ≈ 0.001

The binomial distribution in the table is graphed at the right. Closely examine this 11-point graph of $P(x) = \frac{1}{1024}\binom{10}{x}$, along with the table of values. The individual probabilities are all less than $\frac{1}{4}$. Notice how unlikely it is to get 0 heads or 10 heads in a row. (The probability for each is less than $\frac{1}{1000}$.)

Even for 9 heads in 10 tosses, the probability is less than $\frac{1}{1000}$. Like the graph of the 6-point probability function $P(n) = \frac{1}{32}\binom{5}{n}$ on the previous page, this 11-point graph has a vertical line of symmetry.

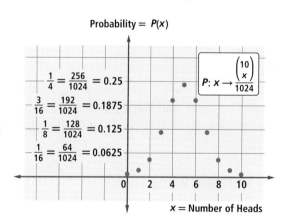

Normal Distributions

As the number of tosses of a fair coin is increased, the points on the graph more closely outline a curve shaped like a bell. On the next page, this bell-shaped curve is positioned so that it is reflection-symmetric to the y-axis and its equation is simplest.

Its equation is $y = \frac{1}{\sqrt{2\pi}}e^{\left(\frac{-x^2}{2}\right)}$.

The function that determines this graph is called a **normal distribution**, and the curve is called a **normal curve**. Notice that its equation involves the famous constants $e \approx 2.718$ and $\pi \approx 3.14$. Every normal curve is the image of the graph at the right under a composite of translations and scale changes. Thus, the graph of $y = \dfrac{1}{\sqrt{2\pi}} e^{\left(\frac{-x^2}{2}\right)}$ is sometimes called the **standard normal curve**.

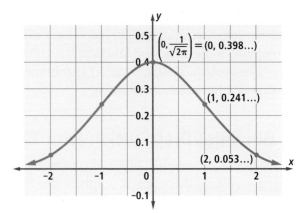

 QY1

In any distribution whose graph is a normal curve, the values fall in certain intervals based on the mean μ and standard deviation σ as shown in the graph below. For example, 34.1% of the function values fall between the mean and 1 standard deviation above the mean.

Normal Distribution Percentages

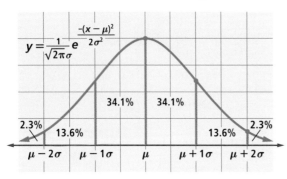

Normal curves are models for many natural phenomena. If you know the mean and standard deviation, you can determine other information about a normally distributed data set.

> ▶ **QY1**
>
> Find the y-values on the standard normal curve when $x = -1$ and $x = -2$.

Example 2

The heights of men in the United States are approximately normally distributed with a mean of 69.2 inches and a standard deviation of 2.8 inches. What percent of men in the U.S. are taller than 72.0 inches?

Solution The difference between the mean height 69.2 inches and the given height 72.0 inches is 2.8 inches. This value is exactly one standard deviation for these data. So, 72.0 inches is 1 standard deviation above the mean, or $\mu + 1\sigma$. From the graph above, notice that to the right of $\mu + 1\sigma$ are 13.6% + 2.3% = 15.9% of the normally-distributed heights. So, about 15.9% of men are taller than 72.0 inches.

 QY2

> ▶ **QY2**
>
> It is estimated that 2.3% of the men in the U.S. are less than a certain height in inches. What is that height?

Normal curves are often good mathematical models for the distribution of scores on an exam. The graph at the right shows an actual distribution of scores on a 40-question test given to 209 geometry students. (It was a hard test!) A possible corresponding normal curve is shown.

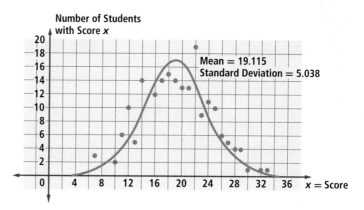

Number of Students with Score x

Mean = 19.115
Standard Deviation = 5.038

x = Score

On some tests, scores are **standardized**. This means that a person's score is not the number of correct answers, but is converted so that it lies in a normal distribution with a predetermined mean and standard deviation. Standardized tests make it easy to evaluate an individual score relative to the mean, but not to other individual scores.

Example 3

SAT scores are standardized to a historic mean of 500 and a standard deviation of 100. What percent of the scores are expected to be between 400 and 700?

Solution Find out how many standard deviations each score, 400 and 700, is from the mean, 500.

$$500 - 400 = 100 \text{ and } 700 - 500 = 200$$

So, a score of 400 is 1 standard deviation below the mean and a score of 700 is 2 standard deviations above the mean.

Refer to the Normal Distribution Percentages graph on the previous page. The percent of scores between $\mu - 1\sigma$ and $\mu + 2\sigma$ is $34.1\% + 34.1\% + 13.6\% = 81.8\%$.

About 81.8% of SAT scores are expected to be between 400 and 700.

The normal distribution is an appropriate topic with which to end this book, because it involves so many of the ideas you have studied in it. The distribution is a function. Its equation involves squares, square roots, π, e, and negative exponents. Its graph is the composite of a translation and scale change image of the curve with equation $y = \frac{1}{\sqrt{2\pi}} e^{\left(\frac{-x^2}{2}\right)}$. It models real data and shows a probability distribution that is used on tests that help to determine which colleges some people will attend. It shows how interrelated and important the ideas of mathematics are.

Questions

COVERING THE IDEAS

1. Let $P(n) = \frac{1}{32} \cdot \binom{5}{n}$.
   a. What kind of function is P?
   b. Find $P(4)$ and describe what it could represent.

2. What are the domain and range of the function $P: x \rightarrow \frac{1}{1024} \cdot \binom{10}{x}$?

3. If a fair coin is tossed 10 times, what is the probability of getting exactly 4 heads?

4. Let $P(n) =$ the probability of getting n heads in 7 tosses of a fair coin.
   a. Make a table of values for P.   b. Graph P.

5. Write an equation for the standard normal curve.

6. Describe one application of normal curves.

7. a. What does it mean for scores to be standardized?
   b. What is one advantage of doing this?

8. Approximately what percent of scores on a normal curve are within 2 standard deviations of the mean?

9. Refer to Example 2. What percent of men in the U.S. are between 66.4 and 72.0 inches tall?

10. Approximately what percent of people score below 300 on an SAT test with mean 500 and standard deviation 100?

APPLYING THE MATHEMATICS

11. Graph the normal curve with mean 50 and standard deviation 10.

12. If you repeatedly toss 8 fair coins, about what percent of the time do you expect to get from 4 to 6 heads?

13. Some tests are standardized so the mean is the grade level at which the test is taken and the standard deviation is 1 grade level. So, for students who take a test at the beginning of 10th grade, the mean is 10.0 and the standard deviation is 1.0.
   a. On such a test taken at the beginning of 10th grade, what percent of students are expected to score below 9.0 grade level?
   b. If a test is taken in the middle of 8th grade (grade level 8.5), what percent of students are expected to score between 7.5 and 9.5?

14. For the graduating class of 2004, total ACT scores for seniors ranged from 1 to 36 with a mean of 20.9 and a standard deviation of 4.8. What percent of students had an ACT score above 25?

15. **Fill in the Blanks** Assume that the lengths of jumps of boys on a track team are normally distributed with a mean of 18 feet and a standard deviation of 1 foot. In a normal distribution, 0.13% of the jumps lie more than 3 standard deviations away from the mean in each direction. This implies that about 1 out of __?__ boys will have a jump over __?__.

16. Let $y = \frac{1}{\sqrt{2\pi}} e^{\left(\frac{-x^2}{2}\right)}$. Estimate y to the nearest thousandth when $x = 1.5$.

17. A game has prizes distributed normally with a mean of $60 and a standard deviation of $20. If you play the game 10 times, what is the probability that you win more than $100 at least 3 times?

REVIEW

18. In the Texas Lotto, a participant picks 6 numbers from 1 to 54 to win. What is the probability of *not* winning the Texas Lotto? (**Lesson 13-8**)

19. **True or False** The probability of getting exactly 50 heads in 100 tosses of a fair coin is less than 5%. Justify your answer. (**Lesson 13-7**)

20. Evaluate $_nC_0$. (**Lesson 13-6**)

21. Expand $\left(6 - \frac{x}{2}\right)^5$. (**Lesson 13-6**)

22. A hot air balloon is sighted from two points on level ground at the same elevation on opposite sides of the balloon. From point P the angle of elevation is $21°$. From point Q the angle of elevation is $15°$. If points P and Q are 10.2 kilometers apart, how high is the balloon? (**Lessons 10-7, 10-1**)

23. Simplify $\sqrt{16} \cdot \sqrt{25} + \sqrt{-16} \cdot \sqrt{25} + \sqrt{-16} \cdot \sqrt{-25} + \sqrt{16} \cdot \sqrt{-25}$. (**Lessons 6-8, 6-2**)

24. Give an equation for the right angle AOB graphed at the right. (**Lesson 6-2**)

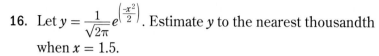

$A = (-25, 25)$

$B = (-25, -25)$

EXPLORATION

25. Together with some other students or using a random number generator, simulate the tossing of 12 coins and count the number of heads. Run the simulation at least 200 times. Let $P(h) = $ the number of times h heads appear out of 12.

   a. Graph the points $(h, P(h))$. How close is $P(h)$ to a normal distribution?

   b. What is the mean number of heads of the distribution?

   c. Estimate the standard deviation of the distribution.

QY ANSWERS

1. ≈ 0.24; ≈ 0.05

2. 2.3% of men are two standard deviations below the mean, and because one standard deviation equals 2.8 in., then the height is $69.2 - 5.6 = 63.6$ in.

Chapter 13 Projects

1 A Skewed Probability Distribution

Use random numbers and technology to simulate the tossing of a fair six-sided die. Call an outcome of 6 a success and an outcome of any number less than 6 a failure.

a. Simulate tossing the die 100 times. How many outcomes were successful?

b. Repeat Part a 100 times. Record the number successes in each set of 100 tosses.

c. Calculate the relative frequency of each number of successes. For example, if you had 30 successes in 12 of the trials, the relative frequency of 30 successes would be $\frac{12}{100}$ or 0.12.

d. Plot the points (number of successes, relative frequency of that number of successes) and draw a smooth curve that roughly models all the points.

e. Compute the mean μ and standard deviation σ of your data set from Part b.

f. Substitute your mean and standard deviation into the equation $y = \frac{1}{\sigma\sqrt{2\pi}} e^{\frac{-(x-\mu)^2}{2\sigma^2}}$.

This is the equation for the normal curve for your data set. Graph this equation on a graphing utility and sketch the graph onto your distribution from Part d. Compare the curves. How close did you come to sketching the normal curve?

2 Convergent and Divergent Geometric Series

In Lesson 13-2, you were given a formula for the sum of a geometric series.

a. Choose five different values for r such that $|r| > 1$. For each of these ratios, consider the geometric series with constant ratio r and first term $g_1 = 2$. Determine the value of each of these geometric series for 10 terms, 50 terms, and 100 terms. Organize you data into a chart.

b. Repeat Part a for five different values for r such that $|r| < 1$.

c. A sequence whose elements approach or draw near to a particular value is said to *converge*. For example, the sequence p with $p_n = \frac{1}{n}$ converges to 0. The word *diverge* means "does not converge." For example, the sequence $S_n = n^2$ diverges, as n^2 becomes larger than any particular value. Using your answers to Parts a and b, make a conjecture about the values of r for which a geometric series will converge.

d. Think about the formulas for geometric series. Write a mathematical argument to support your conjecture in Part c.

3 The Koch Curve

The snowflake curve called a *Koch curve* results from the following recursive process.

Begin with an equilateral triangle. To create the $(n + 1)$st figure, take each segment on the nth figure, trisect it, and replace it with the four congruent segments shown at the right below.

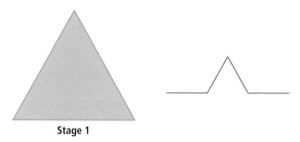

Stage 1

a. Draw the first four stages of the Koch snowflake curve. (*Hint:* The second stage looks like a six-pointed star.)

b. Count the number of segments in the figure at each stage and record this number in a table like the one below. Imagine repeating this process forever. How many segments would the nth figure have? Visualize and describe how the figure changes at each stage.

Stage	1	2	3	4	...	n
Number of Segments	3	?	?	?	...	?

c. Let the length of a side of the first triangle be 1. Find the perimeter of the Koch curve at each stage and record it in a table like the one below. Generalize the pattern to find the perimeter of the figure at the nth stage.

Stage	1	2	3	4	...	n
Perimeter	3	?	?	?	...	?

d. How does the area of the snowflake grow? Fill in a table like the one below to find out. To be able to generalize the pattern, leave the answers in radical form.

Stage	1	2	3	4	...	n
Area	$\frac{\sqrt{3}}{4}$	?	?	?	...	?

e. Suppose the process for creating the Koch curve is repeated infinitely many times. Is the perimeter of the snowflake finite or infinite? Why? Is the area finite or infinite? Why?

f. The Koch curve is an example of a *fractal*. Look in other books and copy an example of another fractal.

4 Famous Mathematical Triangles

Pascal's Triangle is not the only interesting mathematical triangular array. Two others are Euler's Triangle and the Leibniz Harmonic Triangle. Research each triangle. Write the first five rows of each triangle and find out its rule for writing additional rows. Compare and contrast each triangle to Pascal's Triangle.

Chapter 13 Summary and Vocabulary

○ Some sums and products are denoted by special symbols. For instance, the sum $x_1 + x_2 + \ldots + x_n$ is represented by $\sum_{i=1}^{n} x_i$. The product $n(n-1)(n-2) \cdot \ldots \cdot 2 \cdot 1$ is represented by **$n!$ (n factorial)**.

○ A **series** is an indicated sum of terms of a sequence. Values of finite arithmetic or geometric series may be calculated from the following formulas:

For an **arithmetic sequence** $a_1, a_2, \ldots, a_n$ with common difference d:

$$\sum_{i=1}^{n} a_i = \frac{1}{2}n(a_1 + a_n) = \frac{n}{2}(2a_1 + (n-1)d).$$

For a finite **geometric sequence** $g_1, g_2, \ldots, g_n$ with common ratio r:

$$\sum_{i=1}^{n} g_i = g_1 \frac{(1-r^n)}{1-r} = g_1 \frac{(r^n - 1)}{r-1}.$$

○ The **mean absolute deviation** and the **standard deviation** of a data set are measures of the spread, or dispersion, of the data in the set. For the data set $\{x_1, \ldots, x_n\}$, the mean μ is $\frac{1}{n}\sum_{i=1}^{n} x_i$, the mean absolute deviation is $\frac{1}{n}\sum_{i=1}^{n} |x_i - \mu|$, and the standard deviation is $\sqrt{\frac{1}{n}\sum_{i=1}^{n} (x_i - \mu)^2}$.

○ **Pascal's Triangle** is a 2-dimensional sequence. The $(r+1)$st element in row n of Pascal's Triangle is denoted by $\binom{n}{r} = \frac{n!}{r!(n-r)!}$. The expression $\binom{n}{r}$, also denoted $_nC_r$, appears in several other important applications. It is the coefficient of $a^{n-r}b^r$ in the binomial expansion of $(a+b)^n$. It is the number of subsets, or **combinations**, with r elements taken from a set with n elements. If a situation consists of n trials with two outcomes, and the probability of one of these outcomes is p, then the probability of that outcome occurring exactly r times is $\binom{n}{r}p^r(1-p)^{n-r}$. This is a **binomial probability**.

◐ The number of **permutations** of n objects is $n!$. By using permutations and combinations, the probabilities of winning many games of pure chance, such as **lotteries**, can be calculated.

◐ Distributions of binomial probabilities are related to Pascal's Triangle. As the number of the row of Pascal's Triangle increases, the graph of the distribution takes on a shape more and more like a **normal curve**. Some tests are standardized so that graphs of their scores fit that shape. In a **normal distribution**, about 68% of the data are within 1 standard deviation of the mean, and about 95% are within 2 standard deviations. The equation $y = \dfrac{1}{\sigma\sqrt{2\pi}}e^{\frac{-(x-\mu)^2}{2\sigma^2}}$ for a normal distribution with mean μ and standard deviation σ combines many of the ideas of this book in one place.

Theorems

Arithmetic Series Formula (p. 871)
Finite Geometric Series Formula (p. 878)
Number of Permutations Theorem (p. 890)
Factorial Product Theorem (p. 890)
Combination Counting Formula (p. 892)
Pascal's Triangle Explicit Formula (p. 899)
Binomial Theorem (p. 905)
Binomial Probability Theorem (p. 913)

Chapter

13 Self-Test

Take this test as you would take a test in class. You will need a calculator. Then use the Selected Answers section in the back of the book to check your work.

1. Write using summation notation:
$1^4 + 2^4 + 3^4 + \ldots + 17^4$.

2. Evaluate $\displaystyle\sum_{j=0}^{100} (5j + 1)$.

In 3 and 4, seven members of the Stern Rowing Team weigh in before a race. In kilograms, their weights are 68, 69, 70, 71, 71, 74, 74.

3. Find the mean absolute deviation and the standard deviation of the Stern Team's weights.

4. Suppose the Jolly Rowing Team has seven members with a mean team weight identical to the Stern Team's. However, Jolly's standard deviation is 0. What are the seven weights of Jolly's team members?

5. Expand $(x^2 + 2)^5$ using the Binomial Theorem.

6. Find the coefficient of x^3 in the expansion of $(x + 3)^4$.

7. To celebrate the end of the school year, you buy a bowl with three scoops of ice cream from a shop that sells 17 different flavors. How many combinations of three different flavors are possible?

8. Tyler buys 9 different textbooks his freshman year of college. In how many ways can they be arranged on his bookshelf?

9. Evaluate.
   a. $\dbinom{7}{4}$ b. $_{164}C_4$

10. Calculate $_{37}C_{36}$ and explain why your answer makes sense.

11. Rewrite $117 \cdot 116 \cdot 115 \cdot 114 \cdot 113$ as a quotient of factorials.

12. a. Calculate $\dbinom{6}{3}$ and describe its position in Pascal's Triangle.

    b. Calculate the coefficient of p^5q^2 in the expansion of $(p + q)^7$ and describe its position in Pascal's Triangle.

13. Francisco gets a summer job on a trial basis. The first day he is paid $30, the second day he is paid $32, and he continues to get a $2 raise each day. How much will Francisco be paid for 30 days of work?

14. a. Write the arithmetic series $3 + 7 + 11 + \ldots + 87$ using Σ-notation.

    b. Calculate the sum in Part a.

15. Find the sum of the integer powers of 5 from 5^0 to 5^{19}.

16. Find the sum of the first 20 terms of the sequence $\begin{cases} a_1 = 50 \\ a_n = \frac{4}{5}a_{n-1}, \text{ for integers } n \geq 2 \end{cases}$
to the nearest hundredth.

17. A fair coin is flipped five times. Find the probability of each outcome: 0 heads, 1 head, 2 heads, and so on to 5 heads.

18. In Lottery A, you need to match six numbers picked at random from the integers 1 to 50. In Lottery B, you need to match five numbers picked at random from the integers 1 to 70. Order does not matter in either lottery. In which lottery do you have a higher probability of winning? Explain your reasoning.

19. On a recent administration of the SAT test, the mean mathematics score was 515, and the standard deviation was 114.

    a. If test scores are normally distributed, about what percent of scores are within 1 standard deviation of the mean?

    b. To what range of scores on this test does your answer in Part a correspond?

    c. About what percent of scores are at or above 743?

20. Suppose a coin is biased so that there is a 65% chance that the coin will show tails when tossed. Find the probability that there will be 2 heads in 3 tosses.

21. Let $P(n) = \frac{1}{2^6}\binom{6}{n}$.

    a. Make a table of values for this function.

    b. Graph this function.

    c. Describe in words what $P(n)$ represents in the context of a coin toss.

22. Find a value of x such that $\binom{n}{2} = \binom{n}{n-x}$ for all integers $n \geq 2$.

Chapter 13 Chapter Review

SKILLS Procedures used to get answers

OBJECTIVE A Calculate values of arithmetic series. (Lesson 13-1)

In 1–4, evaluate the arithmetic series.

1. $1 + 2 + 3 + \dots + 123$

2. $2 + 8 + 14 + \dots + 92$

3. the sum of the smallest 49 positive integers that are divisible by 7

4. the sum of the first 9 terms of the sequence
$$\begin{cases} a_1 = 120 \\ a_n = a_{n-1} - 6 \text{ for integers } n \geq 2 \end{cases}$$

5. If $1 + 2 + 3 + \dots + k = 1653$, what is the value of k?

OBJECTIVE B Calculate values of finite geometric series. (Lesson 13-2)

In 6–9, evaluate the geometric series.

6. $7 + 2.1 + 0.63 + \dots + 7(0.3)^7$

7. $3 - 12 + 48 - 192 + \dots + 196{,}608$

8. the sum of integer powers of 4 from 4^0 to 4^{17}

9. the sum of the first 11 terms of the sequence $\begin{cases} g_1 = 21 \\ g_n = \frac{3}{7}g_{n-1}, \text{ for integers } n \geq 2 \end{cases}$

10. A geometric series has 18 terms. The constant ratio is 1.037, and the first term is 1313. Estimate the value of the series to the nearest integer.

OBJECTIVE C Use summation (Σ) and factorial (!) notation. (Lesson 13-1, 13-2, 13-3, 13-4)

In 11 and 12, write the terms of the series, and then evaluate the series.

11. $\displaystyle\sum_{n=1}^{5} (3n - 6)$

12. $\displaystyle\sum_{i=-3}^{2} 5 \cdot 7^i$

13. **Multiple Choice** Which equals the sum $1 + 8 + 27 + \dots + 1{,}000{,}000$?

A $\displaystyle\sum_{n=1}^{10} n^6$

B $\displaystyle\sum_{n=1}^{100} 3^n$

C $\displaystyle\sum_{n=1}^{100} n^3$

D $\displaystyle\sum_{n=1}^{1000} n^2$

14. Suppose $a_1 = 23$, $a_2 = 24$, $a_3 = 25$, $a_4 = 26$, $a_5 = 27$. Evaluate $\dfrac{1}{5}\displaystyle\sum_{i=1}^{5} a_i$.

In 15 and 16, rewrite using $\sum$-notation.

15. $3 + 6 + 9 + \dots + 123$

16. $\mu = \dfrac{y_1 + y_2 + y_3 + \dots + y_n}{n}$

17. If $g(n) = n! - n$, calculate $g(3) - g(8)$.

18. Rewrite $41 \cdot 42 \cdot 43 \cdot 44$ as a quotient of factorials.

19. **Multiple Choice** $\dfrac{(n-1)!}{n!} =$

A -1

B n

C $n - 1$

D $\dfrac{1}{n}$

OBJECTIVE D Calculate permutations and combinations. (Lessons 13-4, 13-5)

20. Interpret the symbol $\binom{n}{r}$ in terms of Pascal's Triangle.

21. **Multiple Choice** Which of the following is equal to $\frac{13!}{10! \cdot 3!}$?

    A $_{10}C_3$

    B $\binom{13}{3}$

    C $\binom{10}{3}$

    D $13 \cdot 12 \cdot 11$

In 22 and 23, consider the set {V, E, R, T, I, C, A, L}.

22. How many permutations of the letters in VERTICAL are possible?

23. a. How many subsets have 3 elements?

    b. What is the total number of subsets that can be formed?

In 24–27, evaluate.

24. $\binom{13}{6}$

25. $\binom{432}{432}$

26. $_7C_5$

27. $_{532}C_{531}$

OBJECTIVE E Use the Binomial Theorem to expand binomials. (Lesson 13-6)

In 28–31, expand using the Binomial Theorem.

28. $(x + y)^4$

29. $(t - 3)^5$

30. $(2a^2 - 3)^3$

31. $\left(\frac{p}{2} + 2q\right)^4$

True or False In 32 and 33, if the statement is false, change the statement to make it true.

32. One term of the binomial expansion of $(17x + z)^8$ is $17x^8$.

33. One term of the binomial expansion of $(43a - b)^{15}$ is $\binom{15}{2}(43a)^{13}(-b)^2$.

34. **Multiple Choice** Which equals $\sum_{r=0}^{n} \binom{n}{r} x^{n-r} 7^r$?

    A $(x + n)^7$

    B $(x + r)^n$

    C $(x + 7)^r$

    D $(x + 7)^n$

PROPERTIES Principles behind the mathematics

OBJECTIVE F Recognize properties of Pascal's Triangle. (Lesson 13-5, 13-6)

In 35 and 36, consider the top row in Pascal's Triangle to be row 0.

	row
1	0
1 1	1
1 2 1	2

35. Write row 7 of Pascal's Triangle.

36. What is the sum of the numbers in row n?

37. Find a solution to $\binom{10}{6} + \binom{10}{7} = \binom{x}{y}$

38. **True or False** For all positive integers n, $\binom{n}{1} = \binom{n}{n-1}$. Justify your answer.

39. Describe the coefficient of $r^2 s^{13}$ in the expansion of $(r + s)^{15}$ in terms of Pascal's Triangle.

USES Applications of mathematics in real-world situations

OBJECTIVE G Solve real-world problems using arithmetic or geometric series. (Lessons 13-1, 13-2)

40. A bank stacks rolls of quarters in the following fashion: one roll on top, two rolls in the next layer, three rolls in the next layer, and so on.

    a. If there are 10 layers of quarters, how many rolls are in the stack?

    b. If you want to stack 120 rolls as described above, how many rolls will you need to put on the bottom layer?

41. In a non-leap year, Kevin saved $1 on January 1, $2 on January 2, and $3 on January 3. Each day Kevin saved one dollar more than the previous day.

a. How much did Kevin save on February 15?

b. How much did he save in total by February 15?

c. How many days will it take Kevin to save a total of $10,000?

42. A ball is dropped from a height of 2 meters and bounces to 90% of its previous height on each bounce. When it hits the ground the eighth time, how far has it traveled?

43. A concert hall has 30 rows. The first row has 12 seats. Each row thereafter has 2 more seats than the preceding row. How many seats are in the concert hall?

OBJECTIVE H Solve real-world counting problems involving permutations or combinations. (Lesson 13-4)

44. The visible spectrum is associated with the acronym ROY G BIV (red, orange, yellow, green, blue, indigo, violet). How many ways can these letters be rearranged (ignoring spaces)?

45. A used car dealer has 10 cars that he can line up next to the street. How many different ways can he arrange his cars?

46. There are 25 students in a class. How many handshakes will take place if every student shakes hands with everyone else exactly once?

In 47 and 48, use the fact that the Senate of the 110th Congress had 49 Republicans, 49 Democrats, and 2 Independents.

47. How many choices were there for forming a 5-member committee of Senators?

48. How many 7-member committees could be formed with Independents and Democrats?

OBJECTIVE I Use measures of central tendency or dispersion to describe data or distributions. (Lessons 13-3, 13-9)

In 49 and 50, consider these 2007 profits of the ten largest companies in the United States.

Company	Profit (millions of $)
Wal-Mart Stores	12,731
Exxon Mobil	40,610
Chevron	18,688
General Motors	−38,732
ConocoPhillips	11,891
General Electric	22,208
Ford Motor	−2,723
Citigroup	3,617
Bank of America Corp.	14,982
AT&T	11,951

49. Find the mean absolute deviation for this data set.

50. Find the standard deviation.

In 51–53, use this information: Johns Hopkins University compared the SAT math scores for the incoming 1989 freshman class to the scores for the incoming 2006 freshman class. Some of the data are presented below.

Year	Number of Students in Class	Mean	Standard Deviation
1989	831	662.6	68.2
2006	1211	664.9	62.5

51. Which class shows a greater spread of scores?

52. Assume that the scores for the 2006 class are normally distributed. Within what interval would you expect the middle 68% of the class scores to occur?

53. When the scores for both classes are pooled into one data set, the mean SAT math score is 664.0, which is not the average of the means of the two classes when considered separately. Explain why.

54. Consider the test scores {93, 71, 78, 83, 93, 72, 99, 85}. Give the range of possible values for the mean if a ninth score ranging from 50 to 100 is added to the data set.

OBJECTIVE J Solve problems using combinations and probability.
(Lessons 13-7, 13-8)

In 55 and 56, suppose that a fair coin is tossed 6 times. Calculate the probability of each event.

55. getting exactly 1 head

56. getting exactly 3 heads

In 57–59, suppose a coin is biased so there is a 70% chance that the coin shows tails when tossed. Find the probability of each event to the nearest thousandth.

57. There are 3 heads in 3 tosses.

58. When tossed twice, the coin shows heads the first time and tails the second time.

59. When tossed twice, the coin shows heads once and tails once.

60. If 7 out of 10 is a passing score on a true-or-false quiz with 10 questions and a student guessed at every answer, what is the probability that he passed the test?

61. If 70 out of 100 is a passing score on a true-or-false test and a student guesses at every answer, will he have the same probability of passing the test as the quiz in Question 60? Explain your reasoning.

62. In Illinois' Lotto game, a participant chooses six numbers from 1 to 52. To win the jackpot, the participant must match all six winning numbers. (Order does not matter.) What is the probability of this occurring?

63. The Virginia "Pick 3" lottery requires that a participant choose three numbers, each a digit from 0 to 9. To win the grand prize, the participant must match all three numbers in the order drawn. What is the probability of winning this lottery?

REPRESENTATIONS Pictures, graphs, or objects that illustrate concepts

OBJECTIVE K Graph and analyze binomial and normal distributions. (Lesson 13-9)

64. Consider the function $P(n) = \frac{1}{2^8}\binom{8}{n}$.
   a. Evaluate $P(n)$ for integers 0, 1, ..., 8.
   b. Graph this function.
   c. What name is given to this function?

65. Below is pictured a normal distribution with mean μ and standard deviation σ.

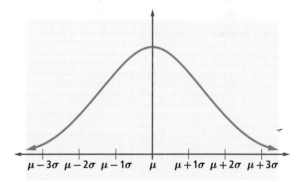

   a. What percent of the data are greater than or equal to μ?
   b. About what percent of the data are between $\mu - 1\sigma$ and $\mu + 1\sigma$?
   c. About what percent of the data are more than 2 standard deviations away from μ?

Properties

Algebra Properties from Earlier Courses

Selected Properties of Real Numbers

For any real numbers a, b, and c:

Postulates of Addition and Multiplication (Field Properties)

	Addition	*Multiplication*
Closure property	$a + b$ is a real number.	ab is a real number.
Commutative property	$a + b = b + a$	$ab = ba$
Associative property	$(a + b) + c = a + (b + c)$	$(ab)c = a(bc)$
Identity property	There is a real number 0 with $0 + a = a + 0 = a$.	There is a real number 1 with $1 \cdot a = a \cdot 1 = a$.
Inverse property	There is a real number $-a$ with $a + -a = -a + a = 0$.	If $a \neq 0$, there is a real number $\frac{1}{a}$ with $a \cdot \frac{1}{a} = \frac{1}{a} \cdot a = 1$.
Distributive property	$a(b + c) = ab + ac$	

Postulates of Equality

Reflexive property	$a = a$
Symmetric property	If $a = b$, then $b = a$.
Transitive property	If $a = b$ and $b = c$, then $a = c$.
Substitution property	If $a = b$, then a may be substituted for b in any arithmetic or algebraic expression.
Addition property	If $a = b$, then $a + c = b + c$.
Multiplication property	If $a = b$, then $ac = bc$.

Postulates of Inequality

Trichotomy property	Either $a < b$, $a = b$, or $a > b$.
Transitive property	If $a < b$ and $b < c$, then $a < c$.
Addition property	If $a < b$, then $a + c < b + c$.
Multiplication property	If $a < b$ and $c > 0$, then $ac < bc$. If $a < b$ and $c < 0$, then $ac > bc$.

Postulates of Powers

For any nonzero bases a and b and integer exponents m and n:

Product of Powers property	$b^m \cdot b^n = b^{m + n}$
Power of a Power property	$(b^m)^n = b^{mn}$
Power of a Product property	$(ab)^m = a^m b^m$
Quotient of Powers property	$\frac{b^m}{b^n} = b^{m - n}$
Power of a Quotient property	$\left(\frac{a}{b}\right)^m = \frac{a^m}{b^m}$

Selected Theorems of Graphing

The set of points (x, y) satisfying $Ax + By = C$, where A and B are not both 0, is a line.

The line with equation $y = mx + b$ has slope m and y-intercept b.

Two nonvertical lines are parallel if and only if they have the same slope.

Two nonvertical lines are perpendicular if and only if the product of their slopes is –1.

The set of points (x, y) satisfying $y = ax^2 + bx + c$ is a parabola.

Selected Theorems of Algebra

For any real numbers a, b, c, and d (with denominators of fractions not equal to 0):

Multiplication Property of Zero	$0 \cdot a = 0$
Multiplication Property of –1	$-1 \cdot a = -a$
Opposite of an Opposite Property	$-(-a) = a$
Opposite of a Sum	$-(b + c) = -b + -c$
Distributive Property of Multiplication over Subtraction	$a(b - c) = ab - ac$
Addition of Like Terms	$ac + bc = (a + b)c$
Addition of Fractions	$\frac{a}{c} + \frac{b}{c} = \frac{a + b}{c}$
Multiplication of Fractions	$\frac{a}{b} \cdot \frac{c}{d} = \frac{ac}{bd}$
Equal Fractions	$\frac{ac}{bc} = \frac{a}{b}$
Means-Extremes	If $\frac{a}{b} = \frac{c}{d}$, then $ad = bc$.
Binomial Square	$(a + b)^2 = a^2 + 2ab + b^2$
Extended Distributive Property	To multiply two polynomials, multiply each term in the first polynomial by each term in the second, and then add the products.
Zero Exponent	If $b \neq 0$, then $b^0 = 1$.
Negative Exponent	If $b \neq 0$, then $b^{-n} = \frac{1}{b^n}$.
Zero Product	$ab = 0$ if and only if $a = 0$ or $b = 0$.
Absolute Value-Square Root	$\sqrt{a^2} = |a|$
Product of Square Roots	If $a \geq 0$ and $b \geq 0$, then $\sqrt{ab} = \sqrt{a} \cdot \sqrt{b}$.
Quadratic Formula	If $ax^2 + bx + c = 0$ and $a \neq 0$, then $x = \frac{-b \pm \sqrt{b^2 - 4ac}}{2a}$.

Geometry Properties from Earlier Courses

In this book, the following symbols are used:

a, b, c	sides	C	circumference	n	number of sides
A	area	d	diameter	p	perimeter
B	area of base	d_1, d_2	diagonals	r	radius
b_1, b_2	bases	h	height	s	side
		ℓ	length	S.A.	surface area
		ℓ	slant height (in conics)	V	volume
		L.A.	lateral area	w	width

Two-Dimensional Figures

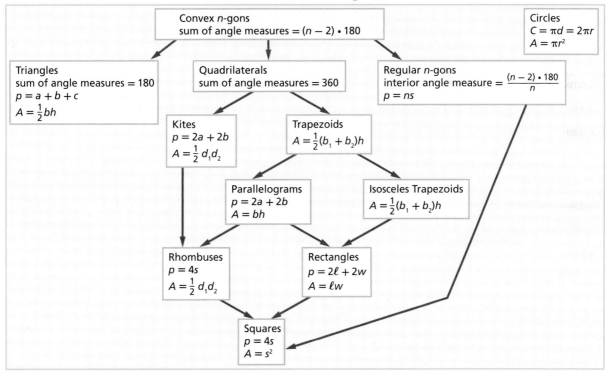

Three-Dimensional Figures

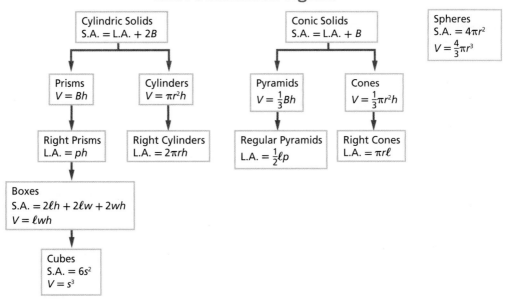

Selected Theorems of Geometry

Parallel Lines

Two lines are parallel if and only if:

1. corresponding angles have the same measure.

2. alternate interior angles are congruent.

3. alternate exterior angles are congruent.

4. they are perpendicular to the same line.

Triangle Congruence

Two triangles are congruent if:

SSS   three sides of one are congruent to three sides of the other.

SAS   two sides and the included angle of one are congruent to two sides and the included angle of the other.

ASA   two angles and the included side of one are congruent to two angles and the included side of the other.

AAS   two angles and a nonincluded side of one are congruent to two angles and the corresponding nonincluded side of the other.

SsA   two sides and the angle opposite the longer of the two sides of one are congruent to two sides and the angle opposite the corresponding side of the other.

Angles and Sides of Triangles

Triangle Inequality

The sum of the lengths of two sides of a triangle is greater than the length of the third side.

Isosceles Triangle

If two sides of a triangle are congruent, the angles opposite those sides are congruent.

Unequal Sides

If two sides of a triangle are not congruent, then the angles opposite them are not congruent, and the larger angle is opposite the longer side.

Unequal Angles

If two angles of a triangle are not congruent, then the sides opposite them are not congruent, and the longer side is opposite the larger angle.

Pythagorean Theorem

In any right triangle with legs a and b and hypotenuse c, $a^2 + b^2 = c^2$.

30-60-90 Triangle

In a 30-60-90 triangle, the sides are in the extended ratio $x : x\sqrt{3} : 2x$.

45-45-90 Triangle

In a 45-45-90 triangle, the sides are in the extended ratio $x : x : x\sqrt{2}$.

Parallelograms

A quadrilateral is a parallelogram if and only if:

1. one pair of sides is both parallel and congruent.

2. both pairs of opposite sides are congruent.

3. both pairs of opposite angles are congruent.

4. its diagonals bisect each other.

Quadrilateral Hierarchy

If a figure is of any type in the hierarchy pictured at the right, it is also of all types above it to which it is connected.

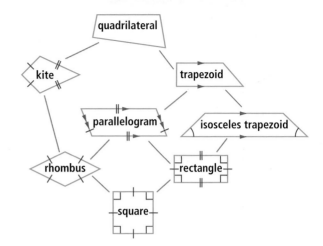

Properties of Transformations

A-B-C-D

Every isometry preserves angle measure, betweenness, collinearity, and distance.

Two-Reflection for Translations

If $m \parallel n$, the translation $r_n \circ r_m$ has magnitude two times the distance between m and n in the direction from m perpendicular to n.

Two-Reflection for Rotations

If m intersects ℓ, the rotation $r_m \circ r_\ell$ has a center at the point of intersection of m and ℓ, and has a magnitude twice the measure of an angle formed by these lines, in the direction from ℓ to m.

Isometry

Every isometry is a transformation that is a reflection or a composite of reflections.

Size-Change

Every size change with magnitude k preserves angle measure, betweenness, and collinearity; a line is parallel to its image; distance is multiplied by k.

Fundamental Theorem of Similarity

If two figures are similar with ratio of similitude k, then:

1. corresponding angle measure are equal.
2. corresponding lengths and perimeters are in the ratio k.
3. corresponding areas and surface areas are in the ratio k^2.
4. corresponding volumes are in the ratio k^3.

Triangle Similarity

Two triangles are similar if:

1. three sides of one are proportional to three sides of the other (SSS).
2. the ratios of two pairs of corresponding sides are equal and the included angles are congruent (SAS).
3. two angles of one are congruent to two angles of the other (AA).

Coordinate Plane Formulas

For all $A = (x_1, y_1)$ and $B = (x_2, y_2)$:

    Distance formula $AB = \sqrt{(x_2 - x_1)^2 + (y_2 - y_1)^2}$

    Midpoint formula   The midpoint of $\overline{AB}$ is $\left(\frac{x_1 + x_2}{2}, \frac{y_1 + y_2}{2}\right)$.

For all points (x, y):

reflection over the x-axis	$(x, y) \rightarrow (x, -y)$
reflection over the y-axis	$(x, y) \rightarrow (-x, y)$
reflection over $y = x$	$(x, y) \rightarrow (y, x)$
size change of magnitude k, center $(0, 0)$	$(x, y) \rightarrow (kx, ky)$
translation h units horizontally, k units verically	$(x, y) \rightarrow (x + h, y + k)$

Theorems

Advanced Algebra Theorems

Chapter 1

Rules for Order of Operations *(Lesson 1-1, p. 8)*

1. Perform operations within parentheses or other grouping symbols from the innermost group out.
2. Within grouping symbols or if there are no grouping symbols:
   a. Take powers from left to right.
   b. Multiply and divide in order from left to right.
   c. Add and subtract in order from left to right.

Distributive Property *(Lesson 1-6, p. 41)*

For all real numbers a, b, and c, $c(a + b) = ca + cb$.

Opposite of a Sum Theorem *(Lesson 1-6, p. 44)*

For all real numbers a and b, $-(a + b) = -a + -b = -a - b$.

Chapter 2

The Fundamental Theorem of Variation

(Lesson 2-3, p. 88)

1. If $y = kx^n$, that is, y varies *directly* as x^n, and x is multiplied by c, then y is multiplied by c^n.
2. If $y = \dfrac{k}{x^n}$, that is, y varies *inversely* as x^n, and x is multiplied by a nonzero constant c, then y is divided by c^n.

Slope of $y = kx$ Theorem *(Lesson 2-4, p. 95)*

The graph of the direct-variation function with equation $y = kx$ has constant slope k.

Converse of the Fundamental Theorem of Variation *(Lesson 2-8, p. 122)*

a. If multiplying every x-value of a function by c results in multiplying the corresponding y-values by c^n, then y varies directly as the nth power of x, that is, $y = kx^n$.

b. If multiplying every x-value of a function by c results in dividing the corresponding y-values by c^n, then y varies inversely as the nth power of x, that is, $y = \dfrac{k}{x^n}$.

Chapter 3

Parallel Lines and Slope Theorem

(Lesson 3-1, p. 154)

Two non-vertical lines are parallel if and only if they have the same slope.

Standard Form of an Equation of a Line Theorem *(Lesson 3-3, p. 163)*

The graph of $Ax + By = C$, where A and B are not both zero, is a line.

Point-Slope Theorem *(Lesson 3-4, p. 170)*

If a line contains the point (x_1, y_1) and has slope m, then it has the equation $y - y_1 = m(x - x_1)$.

nth Term of an Arithmetic Sequence Theorem *(Lesson 3-8, p. 197)*

The nth term a_n of an arithmetic (linear) sequence with first term a_1 and constant difference d is given by the explicit formula $a_n = a_1 + (n - 1)d$.

Constant-Difference Sequence Theorem

(Lesson 3-8, p. 198)

The sequence defined by the recursive formula

$$\begin{cases} a_1 \\ a_n = a_{n-1} + d, \text{ for integers } n \geq 2 \end{cases}$$

is the arithmetic sequence with first term a_1 and constant difference d.

Chapter 4

Size Change Theorem *(Lesson 4-4, p. 244)*

$\begin{bmatrix} k & 0 \\ 0 & k \end{bmatrix}$ is the matrix for S_k.

Pythagorean Distance Formula *(Lesson 4-4, p. 245)*

If $A = (x_1, y_1)$ and $B = (x_2, y_2)$, then

$$AB = \sqrt{\left|x_2 - x_1\right|^2 + \left|y_2 - y_1\right|^2}.$$

Scale Change Theorem *(Lesson 4-5, p. 251)*

$\begin{bmatrix} a & 0 \\ 0 & b \end{bmatrix}$ is a matrix for $S_{a,b}$.

Matrix for r_y Theorem *(Lesson 4-6, p. 256)*

$\begin{bmatrix} -1 & 0 \\ 0 & 1 \end{bmatrix}$ is the matrix for r_y.

Matrix Basis Theorem *(Lesson 4-6, p. 257)*
Suppose A is a transformation represented by a 2×2 matrix. If $A : (1, 0) \to (x_1, y_1)$ and $A : (0, 1) \to (x_2, y_2)$,

then A has the matrix $\begin{bmatrix} x_1 & x_2 \\ y_1 & y_2 \end{bmatrix}$.

Matrices for r_x, $r_{y=x}$, and $r_{y=-x}$ Theorem
(Lesson 4-6, p. 259)

1. $\begin{bmatrix} 1 & 0 \\ 0 & -1 \end{bmatrix}$ is the matrix for r_x.

2. $\begin{bmatrix} 0 & 1 \\ 1 & 0 \end{bmatrix}$ is the matrix for $r_{y=x}$.

3. $\begin{bmatrix} 0 & -1 \\ -1 & 0 \end{bmatrix}$ is the matrix for $r_{y=-x}$.

Matrices and Composites Theorem
(Lesson 4-7, p. 265)
If $M1$ is the matrix for transformation T_1, and $M2$ is the matrix for transformation T_2, then $M2M1$ is the matrix for $T_2 \circ T_1$.

Composite of Rotations Theorem
(Lesson 4-8, p. 270)
A rotation of $b°$ following a rotation of $a°$ with the same center results in a rotation of $(a + b)°$. In symbols, $R_b \circ R_a = R_{a+b}$.

Matrix for R_{90} Theorem *(Lesson 4-8, p. 270)*
$\begin{bmatrix} 0 & -1 \\ 1 & 0 \end{bmatrix}$ is the matrix for R_{90}.

Perpendicular Lines and Slopes Theorem
(Lesson 4-9, p. 276)
Two lines with slopes m_1 and m_2 are perpendicular if and only if $m_1 m_2 = -1$.

Parallel Lines and Translations Theorem
(Lesson 4-10, p. 282)
Under a translation, a preimage line is parallel to its image.

Chapter 5

Addition Property of Inequality *(Lesson 5-1, p. 303)*
For all real numbers a, b, and c, if $a < b$, then $a + c < b + c$.

Multiplication Property of Inequality
(Lesson 5-1, p. 303)
For all real numbers a, b, and c, if $a < b$ and $c > 0$, then $ac < bc$; if $a < b$ and $c < 0$, then $ac > bc$.

Substitution Property of Equality
(Lesson 5-3, p. 314)
If $a = b$, then a may be substituted for b in any arithmetic or algebraic expression.

Inverse Matrix Theorem *(Lesson 5-5, p. 329)*

If $ad - bc \neq 0$ and $M = \begin{bmatrix} a & b \\ c & d \end{bmatrix}$,

then $M^{-1} = \begin{bmatrix} \dfrac{d}{ad - bc} & \dfrac{-b}{ad - bc} \\ \dfrac{-c}{ad - bc} & \dfrac{a}{ad - bc} \end{bmatrix}$.

System-Determinant Theorem *(Lesson 5-6, p. 338)*
An $n \times n$ system of linear equations has exactly one solution if and only if the determinant of the coefficient matrix is *not* zero.

Linear-Programming Theorem *(Lesson 5-9, p. 356)*
The feasible region of every linear-programming problem is convex, and the maximum or minimum quantity is determined at one of the vertices of this feasible region.

Chapter 6

Binomial Square Theorem *(Lesson 6-1, p. 376)*
For all real numbers x and y,
$(x + y)^2 = x^2 + 2xy + y^2$ and
$(x - y)^2 = x^2 - 2xy + y^2$.

Absolute Value-Square Root Theorem
(Lesson 6-2, p. 381)
For all real numbers x, $\sqrt{x^2} = |x|$.

Graph-Translation Theorem *(Lesson 6-3, p. 387)*

In a relation described by a sentence in x and y, the following two processes yield the same graph:

1. replacing x by $x - h$ and y by $y - k$;
2. applying the translation $T_{h,k}$ to the graph of the original relation.

Parabola-Translation Theorem *(Lesson 6-3, p. 388)*

The image of the parabola with equation $y = ax^2$ under the translation $T_{h,k}$ is the parabola with the equation

$$y - k = a(x - h)^2$$

or

$$y = a(x - h)^2 + k.$$

Parabola Congruence Theorem *(Lesson 6-4, p. 395)*

The graph of the equation $y = ax^2 + bx + c$ is a parabola congruent to the graph of $y = ax^2$.

Completing the Square Theorem *(Lesson 6-5, p. 402)*

To complete the square on $x^2 + bx$, add $\left(\frac{b}{2}\right)^2$.

Quadratic Formula Theorem *(Lesson 6-7, p. 414)*

If $ax^2 + bx + c = 0$ and $a \neq 0$, then $x = \frac{-b \pm \sqrt{b^2 - 4ac}}{2a}$.

Square Root of a Negative Number Theorem *(Lesson 6-8, p. 422)*

If $k < 0$, $\sqrt{k} = i\sqrt{-k}$.

Properties of Complex Numbers Postulate *(Lesson 6-9, p. 428)*

In the set of complex numbers:

1. Addition and multiplication are commutative and associative.
2. Multiplication distributes over addition and subtraction.
3. $0 = 0i = 0 + 0i$ is the additive identity; $1 = 1 + 0i$ is the multiplicative identity.
4. Every complex number $a + bi$ has an additive inverse $-a + -bi$ and a multiplicative inverse $\frac{1}{a + bi}$ provided $a + bi \neq 0$.
5. The addition and multiplication properties of equality hold.

Discriminant Theorem *(Lesson 6-10, p. 437)*

Suppose a, b, and c are real numbers with $a \neq 0$. Then the equation $ax^2 + bx + c = 0$ has:

(i)   two real solutions if $b^2 - 4ac > 0$.
(ii) one real solution if $b^2 - 4ac = 0$.
(iii) two complex conjugate solutions if $b^2 - 4ac < 0$.

Chapter 7

Probability of Repeated Independent Events *(Lesson 7-1, p. 454)*

If an event has probability p, and if each occurrence of the event is independent of all other occurrences, then the probability that the event occurs n times in a row is p^n.

Product of Powers Postulate *(Lesson 7-2, p. 459)*

For any nonnegative base b and nonzero real exponents m and n, or any nonzero base b and integer exponents m and n, $b^m \cdot b^n = b^{m+n}$.

Quotient of Powers Theorem *(Lesson 7-2, p. 460)*

For any positive base b and real exponents m and n, or any nonzero base b and integer exponents m and n, $\frac{b^m}{b^n} = b^{m-n}$.

Power of a Product Postulate *(Lesson 7-2, p. 460)*

For any nonnegative bases a and b and nonzero real exponent m, or any nonzero bases a and b and integer exponent m, $(ab)^m = a^m b^m$.

Power of a Quotient Theorem *(Lesson 7-2, p. 461)*

For any positive bases a and b and real exponent n, or any nonzero bases a and b and integer exponent n, $\left(\frac{a}{b}\right)^n = \frac{a^n}{b^n}$.

Zero Exponent Theorem *(Lesson 7-2, p. 461)*

If b is a nonzero real number, $b^0 = 1$.

Power of a Power Postulate *(Lesson 7-2, p. 462)*

For any nonnegative base b and nonzero real exponents m and n, or any nonzero base b and integer exponents m and n, $(b^m)^n = b^{mn}$.

Negative Exponent Theorem *(Lesson 7-3, p. 467)*

For any positive base b and real exponent n, or any nonzero base b and integer exponent n, $b^{-n} = \frac{1}{b^n}$.

Annual Compound Interest Formula *(Lesson 7-4, p. 473)*

Let P be the amount of money invested at an annual interest rate r compounded annually. Let A be the total amount after t years. Then $A = P(1 + r)^t$.

General Compound Interest Formula *(Lesson 7-4, p. 473)*

Let P be the amount invested at an annual interest rate r compounded n times per year. Let A be the amount after t years. Then $A = P\left(1 + \frac{r}{n}\right)^{nt}$.

Recursive Formula for a Geometric Sequence
(Lesson 7-5, p. 479)
Let r be a nonzero constant. The sequence g defined by
the recursive formula $\begin{cases} g_1 = x \\ g_n = rg_{n-1}, \text{ for integers } n \geq 2 \end{cases}$
is the geometric, or exponential, sequence with first
term x and constant multiplier r.

Explicit Formula for a Geometric Sequence
(Lesson 7-5, p. 481)
In the geometric sequence g with first term g_1 and
constant ratio r, $g_n = g_1(r)^{n-1}$, for integers $n \geq 1$.

Number of Real Roots Theorem *(Lesson 7-6, p. 488)*
Every positive real number has:
  2 real nth roots when n is even.
  1 real nth root when n is odd.
Every negative real number has:
  0 real nth roots when n is even.
  1 real nth root when n is odd.
Zero has:
  1 real nth root.

$\frac{1}{n}$ Exponent Theorem *(Lesson 7-6, p. 489)*
When $x \geq 0$ and n is an integer greater than 1,
$x^{\frac{1}{n}}$ is an nth root of x.

Rational Exponent Theorem *(Lesson 7-7, p. 493)*
For any nonnegative real number x and positive
integers m and n,
$x^{\frac{m}{n}} = \left(x^{\frac{1}{n}} \right)^m$, the mth power of the positive nth root of x, and
$x^{\frac{m}{n}} = (x^m)^{\frac{1}{n}}$, the positive nth root of the mth power of x.

Chapter 8

Inverse-Relation Theorem *(Lesson 8-2, p. 524)*
Suppose f is a relation and g is the inverse of f. Then:
1. If a rule for f exists, a rule for g can be found by
   switching x and y in the rule for f.
2. The graph of g is the reflection image of the graph
   of f over the line with equation $y = x$.
3. The domain of g is the range of f, and the range of g
   is the domain of f.

Inverse Functions Theorem *(Lesson 8-3, p. 530)*
Two functions f and g are inverse functions if and only if:
1. For all x in the domain of f, $g(f(x)) = x$, and
2. For all x in the domain of g, $f(g(x)) = x$.

Power Function Inverse Theorem
(Lesson 8-3, p. 532)
If $f(x) = x^n$ and $g(x) = x^{\frac{1}{n}}$ and the domains of f and g are
the set of *nonnegative* real numbers, then f and g are
inverse functions.

Root of a Power Theorem *(Lesson 8-4, p. 539)*
For all positive integers $m \geq 2$ and $n \geq 2$,
$\sqrt[n]{x^m} = \left(\sqrt[n]{x} \right)^m = x^{\frac{m}{n}}$ when $x \geq 0$.

Root of a Product Theorem *(Lesson 8-5, p. 546)*
For any nonnegative real numbers x and y, and any
integer $n \geq 2$, $(xy)^{\frac{1}{n}} = x^{\frac{1}{n}} \cdot y^{\frac{1}{n}}$ (power form) and
$\sqrt[n]{xy} = \sqrt[n]{x} \cdot \sqrt[n]{y}$ (radical form).

nth Root of a Product Theorem *(Lesson 8-7, p. 559)*
When $\sqrt[n]{x}$ and $\sqrt[n]{y}$ are defined and are real numbers,
then $\sqrt[n]{xy}$ is also defined and $\sqrt[n]{xy} = \sqrt[n]{x} \cdot \sqrt[n]{y}$

Chapter 9

Exponential Change Model *(Lesson 9-2, p. 589)*
If a positive quantity a is multiplied by b ($b > 0, b \neq 1$)
in each unit period, then after a period of length x, the
amount of the quantity is ab^x.

Continuously Compounded Interest Formula
(Lesson 9-3, p. 597)
If an amount P is invested in an account paying an
annual interest rate r compounded continuously, the
amount A in the account after t years is $A = Pe^{rt}$.

Log_b of b^n Theorem *(Lesson 9-9, p. 636)*
For every positive base $b \neq 1$, and any real number n,
$\log_b b^n = n$.

Logarithm of 1 Theorem *(Lesson 9-9, p. 636)*
For every positive base $b \neq 1$, $\log_b 1 = 0$.

Logarithm of a Product Theorem *(Lesson 9-9, p. 637)*
For any positive base $b \neq 1$ and positive real numbers x
and y, $\log_b(xy) = \log_b x + \log_b y$.

Logarithm of a Quotient Theorem *(Lesson 9-9, p. 638)*
For any positive base $b \neq 1$ and for any positive real
numbers x and y, $\log_b\left(\frac{x}{y}\right) = \log_b x - \log_b y$.

Logarithm of a Power Theorem *(Lesson 9-9, p. 639)*
For any positive base $b \neq 1$ and for any positive real
number x and any real number n, $\log_b(x^n) = n \log_b x$.

Change of Base Theorem *(Lesson 9-10, p. 646)*
For all positive real numbers a, b, and t, $b \neq 1$ and $t \neq 1$,
$\log_b a = \frac{\log_t a}{\log_t b}$.

Chapter 10

Pythagorean Identity Theorem *(Lesson 10-5, p. 688)*
For all θ, $(\cos \theta)^2 + (\sin \theta)^2 = 1$.

Supplements Theorem *(Lesson 10-7, p. 701)*
For all θ in degrees, $\sin \theta = \sin(180° - \theta)$.

Law of Sines Theorem *(Lesson 10-7, p. 702)*
In any triangle ABC, $\frac{\sin A}{a} = \frac{\sin B}{b} = \frac{\sin C}{c}$.

Law of Cosines Theorem *(Lesson 10-8, p. 706)*
In any triangle ABC, $c^2 = a^2 + b^2 - 2ab \cos C$.

Chapter 11

Extended Distributive Property *(Lesson 11-2, p. 739)*
To multiply two polynomials, multiply each term in the first polynomial by each term in the second and add the products.

Difference of Squares Factoring Theorem
(Lesson 11-3, p. 747)
For all a and b, $a^2 - b^2 = (a + b)(a - b)$.

Binomial Square Factoring Theorem
(Lesson 11-3, p. 747)
For all a and b, $a^2 + 2ab + b^2 = (a + b)^2$ and
$a^2 - 2ab + b^2 = (a - b)^2$.

Zero-Product Theorem *(Lesson 11-4, p. 752)*
For all a and b, $ab = 0$ if and only if $a = 0$ or $b = 0$.

Factor Theorem *(Lesson 11-4, p. 753)*
$x - r$ is a factor of $P(x)$ if and only if $P(r) = 0$, that is, r is a zero of P.

Rational-Root (or Rational-Zero) Theorem
(Lesson 11-5, p. 761)
Suppose that all the coefficients of the polynomial function described by $f(x) = a_n x^n + a_{n-1} x^{n-1} + \dots + a_2 x^2 + a_1 x + a_0$ are integers with $a_n \neq 0$ and $a_0 \neq 0$. If $\frac{p}{q}$ is a root of $f(x)$ in lowest terms, then p is a factor of a_0 and q is a factor of a_n.

The Fundamental Theorem of Algebra
(Lesson 11-6, p. 766)
Every polynomial equation $P(x) = 0$ of any degree ≥ 1 with complex number coefficients has at least one complex number solution.

Number of Roots of a Polynomial Equation Theorem *(Lesson 11-6, p. 768)*
Every polynomial equation of degree n has exactly n roots, provided that multiple roots are counted according to their multiplicities.

Polynomial-Difference Theorem *(Lesson 11-7, p. 773)*
$y = f(x)$ is a polynomial function of degree n if and only if, for any set of x-values that form an arithmetic sequence, the nth differences of corresponding y-values are equal and the $(n - 1)$st differences are not equal.

Chapter 12

Focus and Directrix of a Parabola Theorem
(Lesson 12-1, p. 801)
For any nonzero real number a, the graph of $y = ax^2$ is the parabola with focus at $\left(0, \frac{1}{4a}\right)$ and directrix at $y = -\frac{1}{4a}$.

Circle Equation Theorem *(Lesson 12-2, p. 806)*
The circle with center (h, k) and radius r is the set of points (x, y) that satisfy $(x - h)^2 + (y - k)^2 = r^2$.

Interior and Exterior of a Circle Theorem
(Lesson 12-3, p. 812)
Let c be the circle with center (h, k) and radius r. Then the interior of c is described by $(x - h)^2 + (y - k)^2 < r^2$ and the exterior of c is described by $(x - h)^2 + (y - k)^2 > r^2$.

Equation for an Ellipse Theorem
(Lesson 12-4, p. 818)
The ellipse with foci $(c, 0)$ and $(-c, 0)$ and focal constant $2a$ has equation $\frac{x^2}{a^2} + \frac{y^2}{b^2} = 1$, where $b^2 = a^2 - c^2$.

Length of Axes of an Ellipse Theorem
(Lesson 12-4, p. 820)
In the ellipse with equation $\frac{x^2}{a^2} + \frac{y^2}{b^2} = 1$, $2a$ is the length of the horizontal axis and $2b$ is the length of the vertical axis.

Circle Scale-Change Theorem *(Lesson 12-5, p. 826)*
The image of the unit circle with equation $x^2 + y^2 = 1$ under $S_{a,b}$ is the ellipse with equation $\left(\frac{x}{a}\right)^2 + \left(\frac{y}{b}\right)^2 = 1$.

Graph Scale-Change Theorem *(Lesson 12-5, p. 827)*
In a relation described by a sentence in x and y, the following two processes yield the same graph:
1. replacing x by $\frac{x}{a}$ and y by $\frac{y}{b}$;
2. applying the scale change $S_{a,b}$ to the graph of the original relation.

Area of an Ellipse Theorem *(Lesson 12-5, p. 828)*
An ellipse with semimajor and semiminor axes of lengths a and b has area $A = \pi ab$.

Equation for a Hyperbola Theorem
(Lesson 12-6, p. 832)
The hyperbola with foci $(c, 0)$ and $(-c, 0)$ and focal constant $2a$ has equation $\frac{x^2}{a^2} - \frac{y^2}{b^2} = 1$, where $b^2 = c^2 - a^2$.

Asymptotes of a Hyperbola Theorem
(Lesson 12-6, p. 834)
The asymptotes of the hyperbola with equation $\frac{x^2}{a^2} - \frac{y^2}{b^2} = 1$ are $\frac{y}{b} = \pm\frac{x}{a}$, or $y = \pm\frac{b}{a}x$.

Attributes of $y = \frac{k}{x}$ Theorem *(Lesson 12-7, p. 841)*
The graph of $y = \frac{k}{x}$ or $xy = k$ is a hyperbola. When $k > 0$, this hyperbola has vertices $\left(\sqrt{k}, \sqrt{k}\right)$ and $\left(-\sqrt{k}, -\sqrt{k}\right)$, foci $\left(\sqrt{2k}, \sqrt{2k}\right)$ and $\left(-\sqrt{2k}, -\sqrt{2k}\right)$, and focal constant $2\sqrt{2k}$. The asymptotes of the graph are $x = 0$ and $y = 0$.

Chapter 13

Arithmetic Series Formula *(Lesson 13-1, p. 871)*
In an arithmetic sequence $a_1, a_2, a_3, ..., a_n$ with constant difference d, $\sum_{i=1}^{n} a_i = \frac{n}{2}(a_1 + a_n) = \frac{n}{2}(2a_1 + (n-1)d)$.

Finite Geometric Series Formula
(Lesson 13-2, p. 878)
Let S_n be the sum of the first n terms of the geometric sequence with first term g_1 and constant ratio $r \neq 1$. Then $S_n = \frac{g_1(1 - r^n)}{1 - r}$.

Number of Permutations Theorem
(Lesson 13-4, p. 890)
There are $n!$ permutations of n distinct objects.

Factorial Product Theorem *(Lesson 13-4, p. 890)*
For all $n \geq 1$, $n! = n \cdot (n-1)!$.

Combination Counting Formula *(Lesson 13-4, p. 892)*
The number $_nC_r$ of subsets, or combinations, of r elements that can be formed from a set of n elements is given by the formula $_nC_r = \frac{n!}{r!(n-r)!}$.

Pascal's Triangle Explicit Formula
(Lesson 13-5, p. 899)
If n and r are integers with $0 \leq r \leq n$, then $\binom{n}{r} = {_nC_r} = \frac{n!}{r!(n-r)!}$.

Binomial Theorem *(Lesson 13-6, p. 905)*
For all complex numbers a and b, and for all integers n and r with $0 \leq r \leq n$, $(a + b)^n = \sum_{r=0}^{n} \binom{n}{r} a^{n-r} b^r$.

Binomial Probability Theorem *(Lesson 13-7, p. 913)*
Suppose an experiment has an outcome with probability p, so that the probability the outcome does not occur is $q = 1 - p$. Then in n independent repetitions of the experiment, the probability that the outcome occurs r times is $\binom{n}{r} p^r q^{n-r}$.

CAS Commands

The Computer Algebra System (CAS) commands used in this course and examples of their use are given below. Each command must be followed by a number, variable, expression, or equation, usually enclosed in parentheses.

Command	Description	Example
Define	A rule for a function is stored under the name indicated. Values of that function can then be calculated by entering the function's name followed by the value of the independent variable in parentheses.	Define $f(n)=2000 \cdot n - 1400$ *Done* $f(4)$ 6600
| (such that)	Variable values that appear after the symbol are substituted into an expression, inequality, or equation that appears before the symbol.	$r=\dfrac{\sqrt{v}}{\sqrt{h \cdot \pi}} \mid v=500000$ $r=\dfrac{500 \cdot \sqrt{2}}{\sqrt{h \cdot \pi}}$
solve	An equation, inequality, or system is solved for an indicated variable or variables. All real solutions are given.	$\text{solve}\left(y=\dfrac{1}{2} \cdot x - 5 \text{ and } y = 2 \cdot x - 1, \{x,y\}\right)$ $x=\dfrac{-8}{3}$ and $y=\dfrac{-19}{3}$
expand	The Distributive Property is applied to products and powers of mathematical expressions.	$\text{expand}((72+w) \cdot (24+2 \cdot w))$ $2 \cdot w^2 + 168 \cdot w + 1728$
DelVar	Any stored values for the indicated variable are deleted from memory.	DelVar a *Done*
cSolve	An equation or inequality is solved for an indicated variable. All complex solutions are given.	$\text{cSolve}(z^4=81,z)$ $z=3 \cdot i$ or $z=-3 \cdot i$ or $z=-3$ or $z=3$
factor	A polynomial is factored over the rational numbers. On some CAS, if ",x" is added to the end of the polynomial, it is factored over the real numbers.	$\text{factor}(x^4 - 14 \cdot x^2 + 45)$ $(x-3) \cdot (x+3) \cdot (x^2-5)$ $\text{factor}(x^4 - 14 \cdot x^2 + 45, x)$ $(x-3) \cdot (x+3) \cdot (x+\sqrt{5}) \cdot (x-\sqrt{5})$
cFactor	A polynomial is factored over the complex numbers.	$\text{cFactor}(x^2+36, x)$ $(x+-6 \cdot i) \cdot (x+6 \cdot i)$
rfactor	On some CAS, a polynomial is factored over the real numbers.	$\text{rfactor}(x^2-14)$ $\left(x+\sqrt{14}\right) \cdot \left(x-\sqrt{14}\right)$

Selected Answers

Chapter 1

Lesson 1-1 (pp. 6–13)
Mental Math a. 49 in^2 **b.** 24 cm^2 **c.** 9 ft **d.** 40 mm
Guided Example 2 Step 1: 1; –1; 3; –4; 3 **Step 2:** 1; 49; 3
Step 3: –6; 6 **Step 4:** –1 **Guided Example 3 a.** 33; 9.53;
$\frac{6}{100}$ or 0.06; 0.06; 33; 9.53; 18.87; 18.87 **b.** $\frac{ptu}{1000}$
Questions
1. 1975 **3.** Answers vary. Sample: An equation must
include an equal sign, but an algebraic expression does
not include a verb. **5.** Answers vary. Sample: e^{rt}
7. a. Yes; it is an algebraic sentence that includes an
equal sign. **b.** Answers vary. Sample: Yes; it can be
considered a formula solved for c^2. **9.** $306 - 60m$
11. $28.02 **13. a.** $C + 372t$ **b.** $C + 12bt$ **15.** 4.2
17. –48.1 **19.** (e) **21.** (b) **23.** (d) **25.** C
27. a. $x = -9$ **b.** $2(-9) = 4(-9) + 18$; $-18 = -36 + 18$;
$-18 = -18$. The answer checks.

Lesson 1-2 (pp. 14–19)
Mental Math a. x^9 **b.** a^5 **c.** not possible **d.** n^{14}
Questions
1. Answers vary. Sample: A function is a set of
ordered pairs where no two pairs have the same first
coordinate. **3.** Answers vary. Sample: A mathematical
model is a description of a real situation using the
language and concepts of mathematics. For example,
the equation $x + y + z = 180$ models the relationship
among the angle measures in a triangle. **5.** Yes; for every
ordered pair (r, C), each value of r determines only one
value for C. **7.** No; one x-value generates two different
y-values. **9. a.** D is the dependent variable.
You cannot know the measure of the interior vertex
angles if you do not know how many sides the polygon
has. **b.** 120 **c.** Answers vary. Sample:
A regular hexagon can be split
into six equilateral triangles.
Each of the triangles has
interior angles that sum
to 180°. Since they are
equilateral, they are also
equiangular. So, each
interior angle of each triangle
has measure 60°. Therefore, each
interior angle of the hexagon has measure 120°.

11. $p = 75$ **13.** No; it is not the graph of a function
because there are two y-values for every x-value except –1
and 1. **15.** $\frac{b}{m}$dollars **17.** KL outfits **19.** $\frac{g}{m}n$ eggs

Lesson 1-3 (pp. 20–25)
Mental Math a. $8.85 **b.** $88.50 **c.** $8.85 **d.** $26.55
Activity Step 1: $y = \sqrt{5^2 - x^2}$ **Step 3:** 3; 4.03702; $\sqrt{25 - \pi^2}$;
$\sqrt{25 - 9 \cdot c^2}$
Step 4: Answers vary. Sample: Error: Nonreal answer.
$25 - 36 = -11$, for which there exists no real square root.
Questions
1. f of x **3.** A of x equals $\frac{1}{2}x$ times the quantity 5 minus x.
5. 4.80 **7.** 29.85 **9.** 186.8 **11.** Answers vary.
Sample: define $t(c) = 1.8c + 32$ **13. a.** –12 **b.** 18
15. a. r **b.** 288π **c.** $36\pi R^3$
17. a.

X	h(x)	g(x)	h(x) – g(x)
–4	16	0.06	15.94
–2	4	0.25	3.75
–0.5	0.25	0.71	–0.46
0.5	0.25	1.41	–1.16
2.5	6.25	5.66	0.59
3	9	8	1
3.5	12.25	11.31	0.94
4.5	20.25	22.63	–2.38
10	100	1024	–924

b. No; the table shows, for example, that $h(-0.5) < g(-0.5)$.
c. Answers vary. Sample: The var key should display
the currently stored variables and functions in your
CAS memory. **d.** Answers vary. Sample: Yes; the calculator
uses a different icon to represent stored functions and
stored variables. **e.** The calculator does not evaluate
the function. **19.** 2000, 2003, 2004, 2005, 2006 **21. a.** 12
b. –212 **c.** 2.42916 **d.** $|4b| + (4b - 2)^3$ **23.** No; because
each brother has more than one sibling. **25. a.** $y = 6$
b. $x = -4$ **c.** $y = k, x = h$

Lesson 1-4 (pp. 26–32)
Mental Math a. 2 units **b.** 1 square unit **c.** 5 units
Activity Step 4: (70, 478.33) **Step 5:** $C(75) = 356.25$,
$S(75) = 543.75$; new window: $0 \le x \le 75$, $0 \le y \le 545$
Step 7: No; you do not need to consider negative values
because cars and SUVs cannot travel at a negative speed
or stop in a negative number of feet. **Step 9:** Minimum
value of $C \approx -5$; minimum value of $S \approx -3$

Questions

1. $\{x \mid 0 \leq x \leq 70\}$ 3. Use the TRACE function on your graphing utility to estimate $C(45)$.   5. Answers vary. Sample: $0 \leq x \leq 55, 0 \leq y \leq 310$ 7. $\{x \mid 0 \leq x \leq 8\};$ $\{y \mid 2 \leq y \leq 5\}$ 9. integer, rational number, real number 11. irrational number, real number
13. Answers vary. Sample: –29
15. a. $\{y \mid 0 \leq y \leq 24\}$ b. 17 years c. $7 \leq d \leq 12$
d. yes

Tree Age vs. Trunk Diameter

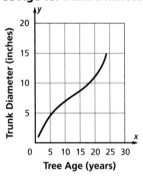

17. a. 5 b. $\{x \mid x \text{ is a real number}\}; \{y \mid y = 5\}$ 19. a. 4
b. $x = 7$ 21. a.

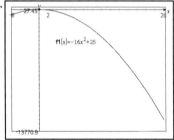

b. 5.64 ft   23. 5050   25. $r = 261; \frac{1}{3}(261) - 27 = 60$

Lesson 1-5 (pp. 33–39)

Mental Math a. 70 mph **b.** 315 miles **c.** 3 hours **d.** 5:30 PM
Guided Example 1 a. $d; t$ **b.** 0; 210; 0; 12 **c.** 6; 6; 60; 6
d. 90; 90 **e.** 45; 60; 45; 60; t
Activity Step 4: t days; $D(t)$ inches; no; negative values of x do not make sense because x represents elapsed time.
Step 5: $D(6) = 3.75$ in. **Step 6:** $D(2) = 5.75$ in.
Questions
1. 0 miles; The graph shows this because the rightmost point is (210, 0), and the coordinate 0 indicates distance from home.   3. 12   5. a. 4 b. -2   7. a. 5.19 in.
b. After 5.6 days   9. (1, 8); (1.5, 6.75); (2, 5); (2.5, 2.75); (3, 0)   11. Yes; domain: {-3, -2, -1, 0, 1, 2, 3}; range: {-0.1, -0.06, -0.03, 0, 0.03, 0.06, 0.1}   13. Yes; domain: {-6, -4, -2, 0, 2, 4, 6}; range: {0, 1, 2, 3}   15. 2,160,000
17. a. -5, 0, 3, 4, 3, 0, -5

b.

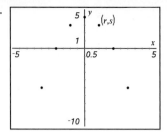

Yes; each value of r is paired with a single value of s.

c.

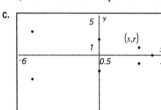

No; certain values of s are paired with more than one value of r.   19. $\{x \mid 0 \leq x \leq 100\}; \{G(x) \mid 0 \leq G(x) \leq 20\}$
21. Yes; $f: x \rightarrow 2x - \pi x^2$ 23. 7,698,000,000 km^2

Lesson 1-6 (pp. 40–46)

Mental Math a. 18 **b.** 120 **c.** $28xy$
Guided Example 2 465; 0.93d; 0.28d; 465; 0.28d; 85; 303.57; 303.57; 196.43; \$550; \$303.57; \$196.43
Guided Example 4 -5a; -7; 5a; -8a; 19; 8a; 2.375
Questions
1. 45,600 in^3 3. 13.9; if she has 122 nickels, her collection is worth \$13.90.   5. $y = 8$ 7. $x = 1.5$ 9. Answers vary.
Sample: 24; $x = \frac{120}{19}$ 11. a. $x = 4$ b. $3(4) - (4 + 1) = 7$
13. 52   15. $0 < A < 200$ 17. $t = \frac{72}{5}$ 19. $z = 10$
21. a. $\frac{17}{20}$ b. $n = 47$ 23. about \$1,000,000 million
25. 1990 through 2000   27. 12.0 cm

Lesson 1-7 (pp. 47–52)

Mental Math a. 4 **b.** 4 **c.** 5
Questions
1. $\approx 1,963.5$ cm^3 3.

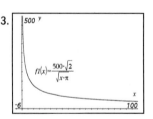

5. $V; B$ and $h.$ 7. $r = \frac{65t}{c} + 65 = 65\left(\frac{t}{c} + 1\right)$ 9. B
11. a. $S = \frac{R}{0.5P}$ b. 40 feet   13. $k = 0.75$ 15. $b_1 = \frac{2A}{h} - b_2$
17. a. $x = 16$ b. $x = -32$ c. $x = \frac{-480}{7}$

19. a.

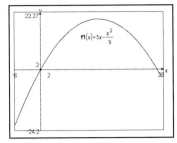

b. all real numbers; all real numbers ≤ 20.25
c. $\{f(x) \mid \frac{50}{9} \leq f(x) \leq 14\}$

Lesson 1-8 (pp. 53–59)

Mental Math a. 0 **b.** 0 **c.** 36 **d.** 36 **e.** 81
Activity Step 1:

Number of Dots
1
3
6
10
15

Step 2:

Number of Dots
21
28
36
45

Step 3: For each term number there corresponds only one number of dots.
Guided Example 2 Solution 1 a. 28,700; 29,876.70; 31,101.64; 32,376.81; 33,704.26 **b.** 1,532,865.74 **c.** 1,532,865.74; $1,532,865.74 **Solution 2 a.** 1; 1; 5 **b.** 1,532,865.74
c. $1,532,865.74
Questions
1. a. term **b.** 21 **3.** 210 **5. a.**
b. 1, 4, 9, 16, 25 **c.** $S_n = n^2$
7. 4.3, 1.3, -1.7, -4.7

9. a.

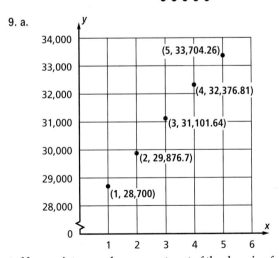

b. No; non-integer values are not part of the domain of the function.

11. a.

Term Number	Value of term
4	$C_4 = \frac{4^2(4+1)^2}{4} = 100$
5	$C_5 = \frac{5^2(5+1)^2}{4} = 225$
6	$C_6 = \frac{6^2(6+1)^2}{4} = 441$
7	$C_7 = \frac{7^2(7+1)^2}{4} = 784$

b. the sum of the first seven cubes **13. a.** (1, 2), (2, 4), (3, 6), (4, 10), (5, 18), (6, 34) **b.** $S_n = 2 + 2^{n-1}$, for all $n > 1$
15. $s = \frac{d+13}{7}$ **17. a.** $r = \frac{7}{3}$ **b.** $5\left(\frac{7}{3}\right) - \left(2\left(\frac{7}{3}\right) + 1\right) = 6$
19. a. t must be a whole number. **b.** 12.5, 14, 15.5, 17
c.

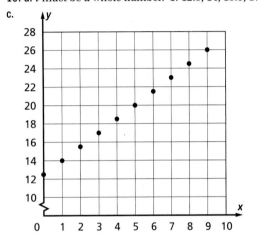

Self-Test (64–65)

1. $f(-1.5) = (-1.5)^2 - 4^{-1.5} = 2.25 - 0.125 \approx 2.13$ **2.** The first statement says that the value of the function $f(x)$ evaluated at $x = a$ is 7, whereas the second statement says that the value of the function $f(x)$ evaluated at $x = 7$ is a. **3.** $g(-2) = 17$ and $g(2) = 3$. **4.** domain: $\{x \mid -2 \leq x \leq 6\}$; range: $\{y \mid 2 \leq y \leq 6\}$. **5. a.** 5 **b.** $x = -1.3$ and $x = 2$. **6.** 135 **7.** B **8.** $s = \frac{m}{4}$ **9.** $5(7x - 4) = 50$; $7x - 4 = 10$; $7x = 14$; $x = 2$ **10.** $3.2a = 0.75 + 1.2a$; $2a = 0.75$; $a = 0.375$
11. $p + 0.2(1200 - p) = 244.4p$; $p + 0.2 \cdot 1200 - 0.2p = 244.4p$; $240 = 243.6p$; $p \approx 0.985$ **12.** 101cm^3 **13.** $\{V \mid 0 < V \leq 200\pi\}$.
14. a. $V = \frac{1}{3}\pi r^2 h$; $h = \frac{V}{\left(\frac{1}{3}\right)\pi r^2}$; $h = \frac{3V}{\pi r^2}$ **b.** $V = \frac{1}{3}\pi r^2 h$; $r^2 = \frac{3V}{\pi h}$; $r = \sqrt{\frac{3V}{\pi h}}$ **15.** domain: $\{0, 4, 8, 16\}$; range: $\{7, 8, 9, 11\}$.
16. B, C **17.** No; it is not a function because $x = 9$ corresponds to both $y = 7$ and $y = 4$. **18.** Yes; it is a function; each x-value defines just one y-value.
19. a. $a_1 = -4^1 + 2 = -2$; $a_2 = -4^2 + 2 = -14$; $a_3 = -4^3 + 2 = -62$; $a_4 = -4^4 + 2 = -254$; $a_5 = -4^5 + 2 = -1022$. **b.** -16,382

20.

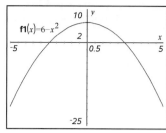

$f1(x)=6-x^2$

a. $\{f(t) \mid -19 \le f(t) \le 6\}$. b. $x \approx -2.45$ and $x \approx 2.45$.

21.

x	-6	-4	-2	0	2
y	-192	-48	-12	-3	-0.75

a. (-6, -192), (-4, -48), (-2, -12), (0, -3), (2, -0.75). b. $x = 2$ **22. a.** $W(10) = -4$; the wind-chill index with a 10 mph wind at $10°$ F is $-4°$ F. **b.** $A = -20$

23. a. (-1, 0), (0, 2), (1, 3), (2, 2), (3, 0). **b.** Yes; there is only one y-value for each x-value. **c.** domain: {-1, 0, 1, 2, 3}; range:{0, 2, 3}. **24.** about 149,000 more people **25.** C
26. about 0.000772 **27.** $T = \frac{1}{4}T + \frac{1}{10}T + 195$; $T = \frac{7}{20}T + 195$; $\frac{13}{20}T = 195$; $T = \frac{20}{13} \cdot 195 = 300$ tickets **28.** The value of c does not change because $a^2 + b^2 = b^2 + a^2 = c^2$. This corresponds to the geometric situation of a reflection, where the orientation of the triangle shifts, but its lengths do not change.

Self-Test Correlation Chart

The chart below keys the **Self-Test** questions to the objectives in the **Chapter Review** at the end of the chapter. This will enable you to locate those **Chapter Review** questions that correspond to questions missed on the **Self-Test**. The lesson where the material is covered is also indicated on the chart.

Question	1	2	3	4	5	6	7	8	9	10
Objective(s)	B	B	B	H	H	B	D	D	C	C
Lesson(s)	1-3	1-3	1-3	1-4, 1-5	1-4, 1-5	1-3	1-7	1-7	1-6	1-6
Question	11	12	13	14	15	16	17	18	19	20
Objective(s)	C	A	H	D	H	G	G	G	E	H, L
Lesson(s)	1-6	1-1	1-4, 1-5	1-7	1-4, 1-5	1-2, 1-5	1-2, 1-5	1-2, 1-5	1-8	1-4, 1-5, 1-8
Question	21	22	23	24	25	26	27	28			
Objective(s)	L	J	G, H	J	F	A	K	I			
Lesson(s)	1-4, 1-5, 1-8	1-3, 1-4	1-2, 1-4, 1-5	1-3, 1-4	1-7	1-1	1-6	1-7			

Chapter Review (pp. 66–69)

1. 119.1016 **3.** $d = 44.1$ m **5.** -15 **7. a.** $h:x \rightarrow 12x - 2\sqrt{x}$
b. 102 **9. a.** $f(x) = 4 - 27x$ **b.** -212 **11.** $x = \frac{1}{15}$; $12\left(\frac{1}{15}\right) = \frac{4}{5}$
13. $v = 3; -\frac{5}{3} = \frac{4}{3} - 3$ **15.** $s = -3; 3 + 5(-3) = 4(2(-3) + 3)$
17. $m = \frac{48}{79}; \left(\frac{48}{79 \cdot 8}\right) + \left(\frac{48}{79 \cdot 6}\right) - 2 = -3\left(\frac{48}{79}\right)$ **19.** $t = \frac{12 - x}{6}$
21. $h = \frac{3A}{\pi(r_1^2 - r_2^2)}$ **23.** Answers vary. Sample: It is not solved for s because s appears on both sides of the equal sign.
25. -9, -15, -21, -27, -33 **27.** 18 **29.** $r = -\frac{\sqrt{3-v}}{\sqrt{h\pi}}$ or $r = \frac{\sqrt{3-v}}{\sqrt{h\pi}}$; $r = \frac{\sqrt{3-v}}{\sqrt{h\pi}}$ **31.** B **33.** no **35.** Yes; each x-value defines a single y-value. **37.** No; an x-value defines two y-values. **39.** domain: {-2,-1,0,1,2}; range: {-1}
41. domain: all real numbers; range: all real numbers
43. domain: all real numbers; range: all nonnegative real numbers **45.** $g(-3) = 1$ **47.** If m_1 and m_2 are switched, F has the same value. This implies that the magnitude of the gravitational force that m_2 exerts on m_1 is equal to that exerted by m_1 on m_2. **49.** D **51.** 240 ft

53. a.

Time (months)	Balance (dollars)
1	200
2	235
3	270
4	305
5	340
6	375

b. $B(t) = 200 + 35(t - 1)$ **55.** 629,672; the difference between the population of Wisconsin and the population of Minnesota in 1980 **57. a.** $d = \frac{W - 4000}{200}$
b. 50 days
59. 10 hours

61. a.

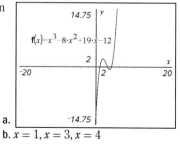

$f(x)=x^3-8\cdot x^2+19\cdot x-12$

b. $x = 1, x = 3, x = 4$

63.

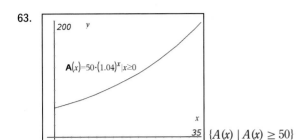

$\{A(x) \mid A(x) \geq 50\}$

65. a.

x	f(x)
1	1
1.2	0.757
1.4	0.330
1.6	−0.359
1.8	−1.419

b. 1.6 **c.** C

Chapter 2

Lesson 2-1 (pp. 72–78)

Mental Math **a.** $\frac{7}{9}$ **b.** $\frac{4}{7}$ **c.** $\frac{1}{3} - \frac{1}{4}$ **d.** $\frac{1}{2} + \frac{1}{4} + \frac{1}{6}$

Guided Example 2 b; g; 729; 2; 729; 64; $\frac{729}{64}$; b; $\frac{729}{64}$; g; $\frac{729}{64}$; 10; 11,390,625; 11,390,625

Questions

1. Answers vary. Sample: The weekly salary of a person varies directly with how many hours that person works that week. **3.** y; x^2; 5 **5.** 118.1 mm **7. a.** B **b.** B
9. 2.09 cm **11.** 7,235.1744 **13. a.** $d = kt$ **b.** 2.4 miles
15. a.

s	1	2	3	4	5	6	7	8	9	10
A	6	24	54	96	150	216	294	384	486	600

b. 4 **c.** 4 **d.** 4 **e.** quadruples **f.** increases by a factor of 9
17. a. 784 feet **b.** 2.28 seconds **19.** No; $x = -1$ is paired with two different y-values.
21. a. Answers vary. Sample:

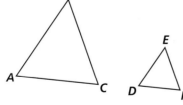

b. They are equal.
23. 5^6

Lesson 2-2 (pp. 79–85)

Mental Math **a.** 8 **b.** 0.32 **c.** $18n^2$ **d.** $18n^2 - 6$

Guided Example 3 d^2; 4; 40; 40; 4^2; 40; 16; 640; 640; 640; 6^2; 17.8

Questions

1. number of workers **3.** A **5.** 3.28 m **7.** 7.9 lumens; it is approximately 0.44 times as intense (or, it is 2.25 times softer). **9.** 187.5 **11. a.**

w	2	3	4	5	6	7	8	9	10	11	12	15	20
t	24	16	12	9.6	8	6.9	6	5.3	4.8	4.4	4	3.2	2.4

b. is halved **c.** is divided by 3 **d.** $t = \frac{48}{w}$, so $\frac{\frac{48}{3w}}{\frac{48}{w}} = \frac{48 \cdot w}{48 \cdot 3w} = \frac{1}{3}$
13. inversely **15.** inversely **17.** 137.9 lb
19. x^7; Quotient of Powers Property: $\frac{x^m}{x^n} = x^{m-n}$
21. a.

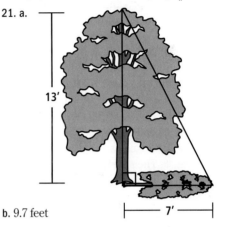

13′

7′

b. 9.7 feet

Lesson 2-3 (pp. 86–92)

Mental Math **a.** 1200 ft³ **b.** 80 ft **c.** 20

Activity Step 1.

Size	Mean Radius r (cm)	r³ (cm³)	Ratio of Radius to Mini's Radius	Ratio of r³ to Mini's r³	Production Cost Estimate
Mini	0.7	0.343	1:1	1:1	$\frac{1}{2}$ cent
Regular	1.4	2.744	2:1	8:1	
Jumbo	2.1	9.261	3:1	27:1	

Step 2. a. 2; 8; 8; 4 **b.** 27; 13.5; 3 **c.** 32 cents; The radius ratio is 4, so the cost ratio is 4^3.
$4^3\left(\frac{1}{2}\text{ cent}\right) = 64\left(\frac{1}{2}\text{ cent}\right) = 32$ cents

Size	Mean Radius r (cm)	r³ (cm³)	Ratio of Radius to Mini's Radius	Ratio of r³ to Mini's r³	Production Cost Estimate
Super Jumbo	2.8	21.952	4:1	64:1	32 cents

Step 3. c^3 Step 4. inversely a.

Size	Radius r (cm)	Ratio of r^3 to Mini's r^3	Number of Pops n in a Carton
Mini	0.7	1:1	270
Regular	1.4	8:1	33
Jumbo	2.1	27:1	10
Super Jumbo	2.8	64:1	4

b. decrease; 8; 8; 33 c. dividing; 3; third; dividing; 4; third
Step 5. c^3
Guided Example 2 Solution 1. 18; 18; 5832; 5832
Solution 2. $18d$; 18^3; kd^3; 5832; 5832
Questions
1. 13.5 cents 3. 4 5. divided by c^n 7. The ancient
rodent was $3.2^3 = 32.8$ times as heavy. 9. When x is
doubled, y is multiplied by 16. This is because $16 = 2^4$.
11. a. The volume is multiplied by 512. b. The volume
is divided by 8. 13. a. Their weight would be about 12^3,
or 1728, times as great. b. Their surface area would
be about 12^2, or 144, times as great. 15. 5:3
17. Let y_1 = original value before multiplying x by c.
Let y_2 = value when x is multiplied by c. To find y_2,
x must be multiplied by c.

$y_1 = \frac{k}{x^n}$ Definition of inverse variation

$y_2 = \frac{k}{(cx)^n}$ Definition of y_2

$y_2 = \frac{k}{c^n x^n}$ Product of Powers Postulate

$y_2 = \frac{1}{c^n} \cdot \frac{k}{x^n}$ Associative and Commutative Properties of Multiplication

$y_2 = \frac{1}{c^n} \cdot y_1$ Substitution of y_1 for $\frac{k}{x^n}$

$y_2 = \frac{y_1}{c^n}$ Definition of division

19. C 21. a. cn cents b. $\frac{d}{2}m$ cents 23. 10.85

Lesson 2-4 (pp. 93–99)
Mental Math a. 0.0145 **b.** $-\sqrt{2}$ **c.** $\frac{7}{11}$ and 0.0145
d. 17 and 512
Activity Step 1. The value of y increases with respect
to x. The graph becomes steeper. **Step 2.** The value
of y decreases with respect to x. The graph becomes
steeper in a downward direction. **Step 3.** The graph is
horizontal. **Step 4.** Answers vary. Sample: Let $k = 3$.
Then $y_1 = 3$. Slope = 3. They are all the same number.
Questions
1. (x_1, y_1) and (x_2, y_2) 3. dependent; independent
5. For every 5 seconds it takes to hear thunder, you
are 1 mile farther away from where the lightning struck.
7. positive; negative 9. line; k; $(0, 0)$ 11. 60 miles per hour
13. Answers vary. Sample: For $k = 3$, $A = (1, 3)$. The point

$B = (2, 6)$ is also on the line $y = 3x$. Using A and B, the
slope is $\frac{6-3}{2-1} = 3$. This is the same slope found using A and
O, and it is equal to k. 15. No; the rate of change between
the first two points is $\frac{1}{3}$, and the rate of change between
second and third points is 1.

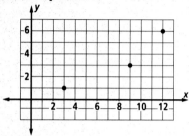

17. a: $y = -4x$; b: $y = 4x$; c: $y = \frac{1}{4}x$; d: $y = -\frac{1}{4}x$
19. a. $\frac{41}{21}$ b. $-\frac{17}{3}$ 21. 9:25 23. direct 25. neither

Lesson 2-5 (pp. 100–105)
Mental Math a. 22 cents **b.** $1.88 **c.** $2.20 **d.** $4.15
Activity Step 1. $y = k(1)^2$, so $y = k$. **Step 3.** It becomes
narrower. **Step 4.** It becomes the x-axis. $A = (1, 0)$.
Step 5. It becomes narrower. When k is negative, the
graph is reflected over the x-axis.
Questions
1. a. 1.5; As speed increases from 10 to 20 miles per hour,
braking distance increases by 1.5 ft for each mph,
on average. b. 5.5; As speed increases from 50 to 60 miles
per hour, braking distance increases by 5.5 ft for each
mph, on average. c. $\frac{d-b}{c-a}$; As speed increases from a to
c miles per hour, braking distance increases by $\frac{d-b}{c-a}$ ft for
each mph, on average. 3. parabola 5. domain: set of
all real numbers, range: $\{y \mid y \leq 0\}$ 7. a. $k > 0$ b. $k < 0$
9. a. A, s b. $\frac{\sqrt{3}}{4}$ 11. a. Substituting $(0, 0)$ into $y = kx^2$
gives the true statement $0 = 0$ for all values of k.
b. Substitute 1 for x: $f(1) = k(1)^2 = k$, so $k = f(1)$ in this
equation. 13. Yes; each 4-inch beam will be half as
strong as the 8-in. beam because strength is directly
proportional to width.
15. a.

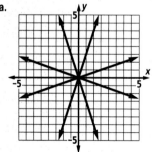

b. 3; $\frac{1}{3}$; $-\frac{1}{3}$; -3 c. $y = \frac{1}{3}x$; $y = -3x$ or $y = -\frac{1}{3}x$; $y = 3x$
17. a. The domain of any sequence is the set of positive
numbers. b. 3818, 4035, 4258, 4487, 4722 19. a. Answers
vary. Samples: A, B, C, D, E, K, M, T, U, V, W, Y
b. Answers vary. Samples: H, I, O, X

Lesson 2-6 (pp. 106–113)

Mental Math a. 108 **b.** 18 **c.** 72 **d.** 108
Activity 1 Step 2. y approaches 0. **Step 3.** y approaches 0.
Step 4. No; the numerator is 10, thus y could never equal 0.
Step 5. No; it is not possible to divide by 0, so 0 is not in
the domain of the function. **Step 6.** Yes, the graphs have
the same behavior. None of the graphs intersect
the x- or y-axis.

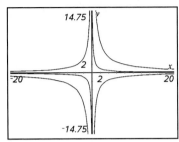

Activity 2 Step 1. It is not possible for the function values
to be negative, since the denominator must be positive.
Step 2. y approaches 0. **Step 3.** y approaches 0.
Step 4. No; the numerator is 10, thus y could never
equal 0. **Step 5.** No; it is not possible to divide by 0.
Step 6. It goes to infinity. The denominator keeps getting
smaller, causing the function values to become greater
and greater.

Questions
1. It is not possible to divide by 0. **3.** hyperbola **5. a.** $-\frac{3}{2}$; 1
b.

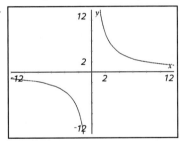

7. a. no **b.** yes **9.** The y-coordinates get larger and larger.
11. a. ii **b.** iii **c.** i **d.** iv **13. a.** 4; 2; $\frac{12}{8} = 1.5$; $\frac{12}{10} = 1.2$
b. $y = \frac{12}{x}$, x = number of kids, y = number of cookies each
kid gets **c.** There cannot be a fractional number of kids.
15. a.

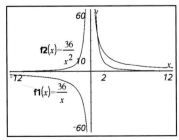

Answers vary. Both
graphs have two
branches, but the
branches of the graph
of f are in Quadrants
I and III, whereas the
branches of the graph
of g are in Quadrants I
and II.

b. –6, –7; the average rate of change of g is more negative
than the average rate of change of f. **c.** $-\frac{1}{2}$, $-\frac{1}{8}$; the average
rate of change of g is less negative than the average
rate of change of f. **17.** $y = -3x^2$ **19. a.** $\frac{4.75}{3} \approx 1.58$
b. $1.58^3 \approx 3.94$ **c.** According to the Fundamental Theorem
of Variation, the answer to Part b is the cube of the answer
to Part a. **21.** C **23. a.** independent variable: time;
dependent variable: temperature
b.

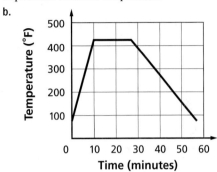

c. domain: 0 to 57; range: 75° to 425°

Lesson 2-7 (pp. 114–120)

Mental Math a. 2 **b.** 4 **c.** $10.12 **d.** no
Guided Example 1 Solution 4.9; 1; 4.9; 4.9 **Check** 4.9; 11.0;
19.6; 30.6; 44.1
Activity Step 1. $k = 10{,}115$; $d = \frac{10{,}115}{w^2}$ **Step 2.** No; only
$w = 85$ lb has the same value, but the other points do not
fit very well. **Step 3.** $k = 119$; $d = \frac{119}{w}$ **Step 4.** The values
are not exactly the same, but they are closer than the first
equation. **Step 5.** Yes; the values that the second equation
gives are much closer to the experimental results.
Questions
1. After it is dropped from the top of a cliff, the
distance d a ball has fallen varies directly as the square
of the time t. **3.** about 10.1 sec **5.** Answers vary.
Sample: Jenna's data shows that when $w = 80$, $d = 1.5$,
but the model gives $d = 1.58$. **7.** 5.95 yd; no, this is
too long for a seesaw to be safe **9. a.** i **b.** $k = 14.3$
c. $V = 200.2$ in^3 **11. a.** iv **b.** $k = 12$ **c.** No; if this were true,
I would have to increase with d, which is not what the data
indicate. **d.** $I = 0.12$ **13.** Matt will need to sit 2.5 times as
far from the pivot as Pat sits. Sample: Matt weighs 80 lb,
Pat weighs 200 lb, and Pat is 3 feet from the pivot. Since
$d = \frac{k}{w}$, $3 = \frac{k}{200}$ and $k = 600$. Then Matt's distance is
$d = \frac{600}{80} = 7.5$, and $3 \cdot 2.5 = 7.5$. **15.** E **17.** C
19. The value of y is multiplied by $3^{23} = 94{,}143{,}178{,}827$.
21. a. $f(1999) = 8{,}698{,}000$ **b.** 2002

Lesson 2-8 (pp. 121–128)

Mental Math a. (2, 7) **b.** (4, 9.5) **c.** (5, 3)

Guided Example 1 Answers vary. Sample: (1, 9) 9; 1; 9; M; 9; w; Answers vary. Sample:

w	1	2	3	4	5	6
M	9	18	27	36	45	54

directly; w

Guided Example 3 Answers vary. Sample answers are given.
1, 212; 2, 106; 3, 71; $\frac{1}{2}$; 2; about $\frac{1}{3}$; 3; inversely; d; d; 212; 1; 212; $M = \frac{212}{d}$;

d	1	2	3	4	5	6
M	212	106	70.7	53	42.4	35.3

inversely; d

Questions
1. w, t, d **3.** 32 lb **5.** a straight line **7.** When d doubles, M is divided by 2, not 4. **9.** numerator **11. a.** P varies directly as w. When w is multiplied by about $\frac{4}{3}$ to go from $w = 75$ to $w = 99$, P is multiplied by $\frac{12.9}{9.8} \approx \frac{4}{3}$. **b.** P varies inversely as h^2. **c.** 17.6 psi; by the table in Part b, $P = 70.2$ when $w = 1.5$, so when w is doubled to 3, $P = \frac{70.2}{4} \approx 17.6$ by the Fundamental Theorem of Variation. **d.** $P = \frac{kw}{h^2}$
13 a. none **b.** none **c.** $x = 0, y = 0$ **d.** $x = 0, y = 0$
15. $r = \frac{512}{729} \approx 0.7$

Lesson 2-9 (pp. 129–135)

Mental Math a. 20 **b.** 10 **c.** 16

Guided Example 3 a. hr^2 **b.** 6; 2; 24; 3.142 **c.** π **d.** $3.142hr^2$ or πhr^2 **e.** cylinder

Questions
1. a. Combined variation is a situation in which both direct and inverse variations occur together. **b.** In joint variation a quantity varies directly as a power of two or more independent variables, but in combined variation there is both direct and inverse variation. **3.** 13.7 mph
5. a. $k \approx 3.4$ **b.** No; she used $k = 3.4$ instead of 3.14 or π.
7. a. F varies jointly as m and a. **b.** $k = 1$ **9. a.** $V = khw\ell$
b. rectangular **c.** $k = 1$ **11. a.** $C = kL(R_o^2 - R_i^2)$
b. $k = 165.33$ **c.** $C = 165.33L(R_o^2 - R_i^2)$ **d.** \$32.40
e. dollars per cubic foot
13. a.

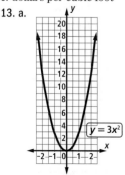

$y = 3x^2$

b. parabola **c.** 9 **d.** No; the rate of change between x_1 and x_2 increases as the values of x_1 and x_2 increase.
15. $k = \frac{wy}{xz}$ **17.** 118.7

Self-Test (pp. 140–142)

1. Direct variation means that n and ℓ increase together. n is the dependent variable and ℓ is the independent variable, so the equation is $n = k\ell$. **2.** Direct variation means that w increases as d^4 increases, and inverse variation means that w decreases as L^2 increases. w is the dependent variable and d and L are the independent variables, so the equation is $w = \frac{kd^4}{L^2}$. **3.** By the form of the equation, we know that s is the dependent variable, p is the independent variable, and k is the constant of variation. Because p^4 is in the denominator, this means that as p^4 increases, s decreases. So, to express this equation, we say: s varies inversely as p^4. **4.** The equation described is $T = ks^3w^2$. Given $s = 2$ and $w = 1$, $T = 10$, thus $10 = k(2^3)(1^2)$ and $k = \frac{5}{4}$. Therefore we can rewrite the equation as $T = \frac{5}{4}s^3w^2$. So for $s = 8$ and $w = \frac{1}{2}$, $T = \frac{5}{4}(8)^3\left(\frac{1}{2}\right)^2 = 160$. **5.** From the equation we can see that y varies inversely as x^2. Therefore, by the Fundamental Theorem of Variation, doubling the x-value divides the y-value by 4. For example, the points $(1, 5)$ and $\left(2, \frac{5}{4}\right)$ satisfy the equation. **6.** If y is multiplied by 8 when x is doubled, this implies that y and x are directly related. As the cube of 2 is 8, we see that y varies directly as the cube of x. A general equation to express this relationship is $y = kx^3$. **7.** The average rate of change can be found by plugging the points into the formula $m = \frac{y_2 - x_2}{x_2 - x_1}$. Here we get $m = \frac{16 - 4}{4 - 2} = \frac{12}{2} = 6$. **8. a.** False; for inverse variation equations $y = \frac{k}{x^n}$, there is no value for y when $x = 0$. **b.** False; if y varies directly with x^n, where $n > 1$, the rate of change will not be constant.
9. a. parabola; $k > 0$ **b.** The opening of the parabola grows wider. **10.** $\{d \mid d \neq 0\}$. Since 0 cannot be in the denominator of a fraction, d^2 cannot equal 0, and thus $d \neq 0$. d is defined for all other real numbers.
11. a. Directly; the volume of the sphere will increase as the radius increases. **b.** Inversely; the more expensive the stock, the fewer shares of stock you can buy.
12. a. Answers vary. Sample:

x	y
0	0
1	−0.5
2	−1
3	−1.5
4	−2
5	−2.5

b.

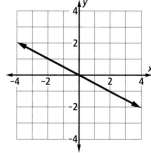

c. We can use two points from the table to find the slope. Because the slope is constant, any two points will work. Therefore, $m = \frac{-2-(-1)}{4-2} = -\frac{1}{2}$.

13. a.

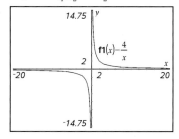

b. $x = 0$, $y = 0$ **14. a.** B; the answer cannot be A or D because both of these graphs would have constant slope. The answer cannot be C because the y-values of this function must be positive, but the graph shown does not have this property.   **b.** Because you cannot have 0 in the denominator, $x \neq 0$, but all other values of x are in the domain. Likewise $y \neq 0$, but all other values of y are in the range.

15. a.

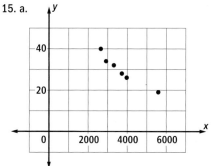

b. $F = \frac{k}{w}$. As w doubles, F is halved. Find the constant of variation by plugging in a pair of values. Answers vary. Sample: $40 = \frac{k}{2655}$, $k = 106{,}200$.   **c.** Plug in 5900 for the weight to find that according to the model, $F = \frac{106{,}200}{5900} = 18$ mpg.   **d.** The model shows that as vehicle weight increases, fuel economy decreases.   **16.** V is the dependent variable, and because its graph with g is a parabola, it is related to g as follows: $V = kg^2$. Its graph with h is a line, so it is related to h as follows: $V = kh$. Because V varies jointly with g and h, we can combine these two equations into the general equation $V = kg^2h$.   **17. a.** Because height is inversely proportional to the square of the radius by the Fundamental Theorem of Variation, as r doubles, h will be divided by 4. So when r increases from 5 cm to 10 cm, h will decrease from 10 cm to $\frac{10}{4} = 2.5$ cm.   **b.** In general, if two cylinders have the same volume and the radius of one is double the radius of the other, their heights are related in the ratio 1:4.   **18. a.** If Paula is right, then the equation that fits is $m = kd^3$. Therefore, plugging in the given points, $k = \frac{6.0 \cdot 10^{24}}{(12{,}700)^3} = 2.9 \cdot 10^{12}$. Thus the equation would be $m = 2.9 \cdot 10^{12}d^3$.   **b.** The model from Part a predicts Jupiter's mass to be 1331 times the mass of Earth, because $11^3 = 1331$. Thus Jupiter's mass would be about $8.0 \cdot 10^{27}$ kg.   **c.** Answers vary. Sample: Not all planets are made of the same materials. Jupiter is less dense than Earth; therefore, their masses cannot vary directly as their volumes alone.   **19. a.** $d = kt$; the data appear to lie on a line, so d varies directly with time, not the square of time.   **b.** Answers vary slightly, depending on the data point chosen. Sample: $41 = 30k$, $k = \frac{41}{30}$, $d = \frac{41}{30}t$.   **c.** Answers vary slightly. Sample: $d = \frac{41}{30}(180) = 246$ in.   **d.** Answers vary slightly. Sample: 50 ft = 600 in., $600 = \frac{41}{30}t$, $t \approx 439$ minutes; the water can run for 439 minutes or 7 hours, 19 minutes.

Self-Test Correlation Chart

Question	1	2	3	4	5	6	7	8	9	10
Objective(s)	A	A	A	B	D	D	C	E	E	E
Lesson(s)	2-1, 2-2, 2-9	2-1, 2-2, 2-9	2-1, 2-2, 2-9	2-1, 2-2, 2-9	2-3, 2-8	2-3, 2-8	2-4, 2-5	2-4, 2-5, 2-6	2-4, 2-5, 2-6	2-4, 2-5, 2-6

Question	11	12	13	14	15	16	17	18	19
Objective(s)	F	C, I	E, I	E, J	G, H, I	H	F	F, G	F, G, H
Lesson(s)	2-1, 2-2	2-4, 2-5, 2-6	2-4, 2-5, 2-6	2-4, 2-5, 2-6	2-1, 2-2, 2-4, 2-5, 2-6, 2-7, 2-8, 2-9	2-7, 2-8	2-1, 2-2	2-1, 2-2, 2-9	2-1, 2-2, 2-7, 2-8, 2-9

Chapter Review (pp. 143–147)

1. $y = kx$ **3.** $z = kxt$ **5.** directly; x; inversely; the square of v **7.** y varies inversely as the third power of w.
9. $y = 49$ **11.** $y = \frac{1}{3}$ **13.** $\frac{32}{7}$ **15.** 8 **17.** $-\frac{1}{40}$ **19.** y is tripled. **21.** y is multiplied by 3^n. **23.** divided by c^n
25. y is multiplied by $\frac{a^n}{b^n}$. **27.** parabola **29.** B, D **31.** true
33. $x = 0$ **35.** $n = k\ell^2$ **37.** $F = \frac{k}{d^2}$ **39.** directly
41. inversely **43.** only 0.0225 min, or about 1.35 sec
45. 500,000,000 seconds, or about 16 yr **47.** ≈ 944.58 lb
49. a.

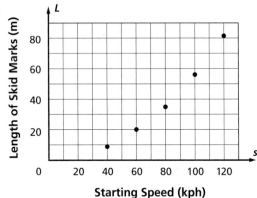

b. $L = kS^2$ **c.** $k = \frac{9}{1600}$; $L = \left(\frac{9}{1600}\right)S^2$ **d.** $L \approx 126.6$ m
51. $P = kdD$; because the points on both graphs lie on straight lines, P varies directly D, and also as d. Therefore, P varies jointly as D and d.
53. a.

q	p
0	undef
1	2
2	5
3	0.222
4	0.125
5	0.08
6	0.056

b.

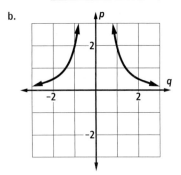

55.

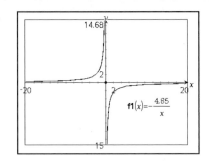

57. The branches pull away from the y-axis.
59. C **61.** negative

Chapter 3

Lesson 3-1 (pp. 150–156)

Mental Math a. 31 or 32 **b.** 17 **c.** 14
Guided Example 2 a. 26; $8t$ **b.** $26 - 8t$; $\frac{13}{4}$; $3\frac{1}{4}$ **c.** 26; –8
Questions
1. m; b **3.** 1 minute, 45 seconds **5.** y-intercept; slope
7. $\frac{1}{2}$ **9.** $y = \frac{2}{3}x - 3$ **11.** a, c, and e; b and d
13. a. $y = 0.03x + 29.99$ **b.** 0.03; 29.99; The slope is the constant rate per minute, the y-intercept is the initial condition, the monthly service fee. **c.** $34.07 **15.** 1
17.

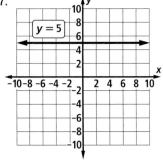

19. a. Answers vary. Sample: $(1, -180)$, $(4, 180)$, $(5, 300)$, $(7, 540)$ **b.** $T = 120D - 300$ **c.** $28,080 **21.** joint
23. combined **25. a.** inversely **b.** $n = 1$ **27.** $y = 48 - 6x$

Lesson 3-2 (pp. 157–162)

Mental Math a. $-12 < x < 12$ **b.** $-6 < x < 6$ **c.** $-4 < x < 8$
d. $x \le -4$ or $x \ge 8$
Activity Step 1. $2.5b + 2d = 30$ **Step 2.** $d = 15 - 1.25b$
Step 4. $(0, 15)$, $(4, 10)$, $(8, 5)$, $(12, 0)$
Guided Example 1 a. 30: 3,:4, continuous **b.** $6: $3: $30: buying 3 adult tickets and 4 child tickets; discrete
Questions
1. a. linear combination **b.** Answers vary. Sample: Biking 600 miles, with P hours at 15 mph and L hours at 12 mph **3.** $0.5T + 0.75S + 1.25K$
5. a. $b + t = 12$ **b.** discrete **7. a.** $5.2x$ **b.** $7.8y$ **c.** $5.2x + 7.8y$
d. $5.2x + 7.8y = 3.6$ **e.** approximately 0.19 L
9. a. nonnegative integers **b.** $190 **c.** $25L + 15S = 225$

d.

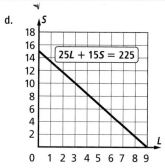

e.

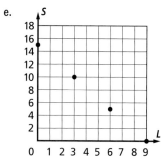

f. $(0, 15)$, $(3, 10)$, $(6, 5)$, $(9, 0)$ **11.** $\frac{5}{3}$ lb   **13.** $t = ksr$ and $r = pm$, where k and p are constants. So then by substitution, $t = ks(pm) = kpsm$. Let $a = kp$. Then a is a constant and $t = asm$, so t varies jointly with s and m.
15. a. $w = kh^3$ **b.** $\frac{1}{173} = \frac{1}{4913}$

Lesson 3-3 (pp. 163–168)

Mental Math a. $c = kn$ **b.** $p = klw$ **c.** $t = \frac{k}{n}$ **d.** $d = \frac{kt}{a^2}$
Guided Example 3 $-\frac{8}{3}x + 8$; $-\frac{8}{3}x + 8$; $-\frac{8}{3}x + 4$; $-\frac{8}{3}x + 2$; (1) and (2); (3) and (4); (1); (2); (1); (2)
Questions
1. a. line   **b.** vertical   **c.** horizontal   **3.** It is undefined.
5. a. 9   **b.** 4
c.

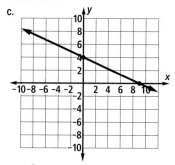

7. a. $\frac{C}{A}$; there is no x-intercept.   **b.** $\frac{C}{B}$; there is no y-intercept
9. 42   **11. a.**

$0x + 4y = 14$

b. Answers vary. Sample: $(0, 3.5)$, $(5, 3.5)$ **c.** 0
13. $x = 17$ **15. a.** $y = -\frac{A}{B}x + \frac{C}{B}$; $y = -\frac{A}{B}x + \frac{D}{B}$
b. The slopes are equal, so the lines are parallel or are the same line.   **c.** The intercepts are different, so the lines cannot be the same line and are thus parallel.
d. If equations for two lines are written in standard form with the same coefficients of x and y but different right sides, then the lines are unique and parallel.
e. Answers vary. Sample: $16x - 13y = 15$, $16x - 13y = 20$
17. $0.50P + H$ **19. a.** $400 - 60t = d$ **b.** constant decrease
21. Answers vary. Samples: **a.** $\{h \mid 0 \le h \le 16\}$ **b.** $\{d \mid 0 < d\}$.
c. $\{t \mid -20 \le t \le 50\}$.

Lesson 3-4 (pp. 169–175)

Mental Math a. $\frac{2}{3}$ **b.** $\frac{3}{11}$ **c.** $\frac{b}{b+r+w}$
Guided Example 2 $-\frac{9}{2}$; 7; $-\frac{9}{2}$; 4; $y = -\frac{9}{2}x + 25$
Questions
1. 2   **3.** false   **5.** Compute the slope and then use either point in the point slope form $y - y_1 = m(x - x_1)$.
7. $y = 6x + \sqrt{3}$ **9.** $y + 3 = \frac{12}{7}(x + 2)$ **11. a.** $c = 1.45s + 30$
b. $30   **c.** $175   **13. a.** 0.759398
b. Define $\texttt{ptslope(a,b,m)} = \texttt{m} \cdot \texttt{(x-a)+b}$
c. $y = 0.759398x - 10.7376$ **d.** Define $\texttt{line2pt}$ $\texttt{(a,b,c,d)} = \texttt{ptslope(a,b,slope(a,b,c,d))}$
e. yes   **15. a.** $w = f(x) = \begin{cases} x + 10, \text{ for } 3 \le x \le 12 \\ \frac{5}{12}x + 17, \text{ for } 12 < x \le 60 \end{cases}$
b. 29.5 pounds   **17.** not 0; 0   **19.** D   **21. a.** $r_3 = 1700$
b. r sub seven equals negative one thousand three hundred twenty-seven.

Lesson 3-5 (pp. 176–181)

Mental Math a. 2.5   **b.** 0   **c.** 0.144   **d.** –0.8
Activity 1 Step 1. Answers vary. Sample:

Number of Squares	Score
240	5083
45	859
12	366
165	3115
174	3407
216	4257
165	3269
69	1409
225	4612
209	4097
225	4524
315	6340
210	4388
248	4807
150	2794

Step 2. Answers vary. Sample:

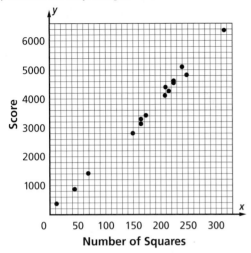

Number of Squares

Yes, the data points appear to have a linear pattern.

Step 3. Answers vary. Sample:

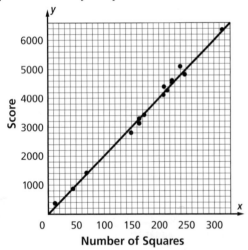

Number of Squares

Step 4. Answers vary. Sample: (200, 4000), (300, 6000). In this case the equation is $y = 20x$. **Step 5.** Answers vary. Sample: Using the equation in the previous step would give a score of 2000.

Activity 2 Step 3. Answers vary. Sample: $y = 0.4829x - 29.90$ **Step 4.** Answers vary. Sample: It is very close to the regression line.

Step 5. Answers vary. Sample: At $x = 20$, the predictions differ by 0.01242. When $x = 80$, the prediction from the movable line differs from the value in the table by ≈ 0.73, and the prediction from the regression line differs from the value in the table by ≈ 0.74.

Questions

1. The student captured 48 squares, for a score of 985 points.

3. a. 19 **b.** 19 points are scored for every square captured.

c. 35 **d.** The player starts with 35 points. **e.** no

5. a. $y = -\frac{4}{3}x + \frac{16}{3}$ **b.** $y = -\frac{4}{3}x + \frac{16}{3}$

7.

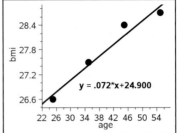

9. $y = 11.74x + 204.3$ **11. a.** $-\frac{5}{2}x + 3$ **b.** $-\frac{5}{2}x - \frac{9}{2}$ **13. a.** $-6, 7, -8, 9$ **b.** negative **15. a.** $-\frac{11}{9}$ **b.** $\frac{5}{7}$ **c.** $\frac{3a+5}{7-8a}$ **d.** $-\frac{200}{897} \approx -0.22$

Lesson 3-6 (pp. 182–188)

Mental Math a. 31.5 **b.** 5007 **c.** 22 **d.** the set of positive integers

Guided Example 1 113; 154; 154; 195; 195;236; 154; 195; 236

Guided Example 2 10,000; 400; 1

Guided Example 5 A1; $= \frac{A1}{2}$; A2; A3; A6; 64; 32; 16; 8; 4; 2

Questions

1. The answer of the last input **3.** a set of statements that indicate the first (or first few) terms and tell how the next term is calculated from the previous term or terms.

5. $\begin{cases} b_1 = 64 \\ b_{n+1} = \frac{1}{2}b_n, \text{ for integers } n = \{1, 2, 3, 4, 5\} \end{cases}$

7. a. $-2, 6, 14, 22, 30$ **b.** $\begin{cases} a_1 = -2 \\ a_n = a_{n-1} + 8, \text{ for } n \geq 2 \end{cases}$

9. Answers vary. Sample: If the formula did not have two parts, you would be missing either the initial condition or the instructions for how to calculate the following terms.

11. $\begin{cases} t_1 = -2 \\ t_{n+1} = t_n - 2, \text{ for } n \geq 1 \end{cases}$

13.a. $\begin{cases} S_1 = 8 \\ S_n = S_{n-1} + 2, \text{ for } 2 \leq n \leq 25 \end{cases}$ **b.** 34 seats

15. Yes; for each n there is a unique S_n. **17. a.** The initial term of the sequence is 6.42, and each subsequent term in the sequence is 2.73 less than the previous term.

b. $\begin{cases} a_1 = 6.42 \\ a_n = a_{n-1} - 2.73, \text{ for } n \geq 2 \end{cases}$ **c.** -12.69

19. Yes; there appears to be a linear relationship between attendance and payroll.

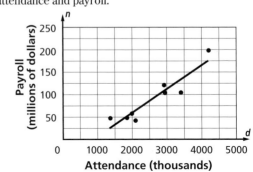

21. a.

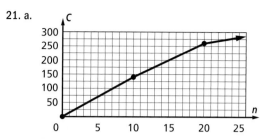

b. $C = \begin{cases} 14n \text{ for } 0 \leq n \leq 10 \\ 12n + 20, \text{ for } 10 < n \leq 20 \\ 10n + 60, \text{ for } 20 < n \end{cases}$

23. a. After 3 seconds, the object has fallen 44.1 feet.
b. No; the object cannot fall for a negative amount of time.

Lesson 3-7 (pp. 189–195)

Mental Math a. $21x$ **b.** $21x$ **c.** $6x - 18y$

Guided Example 1 a.

n	a_n
1	–2
2	2
3	6
4	10
5	14
6	18

b.

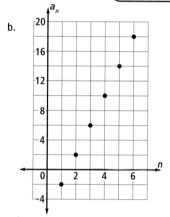

Questions

1. a. 34 **b.** $\begin{cases} a_1 = -2 \\ a_{n+1} = a_n + 4, \text{ for } n \geq 1 \end{cases}$ **3.** D

5. a. 55 **b.** 5 **c.** 5 **7. a.** $\begin{cases} b_1 = 10 \\ b_{n+1} = 1.2b_n, \text{ for } n \geq 1 \end{cases}$

b.

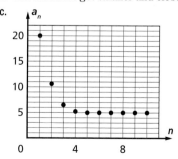

c. 22 days **9. a.**

n	b_n
1	$500
2	$575
3	$650
4	$725
5	$800
6	$875

b. $\begin{cases} b_1 = 500 \\ b_n = b_{n-1} + 75, \text{ for } n \geq 2 \end{cases}$

c.

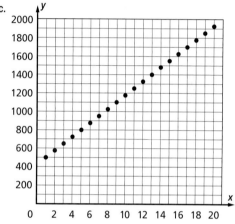

d. Yes; All the points lie on the line $y = 500 + 75n$ because Kamilah saves the same amount every week.

11. a.

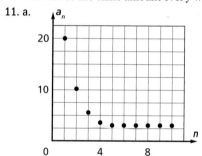

b. The numbers get smaller and closer to 3.

c.

The numbers get smaller and closer to 5. The only change in the answer between the two formulas is in the numbers that the sequences approached.

13. a. one third of **b.** $\begin{cases} b_1 = 81 \\ b_n = \frac{1}{3}b_{n-1}, \text{ for } n \geq 2 \end{cases}$

c. $1, \frac{1}{3}, \frac{1}{9}, \frac{1}{27}$ **15.** E

17. a.

n	s_n
1	1
2	2
3	4
4	7
5	12
6	20
7	33
8	54
9	88
10	143
11	232
12	376
13	609
14	986

b. $s_n = F_{n+2} - 1$ **c.** Answers vary. Sample: If we let t_n be the sum of the squares of the first n Fibonacci numbers, then a table for t_n is as follows:

n	t_n
1	1
2	2
3	6
4	15
5	40
6	104
7	273
8	714
9	1870
10	4895
11	12,816
12	33,552
13	87,841
14	229,970

An equation for t_n is $t_n = F_n \cdot F_{n+1}$.

Lesson 3-8 (pp. 196–202)

Mental Math a. parabola **b.** line **c.** none of these **d.** hyperbola

Questions

1. 151 **3.** $a_n = a_1 + (n-1)d$ **5. a.** $a_n = 12 + (n-1)4$

b. $\begin{cases} a_1 = 12 \\ a_n = a_{n-1} + 4, \text{ for integers } n \geq 2 \end{cases}$ **c.** 196

7. $\begin{cases} a_1 = 1.3 \\ a_n = a_{n-1} + 0.3, \text{ for integers } n \geq 2 \end{cases}$

9. a. $\begin{cases} t_1 = 12 \\ t_n = t_{n-1} + 7, \text{ for integers } n \geq 2 \end{cases}$ **b.** 544

11. $\begin{cases} a_1 = -70 \\ a_{n+1} = a_n + 22.5, \text{ for integers } n \geq 1 \end{cases}$

13. $a_n = 13 - 4(n-1)$

15. \$400 **17. a.** 15.85, 15.71, 15.56

b. $r_n = 15.85 - 0.14545(n-1)$, or $r_n = 16 - 0.14545n$

19. $T_1 = \frac{1(1+1)}{2} = 1; T_2 = \frac{2(2+1)}{2} = 3 = 1 + 2;$
$T_3 = \frac{3(3+1)}{2} = 6 = 3 + 3$ **21.** E **23. a.** $k = -\frac{1}{2}$
b. k is the slope of the line.

Lesson 3-9 (pp. 203–209)

Mental Math a. 10, 12 **b.** –1, –7 **c.** 1, –1

Guided Example 1 a. 5 **b.** greatest integer; –5 **c.** 3.1416; π;
4; **d.** smallest integer; 13; 13

Guided Example 4 6; 0.18; 18

Questions

1. \$1.38 **3.** Answers vary. Sample: $\lceil x \rceil$ means the smallest integer greater than or equal to x. It is called the ceiling function because the ceiling of a room is the *smallest* (or *next up*) level of the building that is above that room.

5. 12 **7.** 7 **9. a.** domain: set of all real numbers; range: set of all integers **b.** domain: set of all positive real numbers; range: {2.50, 2.90, 3.30, ...} **11. a.** 1 **b.** 2 **c.** r rounds to the greatest integer less than or equal to $x + 0.5$.

13. $750 - g\lfloor \frac{750}{g} \rfloor$ **15. a.** yes **b.** $5.10 + 1.15\lceil \frac{a}{100} \rceil$

17. a. Answers vary. Sample: Monday **b.** Thursday

c. Monday **19. a.** $\begin{cases} a_1 = \sqrt{2} \\ a_n = a_{n-1} + 2\sqrt{3}, \text{ for } n \geq 2 \end{cases}$

b. $a_n = \sqrt{2} + 2\sqrt{3}(n-1)$ **c.** $\sqrt{2} + 200\sqrt{3}$

21. a. $y - 2 = 2.5(x - 2)$ **b.** $y + 3 = 2.5x$
c. $y = 2.5(x - 2) + 2 = 2.5x - 5 + 2 = 2.5x - 3$

Self-Test (pp. 213–214)

1.

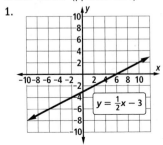

$y = \frac{1}{2}x - 3$

2. a. $\frac{5}{3}$, since $-3y = 10 - 5x$, $y = -\frac{10}{3} + \frac{5}{3}x$ **b.** *x*-intercept: 2, *y*-intercept: $-\frac{10}{3}$; $5x - 3(0) = 10$, $x = 2$; $5(0) - 3y = 10$, $y = -\frac{10}{3}$ **3.** constant decrease; as the *x*-value increases, the *y*-value decreases **4.** $m = \frac{7-3}{-2-5} = -\frac{4}{7}$, so an equation in point-slope form is $y - 7 = -\frac{4}{7}(x + 2)$. **5.** The line will have the same slope, $-\frac{2}{3}$. Use point-slope form: $y - 3 = -\frac{2}{3}(x + 6)$. **6. a.** vertical lines, division by zero is undefined **b.** horizontal lines, there is no vertical change **7.** H hot dogs cost 2.25*H*, and T tacos cost 3.75*T*, so the total amount is $2.25H + 3.75\,T$. **8.** The constant change is $-\frac{1}{3}\frac{ft}{sec}$ and the initial condition is 310; $h = 310 - \frac{1}{3}t$ **9. a.** Explicit; you can find the value of a_n directly by plugging in the value of *n*. **b.** $a_1 = 50 - 3(1 - 1) = 50$; $a_2 = 50 - 3(2 - 1) = 47$; $a_3 = 50 - 3(3 - 1) = 44$; $a_4 = 50 - 3(4 - 1) = 41$; $a_5 = 50 - 3(5 - 1) = 38$ **c.** Yes; there is a constant difference of 3 between consecutive terms **10. a.** $C = 23.50 + 0.08(m - 200)$; $C = 23.50 + 0.08(453 - 200)$;

$C = \$43.74$ **b.** $C = \begin{cases} 23.50, 0 \le m \le 200 \\ 23.50 + 0.08(m - 200), m > 200 \end{cases}$

11. a. 29, 47, 76, 123 since $11 + 18 = 29$, $18 + 29 = 47$, $29 + 47 = 76$, $47 + 76 = 123$

b. $\begin{cases} L_1 = 1 \\ L_2 = 3 \\ L_{n+2} = L_n + L_{n+1}, n \ge 1 \end{cases}$ **12.** $m = \frac{305.4 - 273.2}{90 - 32} = \frac{32.2}{58} = 0.56$;

$K - 273.2 = 0.56(F - 32)$; $K - 273.2 = 0.56F - 17.9$; $K = 0.56F + 255.3$ **13.** $a_n = -25 - 20(n - 1)$, because -20 is the common difference between terms and -25 is the initial term. **14.** B; each piece has a steeper slant than the previous one.

15.

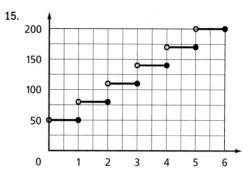

16. a. D; you must add to the initial 50 cent cost, so it cannot be A or C, and $\lceil 1 - w \rceil$ will be negative for $w > 2$, so it cannot be B. **b.** $c = 0.50 + 0.30\lceil 3.2 - 1 \rceil = \$1.40$

17.

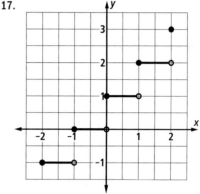

18. a. $y \approx -0.875x + 75.9$ **b.** $y \approx -0.874\,(42) + 75.9 \approx 39.2$ **c.** (80, 8.9); It means that the regression line gives a less accurate prediction for an 80-year-old. **d.** Answers vary. People who are ages 80–100 have many more health

Self-Test Correlation Chart

Question	1	2	3	4	5	6	7	8	9	10
Objective(s)	M	A	G	M	B	E	H	G	D, F	K
Lesson(s)	3-1, 3-3	3-1, 3-3	3-1, 3-8	3-1, 3-3	3-4	3-1, 3-2, 3-3	3-2	3-1, 3-8	3-6, 3-8	3-4, 3-9

Question	11	12	13	14	15	16	17	18
Objective(s)	D, L	I	D	N	N	C, K	N	J
Lesson(s)	3-6, 3-7, 3-8	3-4	3-6, 3-8	3-4, 3-7, 3-9	3-4, 3-7, 3-9	3-4, 3-9	3-4, 3-7, 3-9	3-5

factors affecting their life expectancy than people who are younger.

Chapter Review (pp. 215–219)

1. a. 3 **b.** 4 **c.** -12 **3. a.** 0 **b.** no *x*-intercept **c.** -17 **5. a.** $\frac{6}{5}$ **b.** $-\frac{1}{3}$ **c.** $\frac{2}{5}$ **7.** $y - 10 = \frac{2}{5}(x + 5)$ **9.** $y = -(x + 3) + 4$, or $y = -x + 1$ **11. a.** 13 **b.** -13 **13.** 8 **15. a.** $a_n = 37 + (n - 1)(-21) = -21n + 58$ for integers $n \ge 1$

b. $\begin{cases} a_1 = 37 \\ a_n = a_{n-1} - 21, \text{ for integers } n \ge 2 \end{cases}$ **c.** -26, -47, -68, -89, -110 **17. a.** $a_n = -\frac{11}{12} + \frac{n}{4}$ for integers $n \ge 1$ **b.** $-\frac{2}{3}, -\frac{5}{12}$, $-\frac{1}{6}, \frac{1}{12}, \frac{1}{3}$ **19.** true **21.** $-\frac{b}{m}$ **23.** Answers vary. Sample: $(x + 4, y - 3)$ **25.** It is a set of collinear points. **27.** no **29.** yes **31.** 0.8 **33. a.** $p = 44 - 2t$ **b.** 9 hours **35. a.** $A + B$ **b.** $2.5A + 6.25B$ **c.** $0.75 = 2.5A + 6.25B$ **d.** Answers vary. Sample: (0.3, 0), (0, 0.12), (0.25, 0.02) **37.** 70 mL

39. a.

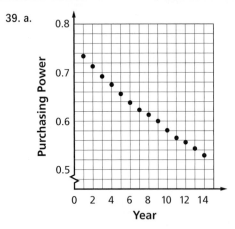

b. $y \approx -0.016x + 0.790$ **c.** Yes; the decrease in the purchasing power is relatively stable from year to year.

41. $C = \begin{cases} 0.99d \text{ for } d \le 20 \\ 19.80 + 0.89(d - 20) \text{ for } d > 20 \end{cases}$ **43. a.** 3 mph

b. 3 mph **c.** $d = \begin{cases} 3t, \text{ for } t \le 0.5 \\ 1.5, \text{ for } 0.5 < t \le 6.5 \\ -3t + 21, \text{ for } 6.5 < t \le 6.75 \\ 0.75, \text{ for } 6.75 < t \le 7.5 \\ -3t + 23.25, \text{ for } 7.5 < t \le 7.75 \end{cases}$

d. Her speed walking home was the same. We can use the graph to calculate the rate of change of d with respect to t to see this. In both cases she walked at 3 mph.

45. a. $\begin{cases} a_1 = 2300 \\ a_n = a_{n-1} - 26, \text{ for integers } n \ge 2 \end{cases}$

b. $a_n = 2326 - 26n$ for integers $n \ge 1$ **c.** Yes; she will have 246 nuts left at the end of the winter.

47.

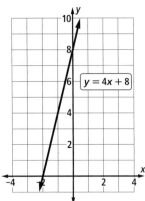

$y = 4x + 8$

49.

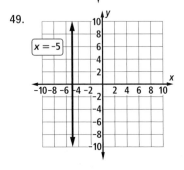

$x = -5$

51. negative   **53.** Answers vary. Sample: $y = -\frac{1}{3}x - 1$

55. a.

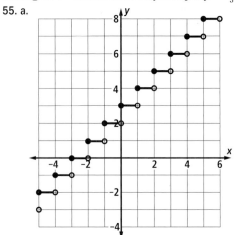

b. Domain: set of all real numbers; range: set of all integers

57. $a_n = 6n - 4$

Chapter 4

Lesson 4-1 (pp. 222–227)

Mental Math a. $y = 4x + 2.5$ **b.** $x = -7$ **c.** $y = \frac{1}{3}x + \frac{9}{10}$
d. $y = 12$

Guided Example a. 3.5; 1.9; 3.4; 1.9; 3.9; 2.8; 3.5; 3.4; 3.9; 1.9; 1.9; 2.8; **b.** 3; 2; 3 × 2; 2; 3; 2 × 3

Activity Step 1. Answers vary.

Sample: $\begin{bmatrix} 3 & 6 & 4 & -2.5 & -5 \\ -5 & -1 & 5 & 4 & -0.75 \end{bmatrix}$

Step 2. Answers vary.

Sample: $\begin{bmatrix} 6 & 4 & -2.5 & -5 & 3 \\ -1 & 5 & 4 & -0.75 & -5 \end{bmatrix}$

and $\begin{bmatrix} -2.5 & -5 & 3 & 6 & 4 \\ 4 & -0.75 & -5 & -1 & 5 \end{bmatrix}$

Step 4.

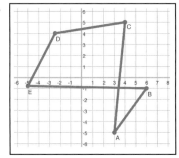

$BACDE$ is not a polygon because $\overline{AC}$ intersects $\overline{BE}$.

Questions

1. A matrix is a rectangular arrangement of objects.

3. Answers vary. Sample: Find the matrix command and enter the dimensions.

5.

	1980	1985	1990	2000
Males	3.5	3.3	3.4	3.9
Females	1.9	1.8	1.9	2.8

7. a. $\begin{bmatrix} x \\ y \end{bmatrix}$ **b.** a point matrix

9. $\begin{array}{ccc} A & B & C \\ \begin{bmatrix} -3 & 5 & -2 \\ 4 & 1 & -2 \end{bmatrix} \end{array}$ **11. a.** 4×2 **b.** the number of law

degrees earned in 2000 **c.** the total number of professional degrees in Medicine, Dentistry, Law, and Theology earned by women in 2000 **13.** -3; 0.1

15. a.
$\begin{array}{cccccc} & A & E & I & O & U \\ \text{English} & \begin{bmatrix} 0.08 & 0.13 & 0.07 & 0.08 & 0.03 \\ \text{SCRABBLE} & 0.09 & 0.12 & 0.09 & 0.08 & 0.04 \end{bmatrix} \end{array}$

b. Answers vary. Sample: Go to the matrix prompt and then specify the 2 by 5 dimensions and enter the elements in their appropriate location.

17.

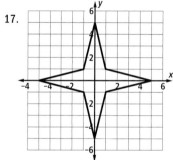

This is a nonconvex polygon. **19.** $a_n = 50 - 2n$
21. The lengths of the edges of A are 3 times as long as the lengths of the edges of B.

Lesson 4-2 (pp. 228–234)

Mental Math a. $11.00 **b.** $12.50 **c.** $10.75 **d.** $4.5p + 3.25s + 3c$

Activity Step 2. $\begin{bmatrix} 2a & 2b & 2c \\ 2d & 2e & 2f \end{bmatrix}$ **Step 3.** $\begin{bmatrix} 3a & 3b & 3c \\ 3d & 3e & 3f \end{bmatrix}$

Step 4. $\begin{bmatrix} 4a & 4b & 4c \\ 4d & 4e & 4f \end{bmatrix}$; The coefficients seem to be increasing by 1 each time I add the original matrix.

Step 5. $\begin{bmatrix} 2a & 2b & 2c \\ 2d & 2e & 2f \end{bmatrix}$ **Step 6.** $\begin{bmatrix} 3a & 3b & 3c \\ 3d & 3e & 3f \end{bmatrix}$

Step 7. $\begin{bmatrix} 4a & 4b & 4c \\ 4d & 4e & 4f \end{bmatrix}$; Again, the coefficients seem to be increasing by 1. **Step 8.** Answer vary. Sample: Yes; Steps 4 and 7 show that the result of adding a matrix to itself a certain number of times is the same as multiplying the original matrix by the number of times you added it.

Questions

1. No; in order to add two matrices, they must have the same dimensions. The two matrices in question have different dimensions. **3.** $\begin{bmatrix} 5 & -15 \\ -6 & 3 \end{bmatrix}$ **5.** $\begin{bmatrix} -1 & -3 \\ -\frac{9}{2} & 10 \end{bmatrix}$

7. $\begin{bmatrix} 6 & 8 & 4 & 3 & 1 \\ 7 & 8 & 5 & 2 & 5 \\ 13 & 1 & 1 & 4 & 4 \\ 12 & 1 & 9 & 4 & 2 \end{bmatrix}$; The matrix represents the cars at

Rusty's Car Dealership that were not damaged by the hailstorm. **9. a.** false **b.** $B - C = \begin{bmatrix} -1 & -1 \\ -1 & -1 \end{bmatrix}$;

$C - B = \begin{bmatrix} 1 & 1 \\ 1 & 1 \end{bmatrix}$ **11. a.** $M = \begin{bmatrix} 2 & -2 & 6 & -101 \\ -3 & 3 & -36 & 57 \\ -2 & 2 & -12 & 113 \\ -4 & 4 & -76 & 47 \end{bmatrix}$

b. the difference in points scored against each team in 2006 from 2005 **c.** 2 means New England had 2 more wins in 2006 compared to 2005; -2 represents the 2 fewer losses New England had in 2006 compared to 2005; 6 represents how many more points New England scored in 2006 compared to 2005; -101 represents how many fewer points against New England were incurred in 2006 compared to 2005 **13. a.** $k = \frac{1}{3}$ **b.** $\ell = 3$ **15.** $a = 4, b = 63, c = -29,$

$d = -5$ **17.** $\begin{bmatrix} 0 & 0 & -1 \\ 0 & 4 & 0 \end{bmatrix}$

19.

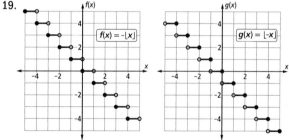

Answers vary. Sample: $f(x) = -\lfloor x \rfloor$ is always 1 greater than $g(x) = \lfloor -x \rfloor$ for all values of x, except when x is an integer, in which case $f(x) = g(x)$. **21. a.** Area $= 4x \cdot 0.5x = 2x^2$

b.

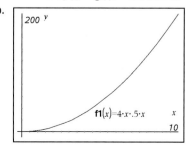

c. $\{x \mid x > 0\}$ **d.** the set of real numbers

e.

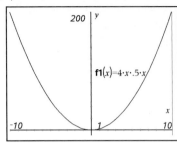

Lesson 4-3 (pp. 235–241)

Mental Math a. $2.50 **b.** $4.80 **c.** $14.80

Guided Example 2 2; 3; 6; $3 \cdot -1$; $1 \cdot 2 + 3 \cdot 4$; $-2 \cdot 5$;
$-2 \cdot 2 + 4 \cdot 4$; 2; 14; -14; 12

Activity Step 2. [358] **Step 3.** [358] **Step 4.** The Associative
Property of Multiplication

Questions

1. 3   **3.** $\$11 \cdot 1 + \$17 \cdot 5 + \$18 \cdot 3 = \$150$ **5.** [−23]
7. a. No; the number of elements in a row of B is not
the same as the number of elements in a column of A.
b. Matrix multiplication is not commutative.

9.
$$\begin{bmatrix} 4 & 6 \\ -1 & 8 \end{bmatrix} \cdot \begin{bmatrix} 7 & 3 \\ 5 & 1 \end{bmatrix} = \begin{bmatrix} 58 & 18 \\ 33 & 5 \end{bmatrix}$$

$$\begin{bmatrix} 7 & 3 \\ 5 & 1 \end{bmatrix} \cdot \begin{bmatrix} 4 & 6 \\ -1 & 8 \end{bmatrix} = \begin{bmatrix} 25 & 66 \\ 19 & 38 \end{bmatrix}$$

11.
$$\begin{array}{c} \text{Wt.} \\ \text{Vol.} \end{array}\begin{array}{ccc} \text{TTB} & \text{TB} & \text{TB} \\ \begin{bmatrix} 1.5 & 5 & 12 \\ 1 & 1 & 0.6 \end{bmatrix} \end{array} \cdot$$

$$\begin{array}{c} \text{TTB} \\ \text{TB} \\ \text{BB} \end{array}\begin{array}{cc} \text{Store 1} & \text{Store 2} \\ \begin{bmatrix} 20 & 60 \\ 50 & 10 \\ 12 & 15 \end{bmatrix} \end{array} =$$

$$\begin{array}{c} \text{Wt.} \\ \text{Vol.} \end{array}\begin{array}{cc} \text{Store 1} & \text{Store 2} \\ \begin{bmatrix} 424 & 320 \\ 77.2 & 19 \end{bmatrix} \end{array}$$

13. a. $\begin{bmatrix} x+y \\ x+y \end{bmatrix}$; A is moved to the right y units and
up x units to B.

b. $\begin{bmatrix} x \\ y \end{bmatrix}$; $B = A$ **c.** $\begin{bmatrix} y \\ x \end{bmatrix}$; A's coordinates are switched to get B.

d. $\begin{bmatrix} -y \\ x \end{bmatrix}$; A's coordinates are switched, and y is changed to
its opposite for B.

and then the coordinates are switched to get B. **e.** $\begin{bmatrix} 2x \\ 2y \end{bmatrix}$;
A's coordinates are doubled to get B. **15.** $x = 3$

17. $\begin{bmatrix} 0 & 0 \\ 0 & 0 \end{bmatrix}$ **19.** 4×3 **21. a.** square **b.** $K' = (0.5, 1.5)$;
$L' = (-2, 1.5)$; $M' = (-2, 4)$; $N' = (0.5, 4)$ **c.** Square; it is a
size-change image of $KLMN$, and size-change images are
similar to their preimages.

Lesson 4-4 (pp. 242–248)

Mental Math a. true **b.** true **c.** false **d.** true

Activity Step 1. $\begin{bmatrix} 4 & 6 & 6 & 3 & 4 \\ 2 & 2 & 6 & 4 & 4 \end{bmatrix}$

Step 2. $\begin{bmatrix} 2 & 0 \\ 0 & 2 \end{bmatrix}\begin{bmatrix} 4 & 6 & 6 & 3 & 4 \\ 2 & 2 & 6 & 4 & 4 \end{bmatrix} = \begin{bmatrix} 8 & 12 & 12 & 6 & 8 \\ 4 & 4 & 12 & 8 & 8 \end{bmatrix}$;

$$\begin{bmatrix} \frac{1}{3} & 0 \\ 0 & \frac{1}{3} \end{bmatrix}\begin{bmatrix} 4 & 6 & 6 & 3 & 4 \\ 2 & 2 & 6 & 4 & 4 \end{bmatrix} = \begin{bmatrix} \frac{4}{3} & 2 & 2 & 1 & \frac{4}{3} \\ \frac{2}{3} & \frac{2}{3} & 2 & \frac{4}{3} & \frac{4}{3} \end{bmatrix}$$

Step 3. The coordinates of $N'U'M'E'R'$ are each 2 (or $\frac{1}{3}$) times
the corresponding coordinates of the original $NUMER$.

Step 4.

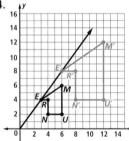 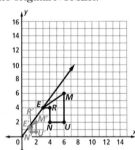

Step 5. $\frac{OE'}{OE} = 2\left(\text{or } \frac{1}{3}\right)$ **Step 6.** $\frac{OR'}{OR} = 2\left(\text{or } \frac{1}{3}\right)$; $\frac{ON'}{ON} = 2\left(\text{or } \frac{1}{3}\right)$; $\frac{OM'}{OM}$
$= 2\left(\text{or } \frac{1}{3}\right)$; $\frac{OU'}{OU} = 2\left(\text{or } \frac{1}{3}\right)$ **Step 7.** The vertices of $N'U'M'E'R'$
are k times the vertices of the original $NUMER$.

Guided Example 2 Missing numbers along top rows of
matrices, left to right: 4, −1, 1, 4; Missing numbers along
bottom rows of matrices, left to right: 0, 4, 8, −12, 8, 4

Questions

1. a. $N'U'M'E'R'$ is similar to $NUMER$, with sides four times
as long. **b.** $N'U'M'E'R'$ is similar to $NUMER$, with sides
half as long. **c.** $N'U'M'E'R'$ is identical to $NUMER$.
3. The size change of magnitude 1.2 and center $(0, 0)$
maps $(4, 3.4)$ onto $(4.8, 4.08)$.   **5.** C   **7.** $\begin{bmatrix} \frac{1}{4} & 0 \\ 0 & \frac{1}{4} \end{bmatrix}$
9. always **11.** always **13.** $\begin{bmatrix} 2768 & 0 \\ 0 & 2768 \end{bmatrix}$ **15. a.** $\begin{bmatrix} \frac{5}{2} & 0 \\ 0 & \frac{5}{2} \end{bmatrix}$

b. 1 **c.** 1
d. yes, by the
Parallel Lines
and Slope Theorem **17. a.**

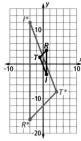

b. $\triangle T^*R^*I^*$ is the rotation image of $\triangle T'R'I'$ under a half turn.

19. a. $\begin{bmatrix} 2 \\ 1 \end{bmatrix}$; $\begin{bmatrix} 4 \\ 1 \end{bmatrix}$; $\begin{bmatrix} 1 \\ 2 \end{bmatrix}$; $\begin{bmatrix} 1 \\ 4 \end{bmatrix}$

b.

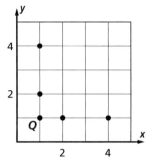

21. a. $\ell = \frac{V}{wh}$ **b.** $V; w; h$ **c.** $\frac{5}{18}$

Lesson 4-5 (pp. 249–254)

Mental Math a. I and II **b.** II and IV **c.** III and IV **d.** I and III
Guided Example 1 4; 0.5; 4; 0.5; 20; 3; 24; 0
Activity Step 1. See first graph in Guided Example 1.

Step 2. $S1 \cdot M = \begin{bmatrix} 0 & 20 & 24 \\ 2 & 3 & 0 \end{bmatrix}$ **Step 3.** See second graph

in Example 1. $S1$ represents the scale change $S_{4,0.5}$.

Step 4. $S2 \cdot M = \begin{bmatrix} 0 & 6 & 7.2 \\ 12 & 18 & 0 \end{bmatrix}$

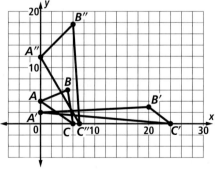

Step 5. $S2$ represents the scale change $S_{1.2,3}$.

Step 6. The matrix $\begin{bmatrix} a & 0 \\ 0 & b \end{bmatrix}$ represents the scale change $S_{a,b}$.

Questions
1. $\left(0.8x, \frac{4}{7}y\right)$ **3. a.** (12, 15) **b.** $S_{4,2}$ stretches horizontally by a factor of 4 and stretches vertically by a factor of 2.
5. $A' = (-6, 10)$, $B' = (0, 14)$, $C' = (12, 2)$, $D' = (-12, -2)$
7. false **9.** size; S_8 **11.** Answers vary. Sample: Move B to make both dimensions 1.6 times as long.
13. a.

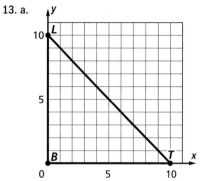

$\triangle BLT$ is an isosceles right triangle. **b.** $\triangle B'L'T' = \begin{bmatrix} 0 & 0 & 40 \\ 0 & 12 & 0 \end{bmatrix}$
c.

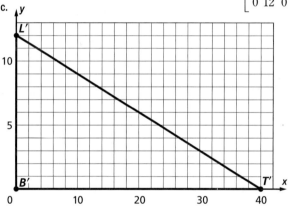

$\triangle B'L'T'$ is also a right triangle but it is scalene, not isosceles. **15. a.** $S_{5,7}$ **b.** $F = (5,2)$, $O = (1,4)$, $O = (4,-1)$
c. $\frac{F'O'}{FO} \approx 5.5$, $\frac{O'R'}{OR} \approx 6.5$ **d.** No; the transformation given here is a scale change that is not a size change. The figure and its image are not similar triangles, so the ratios will

not be the same. **17. a.** 4 **b.** $\frac{1}{2}$ **c.** $\begin{bmatrix} 2 & 0 \\ 0 & 2 \end{bmatrix}$; 2

19. $a_n = a_1 + d(n - 1)$ **21. a.** Answers vary. Sample: distance, angle measure, and collinearity **b.** An isometry is a transformation that preserves distance. Answers vary. Sample: rotation and translation

Lesson 4-6 (pp. 255–261)

Mental Math a. no **b.** no **c.** yes **d.** yes

Guided Example 2 0; 1; 0; 1; 1; 0 $\begin{bmatrix} 0 & 1 \\ 1 & 0 \end{bmatrix}$

Activity Step 4. The corresponding points have the same *x*-coordinate and opposite *y*-coordinates; reflection over the *x*-axis. **Step 5.** Corresponding points have *x*- and *y*-coordinates switched; reflection over the line $y = x$; corresponding points have *x*- and *y*-coordinates switched, and each coordinate is multiplied by –1; reflection over the line $y = -x$.

Questions

1. (–30, –150)   **3.** $\overline{AA'}$ **5.** r_x: $(x, y) \rightarrow (x, -y)$; $r_x(x, y) = (x, -y)$
7. D   **9.** B   **11.** true   **13.** (4, 6); $y = x$; (6, 4)

15. $\begin{bmatrix} 0 & -1 \\ -1 & 0 \end{bmatrix} \begin{bmatrix} x \\ y \end{bmatrix} = \begin{bmatrix} 0(x) + -1(y) \\ -1(x) + 0(y) \end{bmatrix} \begin{bmatrix} -y \\ -x \end{bmatrix}$

17. a. (1, –6)   **b.** $\begin{bmatrix} 0.5 & 0 \\ 0 & 2 \end{bmatrix}$ **c.** a scale change in which

horizontal distances are halved and vertical distances are doubled   **19. a.** $\sqrt{|4 + 2|^2 + |4 - 3|^2} = \sqrt{37}$

b. $\sqrt{|2 + 1|^2 + |2 - \frac{3}{2}|^2} = \frac{\sqrt{37}}{2}$ **c.** no

21. a. $A = 6x^2$ **b.** $x = \sqrt{\frac{A}{6}}$

Lesson 4-7 (pp. 262–268)

Mental Math a. 10 **b.** 5 **c.** 20

Activity Step 4. a counterclockwise rotation 90° about (0, 0)

Step 5. $M = \begin{bmatrix} 0 & -1 \\ 1 & 0 \end{bmatrix}$

Step 6.

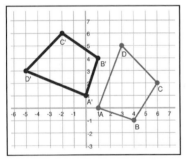

The image of *ABCD* after applying the transformation described by *M* is the same as *A″B″C″D″* from Step 3.

Questions

1. a. 2×2 **b.** closure under multiplication   **3.** $\begin{bmatrix} 5 & 2 \\ \sqrt{8} & \pi \end{bmatrix}$

5. a. yes **b.** yes **c.** Answers vary.

Sample: $\begin{bmatrix} 1 & 2 \\ 3 & 4 \end{bmatrix} \begin{bmatrix} 5 & 6 \\ 7 & 8 \end{bmatrix} \neq \begin{bmatrix} 5 & 6 \\ 7 & 8 \end{bmatrix} \begin{bmatrix} 1 & 2 \\ 3 & 4 \end{bmatrix}$ **7.** B   **9.** A

11. He changed the order of the matrices being multiplied, and matrix multiplication is not always commutative.
13. No; one is a rotation of 90°, the other of –90°. You can also show this with matrices:

$\begin{bmatrix} 1 & 0 \\ 0 & -1 \end{bmatrix} \begin{bmatrix} 0 & 1 \\ 1 & 0 \end{bmatrix} = \begin{bmatrix} 0 & 1 \\ -1 & 0 \end{bmatrix} \neq \begin{bmatrix} 0 & -1 \\ 1 & 0 \end{bmatrix} = \begin{bmatrix} 0 & 1 \\ 1 & 0 \end{bmatrix} \begin{bmatrix} 1 & 0 \\ 0 & -1 \end{bmatrix}$

15. a. $\begin{bmatrix} 0 & 1 \\ -1 & 0 \end{bmatrix}$ **b.** $\begin{bmatrix} 0 & -6 & -6 \\ -1 & -1 & -2 \end{bmatrix}$ **c.** They are the same.

A –90° rotation is equivalent to a 270° rotation.

17. a. $\begin{bmatrix} -\frac{1}{2} & -2 & -3 & -\frac{3}{2} \\ 0 & -\frac{1}{2} & 1 & \frac{5}{2} \end{bmatrix}$ **b.** no **c.** yes

19. a. $AB = \begin{bmatrix} ae + bg & af + bh \\ ce + dg & cf + dh \end{bmatrix}$; $BA = \begin{bmatrix} ae + cf & be + df \\ ag + ch & bg + dh \end{bmatrix}$

b. No, it is not. For instance, $ae + bg$ does not necessarily equal $bg + dh$.   **21.** Answers vary. Sample: (0, 0), (1, 1), and (2, 2)   **23.** $\begin{bmatrix} \frac{3}{2} & 0 \\ 0 & \frac{3}{2} \end{bmatrix}$ **25. a.** $9s + 15r = 78$

b.

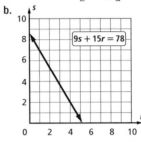

2 sale movies and 4 regular movies or 7 sale movies and 1 regular movie

Lesson 4-8 (pp. 269–273)

Mental Math a. $h = \frac{3v}{\pi r^2}$ **b.** $t = \sqrt{\frac{h}{4.9}}$ **c.** $k = \frac{Md}{wt^2}$ **d.** $x = \frac{y - 12}{4}$

Activity Step 2. 70° **Step 3.** 82° **Step 4.** 152° **Step 5.** –152°

Guided Example a. $\begin{bmatrix} -1 & 0 \\ 0 & -1 \end{bmatrix}$ **b.** 180; $\begin{bmatrix} -1 & 0 \\ 0 & -1 \end{bmatrix}$; $\begin{bmatrix} 0 & 1 \\ -1 & 0 \end{bmatrix}$

Questions

1. clockwise   **3. a.** 135° **b.** $R_{90} \circ R_{45} = R_{135}$

5. R_{90}: $\begin{bmatrix} 0 & -1 \\ 1 & 0 \end{bmatrix}$; R_{180}: $\begin{bmatrix} -1 & 0 \\ 0 & -1 \end{bmatrix}$; R_{270}: $\begin{bmatrix} 0 & 1 \\ -1 & 0 \end{bmatrix}$; R_{360}: $\begin{bmatrix} 1 & 0 \\ 0 & 1 \end{bmatrix}$

7. $\begin{bmatrix} -2 & -8 & -8 & -2 \\ -3 & -3 & -9 & -9 \end{bmatrix}$ **9. a.** $\begin{bmatrix} 0 & 1 \\ 1 & 0 \end{bmatrix}$ **b.** $\begin{bmatrix} 0 & -1 \\ -1 & 0 \end{bmatrix}$

c. Matrix multiplication is not commutative; composition of transformations is not commutative.   **11. a.** $\begin{bmatrix} 4 & 0 \\ 0 & 0.25 \end{bmatrix}$

b. a horizontal stretch of magnitude 4 and a vertical shrink of magnitude 0.25   **13.** false   **15.** about 10,920,000 people

Lesson 4-9 (pp. 274–279)

Mental Math a. 70 **b.** 20 **c.** $\frac{950 - 5n}{10}$

Activity Step 1. $\begin{bmatrix} -8 & 5 \\ 3 & -1 \end{bmatrix}$ **Step 2.** $R = \begin{bmatrix} 0 & -1 \\ 1 & 0 \end{bmatrix}$

Step 3. Answers vary. Sample:
The segments
look perpendicular.

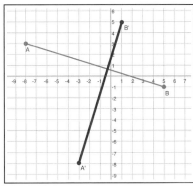

Step 4. The slope of $\overline{AB} = -\frac{4}{13}$; the slope of $\overline{A'B'} = \frac{13}{4}$;
the slopes are negative reciprocals of each other.

Guided Example $\frac{2}{5}; -\frac{5}{2}; y + 3 = -\frac{5}{2}(x - 4)$

Questions

1. $\begin{bmatrix} x_1 & x_2 \\ y_1 & y_2 \end{bmatrix}$ 3. true   5. false   7. $y = -\frac{5}{3}x + 11$

9. $y - 10 = -\frac{1}{4}(x + 5)$ 11. Answers vary. Sample: Lines
with slopes of zero will never satisfy the condition $m_1 m_2 = -1$.

13. $x = 6$ 15. a. $90°$ b. $A' = (2, -6)$ and $B' = (0, 5)$

c. slope of $\overleftrightarrow{AB} = \frac{2}{11}$; slope of $\overleftrightarrow{A'B'} = -\frac{11}{2}$ d. (slope of $\overleftrightarrow{AB}$)·
(slope of $\overleftrightarrow{A'B'}$) = -1, so the two lines are perpendicular

17. a. || b. Answers vary. Sample:

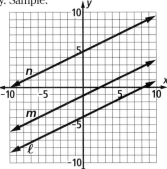

19. a. ⊥ b. Answers vary. Sample:

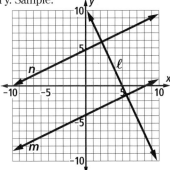

21. a. perpendicular b. neither c. parallel d. parallel

e. parallel f. neither   23. $\begin{bmatrix} 1 & 0 \\ 0 & 1 \end{bmatrix}$ 25. B

Lesson 4-10 (pp. 280–285)

Mental Math a. r_x b. $r_{y=-x}$ c. R_{-90} or R_{270}

Activity Step 1. $D = \begin{bmatrix} 4 & 4 & 4 \\ 3 & 3 & 3 \end{bmatrix}$ **Step 2.** the difference

in the x-coordinates of $\triangle ABC$ and $\triangle A'B'C'$ **Step 3.** the
difference in the y-coordinates of $\triangle ABC$ and $\triangle A'B'C'$

Step 4. $D + M = \begin{bmatrix} 5 & 5 & 7 \\ 1 & 5 & 5 \end{bmatrix}$; $D + M$ represents $\triangle A'B'C'$, the

image of $\triangle ABC$ shifted 4 units to the right and 3 units up.

Step 5. $(x + 4, y + 3)$

Guided Example 1 a. $6 - 8$; $2 + 2$; $8 - 8$; $4 + 2$; 0; 6; $5 - 8$;

$5 + 2$; -3; 7; $\begin{bmatrix} -8 & -8 \\ 2 & 2 \end{bmatrix}$; $\begin{bmatrix} 8 & 5 \\ 4 & 5 \end{bmatrix}$; $\begin{bmatrix} 0 & -3 \\ 6 & 7 \end{bmatrix}$

b.

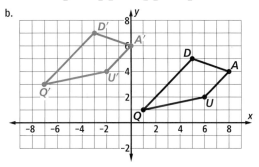

Guided Example 2 3; 2; -6; 3; 2; -6; 3; -12

Questions

1. a. $T_{4,3}(x, y) = (x + 4, y + 3)$ b. $T_{-4,-3}(x, y) = (x - 4, y - 3)$

3. $h; k$ 5. $\begin{bmatrix} -3 & 5 & -12 \\ -5 & -2 & 5 \end{bmatrix}$ 7. a. $\frac{3}{7}$ b. $\frac{3}{7}$

c. Yes; their slopes are equal so the two lines are parallel.

9. a. $T_{2,-4}$ b. slope of $\overline{BC}$ = slope of $\overline{B'C'} = -\frac{2}{3}$

c. $BC = B'C' = \sqrt{13}$ d. parallelogram   11. a. $T_{5,8}$ b. yes

c. yes, because matrix addition is commutative

13. $T_{-2,-1}, T_{-2,1}, T_{1,2}, T_{1,-2}, T_{-1,-2}, T_{2,-1}$ 15. a. $(-y + h, x + k)$;
$(-y - k, x + h)$ b. No, the two images in Part a are not equal.

17. a. $\begin{bmatrix} -1 & 0 \\ 0 & -1 \end{bmatrix}$ b. $\begin{bmatrix} -1 & 0 \\ 0 & -1 \end{bmatrix}$; yes

19. a.

	blue	gray	tan
S	8	8	8
M	12	12	12
L	8	8	8

b. $\begin{bmatrix} 0 & 0 & 1 \\ 2 & 1 & 0 \\ 0 & 1 & 0 \end{bmatrix}$ c. $\begin{bmatrix} 8 & 8 & 7 \\ 10 & 11 & 12 \\ 8 & 7 & 8 \end{bmatrix}$

Self-Test (pp. 291–292)

1. $\begin{array}{cccc} H & U & L & K \\ \begin{bmatrix} -2 & 5 & -2 & -4 \\ 4 & 1 & -2 & 2 \end{bmatrix} \end{array}$

2. a.

	Jicamaport	Okraville	Potatotown
first-class	16	4	2
economy	107	180	321

b.

	first-class	economy
Jicamaport	16	107
Okraville	4	180
Potatotown	2	321

3. The product will exist when the number of columns in the left matrix is equal to the number of rows in the right matrix. This the the case for AB, BA, AC, and CB.

4. $BA = \begin{bmatrix} (5 \cdot 2) + (-2 \cdot 4) + \left(-1 \cdot -\frac{1}{2}\right) & (5 \cdot 0) + (-2 \cdot (-2)) + \left(-1 \cdot \frac{1}{2}\right) \\ \left(\frac{1}{8} \cdot 2\right) + (6 \cdot 4) + \left(5 \cdot -\left(\frac{1}{2}\right)\right) & \left(\frac{1}{8} \cdot 0\right) + (6 \cdot (-2)) + \left(5 \cdot \frac{1}{2}\right) \end{bmatrix}$

$= \begin{bmatrix} 2.5 & 3.5 \\ 21.75 & -9.5 \end{bmatrix}$

5. not possible; A and C have different dimensions

6. $\frac{1}{3}B = \begin{bmatrix} \frac{1}{3} \cdot 5 & \frac{1}{3} \cdot (-2) & \frac{1}{3} \cdot (-1) \\ \frac{1}{3} \cdot \frac{1}{8} & \frac{1}{3} \cdot 6 & \frac{1}{3} \cdot 5 \end{bmatrix} = \begin{bmatrix} \frac{5}{3} & -\frac{2}{3} & -\frac{1}{3} \\ \frac{1}{24} & 2 & \frac{5}{3} \end{bmatrix}$

7. Answers vary. Sample: Multiplication by $\begin{bmatrix} 1 & 0 \\ 0 & 1 \end{bmatrix}$ maps each point $\begin{bmatrix} x \\ y \end{bmatrix}$ onto itself. **8.** No; $m_1 = 5$, $m_2 = \frac{1}{5}$, and $m_1 m_2 = 5 \cdot \frac{1}{5} = 1$. For perpendicular lines, $m_1 m_2 = -1$.

9. $m \cdot \frac{1}{7} = -1$, $m = -7$, $y - (-1) = -7\left(x - \frac{1}{4}\right)$; $y = -7x + \frac{3}{4}$

10. The matrix for $r_y \circ R_{270}$ is $\begin{bmatrix} -1 & 0 \\ 0 & 1 \end{bmatrix}\begin{bmatrix} 0 & 1 \\ -1 & 0 \end{bmatrix} =$

$\begin{bmatrix} (-1 \cdot 0) + (0 \cdot (-1)) & (-1 \cdot 1) + (0 \cdot 0) \\ (0 \cdot 0) + (1 \cdot (-1)) & (0 \cdot 1) + (1 \cdot 0) \end{bmatrix} = \begin{bmatrix} 0 & -1 \\ -1 & 0 \end{bmatrix}$

11. $P = (3, -5)$ maps onto $P' = (8, -10)$, so the translation is $T_{5,-5}$.

12.

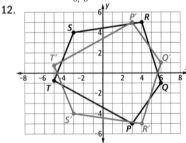

13. $\begin{bmatrix} (38 \cdot 4) + (135 \cdot 6.5) + (169 \cdot 5) \\ (84 \cdot 4) + (101 \cdot 6.5) + (152 \cdot 5) \\ (84 \cdot 4) + (118 \cdot 6.5) + (135 \cdot 5) \\ (67 \cdot 4) + (236 \cdot 6.5) + (34 \cdot 5) \end{bmatrix} = \begin{array}{l} \text{Movie 1} \\ \text{Movie 2} \\ \text{Movie 3} \\ \text{Movie 4} \end{array} \begin{bmatrix} 1874.5 \\ 1752.5 \\ 1778 \\ 1972 \end{bmatrix}$

14.

$\begin{bmatrix} 42+5 & 15+2 \\ 10+3 & 2+1 \\ 7+0 & 1+3 \\ 4+2 & 0+0 \end{bmatrix} = \begin{array}{l} \text{Paperbacks} \\ \text{Hardbacks} \\ \text{Audiotapes} \\ \text{Audio CDs} \end{array}$

	fiction	non-fiction
Paperbacks	47	17
Hardbacks	13	3
Audiotapes	7	4
Audio CDs	6	0

15. $\begin{bmatrix} 2 & 0 \\ 0 & 2 \end{bmatrix}\begin{bmatrix} 2 & -9 \\ 7 & 5 \end{bmatrix} = \begin{bmatrix} (2 \cdot 2) + (0 \cdot 7) & (2 \cdot -9) + (0 \cdot 5) \\ (0 \cdot 2) + (2 \cdot 7) & (0 \cdot -9) + (2 \cdot 5) \end{bmatrix}$

$= \begin{bmatrix} 4 & -18 \\ 14 & 10 \end{bmatrix} = 2\begin{bmatrix} 2 & -9 \\ 7 & 5 \end{bmatrix}$ **16.** $\begin{bmatrix} 0 & -1 \\ 1 & 0 \end{bmatrix}$

17. The reflection of $\triangle XYZ$ over the y-axis gives a triangle with vertices $X' = (4, 5)$, $Y' = (-2, 6)$, and $Z' = (3, 1)$.
18. $R_{270}(x) = (5, 4)$, $R_{270}(y) = (6, -2)$, $R_{270}(z) = (1, 3)$

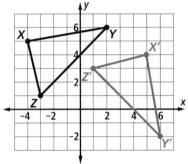

19. a. $\begin{bmatrix} 4 & 0 \\ 0 & 3 \end{bmatrix}\begin{bmatrix} 2.5 & 1 & -6 \\ -5 & 4 & -12 \end{bmatrix} = \begin{bmatrix} 2.5 \cdot 4 & 1 \cdot 4 & -6 \cdot 4 \\ -5 \cdot 3 & 4 \cdot 3 & -12 \cdot 3 \end{bmatrix} =$

$\begin{bmatrix} 10 & 4 & -24 \\ -15 & 12 & -36 \end{bmatrix}$

b. $\begin{bmatrix} 2.5 & 1 & -6 \\ -5 & 4 & -12 \end{bmatrix} + \begin{bmatrix} 2.5 & 2.5 & 2.5 \\ -1 & -1 & -1 \end{bmatrix} =$

$\begin{bmatrix} 2.5+2.5 & 1+2.5 & -6+2.5 \\ -5-1 & 4-1 & -12-1 \end{bmatrix} = \begin{bmatrix} 5 & 3.5 & -3.5 \\ -6 & 3 & -13 \end{bmatrix}$

20. true **21.** False; Examples vary. Sample: $y = x$ rotated $135°$ is the line $y = 0$. $m_1 m_2 = (1 \cdot 0) = 0 \neq -1$.

22. a. $S_7(a, b) = (7a, 7b)$ **b.** $PQ = \sqrt{|3 - 2|^2 + |9 - 5|^2} = \sqrt{17}$; $S_7(2, 5) = (7 \cdot 2, 7 \cdot 5) = (14, 35)$; $S_7(3, 9) = (7 \cdot 3, 7 \cdot 9) = (21, 63)$; $P'Q' = \sqrt{|21 - 14|^2 + |63 - 35|^2} = \sqrt{833}$; $\frac{\sqrt{833}}{\sqrt{17}} = \sqrt{\frac{833}{17}} = \sqrt{49} = 7$. Thus, $P'Q' = 7PQ$.

23. Slope of $\overline{PQ} = \frac{9 - 5}{3 - 2} = 4$; so the slope of the perpendicular bisector $= -\frac{1}{4}$; midpoint $= \left(\frac{2+3}{2}, \frac{5+9}{2}\right) = \left(\frac{5}{2}, 7\right)$; $y - 7 = -\frac{1}{4}\left(x - \frac{5}{2}\right)$

Self-Test Correlation Chart

Question	1	2	3	4	5	6	7	8
Objective(s)	A	I	E	C	B, E	B	E	H
Lesson(s)	4-1	4-1	4-2, 4-3, 4-7	4-3	4-2, 4-3, 4-7	4-2	4-2, 4-3, 4-7	4-9, 4-10

Question	9	10	11	12	13	14	15	16
Objective(s)	D	G	F	K	J	J	B	G
Lesson(s)	4-9	4-4, 4-5, 4-6, 4-7, 4-8, 4-10	4-4, 4-5, 4-6, 4-8, 4-9, 4-10	4-4, 4-5, 4-6, 4-7, 4-8, 4-10	4-2, 4-3	4-2, 4-3	4-2	4-4, 4-5, 4-6, 4-7, 4-8, 4-10

Question	17	18	19	20	21	22	23
Objective(s)	G	K	G	F	F	F	D
Lesson(s)	4-4, 4-5, 4-6, 4-7, 4-8, 4-10	4-4, 4-5, 4-6, 4-7, 4-8, 4-10	4-4, 4-5, 4-6, 4-7, 4-8, 4-10	4-4, 4-5, 4-6, 4-8, 4-9, 4-10	4-4, 4-5, 4-6, 4-8, 4-9, 4-10	4-4, 4-5, 4-6, 4-8, 4-9, 4-10	4-9

Chapter Review (pp. 293–297)

1. a. $\begin{bmatrix} 4 \\ 5 \end{bmatrix}$ **b.** $\begin{bmatrix} 3 \\ -4 \end{bmatrix}$ **c.** $\begin{bmatrix} -4 \\ 1 \end{bmatrix}$ **d.** $\begin{bmatrix} -2 \\ -4 \end{bmatrix}$ **e.** $\begin{bmatrix} 3 \\ 1 \end{bmatrix}$

3. $\begin{bmatrix} 4 & -2 & -2 \\ 3 & 1 & 3 \\ -4 & -4 & 7 \end{bmatrix}$ **5.** $\begin{bmatrix} -6 & 0 & -3 \\ -9 & -5 & -1 \\ 4 & 7 & -4 \end{bmatrix}$ **7.** $p = 3; q = 3$

9. $\begin{bmatrix} 1 & 12 \\ 15 & 0 \end{bmatrix}$ **11.** $\begin{bmatrix} 22 & 19 \\ -8 & 34 \end{bmatrix}$ **13.** $p = 4; q = 2$ **15.** $y = -3x + 1$

17. $y = -\frac{5}{3}x + \frac{7}{3}$ **19. a.** true **b.** Answers vary.

Sample: $\left(\begin{bmatrix} 4 & 3 \\ 7 & 1 \end{bmatrix} \begin{bmatrix} 2 & 9 \\ -1 & -3 \end{bmatrix} \right) \begin{bmatrix} -4 \\ -4 \end{bmatrix} = \begin{bmatrix} 4 & 3 \\ 7 & 1 \end{bmatrix} \left(\begin{bmatrix} 2 & 9 \\ -1 & -3 \end{bmatrix} \begin{bmatrix} -4 \\ -4 \end{bmatrix} \right)$

21. a. false **b.** Answers vary. Sample:

$\begin{bmatrix} 1 & 2 \\ 3 & 4 \end{bmatrix} - \begin{bmatrix} 5 & 6 \\ 7 & 8 \end{bmatrix} \neq \begin{bmatrix} 5 & 6 \\ 7 & 8 \end{bmatrix} - \begin{bmatrix} 1 & 2 \\ 3 & 4 \end{bmatrix}$

23. $m = 1; n = 7$ **25.** $\begin{bmatrix} 1 & 0 \\ 0 & 1 \end{bmatrix}$ **27.** C

29. $\sqrt{(2x_2 - 2x_1)^2 + (2y_2 - 2y_1)^2} =$

$\sqrt{4(x_2 - x_1)^2 + 4(y_2 - y_1)^2} = 2\sqrt{(x_2 - x_1)^2 + (y_2 - y_1)^2}$

31. $2x + 3y = 67$ **33.** $\begin{bmatrix} 1 & 0 \\ 0 & 1 \end{bmatrix}$; the identity transformation

35. a. $\begin{bmatrix} -1 & 0 \\ 0 & 1 \end{bmatrix}$ **b.** r_y **37. a.** S_2 **b.** Multiply the matrix of the points for *FIG* by the size change matrix $\begin{bmatrix} 2 & 0 \\ 0 & 2 \end{bmatrix}$.

39. F **41.** $\begin{bmatrix} 2 & -1 & 4 & 3 \\ -0.7 & 0 & 1 & -5 \end{bmatrix}$ **43.** $\begin{bmatrix} 0 & -4 & 3 \\ 2 & 0 & 2 \end{bmatrix}$

45. No; their slopes are not equal. **47.** Answers vary.

49. Sample: (1, 8)

$\begin{array}{c} \\ 2005 \\ 2006 \end{array} \begin{bmatrix} \text{CC/DC} & \text{non-} & \text{investment} \\ \text{Fraud} & \text{delivery} & \text{fraud} \\ 240 & 410 & 2000 \\ 427.50 & 585 & 2694.99 \end{bmatrix}$

51. The element in the second row and first column. (1:01)

53. a. $\begin{bmatrix} 822 & 456 & 977 \\ 784 & 739 & 879 \end{bmatrix}$ **b.** *Harry Potter* (original)

c. *The Matrix* sequel **55.** $B = \begin{bmatrix} 9 & 11 & 2 \\ 5 & 7 & 5 \end{bmatrix}$; $P = \begin{bmatrix} 1 \\ 2 \\ 3 \end{bmatrix}$; $BP\begin{bmatrix} 37 \\ 34 \end{bmatrix}$;

Brenda scored 37 points and Marisa scored 34 points.

57. a.

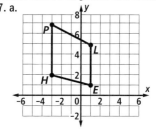

b. $\begin{bmatrix} -6 & 2 & 2 & -6 \\ 6 & 3 & 15 & 21 \end{bmatrix}$

c.

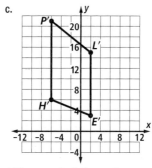

d. No; magnitude is not the same for the horizontal and vertical directions.

59. a.

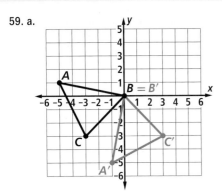

b. R_{90} **61. a.** $\begin{bmatrix} -1 & 0 & 3 & 3 & 2 \\ 1 & -4 & -2 & 0 & 5 \end{bmatrix}$;

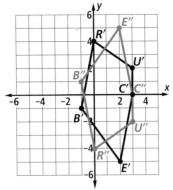

b. Yes; reflection is a congruence transformation.
c. Yes; reflections preserve angle measure and slope.

Chapter 5

Lesson 5-1 (pp. 300–306)
Mental Math a. $\{7,11\}$ **b.** the set of all integers **c.** Ø **d.** the set of all real numbers
Activity Step 1.

Compound Sentence Entry	CAS Output
$x > 4$ and $x \le 8$	$4 < x \le 8$
$x > 4$ or $x \le 8$	true
$a < 5$ and $a < 20$	$a < 5$
$a < 5$ or $a < 20$	$a < 20$
$n < 0$ and $n > 5$	false
$n < 0$ or $n > 5$	$n < 0$ or $n > 5$

Step 2. *True* is displayed when all real numbers satisfy the compound sentence, and *false* is displayed when no real numbers satisfy the compound sentence.
Step 3. Compound sentences involving the word *or* include more values in their solution set than compound sentences involving the word *and* because solutions to *or* sentences have to satisfy only part of the sentence.

Step 4.

Stored Variable	Compound Sentence Entry	CAS Output
$3 \rightarrow x$	$x > 4$ and $x \le 8$	false
$3 \rightarrow x$	$x > 4$ or $x \le 8$	true
$1 \rightarrow a$	$a < 5$ and $a < 20$	true
$1 \rightarrow a$	$a < 5$ or $a < 20$	true
$2 \rightarrow n$	$n < 0$ and $n > 5$	false
$2 \rightarrow n$	$n < 0$ or $n > 5$	false

Step 5. When the stored value satisfies the compound sentence, the output is *true*; when the stored value does not satisfy the compound sentence, the output is *false*.
Questions
1. $0 < h < 4$

3. $9\frac{29}{60} \le h \le 14\frac{48}{60}$

5. B   **7. a.** intersection **b.** union

9.

11. Steel Dragon 2000   **13.** multiply; divide; reverse

15. $m < -0.28$

17. a. $0 \le r \le 61$ **b.** $0 \le r \le 6$ **19.** $\{m \mid -\frac{1}{3} \le m < 3\}$

21. Answers vary. Sample: $T_{0,8}$
23. C

Lesson 5-2 (pp. 307–313)
Mental Math a. 9 times as much **b.** 1.44 times as much
c. 2 times as tall
Guided Example 3 Solution 1. $-3x$; -3; 3; -5; 15; -1.75; 12.25; 1; 4
Solution 2. -1.75; 12.25; 1; 4
Questions
1. A system is a set of conditions joined by the word *and*.
3. a.

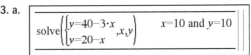

b.

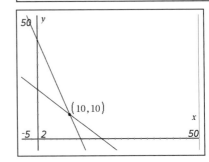

c. Answers vary. The "solve" command because it involved fewer steps.   **5. a.** $x = -3, y = -3$

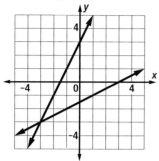

b. Answers vary. Sample: The lines have their x- and y-intercepts exchanged.

c.

$$\text{solve}\left(y = \frac{1}{2} \cdot x - 1.5 \text{ and } y = 2 \cdot x + 3, x, y\right)$$
$$x = -3. \text{ and } y = -3.$$

7. a. Answers vary. Sample: The graphing utility shows more decimal places for the y-coordinate, so 11.2 must be a rounded value. **b.** Answers vary. Sample: No, this is also an estimate. Since the solution involves the irrational number π, no exact decimal answer can be written.
9. a. 2 **b.** $(0.85, 1.45)$, $(-2.35, 11.05)$ **c.** $2(0.85)^2 \approx 1.45$; $-3(0.85) + 4 = 1.45$; $2(-2.35)^2 \approx 11.05$; $-3(-2.35) + 4 = 11.05$
11. a. 2 **b.** $(-0.76, 7)$, $(0.76, 7)$ **c.** $(0.76)^2 \cdot 7 \approx 4$; $(-0.76)^2 \cdot 7 \approx 4$
13. a

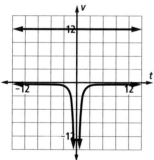

b. 0 c. no solutions

15. yes; The graphs of $xy = 1740$ and $x + y = 89$ intersect in two points, $(29,60)$ and $(60,29)$.

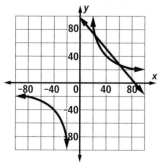

17. $x > -\frac{3}{4}$

![number line with open circle at -3/4, shaded right]

19. $\begin{bmatrix} -4 & -1 \\ 1 & -4 \end{bmatrix}$ **21.** $t = \frac{ea - f}{1.5} = \frac{2}{3}(ea - f)$

Lesson 5-3 (pp. 314–320)

Mental Math a. $4x + 4$ **b.** $8 + \sqrt{34}$ **c.** $2L + 10$ **d.** $\frac{n\sqrt{3}}{2}$

Activity Step 1. $\begin{cases} H + V = 600 \\ H = 4V \end{cases}$ **Step 2.** $H = 4V$; $4V + V = 600$; $V = 600 - H$; $H = 4(600 - H)$; **Step 3.** $V = 120$; $H = 480$; the home team gets 480 tickets and the visiting team gets 120 tickets. **Step 4.** Answers vary. Sample: Yes; one substitution was easier because it required less arithmetic. **Step 5.** The CAS command substitutes $4V$ for H in the first equation. **Step 6.** $V = 120$ **Step 7.** $H = 480$

$$h = 4 \cdot v \mid v = 120 \qquad\qquad h = 480$$

Guided Example 1 $2M$; $5M$; $2M$; $5M$; 20; 40; 100; 40; 20; 100; yes; yes; yes

Questions
1. If $a = b$, then a may be substituted for b in any arithmetic or algebraic expression. **3.** one **5.** $-12 = 4(-3)$; $(-12)(-3) = 36$; the original equations are both true when the values are substituted. **7.** $x = 3, y = -2, z = \frac{1}{2}$
9. a. no solution **b.** inconsistent
c.

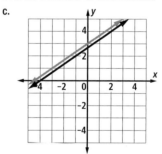

11. There are either no solutions, one solution, or infinitely many solutions. **13. a.** There are no solutions.
b. inconsistent
c.

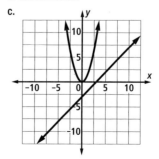

15. a. Answers vary. Sample: $\begin{cases} S = 2C \\ G = 4C \\ S + G + C = 50 \end{cases}$

b. $C = \frac{50}{7}$, $G = \frac{200}{7}$, $S = \frac{100}{7}$; use about 14.3 yd³ of sand, 28.6 yd³ of gravel, and 7.1 yd³ of cement.
17. a. $0.09x \leq 20$ **b.** $x \leq 222.2$ **c.** 222 one page flyers
19. a. domain: all real numbers except 0; range: all real numbers except 0 **b.** domain: all real numbers; range: all real numbers less than or equal to 0 **c.** domain: all real numbers except 0; range: all positive real numbers
d. domain: all real numbers; range: all real numbers

Lesson 5-4 (pp. 321–327)

Mental Math a. $39p + 4s$ dollars **b.** $3d_1 + 6d_2$ dollars
c. $xd_1 + yd_2$ dollars
Activity Step 2. $9c + 2p = 50$ and $65c + 45p = 800$
Step 3. Answers vary. Sample: To eliminate c, multiply the second equation by $\frac{-1}{5}$ and multiply the first equation by $\frac{13}{9}$. To eliminate p, multiply the first equation by –45 and the second equation by 2. **Step 4.** Answers vary.
Sample: $\frac{-55}{9p} = \frac{-790}{9}$; $-275c = -650$ **Step 5.** $p = \frac{158}{11}$; $c = \frac{26}{11}$; she should eat about 14.4 ounces of pasta and 2.4 ounces of chicken for dinner. **Step 6.** Answers vary. Sample: The easier variable to eliminate is usually the one where the coefficients have factors in common. Using a CAS can make this easier.
Guided Example 2 $7x$; $2y$; $-6x$; $6y$; -54; $-4x$; $7y$; $7x$; $2y$; 31; $49y$; -238; $28x$; 124; $57y$; -2; -20; 5; 12; 3; 5; -2; 3; 5; -2; 3
Questions
1. Addition and Multiplication Properties of Equality, Distributive Property **3. a.** p **b.** $p = 14.36$; $c = 2.36$
c. (2.36, 14.36) **5. a.** $\begin{cases} 5h + r = 50 \\ 80h + 95 = 800 \end{cases}$ **b.** 10 oz of chicken, 0 oz of fries **7.** $u = 5$ and $v = -2$
9. $x = 6$, $y = 2$, and $z = -3$ **11.** inconsistent **13.** $4.25
15. a. Answers vary. Sample: Multiply the first equation by –3 and the second equation by 2. **b.** Answers vary. Sample: Multiply the first equation by –2. **c.** $y = -2$, $z = 3$, and $x = 5$ **d.** true **17.** (–14, –10, 1)
19. $\begin{bmatrix} 4x + 3y \\ -2x + 6y \end{bmatrix}$ **21. a.** $3\begin{bmatrix} 4 & -6 & 5.5 \\ 0 & 3 & 0 \end{bmatrix} - 2\begin{bmatrix} 7 & 2 & 0 \\ 0 & 0 & -1 \end{bmatrix}$
b. $\begin{bmatrix} -2 & -22 & 16.5 \\ 0 & 9 & 2 \end{bmatrix}$

Lesson 5-5 (pp. 328–335)

Mental Math a. $\frac{1}{6}$ **b.** $\frac{1}{2}$ **c.** $\frac{1}{2}$ **d.** 1

Activity 1 Step 2. $\begin{bmatrix} 5.5 & 2.5 \\ 3.5 & 1.5 \end{bmatrix}$ **Step 3.** $\begin{bmatrix} 1 & 0 \\ 0 & 1 \end{bmatrix}$ **Step 4.** $\begin{bmatrix} 1 & 0 \\ 0 & 1 \end{bmatrix}$

Guided Example a. -2; -4; -1; 6; -2; 6; 4; -1; -8; $\frac{6}{-8}$, $\frac{-4}{-8}$, $\frac{1}{-8}$, $\frac{-3}{4}$, $\frac{1}{2}$, $\frac{-1}{8}$, $\frac{1}{4}$
b. 5; 10; 3; 6; 5; 6; 10; 3; 0; 0; no inverse
Activity 2 Step 1. 13, 5, 27, 9, 14, 27, 20, 8, 5, 27, 16, 1, 18, 11
Step 3. 5, 9, 27, 8, 27, 1, 11, 13, 27, 14, 20, 5, 16, 18, 27

Step 5. 1, 51, 145, 110, 23, 119, 201, 145, 85, 88, 79, 154, 16, 41, 47, 134, 141, 92, 183, 157

Step 6. $e^{-1} = \begin{bmatrix} -\frac{5}{31} & \frac{8}{31} \\ \frac{2}{31} & \frac{3}{31} \end{bmatrix}$; $e^{-1} \cdot V =$

$\begin{bmatrix} 13 & 5 & 27 & 5 & 9 & 27 & 8 & 27 & 1 & 11 \\ 5 & 20 & 13 & 27 & 14 & 20 & 5 & 16 & 18 & 27 \end{bmatrix}$ **Step 7.** PASS MATH

Questions
1. a. 1 **b.** 0 **c.** $\begin{bmatrix} 1 & 0 \\ 0 & 1 \end{bmatrix}$ **3. a.** $\begin{bmatrix} 1 & 0 \\ 0 & 1 \end{bmatrix}$ **b.** $\begin{bmatrix} 1 & 0 \\ 0 & 1 \end{bmatrix}$

5. a. when $ad - bc \neq 0$ **b.** $\begin{bmatrix} \frac{d}{ad-bc} & \frac{-b}{ad-bc} \\ \frac{-c}{ad-bc} & \frac{a}{ad-bc} \end{bmatrix}$ **7.** 75, $\begin{bmatrix} \frac{2}{25} & -\frac{3}{25} \\ \frac{7}{75} & \frac{2}{75} \end{bmatrix}$

9. $\begin{bmatrix} \frac{d}{ad-bc} & \frac{-b}{ad-bc} \\ \frac{-c}{ad-bc} & \frac{a}{ad-bc} \end{bmatrix}\begin{bmatrix} a & b \\ c & d \end{bmatrix} = \begin{bmatrix} \frac{da}{ad-bc} + \frac{-bc}{ad-bc} & \frac{db}{ad-bc} + \frac{-bd}{ad-bc} \\ \frac{-ca}{ad-bc} + \frac{ac}{ad-bc} & \frac{-cb}{ad-bc} + \frac{ad}{ad-bc} \end{bmatrix}$

$= \begin{bmatrix} \frac{ad-bc}{ad-bc} & \frac{0}{ad-bc} \\ \frac{0}{ad-bc} & \frac{ad-bc}{ad-bc} \end{bmatrix} = \begin{bmatrix} 1 & 0 \\ 0 & 1 \end{bmatrix}$

11. 54, 190, 35, 127, 42, 150, 31, 114, 62, 221, 33, 129, 15, 54, 26, 96, 49, 185 **13.** Answers vary. Sample: $\begin{bmatrix} 2 & 6 \\ 1 & 3 \end{bmatrix}$

15. a. $\begin{bmatrix} 0 & -1 \\ 1 & 0 \end{bmatrix}\begin{bmatrix} 0 & 1 \\ -1 & 0 \end{bmatrix} = \begin{bmatrix} 1 & 0 \\ 0 & 1 \end{bmatrix}$;

$\begin{bmatrix} 0 & 1 \\ -1 & 0 \end{bmatrix}\begin{bmatrix} 0 & -1 \\ 1 & 0 \end{bmatrix} = \begin{bmatrix} 1 & 0 \\ 0 & 1 \end{bmatrix}$ **b.** They are inverses.

If you rotate an object 90° and then 270° or vice versa, you end up in the same place where you started. **c.** It is the matrix for R_{180}, $\begin{bmatrix} -1 & 0 \\ 0 & -1 \end{bmatrix}$, because $R_{180} \cdot R_{180} = R_{360}$.

17. a. $\begin{cases} 50e + 30d = 160 \\ 10e = 20 \\ 0.75d + 0.75b = 3 \end{cases}$ **b.** $e = 2$; $d = 2$; $b = 2$;
2 energy bars, 2 energy drink bottles, and 2 bottles of water **c.** Answers vary. Sample: I used the substitution method because it was easy to solve the second equation for e. **19. a.** $d = 60t$
b.

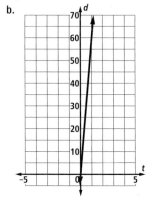

Lesson 5-6 (pp. 336-341)

Mental Math **a.** 4 **b.** 2 **c.** 11 **d.** $ad - bc$

Guided Example 1 $\begin{bmatrix} \frac{2}{7} & \frac{1}{14} \\ -\frac{1}{7} & \frac{3}{14} \end{bmatrix}$; $\begin{bmatrix} \frac{2}{7} & \frac{1}{14} \\ -\frac{1}{7} & \frac{3}{14} \end{bmatrix}$; $\begin{bmatrix} \frac{2}{7} & \frac{1}{14} \\ -\frac{1}{7} & \frac{3}{14} \end{bmatrix}$; $\begin{bmatrix} 1 & 0 \\ 0 & 1 \end{bmatrix}$;

$\begin{bmatrix} 4 \\ -2 \end{bmatrix}$; $\begin{bmatrix} 4 \\ -2 \end{bmatrix}$; 4, –2

Questions

1. a. $\begin{bmatrix} 0.5 & -1 \\ 3 & 8 \end{bmatrix} \begin{bmatrix} x \\ y \end{bmatrix} = \begin{bmatrix} 1.75 \\ 5 \end{bmatrix}$ **b.** about (2.71, –0.39)

3. zero; The determinant of the coefficient matrix is 0, and (2, 2) satisfies the first equation but not the second.
5. infinitely many; the determinant of the coefficient matrix is 0 and (2, 2) satisfies both equations.

7. $\begin{bmatrix} x \\ y \\ z \end{bmatrix} = \begin{bmatrix} -1 \\ 0 \\ 8 \end{bmatrix}$ **9.** $\begin{bmatrix} 3 & 5 \\ 1 & 1 \end{bmatrix} \begin{bmatrix} s \\ n \end{bmatrix} = \begin{bmatrix} 3943 \\ 937 \end{bmatrix}$; $\begin{bmatrix} s \\ n \end{bmatrix} = \begin{bmatrix} 371 \\ 566 \end{bmatrix}$;

371 student tickets and 566 non-student tickets **11.** No; you need three equations to find three unknowns.

13. $n = -12$ **15.** $(w, x, y, z) = (-1, 1, 2, 4)$ **17. a.** $\begin{bmatrix} 1 & -1 \\ 0 & 1 \end{bmatrix}$

b. $\begin{bmatrix} 1 & -1 \\ 0 & 1 \end{bmatrix} \begin{bmatrix} 1 & 1 \\ 0 & 1 \end{bmatrix} = \begin{bmatrix} 1 & 1 \\ 0 & 1 \end{bmatrix} \begin{bmatrix} 1 & -1 \\ 0 & 1 \end{bmatrix} = \begin{bmatrix} 1 & 0 \\ 0 & 1 \end{bmatrix}$

19. $y = -\frac{5}{8}x$

Lesson 5-7 (pp. 342-347)

Mental Math **a.** $8 + 0.5m$ in. **b.** 9.5 in. **c.** 6.5 in. **d.** 9 in.
Guided Example 2 Answers vary. Sample: 0; 0; above; Answers vary. Sample: 0; 0; Answers vary. Sample: does not; below

Questions

1. half-planes; boundary **3.** below **5. a.** ii **b.** i **c.** iii

7.

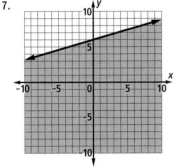

9. 8 **11.** 3 **13.** $-3x + 2y \le 12$ **15. a.** yes **b.** yes **c.** no
17. No; a single x value is paired with multiple y values.
19. Yes; it has exactly one solution.
21. $\{w \mid 2 < w \le 4.5\}$

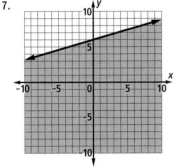

Lesson 5-8 (pp. 348-354)

Mental Math **a.** $p \le 12$ **b.** $c > -3$ **c.** $23 < x \le 31$ **d.** $t < 0.7$ or $t > 0.9$
Activity Step 1. $2T + C \le 12$ **Step 2** $2T + 2C \le 16$

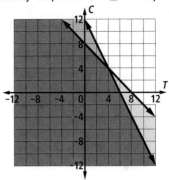

Step 3.

(T, C)	Satisfies Long Block Inequality	Satisfies Square Block Inequality
(2, 4)	yes	yes
(6, 2)	no	yes
(2, 7)	yes	no
(5, 4)	no	no

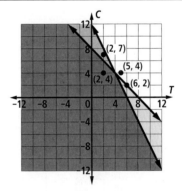

Step 4. (2, 4); It appears in the region where the solutions to the two inequalities overlap.
Guided Example 1 **a.** 2000; L; 2000L; 1500R; 1500R; 2000L; 1500R; 2000L; 15R; 20L **b.** 30; 40; 50; 20; 30; 40; 50; 20

Questions

1. intersection **3. a.** not possible; not enough long blocks or square blocks **b.** possible **c.** possible **d.** not possible; not enough square blocks **5. a.** $\begin{cases} 2T + C = 12 \\ 2T + 2C = 16 \end{cases}$ **b.** (4, 4)
7. a. \$160,000 **b.** \$100,000 **c.** \$107,500

9. a.

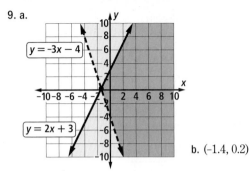

$y = -3x - 4$
$y = 2x + 3$

b. $(-1.4, 0.2)$

11. a. $0; 0; 40; 1.25S + 2.5L$

b.

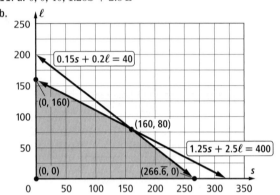

$0.15s + 0.2\ell = 40$
$(0, 160)$
$(160, 80)$
$1.25s + 2.5\ell = 400$
$(0, 0)$
$(266.\overline{6}, 0)$

13. a.
$$\begin{cases} x \leq 40 \\ 2x + 2y \leq 100 \\ 2.5x + 3.5y \leq 150 \\ x \geq 0 \\ y \geq 0 \end{cases}$$

b.

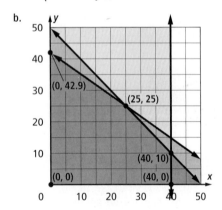

$(0, 42.9)$
$(25, 25)$
$(40, 10)$
$(0, 0)$
$(40, 0)$

15.

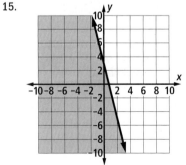

17. false   **19.** $52x$ to $53x + 265$ minutes

Lesson 5-9 (pp. 355–361)

Mental Math a. $45 **b.** $45 **c.** $75 **d.** $95

Activity Step 1. Numbers for empty cells of table.
Top row table: $5x$; $3y$; Middle row table: $2x$; $6y$; 130;

Bottom row table: $2.5x$; $4y$; 120;
$$\begin{cases} 5x + 3y \leq 115 \\ 2x + 6y \leq 130 \\ 2.5x + 4y \leq 120 \\ x \geq 10 \\ y \geq 1 \end{cases}$$

Step 2.

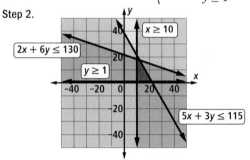

$2x + 6y \leq 130$
$y \geq 1$
$x \geq 10$
$5x + 3y \leq 115$

$2x + 6y = 130$; $5x + 3y = 115$; $2x + 6y = 130$; $5x + 3y = 115$; $2.5x + 4y \leq 120$ does not help define the feasible region.
Step 3. $(10, 1)$, $(10, \frac{55}{3})$, $(22.4, 1)$, $(12.5, 17.5)$
Step 4. $R = 20x + 15y$ **Step 5.** $R_1 = 20(10) + 15(1) = 215$; $R_2 = 20(10) + 15(18) = 470$; $R_3 = 20(22) + 15(1) = 455$; $R_4 = 20(12) + 15(17) = 495$; **Step 6.** To maximize profits, 12 centerpieces and 17 topiaries should be made. $495 would be made from the sales.

Questions
1. a. revenue, $R = 900T + 600C$ **b.** $5100 **c.** possible combinations of tables and chairs **d.** true **3.** He proved the Linear-Programming Theorem. **5.** $(12, 17)$; It means that they should make 12 centerpieces and 17 topiaries. **7.** a procedure for solving large linear programming problems **9. a.** $(7, 10)$ **b.** $(1, 2)$
11. a. Let M = trays of muffins, N = loaves of nut bread
$$\begin{cases} \frac{9}{16}M + \frac{1}{2}N \leq 50 \\ \frac{3}{4}M + \frac{3}{4}N \leq 48 \\ \frac{3}{4}M + \frac{1}{6}N \leq 15 \\ M \geq 0 \\ N \geq 0 \end{cases}$$

b.

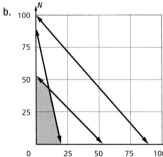

The feasible set is the set of lattice points in the shaded region. $\frac{9}{16}M + \frac{1}{2}N = 50$ is not a boundary of the feasible region.
c. 53 loaves of nut bread and 0 trays of muffins

13.

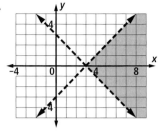

15. $y \geq -\frac{2}{3}x + 2$ **17.** $(s, A) = (2, 4)$ or $(-4, 16)$

Self-Test (pp. 365–366)

1. $-4n + 18 < 30$; $-4n < 12$; $n > -3$ **2.**

3. Estimating the intersection points from the graph, the solutions are approximately $(-1.7, -2.8)$ and $(1.2, -1.4)$.
4. This system is inconsistent because there is no solution. The system describes two lines that do not intersect because they are parallel.   **5.** Substituting the first two equations into the third yields $3(q + 4) - 2(3q) = 3$, or $-3q + 12 = 3$, so $q = 3$. Substituting this value back into the first two equations gives $p = 3(3) = 9$ and $r = 3 + 4 = 7$. Thus, the solution to this system is $p = 9$, $q = 3$, $r = 7$. Substituting these values back into system of equations confirms this solution: $9 = 3 \cdot 3$; $7 = 3 + 4$; $3 \cdot 7 - 2 \cdot 9 = 3$.   **6.** Multiplying the first equation by 2, multiplying the second by 7, and subtracting yields $-53y = -74 \Rightarrow y = \frac{74}{53}$. Substituting this into either equation yields $x = \frac{38}{53}$. We can substitute to check: $7\left(\frac{38}{53}\right) + 5\left(\frac{74}{53}\right) = 12$ and $2\left(\frac{38}{53}\right) + 9\left(\frac{74}{53}\right) = 14$.

7. a. $\begin{cases} H + P = 20 \\ 7.83H + 11.50P = 200 \end{cases}$; H is the number of pounds of organic hazelnuts and P is the number of pounds of organic pecans. **b.** Solving the first equation for H and substituting into the second yields $7.83(20 - P) + 11.50P = 200$, $P \approx 11.8$. The first equation then implies $H = 20 - P \approx 8.2$. Thus, about 11.8 pounds of organic pecans must be mixed with about 8.2 pounds of organic hazelnuts to obtain 20 pounds of mixed nuts worth \$10 per pound.

8. a. $\begin{bmatrix} 3 & 2 \\ -2 & 7 \end{bmatrix}$ **b.** Because $(3)(7) - (2)(-2) = 25 \neq 0$, the Inverse Matrix Theorem says that the inverse of the coefficient matrix is $\begin{bmatrix} \frac{7}{25} & \frac{-2}{25} \\ \frac{2}{25} & \frac{3}{25} \end{bmatrix}$.

c. $\begin{bmatrix} 3 & 2 \\ -2 & 7 \end{bmatrix}\begin{bmatrix} x \\ y \end{bmatrix} = \begin{bmatrix} 24 \\ 39 \end{bmatrix}$; $\begin{bmatrix} \frac{7}{25} & \frac{-2}{25} \\ \frac{2}{25} & \frac{3}{25} \end{bmatrix}\begin{bmatrix} 3 & 2 \\ -2 & 7 \end{bmatrix}\begin{bmatrix} x \\ y \end{bmatrix} =$

$\begin{bmatrix} \frac{7}{25} & \frac{-2}{25} \\ \frac{2}{25} & \frac{3}{25} \end{bmatrix}\begin{bmatrix} 24 \\ 39 \end{bmatrix}$; $\begin{bmatrix} 1 & 0 \\ 0 & 1 \end{bmatrix}\begin{bmatrix} x \\ y \end{bmatrix} = \begin{bmatrix} \frac{18}{5} \\ \frac{33}{5} \end{bmatrix}$; $x = \frac{18}{5}$, $y = \frac{33}{5}$

9. a. Answers vary. Sample: $\begin{bmatrix} 1 & 1 \\ 1 & 1 \end{bmatrix}$ **b.** The determinant of the matrix is zero: $(1)(1) - (1)(1) = 0$.
10. First, graph the boundary $y = -\frac{5}{2}x - 2$ with a dotted line, then shade above it.

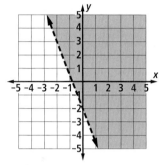

11. $y < -x$ and $y \geq -2$
12. The points are solutions if they are located in the shaded region. **a.** no **b.** yes **c.** no

13. $\begin{cases} \frac{1}{3}e + \frac{1}{6}c \leq 48 \\ \frac{1}{6}e + \frac{1}{3}c \leq 64 \\ e \geq 0 \\ c \geq 0 \end{cases}$

b. vertices: $(0, 0)$; $(0, 144)$; $(160, 64)$; $(192, 0)$

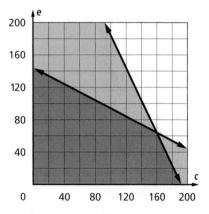

c. The Linear-Programming Theorem states that one of the vertices from Part b will maximize $15e + 12c = profits$. Substituting each vertex's values into this profit expression, we find that the company maximizes its profits at \$2880 by making 160 Cool Contemporary shelves and 64 Olde English shelves.

Self-Test Correlation Chart

Question	1	2	3	4	5	6	7	8
Objective(s)	H	H	I	D	A	A	F	B, C
Lesson(s)	5-1	5-1	5-2	5-2, 5-3, 5-4, 5-6	5-3, 5-4	5-3, 5-4	5-2, 5-3, 5-4, 5-6	5-5, 5-6

Question	9	10	11	12	13
Objective(s)	B	J	K	E	G
Lesson(s)	5-5	5-7	5-8	5-8, 5-9	5-9

Chapter Review (pp. 367–371)

1. D **3.** $x = -12, y = 4$ **5.** $x = \frac{47}{2}, y = -9$ **7.** $a = -\frac{8}{5}, b = \frac{6}{5}$, $c = -\frac{2}{5}$ **9. a.** 4 **b.** $\begin{bmatrix} \frac{1}{4} & 0 \\ 0 & 1 \end{bmatrix}$ **11. a.** 0 **b.** does not exist

13. $\begin{bmatrix} \frac{3}{13} & \frac{-2}{13} \\ \frac{5}{13} & \frac{1}{13} \end{bmatrix}$ **15.** The determinant equals 0.

17. $\begin{bmatrix} 3 & -4 \\ 4 & -6 \end{bmatrix}\begin{bmatrix} x \\ y \end{bmatrix} = \begin{bmatrix} 12 \\ 48 \end{bmatrix}; \begin{bmatrix} x \\ y \end{bmatrix} = \begin{bmatrix} -60 \\ -48 \end{bmatrix}$

19. $\begin{bmatrix} 2 & -5 \\ 3 & -6 \end{bmatrix}\begin{bmatrix} m \\ n \end{bmatrix} = \begin{bmatrix} 4 \\ -3 \end{bmatrix}; \begin{bmatrix} m \\ n \end{bmatrix} = \begin{bmatrix} -13 \\ -6 \end{bmatrix}$

21. No; they have different solutions. **23.** an inconsistent system **25. a.** consistent **b.** 1 **27. a.** consistent **b.** 2 **29.** Answers vary. Sample: $t = 1$ **31.** false **33. a.** Yes; this is a convex region. **b.** No; the region is not convex. **c.** Yes; the region is convex. **d.** No; the region is not convex. **35.** Yes; it is a vertex. **37.** The tank must contain 60% from the first source and 40% from the second source. **39.** 7 oz of meat and 4 oz of vegetables

41. a. $\begin{cases} c \geq 0 \\ r \geq 0 \\ c \leq 30{,}000 \\ c + r \leq 40{,}000 \\ 185c + 225r \leq 8{,}500{,}000 \end{cases}$

b.

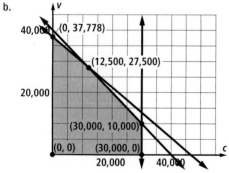

$(0, 0), (30{,}000, 10{,}000), (12{,}500, 27{,}500), (0, 37{,}777.7)$
c. $(12{,}500, 27{,}500)$

43. a.

b. $\{t \mid 48 < t \leq 99\}$

45. $x < -4$

47. $t \leq 5$

49.

51.

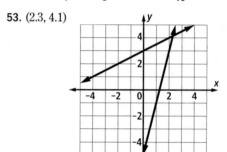

53. $(2.3, 4.1)$

55. no solutions

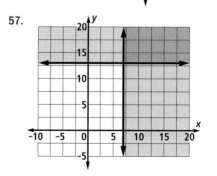

57.

59.

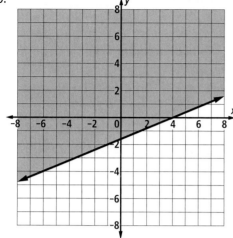

61.

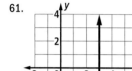

63.

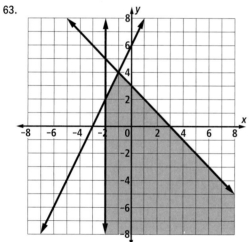

65. C

Chapter 6

Lesson 6-1 (pp. 374–379)

Mental Math a. 21 ft^2 **b.** 58.5 ft^2 **c.** 37.5 ft^2 **d.** 44 ft^2
Guided Example 3 a. $15 - s$; $15 - s$; 15; s, 15; 15; 15; 30; 225; $15 - s$; s^2 **b.** $15 - s$; 225; $30s$; s^2; $30s\pi$; $s^2\pi$; $30\pi s - \pi s^2$
Questions
1. B; it is not a second-degree equation. **3.** Answers vary. Sample: square, circle **5. a.** $6x^2 + 31xy + 35y^2$

b.

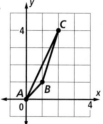

7. a. $6y^2 - 7y + 2$

b.

9. 169 **11.** $p^2 + 2pw + w^2$ **13.** $(x - y)^2 = (x - y)(x - y)$, Definition of second power; $= (x - y)x - (x - y)y$, Distributive Property; $= x^2 - yx - xy + y^2$, Distributive Property; $= x^2 - 2xy + y^2$, Commutative and Associative Properties **15.** $p^2 - 2p + 1$ **17. a.** $4w^2 + 224w + 1200$ **b.** $4w^2 + 224w$ **19.** $4t^2 - \frac{4}{3}kt + \frac{1}{9}k^2$ **21. a.** $x^2 + 12x + 32$ **b.** about 189 tiles **c.** about 16% **d.** about 23% **23.** $h = 10$ **25. a.** -5.25 million visits/yr **b.** 0.129 billion dollars/yr **27. a.**

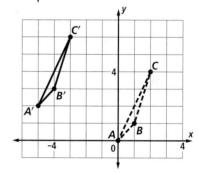

b.

c. The transformation translates $\triangle ABC$ 5 units to the left and 2 units up.

Lesson 6-2 (pp. 380–385)

Mental Math a. $3000 **b.** $3400 **c.** $2500
Activity Step 1. Answers vary. Sample:

	x	$f(x)$	$g(x)$
$x < -10$	-15	15	15
$-10 \leq x \leq -1$	-6	6	6
$-1 < x < 0$	$-\frac{1}{2}$	$\frac{1}{2}$	$\frac{1}{2}$
$x = 0$	0	0	0
$0 < x < 1$	$\frac{1}{3}$	$\frac{1}{3}$	$\frac{1}{3}$
$1 \leq x \leq 10$	7	7	7
$x > 10$	12	12	12

Step 2. $f(x) = g(x)$ for all real numbers x.

Step 3.

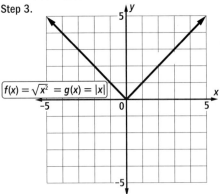

$f(x) = \sqrt{x^2} = g(x) = |x|$

Step 4. The graphs are the same.

Questions

1. a. 17.8 b. 17.8 c. -17.8 d. -17.8   3. No; consider $x \neq 0$.
Then $|x| > 0$, so $-|x| < 0$ and $|x| \neq -|x|$.   5. $|x| = 90$
7. $n = -1$ or $n = -6$ 9. A   11. ± 27 13. 4.51
15. irrational   17. rational, 6   19. irrational
21. a. $1.5 \geq |p - 442.5|$ b. 444 g   23. a pizza with about
a 13.4" diameter   25. Answers vary. Sample: Each small
square with dimensions $\sqrt{3} \times \sqrt{3}$ has an area of $(\sqrt{3})^2 = 3$.
So the large square, which is made up of four small squares,
has an area of $4 \cdot 3 = 12$. You can also find the area of the
large square by squaring its side length: $(\sqrt{3} + \sqrt{3})^2 = (2\sqrt{3})^2$.
So $(2\sqrt{3})^2 = 12$. Taking the square root of both sides gives
$2\sqrt{3} = \sqrt{12}$. (Since the side length must be positive, the
absolute value can be ignored.)   27. $64 - x^2$

29. a.

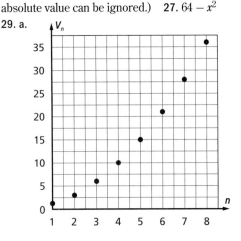

b. $v_{n+1} = v_n + (n + 1)$, for integers $n \geq 1$

Lesson 6-3 (pp. 386–392)

Mental Math a. 16 b. 100 c. 63 d. 441

Activity Step 1. Group B

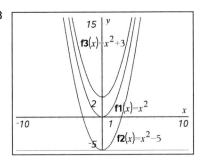

Group C

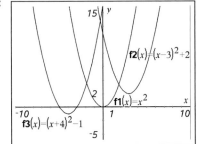

Step 2. Group A: $f2$ is a translation of $f1$ four units to the
right; $f3$ is a translation of $f1$ two units to the left; Group B:
$f2$ is a translation of $f1$ five units down; $f3$ is a translation
of $f1$ three units up; Group C: $f2$ is a translation of $f1$
three units to the right and two units up; $f3$ is a translation
of $f1$ four units to the left and one unit down.

Step 3. a. translation 5 units to the left b. translation 1
unit down c. translation 2 units to the left and 5 units
down Step 4. a. translation h units horizontally (to the
right if h is positive, to the left if h is negative)
b. translation k units vertically (up if k is positive, down
if k is negative) c. translation h units horizontally and
k units vertically Step 5. The conjecture is true.

Guided Example 2 a. $x + 3$; $y - 5$; $T_{-3,5}$; -3; 5 b. $x = 0$; -3
c. -2; -2; 3; 3; (-3, 5); 3; -3

Questions

1. The translation $T_{5,0}$ maps the first graph onto the second.
3. a. $(x + 7, y)$ b. $y = (x - 7)^2$ 5. $y = \frac{7}{5}(x - 4)^2 - 7$
7. true   9. a. $x = -5$ b. $x = h$ 11. a. (4, 5) b. $x = 4$ c. down
d.

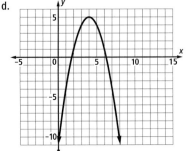

13. $y = -(x - 3)^2 - 2$ 15. a. $y = -4x^2 + 2$ b. $y = -4(x - 2)^2$
17. $x = 9$ 19. $y = mx$; x_1; y_1 21. $d = 2$ or $d = -5$;
$3|2(2) + 3| = 3|7| = 21$ and $3|2(-5) + 3| = 3|-7| = 21$
23. a. $d^2 + 12d + 36$ b. 196 m²; 256 m²   25. $\frac{x^4}{y}$

Lesson 6-4 (pp. 393–400)

Mental Math a. Answers vary. Sample: $y = 3x$ **b.** Answers vary. Sample: $y = \frac{3}{x}$ **c.** Answers vary. Sample: $y = 3x^2$
d. Answers vary. Sample: $y = \frac{3}{x^2}$
Activity Answers vary. Sample: **Steps 1–5.**

	Initial Height h_0 (m)	Elapsed Time Trial 1 (sec)	Elapsed Time Trial 2 (sec)	Elapsed Time Trial 3 (sec)	Elapsed Time Average (t) (sec)
Partner 1	1.2	1.75	1.6	1.9	1.75
Partner 2	1	2.05	1.93	2.02	2.00

Step 6. Answers vary. Sample : For partner 1, $v_0 \approx 7.89 \frac{m}{sec^2}$; $h = -4.9t^2 + 7.89t + 1.2$. For partner 2, $v_0 = 9.3 \frac{m}{sec^2}$; $h = -4.9t^2 + 9.3t + 1$.

Step 7. Answers vary. Sample: for partner 1, about 4.4 m; for partner 2, about 5.4 m

Questions
1. $y = ax^2 + bx + c$ **3.** $y = -3x^2 - 24x - 53$
5. a. height **b.** acceleration due to gravity **c.** time
d. initial velocity **e.** initial height **7.** 0 **9.** 0 **11.** $-5 \frac{ft}{sec}$
13.

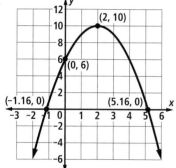

15. a. $h = -16t^2 + 1474$
b.

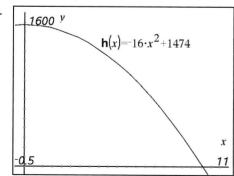

c. about 9.6 sec **d.** about 7.5 sec longer than the equation predicts **17.** $y = -\frac{1}{4}x^2 + 2x - 2$

19. a.

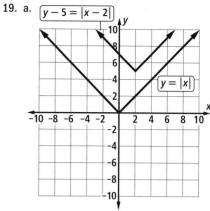

b. The graph of $y - 5 = |x - 2|$ is the image of $y = |x|$ under the translation $T_{2,5}$. **21.** $y \leq \frac{2}{3}x + 4$ **23.** $n = 5$ **25.** $n = 6$

Lesson 6-5 (pp. 401–407)

Mental Math a. 1 **b.** 1 **c.** an infinite number **d.** 0
Guided Example 2 a. 12; 36; 36; 36; 33; 36; 33; 6 **b.** -6; -33
Questions
1. a. $x^2 + 16x$ **b.** 64 **c.** 16; 64; 8 **3.** $\frac{1}{4}$ **5.** $\frac{1}{9}$
7. a. $y + 9.5 = -2(x + 1.5)^2$ **b.** (-1.5, -9.5)
9. a. $h = -16t^2 + 64t + 8$ **b.** 56 ft **c.** 72 ft
d.

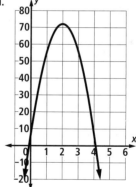

e. about 4.1 sec **11.** $y - 14.8 = 5(x - 0.2)^2$
13. The parabola opens down and thus, has no minimum.
15. 23.2 $\frac{m}{sec}$ **17. a.** $T_{1,4}$ **b.** $y = -(x - 1)^2 + 4$
19. a.

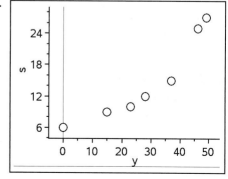

b.

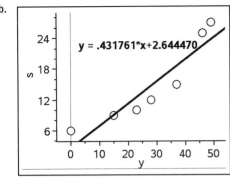

c. about 46 states   **21.** $x = 5$

Lesson 6-6 (pp. 408–413)

Mental Math a. $-\frac{3}{2}$ **b.** $\frac{2}{3}$ **c.** $\frac{2}{3}$ **d.** $-\frac{2}{3}$

Activity: Answers vary. Samples are given.

Step 2.

Drop #	I	M	O
1	0	0	0
2	1	0	0
3	1	1	0
4	1	1	0
5	1	1	0
6	1	0	0
7	1	1	0
8	1	1	0
9	1	1	0
10	1	1	0
11	1	0	0
12	1	1	1
13	0	0	0
14	1	1	0
15	0	0	0
16	1	0	0
17	1	1	0
18	1	1	0
19	1	0	0
20	1	1	0
21	1	0	0
22	1	1	0
23	1	1	1
24	1	0	0
25	1	0	0

Step 3. $I: \frac{22}{25}$; $M: \frac{14}{25}$; $O: \frac{2}{25}$

Step 4.

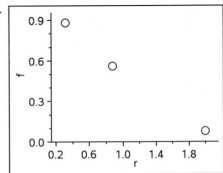

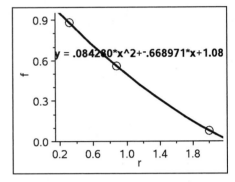

Step 5. Answers vary. Sample: This model seems to pass through all three of my data points. **Step 6.** Answers vary. Sample: Quarter radius $= \frac{15}{32}$ in.; prediction ≈ 0.786; relative frequency $= \frac{19}{25} = 0.76$; This value is close to the predicted value. The value from the combined data is even closer to the predicted value.

Questions

1. a. Answers vary. Sample: $y = 0.084x^2 - 0.669x + 1.081$
b. Answers vary. Sample: 0.236   **3.** $y = x^2 - 8x + 10$
5. $y \approx 0.194x^2 - 1.14x + 3.7$ **7. a.** $y = -0.042x^2 + 0.59x + 10$
b.

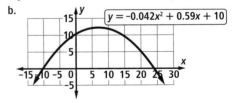

$y = -0.042x^2 + 0.59x + 10$

9. a.

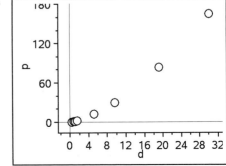

b. $p \approx 0.1089d^2 + 2.281d - 1.535$

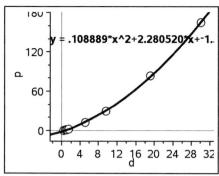

Yes; the model seems to pass through all the data points. **c.** The model predicts about 258 yr for Pluto, so Pluto fits the model fairly well. **d.** about 1,086 yr **11.** $y - 2.25 = 3(x - 1.5)^2$ **13.** $y = -2x^2 - 16x - 32.5$ **15. a.** about 1.37 sec **b.** about 0.56 sec

Lesson 6-7 (pp. 414–419)

Mental Math a. $a^2 - 2ab + b^2$ **b.** $a^2 - 6ab + 9b^2$
c. $25a^2 - 30ab + 9b^2$ **d.** $25a^8 - 30a^4b^6 + 9b^{12}$
Guided Example 2 5; 13; -6; 13; 13; 5; -6; 5; -13; 289; 10; -13; 17; 10; -13; 17; 10; $\frac{2}{5}$; -3
Questions
1. $x = \frac{-b + \sqrt{b^2 - 4ac}}{2a}$ or $x = \frac{-b - \sqrt{b^2 - 4ac}}{2a}$. **3.** to make the coefficient of x^2 1 in order to complete the square
5. $z = -1$ or $z = -\frac{3}{10}$ **7.** x represents the distance along the ground in feet of the ball from home plate, and h represents the height in feet of the ball at that distance. **9.** about 60.25 ft **11.** The solutions are the opposites of the solutions from Example 2; $x = -\frac{2}{5}$ or $x = 3$.
13. $t_n = \frac{n(n+1)}{2} = 10,608$; $n(n + 1) = 21,216$; $n^2 + n = 21,216$; $n^2 + n - 21,216 = 0$; $n = \frac{-1 \pm \sqrt{1^2 - 4(1)(-21,216)}}{2(1)}$; $n = \frac{-1 \pm \sqrt{84,865}}{2}$; n is not an integer, therefore 10,608 cannot be a triangular number. **15. a.** Q_1: $x = -2$ or $x = -3$; Q_2: $x = -3$ or $x = -4$; Q_3: $x = -4$ or $x = -5$; Q_4: $x = -5$ or $x = -6$; Q_5: $x = -6$ or $x = -7$ **b.** Q_1: 6, -5; Q_2: 12, -7; Q_3: 20, -9; Q_4: 30, -11; Q_5: 42, -13 **c.** The sum of the solutions is equal to $-b$, and the product of the solutions is equal to c. **17.** $\left(-\frac{2}{3}, 4\right)$ and $\left(\frac{1}{2}, 4\right)$
19. $y = -2x^2 + 8x + 3$ **21.** $x = -24$ **23. a.** $g_1 = 0$; $g_2 = 2$; $g_3 = 0$; $g_4 = 2$ **b.** $g_n = 0$ for all odd integers n, and $g_n = 2$ for all even integers n. **c.** 2

Lesson 6-8 (pp. 420–426)

Mental Math a. $x = 4$ or $x = -4$ **b.** $x = 7$ or $x = -1$
c. $x = 3.5$ or $x = -0.5$
Activity Step 1. -100; $-i$; $-3i$; $\frac{2i}{5}$; $10i$; -10 **Step 2.** Addition and subtraction of imaginary numbers, and multiplication of an imaginary number by a real scalar, result in an imaginary number, but multiplication of two imaginary numbers results in a real number. **Step 3.** 10; 10; yes; $4 \geq 0$ and $25 \geq 0$, and $\sqrt{4} \cdot \sqrt{25} = \sqrt{100} = 10$.

Step 4. $10i$; 10; yes; If $a \geq 0$ and $b < 0$, then $\sqrt{a} \cdot \sqrt{b} = \sqrt{a} \cdot i\sqrt{-b}$, and $ab < 0$ so $\sqrt{ab} = i\sqrt{a(-b)} = \sqrt{a} \cdot i\sqrt{-b}$. No; $-4 < 0$ and $-25 < 0$, and $\sqrt{-4} \cdot \sqrt{-25} \neq \sqrt{-4 \cdot -25}$, so this is a counterexample.
Questions
1. true **3.** C **5.** Euler **7.** true **9. a.** $x = \sqrt{-16}$ or $x = -\sqrt{-16}$ **b.** $x = 4i$ or $x = -4i$ **11.** $(-6i)^2 = (-6)^2i^2 = 36(-1) = -36$ **13.** $i\sqrt{53}$ **15.** $-6\sqrt{2}$ **17.** $7i$ **19.** $20i$
21. 0 **23.** $-10\sqrt{3}$ **25. a.** yes **b.** no **27.** $691i$ **29.** False; $2i + -2i = 0$, 0 is a real number. **31.** $x = \frac{0 \pm \sqrt{0 - 4(1)(4)}}{2(1)} = \pm\frac{i\sqrt{16}}{2} = \pm 2i$ **33.** $a = \pm i\sqrt{7}$ **35.** $x = \frac{3}{2}$ or $x = -\frac{5}{4}$
37. $y = 0.25x^2 + 300x - 77.7$ **39.** domain: all real numbers; range: all real numbers

Lesson 6-9 (pp. 427–433)

Mental Math a. $w < 30$ **b.** $7 \leq s \leq 10$ **c.** $h \geq 54$
Activity Step 1. $8 + 12i$; $11 + 2i$; $-12 - 3i$ **Step 2.** $-2 - 8i$; $-9 - 8i$; $7 - 3i$ **Step 3.** You can use the Commutative and Associative Properties to add or subtract the real parts and then add or subtract the imaginary parts; $(a + bi) + (c + di) = (a + c) + (b + d)i$ or $(a + bi) - (c + di) = (a - c) + (b - d)i$

Step 4.

$a + b \cdot i + c + d \cdot i$	$a + c + (b + d) \cdot i$
$a + b \cdot i - (c + d \cdot i)$	$a - c + (b - d) \cdot i$

Guided Example 3 $9 + 18i + 6i + 12i^2$; $9 - 6i + 6i - 4i^2$; $9 + 24i + 12i^2$; $9 + 24i + 12(-1)$; -1; -3; 24
Guided Example 4 $3 + 2i$; $5 - 4i$; 23; -2; 8; 2; numerators: 23, -2, 8, 2; denominators: 8, 2, 8, 2; $188 + 30i$; $188 + 30i$
Questions
1. real **3.** real part: -4; imaginary part: $\sqrt{5}$ **5.** $3 + 5i$
7. 29 **9.** $a - bi$ **11. a.** $5 - 2i$ **b.** $-3i$ **c.** $-2 + 3i$ **d.** 4
13. $2 - 3i$ **15.** true **17.** $5i$ or $0 + 5i$ **19. a.** Answers vary. Sample: $3 + 2i$, $8 - 2i$ **b.** Answers vary. Sample: $3i$, $5i$
21. a. $3 + 2i$, $3 - 2i$ **b.** They are complex conjugates.
23. $a^2 + b^2$; Because a and b are real numbers, $a^2 + b^2$ is a real number.
25. $a = i\sqrt{5}$ or $-i\sqrt{5}$ **27.** Answers vary. Sample: By the definition of absolute value, the distance between x and $\frac{-b}{2a}$ is $\sqrt{\frac{b^2 - 4ac}{4a^2}}$, and this means that the difference between x and $\frac{-b}{2a}$ could be $\sqrt{\frac{b^2 - 4ac}{4a^2}}$ or its opposite. So, when absolute value signs are removed, a $\pm$ sign must be added. **29.** $R = \frac{km^2}{P}$

Lesson 6-10 (pp. 434–440)

Mental Math a. S_3 **b.** $\begin{bmatrix} 4.2 & 0 \\ 0 & 4.2 \end{bmatrix}$ **c.** $S_{0.5,6}$ **d.** $\begin{bmatrix} 7 & 0 \\ 0 & 2 \end{bmatrix}$

Activity

$y = ax^2 + bx + c$	Number of x-Intercepts of Graph	Solutions to $ax^2 + bx + c = 0$	Number of Real Solutions to $ax^2 + bx + c = 0$	Value of $b^2 - 4ac$
a. $y = 4x^2 - 24x + 27$	2	$x = 1.5, 4.5$	2	144
b. $y = 4x^2 - 24x + 36$	1	$x = 3$	1	0
c. $y = 4x^2 - 24x + 45$	0	$x = 3 \pm \frac{3}{2}i$	0	-144
d. $y = -6x^2 + 36x - 54$	1	$x = 3$	1	0
e. $y = -6x^2 + 36x - 48$	2	$x = 2, 4$	2	144
f. $y = -6x^2 + 36x - 60$	0	$3 \pm i$	0	-144

Step 3. Answers vary. Sample: the number of x-intercepts is equal to the number of real solutions. **Step 4.** Answers vary. Sample: When the discriminant is positive, there are two real solutions. When the discriminant is zero, there is one real solution. When the discriminant is negative, there are no real solutions. **Step 5a-b.** Answers vary. Check students' work. **Step 5c.** yes

Guided Example Solution a. 1 **b.** –364; 0 **c.** 3; 5; –14, 193; 2

Check a. 1; 1 **b.** 0; 0 **c.** 2; 2

Questions

1. They are equal. **3. a.** $b^2 - 4ac$ **b.** $\frac{-b \pm \sqrt{b^2 - 4ac}}{2a}$ **5. a.** v **b.** iv **7. a.** $x^2 - 10x + 9 = 0$; $b^2 - 4ac = 64$, so there are 2 real roots. **b.** $x^2 - 10x + 9 = y$; 2 x-intercepts
9. Answers vary. Sample:

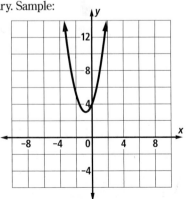

11. a. positive **b.** 2 **13.** Yes; the discriminate D of $0 = -0.4x^2 + 3x - 5$ is 1. Because D is positive there are real solutions, and the ball will reach the height of the rim. **15.** True; the graph intersects the x-axis in two points, so there are two real roots. **17.** No; Explanations vary. Sample: If one root is rational, then by the quadratic formula $x = \frac{a \pm \sqrt{b}}{c}$, where a, $\sqrt{b}$, and c are rational numbers. The other root, $\frac{a - \sqrt{b}}{c}$ would have to be rational because addition, subtraction, and division of rational numbers result in a rational number. **19.** $x = 4 + 3i$ or $x = 4 - 3i$
21. a. $p = -16t^2 + 1200$ **b.** $m = 300 + 30t$
c. $t \approx 6.62$ seconds

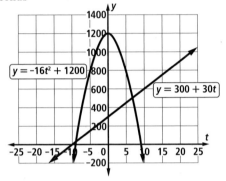

Self-Test (p. 445)

1. $y - 21 = x^2 - 10x$; $y - 21 + 25 = x^2 - 10x + 25$; $y + 4 = (x - 5)^2$ **2.** $(5, -4)$ **3.** $0 = x^2 - 10x + 21$; $x = \frac{10 \pm \sqrt{(-10)^2 - 4(1)(21)}}{2} = \frac{10 \pm 4}{2}$; $x = \frac{10 + 4}{2} = 7$ or $x = \frac{10 - 4}{2} = 3$. The x-intercepts are 7 and 3. **4.** $5i \cdot i = 5i^2 = -5$
5. $\sqrt{-12} \cdot \sqrt{-3} = i\sqrt{12} \cdot i\sqrt{3} = i^2\sqrt{36} = -6$ **6.** $\frac{6 + \sqrt{-64}}{3} = \frac{6 + i\sqrt{64}}{3} = \frac{6 + 8i}{3} = 2 + \frac{8}{3}i$ **7.** $(1 + 5i)(-4 + 3i) = -4 - 20i + 3i + 15i^2 = -19 - 17i$ **8.** $6 - 5i - (2 + 5i) = 6 - 5i - 2 - 5i = 4 - 10i$ **9.** C
10. $y - 3 = -(x + 4)^2$ is in vertex form, so the vertex is $(-4, 3)$. Because $a = -1 < 0$, the graph will open down, it is symmetric about $x = -4$, and the y-intercept is $y = -16 + 3 = -13$.

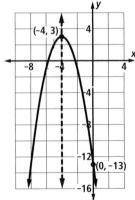

11. From the vertex form of the equation, the vertex is $(-4, 3)$. Since $a < 0$, the parabola opens down. So, the domain is all real numbers and the range is $\{y \mid y \leq 3\}$.

12. $t + 12 = 21$ or $t + -12 = -21$; $t = 9$ or $t = -33$
13. $\sqrt{(6s+5)^2} = 9 \Rightarrow |6s+5| = 9$; $6s + 5 = 9$, $s = \frac{2}{3}$; or
$(6s+5) = -9$, $s = -\frac{7}{3}$ **14.** $17 = y^2 + 36y + 324$;
$0 = y^2 + 36y + 307$; $y = \frac{-36 \pm \sqrt{1296 - 4 \cdot 1 \cdot 307}}{2}$; $y = \frac{-36 \pm \sqrt{68}}{2}$;
$y = -18 \pm \sqrt{17}$ **15.** $3x^2 + 14x - 5 = 0$;
$x = \frac{-14 \pm \sqrt{196 - 4 \cdot 3 \cdot (-5)}}{6}$; $x = \frac{-14 \pm \sqrt{196 + 60}}{6}$; $x = \frac{-14 \pm 16}{6}$; $x = \frac{1}{3}$, $x = -5$
16. By the Absolute Value-Square Root Theorem, $\sqrt{x^2} = |x|$
for all real values of x.   **17.** $4a^2 - 20a + 25 + 4a^2 + 20a + 25 = 8a^2 + 50$ **18.** The solutions are not real.   **19. a.** 0
b. 1 **c.** 2   **20.** $h = -16(1.5)^2 + 28(1.5) + 6$ $h = -36 + 42 + 6$ $h = 12$ feet   **21.** $0 = -16t^2 + 28t + 6$; $t = \frac{-28 \pm \sqrt{784 - 4 \cdot (-16) \cdot 6}}{-32}$;
$t = \frac{-28 \pm \sqrt{1168}}{-32}$; $t \approx -0.193$, $t \approx 1.943$; In this situation, only
$t = 1.943$ seconds makes sense. So the ball hit the ground
after about 1.9 sec.   **22.** $(16 + 2x)(12 + 2x) - 16 \cdot 12 = 192 + 56x + 4x^2 - 192 = 4x^2 + 56x$

23. B; the coefficient of x^2 in B is 4, but the coefficient of x^2
in A, C, and D is 1, so B has a narrower graph.

24. a.

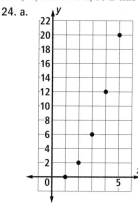

b. Solve the system
$$\begin{cases} a + b + c = 0 \\ 4a + 2b + c = 0 \\ 9a + 3b + c = 6 \end{cases} \text{ to get}$$
$a = 1$, $b = -1$, $c = 0$.
So a quadratic model
for the data is $f(n) = n^2 - n$ **c.** $f(17) = 17^2 - 17 = 272$ points

b. $(n) = n^2 - n$ **c.** $f(17) = 17^2 - 17 = 272$ points

Self-Test Correlation Chart

Question	1	2	3	4	5	6	7	8
Objective(s)	B	B	C	E	E	E	E	E
Lesson(s)	6-4, 6-5	6-4, 6-5	6-2, 6-7, 6-8, 6-10	6-8, 6-9	6-8, 6-9	6-8, 6-9	6-8, 6-9	6-8, 6-9

Question	9	10	11	12	13	14	15	16
Objective(s)	G	K	G	D	C	C	C	F
Lesson(s)	6-3, 6-4	6-2, 6-3, 6-4, 6-10	6-3, 6-4	6-2	6-2, 6-7, 6-8, 6-10	6-2, 6-7, 6-8, 6-10	6-2, 6-7, 6-8, 6-10	6-2

Question	17	18	19	20	21	22	23	24
Objective(s)	A	H	L	I	I	I	G	J
Lesson(s)	6-1	6-10	6-10	6-1, 6-4, 6-5, 6-7	6-1, 6-4, 6-5, 6-7	6-1, 6-4, 6-5, 6-7	6-3, 6-4	6-6

Chapter Review (pp. 446–449)

1. $7p^2 - 19p - 6$ **3.** $36w^2 - 6w - 12$ **5.** $a^2 + 2ab + b^2$
7. $13y^2 - 78y + 117$ **9.** $y = 5x^2 + 10x + 5$ **11.** $y + \frac{101}{4} = \left(x + \frac{9}{2}\right)^2$ **13.** C   **15.** $r = \pm\sqrt{26}$ **17.** $y = \pm 5i$
19. $s = \frac{-5 \pm \sqrt{13}}{2}$ **21.** $k = 3$ **23.** $y = -4$ or $y = 26$ **25.** $x = 1$
or $x = -\frac{7}{3}$ **27.** $t = 5$ **29.** 1   **31.** -15   **33.** -6   **35.** $8 + 3i$
37. $17 + 19i$ **39.** $12 - 5i$ **41.** $\frac{11 - 23i}{13}$ **43.** $x < 0$ **45.** 16
47. 2   **49.** The graph of $y = |x+2|$ is the image of the graph
of $y = |x|$ under $T_{-2, 0}$.   **51. a.** $T_{0.9, 0.6}$ **b.** $y - 0.6 = \frac{3}{4}(x - 0.9)^2$
53. The solutions to $(k + 2)^2 = 25$ are the solutions to $k^2 = 25$
under a horizontal shift of two units to the left.   **55. a.** 57
b. 2 **c.** irrational   **57. a.** 15 **b.** 0 **c.** nonreal   **59. a.** 26.39 ft
b. after 1.16 sec **c.** −15 ft   **61. a.** Answers vary. Sample: $y \approx x^2 + 23.2x + 11.7$ **b.** about 121 frogs **c.** The model does
not take into account the initial number of frogs released.
If the model's prediction of the twentieth ring were
correct, there would be more than 1000 frogs found.
63. a. $y = 76.68x^2 + 2254x + 314{,}255$, where y is the
population x years after 1980. **b.** 450,887 **c.** 2061

65.

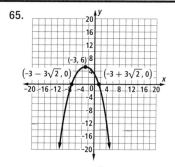

67.

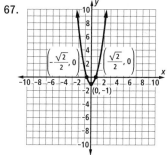

69. A   **71.** 2
73. Yes; the vertex of
the parabola is $\left(\frac{3}{2}, -\frac{27}{4}\right)$,
which is below the line
$y = -1$. The parabola
opens up, so the
parabola intersects the
horizontal line $y = -1$.

Chapter 7

Lesson 7-1 (pp. 452–458)

Mental Math **a.** Answers vary. Sample: –13 **b.** does not exist
c. Answers vary. Sample: $12 + 6i$ **d.** does not exist

Activity Step 1: For even powers, see graph in Activity. Odd powers: functions are in Quadrants I and III and they are symmetric about the origin.

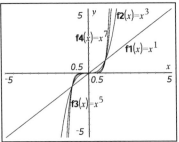

Step 2: Even powers: The domain is the set of all real numbers and the range is $\{y \mid y \geq 0\}$; Odd powers: The domain is the set of all real numbers and the range is the set of all real numbers. **Step 3:** Answers vary. Sample: The graphs of even-power functions are in Quadrants I and II, and they are symmetric to the y-axis. The graphs of odd-power functions are in Quadrants I and III, and they are symmetric about the origin.

Questions
1. 2048; 4096 **3.** the probability of an independent event with probability $p = 0.3$ happening four times in a row is 0.0081. **5. a.** 0.00098 **b.** 0.00000095 **c.** p^{10}

7. a.

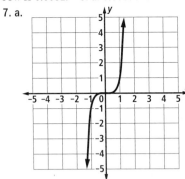

b. domain: all real numbers; range: all real numbers
c. rotation symmetry about the origin

9. $\{y \mid y \geq 0\}$; I; II

11. B; The graph should pass through the origin and be symmetric over the y-axis. B is the only graph that meets both of these requirements. **13. a.** 0.5 **b.** 0.25
c. about 2.98×10^{-8} or 0.0000000298 **15. a.** $x > 0; x < 0$
b. all real numbers x such that $x \neq 0$; none **17.** $y = x^{13}$
19. a.

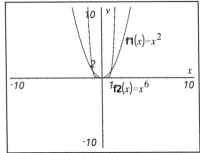

b. $x = -1, 0, 1$ **c.** $\{x \mid -1 < x < 1, x \neq 0\}$ **d.** The difference between $f(x)$ and $g(x)$ increases, then decreases as x increases from 0 to 1. **21. a.** 2 **b.** $4 + 6i$
23. $x = \frac{3}{4}, y = -\frac{27}{8}, z = \frac{1}{2}$ **25.** $\frac{3}{2}x^3$

Lesson 7-2 (pp. 459–465)

Mental Math **a.** false **b.** true **c.** true **d.** true
Activity 1 Step 1: **a.** x^2 **b.** x^{10} **c.** x^{15} **Step 2:** x^{m+n}
Step 3: **a.** $x^5 \cdot y^5$ **b.** $x^2 \cdot y^2$ **c.** $x^{15} \cdot y^{15}$ **Step 4:** $x^m \cdot y^m$
Guided Example 1 3; 10; 7
Activity 2 Step 1: **a.** x^6 **b.** y^{20} **c.** z^{18} **d.** 1 **Step 2:** x^{mn}
Guided Example 5 $2r$; $4r^2$; $4r^2$; 4; 0.785; 78.5
Questions
1. Answers vary. Sample: $3^2 \cdot 3^3 = 3^5$ **3.** Answers vary. Sample: $(7^2)^3 = 7^6$ **5.** Answers vary. Sample: $\left(\frac{3}{5}\right)^4 = \frac{3^4}{5^4}$
7. 5^6; $(5^2)^3 = (5 \cdot 5)^3 = (5 \cdot 5)(5 \cdot 5)(5 \cdot 5) = 5 \cdot 5 \cdot 5 \cdot 5 \cdot 5 \cdot 5 = 5^6$
9. Power of a Quotient Theorem **11.** n^{70} **13.** $27x^{12}$
15. v^a **17.** 6 **19.** about 5417 times
21. a.

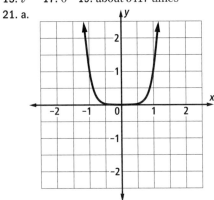

b. 6 **23.** $5^5 \cdot 7^4$ **25.** Answers vary.
Sample: $x^2 \cdot x^5$; $x^1 \cdot x^6$; $x^0 \cdot x^7$ **27.** $\frac{-2}{3x^2}$ **29.** 1 **31.** even

Lesson 7-3 (pp. 466–471)

Mental Math **a.** 4 **b.** 11 **c.** $\left(\frac{b}{2}\right)^2$ **d.** $\left(\frac{b}{2}\right)^2 - c$
Activity Step 1: **a.** $\frac{1}{x^3}$ **b.** $\frac{1}{x^8}$ **c.** $\frac{1}{x^{11}}$ **Step 2:** $\frac{1}{x^n}$
Guided Example 1 2^5; 0.03125
Guided Example 3 **a.** 5; 12; –7 **b.** –3; 5; 2 **c.** 4; 6; 6; 6; 6; 4
Questions

1. t^{-7} **3.** $\frac{1}{x^4}$ **5.** 7 **7. a.** 1 **b.** $\frac{5}{9}$ **c.** $\frac{25}{81}$ **9. a.** 0.001953125
b. $1.953125 \cdot 10^{-3}$ **11.** $\frac{2y^3}{3}$ **13.** $\frac{2}{a^3b^4}$ **15.** 1 **17.** D
19. $A = khg^{-5}$ **21.** $\frac{4x^{38}}{9y^{26}z^{34}}$ **23. a.** $\frac{28}{10^5}; \frac{49}{10^5}; \frac{126}{10^5}$ **b.** $2.8 \cdot 10^{-4}$;
$4.9 \cdot 10^{-4}$; $1.26 \cdot 10^{-3}$ **25.** $4x^2y^{-3}$ **27.** $20a^4$ **29.** A; the range contains no negative numbers. **31. a.** domain: all real numbers except 0; range: all negative real numbers **b.** domain: all real numbers except 0; range: all real numbers except 0

Lesson 7-4 (pp. 472–478)

Mental Math a. $\frac{1}{81}$ b. $\frac{1}{8}$ c. $\frac{1}{648}$

Guided Example 2 a. 0.055; 1.0561; 1.0561; 0.0561; 5.61
b. 0.055, 365; 365 · 1, 1.0565; 1.0565, 0.0565; 5.65

Questions

1. $2593.79; $2691.10; $2792.06; $2896.81; $3005.49
3. a. $A = P\left(1 + \frac{r}{12}\right)^{12t}$ b. $A = P\left(1 + \frac{r}{366}\right)^{366t}$ 5. 4.08%
7. False; because the starting balance gets larger each year, the interest earned each year increases.
9. a. $3405 b. $3423.49 c. No; when interest is compounded, the starting balance increases each year and thus the amount of interest earned each year also increases, but with simple interest, the principal used to calculate the amount of interest earned does not increase during the investment period.
11. $91.36 13. $\frac{1}{729}$ 15. False; Answers vary. Sample: $(2^3)^5$ is not the same as 2^{3^5}. $(2^3)^5 = 2^{15}$ but $2^{3^5} = 2^{243}$.
17. $T_{-3,\,-8}$ 19. a. 4, 12, 36, 108, 324 b. No; the terms do not increase by a constant amount.

Lesson 7-5 (pp. 479–485)

Mental Math a. no b. yes; 11 c. yes; –0.1 d. no

Guided Example 2 1; 4; 5 – 1; 81; 324; $4(-3)^{10-1}$; 4(–19,683)

Questions

1. adding; multiplying 3. no 5. a. 200, 0.02, $2 \cdot 10^{-17}$, $2 \cdot 10^{-47}$ b. The first 15 terms are 200, 20, 2, 0.2, 0.02, 0.002, 0.0002, $2 \cdot 10^{-4}$, $2 \cdot 10^{-5}$,..., $2 \cdot 10^{-12}$.

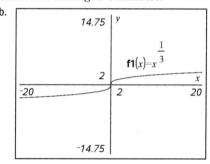

7. a. $r = \frac{g_{n+1}}{g_n}$ b. $g_{n+1} = x(r)^n$ 9. a. i. 0.38 ii. 0.38 iii. 0.38 iv. 0.38 b. They are all equal. 11. a. $g_n = 6.6(0.6)^{n-1}$
b. 1 in. 13. a. $g_t = 5000(1.0225)^t$ b. $7802.54 c. No; there is not a constant ratio between amounts.
15. 224, 448; $g_n = 14(2)^{n-1}$
17. a. $\begin{cases} g_1 = 23 \\ g_n = -g_{n-1}, \text{ for integers } n \geq 2 \end{cases}$
b. $g_n = 23(-1)^{n-1}$ 19. a. $\frac{1}{8}$ b. $A_n = \frac{1}{8}\left(\frac{1}{2}\right)^n$
21. $N = kab^{-2}c^{-3}$ 23. longer

Lesson 7-6 (pp. 486–492)

Mental Math a. positive, two b. positive, two
c. negative, none d. zero, one

Guided Example 2 5th; 77; $\frac{1}{5}$; 2.384; 2.384; 2.384; odd

Questions

1. $2^{\frac{1}{12}}$ 3. 228.6 Hz 5. 3 7. 2 9. 1 11. 1.26
13. $x = 12, -12, 12i, -12i$ 15. false 17. D; The exponent is negative, so it cannot be an nth root.
19. 6% 21. Answers vary. Sample: $x = \frac{1}{4}$
23. > 25. $\frac{1}{4}$ 27. geometric; $g_n 5 \cdot 2^{n-1}$ 29. true

Lesson 7-7 (pp. 493–499)

Mental Math a. 2 qt b. 4 qt

Activity 1 Step 1: $8^{0.5}$; $8^{1.25}$; $8^{1.5}$; $8^{1.\overline{6}}$; 2; 2.83; 4.76; 13.45; 22.63; 32 **Step 2:** It is between 1 and 8.
Step 3: It is between 8 and 64. **Step 4:** It gets larger.
Step 5: Answers vary. Sample: It will be between 32 and 64, since the exponent is between $\frac{5}{3}$ and 2.

Activity 2 Step 1: a. No; because there is no real 6th root of a negative number, so the first graph will only be defined for nonnegative numbers.

b.

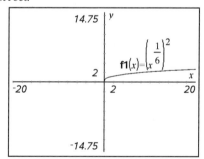

c. Answers may vary. Sample: Yes my prediction was correct.

Step 2: aa. No; the 5th power of a negative number is negative, but the square root of the 10th power of a negative number is a positive number.

ab.

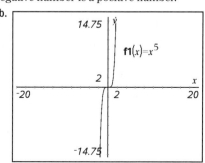

ac.

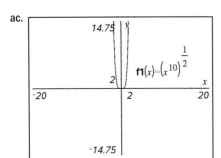

$f1(x) = \left(x^{10}\right)^{\frac{1}{2}}$

b. Answers vary. Sample: I predict the graph will look like the sections of the graphs in Part a that are in Quadrant I.

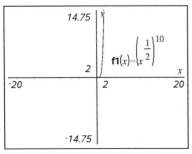

$f1(x) = \left(x^{\frac{1}{2}}\right)^{10}$

Step 3: Answers vary. Sample: I predict the graph will be a ray in Quadrant I.

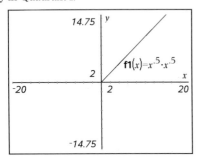

$f1(x) = x^{.5} \cdot x^{.5}$

Step 4: The Power of a Power and Power of a Product Postulates do not hold for noninteger rational exponents with negative bases.

Questions

1. $x^{\frac{3}{7}}$ **3. a.** $\left(10,000,000^4\right)^{\frac{1}{7}}, \left(10,000,000^{\frac{1}{7}}\right)^4$ **b.** $\left(10,000,000^{\frac{1}{7}}\right)^4$
c. 10,000 **5.** 32 **7.** 27 **9.** 2197 **11.** $c^{\frac{25}{12}}$ **13.** $R = 100,000$
15. $s = 10^{\frac{7}{4}}$ **17.** $<$

19. a.

Rational Power	$\left(\frac{1}{4}\right)^0$	$\left(\frac{1}{4}\right)^{\frac{1}{4}}$	$\left(\frac{1}{4}\right)^{\frac{1}{2}}$	$\left(\frac{1}{4}\right)^{\frac{2}{3}}$	$\left(\frac{1}{4}\right)^1$
Decimal Power	$\left(\frac{1}{4}\right)^0$	$\left(\frac{1}{4}\right)^{0.25}$	$\left(\frac{1}{4}\right)^{0.5}$	$\left(\frac{1}{4}\right)^{0.6}$	$\left(\frac{1}{4}\right)^1$
Value	1	0.707	0.5	0.397	0.25
Rational Power	$\left(\frac{1}{4}\right)^{\frac{4}{3}}$	$\left(\frac{1}{4}\right)^{\frac{3}{2}}$	$\left(\frac{1}{4}\right)^{\frac{7}{4}}$	$\left(\frac{1}{4}\right)^2$	
Decimal Power	$\left(\frac{1}{4}\right)^{1.\overline{3}}$	$\left(\frac{1}{4}\right)^{1.5}$	$\left(\frac{1}{4}\right)^{1.75}$	$\left(\frac{1}{4}\right)^2$	
Value	0.157	0.125	0.088	0.063	

b. $\left(\frac{1}{4}\right)^n$ gets smaller as n gets larger. **c.** Answers vary. Sample: Since $\frac{3}{2} < \frac{5}{3} < \frac{7}{4}$, $0.125 < \left(\frac{1}{4}\right)^{\frac{5}{3}} < 0.088$. **d.** smaller
21. 2^n **23.** $\frac{8}{27}$ **25.** 0.008 **27.** 1374.6 times as heavy; it is less than the original estimate. **29.** $\left(-1 - \sqrt{3}i\right)^3$
31. a. nonnegative numbers **b.** nonpositive numbers
33. $T = s_{\frac{2}{3}, \frac{1}{12}}$

Lesson 7-8 (pp. 500–504)

Mental Math a. 64 **b.** 8 **c.** 8
Guided Example 1 1. $\frac{625}{16}, \frac{5}{2}; \frac{125}{8}, \frac{2}{5}; \frac{8}{125}, \frac{125}{8}; \frac{4096}{244,140,625}, \frac{8}{125}; \frac{125}{8}$
Questions
1. $\frac{1}{10}$ **3.** $\frac{27}{125}$ **5.** 0.115 **7.** 0.890 **9.** about 4165 kg
11. $w = \frac{1}{32}$ **13.** $x \approx 0.010$ **15.** 4.5% **17.** True; $\left(x^{-\frac{1}{5}}\right)^{-5} = x$.
19. $t = \frac{1}{2}$ **21.** x; Let $x = 2$: $\left(2^{-4}\right)^{-\frac{1}{4}} = \left(\frac{1}{16}\right)^{-\frac{1}{4}} = 16^{\frac{1}{4}} = 2$.
23. $\frac{1}{2}x^{\frac{8}{3}}$; Let $x = 27$: $\frac{27}{6(27)^{-\frac{4}{3}}}\left(3(27)^{\frac{1}{3}}\right) = \frac{27}{6 \cdot \frac{1}{81}}(9) = \frac{6561}{2}, \frac{1}{2}(27)^{\frac{8}{3}} =$
$\frac{1}{2} \cdot 3^8 = \frac{6561}{2}$. **25.** 100,000 **27.** 0.16 **29.** Since $y = k$ does not intersect $y = x^n$ for $k < 0$ and n even, there are no real solutions.

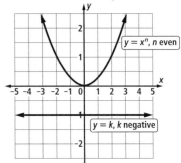

31. $\frac{a^3}{4b^6}$

Self-Test (p. 509)

1. Exponents do not distribute over addition.
$(3 + 4)^{\frac{1}{2}} = 7^{\frac{1}{2}} \approx 2.65$, while $3^{\frac{1}{2}} + 4^{\frac{1}{2}} \approx 3.73$. **2.** $7^{-2} = \frac{1}{7^2} = \frac{1}{49}$
3. $(214,358,881)^{\frac{1}{8}} = 11$ because $11^8 = 214,358,881$.
4. $\left(\frac{343}{27}\right)^{-\frac{4}{3}} = \left(\frac{27}{343}\right)^{\frac{4}{3}} = \left(\frac{3}{7}\right)^4 = \frac{81}{2041}$ **5.** $\frac{7.3 \cdot 10^3}{10^{-4}} = 7.3 \cdot 10^3 \cdot 10^4 =$
$7.3 \cdot 10^7 = 73,000,000$ **6.** $\left(1728x^9y^{27}\right)^{\frac{1}{3}} = 1728^{\frac{1}{3}}x^{\frac{9}{3}}y^{\frac{27}{3}} = 12x^3y^9$
7. $\frac{84x^{21}y^5}{6x^3y^7} = \frac{84}{6}x^{21-3}y^{5-7} = 14x^{18}y^{-2}$, or $\frac{14x^{18}}{y^2}$ **8. a.** True; $-5^{\frac{1}{3}}$
is negative and $5^{-\frac{1}{3}}$ is positive. **b.** False; $3^{-6.4} = \frac{1}{3^{6.4}}$, $3^{-6.5} =$
$\frac{1}{3^{6.5}}$, and $\frac{1}{3^{6.4}} > \frac{1}{3^{6.5}}$. **9.** $\left(x^{\frac{3}{2}}\right)^{\frac{2}{3}} = 0.8^{\frac{2}{3}}$; $x = 0.8^{\frac{2}{3}}$; $x \approx 0.862$ **10.** 1.27
11. n is even because the graph has reflection symmetry about the y-axis. **12. a.** $h_1 = 5$; $h_2 = 0.85 \cdot 5 = 4.25$;
$h_3 = 0.85 \cdot 4.25 \approx 3.61$; $h_4 \approx 0.85 \cdot 3.61 \approx 3.07$.
b. $h_n = 5(0.85)^{n-1}$ **13.** False; the range of $y = x^2$ is
all real numbers greater than or equal to zero, so $y = x^2$ is a
counterexample. **14. a.** $r = \frac{0.6}{3} = 0.2$; $r = 0.2$; $0.024 \cdot 0.2 =$
0.0048, $0.0048 \cdot 0.2 = 0.00096$
b. $\begin{cases} h_1 = 3 \\ h_n = 0.2h_{n-1}, \text{ for integers } n \geq 2 \end{cases}$ **c.** $h_n = 3(0.2)^{n-1}$
15. a. $r = \frac{-12}{-4} = 3$; $-108 \cdot 3 = -324$, $-324 \cdot 3 = -972$
b. $\begin{cases} h_1 = -4 \\ h_n = 3h_{n-1}, \text{ for integers } n \geq 2 \end{cases}$ **c.** $h_n = -4(3)^{n-1}$

16. $\left(\frac{1}{3}\right)^{10} \approx 0.000017$ **17.** $1.23 \cdot (1.2)^2 \approx 1.77$ minutes = 1 minute 46.2 seconds **18.** Solve $x^{10} = 1024$ for x. A CAS in real-number mode shows that $x = 2$ or $x = -2$, so the real 10th roots of 1024 are 2 and -2. **19.** False; $625^{\frac{1}{4}}$ represents only the positive 4th root of 625, which is 5.

20. a. $500(1 + 0.0425)^6 \approx \$641.83$ **b.** $\left(1 + \frac{0.0425}{4}\right)^4 \approx 1.0432$, so the APY is about 4.32%.
21. $7854 = P\left(1 + \frac{0.068}{12}\right)^{12 \cdot 7}$, so $P \approx \$4,885.96$.

Self-Test Correlation Chart

Question	1	2	3	4	5	6	7
Objective(s)	E	A	A	A	B	B	B
Lesson(s)	7-2, 7-6, 7-7, 7-8	7-2, 7-3, 7-6, 7-7, 7-8	7-2, 7-3, 7-6, 7-7, 7-8	7-2, 7-3, 7-6, 7-7, 7-8	7-2, 7-3, 7-7, 7-8	7-2, 7-3, 7-7, 7-8	7-2, 7-3, 7-7, 7-8

Question	8	9	10	11	12	13	14
Objective(s)	E	D	A	I	H	I	C
Lesson(s)	7-2, 7-6, 7-7, 7-8	7-6, 7-7, 7-8	7-2, 7-3, 7-6, 7-7, 7-8	7-1	7-5	7-1	7-5

Question	15	16	17	18	19	20	21
Objective(s)	C	F	F	D	E	G	G
Lesson(s)	7-5	7-1, 7-6, 7-7, 7-8	7-1, 7-6, 7-7, 7-8	7-6, 7-7, 7-8	7-2, 7-6, 7-7, 7-8	7-4	7-4

Chapter Review (pp. 510–513)

Questions
1. 1 **3.** $\frac{3}{5}$ **5.** 16 **7.** 24.73 **9.** 18.52 **11.** $y = 9$
13. $4x^6$ **15.** $\frac{125p^4}{27q^4}$ **17.** $-4x^6y^2$ **19. a.** $g_n = 4(-3)^{n-1}$
b. $\begin{cases} g_1 = 4 \\ g_n = -3g_{n-1}, \text{ for } n \geq 2 \end{cases}$ **c.** $g_{16} = -57,395,628$
21. a. $4(1.075)^{24}$ **b.** 22.691 **23.** 7, 21, 63, 189 **25.** 6, 4, $\frac{8}{3}, \frac{16}{9}$
27. $x = \pm 6$ **29.** $y = \pm 2$ **31.** $y = 216$ **33.** $x \approx 268.01$
35. $m \approx 13.87$ **37.** False; $\pi^{5.4} > \pi^{5.3}$, thus $\frac{1}{\pi^{5.4}} < \frac{1}{\pi^{5.3}}$.
39. a. ± 11 **b.** 11 **41.** true **43.** A **45.** B, C
47. a. odd numbers **b.** 2 **49.** $\frac{1}{n}$th **51.** 61% **53.** $\left(\frac{T}{t}\right)^{\frac{2}{3}}$
55. ≈ 0.825 **57.** \$3465.09 **59. a.** \$3357.34 **b.** \$629.15
61. \$436.45 **63. a.** 18.1 cm × 22.6 cm **b.** 15.2 cm × 12.2 cm
65. $y = x^8$ **67. a.**

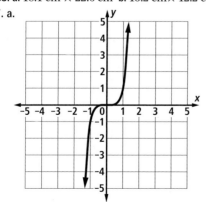

b. domain: all real numbers; range: all real numbers
c. rotation symmetry about the origin

69.

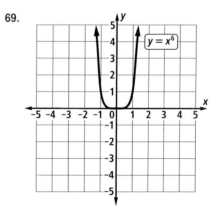

The range $y \geq 0$ does not include $y = -5$.

Chapter 8

Lesson 8-1 (pp. 516–521)
Mental Math a. $\frac{3}{2} + i$ **b.** $-10 - 5i$ **c.** 13 **d.** $k^2 + r^2$
Activity Step 1 \$27,040 **Step 2** \$27,280 **Step 3** discount first and then rebate; \$240
Guided Example 2 a. $3x^2; 3x^2$ **b.** $x - 2; x - 2; x^2 - 4x + 4$; $3x^2 - 12x + 12$
Questions
1. a. \$20,240 **b.** \$20,000 **3. a.** -21,296 **b.** -21,296 **5.** Yes; $f(g(x)) = (x - 3) + 1 = x - 2, g(f(x)) = (x + 1) - 3 = x - 2.$
7. a. $-4a^2 - 4a - 4$ **b.** $16a^2 - 4a + 1$ **9. a.** the set of real numbers other than 0 **b.** no **11.** the set of real numbers other than 0 **13.** $f(x) = 0.6x; g(x) = x - 300;$ $f(g(x)) = 0.6(x - 300) = 0.6x - 180; g(f(x)) = 0.6x - 300;$

He should take Executive Rental's offer. **15. a.** $w = \frac{k_1}{z}$;
$f: z \to \frac{k_1}{z}$ **b.** $z = k_2 x^4$; $g: x \to k_2 x^4$ **c.** $w = \frac{k_1}{k_2 x^4} = \frac{k_3}{x^4}$;
$h: x \to \frac{k_3}{x^4}$ **d.** w varies inversely as the fourth power of x.
e. $h = f \circ g$ **17. a.** x^4 **b.** when $x = 0$

19. $\begin{bmatrix} \frac{1}{a} & \frac{7}{18a} \\ 0 & \frac{1}{18} \end{bmatrix}$ **21.** false **23. a.** $\frac{13}{4}$ **b.** $-\frac{4}{13}$

Lesson 8-2 (pp. 522–528)

Mental Math a. $\frac{1}{100}$ **b.** $\frac{20}{100}$ or $\frac{1}{5}$ **c.** $\left(\frac{4}{10}\right)^n \cdot \left(\frac{5}{10}\right)^m$, or $\left(\frac{2}{5}\right)^n \cdot \left(\frac{1}{2}\right)^m$

Guided Example 2 a. x; y^2 **b.** $y = x^2$; $y = x$

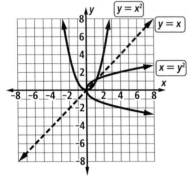

c. is not; Answers vary. Sample: The graph of the inverse contains the points $(1, 1)$ and $(1, -1)$, which have the same first coordinate.

Questions

1. by switching the coordinates of the points in the relation
3. a. $g = \{(-9, -4), (-7, -3), (-1, 0), (5, 3), (7, 4), (11, 6)\}$
b.

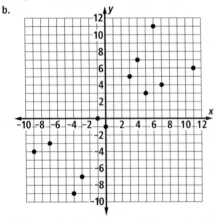

c. The graph of the inverse is the graph of f reflected over the line with equation $y = x$. **d.** $f(x) = 2x - 1$
e. $g(x) = \frac{x+1}{2}$ **5.** The graph of the inverse of any relation f is the reflection of the graph of f over the line with equation $y = x$. **7.** $y = 5x + 10$ **9.** The inverse of q is not a function because the graph of q does not pass the horizontal-line test. **11.** No; the graph of the function does not pass the horizontal-line test. **13.** $H: x \to \frac{x}{2}$

15. a, c.

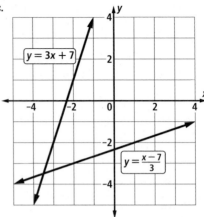

b. $y = \frac{x-7}{3}$
d. Yes; the graph of the inverse satisfies the vertical-line test. **e.** The slope of the inverse function is the reciprocal of the slope of the function. **17. a.** $U = \frac{25}{16}E$ **b.** $E = 0.64U$
c. Yes; the points in the first function are of the form (price in euros, price in dollars), and the points in the second function are of the form (price in dollars, price in euros). **19. a.** 8 **b.** 4 **c.** x **d.** the identity function
21. False; $|-3(-1)| = 3$, but $3(-1) = -3$. **23.** $BC = 9$; $CE = 8$

Lesson 8-3 (pp. 529–536)

Mental Math a. $y - 3 = -2(x - 1)^2$ **b.** $y + 2 = -2(x + 2)^2$
c. $y - 5.4 = 2(x - 0.3)^2$
Activity 1 Step 1 $g(x) = 12x - 4$; $g(x) = \frac{1}{3} + \frac{x}{12}$
Step 2 $f(g(x)) = x$; $g(f(x)) = x$ **Step 3** The results are the same.
Guided Example $8x$, 11; $\frac{8x-11}{3}$; $\frac{8y-11}{3}$; $8y$, 11; 11, $8y$; $\frac{3x+11}{8}$, $\frac{3x+11}{8}$
Activity 2 f^{-1} is a function if the domain of f is $x \geq 0$, or the domain of f is $x \leq 0$.
Questions
1. the inverse of f; f inverse **3.** Yes; $f(g(x)) = 5 - 2\left(\frac{x-5}{-2}\right) = 5 + x - 5 = x$; $g(f(x)) = \frac{(5-2x)-5}{-2} = \frac{-2x}{-2} = x$
5. a. $A(x) = x + 15$ **b.** $A^{-1}(x) = x - 15$ **c.** subtracting 15
d. 63 **7.** No; an equation for the inverse would be $x = 21$, which does not represent a function. **9.** $f^{-1}(x) = x^{\frac{1}{7}}$
11. a. $f^{-1}(x) = -12 - (2x)^{\frac{1}{3}}$ **b.** yes; all real numbers
13. CODES ARE COOL
15. a.

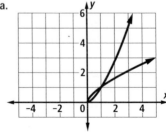

b. The graph of h is the graph of g reflected over the line with equation $y = x$. **c.** Each is equal to x. **d.** $64 = 512^{\frac{2}{3}}$

e. Yes, over the restricted domain $x \geq 0$, they are inverses because the graph of g is the reflection image of the graph of h over the line with equation $y = x$. **17.** $y = 2x - 14$ **19.** f; the inverse of the inverse of a function is the original function. **21.** B **23. a.** $\frac{11}{2}$ **b.** 1 **c.** $2x^{-2} + 5$ **d.** $(2x + 5)^{-2}$ **25. a.** $\frac{1}{4}$ **b.** all real numbers

Lesson 8-4 (pp. 537–544)
Mental Math a. $\$4.50$ **b.** $\$6.75$ **c.** 12 **d.** 20
Guided Example 2 a. 6561; positive; 4th; 9; 9; 9
b. $x^4 = 6561$; 9; -9; 9i; -9i; 4th
Questions
1. René Descartes **3.** false **5.** about 3.77 **7.** 1.69
9. $\sqrt[n]{x^m}$, or $\left(\sqrt[n]{x}\right)^m$ **11. a.** 11, -11, 11i, -11i **b.** 11
13. c^3 **15.** $x^{\frac{1}{16}}$ **17.** A **19. a.** 1.62, 1.43, 2.78, 1.68, 1.51
b. about 1.856; about 85.6% **21. a.** $r = \sqrt[3]{\frac{3}{4\pi}V}$ **b.** $\left(\frac{3}{4\pi}V\right)^{\frac{1}{3}}$
23. a. $\left(43{,}046{,}721z^{48}\right)^{\frac{1}{16}}$ **b.** $\sqrt[16]{43{,}046{,}721z^{48}}$ **c.** 24
25. $\frac{\sqrt[4]{k}}{\sqrt[4]{k}} = 1$ **27. a.** $h(w(x))$ **b.** $h(w(x)) = -0.0010692x^2 + 0.4905x + 4$ **c.** about 420.1 ft **d.** about 466.8 ft **29.** -125
31. $t = -1, 2$, and -3 are not in the domain of g. **33. a.** 1.55
b. 1.60 **c.** 1.61 **d.** $\frac{\sqrt{5}+1}{2}$ **e.** Answers vary. Sample: As the approximations of $\sqrt{1 + \sqrt{1 + \ldots}}$ get more exact, they get closer to $\frac{\sqrt{5}+1}{2}$.

Lesson 8-5 (pp. 545–550)
Mental Math a. w^5 **b.** x^2 **c.** $\frac{36}{y^5}$ or $36y^{-5}$ **d.** $8w^4z^2 + \frac{wz^3}{2}$

Activity Step1: a.
$$\frac{(a \cdot b)^{\frac{1}{3}} - a^{\frac{1}{3}} \cdot b^{\frac{1}{3}} \mid a>0 \text{ and } b>0}{\qquad\qquad\qquad\qquad 0}$$

b.
$$\frac{(a \cdot b)^{\frac{1}{12}} - a^{\frac{1}{12}} \cdot b^{\frac{1}{12}} \mid a>0 \text{ and } b>0}{\qquad\qquad\qquad\qquad 0}$$

c.
$$\frac{a^{\frac{1}{4}} \cdot b^{\frac{1}{4}} - (a \cdot b)^{\frac{1}{4}} \mid a>0 \text{ and } b>0}{\qquad\qquad\qquad\qquad 0}$$

d.
$$\frac{a^{\frac{1}{7}} \cdot b^{\frac{1}{7}} - (a \cdot b)^{\frac{1}{7}} \mid a>0 \text{ and } b>0}{\qquad\qquad\qquad\qquad 0}$$

Step 2: 0 **Step 3:** $a^{\frac{1}{n}} \cdot b^{\frac{1}{n}}$
Guided Example 2 Solution 1 $625x^4$; 625, x^4; $5x$
Solution 2 $125x^3$; $625x^4$; x^4
Questions
1. For any nonnegative real numbers x and y, and any integer $n \geq 2$, $(xy)^{\frac{1}{n}} = x^{\frac{1}{n}} \cdot y^{\frac{1}{n}}$. **3.** 100 **5.** Answers vary.
Sample: $\sqrt[3]{25} \cdot \sqrt[3]{10}$; $\sqrt[3]{50} \cdot \sqrt[3]{5}$; $\sqrt[3]{125} \cdot \sqrt[3]{2}$ **7.** $a = 27, b = 3$
9. $14\sqrt{7}$ **11. a.** $4y^6$ **b.** $\frac{y^8}{2}\sqrt[3]{y}$ **13.** 6 **15.** $\sqrt[3]{5} + \sqrt[3]{5}$

17. a. The Root of a Product Theorem does not apply when multiplying two different nth roots. **b.** $n = 80$
19. $2\sqrt{x^2 + y^2}$ **21. a.** about 0.189 **b.** about 0.040
23. a. $L^{-1}(x) = \frac{x-b}{m}$ **b.** They are reciprocals.
c. when $m = 0$ **25.** $-\frac{4}{85} + \frac{33}{85}i$

Lesson 8-6 (pp. 551–555)
Mental Math a. 12 ft **b.** 4.5 ft **c.** 6 ft
Activity Step1: a. $\frac{\sqrt{2}}{2}$ **b.** $\frac{3\sqrt{5}}{5}$ **c.** $-\frac{13\sqrt{7}}{7}$ **Step 2:** yes **Step 3:** $\frac{\sqrt{a}}{a}$
Guided Example 1 Solution 1 $\sqrt{3x}$; $\sqrt{3x}$; 4; x; 2; $3x$; $\sqrt{3x}$; $2x$
Solution 2 $\sqrt{12x}$; $3\sqrt{12x}$; $\sqrt{3x}$; $\sqrt{12x}$; $2x$
Questions
1. a. 316,800 inches $\cdot \frac{1 \text{ foot}}{12 \text{ inches}} \cdot \frac{1 \text{ yard}}{3 \text{ feet}} = 8800$ yards
b. 8800 yards $\cdot \frac{1 \text{ mile}}{1760 \text{ yards}} = 5$ miles **3.** 0.7 **5.** $\frac{b\sqrt{a}}{a}$
7. $\frac{11\sqrt{3}}{3}$ **9. a.** $\frac{3+\sqrt{5}}{3+\sqrt{5}}$ **b.** $\frac{21}{4} + \frac{7\sqrt{5}}{4}$ **11.** $\frac{\sqrt{3}}{2}$ **13.** $\frac{\sqrt{x}}{x^2}$
15. $\frac{4\sqrt{n}+24}{n-36}$ **17. a.** $\frac{16\sqrt{3}}{3}$ **b.** $\frac{a\sqrt{3}}{3}$; $\frac{2a\sqrt{3}}{3}$ **c.** $\frac{5\sqrt{3}}{3}$ **19.** $3a^2\sqrt[3]{2a}$
21. True; by the Power of a Power Postulate, each side is equivalent to $x^{\frac{1}{2}}y^{\frac{1}{8}}$. **23. a.** $g^{-1}(x) = x^3$ **b.** $(2x)^{\frac{1}{3}}$ **25.** No; the inverse contains two points, (5, 3) and (5, 8), with the same first coordinates. **27.** $\frac{1}{p^{\frac{5}{3}}}$ **29.** Gd

Lesson 8-7 (pp. 556–561)
Mental Math a. 10 min **b.** 9 min, 20 sec **c.** 8 min, 40 sec
Activity Step 1:

x	f1(x):... ▾	f2(x):... ▾	▾
	(-2)^x	(-7)^x	
2	4.	49.	
4	16.	2.4ᴇ3	
6	64.	1.18ᴇ5	
8	256.	5.76ᴇ6	
10	1.02ᴇ3	2.82ᴇ8	
12	4.1ᴇ3	1.38ᴇ10	
2			

x	f1(x):... ▾	f2(x):... ▾	▾
	(-6)^x	(-5)^x	
2	36.	25.	
4	1.3ᴇ3	625.	
6	4.67ᴇ4	1.56ᴇ4	
8	1.68ᴇ6	3.91ᴇ5	
10	6.05ᴇ7	9.77ᴇ6	
12	2.18ᴇ9	2.44ᴇ8	
2			

Step 2: Answers vary. Sample: Every even power of a negative number is positive; every even power of an even number is even; every even power of an odd number is odd.

Step 3:

x	f1(x):...▼ (-2)^x	f2(x):...▼ (-7)^x	▼
1	-2.	-7.	
3	-8.	-343.	
5	-32.	-1.68E4	
7	-128.	-8.24E5	
9	-512.	-4.04E7	
11	-2.05E3	-1.98E9	

1

x	f1(x):...▼ (-6)^x	f2(x):...▼ (-5)^x	▼
1	-6.	-5.	
3	-216.	-125.	
5	-7.78E3	-3.13E3	
7	-2.8E5	-7.81E4	
9	-1.01E7	-1.95E6	
11	-3.63E8	-4.88E7	

1

Every odd power of a negative number is negative; every odd power of an even number is even; and every odd power of an odd number is odd. **Step 4:** Even; positive; odd; negative

Guided Example 1 $5; -2; 3; -27$

Questions

1. a. Answers vary. Sample: 2, 4, 6 **b.** Answers vary. Sample: 1, 3, 5 **3. a.** positive **b.** negative **c.** negative **d.** negative **5.** $(-2)^3$ **7.** -6 **9.** -1 **11.** $-\frac{4}{3}y^9$ **13.** $2q^3\sqrt[3]{10q^2}$ **15.** The intersection point is $(-3, -27)$. This shows that $(-3)^3 = -27$, or $\sqrt[3]{-27} = -3$. **17.** Answers vary. Sample: If $x < 0$, then $\sqrt[18]{x^6} = -\sqrt[3]{x}$. **19. a.** $(2 + 2i)^4 = (2(1 + i))^4 = 2^4((1 + i)^2)^2 = 2^4 \cdot (2i)^2 = 16 \cdot (-4) = -64$ **b.** $(-2 - 2i)^4 = (-(2 + 2i))^4 = (-1)^4(2 + 2i)^4 = 1(-64) = -64$ **c.** $(2 - 2i)^4 = (2(1 - i))^4 = 2^4((1 - i)^2)^2 = 2^4 \cdot (-2i)^2 = 16 \cdot (-4) = -64$ **d.** The other fourth root is $-2 + 2i$. $(-2 + 2i)^4 = (-(2 - 2i))^4 = (-1)^4(2 - 2i)^4 = 1(-64) = -64$ **e.** $\sqrt[4]{-64}$ is not defined because none of the 4th roots of -64 are real numbers. **21.** $7 - 4\sqrt{3}$ **23.** 4 **25.** $r = 512$ **27.** 85

Lesson 8-8 (pp. 562–567)

Mental Math a. $y = \frac{1}{7}x$; yes **b.** $x + y = 4.5$; yes **c.** $x = 13$; no
Guided Example 2 Solution 1 $-\sqrt[6]{x}, -5; -\sqrt[6]{x}; 729, 6th$
Check 729; 729; 3; $x = 729$; no **Solution 2** $\sqrt[6]{x} \geq 0$; no

Questions

1. 1.28 m **3.** $m = 1296$ **5.** $w = 16,777,216$ **7.** $x = 2.375$ **9.** $(1 - \sqrt{15}, -2), (1 + \sqrt{15}, -2)$ **11. a.** about 1.62 ft **b.** 1.275 ft **c.** $18.2 = 4hs + 2s^2$; $18.2 = 2(2hs + s^2)$; $9.1 = 2hs + s^2$; $9.1 + h^2 = 2hs + s^2 + h^2$; $\sqrt{9.1 + h^2} = s + h$; $s = -h + \sqrt{9.1 + h^2}$ **13. a.** 36, 226, 52, 38; 1.36, 3.26, 1.52, 1.38

b. The geometric mean of the size-change factors is about 1.70, indicating an average increase of about 70% per decade. **c.** about \$157,575 **15.** $x = -1$ **17.** $-2y^3$ **19.** Answers vary. Sample: $\left((-1)^4\right)^{\frac{1}{4}} = 1 \neq -1$ **21.** true; $\frac{\sqrt{3t}}{\sqrt{t}} = \frac{\sqrt{3t}}{\sqrt{t}} \cdot \frac{\sqrt{3}}{\sqrt{3}} = \frac{3\sqrt{t}}{\sqrt{3t}}$ **23. a.** $u(v(x)) = \sqrt[3]{x^6} = x^2$; domain: all real numbers **b.** $v(u(x)) = (\sqrt[3]{x})^6 = x^2$; domain: all real numbers

25. $\begin{cases} x \geq -2 \\ y \geq -4 \\ y \leq 4 \\ y \leq 4 - 2x \end{cases}$

Self-Test (pp. 572–573)

1. $4 - (3(-1) + 7)^2 = 4 - 16 = -12$
2. $f(g(x)) = 4 - (3x + 7)^2 = -9x^2 - 42x - 45$
3. $g \circ f(x) = 3(4 - x^2) + 7 = 19 - 3x^2$
4. no; from Question 3, $g(f(x)) = 19 - 3x^2 \neq x$
5. a. $x = \frac{1}{2}y + 3$; $x - 3 = \frac{1}{2}y$; $y = 2x - 6$
b. Yes; Answers vary. Sample: It is a line in slope-intercept form.
6.

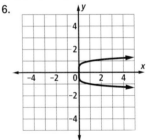

7. Restrict it to either only positive numbers or only negative numbers. That way, the graph of the original function will pass the horizontal-line test. **8.** False; While $(-2)^4 = 16$, the use of the radical symbol indicates the positive real 4th root of 16, which is 2. **9.** Substitute $r = 12$ into the formula and solve for S: $12 = \sqrt{\frac{S}{4\pi}}$; $144 = \frac{S}{4\pi}$; $S = 1810$ cm^2 **10.** $\sqrt[3]{-216x^{12}y^3} = \sqrt[3]{-216} \cdot \sqrt[3]{x^{12}} \cdot \sqrt[3]{y^3} = -6x^4y$ **11.** $\sqrt[4]{\frac{32x^{12}}{x^4}} = \frac{\sqrt[4]{16} \cdot \sqrt[4]{2} \cdot \sqrt[4]{x^{12}}}{\sqrt[4]{x^4}} = \frac{2x^3\sqrt[4]{2}}{x} = 2x^2\sqrt[4]{2}$

12. $\frac{3x^2}{\sqrt{81x}} = \frac{3x^2}{9\sqrt{x}} \cdot \frac{\sqrt{x}}{\sqrt{x}} = \frac{3x^2\sqrt{x}}{9x} = \frac{x\sqrt{x}}{3}$ **13.** Deon and Carmen; Construct an isosceles right triangle with legs of length h and a hypotenuse of length 15. By the Pythagorean Theorem, $2h^2 = 15^2$. So $h = \sqrt{\frac{225}{2}} = \frac{15\sqrt{2}}{2}$. **14.** $\sqrt[4]{\sqrt[3]{13}} = \left(13^{\frac{1}{3}}\right)^{\frac{1}{5}} = 13^{\frac{1}{15}}$ **15.** $\frac{24}{8 - 2\sqrt{3}} \cdot \frac{8 + 2\sqrt{3}}{8 + 2\sqrt{3}} = \frac{192 + 48\sqrt{3}}{52} = \frac{48 + 12\sqrt{3}}{13}$

16. $r = \sqrt[3]{\frac{3(268)}{4\pi}} = \sqrt[3]{\frac{201}{\pi}} \approx 4.00$ in. **17.** $25 = \sqrt[4]{7x}$; $\left(\frac{25}{4}\right)^2 = 7x$; $x = \frac{625}{112}$ **18.** $17 + \sqrt[3]{2x + 7} = 20$; $\sqrt[3]{2x + 7} = 3$; $2x + 7 = 27$; $x = 10$ **19.** odd integers **20.** domain: $x \geq 0$; range: $y \geq 0$ **21.** $\sqrt[6]{1092 \cdot 597 \cdot 46 \cdot 144 \cdot 32 \cdot 27} \approx 125$ bacteria per mL **22.** None; The price when the coupon is used first is $0.7(P - 45) = 0.7P - 31.5$, and the price when the discount is taken first is $0.7P - 45$. For all values of P, $0.7P - 31.5 > 0.7P - 45$.

Self-Test Correlation Chart

Question	1	2	3	4	5	6	7	8
Objective(s)	A	A	A	F	B	J	J	G
Lesson(s)	8-1	8-1	8-1	8-2, 8-3	8-2, 8-3	8-2, 8-3	8-2, 8-3	8-4, 8-5, 8-7
Question	9	10	11	12	13	14	15	16
Objective(s)	I	D	D	D	I	D	D	C
Lesson(s)	8-4, 8-8	8-5, 8-6, 8-7	8-5, 8-6, 8-7	8-5, 8-6, 8-7	8-4, 8-8	8-5, 8-6, 8-7	8-5, 8-6, 8-7	8-4, 8-7
Question	17	18	19	20	21	22		
Objective(s)	E	E	F	J	I	H		
Lesson(s)	8-8	8-8	8-2, 8-3	8-2, 8-3	8-4, 8-8	8-1		

Chapter Review (pp. 574–577)

1. f **3. a.** 25 **b.** $x^2 + x - 5$ **5.** yes; $f \circ g(x) = x = g \circ f(x)$
7. $h(h(x)) = x^{\frac{4}{9}}$ **9.** $y = \frac{1}{6}x + \frac{1}{2}$ **11.** $\frac{x-13}{7}$ **13.** $-\sqrt{-t}$ **15.** 3
17. $\frac{16}{49}$ **19.** 1.44   **21.** -16.87   **23.** b^4 **25.** 2   **27.** $-u^2v^4$
29. $\sqrt[8]{k}$ **31.** $\sqrt{13}$ **33.** $\frac{7\sqrt{3}+7}{2}$ **35.** $y = 15.625$ **37.** $y = -2288$
39. no solution   **41.** true   **43.** domain: $\{x \mid x \geq -6\}$;
range: $\{y \mid y \leq 0\}$ **45.** The range of f is the set of all positive
integers.   **47.** D   **49.** all real numbers   **51.** False; the
Root of a Product Theorem does not hold for negative a
and b and even n.   **53. a.** \$981 **b.** \$981 **c.** The deal is the
same because both the discount and the sales tax involve
multiplication, which is commutative.   **55.** about 1.33
times farther   **57.** 151.25 ft   **59.** the diameter of the first
balloon is about 1.91 times larger

61. a.

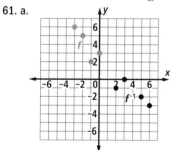

b. reflection over the line with equation $y = x$

63. a.

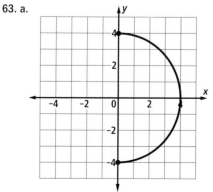

b. No; Explanations vary. Sample: The graph of the
original function does not pass the horizontal-line test.

65. a.

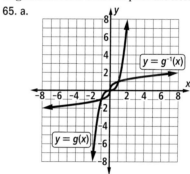

b. the set of all real numbers

67. a.

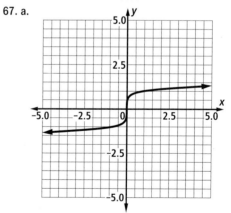

b. domain: all real numbers; range: all real numbers

Chapter 9

Lesson 9-1 (pp. 580–586)

Mental Math 1 hr 21 min
Guided Example 2 a. 280.6; 1.78; 280.6(1.78)^x **b.** 280.6; 1.78;
1.2; 1.2 **c.** 280.6(1.78)^x; 2.4; 2.4; 1.2 1.2
Questions
1. 56,680 rabbits   **3.** 5252 rabbits   **5.** C

7. a.

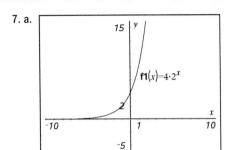

$f1(x)=4\cdot2^x$

b. 1.52 **c.** True; the function satisfies the definition of an exponential function. **9. a.** 0.88 **b.** the speed in teraflops of the fastest computer in 1995 **11. a.** independent: t; dependent: A **b.** $(1 + r)$ **13. a.** $a = 31.1$, $b = 1.0116$, $c = 30.2$, $d = 1.0238$; $C(x) = 31.1(1.0116)^x$, $M(x) = 30.2(1.0238)^x$ **b.** 29.36 million, 26.85 million

c.

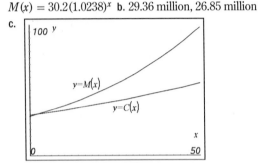

$y=M(x)$

$y=C(x)$

d. Answers vary. Sample: Morocco's population will outgrow Canada's population in the long run. **15. a.** 1 **b.** 1.94 **c.** 1,106,207 bighead carp

17. a. $\begin{bmatrix} 4 & 4 & -2 \\ -3 & -4 & 1 \end{bmatrix}$

b.

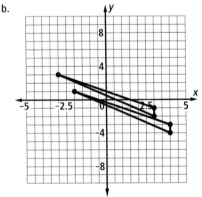

19. a. $22,960 **b.** $22,400 **c.** $28,000\frac{(100-r)}{100}$

Lesson 9-2 (pp. 587–594)

Mental Math a. 0 **b.** 1 **c.** infinitely many
d. infinitely many **e.** 1
Guided Example 1 a. 88; 36,025; 0.88; 36,025$(0.88)^t$ **b.** 4; 4; 36,025; 4; 21,604.03; $21,604.03
Guided Example 3 a. 100; 1; 1; $\left(\frac{1}{2}\right)$ **b.** 0.8; 0.3; 0.3; 1700

Activity Step 1: a. As b is increases, the steepness at the right side of the graph increases. **b.** All of these functions are increasing over their whole domain.
Step 2: a. As b decreases, the graphs drop more sharply.
b. All of these functions are decreasing over their whole domain **Step 3: a.** For $0 < b < 1$, the graph decreases for all x; for $b > 1$, the graph increases for all x. **b.** The shapes of the two kinds of graphs are reflections of each other over the y-axis.
Step 4: The y-intercept changes with a because $y = a$ at $x = 0$.

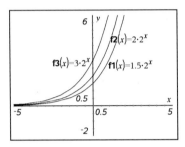

$f2(x)=2\cdot2^x$
$f3(x)=3\cdot2^x$
$f1(x)=1.5\cdot2^x$

Step 5:

Property	Exponential Decay: $a>0, 0<b<1$	Exponential Growth: $a>0, b>1$
Domain	all real numbers	all real numbers
Range	$y > 0$	$y > 0$
y-intercept	a	a
x-intercept	none	none
Horizontal asymptote	$y = 0$	$y = 0$
As x increases, y ...	decreases	increases

Step 6:

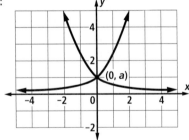

$(0, a)$

Questions
1. $10,033.00 **3.** The *half-life* of a substance is the amount of time it takes for half of the atoms in that substance to decay into other matter. **5.** about 7400 BCE

7. a.

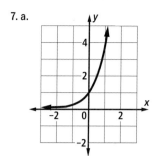

b.

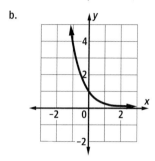

c. The graph of $y = 4^x$ is the reflection of the graph of $y = \left(\frac{1}{4}\right)^x$ over the y-axis, and vice versa.
9. True; $ab^x \neq 0$ for all x, when $a \neq 0$ and $b > 0$.
11. a.

Year (t)	Value (N)
1	$18,000
2	$16,200
3	$14,580
4	$13,122

b.

Year (t)	Value (N)
1	$18,000
2	$16,000
3	$14,000
4	$12,000

c. With the exponential model, $N = 20{,}000(0.9)^t$, so after 6 months, the car is worth $20{,}000(0.9)^{\frac{1}{2}} \approx \$18{,}974$. With the linear model, $N = 20{,}000 - 2000t$, so after 6 months, the car is worth $20{,}000 - 2000\left(\frac{1}{2}\right) = \$19{,}000$. Thus, the exponential depreciation is preferred for this situation.
13. a. $P(x) = 340\left(\frac{1}{2}\right)^x$ **b.** $1360; $680; $340; $170; $85; $42.50
c. 32.5 half-lives; $P(32) \approx \$0.06$ per terabyte (1 terabyte = 1,000,000 megabytes) **d.** after 25 half-lives, that is, in 2011
e. Answers vary. Sample: Yes; this just means that a larger quantity of memory, such as the terabyte, would become the basis for the price of memory. **15. a.** D
b. all real numbers **c.** $\{y \mid y > 0\}$ **17. a.** about 50
b. $g = 12(1.33)^{\frac{w}{2}}$ **19. a.** 125 **b.** 55.9 **c.** 70.7 **21.** tr^n

Lesson 9-3 (pp. 595–601)

Mental Math a. 36π **b.** 8
Activity $1\left(1 + \frac{1}{4}\right)^4$; 12, $1\left(1 + \frac{1}{12}\right)^{12}$, $2.61304; 365, $1\left(1 + \frac{1}{365}\right)^{365}$, $2.71457; 8760, $1\left(1 + \frac{1}{8760}\right)^{8760}$, $2.71813; 525,600, $1\left(1 + \frac{1}{525,600}\right)^{525,600}$, $2.71828
Guided Example 2 0.448; –10.0055; 0.448, –10.0055, 0.08; 0.448, 0.449; 0.998; 99.8
Questions
1. irrational **3. a.** about $2.7146 **b.** e **5. a.** $1061.83
b. $1233.67 **7.** about 81.6% **9. a. i.** increasing
ii. increasing **iii.** decreasing
b. i.

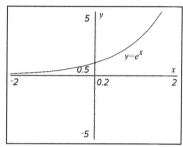

ii.

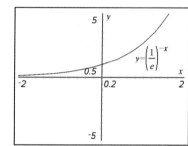

iii.

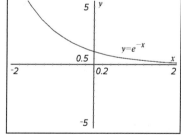

c. The graphs of y_1 and y_2 coincide because $y_2 = \left(\frac{1}{e}\right)^{-x} = (e^{-1})^{-x} = e^x = y_1$. **d.** i, ii **e.** iii **11. a.** about 29.5%
b. about $364,968 **13. a.** about 352.2 million **b.** about 862.4 million **c.** The answer for Part b; Reasons vary. Sample: The newer model uses a faster rate of population growth. **d.** Answers vary. Sample: As of July 2007, the U.S. population was about 301 million. The first model predicts the 2007 U.S. population would be 280.1 million, and the second model predicts 304.3 million. **15. a.** $\frac{1}{4096}$; 1; 64
b. domain: all real numbers; range: all positive real numbers **c.** $x = 16^y$ **17.** $k = \frac{3}{\sqrt[3]{5}}$ **19.** $a > \frac{16}{7}$
21. a. $y = 6.6x + 73.6$ **b.** 152.8 cm

Lesson 9-4 (pp. 602–607)

Mental Math a. yes b. no c. yes
Activity Step 1: $y = 991.709 \cdot (0.890071)^x$
Step 2: $y = 992 \cdot (0.890)^x$ **Step 3:** The model from Step 2 yields 553.939; the one from Example 1 yields 539.835. Therefore, the Step 2 model fits better.
Step 4.

Hours Since Administration	Amount of Drug in Body	Example 1	Activity Step 2
0	1000	1000	992
5	550	539.8	553.9
10	316	291.4	309.3
15	180	157.3	172.7
20	85	84.9	96.5
25	56	45.8	53.9
30	31	24.7	30.1

The model from Step 2 appears more accurate overall, especially as the number of hours since the drug was taken increase.

Questions
1. a. $y = 1000 \cdot (0.891)^x$ b. after approximately 39.9 hr
3. No, the initial condition is not given.
5. a. Answers vary. Sample: $y = 126.8 \cdot (1.127)^x$, where x is the number of years after 2001. b. 2009
7. a. $y = 3887 \cdot (0.4934)^x$ b. (2, 946), (6, 56) c. Tyler is correct because approximately half of the coins land heads on every drop.
9. a. (–1, 0.007), (0, 1), (0.25, 3.49)

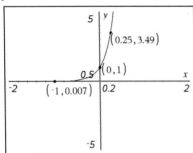

b. domain: all real numbers; range: all positive real numbers
11. a. $5428.72 b. $5439.08 c. $5439.11 13. 16 15. $4xy$

Lesson 9-5 (pp. 608–614)

Mental Math $\sqrt[3]{4.8}$
Guided Example 1 a. 2; 2 b. –5; –5 c. 1; 0
Activity Step 1: 0.01, 0.1, 0.178, 0.316, 0.562, 1, 1.778, 3.162, 5.623, 10, 100

Step 4:

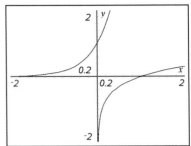

These graphs are the same as those from Steps 2 and 3.
Questions
1. $10^m = n$ 3. 7 5. –7 7. –1 9. 8 and 9 11. –2.510
13. 100,000 15. a. x-intercept: 1; there is no y-intercept.
b. Answers vary. Sample: (1, 0), (10, 1), (100, 2)
c. Answers vary. Sample: (0, 1), (1, 10), (2, 100) 17. $\log_{10} x$
19. 1 and 2 21. 3.765 23. x; f and g are inverse functions.
25. about 64% 27. $1,632,986 29. 13^{-6}

Lesson 9-6 (pp. 615–621)

Mental Math a. 10^4 or 10,000 b. 10^{11} c. $2 \cdot 10^9$ d. $\frac{a}{b} \cdot 10^{m-n}$
Guided Example 4 6.9, 9.3; 9.3, 6.9, 2.4; 2.4, 251.189, 251, 1989
Questions
1. 73.89 dB 3. between 10^{-4} and 10^{-3} 5. 100 times as intense 7. about 15.8 times as loud 9. 31.6 times as much force 11. 7.69 13. 19.95 15. 5.5 17. –8.14
19. $1,000,000 = 10^6$; thus, $\log_{10} 1,000,000 = 6$.
21. $x = 10,000,000$ 23. $y = \log x$, $x = 10^y$
25. exponential decay; Explanations vary. Sample: y will decrease as x increases. 27. $\frac{p^2 q^4}{m^3}$

Lesson 9-7 (pp. 622–628)

Mental Math a. (–3, 5) b. (9, 15) c. (–6, –5) d. (1, 4)
Guided Example 2 a. 2; 2 b. definition of logarithm; 1; $3x$, 1; $3x$, 1; $\frac{1}{3}$, Division Property of Equality; $\frac{1}{3}$ c. 4^x; 4^x, –3; x, –3; –3
Activity Step 1: Answers vary. Sample: $b = 2$;

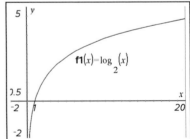

Step 2:

b	Domain	Range	x-intercepts	y-intercepts	Asymptotes
2	all positive real numbers	all real numbers	1	none	$x = 0$

Step 3:

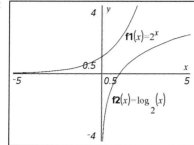

They are reflection images of one another over the line $y = x$.
Step 4: Answers vary.
Step 5: Answers vary. Sample: The functions all have an x-intercept of 1 and an asymptote of $x = 0$. They all have the same domain and range.

Questions
1. $b^x = a$ when b is any positive value other than 1
3. $\log_{6\sqrt{3}} 136,048,896 = 8$ 5. $b^c = a$ 7. domain: $x > 0$; range: all real numbers
9.

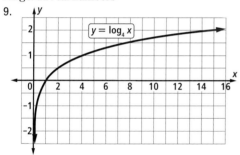

11. $\left(\frac{1}{4}\right)^y = \frac{1}{64}$; $y = 3$ 13. $a^{\frac{1}{3}} = 8$; $a = 512$ 15. $47^{-0.2} = d$; $d \approx 0.463$ 17. $w^1 = w$; $w > 0, w \neq 1$ 19. a. 1 b. 2 c. about 0.203   21. a. 3; $\frac{1}{3}$ b. 2; $\frac{1}{2}$ c. $\log_a b = \frac{1}{\log_b a}$ 23. 10^{10}
25. $4.5 \cdot 10^{-9}$ 27. -4 29. a. x^{12} b. x^{-6} c. x^{27} d. x^3

Lesson 9-8 (pp. 629-634)
Mental Math a. x^8 b. x^8 c. $\sqrt{6x-3}$ d. $6\sqrt{x}-3$
Guided Example 2 6; -34, 6; 17.4; 17.4
Questions
1. natural logarithms   3. $e^0 = 1$ 5. $\ln 148.41 \approx 5$
7. $\ln 1000$; because $e < 10$, e must be raised to a greater power to get 1000.   9. about 27.1 ft   11. 1   13. -2
15. about 20 minutes longer   17. about 26 decibels
19. a. no y-intercept b. no y-intercept c. Graphs of equations of the form $y = \log_b x$, where $b > 0$ and $b \neq 1$ have no y-intercept. This is because $b^y \neq 0$ for all y.
21. 0.896   23. $x = \sqrt{37}$ 25. B; this equation can be written in the the form $f(x) = ab^x$, where $a = 1$ and $b = \frac{1}{7}$. Since $0 < b < 1$, this equation describes exponential decay.

Lesson 9-9 (pp. 635-642)
Mental Math a. $2 \cdot 10, 10^2, 2^{10}$ b. $e^{16}, 3^{16}, \pi^{16}$
c. $0.5, \sqrt{0.5}, \sqrt[3]{0.5}$

Activity 1 Step 1: a. 0; 0; 0; 0 b. 1; 1; 1; 1 c. -2, -1, 0, 1
Step 2: For set a, for any base $b > 0$, $b \neq 1$, $\log_b 1 = 0$. For set b, for any base $b > 0$, $b \neq 1$, $\log_b b = 1$. For set c, for any base $b > 0$, $b \neq 1$, and any real number n, $\log_b b^n = n$.
Activity 2 Step 1: $\log 6$, $\log 6$; $2 \log 6$, $2 \log 6$; $\log_5\left(\frac{175}{2}\right)$, $\log_5\left(\frac{175}{2}\right)$; $7 \ln 2, 7 \ln 2$ **Step 2:** $\log_b x + \log_b y$
Guided Example 3 N, 10^{-12}; N, 12; N, 12
Activity 3 Step 1: a. 0.477 b. 0.954 c. 1.431 d. 1.908 e. 2.303 f. 4.605 g. 6.908 h. 9.210 **Step 2:** $n \cdot \log_b(x)$
Step 3: The Product Property of Logarithms supports this conclusion.
Questions
1. 15.4   3. 0   5. a. $\log_2 7$ b. Logarithm of a Quotient Theorem   7. a. 3 b. Answers vary. Sample: Logarithm of a Power Theorem   9. $\log(64) = \log(2^6) = 6 \log 2$
11. $\log_b\left(\frac{x\sqrt{z}}{y}\right)$ 13. true   15. False; Answers vary. Sample: For $M = 3$ and $N = 7$, $\log(10) \neq \log 3 + \log 7$; $\log(MN) = \log M + \log N$.   17. true   19. $y = \left(\frac{x}{w}\right)^{\frac{1}{4}}$ 21. $\log_{10} 5 + \log_{10} 2 = \log_{10}(5 \cdot 2) = \log_{10} 10 = 1$ and $\log_{10}(a \cdot b) = \log_{10} a + \log_{10} b$; therefore, $\frac{\log_{10} a}{\log_{10} 5 + \log_{10} 2} + \frac{\log_{10} b}{\log_{10} 5 + \log_{10} 2} = \frac{\log_{10}(a \cdot b)}{\log_{10} 5 + \log_{10} 2} = \log_{10}(a \cdot b)$
23. a. b^m b. $(b^m)^n$ c. b^{mn} d. b^{nm} e. nm 25. $x = 12$
27. a.

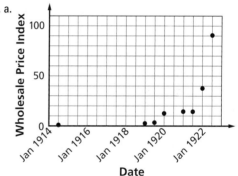

b. Answers vary. Sample: The data seem to roughly lie on an exponential curve. c. Answers vary. Sample: $y \approx 0.533 \cdot 1.30^x$ d. 46.11 e. under predicts
29. a. $\frac{19}{3}$ b. $\frac{19}{3}$ is the slope of the line segment connecting the points $(-3, -9)$ and $\left(-2, -\frac{8}{3}\right)$.

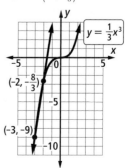

Lesson 9-10 (pp. 643–649)

Mental Math a. $x = 4$ b. $y = 3$ c. $z = 2$ d. $w = 2$

Guided Example 2 $Ce^{-\frac{\ln 2}{5730}t}$; $e^{-\frac{\ln 2}{5730}t}$; $\ln\left(e^{-\frac{\ln 2}{5730}t}\right)$; $-\frac{\ln 2}{5730}t$; $-\frac{5730}{\ln 2}$; 2182.11; 2182

Questions

1. a. 5 b. 5 c. 5 3. about 3.565; $4^{3.565} \approx 140$
5. $r = \frac{\log 2}{12\log e} \approx 0.0578$ 7. about 6067 yr old 9. –6.322; $0.5^{-6.322} \approx 80$ 11. about 5.61 yr 13. $y \approx 0.879$
15. a. about 1.27% b. in 2011; $5{,}995{,}000{,}000 \cdot (1.0127)^{12.28} \approx 7{,}000{,}000{,}000$ c. in 2054 d. Answers vary. Sample: 2008: 6,706,992,932; 2007: 6,627,548,985; growth rate: 1.20; in 2057
17. 2 and 3 19. –3 and –2 21. $x = 3$ 23. about $3409.7\frac{m}{sec}$
25. –2
27. a.

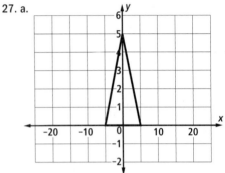

b. isosceles c. $\begin{bmatrix} -20 & 20 & 0 \\ 0 & 0 & 5 \end{bmatrix}$

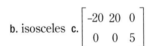

d.

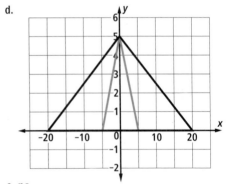

e. no

Self-Test (pp. 654–655)

1. domain: $x > 0$; range: all real numbers 2. The fraction of uranium remaining in an artifact after x half-lives is $U(x) = \left(\frac{1}{2}\right)^x$. Since in this case one half-life is about a week, the fraction remaining after three half-lives is $U(3) = 0.125$. So, there is 12.5% of the original amount of uranium remaining in an artifact after 3 weeks.

3. a.

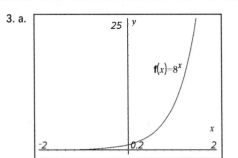

b. $f(\pi) = 8^\pi \approx 687.3$ c. yes; $y = 0$ 4. a. any b with $0 < b < 1$; example: $b = 0.5$ b. domain: all real numbers; range: $y > 0$ if $a > 0$, $y < 0$ if $a < 0$ 5. $\log 100{,}000{,}000 = \log(10^8) = 8$ by the Log_b of b^n Theorem. 6. $\log_2\left(\frac{1}{16}\right) = \log_2 2^{-4} = -4$ by the Log_b of b^n Theorem. 7. $\ln e^{-4} = -4$ because $\ln e^x = x$. 8. $\log a + 2\log t - \log s = \log a + \log(t^2) - \log s = \log(at^2) - \log s = \log\left(\frac{at^2}{s}\right)$ 9. The initial population is 50 and the yearly growth factor is 1.12; so the number of birds after t years is given by $N(t) = 50(1.12)^t$. $N(3) = 70.2464$; thus, after 3 years, there will be about 70 birds. 10. $\log_x 27 = \frac{3}{4}$; $x^{\frac{3}{4}} = 27$; $x = 27^{\frac{4}{3}}$; $x = 81$ 11. $5^y = 40$; $y\log 5 = \log 40$; $y = \frac{\log 40}{\log 5} \approx 2.29$ 12. $\ln(7z) = \ln 3 + \ln 21$; $\ln(7z) = \ln(3 \cdot 21) = \ln 63$; $7z = 63$; $z = 9$ 13. False; if $a = 5$ and $b = 3$, $\log_5 a + \log_3 b = 1 + 1 = 2$, but $\log_{15}(ab) = 1$. 14. The inverse of a logarithmic function is an exponential function with the same base, so an equation for the inverse is $y = 4^x$. 15. $7.3 = 6.1 + \log\left(\frac{23}{C}\right)$; $1.2 = \log 23 - \log C$; $\log C = \log 23 - 1.2 \approx 0.162$; $C \approx 10^{0.162} \approx 1.45$. 16. $\log_{14} 24.72 = \frac{\log 24.72}{\log 14} = 1.215$ 17. $\log_b b^n = n$, $b > 0$, $b \neq 1$ 18. a. Answers vary. Sample: $(1, 0)$; $(5, 1)$; $(25, 2)$ b. Powers of 5 must be positive, so the domain is $x > 0$; all real numbers can be exponents, so the range is all real numbers.

c.

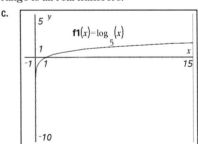

d. $y = 5^x$ e.

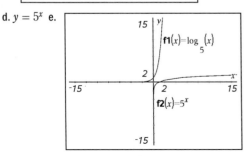

19. A; graphs of exponential growth have this shape and a y-intercept of 1. **20. a.** In 2025, $t = 18$, and so $P = 302{,}200{,}000\, e^{0.006(18)} \approx 336{,}665{,}229 \neq 349{,}400{,}000$ **b.** $349{,}400{,}000 = 302{,}200{,}000 e^{18r}$; $1.156188 = e^{18r}$; $18r = \ln 1.156188 \approx 0.145128$; $r = 0.00806$. So a growth rate of about 0.8% would give the reported projected population. **21. a.** $27^x = 14$; $x \cdot \ln 27 = \ln 14$; $x = \frac{\ln 14}{\ln 27}$ **b.** $x = 0.801$ **22. a.** Exponential regression shows that an exponential model for these data is $y = 572.6123(1.022)^x$, where x is the number of months after March 2007 and y is the number of articles in thousands. **b.** $1000 = 572.6123(1.022)^x$; $1.74638 = 1.022^x$; $x = \ln 1.022 = \ln 1.74638$, $x = \frac{\ln 1.74638}{\ln 1.022} \approx 25.6$. This model predicts that there will be 1 million German *Wikipedia* articles about 26 months after March 2007, or in April 2009.

Self-Test Correlation Chart

Question	1	2	3	4	5	6	7	8
Objective(s)	E	G	D, J	D	A	A	A	F
Lesson(s)	9-5, 9-7, 9-8	9-1, 9-2, 9-3, 9-10	9-1, 9-2, 9-3	9-1, 9-2, 9-3	9-5, 9-7, 9-8, 9-10	9-5, 9-7, 9-8, 9-10	9-5, 9-7, 9-8, 9-10	9-9, 9-10

Question	9	10	11	12	13	14	15	16
Objective(s)	G	C	B	C	F	E	I	A
Lesson(s)	9-1, 9-2, 9-3, 9-10	9-5, 9-6, 9-7, 9-9	9-10	9-5, 9-6, 9-7, 9-9	9-9, 9-10	9-5, 9-7, 9-8	9-6, 9-8	9-5, 9-7, 9-8, 9-10

Question	17	18	19	20	21	22
Objective(s)	F	A, E, J, K	J	G	B	H
Lesson(s)	9-9, 9-10	9-1, 9-2, 9-5, 9-7, 9-8, 9-10	9-1, 9-2	9-1, 9-2, 9-3, 9-10	9-10	9-4

Chapter Review (pp. 656–659)

1. 4 **3.** 8 **5.** 15 **7.** –3 **9.** 4.99 **11.** undefined **13.** 1.65 **15.** $t = 0.90$ **17.** $n = 18.85$ **19.** $z = 3.78$ **21.** $a = 1.89$ **23.** $x = 11$ **25.** $z = 10^{18}$ **27.** $x = 125$ **29.** $x = 7.03$ **31.** domain: all real numbers; range: positive real numbers **33.** when $a > 1$ **35.** $y = 0$ **37.** $f^{-1}(x) = -\ln x$ **39.** true **41.** $3^{-5} = \frac{1}{243}$ **43.** $10^n = m$ **45.** $\log 0.01585 \approx -1.8$ **47.** $\log_x z = y, x > 0, x \neq 1$ **49.** $\ln(204)$; Logarithm of a Product Theorem **51.** $\log_{12}\left(\frac{1}{121}\right)$; Logarithm of a Power Theorem **53.** 79; Logarithm of a Power Theorem, $\log_b$ of b^n Theorem **55.** $\log\left(\frac{a}{b^3}\right)$; Logarithm of a Power Theorem, Logarithm of a Quotient Theorem **57.** about 37.675 million **59.** 2048 **61.** 16.53% **63.** $11,327.88 **65. a.**

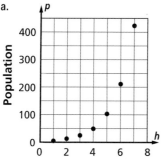

b. $P = 2.72(2.06)^h$ **c.** about 771,100 bacteria **67.** 215.89 decibels **69.** 10,000 **71.** 206.96°F

73.

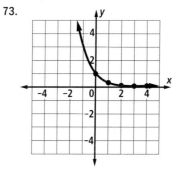

75. a. $y = 3^x$ **b.** $y = 2^x$ **c.** The graph of e^x shares the same y-intercept and falls between the graphs of f and g for all other values of x.

77. a.

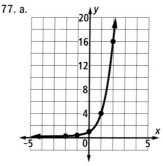

b.

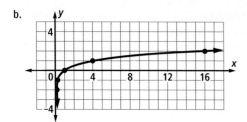

79. a.

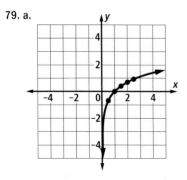

b. $y = e^x$ **81.** 1

Chapter 10

Lesson 10-1 (pp. 662–669)

Mental Math a. yes **b.** no **c.** yes

Activity Steps 1–2:

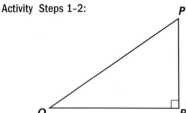

Step 3: Answers vary. Sample: $PQ = 8.4$ cm, $QR = 6.5$ cm, $RP = 5.3$ cm **Step 4:** $\sin 39° \approx 0.631$, $\cos 39° \approx 0.774$, $\tan 39° \approx 0.815$ **Step 5:** $\sin 39° \approx 0.629$, $\cos 39° \approx 0.777$, $\tan 39° \approx 0.810$ **Step 6:** 0.3%, −0.3%, 0.6%

Guided Example 2 16.7°; 490; 490, $\sin 16.7°$; 1705.2; 1705; 1705, 490, 1633; 16.7°, 0.958, 0.958

Questions

1. D **3. a.** $\frac{a}{c}$ **b.** $\frac{b}{c}$ **c.** $\frac{a}{b}$ **5.** about 98.9 m **7.** 4.58 cm
9. a. about 24.1 ft **b.** about 24.0 ft

11. a.

$\sin 30° = 0.5$	$\sin 40° = 0.643$
$\cos 60° = 0.5$	$\cos 50° = 0.643$
$\sin 15° = 0.259$	$\sin 90° = 1$
$\cos 75° = 0.259$	$\cos 0° = 1$

b. They are complementary. **c.** They are equal. **d.** The cosine of an angle is equal to the sine of its complement (and, equivalently, the sine of an angle is equal to the cosine of its complement). **e.** Answers vary.

Sample: $\cos 85° = \sin 5° \approx 0.087$ **13.** about 1.85 units
15. a. $\sin C = \frac{h}{a}$; $h = a \sin C$; $A = \frac{1}{2}bh = \frac{1}{2}ba \sin C$
b. 936.73 **17.** $z = 32$ **19. a.** π **b.** π **21.** false, $\angle 1 \cong \angle 5$
23. a. $(90 - x)°$ **b.** $x°$

Lesson 10-2 (pp. 670–674)

Mental Math a. 7 and 8 **b.** 8 and 9 **c.** 0 and 1 **d.** 5 and 6
Guided Example 2 $\angle A$; Answers vary. Sample: sin, 5, 13; sin; sin, sin, sin, 5, 13; 22.62°; 22.62°; 67.38°
Guided Example 4 elevation β; cosine; β, 110.56, 125.75; β, 0.8792; 28.45°; 28.45

Questions

1. Answers vary. Sample: 2nd SIN .475) ENTER **3.** 45°
5. a. $\tan^{-1}\left(\frac{5}{12}\right) \approx 22.62°$ **b.** $\sin^{-1}\left(\frac{12}{13}\right) \approx 67.38°$ **7.** horizontal
9. The line of sight is a transversal for parallel horizontal lines, so the angles of elevation and depression are alternate interior angles and therefore congruent.
11. a. 0.015 **b.** 0.015 **c.** about 0.86° **13.** about 23.20°
15. about 16.33° **17.** $\frac{11\sqrt{2}}{4}$ **19.** $x = \frac{-s \pm \sqrt{s^2 - 4rt}}{2r}$
21. $\sqrt{13} \approx 3.606$

Lesson 10-3 (pp. 675–681)

Mental Math a. isometry **b.** neither **c.** isometry
Activity 1 Step 1: Answers vary. Sample: My thumb moves more than my partner. The background doesn't seem to jump. After stepping back, only my thumb seems to jump.
Step 2: Apparent position of star from A: 6; apparent position of star from B: 4
Step 3: a. Answers vary. Sample:

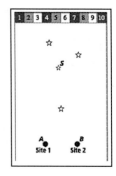

b.

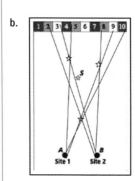

c. Answers vary. Sample: 3, 5; 8, 9; 2, 10; the larger the jump, the larger the angle. **d.** Close objects seem to jump more than distant objects because the angle between the two segments connecting the object to both eyes is greater when the object is closer.

Activity 2 Step 2: Answers vary. Sample: 3 spaces
Step 3: Answers vary. Sample: Par $\angle$: 7.5°
Step 4: Answers vary. Sample: 7.5°, 82.5°, 82.5°
Step 5: Answers vary. Sample: 15.2 feet
Activity 3 Step 1: 89.89°, 89.89° **Step 2:** 179,700 miles
Activity 4 Step 1:

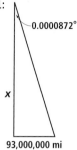
0.0000872°

x

93,000,000 mi

$6.11 \cdot 10^{13}$ miles **Step 2:** about 10.4 light-years
Activity 5
Step 1.

θ

10 m

1.27 m

$\theta = 7.24°$

Step 2:

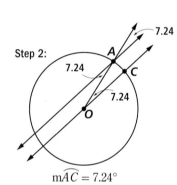
7.24

A

7.24

C

7.24

O

$m\widehat{AC} = 7.24°$

Step 3: 797.45 km **Step 4:** $x \approx 39,652$ km
Questions
1. the illusion of motion created by changing viewing positions 3. We assumed the Moon was equidistant from the two cities. This assumption is justified because the variation is very small compared to the distances being measured. 5. 14.32 ft 7. 1708.6 meters
9. a. $\frac{s}{5r}$ b. $\frac{1}{5}$ c. $\frac{s}{r}$ 11. a. (0, 1) b. (–1, 0) c. (0, –1) d. (0, –1)
13. a. III b. II c. I, III

Lesson 10-4 (pp. 682–686)
Mental Math a. 90 **b.** 16 ft **c.** no
Activity Step 1: Check students' work. **Step 2: a.** Check students' work. $A_{20} = (\cos 20°, \sin 20°)$ **b.** Answers vary. Sample: $A_{20} \approx (0.95, 0.35)$ **c.** $A_{20} = (\cos 20°, \sin 20°) \approx$ (0.940, 0.342) **Step 3: a.** Check students' work. $A_{40} =$ (cos 40°, sin 40°) **b.** Answers vary. Sample: $A_{40} \approx$ (0.77, 0.64) **c.** $A_{40} = (\cos 40°, \sin 40°) \approx (0.766, 0.643)$
Step 4: a. Check students' work. $A_{75} = (\cos 75°, \sin 75°)$
b. Answers vary. Sample: $A_{75} \approx (0.97, 0.26)$
c. $A_{75} = (\cos 75°, \sin 75°) = (0.966, 0.259)$

Step 5: a. For any θ, the coordinates of a rotation R_θ centered at (0, 0) of magnitude θ is (cos θ, sin θ). **b.** Answers vary. Sample: $A_{61} \approx (0.49, 0.68)$ **c.** Answers vary. Sample: cos 61° $\approx$ 0.48; $\frac{0.49 - 0.48}{0.48} \approx 0.02$; sin 61° $\approx$ 0.87; $\frac{0.86 - 0.87}{0.87} \approx -0.01$; both were within 3% of the actual values.
Questions
1. a. x b. y 3. (1, 0); 1; 0 5. 0 7. 0 9. a. a rotation of 180°
b. (–1, 0) c. 0 11. C 13. 0; 1 15. c 17. c 19. a
21. $x = 32$°; cos 392° = cos 32° ≈ 0.848 23. a. 1 b. –1
25. sin(–270°) = sin 450° = 1 27. Because stars are extremely far away, you need to use very distant sites in order to have manageable parallax angles.
29. about 3003 ft away

Lesson 10-5 (pp. 687–692)
Mental Math a. 32 **b.** 7 **c.** 0
Activity Step 1:

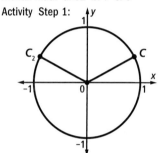
C_2

C

$C_2 \approx (-0.883, 0.469)$ **Step 2:** The acute angles that C and C_2 make with the x-axis are congruent, so they both measure 28°. Thus, the magnitude of the rotation that maps (1, 0) onto C_2 is 180° – 28° = 152°. cos 152° ≈ –0.883, sin 152° ≈ 0.469
Step 3:

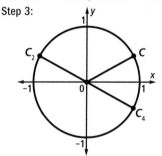
C_2

C

C_4

$C_4 \approx (0.883, -0.469)$ **Step 4:** $\theta = -28° = 332$°; sin 332° ≈ –0.469, cos 332° ≈ 0.883
Step 5:

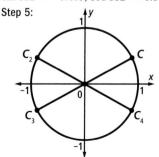
C_2

C

C_3

C_4

$C_3 = (\cos 208°, \sin 208°) \approx (-0.883, -0.469)$

Questions

1. is always negative   3. is always positive
5. negative

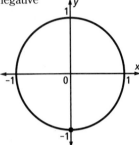

7. a. $\frac{12}{13}, -\frac{12}{13}$ b. $\sin x$ is the y-coordinate of $R_x(1, 0)$, and $\sin(-x)$ is the y-coordinate of $R_{-x}(1, 0)$. Because $R_{-x}(1, 0)$ is the reflection image of $R_x(1, 0)$ over the x-axis, the y-coordinates of these points are oppostites. Therefore, $\sin(-x) = -\frac{5}{13}$.

9.

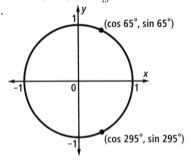

Since $(\cos 295°, \sin 295°)$ is the image of $(\cos 65°, \sin 65°)$ reflected over the y-axis, we know $\sin 295° = -\sin 65°$.
$\sin 295° = -\sin 65° \approx -0.906$ 11. $\frac{\sqrt{2}}{2}$ 13. $-\frac{\sqrt{3}}{2}$ 15. 233°
17. a. $\sin 300° \approx -0.866$; $\cos 300° = 0.5$; $\tan 300° \approx -1.732$
b. $\frac{-0.866}{0.5} = -1.732$ 19. 0   21. The y-coordinates of $R_\theta(1, 0)$ and $R_{180°-\theta}(1, 0)$ are the same, and $\sin \theta$ is the y-coordinate of points on the unit circle.   23. false
25. a. about 10 degrees b. Answers vary. Sample: The stairs will now be less steep.   27. $x = 9$ or $x = 1$

Lesson 10-6 (pp. 693–699)

Mental Math a. the shortest diagonal **b.** the other leg
c. the diagonal **d.** They are the same length.

Activity Step 1:

θ	$\cos \theta$
0°	1.00
15°	0.97
30°	0.87
45°	0.71
60°	0.50
75°	0.26
90°	0.00
105°	−0.26
120°	−0.50
135°	−0.71
150°	−0.87
165°	−0.97
180°	−1.00
195°	−0.97
210°	−0.87
225°	−0.71
240°	−0.50
255°	−0.26
270°	0.00
285°	0.26
300°	0.50
315°	0.71
330°	0.87
345°	0.97
360°	1.00

Step 2:

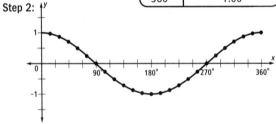

Step 3: The graph is bounded such that $-1 \leq \cos \theta \leq 1$ and has intercepts that are odd multiples of 90°.
Step 4: The largest value is 1 and the smallest value is −1.
Questions
1. the cosine function   3. a. 1; 0 b. 0; −1 c. increase 5. 8
7. −180°, 0°, 180°   9. false   11. true   13. a. Answers vary. Sample: −90°, 270° b. Answers vary. Sample: 45°, 225°
15. yes; 360°   17. No; this is a polynomial function.
19. a. Answers vary. Sample: Temperatures are annually affected by the seasons; therefore, the data are periodic with period 1 year. b. domain: all real numbers; range: $66 \leq T \leq 105$

21. a. Yes; period: 360°

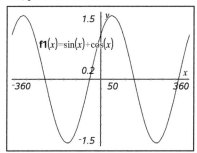

b. domain: all real numbers; range: $-\sqrt{2} \le y \le \sqrt{2}$ **23.** 0
25. a. $h = -1.86t^2 + 15t + 47$
b.

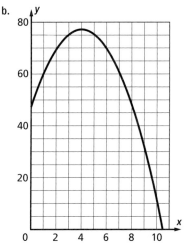

c. 77 m **d.** 10.5 sec **27. a.** $y = -\frac{1}{9}x$ **b.** direct

Lesson 10-7 (pp. 700–705)
Mental Math a. 1 **b.** 0 **c.** infinitely many (any line parallel to $y = x$)
Activity Step 1:

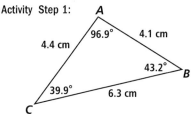

Step 2: $\frac{\sin A}{a} = \frac{\sin B}{b} = \frac{\sin C}{c} \approx 0.158$ **Step 3:** The ratios of the sines of $\triangle ABC$'s angles to the lengths of its sides opposite them are equal, but these ratios are not equal to the corresponding ratios of my partner's triangle.
Guided Example 3 $\frac{\sin 55°}{7}$, $\frac{\sin 17°}{g}$; 2.5
Questions
1. For any value of k between 0 and 1 and for $0° < \theta < 180°$, the graph of $y = k$ intersects $y = \sin \theta$ in two points.
3. 135° **5.** In any triangle, the ratios of the sines of its angles to the lengths of the sides opposite them are equal.
7. $r \approx 6.63$ **9. a.** The statement is incorrect because $\triangle BSG$ is not a right triangle. **b.** No, the Law of Sines can

be used only when either the AAS condition or the ASA condition is satisfied. **11. a.** about $1.24 \cdot 10^8$ m/sec
b. about 2.42 **c.** about $1.36 \cdot 10^8$ m/sec **13.** Transitive Property of Equality; $\triangle ABC$; $ab \sin C = ac \sin B = bc \sin A$; $\frac{ab \sin C}{abc} = \frac{ac \sin B}{abc} = \frac{bc \sin A}{abc}$ **15.** $90°, 270°, 450°$ **17.** $\pm\sqrt{1 - \sin^2\theta}$

Lesson 10-8 (pp. 706–711)
Mental Math a. true **b.** false **c.** true
Guided Example 3 14, 12, 20; 20, 14, 12, 14, 12; 400, 340, 336; 60, -336; -0.1786; 100.29°; 100°
Questions
1. q^2 **3.** $p^2 + q^2 - 2pq \cos R$
5. $BC = \sqrt{x^2 + 100 - 20x \cos 27.5°}$
7. Step 1 $m\angle C = \sin^{-1}\left(\frac{c}{a}\sin A\right) \approx 22.02°$
Step 2 $m\angle B \approx 180° - 30° - 22.02° = 127.98°$
Step 3 $b = \frac{\sin B}{\sin A} \cdot a \approx 12.6$ cm **9.** Consider $\triangle ABC$ with $\angle C$ acute. By the Law of Cosines, $c^2 - a^2 - b^2 = -2ab \cos C$. Since C is acute, $\cos C > 0$, so $-2ab \cos C < 0$. Thus, $c^2 - a^2 - b^2 < 0$, and $c^2 < a^2 + b^2$.
11. $\cos A = \frac{a^2 - b^2 - c^2}{-2bc}$ **13. a.** ≈ 54.35 m **b.** $\approx 12.51°$
15. $59.49°$ **17.** 50.12 m **19.** Yes. Equations vary. Sample: $x = 90°$. **21.** By the Pythagorean Identity, we have $D = \sqrt{r^2 \cos^2\theta + r^2 \sin^2\theta} = \sqrt{r^2(\cos^2\theta + \sin^2\theta)} = \sqrt{r^2} = |r|$.
23. $\frac{85}{360} = \frac{17}{72}$

Lesson 10-9 (pp. 712–717)
Mental Math a. 24 **b.** 11.33 **c.** 16
Guided Example 2 a. $\frac{\pi \text{ radians}}{180 \text{ degrees}}$, $\frac{45\pi}{180}$, $\frac{\pi}{4}$, $\frac{\pi}{4}$
b. $\frac{180 \text{ degrees}}{\pi \text{ radians}}$, $\frac{360}{3}$ degrees; 120°; 120°
Guided Example 3 2, 2, -0.416; $\frac{3\pi}{4}$, $\frac{3\pi}{4}$, -1; 57.3°, 114.6°; 114.6°, -0.416; 135°, 135°, -1
Questions
1. a. 2π **b.** π **c.** $\frac{\pi}{2}$ **d.** $\frac{11\pi}{2}$ **3.** 1440° **5.** 56° **7.** $\frac{\pi}{2}$ **9.** $\frac{5\pi}{4}$
11. a. $\sin 6$ is the sine of 6 radians; $\sin 6°$ is the sine of 6 degrees. **b.** -0.279, 0.105 **13. a.** 0 **b.** $\cos 270° = 0$
15. 2π **17.** B and D **19.** 1 **21. a.** $\frac{5\pi}{2}$ **b.** 5π **23.** $\frac{16\pi}{3}$ feet
25. $\sin \theta = \sin(180° - \theta)$; thus, if you use the Law of Sines, you know $\theta \approx 49.46°$ or 130.54°. **27.** 1
29. The stones form a right triangle, with legs of 2 and 1, and $\tan^{-1}\left(\frac{1}{2}\right) = 26.57°$. Thus, using this technique will always give a 26.57° angle.

Self-Test (pp. 722–723)
1. a. $\cos(17.4°) \approx 0.954$ **b.** $\sin(17.4°) \approx 0.299$
2. $\tan^{-1}(\tan \theta) = \tan^{-1}(0.64)$; $\theta \approx 32.619°$ **3. a.** $\cos \theta = \frac{\text{adj.}}{\text{hyp.}} = \frac{40}{41}$ **b.** $\tan \theta = \frac{\text{opp.}}{\text{adj.}} = \frac{9}{40}$ **4.** $1 - \sin^2 \theta = \cos^2 \theta$; $1 - (0.280)^2 = \cos^2 \theta$; $\cos^2 \theta = 0.9216$; $\cos \theta = \pm 0.96$. Since $90° < \theta < 180°$, $\cos \theta = -0.96$. **5. a.** $(\cos(-423°), \sin(-423°))$ **b.** According to the unit-circle definitions of cosine and sine, for any θ, the point $(\cos \theta, \sin \theta)$ is the image of $(1, 0)$ under R_θ. **6.** $x = 163°$; since $\sin \theta = \sin(180° - \theta)$, $\sin(180° - 17°) = \sin 163° = \sin 17°$.

7. *a*, because the angle that the radius through (a, b) makes with the *x*-axis measures about 70°, and cosine is the *x*-coordinate of points on the unit circle.   **8.** *f*, because (e, f) is in the third quadrant, and sine is the *y*-coordinate of points on the unit circle.   **9.** According to the Law of Cosines, $c^2 = a^2 + b^2 - 2ab \cos C$. Therefore, $c^2 = 250^2 + 175^2 - 2(250)(175) \cos 35°$; $c^2 \approx 21{,}449.2$; $c \approx 146.455$; he is about 146 m away.   **10.** According to the Law of Sines, $\frac{\sin S}{s} = \frac{\sin L}{\ell}$. Therefore, $\frac{\sin S}{425} = \frac{\sin 12°}{321}$; $m\angle S = \sin^{-1}\left(425 \cdot \left(\frac{\sin 12°}{321}\right)\right) \approx 16.0°$ **11.** $\sin(\theta + 360°) = \sin\theta$ for all values of θ; therefore, the period is 360°.   **12.** Answers vary. Sample: $-360° < \theta < -270°$ and $-90° < \theta < 90°$.   **13.** $15^2 = 20^2 + 30^2 - 2(20)(30)\cos\theta$; $\cos\theta = \frac{43}{48}$;

$\cos^{-1}\left(\frac{43}{48}\right) = 26.38°$; The sum of the measures of the angles of a parallelogram is 360°, and opposite angles are equal. Therefore, $360 - 2(26.38) = 2\theta$; $\theta = 153.62$. The angles of the parallelogram have measures 153.62°, 26.38°, 153.62°, and 26.38°.   **14.** Convert 0.008 arc-second to degrees: 0.008 arc-second $\cdot \frac{1 \text{ degree}}{3600 \text{ arc-seconds}} \approx 0.000002°$. Let d be the distance from Earth to Betelgeuse. $\tan(0.000002°) = \frac{93{,}000{,}000}{d}$; $d \approx 2.66 \cdot 10^{15}$ miles. Convert miles to light-years. $d \approx 2.66 \cdot 10^{15}$ miles $\cdot \frac{1 \text{ light-year}}{5.88 \cdot 10^{12} \text{ mi}} \approx 452.4$ light-years   **15. a.** $60° \cdot \frac{\pi}{180°} = \frac{\pi}{3}$ **b.** $25° \cdot \frac{\pi}{180°} = \frac{5\pi}{36}$ **16. a.** $4\pi \cdot \frac{180°}{\pi} = 720°$ **b.** $\frac{\pi}{7} \cdot \frac{180°}{\pi} \approx 25.7°$ **17. a.** $\sin 36° = \frac{1500}{d}$; $d = \frac{1500}{\sin 36°} \approx 2552$ ft **b.** $\tan 36° = \frac{1500}{a}$; $a = \frac{1500}{\sin 36°} \approx 2065$ ft

Self-Test Correlation Chart

Question	1	2	3	4	5	6	7	8	9
Objective(s)	A	B	E	F	E	F	I	I	H
Lesson(s)	10-1, 10-9	10-2, 10-7	10-1, 10-4, 10-5	10-5, 10-7	10-1, 10-4, 10-5	10-5, 10-7	10-4, 10-5	10-4, 10-5	10-7, 10-8

Question	10	11	12	13	14	15	16	17
Objective(s)	C	J	J	C	G	D	D	G
Lesson(s)	10-7, 10-8	10-6	10-6	10-7, 10-8	10-1, 10-2, 10-3	10-9	10-9	10-1, 10-2, 10-3

Chapter Review (pp. 724–727)

1. 0.559   **3.** -0.383   **5.** -0.951   **7.** $\theta = 30°$ or $150°$
9. $\theta = 90°$ **11.** $a = 13.4$ **13.** $c = 86.9°$ **15.** $e = 14.1$
17. $-\frac{7\pi}{12}$ **19.** $\frac{9\pi}{4}$ **21.** $270°$ **23.** $210°$ **25.** D   **27.** C
29. $\frac{5}{13}$ **31.** true; $\tan x \cdot \cos x = \frac{\sin x}{\cos x} \cdot \cos x = \sin x$ as long as x is not an odd multiple of $\frac{\pi}{2}$.   **33.** -0.292; -0.956
35. false; $\cos\theta - \cos(180° - \theta) = \cos\theta + \cos\theta \neq 0$
37. $\cos x \approx \pm 0.968$ **39.** 4.8°   **41.** 11,000 km
43. about 309 cm   **45.** 84°

47. $\frac{\sqrt{2} - \sqrt{6}}{4}$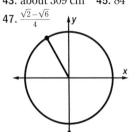

49. $\frac{-\sqrt{2} - \sqrt{6}}{4}$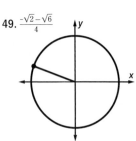

51. $\frac{\sqrt{6} - \sqrt{2}}{4}$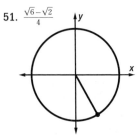

53. 4 radians   **55.** *e*

57. a.

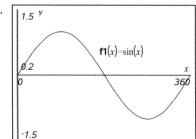

b. domain: all real numbers; range: $-1 \leq y \leq 1$
59. Answers vary. Sample: $T : (x, y) \to (x - 90°, y)$
61. No; this graph cannot be mapped onto the graph of the sine function $s : \theta \to \sin\theta$ by any composite of translations, scale changes, or reflections.

Chapter 11

Lesson 11-1 (pp. 730–737)

Mental Math a. 0 **b.** Answers vary. Sample: $\begin{bmatrix} 2 & 1 \\ 6 & 3 \end{bmatrix}$

c. does not exist **d.** Answers vary. Sample: $f : x \to 1$
Guided Example 1 a. 1; 3; 16; 96; 4; 16; 96; -16; 0; 4; 0; 96
b. 4; 4; -16; -16 **c.** -16; 4; 96
Activity 1 Step 1: a. $f(x) + g(x) = 5x^3 + 2x^2 + 3x + 5$; degree 3 **b.** $f(x) - g(x) = -5x^3 + 2x^2 + 3x + 3$; degree 3 **Step 2: a.** 5 **b.** 10 **Step 3: a.** $10x^5 + 15x^4 + 20x^3 + 2x^2 + 3x + 4$ **b.** Answers vary. They are the same.
Step 4: sum; product

Activity 2 Step 1: $f(g(x)) = 2(5x^3 + 1)^2 + 3(5x^3 + 1) + 4$
a. Answers vary. Sample: 6 **b.** Answers vary. Sample: 50
Step 2: a. $50x^6 + 35x^3 + 9$ **b.** Answers vary. Sample:
They are the same. **Step 3:** Answers vary. Sample: The
degree of the composite of two polynomial functions is the
product of the degrees of the polynomial functions. The
leading coefficient of $f(g(x))$ is the leading coefficient of
$g(x)$ to the power of the degree of $f(x)$ multiplied by the
leading coefficient of $f(x)$.
Guided Example 3 a. 2; 1; 3 **b.** 3; 4; 12 **c.** 2; 1; 2 **d.** 3; 4; 48
Questions
1. Yes; its degree is 1 and its leading coefficient is 8.
3. Yes; its degree is 5 and its leading coefficient is 6.
5. $a_4t^4 + a_3t^3 + a_2t^2 + a_1t + a_0$ **7. a.** –10
b.

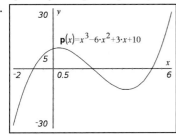

9. **a.** 8 **b.** 6 **c.** $6x^8 + 2x^4 - \frac{5}{2}$ **11. a.** 5 **b.** 8 **c.** 15
13. **a.** 2 **b.** $-\frac{1}{8}g$ or –16 **c.** 47.04 ft **15. a.** $40,000 + 5000 +$
$700 + 2 = 45,702$ **b.** 19,394 **17. a.** $\frac{1}{49}, \frac{1}{343}; 3, -8$
b. $A_n = 49 \cdot \left(\frac{1}{7}\right)^{n-1}$ **c.** $B_n = 47 - 11(n - 1)$ **d.** A

19.

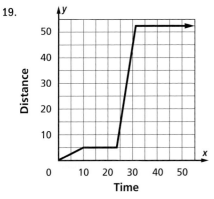

21. B **23. a.** 3 **b.** 128°

Lesson 11-2 (pp. 738–744)
Mental Math $2\sqrt{2} \approx 2.8$ blocks
Guided Example 3 $5x^2$; $4x^2$; $11x$; $220x^5$; 2; 8; –3; –48; $4x^2$;
$11x$; 440; 88
Questions
1. **a.** trinomial **b.** 8 **3. a.** binomial **b.** 3 **5. a.** monomial
b. 0 **7.** Answers vary. Sample: $2xy^4 + y$ **9.** $10b^4 - 15b^3 +$
$10b^2 - 15b$ **11. a.** 384; 384; 300; 84; 0; 0 **b.** $x = 3$
13. **a.** $x^4 - 3x^2 + 9$ **b.** $x^4 - 8x^2 + 16$ **c.** $x^4 - 15x^2 + 25$
d. $x^4 - 24x^2 + 36$ **e.** $x^4 - (n^2 - 2n)x^2 + n^2$ **15. a.** $V(x) =$
$\frac{1}{2}x(15 - 2x\sqrt{3})(\frac{15}{2}\sqrt{3} - 3x)$ **b.** $V(x) = 3\sqrt{3}x^3 - 45x^2 + \frac{225\sqrt{3}}{4}$

c. $V(4) = 417\sqrt{3} - 720 \approx 2.27$ **d.** 62.5 in³ **17.** $(75 - 2b) \cdot$
$(150 - a - c) = 11,250 + 2ab + 2bc - 75a - 300b - 75c$
19. $2y + 1$ and $y + 7$ **21. a.** 5 **b.** $5119.36
23. **a.**

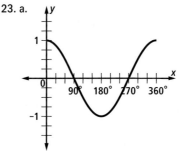

b. $x = 90°, x = 270°$ **25.** $x^2 + 14x + 49$

Lesson 11-3 (pp. 745–751)
Mental Math a. –0.15 **b.** 0.15 **c.** –0.15 **d.** 0.15
Activity Step 1: a.

| $\text{factor}(x^2 - 24 \cdot x + 144)$ | $(x - 12)^2$ |

b.

| $\text{factor}(x^2 - 49)$ | $(x - 7) \cdot (x + 7)$ |

c.

| $\text{factor}(x^2 - 64)$ | $(x - 8) \cdot (x + 8)$ |

d.

| $\text{factor}(x^2 + 8 \cdot x + 16)$ | $(x + 4)^2$ |

e.

| $\text{factor}(4 \cdot x^2 - 81)$ | $(2 \cdot x - 9) \cdot (2 \cdot x + 9)$ |

f.

| $\text{factor}(x^2 - 6 \cdot x + 9)$ | $(x - 3)^2$ |

g.

| $\text{factor}(x^2 + 10 \cdot x + 25)$ | $(x + 5)^2$ |

h.

| $\text{factor}(36 \cdot x^2 - 1)$ | $(6 \cdot x - 1) \cdot (6 \cdot x + 1)$ |

Step 2: (d) and (g) have the form $(a + b)^2$, (a) and (f)
have the form $(a - b)^2$, and (b), (c), (e), and (h) have the
form $(a - b)(a + b)$. **Step 3: a.** $(x + y)(x - y)$ **b.** $(x - y)^2$
c. $(x + y)^2$
Guided Example 2 a. $3x^3$; $10y$; $3x^3$; $10y$; $3x^3$; $10y$ **b.** p; 9; 9; $p - 9$
Questions
1. $3m^2$; $5m$; $-2n$ **3.** $90xy^2(x^2 + 3 + 2x)$ **5. a.** binomial
square **b.** $(x - y)^2$ **7. a.** difference of squares
b. $(7x - 12b)(7x + 12b)$ **9. a.** neither **b.** not possible
11. **a.** $12t(t + 1)(t - 1)$ **b.** $(12t^2 + 12t)(t - 1) = 12t^3 -$
$12t^2 + 12t^2 - 12t = 12t^3 - 12t$ **13.** $(x + 3)(x - 3) \cdot$
$(x + 1)(x - 1)$; For $x = -3$, $(-3)^4 - 10(-3)^2 + 9 = 0$ and
$(-3 + 3)(-3 - 3)(-3 + 1)(-3 - 1) = 0(-6)(-2)(-4) = 0$.
15. $-1(1 + 5y)^2$; common monomial factoring, binomial
square factoring **17.** $3x(2x + 5)(2x - 5)$;

common monomial factoring, difference of squares
19. a. yes **b.** no; $(x - \sqrt{30})(x + \sqrt{30})$ **21.** A **23.** $4x + 7$
25. $2(q + 1)(3q - 2)$ **27. a.** $5x(x^2 - 2x - 35) =$
$5x(x - 7)(x + 5)$ **b.** $(5x^2 - 35x)(x + 5) = 5x^3 + 25x^2 -$
$35x^2 - 175x = 5x^3 - 10x^2 - 175x$ **29. a.** $\frac{(2x + 3)(2x - 3)}{(2x + 3)(x - 1)}$
b. $\frac{(2x - 3)}{(x - 1)}$ **c.** The domain includes all real numbers x
such that $x \neq -1.5$ and $x \neq 1$. If x took these values, the
denominator of the expression would equal zero.
31. $4h^3 + 14h^2 + 12h$ **33. a.** $V = \frac{1}{3}\pi r^2 h$ **b.** $r = \sqrt{81 - h^2}$
c. $V = \frac{1}{3}\pi h\left(\sqrt{81 - h^2}\right)^2 = \frac{1}{3}\pi(81h - h^3)$ **d.** true **35.** –1

Lesson 11-4 (pp. 752–759)

Mental Math a. never **b.** never **c.** always **d.** never
Activity 1 Step 1:

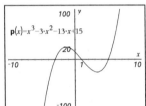
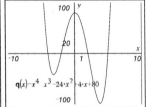

Step 2: $p(x) = x^3 - 3x^2 - 13x + 15$: –3, 1, 5;
$q(x) = x^4 - x^3 - 24x^2 + 4x + 80$: –4, –2, 2, 5
Step 3:

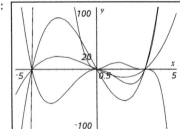

Step 4: If r is an x-intercept of the graph of $P(x)$, then $x - r$
is a factor of $P(x)$. **Step 5:** Answers vary. Sample:
$r(x) = (x + 1)(x - 4)(x - 5) = x^3 - 8x^2 + 11x + 20$
Guided Example 3 $x - (-2); x - (-1); x + 2; x + 1$; **Check:**
Graph should match the one in Guided Example 3.
Activity 2 Step 1: a. –4, 0, 3 **b.** –4, 0, 3 **c.** –4, 0, 3 **d.** –4, 0, 3
Step 2:

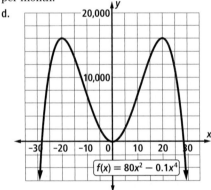

The four graphs have the same x-intercepts. The function
paths between the x-intercepts are different.
Step 3: Answers vary. Sample: $m(x) = x(x + 4)^2(x - 3)$

Questions
1. For all a and b, $ab = 0$ if and only if $a = 0$ or $b = 0$.
3. $-\sqrt{2}, -1, \sqrt{2}, 20$ **5.** C **7. a.** $r(t) = t(t + 3)(2t - 7)$
b. $-3, 0, \frac{7}{2}$ **9. a.** $p(x) = k(x - 5.6)^a(x + 2.9)^b$,
where $k \neq 0$ and a and b are positive integers.
b. Answers vary. Sample:

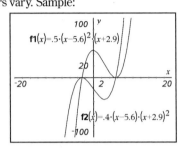

11. $y = \frac{1}{2}(x + 2)(x - 1)(x - 3)^2$ **13. a.** $P(x) = x(x^2 + 5x + 3)$
b. $x = 0, x = \frac{-5 \pm \sqrt{13}}{2}$ **c.** $P(x) = x\left(x + \frac{5 - \sqrt{13}}{2}\right)\left(x + \frac{5 + \sqrt{13}}{2}\right)$;
the second factor in Part a has been factored.
15. $x^2 + 1 \neq 0$ for all real numbers x. **17. a.** –3, 2 **b.** $k = \frac{1}{4}$
19. a. $f(n) = -0.1n^2(n - \sqrt{800})(n + \sqrt{800})$ **b.** $-\sqrt{800} \approx$
$-28.2843, 0, \sqrt{800} \approx 28.2843$ **c.** The zeros represent the
number of employees that results in 0 units produced
per month.
d.

A reasonable domain for this situation is $0 \leq n \leq 28.2843$.
21. No; this expression contains a negative integer
power of x. **23.** yes; 2 **25.** 1; 0 **27. a.** True;
$\sqrt{12} \cdot \sqrt{3} = \sqrt{36} = \sqrt{2} \cdot \sqrt{18}$ by the Root of a Product
Theorem. **b.** False; $\sqrt{-12} \cdot \sqrt{-3} = i\sqrt{12} \cdot i\sqrt{3} = i^2\sqrt{36} =$
$-\sqrt{36} \neq \sqrt{36} = \sqrt{2} \cdot \sqrt{18}$.

Lesson 11-5 (pp. 760–765)

Mental Math a. 12-oz box **b.** $10-off coupon **c.** before tax
d. They pay the same amount.
Activity Step 1: $x = \frac{3}{2}$ or $x = -\frac{4}{9}$; $x = \frac{3}{2}, x = -\frac{4}{9}$, or $x = -\frac{7}{5}$
Step 2: $Q(x)$: $18x^2$, –12; $P(x)$: $90x^3$, –84 **Step 3:** For both
$Q(x)$ and $P(x)$, the denominators of the roots of the
equation are factors of the coefficient of the leading term.
Step 4: For both $Q(x)$ and $P(x)$, the numerators of the
roots of the equation are factors of the constant term.

Questions
1. 5 is a factor of the leading coefficient, and 7 is a factor of the constant term. 3. a. $\pm 24, \pm 12, \pm 8, \pm 6, \pm 4, \pm 3,$ $\pm 2, \pm \frac{3}{4}, \pm 1, \pm \frac{3}{4}, \pm \frac{1}{2}, \pm \frac{3}{8}, \pm \frac{1}{4}, \pm \frac{1}{8}$ b. $-\frac{1}{2}$ 5. a. $\pm 14, \pm 7, \pm 2,$ $\pm 1, \pm \frac{2}{7}, \pm \frac{1}{7}$ b. There are no rational roots. If we look at the graph, we see that the roots must lie in the ranges $-0.5 < x < 0$ and $3 < x < 3.5$. By testing $-\frac{2}{7}$ and $-\frac{1}{7}$, we find that neither of these values are zeros. 7. a. $\pm 12, \pm 6, \pm 4,$ $\pm 3, \pm 2, \pm \frac{3}{2}, \pm 1, \pm \frac{1}{2}$

b.
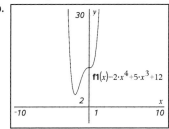

The function does not intersect the x-axis. 9. a. Answers vary. Sample: $y = x^2 - 2ax + (a^2 - p)$ b. Assume that $a + \sqrt{p}$ is rational. Therefore, it should be able to be expressed as $\frac{m}{n}$ and with $n \neq 1$ since $a + \sqrt{p}$ is not an integer. However, according to the Rational-Root Theorem, n is a factor of the leading coefficient, which is 1. Thus, $a + \sqrt{p}$ cannot be rational. 11. a. $0 \leq x \leq 9$ b. 0, 9, 15 c. The beam has no sag when $x = 0$ (the point is on the wall), or $x = 9$ (the point is on the support). $x = 15$ is a zero for the equation, but this value is not in the domain of x in this problem. 13. about $4.8°$

15. a.
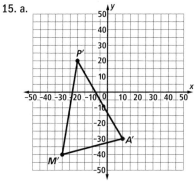

b.
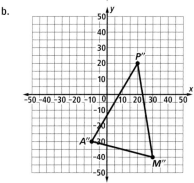

c. a rotation of $180°$ about the origin

Lesson 11-6 (pp. 766-771)

Mental Math a. 2 **b.** 2 **c.** 1 **d.** infinitely many
Activity Step 1: $P(3) = 3^3 + 2(3)^2 - 14(3) - 3 = 0$; the Factor Theorem **Step 2:** 2 **Step 3:** 2 roots; If $Q(x) = 0$, then $P(x) = (x - r) \cdot 0 = 0$. Thus, the roots of a factor of a polynomial are also the roots of the polynomial itself.
Step 4:

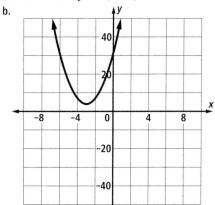

Questions
1. Every polynomial equation $P(x) = 0$ of any degree ≥ 1 with complex number coefficients has at least one complex number solution. 3. C 5. $x = -\frac{b}{a}$ 7. $x = -8$, rational; $x = -8$ multiplicity 2 9. $z = 3$ and $z = \frac{22}{7}$, rational; $z = 3 + 4i$ and $z = 3 - 4i$, nonreal; $x = 3$ multiplicity 4 11. 5 13. $x = 4 - 2i$ 15. a. The graph will be translated vertically when k is changed. b. $k = 40$ c. $k > 40$ 17. $p(x) = 3x^3 - 5x^2 - 58x + 40$
19. a. $V(w) = 4w^3 - 30w^2 + 50w$ b. no c. $w \approx 1.06$
21. 100 23. a. $y = 3(x + 3)^2 + 4$
b.

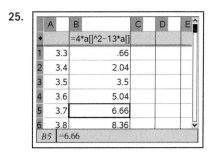

25.

	A	B		C	D	E
		=4*a[]^2−13*a[]				
1	3.3	.66				
2	3.4	2.04				
3	3.5	3.5				
4	3.6	5.04				
5	3.7	6.66				
6	3.8	8.36				
B5	=6.66					

$y = 6.66$

Lesson 11-7 (pp. 772-777)

Mental Math a. $(c^7 + 10)(c^7 - 10)$ **b.** $(r - 5)^2$
c. $\pi x^3 (x - 13)$ **d.** $(2p + 3q)^2$

Activity 1 Step 1: a-c.

A x	B	C	D	E
	=4*x^2−5*x−3			
1	1	-4		
2	2	3	7	
3	3	18	15	8
4	4	41	23	8
5	5	72	31	8
6	6	111	39	8

D3 $=c3-c2$

Step 2: a-b.

A x	B	C	D	E	F
	='x^3				
1	1	-3			
2	2	-1	2		
3	3	7	8	6	
4	4	27	20	12	6
5	5	65	38	18	6
6	6	127	62	24	6

E4 $=d4-d3$

Step 3: Answers vary. Sample: The fourth set of differences will all be the same.

Activity 2 Steps 1 and 2:

A x	B	C	D	E	F
	=3^x				
1	1	3			
2	2	9	6		
3	3	27	18	12	
4	4	81	54	36	24
5	5	243	162	108	72
6	6	729	486	324	216

C2 $=b2-b1$

Step 2: Answers vary. Sample: The method of finite differences does not produce a set of constant differences. Instead, the pattern of differences seems to repeat in each column.

Guided Example 2 a. 15; 31; 63; 127 **b.** first row: 15, 31, 63, 127; second row: 16, 32, 64; third row: 16, 32; will not; is not; polynomial

Questions

1. 3 **3.** yes; 12 **5.** Tanner is not using x-values that form an arithmetic sequence. **7. a.** yes **b.** 4 **9.** yes (degree $= 4$) **11. a.** b; $m + b$; $2m + b$; $3m + b$; $4m + b$ **b.** m, m, m, m; this is a linear function, thus, it has degree 1 and its first differences are equal. **13. a.** 4 by the Number of Roots of a Polynomial Equation Theorem **b.** $\pm1, \pm2, \pm5, \pm10, \pm\frac{1}{3}, \pm\frac{2}{3}, \pm\frac{5}{3}, \pm\frac{10}{3}$ **c.** $x \approx -1.33 + 0.78i, x \approx -1.33 - 0.78i, x \approx 0.71, x \approx 1.96$ **15. a.** $P(x) = -x^3 + 4x^2 - 3$ **b.** 3; $\frac{\sqrt{21} + 3}{2}$

17.

$$\text{solve}\left(\begin{bmatrix} 4 & -2 & 3 \\ 8 & -3 & 5 \\ 7 & -2 & 4 \end{bmatrix} \cdot \begin{bmatrix} x \\ y \\ z \end{bmatrix} = \begin{bmatrix} 1 \\ 4 \\ 5 \end{bmatrix}, x,y,z\right)$$

$$x{=}1 \text{ and } y{=}3 \text{ and } z{=}1$$

$x = 1, y = 3, z = 1$

Lesson 11-8 (pp. 778-785)

Mental Math a. Answers vary. Sample: 2 **b.** Answers vary. Sample: -8 **c.** 5 or -5 **d.** 4 or -4

Guided Example 2 9, 16, 25, 36; 5, 7, 9, 11; 2, 2, 2; 3; $\frac{1}{3}, \frac{1}{2}, \frac{1}{6}$; 0; $\frac{1}{3}n^3 + \frac{1}{2}n^2 + \frac{1}{6}n$

Questions

1. They used the method of finite differences. **3.** 190 **5.** 81 **7.** $12 = 4a + 2b + c$, $60 = 25a + 5b + c$, $162 = 64a + 8b + c$ **9.** $a = \frac{1}{2}, b = -\frac{1}{2}, c = 0$ **11.** $y = \frac{1}{2}x^2 + \frac{1}{2}x + 1$ **13. a.** $H(x) \approx -0.001x^4 + 0.019x^3 - 0.028x^2 - 0.313x$

b.

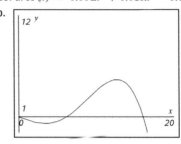

The graph does not have the same shape as the picture due to rounding error and using a limited number of points.

15. a. yes **b.** 3 **17.** \$40,784.87 **19. a.** $\log_7 6$ **b.** $\log 81$ **c.** $\log 2$

Self-Test (pp. 790-791)

1. To calculate the volume, we take the width of the box as $40 - 2x$, the length of the box as $30 - 2x$, and the height of the box as x. Since $V = \ell w h$, $V(x) = (30 - 2x) \cdot (40 - 2x)x = 4x^3 - 140x^2 + 1200x$. **2.** A; this is the only choice that contains a single term from all three factors. **3.** Use the Extended Distributive Property: $(2 + 5)(2^2 - 3z + 1) = (2 + 5)(2^2) + (2 + 5)(-3z) + (2 + 5)(1) = z^3 - 3z^2 - 15z + z + 5$. Now combine like terms and write the resulting terms in descending order: $z^3 + 2z^2 - 14z + 5$. **4.** The degree is the largest sum of the exponents of the variables in any term. A binomial is a polynomial with two terms. Answers vary. Sample: $3xy + z^4$ **5.** $680x^4 + 850x^3 + 1020x^2 + 1105x + 935$; the degree of x in each term is the number of years Francesca has held that summer's savings. **6.** Substitute 0.051 for $r, x = 1.051$; $680(1.051)^4 + 850(1.051)^3 + 1020(1.051)^2 + 1105(1.051) + 935 \approx \$5039.54$ **7. a.** The degree is the largest sum of the exponents of the variables in any term. The degree of $P(x)$ is 6. **b.** A polynomial of degree n has n complex roots. Thus $P(x)$ has 6 complex roots. **8.** According to the Rational-Root Theorem, if $\frac{a}{b}$ is a rational root of a polynomial function, then a is a factor of the polynomial's constant term and b is a factor of the coefficient of the leading term. Here, a is a factor of 11 and b is a factor of 4. Therefore, the possible rational roots are $\pm11, \pm\frac{11}{2}, \pm\frac{11}{4}, \pm1, \pm\frac{1}{2}$, and $\pm\frac{1}{4}$. **9. a.** The graph intersects the x-axis 3 times, so there are 3 real zeros. **b.** The function has a total of 5 complex zeros and 3 real zeros, so there are 2 nonreal zeros.

10. a. According to the Rational-Root Theorem, if $\frac{a}{b}$ is a rational root of a polynomial function, then a is a factor of the polynomial's constant term and b is a factor of the coefficient of the leading term. Here, a is a factor of 12, and b is a factor of 1. Therefore, the possible rational roots are $\pm 12, \pm 6, \pm 4, \pm 3, \pm 2$, and ± 1. **b.** Looking at the graph, there are two real zeros between consecutive integers that must be irrational because the rational roots are all integer values. The irrational zeros must lie between -1 and 0 and 3 and 4. **11.** Set each factor equal to 0 and solve for x: $(x - 2)^3 = 0, x - 2 = 0, x = 2$; $(11x + 37) = 0, 11x = -37$, $x = -\frac{37}{11}$; $x^2 - 7 = 0, x^2 = 7, x = \pm\sqrt{7}$. So the zeros of g are $2, -\frac{37}{11}, \sqrt{7}$, and $-\sqrt{7}$. **12.** The multiplicity of a zero r is the highest power of $x - r$ that appears as a factor of the polynomial. The multiplicity of 2 is 3, and the multiplicity of each other zero is 1. **13.** D; D is only true when $x = r$ or $x = s$. The rest are true for all x by the Factor Theorem and the Number of Roots of a Polynomial Equation Theorem. **14.** $\sqrt{21}$ is a solution to $x^2 = 21$ and a root of $x^2 - 21 = 0$. According to the Rational-Root Theorem, if $\frac{a}{b}$ is a rational root of a polynomial, then a is a factor of the polynomial's constant term and b is a factor of the coefficient of the leading term. Here, a is a factor of –21, and b is a factor of 1. Therefore, the possible rational roots are $\pm 21, \pm 7, \pm 3$, and ± 1. If we test these values, none are solutions for the equation $x^2 - 21 = 0$. Thus, $\sqrt{21}$ is irrational. **15. a.** Use a CAS: $(y - 3)(y^2 - 5)(y^2 + 3y + 9)$ **b.** Use a CAS or use difference of squares factoring and the quadratic formula on the quadratic factors you found in Part a: $(y - 3)(y + \sqrt{5})(y - \sqrt{5})(y + \frac{3}{2} + \frac{3}{2}i\sqrt{3}) \cdot (y + \frac{3}{2} - \frac{3}{2}i\sqrt{3})$ **16.** By the Factor Theorem, a is a zero

of r if $y - a$ is a factor of $r(y)$. So by your answer to 15b, the zeros are $3, \sqrt{5}, -\sqrt{5}, -\frac{3}{2} + \frac{3\sqrt{3}}{2}i$, and $-\frac{3}{2} - \frac{3\sqrt{3}}{2}i$
17. Factor out the greatest common monomial factor: $2vw(5uv + 12)$; these two factors are prime, so the polynomial is factored completely. **18.** For a polynomial to have a single root at $x = -2$ and a double root at $x = 6$, $x + 2$ must appear as a factor once and x - 6 must appear as a factor twice. Answers vary.
Sample: $p(x) = (x + 2)(x - 6)^2 = x^3 - 10x^2 + 12x + 72$.

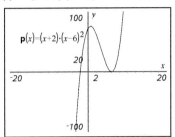

19. a. Yes, the 3rd differences are equal; therefore, f can be modeled by a polynomial. **b.** 3; the 3rd differences are equal, and the 2nd differences are not. **20.** From using the method of finite differences, we find that the 2nd set of differences is constant; therefore, these data can be modeled by a quadratic function. The function has the form $r = at^2 + bt + c$. To solve for the coefficients and constant term, we pick three data points and solve. For example, use (-2, 12), (0, 0) and (3, 12).

Solving the system $\begin{cases} 12 = 4a - 2b + c \\ 0 = c \\ 12 = 9a + 3b + c \end{cases}$ gives $a = 2, b = -2$,

$c = 0$. So, the equation is $r = 2t^2 - 2t$.

Self-Test Correlation Chart

Question	1	2	3	4	5	6	7	8	9	10
Objective(s)	I	A	A	E	H	H	E, F	G	K	G, K
Lesson(s)	11-2	11-2	11-2	11-1, 11-2, 11-3	11-1, 11-8	11-1, 11-8	11-1, 11-2, 11-3, 11-4, 11-6	11-5	11-4, 11-5	11-4, 11-5

Question	11	12	13	14	15	16	17	18	19	19
Objective(s)	C	C	F	G	B	C	B	J	D	D
Lesson(s)	11-4, 11-6	11-4, 11-6	11-4, 11-6	11-5	11-3	11-4, 11-6	11-3	11-1, 11-4, 11-5	11-7, 11-8	11-7, 11-8

Chapter Review (pp. 792–795)

1. $x^3 - x^2 + x + 3$ **3.** $x^3 - x^2y - xy^2 + y^3$ **5.** $5a^7b^6$; $8a^6$ **7.** $(x - 3)^2$ **9.** $(pq + 2)(pq - 2)(p^2q^2 + 4)$
11. not possible **13.** $(t - \frac{3}{2} + \frac{\sqrt{157}}{2})(t - \frac{3}{2} - \frac{\sqrt{157}}{2})$
15. $n = 0, n = -3, n = \frac{7}{8}$; no multiple roots **17.** $x = 0$, $x = -6$; $x = -6$ is a double root. **19.** 0, 5, -5 **21.** No; there are no constant differences in the 1st through 3rd

differences. **23.** No; there are no constant differences in the 1st through 3rd differences. **25. a.** 6 **b.** 18
27. B **29.** D **31. a.** yes **b.** no; $(x + \sqrt{42})(x - \sqrt{42})$
33. The left side of the equation is not a product.
35. true **37. a.** 3 **b.** 2 **39. a.** $\pm 3, \pm\frac{3}{2}, \pm 1, \pm\frac{1}{2}$ **b.** $x = 3$, $x = \frac{1}{2}$ **41. a.** $\pm 6, \pm 3, \pm 2, \pm 1, \pm\frac{3}{2}, \pm\frac{6}{5}, \pm\frac{3}{5}, \pm\frac{1}{2}, \pm\frac{2}{5}, \pm\frac{3}{10}$, $\pm\frac{1}{5}, \pm\frac{1}{10}$

b.

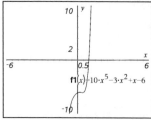

The range of possible rational roots is from $x = -6$ to $x = 6$. By restricting the graph to this range, we see that the graph of f only intersects the x-axis once in this interval, between $\frac{1}{2}$ and 1. The only possible rational root in this range is $\frac{3}{5}$, and $f(\frac{3}{5}) \neq 0$. Therefore, f has exactly one irrational zero. **43. a.** $G = n^2 - n$ **b.** 552 **45. a.** $250x^5$ $+ 250x^4 + 250x^3 + 250x^2 + 250x + 250$ **b.** $1743.83 **c.** about 4.99% **47.** $V(s) = 4s^3 - 16s^2 + 15s$ **49.** $v(x) = 4x^3 - 260x^2 + 4000x$

51. a.

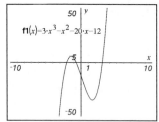

b. $f(x) = (x - 3)(x + 2)(3x + 2)$ **53. a.** Answers vary. Sample: $h(x) = x(x + 1)(x - 1)(x - 3)$ **b.** $h(x) = 3x^4 - 9x^3 - 3x^2 + 9x$

c.

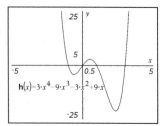

55. $-4, -3, -3, -1$ **57. a.** $\pm 6, \pm 3, \pm 2, \pm\frac{3}{2}, \pm 1, \pm\frac{1}{2}$ **b.** You should test ± 1 and $-\frac{3}{2}$ because the graph indicates that the function does not cross the x-axis around the other values. **c.** There are no rational roots.

Chapter 12

Lesson 12-1 (pp. 798–803)

Mental Math a. $\log_{16} 1742$ **b.** $\log_3 20$ **c.** $r = 3e^t$ **d.** All are equivalent.

Activity Step 1:

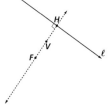

Step 2:

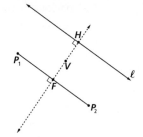

Step 3: a.

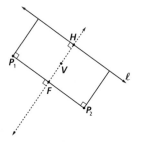

b. FP_1 = perpendicular distance from P_1 to ℓ. P_1 is on the parabola with focus F and directrix ℓ. **c.** FP_2 = perpendicular distance from P_2 to ℓ. P_2 is also on the parabola.

Step 4:

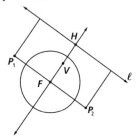

Step 5: a.

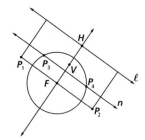

b. Yes; they are equidistant from F and ℓ. (Each distance is r.)

Step 6:

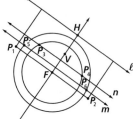

Step 7:

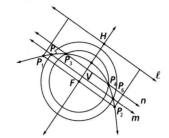

Questions

1. a. No; the focus is defined as a point not on the directrix.
b. No; the vertex must be equidistant from the focus and directrix, and a point on the directrix has distance zero from the directrix. **3.** false

5. a.

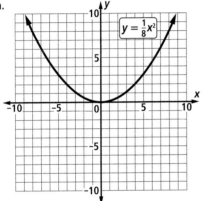

b. focus: $(0, 2)$; vertex: $(0, 0)$; directrix: $y = -2$
c. The distance from $(5, 3.125)$ to $(0, 2)$
is $\sqrt{(5 - 0)^2 + (3.125 - 2)^2} = 5.125$, and the distance from $(5, 3.125)$ to $y = -2$ is $3.125 - (-2) = 5.125$.
d. Answers vary. Sample: $(8, 8)$; $\sqrt{(8 - 0)^2 + (8 - 2)^2} = 10 = 8 - (-2)$ **7.** $y = -\frac{1}{12}x^2$ **9.** $\left(0, -\frac{1}{8}\right)$; $y = \frac{1}{8}$
11. $\left(-5, \frac{25}{4}\right)$; $y = \frac{23}{4}$ **13. a.** down **b.** $\left(0, -\frac{1}{20}\right)$ **15. a.** $x = \frac{1}{20}y^2$
b. Answers vary. Sample: $(0, 0)$, $(20, 20)$, $(20, -20)$
17. a. $y = \pm\sqrt{13}$ **b.** $y = -3 \pm \sqrt{13}$ **c.** $y = 169$ **d.** $y = 166$
19. a. $\frac{4}{7}$ **b.** 1000 **21.** C

Lesson 12-2 (pp. 804–809)

Mental Math a. upward **b.** $4.5 \frac{m}{sec}$ **c.** 25 m **d.** yes
Guided Example 2 $12; \frac{17}{2}, \frac{27}{2}; 12; \frac{17}{2}, \frac{729}{4}$
Guided Example 5 $1; 1; 24; 25; y - 5; 24; 5; 2\sqrt{6}; 5; -2\sqrt{6};$
$5 + 2\sqrt{6}; 5 - 2\sqrt{6}; 5$

Questions

1. $x^2 + y^2 = 230{,}400$ **3.** 34 **5.** Answers vary.
Sample: $x^2 + y^2 = 25$, $x^2 + y^2 = 16$

7. center: $(0, 0)$; radius: 11

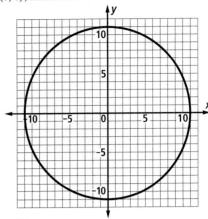

$y = \pm\sqrt{121 - x^2}$ **9. a.** There are no points on the circle where $x = -6$. **b.** There are no points on the circle where $x = 800$. **11. a.** $x^2 + y^2 = 6.25$ **b.** 47,901.4 mph
13. a. $(x - 4)^2 + (y + 1)^2 = 25$ **b.** Answers vary.
Sample: $(4, 3)$; image is $(8, 2)$, $(8 - 4)^2 + (2 + 1)^2 = 25$
15. a. $(0, 0)$; maximum **b.** $y = -\frac{1}{4}x^2$ **c.** $x = 0$
17. $-4, -1, 1, 4$ **19.** $y = -\frac{7}{3}x$

Lesson 12-3 (pp. 810–815)

Mental Math a. 4 **b.** $-2, 7, -\frac{3}{2}$ **c.** 7
Guided Example 3 a. $(0, 0)$; 2.5; $x^2 + y^2 = 6.25$; $(0, 0)$; 9;
$x^2 + y^2 = 81$; $x^2 + y^2$; 6.25; $x^2 + y^2$; 81 **b.** 6.25π; 81π; 81π;
6.25π; 74.75π; 234.83

Questions

1. a.

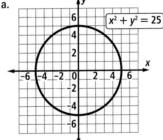

b.

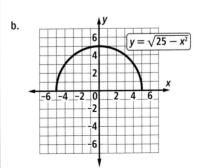

c.

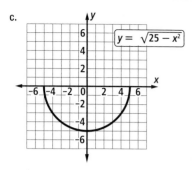

$y = \sqrt{25 - x^2}$

3. 9 ft 8 in.   **5.** the interior of the circle C

7. $(x - 2)^2 + (y - 3)^2 > 9$ **9. a.** about 15 ft 5 in.

b. about 14 ft 10 in.

11. $\begin{cases} x < 3.35 \\ x > -3.35 \\ y < 3.35 \\ y > -3.35 \\ x^2 + y^2 > 2.275^2 \end{cases}$

13. For a line ℓ and a point F not on ℓ, a parabola is the set of all points in the plane of ℓ and F equidistant from F and ℓ.

15. a. 13 **b.** $x^2 + y^2 = 169$ **17. a.** $x^2 + 16x + y^2 + 64$

b. $x^2 + 2xy + y^2 + 64$ **c.** $x^2 + 2xy + 16x + y^2 + 16y + 64$

19. a. $0 \le x \le 8$ **b.** $0 \le y \le 4$ **21.** No; the pieces do not satisfy the Triangle Inequality. To make a triangle, the sum of the measures of any two sides of the triangle should always be greater than the measure of the third side. Here, $1.5 + 2.3 < 4$.

Lesson 12-4 (pp. 816–823)

Mental Math a. 17 **b.** 11 **c.** 9

Activity Steps 1–3:

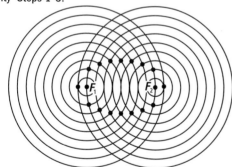

Step 4:

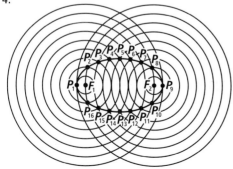

Guided Example 1 a. ± 3; 0; ± 2 **b.** 1; -2; 2 **c.** y-axis; $(x, -y)$;

y; $-y$; $\dfrac{x^2}{9} + \dfrac{(-y)^2}{4} = 1$; x-axis

Guided Example 2 10; 5; 6; 3; -5; 5; -3; 3; -4; 4; 25; 9

Questions

1. 10   **3. a.** I, J **b.** $\overline{AB}$ **c.** $\overline{CD}$ **5.** major axis: $(0, -b)$, $(0, b)$;

minor axis: $(-a, 0)$, $(a, 0)$ **7.** $-a \le x \le a$; $-b \le y \le b$

9.

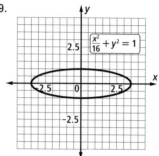

$\dfrac{x^2}{16} + y^2 = 1$

11. a. $\dfrac{x^2}{64} + \dfrac{y^2}{16} = 1$ **b.** $\dfrac{x^2}{64} + \dfrac{y^2}{16} < 1$ **13. a.** both foci at $(0, 0)$;

focal constant 20 **b.** The ellipse is a circle.

15. a. about 43.4 million mi **b.** about 28.6 million mi

c. about 14.8 million mi

17. a.

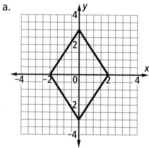

b. major axis: y-axis from -3 to 3; minor axis: x-axis from -2 to 2 **c.** x-intercepts: -2, 2; y-intercepts: -3, 3

d.

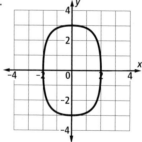

major axis: y-axis from -3 to 3; minor axis: x-axis from -2 to 2 **e.** Answers vary. Sample: They have the same x- and y-intercepts and endpoints of the major and minor axes. The graph from Part a is a polygon made of line segments, whereas the graph from Part d has smooth curves.

19. a. $x^2 + y^2 = 9$ **b.** $\left(x - \dfrac{3\sqrt{2}}{2}\right)^2 + \left(y - \dfrac{3\sqrt{2}}{2}\right)^2 = \dfrac{9}{2}$

c. $x^2 + y^2 < 9$ **d.** A: 6π; B: $3\pi\sqrt{2}$ **e.** A: 9π; B: $\dfrac{9\pi}{2}$

21. $b \approx 35.9$ **23.** $(7x, 8y)$

Lesson 12-5 (pp. 824–830)

Mental Math a. $T_{7,-2}$ **b.** $S_{3,2.5}$ **c.** $T_{7,-2} \circ S_{3,2.5}$
d. $r_{y=x} \circ T_{7,-2} \circ S_{3,2.5}$

Guided Example $2y$; $\frac{x'}{4}$; $\frac{y'}{2}$; $\frac{x'}{4}$; $\frac{y'}{2}$; $\frac{x}{4}$; $\frac{y}{2}$; 4; 8; 4^2; 2^2; 1; 0; 1; 4^2; 2^2; 0; 1; 1

Questions

1. C **3.** true **5. a.** $(5,0)$ **b.** $\left(0, \frac{1}{2}\right)$ **c.** $\left(\frac{x}{5}\right)^2 + (2y)^2 = 1$
7. a. $S_{4,8}$ **b.** $\left(\frac{x}{4}\right)^2 + \left(\frac{y}{8}\right)^2 = 1$ **c.** 32π **9.** Answers vary.
Samples: **a.** $S_{2,2}$ **b.** $S_{1,2}$ **11.** $19{,}973\pi \approx 62{,}746$ million mi^2
13. $2b = 8\sqrt{2} \approx 11.31$ **15.** Answers vary. Samples: $\left(\frac{x}{24}\right)^2 +$
$y^2 = 1$; $\left(\frac{x}{12}\right)^2 + \left(\frac{y}{2}\right)^2 = 1$; $\left(\frac{x}{6}\right)^2 + \left(\frac{y}{4}\right)^2 = 1$; $x^2 + \left(\frac{y}{24}\right)^2 = 1$
17. iii **19.** ii **21.** i **23.** $n = -0.25$ or $n = -0.75$
25. $y = -\frac{3}{7}x + \frac{29}{7}$

Lesson 12-6 (pp. 831–837)

Mental Math a. -1 **b.** $\frac{1}{4}$ **c.** no

Activity Step 1: 6; 6; 4; 4; 2; 2; yes

Step 2:

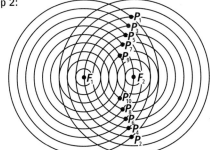

Step 3:

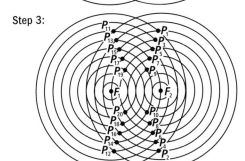

Step 4:

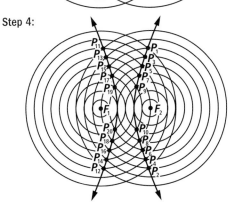

Questions

1. $\frac{x^2}{a^2} - \frac{y^2}{b^2} = 1$ where $b^2 = c^2 - a^2$; $(a, 0)$; $(-a, 0)$
3. $(1, 0)$, $(-1, 0)$; $(-\sqrt{2}, 0)$, $(\sqrt{2}, 0)$; $y = x$, $y = -x$
5. false **7.** Answers vary. Sample: If x is very large, then
the difference between $\sqrt{x^2 - 1}$ and $\sqrt{x^2}$ is very small.
9. Answers vary. Sample: It does not include the
asymptotes for the parts of the hyperbola in the second
and third quadrants. **11. a.** about 6.83 **b.** $\frac{x^2}{11.7} - \frac{y^2}{4.3} = 1$
c.

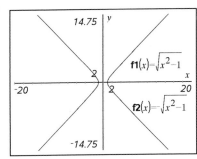

13. $y = \pm\sqrt{x^2 - 1}$

15.

17. a. no closer than 2 ft **b.** $2400\pi \approx 7539.82$ ft^3

Lesson 12-7 (pp. 838–843)

Mental Math a. $45°$ **b.** $60°$ **c.** $72°$
Guided Example 3 $-\sqrt{2k}$; $-\sqrt{2k}$; 8; 8; 8; 64; 32; 32; $\frac{32}{x}$

Questions

1. yes; $x^2 + y^2 - 6x + 14y - 60 = 0$ **3.** yes; $x^2 - 6xy +$
$y^2 + \pi y - \sqrt{11} = 0$ **5. a.** $\left(\sqrt{238}, \sqrt{238}\right)$, $\left(-\sqrt{238}, -\sqrt{238}\right)$
b. $2\sqrt{238} \approx 30.85$ **7.** $w = 0$, $d = 0$

9.

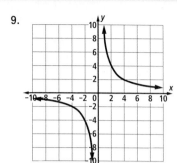

foci: (4, 4), (–4, –4); vertices: $\left(2\sqrt{2}, 2\sqrt{2}\right), \left(-2\sqrt{2}, -2\sqrt{2}\right)$; asymptotes: $x = 0$, $y = 0$; focal constant: 8
11. $\frac{1}{4}x^2 - \frac{1}{9}y^2 - 1 = 0$ (or $9x^2 - 4y^2 - 36 = 0$); $A = \frac{1}{4}$, $B = 0$, $C = \frac{1}{9}$, $D = 0$, $E = 0$, $F = -1$ (or $A = 9$, $B = 0$, $C = 4$, $D = 0$, $E = 0$, $F = -36$) **13. a.** Yes; the asymptotes $x = 0$ and $y = 0$ are perpendicular. **b.** Yes; the asymptotes $y = x$ and $y = -x$ are perpendicular. **c.** No; the asymptotes $y = \frac{3}{2}x$ and $y = -\frac{3}{2}x$ are not perpendicular.

15. a.

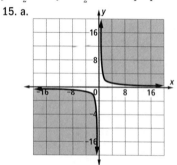

b.

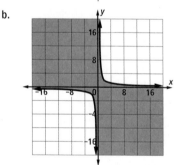

17. C **19.** B **21.** about 401 ft **23.** $0.90

Lesson 12-8 (pp. 844–849)

Mental Math a. (0, 0) **b.** (–2, 4) and (2, 4) **c.** (0, 0) and (1, 1) **d.** no intersections

Activity Step 1:

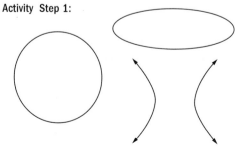

Step 2:

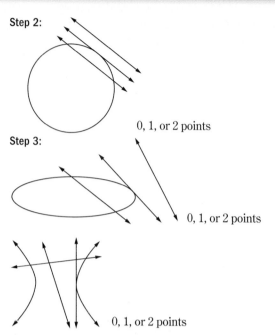

0, 1, or 2 points

Step 3:

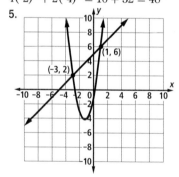

0, 1, or 2 points

0, 1, or 2 points

Step 4: 0, 1, or 2; 0, 1, or 2; 0, 1, or 2
Guided Example 3 0; x; x^2; x; 2; 1; –7; 2; 0
Questions
1. 0, 1, or 2 **3. a.** 2 **b.** (–2, –4), (2, 4) **c.** $4 = 2 \cdot 2$; $-4 = 2 \cdot -2$; $4(2)^2 + 2(4)^2 = 16 + 32 = 48$; $4(-2)^2 + 2(-4)^2 = 16 + 32 = 48$
5.

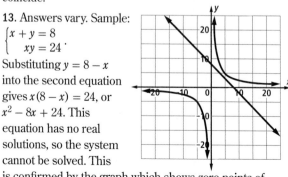

estimated solutions: (–3, 2), (1, 6); exact solutions: (–3, 2), (1, 6) **7.** (1, –2), $\left(-\frac{1}{2}, -\frac{7}{2}\right)$ **9.** $-16 = (-6)^2 + 7 \cdot -6 - 10$; $-16 = -5(-6) - 46$ **11.** A quadratic-linear system can have at most 2 solutions because a straight line and a quadratic curve cannot coincide.

13. Answers vary. Sample: $\begin{cases} x + y = 8 \\ xy = 24 \end{cases}$.
Substituting $y = 8 - x$ into the second equation gives $x(8 - x) = 24$, or $x^2 - 8x + 24$. This equation has no real solutions, so the system cannot be solved. This is confirmed by the graph which shows zero points of intersection. **15.** (3.79, 1.90), (–3.79, –1.90) **17.** $y = \frac{50}{x}$

19. a. difference of squares **b.** $(r^{10} + t)(r^{10} - t)$
21. a. sum of squares **b.** not possible **23.** $-\frac{7}{2} + 8i$

Lesson 12-9 (pp. 850–855)

Mental Math a. $n > 4t$ **b.** $\ell \geq 30$ **c.** $85.9 \leq s \leq 86.9$,
or $|86.4 - s| \leq 0.5$ **d.** $a < 7$

Activity

	Parabola	Circle	Ellipse	Hyperbola
Parabola	0, 1, 2, 3, 4, or infinite	0, 1, 2, 3, or 4	0, 1, 2, 3, or 4	0, 1, 2, 3, or 4
Circle		0, 1, 2, or infinite	0, 1, 2, 3, 4, or infinite	0, 1, 2, 3, or 4
Ellipse			0, 1, 2, 3, 4, or infinite	0, 1, 2, 3, or 4
Hyperbola				0, 1, 2, 3, 4, or infinite

Guided Example 2 ellipse; hyperbola; 9; 3; 15; 13; 52; 4; 2; 2; 2; $\sqrt{7}$; -2; -2; $\sqrt{7}$; 2; $\sqrt{7}$; 2; -$\sqrt{7}$; -2; $\sqrt{7}$; -2; -$\sqrt{7}$

Questions

1. 4 **3.** (0, 3), (0, -3) **5. a.** 2 **b.** (-1.59, 2.54) and (1.59, 2.54)
7. $42 in the first month and $60 in the second month
9. (-4, 3), (4, 3), (-3, -4), (3, -4) **11.** (-1, -6)

13. a.
$$\begin{cases} x^2 + y^2 = 1{,}000{,}000 \\ (x - 200)^2 + (y - 800)^2 = 160{,}000 \\ (x - 1000)^2 + (y - 1100)^2 = 250{,}000 \end{cases}$$

b. $x = 600$, $y = 800$; The net is 600 m east and 800 m north of the first boat.

15. $\begin{cases} x + y = 37 \\ xy = 300 \end{cases}$; The solutions are $x = 12$ and $y = 25$
or $x = 25$ and $y = 12$. Mr. Hwan's age is 25 yr, and his
baby's is 12 mo. **17.** yes; $13y^2 - \frac{1}{2}x + \sqrt{7}y = 0$
19.

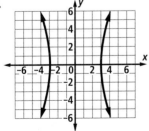

21. Answers vary. Sample: Transformations map each
point of a preimage onto a unique point in the image.

23. $\begin{bmatrix} -9 & 14 \\ -8 & 14 \end{bmatrix}$

Self-Test (pp. 860–861)

1. a. $\frac{1}{25}x^2 - \frac{1}{144}y^2 - 1 = 0$ (or $144x^2 - 25y^2 - 3600 = 0$)
b. $A = \frac{1}{25}, B = 0, C = -\frac{1}{144}, D = 0, E = 0, F = -1$
(or $A = 144, B = 0, C = -25, D = 0, E = 0, F = -3600$)
c. The equation is of the form $\frac{x^2}{a^2} - \frac{y^2}{b^2} = 1$, so it represents
a hyperbola. **2.** The interior is the set of points whose
distance from (3, -5) is less than 7, so the inequality is
$(x - 3)^2 + (y + 5)^2 < 49$. **3.** The circle's center shifts to
(-2, 5) under the transformation. By the Graph-Translation
Theorem, the equation for the image is $(x + 2)^2 +$
$(y - 5)^2 = 1$. **4. a.** $\left(0, \frac{1}{4 \cdot \frac{1}{11}}\right) = \left(0, \frac{1}{\frac{4}{11}}\right) = \left(0, \frac{11}{4}\right)$ **b.** (0, 0)
c. $y = -\frac{1}{4 \cdot \frac{1}{11}} = -\frac{1}{\frac{4}{11}} = -\frac{11}{4}$ **5.** The major and minor axes of a
circle will be the same length, and the foci will coincide at
a single point (the center of the circle).

6. a. $\begin{cases} (n + 70)(p - 2) = 5880 \\ np = 5600 \end{cases}$; $np + 70p - 2n - 140 = 5880$;
$5600 + 70p - 2\frac{5600}{p} - 140 = 5880$; $70p - \frac{11{,}200}{p} = 420$;
$70p^2 - 420p - 11{,}200 = 0$; $p^2 - 6p - 160 = 0$; $p = -10$
or $p = 16$, but since price cannot be negative, the original
price was $p = 16$. The new price is $14. **b.** 420; $n = \frac{5600}{p} =$
$\frac{5600}{16} = 350$; this year he sold $n + 70 = 420$ pairs. **7.** $a = 10$
and $b = 7$, so an equation is $\frac{x^2}{10^2} + \frac{y^2}{7^2} = \frac{x^2}{100} + \frac{y^2}{49} = 1$.
8. $A = \pi ab = \pi(10 \cdot 7) = 70\pi$ **9.** $S_{10,7}$; The image of
the circle with equation $x^2 + y^2 = 1$ under $S_{10,7}$
is $\frac{x^2}{100} + \frac{y^2}{49} = 1$ by the Circle Scale-Change Theorem.
10.

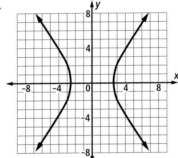

The equation represents a hyperbola with foci at $(\sqrt{41}, 0)$,
and $(-\sqrt{41}, 0)$, a focal constant of 10, and vertices (-5, 0)
and (5, 0). The asymptotes are $y = \pm\frac{4}{5}x$.
11.

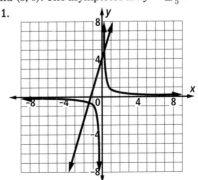

Substitute $2x + 5$ for y in the first equation: $x(2x + 5) = 2$; $2x^2 + 5x - 2 = 0$; Use the quadratic formula to solve for x: $x = \frac{-5 \pm \sqrt{41}}{4}$, or $x \approx 0.35$ or $x \approx -2.85$. Substitute these values into $xy = 2$ to find the y-values: $y \approx 5.70$ or $y = -0.70$. So, the points of intersection are $(0.35, 5.70)$ and $(-2.85, -0.70)$.

12. Substitute $x^2 - 4x + 3$ for y in the second equation: $x^2 + (x^2 - 4x + 3)^2 = 9$; $x^2 + x^4 - 8x^3 + 22x^2 - 24x + 9 = 9$; $x^4 - 8x^3 + 23x^2 - 24x = 0$; $x(x^3 - 8x^2 + 23x - 24) = 0$; $x(x - 3)(x^2 - 5x + 8) = 0$; $x = 0$ or $x = 3$ by the Zero-Product Theorem because $x^2 - 5x + 8 = 0$ has no real solutions. When $x = 0$, $y = 0^2 - 4(0) + 3 = 3$. When $x = 3$, $y = 3^2 - 4(3) + 3 = 9 - 12 + 3 = 0$. So, the solutions to the system are $(0, 3)$ and $(3, 0)$. **13. a.** hyperbola
b. The foci of the hyperbola with equation $xy = 15$ are $\left(\sqrt{2 \cdot 15}, \sqrt{2 \cdot 15}\right) = \left(\sqrt{30}, \sqrt{30}\right)$ and $\left(-\sqrt{2 \cdot 15}, -\sqrt{2 \cdot 15}\right) = \left(-\sqrt{30}, -\sqrt{30}\right)$, and the focal constant is $2\sqrt{30}$. The graph of $xy = -15$ is the reflection image of the graph of $xy = 15$ over the x-axis, so the foci for $xy = -15$ are $\left(\sqrt{30}, -\sqrt{30}\right)$ and $\left(-\sqrt{30}, \sqrt{30}\right)$, and the focal constant is still $2\sqrt{30}$. **c.** Yes; the hyperbola is symmetric over $y = \pm x$. **14.** Area of pool: $A = \pi\left(\frac{15}{2} \cdot 4\right) = 30\pi$; Area of walkway and pool: $A = \pi\left(9 \cdot \frac{11}{2}\right) = 49.5\pi$; Area of walkway: $49.5\pi - 30\pi = 19.5\pi \approx 61.26 \text{ m}^2$

15. Solve the system $\begin{cases} \ell w = 104{,}094 \\ 2\ell + 2w = 1320 \end{cases}$ by substitution.

$w = \frac{104{,}094}{\ell}$; $2\ell + 2\left(\frac{104{,}094}{\ell}\right) = 1320$; $2\ell^2 + 208{,}188 = 1320\ell$; $\ell^2 - 660\ell + 104{,}094 = 0$; $\ell \approx 260.7$ or 399.3. Since addition and multiplication are commutative, solving for w would give the same answers. Colorado is longer than it is wide, so $\ell = 399.3$ and $w = 260.7$. The dimensions of Colorado are approximately 399.3 mi wide by 260.7 mi long.
16. $(5.3 \cdot 10^9) + (9 \cdot 10^7) = 5.39 \cdot 10^9 \text{ km}$ **17.** Answers vary. Sample: A parabola will intersect a circle three times if its vertex is on the highest point of the circle and it opens down. This is the case for the circle and the parabola described by $x^2 + y^2 = 1$ and $y = 1 - x^2$.

18. This set of points is a circle with radius 8 and center $(3, 2.5)$ by the definition of a circle. It has equation $(x - 3)^2 + (y - 2.5)^2 = 64$ because (h, k) is $(3, 2.5)$ and $r = 8$.

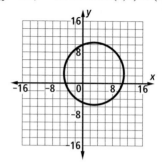

19. If the opening has center $(0, 0)$ and a radius of 7, we can describe the arch with the equation $y = \sqrt{49 - x^2}$. Substitute the spelunker's height for y to find the place farthest from the center that he can stand: $5\frac{8}{12} = 5\frac{2}{3} = \sqrt{49 - x^2}$; $\frac{289}{9} = 49 - x^2$; $x^2 = 49 - \frac{289}{9}$; $x = \sqrt{\frac{152}{9}} \approx \pm 4.1$. The spelunker can stand at most 4.1 ft away from the center.

20.

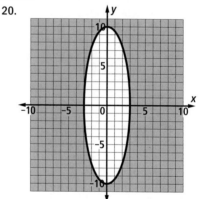

Self-Test Correlation Chart

Question	1	2	3	4	5	6	7
Objective(s)	A, F	B	B	E	F	H	I
Lesson(s)	12-1, 12-2, 12-4, 12-5, 12-6, 12-7	12-1, 12-2, 12-3, 12-4, 12-6, 12-7	12-1, 12-2, 12-3, 12-4, 12-6, 12-7	12-1, 12-2, 12-4, 12-6, 12-7	12-1, 12-2, 12-4, 12-5, 12-6	12-8, 12-9	12-2, 12-4, 12-6, 12-7

Question	8	9	10	11	12	13	14
Objective(s)	C	F	E, I	K	D	E, F	G
Lesson(s)	12-5	12-1, 12-2, 12-4, 12-5, 12-6	12-1, 12-2, 12-4, 12-6, 12-7	12-8, 12-9	12-8, 12-9	12-1, 12-2, 12-4, 12-5, 12-6, 12-7	12-2, 12-3, 12-4, 12-5

Question	15	16	17	18	19	20
Objective(s)	H	G	K	B, L	G	J
Lesson(s)	12-8, 12-9	12-2, 12-3, 12-4, 12-5	12-8, 12-9	12-1, 12-2, 12-3, 12-4, 12-6, 12-7	12-2, 12-3, 12-4, 12-5	12-3

Chapter Review (pp. 862–865)

1. $x^2 + y^2 - 14x + 4y - 28 = 0$; $A = 1, B = 0, C = 1$, $D = -14, E = 4, F = -28$ **3.** $\frac{1}{36}x^2 - \frac{1}{16}y^2 - 1 = 0$ (or $16x^2 - 36y^2 - 576 = 0$); $A = \frac{1}{36}, B = 0, C = -\frac{1}{16}$, $D = 0, E = 0, F = -1$ (or $A = 16, B = 0, C = -36, D = 0$, $E = 0, F = -576$) **5.** $x^2 + y^2 - 169 = 0$; $A = 1, B = 0$, $C = 1, D = 0, E = 0, F = -169$ **7.** $x^2 + y^2 = 144$

9. $b = \sqrt{35 - a^2}$ **11.** $x^2 + y^2 > 64$ **13.** $\frac{x^2}{16} + \frac{y^2}{49} = 1$

15. $y = \frac{4}{x}$ **17.** $70\pi \approx 219.91$ **19.** the ellipse by 5π or 15.71 units^2 **21.** $(0, 8)$ or $\left(\frac{7}{2}, \frac{81}{4}\right)$ **23.** no solutions

25. $a = \frac{-4}{3}$ and $b = -3$ or $a = 1$ and $b = 4$

27. a. $\begin{cases} ab = 1073 \\ (a + 3)(b - 7) = 960 \end{cases}$

b. $-\frac{111}{7}$ and $-\frac{203}{3}$ or 29 and 37 **29.** $(0, 0)$; 19 **31. a.** $\left(0, \frac{7}{4}\right)$

b. $(0, 0)$ **c.** $y = -\frac{7}{4}$ **33. a.** $\left(0, 2\sqrt{21}\right), \left(0, -2\sqrt{21}\right)$ **b.** 20

35. ellipse **37. a.** hyperbola **b.** parabola **c.** ellipse **d.** circle

39. true **41. a.** $\frac{x^2}{25} + \frac{y^2}{64} = 1$ **b.** ellipse

43. $\begin{cases} x^2 + y^2 \le 250{,}000 \\ x^2 + y^2 \ge 235{,}225 \end{cases}$ **45. a.** If the moving van is in a lane on a two-way road, it will not be able to pass through the tunnel because the tunnel is only 11.5 ft tall 6 ft from the center. The outer corner of the truck would not fit. **b.** Yes, the tunnel is 12.6 ft tall 3 ft from the center, so if the truck drives in the center of the roadway, it will be able to fit through the tunnel with more than 0.5 ft to spare.

47. a. $\begin{cases} x^2 + y^2 \le 5625 \\ (x + 30)^2 + (y - 50)^2 = 5625 \end{cases}$

b. $(-74.25, -10.55)$ or $(44.25, 60.55)$ **49.** The price was $60 in the first month and $80 in the second month.

51.

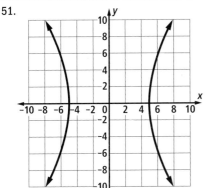

53.

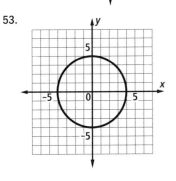

55. $\frac{x^2}{25} + \frac{y^2}{100} = 1$ **57.** B

59.

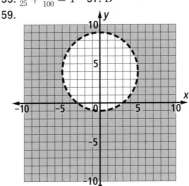

61. Answers vary. Sample: $(x + 1)^2 + y^2 = 9, \frac{x^2}{4} - y^2 = 1$

63. The system is: $\begin{cases} x + y = 17 \\ xy = 73 \end{cases}$. It appears on the graph of the system that the functions come very close to each other and might intersect. But, in fact, they do not intersect because there is no real solution to the system. Substituting $y = 17 - x$ into the second equation gives $x(17 - x) = 73$, or $x^2 - 17x + 73 = 0$, which has no real solution. So, the situation is impossible.

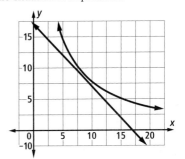

65.

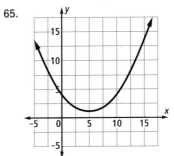

67.

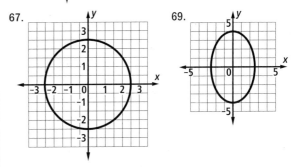

69.

71.

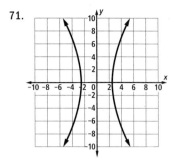

Chapter 13

Lesson 13-1 (pp. 868–875)

Mental Math a. –9; –6; –3 b. –18 c. 291; 294; 297 d. 882
Guided Example 1 12; 12; 78; 78
Guided Example 3 a. 1; 2; 6; $(4 \cdot 3 + 6)$; $(4 \cdot 500 + 6)$; 10; 14; 18; 2006 b. 10; 4; 500; 500; 10; 500; 4; 504,000

Questions

1. arithmetic series   3. Add the whole numbers from 1 to 100; 5050   5. sigma   7. a. $a_n = a_1 + (n-1)d$

b. $\sum_{i=1}^{n} a_i = \frac{n}{2}(2a + (n-1)d)$ c. $\sum_{i=1}^{n} a_i = \frac{n}{2}(a_1 + a_n)$

9. $\frac{500}{2}(10 + 2006) = 504,000$ 11. C   13. 1850

15. the number of terms   17. $2850   19. a. $\sum_{i=1}^{100} i^2$

b. 338,350   21. $\frac{1}{n}\sum_{i=1}^{n} a_i$ 23. a. –2.5 b. 625

c. $a_n = 16(-2.5)^{n-1}$ 25. $y = -\frac{11}{8}x^2 + 7x - \frac{5}{8}$

Lesson 13-2 (pp. 876–882)

Mental Math a. yes b. no c. yes d. no
Activity Steps 1–2:

In Step 2, half of the square has been shaded.
Step 3: This region represents $\frac{1}{4}$ of the original square, so $\frac{3}{4}$ of the original square has been shaded.
Step 4:

$\frac{1}{2} + \frac{1}{4} + \frac{1}{8} + \frac{1}{16}$;
$\frac{1}{2} + \frac{1}{4} + \frac{1}{8} + \frac{1}{16} + \frac{1}{32}$;
$\frac{7}{8}, \frac{15}{16}, \frac{31}{32}$

Step 5: $\frac{63}{64}$; the nth term of the series is $\frac{1}{2^n}$, and

$S_n = \sum_{i=1}^{n}\left(\frac{1}{2}\right)^n = \frac{2^n - 1}{2^n}$. **Step 6:** The original square will never be entirely shaded because $S_n \neq 1$ for any finite value of n.

Guided Example 1 a. $\frac{1}{16}; \frac{1}{32}; \frac{1}{64}; \frac{1}{128}; \frac{1}{256}; \frac{1}{512}; \frac{1}{1024}$

b. 10; $k = 1$; $\left(\frac{1}{2}\right)^k$ c. $\frac{1023}{1024}$; 0.999

Guided Example 2 a. $-\frac{1}{3}; -\frac{1}{3}; -\frac{1}{3}; 2; -\frac{1}{3}; 3; 27; -\frac{1}{3}; 4; 27; -9; 3;$

$-1; \frac{1}{3}$ b. 27; $-\frac{1}{3}; 5; -\frac{1}{3}; \frac{61}{3}$ **Check:** 27; –9; 3; –1; $\frac{1}{3}; \frac{61}{3}$

Questions

1. $\frac{255}{256}$ 3. a. 8 b. $\sum_{k=1}^{8} 5(2)^{k-1}$ c. 1275   5. a. 6

b. $\sum_{k=1}^{6} 170\left(-\frac{1}{10}\right)^{k-1}$ c. ≈ 154.55 7. a. i. 18 ii. 21 iii. 20

iv. $\frac{61}{3}$ b.

12 14 16 18 20 22 24 26 28 30

S_n is between S_{n-1} and S_{n-2}.

9. a. $500(1.04) + 500(1.04)^2 + \ldots + 500(1.04)^7$

b. $\sum_{k=1}^{7} 500(1.04)^k$ c. $4107.11

11. a. $r^2 - 1$; $(r-1)(r^2 + r + 1)$;
$(r-1)(r+1)(r^2 + 1)$; $(r-1)(r^4 + r^3 + r^2 + 1)$;
$(r-1)(r+1)(r^2 + r + 1)(r^2 - r + 1)$
b. 1; 0; Suppose $P(r) = r^n - 1$, then $P(1) = 1^n - 1 = 0$ for all n. Thus, by the Factor Theorem, $r - 1$ is a factor of $P(r) = r^n - 1$ for all n. c. $n = 2: \frac{r^n - 1}{r - 1} = r + 1$; $n = 3$:
$\frac{r^n - 1}{r - 1} = r^2 + r + 1$; $n = 4: \frac{r^n - 1}{r - 1} = (r+1)(r^2 + 1)$;
$n = 5: \frac{r^n - 1}{r - 1} = r^4 + r^3 + r^2 + 1$; $n = 6: \frac{r^n - 1}{r - 1} =$
$(r+1)(r^2 + r + 1)(r^2 - r + 1)$ 13. a. x^5

b. $S_{15} = \frac{x^7(1 - x^{75})}{1 - x^5}$

c.

expand $\left(\dfrac{x^7 \cdot \left(1 - x^{75}\right)}{1 - x^5}\right)$

$x^{77} + x^{72} + x^{67} + x^{62} + x^{57} + x^{52} + x^{47} + x^{42} \blacktriangleright$

15. 15,050   17. 4230   19. a. $f(-3) = 27^{-3} = \frac{1}{19,683}$;

$f(0) = 27^0 = 1$; $f\left(\frac{2}{3}\right) = 27^{\frac{2}{3}} = 9$ b. domain: set of all real numbers; range: set of positive real numbers
c. $f(x) = \log_{27} x$ 21. $y - 5 = -\frac{7}{2}(x - 6)$

Lesson 13-3 (pp. 883–888)

Mental Math a. 96 b. 36 c. 111
Activity Step 1:

	Number of Words	
Sentence Number	Paragraph 1	Paragraph 2
1	12	27
2	11	16
3	10	19
4	11	31
5	9	14

Step 2: 10.6; 21.4 **Step 3:** 1.96, 0.16, 0.36, 0.16, 2.56; 31.36, 29.16, 5.76, 92.16, 54.76 **Step 4:** 1.04, 42.64; 1.02, 6.53
Step 5: Paragraph one: it has fewer words per sentence on average, and since the standard deviation is much lower, each sentence does not differ much from the average sentence length.

Questions
1. 1.44   3. about 5.43   5. the sum of all of the scores
7. the average deviation from the mean   9. the standard deviation   11. The results are more spread out in the first option. It has a lower $m.a.d.$ and the chances of winning something close to the mean are greater.   13. a. 11, 5.74
b. 12, 5.74   15. Answers vary. Sample: {0, 150, 300} and {100, 150, 200} 17. a. $10.63 b. $2487.05   19. 2.33 m
21. $2i$ 23. 0   25. a. 95 b. at least 90 c. 90

Lesson 13-4 (pp. 889–895)
Mental Math a. the set of all real numbers **b.** $\{p \mid p$ is divisible by 6$\}$ **c.** the set of all ellipses **d.** Ø
Activity Step 1: Top row: 6; 24; 720; 5040; 40,320; 3,628,800;
Middle row: 6; 24; 120; 5040; 40,320; 362,880;
Bottom row: 3; 4; 5; 6; 7; 8; 9; 10

Step 2: The pattern is $n = \dfrac{n!}{(n-1)!}$. $\begin{cases} f_1 = 1 \\ f_n = n \cdot f_{n-1}, n \geq 2 \end{cases}$

Guided Example 4 13; 11; 13!; 11!; 2!; 78
Questions
1. **a.** PRM, PMR, RPM, RMP, MPR, MRP **b.** PRM
3. 1, 2, 6, 24, 120, 720, 5040   5. Answers vary. Sample: A permutation is a choice of objects where order matters. A combination is a choice of objects where order does not matter.   7. $\dfrac{n!}{r!(n-r)!}$ 9. 10   11. 21   13. a. 3,921,225
b. 1,471,429,260 c. about 5.770 quadrillion
15. a. 177,100; she offers more dye combinations than advertised. b. 127,512,000   17. The scores in the first set are more spread out.   19. a. $y = \frac{1}{m}x - \frac{b}{m}$ b. They are reciprocals. c. when $m = 0$ 21. Answers vary. Sample: {8, 12, 14, 14}

Lesson 13-5 (pp. 896–902)
Mental Math a. Answers vary. Samples: $(x^2 + 3, 2)$, $(h + hk^3, 4)$, $(\pi, 0)$ **b.** yes **c.** the set of nonnegative integers **d.** no
Activity Step 1: $_4C_0 = 1$, $_4C_1 = 4$, $_4C_2 = 6$, $_4C_3 = 4$, $_4C_4 = 1$
Step 2: These are the 5 elements in row 4 of Pascal's Triangle.
Step 3: $_7C_0 = 1$, $_7C_1 = 7$, $_7C_2 = 21$, $_7C_3 = 35$, $_7C_4 = 35$, $_7C_5 = 21$, $_7C_6 = 7$, $_7C_7 = 1$ **Step 4:** These are the 8 elements in row 7 of Pascal's Triangle.
Step 5: $_nC_r$ is the $(r + 1)$st element in row n of Pascal's Triangle.
Guided Example 3 Solution 1 8; 5; 8; 5; 56 **Solution 2** 6; 8; 56

Questions
1. The array that later became known as Pascal's Triangle first appeared in the 11th century in the works of Abu Bakr al-Karaji, a Persian mathematician, and Jia Xian, a Chinese mathematician.   3. Pascal's triangle is generated as follows: The only element in the top row is 1, and every other row begins and ends with a 1. If x and y are adjacent elements in a row, then $x + y$ is the element immediately beneath and between them.   5. $\binom{5}{0}$ $\binom{5}{1}$ $\binom{5}{2}$ $\binom{5}{3}$ $\binom{5}{4}$ $\binom{5}{5}$ 7. 15; 5th element, row 6
9. 1; 1st element, row 10   11. 21; 6th element, row 7
13. 4950   15. $\binom{y}{x} = \binom{10}{5}$ 17. a. 2, 4, 8, 16, 32, 64 b. 128
c. 2^n 19. This sequence is formed by the 3rd elements of consecutive rows, beginning with row 2.   21. 495
23. 58   25. about 3.99   27. $\det \begin{bmatrix} 3 & 1 \\ 6 & 2 \end{bmatrix} = 0$, so its inverse is undefined.   29. $y = \frac{12}{5}$

Lesson 13-6 (pp. 903–908)
Mental Math a. $r^2 - 2rs + s^2$ **b.** $49p^2 + 126p + 81$
c. $4a^2b^2 - 12abc + 9c^2$ **d.** $25m^4k^4 + 10m^2n^3k^4 + n^6k^4$
Activity Steps 1–2:

$a^3 + 3a^2b + 3ab^2 + b^3$	3
$a^4 + 4a^3b + 6a^2b^2 + 4ab^3 + b^4$	4
$a^5 + 5a^4b + 10a^3b^2 + 10a^2b^3 + 5ab^4 + b^5$	5
$a^6 + 6a^5b + 15a^4b^2 + 20a^3b^3 + 15a^2b^4 + 6ab^5 + b^6$	6

Step 2. $n(n + 1)$
Steps 3 and 6.

1, 3, 3, 1	3, 2, 1, 0	0, 1, 2, 3
1, 4, 6, 4, 1	4, 3, 2, 1, 0	0, 1, 2, 3, 4
1, 5, 10, 10, 5, 1	5, 4, 3, 2, 1, 0	0, 1, 2, 3, 4, 5
1, 6, 15, 20, 15, 6, 1	6, 5, 4, 3, 2, 1, 0	0, 1, 2, 3, 4, 5, 6

Step 4: The coefficients form Pascal's triangle.
Step 5: $\binom{5}{0}$ $\binom{5}{1}$ $\binom{5}{2}$ $\binom{5}{3}$ $\binom{5}{4}$ $\binom{5}{5}$ **Step 7:** The exponents of a start at n and decrease by 1 with each term, and the exponents of b start at 0 and increase by 1 with each term.
Guided Example 3 1; 4; 4; 3; 1; 6; 2; 2; 4; 1; 3; 1; 4; 16; 8; 32; 6; 24; 4; 8; 2; 1
Questions
1. **a.** $x^2 + 2xy + y^2$ **b.** 1, 3, 3, 1   3. $\frac{1}{16} - \frac{1}{2}m^2 + \frac{3}{2}m^4 - 2m^6 + m^8$ 5. $210x^6y^4$ 7. $11{,}520n^8$ 9. $(x + 2)^{14}$
11. $(a + b)(a^3 + 3a^2b + 3ab^2 + b^3) = a^4 + 3a^3b + 3a^2b^2 + ab^3 + a^3b + 3a^2b^2 + 3ab^3 + b^4 = a^4 + 4a^3b + 6a^2b^2 + 4ab^3 + b^4 = \sum_{r=0}^{4} \binom{4}{r}a^{4-r}b^r$ 13. $(1 + 0.003)^3 \approx 1^3 + 3 \cdot 1^2 \cdot (0.003) = 1 + 0.009 = 1.009$

15. a. 1; 11; 121; 1331; 14,641; The digits in each term form a row of Pascal's triangle.

b. $\sum_{r=0}^{4}\binom{4}{r}10^{4-r}1^{r} = 10,000 + 4000 + 600 + 40 + 1 = 14,641$

c. $\sum_{r=0}^{5}\binom{5}{r}10^{5-r}1^{r} = 100,000 + 50,000 + 10,000 + 1000 + 50 + 1 = 161,051$ 17. True; $\frac{n!}{(n-2)!} = \frac{n(n-1)(n-2)!}{(n-2)!} = n^2 - n$;

this is the difference of two integers, which will be an integer.   19. a. $2\sqrt{26}$ b. $(0, -\sqrt{26})$, $(0, \sqrt{26})$; $(-\sqrt{15}, 0)$, $(\sqrt{15}, 0)$, c. $(0, \sqrt{11})$, $(0, -\sqrt{11})$ d. $2\sqrt{26}$

e. $\pi\sqrt{390} \approx 62.04$ 21. $y = 169$ 23. $z = 100,000$

Lesson 13-7 (pp. 909-916)

Mental Math a. 1000 b. 1 c. 1000 d. 499,500
Activity 1 Steps 1–2: Answers vary. Sample:

Trial	Sequence	Number of Heads
1	HHTH	3
2	HHTT	2
3	HTHH	3
4	TTHH	2
5	TTTH	1
6	THHT	2
7	TTTT	0
8	HTHH	3
9	TTTT	0
10	HTHH	3

Number of occurrences of: 0 heads: 2, 1 head: 1, 2 heads: 3, 3 heads: 4, 4 heads: 0 Step 3: Answers vary. Sample: The relative frequency for getting 2 heads in 4 flips is 0.39.

**Relative Frequency
of the Number of Heads
in 4 Flips of a Coin**

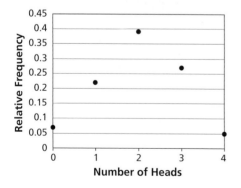

Activity 2 Step 1:

	n	0	1	2	3	4
Sequences of 4 Tosses with n Heads		TTTT	HTTT THTT TTHT TTTH	HHTT HTHT HTTH THHT THTH TTHH	HHHT HHTH HTHH THHH	HHHH
y = Number of Sequences with n Heads		1	4	6	4	1

Step 2:   **Number of Sequences y
with n Heads**

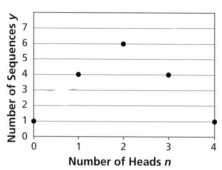

Step 3: The two graphs are nearly identical in shape.

Guided Example 3 $\frac{1}{6}$, $\frac{1}{36}$, $\frac{1}{36}$, $\frac{1}{36}$, $\frac{35}{36}$, $\left(\frac{35}{36}\right)^3$, 0.919; 2, 3 $\cdot \frac{1}{36} \cdot \left(\frac{35}{36}\right)^2 \approx$ 0.0788; $\binom{3}{2}$, 2, 1; 3 $\cdot \left(\frac{1}{36}\right)^2 \cdot \frac{35}{36} \approx$ 0.00225; 0; $\binom{3}{3}$ $p^3 q^0$; $\left(\frac{1}{36}\right)^3 \approx$ 0.0000214

Questions
1. a. 10 b. 10 c. $\frac{10}{32} = 0.3125$ 3. $\binom{4}{1}\left(\frac{1}{35}\right)\left(\frac{34}{35}\right)^3 \approx 0.1048$
5. $\frac{35}{128} \approx 0.2734$ 7. a. 0.45 b. 0.2475 c. 0.495
9. about 0.02799   11. about 0.0016   13. about 0.0231
15. RRRR, RRRW, RRWR, RWRR, WRRR, RRWW, RWRW, RWWR, WRRW, WRWR, WWRR, RWWW, WRWW, WWRW, WWWR, WWWW
17. $a^5 + 5a^4b + 10a^3b^2 + 10a^2b^3 + 5ab^4 + b^5$
19. 30, 60 or −45, 135   21. $t^4 - r^4$ 23. $\frac{y}{x}$

Lesson 13-8 (pp. 917-921)

Mental Math a. −0.788 b. −0.788 c. 0.788 d. 0.788
Guided Example 1 5; 5!; 51!; 3,819,816; 3,819,816; 175,711,536; 175,711,536; 175,711,536; 5.69 · 10^{-9}
Questions
1. a game or procedure in which prizes are distributed by pure chance   3. $\frac{5}{19,523,504} \approx 0.00000026 \approx \frac{1}{3,846,154}$ 5. $\frac{1}{1000}$
7. $\frac{1}{76,275,360}$ 9. $\frac{5}{324,632} \approx 0.0000154 \approx \frac{1}{64,935}$ 11. $\frac{35}{128} \approx 0.2734$
13. $\frac{1}{27,405}$ 15. a. 5 b. 0   17. B; it is equivalent to the equation $f(x) = \left(\frac{1}{5}\right)^x$, which has a growth factor between 0 and 1.   19. $\frac{8}{1}$

Lesson 13-9 (pp. 922-927)

Mental Math a. no **b.** yes **c.** no **d.** yes
Guided Example 1 a. 1; $\frac{5}{32}, \frac{5}{16}, \frac{5}{16}, \frac{5}{32}, \frac{1}{32}$
b. from left to right: $\frac{5}{32}, \frac{5}{16}, \frac{5}{16}, \frac{5}{32}, \frac{1}{32}$

Questions

1. a. binomial probability distribution **b.** $\frac{5}{32}$; Answers
vary. Sample: $P(4)$ this could represent the probability of
getting 4 heads in 5 tosses of a fair coin. **3.** $\frac{210}{1024} \approx 0.21$
5. $y = \frac{1}{\sqrt{2\pi}} e^{\left(\frac{-x^2}{2}\right)}$ **7. a.** Standardized scores are adjusted
so they lie in a normal distribution with a predetermined
mean and standard deviation. **b.** Standardized scores
make it easy to evaluate an individual score relative to the
mean. **9.** 68.2%

11.

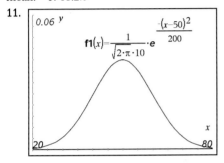

$$f1(x) = \frac{1}{\sqrt{2 \cdot \pi \cdot 10}} \cdot e^{\frac{-(x-50)^2}{200}}$$

13. a. 15.9% **b.** 68.2% **15.** 770; 21 ft **17.** 0.00129
19. false; $\binom{100}{50}\left(\frac{1}{2}\right)^{100} \approx 0.0796 \approx 8\%$ **21.** $7776 - 3240x +$
$540x^2 - 45x^3 + \frac{15}{8}x^4 - \frac{1}{32}x^5$ **23.** $40i$

Self-Test (pp. 932-933)

1. This is a series with 17 terms where each term
is the fourth power of the index variable: $\sum_{i=1}^{17} i^4$.

2. $\sum_{j=0}^{17}(5j+1) = \frac{n}{2}(a_1 + a_n) = 50.5(1 + 501) = 25{,}351$

3. $\mu = \frac{68+69+70+71+71+74+74}{7} = 71$; $m.a.d = \frac{1}{7}\sum_{i=1}^{7}|x_i - \mu| =$
$\frac{1}{7}(|68-71| + |69-71| + |70-71| + |71-71| + |71-71| +$
$|74-71| + |74-71|) = \frac{12}{7}$; $\sigma = \sqrt{\frac{1}{n}\sum_{i=1}^{n}(x_i - 71)^2} =$
$\sqrt{\frac{1}{7}(9+4+1+0+0+9+9)} = \sqrt{\frac{32}{7}} \approx 2.14$

4. The mean team weight is 71 kg. If the standard
deviation is 0, then $\sqrt{\frac{1}{7}\sum_{i=1}^{n}(x_i - 71)^2} = 0$, which means
that each of the seven team members weighs 71 kg.

5. $(x^2 + 2)^5 = \sum_{r=0}^{n=5}\binom{n}{r}(x^2)^{n-r}2^r = x^{10} + 10x^8 + 40x^6 +$
$80x^4 + 80x^2 + 32$ **6.** The second term in the expansion
is $\binom{4}{1}x^3(3)^1$, so the coefficient is $\binom{4}{1}3 = 12$.

7. $\binom{17}{3} = \frac{17!}{3!14!} = 680$ combinations **8.** By the Number of
Permutations Theorem, there are $9! = 362{,}880$ arrangements.
9. a. $\binom{7}{4} = \frac{7!}{4!(7-4)!} = \frac{7!}{4!3!} = 35$ **b.** $_{164}C_4 = \binom{164}{4}$
$= \frac{164!}{4!160!} = 29{,}051{,}001$ **10.** $_{37}C_{36} = \binom{37}{36} = \frac{37!}{36!1!} = 37$.
This makes sense because each subset of 36 elements
is created by removing one element, and there are 37
possibilities. **11.** $\frac{117 \cdot 116 \cdot 115 \cdot 114 \cdot 113}{1} \cdot \frac{112!}{112!} = \frac{117!}{112!}$
12. a. $\binom{6}{3} = \frac{6!}{3!(6-3)!} = \frac{6!}{3!3!} = \frac{6 \cdot 5 \cdot 4}{6} = 20$; This number is the
4th element in row 6 of Pascal's triangle. **b.** This is the
3rd term of the expansion with eight terms, so its
coefficient is $\binom{7}{2} = \frac{7!}{2!(7-2)!} = \frac{7!}{2!5!} = \frac{7 \cdot 6}{2} = 21$. This value is
the 3rd element in row 7 of Pascal's Triangle. **13.** This is
an arithmetic series with first term 30 and constant
difference 2. So, the sum will be $\sum_{i=1}^{30} 30 + 2(i-1) =$
$\frac{30}{2}(30 + 88) = \$1770$ **14. a.** This is an arithmetic series
with first term 3 and constant difference 4. Since we know
the final term is 87, to find n solve $3 + 4(n-1) = 87$.
$4n - 1 = 87$; $n = 22$. The series can be expressed in sigma
notation as $\sum_{i=1}^{22}(4i-1)$. **b.** $\sum_{i=1}^{22}(4i-1) = \frac{22}{2}(3+87) = 990$
15. This is a geometric series with $n = 20$ terms, first term
$5^0 = 1$, and constant ratio 5: $\sum_{i=1}^{20} 5^{i-1} = \frac{a(1-r^n)}{1-r} = \frac{1(1-5^{20})}{1-5} =$
$23{,}841{,}857{,}910{,}156$ **16.** $\sum_{i=0}^{19} 50\left(\frac{4}{5}\right)^i = \frac{50\left(1-\left(\frac{4}{5}\right)^{20}\right)}{1-\frac{4}{5}} =$
$\frac{188{,}535{,}840{,}025{,}698}{762{,}939{,}453{,}125} \approx 247.12$ **17.** There are $2^5 = 32$ possible
outcomes. There are $\binom{5}{0} = 1$ possibility for 0 heads,
$\binom{5}{1} = 5$ possibilities for 1 head, $\binom{5}{2} = 10$ possibilities
for 2 heads, $\binom{5}{3} = 10$ possibilities for 3 heads, $\binom{5}{4} =$
5 possibilities for 4 heads, and $\binom{5}{5} = 1$ possibility for
5 heads. Therefore, the respective probabilities are
$\frac{1}{32}, \frac{5}{32}, \frac{5}{16}, \frac{5}{16}, \frac{5}{32}$, and $\frac{1}{32}$. **18.** The number of possible
combinations of 6 numbers in Lottery A $= \binom{50}{6} =$
$15{,}890{,}700$. Since the probability of picking each
combination is the same, the probability of winning
is $\frac{1}{15{,}890{,}700} \approx 6.29 \cdot 10^{-8}$. The number of possible
combinations of 5 numbers in Lottery B $= \binom{70}{5} =$
$12{,}103{,}014$, so similarly, the probability of winning
is $\frac{1}{12{,}103{,}014} \approx 8.26 \cdot 10^{-8}$. You have a higher probability of
winning in Lottery B. **19. a.** According to the Normal
Distribution Percentages graph, $34.1\% + 34.1\% = 68.2\%$ of
scores are in this range. **b.** $\mu - \sigma = 515 - 114 = 401$ and μ
$+ \sigma = 515 + 114 = 629$. So, the scores range from 401 to
629. **c.** $743 = 515 + 228 = \mu + 2\sigma$; therefore, about 2.3%
of scores are at or above 743.

20. The probability of r heads in n tosses is $\binom{n}{r}p^r q^{n-r}$. Here, $p = 0.35$, $q = 0.65$, $n = 3$, and $r = 2$; $\binom{3}{2}(0.35)^2(0.65)^1 \approx 0.239$.

21. a.

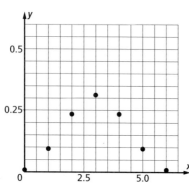

n	$P(n)$
0	$\frac{1}{64}$
1	$\frac{3}{32}$
2	$\frac{15}{64}$
3	$\frac{5}{16}$
4	$\frac{15}{64}$
5	$\frac{3}{32}$
6	$\frac{1}{64}$

c. $P(n)$ represents the probability of a certain number of outcomes in 6 tosses. For example, $P(5)$ represents the probability that there will be exactly 5 of either heads or tails in 6 fair coin tosses.   22. $\binom{n}{2} = \binom{n}{n-2}$ because

$$\binom{n}{2} = \frac{n!}{2!(n-2)!} = \frac{n!}{(n-2)!2!} = \frac{n!}{(n-2)!(n-(n-2))!} = \binom{n}{n-2}.$$

Therefore, $x = 2$. This value of x can also be deduced from the symmetry of Pascal's Triangle.

Self-Test Correlation Chart

Question	1	2	3	4	5	6	7	8
Objective(s)	C	C	I	I	E	E	H	H
Lesson(s)	13-1, 13-2, 13-3, 13-4	13-1, 13-2, 13-3, 13-4	13-3, 13-9	13-3, 13-9	13-6	13-6	13-4	13-4

Question	9	10	11	12	13	14	15	16
Objective(s)	D	D	C	D, F	G	A, C	B	B
Lesson(s)	13-4, 13-5	13-4, 13-5	13-1, 13-2, 13-3, 13-4	13-4, 13-5, 13-6	13-1, 13-2	13-1, 13-2, 13-3, 13-4	13-2	13-2

Question	17	18	19	20	21	22
Objective(s)	J	J	I	J	K	F
Lesson(s)	13-7, 13-8	13-7, 13-8	13-3, 13-9	13-7, 13-8	13-9	13-5, 13-6

Chapter Review (pp. 934–937)

1. 7626   3. 8575   5. 57   7. 157,287   9. ≈ 36.75

11. $-3, 0, 3, 6, 9; 15$ 13. C   15. $\sum\limits_{i=1}^{41} 3i$ 17. $-40,309$ 19. D

21. B   23. a. 56 b. 256   25. 1   27. 532   29. $t^5 - 15t^4 + 90t^3 - 270t^2 + 405t - 243$ 31. $\frac{p^4}{16} + p^3 q + 6p^2 q^2 + 16pq^3 + 16q^4$ 33. true   35. 1 7 21 35 35 21 7 1

37. $\binom{x}{y} = \binom{11}{7}$ 39. The coefficient is the 14th element of row 15 of Pascal's Triangle.   41. a. \$46 b. \$1081 c. 141 days   43. 1230 seats   45. 3,628,800   47. 75,287,520

49. 13,280.98   51. the 1989 freshman class   53. Answers vary. Sample: The two classes have different numbers of students, which means that the mean score of incoming 1989 freshmen is weighted less than the mean score of 2006 freshmen.   55. $\frac{3}{32} = 0.09375$ 57. 0.027

59. 0.42   61. No; $\left(\frac{1}{2}\right)^{100}\sum\limits_{k=70}^{100} {}_{100}C_k \neq \left(\frac{1}{2}\right)^{10}\sum\limits_{i=7}^{10} {}_{10}C_i$.

63. $\frac{1}{1000} = 0.001$ 65. a. 50% b. 68.2% c. 4.6%

Glossary

A

absolute deviation The difference $|x_i - \mu|$, where x_i is an element of a data set and μ is the mean of the set. **(884)**

absolute value The operation or function defined by $|x| = \begin{cases} x \text{ when } x \geq 0 \\ -x \text{ when } x < 0 \end{cases}$; geometrically, the distance of x from 0 on a number line. **(380)**

absolute-value function The function with equation $f(x) = |x|$. **(380)**

acceleration The rate of change of the velocity of an object. **(12)**

algebraic expression See *expression*. **(6)**

algebraic sentence A sentence in which expressions are related by equality or inequality. **(6)**

angle of depression The angle between the line of sight and the horizontal when the line of sight points down. **(672)**

angle of elevation The angle between the horizontal base of an object and the observer's line of sight to the object when the line of sight points up. **(665)**

annual percentage yield, or APY The rate of interest earned after all the compoundings have taken place in one year. Also called *effective annual yield* or *yield*. **(475)**

argument of a function The domain variable in a function. **(21)**

arithmetic mean See *mean of a data set*. **(883)**

arithmetic sequence A sequence with a constant difference between consecutive terms. Also called *linear sequence*. **(196)**

arithmetic series An indicated sum of successive terms of an arithmetic sequence. **(869)**

asymptote A line approached by the graph of a function. **(581)**

asymptotes of a hyperbola The two lines that are approached by the points on the branches of a hyperbola as the points get farther from the foci. **(107)**

axis of symmetry of a parabola The line through the focus of a parabola perpendicular to the directrix. **(388, 798)**

B

base x in the expression x^n. **(452)**

bel A unit of sound intensity; 10 bels is a decibel. **(616)**

binomial An expression with two terms. A polynomial with two terms. **(375, 738)**

binomial coefficients The coefficients of terms in the expansion of $(a + b)^n$, often displayed as the numbers in Pascal's Triangle. **(906)**

binomial distribution A probability function P resulting from calculations of binomial probabilities. Also called *binomial probabilty distribution*. **(923)**

binomial expansion The result of writing the power of a binomial as a sum. **(903)**

binomial experiment A situation in which n independent trials occur, and each trial has exactly two mutually exclusive outcomes. **(912)**

binomial probability distribution See *binomial distribution*. **(923)**

boundary of a half plane The line in a plane that separates the half-plane from another mutually exclusive half-plane. **(342)**

branches of a hyperbola The two separate parts of the graph of a hyperbola. **(107)**

C

ceiling function The step function $f(x) = \lceil x \rceil$, the least integer greater than or equal to x. Also called *least-integer function* or *rounding-up function*. **(204)**

ceiling symbol The symbol $\lceil x \rceil$ indicating the least integer greater than or equal to x. **(204)**

center of a circle The fixed point from which the set of points of the circle are at a given distance. **(804)**

center of a rotation In a rotation, a point that coincides with its image. **(269)**

center of a size change The point in a size change of magnitude $k \neq 1$ that coincides with its image. **(242)**

center of an ellipse The intersection of the axes of an ellipse. **(819)**

circle The set of all points in a plane at a given distance from a fixed point. **(804)**

closure The property of an operation on elements of a set S when all the results of performing the operation are elements of S. **(262)**

coefficient matrix A matrix that contains the coefficients of the variables in a system of equations. **(336)**

coefficients of a polynomial The numbers a_n, a_{n-1}, a_{n-2}, ..., a_0 in the polynomial $a_n x^n + a_{n-1}x^{n-1} + a_{n-2}x^{n-2} + ... + a_0$. **(730)**

combination Any choice of r objects from n objects when the order of the objects does not matter. **(892)**

combined variation A situation in which direct and inverse variations occur together. **(129)**

common logarithm A logarithm to the base 10. Also called *common log*. **(609)**

common logarithm function The function that maps x onto $\log_{10}x$ for all positive numbers x. Also called the *logarithm function to the base 10*. **(611)**

completing the square A technique used to transform a quadratic from $ax^2 + bx + c$ form to $a(x - h)^2 + k$ form. **(401)**

complex conjugate For any complex number $a + bi$, the difference $a - bi$. **(429)**

complex number A number that can be written in the form $a + bi$, where a and b are real numbers and $i = \sqrt{-1}$. **(427)**

composite $g \circ f$ For two functions f and g, the function that maps x onto $g(f(x))$ and whose domain is the set of all values in the domain of f for which $f(x)$ is in the domain of g. **(517)**

composite of two transformations Given transformation T_1 that maps figure G onto figure G' and transformation T_2 that maps figure G' onto figure G'', the transformation that maps G onto G'', the composite of T_1 and T_2, written $T_2 \circ T_1$. **(263)**

composition See *function composition*. **(517)**

compound sentence A sentence in which two clauses are connected by the word *and* or by the word *or*. **(300)**

compounding The process of earning interest on the interest of an investment. **(473)**

concentric circles Circles with the same center but different radii. **(804)**

conic section The intersection of a plane and a double cone. **(838)**

conjugate For any expression of the form $a + \sqrt{b}$, the expression $a - \sqrt{b}$. **(553)**

consistent system A system that has one or more solutions. **(316)**

constant-decrease situation A situation in which a quantity y decreases by a constant amount for every fixed increase in x. **(150)**

constant difference In an arithmetic sequence, the difference of two consecutive terms. **(196)**

constant-increase situation A situation in which a quantity y increases by a constant amount for every fixed increase in x. **(150)**

constant of variation The nonzero real constant k in the equation $y = kx^n$ or $y = \frac{k}{x^n}$ or other equation of a function of variation. **(72)**

constant matrix A matrix that contains the constants on the right sides of the equations in a system. **(336)**

constant multiplier The constant in a geometric sequence. Also called *constant ratio*. **(479)**

constant ratio In a geometric sequence, the ratio of successive terms. Also called *constant multiplier*. **(480)**

constant term, constant In the polynomial $a_n x^n + a_{n-1}x^{n-1} + a_{n-2}x^{n-2} + ... + a_0$, the number a_0. **(730)**

constraint A restriction on a variable or variables in a situation. **(299)**

continuous change model The equation $N(t) = N_0 e^{rt}$, where N_0 is the initial amount and r is the growth factor over a time t. **(598)**

continuous compounding The limit of the process of earning interest with periods of compounding approaching zero. Also called *instantaneous compounding*. **(596)**

convex region A region of the plane in which any two points of the region can be connected by a line segment which lies entirely in the region. **(354)**

coordinate plane A plane in which there is a one-to-one correspondence between the points in the plane and the set of ordered pairs of real numbers. **(33)**

corollary A theorem that follows immediately from another theorem. (388)

correlation coefficient A number between -1 and 1 that indicates how well a linear equation fits data. (181)

cosine function The correspondence $\theta \rightarrow \cos \theta$ that maps a number θ onto its cosine. (663)

cosine of θ (cos θ) In a right triangle with acute angle θ, $\cos \theta = \dfrac{\text{length of leg adjacent to } \theta}{\text{length of hypotenuse}}$, the x-coordinate of the image of $(1, 0)$ under R_θ, the rotation with center $(0, 0)$ and magnitude θ. (663, 684)

counterexample An instance which shows a generalization to be false. (424)

counting numbers See *natural numbers*. (29)

cos^{-1} x The number between 0 and 180°, or between 0 and π, whose cosine is x. (671)

cube of x The third power of x, x^3. (454)

cube root A cube root t of x, denoted $\sqrt[3]{x}$, is a solution to the equation $t^3 = x$. (487)

cubic polynomial A polynomial of a single variable with degree 3, such as $ax^3 + bx^2 + cx + d$. (731)

cubing function A function f with equation $f(x) = x^3$. (454)

D

data set A collection of elements in which an element may appear more than once. (883)

decade growth factor The ratio of an amount in a specific year to the amount ten years earlier. (542)

decibel (dB) A measure of relative sound intensity; $\frac{1}{10}$ of a bel. (615)

default window The window that is set on an automatic grapher by the manufacturer. (28)

degree of a polynomial in a single variable The largest exponent of the variable in a polynomial. (730)

degree of a polynomial in several variables The largest sum of the exponents of the variables in any term in a polynomial expression. (738)

dependent variable A variable whose value always depends on the value(s) of other variable(s). (15)

depreciation The decrease in value over time of manufactured goods. (587)

determinant of a 2 × 2 matrix For the matrix $M = \begin{bmatrix} a & b \\ c & d \end{bmatrix}$, the number $ad - bc$. (331)

deviation For each value of an independent variable, the difference $p - a$ between the value p of the dependent variable predicted by a model and the actual value a. (178)

difference of two matrices For two matrices A and B with the same dimensions, their difference $A - B$ is the matrix in which each element is the difference of the corresponding elements in A and B. (230)

dimensions of a matrix A matrix with m rows and n columns has dimensions $m \times n$. (222)

direct-variation equation An equation in which one variable is a multiple of another variable or product of variables. (72)

direct-variation function A function that can be described by a formula of the form $y = kx^n$, $k \neq 0$ and $n > 0$. (73)

directly proportional to The situation in which y varies directly as x^n. Also called *varies directly as*. (73)

directrix of a parabola The line such that the distance from it to any point on the parabola is equal to the distance from that point to the focus. (798)

discrete function A function whose domain can be put into one-to-one correspondence with a finite or infinite set of integers, with gaps or intervals between successive values in the domain. (57)

discrete graph A graph that is made up of unconnected points. (57)

discrete set A set in which there is a positive distance greater than some fixed amount between any two elements of the set. (108)

discriminant of a quadratic equation For the equation $ax^2 + bx + c = 0$, the value of the expression $b^2 - 4ac$. (435)

domain of a function The set of values which are allowable substitutions for the independent variable. (27)

double cone The surface generated by a line rotating about an axis that contains a point on the line. (797)

double inequality A sentence that has two inequality symbols. (301)

double root The root of a quadratic equation when the discriminant is 0; the root of a quadratic equation that has only one solution; a root r of a polynomial in x for which $(x - r)^2$ is a factor, but not $(x - r)^3$. Also called a *root with multiplicity 2*. (767)

E

e The constant 2.7182818459... that the sequence of numbers of the form $\left(1 + \frac{1}{n}\right)^n$ approaches as n increases without bound; the base of natural logarithms. (596)

element of a matrix Each object in a matrix. (222)

ellipse Given two points F_1 and F_2 (the *foci*) and a positive number d, the set of points P in a plane for which $PF_1 + PF_2 = d$. (817)

equal complex numbers Two complex numbers with equal real parts and equal imaginary parts. (427)

equal matrices Two matrices that have the same dimensions and in which corresponding elements are equal. (223)

equation A sentence stating that two expressions are equal. (9)

equivalent formulas Two or more formulas for which all values of the variables that satisfy one formula satisfy the other(s). (48)

Euler's f(x) notation Notation in which $f(x)$ represents the value of a function f with argument x. (20)

evaluating an expression Substituting numbers for the variables in an expression and calculating a result. (8)

expanding a polynomial Writing a power of a polynomial or the product of polynomials as a sum. (739)

explicit formula for nth term A formula that describes any term in a sequence in terms of its position in the sequence. (54)

exponent n in the expression b^n. (452)

exponential curve The graph of an exponential function. (582)

exponential decay A situation described by an exponential function in which the growth factor is between zero and 1, in which $f(x)$ decreases as x increases. (587)

exponential function A function f with the equation $f(x) = ab^x$ ($a \neq 0$, $b > 0$, $b \neq 1$). (582)

exponential growth A situation described by an exponential function where the growth factor is greater than one. (582)

exponential growth model If a quantity a has growth factor b for each unit period, then after a period of length x, there will be ab^x. (580)

exponential sequence See *geometric sequence*. (479)

exponentiation See *powering*. (452)

expression A combination of numbers, variables, and operations that stands for a number. Also called *algebraic expression*. (6)

exterior of a circle The region outside a circle; the set of points whose distance from the center of the circle is greater than the radius. (811)

extraneous solution A possible solution that is obtained from an equation-solving procedure but that does not check in the original equation. (563)

F

f(x) notation The notation used to describe functions, read "*f* of *x*." (20)

factor A number or expression which evenly divides a given expression. (745)

factored form of a polynomial A polynomial written as a product of two or more factors. (745)

factorial For any integer $n \geq 2$, the product of the integers from 1 through n, denoted by $n!$. (890)

factorial symbol, ! The symbol used to represent the product of the integers n through 1. (890)

factoring a polynomial The process of rewriting a polynomial as a product of two or more factors. (745)

fair coin, fair die A coin or die that has an equal probability of landing on each of its sides. Also called *unbiased coin* or *unbiased die*. (910)

feasible region The set of solutions to a system of linear inequalities. Also called *feasible set*. (349)

feasible set See *feasible region*. (349)

Fibonacci sequence The sequence 1, 1, 2, 3, 5, 8, 13, ... A recursive definition is

$$\begin{cases} F_1 = 1 \\ F_2 = 1 \\ F_n = F_{n-1} + F_{2n-2} \text{ for } n \geq 3 \end{cases} \quad \text{(192)}$$

field properties The assumed properties of addition and multiplication of real numbers. (S1)

floor function The step function $f(x) = \lfloor x \rfloor$, indicating the greatest integer less than or equal to x. Also called *greatest-integer function, int function,* or *rounding-down function.* (204)

floor symbol The symbol $\lfloor x \rfloor$ indicating the greatest integer less than or equal to x. (204)

focal constant The constant sum of the distances from any point P on an ellipse to the foci. For a hyperbola, the absolute value of the difference of the distances from a point on a hyperbola to the two foci of the hyperbola. (817, 832)

focus, plural foci For a parabola, the point along with the directrix from which a point is equidistant. The two points from which the sum (ellipse) or difference (hyperbola) of distances to a point on the conic section is constant. (798, 817, 832)

formula A sentence stating that a single variable is equal to an expression with one or more different variables on the other side. (9)

function A set of ordered pairs (x, y) in which each first component x of the pair is paired with exactly one second component y. A relation in which no two ordered pairs have the same first component x. (15)

function composition The operation that results from first applying one function, then another; denoted by the symbol ∘. (517)

G

general form of a quadratic relation An equation of the form $Ax^2 + Bxy + Cy^2 + Dx + Ey + F = 0$, where $A, B, C, D, E,$ and F are real numbers and at least one of $A, B,$ or C is not zero. (796)

geometric mean The nth root of the product of n numbers. (537)

geometric sequence A sequence in which each term after the first is found by multiplying the previous term by a constant. Also called *exponential sequence.* (479)

geometric series An indicated sum of successive terms of a geometric sequence. (877)

gravitational constant The acceleration of a moving object due to gravity, often denoted by g. Near Earth's surface, $g \approx 32 \frac{\text{ft}}{\text{sec}^2} \approx 9.8 \frac{\text{m}}{\text{sec}^2}$. (12)

greatest common monomial factor The monomial with the greatest coefficient and highest degree that is a factor of all the terms of a polynomial. (746)

greatest integer function See *floor function.* (204)

growth factor In the exponential function $y = ab^x$ with $a > 0$, the base b. (582)

H

half-life The amount of time required for a quantity in an exponential decay situation to decay to half its original value. (588)

half-plane Either of the two sets of points, or regions, separated by a line in a plane. (342)

hierarchy A diagram that shows how various ideas are related, with a direction that moves from more specific to more general. (S5)

horizontal asymptote A horizontal line that is approached by a graph as the values of x get very large or very small. (107)

horizontal line A line with an equation of the form $y = b$. (153)

horizontal-line test The inverse of a function is itself a function if and only if no horizontal line intersects the graph of the function in more than one point. (525)

horizontal magnitude The number a in the scale change that maps (x, y) onto (ax, by). (249)

horizontal scale change The stretching or shrinking of a figure in only the horizontal direction; a transformation which maps (x, y) onto (kx, y). (249)

hyperbola The graph of a function with equation of the form $y = \frac{k}{x}$, where $k \neq 0$. For any two points F_1 and F_2 (the *foci* of the hyperbola) and d (the *focal constant* of the hyperbola) with $0 < d < F_1F_2$, the set of points P in a plane that satisfy $|PF_1 - PF_2| = d$. (107, 832)

I

i One of the two square roots of –1, denoted by $\sqrt{-1}$. (421)

identity A relationship that is true for all values of variables in a domain. (688)

identity function The function with equation $f(x) = x$. (454)

2 × 2 identity matrix The matrix $\begin{bmatrix} 1 & 0 \\ 0 & 1 \end{bmatrix}$, which maps each point $\begin{bmatrix} x \\ y \end{bmatrix}$ of a figure onto itself. **(244)**

3 × 3 identity matrix The matrix $\begin{bmatrix} 1 & 0 & 0 \\ 0 & 1 & 0 \\ 0 & 0 & 1 \end{bmatrix}$, which maps each point $\begin{bmatrix} x \\ y \\ z \end{bmatrix}$ of a figure onto itself. **(268)**

identity transformation The size-change transformation of magnitude 1; the transformation in which each point coincides with its image. **(244)**

image The result of applying a transformation to a preimage. **(242)**

imaginary number A number that is the square root of a negative real number. **(421)**

imaginary part Of the complex number $a + bi$, bi. **(427)**

imaginary unit The complex number i. **(421)**

inconsistent system A system that has no solutions. **(316)**

independent events Two or more events in which the occurrence of one event does not affect the probabilities the other events occur. **(453)**

independent variable In a formula, a variable upon whose value other variables depend. **(15)**

index The subscript used for a term in a sequence indicating the position of the term in the sequence. **(54)**

inequality A sentence containing one of the symbols $<, >, \le, \ge, \ne$, or $\approx$. **(303)**

index variable The variable i under the Σ sign in summation notation. Also called *index*. **(871)**

input A value of the independent variable in a function. **(15)**

interval The set of numbers x such that $x \le a$ or $a \le x \le b$, where the $\le$ can be replaced by $<, >,$ or $\ge$. **(301)**

int function See *floor function*. **(204)**

integers The set of numbers {... , –3, -2, –1, 0, 1, 2, 3, ...}; the set of natural numbers and their opposites. **(29)**

interior of a circle The region inside a circle; the set of points whose distance from the center of the circle is less than the radius. **(811)**

intersection of two sets The set consisting of those values common to both sets. **(301)**

inverse cosine function, cos⁻¹ For the function that maps x onto $\cos x$, restricted to the domain $0 \le x \le \pi$ or $0 \le x \le 180°$, the function that maps $\cos x$ onto x. **(671)**

inverse function, f^{-1} The inverse relation formed by switching the coordinates of each ordered pair of f, when this inverse is itself a function. A function that when composed with another function gives the identity function. **(529, 531)**

inverse matrices Two matrices whose product is the identity matrix. **(328)**

inverse of a relation The relation obtained by switching the coordinates of each ordered pair in the relation. **(523)**

inverse sine function, sin⁻¹ For the function that maps x onto $\sin x$, restricted to the domain $-\frac{\pi}{2} \le x \le \frac{\pi}{2}$, the function that maps $\sin x$ onto x. **(670)**

inverse-square curve The graph of $y = \frac{k}{x^2}$. **(108)**

inverse tangent function, tan⁻¹ For the function that maps x onto $\tan x$, restricted to the domain $-\frac{\pi}{2} \le x \le \frac{\pi}{2}$, the function that maps $\tan x$ onto x. **(671)**

inverse-variation function A function that can be described by a formula of the form $y = \frac{k}{x^n}$, for $k \ne 0$ and $n > 0$. **(79)**

inversely proportional to The situation in which y varies indirectly with x^n. Also called *varies inversely as*. **(80)**

irrational number A real number that is not rational, that is, cannot be expressed as a simple fraction or ratio of the form $\frac{a}{b}$, where a and b are integers and $b \ne 0$. **(29, 383)**

irreducible polynomial See *prime polynomial*. **(748)**

J

joint variation A situation in which one quantity varies directly as the product of two or more independent variables, but not inversely as any variable. **(132)**

L

lattice point A point in a solution set whose coordinates are both integers. **(344)**

leading coefficient In the polynomial $a_n x^n + a_{n-1} x^{n-1} + a_{n-2} x^{n-2} + ... + a_0$, the number a_n. **(730)**

line of reflection The line over which a figure is reflected. (255)

line of sight An imaginary line from one position to another, or in a particular direction. (672)

line of best fit A line that best fits a set of data, found by using regression. Also called *least-squares line* or *regression line*. (177)

line of reflection See *reflecting line*. (255)

line of symmetry For a figure F, a line m such that the reflection image of F over m is F itself. (102)

linear combination The sum of the multiples of two or more variables. (157)

linear-combination method A method of solving systems that involves adding multiples of the given equations. (321)

linear function A function whose graph is a line or part of a line. A function f with the equation $f(x) = ax + b$, where a and b are real numbers. (151)

linear inequality An inequality in which both sides are linear expressions. (342)

linear polynomial A polynomial of the first degree. (731)

linear-programming problem A problem of maximizing or minimizing a quantity based on solutions to a system of linear inequalities. (356)

linear regression A method that uses all the data points to find the line of best fit for those points. (177)

linear scale A scale with units spaced so that the distance between successive units is constant. (617)

linear sequence See *arithmetic sequence*. (196)

log of *x* to the base 10 See *logarithm of x to the base 10*. (608)

logarithm function to the base 10 See *common logarithm function*. (611)

logarithm function with base *b*, $\log_b x$ The inverse of the exponential function with base b, $f(x) = b^x$; the function that maps x onto $\log_b x$ for all positive numbers x. (624)

logarithm of *x* to the base 10 y is the logarithm of x to the base 10, written $y = \log_{10} x$, if and only if $10^y = x$. Also called *log of x to the base 10* or *log base 10 of x*. (608)

logarithm of *a* to the base *b* Let $b > 0$ and $b \neq 1$. Then x is the logarithm of a to the base b, written $x = \log_b a$, if and only if $b^x = a$. (622)

logarithmic curve The graph of a function of the form $y = \log_b x$. (611)

logarithmic equation An equation of the form $y = \log_b x$. (610)

logarithmic scale A scale in which the scale values are the exponents of the powers. (615)

M

matrix subtraction If two matrices A and B have the same dimensions, their difference $A - B$ is the matrix whose element in each position is the difference of the corresponding elements in A and B. (230)

magnitude of a size change In the size change that maps (x, y) onto (kx, ky), the number k. Also called *size-change factor*. (242)

magnitude of a rotation In a rotation, the amount that the preimage is turned about the center of rotation, measured in degrees from $-180°$ (clockwise) to $180°$ (counterclockwise), m$\angle POP'$, where P' is the image of P under the rotation and O is its center. (269)

major axis of an ellipse The segment that contains the foci of the ellipse and has two vertices of the ellipse as its endpoints. (819)

mapping A synonym for *function*. Also called *map*. (22)

mapping notation The notation $f: x \rightarrow y$ for a function f. (22)

mathematical model A mathematical graph, sentence, or idea that parallels some or all of the structure of a real situation. (16)

matrix A rectangular arrangement of objects or numbers, its *elements*. (222)

matrix addition If two matrices A and B have the same dimensions, their sum $A + B$ is the matrix in whose element in each position is the sum of the corresponding elements in A and B. (228)

matrix form of a system A representation of a system using matrices. The matrix form for $\begin{cases} ax + by = e \\ cx + dy = f \end{cases}$ is $\begin{bmatrix} a & b \\ c & d \end{bmatrix} \cdot \begin{bmatrix} x \\ y \end{bmatrix} = \begin{bmatrix} e \\ f \end{bmatrix}$. (336)

matrix multiplication If A is an $m \times n$ matrix and B is an $n \times p$ matrix, the product $A \cdot B$, or AB, is the $m \times p$ matrix whose element in row i and column j is the product of row i of A and column j of B. **(237)**

matrix product The result of matrix multiplication. **(237)**

maximum The greatest value of a data set or function. **(388)**

mean absolute deviation, m.a.d. The mean of the absolute deviations in a data set. **(884)**

mean of a data set The result of dividing the sum of the numbers in a data set by the number of numbers in the set. **(883)**

measure of center A number which in some sense is at the "center" of a data set; the mean or median of a data set. Also called *measure of central tendency*. **(884)**

measure of spread A number, like standard deviation, which describes the extent to which elements of a data set are dispersed or spread out. **(884)**

median When the terms of a data set are placed in increasing order, if the set has an odd number of terms, the middle term; if the set has an even number of terms, the average of the two terms in the middle. **(537)**

method of finite differences The use of successive calculations of differences of values of polynomial functions to determine whether a polynomial function of a particular degree can be an exact model for a set of points. **(774)**

minimum The least value of a set or function. **(388)**

minor axis of an ellipse The segment that has two vertices of the ellipse as its endpoints and does not contain the foci. **(819)**

mode The number or numbers which occur most often in a data set. **(608)**

model for an operation A pattern that describes many uses of that operation. **(7)**

monomial A polynomial with one term. **(738)**

multiplicity of a root For a root r in an equation $P(x) = 0$, the highest power of the factor $x - r$. **(767)**

N

natural logarithm of m n is the natural logarithm of m, written $n = \ln m$, if and only if $m = e^n$. **(629)**

natural numbers The set of numbers $\{1, 2, 3, 4, 5, ...\}$, sometimes also including 0. Also called the *counting numbers*. **(29)**

normal curve The graph of a normal distribution. **(924)**

normal distribution A function whose graph is the image of the graph of $y = \frac{1}{\sqrt{2\pi}} e^{\left(\frac{-x^2}{2}\right)}$ under a composite of translations or scale transformations. **(924)**

normalized scores See *standardized scores*. **(925)**

nth-power function The function defined by $f(x) = x^n$, where n is a positive integer. **(454)**

nth root Let n be an integer greater than 1. Then b is an nth root of x if and only if $b^n = x$. **(486)**

nth term The term occupying the nth position in the listing of a sequence. The general term of a sequence. **(197)**

O

oblique line A line that is neither horizontal nor vertical.

one-to-one correspondence A mapping in which each member of one set is mapped to a distinct member of another set, and vice versa. **(242)**

order of operations A set of rules used to evaluate expressions, specifically: 1. Perform operations within grouping symbols from inner to outer; 2. Take powers from left to right; 3. Do multiplications or divisions from left to right; 4. Do additions or subtractions from left to right. **(8)**

output A value of the dependent variable in a function. **(15)**

P

parabola For a line ℓ and a point F not on ℓ, the set of all points in the plane of ℓ and F equidistant from F and ℓ. **(102, 798)**

paraboloid A 3-dimensional figure created by rotating a parabola in space around its axis of symmetry. **(801)**

parallax angle for a star Let P and Q be two positions on Earth and let S be the position of a star. Then the parallax angle θ is half the measure of $\angle PSQ$. **(676)**

Pascal's Triangle The sequence satisfying

1. $\binom{n}{0} = \binom{n}{n} = 1$ for all integers $n \geq 0$ and

2. $\binom{n+1}{r+1} = \binom{n}{r} + \binom{n}{r+1}$ for all integers $0 \leq r \leq n$.

The triangular array

$$
\begin{array}{ccccccccc}
 & & & & 1 & & & & \\
 & & & 1 & & 2 & & 1 & \\
 & & 1 & & 3 & & 3 & & 1 \\
 & 1 & & 4 & & 6 & & 4 & & 1 \\
1 & & 5 & & 10 & & 10 & & 5 & & 1 \\
 & & & & \vdots & & & &
\end{array}
$$

where if x and y are located next to each other on a row, the element just below and directly between them is $x + y$. **(898, 867)**

perfect-square trinomial A trinomial of the form $a^2 + 2ab + b^2$ or $a^2 - 2ab + b^2$; the square of a binomial. **(401)**

period The horizontal translation of smallest positive magnitude that maps the graph of a function onto itself. **(694)**

periodic function A function whose graph can be mapped to itself by a horizontal translation. **(694)**

permutation An arrangement of objects in which order matters. **(889)**

pH scale A logarithmic scale used to measure the acidity of a substance. **(618)**

piecewise linear Relating to a function or graph that is described as a union of segments or other subsets of lines. **(171)**

pitch The measure of the steepness of the slant of a roof. **(51)**

point matrix A 2×1 matrix. **(224)**

point-slope form of a linear equation For a line, an equation of the form $y - y_1 = m(x - x_1)$, where (x_1, y_1) is a point on the line with slope m. **(170)**

polynomial equation An equation of the form $y = a_n x^n + a_{n-1} x^{n-1} + \ldots + a_1 x^1 + a_0$, where n is a positive integer and $a_n \neq 0$. **(729)**

polynomial function A function of the form $P: x \rightarrow P(x)$, where $P(x)$ is a polynomial. **(731)**

polynomial in x of degree n An expression of the form $a_n x^n + a_{n-1} x^{n-1} + \ldots + a_1 x^1 + a_0$, where n is a positive integer and $a_n \neq 0$. **(730)**

polynomial model A polynomial equation which fits a data set. **(772)**

power The expression x^n; the result of the operation of powering, or exponentiation. **(452)**

powering An operation by which a variable is raised to a power. Also called *exponentiation*. **(452)**

preimage An object to which a transformation is applied. **(242)**

prime polynomial Over a set of numbers, a polynomial that cannot be factored into polynomials of lower degree whose coefficients are in the set. Also called *irreducible polynomial*. **(748)**

principal The original amount of money invested. **(473)**

probability distribution A function that maps a set of events onto their probabilities. Also called *probability function*. **(923)**

probability of an event If a situation has a total of t equally likely outcomes and e of these outcomes satisfy conditions for a particular event, then the probability of the event is $\frac{e}{t}$. **(453)**

pure imaginary numbers Multiples of the complex number i. **(422)**

Q

quadratic equation An equation that involves quadratic expressions. **(374)**

quadratic equation in two variables An equation of the form $Ax^2 + Bxy + Cy^2 + Dx + Ey + F = 0$, where A, B, C, D, E, and F are real numbers and at least one of A, B, or C is not zero. **(374)**

quadratic expression An expression that contains one or more terms in its variables, such as x^2, y^2, or xy, but no higher powers of x and y. **(374)**

quadratic form An expression of the form $Ax^2 + Bxy + Cy^2 + Dx + Ey + F$. **(374)**

Quadratic Formula If $ax^2 + bx + c = 0$ and $a \neq 0$, then $x = \dfrac{-b \pm \sqrt{b^2 - 4ac}}{2a}$. **(414)**

quadratic function A function f with an equation of the form $f(x) = ax^2 + bx + c$. **(374)**

quadratic polynomial A polynomial of a single variable with degree 2. **(731)**

quadratic regression A process of finding the model with the least sum of squares of differences from the given data points to the values predicted by the model. (**408**)

quadratic system A system that involves polynomial sentences of degrees 1 and 2, at least one of which is a quadratic sentence. (**844**)

quadratic-linear system A quadratic system with at least one linear sentence. (**844**)

quadratic-quadratic system A system that involves two or more quadratic sentences. (**851**)

quartic equation A fourth degree polynomial equation. (**766**)

quartic polynomial A polynomial of a single variable with degree 4. (**731**)

quintic equation A fifth degree polynomial equation. (**766**)

R

radian A measure of an angle, arc, or rotation such that π radians $=180$ degrees. (**712**)

radical symbol The symbol $\sqrt{}$, as in $\sqrt{2x}$ or $\sqrt[3]{9}$. (**537**)

radius The distance between any point on a circle and the center of the circle. (**804**)

random numbers Numbers which have the same probability of being selected. (**928**)

range of a function The set of values of the dependent variable that can result from all possible substitutions for the independent variable. (**27**)

rate of change See *slope*. (**94**)

ratio of similitude In two similar figures, the ratio between a length in one figure and the corresponding length in the other. (**245**)

rational number A number that can be represented as a simple fraction or ratio of the form $\frac{a}{b}$, where a and b are integers and $b \neq 0$. (**29, 383**)

rationalizing the denominator The process of rewriting a fraction so that its denominator is a rational number. (**551**)

real function A function whose independent and dependent variables stand for only real numbers. (**26**)

real numbers Those numbers that can be represented by finite or infinite decimals. (**29**)

real part Of the complex number $a + bi$, the real number a. (**427**)

rectangular hyperbola A hyperbola with perpendicular asymptotes. (**842**)

recursive formula A set of statements that indicates the first term (or first few terms) of a sequence and tells how the next term is calculated from the previous term or terms. Also called *recursive definition*. (**183**)

reflecting line The line over which a point is reflected. Also called *line of reflection*. (**255**)

reflection A transformation under which the image of a point P over a reflecting line m is (1) P itself, if P is on m; (2) the point P' such that m is the perpendicular bisector of the segment connecting P with P' that maps a figure to its reflection image. (**255**)

reflection image of a point A over a line m The point A if A is on m and the point A' such that m is the perpendicular bisector of $\overline{AA'}$ if A is not on m. (**255**)

reflection-symmetric Coinciding with a reflection image of itself. (**102**)

regression line See *line of best fit*. (**177**)

relation Any set of ordered pairs. (**15**)

root of a polynomial For a polynomial function P, a zero of the equation $P(x) = 0$. Also called *zero of a polynomial*. (**752**)

root with multiplicity 2 See *double root*. (**767**)

roots of an equation Solutions to an equation. (**437**)

rotation A transformation with a center O under which the image of O is O itself and the image of any other point P is the point P' such that $m\angle POP'$ is a fixed number (its *magnitude*). (**269**)

rounding-down function See *floor function*. (**204**)

rounding-up function See *ceiling function*. (**204**)

row A horizontal list in a table, rectangular array, or spreadsheet. (**222**)

row-by-column multiplication The process of obtaining an element in the product of two matrices in which a row matrix is multiplied by a column matrix. (**236**)

S

scalar multiplication An operation leading to the product kA of a scalar k and a matrix A, in which each element of kA is k times the corresponding element in A. (230)

scalar product The result of scalar multiplication. (230)

scale change The stretching or shrinking of a figure in either a horizontal direction only, in a vertical direction only, or in both directions. A horizontal scale change of magnitude a and a vertical scale change of magnitude b maps (x, y) onto (ax, by), and is denoted by $S_{a,b}$. (249)

scatterplot A plot with discrete points used to display a data set. (178)

semicircle A half-circle. (806)

semimajor axis Half the major axis of an ellipse. (820)

semiminor axis Half the minor axis of an ellipse. (820)

sequence A function whose domain is the set of all positive integers or the set of positive integers from a to b. (53)

set-builder notation The notation $\{x \mid \ldots\}$ read "the set of all x such that ... ." Also written $\{x: \ldots\}$ (27)

series An indicated sum of terms in a sequence. (869)

shrink A scale change in the horizontal (or vertical) direction in which the absolute value of the magnitude is less than 1. (249)

sigma, Σ The Greek letter that indicates a sum. (871)

sigma notation, Σ-notation A shorthand notation used to restate a series. Also called *summation notation*. (871)

similar figures Two figures such that one is the image of the other under a composite of isometries (reflections, rotations, translations, glide reflections) and size changes. (245)

simple fraction A fraction of the form $\frac{a}{b}$, where a and b are integers and $b \neq 0$. (383)

simple interest The amount of interest I earned when calculated using the formula $I = Prt$, where P is the principal, r is the rate, and t is the time. (477)

sine function The correspondence $\theta \to \sin \theta$ that maps a number onto its sine. (663)

sine of θ (sin θ) In a right triangle with acute angle θ, $\sin \theta = \frac{\text{length of leg opposite } \theta}{\text{length of hypotenuse}}$; the y-coordinate of the image of $(1, 0)$ under R_θ, the rotation with center $(0, 0)$ and magnitude θ. (663, 684)

sine wave A curve that is the image of the graph of the sine function $s: \theta \to \sin \theta$ by any composite of translations, scale changes, or reflections. (695)

singular matrix A matrix whose multiplicative inverse does not exist. (330)

sinusoidal Pertaining to sine waves. (695)

size change The transformation that maps the point (x, y) onto (kx, ky); a transformation with center O such that the image of O is O itself and the image of any other point P is the point P' such that $OP' = k \cdot OP$ and P' is on ray OP if k is positive, and on the ray opposite ray OP if k is negative. (242)

sin⁻¹ x The number between -90° and 90°, or between $-\frac{\pi}{2}$ and $\frac{\pi}{2}$, whose sine is x. (670)

size change factor *See magnitude of a size change.* (242)

slope The slope of a line through two points (x_1, y_1) and (x_2, y_2) is the quantity $\frac{y_2 - y_1}{x_2 - x_1}$. Also called *rate of change*. (94)

slope-intercept equation of a line An equation of the form $y = mx + b$, where m is the slope of the line and b is its y-intercept. (151)

solution set of a system The intersection of the solution sets of the individual sentences in a system. (307)

solving a sentence Finding all solutions to a sentence. (40)

solving a triangle Using theorems from geometry and trigonometry to find all the missing measures of sides and angles of a triangle. (700)

square matrix A matrix with the same number of rows and columns. (328)

square root A square root x of t is a solution to $x^2 = t$. The positive square root of a positive number x is denoted $\sqrt{x}$. (381)

square root function The function f with equation $f(x) = \sqrt{x}$, where x is a nonnegative real number. (537)

squaring function The function f with equation $f(x) = x^2$. (454)

standard deviation, s.d. Let S be a data set of n numbers $\{x_1, x_2, ..., x_n\}$. Let μ be the mean of S. Then the standard deviation, s.d., of S is given by

$$s.d. = \sqrt{\frac{\sum\limits_{i=1}^{n}(x_i - \mu)^2}{n}}.\ \textbf{(885)}$$

standard form for an equation of a line An equation for a line in the form $Ax + By = C$, where A and B are not both zero. **(159)**

standard form of an equation for a parabola An equation for a parabola in the form $y = ax^2 + bx + c$, where $a \neq 0$. **(374)**

standard form of a polynomial A polynomial written in the form $a_n x^n + a_{n-1} x^{n-1} + ... + a_1 x^1 + a_0$, where n is a positive integer and $a_n \neq 0$. **(730)**

standard form of a quadratic equation An expression of the form $ax^2 + bx + c = 0$, where $a \neq 0$. **(374)**

standard form of an equation for a hyperbola An equation for a hyperbola in the form $\frac{x^2}{a^2} - \frac{y^2}{b^2} = 1$, where $b^2 = c^2 - a^2$, the foci are $(c, 0)$ and $(-c, 0)$ and the focal constant is $2a$. **(833)**

standard form of an equation for a quadratic relation An equation of the form $Ax^2 + Bxy + Cy^2 + Dx + Ey + F = 0$, where $A, B, C, D, E,$ and F are real numbers and at least one of $A, B,$ or C is nonzero. **(839)**

standard form of an equation for an ellipse An equation for an ellipse in the form $\frac{x^2}{a^2} + \frac{y^2}{b^2} = 1$, where $b^2 = a^2 - c^2$, the foci are $(c, 0)$ and $(-c, 0)$ and the focal constant is $2a$. **(818)**

standard normal curve The graph of a normal distribution. **(924)**

standard position for an ellipse The position of an ellipse centered at the origin with its foci on an axis. **(818)**

standard window The default window of a grapher that shows all four quadrants at a reasonably close scale. **(28)**

standardized scores Scores whose distribution is normal with a predetermined mean and standard deviation. **(925)**

statistical measure A single number which is used to describe an entire set of numbers.

step function A piecewise function whose graph looks like a series of steps, such as the graph of the function with equation $y = \lfloor x \rfloor$. **(204)**

stretch A scale change in the horizontal (or vertical) direction in which the absolute value of the magnitude is greater than 1; that is, $|a| > 1$ (or $|b| > 1$). **(249)**

subscript A number or variable written below and to the right of a variable. **(54)**

subset A set whose elements are all from a given set. **(891)**

substitution method A method of solving systems in two or more variables by solving one equation for one variable, substituting the expression for that variable into the other equation(s), and solving the resulting equation(s). **(314)**

subtraction of matrices Given two matrices A and B having the same dimensions, their difference $A - B$ is the matrix whose element in each position is the difference of the corresponding elements in A and B. **(230)**

sum of two matrices For two matrices A and B with the same dimensions, the matrix $A + B$ in which each element is the sum of the corresponding elements in A and B. **(228)**

summation notation A shorthand notation used to denote a series. Also called *sigma notation* or *Σ-notation*. **(871)**

system A set of conditions joined by the word "and"; a special kind of compound sentence. **(307)**

T

$\tan^{-1} x$ The number between $0°$ and $180°$, or between 0 and π whose tangent is x. **(671)**

tangent function The correspondence that maps a number x onto its tangent. **(663)**

tangent line A line that intersects a circle or ellipse in exactly one point. **(846)**

tangent of θ (tan θ) In a right triangle with acute angle θ, $\tan \theta = \frac{\text{length of leg opposite } \theta}{\text{length of leg adjacent to } \theta}$; $\tan \theta = \frac{\sin \theta}{\cos \theta}$, provided $\cos \theta \neq 0$. **(663)**

term of a polynomial Any one of the separate addends in a polynomial. **(730)**

term of a sequence An element of a sequence. **(53)**

theorem In a mathematical system, a statement that has been proved. **(88)**

transformation A one-to-one correspondence between the points of a preimage and the points of an image. **(242)**

translation A transformation for all x and y that maps (x, y) onto $(x + h, y + k)$ denoted by $T_{h,k}$. **(281)**

trial One repetition of an experiment. **(912)**

triangular number An element of the sequence 1, 3, 6, 10, ..., whose nth term is $\frac{n(n + 1)}{2}$. **(54)**

triangulation The process of determining the location of points using triangles and trigonometry. **(703)**

trigonometric ratios The ratios of the lengths of the sides in a right triangle. **(663)**

trinomial A polynomial with three terms. **(738)**

U

union of two sets The set consisting of those elements in either one or both sets. **(302)**

unit circle The circle with center at the origin and radius 1 unit. **(683)**

unit fraction A simple fraction with 1 in its numerator. **(918)**

V

value of a function For a function f, if $y = f(x)$, the value of y. **(21)**

variable A symbol that can be replaced by any one of a set of numbers or other objects. **(6)**

varies directly as See *directly proportional to*. **(72)**

varies inversely as See *inversely proportional to*. **(80)**

velocity The rate of change of distance with respect to time. **(52)**

vertex form of an equation of a parabola An equation of the form $y - k = a(x - h)^2$ where (h, k) is the vertex of the parabola. **(388)**

vertex of a parabola The intersection of a parabola with its axis of symmetry. **(102, 798)**

vertical asymptote A vertical line that is approached by the graph of a relation as the values of x approach a particular real number. **(107)**

vertical line A line in the plane with an equation of the form $x = b$. **(153)**

vertical magnitude The number b in the scale change that maps (x, y) onto (ax, by). **(249)**

vertical scale change A transformation that maps (x, y) onto (x, by). **(249)**

vertices of an ellipse The endpoints of the major and minor axes of the ellipse. **(819)**

vertices of a hyperbola The points of intersection of a hyperbola and the line containing its foci. **(832)**

vinculum The bar in a fraction or a radical symbol. **(8)**

W

whole numbers The set of numbers {0, 1, 2, 3, 4, 5, ... }. **(29)**

window The part of the coordinate grid shown on the screen of an automatic grapher. **(27)**

X

x-axis The line in the coordinate plane in which the second coordinates of points are 0. **(152)**

x-intercept The x-coordinate of the point at which a graph crosses the x-axis. **(152)**

Y

y-axis The line in the coordinate plane in which the first coordinates of points are 0. **(255)**

y-intercept The y-coordinate of the point at which a graph crosses the y-axis. **(151)**

yield See *annual percentage yield*. **(475)**

Z

zeros of a function For a function f, a value of x for which $f(x) = 0$. **(437)**

zero of a polynomial See *root of a polynomial*. **(752)**

zoom A feature on an automatic grapher which enables the window of a graph to be changed without keying in interval endpoints for x and y. **(32)**

Index

R

rad, abbreviation, 714
radian measure, 712–715
radical sign, 515, 537, 538–539
radicals
   and inverses, 514–515
   multiplying, 545–547
   and nth roots, 537–540
   rationalizing the denominator,
     551–553
   simplifying, 547–548
   solving equations with, 562–565
radioactive decay, 588–589, 644
radius of circle, 804, 806
random number generator, 927
range of a function, 27
   of direct variation, 102
   exponential, 624
   in inverse relations, 522–524
   of inverse-square, 109
   logarithmic, 624
   for positive integer power,
     454–456, 458
   in slope of line, 94, 96
rate of change, 94, 100–101, 137
Rational Exponent Theorem,
   493–494, 497
rational numbers, 29, 383
   as exponents, 493–497, 500–502
   as roots of polynomials, 760–762
Rational-Root Theorem, 761–762
Rational-Zero Theorem, 761
rationalizing the denominator,
   551–553
readability index, 885
real function, 26
real numbers, 29
   real nth roots of, 488–489
real part, of complex number, 427
real solutions to quadratic equation,
   435–437
reciprocals, 328
   exponents of same base, 466–467
rectangles, 375, 600
rectangular box
   base-side length formula, 566
   surface area of, 77, 261
   volume of, 68, 740, 741–742
rectangular hyperbola, 842, See also
   hyperbola(s).
recursive definitions, 183
   of Pascal's triangle, 897–899
   of powers, 452
   spreadsheets and, 185–186
Recursive Formula for a Geometric
   Sequence, 479–480

recursive formula(s), 183, 189, See
   also *formula(s)*; *sequence(s)*.
   for geometric sequences,
     479–480, 487, 580
   graphing a sequence using,
     191–192
   hailstone numbers, 202
   notation for, 184–185
   translation from explicit
     formula, 199
   triangular numbers, 200, 201
recursive notation, for arithmetic
   sequences, 198–199
reflecting line, 255
reflection image
   of exponentiation and
     logarithmic functions, 630
   of hyperbola over axes, 834
   of point over line, 255
   in solutions to sine and
     cosine, 701
reflection properties of conics, 856
reflection-symmetry
   of circles, 688
   of hyperbola, 840
   of parabola, 102, 388, 798
   to y-axis, 388
reflections, 255
   composites of, 264
   over other lines, 257–259
   over y-axis, 255–256
refraction of light, 536, 704, 794
refractive index, 704
Regiomontanus, 711
regression
   exponential, 603–604
   linear, 177–178
   polynomial, 782
   quadratic, 408, 410
regression lines, 177–178, 407
relations, 15
   functions and, 14–17
   inverse, 522–525
repeated multiplication of
   powers, 452
residual squares, 210
Review questions, 12–13, 19, 25, 32,
   39, 46, 52, 59, 78, 85, 92, 99,
   105, 112–113, 120, 128,
   134–135, 156, 162, 167–168,
   175, 181, 188, 194–195,
   201–202, 209, 227, 234, 241,
   247–248, 253–254, 261, 268,
   272–273, 279, 284–285, 306,
   312–313, 320, 327, 335, 341,
   346–347, 354, 361, 379, 385,
   392, 400, 407, 412–413, 419,

   426, 433, 440, 458, 465,
   470–471, 478, 485, 492,
   499, 503–504, 521, 528,
   535–536, 543–544, 549–550,
   555, 561, 567, 586, 593–594,
   600–601, 607, 613–614,
   620–621, 627–628, 634,
   641–642, 648–649, 669, 674,
   680–681, 686, 691–692,
   698–699, 705, 711, 717, 737,
   744, 751, 758–759, 764–765,
   771, 777, 785, 803, 809,
   814–815, 823, 829–830, 837,
   843, 849, 855, 875, 882, 888,
   895, 901–902, 908, 915–916,
   921, 927, See also *Chapter
   Review*.
rhombus, 600
Richardson, L.F., 621
Richter, Charles, 617
Richter scale, 617–618
right-triangle trigonometry
   3-4-5, altitude, 855
   5-12-13, angles, 671
   30-60-90, side ratios, 554
   30-60-90, sines and cosines,
     666–667
   45-45-90, sines and cosines,
     666–667
   angle of depression, 671–673
   angle of elevation, 665, 671–673,
     686
   auxiliary lines to create, 666
   definitions of sine, cosine, and
     tangent, 663
   finding sides of right triangles,
     664–665
Root of a Power Theorem, 539
Root of a Product Theorem, 546
root of multiplicity 2, 767
roots
   of an equation, 437
   of negative numbers, 558–559
   of a polynomial, 752, 767–769
   of positive numbers
     graphs of, 568
     nth, 486–460
     powers and, 489–490
     radical notation for, 537–539
   of roots, 540
*Rosencrantz & Guildenstern are
   Dead* (Stoppard), 453, 457
rotation-symmetry of hyperbola, 840
rotations, 264, 269
   of 90°, and translations, 285
   of 180°, 285
   composite of two, 269–270

Photo Credits

Photo Credits

Chapters 1–6

Cover, back: ©Pavel Trotsenko/paxtyvisuals; **vi** (l) ©tforgo/iStock, (r) ©P_Wei/iStock; **vii** (l) ©Christopher Futcher/iStock, (r) ©fotoVoyager/iStock; **viii** (l) ©ilbusca/iStock, (r) ©Wallace Weeks/Shutterstock; **ix** (l) ©bereta/iStock, (r) ©John Vachon/ Library of Congress, Prints and Photographs Division [LC-DIG-fsac-1a34273]; **x** (l) ©elkor/iStock, (r) ©dtkindler/iStock; **xi** (l) ©ChameleonsEye/Shutterstock, (r) ©MarcelC/iStock; **xii** ©David Madison/Getty Images; **3** ©Zoran Kolundzija/iStock; **4–5** ©tforgo/iStock; **7** Courtesy of the Rauner Special Collections Library, Dartmouth University; **12** NASA/JPL/UA/ Lockheed Martin; **16** ©michaeljung/iStock; **19** ©GoodDween/iStock; **20** ©Georgios Kollidas/Shutterstock; **21** ©rzelich/iStock; **25** ©carterdayne/iStock; **27** ©Sharpshot/Dreamstime; **32** ©blogs.swa-jkt.com; **36** ©PeopleImages/iStock; **39** ©svetikd/ iStock; **40** ©Pamela Moore/iStock; **42** Courtesy of the National Baseball Hall of Fame and Museum; **43** ©Keith Allison/ Wikimedia Commons; **45** ©ozgurcankaya/iStock; **51** ©citizenfresh/iStock; **56** ©Jeff Greenberg/PhotoEdit; **58** *Anno's Magic Seeds* by Mitsumasa Anno. Used by permission of Penguin Random House LLC. All rights reserved. **59** ©luismmolina/ iStock **60** (l) ©Galleria Nazionale d'Arte Moderna, Rome, Italy/De Agostini Picture Library/A. Dagli Orti/Bridgeman Images, (r) ©STILLFX/iStock; **61** ©og-vision/iStock; **70–71** ©P_Wei/iStock; **71** (b) ©PeopleImages/iStock; **77** ©Johannes Gerhardus Swanepoel/iStock; **85** ©Steve Lipofsky/Corbis; **86** ©spet//iStock; **89** ©Science/Illustration Carin L. Cain; **92** ©hanibaram/iStock; **93** ©skystardream/iStock; **104** ©yozks/iStock; **112** Courtesy NASA; **117** ©daphotovideo/iStock; **119** ©LocalTravelPhotos/Wikimedia/Creative Commons;**130** (t,b) ©RobertDodge/iStock; **136** (l) ©HultonArchive/Shutterstock, (r) ©Merts/iStock; **137** ©Museo Galileo; **148–149** ©Christopher Futcher/iStock; **150** ©www.RoadsideArchitecture.com; **161** ©traveller1116/iStock; **168** ©Kubrak78/iStock; **181** ©gerenme/iStock; **185** ©Aspen Photo/Shutterstock; **194** ©GlobalP/ iStock; **202** ©LAWaterhousePhotography/iStock; **206** ©kropic/iStock; **210** ©Michael Westhoff/iStock; **211** ©han2617/ iStock; **220–221** ©fotoVoyager/iStock; **225** ©dwhob/iStock; **226** ©LawrenceSawyer/iStock; **228** ©gary yim/Shutterstock; **233** ©Public Domain/Wikimedia Commons; **235** ©doram/iStock; **240** ©jeffhochstrasser/iStock; **246** ©James Anderson/ iStock; **249** ©Delpixart/iStock; **259** ©Zoran Kolundzija/iStock; **265** ©Michael Krinke/iStock; **267** ©Samot/Shutterstock; **273** ©Davel5957/iStock; **278** (t) ©Theodore Trimmer/Shutterstock, (c) ©Thanakorn Thaneewach/Shutterstock, (b) ©Stephensavage/iStock; **279** ©Kokkai Ng/iStock; **282** ©Mario Savoia/Shutterstock; **285** Courtesy University of Chicago; **286** (l) Courtesy Library of Congress, (c) ©Wellcome Images/Wikimedia Commons, (r) ©Mordolff/iStock; **287** NOAA; **298–299** ©ilbusca/iStock; **298** (l) ©Nerthuz/iStock; **302** ©Coprid/iStock; **305** ©Wikimedia Commons; **312** ©Susan Chiang/ iStock; **313** ©varela/iStock; **319** ©TERADAT SANTIVIVUT/iStock; **322** ©SVLIET/iStock; **326** ©serpia/iStock; **331** Public Domain/Wikimedia Commons; **335** ©enviromantic/iStock; **340** ©SomkiatFakmee/iStock; **347** ©StephanHoerold/iStock; **352** ©ozgurdonmaz/iStock; **353** ©alice-photo/iStock; **356** ©David Monniaux/Wikimedia Commons; **362** (l) ©Jasmina007/iStock; (r) ©Everett Historical/Shutterstock; **363** ©apomares/iStock; **372–373** ©Wallace Weeks/Shutterstock; **373** (b) ©Jitalia17/ iStock; **377** ©SeanShot/iStock; **378** ©OJO Images/iStock; **390** ©kcconsulting/iStock; **391** ©Thomas Barrat/Shutterstock; **392** ©Kemter/iStock; **395** ©GeorgiosArt/iStock; **397** ©Air Images/Shutterstock; **399** ©Holger Mette/iStock; **411** ©Mike Morbeck/Wikimedia Commons; **412** Courtesy R. Albrecht/ESA/ESO Space Telescope European Coordinating Facility/ NASA; **418** ©Christopher Penler/Shutterstock; **421** ©Public Domain/Wikimedia Commons; **426** ©PJPhoto69/iStock; **431** (t, b) ©Richard Megna/Fundamental Photographs, NYC; **438** ©Purdue9394/iStock; **441** ©FooTToo/iStock; **442** ©terminator1/ iStock

Chapters 7–13

450–451 ©bereta/iStock; **451** (b) ©Ivenks/iStock; **453** ©alexeys/iStock; **458** ©shgmom56/Wikimedia Commons; **463** Courtesy NASA/ESA/H. Weaver (JHU/APL), A. Stern (SwRI), and the HST Pluto Companion Search Team; **464** ©Royalty Free/Corbis; **470** (t) ©JC559/iStock; (b) ©Roberto A Sanchez/iStock; **474** ©YinYang/iStock; **475** ©Robert Brenner/PhotoEdit; **476** ©richcano/iStock; **478** ©ngirish/iStock; **479** ©Peter Frunch/LEVINE/LEAVITT; **487** bbossom/iStock; **491** ©bereta/iStock; **496** ©muldoon/iStock; **498** ©nstanev/iStock; **501** ©kupicoo/iStock; **502** (t) ©Wavebreak/iStock; (b) ©fotoon/iStock; **503** ©GlobalP/iStock; **504** ©GeorgiosArt/iStock; **505** (l) ©janniswerner/iStock; (r) ©micro_photo/iStock; **506** ©jonya/iStock; **514–515** John Vachon/Library of Congress, Prints and Photographs Division [LC-DIG-fsac-1a34273]; **516** ©IS_ImageSource/iStock; **522** Debra Meloy Elmegreen (Vassar College) et al./Hubble Heritage Team/AURA/STScI/NASA); **526** W. Keel (U. Alabama)/NASA/ESA/Hubble Heritage Team/STScI/AURA); **536** ©underworld/Shutterstock; **538** ©luckyraccoon/iStock; **541** NASA/JPL-Caltech/Cornell University; **544** ©Pavel_Markevych/Shutterstock; **549** ©2dnberty/iStock; **561** bulentozber/iStock; **566** Collection of the Australian National Maritime Museum; **568** ©Maxim Anisimov/iStock; **569** ©piccaya/iStock; **578–579** ©elkor/iStock; **580** ©alexmak72427/iStock; **583** Courtesy Lawrence Livermore National Laboratory; **585** ©leonori/iStock; **586** ©mikvivi/iStock; **589** ©bigapple/iStock; **592** ©tomlamela/iStock; **594** ©mtreasure/iStock; **597** ©Bratila Andrei/iStock; **602** ©mphillips007/iStock; **607** ©Coast-to-Coast/iStock; **613** ©Anna Vignet/Wikimedia Commons; **616** Harris & Ewing/Library of Congress Prints and Photographs Division [LC-H25-11186-F]; **618** ©J. K. Nakata/Wikimedia Commons; **619** ©lev radin/Shutterstock; **620** ©Rogelio Bernal Andreo/NASA; **621** ©Juanmonino/iStock; **628** ©Asim Patel/Wikimedia/Creative Commons; **629** Image courtesy the Old Calculator Museum; **631** ©stephankerkhofs/iStock; **633** ©Anne Clark/iStock; **634** ©Science Museum/Science and Society Picture Library; **638** ©Blulz60/iStock; **642** (l, r) delectus/iStock; **644** ©microgen/iStock; **648** ©bonnie jacobs/iStock; **649** (l) simongurney/iStock, (r) ©BanksPhotos/iStock; **650** (l) ©Cornelia Schaible/iStock, (r) ©tomeng/iStock; **651** ©deyangeorgiev/iStock; **660–661** ©dtkindler/iStock; **665** ©peplow/iStock; **668** (t) ©FOTOGRAFIA INC./iStock, (b) ©markgoddard/iStock; **673** ©slobo/iStock; **678** ©Ny Carlsberg Glyptotek/Public Domain/Wikimedia Commons; **680** ©North Wind/North Wind Picture Archives; **686** ©PKM1/iStock; **692** ©Lifesizeimages/iStock; **695** ©University of Chicago Photographic Archive [apf3-03150], Special Collections Research Center, University of Chicago Library; **696** ©tgellan/iStock; **698** Courtesy NASA/JPL-Caltech/Cornell; **705** ©blyjak/iStock; **710** ©Alina Solovyova-Vincent/iStock; **718** ©Jorge Salcedo/Shutterstock; **719** (l) ©Difydave/iStock, (r) ©Neftali/Shutterstock; **728–729** ©ChameleonsEye/Shutterstock; **728** (b) ©smaehl/iStock; **735** ©Yuri_Arcurs/iStock; **736** ©OJO Images/iStock; **737** ©Antagain/iStock; **743** ©IvonneW/iStock; **758** ©gerenme/iStock; **766** ©traveler1116/iStock; **781** ©MACIEJ NOSKOWSKI/iStock; **784** ©Grafissimo/iStock; **786** ©spastonov/iStock; **787** ©vgajic/iStock; **796–797** ©MarcelC/iStock; **801** ©jif2003/iStock; **802** ©Jim.henderson/Wikimedia Commons; **808** ©Allthesky.com Astrophotography; **809** ©Cylonphoto/iStock; **810** ©webkojak/iStock; **812** ©Public Domain/Wikimedia Commons; **822** Courtesy NASA/Johns Hopkins University Applied Physics Laboratory/Carnegie Institution of Washington; **823** ©David S. Gunderson; **829** ©CEMSE/University of Chicago; **831** (l) ©tupungato/iStock, (r) ©Franck-Boston/iStock; **837** ©RedLeash/iStock; **842** ©ciapix/Shutterstock; **848** ©MaxyM/Shutterstock; **849** ©Paul Velgos/iStock; **852** ©blakisu/iStock; **854** ©igabriela/iStock; **856** (l) ©FroukjeBrouwer/iStock, (r) ©tella_db/iStock; **857** ©ChrisGorgio/iStock; **866–867** ©David Madison/Getty Images; **874** ©kickstand/iStock; **879** ©monkeybusinessimages/iStock; **881** ©Chris Fisher/iStock; **886** ©Keith Allison/Wikimedia Commons; **887** ©Everett Historical/Shutterstock; **889** ©Joseph Sohm/Shutterstock; **892** ©Rawpixel Ltd/iStock; **894** ©makluk/iStock; **901** ©pkline/iStock; **904** ©Public Domain/Wikimedia Commons; **909** ©mattesimages/iStock; **911** ©2008 Associated Press; **913** ©mikdam/iStock; **914** ©olaser/iStock; **915** ©nayladen/iStock; **916** ©alexh/iStock; **919** ©Talaj/iStock; **921** ©Maica/iStock; **927** ©technotr/iStock; **928** ©RapidEye/iStock; **929** Public Domain/Wikimedia Commons

Symbols

$\{x \mid x > n\}$	the set of all x such that x is greater than n
$-x$	opposite of x
$\cap$	intersection
$\cup$	union
$f(x)$	function notation read "f of x"
$f : x \rightarrow y$	function notation read "f maps x onto y"
A'	image of A
S_k	size change of magnitude k
$S_{a,b}$	scale change with horizontal magnitude a and vertical magnitude b
r_x	reflection over the x-axis; transformation with matrix $\begin{bmatrix} 1 & 0 \\ 0 & -1 \end{bmatrix}$
r_y	reflection over the y-axis; transformation with matrix $\begin{bmatrix} -1 & 0 \\ 0 & 1 \end{bmatrix}$
$r_{y=x}$	reflection over the line $y = x$; transformation with matrix $\begin{bmatrix} 0 & 1 \\ 1 & 0 \end{bmatrix}$
r_m	reflection over the line m
R_θ	rotation of magnitude θ *counterclockwise* with center at the origin
R_{180}	rotation of magnitude $180°$; transformation with matrix $\begin{bmatrix} -1 & 0 \\ 0 & -1 \end{bmatrix}$
R_{90}	rotation of magnitude $90°$; transformation with matrix $\begin{bmatrix} 0 & -1 \\ 1 & 0 \end{bmatrix}$
$T_2 \circ T_1$	composite of transformations T_1 and T_2
$T_{h,k}$	translation of h units horizontally and k units vertically
$\parallel$	parallel
$\perp$	perpendicular
$\begin{bmatrix} a & b \\ c & d \end{bmatrix}$	2×2 matrix
M^{-1}	inverse of matrix M
$\det M$	determinant of matrix M

$\sqrt{}$	radical sign; square root
$\sqrt[n]{x}$	the largest real nth root of x
i	$\sqrt{-1}$
$\sqrt{-k}$	a solution of $x^2 = -k$, $k > 0$
$a + bi$	a complex number, where a and b are real numbers
$g \circ f$	composite of functions f and g
$g(f(x))$	value at x of the composite of functions f and g
$\lvert x \rvert$	absolute value of x
$\lfloor x \rfloor$	greatest integer less than or equal to x
$\lceil x \rceil$	least integer greater than or equal to x
f^{-1}	inverse of a function f read "f inverse"
a^b, $a^\wedge b$	the bth power of a
$\log_b a$	logarithm of a to the base b
e	$2.71828\ldots$
$\ln x$	natural logarithm of x
$m\angle ABC$	measure of angle ABC
$\sin \theta$	sine of θ
$\cos \theta$	cosine of θ
$\tan \theta$	tangent of θ
rad	radian
a_n	"a sub n"; the nth term of a sequence
$\displaystyle\sum_{i=1}^{n} i$	the sum of the integers from 1 to n
S_n	the sum of the first n terms of a sequence
$x!$	x factorial
$\dbinom{n}{r}$, $_nC_r$	the number of ways of choosing r objects from n objects
`int`	greatest integer calculator command
`seq`	sequence calculator command
`det`	determinant of a matrix calculator command
`fMax`, `fMin`	maximum or minimum function value calculator command
`stDevPop`	standard deviation calculator command
`nCr`	combination calculator command